9-5-06

Advanced
University Physics
Second Edition

Advanced
University Physics
Second Edition

Mircea S Rogalski
Instituto Superior Técnico, Portugal

Stuart B Palmer
University of Warwick, UK

Chapman & Hall/CRC
Taylor & Francis Group
Boca Raton   London   New York   Singapore

Cover photograph of a Wigner function plot courtesy of The University of Nottingham School of Physics and Astronomy.

Published in 2006 by
CRC Press
Taylor & Francis Group
6000 Broken Sound Parkway NW, Suite 300
Boca Raton, FL 33487-2742

© 2006 by Taylor & Francis Group, LLC
CRC Press is an imprint of Taylor & Francis Group

No claim to original U.S. Government works
Printed in the United States of America on acid-free paper
10 9 8 7 6 5 4 3 2

International Standard Book Number-10: 1-58488-511-4 (Hardcover)
International Standard Book Number-13: 978-1-58488-511-5 (Hardcover)
Library of Congress Card Number 2005045510

This book contains information obtained from authentic and highly regarded sources. Reprinted material is quoted with permission, and sources are indicated. A wide variety of references are listed. Reasonable efforts have been made to publish reliable data and information, but the author and the publisher cannot assume responsibility for the validity of all materials or for the consequences of their use.

No part of this book may be reprinted, reproduced, transmitted, or utilized in any form by any electronic, mechanical, or other means, now known or hereafter invented, including photocopying, microfilming, and recording, or in any information storage or retrieval system, without written permission from the publishers.

For permission to photocopy or use material electronically from this work, please access www.copyright.com (http://www.copyright.com/) or contact the Copyright Clearance Center, Inc. (CCC) 222 Rosewood Drive, Danvers, MA 01923, 978-750-8400. CCC is a not-for-profit organization that provides licenses and registration for a variety of users. For organizations that have been granted a photocopy license by the CCC, a separate system of payment has been arranged.

Trademark Notice: Product or corporate names may be trademarks or registered trademarks, and are used only for identification and explanation without intent to infringe.

Library of Congress Cataloging-in-Publication Data

Rogalski, Mircea S.
   Advanced university physics / by Mircea S. Rogalski and Stuart B. Palmer.-- 2nd ed.
      p. cm.
      Palmer's name appears first on earlier edition.
   Includes bibliographical references and index.
   ISBN 1-58488-511-4 (alk. paper)
   1. Physics--Textbooks. I. Palmer, Stuart B. II. Title.

QC21.3.P35 2005
530--dc22
                                     2005045510

Taylor & Francis Group
is the Academic Division of T&F Informa plc.

Visit the Taylor & Francis Web site at
http://www.taylorandfrancis.com
and the CRC Press Web site at
http://www.crcpress.com

CONTENTS

Preface		xiii
Chapter 1	INTRODUCTION	1
	1.1. SI Units	1
	1.2. Dimensional Analysis	3
	1.3. Fundamental Physical Constants	9
Chapter 2	PRINCIPLES OF NEWTONIAN MECHANICS	11
	2.1. Newton's Laws of Motion	11
	2.2. Conservation of Momentum	23
	2.3. Conservation of Energy	24
	PROBLEMS	28
Chapter 3	ANGULAR MOMENTUM	40
	3.1. Conservation of Angular Momentum	40
	3.2. Motion under a Central Force	42
	3.3. Kepler's Problem	47
	3.4. Kepler's Laws	50
	PROBLEMS	53
Chapter 4	PRINCIPLES OF SPECIAL RELATIVITY	65
	4.1. Postulates of Special Relativity	65
	4.2. The Lorentz Transformation	67
	4.3. Spacetime Diagrams	73
	4.4. Relativistic Invariance	77
	PROBLEMS	81
Chapter 5	RELATIVISTIC LAWS OF MOTION	85
	5.1. Relativistic Momentum and Energy	85
	5.2. The Relativistic Equation of Motion	89
	5.3. The Equivalence Principle	93
	PROBLEMS	97

Chapter 6   CONTINUUM MECHANICS   105

6.1. Strain in a Continuum   105
6.2. Stress in a Continuum   109
6.3. Elastic Solids   113
6.4. Fluid Mechanics   116
PROBLEMS   121

Chapter 7   THE LAWS OF THERMODYNAMICS   129

7.1. Zeroth Law of Thermodynamics   129
7.2. First Law of Thermodynamics   132
7.3. Second Law of Thermodynamics   135
PROBLEMS   141

Chapter 8   THERMODYNAMIC FUNCTIONS   149

8.1. Clausius's Theorem   149
8.2. Entropy   151
8.3. Thermodynamic Potentials   153
8.4. Hydrostatic Systems   157
8.5. Heat Capacities   160
PROBLEMS   162

Chapter 9   PERFECT AND REAL GASES   168

9.1. Perfect Gas Laws   168
9.2. Thermodynamic Functions of a Perfect Gas   172
9.3. van der Waals Gas   175
PROBLEMS   180

Chapter 10   PHASE TRANSITIONS   187

10.1. Equilibrium Between Phases   187
10.2. First Order Phase Transitions   190
10.3. Higher Order Phase Transitions   192
10.4. The Phase Rule   193
10.5. Third Law of Thermodynamics   195
PROBLEMS   198

Chapter 11   ELECTROSTATICS   204

11.1. Electric Field   204
11.2. Electrostatic Potential   208
11.3. Polarization   213
11.4. Electrostatic Energy   216
PROBLEMS   218

| **Chapter 12** | MAGNETIC INDUCTION | 227 |

12.1. Current Flow — 227
12.2. Magnetic Effects — 231
12.3. Magnetic Vector Potential — 239
PROBLEMS — 242

| **Chapter 13** | MAGNETIC FIELDS | 247 |

13.1. Magnetization — 247
13.2. Faraday's Law — 252
13.3. Energy in Magnetic Fields — 257
PROBLEMS — 259

| **Chapter 14** | MAXWELL'S EQUATIONS | 265 |

14.1. Maxwell's Field Equations — 265
14.2. Electromagnetic Energy — 270
14.3. Potential Equations — 273
PROBLEMS — 276

| **Chapter 15** | WAVE EQUATIONS | 280 |

15.1. Equation of Wave Motion — 280
15.2. Elastic Waves on a String — 284
15.3. Sound Waves in Fluids — 288
15.4. Electromagnetic Waves in Isotropic Dielectrics — 292
PROBLEMS — 295

| **Chapter 16** | WAVE PROPAGATION | 300 |

16.1. Harmonic Waves — 300
16.2. Wave Propagation in Three Dimensions — 304
16.3. Stationary Waves — 309
16.4. Continuous Waves — 313
16.5. Wave Dispersion — 315
PROBLEMS — 317

| **Chapter 17** | WAVE ENERGY | 326 |

17.1. Energy Density — 326
17.2. Energy Flow — 329
17.3. Wave Momentum — 334
17.4. Attenuation of Waves — 336
17.5. Wave Energy at an Interface — 338
PROBLEMS — 341

| **Chapter 18** | INTERFERENCE | 348 |

18.1. Interference of Two Monochromatic Waves — 348

	18.2. Interference with Multiple Beams	354
	PROBLEMS	359
Chapter 19	DIFFRACTION	366
	19.1. Diffraction of Scalar Waves	366
	19.2. The Linear Approximation of Diffraction	370
	19.3. The Diffraction Grating	376
	PROBLEMS	379
Chapter 20	POLARIZATION	387
	20.1. The Transverse Nature of Electromagnetic Waves	387
	20.2. Intensity of Electromagnetic Waves	390
	20.3. The State of Polarization	393
	20.4. Alternative Descriptions of Polarization	401
	PROBLEMS	404
Chapter 21	REFLECTION AND REFRACTION	411
	21.1. Electromagnetic Waves at an Interface	411
	21.2. Fresnel's Equations	414
	PROBLEMS	422
Chapter 22	WAVES IN ANISOTROPIC MEDIA	429
	22.1. The Plane Wave Equation	429
	22.2. Optical Anisotropic Media	433
	22.3. Ray and Phase Velocity	437
	PROBLEMS	441
Chapter 23	ABSORPTION AND DISPERSION	450
	23.1. Optical Properties of Conducting Media	450
	23.2. Origin of Complex Constitutive Parameters	457
	PROBLEMS	463
Chapter 24	GENERALIZED MECHANICS	468
	24.1. d'Alembert's Principle	468
	24.2. Hamilton's Principle	471
	24.3. Lagrange's Equations	473
	24.4. The Canonical Equations	475
	24.5. The Poisson Brackets	478
	24.6. The Hamilton-Jacobi Equation	480
	PROBLEMS	483

Chapter 25	**PRINCIPLES OF STATISTICAL MECHANICS**	490
	25.1. Liouville's Theorem	490
	25.2. Basic Statistical Assumptions	494
	25.3. Boltzmann's Principle	498
	25.4. Microcanonical Ensemble	504
	25.5. Canonical Ensemble	506
	25.6. Grand Canonical Ensemble	511
	PROBLEMS	515
Chapter 26	**STATISTICAL THERMODYNAMICS**	522
	26.1. Equipartition Theorem	522
	26.2. Statistics of the Perfect Gas	525
	26.3. The Maxwell-Boltzmann Distribution Law	529
	26.4. The Gibbs Paradox	534
	PROBLEMS	538
Chapter 27	**THERMAL RADIATION**	544
	27.1. Thermodynamics of Radiation	544
	27.2. Statistics of Radiation	549
	27.3. Planck's Radiation Formula	553
	PROBLEMS	556
Chapter 28	**WAVE MECHANICS**	561
	28.1. The Corpuscular Nature of Radiation	561
	28.2. The Old Quantum Theory	563
	28.3. The Wave Nature of Particles	569
	28.4. The Uncertainty Principle	573
	PROBLEMS	575
Chapter 29	**POSTULATES OF QUANTUM MECHANICS**	581
	29.1. Postulate 1: Wave Functions	581
	29.2. Postulate 2: Operators	584
	29.3. Postulate 3: Eigenvalues	587
	29.4. Postulate 4: Commutation Relations	591
	PROBLEMS	595
Chapter 30	**THE SCHRÖDINGER PICTURE**	601
	30.1. Postulate 5: The Time-Dependent Schrödinger Equation	601
	30.2. The Time-Independent Schrödinger Equation	604
	30.3. Unbound States. Probability Current Density	606
	30.4. Bound States. The Harmonic Oscillator	609
	PROBLEMS	613

Chapter 31 ORBITAL ANGULAR MOMENTUM — 619

- 31.1. Orbital Angular Momentum Operators — 619
- 31.2. Eigenvalue Equations in a Central Field — 622
- 31.3. Quantization of Orbital Angular Momentum — 625
- PROBLEMS — 630

Chapter 32 ONE-ELECTRON ATOMS — 636

- 32.1. The Radial Equation for One-Electron Atoms — 636
- 32.2. The Hydrogen Atom — 639
- 32.3. The One-Electron Atom in an External Magnetic Field — 643
- PROBLEMS — 646

Chapter 33 MATRIX MECHANICS OF ANGULAR MOMENTUM — 653

- 33.1. Matrix Representations — 653
- 33.2. Angular Momentum Matrices — 656
- 33.3. Spin Angular Momentum — 661
- PROBLEMS — 667

Chapter 34 THE SPINNING ELECTRON — 673

- 34.1. Addition of Angular Momenta — 673
- 34.2. Spin-Orbit Interaction — 679
- 34.3. The Spinning Electron in External Magnetic Field — 682
- PROBLEMS — 686

Chapter 35 SYSTEMS OF IDENTICAL PARTICLES — 692

- 35.1. Many-Particle Systems — 692
- 35.2. The Pauli Exclusion Principle — 695
- 35.3. Distribution Laws for Identical Particles — 698
- PROBLEMS — 702

Chapter 36 MULTIELECTRON ATOMS — 709

- 36.1. Stationary State Perturbation Theory — 709
- 36.2. The Helium Atom — 713
- 36.3. The Central-Field Approximation — 718
- PROBLEMS — 723

Chapter 37 ATOMIC RADIATION — 730

- 37.1. Time-Dependent Perturbation Theory — 730
- 37.2. Fermi's Golden Rule — 733
- 37.3. Emission and Absorption of Radiation — 735
- 37.4. Spontaneous Emission — 739
- PROBLEMS — 742

Chapter 38	SYSTEMS OF ATOMS	747
	38.1. The Adiabatic Approximation	747
	38.2. Linear Lattice Vibrations	750
	38.3. Phonons	756
	38.4. Lattice Heat Capacity	759
	PROBLEMS	763
Chapter 39	STRUCTURE OF SOLIDS	770
	39.1. The Crystal Lattice	770
	39.2. The Reciprocal Lattice	777
	39.3. Structure Determination	781
	PROBLEMS	785
Chapter 40	FREE ELECTRONS IN SOLIDS	790
	40.1. Free-Electron Approximation	790
	40.2. Free-Electron Thermodynamic Functions	796
	40.3. Electron Spin Paramagnetism	799
	40.4. Electrical Conduction	801
	PROBLEMS	804
Chapter 41	ELECTRONIC ENERGY BANDS	810
	41.1. Bloch Waves	810
	41.2. The Weak-Binding Approximation	814
	41.3. The Tight-Binding Approximation	820
	PROBLEMS	823
Chapter 42	SEMICONDUCTOR PHYSICS	829
	42.1. Free Charge Carriers	829
	42.2. Intrinsic Semiconductors	834
	42.3. Impurity Semiconductors	837
	PROBLEMS	845
Chapter 43	SOLID STATE ELECTRONICS	850
	43.1. Carrier Transport Phenomena	850
	43.2. The *pn* Junction	855
	43.3. The Junction Transistor	861
	PROBLEMS	865
Chapter 44	SOLID STATE MAGNETISM	870
	44.1. Diamagnetism	870
	44.2. Paramagnetism	872
	44.3. Ferromagnetism	875
	44.4. Antiferromagnetism	883

	44.5. Ferrimagnetism	885
	PROBLEMS	888
Chapter 45	SUPERCONDUCTIVITY	893
	45.1. The Superconducting State	893
	45.2. Cooper Pairs	898
	PROBLEMS	904
Chapter 46	NUCLEAR STRUCTURE	910
	46.1. Semiclassical Models of the Nucleus	910
	46.2. The Shell Model of the Nucleus	920
	PROBLEMS	925
Chapter 47	NUCLEAR DYNAMICS	931
	47.1. Radioactive Decay Law	931
	47.2. Alpha Decay	936
	47.3. Beta Decay	939
	PROBLEMS	944
Appendix I	VECTOR CALCULUS	949
Appendix II	MATRICES	954
Appendix III	SOME PROPERTIES OF PARTIAL DERIVATIVES	960
Appendix IV	EVALUATION OF SOME INTEGRALS	963
Appendix V	GLOSSARY OF SYMBOLS	967
PROBLEM HINTS		974
ANSWERS		988
Index		997

PREFACE TO THE SECOND EDITION

The goal of this edition is to provide the students with the opportunity to increase the depth of their knowledge of physics in one and two dimensions. We have built upon the strengths of the first edition, aiming to strike a good balance between theory, which has been reduced to 47 chapters, and relevant problems, complete with detailed solutions, which complement every chapter. There is a selection of 233 solved problems designed to encourage the student to develop the ability and desire to master the methods and techniques for a solid understanding of university physics. Problems the student may attempt, as a proof of learning, are also included at the end of each chapter. Hints, facilitating the analysis and solution, and answers to that list of 232 problems are given at the end of the book.

We wish to acknowledge the criticism and advice from several of our colleagues and graduate students, resulting in significant improvements of presentation. We are also grateful to the editorial staff of Taylor & Francis Group for their continuing encouragement and support.

MIRCEA S. ROGALSKI   STUART B. PALMER

PREFACE TO THE FIRST EDITION

The purpose of this book is to bridge the gap between the mainly descriptive treatment of phenomena, given in compendium University texts, and the highly theoretical accounts found in graduate level texts. The former seldom satisfy the serious student or the teacher, while the latter can often be intimidating without an appropriate introduction. The present book is intended to be used as a realistic introduction to both macroscopic and microscopic physics, at an intermediate level allowing the reader to step from empirical books to advanced texts.

There are several good textbooks which contain suitable advanced material for each particular branch of physics, but no one book covers the entire syllabus in a unified way. The texts that are recommended for specialized topics devote considerable space to the development of the necessary background and the application of the derived results to further topics. The present book avoids this and presents a concise, condensed sequence of physical principles linking macroscopic mechanical, thermal and electrical phenomena to their microscopic quantum origin, statistical description and atomic structure. It is written firstly as a book for learning and revising where the advanced undergraduates will find a coherent summary and the logical connections with earlier work to supplement and support more specialized books. In addition it also provides a reference material for professional physicists in both university and industry, who may find it useful for recapping aspects of physics related to their specialist activity.

The text is designed to encourage the student of physics, engineering or materials science to use it continually, and to make it the basis of his or her course. A rigorous treatment is applied to each topic by starting from first principles and deriving the basic laws and the significant consequences. We assume that the mathematical background of the student includes at least a year's course in calculus, and we aim to develop the student's facility with applied mathematics by gradually increasing the mathematical sophistication as the chapters progress and by the use of several Appendices.

The book is divided into two parts, although this division is not formal. The first part deals with macroscopic physics, and the second part (starting with Chapter 28) is concerned with microscopic physics. We start with a discussion of the basic concepts of momentum, energy and angular momentum (Chapters 2 and 3) followed by two chapters on special relativity. Since we have supposed that much of this is familiar to the reader, the treatment is rather abbreviated in some sections. Chapter 6, which is dedicated to the continuum limit of mechanics, is intended to establish a connection with thermodynamics, whose basic laws and applications to continuum systems are discussed in Chapters 7 to 10. The basic principles of electromagnetism, starting from the fundamental experimental observations and building up to Maxwell's equations, are introduced in Chapters 11 to 14, allowing the properties of electromagnetic waves to be deduced. The generality of concepts and techniques concerning waves is emphasized in Chapters 15 to 22, where we use the wave equation to discuss wave propagation and the phenomena of interference, coherence and diffraction. The techniques of Fourier analysis are employed where necessary. Chapters 23 to 26 deal with the transverse nature of electromagnetic waves and optical phenomena in anisotropic media. Chapter 27 describes the macroscopic phenomena of absorption and dispersion and demonstrates the need for a microscopic description.

In approaching microscopic physics, we first present the generalized mechanics appropriate to many-particle systems (Chapter 28) and then introduce the principles and methods of statistical mechanics and thermodynamics (Chapters 29 to 31). Thermal radiation (Chapter 32) and the phenomena and hypotheses related to wave-particle duality (Chapter 33) are discussed with the aim of making the transition to quantum physics more transparent. The postulates of quantum mechanics lie at the heart of microscopic physics (Chapter 34). They are used as the basis of our discussion of orbital angular momentum and one-electron atoms using the Schrödinger picture (Chapters 35 to 37), the spinning electron in the matrix formalism (Chapters 38 and 39), multielectron atoms as systems of identical particles (Chapters 40 and 41) and the semi-classical theory of radiation (Chapter 42). A description of the systems of atoms and the crystal lattice (Chapters 43 and 44) prepares the reader to cope with the problems of solid state physics. Here we cover free electrons in metals (Chapter 45), electronic energy bands, semiconductor physics and physical electronics (Chapters 46 to 48), lattice dynamics (Chapter 49), magnetism and superconductivity (Chapters 50 and 51). The last two chapters on nuclear structure and dynamics show how classical and quantum concepts find applicability in the domain of very short distances.

Teachers should appreciate that each chapter is a suitable core for a university lecture course at the second, third or fourth year level. We have tried to provide coverage of the subject matter, using simple models yet still allowing contact with the behaviour of real phenomena, real substances and aspects of present-day research. Several relevant examples are included in every chapter. However needs differ and each University department will probably find itself teaching material which is not in the text. Some specialist topics reserved for graduate courses are added as notes. Various applications are discussed for the purpose of illustrating the core topics. All formulae are presented in SI form and should be used with SI units. Rather than scattering the references among the text, some titles of books are collected at the end of each chapter, with no intention other than to provide supplementary reading. In many cases a range of problems help to clarify material in the text in addition to dealing with related areas that have not been included. For these reasons a companion volume of worked examples, applications and problems is in preparation.

Parts of the manuscript have been read by Professors M. Gunn, E. J. S. Lage, D. Iordache and A. Lupascu, Doctors N. Appleyard, T. J. Jackson and A. Dorobantu. They are to be thanked for numerous improvements and corrections. We, of course, are responsible for all errors, omissions and faults of presentation which, despite our best efforts, will no doubt remain. We also thank the editors at Gordon and Breach Science Publishers for their continuing support and encouragement.

STUART B. PALMER

MIRCEA S. ROGALSKI

1. INTRODUCTION

1.1. SI UNITS
1.2. DIMENSIONAL ANALYSIS
1.3. FUNDAMENTAL PHYSICAL CONSTANTS

1.1. SI Units

The laws of physics are concerned with the physical causes of observable effects. They are formulated as relationships between the magnitudes of physical quantities. It is possible to adopt for each physical quantity Q a standard amount $[Q]$, called the *unit*. An experimental measurement, which consists of a comparison of Q with $[Q]$, can then be expressed as

$$Q = n[Q] \tag{1.1}$$

where n is a number which represents the ratio of Q to $[Q]$. A physical equation, which shows how the magnitude of one quantity depends on the magnitudes of other quantities, is expressed in terms of symbols which are associated with both the numerical magnitudes and the units. In other words each symbol represents the number which is the measure of the physical quantity in terms of a given unit. Physical theories can be developed without any consideration of units, although these must be specified when considering practical problems. A particular set of equations will be satisfied by replacing the symbols by the measured quantities expressed in their appropriate units. Such a system of units, for which the equations hold when the symbols are replaced by numerical magnitudes, is called a *coherent* system.

It is common practice to consider the magnitudes of physical quantities as either *fundamental*, if their definition can be formulated without reference to the magnitude of any other quantity or *derived*, if defined in terms of the magnitudes of other quantities. Units of fundamental quantities, known as *base units*, are defined with reference to appropriately recognised standards, whereas units of derived quantities, for which comparison with a standard is difficult, are defined in terms of the base units and are referred to as *derived units*. Since the classification of physical quantities as fundamental and derived is to a large extent arbitrary, base units can be chosen for reasons of practical convenience. The system we have adopted throughout this text is the International Metric System (abbreviated as SI) which introduces units of seven base quantities

- length : metre (m)
- mass : kilogram (kg)
- time : second (s)
- temperature : kelvin (K)
- electric current : ampere (A)
- luminous intensity : candela (cd)
- amount of substance : mole (mol)

The units of length, mass and time are the base units of mechanics. The unit of mass, the kilogram, is equal to the mass of the international prototype of the kilogram, which originally was made to correspond with the mass of one cubic decimetre of pure water at a standard temperature and pressure. The second is defined in terms of an atomic standard as the duration of 9 192 631 770 periods of the radiation corresponding to the transition between the two hyperfine levels of the ground state of the cesium-130 atom. The unit of length, the meter, is equal to the path length travelled by light in vacuum during a time interval of 1/299 792 458 of a second. Some derived SI units encountered in mechanics have the following recommended names and symbols

- frequency : hertz (Hz) $1 Hz = 1 s^{-1}$
- force : newton (N) $1 N = 1 kg \cdot m \cdot s^{-2}$
- pressure : pascal (Pa) $1 Pa = 1 N \cdot m^{-2}$
- energy : joule (J) $1 J = 1 N \cdot m$
- power : watt (W) $1 W = 1 J \cdot s^{-1}$

Heat is, of course, a form of energy, and so the unit of heat is the same as that of mechanical energy, the joule. Units derived from only length, mass and time are, however, insufficient for expressing the magnitudes of all the quantities related to thermal phenomena. The additional SI base unit is the unit of temperature, called the kelvin, which is defined as the fraction 1/273.16 of the thermodynamic temperature of the triple point of water. No names have yet been proposed for the derived SI units in thermodynamics which are expressed in terms of the kelvin, such as

- entropy : $J \cdot K^{-1}$
- heat capacity : $J \cdot K^{-1}$
- coefficient of expansion : K^{-1}

Another additional base unit is needed for defining the magnitude of quantities involved in electric and magnetic phenomena. This was chosen to be the SI unit of electric current, the ampere, defined to be the constant current which, if maintained in each of two straight parallel conductors of infinite length and negligible circular cross section separated by a distance of one metre in a vacuum, produces between these two conductors a force of 2×10^{-7} newton per metre of length. All the other electric and magnetic units can be derived in terms of the four base units m, kg, s and A as

- electric charge   :   coulomb (C) $1C = 1A \cdot s$
- electric potential   :   volt (V) $1V = 1W \cdot A^{-1}$
- electric resistance   :   ohm (Ω) $1\Omega = 1V \cdot A^{-1}$
- electric conductance   :   siemens (S) $1S = 1\Omega^{-1}$
- electric capacitance   :   farad (F) $1F = 1C \cdot V^{-1}$
- magnetic flux   :   weber (Wb) $1Wb = 1V \cdot s$
- magnetic induction   :   tesla (T) $1T = 1Wb \cdot m^{-2}$
- inductance   :   henry (H) $1H = 1Wb \cdot A^{-1}$

The magnitude of quantities involved in optical phenomena can only be expressed using an additional base unit, related to the intensity of light. The SI unit of luminous intensity is the candela, defined as the luminous intensity, in a given direction, of a source that emits monochromatic radiation of frequency 540×10^{12} hertz and that has a radiant intensity in that direction of (1/683) watt per steradian. Note that steradian (sr) is the SI supplementary unit for solid angle. Two derived units are given a name and a recommended symbol

- luminous flux   :   lumen (lm) $1lm = 1cd \cdot sr$
- illumination   :   lux (lx) $1lx = 1lm \cdot m^{-2}$

The SI base unit for the amount of a substance, the mole, is defined as the amount of substance of a system that contains as many elementary entities (molecules, atoms, ions, electrons and so on) as there are atoms in 0.012 kg of the pure carbon-12.

1.2. DIMENSIONAL ANALYSIS

The *dimensional formula* of a physical quantity shows how its magnitude is defined in terms of fundamental magnitudes, which are usually represented by symbols: $[L]$ for a unit of length, $[M]$ for a unit mass, $[T]$ for a unit interval of time, $[\theta]$ for a unit of temperature, $[I]$ for a unit of electric current and so on. For each particular quantity the dimensional formula is obtained by means of the defining equation. For instance, using Newton's second law as the definition of force

$$F = \frac{d}{dt}(mv) = m\frac{d^2r}{dt^2}$$

the magnitude of length, mass and time interval can be expressed by $n'[L]$, $n''[M]$ and $n'''[T]$, where n', n'' and n''' are arbitrary positive numbers, and so the magnitude of force is given by

$$F = n\frac{[M][L]}{[T]^2} = n\left[MLT^{-2}\right]$$

where $n = n'n''/(n''')^2$ is a ratio of numbers, or

$$[F] = \left[MLT^{-2}\right] \qquad (1.2)$$

Equation (1.2) represents the dimensional formula for force, and the power to which each fundamental unit is raised in the expression on the right hand side is said to be the *dimension* of force in respect to that fundamental magnitude. It is clear that a dimensional formula for the magnitude of a particular quantity depends on the choice of the fundamental magnitudes and therefore is, to that extent, arbitrary. If the fundamental magnitudes are chosen to be those corresponding to the SI base units, the dimensional formulae of the quantities commonly appearing in mechanics have the form

$$[Q] = \left[M^\alpha L^\beta T^\gamma\right] \qquad (1.3)$$

Any magnitude which can be expressed by a number only, irrespective of the base units, such as an angle, has no dimensions and corresponds to a *dimensionless* quantity, which finds no place in a dimensional formula.

An equation which describes a particular physical situation is always a statement of the equality of quantities, which have the same dimensional formula for a given choice of fundamental magnitudes. It follows that if the equation contains a pair of terms $\left[M^\alpha L^\beta T^\gamma\right]$ and $\left[M^{\alpha'} L^{\beta'} T^{\gamma'}\right]$ we must have

$$\left[M^\alpha L^\beta T^\gamma\right] \cong \left[M^{\alpha'} L^{\beta'} T^{\gamma'}\right] \qquad (1.4)$$

for all the independent base units $[M], [L], [T]$, which yields the requirement of *dimensional homogeneity*

$$\alpha = \alpha' \quad , \quad \beta = \beta' \quad , \quad \gamma = \gamma' \qquad (1.5)$$

As Eq.(1.4) must be valid for any pair of terms in a given equation, it is an expression of the so-called *principle of homogeneity* which states that *all terms in a physical equation must be dimensionally homogenous*. Any lack of dimensional homogeneity indicates an error in the derivation of a particular equation. A special case is that the arguments of mathematical functions which can be developed as power series, such as e^x, $\ln x$, $\cos x$, $\sin x$, $\tan x$, $\sinh x$, $\tanh x$ and so on, must be dimensionless, otherwise the terms in the series cannot have the same dimensions, as required by Eq.(1.5).

The principle of dimensional homogeneity provides us with a means to determine the form of the relationship between the magnitudes of the quantities involved in a physical problem. We can write

$$Q = \Phi(Q', Q'', Q''', \ldots) = \sum C(Q')^x C(Q'')^y C(Q''')^z \ldots \qquad (1.6)$$

where Q', Q'', Q''' and so on must be either constant or independent quantities, which could be changed in magnitude without affecting the others. The function Φ in Eq.(1.6) is assumed continuous and therefore can always be approximated by a series with numerical coefficients and exponents. Since the dimensional formula for Q must be the same as the dimensional formula for each term of the series, we may write

$$[Q] = [Q']^x [Q'']^y [Q''']^z \qquad (1.7)$$

The Rayleigh method of *dimensional analysis* consists of equating the dimensions with respect to each fundamental magnitude on the two sides of Eq.(1.7), provided the dimensional formulae are written in terms of the same fundamental magnitudes. Substituting Eq.(1.3), Eq.(1.7) becomes

$$\left[M^\alpha L^\beta T^\gamma \right] = \left[M^{\alpha'} L^{\beta'} T^{\gamma'} \right]^x \left[M^{\alpha''} L^{\beta''} T^{\gamma''} \right]^y \left[M^{\alpha'''} L^{\beta'''} T^{\gamma'''} \right]^z$$

and we obtain three separate equations

$$\begin{aligned} \alpha' x + \alpha'' y + \alpha''' z + \ldots &= \alpha \\ \beta' x + \beta'' y + \beta''' z + \ldots &= \beta \\ \gamma' x + \gamma'' y + \gamma''' z + \ldots &= \gamma \end{aligned} \qquad (1.8)$$

with unknowns x, y, z and so on. It is clear that, in this case, a complete solution cannot be obtained if there are more than three unknowns. Such a method yields no information about the numerical coefficient C or the form of the function Φ, and therefore, a complete determination of the relationship (1.6) is never possible. However, dimensional analysis can provide information about the effect of some particular magnitudes, which permits considerable simplifications in investigating physical phenomena.

EXAMPLE 1.1. Vibrations of a liquid drop

The equilibrium shape of a liquid drop may to a first approximation be considered as a sphere, if gravitational force is neglected and only surface tension is taken into account. To apply dimensional analysis it is sufficient to recall that in order to increase the surface of a liquid it is necessary to apply to each element of its contour a force which is proportional to the length of the element, $F = fl$, where f is a characteristic constant of the liquid, called the surface tension. We may assume that the frequency of vibration ν, due to the external perturbation of the equilibrium shape, depends only on the radius r, density ρ and surface tension σ of the liquid drop in the form

$$v = \Phi(r, \rho, f)$$

Substituting the dimensional formulae of these independent magnitudes

$$[v] = [T^{-1}], \qquad [r] = [L], \qquad [\rho] = [m/V] = [ML^{-3}], \qquad [f] = [F/l] = [MT^{-2}]$$

Eq.(1.7) becomes

$$[T^{-1}] = [L]^x [ML^{-3}]^y [MT^{-2}]^z$$

Solving the simultaneous equations

$$y + z = 0, \qquad x - 3y = 0, \qquad -2z = -1$$

yields an expression for v which is difficult to obtain by a direct application of the theory

$$v = C\sqrt{\frac{f}{\rho r^3}}$$

If the liquid drop is placed on a horizontal plane under the influence of gravity, its equilibrium shape becomes symmetrical about the vertical axis, and r can be regarded as the radius of the maximum horizontal section. The frequency of vibration will now depend on the specific weight ρg rather than the density

$$v = \Phi(r, \rho, f, g)$$

where $[g] = [F/m] = [LT^{-2}]$. Hence from Eq.(1.7) we obtain

$$[T^{-1}] = [L]^x [ML^{-3}]^y [MT^{-2}]^z [LT^{-2}]^w$$

which gives three separate equations

$$y + z = 0, \qquad x - 3y + w = 0, \qquad -2z - 2w = -1$$

It is now only possible to determine three of the exponents in terms of the fourth. If we select w as arbitrary, we have

$$v = Cr^{-3/2+2w} \rho^{-1/2+w} f^{1/2-w} g^w = \sqrt{\frac{f}{\rho r^3}} \left[C \left(\frac{\rho g r^2}{f} \right)^w \right]$$

It follows that the frequency of vibration can be written in the form (1.6) as

$$v = \sqrt{\frac{f}{\rho r^3}} \sum C \left(\frac{\rho g r^2}{f} \right)^w = \sqrt{\frac{f}{\rho r^3}} \Phi\left(\frac{\rho g r^2}{f} \right)$$

where the function Φ remains unspecified. The function $\rho g r^2 / f$ is dimensionless, so that any value for w is consistent with the requirement of dimensional homogeneity.

1. INTRODUCTION

The magnitudes of thermal quantities must be defined in terms of four fundamental magnitudes

$$[Q] = \left[M^\alpha L^\beta T^\gamma \theta^\delta \right] \tag{1.9}$$

where the fourth dimension δ is defined with respect to the magnitude $[\theta]$ of temperature. Dimensional formulae for some thermal quantities in terms of $[M], [L], [T]$ and $[\theta]$ are obtained from the defining equations as follows

- quantity of heat $\quad [Q] = \left[ML^2 T^{-2} \right]$
- enthalpy $\quad [H] = \left[ML^2 T^{-2} \right]$
- entropy $\quad [S] = \left[ML^2 T^{-2} \theta^{-1} \right]$ $\quad (1.10)$
- specific heat $\quad [c_V] = [c_p] = \left[L^2 T^{-2} \theta^{-1} \right]$

EXAMPLE 1.2. Thermal conductivity of a gas

The theory of heat conduction in an isotropic medium is based on experimental facts which establish a proportionality between the rate of heat transfer dQ/dt across a surface A and the temperature gradient which $d\theta/dx$, can be written as:

$$\frac{dQ}{dt} = -KA \frac{d\theta}{dx} \tag{1.11}$$

where K is a coefficient characteristic of the medium and called the thermal conductivity. It follows that the dimensional formula for thermal conductivity is $[K] = \left[MLT^{-3} \theta^{-1} \right]$.

We may assume that the thermal conductivity of a gas is determined by the magnitude of four independent quantities: the mean free path λ of the gas molecules, which is the mean distance travelled by molecules between collisions with other molecules, the average molecular velocity $\overline{v}$, the gas density ρ and the specific heat at constant volume c_V

$$K = \Phi(\lambda, \overline{v}, \rho, c_V)$$

Substituting the dimensional formulae into Eq.(1.7) gives

$$\left[MLT^{-3} \theta^{-1} \right] = C[L]^x \left[LT^{-1} \right]^y \left[ML^{-3} \right]^z \left[L^2 T^{-2} \theta^{-1} \right]^w$$

or

$$z = 1, \quad x + y - 3z + 2w = 1, \quad -y - 2w = -3, \quad -w = -1$$

The analysis thus yields the following definite dimensions

$$x = y = z = w = 1$$

which leads to

$$K = C\rho c_V \lambda \bar{v}$$

and this is the same as the result which can be derived from the kinetic theory, with $C = 1/3$.

With $[M], [L], [T]$ and $[I]$ as fundamental magnitudes, the dimensional formulae for various electric and magnetic quantities take the form

$$[Q] = \left[M^\alpha L^\beta T^\gamma I^\delta \right] \quad (1.12)$$

Note, however, that the electric charge is often regarded as having a fundamental magnitude $[q]$, so that the dimensional formulae

$$[Q] = \left[M^\alpha L^\beta T^\gamma q^\delta \right] \quad (1.13)$$

can be equally adopted. The dimensional formulae of some relevant electric and magnetic quantities can be readily derived from their defining equations

- electric field $\quad [E] = [F/q] = \left[MLT^{-2} q^{-1} \right] = \left[MLT^{-3} I^{-1} \right]$
- electric potential $\quad [V] = [EL] = \left[ML^2 T^{-2} q^{-1} \right] = \left[ML^2 T^{-3} I^{-1} \right]$
- permittivity $\quad [\varepsilon] = \left[q^2 / (FL^2) \right] = \left[M^{-1} L^{-3} T^2 q^2 \right] = \left[M^{-1} L^{-3} T^4 I^2 \right]$
- magnetic induction $\quad [B] = [F/(IL)] = \left[MT^{-1} q^{-1} \right] = \left[MT^{-2} I^{-1} \right] \quad (1.14)$
- magnetic field $\quad [H] = [I/L] = \left[L^{-1} T^{-1} q \right] = \left[L^{-1} I \right]$
- permeability $\quad [\mu] = [B/H] = \left[ML q^{-2} \right] = \left[MLT^{-2} I^{-2} \right]$

EXAMPLE 1.3. Intensity of electromagnetic radiation

The intensity of electromagnetic radiation in an isotropic medium is defined as the power transported by the radiation per unit area and denoted by $S = P/A$. Let us assume that it depends on the characteristic permittivity ε and permeability μ of the medium and on the electric field E and magnetic flux density (or magnetic induction) B of the electromagnetic wave. In other words we assume that

$$S = \Phi(\varepsilon, \mu, E, B)$$

where $[S] = \left[P/L^2 \right] = \left[MT^{-3} \right]$. It follows that

$$\left[MT^{-3} \right] = C \left[M^{-1} L^{-3} T^4 I^2 \right]^x \left[MLT^{-2} I^{-2} \right]^y \left[MLT^{-3} I^{-1} \right]^z \left[MT^{-2} I^{-1} \right]^w$$

or

$$-x+y+z+w=1, \quad -3x+y+z=0, \quad 4x-2y-3z-2w=-3, \quad 2x-2y-z-w=0$$

which gives

$$x=0, \quad y=-1, \quad z=1, \quad w=1$$

We then obtain

$$S = CEB/\mu = CEH$$

which is the same as the magnitude of the Poynting vector for a plane electromagnetic wave, provided that $C=1$.

1.3. FUNDAMENTAL PHYSICAL CONSTANTS

Many laws of physics are formulated in terms of proportionality relationships between the physical quantities involved in particular phenomena. In order to obtain the corresponding equations between the magnitudes of these physical quantities, it is necessary to introduce a proportionality factor, such as K in Eq.(1.11). The constant factor can always be regarded as a derived quantity, whose magnitude may be determined knowing the base units and the dimensions in each particular case. If the magnitude of the constant factor depends on the nature of the physical system to which an equation is applied, it is called a *characteristic constant*. Values of characteristic constants, determined by experimental measurements, are found in tables of numerical data. If a physical equation remains invariant whatever the nature of the system to which it is applied, the proportionality factor is called a *universal constant*. There is at present a list of more than 100 different universal constants, each corresponding to a particular law, which are classified according to various criteria. Basic constants are always considered to be those whose values are determined by experiment. Approximate SI values are listed below for some experimentally evaluated constants:

- ◆velocity of light in vacuum $c = 2.997 \times 10^8$ $(\text{m} \cdot \text{s}^{-1})$
- ◆gravitational constant $G = 6.672 \times 10^{-11}$ $(\text{N} \cdot \text{m}^2 \cdot \text{kg}^{-2})$
- ◆Planck constant $h = 6.626 \times 10^{-34}$ $(\text{J} \cdot \text{s})$
- ◆Avogadro constant $N_A = 6.022 \times 10^{26}$ (kmol^{-1})
- ◆Boltzman constant $k = 1.380 \times 10^{-23}$ $(\text{J} \cdot \text{K}^{-1})$
- ◆electron charge $e = 1.602 \times 10^{-19}$ (C)
- ◆electron rest mass $m_e = 9.109 \times 10^{-31}$ (kg)
- ◆proton rest mass $m_p = 1.672 \times 10^{-27}$ (kg)
- ◆neutron rest mass $m_n = 1.674 \times 10^{-27}$ (kg)

There are physical constants for which standard SI values have been fixed by definition, such as

- standard acceleration due to gravity $\quad g = 9.80665 \quad (m \cdot s^{-1})$
- standard temperature $\quad 273.16 \quad (K)$
- standard pressure $\quad 1.013250 \times 10^5 \quad (N \cdot m^{-1})$
- permeability of free space $\quad \mu_0 = 4\pi \times 10^{-7} \quad (H \cdot m^{-1})$
- permittivity of free space $\quad \varepsilon_0 = 8.85 \times 10^{-12} \quad (F \cdot m^{-1})$
- atomic mass unit $\quad u = 1.6605655 \times 10^{-27} \quad (kg)$

Most of the other fundamental constants can be evaluated from their defining equations, written in terms of the previous basic and defined constants.

FURTHER READING
1. G. W. C. Kaye, T. H. Laby - TABLES OF PHYSICAL AND CHEMICAL CONSTANTS AND SOME MATHEMATICAL FUNCTIONS, Wiley, New York, 1995.
2. H. L. Anderson - PHYSICIST'S DESK REFERENCE, Springer-Verlag, New York, 1995.
3. H. G. Jerrard, D. B. McNeill - DICTIONARY OF SCIENTIFIC UNITS, Chapman & Hall, London, 1992.

2. Principles of Newtonian Mechanics

2.1. NEWTON'S LAWS OF MOTION
2.2. CONSERVATION OF MOMENTUM
2.3. CONSERVATION OF ENERGY

2.1. Newton's Laws of Motion

The principal concepts of Newtonian mechanics, which form the basis of all dynamics, are those of *space* (position), *time* (time-interval) and *momentum* (quantity of motion). The *physical space* for the dynamical description of phenomena is three-dimensional Euclidean space, where the *position* of a particle can be specified by its *position vector* $\vec{r}$ relative to a *reference frame*, written as

$$\vec{r} = r\vec{e}_r$$
$$\vec{r} = x\vec{e}_x + y\vec{e}_y + z\vec{e}_z$$
(2.1)

It is convenient to define a Cartesian frame by the orthogonal unit vectors $\vec{e}_x$, $\vec{e}_y$ and $\vec{e}_z$, where the position of a particle is determined by both the magnitude r and the direction $\vec{e}_r$ of the vector $\vec{r}$. The position is specified by the projections x, y and z of the position vector on the three axes, called Cartesian coordinates. The choice of origin and direction of the Cartesian axes in a particular reference frame is entirely a matter of convenience in Newtonian mechanics. The arbitrary nature of the location of the origin and of the direction of the axes implies respectively that space is *homogenous*, that is, the properties of space are the same in different places, and *isotropic*, i.e., the properties of space are the same in all directions.

The concept of time in classical physics is *absolute*. Time flows continuously and uniformly. As a consequence, the *time-interval* between two events at one position in space will be the same in any reference frame, and if two events taking place at two different places are *simultaneous* in one reference frame, they will also be simultaneous in any other reference frame.

The absolute simultaneity of time allows the measurement of *length* between two positions by laying a meter stick alongside and reading off the two end marks simultaneously. The length is therefore also absolute, i.e., the same in all the reference frames.

Using the concepts of length and time-interval, the rate at which the length is traversed is given by the *velocity* $\vec{v}$ and *acceleration* $\vec{a}$ of a particle, that is by the first and second derivatives of the position vector with respect to time, written as

$$\vec{v} = \dot{\vec{r}} = \frac{d\vec{r}}{dt} = \dot{x}\vec{e}_x + \dot{y}\vec{e}_y + \dot{z}\vec{e}_z = v_x\vec{e}_x + v_y\vec{e}_y + v_z\vec{e}_z$$

$$\vec{a} = \dot{\vec{v}} = \frac{d\vec{v}}{dt} = \ddot{x}\vec{e}_x + \ddot{y}\vec{e}_y + \ddot{z}\vec{e}_z = a_x\vec{e}_x + a_y\vec{e}_y + a_z\vec{e}_z$$

(2.2)

EXAMPLE 2.1. Planar motion

Two parameters are needed to specify the position of a particle in a two dimensional space. In Cartesian coordinates they are x and y. The time derivative of the position vector (2.1) can also be written as

$$\vec{v} = \dot{\vec{r}} = \dot{r}\vec{e}_r + r\dot{\vec{e}}_r \qquad (2.3)$$

where the first term on the right is the component of the velocity directed radially outward from the origin. The second term gives the component of velocity in the direction perpendicular to $\vec{e}_r$ (Figure 2.1). If we allow a change $\Delta\varphi$ in the direction of the unit vector $\vec{e}_r$, in a time interval Δt, we can write

$$\dot{\vec{e}}_r = \frac{d\vec{e}_r}{dt} = \lim_{\Delta t \to 0} \frac{\Delta \vec{e}_r}{\Delta t} = \lim_{\Delta t \to 0} \frac{\Delta \varphi}{\Delta t}\vec{e}_\varphi = \dot{\varphi}\vec{e}_\varphi \qquad (2.4)$$

Thus, for a planar motion, the position of a particle may be defined by two polar coordinates r, φ and its velocity is given by

$$\vec{v} = \dot{r}\vec{e}_r + r\dot{\varphi}\vec{e}_\varphi = v_r\vec{e}_r + v_\varphi\vec{e}_\varphi \qquad (2.5)$$

where v_r is known as the *radial* velocity, and v_φ is called the *tangential* velocity. We then differentiate $\vec{v}$ to find the acceleration as

$$\vec{a} = \frac{d\vec{v}}{dt} = \frac{d}{dt}(\dot{r}\vec{e}_r + r\dot{\varphi}\vec{e}_\varphi) = \ddot{r}\vec{e}_r + \dot{r}\dot{\vec{e}}_r + \dot{r}\dot{\varphi}\vec{e}_\varphi + r\ddot{\varphi}\vec{e}_\varphi + r\dot{\varphi}\dot{\vec{e}}_\varphi$$

where the time derivative $\dot{\vec{e}}_\varphi$ can be evaluated, using Figure 2.1, as

$$\dot{\vec{e}}_r = \frac{d\vec{e}_\varphi}{dt} = \lim_{\Delta t \to 0} \frac{\Delta\vec{e}_\varphi}{\Delta t} = \lim_{\Delta t \to 0} \frac{\Delta\varphi}{\Delta t}(-\vec{e}_r) = -\dot{\varphi}\vec{e}_r \qquad (2.6)$$

We finally obtain

$$\vec{a} = \ddot{\vec{r}} = \ddot{r}\vec{e}_r + \dot{r}\dot{\varphi}\vec{e}_\varphi + \dot{r}\dot{\varphi}\vec{e}_\varphi + r\ddot{\varphi}\vec{e}_\varphi - r\dot{\varphi}^2\vec{e}_r = (\ddot{r} - r\dot{\varphi}^2)\vec{e}_r + (r\ddot{\varphi} + 2\dot{r}\dot{\varphi})\vec{e}_\varphi = a_r\vec{e}_r + a_\varphi\vec{e}_\varphi \qquad (2.7)$$

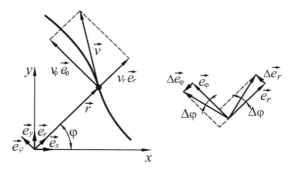

Figure 2.1. Cartesian and polar coordinates of a planar motion

The radial component a_r includes a linear acceleration $\ddot{r}$ in the radial direction, due to a change in radial speed and a *centripetal* acceleration $-r\dot{\varphi}^2$. Similarly, the tangential component a_φ contains two terms: a linear acceleration $r\ddot{\varphi}$ in the tangential direction, due to a change in the magnitude of the angular velocity, and an acceleration $2\dot{r}\dot{\varphi}$, arising when r and φ both change with time, and this is called the *Coriolis* acceleration.

The *momentum* of a particle is a vector parallel to its velocity, whose magnitude is zero when the particle is at rest and increases with the magnitude of velocity. *Inertial mass* is the scalar coefficient of velocity in the expression for the momentum

$$\vec{p} = m\vec{v} \tag{2.8}$$

and is a measure of the ability of a particle to resist a change in its motion, called *inertia*. Experiments show that inertial mass is equivalent to the gravitational mass of a particle, which measures its response to gravitational fields, so that normally the term *mass* is used in both cases. Mass was defined by Newton as the quantity of matter present, given by the product of density and volume. It is additive, the mass of a system is the sum of the masses of the isolated parts. Mass is a secondary concept, derived from space, time and momentum. It is a convenient term that provides, in Newtonian mechanics, a quantitative measure of inertia. Momentum plays a central role in the formulation of the laws of motion.

Newton's first law: *An isolated particle, independent of any external interactions, remains indefinitely in its state of rest or of uniform motion in a straight line.*

In other words the quantity of motion or momentum of an isolated particle is constant in time

$$m\vec{v} = m\dot{\vec{r}} = m\dot{\vec{r}}_0 = m\vec{v}_0$$

where $\vec{v}$ is the velocity at any time t, and $\vec{v}_0$ is the initial velocity. After integration this gives the position $\vec{r}$ as

$$\vec{r} = \vec{v}_0 t + \vec{r}_0$$

so that the isolated particle remains at rest when $\vec{v}_0 = 0$ and moves uniformly in a straight line when $\vec{v}_0 \neq 0$.

Newton's second law: *The rate of change of momentum is proportional to the applied force and is parallel to the direction in which the force acts*

$$\frac{d}{dt}(m\vec{v}) = \frac{d\vec{p}}{dt} = \vec{F} \qquad (2.9)$$

where the proportionality factor is unity if a set of consistent units is employed. The vector equation of the second law can be regarded as the definition of force, provided mass was previously defined. Hence, force is another secondary concept, although very convenient for describing how fields affect the particle motion. The equation of motion (2.9) gives the analytical expression of both the first and second laws of motion, since if $\vec{F} = 0$ the momentum $\vec{p}$ will be a constant.

Since in Newtonian mechanics mass does not change with time, we can write

$$m\ddot{\vec{r}} = m\vec{a} = \vec{F} \qquad (2.10)$$

and hence, Eq.(2.10) can be expressed in terms of scalar component equations. In planar motion, for instance, we can use either the Cartesian coordinates

$$m\ddot{x} = F_x$$
$$m\ddot{y} = F_y \qquad (2.11)$$

or the plane polar coordinates

$$m(\ddot{r} - r\dot{\varphi}^2) = F_r$$
$$m(r\ddot{\varphi} + 2\dot{r}\dot{\varphi}) = F_\varphi \qquad (2.12)$$

where the components of the acceleration given by Eq.(2.7) have been inserted. The system of second order differential equations, which describe a given motion of a particle, can also be expressed in a form independent of the choice of coordinates, called the Lagrangian formulation, which will be introduced in Chapter 24.

Newton's third law: *The forces that two particles exert upon each other are equal, opposite, and lie along the line joining them.*

The third law is a statement of the way in which particles interact with each other, and it is true not only for interactions transmitted by some mechanical contact, but also for the electrostatic and gravitational forces, transmitted across a vacuum by appropriate fields.

The solution of the equation of motion (2.10) is in the form of a trajectory $\vec{r} = \vec{r}(t)$, which gives a description of the system in terms of *space-time*. Since it is a vector differential equation of second order, when the equation of motion is integrated there are two vector constants which must be chosen to fit the physical (boundary) conditions. A basic assumption of Newtonian mechanics is the possibility of the simultaneous knowledge of position $\vec{r}$ and momentum $\vec{p}$, so that it is convenient to assume that the initial values $\vec{r}_0$ and $\vec{p}_0$ at time zero, called the *initial conditions*, are known.

EXAMPLE 2.2. Motion under a constant force

For all constant forces, the equation of motion (2.10) can be integrated to give

$$m\dot{\vec{r}} = \vec{p} = \vec{F}t + \vec{p}_0$$

where $\vec{p}_0$ is the initial momentum, and $\vec{F}$ = const. Then we can perform a second integration, which yields

$$m\vec{r} = \frac{1}{2}\vec{F}t^2 + \vec{p}_0 t + m\vec{r}_0$$

provided $\vec{r}_0$ is the initial position. It follows that the trajectory becomes

$$\vec{r} = \frac{1}{2}\frac{\vec{F}}{m}t^2 + \frac{\vec{p}_0}{m}t + \vec{r}_0$$

Since the first term will dominate the solution after a sufficiently long time, the motion will then be in the direction of the force $\vec{F}$, whatever the initial conditions.

The reference frames in which Newton's laws hold are called *inertial frames*. Assuming the existence of a fixed inertial frame in which Eq.(2.10) provides an exact description of the motion of a particle, let us examine how it transforms under a uniform translation of the reference frame. As illustrated in Figure 2.2 the same event specified by the position vector $\vec{r}(x, y, z)$ and t in the fixed frame is specified in the moving frame by $\vec{r}'(x', y', z')$ and t'. In Newtonian mechanics the common sense view is accepted that the relative motion does not affect timekeeping or geometry, so that $t' = t$ and the coordinates in the two frames are related by the law of vector addition in a common space, which reads

$$\vec{r}' = \vec{r} - \vec{R} = \vec{r} - \vec{v}t \quad \text{or} \quad \begin{aligned} x' &= x - v_x t \\ y' &= y - v_y t \\ z' &= z - v_z t \end{aligned} \qquad (2.13)$$

which is known as the *Galilean transformation*.

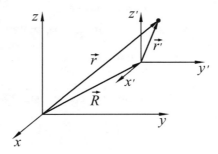

Figure 2.2. Galilean transformation of coordinates. Cartesian reference frame $x'y'z'$ is in motion relative to xyz

For a uniform translation of the moving frame with respect to the fixed one, with a velocity $\vec{v} = \text{const}$, we obtain

$$\begin{aligned} \dot{x}' &= \dot{x} - v_x & \ddot{x}' &= \ddot{x} \\ \dot{y}' &= \dot{y} - v_y \quad \text{or} \quad & \ddot{y}' &= \ddot{y} \\ \dot{z}' &= \dot{z} - v_z t & \ddot{z}' &= \ddot{z} \end{aligned} \qquad (2.14)$$

in other words the acceleration of a particle is invariant under a uniform translation of the reference frame.

Since in Newtonian mechanics the inertial mass is independent of velocity in any particular reference frame $m' = m$, the equation of motion (2.10) is invariant under the Galilean transformation if the interactions on a particle are the same, $\vec{F} = \vec{F}'$, in all the inertial frames. This is consistent with Newton's formulation of *Galileo's principle of relativity*: *in any frame the laws of physics have the same form and the constants appearing in the equations have the same numerical values.*

Newtonian mechanics assumes the existence of a privileged inertial frame, at absolute rest. A system of axes at rest with respect to the earth approximates such a situation for most practical purposes, as our understanding of mechanics necessarily comes from experiments performed in earth laboratories. A better approximation is provided by a frame at rest with respect to the average positions of the stars. But as we shall show in Chapter 4, the concept of absolute rest becomes meaningless in the relativistic mechanics, which extends the validity of the principle of relativity for all physical phenomena, including those involving electromagnetic fields.

EXAMPLE 2.3. Oscillatory motion in one dimension

A bound motion in one dimension is of necessity oscillatory and is approximated by all systems in stable equilibrium, which are carrying out small motions about their equilibrium configurations. An analysis of the oscillatory motion in terms of a variable x, which describes motion in a straight line, provides results valid for any oscillation that can be described by a single variable. A bound motion of small amplitude is described by a *harmonic* force of the form

$$F = -kx$$

where k is a positive constant, so that the equation of motion (2.10) becomes

$$m\ddot{x} = -kx \quad \text{or} \quad \ddot{x} + \omega_0^2 x = 0$$

where $\omega_0 = \sqrt{k/m}$ is a real number called the *angular frequency*, and is a starting point for describing *simple harmonic motion*. It is a characteristic of systems demonstrating simple harmonic motion that their angular frequency is only determined by mechanical properties, such as the strength k of the force and the inertia m of the response to it. It follows that the differential equation of motion (2A.3) is *linear* with constant coefficients, and therefore, the *principle of superposition* holds: *if two particular solutions of the equation of motion are found, their linear combination is also a solution of the equation.*

A convenient way of solving the equation of motion is the trial substitution

$$x = e^{\lambda t} \quad \text{i.e.,} \quad \dot{x} = \lambda e^{\lambda t}, \quad \ddot{x} = \lambda^2 e^{\lambda t}$$

where λ is quantity to be determined from the *characteristic equation*

$$\lambda^2 + \omega_0^2 = 0$$

which must be satisfied if $e^{\lambda t}$ is to be a solution. Using the two roots $\lambda_1 = +i\omega_0$, $\lambda_2 = -i\omega_0$ we may construct the general solution to be

$$x = C_1 e^{\lambda_1 t} + C_2 e^{\lambda_2 t} \quad \text{or} \quad x = C_1 e^{i\omega_0 t} + C_2 e^{-i\omega_0 t}$$

where C_1 and C_2 are arbitrary complex constants. The condition to obtain a real solution, which is $x = x^*$, may be written as

$$C_1^* e^{-i\omega_0 t} + C_2^* e^{i\omega_0 t} = C_1 e^{i\omega_0 t} + C_2 e^{-i\omega_0 t}$$

so that the constants C_1, C_2 must be complex conjugates, $C_1^* = C_2$. Choosing appropriate forms

$$C_1 = \frac{A}{2} e^{i\varphi_0} \quad , \quad C_2 = \frac{A}{2} e^{-i\varphi_0}$$

the general solution becomes

$$x = \frac{A}{2}\left[e^{i(\omega_0 t + \varphi_0)} + e^{-i(\omega_0 t + \varphi_0)}\right] = A\cos(\omega_0 t + \varphi_0)$$

and the physical motion is given by the real part of a complex solution

$$x = \text{Re}\left[Ae^{i(\omega_0 t + \varphi_0)}\right]$$

The arbitrary constants A and φ_0 are respectively called the *amplitude* (which gives the maximum displacement from the equilibrium position, $x_{\max} = A$) and the *phase constant* (which gives the initial phase $\omega_0 t + \varphi_0 = \varphi_0$ at time zero) of the oscillatory motion. These two constants can be expressed in terms of the initial conditions for position x_0 and momentum p_0 as

$$x = A\cos(\omega_0 t + \varphi_0) \qquad \text{i.e.,} \qquad x_0 = A\cos\varphi_0$$

$$\dot{x} = -A\omega_0 \sin(\omega_0 t + \varphi_0) \qquad \text{i.e.,} \qquad p_0 = -Am\omega_0 \sin\varphi_0$$

and this gives

$$A = \sqrt{x_0^2 + \frac{p_0^2}{mk}} \quad , \qquad \tan\varphi_0 = -\frac{p_0}{x_0\sqrt{mk}}$$

In practice, apparently free oscillations do always decrease in amplitude with time, and the representation of such a motion requires the addition of a *damping* force, which acts against the velocity $\dot{x}$ slowing down the oscillator. A simple description of the damping force is $-\alpha\dot{x}$, where α is a positive constant, and when this is added to the simple harmonic force the equation of motion (2.10) becomes

$$m\ddot{x} = -kx - \alpha\dot{x} \qquad \text{or} \qquad \ddot{x} + 2\beta\dot{x} + \omega_0^2 x = 0$$

where ω_0 is the angular frequency of the undamped motion, and β stands for $\alpha/2m$. The motion described by this equation is known as *damped harmonic motion*. Assuming a particular solution $x = e^{\lambda t}$, we obtain the characteristic equation

$$\lambda^2 + 2\beta\lambda + \omega_0^2 = 0$$

for which we have the two roots

$$\lambda_1 = -\beta + \sqrt{\beta^2 - \omega_0^2} \qquad \text{and} \qquad \lambda_2 = -\beta - \sqrt{\beta^2 - \omega_0^2}$$

We then can write the general solution as

$$x = C_1 e^{\lambda_1 t} + C_2 e^{\lambda_2 t}$$

Since β and ω_0 are positive numbers, there are three general cases to be considered.

Underdamped case: $\beta^2 < \omega_0^2$. For a small damping force we can define $\omega = \sqrt{\omega_0^2 - \beta^2}$ so that the roots become

$$\lambda_1 = -\beta + i\omega \quad , \quad \lambda_2 = -\beta - i\omega$$

and the solution takes the form

$$x = C_1 e^{(-\beta + i\omega)t} + C_2 e^{(-\beta - i\omega)t} = e^{-\beta t}\left(C_1 e^{i\omega t} + C_2 e^{-i\omega t}\right)$$

If the constants C_1 and C_2 are defined as before, we then obtain

$$x = A_0 e^{-\beta t} \cos(\omega t + \varphi_0)$$

The constants A_0 and φ_0 are arbitrary, and ω defines the *natural frequency* of the system, which is less than the undamped frequency ω_0. Thus, the displacements will lie between the two curves $A_0 e^{-\beta t}$ and $-A_0 e^{-\beta t}$, represented by dotted lines in Figure 2.3. The solid line shows that the amplitude $A_0 e^{-\beta t}$ decays with time. It diminishes by the factor of $1/e$ over any time interval of $1/\beta$, which is known as the decay time τ, that is

$$e^{-1} = e^{-\beta t} \quad \text{or} \quad \tau = \beta^{-1}$$

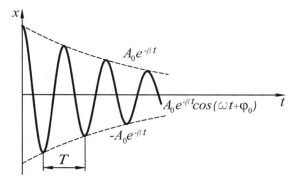

Figure 2.3. Underdamped oscillatory motion

If $T = 2\pi/\omega$ is the period of the damped oscillation, the ratio between the amplitudes of two successive maxima is

$$\frac{A(t)}{A(t+T)} = \frac{A_0 e^{-\beta t}}{A_0 e^{-\beta(t+T)}} = e^{\beta T}$$

The logarithm of this ratio is called the *logarithmic decrement* of the motion, given by

$$\delta = \ln \frac{A(t)}{A(t+T)} = \beta T$$

so that the decay of the amplitude takes the form

$$A(t) = A_0 e^{-\delta t/T}$$

The amplitude diminishes by a factor of $1/e$ during the decay time τ, and so we can write $e^{-\delta t/T} = e^{-1}$. If N_e stands for the number of damped oscillations over a time interval equal to the decay time τ, then $N_e = \tau/T$ and $\delta \tau/T = \delta N_e = 1$ and therefore $\delta = N_e^{-1}$. The period of the damped oscillation increases with the damping factor β as

$$T = \frac{2\pi}{\sqrt{\omega_0^2 - \beta^2}}$$

Critically damped case: $\beta^2 = \omega_0^2$. As the damping force becomes sufficiently large as compared to the restoring force in the system, the condition of critical damping is reached, at which $T \to \infty$ and no oscillation occurs. The solution takes the form

$$x = (C_1 + C_2)e^{-\beta t} = Ce^{-\beta t}$$

which contains only one adjustable constant and therefore has to be treated differently from the general solution.

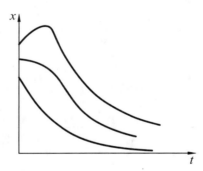

Figure 2.4. Critically damped motions

Direct substitution shows that $Bte^{-\beta t}$ is an independent solution. Therefore the motion of the critically damped oscillator is described by

$$x = (C + Bt)e^{-\beta t}$$

and is represented in Figure 2.4, where the oscillator moves back from the displaced position toward equilibrium, with a decay time $\tau = \beta^{-1}$ as before.

Overdamped case: $\beta^2 > \omega_0^2$. For this situation, the general solution consists of real exponentials of the form

$$x = C_1 e^{-\left(\beta - \sqrt{\beta^2 - \omega_0^2}\right)t} + C_2 e^{-\left(\beta + \sqrt{\beta^2 - \omega_0^2}\right)t}$$

The system also returns to equilibrium in a longer time than that for the critically damped case, since the first term has a decay time larger than $1/\beta$.

Let us now investigate the effect of an applied time varying force $F(t)$ on a damped harmonic oscillator. When such a force is added, the equation of motion becomes

$$\ddot{x} + 2\beta\dot{x} + \omega_0^2 x = \frac{1}{m} F(t)$$

where we consider only a harmonic driving force, of amplitude F_0 and *forcing* angular frequency ω_f, and this gives

$$\frac{1}{m} F(t) = \frac{F_0}{m} \cos\omega_f t = f_0 \cos\omega_f t$$

The complete solution of the equation of *forced damped* harmonic motion, which is written as

$$\ddot{x} + 2\beta\dot{x} + \omega_0^2 x = f_0 \cos\omega_f t$$

consists of the already found general solution to the homogeneous equation plus a particular solution which satisfies the inhomogeneous equation, which is conveniently written in the complex form

$$\ddot{x} + 2\beta\dot{x} + \omega_0^2 x = f_0 e^{i\omega_f t}$$

Since we might expect the particle to oscillate at the forcing angular frequency, a trial solution will be of the form $x = Ae^{i\omega_f t}$, where x and A are complex quantities

$$\dot{x} = i\omega_f A e^{i\omega_f t} \quad , \quad \ddot{x} = -\omega_f^2 A e^{i\omega_f t}$$

Substitution yields

$$-\omega_f^2 A + 2i\beta\omega_f A + \omega_0^2 A = f_0 \quad \text{or} \quad A = \frac{f_0}{(\omega_0^2 - \omega_f^2) + 2i\beta\omega_f}$$

The complex denominator may be denoted as $(\omega_0^2 - \omega_f^2) + 2i\beta\omega_f = \rho e^{-i\varphi}$, where it is apparent that

$$\rho = \sqrt{(\omega_0^2 - \omega_f^2)^2 + 4\beta^2\omega_f^2} \quad , \quad -\varphi = \arctan\frac{2\beta\omega_f}{\omega_0^2 - \omega_f^2}$$

and this results in

$$A = \frac{f_0}{\rho} e^{i\varphi} \quad \text{or} \quad x = \frac{f_0}{\rho} e^{i(\omega_f t + \varphi)}$$

The particular solution for the physical motion is then given by the real part of x, which is

$$x = \frac{f_0}{\rho} \cos(\omega_f t + \varphi)$$

In order to find the complete solution, this expression should be added to the general solution of the unforced equation

$$x = A_0 e^{-\beta t}\cos(\omega t + \varphi_0) + \frac{f_0}{\rho}\cos(\omega_f t + \varphi)$$

The first part, called the *transient* solution, oscillates with the natural frequency ω and becomes negligible at large values of time, since its amplitude decays with a time constant of $1/\beta$. The second and remaining part, called the *steady-state* solution, has the form

$$x = \frac{F_0/m}{\sqrt{\left(\omega_0^2 - \omega_f^2\right)^2 + 4\beta^2\omega_f^2}}\cos\left(\omega_f t - \arctan\frac{2\beta\omega_f}{\omega_0^2 - \omega_f^2}\right)$$

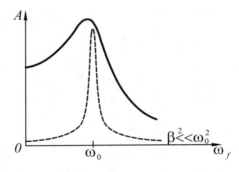

Figure 2.5. Amplitude resonance. The dashed line corresponds to $\beta^2 \ll \omega_0^2$

In the steady-state the particle oscillates at the forcing frequency, although the phase constant φ shows that the displacement x is shifted in phase with respect to the applied force. If the applied force has amplitude F_0, the amplitude of the oscillation is a maximum for a particular forcing angular frequency called the *resonance frequency* ω_{res}, which is given by

$$-4\left(\omega_0^2 - \omega_f^2\right)\omega_f + 8\beta^2\omega_f = 0 \quad \text{or} \quad \omega_f = \sqrt{\omega_0^2 - 2\beta^2} = \omega_{res}$$

so that the maximum value A_{res} of the amplitude results as

$$A_{res} = \frac{F_0/m}{2\beta\sqrt{\omega_0^2 - \beta^2}}$$

This increase in amplitude, when a particle is driven at a certain frequency, is known as *resonance* and is illustrated in Figure 2.5, where the light damping case $\left(\beta^2 \ll \omega_0^2\right)$ is represented by the dashed line.

2.2. CONSERVATION OF MOMENTUM

The most important conclusions of Newtonian mechanics can be expressed in the form of conservation laws, which state under what conditions various mechanical quantities are independent of time.

We have pointed out that the conservation law for the momentum $\vec{p}$ of an isolated particle may be regarded as equivalent to Newton's first law. The result can be generalized to many-particle systems, however we must distinguish between the *external forces* $\vec{F}_{ei}$ and the *internal forces* $\vec{F}_{ij}$ arising because of the presence of the other particles acting on a given particle i. According to Newton's third law, the sum of internal forces over all particles vanishes according to

$$\sum_i \sum_{j \neq i} \vec{F}_{ij} = 0$$

since it consists of pairs of terms which mutually cancel one another $\left(\vec{F}_{ij} + \vec{F}_{ji} = 0 \right)$. The equation of motion (2.10) for the ith particle takes the form

$$m_i \ddot{\vec{r}}_i = \vec{F}_{ei} + \sum_{j \neq i} \vec{F}_{ij} \qquad (2.15)$$

so that by summing over all particles we obtain

$$\sum_i m_i \ddot{\vec{r}}_i = \sum_i \vec{F}_{ei} + \sum_i \sum_{j \neq i} \vec{F}_{ij} = \sum_i \vec{F}_{ei} = \vec{F}_e \qquad (2.16)$$

where $\vec{F}_e$ stands for the total external force. It is now convenient to introduce a position vector $\vec{R}$ which defines the point called *centre of mass* of the system. $\vec{R}$ is the average of the position vectors of the particles, weighted in proportion to their mass

$$\vec{R} = \frac{\sum_i m_i \vec{r}_i}{\sum_i m_i} = \frac{\sum_i m_i \vec{r}_i}{M} \qquad (2.17)$$

where M denotes the mass of the system. Hence Eq.(2.16) becomes

$$M \ddot{\vec{R}} = \sum_i m_i \ddot{\vec{r}}_i = \vec{F}_e \qquad (2.18)$$

and shows that the centre of mass moves as a single particle with mass M under the action of the total external force. The definition (2.17) transforms to $\vec{P} = M\dot{\vec{R}}$, if we define the *total momentum* of the many-particle system as

$$\vec{P} = \sum_i \vec{p}_i = \sum_i m_i \dot{\vec{r}}_i \qquad (2.19)$$

Therefore the equation of motion for the centre of mass (2.18) reduces to

$$\frac{d\vec{P}}{dt} = \vec{F}_e \qquad (2.20)$$

When $\vec{F}_e = 0$ we obtain $\vec{P}$ = const, which means that *the total momentum of an isolated system of particles is constant in time*. This result is known as the *law of conservation of momentum* for a many-particle system. Formerly regarded as a specific property of certain interactions, this conservation law was later interpreted as a manifestation of the homogeneity of physical space.

2.3. Conservation of Energy

Newtonian mechanics allows us to describe a system not only in terms of space-time but also in terms of *momentum-energy*, which is appropriate for the representation of interactions where we know how a force varies with position but not how it varies with time. The equation of motion of a particle (2.9) transforms according to

$$\frac{1}{m}\vec{p} \cdot \frac{d\vec{p}}{dt} = \vec{F}(\vec{r},t) \cdot \vec{v} \quad \text{or} \quad \frac{d}{dt}\left(\frac{p^2}{2m}\right) = \vec{F}(\vec{r},t) \cdot \frac{d\vec{r}}{dt} \qquad (2.21)$$

where a scalar quantity, called the *kinetic energy T*, can be defined as

$$T = \frac{p^2}{2m} = \frac{1}{2}mv^2 \qquad (2.22)$$

The right hand side of Eq.(2.21) takes the form

$$\vec{F}(\vec{r}) \cdot \frac{d\vec{r}}{dt} = \frac{d}{dt} \int_{\vec{r}_i}^{\vec{r}} \vec{F}(\vec{r}) \cdot d\vec{r}$$

where the integral gives the *work W* done on the particle by an external force field $\vec{F}(\vec{r})$ when it moves from the initial position $\vec{r}_i$ to position $\vec{r}$. If the work is independent of the path of integration between the points $\vec{r}_i$ and $\vec{r}$, it can be expressed as a change in a quantity called *potential energy* $U(\vec{r})$ due to a change in the position $\vec{r}$ with respect to a reference position $\vec{r}_i$. The potential energy is defined by

$$U(\vec{r}) = -\int_{\vec{r}_i}^{\vec{r}} \vec{F}(\vec{r}) \cdot d\vec{r} \qquad (2.23)$$

Therefore Eq.(2.21) reduces to

$$\frac{d}{dt}\left[T - \int_{\vec{r}_i}^{\vec{r}} \vec{F}(\vec{r}) \cdot d\vec{r}\right] = 0 \qquad (2.24)$$

which states *that the total energy of the particle*

$$T - \int_{\vec{r}_i}^{\vec{r}} \vec{F}(\vec{r}) \cdot d\vec{r} = T + U(\vec{r}) \qquad (2.25)$$

is constant in time, if the forces acting on the particle are conservative. This is the *law of conservation of energy* for a particle. A force field is called conservative provided Eq.(2.23) is valid. In other words, in a conservative force field, for a path length element $d\vec{r}$ we can write

$$dU(\vec{r}) = -\vec{F}(\vec{r}) \cdot d\vec{r}$$

Inserting here the Cartesian coordinate form, given by

$$dU = \frac{\partial U}{\partial x} dx + \frac{\partial U}{\partial y} dy + \frac{\partial U}{\partial z} dz = \nabla U \cdot d\vec{r} = grad U \cdot d\vec{r}$$

shows that a conservative force field $\vec{F}(\vec{r})$ is the gradient of the potential energy, that is

$$\vec{F}(\vec{r}) = -\nabla U(\vec{r}) = -grad U(\vec{r}) \qquad (2.26)$$

and this can be written by components as

$$F_x = -\frac{\partial U}{\partial x} \quad , \quad F_y = -\frac{\partial U}{\partial y} \quad , \quad F_z = -\frac{\partial U}{\partial z} \qquad (2.27)$$

Thus, the equilibrium condition for a particle $(F_x = F_y = F_z = 0)$ becomes an extremum condition on its potential energy, of the form

$$\frac{\partial U}{\partial x} = \frac{\partial U}{\partial y} = \frac{\partial U}{\partial z} = 0$$

As illustrated in Figure 2.6 the stable equilibrium position corresponds to the minimum of the potential energy U. The equilibrium is not stable where U has a maximum. If higher derivatives of U also vanish, so that U is a constant, the equilibrium is said to be neutral.

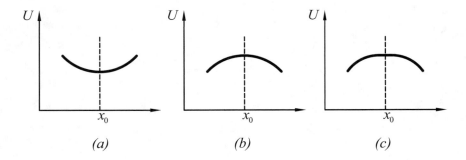

Figure 2.6. Stability of equilibrium: (a) stable, (b) unstable, (c) neutral

The energy equation for a system of particles can be derived, as in the case of a single particle, by transforming the equation of motion (2.15) for the ith particle of the system and then summing over all particles

$$\sum_i \frac{d}{dt}\left(\frac{p_i^2}{2m_i}\right) = \sum_i \left[\vec{F}_{ei}(\vec{r}_i,t)\cdot\frac{d\vec{r}_i}{dt} + \sum_{j\neq i}\vec{F}_{ij}(\vec{r}_i,t)\cdot\frac{d\vec{r}_i}{dt}\right] \qquad (2.28)$$

where the *total kinetic energy* T of the many-particle system is defined by extension of Eq.(2.22) as

$$T = \sum_i \frac{p_i^2}{2m_i} = \frac{1}{2}\sum_i m_i v_i^2 \qquad (2.29)$$

If the external forces are conservative, as given by Eq.(2.26), the first term in the right-hand side becomes

$$\sum_i \vec{F}_{ei}\cdot\frac{d\vec{r}_i}{dt} = -\sum_i \frac{\nabla_i U_i \cdot d\vec{r}_i}{dt} = -\sum_i \frac{dU_i}{dt} = -\frac{d}{dt}\sum_i U_i \qquad (2.30)$$

where ∇_i only acts on the components of $\vec{r}_i$. If the internal forces $\vec{F}_{ij}$ can also be derived from a potential energy U_{ij}, this must be a function of the distance $r_{ij} = |\vec{r}_i - \vec{r}_j| = |\vec{r}_{ij}|$ between the particles only, $U_{ij} = U_{ij}(r_{ij})$, so that any pair of mutual forces obeys Newton's third law

$$\vec{F}_{ij} + \vec{F}_{ji} = -\nabla_i U_{ij} - \nabla_j U_{ij} = -\nabla_{ij} U_{ij} + \nabla_{ij} U_{ij} = 0$$

where ∇_i, ∇_j and ∇_{ij} stand for the gradients with respect to $\vec{r}_i$, $\vec{r}_j$ and $\vec{r}_{ij} = \vec{r}_i - \vec{r}_j$, respectively. It is now convenient to write the second term in the right-hand side of Eq.(2.28), for conservative internal forces, as a sum over pairs of particles

$$\sum_i \sum_{j \neq 1} \vec{F}_{ij} \cdot \frac{d\vec{r}_i}{dt} = -\frac{1}{2} \sum_i \sum_{j \neq 1} \left(\nabla_i U_{ij} \cdot \frac{d\vec{r}_i}{dt} + \nabla_j U_{ij} \cdot \frac{d\vec{r}_j}{dt} \right)$$

$$= \frac{1}{2} \sum_i \sum_{j \neq 1} \left[\nabla_i U_{ij} \cdot \frac{d_i}{dt} (\vec{r}_i - \vec{r}_j) \right] \tag{2.31}$$

$$= -\frac{1}{2} \sum_i \sum_{j \neq 1} \frac{\nabla_{ij} U_{ij} \cdot d\vec{r}_{ij}}{dt} = -\frac{1}{2} \sum_i \sum_{j \neq 1} \frac{dU_{ij}}{dt} = -\frac{d}{dt}\left[\frac{1}{2} \sum_i \sum_{j \neq 1} U_{ij} \right]$$

Substituting Eqs.(2.29), (2.30) and (2.31), Eq.(2.28) becomes

$$\frac{d}{dt}\left[T + \sum_i U_i + \frac{1}{2} \sum_i \sum_{j \neq 1} U_{ij} \right] = 0 \tag{2.32}$$

where T is given by Eq.(2.29), and a *total potential energy U* of the system is defined as

$$U = \sum_i U_i + \frac{1}{2} \sum_i \sum_{j \neq 1} U_{ij} \tag{2.33}$$

Equation (2.32) provides the *energy conservation law* for a system of particles: *the total energy $T + U$ of a system is conserved, if both the external and internal forces are conservative.*

FURTHER READING
1. H. Goldstein, C. Pool, J. Safko - CLASSICAL MECHANICS, Pearson Education, Boston, 2001.
2. M. W. McCall - CLASSICAL MECHANICS: A MODERN INTRODUCTION, Wiley, New York, 2000.
3. J. L. McCauley - CLASSICAL MECHANICS, Cambridge University Press, Cambridge, 1997.

PROBLEMS

2.1. A mass m is released from rest on the frictionless inclined face of a wedge of mass M and wedge of angle α. Determine its acceleration relative to the wedge, ignoring friction between the wedge and the lower block.

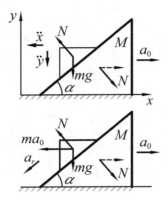

Solution

Let us put $\vec{a}_0$ for the horizontal acceleration of the wedge and $\vec{a}_r$ for the acceleration of m relative to its inclined face. In terms of an inertial frame at rest, the acceleration of m is $\vec{a} = \vec{a}_0 + \vec{a}_r$. Using the free-body diagram for each mass, Eqs.(2.10) and (2.11) can be written as

$$Ma_o = N \sin \alpha$$

$$m\ddot{x} = m(a_0 - a_r \cos \alpha) = -N \sin \alpha$$

$$m\ddot{y} = m(-a_r \sin \alpha) = N \cos \alpha - mg$$

where N stands for some unknown normal force by which m pushes the wedge. Solving for $\vec{a}_r$, one obtains

$$a_r = \frac{(M+m)g \sin \alpha}{M + m \sin^2 \alpha}$$

There is an easier way of showing this, in the frame of reference which is accelerated at the same rate as the wedge, where, in view of Eq.(2.10), the law equation of mass m reads

$$m\vec{a}_r = m\vec{a} - m\vec{a}_0 = \vec{F} - m\vec{a}_0$$

The description of relative motion in this noninertial frame requires the introduction of a friction force, in order to account for the apparent acceleration of mass m. Hence, the free-body diagram for mass m yields

$$ma_r = mg \sin \alpha + ma_0 \cos \alpha$$

$$N = mg \cos \alpha - ma_0 \sin \alpha$$

and it follows that $Ma_0 = N \sin \alpha = (mg \cos \alpha - ma_0 \sin \alpha)\sin \alpha$, which can be rearranged to give

$$a_0 = \frac{mg \sin \alpha \cos \alpha}{M + m \sin^2 \alpha}$$

Substituting this value of a_0 into the equation of relative motion of mass m, one obtains the same expression for a_r as before.

2.2. A ball of mass m is propelled vertically upward with initial speed v_0. Assuming that its motion is affected by air friction $F = -kv^2$, determine how high the ball will rise, and what is its speed of arrival back at the starting level.

Solution

The ball is moving up in one dimension, under the influence of a constant gravitational field and air friction, so that Eq.(2.11) reduces to

$$m\ddot{y} = -mg - k\dot{y}^2 \quad \text{or} \quad m\frac{dv}{dt} = -mg - kv^2$$

since $\dot{y} = v$ and $\ddot{y} = dv/dt$. Rearranging results in

$$dt = -\frac{dv}{g(1 + kv^2/mg)} = -\frac{dv}{g(1 + v^2/c^2)}$$

where $c^2 = mg/k$ is a constant. It follows that

$$dy = vdt = -\frac{c^2}{2g}\frac{2vdv}{c^2 + v^2} \quad \text{or} \quad y = \frac{c^2}{2g}\ln\frac{c^2 + v_0^2}{c^2 + v^2}$$

in view of the initial conditions $v = v_0$, when $y = 0$. The ball attains its maximum height $y = H$ when $v = dy/dt = 0$, and this gives

$$H = \frac{c^2}{2g}\ln\left(1 + \frac{v_0^2}{c^2}\right)$$

When the ball is moving down, from $y = H$, air friction will be positive, and then

$$m\ddot{y} = -mg + k\dot{y}^2 \quad \text{or} \quad m\frac{dv}{dt} = mg - kv^2$$

where $\dot{y} = -v$ and $\ddot{y} = -dv/dt$. If we set $dv/dt = 0$, it becomes apparent that $c = \sqrt{mg/k}$ stands for the terminal speed during the fall with air friction. Rearranging gives $dt = c^2 dv/g(c^2 - v^2)$, so that

$$dy = -vdt = \frac{c^2}{2g}\frac{(-2vdv)}{c^2 - v^2} \quad \text{or} \quad y = H + \frac{c^2}{2g}\ln\left(1 - \frac{v^2}{c^2}\right)$$

where $v = 0$, when $y = H$. The return speed at the starting level is obtained for $y = 0$, after putting in the value for H, and it is obviously less than v_0, namely

$$v_f = \frac{v_0 c}{\sqrt{v_0^2 + c^2}}$$

2.3. A projectile of mass m is launched at an angle α above the horizontal, with initial speed v_0. Assuming that its motion experiences air friction $\vec{F} = -k\vec{v}$, determine the range and the maximum height reached by the projectile along its trajectory.

Solution

If we set up a coordinate system in which the x-axis is horizontal, the y-axis is vertical, and the origin is located at the point from which the projectile is launched, the planar motion in a constant gravitational field can be describes in terms of Eq.(2.11) as

$$m\ddot{x} = -k\dot{x}, \qquad m\ddot{y} = -mg - k\dot{y}$$

Rearranging the first equation, where $\dot{x} = v_x$ and $\ddot{x} = dv_x/dt$, one obtains

$$\frac{dv_x}{dt} + \frac{k}{m}v_x = 0 \quad \text{or} \quad \frac{dv_x}{v_x} = -\frac{k}{m}dt$$

If the projectile is launched at time $t = 0$ with speed v_0 at an angle α to the horizontal, integrating dt between the limits 0 and t, and dv_x between the corresponding limits $v_0 \cos\alpha$ and v_x, yields

$$v_x = v_0 \cos\alpha \, e^{-kt/m}$$

Since $v_x = dx/dt$ and $x = 0$, when $t = 0$, integrating the above equation results in

$$x = \frac{mv_0 \cos\alpha}{k}\left(1 - e^{-kt/m}\right)$$

The equation of vertical motion, where $\dot{y} = v_y$ and $\ddot{y} = dv_x/dt$, takes the form

$$\frac{dv_y}{dt} + \frac{k}{m}v_y = -g$$

Since its associated homogeneous equation is the same as the equation of horizontal motion, we look for a general solution

$$v_y = Ae^{-kt/m} - \frac{mg}{k}$$

where $v_y = -mg/k$ is an obvious particular solution. The initial conditions $v_y = v_0 \sin \alpha$, when $t = 0$, determine the constant A, such that

$$v_y = \left(v_0 \sin \alpha + \frac{mg}{k}\right) e^{-kt/m} - \frac{mg}{k}$$

Integrating the above equation for $v_y = dy/dt$ to get a relation between y and t, where $y = 0$ when $t = 0$, yields

$$y = \frac{m}{k}\left(v_0 \sin \alpha + \frac{mg}{k}\right)\left(1 - e^{-kt/m}\right) - \frac{mg}{k} t$$

The projectile attains its maximum height H when

$$\frac{dy}{dt} = \left(v_0 \sin \alpha + \frac{mg}{k}\right) e^{-kt/m} - \frac{mg}{k} = 0 \quad \text{or} \quad t = \frac{m}{k} \ln\left(1 + \frac{kv_0 \sin \alpha}{mg}\right)$$

Substituting this value of t into the equation for y gives

$$H = \frac{mv_0 \sin \alpha}{k} - \frac{m^2 g}{k^2} \ln\left(1 + \frac{kv_0 \sin \alpha}{mg}\right)$$

By eliminating t between the equations of the two coordinates, the trajectory of the projectile in the vertical plane becomes

$$y = \frac{kv_0 \sin \alpha + mg}{kv_0 \cos \alpha} x + \frac{m^2 g}{k^2} \ln\left(1 - \frac{kx}{mv_0 \cos \alpha}\right)$$

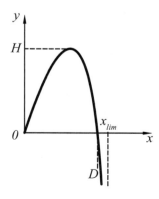

It is clear that $y = 0$ either when $x = 0$, at the point from which the projectile was launched, or when $x = D$, at the point where it strikes the ground. The range D is thus given by

$$\frac{kv_0 \sin\alpha + mg}{kv_0 \cos\alpha} D + \frac{m^2 g}{k^2} \ln\left(1 - \frac{kD}{mv_0 \cos\alpha}\right) = 0$$

As there is no solution to this equation in terms of elementary functions, a convenient method to obtain a qualitative picture is the graphical solution. For $x > D$, an asymptotic behaviour is apparent, as $y \to -\infty$ when x tends to the limiting value $x_{\lim} = mv_0 \cos\alpha / k$.

If the air friction coefficient k is small enough to consider $D \ll x_{\lim}$, that is, $kx \ll mv_0 \cos\alpha$, we may use the Taylor series expansion $\ln(1+z) = z - z^2/2 +$ terms in higher powers of z, where $z = -kx / mv_0 \cos\alpha$, such that the trajectory reduces to a parabola given by

$$y = x\tan\alpha + \frac{mg}{v_0 \cos\alpha}x - \frac{m^2 g}{k^2}\left(\frac{kx}{mv_0 \cos\alpha} + \frac{k^2 x^2}{2m^2 v_0^2 \cos^2\alpha}\right) = x\tan\alpha - \frac{gx^2}{2v_0^2 \cos^2\alpha}$$

To this approximation, we may ignore the effects of air resistance, and hence, one obtains

$$H = \frac{v_0^2 \sin^2\alpha}{2g}, \qquad D = \frac{v_0^2 \sin 2\alpha}{g}$$

2.4. A mass m, connected to a fixed point by a light horizontal spring of spring constant k, is displaced X_0 from its equilibrium position and then released along a horizontal surface. Assuming a coefficient of friction μ between mass and surface, describe its damped bound motion.

Solution

Because of the reversal of the friction force, which is $+\mu mg$ for the odd-numbered half-cycles, when $v = \dot{x} < 0$, and $-\mu mg$ for the alternate half-cycles, when $v = \dot{x} > 0$, the equation of motion (2.10) reads

$$m\ddot{x} = -kx \pm \mu mg \qquad \text{or} \qquad \ddot{x} + \frac{k}{m}\left(x \mp \frac{\mu mg}{k}\right) = 0$$

It is convenient to put $\xi = x \mp \mu mg / k$, so that $\ddot{\xi} = \ddot{x}$, and the equation reduces to that of a frictionless harmonic motion of angular frequency $\omega_0 = \sqrt{k/m}$. However, friction will affect the amplitude of motion, and hence, the solution must be written for a half-cycle at a time, in the form

$$\xi = A\cos\omega_0 t \qquad \text{or} \qquad x(t) = \pm\frac{\mu mg}{k} + A\cos\omega_0 t$$

The initial phase is $\varphi_0 = 0$, because $p_0 = m\dot{x}(n\pi/\omega_0) = 0$. The corresponding values for A read

$$A = \left|\xi(n\pi/\omega_0)\right| = \left|x(n\pi/\omega_0) - (-1)^n \frac{\mu mg}{k}\right|$$

For the first cycle, that is from $\omega_0 t = 0$ to $\omega_0 t = \pi$, this gives

$$A = X_0 - \frac{\mu mg}{k} \quad \text{or} \quad X = \frac{\mu mg}{k} + \left(X_0 - \frac{\mu mg}{k}\right)\cos\omega_0 t$$

The amplitude of the bound motion diminishes by $2\mu mg/k$, from $X_0 = x(0)$ to $X_1 = -x(\pi/\omega_0) = X_0 - 2\mu mg/k$. For the second half-cycle, from $\omega_0 t = \pi$ to $\omega_0 t = 2\pi$, one obtains

$$A = X_0 - 3\frac{\mu mg}{k} \quad \text{or} \quad X = -\frac{\mu mg}{k} + \left(X_0 - 3\frac{\mu mg}{k}\right)\cos\omega_0 t$$

and the amplitude $X_2 = x(2\pi/\omega_0) = X_0 - 4\mu mg/k$, a value which is $2\mu mg/k$ less than the starting amplitude X_1. By a similar development, if n is the number of half-cycles already completed, the motion during the next half-cycle is described by

$$X = (-1)^n \frac{\mu mg}{k} + \left[X_0 - (2n+1)\frac{\mu mg}{k}\right]\cos\omega_0 t$$

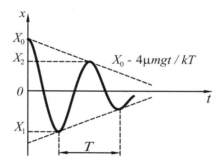

The decrease in amplitude, which follows an arithmetic progression of the form

$$X_n = X_0 - 2n\frac{\mu mg}{k}$$

can be understood in terms of the work being done against friction. This can be expressed as a decrease in the potential energy U of the spring, given by

$$-kx = -\frac{\partial U}{\partial x} \quad \text{or} \quad U(x) = \frac{1}{2}kx^2$$

where use has been made of Eq.(2A.1) and Eq.(2.27). Thus, we have

$$\frac{1}{2}kX_1^2 - \frac{1}{2}kX_0^2 = -\mu mg(X_0 + X_1) \quad \text{or} \quad X_1 = X_0 - 2\frac{\mu mg}{k}$$

for the first half-cycle,

$$\frac{1}{2}kX_2^2 - \frac{1}{2}kX_1^2 = -\mu mg(X_1 + X_2) \quad \text{or} \quad X_2 = X_1 - 2\frac{\mu mg}{k} = X_0 - 4\frac{\mu mg}{k}$$

for the second half-cycle, and so on, as shown before.

2.5. A body mass M is connected to a fixed point by a horizontal spring of mass m and spring constant k. Determine its period of oscillation along a frictionless horizontal surface.

Solution

Assuming a horizontal displacement x of the mass M, each spring element of mass $dm = (dr/L)m$ and position r will describe linear oscillation given by

$$\xi = a\cos(\omega t + \varphi)$$

where $\xi = rx/L$ and $a = rA/L$. The angular frequency can be written in terms of two constants, $\omega^2 = q/p$, where q has the dimensions of force, $q = F/\xi$, and p those of the inertia of the response to it, $p = F/\ddot{\xi}$. We can write the differential equation of motion as

$$\ddot{\xi} + \omega^2 \xi = 0 \quad \text{or} \quad p\ddot{\xi} + q\xi = 0$$

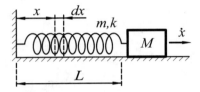

An alternative method of interpreting p and q is to consider the energy of the system, by transforming the equation of motion according to

$$p\dot{\xi}\ddot{\xi} + q\xi\dot{\xi} = 0 \quad \text{or} \quad p\dot{\xi}\frac{d\dot{\xi}}{dt} + q\xi\frac{d\xi}{dt} = 0$$

Rearranging gives

$$\frac{d}{dt}\left(\frac{1}{2}p\dot{\xi}^2 + \frac{1}{2}q\xi^2\right) = 0 \quad \text{or} \quad \frac{d}{dt}(T+U) = \frac{dE}{dt} = 0$$

where E stands for the total energy of the system, which is a constant. Thus, if the energy of an oscillating system can be written in the form

$$E = \frac{1}{2}p\dot{\xi}^2 + \frac{1}{2}q\xi^2$$

it follows that its motion is simple harmonic with angular frequency $\omega^2 = q/p$. In our case, the potential energy is that of the spring, $U = kx^2/2$, whereas the kinetic energy is given by

$$T = \frac{1}{2}M\dot{x}^2 + \frac{1}{2}\int_0^M \dot{\xi}^2 dm = \frac{1}{2}M\dot{x}^2 + \frac{1}{2}\int_0^L \left(\frac{r}{L}\dot{x}\right)^2 \frac{m}{L}dr$$

and hence

$$E = T + U = \frac{1}{2}\left(M + \frac{m}{3}\right)\dot{x}^2 + \frac{1}{2}kx^2$$

This finally provides

$$T = \frac{2\pi}{\omega} = 2\pi\sqrt{\frac{p}{q}} = 2\pi\sqrt{\frac{M + m/3}{k}}$$

2.6. A body of mass m_1 and momentum $\vec{p}_0$ collides with a mass m_2 at rest, and it is observed that after the collision m_1 and m_2 move with momenta $\vec{p}_1$ and $\vec{p}_2$, respectively. Show that the total momentum is conserved, that is $\vec{p}_0 = \vec{p}_1 + \vec{p}_2$, assuming that in any inertial frame the energy equation

$$\frac{\vec{p}_0^{\,2}}{2m_1} = \frac{\vec{p}_1^{\,2}}{2m_1} + \frac{\vec{p}_2^{\,2}}{2m_2} + Q$$

holds, where Q stands for the change in the internal energy. Provided that Q is a small percentage f of the original kinetic energy, discuss the post-collision motion of the two bodies in terms of their mass ratio $A = m_1/m_2$.

Solution

In view of the Galilean transformation (2.13), under a uniform translation of the reference frame with velocity $\vec{V}$ = const one obtains

$$\vec{v}_i' = \vec{v}_i - \vec{V} \quad \text{or} \quad \vec{p}_i' = \vec{p}_i - m_i\vec{V}$$

where $i = 0, 1, 2$. In the new frame, the energy equation reads

$$\frac{(\vec{p}_0' - m_1\vec{V})^2}{2m_1} + \frac{(-m_2\vec{V})^2}{2m_2} = \frac{(\vec{p}_1 - m_1\vec{V})^2}{2m_1} + \frac{(\vec{p}_2 - m_2\vec{V})^2}{2m_2} + Q$$

which, by rearranging, results in

$$(\vec{p}_0 - \vec{p}_1 - \vec{p}_2)\cdot\vec{V} = 0$$

This is true for any $\vec{V}$, so that the momentum conservation follows as required, since no external forces act on the system. Providing there is no appreciable friction at the interface during impact, the struck object m_2 always leaves the collision along the direction θ of the line of centers at impact, $\sin\theta = \delta/(r_1 + r_2)$, where δ stands for the impact parameter. By combining the momentum conservation, written as

$$\vec{p}_1 = \vec{p}_0 - \vec{p}_2 \quad \text{or} \quad p_1^2 = p_0^2 + p_2^2 - 2p_0 p_2 \cos\theta$$

and the energy equation

$$(1-f)\frac{p_0^2}{2m_1} = \frac{p_1^2}{2m_1} + \frac{p_2^2}{2m_2} \quad \text{or} \quad p_1^2 = (1-f)p_0^2 - Ap_2^2$$

we find that

$$p_2^2(1+A) - 2p_0 p_2 \cos\theta + fp_0^2 = 0$$

For small f, this gives

$$p_2 = \frac{p_0 \cos\theta}{1+A}\left(1 + \sqrt{1 - \frac{f(1+A)}{\cos^2\theta}}\right) \approx \frac{2p_0 \cos\theta}{1+A}\left(1 - f\frac{1+A}{4\cos^2\theta}\right)$$

and p_1 follows from the energy equation. If m_1 is observed to move after collision along a line at φ with its original direction, we have

$$p_2 \sin\vartheta = p_1 \sin\varphi \quad \text{or} \quad \sin\varphi = \frac{p_1}{p_2}\sin\vartheta$$

(a)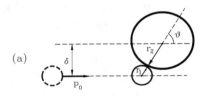

(b)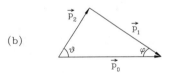

When $f \to 0$, the collision becomes elastic, and hence

$$p_2 = \frac{2p_0 \cos\theta}{1+A}, \quad p_1 = p_0\sqrt{1 - \frac{4A}{(1+A)^2}\cos^2\theta}$$

so that m_1 moves after collision at an angle defined by

$$\sin\varphi = \frac{\sin 2\theta}{\sqrt{(1+A)^2 - 4A\cos^2\theta}}$$

If $A = 1$ ($m_1 = m_2$), it is apparent that $\sin\varphi = \cos\theta$, which means that the two bodies move at right angles to each other ($\theta + \varphi = \pi/2$). When $A \ll 1$ ($m_1 \ll m_2$), we obtain to a first approximation that $\sin\varphi \approx \sin 2\theta$, and hence $2\theta + \varphi \approx \pi$. To find a maximum possible value of φ, we have to differentiate our expression for $\sin\varphi$ and set this equal to zero, that is

$$\frac{d\sin\varphi}{d\theta} = 0 \quad \text{or} \quad 4A\cos^4\theta - 2(1+A)^2\cos^2\theta + (1+A)^2 = 0$$

Using some algebra, one obtains

$$\cos^2\theta = \frac{1+A}{2A}, \quad \sin^2\theta = \frac{A-1}{2A} \quad \text{or} \quad \sin\varphi_{max} = \frac{1}{A}$$

Thus, there is a maximum value for φ provided $A > 1 (m_1 > m_2)$.

2.7. Three identical smooth metal cylinders, touching each other, are resting freely on a frictionless platform that is inclined θ with the horizontal. Find the θ values corresponding to a stable equilibrium condition.

Solution

Let us assume that, at a given θ, a force F pushing along the x-axis is required in order to maintain the equilibrium. It can be determined by considering the virtual work $\vec{F}\cdot\delta\vec{r}$ induced by a virtual displacement $\delta\vec{r}$, which is consistent with the conditions of constraint, during an infinitesimal dislocation. In view of Eq.(2.26), the work performed by the force F undergoing an infinitesimal displacement

$$\vec{F}\cdot\delta\vec{r} = -\delta U$$

must equal the work of the conservative forces resulting from that displacement.

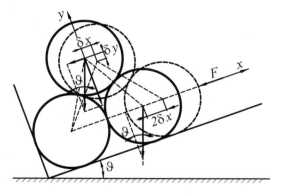

Assuming virtual displacements δx and δy for the centre of mass of the top cylinder, one of the bottom cylinders will be displaced $2\delta x$, and hence, the equation reduces to

$$-F(2\delta x) = mg\cos\theta\,\delta y + mg\sin\theta\,\delta x + mg\sin\theta(2\delta x)$$

Given the constraints $x = 2R\cos\alpha$ and $y = 2R\sin\alpha$ on the coordinates, where R is the radius of each cylinder and $\alpha = \pi/3$, one obtains

$$\delta x = \frac{dx}{d\alpha}\delta\alpha = -2R\sin\alpha\,\delta\alpha, \quad \delta y = \frac{dy}{d\alpha}\delta\alpha = 2R\cos\alpha\,\delta\alpha$$

and it follows that

$$F = \frac{mg}{2}(\cos\theta\cot\alpha - 3\sin\theta) = \frac{mg}{2}\left(\frac{\cos\theta}{\sqrt{3}} - 3\sin\theta\right)$$

It is apparent that $F \leq 0$ when $\theta \geq \tan^{-1}(1/3\sqrt{3})$, and this is a condition for stable equilibrium. However, the rotational equilibrium of the top cylinder about the fixed axis of the bottom cylinder requires that

$$mg\sin\theta(2R\sin\alpha) \leq mg\cos\theta(2R\cos\alpha) \quad \text{or} \quad \tan\theta \leq \cot\alpha = \frac{1}{\sqrt{3}}$$

and hence, the stable equilibrium can be reached for θ values in the range

$$\tan^{-1}\left(\frac{1}{3\sqrt{3}}\right) \leq \theta \leq \frac{\pi}{6}$$

2.8. A body of mass M is free to slide on a frictionless horizontal groove. A mass m, connected to the wall by a flexible, inextensible cord passing over a pulley of negligible mass, is pulled until the cord is at an angle α with the vertical, and then it is released. Find the mass m provided that α is not altered during its motion.

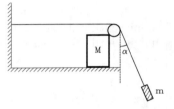

2.9. A mass m hanging from a spring of spring constant k through an inextensible and perfectly flexible string is pulled down a distance D and then released. What is the maximum value of D for the string to stay stretched during oscillation?

2.10. A ball is propelled towards an incline which slopes at an angle φ to the horizontal. Assuming that the ball collides elastically with the face of the incline, find the direction α in which it should be aimed to return to the starting point.

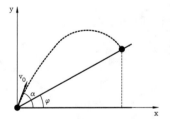

2.11. A mass m is sliding with velocity v_0 along a smooth horizontal plane and then on the frictionless surface of a block of mass M and height H that is initially at rest. Neglecting any friction, find the final velocities u and v of the bodies m and M, assuming that they will move in opposite directions.

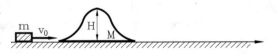

2.12. Two balls of masses m and $2m$ are placed together on the same plane, being connected by a flexible, inextensible string of length l. If the ball $2m$ is propelled vertically upward with initial speed v_0 and $l < v_0^2/2g$, find how high it will rise and how long it takes until colliding with the ball m.

2.13. A particle of mass m_1 and momentum $\vec{p}_0$ collides elastically with a stationary mass m_2 and after collision the particles move with momenta $\vec{p}_1$ and $\vec{p}_2$, respectively. Show that the curve traced by the tip of vector $\vec{p}_2$ is a circle of radius $p_2/(1+A)$, where $A = m_1/m_2$, passing through the origin from which the vector $\vec{p}_2$ is directed.

2.14. A framework of total mass m, made up of articulated uniform rods that can freely rotate about the connecting pins, is maintained in static equilibrium by a flexible, inextensible wire. What is the tension T in the wire?

3. ANGULAR MOMENTUM

3.1. CONSERVATION OF ANGULAR MOMENTUM
3.2. MOTION UNDER A CENTRAL FORCE
3.3. KEPLER'S PROBLEM
3.4. KEPLER'S LAWS

3.1. CONSERVATION OF ANGULAR MOMENTUM

If the momentum of a particle is denoted by $\vec{p}$, then by definiton its *angular momentum* $\vec{L}$ about a given point O is

$$\vec{L} = \vec{r} \times \vec{p} = \vec{r} \times m\dot{\vec{r}} = m(\vec{r} \times \dot{\vec{r}}) \tag{3.1}$$

where $\vec{r}$ is the position vector from the origin O to the particle. For a particle subject to a force $\vec{F} = m\ddot{\vec{r}}$, Eq.(2.10), we can take the moment about the point O and obtain

$$\vec{G} = \vec{r} \times \vec{F} = \vec{r} \times m\ddot{\vec{r}}$$

where $\vec{G}$ is the torque of force $\vec{F}$ about O. The right-hand side can be rewritten by using the vector identity

$$\vec{r} \times m\ddot{\vec{r}} = \frac{d}{dt}(\vec{r} \times m\dot{\vec{r}}) - \dot{\vec{r}} \times m\dot{\vec{r}} = \frac{d}{dt}(\vec{r} \times m\dot{\vec{r}}) = \frac{d\vec{L}}{dt}$$

where the cross product of the vector $\dot{\vec{r}}$ with itself obviously vanishes. Hence we have

$$\frac{d\vec{L}}{dt} = \vec{G}$$

The torque $\vec{G}$ is zero either if $\vec{F}$ is zero or if the line of action of $\vec{F}$ is parallel to that of $\vec{r}$. In the latter situation $\vec{F}$ is called a *central force*. This yields the *conservation law for the angular momentum* of a particle: *the angular momentum $\vec{L}$ is conserved if there are no forces acting on a particle or if the force acting is central.*

In a similar manner we may derive a conservation law for the total *angular momentum* of a system of particles, which is expressed by summing over the cross products $\vec{r}_i \times \vec{p}_i$ about the same origin

$$\vec{L} = \sum_i \vec{r}_i \times \vec{p}_i = \sum_i m_i \left(\vec{r}_i \times \dot{\vec{r}}_i \right) \tag{3.2}$$

Taking the moment of both sides of Eq.(2.15) with respect to the origin and summing over all particles, we have

$$\sum_i m_i \left(\vec{r}_i \times \ddot{\vec{r}}_i \right) = \sum_i \vec{r}_i \times \vec{F}_{ei} + \sum_i \sum_{j \neq 1} \vec{r}_i \times \vec{F}_{ij}$$

In view of the previously used vector identity, the left-hand side gives the time rate of change of the total angular momentum, so that the equation reduces to

$$\frac{d\vec{L}}{dt} = \vec{G}_e + \sum_i \sum_{j \neq 1} \vec{r}_i \times \vec{F}_{ij}$$

where $\vec{G}_e$ is the total external torque

$$\vec{G}_e = \sum_i \vec{r}_i \times \vec{F}_{ei}$$

and the double sum is the total internal torque, which vanishes if the line of action of all the internal forces $\vec{F}_{ij}$ lie along the lines $\vec{r}_{ij}$ joining the particles, as illustrated in Figure 3.1.

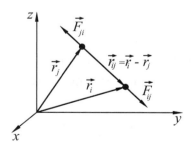

Figure 3.1. Central internal forces

In other words, if the internal forces are all central forces, we have

$$\sum_i \sum_{j \neq 1} \vec{r}_i \times \vec{F}_{ij} = \frac{1}{2} \sum_i \sum_{j \neq 1} \left(\vec{r}_i \times \vec{F}_{ij} + \vec{r}_j \times \vec{F}_{ji} \right)$$

$$= \frac{1}{2} \sum_i \sum_{j \neq 1} \left(\vec{r}_i - \vec{r}_j \right) \times \vec{F}_{ij} = \frac{1}{2} \sum_i \sum_{j \neq 1} \vec{r}_{ij} \times \vec{F}_{ij} = 0$$

Therefore, the time derivative of the total angular momentum about a given point is equal to the total external torque about the same point, that is

$$\frac{d\vec{L}}{dt} = \vec{G}_e$$

The *conservation law for the total angular momentum* can be stated as: *If the total external torque is zero, the total angular momentum of a system of particles, in which the internal forces are all central forces, is constant in time.*

3.2. Motion under a Central Force

The simplest force that can be associated with point-like particles is a mutual *central force* acting along the line joining them, since this line is the only direction the particles can define, as far as they have no intrinsic directional properties and Newton's third law is true. Let us consider two particles, one fixed at the origin and the other with its position specified by the position vector $\vec{r}$. The central force on the second particle which can be written as

$$\vec{F} = f(r) \frac{\vec{r}}{r} = f(r) \vec{e}_r$$

where $f(r)$ gives the magnitude of the interaction and $\vec{e}_r$ is defined in Eq.(2.1), is both isotropic and *conservative*. Its Cartesian components are

$$F_x = \frac{x}{r} f(r), \quad F_y = \frac{y}{r} f(r), \quad F_z = \frac{z}{r} f(r)$$

so that, for a differential path length, we have the relation

$$dU = -\left(F_x dx + F_y dy + F_z dz \right) = -\frac{f(r)}{r} \left(x dx + y dy + z dz \right) = -f(r) dr$$

where use has been made of the identity

$$2rdr = d(r^2) = d(x^2 + y^2 + z^2) = 2(xdx + ydy + zdz)$$

For the motion of a particle in space, which is described by three coordinates, dU is a function of the radial distance r only. It follows that the change in potential energy is independent of the path, and dU is an exact differential. Therefore the central force is indeed conservative, and the central force potential energy is a function of r only

$$U(r) = -\int_{r_i}^{r} f(r) dr \qquad (3.3)$$

where $U(r_i)$ appears as an arbitrary constant of integration. A discussion assuming a fixed centre of force applies with little error when the two interacting particles have very different masses, as in the case of planetary motion or with an electron and an atomic nucleus, provided the more massive particle is considered fixed. In the case where the particles have comparable masses m_1 and m_2 their motion under a mutual central force must be first reduced to an equivalent one-particle motion under a central force having a fixed centre.

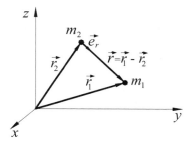

Figure 3.2. Two body configuration

Using the coordinates given in Figure 3.2 the equation of motion (2.10) can be written for two particles in the form

$$m_1 \ddot{\vec{r}}_1 = f(r)\vec{e}_r$$
$$m_2 \ddot{\vec{r}}_2 = -f(r)\vec{e}_r \qquad (3.4)$$

where $f(r) < 0$ if the force is attractive, and $f(r) > 0$ for a repulsive force. A single equation of motion for $\vec{r} = \vec{r}_1 - \vec{r}_2$ is obtained if we divide the first equation (3.4) by m_1 and the second by m_2 and subtract. Denoting the ratio $m_1 m_2 / (m_1 + m_2)$ by μ, known as the *reduced mass* and usually written in the form

$$\frac{1}{\mu} = \frac{1}{m_1} + \frac{1}{m_2} \qquad (3.5)$$

we may reduce Eqs.(3.4) to the equation of motion of a single particle

$$\mu \ddot{\vec{r}} = f(r)\vec{e}_r \qquad (3.6)$$

having a mass μ and a position vector $\vec{r}$ with respect to a fixed centre of force, which will be taken to be the origin of the coordinate system. The central force exerts no torque on the reduced mass μ, hence the angular momentum vector $\vec{L}$ is conserved during the motion. This is formally proved by taking the moment of both sides of Eq.(3.6), which gives

$$f(r)(\vec{r} \times \vec{e}_r) = \mu(\vec{r} \times \ddot{\vec{r}}) = \frac{d}{dt}(\vec{r} \times \mu \dot{\vec{r}}) = \frac{d\vec{L}}{dt} = 0$$

where $\vec{r}$ is parallel with $\vec{e}_r$, and their cross product vanishes. Since the direction of $\vec{L}$ is fixed in space, the position vector $\vec{r}$ can only move in a plane perpendicular to $\vec{L}$ through the centre of force, that is *the motion is planar*. In this case we need only two-dimensional polar coordinates to define the position of the system. Using the components (2.7) of the acceleration $\ddot{\vec{r}}$, the radial and tangential equations of motion become

$$\mu(\ddot{r} - r\dot{\varphi}^2) = f(r) \qquad (\vec{e}_r) \qquad (3.7)$$

$$\mu(r\ddot{\varphi} + 2\dot{r}\dot{\varphi}) = 0 \qquad (\vec{e}_\varphi) \qquad (3.8)$$

The determination of the orbit of a particle moving in a central force field involves the integration of these two equations, which give the description of the system in terms of space-time. However, an alternative description, in terms of angular momentum and energy, which are constants of central force motion, is easier and provides greater physical insight. It provides the basis of a further transition to quantum mechanics where the initial values of the coordinates $r, \dot{r}, \varphi$ and $\dot{\varphi}$ required by the integration of Eq.(3.7) become meaningless, whilst we can still consider the system angular momentum and system energy.

The magnitude L of the angular momentum is obtained by direct integration of Eq.(3.8) written in the form

$$\frac{\mu}{r}\frac{d}{dt}(r^2\dot{\varphi}) = 0$$

which becomes

$$L = \mu r^2 \dot{\varphi} = \text{const} \qquad (3.9)$$

so that L is a constant for central force motion. It can be used to eliminate φ from Eq.(3.7), which takes the form

$$\mu\ddot{r} - \frac{L^2}{\mu r^3} = f(r) \quad \text{or} \quad \mu\ddot{r} = -\frac{d}{dr}\left[U(r) + \frac{L^2}{2\mu r^2}\right] \tag{3.10}$$

where the central force potential energy was inserted, according to Eq.(3.3). Upon multiplication of both sides by $\dot{r}$, we have

$$\mu\ddot{r}\dot{r} = -\frac{d}{dr}\left[U(r) + \frac{L^2}{2\mu r^2}\right]\frac{dr}{dt} = -\frac{d}{dt}\left[U(r) + \frac{L^2}{2\mu r^2}\right]$$

which is equivalent to

$$\frac{d}{dt}\left[\frac{1}{2}\mu\dot{r}^2 + \frac{L^2}{2\mu r^2} + U(r)\right] = 0$$

so that the total energy E is also a constant of central force motion, that is

$$E = \frac{1}{2}\mu\dot{r}^2 + \frac{L^2}{2\mu r^2} + U(r) = \text{const} \tag{3.11}$$

Equations (3.9) and (3.11), known as *first integrals* of the equations of motion (3.8) and (3.7), give the desired description of the motion in a central force field in terms of the angular momentum and energy.

It is convenient to define an *effective potential energy* in the form

$$U_{\text{eff}}(r) = U(r) + \frac{L^2}{2\mu r^2} \tag{3.12}$$

so that the energy conservation law may be written in a form that describes a motion in one dimension as

$$E = \frac{1}{2}\mu\dot{r}^2 + U_{\text{eff}}(r)$$

The second term on the right-hand side of Eq.(3.12) is called the *centrifugal potential energy*, since it corresponds to an additional force in Eq.(3.10), which reads

$$\mu\ddot{r} = f(r) + \frac{L^2}{\mu r^3} = f(r) + \mu r\dot{\varphi}^2$$

and this is known as the centrifugal force. Equation (3.12) involves only the radial motion, whose qualitative features can be discussed by examining a plot of the effective potential energy (3.12) against r. Such a diagram is given in Figure 3.3 for the specific case of an inverse-square force, given by

$$f(r) = -\frac{C}{r^2} \quad \text{i.e.,} \quad U(r) = -\frac{C}{r} \qquad (3.13)$$

where the interaction is attractive for a positive C.

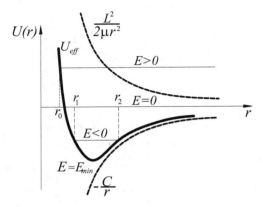

Figure 3.3. Effective potential energy U_{eff} due to an attractive inverse-square force

Since both the gravitational and electrostatic forces have an inverse-square dependence

$$f(r) = -G\frac{m_1 m_2}{r^2} \quad \text{i.e.,} \quad U(r) = -G\frac{m_1 m_2}{r} = -\frac{C}{r}$$

where G, not to be confused with the torque, stands for the gravitational constant, and, respectively

$$f(r) = -\frac{1}{4\pi\varepsilon_0}\frac{Ze^2}{r^2} = -\frac{Ze_0^2}{r^2} \quad \text{i.e.,} \quad U(r) = -\frac{Ze_0^2}{r} = -\frac{C}{r}$$

where $e_0^2 = e^2/4\pi\varepsilon_0$, this case is the most important of all the central force laws. The dotted lines in Figure 3.3 represent the two components of the effective potential energy

$$U_{eff} = -\frac{C}{r} + \frac{L^2}{2\mu r^2} \qquad (3.14)$$

which is given by the solid line.

The radial motion is restricted to regions where the kinetic energy $T \geq 0$. According to the energy conservation law (3.11), we have $T = E - U_{eff}(r)$, and the nature of the motion is determined by the total energy.

$E > 0$ (unbound motion). There is no upper limit to the possible value of r. For $r \to \infty$, we obtain from Eq.(3.11) that $v_r = \dot{r} \to \sqrt{2E/\mu}$, while according to Eq.(3.9)

$v_\varphi = r\dot\varphi = L/\mu r \to 0$. Thus the orbit approaches a straight line, called an *asymptote*, at infinity. By reversing the motion, the particle cannot come closer than a nearest approach distance r_0, where the kinetic energy becomes zero, and the particle is pushed away by the centrifugal barrier and travels back out to infinity.

$E = 0$ (unbound motion). A similar picture of the motion can be applied, also corresponding to an unbound orbit.

$E < 0$ (bound motion). The position r has a lower limit r_1 and also cannot exceed an upper limit r_2 with positive kinetic energy so that the two particles form a bound system. The planar motion is contained between two circles of radius r_1 and r_2.

$E = E_{min}$ (bound motion). The position r is restricted to one value so that motion is possible at only one radius, and the orbit is a circle.

The previous results hold for planar motion under a central force field, which requires a fixed direction of $\vec{L}$ in space. If $\vec{L} = 0$, $\vec{r}$ is parallel to $\dot{\vec{r}}$, and hence, the motion must be along a straight line going through the centre of force. Since there is no centrifugal barrier to hold the particles apart, we obtain a straight line motion on a collision course.

3.3. KEPLER'S PROBLEM

The determination of the orbit of a particle moving in an inverse-square field of force, given by Eq.(3.13), is known as *Kepler's problem*. It is convenient to make use of the angular momentum and energy integrals (3.9) and (3.11) which become

$$E = \frac{\mu}{2}\left(\dot r^2 + r^2 \dot\varphi^2\right) - \frac{C}{r}$$

$$L = \mu r^2 \dot\varphi \tag{3.15}$$

rather than the equations of motion (3.7) and (3.8). The time derivatives

$$\dot\varphi = \frac{L}{\mu r^2}, \qquad \dot r = \frac{dr}{d\varphi}\dot\varphi = \frac{dr}{d\varphi}\frac{L}{\mu r^2}$$

may be eliminated from the energy equation by means of the angular momentum equation, and therefore we have

$$\frac{\mu}{2}\left[\frac{L^2}{\mu^2 r^4}\left(\frac{dr}{d\varphi}\right)^2 + \frac{L^2}{\mu^2 r^2}\right] = E + \frac{C}{r}$$

which may be written as

$$\frac{1}{r^4}\left(\frac{dr}{d\varphi}\right)^2 = \frac{2\mu E}{L^2} + \frac{2\mu C}{L^2 r} - \frac{1}{r^2}$$

A procedure often used in the treatment of central fields is to make the substitution $\rho = 1/r$, so that

$$\frac{d\rho}{d\varphi} = -\frac{1}{r^2}\frac{dr}{d\varphi} \quad \text{and} \quad \left(\frac{d\rho}{d\varphi}\right)^2 = \frac{2\mu E}{L^2} + \frac{2\mu C}{L^2}\rho - \rho^2 \quad (3.16)$$

Upon differentiation we have

$$2\frac{d\rho}{d\varphi}\frac{d^2\rho}{d\varphi^2} = \frac{2\mu C}{L^2}\frac{d\rho}{d\varphi} - 2\rho\frac{d\rho}{d\varphi} \quad \text{i.e.,} \quad \frac{d\rho}{d\varphi}\left[\frac{d^2\rho}{d\varphi^2} + \rho - \frac{\mu C}{L^2}\right] = 0$$

and, for $d\rho/d\varphi \neq 0$, Eq.(3.16) reduces to the so-called differential equation of the *orbit*

$$\frac{d^2\rho}{d\varphi^2} + \rho = \frac{\mu C}{L^2} \quad (3.17)$$

which is similar to the equation of a forced oscillator with a constant forcing term. A particular solution which satisfies this inhomogeneous equation is $\mu C/L^2$, while the general solution of the reduced equation can be written as

$$\frac{d^2\rho}{d\varphi^2} + \rho = 0 \quad \text{i.e.,} \quad \rho = A\cos\varphi + B\sin\varphi$$

where A and B are arbitrary constants. The complete solution of Eq.(3.17) is then given by

$$\rho = \frac{\mu C}{L^2} + A\cos\varphi + B\sin\varphi$$

For convenience we may assume that $\varphi = 0$ where $r = r_{\min}$, that is where $\rho = \rho_{\max}$, which requires that

$$\frac{d\rho}{d\varphi} = -A\sin\varphi + B\cos\varphi = 0$$

This gives $B = 0$ when $\varphi = 0$, and the solution for the orbit simplifies to

$$\rho = \frac{\mu C}{L^2} + A\cos\varphi \quad \text{or} \quad \frac{1}{r} = \frac{\mu C}{L^2} + A\cos\varphi \tag{3.18}$$

and r takes the form

$$r = \frac{L^2/\mu C}{1 + (AL^2/\mu C)\cos\varphi} \tag{3.19}$$

In order to understand Eq.(3.19), let us compare it to the definition of a *conic section*, derived in Problem 3.4 and illustrated in Figure 3.4, which reads

$$r = \frac{p}{1 + e\cos\varphi} \tag{3.20}$$

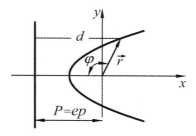

Figure 3.4. Parameters of a conic section

It is obvious that the orbit of the motion under a central force field is a conic, namely a *hyperbola* for $e > 1$, a *parabola* for $e = 1$, an *ellipse* for $e < 1$, or a circle for $e = 0$.

The geometrical parameters of a conic section, namely p and the so-called *eccentricity e*, can be expressed, by comparing Eqs.(3.19) and (3.20) as

$$p = \frac{L^2}{\mu C}, \quad e = \frac{AL^2}{\mu C} \tag{3.21}$$

As $\varphi = 0$ for $r = r_{\min}$, the results (3.21) give upon insertion into Eq.(3.19)

$$r_{\min} = \frac{L^2}{\mu C(1+e)} \tag{3.22}$$

It is convenient to evaluate the energy E, which is a constant of motion, for $r = r_{\min}$, where $\dot{r} = 0$, and its expression (3.15) reduces to

$$E = \frac{L^2}{2\mu r_{\min}^2} - \frac{C}{r_{\min}} = \frac{\mu C^2}{2L^2}\left(e^2 - 1\right) \tag{3.23}$$

so that the eccentricity e, which characterizes the shape of the orbit, can be written as

$$e = \sqrt{1 + \frac{2EL^2}{\mu C^2}}$$

This result allows us to express the solution of Kepler's problem, as given by the conic section (3.20), in terms of the constants of motion L and E

$$r = \frac{L^2/\mu C}{1 + \left(1 + 2EL^2/\mu C^2\right)^{1/2}\cos\varphi} \tag{3.24}$$

The classification of the conic sections according to the range of e corresponds to the previous qualitative discussion of the orbits based on the energy diagram. It follows from Eq.(3.23) that

$$\begin{array}{llll}
e > 1 & \textit{(hyperbola)} & E > 0 & \\
e = 1 & \textit{(parabola)} & E = 0 & \\
0 < e < 1 & \textit{(ellipse)} & E_{\min} < E < 0 & \\
e = 0 & \textit{(circle)} & E = E_{\min} = -\mu C^2/2L^2 &
\end{array} \tag{3.25}$$

3.4. Kepler's Laws

Kepler's three laws for planetary motion were set on an empirical basis and are restricted specifically to the inverse-square force field (3.13), although the second law holds for any central force motion. The derivation of Kepler's laws is an interesting application of the previous discussion about motion in a central force field.

Kepler's first law: *The planets move in elliptical orbits with the sun at one focus.*

This law can be regarded as a particular case of the solution (3.24) to Kepler's problem. The extremum condition for Eq.(3.16), $d\rho/d\varphi = 0$, yields the algebraic equation

$$\rho^2 - 2\frac{\mu C}{L^2}\rho - \frac{2\mu E}{L^2} = 0 \qquad (3.26)$$

with the obvious properties

$$\rho_{max} + \rho_{min} = \frac{2\mu C}{L^2}, \qquad \rho_{max}\rho_{min} = -\frac{2\mu E}{L^2} \qquad (3.27)$$

Since r and therefore ρ are always positive, Eq.(3.27) implies that E must be negative, which from Eq.(3.25) identifies the orbit as an ellipse, represented in Figure 3.5. It has two turning points at $r_{min} = \rho_{max}$, called *aphelion* in planetary motion, and $r_{max} = \rho_{min}$, called *perihelion,* where in both cases the radial component of the velocity is zero, $\dot{r} = 0$.

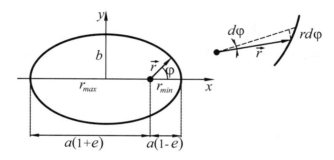

Figure 3.5. Ellipse of the planetary motion and the areal velocity

According to Eq.(3.20) we have

$$r_{max} = \frac{p}{1-e}, \qquad r_{min} = \frac{p}{1+e} \qquad (3.28)$$

so that the semimajor axis a, of the ellipse, which is one half the sum of the two *apsidal* distances r_{min} and r_{max}, is given by

$$a = \frac{1}{2}(r_{min} + r_{max}) = \frac{p}{2}\left(\frac{1}{1-e} + \frac{1}{1+e}\right) = \frac{p}{1-e^2} \qquad (3.29)$$

From Eqs.(3.27) and (3.28) it follows that

$$r_{min}r_{max} = \frac{p^2}{1-e^2} = -\frac{L^2}{2\mu E}$$

and, in view of Eq.(3.29), we can write

$$E = -\frac{L^2}{2\mu}\frac{1-e^2}{p^2} = -\frac{L^2}{2\mu a p}$$

Eqs.(3.27) and (3.28) also give an independent expression of p as

$$\rho_{\min} + \rho_{\max} = \frac{1-e}{p} + \frac{1+e}{p} = \frac{2}{p} = \frac{2\mu C}{L^2} \quad \text{or} \quad p = \frac{L^2}{\mu C} \tag{3.30}$$

We can now combine the last two equations to obtain

$$E = -\frac{C}{2a} \tag{3.31}$$

which is a result of considerable importance in the old quantum theory, as we shall show in Chapter 28, stating that the major axis of elliptic orbits depends solely on the energy. Orbits with the same major axis have the same energy, independent of L.

Kepler's second law: *The radius vector from the sun to a planet describes equal areas in equal times.*

This law is equivalent to the conservation of angular momentum (3.9) in a central force field, which can be written in the form

$$\frac{L}{2\mu} = \frac{1}{2}r^2\dot{\varphi} = \text{const}$$

As illustrated in Figure 3.5 the differential area swept out by the radius vector in time dt is to a first approximation given by

$$dS = \frac{1}{2}r(rd\varphi) \quad \text{or} \quad \frac{dS}{dt} = \frac{1}{2}r^2\dot{\varphi}$$

and is called the *areal velocity*. We then have a constant areal velocity

$$\frac{dS}{dt} = \frac{L}{2\mu} = \text{const} \tag{3.32}$$

for any central force motion.

Kepler's third law: *The square of the period of a planet is proportional to the cube of the semimajor axis of its orbit.*

The period T of a planet may be written as

$$T = \frac{S}{dS/dt} = \frac{\pi ab}{dS/dt} = \frac{\pi ab}{L/2\mu}$$

where b is the semiminor axis of the ellipse and πab is its area. Using the expressions of a and b in terms of e, derived in Problem 3.4, we obtain

$$T^2 = \frac{4\mu^2}{L^2}\pi^2 a^4 (1-e^2)$$

From Eqs.(3.29) and (3.30) it follows that:

$$L^2 = \mu C p = \mu C a (1-e^2)$$

and hence

$$T^2 = \frac{4\pi^2 \mu}{C} a^3 \tag{3.33}$$

which is the mathematical statement of Kepler's third law.

FURTHER READING
1. J. R. Taylor - CLASSICAL MECHANICS, University Science Books, Sausalito, 2004.
2. A. P. Arya - INTRODUCTION TO CLASSICAL MECHANICS, Pearson Education, Boston, 1997.
3. T. W. B. Kibble - CLASSICAL MECHANICS, Addison-Wesley, San Francisco, 1997.

PROBLEMS

3.1. Show that the equation of motion for a rigid body of mass M rotating about a fixed axis can be written as

$$I\frac{d\vec{\omega}}{dt} = (I_0 + Ma^2)\frac{d\vec{\omega}}{dt} = \vec{G}_e$$

where $\vec{\omega}$ is the angular velocity, and I stands for the moment of inertia, given by $I = \sum_i m_i r_i^2$. If one assumes the body to be made up of point masses m_i, at perpendicular distances r_i from the axis of rotation, I_0 represents the moment of inertia with respect to a parallel axis passing through the center of mass, and a is the distance between the two axes.

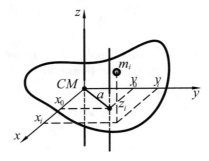

Solution

In view of Eqs.(2.3) and (2.4), the angular momentum of the body with respect to the given axis, Eq.(3.2), can be written in the form

$$\vec{L} = \sum_i m_i \left[\vec{r}_i \times \left(\dot{r}_i \vec{e}_{r_i} + r_i \dot{\vec{e}}_{r_i}\right)\right] = \sum_i m_i r_i^2 \left(\vec{e}_{r_i} \times \dot{\vec{e}}_{r_i}\right) = \dot{\varphi} \sum_i m_i r_i^2 \left(\vec{e}_{r_i} \times \vec{e}_\varphi\right) = \dot{\varphi} \vec{e}_\omega \sum_i m_i r_i^2 = I\vec{\omega}$$

where the vector $\vec{\omega} = \dot{\varphi} \vec{e}_\omega$ is always perpendicular to the plane of motion. Taking the derivative of this angular momentum, which must be equal to the total external torque about the same axis, one obtains

$$I \frac{d\vec{\omega}}{dt} = \vec{G}_e$$

as required. We may set up a Cartesian coordinate system in which the z-axis is parallel to the given axis of rotation, and the origin is located at the centre of mass, so that Eq.(2.17) reads

$$\sum_i m_i x_i = \sum_i m_i y_i = \sum_i m_i z_i = 0$$

Let the axis of rotation pass through a point (x_0, y_0), in the xy-plane through the centre of mass. Hence, we have $a^2 = x_0^2 + y_0^2$. The moment of inertia with respect to the z-axis is equal to

$$I_0 = \sum_i m_i \left(x_i^2 + y_i^2 + z_i^2\right)$$

whereas the moment of inertia with the respect to the axis of rotation reads

$$I = \sum_i m_i \left[(x_i - x_0)^2 + (y_i - y_0)^2 + z_i^2\right]$$

$$= \left(x_0^2 + y_0^2\right)\sum_i m_i + \sum_i m_i \left(x_i^2 + y_i^2 + z_i^2\right) - 2x_0 \sum_i m_i x_i - 2y_0 \sum_i m_i y_i = Ma^2 + I_0$$

This result is known as the *Steiner parallel axes theorem*.

3.2. Find an expression for the moment of inertia of a homogeneous solid of revolution, whose surface is formed by the rotation of a plane curve $R(x)$ about the x-axis, given the density ρ of the material of the body. Calculate the moments of inertia of a cylinder and of a ball, of known mass and radius, with respect to their geometrical axes.

Solution

Let the solid be divided into thin disks of radius $R(x)$, thickness dx and mass $dM = \rho \pi R^2(x) dx$. By further dividing the disks into concentric layers of width dr, all the particles of a layer are at the same distance r from the axis, so that the moment of inertia of such a layer is $dI = r^2 dm = r^2 \rho (2\pi r dr) dx$.

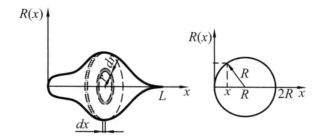

The moment of inertia of the disk can be found by adding together these infinitesimal moments of inertia, and this gives

$$I(x) = \int_0^{R(x)} r^2 dm = 2\pi \rho dx \int_0^{R(x)} r^3 dr = \frac{\pi}{2} \rho dx R^4(x) = \frac{1}{2} R^2(x) dM$$

and hence, the required expression for the moment of inertia of the whole solid of revolution is

$$I = \frac{1}{2} \int_0^M R^2(x) dM = \frac{\pi}{2} \rho \int_0^L R^4(x) dx$$

For a solid cylinder of radius R and mass $M = \pi R^2 L \rho$, we can readily compute the moment of inertia by taking $R(x)=R$, which yields

$$I = \frac{\pi}{2} \rho R^4 L = \frac{1}{2} MR^2$$

For a ball of radius R and mass $M = 4\pi R^3 \rho / 3$, we have

$$R(x) = R^2 - (R-x)^2 = x(x - 2R)$$

By substituting this function, one obtains the moment of inertia of a uniform ball as

$$I = \frac{\pi}{2} \rho \int_0^{2R} x^2 (x - 2R)^2 dx = \frac{8\pi}{15} \rho R^5 = \frac{2}{5} MR^2$$

3.3. A solid spool, which can be considered to be a solid uniform cylinder of radius R and mass M, rolls on a horizontal surface and is connected to a load m by an inextensible thread passing over a pulley, both of negligibly small mass. Find the accelerations of the bodies, assuming a coefficient of friction μ between spool and surface.

Solution

The equation for the load is written in the form (2.10) as

$$ma = mg - F \qquad (i)$$

where F denotes the magnitude of the force of tension in the thread that can only pull the bodies it is attached to. The rolling motion of the spool can be represented as a rotation about an instantaneous axis which is at rest at any given instant. In the case of rolling without sliding, this axis passes through the point of contact C between the spool and the surface. During time dt the spool turns about the instantaneous axis with angular velocity $\vec{\omega}$, so that the velocities of all the other points of the body can be written as $\vec{v} = \vec{\omega} \times \vec{r}$. It follows that the velocity of the load is $v = 2\omega R$, and its acceleration is $a = 2R d\omega / dt = 2R\alpha$.

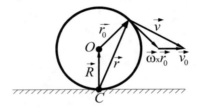

The angular acceleration $\alpha = d\omega / dt$ can be specified, as shown in Problem 3.1, in terms of the torque of the force of tension and the moment of inertia of the cylinder with respect to the given axis, in the form

$$I \frac{d\omega}{dt} = I\alpha = 2FR \qquad \text{or} \qquad \frac{I}{R^2} a = 4F \qquad (ii)$$

On solving the simultaneous equations (i) and (ii), we have

$$a = g \frac{4m}{4m + I/R^2} = g \frac{8m}{8m + 3M} \quad , \quad F = mg \frac{3M}{8m + 3M}$$

where use has been made of the moment of inertia of a cylinder with respect to an axis passing through its generatrix, which reads

$$I = I_0 + MR^2 = \frac{1}{2} MR^2 + MR^2 = \frac{3}{2} MR^2$$

The axis of the spool moves with velocity $v_0 = \omega R$, since the displacement $v_0 dt$ of the axis is equal to the length $\omega R dt$ of the corresponding circular arc traveled by the point of contact between spool and surface during the same time interval. It follows that the acceleration of the translation motion is $a_0 = \alpha R = a/2$ and it is determined by the equation

$$Ma_0 = F + f$$

where f is the adhesive force appearing in a rolling motion without sliding. Like the force of friction at rest, the adhesive force is specified by the magnitude of the external forces

$$f = Ma_0 - F = Mg\frac{m}{8m+3M}$$

and has a maximum value μN, where N is the force of normal pressure exerted by the cylinder on the surface. The spool rolls without sliding when

$$f < \mu Mg \qquad \text{or} \qquad \mu > \frac{m}{8m+3M}$$

For lower μ values, the spool rolls with sliding, that is the point of contact slides with some velocity, and this implies that the instantaneous axis of rotation passes through a point lying either below the surface (if $v_0 > \omega R$) or above it (when $v_0 < \omega R$).

There is an alternative description of the rolling motion, where we let the radius vector of a given point of the spool, issued from the instantaneous axis perpendicular to it, be $\vec{r} = \vec{R} + \vec{r}_0$, where $\vec{r}_0$ is the corresponding radius vector issued from the axis of symmetry. This gives

$$\vec{v} = \vec{\omega} \times (\vec{R} + \vec{r}_0) = \vec{v}_0 + \vec{\omega} \times \vec{r}_0$$

and the rolling motion is described as a contribution from the translation, with velocity $\vec{v}_0$, of the axis of the spool passing through its centre of mass, and the rotation with angular velocity $\vec{\omega}$ about this axis.

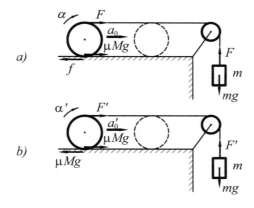

The acceleration a_0 of the centre of mass, Figure (a) is given by

$$Ma_0 = F + f \qquad \text{(iii)}$$

and the angular acceleration is determined in terms of the moment of inertia $I_0 = MR^2/2$ of the cylinder with respect with its central axis as

$$I_0\frac{d\omega}{dt} = I_0\alpha = (F - f)R \qquad \text{(iv)}$$

When the spool rolls without sliding, the angular and linear acceleration are connected by

$$a_0 = R\frac{d\omega}{dt} = R\alpha$$

and the acceleration of the load is

$$a = a_0 + R\alpha = 2a_0$$

On solving the simultaneous equations (i), (iii) and (iv), one obtains

$$a_0 = g\frac{2m}{4m + M + I_0/R^2} = g\frac{4m}{8m + 3M}$$

and then the same expressions for F, f and a as before. The rolling motion with sliding can be described in like manner, Figure (b), substituting f by μMg, and hence, we have

$$ma' = mg - F \quad , \quad Ma'_0 = F - \mu Mg \quad , \quad I_0\alpha' = (F - \mu Mg)R \quad , \quad a' = a'_0 + R\alpha'$$

In this case, it follows that

$$a' = g\frac{3m - \mu M}{3m + M} \quad , \quad a'_0 = g\frac{m + \mu(4m + M)}{3m + M} \quad , \quad F = mg\frac{M(1+\mu)}{3m + M}$$

3.4. Derive the equations of the conic sections, defined by Eq.(3.20), for various values of e, in Cartesian coordinates.

Solution

A conic is the locus of a point that moves so that the ratio of its distance r from a fixed point, called the focus, to its distance d from a fixed line, called the directrix, is a constant. This ratio is called the eccentricity of the conic

$$e = \frac{r}{d} = \text{const}$$

Such a curve, represented in Figure 3.4, can be obtained by taking a plane section through a cone. We can set a parameter $p = eP$ where P, according to Figure 3.4, is given by

$$P = d + r\cos\varphi = \frac{p}{e}$$

so that we obtain the general equation (3.20) of a conic with one focus at the origin, that is

$$r = \frac{p}{1 + e\cos\varphi}$$

The nature of the conic section depends on the magnitude of e. The curves which are generated for various values of e are shown in the figure. Their equations look more familiar in Cartesian coordinates, given by

$$x = r\cos\varphi \quad , \quad y = r\sin\varphi \quad , \quad r = \sqrt{x^2 + y^2}$$

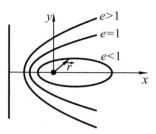

Thus, Eq.(3.20) can be rewritten as

$$p = r + ex \quad \text{or} \quad \sqrt{x^2 + y^2} = p - ex$$

and hence

$$(1-e^2)x^2 + 2pex + y^2 = p^2$$

For $e = 1$, the equation becomes

$$y^2 = -2px + p^2$$

which is the equation of a *parabola*, obtained by cutting a cone parallel to the slant. If $e \neq 1$, the equation can be rewritten as

$$\frac{[x + pe/(1-e^2)]^2}{[p/(1-e^2)]^2} + \frac{y^2}{p^2/(1-e^2)} = 1$$

When $e < 1$, we may set

$$a = \frac{p}{1-e^2} \quad , \quad b = \frac{p}{\sqrt{1-e^2}} \quad \text{or} \quad b = a\sqrt{1-e^2}$$

and the last equation becomes

$$\frac{(x+ea)^2}{a^2} + \frac{y^2}{b^2} = 1$$

which is the equation of an *ellipse*. The linear term in x means that the geometric centre of the ellipse is not at the origin. When $e = 0$ the equation reduces to $x^2 + y^2 = p^2$, and the ellipse degenerates to a circle. When $e > 1$, it is appropriate to set

$$A = \frac{p}{e^2 - 1} \quad , \quad B = \frac{p}{\sqrt{e^2 - 1}} \quad \text{or} \quad B = A\sqrt{e^2 - 1}$$

and we get the equation of a *hyperbola* given by

$$\frac{(x - eA)^2}{A^2} - \frac{y^2}{B^2} = 1$$

3.5. Show that the attractive inverse-square law of force, Eq.(3.13), can be derived from Kepler's laws for planetary motion.

Solution

The law of constant areal velocity shows that there is a central force interaction between the planet and the sun. The determination of the orbit under a central force of magnitude $f(r)$ involves the integration of the two equations of motion (3.7) and (3.8). We have already shown that Eq.(3.8) is equivalent to Eq.(3.9) stating the conservation of angular momentum, so that $\dot{\varphi} = L/\mu r^2$, and it follows that φ can be eliminated from Eq.(3.7) which becomes

$$\ddot{r} - \frac{L^2}{\mu^2 r^3} = \frac{f(r)}{\mu}$$

Using then

$$\dot{r} = \frac{dr}{d\varphi}\dot{\varphi} = \frac{dr}{d\varphi}\frac{L}{\mu r^2} \quad \text{i.e.,} \quad \ddot{r} = \frac{L^2}{\mu^2 r^4}\left[\frac{d^2r}{d\varphi^2} - \frac{2}{r}\left(\frac{dr}{d\varphi}\right)^2\right]$$

the radial equation takes the form

$$\frac{d^2r}{d\varphi^2} - \frac{2}{r}\left(\frac{dr}{d\varphi}\right)^2 - r = \frac{\mu f(r) r^4}{L^2}$$

It is convenient to use again the substitution $\rho = 1/r$, so that

$$\frac{dr}{d\varphi} = -\frac{1}{\rho^2}\frac{d\rho}{d\varphi} \quad \text{or} \quad \frac{d^2r}{d\varphi^2} = \frac{2}{\rho^3}\left(\frac{d\rho}{d\varphi}\right)^2 - \frac{1}{\rho^2}\frac{d^2\rho}{d\varphi^2}$$

and we obtain

$$\frac{d^2\rho}{d\varphi^2} + \rho = -\frac{\mu f(r) r^2}{L^2}$$

From Kepler's second law (3.20) we can express ρ as

$$\rho = \frac{1}{p} + \frac{e}{p}\cos\varphi \quad \text{or} \quad \frac{d^2\rho}{d\varphi^2} + \rho = \frac{1}{p}$$

so that the force can be written as

$$f(r) = -\frac{L^2}{\mu p} \cdot \frac{1}{r^2} = -\frac{C}{r^2}$$

This describes an attractive interaction, following an inverse-square law of force, Eq.(3.13).

3.6. Find the relationship between the impact parameter δ and the scattering angle θ for a particle of mass m and incident velocity v_0 subject to a central force from a stationary mass $M \gg m$. Determine the distance of closest approach between the particle and the mass M.

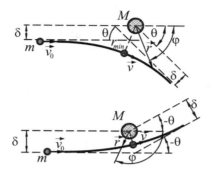

Solution

The orbit of a particle moving in an attractive inverse-square field of force $U(r) = -C/r$ is described by the solutions of the differential equation (3.17), which reads

$$\frac{d^2\rho}{d\varphi^2} + \rho = \frac{mC}{L^2}$$

since the reduced mass $\mu \to m$, when $M \gg m$. The general solution

$$\rho = \frac{1}{r} = \frac{mC}{L^2} + A\cos\varphi + B\sin\varphi$$

is specified by the initial conditions, namely $\varphi = \pi$ at $r \to \infty$, and hence, $A = mC/L^2$. The solution can be written as

$$\frac{1}{r\sin\varphi} = \frac{mC}{L^2}\frac{1+\cos\varphi}{\sin\varphi} + B$$

where $r\sin\varphi \to \delta$ when $\varphi = \pi$, and this gives $B = 1/\delta$. Thus the orbit of the particle follows the path

$$\rho = \frac{1}{r} = \frac{mC}{L^2}(1+\cos\varphi) + \frac{1}{\delta}\sin\varphi$$

where the scattering angle $\varphi = -\theta$ also corresponds to $r \to \infty$, so that

$$\frac{L^2}{mC} = \frac{1+\cos\theta}{\sin\theta} = \cot\frac{\theta}{2}$$

Since $L = mv_0\delta$, one finally obtains

$$\delta \tan\frac{\theta}{2} = \frac{C}{mv_0^2}$$

Substituting $C = GmM$ for the gravitational force, this relationship describes the scattering of a celestial object that is tracked in a hyperbolic orbit about a planet of mass M, where

$$\delta \tan\frac{\theta}{2} = \frac{GM}{v_0^2}$$

Let v be the speed of the particle at its closest distance r_{min} ($\dot{r}=0$) from mass M, so that $mv_0\delta = mvr_{min}$. On substitution into the energy conservation law

$$\frac{mv_0^2}{2} = \frac{(mvr_{min})^2}{2mr_{min}^2} - \frac{C}{r_{min}}$$

we get

$$\frac{\delta^2 - r_{min}^2}{2r_{min}} = \frac{C}{mv_0^2} = \delta \tan\frac{\theta}{2} \quad \text{or} \quad r_{min} = \delta\left(\sqrt{1 + \tan^2\frac{\theta}{2}} - \tan\frac{\theta}{2}\right)$$

In the case where the force between m and M is one of repulsion, that is $U(r) = +C/r$, the orbit is described by

$$\rho = \frac{1}{r} = -\frac{mC}{L^2}(1 + \cos\varphi) + \frac{1}{\delta}\sin\varphi$$

where C is changed to $-C$. However, because to $r \to \infty$ corresponds in this case a scattering angle $\varphi = 0$, the relationship between δ and θ keeps the same form. For instance, when positively charged particles $+ze$ approach a fixed positive charge $+Ze$ of a nucleus (thus $C = zZe_0^2$ for the electrostatic force), they are subjected to the so-called Rutherford scattering in accordance with

$$\delta \tan\frac{\theta}{2} = \frac{zZe_0^2}{mv_0^2} = \frac{r_{min}^2 - \delta^2}{2r_{min}}$$

and thus, the closest distance between the two charges is given by

$$r_{min} = \delta\left(\sqrt{1 + \tan^2\frac{\theta}{2}} + \tan\frac{\theta}{2}\right)$$

3.7. A spacecraft is in a circular orbit of radius $r = 2R_0$ about a planet of radius R_0 and mass M_0. Given the acceleration of gravity g_0 at the surface of the planet, calculate the relative velocity with which a projectile would need to be launched tangentially from the spacecraft, in order to reach the opposite point of the surface of the planet. What is its time of flight?

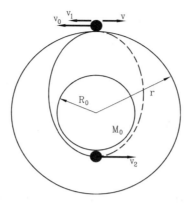

Solution

The trajectory of the projectile should be an ellipse, the centre of the planet being at the focus of the ellipse which is nearer to the landing point, that is the major axis is $r + R_0$. The energy conservation law between the two turning points, where $\dot{r} = 0$, reads

$$\frac{mv_1^2}{2} - G\frac{mM_0}{r} = \frac{mv_2^2}{2} - G\frac{mM_0}{R_0}$$

The mass of the planet M_0 is related to the acceleration of gravity at its surface by

$$G\frac{mM_0}{R_0^2} = mg_0 \quad \text{or} \quad GM_0 = g_0 R_0^2$$

thus

$$\frac{v_1^2}{2} - g_0\frac{R_0}{r} = \frac{v_2^2}{2} - g_0 R_0$$

In order to conserve the angular momentum, Eq.(3.32), we must have $v_1 r = v_2 R_0$, and hence, by eliminating v_2, we have

$$v_1 = R_0\sqrt{\frac{2g_0 R_0}{r(R_0 + r)}} = \sqrt{\frac{g_0 R_0}{3}}$$

using $r = 2R_0$. Since the velocity of motion of the spacecraft in a circular orbit of radius r is given by the condition that the centripetal acceleration must be equal to the acceleration of gravity

$$\frac{v^2}{r} = G\frac{M}{r^2} = g_0\frac{R_0^2}{r^2} \quad \text{or} \quad v = \sqrt{\frac{g_0 R_0}{2}}$$

it follows that the projectile should be launched with a relative velocity given by

$$v_0 = v - v_1 = \sqrt{g_0 R_0}\left(\frac{1}{\sqrt{2}} - \frac{1}{\sqrt{3}}\right)$$

By comparing the period of motion, Eq.(3.33), along the elliptical orbit with major axis of $r+R_0$ with that along the circular orbit with diameter of $4R_0$, we get

$$\left(\frac{T}{T_0}\right)^2 = \left(\frac{r+R_0}{4R_0}\right)^3 \quad \text{or} \quad T = 3\pi\sqrt{\frac{3R_0}{2g_0}}$$

because $T_0 = 4\pi R_0 / v = 4\pi\sqrt{2R_0/g_0}$. The time of flight of the projectile is half its period of revolution along the elliptical trajectory.

3.8. Show that the kinetic energy of a rigid body M in a plane motion is the sum of the kinetic energy of the translation motion with velocity v_0 and the energy of rotational motion with angular velocity ω, that is

$$T = \frac{1}{2}Mv_0^2 + \frac{1}{2}I_0\omega^2$$

where I_0 stands for the moment of inertia of the body with respect to the axis passing through the centre of mass.

3.9. A load m is connected by a flexible, inextensible cord with a spool of equal mass m, radius R and moment of inertia I, over a pulley of negligible mass. When the system is released, what is the acceleration of the centre of mass of the spool? What will be the acceleration in the case where the load is disconnected and one pulls down the cord with a force equal to its weight mg?

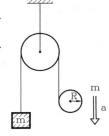

3.10. A hollow cylinder of moment of inertia $I = MR^2$ rolls without sliding down a plane which slopes at an angle α to the horizontal. What is the angular position β, with respect to a direction normal to the plane, of a small mass $m=M/2$ that can move freely inside the cylinder, during its rolling motion.

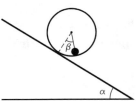

3.11. Find expressions for the first and the second cosmic velocities, defined as the velocity of motion v_{cr} of a satellite of the Earth in a circular orbit of radius R, equal to that of the Earth and, respectively, the velocity v_{par} of a spacecraft that, moving along a parabolic trajectory about the Earth, will never return to the Earth again.

3.12. Determine the solar escape velocity v_{sev}, that should be given to a spacecraft launched from the Earth for it to leave the Solar system, in terms of the Earth's velocity v_0 relative to the Sun and the parabolic velocity v_{par}.

4. PRINCIPLES OF SPECIAL RELATIVITY

4.1. POSTULATES OF SPECIAL RELATIVITY
4.2. THE LORENTZ TRANSFORMATION
4.3. SPACETIME DIAGRAMS
4.4. RELATIVISTIC INVARIANCE

4.1. POSTULATES OF SPECIAL RELATIVITY

The concepts of absolute space and absolute time must be regarded merely as postulates of Newtonian mechanics, where the properties of length were related with the properties of metre sticks and the properties of time intervals were connected with those of clocks. These postulates were determined by the motion of physical systems. Newtonian mechanics follows from these postulates and of course different laws of motion could follow from other different primary assumptions. The revision of the Newtonian concept of space and time was inspired by attempts to extend the validity of Galileo's principle of relativity to electromagnetic phenomena.

A simple representation of the propagation of an electromagnetic wave is obtained using a one-dimensional wave function

$$\Psi(x,t) = f(x \pm ct)$$

where the function $f(x)$ describes the form of the wave travelling along the x-axis with the speed of light c. Setting $\xi = x \pm ct$ we have

$$\frac{\partial \Psi}{\partial x} = \frac{\partial \Psi}{\partial \xi} \quad \text{i.e.,} \quad \frac{\partial^2 \Psi}{\partial x^2} = \frac{\partial^2 \Psi}{\partial \xi^2}$$

$$\frac{\partial \Psi}{\partial t} = \pm c \frac{\partial \Psi}{\partial \xi} \quad \text{i.e.,} \quad \frac{\partial^2 \Psi}{\partial t^2} = c^2 \frac{\partial^2 \Psi}{\partial \xi^2}$$

and the wave equation becomes

$$\frac{\partial^2 \Psi}{\partial x^2} = \frac{1}{c^2}\frac{\partial^2 \Psi}{\partial t^2} \qquad (4.1)$$

The harmonic trial solution of this equation with respect to a fixed inertial frame (x,t) is given by

$$\Psi(x,t) = \Psi_0 \cos k(x-ct) \quad \text{or} \quad \Psi = \Psi_0 e^{ik(x-ct)}$$

In another inertial frame (x',t'), considered to be in uniform translational motion with velocity v, as described by the Galilean transformation (2.13), with respect to the fixed frame, the wave is represented by a wave function $\Psi(x',t')$ related to $\Psi(x,t)$ as

$$\frac{\partial \Psi}{\partial x} = \frac{\partial \Psi}{\partial x'}\frac{\partial x'}{\partial x} + \frac{\partial \Psi}{\partial t'}\frac{\partial t'}{\partial x} = \frac{\partial \Psi}{\partial x'} \quad \text{i.e.,} \quad \frac{\partial^2 \Psi}{\partial x^2} = \frac{\partial^2 \Psi}{\partial x'^2}$$

and

$$\frac{\partial \Psi}{\partial t} = \frac{\partial \Psi}{\partial x'}\frac{\partial x'}{\partial t} + \frac{\partial \Psi}{\partial t'}\frac{\partial t'}{\partial t} = -v\frac{\partial \Psi}{\partial x'} + \frac{\partial \Psi}{\partial t'} \quad \text{i.e.,} \quad \frac{\partial^2 \Psi}{\partial t^2} = v^2\frac{\partial^2 \Psi}{\partial x'^2} - 2v\frac{\partial^2 \Psi}{\partial x'\partial t'} + \frac{\partial^2 \Psi}{\partial t'^2}$$

where use has been made of known properties of partial derivatives. It follows that $\Psi(x',t')$ obeys the following wave equation, with respect to the moving inertial frame

$$(c^2 - v^2)\frac{\partial^2 \Psi}{\partial x'^2} = \frac{\partial^2 \Psi}{\partial t'^2} - 2v\frac{\partial^2 \Psi}{\partial x'\partial t'} \qquad (4.2)$$

which is obviously different from the wave equation (4.1) that holds in the fixed inertial frame. In other words the electromagnetic wave equation is not invariant with respect to a uniform translation of the inertial frame, using a Galilean transformation. Inserting into Eq.(4.2) harmonic solution written as

$$\Psi = \Psi_0 e^{ik(x'-wt')}$$

we obtain

$$(c^2-v^2)(-k^2) = -k^2 w^2 - 2v(k^2 w) \quad \text{i.e.,} \quad w = -v \pm c \quad \text{or} \quad |w| = c \pm v \qquad (4.3)$$

Hence Galileo's transformation predicts that the velocity of light should be different from c, if measured with respect to a moving inertial frame. We therefore have a choice to make before we can fully understand the electromagnetic phenomena. We can either accept Galileo's principle of relativity which states that all inertial frames are equivalent or the Galilean transformation (2.13). We can sacrifice Galileo's principle of relativity by saying that there is a privileged inertial frame, called the *ether*, in which it is supposed that the electromagnetic wave velocity is c as given by Maxwell's equations, although they made no essential reference to the ether. A long series of investigations has been carried out, in order to detect some effect of the motion of the earth through the ether, always with negative results. All of the experiments (the best known being that of Michelson-Morley) have indicated that the speed of light is always the same in all directions and is independent of the relative motion of observer, source and transmitting medium. Einstein has reaffirmed Galileo's principle of relativity on the equivalence of all the inertial frames and, in addition, has assumed as a basic postulate the experimental fact that the speed of light is always constant. As a consequence, the Galilean transformation had to be replaced by another, in order to preserve the same speed of light in all systems. Einstein's basic assumptions are known as the **postulates of special relativity**:

1. *The laws of physics take the same form in all inertial systems*
2. *The speed of light in empty space is a universal constant c and is the same for all observers. It is independent of the state of motion of the emitting body*

Special relativity is only concerned with transformations between inertial frames, as indicated by the restriction contained in the first postulate. *General relativity* covers reference frames in arbitrary relative motion, and is based on the principle of equivalence, confirmed experimentally, which unifies gravitational and inertial forces.

The first of Einstein's postulates implies that it is only possible, by means of physical measurements, to demonstrate that two gravity-free reference frames are moving *relative* to each other. There is no physical basis to infer that any such frame is intrinsically at rest or in uniform motion. The notion of absolute rest is not consistent with the first postulate. The second postulate is in contradiction with the Galilean transformation which would preserve the form of Newton's laws of motion, as illustrated by Eq.(4.3). Therefore, we must find a transformation between two frames moving with uniform relative motion, which is consistent with the second postulate. We must then generalize the laws of Newtonian mechanics in a form which is invariant under that transformation, as required by the first postulate. A fundamental requirement of special relativity is that the laws of Newtonian mechanics are valid in the low speed limit and must be obtained as limiting cases of relativistic formulae for $v \ll c$. This assumption plays, for special relativity, a role analogous to that of Bohr's correspondence principle in quantum mechanics (see Chapter 28).

4.2. THE LORENTZ TRANSFORMATION

Einstein's procedure is to locate an event in a reference frame by its spatial Cartesian coordinates x, y and z and a temporal coordinate t. Consider two frames S and S' moving with uniform relative motion, such that the Cartesian axis system in S', $O'x'y'z'$ coincides with that in S, $Oxyz$, at time $t' = t = 0$. As illustrated in Figure 4.1, the

x and x' axes are parallel to the direction of relative motion: in S, the frame S' moves in the x direction with velocity v, and in S', the frame S moves with velocity $-v$ in the x' direction. A pulse of light emitted from O at time $t=0$ spreads as a spherical wave travelling with speed c, so that the equation of the wave front in S, at a time t will be

$$x^2 + y^2 + z^2 = c^2 t^2 \tag{4.4}$$

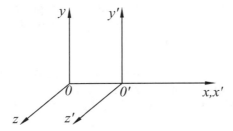

Figure 4.1. Cartesian axes of two inertial frames in uniform relative motion

The second postulate requires that the wave front in S' can be represented by

$$x'^2 + y'^2 + z'^2 = c^2 t'^2 \tag{4.5}$$

In other words an observer in S' also sees a spherical wave travelling from its origin O', if not, any change of the wave front could be used to infer that the system S' is moving uniformly instead of being at rest. The invariance of the expression

$$x^2 + y^2 + z^2 - c^2 t^2 = x'^2 + y'^2 + z'^2 - c^2 t'^2 \tag{4.6}$$

is a basic requirement that the transformation between the two systems must satisfy. The Galilean transformation (2.13), which in our case reads

$$x' = x - vt, \quad y' = y, \quad z' = z, \quad t' = t$$

is not consistent with the invariance property (4.6), as it gives

$$x'^2 + y'^2 + z'^2 - c^2 t'^2 = x^2 + y^2 + z^2 - c^2 t^2 + v^2 t^2 - 2xvt$$

The first postulate brings some simplification to the possible form of the desired transformation. Since a uniform rectilinear motion in S must go over into a uniform rectilinear motion in S', the transformation must be *linear*. A nonlinear transformation would predict acceleration in S' even if the velocity were constant in S. The choice of the origins in space and time has avoided constant terms, and the transformation will be *homogenous*. The directions perpendicular to the relative motion, which are effectively at rest, must be left unchanged by the transformation. We then take the general form

4. PRINCIPLES OF SPECIAL RELATIVITY

$$x' = Ax + Bt$$
$$y' = y$$
$$z' = z \qquad (4.7)$$
$$t' = Dx + Ft$$

Further simplification comes from the argument that the origin O', which has the coordinate $x' = 0$, moves along the x axis with velocity $dx/dt = v$, that is

$$0 = Ax + Bt \quad \text{or} \quad v = -B/A$$

while the origin O, with coordinate $x = 0$, moves along x-axis with velocity $dx'/dt' = -v$. Recalling that $x = 0$ when $t = 0$, we can write

$$x' = A \cdot 0 + B\frac{t' - D \cdot 0}{F} = \frac{B}{F}t' \quad \text{or} \quad -v = \frac{B}{F}$$

so that $A = F$. Inserting now Eqs.(4.7) into Eq.(4.5), we obtain the equation

$$x'^2 + y'^2 + z'^2 - c^2 t'^2 = \left(A^2 - c^2 D^2\right)x^2 + y^2 + z^2 - c^2 t^2 \left(A^2 - \frac{B^2}{c^2}\right) + 2Axt\left(B - c^2 D\right) = 0$$

The identity (4.6) requires that

$$A^2 - c^2 D^2 = 1, \qquad A^2 - \frac{B^2}{c^2} = 1, \qquad B - c^2 D = 0$$

Since $B = vA$, we obtain

$$A = \frac{1}{\sqrt{1 - v^2/c^2}}, \qquad B = \frac{-v}{\sqrt{1 - v^2/c^2}}, \qquad D = \frac{-v/c^2}{\sqrt{1 - v^2/c^2}} \qquad (4.8)$$

so that Eqs.(4.7) now read

$$x' = \frac{x - vt}{\sqrt{1 - v^2/c^2}}$$
$$y' = y$$
$$z' = z \qquad (4.9)$$
$$t' = \frac{t - vx/c^2}{\sqrt{1 - v^2/c^2}}$$

The four equations (4.9) are known as the *Lorentz transformation*, which in this case applies to the relative motion of inertial frames along the x-axis. It is a straightforward matter to show that Eqs.(4.9) are consistent with the invariance relation (4.6), which is thus a sufficient as well as a necessary condition for the Lorentz transformation. With the usual abbreviations

$$\beta = \frac{v}{c} \quad , \quad \gamma = \frac{1}{\sqrt{1-\beta^2}} = \frac{1}{\sqrt{1-v^2/c^2}} \qquad (\gamma \geq 1) \qquad (4.10)$$

the Lorentz transformation simplifies to

$$x' = \gamma(x - \beta ct) \qquad x = \gamma(x' + \beta ct') \qquad \text{a)}$$

$$y' = y \qquad \text{or} \qquad y = y' \qquad \text{b)} \qquad (4.11)$$

$$z' = z \qquad z = z' \qquad \text{c)}$$

$$t' = \gamma\left(t - \frac{\beta}{c}x\right) \qquad t = \gamma\left(t' + \frac{\beta}{c}x'\right) \qquad \text{d)}$$

The right hand side shows the inverse transformations where the primed and unprimed quantities are interchanged and the sign of the relative velocity is reversed, because its direction is the only difference between the systems S and S'. In the low velocity limit $vc \ll 1$, the Lorentz transformation reduces to the Galilean transformation since the only difference consists in the presence of the factor γ and the term $-(\beta/c)x$ in d). It is clear that, in Eq.(4.11a), γ indicates a change of measuring stick length and in Eq.(4.11d) a change in the clock rate in the Lorentz transformation. The term $-(\beta/c)x$ in Eq.(4.11d) shows that the time is dependent on the position of a given event. Therefore, the immediate consequences of the Lorentz transformation require the revision of the usual concepts of simultaneity, time and length.

Simultaneity

Consider two events occuring at the same time $t_1 = t_2 = t$ at two different positions x_1 and x_2 in the unprimed system S. According to Eqs.(4.11) the events are recorded in the primed system S' at different times

$$t_1' = \frac{t - \beta x_1/c}{\sqrt{1-\beta^2}} \quad , \quad t_2' = \frac{t - \beta x_2/c}{\sqrt{1-\beta^2}}$$

so that the apparent time interval in S' is given by

$$\Delta t' = t_2' - t_1' = \frac{\beta(x_1 - x_2)/c}{\sqrt{1-\beta^2}} \neq 0 \qquad (4.12)$$

and hence, it is a function of the relative velocity v. The events will not appear as simultaneous to observers in S'. We are then forced to abandon the intuitive idea of absolute simultaneity, which is based on our experience with objects of low velocity, and accept the relativity of simultaneity, which depends on the coordinate system and applies in the domain of high velocities.

Length contraction

Consider a rod at rest in the primed system S' lying along the x-axis and having the length $\lambda = x'_2 - x'_1$ given by the distance between its ends x'_1 and x'_2 measured at the same instant of time t'. The length in an inertial frame in which the rod is at rest is called its *proper length*. To define length in a frame S in which the rod is moving, an observer must locate the position of both end points, x'_1 and x'_2, at the same time $t_1 = t_2 = t$. From the inverse equations (4.11) we obtain

$$x'_1 = \frac{x_1 - \beta ct}{\sqrt{1-\beta^2}} \quad , \quad x'_2 = \frac{x_2 - \beta ct}{\sqrt{1-\beta^2}} \quad \text{or} \quad x'_2 - x'_1 = \frac{x_2 - x_1}{\sqrt{1-\beta^2}}$$

so that the apparent length $l = x_2 - x_1$ is

$$l = \lambda\sqrt{1-\beta^2} \qquad (4.13)$$

This equation is known as the *Lorentz contraction* and states that a rod parallel to the direction of its motion is shortened by the factor $\sqrt{1-\beta^2}$. It results from the fact that simultaneous events in the rest frame of one rod are not simultaneous in the rest frame of the other. A rod transverse to its direction of motion is unchanged in length. The *proper length* λ is independent of the orientation of the rod, as a manifestation of the isotropy of inertial *rest frame S'*. If the orientation of a rod is neither longitudinal nor transverse to the direction of relative motion, both the length and the inclination are changed in the *laboratory frame S*, as an effect of a contraction in the direction of the relative motion with no change of transverse dimensions.

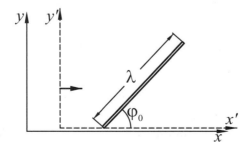

Figure 4.2. Length contraction in a direction of arbitrary orientation φ_0 with respect to the relative direction of motion

Taking $\Delta x' = \lambda \cos\varphi_0$, $\Delta y' = \lambda \sin\varphi_0$ in the rest frame, as illustrated in Figure 4.2, we obtain $\Delta x' = \Delta x / \sqrt{1-\beta^2} = \gamma\Delta x$ and $\Delta y' = \Delta y$, where Δx and Δy are lengths in the laboratory frame, at a time t, where the inclination is given by

$$\tan\varphi = \frac{\Delta y}{\Delta x} = \gamma \tan\varphi_0 \tag{4.14}$$

and the length is

$$l = \sqrt{\Delta x^2 + \Delta y^2} = \lambda\sqrt{(1-v^2/c^2)\cos^2\varphi_0 + \sin^2\varphi_0} = \lambda\sqrt{1-\beta^2\cos^2\varphi_0}$$

Time dilation

Consider two successive events occurring at the same point $x_1' = x_2' = x'$ in the rest frame S. The time interval between the events in the rest frame is called the *proper time interval* $\Delta\tau = t_2' - t_1'$. In the laboratory frame we have

$$t_1 = \frac{t_1' + \beta x'/c}{\sqrt{1-\beta^2}}, \quad t_2 = \frac{t_2' + \beta x'/c}{\sqrt{1-\beta^2}} \quad \text{or} \quad t_2 - t_1 = \frac{t_2' - t_1'}{\sqrt{1-\beta^2}}$$

and the time interval $\Delta t = t_2 - t_1$ is given by

$$\Delta t = \frac{\Delta\tau}{\sqrt{1-\beta^2}} = \gamma\Delta\tau \tag{4.15}$$

It follows that the proper time interval is a minimum. The effect is known as *time dilation* and is equivalent to the slowing down of moving clocks.

Transformation of velocity

The velocity of a particle in the unprimed system S has the components

$$v_x = \frac{dx}{dt}, \quad v_y = \frac{dy}{dt}, \quad v_z = \frac{dz}{dt} \tag{4.16}$$

The corresponding components in the primed system S' are

$$v_x' = \frac{dx'}{dt'}, \quad v_y' = \frac{dy'}{dt'}, \quad v_z' = \frac{dz'}{dt'} \tag{4.17}$$

Differentiation of the Lorentz transformation equations (4.11) yields

$$dx' = \frac{dx - vdt}{\sqrt{1-v^2/c^2}}, \quad dy' = dy, \quad dz' = dz, \quad dt' = \frac{dt - vdx/c^2}{\sqrt{1-v^2/c^2}} \tag{4.18}$$

so that Eqs.(4.17) become

$$v'_x = \frac{dx - vdt}{\sqrt{1-v^2/c^2}\,dx - vdx/c^2} = \frac{v_x - v}{1 - vv_x/c^2} \qquad v_x = \frac{v'_x + v}{1 + vv'_x/c^2}$$

$$v'_y = \frac{v_y\sqrt{1-v^2/c^2}}{1 - vv_x/c^2} \qquad \text{or} \qquad v_y = \frac{v'_y\sqrt{1-v^2/c^2}}{1 + vv'_x/c^2} \qquad (4.19)$$

$$v'_z = \frac{v_z\sqrt{1-v^2/c^2}}{1 - vv_x/c^2} \qquad v_z = \frac{v'_z\sqrt{1-v^2/c^2}}{1 + vv'_x/c^2}$$

giving the transformation of velocity components. For the inverse transformation, see Eqs.(4.11), we must interchange primed and unprimed quantities and reverse the sign of v. The first equation in (4.19) is known as the relativistic *addition law for velocities*, which reduces to the Galilean result for $v \ll c$. We can see that it is not possible to obtain a speed greater than c by changing the reference frame since for the limiting case $v'_x = c$, the velocity in the moving system is

$$v_x = \frac{c + v}{1 + vc/c^2} = c$$

So the speed of light is the same for all inertial frames, as required by the second postulate.

4.3. SPACETIME DIAGRAMS

A geometrical representation of the Lorentz transformation, written as

$$x' = \gamma(x - \beta ct)$$
$$ct' = \gamma(ct - \beta x)$$
$$y' = y \qquad (4.20)$$
$$z' = z$$

was given by Minkovski, by analogy with the transformation of Cartesian coordinates under a spatial rotation, represented in Figure 4.3.

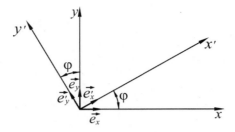

Figure 4.3. Rotation of coordinates through angle φ about the z-axis

For a rotation of Cartesian coordinates through angle φ about the z-axis, we have

$$x'\vec{e}'_x + y'\vec{e}'_y = x\vec{e}_x + y\vec{e}_y$$

where

$$\vec{e}'_x \cdot \vec{e}_x = \cos\varphi, \quad \vec{e}'_x \cdot \vec{e}_y = \cos\left(\frac{\pi}{2} - \varphi\right) = \sin\varphi$$

$$\vec{e}'_y \cdot \vec{e}_x = \cos\left(\frac{\pi}{2} + \varphi\right) = -\sin\varphi, \quad \vec{e}'_y \cdot \vec{e}_y = \cos\varphi$$

so that

$$x' = x(\vec{e}'_x \cdot \vec{e}_x) + y(\vec{e}'_x \cdot \vec{e}_y) \qquad \qquad x' = x\cos\varphi + y\sin\varphi$$
$$\text{or}$$
$$y' = x(\vec{e}'_y \cdot \vec{e}_x) + y(\vec{e}'_y \cdot \vec{e}_y) \qquad \qquad y' = -x\sin\varphi + y\cos\varphi$$

and the coordinates must transform according to the relations

$$x' = x\cos\varphi + y\sin\varphi$$
$$y' = -x\sin\varphi + y\cos\varphi$$
$$z' = z \qquad (4.21)$$
$$t' = t$$

The scalar product of the position vector $\vec{r}(x,y,z)$ with itself, called the *norm* of the vector, is a scalar invariant under any rotation

$$\vec{r}^2 = r^2 = x^2 + y^2 + z^2 = x'^2 + y'^2 + z'^2$$

This is the basic invariance property which can be used to define the rotation transformation in space. In view of the analogous invariance property (4.6) of the

4. PRINCIPLES OF SPECIAL RELATIVITY

Lorentz transformation, it is usual to define a scalar invariant s^2 in spacetime by

$$s^2 = r^2 - c^2 t^2 = r'^2 - c^2 t'^2 \qquad (4.22)$$

In a spacetime diagram, with coordinates (r, ct), a single point at fixed r and fixed t is called an *event*. A line which gives a relation $r = r(t)$ represents a sequence of events (for instance the position of a particle at different times) and is called a *worldline*. The slope of this line with respect to the ct-axis gives the velocity $cdr/d(ct)$, so that a straight line corresponds to a uniform motion and curvature indicates acceleration. A light ray travels in such a diagram along the dotted worldline of equations $r = \pm ct$, as in Figure 4.4. Two frames S and S' moving with uniform relative motion are represented in a spacetime diagram by different coordinate systems, say $S(r, ct)$ and $S'(r', ct')$. Since the same events, that is the same spacetime, are common for the two frames, it is possible to draw the coordinate lines of S', which is moving with velocity v relative to S, on the spacetime diagram drawn for S. The ct'-axis is the locus of events at constant $r' = 0$, that is the locus of the origin O' given by a worldline making an angle $\tan^{-1}(v/c)$ with the ct-axis of S. The r'-axis is the locus of events occuring at $t' = 0$ and making an equal angle with the x-axis of S. The slope of the r'-axis increases and that of the ct'-axis decreases with the velocity of the relative motion up to that of the light ray worldline. The geometrical intuition based upon Euclidean geometry is not appropriate to describe this rotation of axes as a function of the relative velocity v, since the Euclidean invariant under rotation in spacetime is expected to be $r^2 + c^2 t^2$ while the relativistic invariant is $r^2 - c^2 t^2$. The axes r' and ct' of S' can be calibrated in the (r, ct) spacetime diagram of Figure 4.4 using the invariant hyperbolae (4.22) in the (r, ct) plane, which are asymptotic to the light paths $r = \pm ct$.

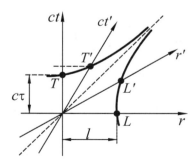

Figure 4.4. Spacetime diagram

Consider an event T on the ct-axis, where $r = 0$, at a distance $c\tau$ from the origin. For $s^2 = -c^2 \tau^2$, the hyperbola through T has the equation

$$r^2 - c^2 t^2 = s^2 \quad \text{or} \quad -c^2 t^2 = -c^2 \tau^2 \quad \text{i.e.,} \quad t = \tau$$

In the same manner, the event T' lying on the ct'-axis has $r' = 0$ and the invariant hyperbola reads

$$r'^2 - c^2 t'^2 = -c^2 \tau^2 \quad \text{or} \quad -c^2 t'^2 = -c^2 \tau^2 \quad \text{i.e.,} \quad t' = \tau$$

that is T and T' correspond to equal time intervals τ from the common origins O and O'. Similarly the hyperbolae (4.22) written for $s^2 = l^2$, calibrate the spatial axis r' such that both events L and L' are located at equal space intervals from the common origin in S and S'. The fact that both T' and L' look to be further from the common origin than T and L, respectively, is related to our Euclidean intuition.

EXAMPLE 4.1. Interpretation of relativistic kinematics in spacetime

Consider a rod at rest in S'. Its ends are the simultaneous events O' and L', and its square length is l^2, see Figure 4.5. The worldlines of the two ends show that the length of the rod in S is the space interval squared s^2 between O and L'', and is shorter than l^2, which is given by the space interval squared OL along the r axis. This illustrates the Lorentz contraction (4.13) which results from the fact that the simultaneous events O' and L' in S' are not simultaneous in S, as $O'L'$ is not parallel to the r axis. Similarly, from Figure 4.4, one may deduce that when a clock moving along the ct' axis reaches T' it has a reading $t' = \tau$. The event T' has in S a coordinate located above T, as determined by a horizontal worldline through T', which corresponds to a time interval longer than τ. In other words, the event T' appears to run slower in S. This is the time dilation (4.15).

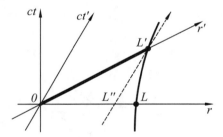

Figure 4.5. Interpretation of the Lorentz contraction in spacetime

---

The *interval squared* between any two events, which are separated by coordinate increments $\Delta x, \Delta y, \Delta z$ and $c\Delta t$ in spacetime, is defined as

$$\Delta s^2 = \Delta r^2 - c^2 \Delta t^2 = \Delta r'^2 - c\Delta t'^2 \qquad (4.23)$$

Whereas the space interval Δr and the time interval Δt may vary from one inertial frame to another, the total interval squared Δs^2 is invariant. Because it is independent of the observer, Δs^2 appears as a property of the two events only, which thus can be used to classify the relation between events. If it is positive, that is

$\Delta r^2 > c^2 \Delta t^2$, the interval is called *spacelike*, and, since the spatial separation of the events is greater than the distance travelled by light during their temporal separation, such events cannot be physically related to one another. If $\Delta s^2 < 0$, the interval is called *timelike* and the events may be causally related, since $c^2 \Delta t^2 > \Delta r^2$ assures that in all inertial frames the time order is the same. For instance, taking the x and x' axes along the line joining the positions of two events, we have $\Delta r = \Delta x$ and $c \Delta t > \Delta x > \beta \Delta x$ since $\beta < 1$. From Eq.(4.20) it then follows that $c \Delta t'$ has the same sign as $c \Delta t$.

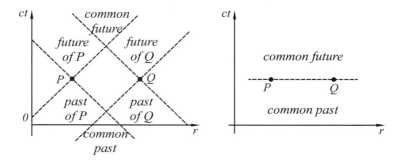

Figure 4.6. Relation between events in spacetime, in special relativity and Newtonian mechanics

A graphical representation of intervals between events is given in Figure 4.6, in a spacetime diagram. The events which are on the same light path have $\Delta s^2 = 0$ and are illustrated by a so-called *light cone*, whose apex is an arbitrary event P or Q, as in Figure 4.6. In the upper half of the light cone, called *absolute future*, the events are related by timelike intervals $c \Delta t > \Delta r$ and all occur after a reference event P or Q, in every inertial frame. For similar reasons, the lower half of the light cone is called *absolute past*. The region outside the cone, where the interval is spacelike, is called *elsewhere*. Here the time order of the events depends on the reference frame used, but the spatial separation always exceeds a minimum value $\Delta s^2 = \Delta r^2 - c^2 \Delta t^2 > 0$. As the interval squared is invariant, all of the inertial frames indicate the same future, past and elsewhere of given events P or Q. As seen from Figure 4.6, in Newtonian mechanics any simultaneous events P and Q have a common future and past, whereas in special relativity each different event has a different future and past.

4.4. RELATIVISTIC INVARIANCE

The covariant formulation of physical laws under a Lorentz transformation is facilitated by writing them in a four-dimensional space, with coordinates $x_1 = x$, $x_2 = y$,

$x_3 = z$, $x_4 = ict$. An event specified by x, y, z and t is regarded as a point in *spacetime* or *world space*, having a position four-vector s_μ of components x_1, x_2, x_3, x_4. The Lorentz transformation, which relates an event in different inertial frames, represents a transformation of the components x_μ from one coordinate system to another. In this notation, Eqs.(4.11) become

$$x_1' = \gamma(x_1 + i\beta x_4) \qquad x_1 = \gamma(x_1' - i\beta x_4')$$
$$x_2' = x_2 \qquad x_2 = x_2'$$
$$x_3' = x_3 \qquad x_3 = x_3' \qquad (4.24)$$
$$x_4' = \gamma(-i\beta x_1 + x_4) \qquad x_4 = \gamma(i\beta x_1' + x_4')$$

and can be written in the compact notation

$$x_\nu' = \sum_{\mu=1}^{4} a_{\nu\mu} x_\mu \quad , \quad x_\mu = \sum_{\nu=1}^{4} a_{\nu\mu} x_\nu' \qquad (4.25)$$

where the coefficients $a_{\nu\mu}$ define the matrix elements of the direct and inverse Lorentz transformations as

$$\begin{pmatrix} \gamma & 0 & 0 & i\beta\gamma \\ 0 & 1 & 0 & 0 \\ 0 & 0 & 1 & 0 \\ -i\beta\gamma & 0 & 0 & \gamma \end{pmatrix} \text{ and } \begin{pmatrix} \gamma & 0 & 0 & -i\beta\gamma \\ 0 & 1 & 0 & 0 \\ 0 & 0 & 1 & 0 \\ i\beta\gamma & 0 & 0 & \gamma \end{pmatrix} \qquad (4.26)$$

The second matrix is obtained simply by transposing the first one (see Appendix II), and the matrix elements satisfy the orthogonality condition

$$\sum_{\nu=1}^{4} a_{\nu\mu} a_{\lambda\mu} = \delta_{\nu\lambda} \qquad (4.27)$$

A four-dimensional vector may be specified in terms of components $A_\mu = (A_x, A_y, A_z, A_t) = (A_1, A_2, A_3, A_4)$ and must obey a transformation rule analogous to (4.24) that reads

$$A_1' = \gamma(A_1 + i\beta A_4) \qquad A_x' = \gamma(A_x + i\beta A_t)$$
$$A_2' = A_2 \qquad A_y' = A_y$$
$$\qquad\qquad\qquad\text{or}\qquad\qquad\qquad (4.28)$$
$$A_3' = A_3 \qquad A_z' = A_z$$
$$A_4' = \gamma(-i\beta A_1 + A_4) \qquad A_t' = \gamma(A_t - i\beta A_x)$$

which also allows the notation

$$A'_\nu = \sum_{\mu=1}^{4} a_{\nu\mu} A_\mu \qquad (4.29)$$

As we expect, the norm of A_μ is a Lorentz invariant

$$\sum_{\mu=1}^{4} A'^2_\mu = \gamma^2\left(A_x^2 + 2i\beta A_x A_t - \beta^2 A_t^2\right) + A_y^2 + A_z^2 + \gamma^2\left(A_t^2 - 2i\beta A_t A_x - \beta^2 A_x^2\right)$$

$$\qquad (4.30)$$

$$= \gamma^2 A_x^2\left(1 - \beta^2\right) + A_y^2 + A_z^2 + \gamma^2 A_t^2\left(1 - \beta^2\right) = \sum_{\mu=1}^{4} A_\mu^2$$

A familiar example is the interval squared (4.23) which is the norm of the four-interval $\Delta s_\mu(\Delta x, \Delta y, \Delta z, ic\Delta t)$ between two events. Its differential form $ds_\mu(dx, dy, dz, icdt)$ defines the Lorentz invariant

$$ds^2 = dx^2 + dy^2 + dz^2 - c^2 dt^2 \qquad (4.31)$$

A related Lorentz invariant is

$$d\tau^2 = -\frac{ds^2}{c^2} = dt^2 - \frac{1}{c^2}\left(dx^2 + dy^2 + dz^2\right) \qquad (4.32)$$

which is the *proper time interval squared*: in the rest frame S' of a particle the space coordinates are constant $(dx = dy = dz = 0)$ and $d\tau = dt$, thus $d\tau$ is the time interval as measured on a clock travelling with the particle.

Lorentz invariants are also called *four-scalars*, in analogy to the ordinary scalars which are invariant under rotations. In a three-dimensional space, dividing a vector by a scalar yields another vector. In a similar manner, dividing a four-vector by a four-scalar yields another four-vector. Along the worldline of a particle it is convenient to divide the components of the differential four-interval ds_μ by the proper time interval $d\tau$ of the particle, defined in (4.28), to obtain the four-vector

$$v_\mu = \frac{ds_\mu}{d\tau} = \left(\frac{dx}{d\tau}, \frac{dy}{d\tau}, \frac{dz}{d\tau}, ic\frac{dt}{d\tau}\right) \qquad (4.33)$$

This is called the *four-velocity* which gives the rate of change of the position four-vector of a particle with respect to its proper time. Using the time-dilation formula (4.15), we find an expression of the four-velocity of a moving particle

$$v_\mu = \gamma\left(\frac{dx}{dt}, \frac{dy}{dt}, \frac{dz}{dt}, ic\right) = \gamma(v_x, v_y, v_z, ic) = \gamma(\vec{v}, ic) \qquad (4.34)$$

The norm of the world four-velocity has a constant magnitude

$$v_\mu^2 = \gamma^2(v_x^2 + v_y^2 + v_z^2 - c^2) = \frac{1}{1-v^2/c^2}(v^2 - c^2) = -c^2 \quad (4.35)$$

where the negative value of the Lorentz invariant indicates that the four-velocity of a particle is a timelike four-vector.

EXAMPLE 4.2. The covariant electromagnetic wave equation

An instructive illustration of the covariant formulation of a physical law in four-dimensional space is to reconsider the electromagnetic wave equation (4.1), which in three dimensions reads

$$\nabla^2 \Psi = \frac{1}{c^2}\frac{\partial^2 \Psi}{\partial t^2} \quad (4.36)$$

and was shown to be not covariant with respect to the Galilean transformation. We can define a *four-gradient* or a *quad* differential operator, with components

$$\frac{\partial}{\partial x_\mu} = \left(\frac{\partial}{\partial x_1}, \frac{\partial}{\partial x_2}, \frac{\partial}{\partial x_3}, \frac{\partial}{\partial x_4}\right) = \left(\frac{\partial}{\partial x}, \frac{\partial}{\partial y}, \frac{\partial}{\partial z}, \frac{1}{ic}\frac{\partial}{\partial t}\right) \quad (4.37)$$

which transforms as

$$\frac{\partial}{\partial x'_\nu} = \sum_{\mu=1}^{4} \frac{\partial x_\mu}{\partial x'_\nu} \frac{\partial}{\partial x_\mu}$$

In view of the inverse Lorentz transformation (4.25), we have $\partial x_\mu / \partial x'_\nu = a_{\nu\mu}$, so that

$$\frac{\partial}{\partial x'_\nu} = \sum_{\mu=1}^{4} a_{\nu\mu} \frac{\partial}{\partial x_\mu}$$

which is the transformation equation (4.29) for the components of a four-vector. The norm of the quad operator, called the *d'Alembertian*, is then a Lorentz invariant of the form

$$\sum_\mu \frac{\partial^2}{\partial x_\mu^2} = \left(\frac{\partial^2}{\partial x^2} + \frac{\partial^2}{\partial y^2} + \frac{\partial^2}{\partial z^2} - \frac{1}{c^2}\frac{\partial^2}{\partial t^2}\right) \quad (4.38)$$

Therefore, the expression (4.36) of the electromagnetic wave equation, written as

$$\left(\frac{\partial^2}{\partial x^2} + \frac{\partial^2}{\partial y^2} + \frac{\partial^2}{\partial z^2} - \frac{1}{c^2}\frac{\partial^2}{\partial t^2}\right)\Psi = 0$$

provides the desired covariant formulation under a Lorentz transformation. It follows that the d'Alembertian (4.37) plays, in special relativity, the same role as that of the Laplacian in classical mechanics.

FURTHER READING
1. W. Rindler - ESSENTIAL RELATIVITY, Springer-Verlag, New York, 2001.
2. J. H. Smith - INTRODUCTION TO SPECIAL RELATIVITY, Dover Publications, New York, 1996.
3. R. Resnick, D. Halliday - BASIC CONCEPTS IN RELATIVITY, Prentice Hall, Englewood Cliffs, 1992.

PROBLEMS

4.1. Find a representation for the Lorentz transformation in a vector form, which is specified by the direction and the magnitude of a velocity vector $\vec{v}$.

Solution

Let $\vec{r} = x\vec{e}_x + y\vec{e}_y + z\vec{e}_z$ and $\vec{r}\,' = x'\vec{e}\,'_x + y'\vec{e}\,'_y + z'\vec{e}\,'_z$ be the position vectors in frames S and S', respectively. In view of Eqs.(4.11), one obtains

$$\vec{r}\,' = \gamma(x - vt)\vec{e}_x + y\vec{e}_y + z\vec{e}_z + x\vec{e}_x - x\vec{e}_x = \vec{r} + (\gamma - 1)x\vec{e}_x - \gamma vt\vec{e}_x$$

It is convenient to make the obvious substitutions $\vec{e}_x = \vec{v}/v$ and $x = \vec{r}\cdot\vec{e}_x = (\vec{r}\cdot\vec{v})/v$, and this leads to the required representation

$$\vec{r}\,' = \vec{r} + (\gamma - 1)\frac{\vec{r}\cdot\vec{v}}{v^2}\vec{v} - \gamma t\vec{v}, \qquad t' = \gamma\left(t - \frac{\vec{r}\cdot\vec{v}}{c^2}\right)$$

where $\vec{r}\,' = \vec{r} = 0$ when $t' = t = 0$. The vector form of the transformation is valid for any direction of the vector $\vec{v}$ and for any point $\vec{r}$. It is apparent that the components of the position vector $\vec{r}$ normal to $\vec{v}$ do not change. For $\gamma \to 1$ the transformations of coordinates and time reduce to the vector form

$$\vec{r}\,' \to \vec{r} - \vec{v}t, \qquad t' \to t$$

which express the Galilean transformation, Eq.(2.13).

4.2. Given two frames S and S', with S' having velocity v along the positive x-axis relative to S, determine a relationship between the direction φ of a ray of light with respect to the x-axis in frame S and its direction φ' relative to the x'-axis in frame S'.

Solution

Suppose that a particle moves with velocity $\vec{v}_0$ in frame S, so that

$$v_x = v_0 \cos\varphi, \qquad v_y = v_0 \sin\varphi$$

and with velocity $\vec{v}'$ in frame S', where we have

$$v'_x = v'\cos\varphi', \qquad v'_y = v'\sin\varphi'$$

Substitution into Eqs.(4.19) for transformation of velocity components gives

$$v_0 \cos\varphi = \frac{v'\cos\varphi' + v}{1 + vv'\cos\varphi'/c^2}, \qquad v_0 \sin\varphi = \frac{v'\sin\varphi'}{\gamma(1 + vv'\cos\varphi'/c^2)}$$

If we have $v_0 = c$ for the speed of light in frame S, then in frame S' we also have $v' = c$, and hence

$$\sin\varphi = \frac{\sin\varphi'}{\gamma(1 + \beta\cos\varphi')}, \qquad \cos\varphi = \frac{\cos\varphi' + \beta}{1 + \beta\cos\varphi'}$$

This change of direction of the rays of light can be put in the form

$$\tan\varphi = \frac{\sin\varphi'}{\gamma(\cos\varphi' + \beta)}$$

and accounts for the observed seasonal change in the position of stars, which is caused by the orbital motion of the Earth and is known as the aberration of starlight.

4.3. To an observer O, a particle moves with the velocity $\vec{v}_0$ of magnitude v_0 and orientation specified by the usual spherical coordinates θ and φ. Find its apparent velocity $\vec{v}'(v',\theta',\varphi')$ for a second observer O' moving with uniform velocity v in the positive x-direction relative to O.

Solution

To the observer O, the velocity $\vec{v}_0$ has components

$$v_x = v_0 \cos\varphi \sin\theta, \qquad v_y = v_0 \sin\varphi \sin\theta, \qquad v_z = v_0 \cos\theta$$

whereas, to the observer O', it appears that

$$v'_x = v'\cos\varphi' \sin\theta', \qquad v'_y = v'\sin\varphi' \sin\theta', \qquad v'_z = v'\cos\theta'$$

Substituting into Eqs.(4.19) gives

$$v'\cos\varphi' \sin\theta' = \frac{v_0 \cos\varphi \sin\theta - v}{1 - v_0 v \cos\varphi \sin\theta/c^2}, \qquad v'\sin\varphi' \sin\theta' = \frac{(v_0/\gamma)\sin\varphi \sin\theta}{1 - v_0 v \cos\varphi \sin\theta/c^2},$$

$$v'\cos\theta' = \frac{(v_0/\gamma)\cos\theta}{1 - v_0 v \cos\varphi \sin\theta/c^2}$$

On dividing the first two equations results in

$$\tan\varphi' = \frac{v_0 \sin\varphi \sin\theta}{\gamma(v_0 \cos\varphi \sin\theta - v)}$$

On squaring and adding together the first two equations, one obtains an expression for $(v'\sin\theta')^2$. Then, on dividing the result by $(v'\cos\theta')^2$ yields

$$\tan\theta' = \frac{\gamma\left(v_0^2 \sin^2\theta + v^2 - \beta^2 v_0^2 \sin^2\varphi \sin^2\theta - 2v_0 v \cos\varphi \sin\theta\right)^{1/2}}{v\cos\theta}$$

Finally, the magnitude v' of the apparent velocity is found by squaring all the three equations and adding them together. After a little algebra, this gives

$$v' = \frac{\left(v_0^2/\gamma^2 + v^2 + \beta^2 v_0^2 \cos^2\varphi \sin^2\theta - 2v_0 v \cos\varphi \sin\theta\right)^{1/2}}{1 - v_0 v \cos\varphi \sin\theta / c^2}$$

Assuming that, to the observer O, the particle moves in the xy-plane, that is $\theta = \pi/2$, it follows that $\tan\theta' \to \infty$, thus $\theta' = \pi/2$ and the motion will be restricted to the $x'y'$-plane for the second observer.

4.4. Derive an expression for the acceleration four-vector that gives the rate of change of the velocity four-vector of a particle with respect to its proper time. What are its components in the case when either the magnitude or the direction of velocity is invariable?

Solution

The acceleration four-vector has components

$$a_\mu = \frac{dv_\mu}{d\tau} = \gamma \frac{dv_\mu}{d\tau} = \gamma\left[\frac{d(\gamma\vec{v})}{dt}, ic\frac{d\gamma}{dt}\right]$$

where use has been made of Eqs.(4.34) and (4.15). In view of Eq.(4.10) one obtains

$$\frac{d\gamma}{dt} = \frac{\gamma^3}{c^2}(\vec{v}\cdot\vec{a})$$

where $\vec{a} = d\vec{v}/dt = \dot{\vec{v}}$. Then we have

$$\frac{d(\gamma\vec{v})}{dt} = \gamma\frac{d\vec{v}}{dt} + \frac{d\gamma}{dt}\vec{v} = \gamma\vec{a} + \frac{\gamma^3}{c^2}(\vec{v}\cdot\vec{a})\vec{v}$$

and this finally gives

$$a_\mu = \gamma^2\left[\vec{a} + \frac{\gamma^2}{c^2}(\vec{v}\cdot\vec{a})\vec{v}, i\frac{\gamma^2}{c}(\vec{v}\cdot\vec{a})\right]$$

For a rotational motion, where $v^2 = \vec{v}\cdot\vec{v} = \text{const}$, we have $\vec{v}\cdot\dot{\vec{v}} = \vec{v}\cdot\vec{a} = 0$ and the acceleration four-vector reduces to

$$a_\mu = \left(\gamma^2 \vec{a}, 0\right)$$

For a rectilinear motion with acceleration $a = |d\vec{v}/dt| = dv/dt$, one obtains

$$a_\mu = \gamma^2\left(a + \gamma^2\beta^2 a, i\gamma^2\beta a\right) = \gamma^4\left(a, i\beta a\right)$$

4.5. Two frames S and S' are in the standard configuration, frame S' having velocity v in the positive x-direction with respect to S. Two light beams are emitted from S' along the positive and the negative y'-axis. What is the angle α observed between the two rays of the light in frame S.

4.6. Find an expression for the magnitude of the relative velocity of two particles, defined as a velocity of one particle in the frame in which the other one is at rest.

4.7. Show that the four-volume $dV = dx_1 dx_2 dx_3 dx_4$ is a Lorentz invariant.

5. RELATIVISTIC LAWS OF MOTION

5.1. RELATIVISTIC MOMENTUM AND ENERGY
5.2. THE RELATIVISTIC EQUATION OF MOTION
5.3. THE EQUIVALENCE PRINCIPLE

5.1. RELATIVISTIC MOMENTUM AND ENERGY

We now need to address the task of writing the laws of Newtonian mechanics in a form which satisfies the postulates of special relativity. One possible approach is to reformulate the conservation laws so they are preserved under a Lorentz transformation.

In order to find a conserved quantity analogous to classical momentum, we define the momentum of a particle as a vector parallel to its velocity $\vec{w}$ given by

$$\vec{p} = m(w)\vec{w}$$

with a magnitude equal to a scalar function of the speed times the speed. The scalar coefficient $m(w)$ is the analogue of Newtonian mass. The form of $m(w)$ can be obtained by applying to an elastic collision between two identical particles the requirement that the vector sum of momenta, for an isolated system of particles, is the same in every inertial frame. The collision is represented in Figure 5.1 (a) as viewed in the rest frame $S_1(x_1, y_1)$ of particle 1 and in Figure 5.1 (b) in the rest frame $S_2(x_2, y_2)$ of particle 2. The axes x_1 and x_2 are chosen to lie along the direction of relative motion. We assume that S_2 moves with a velocity v with respect to S_1 and the y_1 and y_2 axes are coplanar. Note that in their respective rest frames each particle has the same speed v_0 in the y-direction before the collision, this reverses its direction and becomes v' after the collision. The y velocity of the opposite particle in each rest frame is denoted v_y before the collision and v'_y after the collision and is given by Eq.(4.19) of the transformation for the velocity components

$$v_y = \frac{v_0\sqrt{1-v^2/c^2}}{1-vv_x/c^2} = v_0\sqrt{1-\frac{v^2}{c^2}} = \frac{v_0}{\gamma} \quad , \quad v'_y = \frac{v'\sqrt{1-v^2/c^2}}{1-vv_x/c^2} = v'\sqrt{1-\frac{v^2}{c^2}} = \frac{v'}{\gamma}$$

where $v_x = 0$, since the particles collide with the collision diameter perpendicular to the direction of relative motion, so that the x components of the velocities are zero in the rest frame.

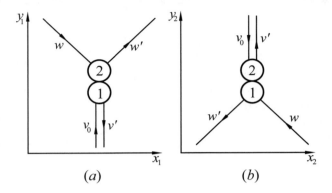

Figure 5.1. Representation of an elastic collision as viewed from the proper frames of the colliding particles. (a) Rest frame of particle 1 (b) Rest frame of particle 2

Therefore the speed of the opposite particle (the particle not in the rest frame), denoted by w before the collision and by w' after the collision is given by

$$w = \left(v^2 + \frac{v_0^2}{\gamma^2} \right)^{1/2} \quad , \quad w' = \left(v^2 + \frac{v'^2}{\gamma^2} \right)^{1/2}$$

Conservation of momentum should hold in S_1. For the x component, it reads

$$m(v_0) \cdot 0 + m(w)v = m(v') \cdot 0 + m(w')v$$

so that $w = w'$ and it follows $v' = v_0$. For the y component, in S_1, we have

$$-m(v_0)v_0 + m(w)\frac{v_0}{\gamma} = m(v_0)v_0 - m(w)\frac{v_0}{\gamma}$$

Thus

$$m(w) = \gamma m(v_0)$$

and this gives the functional form of the dependence of inertial mass on speed. It holds for any value of v_0 and the limit $v_0 \to 0$ gives $m(v_0) \to m(0) = m_0$ which can be taken as the Newtonian mass or *rest mass* of the particle. Hence the inertial mass of the particle depends on its speed as

$$m(v) = \gamma m(0) = \frac{m_0}{\sqrt{1 - v^2/c^2}} \quad (5.1)$$

and the momentum of a particle moving with arbitrary velocity $\vec{v}$ is

$$\vec{p} = \frac{m_0 \vec{v}}{\sqrt{1-v^2/c^2}} = m\vec{v} \tag{5.2}$$

We can now find an energy-velocity relation, as a consequence of Newton's law of motion (2.9), which transforms according to

$$dE = \vec{F} \cdot d\vec{r} = \frac{d\vec{p}}{dt} \cdot d\vec{r} = d\vec{p} \cdot \vec{v} = vdp \tag{5.3}$$

where the parallelism of $\vec{p}$ and $\vec{v}$, see Figure 5.2, allows us to write

$$d\vec{p} \cdot \vec{v} = |d\vec{p}| v \cos\alpha = vdp$$

In other words, a change in energy requires a change in the magnitude of the momentum. A change in just the direction of $\vec{p}$, as given by a transverse force, leaves E constant.

Figure 5.2. Infinitesimal change of the momentum

Using Eq.(5.2), we can evaluate

$$dp = d\left(\frac{m_0 v}{\sqrt{1-v^2/c^2}}\right) = \frac{m_0 dv}{(1-v^2/c^2)^{3/2}}$$

and Eq.(5.3) successively becomes

$$dE = \frac{m_0 v dv}{(1-v^2/c^2)^{3/2}} = -\frac{m_0 c^2}{2} \frac{d(1-v^2/c^2)}{(1-v^2/c^2)^{3/2}} = m_0 c^2 d\left(\frac{1}{\sqrt{1-v^2/c^2}}\right) = m_0 c^2 d\gamma \tag{5.4}$$

Integrating this equation we obtain the increment of energy due to the work done on the particle to bring it from rest (where $\gamma = 1$) to speed v, which is the kinetic energy

$$T = m_0 c^2 \int_1^\gamma d\gamma = m_0 c^2 (\gamma - 1) \tag{5.5}$$

This equation bears little resemblance to the classical expression (2.22) for T. However in the nonrelativistic limit $v \ll c$, it approaches the classical form of the kinetic energy

$$T = m_0 c^2 \left(\frac{1}{\sqrt{1-v^2/c^2}} - 1 \right) \cong m_0 c^2 \left(1 + \frac{1}{2}\frac{v^2}{c^2} - 1 \right) = \frac{1}{2} m_0 v^2$$

Integrating Eq.(5.4) and setting the constant of integration equal to zero, we obtain the *total energy* as

$$E = m_0 c^2 \gamma = \frac{m_0 c^2}{\sqrt{1-v^2/c^2}} \tag{5.6}$$

which, in view of Eq.(5.5), is the sum of the increment of energy from rest and the *rest energy* $m_0 c^2$ that depends only on the rest mass of the particle

$$E = T + m_0 c^2 \tag{5.7}$$

Equation (5.4) can be also written in the form

$$dE = c^2 dm \tag{5.8}$$

known as *energy-inertia relation*, which upon integration becomes

$$\Delta E = c^2 \Delta m \tag{5.9}$$

and means that an energy ΔE added to a particle is associated with an increase of its inertial mass Δm. This is true irrespective of the form of energy and therefore goes beyond the classical conservation law for mechanical energy. An increment ΔE in the energy of a particle increases its velocity by an amount $\Delta \gamma = \Delta E / m_0 c^2$ but, no matter how large γ might become, Eq.(4.10) gives

$$\beta = \sqrt{1 - \frac{1}{\gamma^2}} < 1$$

so that v remains less than c. It follows that the energy of a particle is allowed to increase without limit as $v \to c$.

The energy E can be also expressed by eliminating v between Eq.(5.2), which reads

$$p^2 = \frac{m_0^2 v^2}{1-v^2/c^2} \quad \text{or} \quad \frac{v^2}{c^2} = \frac{p^2}{p^2 + m_0^2 c^2} \quad \text{i.e.,} \quad \gamma = \sqrt{1 + \frac{p^2}{m_0^2 c^2}}$$

and Eq.(5.6), which takes the form

$$E = m_0 c^2 \sqrt{1 + \frac{p^2}{m_0^2 c^2}} \quad \text{or} \quad E^2 = p^2 c^2 + \left(m_0 c^2\right)^2 \qquad (5.10)$$

This *momentum-energy relation*, which is also useful in its differential form

$$EdE = c^2 p dp \qquad (5.11)$$

gives the energy of a particle in terms of its momentum and rest mass.

EXAMPLE 5.1. The photon

The relativistic energy-momentum relation (5.10) allows us to examine the possibility of particles with zero rest mass. Setting $m_0 = 0$ in Eq.(5.10) we see that the momentum and energy of such a particle can be finite and must satisfy the equation

$$p = \frac{E}{c} \qquad (5.12)$$

In order to have nonzero momentum, the expression (5.2) must maintain a finite value in the limit $m_0 \to 0$, which is only possible if $v \to c$. Therefore massless particles must travel at the speed of light. This special class of particles includes *the photon,* associated with the electromagnetic interaction, various kinds of *neutrinos,* associated with weak interactions in the radioactive beta decay and the *graviton,* associated the gravitational interaction. The concept of the photon was first introduced by Einstein, as we shall show in Chapter 28.

5.2. THE RELATIVISTIC EQUATION OF MOTION

Newton's equation of motion (2.9), which is invariant under the Galilean transformation, whose components read

$$\frac{dp_i}{dt} = F_i \qquad (i = 1,2,3) \qquad (5.13)$$

must be generalized to a form invariant under Lorentz transformation, which is expected to be a four-vector equation

$$\frac{dp_\mu}{dt} = F_\mu \qquad (\mu = 1,2,3,4) \qquad (5.14)$$

The spatial component of Eq.(5.14) must reduce to Eq.(5.13) in the low velocity limit.

The four-vector p_μ is expressed as the product of the four-velocity v_μ given by Eq.(4.34) and the rest mass m_0, which is a Lorentz invariant

$$p_\mu = m_0 v_\mu = \gamma(m_0 \vec{v}, im_0 c) = (m\vec{v}, imc) \qquad (5.15)$$

The spacelike component of the *four-momentum* p_μ is the momentum of the particle $\vec{p}$ as defined by (5.2). The timelike component is $p_4 = imc = (i/c)E$, in view of Eq.(5.6), so that the four-momentum

$$p_\mu = \left(\vec{p}, \frac{i}{c}E\right) \qquad (5.16)$$

combines both the momentum and the energy, being also called the *momentum-energy four-vector*. The transformation rule (4.28) reads

$$p'_x = \gamma\left(p_x - \frac{\beta}{c}E\right)$$

$$p'_y = p_y$$

$$p'_z = p_z \qquad (5.17)$$

$$E' = \gamma(E - \beta c p_x)$$

showing that energy and the longitudinal momentum are interrelated, like time and longitudinal position in the Lorentz transformation (4.11). The norm of the four-momentum is a Lorentz invariant given by

$$\sum_\mu p_\mu^2 = p^2 - \frac{E^2}{c^2} = -m_0^2 c^2 \qquad (5.18)$$

a result which is equivalent to the momentum-energy relation (5.10). It can be conveniently represented by a right triangle, as given in Figure 5.3, with p and mc as sides and E/c as hypotenuse. The angle opposite the momentum side of this triangle has the sine given by

$$\frac{p}{E/c} = \frac{m\beta c}{mc} = \beta$$

which varies between the nonrelativistic region $\beta \to 0$ and the extreme relativistic range $\beta \to 1$. The left hand side of the four-vector equation (5.14) takes a covariant form using $dt = \gamma d\tau$, as given by Eq.(4.15), where the proper time interval $d\tau$ is a Lorentz invariant

$$\frac{dp_\mu}{d\tau} = \gamma F_\mu = K_\mu \qquad (5.19)$$

The four-vector K_μ is called the *four-force*. Its spacelike component is γ times the physical force. Its timelike part is given by

$$K_4 = \frac{dp_4}{d\tau} = \frac{i}{c}\frac{dE}{d\tau} = \frac{i\gamma}{c}\frac{dE}{dt} = \frac{i\gamma}{c}\frac{\vec{F}\cdot d\vec{r}}{dt} = \frac{i\gamma}{c}\vec{F}\cdot\vec{v} \qquad (5.20)$$

and is proportional to the rate at which the physical force does work

$$K_\mu = \gamma\left(\vec{F}, \frac{i}{c}\vec{F}\cdot\vec{v}\right) \qquad (5.21)$$

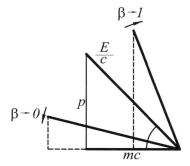

Figure 5.3. Triangle representation of four-momentum

The spatial component of the four-vector equation (5.19) is identical to Newton's law of motion (5.13). Using Eqs.(5.16) and (5.21) the temporal component reads

$$\frac{dE}{d\tau} = \gamma \vec{F}\cdot\vec{v} \quad \text{or} \quad \frac{dE}{dt} = \vec{F}\cdot\vec{v} \quad \text{i.e.,} \quad dE = \vec{F}\cdot d\vec{r}$$

which is the theorem of work. However we must bear in mind that the components of p_μ are related by the Lorentz invariant (5.18), thus the four component equations are not independent. For an isolated particle $\left(\vec{F}=0\right)$ all the components of K_μ vanish, so that both momentum and energy are conserved, since the four-momentum is a constant. In special relativity, momentum and energy which are independent concepts in Newtonian mechanics appear as spacelike and timelike parts of a single four-dimensional concept. The conservation laws for momentum and energy reduce to a single law for conservation of four-momentum.

EXAMPLE 5.2. The equivalence of mass and energy

Let us apply the transformation law of momentum and energy (5.17) to an inelastic collision between two identical particles, with rest mass m_0 moving along the x' axis of their centre of mass frame S' with velocities v' and $-v'$, as illustrated in Figure 5.4. After the collision the particles stick together, resulting a single particle with rest mass $2m_0$, which is at rest in S'.

In the *centre of mass frame* S' we have

- before the collision, a total energy given by

$$E' = 2\frac{m_0 c^2}{\sqrt{1 - v'^2/c^2}} = 2m_0 c^2 + T' \tag{5.22}$$

- after the collision, the total *mechanical* energy of

$$E'' = 2m_0 c^2 \tag{5.23}$$

so that there is lost kinetic energy of $\Delta E' = T'$. The conservation of momentum ($p'_x = 0$ before and after the collision) and the loss of kinetic energy which is converted to heat, are the familiar concepts used in Newtonian mechanics to describe a non-conservative interaction, where conservation of mechanical energy only could not apply.

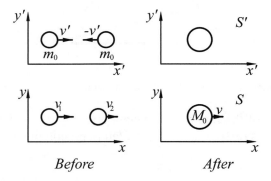

Figure 5.4. Representation of an inelastic collision in the centre of mass frame $S'(x'y')$ and in the laboratory frame $S(x, y)$, with respect to which S' is moving with velocity v

In the *laboratory frame S* we obtain

- before the collision, the momentum and energy are given by the transformation law (5.17), which reads

$$p_x = \gamma\left(p'_x + \frac{\beta}{c}E'\right) = \frac{\gamma\beta}{c}E', \qquad E = \gamma(E' + \beta c p'_x) = \gamma E'$$

where $p'_x = 0$ in the centre of mass frame, and $-v$ is the velocity of the laboratory frame with respect to S'. By inserting E', as given by (5.22), we obtain

$$p_x = \gamma\left(2m_0 + \frac{T'}{c^2}\right)v$$

$$E = \gamma(2m_0 c^2 + T') = \gamma\left(2m_0 + \frac{T'}{c^2}\right)c^2 \quad (5.24)$$

- after the collision, the particle with rest mass $2m_0$ moves with a velocity v in the laboratory frame. However, the four-momentum conservation law requires that its momentum preserves the magnitude

$$p_x = \frac{(2m_0 + T'/c^2)v}{\sqrt{1 - v^2/c^2}} = \frac{M_0 v}{\sqrt{1 - v^2/c^2}}$$

which means, in view of Eq.(5.2), that the final rest mass M_0 must be greater than the initial rest mass $2m_0$. The increase corresponds to the classical loss of kinetic energy. This shows that a change in kinetic energy should be accompanied by a change in the total rest mass. In other words, rest mass can be created from energy and vice-versa. Eq.(5.6) can be used to express the energy after the collision in the laboratory frame, and becomes identical to Eq.(5.23) upon insertion of M_0, so that the rest mass increase is consistent with energy conservation. We must also make the rest mass correction of the energy E'' after the collision in the centre of mass frame.

The distinction between various forms of energy, which allows us, in Newtonian mechanics, to speak about a mechanical energy lost as heat, does not occur in special relativity. Such a lost energy is present as an energy $\Delta E'$ added to a particle and associated to a rest mass increase, according to the energy-inertia relation (5.9). There are many confirmations of this relation, the earliest being the Cockroft-Walton experiment.

5.3. THE EQUIVALENCE PRINCIPLE

Special relativity covers the transformation between inertial frames, which appear as a privileged class of reference frames, allowing the simplest formulation of physical laws. A broader theory must be concerned with reference frames in arbitrary (accelerated) relative motion, and follows from a principle which unifies two apparently independent properties: inertia and gravitation.

The equivalence of the inertial and gravitational masses $m_a = m_g$, which are proportional and are usually made equal by a suitable choice of units, was carefully established by experiments to an accuracy of 10^{-11}. This property implies that all particles experience the same acceleration in a given gravitational field

$$\vec{a} = \frac{\vec{F}}{m_a} = \frac{m_g \vec{g}}{m_a} = \vec{g}$$

Equal acceleration of all particles is precisely the effect expected to occur if the laboratory frame is accelerated. An example of such a frame is an elevator cabin allowed to fall freely under gravity. This is equivalent to an unaccelerated frame, far away from all interacting masses. If we regard the earth as an inertial frame, then Newton's equation of motion $m_a \vec{a}_0 = \vec{F}_e + m_g \vec{g}$ with respect to the earth holds in the cabin. Since $m_a = m_g$ it will take the form $m_a(\vec{a}_0 - \vec{g}) = \vec{F}_e$. One can therefore eliminate a uniform gravitational field $\vec{g}$ by replacing it with the equivalent acceleration of the laboratory frame $\vec{a} = -\vec{g}$ and this yields $m_a(\vec{a}_0 + \vec{a}) = \vec{F}_e$. Alternatively, the effect of acceleration $\vec{a}$ of a frame, with respect to an inertial frame, is the same as that of adding a uniform gravitational field $\vec{g} = -\vec{a}$. The equivalence of a frame which experiences a uniform gravitational field $\vec{g}$ to another which has translational acceleration $-\vec{g}$ and no gravitational field was stated by Einstein, in the form of

The equivalence principle: *All local, freely falling, no rotating frames are fully equivalent for the performance of all physical experiments.*

Each freely falling and no rotating frame represents a *local inertial frame*. Special relativity is concerned with local inertial frames in uniform motion and appears in this context as a local theory. The effect of a gravitational field can be calculated by a kinematical analysis of an equivalent accelerated frame.

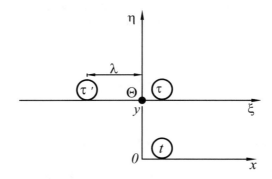

Figure 5.5. Synchronisation of clocks in an accelerated frame

We will assume that Σ is a Cartesian system with origin Θ and coordinates ξ, η which is accelerated in the ξ direction with respect to an inertial frame S with origin O and coordinates x, y. If we take the two frames to be coincident at $\tau = t = 0$, as in Figure 5.5, because the frame Σ is initially at rest, the time intervals measured by the two clocks are identical $\Delta \tau = \Delta t$. A second clock located at $\xi = -\lambda$ in Σ, when arriving at coincidence with the clock at rest in O, is moving with velocity $v = \sqrt{2a(-\lambda)} = \sqrt{2g\lambda}$, so that their proper time intervals are related by the time dilation formula (4.15), which gives

$$\frac{\Delta \tau}{\Delta t} = \frac{1}{\gamma} = \left(1 - \frac{v^2}{c^2}\right)^{1/2} \cong 1 - \frac{g\lambda}{c^2} = 1 + \frac{g\xi}{c^2}$$

It follows that

$$\Delta \tau' \cong \Delta \tau \left(1 + \frac{g\xi}{c^2}\right)$$

a result which is also true, according to the equivalence principle, in an inertial frame Σ, under a uniform gravitational field pointing in the $-\xi$ direction. Hence, $g\xi$ gives the increase $\Delta\varphi$ of the gravitational potential, and we have

$$\Delta \tau' \cong \Delta \tau \left(1 + \frac{\Delta\varphi}{c^2}\right) \qquad (5.25)$$

The time dilation shows that a clock runs faster at higher gravitational potential. In a similar manner, it follows from the Lorentz contraction formula that the distances $\Delta\xi$ and $\Delta\xi'$ between the ends of two rods in Σ, that match with a same rod Δx in S, when Σ is at rest and when Σ is moving with velocity $v = \sqrt{2g\lambda}$ respectively, are related by

$$\frac{\Delta\xi'}{\Delta\xi} = \gamma = \left(1 - \frac{v^2}{c^2}\right)^{-1/2} \cong 1 + \frac{g\lambda}{c^2} = 1 - \frac{g\xi}{c^2}$$

The Lorentz contraction means that the length of a metre stick parallel to the gravitational field increases as the gravitational potential increases

$$\Delta\xi' \cong \Delta\xi \left(1 - \frac{\Delta\varphi}{c^2}\right) \qquad (5.26)$$

EXAMPLE 5.3. The gravitational red shift

The validity of Eq.(5.25) was confirmed by the experiment of Pound and Rebka, using a sharp spectral line of frequency v_0, emitted in the downward direction by a ^{57}Co gamma-ray source, at a height d from the detector. The change of gravitational potential from source to detector is $\Delta\varphi = -gd$, so that the detected frequency v, as given by Eq.(5.25) should be

$$v \cong v_0\left(1 - \frac{\Delta\varphi}{c^2}\right) = v_0\left(1 + \frac{gd}{c^2}\right)$$

In other words there should be an increase in frequency for a falling photon. The experiment measures the photon absorbtion in an iron foil as a function of absorber velocity and shows that the frequency shift along the potential drop $\Delta\varphi$ is exactly compensated by giving the absorber a velocity of g times the photon time of flight. This provides a direct test of the equivalence principle.

The light travelling *against* a gravitational field experiences a decrease in frequency, called a *redshift*. Since Eq.(5.26) only holds in local inertial frames, to find the frequency shift for a light path at an astronomical scale, we must divide the path into a number of portions n each having an equal increase in potential $\Delta\varphi = (\varphi - \varphi_0)/n$. The frequencies at the extremities of successive portions form a series $\nu_0, \nu_1, \nu_2, \ldots, \nu_n = \nu$. We then obtain

$$\frac{\nu}{\nu_0} = \frac{\nu_1}{\nu_0}\frac{\nu_2}{\nu_1}\cdots\frac{\nu}{\nu_{n-1}} = \left(1 - \frac{\varphi - \varphi_0}{nc^2}\right)^n$$

In the limit $n \to \infty$ the frequency shift becomes

$$\frac{\Delta\nu}{\nu_0} \to e^{-(\varphi-\varphi_0)/c^2} - 1 \cong -\frac{1}{c^2}(\varphi - \varphi_0)$$

It is useful to make a comparison between a spectral line ν_0 emitted by an atom which is at a distance r_0 from the centre of the sun and a line emitted by an identical atom on Earth. We will assume the atom at r_0 to be at lower potential in the sun's gravitational field, while the atom on Earth is at a higher potential, as $r \gg r_0$. The line emitted by the latter atom will therefore have a lower frequency ν. The shift toward the red is estimated to be

$$\frac{\Delta\nu}{\nu_0} \cong -\frac{kM}{c^2}\left(\frac{1}{r} - \frac{1}{r_0}\right) \cong \frac{kM}{c^2}\frac{1}{r_0} \cong 2 \cdot 10^{-6}$$

where M, r_0 are the mass and radius of the sun, respectively. This result is in agreement with astronomical data.

Using Eqs.(5.25) and (5.26), in an accelerated frame Σ, or in an inertial frame Σ *under gravitational field*, the differential invariant ds^2, given by Eq.(4.31), becomes

$$ds^2 \cong d\xi^2\left(1 + \frac{2\Delta\varphi}{c^2}\right) + d\eta^2 + d\zeta^2 - c^2 d\tau^2\left(1 - \frac{2\Delta\varphi}{c^2}\right) \qquad (5.27)$$

Equation (5.27) suggests the curvature of space in the presence of mass, illustrated by the bending of a light path in a gravitational field. Consider a light flash emitted in a freely falling cabin, at right angles to its motion. The equivalence principle requires that this flash travels uniformly along a straight line inside the cabin, which is an accelerated frame. It follows that, relative to the Earth, considered as an inertial frame, the light path must be curved parabolically, just like the path of a projectile. The *local Euclidean space* must be then embedded in the *Riemann space* of a broader theory of gravitation called *general relativity*.

FURTHER READING
1. M. S. Longair - THEORETICAL CONCEPTS IN PHYSICS, Cambridge University Press, Cambridge, 2003.
2. W. D. McGlinn - INTRODUCTION TO RELATIVITY, Johns Hopkins University Press, Baltimore, 2002.
3. J. B. Kogut - INTRODUCTION TO RELATIVITY FOR PHYSICISTS AND ASTRONOMERS, Elsevier, Amsterdam, 2000.

PROBLEMS

5.1. Determine the worldline $x = x(ct)$ for the motion in one dimension of a particle of rest mass m_0 under the influence of a constant force F.

Solution

The equation of motion is stated the same way as in Newtonian mechanics, Eq.(2.9), which reads

$$\frac{d}{dt}\left(\frac{m_0 \vec{v}}{\sqrt{1-v^2/c^2}}\right) = \vec{F} \quad \text{or} \quad m_0 \vec{v}\frac{d\gamma}{dt} + m_0 \gamma \frac{d\vec{v}}{dt} = \vec{F}$$

and it follows that the vectors representing the force, the acceleration and the velocity lie in one plane, but the directions of $\vec{F}$ and $\vec{a} = d\vec{v}/dt$ do not coincide in the general case. However, for the motion in one dimension, where $v = dx/dt$, the equation of motion reduces to

$$v\frac{d\gamma}{dt} + \gamma \frac{dv}{dt} = \frac{F}{m_0}$$

where

$$\frac{d\gamma}{dt} = \left(1-\frac{v^2}{c^2}\right)^{-3/2}\frac{v}{c^2}\frac{dv}{dt} = \frac{\gamma^3}{c^2}v\frac{dv}{dt} \quad \text{or} \quad \frac{F}{m_0} = \gamma\frac{dv}{dt}\left(1+\gamma^2\frac{v^2}{c^2}\right) = \gamma^3\frac{dv}{dt}$$

This gives

$$\frac{F}{m_0}t = \int_0^v \frac{dv}{\left(1-v^2/c^2\right)^{3/2}} = \int_0^\varphi \frac{c\cos\varphi\, d\varphi}{\left(1-\sin^2\varphi\right)^{3/2}} = \int_0^\varphi \frac{c\, d\varphi}{\cos^2\varphi} = \tan\varphi = \frac{v}{\sqrt{1-v^2/c^2}}$$

where $\sin\varphi = v/c$, and hence, $\tan\varphi = \gamma v$. Rearranging yields

$$\frac{(F/m_0)t}{\sqrt{1+(F/m_0 c)^2 t^2}} = v = \frac{dx}{dt} \quad \text{or} \quad x = c\int_0^t \frac{(F/m_0 c)t\, dt}{\sqrt{1+(F/m_0 c)^2 t^2}} = \frac{m_0 c^2}{F}\sqrt{1+(F/m_0 c)^2 t^2}$$

Therefore, the worldline is a hyperbola of equation

$$x^2 - (ct)^2 = \left(\frac{m_0 c^2}{F}\right)^2$$

5.2. Find the trajectory of motion of a particle of rest mass m_0, charge q and initial velocity $\vec{v}_0$ directed along the positive x-axis of frame S, in a region occupied by a uniform field of force $\vec{F} = q\vec{E}$ which is parallel to the y-axis.

Solution

Since the direction of the electric force $\vec{F} = q\vec{E}$ is perpendicular to the velocity $\vec{v}_0$, it is convenient to resolve the equation of motion into a component parallel to $\vec{v}_0$, which reads

$$\frac{d}{dt}(m_0 \gamma v_x) = 0 \quad \text{or} \quad m_0 \gamma v_x = m_0 \gamma v_0 = p_0$$

and a perpendicular component, of the form

$$m_0 \gamma \frac{dv_y}{dt} = F \quad \text{or} \quad m_0 \gamma v_y = Ft = p_y$$

In view of Eq.(5.6), one obtains

$$v_x = \frac{p_0}{m_0 \gamma} = p_0 \frac{c^2}{E} \quad \text{and} \quad v_y = \frac{Ft}{m_0 \gamma} = Ft \cdot \frac{c^2}{E}$$

where the total energy (5.10) can be written in terms of the initial energy E_0 of the particle as

$$E = \sqrt{p_0^2 c^2 + F^2 t^2 c^2 + m_0^2 c^4} = \sqrt{E_0^2 + F^2 t^2 c^2}$$

Hence we have

$$\frac{dx}{dt} = \frac{p_0 c^2}{\sqrt{E_0^2 + F^2 t^2 c^2}} \quad \text{or} \quad x = \frac{p_0 c^2}{E_0} \int \frac{dt}{\sqrt{1+(Ftc/E_0)^2}}$$

Let $u = Ftc/E_0$, which gives

$$x = \frac{p_0 c}{F} \int \frac{du}{\sqrt{1+u^2}} = \frac{p_0 c}{F} \sinh^{-1}\left(\frac{Fct}{E_0}\right) \quad \text{or} \quad \frac{Fct}{E_0} = \sinh\left(\frac{Fx}{p_0 c}\right)$$

In a similar way, using

$$dE = \frac{F^2 c^2 t \, dt}{E} \quad \text{or} \quad t \, dt = \frac{E \, dE}{F^2 c^2}$$

we obtain

$$y = Fc^2 \int \frac{tdt}{E} = \frac{1}{F}\int dE = \frac{E_0}{F}\sqrt{1+(Fct/E_0)^2}$$

Substituting $(Fct/E_0) = \sinh(Fx/p_0 c)$ in this relation, provides the required equation for the trajectory as

$$y = \frac{E_0}{F}\sqrt{1+\sinh^2\left(\frac{Fx}{p_0 c}\right)} = \frac{E_0}{F}\cosh\left(\frac{Fx}{p_0 c}\right)$$

When $v \ll c$ we can pass to Newtonian mechanics, by substituting $p_0 = m_0 v_0$ and $E_0 = m_0 c^2$, and the trajectory reduces to a parabola, of equation

$$y \cong \frac{F}{2m_0 v_0^2}x^2 + \text{const}$$

5.3. A particle of rest mass m_1 and momentum $\vec{p}_0$ collides with another particle of rest mass m_2, which is at rest. Discuss the post-collision motion of the two particles, assuming an elastic impact in which the total rest mass of the system of the two particles remains invariable.

Solution

The law of conservation of momentum in the laboratory frame

$$\vec{p}_0 = \vec{p}_1 + \vec{p}_2$$

is valid both in classical mechanics, and in relativistic mechanics, and hence, it is convenient to give the same meaning to the notations already used in Problem 2.6, that is

$$p_1^2 = p_0^2 + p_2^2 - 2p_0 p_2 \cos\theta$$

The energy equation reads

$$E = E_0 + m_2 c^2 = E_1 + E_2$$

In view of Eq.(5.10), we have

$$E_0^2/c^2 = p_0^2 + m_1^2 c^2, \quad E_1^2/c^2 = p_1^2 + m_1^2 c^2, \quad E_2^2/c^2 = p_2^2 + m_2^2 c^2$$

and this gives

$$p_1^2 = E_1^2/c^2 - m_1^2 c^2 = E_1^2/c^2 - E_0^2/c^2 + p_0^2 = E_1^2/c^2 - (E - m_2 c^2)^2/c^2 + p_0^2$$

where

$$E_1 = E - E_2 = E - \sqrt{p_2^2 c^2 + m_2^2 c^4}$$

The substitution of the expression for p_1^2 into the law of conservation of momentum results in

$$\left(E-\sqrt{p_2^2c^2+m_2^2c^4}\right)^2/c^2 = (E-m_2c^2)^2/c^2 + p_2^2 - 2p_0p_2\cos\theta$$

On rearranging and rationalizing we obtain the form

$$E^2 p_2^2 = p_0^2 c^2 (p_2\cos\theta)^2 + 2Em_2 p_0 c^2 (p_2\cos\theta)$$

which can be written in terms of the two components $p_{2x} = p_2\cos\theta$ and $p_{2y} = p_2\sin\theta$ of the momentum of particle m_2 after the impact, and this gives

$$\left(1-\frac{p_0^2 c^2}{E^2}\right)p_{2x}^2 + p_{2y}^2 = 2\frac{m_0 p_0 c^2}{E}p_{2x}$$

This equation describes the curve traced by the tip of the vector $\vec{p}_2$ and is an ellipse passing through its origin

$$\left(\frac{p_{2x}}{a}-1\right)^2 + \frac{p_{2y}^2}{b^2} = 1$$

where

$$a = \frac{m_2 c^2 E}{E^2 - p_0^2 c^2}p_0, \qquad b = \frac{m_2 c^2}{\sqrt{E^2 - p_0^2 c^2}}p_0, \qquad b = a\sqrt{1-\frac{p_0^2 c^2}{E^2}}$$

and hence, a is the major semi-axis, and b is the minor one. It is now convenient to express a in terms of the rest mass ratio $A = m_1/m_2$, and this can done by developing

$$E^2 = E_0^2 + m_2^2 c^4 + 2E_0 m_2 c^2 = (p_0^2 c^2 + m_1^2 c^4) + m_2^2 c^4 + 2E_0 m_2 c^2$$

$$= p_0^2 c^2 + m_1^2 c^4 - m_2^2 c^4 + 2(E_0 + m_2 c^2)m_2 c^2 = p_0^2 c^2 + m_1^2 c^4 - m_2^2 c^4 + 2Em_2 c^2$$

The substitution of $E^2 - p_0^2 c^2$ into the formula for a gives

$$a = \frac{p_0}{2+(A^2-1)m_2 c^2/E}$$

It is apparent that $2a < p_0$ when $A > 1$ and both particles move forwards after the impact, relative to the direction of $\vec{p}_0$, $2a > p_0$ when $A < 1$, and it is possible that after the impact particle m_1 moves backwards, and $2a = p_0$ for $A = 1$, in which case φ varies from 0 to $\pi/2$.

In the limiting case of the Newtonian mechanics, we can neglect $p_0 c$ in comparison with $m_1 c^2$, since $v \ll c$. To this approximation, the total energy reduces to

$$E = E_0 + m_2 c^2 = \sqrt{p_0^2 c^2 + m_1^2 c^4} + m_2 c^2 \approx (m_1 + m_2)c^2$$

and we can also neglect $p_0^2 c^2$ in comparison with E^2, to obtain

$$a \cong \frac{m_2 c^2}{E} p_0 = \frac{m_2}{m_1 + m_2} p_0 = \frac{p_0}{1+A}, \qquad b \cong \frac{m_2 c^2}{E} = \frac{p_0}{1+A}$$

where $A = m_1/m_2$ is the rest mass ratio. The ellipse degenerates into a circle passing through the origin of the vector $\vec{p}_0$, and the particles move after the elastic impact as described before in Problem 2.13.

5.4. Plane wave light emitted in frame S are represented by the solutions of the wave equation in the form

$$\Psi(\vec{r},t) = A e^{i(\vec{k}\cdot\vec{r}-\omega t)}$$

where $\vec{k}$ is the wave vector, of magnitude $k = \omega/c$, directed at an angle θ to the x-axis, and ω is the angular frequency of light. Find the frequency ω' received by an observer in frame S', moving at uniform velocity v in the positive x-direction relative to S.

Solution

The phase $\vec{k}\cdot\vec{r} - \omega t$ of the wave can be interpreted as the scalar product of the wave four-vector

$$k_\mu = \left(\vec{k}, i\frac{\omega}{c}\right)$$

and the position four-vector $S_\mu = (\vec{r}, ict)$. There is proportionality between k_μ and the momentum four-vector p_μ, Eq.(5.16), as $k \sim p$ and $\omega \sim E$ hold for the same scalar coefficient

$$p = \hbar k, \qquad E = \hbar \omega$$

according to the de Broglie hypothesis, see Eq.(28.35), and Planck's hypothesis, see Eq.(28.10). Since the wave four-vector in the two frames are

$$k_\mu = \left(\frac{\omega}{c}\cos\theta, \frac{\omega}{c}\sin\theta, 0, i\frac{\omega}{c}\right), \qquad k'_\mu = \left(\frac{\omega'}{c}\cos\theta', \frac{\omega'}{c}\sin\theta', 0, i\frac{\omega'}{c}\right)$$

substituting into the last transformation equation (4.28), we obtain

$$\frac{\omega'}{c} = \gamma\left(\frac{\omega}{c} - \beta\frac{\omega}{c}\cos\theta\right) \qquad \text{or} \qquad \omega' = \omega\frac{1-\beta\cos\theta}{\sqrt{1-\beta^2}}$$

The wave is observed along the direction θ' given either by the first equation (4.28) as

$$\frac{\omega'}{c}\cos\theta' = \gamma\left(\frac{\omega}{c}\cos\theta - \beta\frac{\omega}{c}\right) \qquad \text{i.e.,} \qquad \cos\theta' = \frac{\cos\theta - \beta}{1-\beta\cos\theta}$$

or, by the second transformation, in the form

$$\frac{\omega'}{c}\sin\theta' = \frac{\omega}{c}\sin\theta \qquad \text{i.e.,} \qquad \sin\theta' = \frac{\sin\theta\sqrt{1-\beta^2}}{1-\beta\cos\theta}$$

Assuming that $\theta = 0$, one obtains $\theta' = 0$, and hence, we have

$$\omega' = \omega\frac{1-\beta}{\sqrt{1-\beta^2}} = \omega\sqrt{\frac{1-\beta}{1+\beta}}$$

which is the longitudinal Doppler effect. Most Doppler measurements are concerned with $d\lambda/\lambda$ values, where

$$\lambda' = \lambda\frac{\omega}{\omega'} \qquad \text{i.e.,} \qquad \frac{d\lambda}{\lambda} = \frac{\lambda'}{\lambda} - 1 = \sqrt{\frac{1+\beta}{1-\beta}} - 1$$

When the relative speed between source and observer is $v \ll c$, that is $\beta \ll 1$, we have

$$\frac{d\lambda}{\lambda} \cong \left(1 + \frac{\beta}{2}\right)^2 - 1 \cong \beta$$

a result established in classical mechanics, where $\omega' = \omega c/(c+v)$, thus

$$\lambda' = \lambda\left(1 + \frac{v}{c}\right) \qquad \text{or} \qquad \frac{\Delta\lambda}{\lambda} = \frac{v}{c} = \beta$$

However, for $\theta = \pi/2$, the wave is observed in S' along a different direction, since $\cos\theta' = -\beta$, and its angular frequency is

$$\omega' = \frac{\omega}{\sqrt{1-\beta^2}}$$

This transversal Doppler effect, which reduces for $\beta \ll 1$ to

$$\omega' \cong \omega\left(1 - \frac{\beta^2}{2}\right)$$

is not predicted in the frame of Newtonian mechanics.

5.5. Consider two identical twins, the first one, A, staying on Earth which is assumed to be an inertial frame, while the second one, B, leaving the Earth with uniform velocity v and, after a long trip, reversing course and returning to Earth with opposite uniform velocity $-v$. In the Earth's frame A observes B's clock, which runs slower due to time dilatation, and concludes that the measured trip duration is shorter by a factor γ than the duration measured by his/her own clock, which means that B is younger than A at the end of the voyage. However, since time dilation only depends on relative motion, the twin B could also conclude that A is younger than him/her when the trip is finished. Find an explanation for this apparent paradox.

Solution

A spacetime diagram readily shows the asymmetry of the two world lines, with the consequence that the roles of the twins are not interchangeable.

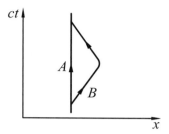

According to the equivalence principle, the acceleration of the reference frame of B, at the reversal from v to $-v$, which takes place in a time interval $\Delta\tau$, has the same effect as an average gravitational field of strength $2v/\Delta\tau$. In this frame the field $\vec{g}$ is always opposite to the acceleration, and points towards A, so that the clock of A is at a higher gravitational potential than that of B, the difference increasing with the distance $v\tau$ between them as

$$\Delta\varphi = gd = \frac{2v}{\Delta\tau}v\tau = 2v^2\frac{\tau}{\Delta\tau}$$

As seen from the B frame, A's clock runs faster than B's clock during reversal, according to Eq.(5.25) which gives

$$\Delta t \cong \Delta\tau\left(1 + \frac{2v^2}{c^2}\frac{\tau}{\Delta\tau}\right) = \Delta\tau + \frac{2v^2}{c^2}\tau$$

However, during the field-free time interval 2τ, A's clock appears from the B frame to be moving uniformly with velocity $\pm v$ and therefore running slower than B's clock, according to

$$2t = \frac{2\tau}{\gamma} = 2\tau\left(1 - \frac{v^2}{c^2}\right)^{1/2} \cong 2\tau - \frac{v^2}{c^2}\tau$$

Thus the total time measured by A's clock, calculated in the B frame, is always longer than the proper time $2\tau + \Delta\tau$ on B's clock, as given by

$$2t + \Delta t = 2\tau + \Delta\tau + \frac{v^2}{c^2}\tau$$

The paradox is therefore resolved if we take into account the principle of equivalence.

5.6. Determine how the mass density of a body depends on its speed.

5.7. Find a representation of the general transformation for momentum and energy in a vector form that is specified by the direction and the magnitude of a velocity vector $\vec{v}$.

5.8. Determine the acceleration of a particle of rest mass m_0 for given values of the applied force $\vec{F}$ and velocity $\vec{v}$. Consider the special case of a particle of charge q under the influence of the Lorentz force $\vec{F} = q(\vec{E} + \vec{v} \times \vec{B})$.

5.9. Find a relationship between the relativistic acceleration of a particle of rest mass m_0 and force, in terms of their components $a_\perp$, $a_\parallel$ and $F_\perp$, $F_\parallel$ which are perpendicular and parallel to the velocity $\vec{v}$ of the particle.

5.10. A particle of rest mass m_1 and velocity $\vec{v}_0$ collides with a stationary particle of rest mass m_2. Show that the rest mass of the newly formed particle in the case of perfectly plastic impact exceeds the sum of m_1 and m_2 and find the velocity of that particle.

6. CONTINUUM MECHANICS

6.1. STRAIN IN A CONTINUUM
6.2. STRESS IN A CONTINUUM
6.3. ELASTIC SOLIDS
6.4. FLUID MECHANICS

6.1. STRAIN IN A CONTINUUM

The macroscopic description of local motion in a deformable medium, where the particles are not assumed to remain a fixed distance apart, is based upon the concept of a *continuum*, which can be indefinitely divided into *elements*. The dimensions of the elements can therefore take any value down to zero. We will consider a typical element as a rectangular parallelepiped, of elementary volume $dV = dxdydz$, located about a given position x, y, z in a Cartesian coordinate system, at a time t. The variation of the properties of the element with size may be neglected, and the physical quantities are defined as continuous functions of position and time only. This is known as the *Eulerian representation* of motion in a continuum, in terms of the time dependence of properties associated to certain points in the medium. The rate of change of any property $F(x, y, z, t)$, associated with a moving element of the continuous medium, is defined by its total time derivative

$$\frac{dF}{dt} = \frac{\partial F}{\partial x}\frac{dx}{dt} + \frac{\partial F}{\partial y}\frac{dy}{dt} + \frac{\partial F}{\partial z}\frac{dz}{dt} + \frac{\partial F}{\partial t}$$

$$= v_x \frac{\partial F}{\partial x} + v_y \frac{\partial F}{\partial y} + v_z \frac{\partial F}{\partial z} + \frac{\partial F}{\partial t} = (\vec{v} \cdot \nabla)F + \frac{\partial F}{\partial t}$$

(6.1)

where $\partial F / \partial t$ is the time rate of change of F at a given position (x, y, z).

A change in the relative position of elements of a continuum is called a *deformation*. If an element, located at x, y, z (O in Figure 6.1) in an unstrained medium at time $t = 0$, has been displaced in the process of deformation, to a point $x + \xi_x, y + \xi_y, z + \xi_z$ (O' in Figure 6.1) at time t, we may define a displacement vector

$\vec{\xi} = (\xi_x, \xi_y, \xi_z)$ which is a function of position and time. If the element is distorted in size or shape, we have a *strain*, determined by the derivatives of ξ_x, ξ_y, ξ_z with respect to x, y, z. The two dimensional case represented in Figure 6.1 shows how to express the strain components in terms of displacements along the coordinate axes. The directions in space of two mutually perpendicular edges dx, dy of an element before deformation are chosen as Cartesian coordinate axes. On deformation these lines rotate from OX and OY to $O'X'$ and $O'Y'$ and are no longer mutually perpendicular. For small strains we may define

- the *dilatational* (or *extensional*) *strain* in the *x*-direction

$$\varepsilon_{xx} = \lim_{dx \to 0} \frac{\xi_x(x+dx, y, z) - \xi_x(x, y, z)}{dx} = \frac{\partial \xi_x}{\partial x}$$

and similarly the dilatational strain in the *y*- and *z*-directions, respectively, as

$$\varepsilon_{yy} = \frac{\partial \xi_y}{\partial y}, \qquad \varepsilon_{zz} = \frac{\partial \xi_z}{\partial z}.$$

- the *shear strain* in the *y*-direction of a point on the *x*-axis, given by the angle α (see Figure 6.1) and denoted by ε_{yx}

$$\varepsilon_{yx} = \alpha = \lim_{dx \to 0} \frac{\xi_y(x+dx, y, z) - \xi_y(x, y, z)}{dx} = \frac{\partial \xi_y}{\partial x}$$

and similarly the shear strain in the *x*-direction of a point on the *y*-axis

$$\varepsilon_{xy} = \beta = \frac{\partial \xi_x}{\partial y}$$

The last two components are not independent, as $\alpha + \beta = \pi/2 - \theta$, so that the decrease of the angle between dx and dy is introduced as a new parameter, called *angle of shear*, in the *xy*-plane and given the symbol γ_{xy} according to

$$\gamma_{xy} = \frac{\pi}{2} - \theta = \varepsilon_{yx} + \varepsilon_{xy} = \frac{\partial \xi_y}{\partial x} + \frac{\partial \xi_x}{\partial y}$$

It is obvious that a rigid rotation of the deformed body about O allows α and β to be varied, so that we may always assume that $\alpha = \beta$ or $\varepsilon_{yx} = \varepsilon_{xy}$. Consequently, we have

$$\varepsilon_{xy} = \frac{1}{2} \gamma_{xy} = \frac{1}{2}\left(\frac{\partial \xi_y}{\partial x} + \frac{\partial \xi_x}{\partial y} \right) \tag{6.2}$$

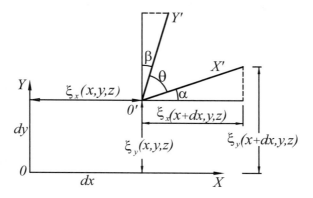

Figure 6.1. Components of strain tensor

Similar results apply to the shear strains in the *yz*- and *xz*-planes. We thus have the following six independent parameters which define the state of strain in terms of the three displacement components:

$$\varepsilon_{xx} = \frac{\partial \xi_x}{\partial x}, \quad \varepsilon_{yy} = \frac{\partial \xi_y}{\partial y}, \quad \varepsilon_{zz} = \frac{\partial \xi_z}{\partial z}$$

$$\varepsilon_{xy} = \varepsilon_{yx} = \frac{1}{2}\left(\frac{\partial \xi_y}{\partial x} + \frac{\partial \xi_x}{\partial y}\right), \quad \varepsilon_{yz} = \varepsilon_{zy} = \frac{1}{2}\left(\frac{\partial \xi_z}{\partial y} + \frac{\partial \xi_y}{\partial z}\right), \quad \varepsilon_{zx} = \varepsilon_{xz} = \frac{1}{2}\left(\frac{\partial \xi_x}{\partial z} + \frac{\partial \xi_z}{\partial x}\right) \quad (6.3)$$

The strain components are positive if the displacement ξ_i corresponds to an axis which has the same sign as the position axis x_i and negative if these signs are different. The matrix

$$\hat{\varepsilon} = \begin{pmatrix} \varepsilon_{xx} & \varepsilon_{xy} & \varepsilon_{xz} \\ \varepsilon_{yx} & \varepsilon_{yy} & \varepsilon_{yz} \\ \varepsilon_{zx} & \varepsilon_{zy} & \varepsilon_{zz} \end{pmatrix} \quad (6.4)$$

is called the *strain matrix* of a given element in the continuous medium. It gives the representation of a *second rank symmetric tensor* $\hat{\varepsilon}_{ij}$, which allows us to write the relationship between the position vector components x_i and those of the displacement vector ξ_i as

$$\xi_i = \sum_j \varepsilon_{ij} x_j \quad (6.5)$$

where $i, j = x, y, z$. Since it is a symmetrical matrix about the principal diagonal, three principal directions always exist, which are orthogonal to one another, with reference to which the matrix is diagonal (see Appendix II). This means that three orthogonal directions exist which remain orthogonal after deformation. The three strain components corresponding to these principal directions are called the *principal strains* and will be denoted by the use of a single suffix, that is ε_x, ε_y and ε_z. When the principal axes are known, they are usually taken as the coordinate axes and the matrix becomes diagonal.

EXAMPLE 6.1. Volume expansion

Let us consider a rectangular element at the origin of the Cartesian coordinate system, whose edges of length dx, dy and dz are parallel to the *principal axes* of strain. On deformation this element becomes again a rectangular parallelepiped, with edges $dx + d\xi_x = dx(1+\varepsilon_x)$, $dy(1+\varepsilon_y)$ and $dz(1+\varepsilon_z)$ in length.

The expansion of a volume $V = dxdydz$ due to strain produced in the medium can be written as

$$dV = dxdydz(1+\varepsilon_x)(1+\varepsilon_y)(1+\varepsilon_z) - dxdydz \cong dxdydz(\varepsilon_x + \varepsilon_y + \varepsilon_z)$$

so that the expansion of the unit volume, called *dilatation* Δ or volume expansion takes the simple form

$$\Delta = \frac{dV}{V} = \varepsilon_x + \varepsilon_y + \varepsilon_z \tag{6.6}$$

The sum of the principal values of a tensor is well known to be invariant, and hence, it is always equal to the sum of its diagonal components in any coordinate system, $Tr(\hat{\varepsilon}) = \varepsilon_{xx} + \varepsilon_{yy} + \varepsilon_{zz}$, which can be written as

$$\Delta = \frac{dV}{V} = \varepsilon_x + \varepsilon_y + \varepsilon_z = \varepsilon_{xx} + \varepsilon_{yy} + \varepsilon_{zz} = \frac{\partial \xi_x}{\partial x} + \frac{\partial \xi_y}{\partial y} + \frac{\partial \xi_z}{\partial z}$$

$$= \left(\vec{e}_x \frac{\partial}{\partial x} + \vec{e}_y \frac{\partial}{\partial y} + \vec{e}_z \frac{\partial}{\partial z}\right) \cdot (\xi_x \vec{e}_x + \xi_y \vec{e}_y + \xi_z \vec{e}_z) = \nabla \cdot \vec{\xi} \tag{6.7}$$

The time rate of change of deformation is obtained by replacing the components of strain by their time derivatives, involving the velocities of the element, which define a vector $\vec{v} = (v_x, v_y, v_z)$. In view of Eq.(6.2) the components of the *rate of strain* are expressed in terms of velocities as

$$e_{xx} = \frac{\partial v_x}{\partial x}, \qquad e_{yy} = \frac{\partial v_y}{\partial y}, \qquad e_{zz} = \frac{\partial v_z}{\partial z}$$

$$e_{xy} = e_{yx} = \frac{1}{2}\left(\frac{\partial v_y}{\partial x} + \frac{\partial v_x}{\partial y}\right), \quad e_{yz} = e_{zy} = \frac{1}{2}\left(\frac{\partial v_z}{\partial y} + \frac{\partial v_y}{\partial z}\right), \quad e_{zx} = e_{xz} = \frac{1}{2}\left(\frac{\partial v_x}{\partial z} + \frac{\partial v_z}{\partial x}\right) \tag{6.8}$$

and form the symmetric matrix

$$\hat{e} = \begin{pmatrix} e_{xx} & e_{xy} & e_{xz} \\ e_{yx} & e_{yy} & e_{yz} \\ e_{zx} & e_{zy} & e_{zz} \end{pmatrix} \qquad (6.9)$$

Substituting the velocities associated with the deformation of an element into Eq.(6.7) yields

$$\frac{1}{V}\frac{dV}{dt} = e_x + e_y + e_z \qquad (6.10)$$

and this describes the volume expansion per unit time. The left-hand side of this equation may be written in terms of density or mass per unit volume ρ of the element, using the conservation of its mass

$$\frac{d}{dt}(\rho V) = 0 = V\frac{d\rho}{dt} + \rho\frac{dV}{dt} \qquad \text{or} \qquad \frac{1}{V}\frac{dV}{dt} = -\frac{1}{\rho}\frac{d\rho}{dt}$$

The right-hand side of Eq.(6.10) is the divergence of velocity

$$e_{xx} + e_{yy} + e_{zz} = \frac{\partial v_x}{\partial x} + \frac{\partial v_y}{\partial y} + \frac{\partial v_z}{\partial z} = \nabla \cdot \vec{v}$$

so that Eq.(6.7) takes the form

$$\frac{1}{\rho}\frac{d\rho}{dt} + \nabla \cdot \vec{v} = 0 \qquad (6.11)$$

called the *equation of continuity* for a fluid of density ρ.

6.2. Stress in a Continuum

A force can be applied on an element of volume V, bounded by a surface S, in two ways. If it acts directly on every particle of the element so that it is proportional to the volume V and the density ρ, the force is called a *body force*. Gravitational and electromagnetic interactions are both examples of a body force. If it acts directly on the surface S and is transmitted to the interior through the internal particles, the force is known as a *contact* or *stress force*. While a body force is a function of position and time, a stress force is a function of position, time, and orientation of the element of surface, of area dS, across which it is exerted. Therefore it is convenient to define a *stress vector*, at a given point on the surface, as

$$\vec{\sigma}_S = \lim_{dS \to 0} \frac{d\vec{F}}{dS}$$

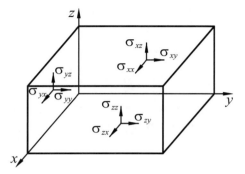

Figure 6.2. Components of stress tensor

For the element with sides dx, dy, dz located about a point (x, y, z), the stress vector at each face has three components, as in Figure 6.2.

The components perpendicular to dS define the *dilatational stress*, and those parallel to dS give the *shear stress*. The stress vectors on the three planes S_x, S_y and S_z which are perpendicular to the x-, y- and z-axes, respectively, have components along the coordinate axes of the form

$$\vec{\sigma}_{S_x} = (\sigma_{xx}, \sigma_{xy}, \sigma_{xz})$$
$$\vec{\sigma}_{S_y} = (\sigma_{yx}, \sigma_{yy}, \sigma_{yz}) \qquad (6.12)$$
$$\vec{\sigma}_{S_z} = (\sigma_{zx}, \sigma_{zy}, \sigma_{zz})$$

which are denoted by two suffixes: the first one specifies the direction of the axis normal to the face on which the stress acts, and the second specifies the axis along which the stress component is directed. The usual sign convention is that the stress component is positive if the directions of both specified axes have the same sign and negative if they have different signs. The nine components define the *stress matrix*

$$\hat{\sigma} = \begin{pmatrix} \sigma_{xx} & \sigma_{xy} & \sigma_{xz} \\ \sigma_{yx} & \sigma_{yy} & \sigma_{yz} \\ \sigma_{zx} & \sigma_{zy} & \sigma_{zz} \end{pmatrix} \qquad (6.13)$$

where the diagonal elements are associated with the normal stress (called *tensile* if positive and *compressive* if negative) and the off-diagonal ones give the shear stress. The matrix and the corresponding second rank stress tensor $\hat{\sigma}_{ij}$, which relates the force vector to the vector denoting area, are symmetrical about the principal diagonal, just like the strain matrix (6.4). This can be shown by considering the equilibrium of our parallelepiped element. Figure 6.3 illustrates the diagram of our isolated element in the *xy*-plane.

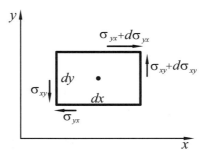

Figure 6.3. Diagram in the *xy*-plane of an isolated differential element

The total torque about an axis passing through the centre point parallel to the *z*-axis must be zero and is given by

$$G = \sigma_{xy} dydz \frac{dx}{2} + \left(\sigma_{xy} + d\sigma_{xy}\right) dydz \frac{dx}{2} - \sigma_{yx}(dxdz)\frac{dy}{2} - \left(\sigma_{yx} + d\sigma_{yx}\right) dxdz \frac{dy}{2}$$

$$\cong \left(\sigma_{xy} - \sigma_{yx}\right) dxdydz = 0$$

It follows that $\sigma_{xy} = \sigma_{yx}$, and two others similar relations can be obtained in the form

$$\sigma_{xy} = \sigma_{yx}, \quad \sigma_{xz} = \sigma_{zx}, \quad \sigma_{yz} = \sigma_{zy} \qquad (6.14)$$

Therefore there are only six independent stress components, in terms of which we can express the *equations of motion* of a continuous medium. Newton's second law requires the total force acting on an element to be equal to the mass times the acceleration. Let us compute all the forces acting at an instant *t*, in a given direction on an element of volume $V = dxdyd$, with one corner at the point (x, y, z). Figure 6.4 illustrates the components of the stress, in the direction $\vec{e}_y$, across a pair of faces *dydz*, which are σ_{xy} and $\sigma_{xy} + \left(\partial \sigma_{xy} / \partial x\right) dx$. According to Newton's third law, the outside medium exerts across this pair of faces a net force corresponding to the difference of the two stress components (see Figure 6.3), that is

$$\left(\frac{\partial \sigma_{xy}}{\partial x} dx\right) dydz$$

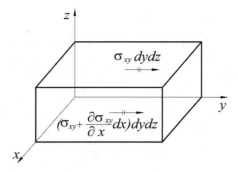

Figure 6.4. Components of stress across a pair of opposite faces of a differential element

In the same way we can find the components of the stress exerted in the same direction on the other two pairs of faces. Their superposition gives the total $\vec{e}_y$ component of stress as

$$\left(\frac{\partial \sigma_{xy}}{\partial x} + \frac{\partial \sigma_{yy}}{\partial y} + \frac{\partial \sigma_{zy}}{\partial z}\right) dxdydz$$

If ρ is the density, dv_y/dt is the $\vec{e}_y$ component of the acceleration and f_y stands for the $\vec{e}_y$-component of the body force per unit mass, we have

$$\rho \frac{dv_y}{dt} = \rho f_y + \frac{\partial \sigma_{xy}}{\partial x} + \frac{\partial \sigma_{yy}}{\partial y} + \frac{\partial \sigma_{zy}}{\partial z} \qquad (6.15)$$

where the volume $dxdydz$ has been divided out of all terms. Equations for the $\vec{e}_x$ and $\vec{e}_z$ -directions can be obtained in a similar manner as

$$\rho \frac{dv_x}{dt} = \rho f_x + \frac{\partial \sigma_{xx}}{\partial x} + \frac{\partial \sigma_{yx}}{\partial y} + \frac{\partial \sigma_{zx}}{\partial z} \qquad (6.16)$$

and

$$\rho \frac{dv_z}{dt} = \rho f_z + \frac{\partial \sigma_{xz}}{\partial x} + \frac{\partial \sigma_{yz}}{\partial y} + \frac{\partial \sigma_{zz}}{\partial z} \qquad (6.17)$$

The general structure of Eqs.(6.15), (6.16) and (6.17) is valid for both the solid and fluid continuous media. It can be converted into the two specific forms using the appropriate expressions for stress, either in terms of strain for elastic media or in terms of rate of strain for fluids.

6.3. ELASTIC SOLIDS

A solid where the strain is completely determined by the stress is called a *perfectly elastic* material. The stress-strain relationships can be derived from the representation of any deformation as a sum of a *dilatation* Δ which causes a change in the volume of an element, but no change in its shape and a *pure shear* which alters the shape while the volume is unchanged. We assume a deformable material obeying *Hooke's law*, which states that *the strain is proportional to the stress*, that is both dilatation and pure shear are linearly dependent on stress. Furthermore we assume that the material is *isotropic*. The slope of the linear relationship between the dilatational stress σ_d and strain Δ, is a constant called the *bulk modulus K*, given by

$$K = \frac{\sigma_d}{\Delta} \qquad (6.18)$$

where the dilatation Δ is expressed in Eq.(6.6). The *shear modulus* μ is defined in a similar way as

$$\mu = \frac{\sigma_s}{\gamma} \qquad (6.19)$$

where σ_s defines a pure shear stress component, and γ is the corresponding angle of shear. Provided the strains are small, the principal axes of strain will be coincident with those of the applied stress. Choosing the coordinate axes along these *principal axes*, Eq.(6.6) takes the form

$$\varepsilon_x + \varepsilon_y + \varepsilon_z = \Delta = 3\varepsilon_d \qquad (6.20)$$

where ε_d is the principal dilatational strain and is a constant in an isotropic solid. The *principal shear* strain components ε_{xs}, ε_{ys}, ε_{zs}, related to the corresponding angles of shear as given by Eq.(6.2), can be then defined as

$$\varepsilon_{xs} = \varepsilon_x - \varepsilon_d, \qquad \varepsilon_{ys} = \varepsilon_y - \varepsilon_d, \qquad \varepsilon_{zs} = \varepsilon_z - \varepsilon_d \qquad (6.21)$$

and obviously obey the equation

$$\varepsilon_{xs} + \varepsilon_{ys} + \varepsilon_{zs} = 0 \qquad (6.22)$$

If compared to Eq.(6.20), this relation shows that the dilatation Δ occuring when a pure shear strain is applied should be zero. We may then similarly write each of the principal stresses $\sigma_x, \sigma_y, \sigma_z$ as a sum of a pure shear and a constant dilatational stress component σ_d, which is proportional to a constant principal strain ε_d, that is

$$\sigma_x = \sigma_{xs} + \sigma_d, \qquad \sigma_y = \sigma_{ys} + \sigma_d, \qquad \sigma_z = \sigma_{zs} + \sigma_d \qquad (6.23)$$

By analogy with shear strain, we can define a pure shear stress as the one for which the sum of the three principal stresses is zero, as given by

$$\sigma_{xs} + \sigma_{ys} + \sigma_{zs} = 0 \tag{6.24}$$

Adding together Eqs.(6.23) yields

$$\sigma_d = \frac{1}{3}(\sigma_x + \sigma_y + \sigma_z) \tag{6.25}$$

that is the dilatational stress component σ_d is equal to the average of the principal stresses. Substituting Eqs.(6.18), (6.19) and (6.20) into Eq.(6.23) we obtain

$$\sigma_x = 3K\varepsilon_d + 2\mu\varepsilon_{xs}$$

$$\sigma_y = 3K\varepsilon_d + 2\mu\varepsilon_{ys} \tag{6.26}$$

$$\sigma_z = 3K\varepsilon_d + 2\mu\varepsilon_{zs}$$

By eliminating the shear stress components between Eqs.(6.21) and (6.26), we obtain

$$\sigma_x = (3K - 2\mu)\varepsilon_d + 2\mu\varepsilon_x$$

$$\sigma_y = (3K - 2\mu)\varepsilon_d + 2\mu\varepsilon_y \tag{6.27}$$

$$\sigma_z = (3K - 2\mu)\varepsilon_d + 2\mu\varepsilon_z$$

It is common practice to introduce a new constant, defined as

$$\lambda = K - \frac{2}{3}\mu \tag{6.28}$$

so that, substituting for ε_d from Eq.(6.20), Eq.(6.27) takes the form

$$\sigma_x = \lambda\Delta + 2\mu\varepsilon_x = (\lambda + 2\mu)\varepsilon_x + \lambda\varepsilon_y + \lambda\varepsilon_z$$

$$\sigma_y = \lambda\Delta + 2\mu\varepsilon_y = \lambda\varepsilon_x + (\lambda + 2\mu)\varepsilon_y + \lambda\varepsilon_z \tag{6.29}$$

$$\sigma_z = \lambda\Delta + 2\mu\varepsilon_z = \lambda\varepsilon_x + \lambda\varepsilon_y + (\lambda + 2\mu)\varepsilon_z$$

The two constants λ and μ, which relate the principal stress and strain in an isotropic elastic material, are called the *Lamé constants*. For *arbitrary axes* the equations relating stress and strain components are obtained as

6. CONTINUUM MECHANICS

$$\sigma_{xx} = \lambda\Delta + 2\mu\varepsilon_{xx}, \qquad \sigma_{yy} = \lambda\Delta + 2\mu\varepsilon_{yy}, \qquad \sigma_{zz} = \lambda\Delta + 2\mu\varepsilon_{zz}$$
$$\sigma_{xy} = 2\mu\varepsilon_{xy}, \qquad \sigma_{yz} = 2\mu\varepsilon_{yz}, \qquad \sigma_{xz} = 2\mu\varepsilon_{xz}$$
(6.30)

where Δ is now expressed by Eq.(6.7). The equations of motion (6.15), (6.16) and (6.17) can then be converted into equations for the displacement, substituting for the stress in terms of strain, as given by Eq.(6.30), which yields

$$\rho\frac{d^2\xi_x}{dt^2} = \rho f_x + (\lambda + \mu)\frac{\partial}{\partial x}\nabla\cdot\vec{\xi} + \mu\nabla^2\xi_x$$

$$\rho\frac{d^2\xi_y}{dt^2} = \rho f_y + (\lambda + \mu)\frac{\partial}{\partial y}\nabla\cdot\vec{\xi} + \mu\nabla^2\xi_y \qquad (6.31)$$

$$\rho\frac{d^2\xi_z}{dt^2} = \rho f_z + (\lambda + \mu)\frac{\partial}{\partial z}\nabla\cdot\vec{\xi} + \mu\nabla^2\xi_z$$

or in compact vector notation

$$\rho\frac{d^2\vec{\xi}}{dt^2} = \rho\vec{f} + (\lambda + \mu)\nabla(\nabla\cdot\vec{\xi}) + \mu\nabla^2\vec{\xi} \qquad (6.32)$$

EXAMPLE 6.2. Elastic wave equations

Consider a vibrating continuum in one dimension, the simplest case being an elastic string, under negligible body forces, $\vec{f} = 0$. If the displacement is a function of time and of distance in one direction, say x, we assume in Eqs.(6.31) that any derivative with respect to y and z is zero. The three component equations reduce to

$$\rho\frac{d^2\xi_x}{dt^2} = (\lambda + 2\mu)\frac{\partial^2\xi_x}{\partial x^2}$$

$$\rho\frac{d^2\xi_y}{dt^2} = \mu\frac{\partial^2\xi_y}{\partial x^2} \qquad (6.33)$$

$$\rho\frac{d^2\xi_z}{dt^2} = \mu\frac{\partial^2\xi_z}{\partial x^2}$$

and define the motion of an elastic solid along the x-axis. Equations (6.33) are known as the elastic plane wave equations in one dimension. The first equation corresponds to a displacement ξ_x in the direction of wave propagation, x, and is called a *longitudinal* wave. The second and third equations describe *transverse* waves where displacements are at right angles to the direction of propagation. Comparison with the wave equation (4.1) shows that the velocity $\sqrt{(\lambda+2\mu)/\rho}$ of longitudinal waves is always greater than that of transverse waves $\sqrt{\mu/\rho}$.

6.4. FLUID MECHANICS

The relationships derived above for elastic solids apply to media undergoing a displacement proportional to the applied stress, which is constant with time and where we assume Hooke's law behaviour. However many solids flow under continued stress, the motion being opposed by a viscous resistance which is proportional to the velocity, rather than to the displacement. These solids approach the properties of a liquid which can only sustain a shear stress. A distinction between a viscous fluid and a solid is that the viscous flow corresponds to large displacements of the particles, whereas in elasticity they are always regarded as small. The equations of fluid mechanics can be derived from a *postulated linear relationship between stress and rate of strain*. The shear stress per unit velocity gradient is called the dynamic *coefficient of viscosity* η, defined as

$$\sigma_{xy} = 2\eta e_{xy} = \eta\left(\frac{\partial v_x}{\partial y} + \frac{\partial v_y}{\partial x}\right)$$

$$\sigma_{yz} = 2\eta e_{yz} = \eta\left(\frac{\partial v_y}{\partial z} + \frac{\partial v_z}{\partial y}\right) \quad (6.34)$$

$$\sigma_{xz} = 2\eta e_{xz} = \eta\left(\frac{\partial v_x}{\partial z} + \frac{\partial v_z}{\partial x}\right)$$

where the rate of shear strain components (6.8) give the angular change per unit time.

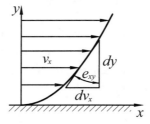

Figure 6.5. Velocity distribution in the *xy*-plane in a viscous fluid flow parallel to *x*

The significance of the first equation (6.34) is illustrated in Figure 6.5 for a viscous flow along the *x*-axis. From Eq.(6.34) it follows that in a fluid which has zero viscosity there are no shear forces $\sigma_{xy} = \sigma_{yz} = \sigma_{xz} = 0$. The normal stress components σ_{xx}, σ_{yy} and σ_{zz} cannot obey equations of the form (6.34) since they must not vanish in the absence of shear strain. Let us consider a fluid at rest where each element is in equilibrium in the absence of shear forces so that the fluid surrounding the element exerts only normal forces across its faces. Taking a tetrahedral element represented in Figure 6.6 and denoting by p_x, p_y, p_z, p_n the forces per unit area across its four faces, we obtain for the component in the $\vec{e}_y$-direction of the vector equation of motion of the element at rest the form

$$0 = p_y \frac{dxdz}{2} - p_n(\vec{e}_n \cdot \vec{e}_y)dS = p_y \frac{dxdy}{2} - p_n dS \cos\alpha = (p_y - p_n)\frac{dxdy}{2}$$

where α is the angle between $\vec{e}_y$ and the outward unit normal $\vec{e}_n$ to dS. It follows that $p_y = p_n = \text{const}$, as the orientation α is arbitrary.

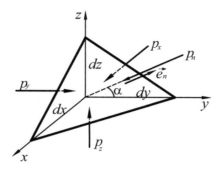

Figure 6.6. Normal forces across the faces of a tetrahedral fluid element at rest

By similar arguments we can define a *scalar* quantity

$$p_x = p_y = p_z = p_n = p$$

for any given element of a fluid at rest, called *pressure p*. Since this stress force is independent of orientation, in a fluid at rest we have

$$\sigma_{xx} = \sigma_{yy} = \sigma_{zz} = -p, \qquad \sigma_{xy} = \sigma_{yz} = \sigma_{xz} = 0$$

Therefore, in a viscous fluid, the direct stress components will be given in terms of the rate of strain components in the form

$$\sigma_{xx} = -p + 2\eta e_{xx} = -p + 2\eta \frac{\partial v_x}{\partial x}$$

$$\sigma_{yy} = -p + 2\eta e_{yy} = -p + 2\eta \frac{\partial v_y}{\partial y} \qquad (6.35)$$

$$\sigma_{zz} = -p + 2\eta e_{zz} = -p + 2\eta \frac{\partial v_z}{\partial z}$$

The dynamic equations with viscous forces are obtained by substituting the foregoing relationships between stress and rate of strain, Eqs.(6.34) and (6.35), into the equations of motion for a continuum. The resulting equation for the *y*-direction follows from Eq.(6.15) as

$$\rho\frac{dv_y}{dt} = \rho f_y - \frac{\partial p}{\partial y} + \eta\frac{\partial}{\partial y}\left(\frac{\partial v_x}{\partial x} + \frac{\partial v_y}{\partial y} + \frac{\partial v_z}{\partial z}\right) + \eta\left(\frac{\partial^2 v_y}{\partial x^2} + \frac{\partial^2 v_y}{\partial y^2} + \frac{\partial^2 v_y}{\partial z^2}\right)$$

$$= \rho f_y - \frac{\partial p}{\partial y} + \eta\frac{\partial}{\partial y}\nabla\cdot\vec{v} + \eta\nabla^2 v_y$$

For an *incompressible fluid* of constant density $d\rho/dt = 0$, the equation of continuity (6.11) takes the form

$$\nabla\cdot\vec{v} = 0 \tag{6.36}$$

and this allows us to simplify the equations of motion under viscous forces. Under this restriction the component equations can be written as

$$\rho\frac{dv_x}{dt} = \rho f_x - \frac{\partial p}{\partial x} + \eta\nabla^2 v_x$$

$$\rho\frac{dv_y}{dt} = \rho f_y - \frac{\partial p}{\partial y} + \eta\nabla^2 v_y \tag{6.37}$$

$$\rho\frac{dv_z}{dt} = \rho f_z - \frac{\partial p}{\partial z} + \eta\nabla^2 v_z$$

or in the more compact form

$$\rho\frac{d\vec{v}}{dt} = \rho\vec{f} - \nabla p + \eta\nabla^2\vec{v} \tag{6.38}$$

The equations (6.37) or (6.38) are known as the *Navier-Stokes equations* for incompressible viscous fluids.

Let us apply Eq.(6.38) to the steady-state flow of a viscous fluid in a horizontal pipe of radius R, parallel to its long axis, which we will take as the x-axis. There are no body forces $f_x = f_y = f_z = 0$ and also $v_y = v_z = 0$. The equation of continuity (6.36) reduces to $\partial v_x/\partial x = 0$, showing that v_x is a function of y and z only, since in a steady-state flow the velocity has not an explicit dependence on time $(\partial v_x/\partial t = 0)$. In view of Eq.(6.1) we then obtain

$$\frac{dv_x}{dt} = v_x\frac{\partial v_x}{\partial x} + v_y\frac{\partial v_x}{\partial y} + v_z\frac{\partial v_x}{\partial z} + \frac{\partial v_x}{\partial t} = 0$$

and Eq.(6.38) reduces to

$$\frac{dp}{dx} = \eta\left(\frac{\partial^2 v_x}{\partial y^2} + \frac{\partial^2 v_x}{\partial z^2}\right) \tag{6.39}$$

since there is no pressure gradient along the y- and z-directions and so $\partial p / \partial y = \partial p / \partial z = 0$. It is now convenient to take the divergence of Eq.(6.38) which, in view of Eq.(6.36), becomes

$$\rho \frac{d}{dt}(\nabla \cdot \vec{v}) = \rho \nabla \cdot \vec{f} - \nabla \cdot (\nabla p) + \eta \nabla^2 (\nabla \cdot \vec{v}) \quad \text{or} \quad \nabla \cdot (\nabla p) = \nabla^2 p = 0$$

that is the pressure gradient is *constantly* decreasing along the pipe, $\partial p / \partial x = $ const. The cylindrical symmetry, see Eq.(I.28) from Appendix I, allows us to write Eq.(6.39) in the form

$$\frac{d^2 v_x}{dr^2} + \frac{1}{r}\frac{dv_x}{dr} - \frac{1}{\eta}\frac{dp}{dx} = 0$$

where r is the distance from the axis of the cylinder. Integration gives the relation

$$v_x = \frac{1}{4\eta}\frac{dp}{dx} r^2 + A \ln r + B$$

where $A = 0$, since v_x has a finite magnitude for $r = 0$. The *no-slip boundary condition* requires that the velocity of a viscous fluid is zero at the solid surface of the pipe, that is $v_x = 0$ for $r = R$, and hence, we obtain

$$v_x = \frac{1}{4\eta}\frac{dp}{dx}\left(r^2 - R^2\right) \tag{6.40}$$

known as *Poiseuille's law*. It describes a steady-state or *laminar* flow, with a velocity whose magnitude varies parabolically across a diameter of the pipe, as in Figure 6.7, where $dp/dx < 0$.

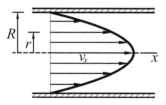

Figure 6.7. Laminar viscous flow in a horizontal pipe

If we assume that the coefficient of viscosity is negligible $(\eta = 0)$ we have defined a *perfect fluid*, which is characterized by the fact that it supports no shear. For this case, the Navier-Stokes equations (6.38) reduce to

$$\rho \frac{d\vec{v}}{dt} = \rho \vec{f} - \nabla p \qquad (6.41)$$

called *Euler's equations of hydrodynamics*, which also take the form

$$\frac{\partial \vec{v}}{\partial t} + \vec{v}(\nabla \cdot \vec{v}) = \vec{f} - \frac{1}{\rho} \nabla p \qquad (6.42)$$

in view of Eq.(6.1). The motion of a perfect fluid follows a *tube of flow* consisting of a family of *streamlines*, defined as imaginary curves in the fluid which are tangential at every point to the velocity vector, that is

$$\vec{v} \times d\vec{r} = 0 \qquad (6.43)$$

as illustrated in Figure 6.8.

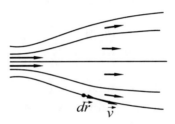

Figure 6.8. A family of streamlines in a perfect fluid

If a perfect fluid is acted on by a conservative body force, such as gravity, that is $\vec{f} = -\nabla \varphi$, where φ denotes the gravitational potential, its steady-state flow $(\partial \vec{v} / \partial t = 0)$ is described by the equation

$$\vec{v}(\nabla \cdot \vec{v}) = -\nabla \left(\frac{p}{\rho} + \varphi \right)$$

where v is the speed of the fluid. In view of a known vector identity, given in Appendix I, Eq.(I.16), this equation can be rewritten as

$$\frac{1}{2} \nabla v^2 - \vec{v} \times (\nabla \times \vec{v}) = -\nabla \left(\frac{p}{\rho} + \varphi \right)$$

Taking the scalar product of this equation with a line element $d\vec{r}$ of a streamline yields

$$d\vec{r} \cdot \nabla \left(\frac{v^2}{2} + \varphi \right) = d\vec{r} \cdot \left[-\nabla \left(\frac{p}{\rho} \right) \right] \quad \text{or} \quad d\left(\frac{v^2}{2} + \varphi \right) = d\vec{r} \cdot \left[-\nabla \left(\frac{p}{\rho} \right) \right] \qquad (6.44)$$

since, according to Eq.(6.43), $d\vec{r} \cdot [\vec{v} \times (\nabla \times \vec{v})] = (\nabla \times \vec{v}) \cdot (d\vec{r} \times \vec{v}) = 0$. The equation

(6.44) states the energy conservation law for a perfect fluid: the change of energy per unit mass of fluid equals the work done by the pressure gradient force per unit mass $-\nabla(p/\rho)$ along a streamline. Its equivalent form, written as

$$d\left(\frac{p}{\rho}+\frac{v^2}{2}+\varphi\right)=0 \quad \text{or} \quad \frac{p}{\rho}+\frac{v^2}{2}+\varphi=\text{const} \tag{6.45}$$

is known as the *Bernoulli equation* along a streamline of a perfect fluid.

FURTHER READING
1. A. J. Spencer - CONTINUUM MECHANICS, Dover Publications, New York, 2004.
2. I. S. Liu - CONTINUUM MECHANICS, Springer-Verlag, New York, 2002.
3. P. Chadwick - CONTINUUM MECHANICS: CONCISE THEORY AND PROBLEMS, Dover Publications, New York, 1999.

PROBLEMS

6.1. For an isotropic elastic material, Hooke's law states that there is a constant ratio E, called Young's ratio, between the linear stress and the linear strain. When a body is given a linear extension, it contracts in directions at right angles to the direction of stress and the Poisson's ratio υ between the relative contraction in the transverse direction and the extensional strain is found to be a constant. Show that the shear modulus μ and the bulk modulus K can be expressed in terms of E and υ by

$$\mu = \frac{E}{2(1+\upsilon)}, \quad K = \frac{E}{3(1-2\upsilon)}$$

Solution

Consider a small cube of edge a with faces perpendicular to the principal stress (and strain) axes, and let $\sigma_x = \sigma_0$. The simple relation $\varepsilon_x = \sigma_x / E$ holds only in the simple case when there are no stresses on the other two pairs of faces. However, tensile stresses σ_y and σ_z, when different from zero, produce contractions along the x-axis, so that ε_x is less than σ_x / E by the magnitude of the resultant contraction

$$\varepsilon_x = \frac{\sigma_x}{E} - \upsilon\left(\frac{\sigma_y}{E} + \frac{\sigma_z}{E}\right)$$

where υ is Poisson's ratio. It can be similarly shown that

$$\varepsilon_y = \frac{\sigma_y}{E} - \upsilon\left(\frac{\sigma_z}{E} + \frac{\sigma_x}{E}\right) \quad \text{and} \quad \varepsilon_z = \frac{\sigma_z}{E} - \upsilon\left(\frac{\sigma_x}{E} + \frac{\sigma_y}{E}\right)$$

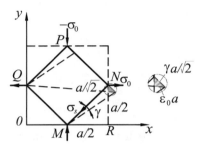

In the special case illustrated in the figure, where $\sigma_y = -\sigma_0$ is a compressive stress, if we assume for simplicity that $\sigma_z = 0$, one obtains

$$\varepsilon_x = \frac{\sigma_0}{E} - \upsilon\frac{(-\sigma_0)}{E} = (1+\upsilon)\frac{\sigma_0}{E} = \varepsilon_0 \quad \text{and} \quad \varepsilon_y = \frac{(-\sigma_0)}{E} - \upsilon\frac{\sigma_0}{E} = -(1+\upsilon)\frac{\sigma_0}{E} = -\varepsilon_0$$

The inner prism with a square base MNPQ of edge $a\sqrt{2}$ is under the action of pure shear stresses σ_s. The equilibrium condition for the forces acting upon the prism with a triangle MNR as base gives

$$\sigma_s\left(\frac{a^2}{\sqrt{2}}\right)\cos\frac{\pi}{4} = \sigma_0\frac{a^2}{2} \quad \text{or} \quad \sigma_s = \sigma_0$$

In view of Eq.(6.19), it follows that the angle of shear is $\gamma = \sigma_0/\mu$, where μ is the shear modulus. From the small shaded triangle, corresponding to the deformation of the inner prism, one obtains

$$\gamma\frac{a}{\sqrt{2}} = \sqrt{2}\varepsilon_0 a \quad \text{or} \quad \gamma = 2\varepsilon_0$$

Substituting ε_0 and γ as found above gives the required shear modulus as

$$\mu = \frac{E}{2(1+\upsilon)}$$

By adding together the three foregoing expressions found for the extensional strains, we have

$$\varepsilon_x + \varepsilon_y + \varepsilon_z = (\sigma_x + \sigma_y + \sigma_z)\frac{1-2\upsilon}{E} \quad \text{or} \quad \Delta = 3\sigma_d\frac{1-2\upsilon}{E}$$

where use has been made of Eqs.(6.20) and (6.25). Finally, in view of Eq.(6.18), the bulk modulus results as

$$K = \frac{E}{3(1-2\upsilon)}$$

6.2. Find an expression for the strain energy density of an isotropic elastic material.

Solution

The density of the potential energy of strain is given by the work that should be expended on a deformation of a volume $dV = dxdydz$ of an elastic material. This can be calculated independently for dilatational and shear strains. Taking into account that the normal stress along a given axis is only dependent on the dilatational strains along all the three axes, as given by Eqs.(6.29), the elementary amounts of work performed by the normal forces have the form

$$(\sigma_x dydz)d\varepsilon_x dx = (2\mu\varepsilon_x + \lambda\Delta)dVd\varepsilon_x$$

$$(\sigma_y dzdx)d\varepsilon_y dy = (2\mu\varepsilon_y + \lambda\Delta)dVd\varepsilon_y$$

$$(\sigma_z dxdy)d\varepsilon_z dz = (2\mu\varepsilon_z + \lambda\Delta)dVd\varepsilon_z$$

and this gives the total elementary work as

$$(2\mu\varepsilon_x d\varepsilon_x + 2\mu\varepsilon_y d\varepsilon_y + 2\mu\varepsilon_z d\varepsilon_z + \lambda\Delta d\Delta)dV$$

where, in view of Eq.(6.20), $d\Delta = d\varepsilon_x + d\varepsilon_y + d\varepsilon_z$. The expended work on bringing the dilatation Δ of the volume dV from zero to Δ is given by

$$dW_d = \left(2\mu\int_0^{\varepsilon_x}\varepsilon_x d\varepsilon_x + 2\mu\int_0^{\varepsilon_y}\varepsilon_y d\varepsilon_y + 2\mu\int_0^{\varepsilon_z}\varepsilon_z d\varepsilon_z + \lambda\int_0^{\Delta}\Delta d\Delta\right)dV$$

Thus the dilatational strain energy density results as

$$U_d = \mu(\varepsilon_x^2 + \varepsilon_y^2 + \varepsilon_z^2) + \frac{1}{2}\lambda\Delta^2$$

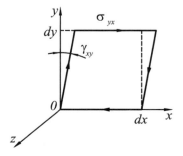

The shear stress is related with the angle of shear as given by Eq.(6.19), and hence, the elementary work about a given axis only depends on the angle of shear in the plane perpendicular to that axis, as shown in the figure, where it reduces to

$$\left(\sigma_{yx}dxdz\right)dyd\gamma_{xy} = \sigma_{yx}d\gamma_{xy}dV = \mu\gamma_{xy}d\gamma_{xy}dV$$

It is apparent that the tangential stresses on the two parallel faces *dydz* are normal to the displacement, and hence, perform no work, whereas there is no displacement of the lower face *dxdz*. It follows that the whole shear work performed on the volume *dV* can be written as

$$dW_s = \mu\left(\int_0^{\gamma_{xy}}\gamma_y d\gamma_y + \int_0^{\gamma_{yz}}\gamma_{yz}d\gamma_{yz} + \int_0^{\gamma_{zx}}\gamma_{zx}d\gamma_{zx}\right)dV$$

and this gives the shear strain energy density as

$$U_s = \frac{1}{2}\mu\left(\gamma_{xy}^2 + \gamma_{yz}^2 + \gamma_{zx}^2\right)$$

6.3. A steel cable of length L, radius r and shear modulus μ is anchored at one end and twisted through an angle θ by a torque G applied to the other end. Show that the torsion constant $\tau = G/\theta$ is proportional to the shear modulus μ. If a coil spring of radius R is made of this cable, show that its spring constant k is proportional to the torsion constant τ of the cable.

Solution

It is convenient to write down Eq.(6.19) for the outer cylindrical shell of the cable, of radius r and thickness dr, for which we have $\sigma_s = F/dA = F/2\pi r dr$, and hence

$$\frac{F}{2\pi r dr} = \mu\gamma$$

where the amount of twist θ at one end, relative to the other end, is obviously related to the angle of shear γ by $r\theta = L\gamma$.

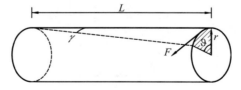

Rearranging to introduce the torque $G = Fr$ causing the twist gives

$$\frac{G}{2\pi r^2 dr} = \mu\frac{r}{L}\theta$$

which is a simple law of direct proportionality between the torsional stress on a cylindrical shell and the strain $r\theta/L$, where the torsion modulus is the same as the shear modulus. A solid cable is a series of adjacent shells each of thickness dr, so that we can move from a shell to a cable in torsion by integrating the shell formula from 0 to r. The torque on the cable is then

$$G = \frac{2\pi\mu\theta}{L}\int_0^r r^3 dr = \mu\frac{\pi R^4}{2L}\theta$$

and hence, the torsion constant $\tau = \mu\left(\pi R^4 / 2L\right)$ is proportional to the shear modulus.

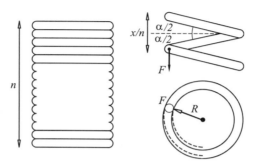

A linear coiled spring of n turns of radius R and unstreched length l obeys the relationship $F = kx$ under a linear force F and extends a distance x beyond length l. However, the statement for evaluating linear stress-strain

$$\frac{F}{A} = E\frac{x}{l} \quad \text{or} \quad F = \frac{EA}{l}x$$

does not apply to springs, which lengthen under stress by twisting and not by linear strain. Each turn of the spring contributes x/n to the extension, although subject to the same total linear force F given by

$$F = kx = (nk)\frac{x}{n}$$

It can be seen that the spring constant for a length l/n is nk. Given the extension

$$\frac{x}{n} = 2R\sin\frac{\alpha}{2} \cong R\alpha = R\frac{\theta}{n}$$

each half-turn is twisted by $\alpha/2$, and hence each turn contributes $\alpha = \theta/n$ to the amount of twist at the end of the cable, given by $\theta = G/\tau = FR/\tau$. The substitution of θ into the foregoing expressions gives

$$k = \frac{\tau}{R^2}$$

and hence, the spring constant k is proportional to the torsion constant τ, as required.

6.4. A horizontal uniform disk of radius R is mounted on the lower end of a vertical elastic rod of torsion constant τ, at its centre of mass. A parallel identical disk a distance h below it is rotating with angular velocity ω about the axis of the rod. Given the dynamic coefficient of viscosity η, find the angle of twist θ of the upper disk.

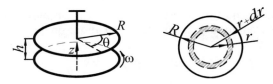

Solution

Let us separate, on the upper disk, a ring of radius r and width dr. The tangential stress produced by viscous forces is proportional to the gradient of velocity in the direction perpendicular to it, as given by the last part of Eq.(6.34), which reduces to

$$\sigma_{xz} = \eta \frac{dv_x}{dz} \quad \text{or} \quad \frac{dF}{dS} = \eta \frac{dv}{dz}$$

This is known as Newton's law for viscous friction between thin layers of fluid parallel to the sliding surfaces. Substituting $dS = 2\pi r dr$ and the normal velocity gradient $dr/dz = r\omega/h$ gives

$$dF = \eta \frac{r\omega}{h} 2\pi r dr$$

The integration of the corresponding torque $dG = r dF$ results in

$$G = \frac{2\pi\eta\omega}{h} \int_0^R r^3 dr = \frac{\pi\eta\omega}{2h} R^4$$

Thus, the angle of twist $\theta = G/\tau$ of the elastic rod is finally given by $\theta = \dfrac{\pi\eta\omega}{2h\tau} R^4$.

6.5. Find the shape of an hourglass timer, assuming a constant velocity v for the free surface of a perfect heavy fluid that is outflowing through the bottom opening of small cross section area S_0.

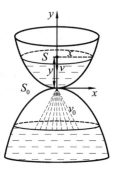

Solution

For a laminar flow of the perfect fluid, the Bernoulli equation (6.45) reduces to

$$\frac{p}{\rho} + \frac{v^2}{2} + gy = \frac{p}{\rho} + \frac{v_0^2}{2}$$

and the condition for the constancy of mass flux reads

$$\rho v S = \rho v_0 S_0$$

Since $S_0 \ll S$, it follows that $v = v_0 S_0 / S \ll 0$, thus we may neglect $v^2/2 \ll v_0^2/2$ in the Bernoulli equation and this gives $v_0 \cong \sqrt{2gy}$. As v is a constant, we obtain the constant ratio

$$\frac{S_0}{v} = \frac{S}{v_0} = \frac{\pi x^2}{\sqrt{2gy}}$$

where $S = \pi x^2$ is the free surface area. It follows that the hourglass shape is given by

$$y = \left(\frac{\pi^2 v^2}{2g S_0^2}\right) x^4$$

6.6. Show that the general relationship between the stress and strain tensors, defined by Eqs.(6.13) and (6.4), can be written, making use of the unit tensor $\hat{I}$, as

$$\hat{\sigma} = 2\mu \hat{\varepsilon} + \lambda \Delta \hat{I}$$

6.7. A uniform spring of mass m and spring constant k has a normal, unstretched length l. Determine its new length $l + x$ when the spring hangs vertically from one end.

6.8. A load m is hung from the end of a flexible, inextensible thread, passing over a pulley of negligible mass. Given the same spring constant k for the two springs, determine the period of simple harmonic motion of the load.

6.9. A spring of spring constant k is attached at either end to equal masses m at rest on a horizontal surface. Given the coefficient of friction μ between masses and surface, find the minimum velocity v_0 with which one mass should be pushed toward the other, in order to remove the left hand mass from the vertical wall.

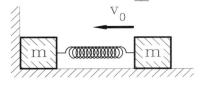

6.10. A spring of unstretched length l and spring constant k is attached at either end to equal masses m at rest on a horizontal frictionless surface. Determine the maximum and minimum length of the spring after a central elastic impact of a ball of the same mass m which approaches as shown with velocity v_0.

6.11. Show that the gradient of velocity in a stationary flow of a compressible gas is related to the variation of the cross-section area of the steam tube by

$$\frac{dv}{v} = \frac{dS/S}{(v/v_S)^2 - 1}$$

where v_S stands for the speed of sound.

6.12. Let h be the height of the free surface of a column of fluid, with density ρ and viscosity coefficient η, above an opening placed in the wall of the vessel and equipped with a pipe directing the outflowing jet horizontally. Given the length l and the radius r of the pipe, find the reaction force F of the fluid exerted on the vessel.

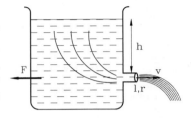

6.13. A vessel containing water slides down a plane that is inclined at angle α above the horizontal, and it is noted that the free surface of water lies parallel to the plane during the motion. The fluid flows out of the vessel with velocity v through a small opening of cross-section area S. Assuming a constant mass m for water and vessel, determine the coefficient μ of sliding friction between vessel and plane.

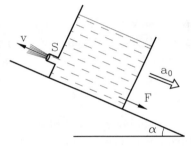

7. THE LAWS OF THERMODYNAMICS

7.1. ZEROTH LAW OF THERMODYNAMICS
7.2. FIRST LAW OF THERMODYNAMICS
7.3. SECOND LAW OF THERMODYNAMICS

7.1. ZEROTH LAW OF THERMODYNAMICS

Thermodynamics is concerned with processes where temperature and heat play an important role. Its laws correlate *macroscopic* parameters, without making any microscopic assumptions, thus ensuring a large degree of generality. The macroscopic parameters can be direct observables (e.g., the pressure or the volume of a fluid) or specific observables, related to the former and following from the laws of thermodynamics: *temperature* from the zeroth law, *internal energy* from the first law and *entropy* from the second law.

Any ensemble in thermodynamics is called a *system*, and it can be either simple or made up of several homogeneous *phases* with well-defined boundaries. All interactions on the system can be performed either by exchange of work, or through heat flow. If heat exchange is forbidden at the boundaries, these are called *adiabatic* boundaries, and the system itself is *thermally isolated*. If the boundary between two systems is not adiabatic, then the two systems are in *thermal contact*. The thermodynamic variables (direct and specific observables) belong to two classes: *intensive variables* X, if of local character (pressure, force, density, etc.) and *extensive variables* x, referring to the system as a whole (mass, volume, internal energy, length). Extensive variables are, in general, proportional to mass. Their values, related to the unit of mass, are called *specific* (e.g., the energy of unit mass is called specific energy). Many direct observables form conjugate pairs, their product Xx having the dimension of energy: pV for a hydrostatic system, BM for a magnetic material or TS for all thermodynamic systems. For that reason, the intensive variable X may be called *force*, and its conjugate extensive variable x is known as *displacement*.

Any quantity which has a unique value for each of the states of a system, is called a *function of state*. A state of the system which lasts long enough (compared, for instance, to the duration of microscopic phenomena), is said to be a state of *thermodynamic equilibrium.* All thermodynamic quantities have constant values in such a state. When two systems in thermal contact have each come to thermodynamic equilibrium, a *thermal equilibrium* is said to be reached between the two systems.

Experimental evidence associated with thermal equilibrium between systems is stated in a form called the **zeroth law of thermodynamics**:

Two systems in thermal equilibrium with a third system, are themselves in thermal equilibrium.

From this law follows the existence of temperature. In this respect let us consider three systems 1, 2 and 3 and let X_i and x_i be the appropriate force and displacement chosen as parameters of state. In addition we will assume system 3 is the reference system. By changing X_1, x_1 we can bring system 1 into equilibrium with system 3. We obtain thus a set of isotherms with temperatures $\theta, \theta', \theta''$ etc., for different reference states of system 3 (Figure 7.1).

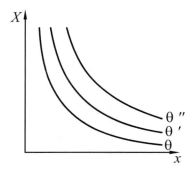

Figure 7.1. Equilibrium isotherms for the thermal contact between hydrostatic systems

We can do the same for system 2. Thus, according to the zeroth law, the two systems 1 and 2 are at equilibrium along each of the isotherms. The property which is common to these systems is called *temperature*. For fixed values of X_3, x_3 a certain value of the force X_1 corresponds to a value of the displacement x_1. Therefore, a relation exists between the four variables

$$F_{13}(X_1, x_1 X_3, x_3) = 0$$

which expresses the equilibrium between systems 1 and 3. A similar relation holds for systems 2 and 3 as

$$F_{23}(X_2, x_2 X_3, x_3) = 0$$

so that, solving for X_3 the two equations, we obtain

$$X_3 = f_{13}(X_1, x_1, x_3) \quad , \quad X_3 = f_{23}(X_2, x_2, x_3) \tag{7.1}$$

Hence, we have

$$f_{13}(X_1, x_1, x_3) = f_{23}(X_2, x_2, x_3) \qquad (7.2)$$

and this yields

$$X_1 = f_{123}(x_1, X_2, x_2, x_3) \qquad (7.3)$$

The equilibrium condition for systems 1 and 2, following from the zeroth law, can be written as

$$F_{12}(X_1, x_1 X_2, x_2) = 0$$

and it follows that

$$X_1 = f_{12}(x_1, X_2, x_2) \qquad (7.4)$$

This allows us to eliminate x_3 from Eq.(7.3). In a similar manner we can eliminate x_3 from Eq.(7.2), which becomes

$$f_1(X_1, x_1) = f_2(X_2, x_2)$$

Thus, for systems at equilibrium, there exists a function which takes the same value in all systems. Denoting it by

$$f(X, x) = \theta \qquad (7.5)$$

this equation is called *the equation of state*, and θ is known as *empirical temperature*.

EXAMPLE 7.1. The temperature scale

To establish a temperature scale, one chooses a thermometric property y and a procedure to obtain a linear relation

$$\theta(y) = ay + b \qquad (7.6)$$

To determine the constants one needs two fixed reference points. For the Celsius scale these are the melting and boiling points of water. The thermometers with the most precise results are those based on the behaviour of gases where X is the pressure p and x the volume V. Fixing one of the parameters (p or V), allows us to take the temperature to be proportional to the second one. *Celsius* temperature, for instance, can be defined in the gas thermometer at either constant pressure or constant volume

$$\theta = \frac{V - V_0}{V_{100} - V_0} \times 100 \quad (p = \text{const}), \qquad \theta = \frac{p - p_0}{p_{100} - p_0} \times 100 \quad (V = \text{const}) \qquad (7.7)$$

In the limit $p \to 0$, all the gases have the same value of temperature, independent of scale, and this defines *the Celsius scale of the ideal gas* in the form

$$t = \lim_{p \to 0} \frac{V - V_0}{V_{100} - V_0} \times 100 \quad (p = \text{const})$$

$$t = \lim_{p \to 0} \frac{p - p_0}{p_{100} - p_0} \times 100 \quad (V = \text{const}) \tag{7.8}$$

By means of the well known equation of state for a mole of ideal gas

$$pV = RT$$

which will be derived later on in Chapter 9, one obtains

$$t = \lim_{p \to 0} \frac{pV/R - pV_0/R}{pV_{100}/R - pV_0/R} \times 100 = T - T_0$$

where T is the absolute temperature, and T_0 is the normal melting point of water, given by

$$T_0 = \lim_{p \to 0} \frac{V_0}{V_{100} - V_0} \times 100 = 273.15$$

The *Kelvin* or *absolute temperature scale* can now be defined as

$$T = t + 273.15 \tag{7.9}$$

7.2. FIRST LAW OF THERMODYNAMICS

The **first law of thermodynamics** is a statement of the experimental evidence concerning the behaviour of thermally isolated systems whose states are changed by adiabatic means only, that is

If a state of a thermally isolated system is changed through the performance of adiabatic work, the amount of work depends only on the performed change and not on the intermediate states through which the system passes from its initial to its final state.

If a certain energy is associated with the change of state under adiabatic conditions, then the change of total energy of the system, called the *internal energy U*, is given by the work performed *on* the system

$$\Delta U = W \tag{7.10}$$

and, because the adiabatic work W is not path-dependent, as stated by the first law, U is a *function of state*. If the system is not thermally isolated, then in addition to work, *heat Q* must be included

$$\Delta U = W + Q \tag{7.11}$$

Q is a measure of the amount of *nonadiabatic* change in the system. Since it is not possible to separate the nonadiabatic contribution, Q, from the adiabatic one, W, to the internal energy change, unless Eq.(7.11) is applied to a particular process, we must consider both Q and W as path-dependent functions. The mathematical formulation of the first law, given by Eq.(7.11), is an expression of the conservation of energy.

For an infinitesimal process, the first law takes the form

$$\Delta U = \delta Q + \delta W \tag{7.12}$$

where δ means that δQ and δW are not exact differentials, as required by the path-dependence of heat Q and work W (see Appendix III).

The work performed reversibly *on* the system can be written as the sum of products of conjugated forces X_k and displacements x_k in the form

$$\delta W = \sum_k X_k dx_k \tag{7.13}$$

so that the heat (7.12) absorbed in a reversible process can be expressed as

$$\delta Q = dU - \sum_k X_k dx_k \tag{7.14}$$

Denoting by f an arbitrary function of state, $\delta Q/df$ is called *heat capacity* and expresses the rate of heat absorption during a change of f. Since heat is path-dependent, one has to specify also the constraints, that is $(n-1)$ constraints for a n-parameter system, of the form

$$C_{f'f''...}^{(f)} = \frac{\delta Q_{f'f''...}}{df} \tag{7.15}$$

where f is the variable, and $f', f''...$ are the constraints. As examples, we have the heat capacities at constant volume and at constant pressure of a hydrostatic system, defined as

$$\frac{\delta Q_V}{dT} = C_V^{(T)} = C_V \quad , \quad \frac{\delta Q_p}{dT} = C_p^{(T)} = C_p \tag{7.16}$$

where the superscript T is normally omitted. Since these are *extensive* quantities, one can also define the corresponding *specific heats* c_V and c_p per unit mass.

EXAMPLE 7.2. Heat capacities for a hydrostatic system

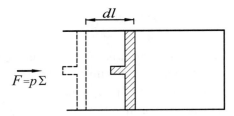

Figure 7.2. Work performed on a hydrostatic system

In a reversible compression of a hydrostatic system, illustrated in Figure 7.2, Eq.(7.13) becomes

$$\delta W = Fdl = p\Sigma dl = -pdV \qquad (7.17)$$

since we have a volume decrease, so that, according to Eq.(7.14), the first law reads

$$\delta Q = dU + pdV \qquad (7.18)$$

where U is a function of any two of p, V and T. Choosing V and T as independent variables, we have

$$dU = \left(\frac{\partial U}{\partial T}\right)_V dT + \left(\frac{\partial U}{\partial V}\right)_T dV$$

and the first law becomes

$$\delta Q = \left(\frac{\partial U}{\partial T}\right)_V dT + \left[p + \left(\frac{\partial U}{\partial V}\right)_T\right] dV$$

Under the constraint that $V = $ const, this reduces to

$$C_V^{(T)} = \frac{\delta Q_V}{dT} = \left(\frac{\partial U}{\partial T}\right)_V = C_V \qquad (7.19)$$

Under the constraint that $T = $ const, one alternatively obtains the *rate of heat* absorbed with a change of volume, along an isotherm

$$C_T^{(V)} = \frac{\delta Q_T}{dV} = p + \left(\frac{\partial U}{\partial V}\right)_T \qquad (7.20)$$

Choosing p and T as independent variables, we must first express

$$dU = \left(\frac{\partial U}{\partial p}\right)_T dp + \left(\frac{\partial U}{\partial T}\right)_p dT \quad , \quad dV = \left(\frac{\partial V}{\partial p}\right)_T dp + \left(\frac{\partial V}{\partial T}\right)_p dT$$

and the first law becomes

$$\delta Q = \left[\left(\frac{\partial U}{\partial p}\right)_T + p\left(\frac{\partial V}{\partial p}\right)_T\right]dp + \left[\left(\frac{\partial U}{\partial T}\right)_p + p\left(\frac{\partial V}{\partial T}\right)_p\right]dT$$

Two heat capacities may be then derived as

$$C_p^{(T)} = \left(\frac{\partial U}{\partial T}\right)_p + p\left(\frac{\partial V}{\partial T}\right)_p = C_p \quad , \quad C_T^{(p)} = \left(\frac{\partial U}{\partial p}\right)_T + p\left(\frac{\partial V}{\partial p}\right)_T \tag{7.21}$$

We can further express C_p in a form similar to that given for C_V in Eq.(7.19), which reads

$$C_p = \left[\frac{\partial}{\partial T}(U + pV)\right]_p = \left(\frac{\partial H}{\partial T}\right)_p \tag{7.22}$$

where the new function of state is called *enthalpy H* and is defined as

$$H = U + pV \tag{7.23}$$

7.3. Second Law of Thermodynamics

If one applies the first law to a *cyclic process* in which a system is brought back to its initial state, one obtains

$$W + Q = 0 \tag{7.24}$$

so that one of the following cases hold

$(a) \quad W = 0 \quad Q = 0$

$(b) \quad W < 0 \quad Q > 0$

$(c) \quad W > 0 \quad Q < 0$

When the case (b) holds one says that the system is a *heat engine*, which absorbs heat and delivers work by performing the cycle. Studying such engines, Carnot was able to show that they cannot absorb heat from only one reservoir, at a given temperature, by so called monothermal processes. This statement may be postulated as **the second law of thermodynamics** in the form

In a monothermal cyclic process one has

$$W \geq 0 \quad , \quad Q \leq 0$$

Provided the cyclic process is reversible, it may be performed in a direction opposite to that of an engine, so that the work W' done on the system and the exchanged heat Q' can be expressed as

$$W' = -W \quad , \quad Q' = -Q$$

Because

$$W' \geq 0 \quad , \quad Q' \leq 0$$

we have

$$-W \geq 0 \quad , \quad -Q \leq 0$$

which are compatible with the second law if, and only if, we have

$$W = 0 \quad , \quad Q = 0 \tag{7.25}$$

Let us further consider a bithermal cyclic, reversible process, in which system 1 exchanges heat Q_M and Q_m with reservoirs at temperatures θ_M and θ_m. After n cycles, the exchanged heat will be

$$nQ_M \quad , \quad nQ_m$$

Considering then another system 2, with k cycles and heat quantities Q'_M and Q'_m, the heat exchanged by systems 1 and 2 with the same two reservoirs will be

$$nQ_M + kQ'_M \quad , \quad nQ_m + kQ'_m$$

Choosing k and n so that the heat exchanged with the reservoir at temperature θ_m is zero

$$nQ_m + kQ'_m = 0$$

one obtains a reversible monothermal process for which, according to Eq.(7.25), one has

$$nQ_M + kQ'_M = 0$$

From the last two equations we can write

$$\frac{Q_M}{Q_m} = \frac{Q'_M}{Q'_m} = f(\theta_M, \theta_m) \tag{7.26}$$

that is *the ratio of heats exchanged with the reservoirs in a cyclic reversible bithermal process depends only on the temperature of the reservoirs and does not depend on the system*. Therefore f is a universal function of θ_M and θ_m. Equation (7.26) is an expression of the *Carnot theorem*, which can be used to define absolute temperatures.

Consider a system 1 extracting heat Q_M from a reservoir at temperature θ_M and rejecting heat Q_0 to a reservoir at temperature θ_0 and a system 2 extracting heat Q_0 at θ_0 and rejecting heat Q_m at θ_m, as illustrated in Figure 7.3. According to the Carnot theorem

$$\frac{Q_M}{Q_0} = f(\theta_M, \theta_0) \quad , \quad \frac{Q_0}{Q_m} = f(\theta_0, \theta_m)$$

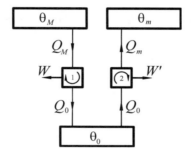

Figure 7.3. Composite engine used to derive the absolute temperature scale from the Carnot theorem

Since the net exchange at θ_0 is zero, the composite system 1 and 2 undergoes a bithermal cyclic process between Q_m and θ_m, so that

$$\frac{Q_M}{Q_m} = f(\theta_M, \theta_m)$$

One thus obtains

$$f(\theta_M, \theta_m) = f(\theta_M, \theta_0) f(\theta_0, \theta_m)$$

Since the left-hand side is independent of θ_0, this temperature must be also eliminated on the right-hand side, requiring

$$f(\theta_M, \theta_m) = \frac{T(\theta_M)}{T(\theta_m)} \tag{7.27}$$

where $T(\theta)$ are functions of the empirical temperature only. Therefore Eq.(7.26) may be written as

$$\frac{Q_M}{Q_m} = \frac{T_M}{T_m} \qquad (7.28)$$

This defines the *Kelvin* or *absolute temperature scale*: the ratio of absolute temperatures of two reservoirs equals the ratio of heats exchanged with a heat engine coupled to them. Such a bithermal, reversible cyclic process is called a *Carnot cycle*. If T_M and T_m are the temperatures of the two reservoirs, the efficiency η of the Carnot cycle is

$$\eta = \frac{-W}{Q_M} = 1 - \frac{Q_m}{Q_M} = 1 - \frac{T_m}{T_M} \qquad (7.29)$$

EXAMPLE 7.3. The Carnot cycle

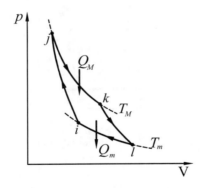

Figure 7.4. The Carnot cycle of a hydrostatic system

For a *real* or a *perfect gas*, the Carnot cycle has a standard representation on a pV diagram given in Figure 7.4, where area inside curve corresponds to work done by the system. The cycle consists of a sequence of four processes:

$i \to j$ reversible adiabatic compression from T_m to T_M
$j \to k$ reversible isothermal expansion at T_M, absorbing heat from a reservoir T_M
$k \to l$ reversible adiabatic expansion from T_M to T_m
$l \to i$ reversible isothermal compression at T_m, yielding heat to a reservoir at T_m until the initial state is obtained

In the Carnot cycle for a *liquid-vapour mixture*, illustrated in Figure 7.5, the isotherm $j \to k$ coincides with the straight line representing the vaporization process at temperature T_M under constant pressure, and the isotherm $l \to i$ corresponds to the converse process of isobaric condensation at temperature T_m. All states the two isotherms are passing through belong to a liquid-vapour mixture.

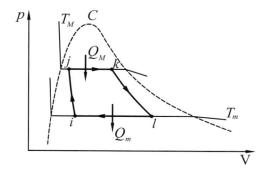

Figure 7.5. The Carnot cycle of a liquid-vapour mixture

For a *paramagnetic substance*, whose magnetization M (the extensive variable) is proportional to the applied magnetic induction B (the intensive variable) and inversely proportional to the temperature T, the Carnot cycle is shown in Figure 7.6 on an appropriate BM diagram. The substance is subjected to magnetization followed by demagnetization processes which require the expenditure of external work, so that the cycle is performed in a direction opposite to that of a heat engine.

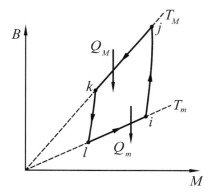

Figure 7.6. The Carnot cycle of a paramagnet

It is immediately apparent that the nature of working substance imposes both the shape and the coordinates used in the representation of the Carnot cycle. We will show in Chapter 8 that the Carnot cycle may be represented as a rectangle, for all working substances if the coordinates of temperature T and entropy S are used.

As illustrated in the last example, the Carnot cycle can be also performed in a direction opposite to that of a heat engine absorbing heat from the cold reservoir T_m and rejecting it to a hotter one, at temperature T_M. However this requires the expenditure of work done on the system, as shown in Figure 7.7, in view of Eq.(7.28) where $T_m < T_M$.

A *refrigerator* is such an engine, that extracts heat from a body colder than the surroundings. Its efficiency is expressed by the *cooling energy ratio* ε_r of the heat extracted from the cold reservoir to the work done on the refrigerant

$$\varepsilon_r = \frac{Q_m}{W} = \frac{T_m}{T_M - T_m} = \frac{1}{\eta} - 1 \qquad (7.30)$$

where η, given by Eq.(7.29), is the efficiency of the Carnot engine.

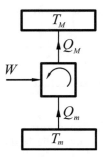

Figure 7.7. Carnot engine that transfers heat from cold to hot reservoir

The *heat pump* is another example that transfers heat to a body at higher temperature than its surroundings, with a *coefficient of performance* ε_p related to the efficiency of the Carnot engine by

$$\varepsilon_p = \frac{Q_M}{W} = \frac{T_M}{T_M - T_m} = \frac{1}{\eta} \qquad (7.31)$$

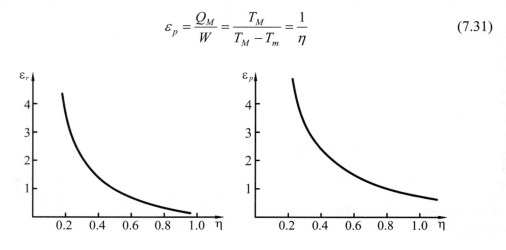

Figure 7.8. Efficiencies $\varepsilon_r, \varepsilon_p$ of a Carnot refrigerator and Carnot heat pump respectively, in terms of the efficiency η of a Carnot engine

Both ε_r and ε_p are plotted in Figure 7.8 as a function of η. When $\eta \to 1$, that is $T_M \to 0$, the cooling ratio $\varepsilon_r \to 0$, so that it becomes increasingly difficult to obtain cooling at low temperatures, and $\varepsilon_p \to 1$ showing that heat delivered by a heat pump equals the expenditure of work. A potential advantage of both devices is apparent for $\eta \to 0$, but practical efficiencies ε_r, ε_p are far below this ideal behaviour.

FURTHER READING
1. D. V. Schroeder - AN INTRODUCTION TO THERMAL PHYSICS, Addison-Wesley, San Francisco, 1999.
2. R. E. Sonntag, C. Borgnakke, G. J. van Wylen - FUNDAMENTALS OF THERMODYNAMICS, Wiley, New York, 1998.
3. C. B. Finn - THERMAL PHYSICS, Chapman & Hall, London, 1994.

PROBLEMS

7.1. A tunnel is bored in a straight line through the Earth, between two points located at latitude θ on its surface, and it is filled with air at normal temperature. Given the radius R_0 of the Earth, assumed to be a uniform-density sphere, the air pressure p_0 and the acceleration of gravity g at the Earth's surface, find the air pressure at the mid-point of the tunnel.

Solution

It is convenient to imagine a second tunnel, perpendicular to the first and passing through the centre of the Earth and to evaluate the pressure at the cross-point, that is, at a radial distance $r = R_0 \sin\theta$ from the Earth's centre, where the equilibrium of any small volume of air reads:

$$dp + \rho g(r) dr = 0 \qquad \text{or} \qquad dp = -\rho g(r) dr$$

According to Gauss's law, see Eq.(11.4) in Chapter 11, which holds for any inverse-square field of force, the gravitational field at the cross point is produced by all the matter inside the sphere of radius r, and hence, it is proportional to the radial distance from the centre of the Earth

$$g(r) = g \frac{r}{R_0}$$

For a given volume of air, the density ρ is proportional to the pressure, provided the temperature is constant, that is

$$pV = \nu RT = \frac{m}{\mu} RT \qquad \text{or} \qquad \rho = \frac{p\mu}{RT}$$

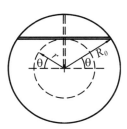

Substituting $g(r)$ and ρ into the equilibrium condition gives

$$\frac{dp}{p} = -\frac{\mu}{RT}\frac{g}{R_0} r\, dr \qquad \text{or} \qquad \ln\frac{p}{p_0} = \frac{\mu g}{2RTR_0}\left(R_0^2 - r^2\right)$$

where $p = p_0$, when $r = R_0$. Thus the air pressure at the mid-point of the tunnel depends on latitude as

$$\ln\frac{p}{p_0} = \frac{\mu g R_0}{2RT}\cos^2\theta$$

7.2. A thermally isolated cylinder is evacuated and fitted with a stopcock. When the stopcock is opened air flows in from the atmosphere, where the temperature is T_0. Determine the equilibrium temperature of the air in the cylinder.

Solution

A quantity of air undergoes an adiabatic expansion in which the change in the internal energy is given by the work performed on the system, Eq.(7.10). Assuming ν moles of atmospheric air in a state defined by the equation

$$p_0 V_0 = \nu R T_0$$

where V_0 is a volume of air equal to that of the cylinder, the adiabatic increase in the internal energy can be written in terms of C_V, Eq.(7.19), in the form

$$\Delta U = \nu C_V (T - T_0)$$

The amount of work done on the system, Eq.(7.13), is $W = p_0 V_0$, and hence, $W = \nu R T_0$. On substitution of ΔU and W into Eq.(7.10) one obtains

$$T = T_0\left(1 + \frac{R}{C_V}\right) = \gamma T_0$$

where $\gamma = C_p / C_V$ stands for the adiabatic exponent, see Eq.(8.45) later on.

7.3. One mole of perfect gas is confined in one compartment of a container with a tightly fitted piston, assumed to be frictionless, which is connected to the opposite wall by a spring of spring constant k. Given that the other compartment is evacuated and the unstretched length of the spring is equal to that of the container, find the heat capacity of the gaseous system.

Solution

The equilibrium condition for a given compression x of the string reads

$$p\Sigma = kx \qquad \text{or} \qquad \Sigma dp = k dx$$

where Σ is the cross-sectional area of the piston, and p is the gas pressure. If the piston moves a distance x, the amount of heat absorbed in the process, Eq.(7.18), is given by

$$\delta Q = dU + p\Sigma dx = C_V dT + kx dx$$

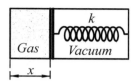

Differentiating the equation of state yields $RdT = d(p\Sigma x) = \Sigma x dp + \Sigma p dx = 2kx dx$, and this gives

$$\delta Q = \left(C_V + \frac{R}{2}\right) dT$$

In view of Eq.(7.15), it follows that the required heat capacity is

$$C = C_V + \frac{R}{2}$$

This indicates that the gaseous system undergoes a polytropic process, see Eq.(9.10) in Chapter 9, with polytropic exponent $n = -1$, so that

$$C = \frac{(-1)C_V - C_p}{(-1) - 1} = \frac{C_V + C_p}{2} = C_V + \frac{R}{2}$$

The polytropic equation has the form

$$p\Sigma = kx = \frac{k}{\Sigma} V \quad \text{or} \quad pV^{-1} = \frac{k}{\Sigma^2} = \text{const}$$

According to the theorem of equipartition of energy, see Eq.(31.7) later on, for one mole of perfect monatomic gas we have $f = 3N_A$ degrees of freedom of translational motion, where N_A stands for the Avogadro constant, and this gives

$$C_V = \frac{3}{2} N_A k_B = \frac{3}{2} R \quad \text{or} \quad C = 2R$$

If we assume a rigid-dumbbell model for a diatomic molecule, this allows two additional degrees of freedom, as the molecule can rotate about two axes. Thus, for one mole of a perfect diatomic gas with $f = 5N_A$ degrees of freedom, we have

$$C_V = \frac{5}{2} R \quad \text{or} \quad C = 3R$$

7.4. A Carnot engine operates between two bodies having the same heat capacity at constant pressure C_p and temperatures T_M and T_m, respectively, until their thermal equilibrium is reached at a certain temperature T_0. Show that T_0 is smaller than the temperature achieved by thermal contact between the two bodies. Calculate the amount of work that is obtained this way and the efficiency of the process. What are the results if the body at lower temperature T_m is a heat reservoir.

Solution

Since the temperatures of the two bodies change during the operation of the engine, it is necessary for the Carnot engine to operate in infinitesimal cycles for the heat transfer to be reversible. Let the heat absorbed at T_M be δQ_M and that absorbed at $T_m < T_M$ be δQ_m, that is

$$\delta Q_M = -C_p dT, \qquad \delta Q_m = -C_p dT'$$

In view of Eq.(7.28), we have

$$\frac{\delta Q_M}{\delta Q_m} = \frac{T}{T'} \qquad \text{or} \qquad C_p \frac{dT}{T} = -C_p \frac{dT'}{T'}$$

On integration results

$$C_p \int_{T_M}^{T_0} \frac{dT}{T} = -C_p \int_{T_m}^{T_0} \frac{dT'}{T'} \qquad \text{or} \qquad C_p \ln \frac{T_0}{T_M} = C_p \ln \frac{T_m}{T_0}$$

and hence, the equilibrium temperature is

$$T_0 = \sqrt{T_M T_m}$$

This is smaller than that achieved by thermal contact, say T_c, since the work interaction may be neglected in this case, and the first law gives

$$C_p(T_M - T_c) + C_p(T_m - T_c) = 0 \qquad \text{or} \qquad T_c = \frac{1}{2}(T_M + T_m) \geq \sqrt{T_M T_m}$$

The work performed by the working substance of the Carnot engine is obtained by applying the first law to one cycle of operation in the form

$$\delta Q_M + \delta Q_m + \delta W = 0 \qquad \text{or} \qquad -\delta W = -C_p dT - C_p dT'$$

On integration one obtains the total amount of work done by the engine in the form

$$-W = -C_p \int_{T_M}^{T_0} dT - C_p \int_{T_M}^{T_0} dT' = C_p(T_M - T_0) + C_p(T_m - T_0) = C_p\left(\sqrt{T_M} - \sqrt{T_m}\right)^2$$

Thus, the efficiency, Eq.(7.29), can be written as

$$\eta = \frac{-W}{Q_M} = \frac{C_p\left(\sqrt{T_M} - \sqrt{T_m}\right)^2}{C_p(T_M - T_0)} = 1 - \sqrt{\frac{T_m}{T_M}}$$

Assuming a heat reservoir of constant temperature T_m and heat capacity $C_p = \infty$, the equilibrium temperature will be $T_0 = T_m$. In this case, Eq.(7.28) reads

$$\frac{\delta Q_M}{\delta Q_m} = \frac{T}{T_m} \qquad \text{or} \qquad \delta Q_m = \frac{T_m}{T}(-C_p dT)$$

On substitution into the first law for one infinitesimal cycle we obtain

$$-\delta W = -C_p dT\left(1 - \frac{T_m}{T}\right)$$

and integration gives the total amount of work performed by the working substance as

$$-W = -C_p \int_{T_M}^{T_m} dT + C_p T_m \int_{T_M}^{T_m} \frac{dT}{T} = C_p\left(T_M - T_m - T_m \ln\frac{T_M}{T_m}\right)$$

The corresponding efficiency is

$$\eta = 1 - \frac{T_m}{T_M - T_m}\ln\left(\frac{T_M}{T_m}\right)$$

7.5. A certain heat engine has one mole of a perfect gas of adiabatic exponent $\gamma = C_p/C_V$ as its working substance and undergoes a cycle of operation made up of the three consecutive steps: a reversible process ij at constant volume V_0, a reversible process represented by a straight line jk and a reversible process ki at constant pressure p_0. Calculate the thermal efficiency of the engine. Derive the efficiency of a Carnot engine operating between reservoirs at T_m and T_M, which are the lowest and, respectively, the highest temperature reached during the given cycle.

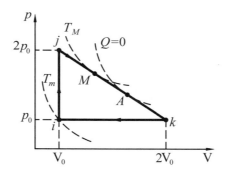

Solution

The perfect gas equation shows that the lowest temperature corresponds to the equilibrium state i, where

$$p_0 V_0 = RT_m$$

while the equilibrium states j and k are at the same temperature $2T_m$. During process jk, heat is absorbed in stage jA and then delivered in stage Ak, where the equilibrium state A is defined by the condition $\delta Q = 0$. By combining the first law, written as

$$\delta Q = dU + pdV = C_V dT + pdV$$

with the differential from of the equation of state

$$pdV + Vdp = RdT$$

one obtains
$$\delta Q = C_V \frac{pdV + Vdp}{R} + pdV = 0$$

The process jk is described by a linear equation $p = a - bV$, verified by the coordinates $(2p_0, V_0)$ and $(p_0, 2V_0)$ of the equilibrium states j and k, respectively, so that

$$p = 3p_0 - \frac{p_0}{V_0}V \quad \text{or} \quad dp = -\frac{p_0}{V_0}dV$$

Since the straight line jk is tangent at A to the curve representing the adiabatic expansion, as shown in the figure, we may substitute these results into the adiabatic equation, and this gives

$$\frac{C_V}{R}\left(p_A - \frac{p_0}{V_0}V\right)dV + p_A dV = 0 \quad \text{or} \quad p_A\left(1 + \frac{R}{C_V}\right) = \frac{p_0}{V_0}V_A$$

where $p_0 V_A / V_0 = 3p_0 - p_A$. It follows, see Eq.(8.45), that in equilibrium state A we have

$$p_A = \frac{3p_0}{2 + R/C_V} = \frac{3}{\gamma + 1}p_0$$

Then, from the linear equation of the process jk and the equation of state one obtains

$$V_A = \frac{3\gamma}{\gamma + 1}V_0, \quad T_A = \frac{9\gamma}{(\gamma + 1)^2}T_m$$

Thus, the work done in stage jA is given by

$$-W_{jA} = \int_{V_0}^{V_A} pdV = \int_{V_0}^{V_A}\left(3p_0 - \frac{p_0}{V_0}V\right)dV = 3p_0 V_0\left(\frac{3\gamma}{\gamma + 1} - 1\right) - \frac{p_0}{2V_0}V_0^2\left[\frac{9\gamma^2}{(\gamma + 1)^2} - 1\right]$$

$$= \frac{1}{2}p_0 V_0 \frac{(2\gamma + 5)(2\gamma - 1)}{(\gamma + 1)^2} = \frac{1}{2}RT_m\frac{(2\gamma + 5)(2\gamma - 1)}{(\gamma + 1)^2}$$

and the increase in the internal energy reads

$$\Delta U_{jA} = C_V(T_A - 2T_m) = C_V T_m\left[\frac{9\gamma}{(\gamma + 1)^2} - 2\right] = \frac{RT_m}{\gamma - 1}\left[\frac{9\gamma}{(\gamma + 1)^2} - 2\right]$$

Thus, the heat absorbed in this stage is obtained, after a little algebra, in the form

$$Q_{jA} = \Delta U_{jA} - W_{jA} = \frac{1}{2}RT_m\frac{(2\gamma - 1)^2}{\gamma^2 - 1}$$

In process ij the work interaction is zero, so that the heat absorbed by the system is given by

$$Q_{ij} = C_V(2T_m - T_m) = C_V T_m = \frac{R}{\gamma - 1} T_m$$

It follows that the heat absorbed in each cycle can be written as

$$Q = Q_{ij} + Q_{jA} = \frac{1}{2} RT_m \frac{4\gamma^2 - 2\gamma + 3}{\gamma^2 - 1}$$

while the work done by the system in each cycle is given by its pressure-volume area

$$-W = \frac{1}{2}(2p_0 - p_0)(2V_0 - V_0) = \frac{1}{2} p_0 V_0 = \frac{1}{2} RT_m$$

Finally, the required thermal efficiency of the engine reads

$$\eta = \frac{-W}{Q} = \frac{\gamma^2 - 1}{4\gamma^2 - 2\gamma + 3}$$

Since the efficiency of the Carnot engine depends on the temperatures of the reservoirs only, Eq.(7.29), and not on the nature of the working substance, we are left to determine the highest temperature T_M, reached during process jk, and this is higher than T_A. If we let M be the equilibrium state where the straight line jk is tangent to the curve representing the isothermal expansion at T_M, we have

$$p = 3p_0 - \frac{p_0}{V_0} V \quad \text{and} \quad pV = RT_M$$

and hence, T_M is a function of volume

$$T_M(V) = \frac{p_0}{R}\left(3V - \frac{V^2}{V_0}\right)$$

with a maximum given by the condition

$$\frac{dT_M(V)}{dV} = \frac{p_0}{R}\left(3 - 2\frac{V}{V_0}\right) = 0 \quad \text{or} \quad V = \frac{3}{2} V_0$$

It follows that $p = 3p_0/2$, so that we obtain $T_M = 9T_m/4$, and this temperature is indeed higher than T_A, since $(\gamma+1)^2 > 4\gamma$ when $\gamma > 1$. The thermal efficiency of the Carnot engine, Eq.(7.29), is given by

$$\eta_c = 1 - \frac{T_m}{T_M} = \frac{5}{9}$$

7.6. One mole of a perfect gas of heat capacity C_V at constant volume undergoes a process described by the equation $p = a\sqrt{V}$, where a is a constant. Find its heat capacity C during the reversible process.

7.7. A certain perfect gas of heat capacity C_V at constant volume undergoes a reversible process in which its heat capacity is given by $C = C_0 + aT^2$, where C_0 and a are constants. Derive the pressure-volume equation of the process.

7.8. A system may change its temperature from an initial equilibrium value T_i to a final equilibrium value T_f by two different reversible processes, *ij* and *ik*. Which one requires a larger amount of heat to be absorbed?

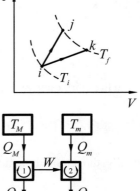

7.9. The work performed by a Carnot engine, operating between reservoirs at temperatures T_M and T_0, with T_M greater than T_0, is expanded on a Carnot heat pump, operating between reservoirs at temperatures T_m and T_0. What is the temperature T_0, provided that the total amount of heat delivered to this reservoir is x times that absorbed from the reservoir at temperature T_m?

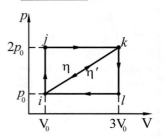

7.10. A certain perfect gas of adiabatic exponent $\gamma = C_p/C_V$ undergoes the reversible cyclic processes *ijki* and *ikli* represented on the pressure p against volume V graph. Derive the ratio η'/η of the efficiencies of the two processes.

8. THERMODYNAMIC FUNCTIONS

8.1. CLAUSIUS'S THEOREM
8.2. ENTROPY
8.3. THERMODYNAMIC POTENTIALS
8.4. HYDROSTATIC SYSTEMS
8.5. HEAT CAPACITIES

8.1. CLAUSIUS'S THEOREM

Consider a system, illustrated in Figure 8.1, performing a polythermal cyclic process, during which it exchanges heat $Q_1, Q_2, \ldots, Q_n$ with a set of n reservoirs at temperatures $T_1, T_2, \ldots, T_n$. The quantities Q_i are positive when absorbed and negative when yielded. Consider, separately, a reference reservoir at T_0 and n reversible, bithermal cyclic processes (n Carnot cycles $C_1, C_2, \ldots, C_n$) operating between $T_1, T_2, \ldots, T_n$ and T_0.

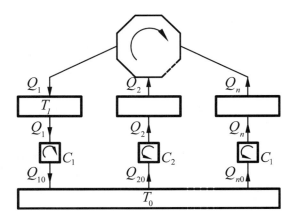

Figure 8.1. System undergoing a cyclic monothermal process

We may choose cycle C_i with the condition that it exchanges at T_i the same amount of heat as the system does, in order to leave each reservoir unchanged. According to the Carnot theorem, if the cycle C_i exchanges Q_{i0} with the reference reservoir T_0, we have

$$\frac{Q_{i0}}{Q_i} = \frac{T_0}{T_i} \quad \text{or} \quad Q_{i0} = \frac{T_0}{T_i} Q_i$$

After summing up the polythermal processes, the cycle undergone by the system does not exchange heat with reservoirs T_i but only with reservoir T_0. Heat transferred to Carnot cycles is therefore

$$Q_0 = \sum_i Q_{i0} = T_0 \sum \frac{Q_i}{T_i} \tag{8.1}$$

so that one deals with a cyclic monothermal process, performing work as a result of heat absorption Q_0. According to the second law of thermodynamics, $Q_0 \leq 0$, and hence

$$T_0 \sum \frac{Q_i}{T_i} \leq 0 \quad \text{or} \quad T_0 \int_{cycle} \frac{\delta Q}{T} \leq 0 \tag{8.2}$$

where rather than simple sums we consider integrals for cycles performed in infinitesimal steps. Since $T_0 > 0$, one has

$$\int_{cycle} \frac{\delta Q}{T} \leq 0 \tag{8.3}$$

Due to its reversibility, the polythermal cycle can be also performed by the system in the opposite direction, so that all heats Q_i change their sign, that is

$$\sum -\frac{Q_i}{T_i} \leq 0 \quad \text{or} \quad \sum \frac{Q_i}{T_i} \geq 0 \tag{8.4}$$

The inequalities (8.2) and (8.4) can be both satisfied over the same cycle only if

$$\sum \frac{Q_i}{T_i} = 0 \quad \text{or} \quad \int_{cycle} \frac{\delta Q}{T} = 0 \tag{8.5}$$

the latter again being used to describe a reversible cycle performed in infinitesimal steps. Equations (8.3) and (8.5) taken together form **Clausius's theorem**: *In any cyclic process one has*

$$\int_{cycle} \frac{\delta Q}{T} \leq 0 \tag{8.6}$$

where the equality holds only for reversible cycles.

8.2. ENTROPY

In an infinitesimal *reversible* process, the *entropy S* is defined as

$$dS = \frac{\delta Q_{rev}}{T} \tag{8.7}$$

so that for a finite reversible process one has

$$S_j - S_i = \int_i^j \frac{\delta Q_{rev}}{T} \tag{8.8}$$

The entropy is a *function of state*. This can be easily proved by means of a reversible cycle performed between the states i and j, illustrated in Figure 8.2 (a), where

$$0 = \int_{iRjR'i} \frac{\delta Q_{rev}}{T} = \int_{iRjR'i} dS = \int_{iRj} dS + \int_{jR'i} dS$$

It follows that

$$\int_{iRj} dS = \int_{iR'j} dS = S_j - S_i \tag{8.9}$$

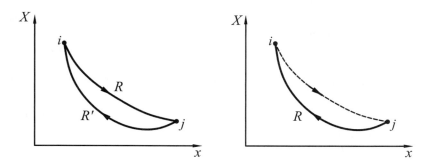

Figure 8.2. Cyclic processes performed between two states i and j: (a) reversibly (b) irreversibly (the dotted line)

Since the entropy change takes the same value for any reversible process from i to j, it is clear from Eq.(8.9) that, up to an integration constant, S is defined in a unique way for every state i or j of the system. It is therefore a function of state. Consequently, dS is an exact differential, whereas δQ_{rev} is not: $1/T$ is said to be an *integrating factor* for δQ_{rev}. Every reversible adiabatic process $(\delta Q = 0)$ follows a path included in a surface of constant entropy (an *isentropic* process).

For an *irreversible* process $i \to j$, one can construct a reversible path from i to j, as done in Figure 8.2 (b), thus forming an irreversible cycle $ijRi$ along which we have

$$\oint_{cycle} \frac{\delta Q}{T} \leq 0 \quad \text{or} \quad \int_i^j \frac{\delta Q}{T}\bigg|_{irrev} + \int_j^i \frac{\delta Q}{T}\bigg|_{rev} \leq 0$$

that is

$$\int_i^j \frac{\delta Q}{T}\bigg|_{irrev} \leq \int_j^i \frac{\delta Q}{T}\bigg|_{rev} = S_j - S_i$$

It follows for an irreversible process that

$$\int_i^j \frac{\delta Q}{T}\bigg|_{irrev} \leq S_j - S_i \quad \text{or} \quad dS \geq \frac{\delta Q}{T}$$

In any infinitesimal process we may therefore write

$$dS \geq \frac{\delta Q}{T} \tag{8.10}$$

where the equality applies to the reversible change. In a thermally isolated system, $\delta Q = 0$, therefore $dS \geq 0$, and this is called the *law of increase of entropy* which can be stated as follows: *the entropy of an isolated system increases whenever an irreversible process takes place*. It allows us to find the equilibrium configuration of an isolated system, given by the maximum value of its entropy. Since the various parts of the universe, defined as an assembly of systems and reservoirs interacting in an adiabatic enclosure, are not in thermodynamic equilibrium, the natural direction of all phenomena is that of increasing entropy. Therefore one has to assert that processes are not reversible in time, as often assumed in the framework of mechanics, but that there is the inevitability of an arrow of time.

The first law (7.14) can be reformulated using

$$\delta Q_{rev} = TdS \tag{8.11}$$

for *reversible* processes, to give the basic relation

$$dU = TdS + \sum_k X_k dx_k \tag{8.12}$$

For a *reversible* process of a hydrostatic system, it takes the form

$$dU = TdS - pdV \tag{8.13}$$

which also holds for *irreversible* processes, since all variables are functions of state.

It is to be noticed that the term TdS has the general form Xdx, where S is the extensive and T is the intensive conjugated variable. Therefore, the first law can be written in a compact form as

$$dU = \sum_k X_k dx_k \tag{8.14}$$

for both reversible and irreversible processes.

The heat transfer $\delta Q_{rev} = TdS$ can be represented in a TS diagram in a similar manner to the pV diagram used for the work performed by the system $-\delta W = pdV$ (Figure 8.3).

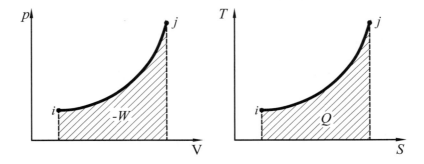

Figure 8.3. Representation of work on a pV diagram and of heat transfer on a TS diagram

Work and heat exchange in a finite process are both given by the areas below the process lines, represented in the pV and TS diagrams, respectively.

8.3. THERMODYNAMIC POTENTIALS

The conditions for equilibrium under various constraints can be determined in terms of functions of state having dimensions of energy and called *thermodynamic potentials*. For systems with two degrees of freedom, four potentials may be defined. They are, for a hydrostatic system, the *internal energy* U and the *enthalpy H*, defined by

$$H = U + pV \tag{8.15}$$

as already introduced by Eqs.(7.11) and (7.23), the *Helmholtz function F*, of the form

$$F = U - TS \tag{8.16}$$

and the *Gibbs function G*, given by

$$G = U - TS + pV \qquad (8.17)$$

The differential forms of the above expressions make their meaning even clearer. We have already seen in Eq.(8.13) that the internal energy may be expressed as

$$dU = TdS - pdV$$

The enthalpy takes the form

$$dH = dU + pdV + Vdp = TdS - pdV + pdV + Vdp$$

so that

$$dH = TdS + Vdp \qquad (8.18)$$

The Helmholtz function may be written as

$$dF = dU - TdS - SdT = TdS - pdV - TdS - SdT$$

that is

$$dF = -SdT - pdV \qquad (8.19)$$

For the Gibbs function we have

$$dG = dU - TdS - SdT + pdV + Vdp$$

which becomes

$$dG = -SdT + Vdp \qquad (8.20)$$

Let us now consider the main features of these potentials functions, which make each of them appropriate for the thermodynamical approach of a different process, performed under given constraints.

Internal energy: U being an exact differential, we have from Eq.(8.13) that

$$T = \left(\frac{\partial U}{\partial S}\right)_V \quad , \quad p = -\left(\frac{\partial U}{\partial V}\right)_S \qquad (8.21)$$

Under *thermal isolation* $(\delta Q = 0)$ and, according to the first law, we have

$$W_S = U_j - U_i$$

8. THERMODYNAMIC FUNCTIONS

which means that the work performed on the system equals the gained internal energy. In a process at *constant volume*, $dU = \delta Q_V$, and one may express

$$C_V = \left(\frac{\partial U}{\partial T}\right)_V = T\left(\frac{\partial S}{\partial T}\right)_V \tag{8.22}$$

Enthalpy: From Eq.(8.18) and since H is an exact differential, we can write

$$T = \left(\frac{\partial H}{\partial S}\right)_p \quad , \quad V = \left(\frac{\partial H}{\partial p}\right)_S \tag{8.23}$$

In an *isobaric processes* we have $dH = \delta Q_p$ or

$$Q_p = H_j - H_i$$

so that a change in enthalpy measures the heat transferred at constant pressure. Hence

$$C_p = \left(\frac{\partial H}{\partial T}\right)_p = T\left(\frac{\partial S}{\partial T}\right)_p \tag{8.24}$$

and enthalpy plays, in isobaric processes, the same role as that of internal energy for isochoric processes.

Helmholtz function: From Eq.(8.19) it follows that

$$S = -\left(\frac{\partial F}{\partial T}\right)_V \quad , \quad p = -\left(\frac{\partial F}{\partial V}\right)_T \tag{8.25}$$

In a reversible *isothermal process*, where $dF = -pdV$, one obtains

$$W_T = F_j - F_i$$

Thus, the work that can be extracted from the system is equal to the lost free energy, which thus plays, for isothermal processes, the same role as that of internal energy for adiabatic processes.

However, the equation $W_S = U_j - U_i$ holds for both reversible and irreversible processes, whereas the equation $W_T = F_j - F_i$ does not apply for irreversible processes. Using $\delta Q \leq TdS$ and taking into account that $\delta Q = dU + pdV$, we have

$$TdS \geq dU + pdV$$

which, in an isothermal process, reads

$$pdV \leq -d(U - TS)$$

The integral form reads

$$W_T \leq F_j - F_i$$

meaning that the lowest limit of free energy represents the highest limit for the work performed in an irreversible *isothermal* process. Taking the time derivatives we obtain

$$T\frac{dS}{dt} \geq \frac{dU}{dt} + p\frac{dV}{dt}$$

which may be written, at constant temperature, as

$$\frac{d(U-TS)}{dt} \leq -p\frac{dV}{dt}$$

If the volume is also maintained constant, we have

$$\frac{d(U-TS)}{dt} = \frac{dF}{dt} \leq 0 \tag{8.26}$$

that is, in irreversible processes *under constant T and V*, the Helmholtz function is decreasing with time and attains its minimum at thermodynamic equilibrium.

Gibbs function: From Eq.(8.20) we have

$$S = -\left(\frac{\partial G}{\partial T}\right)_p \quad , \quad V = \left(\frac{\partial G}{\partial p}\right)_T \tag{8.27}$$

In a similar manner we can derive that, in an irreversible process *under constant T and p*, the time derivatives obey

$$\frac{d(U+pV-TS)}{dt} = \frac{dG}{dt} \leq 0 \tag{8.28}$$

so that the Gibbs function is decreasing with time until its minimum is reached at thermodynamic equilibrium.

8.4. Hydrostatic Systems

Each thermodynamic potential of a hydrostatic system has a pair of proper independent variables: $U = U(S, V)$, $H = H(S, p)$, $F = F(T, V)$ and $G = G(T, p)$. Their differential properties are expressed in the form of the foregoing relations, widely used in applications.

Gibbs-Helmholtz equations

If any one of the thermodynamic potentials is explicitly known, then we have complete information on the system, since all the functions of state can be expressed in terms of this potential. The Helmholtz function F is of particular significance, since it is related to the statistical description of thermodynamic properties. An expression for the internal energy U can be derived, in terms of F, from Eq.(8.16), in the form

$$U = F + TS \quad \text{or} \quad U = F - T\left(\frac{\partial F}{\partial T}\right)_V \tag{8.29}$$

using Eq.(8.25) for entropy. This equation can be rewritten as

$$U = -T^2 \frac{\partial}{\partial T}\left(\frac{F}{T}\right)_V = T^2 \left(\frac{\partial \phi}{\partial T}\right)_V \tag{8.30}$$

where the expression $\phi = -F/T = S - U/T$ is called the *Planck function*. In a similar way one may derive the enthalpy

$$H = U + pV = F - T\left(\frac{\partial F}{\partial T}\right)_V - V\left(\frac{\partial F}{\partial V}\right)_T \tag{8.31}$$

and the Gibbs function

$$G = F + pV = F - V\left(\frac{\partial F}{\partial V}\right)_T = -V^2 \frac{\partial}{\partial V}\left(\frac{F}{V}\right)_T \tag{8.32}$$

in terms of the Helmholtz function. All these equations and others, of the same kind, which express various functions of state in terms of a given thermodynamic potential are called *Gibbs-Helmholtz equations*.

Maxwell relations

The condition of exact differentiability applied to thermodynamic potentials allows us to derive relations between the partial derivatives of thermodynamic variables, as follows

$$dU = TdS - pdV \quad \text{or} \quad \left(\frac{\partial T}{\partial V}\right)_S = -\left(\frac{\partial p}{\partial S}\right)_V \qquad (8.33)$$

$$dH = TdS + Vdp \quad \text{or} \quad \left(\frac{\partial T}{\partial p}\right)_S = \left(\frac{\partial V}{\partial S}\right)_p \qquad (8.34)$$

$$dF = -SdT - pdV \quad \text{or} \quad \left(\frac{\partial S}{\partial V}\right)_T = \left(\frac{\partial p}{\partial T}\right)_V \qquad (8.35)$$

$$dG = -SdT + Vdp \quad \text{or} \quad \left(\frac{\partial S}{\partial p}\right)_T = -\left(\frac{\partial V}{\partial T}\right)_p \qquad (8.36)$$

These four equations are called the *Maxwell relations*.

Reciprocity theorem

Taking p and T as functions of state, one may write their differential forms

$$p = p(V, T) \quad : \quad dp = \left(\frac{\partial p}{\partial V}\right)_T dV + \left(\frac{\partial p}{\partial T}\right)_V dT$$

$$T = T(p, V) \quad : \quad dT = \left(\frac{\partial T}{\partial p}\right)_V dp + \left(\frac{\partial T}{\partial V}\right)_p dV$$

Inserting the second expression into the first, one obtains

$$dp = \left(\frac{\partial p}{\partial V}\right)_T dV + \left(\frac{\partial p}{\partial T}\right)_V \left(\frac{\partial T}{\partial p}\right)_V dp + \left(\frac{\partial p}{\partial T}\right)_V \left(\frac{\partial T}{\partial V}\right)_p dV$$

or

$$dp = \left(\frac{\partial p}{\partial T}\right)_V \left(\frac{\partial T}{\partial p}\right)_V dp + \left[\left(\frac{\partial p}{\partial V}\right)_T + \left(\frac{\partial p}{\partial T}\right)_V \left(\frac{\partial T}{\partial V}\right)_p\right] dV$$

If we then take $dp = 0$ and $dV \neq 0$, it follows that

$$\left(\frac{\partial p}{\partial V}\right)_T = -\left(\frac{\partial p}{\partial T}\right)_V \left(\frac{\partial T}{\partial V}\right)_p$$

or, more generally, one obtains

$$\left(\frac{\partial p}{\partial T}\right)_V \left(\frac{\partial T}{\partial V}\right)_p \left(\frac{\partial V}{\partial p}\right)_T = -1 \qquad (8.37)$$

Such a theorem can be derived relating any three functions of state.

TdS equations

The differential form of entropy in terms of T and V, which reads

$$dS = \left(\frac{\partial S}{\partial T}\right)_V dT + \left(\frac{\partial S}{\partial V}\right)_T dV$$

leads to

$$TdS = T\left(\frac{\partial S}{\partial T}\right)_V dT + T\left(\frac{\partial S}{\partial V}\right)_T dV$$

Using Eqs.(8.22) and (8.35) we can write the *first TdS equation* as

$$TdS = C_V dT + T\left(\frac{\partial p}{\partial T}\right)_V dV \qquad (8.38)$$

If we further consider the differential form of S in terms of T and p, written as

$$dS = \left(\frac{\partial S}{\partial T}\right)_p dT + \left(\frac{\partial S}{\partial p}\right)_T dp \qquad \text{or} \qquad TdS = T\left(\frac{\partial S}{\partial T}\right)_p dT + T\left(\frac{\partial S}{\partial p}\right)_T dp$$

and make use of Eqs.(8.24) and (8.36), we obtain the *second TdS equation* in the form

$$TdS = C_p dT - T\left(\frac{\partial V}{\partial T}\right)_p dp \qquad (8.39)$$

We may eliminate TdS from Eqs.(8.13) and (8.38) to get

$$dU = C_V dT + \left[T\left(\frac{\partial p}{\partial T}\right)_V - p\right] dV$$

where U is expressed in terms of T and V, so that

$$\left(\frac{\partial U}{\partial V}\right)_T = T\left(\frac{\partial p}{\partial T}\right)_V - p \qquad (8.40)$$

and this is sometimes called the *internal energy equation*.

8.5. HEAT CAPACITIES

The main heat capacities are those corresponding to constant values of primary variables. In a hydrostatic system, as expressed by Eqs.(8.22) and (8.24), we have

$$C_V = \frac{\delta Q_V}{dT} = T\left(\frac{\partial S}{\partial T}\right)_V, \qquad C_p = \frac{\delta Q_p}{dT} = T\left(\frac{\partial S}{\partial T}\right)_p$$

The thermodynamical properties of heat capacities, which are now derived, can be either regarded as applications of the characteristic equations or as reference results for further developments in solid state and materials science.

The change of C_p as a function of pressure may be expressed as

$$\left(\frac{\partial C_p}{\partial p}\right)_T = \frac{\partial}{\partial p}\left[T\left(\frac{\partial S}{\partial T}\right)_p\right]_T = T\frac{\partial}{\partial p}\left[\left(\frac{\partial S}{\partial T}\right)_p\right]_T = T\frac{\partial}{\partial T}\left[\left(\frac{\partial S}{\partial p}\right)_T\right]_p$$

and this, in view of Maxwell relations, becomes

$$\left(\frac{\partial C_p}{\partial p}\right)_T = -T\frac{\partial}{\partial T}\left[\left(\frac{\partial V}{\partial T}\right)_p\right]_p = -T\left(\frac{\partial^2 V}{\partial T^2}\right)_p$$

In a similar way, we may derive the change of C_V as a function of volume in the form

$$\left(\frac{\partial C_V}{\partial V}\right)_T = T\left(\frac{\partial^2 p}{\partial T^2}\right)_V$$

To obtain the *difference* of heat capacities $C_p - C_V$ which is an important result, we start with the differential form of $S(T,V)$, which reads

$$dS = \left(\frac{\partial S}{\partial T}\right)_V dT + \left(\frac{\partial S}{\partial V}\right)_T dV$$

and this may be differentiated with respect to T, at constant pressure, to give

$$\left(\frac{\partial S}{\partial T}\right)_p = \left(\frac{\partial S}{\partial T}\right)_V + \left(\frac{\partial S}{\partial V}\right)_T\left(\frac{\partial V}{\partial T}\right)_p$$

From the definitions of C_p and C_V and using Maxwell relations we have

8. THERMODYNAMIC FUNCTIONS

$$C_p - C_V = T\left(\frac{\partial S}{\partial V}\right)_T \left(\frac{\partial V}{\partial T}\right)_p = T\left(\frac{\partial p}{\partial T}\right)_V \left(\frac{\partial V}{\partial T}\right)_p$$

From the reciprocity theorem (8.37) we obtain

$$\left(\frac{\partial p}{\partial T}\right)_V = -\left(\frac{\partial V}{\partial T}\right)_p \left(\frac{\partial p}{\partial V}\right)_T \quad \text{or} \quad C_p - C_V = -T\left(\frac{\partial V}{\partial T}\right)_p^2 \left(\frac{\partial p}{\partial V}\right)_T$$

This relation shows that C_p cannot be smaller than C_V since $(\partial p/\partial V)_T < 0$ for all known substances. It also states that $C_p = C_V$ if either $(\partial V/\partial T)_p = 0$ (e.g., at $4°C$ for water, when its density is at a maximum) or if $T \to 0$. It is convenient to define the coefficients of *isobaric expansion* by

$$\beta_p = \frac{1}{V}\left(\frac{\partial V}{\partial T}\right)_p \tag{8.41}$$

and *isothermal compressibility* as

$$k_T = -\frac{1}{V}\left(\frac{\partial V}{\partial p}\right)_T \tag{8.42}$$

so that the difference of heat capacities may be expressed in the form

$$C_p - C_V = VT\frac{\beta_p^2}{k_T} \tag{8.43}$$

The ratio of heat capacities can be expressed introducing the *adiabatic compressibility* k_S defined as

$$k_S = -\frac{1}{V}\left(\frac{\partial V}{\partial p}\right)_S \tag{8.44}$$

We may then calculate

$$\frac{k_T}{k_S} = \frac{(\partial V/\partial p)_T}{(\partial V/\partial p)_S} = \frac{(\partial V/\partial T)_p (\partial T/\partial p)_V}{(\partial V/\partial S)_p (\partial S/\partial p)_V} = \frac{(\partial S/\partial V)_p (\partial V/\partial T)_p}{(\partial S/\partial p)_V (\partial p/\partial T)_V} = \frac{(\partial S/\partial T)_p}{(\partial S/\partial T)_V} = \frac{C_p}{C_V} = \gamma \tag{8.45}$$

The ratio of heat capacities, denoted by γ is often called the *adiabatic exponent*.

FURTHER READING

1. D. Dunn - FUNDAMENTAL ENGINEERING THERMODYNAMICS, Longman, London, 2001.
2. W. Z. Black, J. G. Hartley - THERMODYNAMICS, Addison-Wesley, San Francisco, 1997.
3. J. P. Holman - THERMODYNAMICS, McGraw-Hill, New York, 1990.

PROBLEMS

8.1. Use the formulation (8.11) of the first law for reversible processes to derive the appropriate results for the heat engines operation.

Solution

Let Q be the absorbed heat from a reservoir at temperature T. The work performed by the engine in a monothermal cycle follows from

$$-W = Q = \int_{cycle} TdS = T\int_{cycle} dS = 0$$

and this result is known as the Kelvin statement of the second law:

> *No work can be performed in a monothermal cyclic process or heat cannot be completely converted into work.*

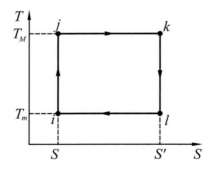

A bithermal cycle exchanging heat with two reservoirs T_M and T_m can be represented on a TS diagram as a rectangle, for any arbitrary performing system. We may express

$$-W = T_M \Delta S_M - T_m \Delta S_m = Q_M - Q_m \quad \text{and} \quad 0 = \Delta S_M - \Delta S_m$$

It follows that

$$\frac{Q_M}{T_M} = \frac{Q_m}{T_m} \quad \text{or} \quad -W = Q_M\left(1 - \frac{T_m}{T_M}\right)$$

Therefore the expression of the Carnot cycle efficiency becomes

$$\eta = \frac{-W}{Q_M} = 1 - \frac{T_m}{T_M}$$

and this is identical to Eq.(7.29), as expected.

8.2. The isobaric expansion of water is described by the coefficient $\beta_p = a\theta - b$, where θ is the empirical temperature, and the two constants are related by $b \cong a$. Show that the temperature $\theta_m = b/a$ of 4°C is not accessible by a reversible adiabatic process.

Solution

A given volume of water depends on the empirical temperature as described by Eq.(8.41), which reads

$$dV = \beta_p V d\theta \quad \text{or} \quad \Delta V = V - V_0 = V_0 \int_0^\theta \beta_p d\theta = V_0 \int_0^\theta (a\theta - b) d\theta$$

so that we obtain

$$\frac{\Delta V}{V_0} = \frac{\theta}{2}(a\theta - 2b)$$

where $V = V_0$ at the melting point of water ($\theta = 0°C$) and at a temperature of 8°C ($\theta = 2b/a$). Thus, $\Delta V/V_0$ is negative between 0°C and $\theta_m = b/a = 4°C$, where $\beta_p = 0$, and the volume is a minimum, as given by

$$\frac{dV}{d\theta} = V_0(a\theta - b) = 0 \quad \text{or} \quad V_{\min} = V_0\left(1 - \frac{b^2}{2a}\right)$$

and it is positive for temperatures above θ_m. The heat transfer in a reversible process is given by Eq.(8.38) as

$$TdS = C_V dT + T\left(\frac{\partial p}{\partial T}\right)_V dV$$

and it is convenient to combine the reciprocity theorem, Eq.(8.37), and the definitions (8.41) and (8.42), which gives

$$\left(\frac{\partial p}{\partial T}\right)_V = -\left(\frac{\partial V}{\partial T}\right)_P \Big/ \left(\frac{\partial V}{\partial p}\right)_T = \frac{\beta_p}{k_T}$$

Substituting this result, one obtains

$$TdS = C_V dT + T\frac{\beta_p}{k_T} dV$$

and hence, in an adiabatic process, we should have

$$dT = -T\frac{\beta_p}{k_T C_V}dV$$

It follows that water can be cooled by adiabatic compression at temperatures between 0°C and 4°C, where $\beta_p < 0$, and hence $dT < 0$. However, the temperature of 4°C is not accessible by reversible adiabatic processes, and this can be proved by considering the Carnot engine illustrated in Problem 8.1, operating between reservoirs at temperatures T_M and $T_m = 277$ K, with T_M greater than T_m. In view of the Clausius theorem, Eq.(8.6), we have

$$0 = \Delta S_{jk} + \Delta S_{li} \quad \text{or} \quad \frac{Q_M}{T_M} = \frac{Q_m}{T_m}$$

During the reversible isothermal compression at $T_m = 277$ K we have $dT = 0$ and $\beta_p = 0$, and this means that $dS = 0$. There is no heat transfer allowed at temperature T_m, thus $Q_m = 0$. The Clausius theorem shows that Q_M should also be equal to zero. In other words, water cannot be cooled down to 4°C by a reversible adiabatic process.

8.3. For a crystalline material at low temperature the heat capacity at constant volume can be written as $C_V = \alpha T^3$, where α depends on pressure and volume. Derive an expression for the difference $C_p - C_V$, where C_p is the heat capacity at constant pressure.

Solution

To obtain the difference of heat capacities in terms of temperature, we start with the equation

$$C_p - C_V = T\left(\frac{\partial p}{\partial V}\right)_V \left(\frac{\partial V}{\partial T}\right)_p$$

which can be written, in view of the Maxwell relations (8.35) and (8.36), in the form

$$C_p - C_V = -T\left(\frac{\partial S}{\partial V}\right)_T \left(\frac{\partial S}{\partial p}\right)_T$$

On substitution of C_V into Eq.(7.19), one obtains the internal energy in the form

$$\left(\frac{\partial U}{\partial T}\right)_V = \alpha T^3 \quad \text{or} \quad U = \frac{\alpha}{4}T^4$$

so that Eq.(8.30) provides the temperature dependence of the Helmholtz function as

$$\frac{\alpha}{4}T^4 = -T^2 \frac{\partial}{\partial T}\left(\frac{F}{T}\right) \quad \text{or} \quad F = -\frac{\alpha}{12}T^4$$

The entropy follows from the definition (8.16), which gives

$$S = \frac{1}{T}(U-F) = \frac{\alpha}{3}T^4$$

We finally obtain the required temperature dependence as

$$C_p - C_V = -\frac{1}{9}\frac{\partial \alpha}{\partial V}\frac{\partial \alpha}{\partial p}T^7$$

8.4. For a perfectly elastic material in the form of a rod of volume V that has a length ℓ under an axial load f at a temperature T, show that the difference between the heat capacity under constant load C_f and that at constant length C_ℓ can be written as

$$C_f - C_\ell = TV\lambda^2 E$$

where Young's modulus E and the linear expansivity λ are defined by the equations

$$E = \frac{\ell}{\Sigma}\left(\frac{\partial f}{\partial \ell}\right)_T, \qquad \lambda = \frac{1}{\ell}\left(\frac{\partial \ell}{\partial T}\right)_f$$

and $\Sigma = V/\ell$ stands for the cross-sectional area of the rod.

Solution

The heat capacity at constant load C_f is analogous to that at constant pressure for a closed hydrostatic system, Eq.(8.24), thus

$$C_f = T\left(\frac{\partial S}{\partial T}\right)_f$$

while the heat capacity at constant length C_ℓ is analogous to that at constant volume, Eq.(8.22), that is

$$C_\ell = T\left(\frac{\partial S}{\partial T}\right)_\ell$$

By considering the entropy as a function of T and ℓ, we obtain

$$dS = \left(\frac{\partial S}{\partial T}\right)_\ell dT + \left(\frac{\partial S}{\partial \ell}\right)_T d\ell \qquad \text{or} \qquad TdS = T\left(\frac{\partial S}{\partial T}\right)_\ell dT + T\left(\frac{\partial S}{\partial \ell}\right)_T d\ell$$

and then, differentiating with respect to T, at constant f, results in

$$T\left(\frac{\partial S}{\partial T}\right)_f = T\left(\frac{\partial S}{\partial T}\right)_\ell + T\left(\frac{\partial S}{\partial \ell}\right)_T\left(\frac{\partial L}{\partial T}\right)_f \qquad \text{or} \qquad C_f - C_\ell = T\left(\frac{\partial S}{\partial \ell}\right)_T\left(\frac{\partial \ell}{\partial T}\right)_f$$

The work done on the system in an infinitesimal reversible process is

$$\delta W = fd\ell \quad \text{and} \quad \delta W = -pdV$$

for an elastic rod in tension and, respectively, for a closed hydrostatic system, Eq.(7.17). Thus, by analogy, Maxwell relations and the reciprocity theorem for an elastic rod may be obtained from those for a closed hydrostatic system by substituting f for p and $-\ell$ for V. Then, Eq.(8.35) becomes

$$\left(\frac{\partial S}{\partial \ell}\right)_f = -\left(\frac{\partial f}{\partial T}\right)_\ell$$

and Eq.(8.37) reads

$$\left(\frac{\partial f}{\partial T}\right)_\ell \left(\frac{\partial T}{\partial \ell}\right)_f \left(\frac{\partial \ell}{\partial f}\right)_T = -1 \quad \text{or} \quad \left(\frac{\partial f}{\partial T}\right)_\ell = -\left(\frac{\partial f}{\partial \ell}\right)_T \left(\frac{\partial \ell}{\partial T}\right)_f$$

Substituting into $C_f - C_\ell$ and using the definitions of E and λ one obtains

$$C_f - C_\ell = T\left(\frac{\partial f}{\partial \ell}\right)_T \left(\frac{\partial \ell}{\partial T}\right)_f^2 = T\frac{\Sigma E}{\ell}\ell^2\lambda^2 = TV\lambda^2 E$$

as required.

8.5. Determine the internal energy U and the Helmholtz function F for an ideal spring that obeys Hooke's law $f = k(T)x$, where the spring constant is a function of temperature.

Solution

Substituting f for p and $-x$ for V, as established in Problem 8.4, the differential form of the Helmholtz function, Eq.(8.19), reads

$$dF = -SdT + fdx$$

and this gives

$$\left(\frac{\partial F}{\partial x}\right)_F = f = k(T)x \quad \text{or} \quad F(T,x) = F(T,0) + k(T)\frac{x^2}{2}$$

Substituting $F(T,x)$ into Eq.(8.25), we have

$$S(T,x) = -\left(\frac{\partial F}{\partial T}\right)_x = -S(T,0) - \left[\frac{\partial k(T)}{\partial T}\right]_x \frac{x^2}{2}$$

and the internal energy is obtained from Eq.(8.16) in the form

$$U(T,x) = F + TS = U(T,0) + k(T)\frac{x^2}{2} - T\left[\frac{\partial k(T)}{\partial T}\right]_x \frac{x^2}{2}$$

It follows that the elastic energy $k(T)x^2/2$ stands for the free energy of the spring during an isothermal expansion and plays the same role as that of internal energy for an adiabatic expansion.

8.6. One mole of a perfect gas of adiabatic exponent $\gamma = C_p/C_V$ undergoes a reversible process described by the equation $p = a - bV$, where a and b are positive constants. What is the temperature of the gas when its entropy is a maximum?

8.7. The coefficients of isobaric expansion β_p and isothermal compressibility k_T for a certain closed hydrostatic system are given by

$$\beta_p = \frac{V-b}{VT}, \qquad k_T = \frac{3(V-b)}{4pV}$$

where b is a constant. Derive an expression for the equation of state of the system.

8.8. A certain perfect gas has a molar heat capacity at constant volume given by

$$C_V = a + bT$$

where a and b are constants. Derive expressions for the Helmholtz function F, the Gibbs function G and the enthalpy H of the gas.

8.9. Derive the following relations for a closed hydrostatic system:

$$\left(\frac{\partial U}{\partial V}\right)_T = \nu(C_p - C_V)\left(\frac{\partial T}{\partial V}\right)_p - p, \qquad \left(\frac{\partial U}{\partial p}\right)_T = -\nu(C_p - C_V)\left(\frac{\partial T}{\partial p}\right)_V - p\left(\frac{\partial V}{\partial p}\right)_T$$

8.10. Show that, during the isobaric expansion of a homogeneous body, the entropy may either increase or decrease, provided the coefficient β_p is either respectively positive or negative.

9. PERFECT AND REAL GASES

9.1. PERFECT GAS LAWS
9.2. THERMODYNAMIC FUNCTIONS OF A PERFECT GAS
9.3. VAN DER WAALS GAS

9.1. PERFECT GAS LAWS

It is usual to postulate a *perfect* or *ideal gas* whose properties are defined by Boyle's law and Joule's law, which are only approximately obeyed at low pressures by a real gas. For that system the laws of the gaseous state become particularly simple.

Boyle's law states that *the product pV is a constant provided the temperature is maintained constant*. Isothermal measurements of real gases can be fitted with a power series, called the *virial expansion*, of the form

$$pV = A + Bp + Cp^2 + \ldots \quad \text{or} \quad V = \frac{A}{p} + B + Cp + \ldots \tag{9.1}$$

where the *virial coefficients* A, B, C are temperature dependent. Boyle's law is observed either at low pressures or if all coefficients but A are zero. At normal pressures, one may retain only the first two terms of the expansion

$$V = \frac{A}{p} + B$$

so that Boyle's law is obeyed if the second virial coefficient B vanishes. This is said to occur at a given *Boyle temperature*. The virial expansion (9.1) is sometimes expressed as

$$p = \frac{A}{V}\left(1 + \frac{B}{V} + \frac{C'}{V^2} + \ldots\right)$$

where the same A and B coefficients appear in both power series.

9. PERFECT AND REAL GASES

Joule's law states that *internal energy is a function of temperature only*. It can be derived from the experimental free expansion of a gas, illustrated in Figure 9.1, in which no work is done and no heat is transferred, so that

$$dU = \delta Q + \delta W = 0$$

The result of the experiment is that the temperature change is too small to be detected, to a first approximation, and can be expressed as

$$\left(\frac{\partial T}{\partial V}\right)_U = -\left(\frac{\partial T}{\partial U}\right)_V \left(\frac{\partial U}{\partial V}\right)_T = 0$$

where the reciprocity theorem (8.37) has been used. Since $(\partial U/\partial T)_V = C_V$ is the heat capacity at constant volume and is finite, it follows that

$$\left(\frac{\partial U}{\partial V}\right)_T = 0 \qquad (9.2)$$

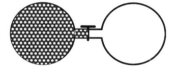

Figure 9.1. Free expansion of a gas in a thermally isolated vessel with rigid walls

We may then write

$$\left(\frac{\partial U}{\partial p}\right)_T = \left(\frac{\partial U}{\partial V}\right)_T \left(\frac{\partial V}{\partial p}\right)_T$$

where $(\partial V/\partial p)_T = -Vk_T$ is a finite quantity, defined in Eq.(8.42), so that Eq.(9.2) becomes

$$\left(\frac{\partial U}{\partial p}\right)_T = 0 \qquad (9.3)$$

Therefore the internal energy U of a perfect gas is independent of p and V, and thus it is a function of T only. Joule's law, as expressed by both Eqs.(9.2) and (9.3), describes the limiting behaviour of any real gas at low pressures.

Equation of state

Inserting Joule's law (9.2) into the equation of internal energy (8.40), we get

$$\left(\frac{\partial U}{\partial V}\right)_T = \left(\frac{\partial p}{\partial T}\right)_V - p = 0$$

so that

$$\left(\frac{\partial p}{\partial T}\right)_V = \frac{p}{T} \quad \text{or} \quad \ln p = \ln T + \ln f(V) \tag{9.4}$$

where the integration constant can be replaced by an arbitrary function $\ln f(V)$, because V is maintained constant during differentiation. It follows that

$$pf(V) = T$$

According to Boyle's law, pV is constant at a given T, thus $f(V) \sim V$, and we obtain the equation of state of an ideal gas in the form

$$pV = RT \tag{9.5}$$

where R is known as the *gas constant*.

Polytropic equation

An ideal gas is said to undergo a *polytropic* process when its heat capacity, defined according to Eq.(7.15) and denoted by C, is held constant. Using Eq.(7.18) written as

$$\delta Q = dU + pdV \quad \text{or} \quad CdT = C_V dT + pdV$$

where $\delta Q = CdT$ and $dU = C_V dT$, and differentiating the equation of state (9.5), which reads

$$pdV + Vdp = RdT$$

we may eliminate dT to obtain

$$pdV + Vdp = R\frac{pdV}{C - C_V} \quad \text{or} \quad (C - C_V - R)pdV + (C - C_V)Vdp = 0 \tag{9.6}$$

The difference of heat capacities for an ideal gas is given by the so-called *Mayer relation* which has the form

9. PERFECT AND REAL GASES

$$C_p - C_V = -T\left(\frac{R}{p}\right)^2\left(-\frac{RT}{V^2}\right) = R \tag{9.7}$$

This result can be inserted into Eq.(9.6), to give

$$(C - C_p)\frac{dV}{V} + (C - C_V)\frac{dp}{p} = 0$$

Integration leads now to

$$(C - C_p)\ln V + (C - C_V)\ln p = \text{const} \tag{9.8}$$

The *polytropic exponent n* is given by the ratio

$$n = \frac{C - C_p}{C - C_V} \tag{9.9}$$

which allows us to write the polytropic equation as

$$pV^n = \text{const} \tag{9.10}$$

We have assumed so far that $C \neq C_V$. When $C = C_V$, Eq.(9.8) reduces to $(C - C_p)\ln V = \text{const}$, so that $V = \text{const}$. It follows that a polytropic process becomes isochoric when $C = C_V$. When $n = 0$, $C = C_p$ and the polytropic process becomes isobaric. In the same way, isothermal and adiabatic processes correspond to $n = 1$ and $n = \gamma$, respectively. Expressing the heat capacity as

$$C = \frac{nC_V - C_p}{n - 1}$$

we can see that in isothermal processes we have $C = \infty$, in agreement with the definition $C = \delta Q / dT$ where $dT = 0$ and $\delta Q \neq 0$. In adiabatic processes $C = 0$, since $\delta Q = 0$ and $dT \neq 0$. The curves representing basic polytropic processes are plotted on a pV diagram in Figure 9.2.

9.2. THERMODYNAMIC FUNCTIONS OF A PERFECT GAS

Using the equation of state of a perfect gas, its entropy can be expressed in terms of any two of the fundamental variables. Taking $S = S(p,T)$ we may differentiate to obtain

$$dS = \left(\frac{\partial S}{\partial p}\right)_T dp + \left(\frac{\partial S}{\partial T}\right)_p dT = -\left(\frac{\partial V}{\partial T}\right)_p dp + \left(\frac{\partial S}{\partial T}\right)_p dT$$

where we made use of Maxwell's relation (8.36). Since

$$p\left(\frac{\partial V}{\partial T}\right)_p = R \quad , \quad C_p = T\left(\frac{\partial S}{\partial T}\right)_p$$

as given by the equation of state (9.5) and by Eq.(8.24), respectively, we obtain the form

$$dS = -\frac{R}{p}dp + \frac{C_p}{T}dT \tag{9.11}$$

Regarding C_p as independent of temperature and integrating, the entropy of a perfect gas becomes

$$S(p,T) = S_0 - R\ln p + C_p \ln T \tag{9.12}$$

In a similar way we may express $S(V,T)$ and $S(p,V)$ as

$$S(V,T) = S'_0 + R\ln V + C_V \ln T \tag{9.13}$$

$$S(p,V) = S''_0 + C_V \ln p + C_p \ln V \tag{9.14}$$

TS diagram
Using $C_V/C_p = \gamma$, $C_V/R = 1/(\gamma-1)$, $C_p/R = \gamma/(\gamma-1)$ we may express Eqs.(9.12) and (9.13) as

$$S - S_0 = C_p\left[\frac{1-\gamma}{\gamma}\ln p + \ln T\right] = C_p \ln\left[Tp^{(1-\gamma)/\gamma}\right] = \frac{R\gamma}{\gamma-1}\ln\left[Tp^{(1-\gamma)/\gamma}\right] \tag{9.15}$$

$$S - S'_0 = C_V\left[(\gamma-1)\ln V + \ln T\right] = C_V \ln\left(TV^{\gamma-1}\right) = \frac{R}{\gamma-1}\ln\left(TV^{\gamma-1}\right)$$

It follows that

$$Tp^{(1-\gamma)/\gamma} = e^{(S-S_0)(\gamma-1)/R\gamma} = \text{const } e^{(\gamma-1)S/R\gamma}$$

If one now takes $p = \text{const}$, the *isobaric equation* results as

$$T = \text{const } e^{(\gamma-1)S/R\gamma} \tag{9.16}$$

In a similar way, we may write

$$TV^{\gamma-1} = e^{(S-S_0')(\gamma-1)/R} = \text{const } e^{(\gamma-1)S/R}$$

so that, taking $V = \text{const}$, we obtain the *isochoric equation* in terms of entropy as

$$T = \text{const } e^{(\gamma-1)S/R} \tag{9.17}$$

The basic polytropic processes may be represented on a TS diagram, as shown in Figure 9.2. Since $\gamma > 1$, it follows that an isochoric curve $(n = \infty)$ has a steeper positive slope than an isobaric curve $(n = 0)$ at the same point.

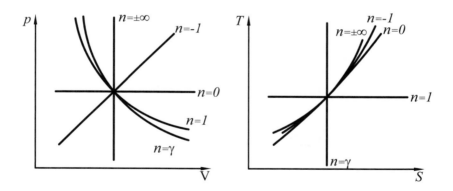

Figure 9.2. Curves representing polytropic processes of a perfect gas on pV and TS diagrams

Energy transfer in polytropic processes

In each reversible polytropic process we may separate the adiabatic contribution W from the nonadiabatic one Q to the internal energy change of a perfect gas. For an *isothermal process* $(T = \text{const})$ we may write

$$W_T = -\int_{V_i}^{V_j} p dV = \int_{V_j}^{V_i} \frac{RT}{V} dV = RT \int_{V_j}^{V_i} \frac{dV}{V} = RT \ln \frac{V_i}{V_j} \tag{9.18}$$

$$Q_T = \int_{S_i}^{S_j} T dS = T(S_j - S_i) \tag{9.19}$$

Using Eq.(9.15) it also follows that

$$Q_T = \frac{RT}{\gamma-1} \ln\left(\frac{V_j}{V_i}\right)^{\gamma-1} = RT \ln \frac{V_j}{V_i} \tag{9.20}$$

For an *adiabatic process* $(S = \text{const})$, where $Q_S = 0$, we have $pV^\gamma = \text{const} = p_i V_i^\gamma$ and therefore

$$W_S = -\int_{V_i}^{V_j} p\,dV = p_i V_i^\gamma \int_{V_j}^{V_i} \frac{dV}{V^\gamma} = \frac{p_i V_i^\gamma}{1-\gamma}\left(V_i^{1-\gamma} - V_j^{1-\gamma}\right) = \frac{p_i V_i}{1-\gamma}\left[1 - \left(\frac{V_j}{V_i}\right)^{1-\gamma}\right]$$

that is

$$W_S = \frac{RT_i}{1-\gamma}\left[1 - \left(\frac{V_j}{V_i}\right)^{1-\gamma}\right] \tag{9.21}$$

For an *isobaric process* $(p = \text{const})$ the work and the heat are given by

$$W_p = -\int_{V_i}^{V_j} p\,dV = p(V_i - V_j) = R(T_i - T_j) \tag{9.22}$$

$$Q_p = \int_{S_i}^{S_j} T dS = \int_{T_i}^{T_j} C_p dT = C_p(T_j - T_i) = \frac{R\gamma}{\gamma-1}(T_j - T_i) \tag{9.23}$$

where use has been made of Eq.(9.11), where $dp = 0$, that is

$$dS = \frac{C_p}{T} dT \quad \text{or} \quad TdS = C_p dT$$

For an *isochoric process* $(V = \text{const})$, where $dV = 0$, we obtain

$$Q_V = \int_{S_i}^{S_j} T dS = \int_{T_i}^{T_j} C_V dT = C_V(T_j - T_i) = \frac{R\gamma}{\gamma-1}(T_j - T_i) \tag{9.24}$$

where we made use of $TdS = C_V dT$, which is valid because $dV = 0$.

Thermodynamic potential functions

Internal energy U. From Joule's law (9.2) and Eq.(7.19) we have

$$dU = \left(\frac{\partial U}{\partial V}\right)_T dV + \left(\frac{\partial U}{\partial T}\right)_V dT = C_V dT$$

It follows that

$$U = C_V T + U_0 \qquad (9.25)$$

where the significance of the integration constant U_0 is assigned in each particular application.

Enthalpy H. We can rewrite the definition of enthalpy (7.23) in the form

$$H = U + pV = C_V T + U_0 + RT$$

so that, using the Mayer relation (9.7), we obtain

$$H = C_p T + U_0 \qquad (9.26)$$

Helmholtz function F can be expresed, using Eq.(9.13), in the form

$$F(V,T) = U - TS = C_V T + U_0 - T(S_0' + R \ln V + C_V \ln T)$$

so that

$$F(V,T) = C_V T(1 - \ln T) - RT \ln V - TS_0' + U_0 \qquad (9.27)$$

Gibbs function G can be written as

$$G(p,T) = U - TS + pV = C_V T + U_0 + pV - T(S_0 - R \ln p + C_p \ln T)$$

where the expression (9.12) for the entropy has been inserted. It follows that

$$G(p,T) = C_p T(1 - \ln T) + RT \ln p - TS_0 + U_0 \qquad (9.28)$$

9.3. VAN DER WAALS GAS

A *real gas* differs from a perfect gas to the extent that its molecules are now considered to have a finite size and to be influenced by forces due to intermolecular interactions. When molecules are far apart, both effects can be neglected and therefore, in the limit of low pressure, a real gas has an ideal behaviour. However except in this special case, the equation of state (9.5) must be amended to describe the molecular behaviour of a real gas. The two most commonly used modifications of the equation of state take the form

$$\left(p+\frac{a}{V^2}\right)(V-b)=RT \tag{9.29}$$

called the *van der Waals equation of state*, and

$$p(V-b)=RTe^{-a/RTV} \tag{9.30}$$

called the *Dieterici equation of state*. The constant b is associated with the finite size of the molecules, and the constant a is related to attractive intermolecular forces. At low pressure, both equations become

$$V=\frac{RT}{p}+b-\frac{a}{RT}$$

provided the second order terms are neglected. This relation is equivalent to the virial expansion (9.1), if the second virial coefficient is taken as

$$B=b-\frac{a}{RT}$$

Therefore, the finite molecular size and the intermolecular forces are linked concepts. If a and b were both zero, the ideal gas behaviour would hold at all temperatures and pressures.

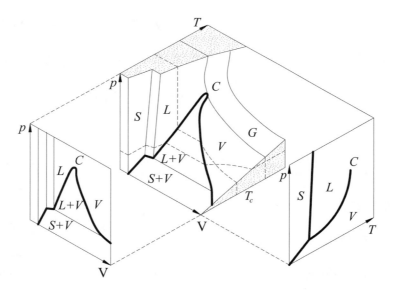

Figure 9.3. The pVT surface of a pure substance and its pT and pV projections

The real behaviour of a pure substance is represented in Figure 9.3 in pVT, pV and pT diagrams. A critical isotherm T_c on the pVT surface separates the gas phase, situated above the critical temperature, from liquid and vapour phases. The *critical point*

C on this isotherm corresponds to the conditions

$$\left(\frac{\partial p}{\partial V}\right)_T = \left(\frac{\partial^2 p}{\partial V^2}\right)_T = 0$$

which define unique values p_c, V_c, T_c, called *critical constants*, for a particular substance. Their experimental values give a ratio $p_c V_c / RT_c$ in the range of 0.25-0.30.

For a fixed volume of substance, the pVT surface reduces to a pT diagram. The vaporisation curve ends at the critical point C and corresponds to the liquid-vapour region of the pVT surface, the sublimation curve corresponds to the solid-vapour region and the *triple point* T_{tr}, a unique temperature at which all phases meet together, is the projection of the triple-point line.

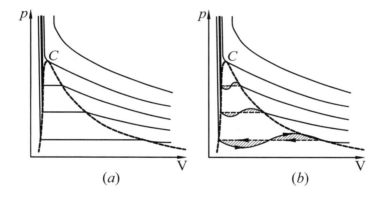

Figure 9.4. Isotherms in a pV diagram: (a) liquefaction of a real gas (b) van der Waals isotherms

For a given temperature, liquefaction proceeds at constant pressure across the liquid-vapour region and corresponds to the horizontal isotherms in the pV diagram illustrated in Figure 9.4 (a). The van der Waals isotherms obtained from Eq.(9.29) are represented in Figure 9.4 (b). Above the critical temperature, they properly describe the real gas behaviour. However, below T_c, the van der Waals isotherms, which are cubic equations of V at a given pressure, show maxima and minima instead of a constant pressure liquid-vapour line. The shaded surfaces situated above and below that line have equal areas. Indeed, for the reversible shaded cycle, marked in Figure 9.4 (b), $W + Q = 0$ and the Clausius theorem gives

$$\int_{cycle} \frac{\delta Q}{T} = 0$$

that is $Q = 0$, because $T = $ const. Since the total work is zero, the positive work, below the line, equals the negative work above the line, and it follows that the two areas are equal.

Equation of corresponding states

The critical parameters p_c, V_c, T_c may be given in terms of the a and b constants of van der Waals equation of state (9.29) which, for given $p = p_c$ and $T = T_c$, reads

$$p_c V^3 - (p_c b + RT_c)V^2 + aV - ab = 0 \tag{9.31}$$

Since the critical isotherm $T = T_c$ has an inflection point for $p = p_c$ and $V = V_c$, the equation must take the form $p_c (V - V_c)^3 = 0$ which, compared to Eq.(9.31), provides

$$V_c^3 = \frac{ab}{p_c} \quad , \quad 3V_c^2 = \frac{a}{p_c} \quad , \quad 3V_c = \frac{p_c b + RT_c}{p_c} \tag{9.32}$$

The solutions of this system of equations are

$$V_c = 3b \quad , \quad p_c = \frac{a}{27b^2} \quad , \quad T_c = \frac{8a}{27Rb} \tag{9.33}$$

It follows that the value of the ratio $p_c V_c / RT_c = 3/8 = 0.375$ falls outside the experimental range of critical parameters, previously mentioned. We can further define the *reduced parameters* p_r, V_r, T_r as

$$V_r = \frac{V}{V_c} \quad , \quad p_r = \frac{p}{p_c} \quad , \quad T_r = \frac{T}{T_c} \tag{9.34}$$

which may be substituted in Eq.(9.33) to obtain

$$V = 3b\, V_r \quad , \quad p = \frac{a p_r}{27b^2} \quad , \quad T = \frac{8a T_r}{27Rb}$$

Therefore, the van der Waals equation of state becomes

$$\left(p_r + \frac{3}{V_r^2}\right)\left(V_r - \frac{1}{3}\right) = \frac{8}{3} T_r \tag{9.35}$$

The states of different substances having the same reduced parameters are called corresponding states, and Eq.(9.35) is known as *equation of corresponding states*.

Entropy of a van der Waals gas

The van der Waals equation (9.29) can be written in terms of pressure as

$$p = \frac{RT}{V-b} - \frac{a}{V^2}$$

9. PERFECT AND REAL GASES 179

The internal energy U, as expressed by Eq.(8.40), is a function of pressure, and so we have

$$\left(\frac{\partial U}{\partial V}\right)_T = T\frac{\partial}{\partial T}\left(\frac{RT}{V-b} - \frac{a}{V^2}\right) - \frac{RT}{V-b} + \frac{a}{V^2} = \frac{a}{V^2}$$

It follows that

$$dU = C_V dT + \frac{a}{V^2}dV \tag{9.36}$$

and, integrating, we obtain

$$U = \int C_V dT - \frac{a}{V} + \text{const} \tag{9.37}$$

where the second term gives the contribution of molecular cohesion to the internal energy. Using Eq.(9.36) we may express the entropy of van der Waals gas as

$$dS = \frac{1}{T}(dU + pdV) = \frac{1}{T}\left(C_V dT + \frac{a}{V^2}dV\right) + \frac{1}{T}\left(\frac{RT}{V-b} - \frac{a}{V^2}\right)dV$$

or

$$dS = C_V \frac{dT}{T} + R\frac{dV}{V-b}$$

so that, assuming that $C_V = \text{const}$, we obtain the form

$$S = S_0 + C_V \ln T + R\ln(V-b) \tag{9.38}$$

which is similar to the entropy of an ideal gas, given by Eq.(9.13) and suggests that b represents the finite volume of the gas molecules.

FURTHER READING
1. Y. A. Cengel, M. A. Boles - THERMODYNAMICS, McGraw Hill, New York, 2001.
2. R. Baierlein - THERMAL PHYSICS, Cambridge University Press, Cambridge, 1999.
3. J. R. Howell, R. O. Buckius - FUNDAMENTALS OF ENGINEERING THERMODYNAMICS, McGraw Hill, New York, 1992.

180 ADVANCED UNIVERSITY PHYSICS

PROBLEMS

9.1. A quantity of perfect gas is contained in a vertical cylinder of cross-section area Σ by means of a well-fitting, frictionless piston of mass m. When equilibrium at outside atmospheric pressure p_0 is achieved, the volume occupied by the gas in the lower part of the cylinder is $V = \Sigma h$. Find the angular frequency of the vertical harmonic motion of the piston, assuming that the gas undergoes a reversible polytropic process.

Solution

For a small vertical displacement x of the piston the gas in the lower part of the cylinder suffers a change in volume $dV = \Sigma x$, and the corresponding change in pressure is obtained by differentiating the polytropic equation (9.10), which gives

$$\frac{dp}{p} + n\frac{dV}{V} = 0 \quad \text{or} \quad dp = -\frac{np}{V}dV = -\frac{np\Sigma}{V}x$$

where the equilibrium pressure is $p = p_0 + mg/\Sigma$. On substitution of p and $V = \Sigma h$ one obtains

$$dp = -\frac{np}{h}x = -\frac{n}{h}\left(p_0 + \frac{mg}{\Sigma}\right)x$$

and hence, the force acting on the piston has the form

$$F = \Sigma dp = -\frac{n}{h}(p_0\Sigma + mg)x = -kx$$

It follows that F is a restoring force driving a harmonic motion of the piston, where

$$k = \frac{n}{h}(p_0\Sigma + mg)$$

The angular frequency of the simple harmonic motion of the piston is then given by

$$\omega = \sqrt{\frac{k}{m}} = \sqrt{\frac{n}{h}\left(\frac{p_0\Sigma}{m} + g\right)}$$

The result holds for polytropic processes where $n > 0$.

9.2. A heat engine has a quantity of perfect gas as its working substance and undergoes a cycle of operation similar to that represented in Figure 7.4, where the reversible adiabatic compression $i \to j$ and expansion $k \to l$, between T_m and T_M, are substituted by polytropic processes of positive heat capacity C. Calculate the thermal efficiency of such an engine.

Solution

The heat capacity of the gas during a reversible polytropic process is a constant determined by the value of its polytropic exponent, Eq.(9.9), in the form

$$C = \frac{nC_V - C_p}{n-1}$$

and hence, it is positive when $n < 1$ and $n > \gamma$. Applying the first law to each stage gives:

for $i \to j$: The polytropic increase in temperature from T_m to T_M requires the heat absorbed and the work done to be written as

$$Q_{ij} = \nu C(T_M - T_m), \qquad -W_{ij} = \nu(C - C_V)(T_M - T_m)$$

for $i \to k$: The heat absorbed in the isothermal expansion at T_M, Eq.(9.18), reads

$$Q_{jk} = -W_{jk} = \nu R T_M \ln \frac{V_k}{V_j}$$

for $k \to l$: During the polytropic process from T_M to T_m we have

$$Q_{kl} = \nu C_V(T_m - T_M), \qquad -W_{kl} = \nu(C - C_V)(T_m - T_M)$$

for $l \to i$: The isothermal compression at T_m results in heat delivered to the surroundings, due to the work done on the system

$$Q_{li} = -W_{li} = \nu R T_M \ln \frac{V_i}{V_l}$$

Using the polytropic equations

$$p_i V_i^n = p_j V_j^n, \quad p_j V_j = p_k V_k, \quad p_k V_k^n = p_l V_l^n, \quad p_l V_l = p_i V_i$$

it is immediately obvious that

$$p_i p_k (V_i V_k)^n = p_j p_l (V_j V_l)^n \qquad \text{and} \qquad p_i p_k (V_i V_k) = p_j p_l (V_j V_l)$$

and hence, we obtain

$$V_i V_k = V_j V_l \qquad \text{or} \qquad \frac{V_k}{V_j} = \frac{V_l}{V_i}$$

The total work done by the working substance in one cycle is given by

$$-W = -W_{ij} - W_{jk} - W_{kl} - W_{li} = \nu R(T_M - T_m) \ln \frac{V_k}{V_j}$$

while the heat absorbed from the surroundings can be written as

$$Q = Q_{ij} + Q_{jk} = \nu C(T_M - T_m) + \nu R T_M \ln \frac{V_k}{V_j}$$

On substitution into Eq.(7.29) one obtains the thermal efficiency in the form

$$\eta = \frac{T_M - T_m}{T_M + \dfrac{C(T_M - T_m)}{R \ln(V_k / V_j)}}$$

This efficiency is smaller than that of a Carnot engine operating between reservoirs at temperatures T_M and T_m, and this can be shown rearranging the expression as

$$\frac{1}{\eta} = \frac{T_M}{T_M - T_m} + \frac{R}{C} \ln \frac{V_k}{V_j} = \frac{1}{\eta_C} + \frac{R}{C} \ln \frac{V_k}{V_j}$$

where η_C is the efficiency of the Carnot engine, Eq.(7.29). Since $C > 0$ and $V_k > V_j$, it follows that $(R/C) \ln(V_k / V_j) > 0$, and thus

$$\frac{1}{\eta} > \frac{1}{\eta_C} \quad \text{or} \quad \eta < \eta_C$$

as required.

9.3. Determine the rate of change of temperature with volume for a mole of gas of heat capacity C_V at constant volume undergoing the free adiabatic expansion illustrated in Figure 9.1. What is the change in temperature and entropy when the volume of the system changes from V_i to V_j.

Solution

In a free adiabatic expansion of a system no work is done, that is $U = $ const. The temperature change expressed by $(\partial T / \partial V)_U$ is called the *Joule coefficient*, and can be derived starting from

$$\left(\frac{\partial T}{\partial V}\right)_U = -\left(\frac{\partial T}{\partial U}\right)_V \left(\frac{\partial U}{\partial V}\right)_T$$

and this, using $C_V = (\partial U / \partial T)_V$ and the internal energy equation (8.40), becomes

$$\left(\frac{\partial T}{\partial V}\right)_U = -\frac{1}{C_V}\left[T\left(\frac{\partial p}{\partial T}\right)_V - p\right] = -\frac{T^2}{C_V} \frac{\partial}{\partial T}\left(\frac{p}{T}\right)_V$$

For a finite increase in the volume, during a Joule expansion, the corresponding temperature change can be obtained by integration as

$$\Delta T = -\int_{V_i}^{V_j} \frac{1}{C_V}\left[T\left(\frac{\partial p}{\partial T}\right)_V - p\right] dV$$

For an ideal gas the Joule coefficient is zero, since $(\partial U / \partial V)_T = 0$. However, a real gas is cooled as a result of a Joule expansion, because, as illustrated in Figure 7.1, in an expansion we have $(\partial U / \partial V)_T > 0$ and $(\partial T / \partial U)_V = 1/C_V > 0$, so that $(\partial T / \partial V)_U < 0$.

The entropy change during a Joule expansion can be derived using the restriction $U = $ const or $dU = TdS - pdV = 0$, so that

$$\Delta S = \int_{V_i}^{V_j} \frac{pdV}{T}$$

Substituting the equation of state (9.5), we obtain for an ideal gas:

$$\Delta S = R \ln \frac{V_j}{V_i}$$

which agrees with Eq.(9.13), since T is constant, during expansion. As $\Delta S = 0$, a Joule expansion is irreversible.

9.4. Derive an expression for the rate of change of temperature with pressure for a fixed mass of gas, of specific heat c_p at constant pressure, undergoing a throttling process under conditions of thermal isolation. What is the change in temperature and entropy when a constant pressure p_i is held on one side of the porous plug and a lower pressure p_j on the other side.

Solution

Let a gas be made to undergo a continuous *throttling* process through a porous plug, under conditions of thermal isolation. A constant pressure can be held on one side of the porous plug and a lower pressure on the other side, as the gas expands and the pressure drops. The steady flow process for the gas is isenthalpic, since no work and heat are exchanged with the surroundings.

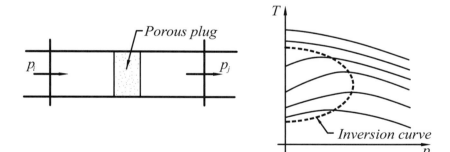

The temperature change with pressure as given by the Joule-Kelvin coefficient μ, which can be derived from the reciprocity theorem (8.37) as

$$\mu = \left(\frac{\partial T}{\partial p}\right)_h = -\left(\frac{\partial T}{\partial h}\right)_p \left(\frac{\partial h}{\partial p}\right)_T$$

where the equations are written for unit mass of ideal gas, in terms of specific extensive variables. Using $dh = Tds + vdp$ we obtain

$$\left(\frac{\partial h}{\partial T}\right)_P = T\left(\frac{\partial s}{\partial T}\right)_P = c_p$$

and

$$\left(\frac{\partial h}{\partial p}\right)_T = T\left(\frac{\partial s}{\partial p}\right)_T + v = -T\left(\frac{\partial v}{\partial T}\right)_p + v$$

where use has been made of the Maxwell relation (8.36). It follows that

$$\mu = \left(\frac{\partial T}{\partial p}\right)_h = \frac{1}{c_p}\left[T\left(\frac{\partial v}{\partial T}\right)_p - v\right] = \frac{T^2}{c_p}\frac{\partial}{\partial T}\left(\frac{v}{T}\right)_p$$

For a finite pressure drop during a Joule-Kelvin expansion, the temperature change can be obtained by integration as

$$\Delta T = \int_{p_i}^{p_j}\frac{1}{c_p}\left[T\left(\frac{\partial v}{\partial T}\right)_p - v\right]dp$$

For a perfect gas $\Delta T = 0$, because $\mu = (1/c_p)[T(R/mp) - v] = 0$. However, a real gas can be cooled or heated, as a result of a Joule-Kelvin expansion, according to the sign of μ. The cooling and heating regimes can be separated by plotting the isenthalpic curves of the real gas in a Tp diagram as given in the figure. The locus of maxima of the isenthalpic curves, at which $\mu = (\partial T/\partial p)_h = 0$, is called the inversion curve. Inside the inversion curve μ is positive (region of cooling), whereas outside it is negative (region of heating).

The change in the entropy during a Joule-Kelvin expansion can be derived using $h = $ const, so that

$$\Delta s = -\int_{p_i}^{p_j}\frac{v\,dp}{T} \quad \text{or} \quad \Delta S = -\int_{p_i}^{p_j}\frac{V\,dp}{T}$$

For a perfect gas

$$\Delta S = R\ln(p_i/p_j)$$

and this agrees with Eq.(9.12) because T is a constant during expansion. Since $\Delta S > 0$, a Joule-Kelvin expansion of the perfect gas is irreversible.

9.5. Derive the critical parameters and the equation of corresponding states for a system that consists of one mole of a gas that obeys the Dieterici equation of state (9.30).

Solution

It has been pointed out that the Dieterici equation of state (9.30) is equivalent to van der Waals equation of state in the limit of low pressure. The critical parameters and the equation of corresponding states for a Dieterici gas can be derived using the critical conditions

$$\left(\frac{\partial p}{\partial V}\right)_T = \left(\frac{\partial^2 p}{\partial V^2}\right)_T = 0$$

where the pressure is given by the equation of state. It is a straightforward matter to show that the two equations can be written in the form

$$\frac{a}{RT_c} = \frac{V_c^2}{V_c - b} = \frac{V_c^3}{2(V_c - b)^2}$$

and it follows that

$$b = \frac{V_c}{2}, \quad a = 2RT_c V_c$$

Therefore, the equation of state reads

$$p\left(V - \frac{V_c}{2}\right) = RTe^{-2T_c V_c / TV}$$

Setting $V = V_c$ and $T = T_c$, we obtain

$$p_c = \frac{2RT_c}{V_c} e^{-2}$$

or $p_c V_c / RT_c = 2e^{-2} = 0.271$. We can see that the value derived from the Dieterici model falls inside the experimental range 0.25-0.30 for the critical parameters. The reduced parameters, as defined by Eq.(9.34), may be then inserted into the equation of state to obtain the Dieterici equation of corresponding states in the form

$$p_r\left(V_r - \frac{1}{2}\right) = \frac{1}{2} e^{2 - 2/T_r V_r}$$

Both the van der Waals and Dieterici forms for the equation of corresponding states are accurately obeyed by several noble gases (Ne, Ar, Kr, Xe) as well as by oxygen, carbon dioxide and methane.

9.6. A fixed mass of perfect gas undergoes the reversible cyclic processes *ijki* and *ikli* represented on the *TS* diagram, where $T_M = xT_m$. Derive the ratio of the efficiencies of the two heat engines.

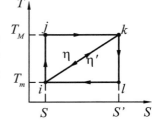

9.7. For a gas of heat capacity C_V at constant volume, isobaric expansivity β_p and isothermal compressibility k_T, which obeys the equation of state $p = p_0(1 + aT - bV)$, where a and b are constants, determine the relationship between pressure and volume in a reversible adiabatic process.

9.8. Find an expression for the Joule-Kelvin coefficient μ for a gas of molar heat capacity C_p at constant pressure that obeys the equation of state $p(V - b) = RT$.

9.9. For a gas obeying the van der Waals equation of state, derive an expression for the difference between the molar heat capacities at constant pressure C_p and at constant volume C_V.

9.10. For a gas obeying the equation of state of the form

$$\left(p + \frac{a}{V^{5/3}}\right)(V - b) = RT$$

derive the critical constants p_c, V_c, T_c, and find an expression for the equation of state at the critical point.

10. PHASE TRANSITIONS

10.1. EQUILIBRIUM BETWEEN PHASES
10.2. FIRST ORDER PHASE TRANSITIONS
10.3. HIGHER ORDER PHASE TRANSITIONS
10.4. THE PHASE RULE
10.5. THIRD LAW OF THERMODYNAMICS

10.1. EQUILIBRIUM BETWEEN PHASES

We first consider a system consisting of one component, having a uniform chemical composition, in contact with its surroundings through the exchange of heat and work. We can write

$$\delta Q \leq TdS \quad , \quad \delta W = -pdV \tag{10.1}$$

where T, p refers to the surroundings and may differ from the actual temperature and pressure of the system. It follows from the first law that

$$dU \leq TdS - pdV \tag{10.2}$$

Therefore the quantity

$$dU + pdV - TdS \leq 0 \tag{10.3}$$

cannot increase, so that, for a given temperature and pressure of the surroundings, the condition for equilibrium is given by

$$dU + pdV - TdS = 0 \tag{10.4}$$

Simpler forms of this general relation can be derived under the condition that two of the thermodynamic coordinates remain constant.

S and V constant: Eq.(10.4) reduces to $dU = 0$, so that the conditions for equilibrium are

$$dS = 0 \quad , \quad dV = 0 \quad , \quad dU = 0 \qquad (10.5)$$

S and p constant: Eq.(10.3) becomes $dU + pdV = dH = 0$, and the set of conditions reads

$$dS = 0 \quad , \quad dp = 0 \quad , \quad dH = 0 \qquad (10.6)$$

T and V constant: Eq.(10.3) can be written as $d(U - TS) = dF = 0$ and the conditions on the system are

$$dT = 0 \quad , \quad dV = 0 \quad , \quad dF = 0 \qquad (10.7)$$

T and p constant: Eq.(10.3) reduces to $d(U + pV - TS) = dG = 0$ and the appropriate set of conditions is

$$dT = 0 \quad , \quad dp = 0 \quad , \quad dG = 0 \qquad (10.8)$$

The four sets of conditions for equilibrium are equivalent. The thermodynamic potentials involved are always coupled through their specified variables. Equilibrium is defined as the minimum of the appropriate potential function depending on which variables are held constant. The analogy with mechanical potential energy shows that we deal with conditions for stable equilibrium.

When a system consists of more than one phase, the thermodynamic functions of the whole may be constructed out of those of the component phases. For a system of two phases, denoted as α and β, we can then write

$$G = m_\alpha g_\alpha + m_\beta g_\beta \qquad (10.9)$$

where the specific Gibbs functions depend on p and T only, while G is a function of p, T and the masses m_α, m_β of each phase. A unique condition for phase equilibrium holds whatever the external constraints, as we shall demostrate below.

At *constant T and p* from Eqs.(10.8) and (10.9) the condition for equilibrium becomes

$$dG = g_\alpha dm_\alpha + g_\beta dm_\beta = 0 \qquad (10.10)$$

As the system is closed, mass is conserved according to $dm_\alpha + dm_\beta = 0$, so that Eq.(10.10) reduces to

$$g_\alpha = g_\beta \qquad (10.11)$$

Therefore *the condition for equilibrium between phases is that their specific Gibbs functions are equal*. The result also holds when more than two phases are present.

For *constant T and V* the proper condition for equilibrium is given by Eq.(10.7) and we can write $F = m_\alpha f_\alpha + m_\beta f_\beta$ so that

$$dF = f_\alpha dm_\alpha + f_\beta dm_\beta + m_\alpha df_\alpha + m_\beta df_\beta = 0 \tag{10.12}$$

We can express the constant volume V in terms of the specific volumes of the two phases, so that

$$m_\alpha v_\alpha + m_\beta v_\beta = \text{const} \quad \text{or} \quad m_\alpha dv_\alpha + m_\beta dv_\beta + v_\alpha dm_\alpha + v_\beta dm_\beta = 0$$

Multiplying the last equation by p and taking into account the condition for equilibrium, we get

$$(f_\alpha + pv_\alpha)dm_\alpha + (f_\beta + pv_\beta)dm_\beta + m_\alpha(df_\alpha + pdv_\alpha) + m_\beta(df_\beta + pdv_\beta) = 0$$

where $df + pdv = -sdT = 0$, since T is assumed constant. Using again $dm_\alpha + dm_\beta = 0$ it follows that

$$f_\alpha + pv_\alpha = f_\beta + pv_\beta \quad \text{or} \quad g_\alpha = g_\chi \tag{10.13}$$

that is the same condition for equilibrium as derived in Eq.(10.11) under different constraints

If S and p are *constant* we may use the enthalpy $H = m_\alpha h_\alpha + m_\beta h_\beta$ and the condition (10.6) for equilibrium

$$dH = h_\alpha dm_\alpha + h_\beta dm_\beta + m_\alpha dh_\alpha + m_\beta dh_\beta = 0$$

under the constraint of constant entropy, which reads

$$m_\alpha s_\alpha + m_\beta s_\beta = \text{const} \quad \text{or} \quad s_\alpha dm_\alpha + s_\beta dm_\beta + m_\alpha ds_\alpha + m_\beta ds_\beta = 0$$

Subtracting the last relation, multiplied by T, from the condition for equilibrium we obtain

$$(h_\alpha - Ts_\alpha)dm_\alpha + (h_\beta - Ts_\beta)dm_\beta + m_\alpha(dh_\alpha - Tds_\alpha) + m_\beta(dh_\beta - Tds_\beta) = 0$$

Using $dh - Tds = Vdp = 0$, since p is assumed to be constant, together with the conservation of total mass, the equation reduces to

$$h_\alpha - Ts_\alpha = h_\beta - Ts_\beta \quad \text{or} \quad g_\alpha = g_\beta \tag{10.14}$$

which is also identical to Eq.(10.11).

Finally, the internal energy $U = m_\alpha u_\alpha + m_\beta u_\beta$ is used if S and V are *constant*. Then, by means of Eq.(10.4), one obtains

$$dU = u_\alpha dm_\alpha + u_\beta dm_\beta + m_\alpha du_\alpha + m_\beta du_\beta = 0$$

and the constant constraints are expressed as

$$m_\alpha v_\alpha + m_\beta v_\beta = \text{const} \quad \text{or} \quad v_\alpha dm_\alpha + v_\beta dm_\beta + m_\alpha dv_\alpha + m_\beta dv_\beta = 0$$

and

$$m_\alpha s_\alpha + m_\beta s_\beta = \text{const} \quad \text{or} \quad s_\alpha dm_\alpha + s_\beta dm_\beta + m_\alpha ds_\alpha + m_\beta ds_\beta = 0$$

Multiplying the two differential relations by p and T respectively, and subtracting them from the condition of equilibrium, we may write

$$(u_\alpha + pv_\alpha - Ts_\alpha) dm_\alpha + (u_\beta + pv_\beta - Ts_\beta) dm_\beta = 0$$

since $du + pdv - Tds = 0$. Taking into account the conservation of total mass, we obtain again that the condition (10.11) for equilibrium holds

$$u_\alpha + pv_\alpha - Ts_\alpha = u_\beta + pv_\beta - Ts_\beta \quad \text{or} \quad g_\alpha = g_\beta \tag{10.15}$$

10.2. First Order Phase Transitions

In a one component system, a phase transition can be best represented in a pT diagram, as it usually occurs at given p and T. At any point on the equilibrium boundary between two phases α and β, illustrated in Figure 10.1, the specific Gibbs functions must be equal. We may consider two phase transitions at p, T and $p + dp, T + dT$ respectively, for which we can write

$$g_\alpha = g_\beta \quad , \quad g_\alpha + dg_\alpha = g_\beta + dg_\beta$$

so that

$$dg_\alpha = dg_\beta \tag{10.16}$$

Since we have

$$dg_\alpha = v_\alpha dp - s_\alpha dT \quad , \quad dg_\beta = v_\beta dp - s_\beta dT$$

it follows that

$$v_\alpha dp - s_\alpha dT = v_\beta dp - s_\beta dT \tag{10.17}$$

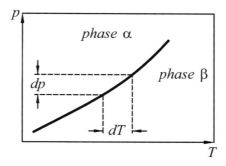

Figure 10.1. Phase equilibrium boundary in a *pT* diagram

Hence, one obtains

$$\frac{dp}{dT}(v_\alpha - v_\beta) = s_\alpha - s_\beta \quad \text{or} \quad \frac{dp}{dT} = \frac{S_\alpha - S_\beta}{V_\alpha - V_\beta} = \frac{\lambda}{T(V_\alpha - V_\beta)} \quad (10.18)$$

Equation (10.18) where λ is the *latent heat* transferred during the phase transition is called the *Clapeyron equation*. It gives the gradient of the phase boundary in a *pT* diagram and holds for *first order* phase transitions, in which there are changes of entropy and volume.

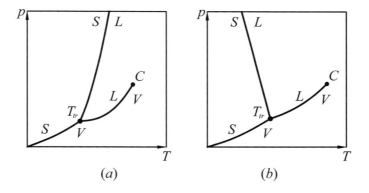

Figure 10.2. The *pT* phase diagrams for: (a) a pure substance that expands on melting (b) a pure substance that contracts on melting

The *pT* phase diagrams of a one component system, given in Figure 10.2, show that melting, vaporization and sublimation satisfy these requirements. The gradients of the vaporization and sublimation curves are positive for all substances. Although the gradient of the fusion curve is positive for most substances, it may also be negative in certain cases, as illustrated in Figure 10.2 (b). Water is one familiar example, since $\Delta V < 0$ and water contracts upon melting. The condition for three phases to coexist, that is $g_\alpha = g_\beta = g_\gamma$, gives the coordinates of the *triple point* T_{tr}.

10.3. HIGHER ORDER PHASE TRANSITIONS

We have seen that the entropy $S = -(\partial G / \partial T)_p$ and the volume $V = (\partial G / \partial p)_T$, which both show discontinuous changes during a first order transition, are the first derivatives of the Gibbs function. Following the scheme suggested by Ehrenfest, all phase transitions may be classified according to the *order of the lowest derivative of the Gibbs function, undergoing a finite change at the transition*. The classification is called the *order of the transition*. Therefore the Clapeyron equation (10.18) can only be applied to transitions of the first order. The definition of higher order changes requires that both ΔS and ΔV should be zero.

In a *second order phase transition* finite changes are to be expected in the second order derivatives of the Gibbs function, which may be expressed in terms of the following experimental quantities

$$\frac{\partial^2 G}{\partial T^2} = \frac{\partial}{\partial T}\left(\frac{\partial G}{\partial T}\right) = -\left(\frac{\partial S}{\partial T}\right)_p = -\frac{C_p}{T} \qquad (10.19)$$

$$\frac{\partial^2 G}{\partial T \partial p} = \frac{\partial}{\partial T}\left(\frac{\partial G}{\partial p}\right) = \left(\frac{\partial V}{\partial T}\right)_p = \beta_p V \qquad (10.20)$$

$$\frac{\partial^2 G}{\partial p^2} = \frac{\partial}{\partial p}\left(\frac{\partial G}{\partial p}\right) = \left(\frac{\partial V}{\partial p}\right)_T = -k_T V \qquad (10.21)$$

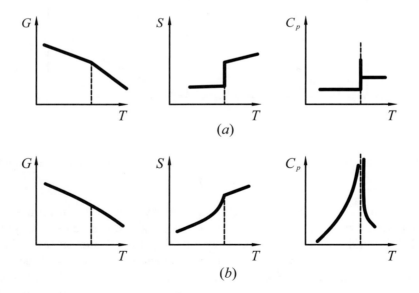

Figure 10.3. Variation of the Gibbs function and its temperature derivatives S and C_p in (a) first order and (b) second order phase transitions

Since V and T remain unchanged, a second-order phase transition is associated with the experimental discontinuity of C_p, β_p or k_T. An example is the superconducting transition in zero magnetic field, which we shall consider later on in the frame of experimental superconductivity. Figure 10.3 illustrates the behaviour of the Gibbs function and its derivatives at both first and second order phase transitions.

Appropriate equations for second order phase transitions can be derived from the constancy of V and S. Equating the volumes along the phase boundary $V_\alpha = V_\beta$ and expanding in terms of T and p, we may write

$$\left(\frac{\partial V_\alpha}{\partial T}\right)_p dT + \left(\frac{\partial V_\alpha}{\partial p}\right)_T dp = \left(\frac{\partial V_\beta}{\partial T}\right)_p dT + \left(\frac{\partial V_\beta}{\partial p}\right)_T dp$$

so that

$$\frac{dp}{dT} = -\frac{(\partial V_\beta / \partial T)_p - (\partial V_\alpha / \partial T)_p}{(\partial V_\beta / \partial p)_T - (\partial V_\alpha / \partial p)_T} = \frac{\Delta\beta_p}{\Delta k_T}$$

Starting from $dS_\alpha = dS_\beta$, and considering expansions in terms of T and p, we get in a similar way

$$\frac{dp}{dT} = -\frac{(\partial S_\beta / \partial T)_p - (\partial S_\alpha / \partial T)_p}{(\partial S_\beta / \partial p)_T - (\partial S_\alpha / \partial p)_T} = \frac{(\partial S_\beta / \partial T)_p - (\partial S_\alpha / \partial T)_p}{(\partial V_\beta / \partial T)_p - (\partial V_\alpha / \partial T)_p} = \frac{1}{VT}\frac{\Delta C_p}{\Delta\beta_p}$$

The gradient at phase boundary in a second order phase transition can thus be written as

$$\frac{dp}{dT} = \frac{\Delta\beta_p}{\Delta k_T} = \frac{1}{VT}\frac{\Delta C_p}{\Delta\beta_p} \tag{10.22}$$

which are called the *Ehrenfest equations*.

10.4. THE PHASE RULE

Consider now a system of c components and ϕ phases. Assuming that every component is present in every phase, there are $2 + c\phi$ variables as follows

$$T, p, m_1^1, m_2^1, \ldots m_c^1$$
$$m_1^2, m_2^2, \ldots m_c^2$$
$$\vdots$$
$$m_1^\phi, m_2^\phi, \ldots m_c^\phi$$

where subscripts denote the components, and superscripts denote the phases. The Gibbs function of the phase denoted by α can be expanded in terms of T, p and the masses of all the components

$$dG^\alpha = \left(\frac{\partial G^\alpha}{\partial T}\right)_{p,m_i} dT + \left(\frac{\partial G^\alpha}{\partial p}\right)_{T,m_i} dp + \sum_{i=1}^{c} \left(\frac{\partial G^\alpha}{\partial m_i}\right)_{T,p} dm_i$$

$$= -S^\alpha dT + V^\alpha dp + \sum_{i=1}^{c} g_i^\alpha dm_i \qquad (10.23)$$

Here g_i^α are intensive variables, called *chemical potentials* of different components in the phase α, corresponding to the specific Gibbs functions of a system of one component. It is convenient for applications to specify the number of moles v_i of a component, instead of its mass m_i, so that we must introduce the *molar chemical potentials* as

$$\mu_i^\alpha = M_i g_i^\alpha \qquad (10.24)$$

and set $dv_i = dm_i / M_i$, so that Eq.(10.23) may be rewritten in the form

$$dG^\alpha = -S^\alpha dT + V^\alpha dp + \sum_{i=1}^{c} \mu_i^\alpha dv_i$$

A mass transfer between two phases α and β of the *i*th component of the system, at constant temperature and pressure, obeys the condition

$$dG = -\mu_i^\alpha dv_i + \mu_i^\beta dv_i = 0 \quad \text{or} \quad \mu_i^\alpha = \mu_i^\beta$$

The result holds for any two phases of every component, so that we obtain a set of c equations of the form

$$\mu_i^1 = \mu_i^2 = \ldots = \mu_i^\phi \quad , \quad i = 1, 2, \ldots, c \qquad (10.25)$$

Hence, there are altogether $c(\phi-1)$ conditions for equilibrium among the molar chemical potentials. We must also take into account one equation for normalization of mole fractions for each phase, since the chemical potential for a component in a phase is an intensive variable, dependent on the composition of that phase but not on its total mass. Therefore the number of intensive degrees of freedom available for the system can be expressed by

$$N = 2 + c\phi - \left[c(\phi-1) + \phi\right] = 2 + c - \phi \tag{10.26}$$

which is called the *Gibbs phase rule*.

The former discussion about one component systems $(c=1)$ illustrates the use of the phase rule. Each phase $(\phi=1)$ has two intensive degrees of freedom, T and p. Along the boundary between two phases α and β in equilibrium $(\phi=2)$ it is one degree of freedom, as expressed by the Clapeyron equation. Three phases in equilibrium $(\phi=3)$ have no degree of freedom, corresponding to the triple point.

10.5. Third Law of Thermodynamics

Experimental investigation of phase transitions experienced by systems of one component at very low temperature provides evidence for an independent principle, which cannot be derived from the second law. Let us consider a first order transition at low temperature, e.g., the melting of solid helium, which obeys an experimental relation of the form

$$\frac{dp}{dT} = 0.245\, T^7$$

and so $dp/dT \to 0$ as $T \to 0$. Since $\Delta V = V_\alpha - V_\beta$ is not zero for a first order transition, from Eq.(10.18) we obtain

$$\lim_{T \to 0} \frac{dp}{dT} = \lim_{T \to 0} \frac{\Delta S}{\Delta V}$$

and it follows that

$$\lim_{T \to 0} (\Delta S)_T = 0 \tag{10.27}$$

which is the **Nernst-Simon statement of the third law**:

The entropy change associated with any isothermal reversible process of a condensed (solid or liquid) system approaches zero as the temperature approaches absolute zero.

If S_0^α and S_0^β are the entropies of two phases in equilibrium at absolute zero, their entropies at temperatures $T > 0$ may be written, according to Eq.(8.24), as

$$S^\alpha = S_0^\alpha + \int_0^T \frac{C_p^\alpha}{T} dT, \qquad S^\beta = S_0^\beta + \int_0^T \frac{C_p^\beta}{T} dT$$

Consider the phase α at a temperature T^β to undergo an isentropic transition to a phase β at a lower temperature T^β, that is

$$S_0^\alpha + \int_0^{T^\alpha} \frac{C_p^\alpha}{T} dT = S_0^\beta + \int_0^{T^\beta} \frac{C_p^\beta}{T} dT \tag{10.28}$$

Assuming that the Nernst-Simon statement holds, that is $S_0^\alpha = S_0^\beta$, we obtain, by taking $T^\beta = 0$, the relation

$$\int_0^{T^\alpha} \frac{C_p^\alpha}{T} dT = 0$$

However, this is impossible, since $C_p^\alpha > 0$, except at absolute zero. Therefore, it is impossible to attain the absolute zero, through an isentropic phase transition, from a higher temperature T^α, provided the Nernst-Simon statement is true. This result is generalized as the **unattainability statement of the third law**:

> *It is impossible, by means of any process to reduce the temperature of a system to absolute zero, in a finite number of steps.*

which appears to be derived from the Nernst-Simon statement.

Let us now assume that the unattainability statement is true. This means that an isentropic phase transition from a phase α at T^α to a phase χ at $T^\beta = 0$ is impossible. Since, for such a transition, Eq.(10.28) reduces to

$$S_0^\beta - S_0^\alpha = \int_0^{T^\alpha} \frac{C_p^\alpha}{T} dT$$

where $C_p^\alpha > 0$, this implies that $S_0^\beta \leq S_0^\alpha$. Similarly, an isentropic transition from a phase β at T^χ to a phase α at $T^\alpha = 0$ is impossible, so that Eq.(10.28), which reduces to

$$S_0^\alpha - S_0^\beta = \int_0^{T^\beta} \frac{C_p^\beta}{T} dT$$

with $C_p^\beta > 0$, implies that $S_0^\alpha \leq S_0^\beta$. Thus, we must conclude that

$$\Delta S = S_0^\beta - S_0^\alpha = 0$$

In other words the Nernst-Simon statement (10.27) follows from the unattainability statement of the third law.

Therefore the two statements of the third law are equivalent, as illustrated in Figure 10.4: if Nernst-Simon statement is violated, a system can be cooled to absolute zero in a finite number of steps, but absolute zero is not attainable in a finite number of steps if $\lim_{T \to 0} (\Delta S)_T = 0$.

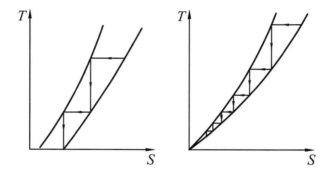

Figure 10.4. Equivalence between the Nernst-Simon and unattainability statements of the third law

EXAMPLE 10.1. Entropy at absolute zero

Consider the entropy change in an irreversible phase transition by which a substance is transformed into another, such as a chemical reaction given by

$$\alpha + \beta \to \alpha\beta$$

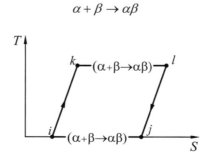

Figure 10.5. Entropy change in an irreversible phase transition

Since the entropy is a function of state, ΔS is independent of the path between the initial and final state. If the phases α and β having been heated from $T = 0$ up to some temperature $T > 0$, are allowed to react and form the new phase $\alpha\beta$, which is then cooled down again to absolute zero, as in Figure 10.5, the entropy change around path *iklj* will be the same as that obtained by direct reaction at $T = 0$ along path *ij*. Eq.(10.27) shows that there is no entropy change in going from one state to another at absolute zero, and this must be equally true for path *iklj*. In other words, each phase has zero entropy at absolute zero. If this is true for one phase and by any reaction this phase is transformed into another with no entropy change, then the new phase must likewise have zero entropy. We may conclude that any crystalline substance has zero entropy at absolute zero.

Applying the third law to either Eq.(8.22) or Eq.(8.24), it follows that a heat capacity C_V or C_p must vanish as the temperature approaches absolute zero, that is

$$\lim_{T \to 0} C = \lim_{T \to 0} T\left(\frac{\partial S}{\partial T}\right) = \lim_{T \to 0}\left(\frac{\partial S}{\partial \ln T}\right) = 0 \qquad (10.29)$$

because $T \to \infty$ and $\Delta S \to 0$ as $T \to 0$. Such a temperature dependence conflicts with the classical theory, which predicts constant heat capacities, as it will be derived later on in the frame of statistical thermodynamics.

FURTHER READING
1. B. Linder - THERMODYNAMICS AND INTRODUCTORY STATISTICAL MACHANICS, Wiley, New York, 2004.
2. M.Sprackling - THERMAL PHYSICS, Springer-Verlag, New York, 1998.
3. R. H. Dittman, M. W. Zemansky - HEAT AND THERMODYNAMICS, McGraw-Hill, New York, 1996.

PROBLEMS

10.1. Derive the relationship between vapour pressure and temperature for a vaporization process at low vapour pressure.

Solution

Assuming an ideal gas behaviour for the vapour, which holds at low pressure, so that the specific volume of the liquid $v_\beta \equiv v_L$ is negligible in comparison with that of the vapour

$v_\alpha \equiv v_V$, we may write $v_\alpha - v_\beta \approx v_\alpha \equiv v_V$, and also $V_V = RT/p$ so that the Clapeyron equation reduces to

$$\frac{dp}{dT} = \frac{\lambda p}{RT^2}$$

An explicit relationship between vapour pressure and temperature can be derived taking into account the temperature dependence of the latent heat given by

$$\frac{\lambda}{T} = (S_V - S_L)$$

where V and L denote the vapour and the liquid phases. It follows that

$$\frac{d}{dT}\left(\frac{\lambda}{T}\right) = \frac{d}{dT}(S_V - S_L) = \left[\left(\frac{\partial S_V}{\partial T}\right)_p - \left(\frac{\partial S_L}{\partial T}\right)_p\right] + \frac{dp}{dT}\left[\left(\frac{\partial S_V}{\partial p}\right)_T - \left(\frac{\partial S_L}{\partial p}\right)_T\right]$$

and using Eqs.(8.24) and (8.36) we obtain

$$\frac{d}{dT}\left(\frac{\lambda}{T}\right) = \frac{1}{T}(C_{pV} - C_{pL}) - \frac{dp}{dT}\left[\left(\frac{\partial V_V}{\partial T}\right)_p - \left(\frac{\partial V_L}{\partial T}\right)_p\right]$$

In view of the negligible volume of liquid phase, substituting dp/dT gives

$$\frac{d}{dT}\left(\frac{\lambda}{T}\right) \approx \frac{\Delta C_p}{T} - \frac{\lambda}{T^2}$$

which implies a linear temperature dependence of λ of the form

$$d\lambda = \Delta C_p dT \quad \text{or} \quad \lambda = \lambda_0 + \Delta C_p T$$

Inserting this result into the Clapeyron equation, we may express dp/p as

$$\frac{dp}{p} = \frac{\lambda_0}{R}\frac{dT}{T^2} + \frac{\Delta C_p}{R}\frac{dT}{T}$$

and the vapour pressure results as

$$\ln p = -\frac{\lambda_0}{RT} + \frac{\Delta C_p}{R}\ln T + \text{const} \quad \text{or} \quad p = \text{const}\, T^{\Delta C_p/R} e^{-\lambda_0/RT}$$

In a small temperature range around the normal boiling point, the standard approximation is to regard λ as a constant, so that we have $\Delta C_p = 0$, and hence

$$p = \text{const}\, e^{-\lambda/RT}$$

This result can also be derived by direct integration of Eq.(10.19), provided λ is a constant.

10.2. Find the relationship between the latent heats of fusion λ_f, vaporization λ_v and sublimation λ_s at the triple point T_{tr} of a pure substance.

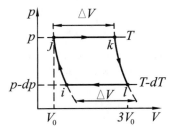

Solution

Consider the pV projection of a pVT surface of a pure substance, illustrated in Figure 9.3. When condensation of the vapour occurs at a temperature T in the range $T_{tr} \leq T \leq T_C$ the condensation is to the liquid phase, while for temperatures less than T_{tr} condensation is directly to the solid phase. A Carnot engine having vapour as its working substance can be operated in an infinitesimal cycle $ijkl$, consisting of an isothermal expansion jk (vaporization) at temperature T, where a latent heat $Q_{jk} = \lambda$ is absorbed and a work $p\Delta V$ is performed by the working substance, followed by an adiabatic expansion kl from p,T to $p-dp, T-dT$, an isothermal compression li (condensation) where an amount of work $(p-dp)\Delta V$ is done on the system and finally, an adiabatic compression ij until the initial parameters are reached. The efficiency of the Carnot engine, Eq.(7.29), reads

$$\eta_c = \frac{-W}{Q_{jk}} = \frac{dp\Delta V}{\lambda} \qquad \text{or} \qquad \eta_c = 1 - \frac{T-dT}{T} = \frac{dT}{T}$$

and this provides λ in terms of dp/dT as given by Eq.(10.18). Applying this result in turn for the three phase transitions yields

$$\frac{dp}{dT} = \frac{\lambda_s}{T(V_V - V_S)}, \qquad \frac{dp}{dT} = \frac{\lambda_v}{T(V_V - V_L)}, \qquad \frac{dp}{dT} = \frac{\lambda_f}{T(V_L - V_S)}$$

Thus, at the triple point T_{tr} one obtains

$$\frac{\lambda_s}{V_V - V_S} = \frac{\lambda_v}{V_V - V_L} = \frac{\lambda_f}{V_L - V_S}$$

and hence, the required relationship can be written as

$$\lambda_s = \lambda_v + \lambda_f$$

10.3. Find a temperature T_0 on the vaporization curve of a pure substance of adiabatic exponent γ that makes it possible, at temperatures below T_0, to produce a condensed phase of the substance by a reversible adiabatic expansion.

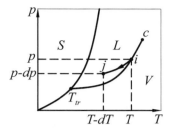

Solution

An adiabatic expansion ij from a point p,T on the vaporization curve to $p-dp, T-dT$ will produce condensation provided it ends in the liquid phase, and hence, the slope of the adiabatic expansion has to be less than that of the vaporization curve. Differentiating the adiabatic equation (9.10), written as

$$Tp^{(1-\gamma)/\gamma} = \text{const} \qquad \text{or} \qquad \frac{dT}{T} + \frac{1-\gamma}{\gamma}\frac{dp}{p} = 0$$

it follows that

$$\left(\frac{dp}{dT}\right)_S = \frac{\gamma}{\gamma-1}\frac{p}{T}$$

This should be less than the slope derived for the vaporization curve in Problem 10.1, namely

$$\frac{dp}{dT} = \frac{\lambda_0}{R}\frac{p}{T^2} + \frac{\Delta C_p}{R}\frac{p}{T}$$

which means that

$$\frac{\gamma}{\gamma-1} < \frac{\lambda_0}{RT} + \frac{C_{pV}-C_{pL}}{R} \qquad \text{or} \qquad T < \frac{\lambda_0}{C_{pL}} = T_0$$

as required.

10.4. Show that the value of any extensive or intensive variable of a closed thermodynamic system becomes independent of temperature as the absolute zero of temperature is approached.

Solution

Let x be an extensive variable and X its conjugate intensive variable, and consider a thermal coefficient defined as

$$\gamma_x(T) = \frac{1}{x}\left(\frac{\partial x}{\partial T}\right)_X$$

In view of a Maxwell relation similar to Eq.(8.36), we also have

$$\gamma_x(T) = -\frac{1}{x}\left(\frac{\partial S}{\partial X}\right)_T$$

and this can be written in terms of C_X, defined as established by Eq.(8.24) for a hydrostatic system

$$C_X = T\left(\frac{\partial S}{\partial T}\right)_X$$

Since the entropy $S(X,T)$ is an exact differential, one obtains

$$\frac{\partial}{\partial T}\left(\frac{\partial S}{\partial X}\right)_T = \frac{\partial}{\partial X}\left(\frac{\partial S}{\partial T}\right)_X \quad \text{or} \quad \left(\frac{\partial S}{\partial X}\right)_T = \int_0^T \frac{\partial}{\partial X}\left(\frac{C_X}{T}\right) dT$$

and it follows that

$$\gamma_x(T) = -\frac{1}{x}\int_0^T \frac{1}{T}\left(\frac{\partial C_X}{\partial X}\right)_T dT$$

In a similar way to that used for a hydrostatic system, we may then derive

$$\left(\frac{\partial C_X}{\partial X}\right)_T = -T\left(\frac{\partial^2 x}{\partial T^2}\right)_X$$

so that we finally get

$$\gamma_x(T) = \frac{1}{x}\int_0^T \left(\frac{\partial^2 x}{\partial T^2}\right)_X dT = \frac{1}{x}\left(\frac{\partial x}{\partial T}\right)_X - \frac{1}{x}\left(\frac{\partial x}{\partial T}\right)_{X,T=0} = \gamma_x(T) - \gamma_x(0)$$

This shows that $\gamma_x(0) = 0$, which means that an extensive variable x becomes independent of temperature when $T \to 0$. The same is true for any extensive variable X, and this can be proved in a similar way, by considering a thermal coefficient

$$\gamma_X(T) = \frac{1}{X}\left(\frac{\partial X}{\partial T}\right)_x$$

Since it is defined in terms of a partial derivative that can be transformed using a Maxwell relation, see Eq.(8.35), the third law, Eq.(10.27), allows us to predict that $\gamma_X(T) \to 0$ as $T \to 0$, and hence, X becomes independent of temperature.

10.5. Show that the isothermal compressibility k_T of a closed hydrostatic system becomes independent of temperature as T approaches the absolute zero.

Solution

The value of k_T at absolute zero cannot be derived from the third law, since it is defined by Eq.(8.42) in terms of $(\partial V/\partial p)_T$, and this is not related to a partial derivative of the entropy. However, we may use Eq.(8.35) to obtain

$$\left[\frac{\partial}{\partial T}\left(\frac{\partial p}{\partial V}\right)_T\right]_V = \left[\frac{\partial}{\partial V}\left(\frac{\partial p}{\partial T}\right)_V\right]_T = \left[\frac{\partial}{\partial V}\left(\frac{\partial S}{\partial V}\right)_T\right]_T$$

and this gives

$$\left[\frac{\partial}{\partial T}\left(\frac{1}{k_T}\right)\right]_V = -V\left[\frac{\partial}{\partial V}\left(\frac{\partial S}{\partial V}\right)_T\right]_T$$

In view of Eq.(10.27), it follows that $(\partial S/\partial V)_T \to 0$ as T approaches absolute zero, thus

$$\lim_{T \to 0}\left[\frac{\partial}{\partial T}\left(\frac{1}{k_T}\right)\right]_V = 0$$

and hence, the value of k_T becomes independent of temperature, as required.

10.6. Derive an approximate expression for the linear dependence of temperature on pressure, in a small temperature range around the normal boiling point p_0, T_0.

10.7. In a Dewar vessel of negligible heat capacity, a mass m_0 of water of specific heat c_0 is cooled down while a mass m of ether, of specific heat c and specific latent heat of vaporization λ, boils. Given the initial temperature θ of the two liquids, calculate the fraction of the ether that must be boiled away before ice starts to appear.

10.8. Derive an expression for the latent heat of a pure substance in the temperature range in which its vaporization curve is given by the equation $p = A\theta^4$, where θ stands for the empirical temperature, and A is a constant.

10.9. Calculate the amount of work that must be supplied to a Carnot refrigerator to convert a quantity of water at room temperature T into ice at a temperature $T_0 = 273$ K, under atmospheric pressure, given the heat capacity C_p at constant pressure and the latent heat of fusion λ.

10.10. Show that, for a closed hydrostatic system, the coefficient of isobaric expansion β_p tends to zero as the temperature tends to zero.

11. Electrostatics

11.1. ELECTRIC FIELD
11.2. ELECTROSTATIC POTENTIAL
11.3. POLARIZATION
11.4. ELECTROSTATIC ENERGY

11.1. Electric Field

Electrostatics is concerned with the interactions between stationary electric charges. It is known experimentally that the force between *point charges*, which can be either *positive* or *negative*, is a central force, acting along their line of centres. The magnitude of the force is given by Coulomb's inverse square law

$$\vec{F} = \frac{1}{4\pi\varepsilon_0} \frac{q_1 q_2}{r^2} \vec{e}_r = f(r)\vec{e}_r \tag{11.1}$$

where the proportionality constant $1/4\pi\varepsilon_0$ corresponds to the choice of SI units and is given in terms of the permittivity of free space ε_0.

The electrostatic forces between pairs of point charges obey the *superposition principle* which states that *the net force on a given charge q is the vector sum of the individual forces found by pairing it with each of the neighbouring point charges*. The ratio of the net force on the charge q to the magnitude of q itself is a quantity independent of the magnitude and is called the *electric field intensity* $\vec{E}$, given by

$$\vec{E} = \frac{\vec{F}}{q} = \frac{f(r)}{q} \vec{e}_r \tag{11.2}$$

The superposition principle requires that an array of charges q_i, which are at distances r_i from a point P, give rise at P to an electric field

$$\vec{E}_P = \frac{1}{4\pi\varepsilon_0} \sum_i \frac{q_i \vec{r}_i}{r_i^3}$$

It is common practice to represent electric fields by *flux lines* which always start at positive charges, terminate at negative charges and never cross, for obvious reasons. The number of lines per unit area gives the strength of an electric field at a given point in space. The analogy with hydrodynamic streamlines is strictly formal, as electrostatics is not associated with flow at all, but illustrates the concept of the *continuous* distribution of the field which describes the electrostatic interaction. In a similar manner it is often convenient to speak in terms of a continuous charge distribution rather than of arrays of point charges and to introduce a charge density $\rho(\vec{r}) = dq/dV$ for a volume element or $\sigma(\vec{r}) = dq/dS$ for a surface element. For a continuous charge distribution in a volume V, the electric field at a point P may be written as an integral

$$\vec{E}_P = \frac{1}{4\pi\varepsilon_0} \int_V \frac{\rho \vec{r}}{r^3} dV \qquad (11.3)$$

where the volume element is located at a distance r with respect to the point P. The integration is over the volume occupied by the charge distribution.

Electrostatics can be formulated in terms of basic properties of electric fields, as expressed by a flux law and a circulation law, which will be derived from Coulomb's experimental inverse square law (11.1). Let us first calculate the total flux of an electric field over a closed surface S of arbitrary shape which is surrounding a single point charge q (Figure 11.1).

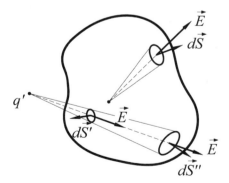

Figure 11.1. Arbitrary closed surface for the proof of Gauss's law

The flux of $\vec{E}$ across an elementary area dS on S is $\vec{E} \cdot d\vec{S}$, so that the total flux of the electric field is given by an integral, which in view of Eq.(11.1) becomes

$$\int_S \vec{E} \cdot d\vec{S} = \int_S E \cos\theta \, dS = \frac{q}{4\pi\varepsilon_0} \int_S \frac{dS \cos\theta}{r^2} = \frac{q}{4\pi\varepsilon_0} \int_S d\Omega$$

where θ is the angle between the normal to dS and the field direction, and $d\Omega = dS\cos\theta/r^2$ is the solid angle of the cone associated with the area dS. The total solid angle around a point is given by $\int_S d\Omega = 4\pi$. Thus

$$\int_S \vec{E}\cdot d\vec{S} = \frac{q}{\varepsilon_0}$$

This result holds independently of the location of the charge within the volume, so that the flux of an arbitrary number of charges inside S is expressed by applying the principle of superposition

$$\int_S \vec{E}\cdot d\vec{S} = \frac{1}{\varepsilon_0}\sum q_i$$

If there is a continuous distribution of charge within S, defined by a charge density ρ, the summation becomes a volume integral over volume V of the form

$$\sum q_i = \int_V \rho dV$$

so that we obtain

$$\int_S \vec{E}\cdot d\vec{S} = \frac{1}{\varepsilon_0}\int_V \rho dV \tag{11.4}$$

Equation (11.4) is known as integral form of *Gauss's law* which states that *the total flux of $\vec{E}$ across a closed surface equals the total charge inside the surface divided by ε_0*. It is noteworthy that for a charge q' located *outside* the volume, any flux cone cuts the surface S twice, as represented in Figure 11.1. The flux *into* the volume across any elementary area dS' is exactly the same as the flux *out* of the volume across dS'' so that the net flux for the surface S is zero.

The total flux of the electric field across a closed surface can be equated to the volume integral of the divergence of $\vec{E}$ over the volume, according to Gauss's divergence theorem given in Appendix I, Eq.(I.3), so that Eq.(11.4) reads

$$\int_S \vec{E}\cdot d\vec{S} = \int_V div\vec{E}\, dV = \int_V \nabla\cdot\vec{E}\, dV = \frac{1}{\varepsilon_0}\int_V \rho dV$$

and because this is true for any volume however small, it can be alternatively written as

$$\nabla\cdot\vec{E} = \frac{1}{\varepsilon_0}\rho \quad \text{or} \quad \frac{\partial E_x}{\partial x} + \frac{\partial E_y}{\partial y} + \frac{\partial E_z}{\partial z} = \frac{1}{\varepsilon_0}\rho \tag{11.5}$$

Equation (11.5) is called the *differential form* of Gauss's law and relates the variation of

$\vec{E}$ at a point in space to the density of charge at the same point. The divergence can be regarded as a measure of the strength of the source field at a point. If the divergence is zero, as required for instance by the equation of continuity (6.36) for an incompressible fluid, there are no sources and no field lines starting or ending at that point.

Like all central force fields, the electric field (11.2) is *conservative*, that is the work done moving a charge q in the field $\vec{E}$ is independent of path. Substituting Eq.(11.2) into Eq.(3.3) and integrating along an arbitrary closed path Γ, as in Figure 11.2, gives

$$\int_\Gamma \vec{E} \cdot d\vec{r} = 0 \tag{11.6}$$

Equation (11.6) is independent of the starting point $\vec{r}_i$ and is known as the *circulation law* for electrostatics: the line integral of $\vec{E}$ vanishes identically for any closed path.

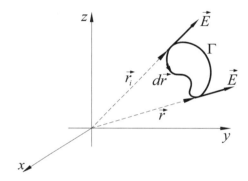

Figure 11.2. Arbitrary closed path for the proof of the circulation law

Applying Stoke's theorem, given in Appendix I, Eq.(I.5), which equates the circulation of $\vec{E}$ along a closed path Γ to the flux of $curl\,\vec{E}$ through any surface S bounded by Γ, Eq.(11.6) reads

$$\int_\Gamma \vec{E} \cdot d\vec{r} = \int_S curl\,\vec{E} \cdot d\vec{S} = \int_S \left(\nabla \times \vec{E}\right) \cdot d\vec{S} = 0$$

and is true for any surface however small, so that

$$\nabla \times \vec{E} = 0 \tag{11.7}$$

which holds at all points for which the conservative field $\vec{E}$ is defined.

Gauss's law and the circulation law, in either integral forms (11.4) and (11.6) or differential forms (11.5) and (11.7), are completely equivalent to Coulomb's inverse square law (11.1), given the principle of superposition, and can be regarded as axioms of the theory of electrostatics.

11.2. ELECTROSTATIC POTENTIAL

The circuit law (11.6), regarded as a statement that the line integral of $\vec{E}$ depends only on the end points of the paths and not on the path itself, allows us to define an *electrostatic scalar potential* $V(r)$ as

$$V(r) = -\int_{r_i}^{r} \vec{E} \cdot d\vec{r} \quad \text{or} \quad dV = -\vec{E} \cdot d\vec{r} \tag{11.8}$$

where the negative sign shows that $\vec{E}$ points toward a decrease in potential. A comparison between the definitions of both the potential energy (2.23) and the electric field (11.2) shows that $V(r)$ gives the electrostatic energy per unit charge. The potential $V(r)$ at a point depends of the choice of a reference $V(r_i) = 0$ from which it is to be measured. We can only measure the *difference* of potential with respect to that reference. A convenient choice in electrostatics is to assign a potential of zero to a point at infinite distance from a charge distribution. In many practical problems the potential of the earth is taken as zero. In Cartesian coordinates we can regard V as a function of space coordinates x, y and z so that Eq.(11.8) reads

$$dV = \frac{\partial V}{\partial x}dx + \frac{\partial V}{\partial y}dy + \frac{\partial V}{\partial z}dz = -(E_x dx + E_y dy + E_z dz)$$

which yields the relations to calculate $\vec{E}$ from $V(r)$ as

$$E_x = -\frac{\partial V}{\partial x} \quad , \quad E_y = -\frac{\partial V}{\partial y} \quad , \quad E_z = -\frac{\partial V}{\partial z} \tag{11.9}$$

In a more compact form we see that the electric field is minus the gradient of the electrostatic potential, that is

$$\vec{E} = -\text{grad}\, V = -\nabla V \tag{11.10}$$

as expressed for all conservative fields by Eq.(2.26).

EXAMPLE 11.1. The electric dipole

An electric dipole is a common source of electric field, consisting of two equal and opposite point charges $+q$ and $-q$ a short distance d apart. If $\vec{d}$ is the position vector of $+q$ with respect to $-q$ the *dipole moment* is then defined by $\vec{p} = q\vec{d}$. The expression for the potential and field due to an electric dipole simplifies considerably if determined at a distance from the charges which is large compared with d, as in the arrangement shown in Figure 11.3.

If we take the origin of the coordinate system at the dipole centre, the electric potential due to the two charges at a point P with position vector $\vec{r}$ may be expressed, by applying Eqs.(11.8), (11.2) and (11.1), as

$$V(r) = -\frac{q}{4\pi\varepsilon_0}\int_0^{r_+}\frac{dr_+}{r_+^2} + \frac{q}{4\pi\varepsilon_0}\int_0^{r_-}\frac{dr_-}{r_-^2} = \frac{q}{4\pi\varepsilon_0}\left(\frac{1}{r_+} - \frac{1}{r_-}\right) \quad (11.11)$$

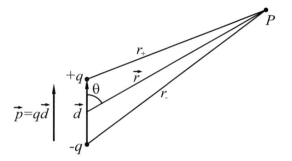

Figure 11.3. Calculation of the potential due to an electric dipole

It is convenient to expand r_+ and r_- into power series of the ratio $d/2r$ (see Figure 11.3) as

$$\frac{1}{r_\pm} = \frac{1}{\sqrt{r^2 + d^2/4 \mp rd\cos\theta}} = \frac{1}{r}\left(1 + \frac{d^2}{4r^2} \mp \frac{d}{r}\cos\theta\right)^{-1/2}$$

$$= \frac{1}{r}\left[1 \pm \frac{d}{2r}\cos\theta + \frac{1}{2}(3\cos^2\theta - 1)\left(\frac{d}{2r}\right)^2 \pm \cdots\right]$$

so that the potential is given in terms of dipole moment $\vec{p}$ by

$$V(r) = \frac{qd\cos\theta}{4\pi\varepsilon_0 r^2} = \frac{p\cos\theta}{4\pi\varepsilon_0 r^2} = \frac{\vec{p}\cdot\vec{r}}{4\pi\varepsilon_0 r^3} \quad (11.12)$$

The electric field due to the electric dipole is obtained by further differentiation since, according to Eq.(11.10), we have

$$\vec{E} = -\nabla\left(\frac{\vec{p}\cdot\vec{r}}{4\pi\varepsilon_0 r^3}\right) = \frac{1}{4\pi\varepsilon_0}\left[\frac{1}{r^3}\nabla(\vec{p}\cdot\vec{r}) + (\vec{p}\cdot\vec{r})\nabla\left(\frac{1}{r^3}\right)\right]$$

We can now write

$$\nabla(\vec{p}\cdot\vec{r}) = \vec{p}$$

and since

$$\frac{\partial r}{\partial x} = \frac{\partial}{\partial x}(x^2 + y^2 + z^2)^{1/2} = \frac{x}{r} \quad , \quad \frac{\partial r}{\partial y} = \frac{y}{r} \quad , \quad \frac{\partial r}{\partial z} = \frac{z}{r}$$

it follows that

$$\nabla\left(\frac{1}{r^3}\right) = -\frac{3}{r^4}\left(\frac{x\vec{e}_x}{r} + \frac{x\vec{e}_y}{r} + \frac{z\vec{e}_z}{r}\right) = -\frac{3\vec{r}}{r^5}$$

Thus, one obtains

$$\vec{E} = \frac{1}{4\pi\varepsilon_0}\left[3\frac{(\vec{p}\cdot\vec{r})\vec{r}}{r^5} - \frac{\vec{p}}{r^3}\right]$$

For a group of point charges q_i, located at distances r_i from a point P, Eq.(11.11) assumes the general form

$$V_P = \frac{1}{4\pi\varepsilon_0}\sum_i \frac{q_i}{r_i}$$

where V is the electrostatic potential at P. With a continuous distribution of charge in a volume V, the density of charge being ρ, we obtain

$$V_P = \frac{1}{4\pi\varepsilon_0}\int_V \frac{\rho}{r}dV \qquad (11.13)$$

where the volume element dV is located at a distance r from P. The integration is over the total volume of the charge distribution. If we eliminate $\vec{E}$, by combining the differential form of Gauss's law (11.5) and the relation (11.10) between electric field and potential, one obtains a relation between the potential and the charge density of the form

$$\nabla\cdot(-\nabla V) = -\left(\vec{e}_x\frac{\partial}{\partial x} + \vec{e}_y\frac{\partial}{\partial y} + \vec{e}_z\frac{\partial}{\partial z}\right)\cdot\left(\vec{e}_x\frac{\partial V}{\partial x} + \vec{e}_y\frac{\partial V}{\partial y} + \vec{e}_z\frac{\partial V}{\partial z}\right)$$

$$= -\left(\frac{\partial^2 V}{\partial x^2} + \frac{\partial^2 V}{\partial y^2} + \frac{\partial^2 V}{\partial z^2}\right) = \frac{1}{\varepsilon_0}\rho$$

This differential equation relating the potential to the charge density at a given point as

$$\nabla^2 V = -\frac{1}{\varepsilon_0}\rho \qquad (11.14)$$

is called *Poisson's equation*. It allows us to find the potential function V given the distribution of charge density ρ at every point. The expression (11.13) is a particular solution to Poisson's equation. For regions which contain no charge density, Eq.(11.14) reduces to

$$\nabla^2 V = 0 \quad \text{or} \quad \frac{\partial^2 V}{\partial x^2} + \frac{\partial^2 V}{\partial y^2} + \frac{\partial^2 V}{\partial z^2} = 0 \tag{11.15}$$

and this is called *Laplace's equation*. The integration of Poisson's and Laplace's equations requires the choice of appropriate boundary conditions. For problems involving spherical symmetry we must use the Laplacian operator in spherical coordinates, as given in Appendix I, Eq.(I.23), so that Laplace's equation becomes

$$\frac{1}{r^2}\frac{\partial}{\partial r}\left(r^2 \frac{\partial V}{\partial r}\right) + \frac{1}{r^2 \sin\theta}\frac{\partial}{\partial \theta}\left(\sin\theta \frac{\partial V}{\partial \theta}\right) + \frac{1}{r^2 \sin\theta}\frac{\partial^2 V}{\partial \varphi^2} = 0 \tag{11.16}$$

For the highest possible symmetry, the electrostatic potential is independent of θ and φ, and Eq.(11.16) reads

$$\frac{\partial}{\partial r}\left(r^2 \frac{\partial V}{\partial r}\right) = 0$$

a solution to which is the potential function $V(r) = 1/r$ which describes the spherically symmetric potential of a point charge. For an electrostatic potential symmetrical about the polar axis, $\partial^2 V / \partial \varphi^2 = 0$ and Eq.(11.16) reduces to

$$\frac{\partial}{\partial r}\left(r^2 \frac{\partial V}{\partial r}\right) + \frac{1}{\sin\theta}\frac{\partial}{\partial \theta}\left(\sin\theta \frac{\partial V}{\partial \theta}\right) = 0 \tag{11.17}$$

A solution can be found by separation of variables, in the form

$$V = R(r)F(\theta) \tag{11.18}$$

which yields

$$\frac{1}{R}\frac{\partial}{\partial r}\left(r^2 \frac{\partial R}{\partial r}\right) + \frac{1}{F \sin\theta}\frac{\partial}{\partial \theta}\left(\sin\theta \frac{\partial F}{\partial \theta}\right) = 0$$

so that

$$\frac{1}{R}\frac{\partial}{\partial r}\left(r^2 \frac{\partial R}{\partial r}\right) = -\frac{1}{F \sin\theta}\frac{\partial}{\partial \theta}\left(\sin\theta \frac{\partial F}{\partial \theta}\right)$$

Since one side is independent of θ and the other is independent of r, they must be separately equal to some constant k. The angular equation

$$\frac{\partial}{\partial \theta}\left(\sin\theta \frac{\partial F}{\partial \theta}\right) = -kF \sin\theta$$

has the solution $F(\theta) = \cos\theta$, provided $k = 2$. The radial equation, written as

$$\frac{\partial}{\partial r}\left(r^2 \frac{\partial R}{\partial r}\right) = 2R$$

has the trial solutions $R(r) = r$ and $R(r) = 1/r^2$, which lead to independent forms of the electrostatic potential V, upon multiplication by $\cos\theta$. We can then express the solution to Eq.(11.17) by the linear combination

$$V = \alpha r \cos\theta + \beta \frac{\cos\theta}{r^2} \tag{11.19}$$

where the constants α and β are determined by the boundary conditions.

EXAMPLE 11.2. The potential around a spherical conductor in a uniform electric field

A sphere of radius a, at the origin of a Cartesian coordinate system, which experiences a uniform electric field $\vec{E}$ along the z-axis (Figure 11.4), must have a symmetrical potential about the field direction, as described by Eq.(11.19). For an uncharged conducting sphere, the potential at its surface can be chosen as a reference potential: $V = 0$ at $r = a$. Assuming that the presence of the sphere has a negligible effect on the uniform field at large distances, its potential must take the asymptotic form $V = -Er\cos\theta$ as $r \to \infty$ which gives the correct behaviour for the field components

$$E_r = -\frac{\partial V}{\partial r} = E\cos\theta \quad , \quad E_\theta = -\frac{1}{r}\frac{\partial V}{\partial \theta} = -E\sin\theta \tag{11.20}$$

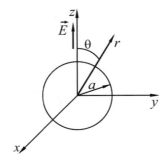

Figure 11.4. A conducting sphere in a uniform electric field

Applying the boundary condition (11.20) as $r \to \infty$ we obtain

$$V \to \alpha r \cos\theta = -Er\cos\theta \quad \text{or} \quad \alpha = -E$$

while the condition for a vanishing surface potential gives

$$0 = -Ea\cos\theta + \beta\frac{\cos\theta}{a^2} \quad \text{or} \quad \beta = Ea^3$$

The potential function (11.19) is finally obtained as

$$V = Er\cos\theta\left(\frac{a^3}{r^3} - 1\right) \tag{11.21}$$

The general solution of Laplace's equation (11.15) will be derived in Chapter 31, in connection with the quantum mechanical approach to the central field problem.

11.3. POLARIZATION

The laws of electrostatics that apply in free space must be modified when neutral matter occupies the space between the electric charges. This is accomplished by adding an independent function to the field and potential, which is determined by the location of charges. It is assumed that neutral matter contains atoms or molecules in positions which are fixed and given by the time average over their motion. We assume it consists of positively charged nuclei $+q$ surrounded by negative electron clouds. We differentiate between *conductors*, inside which electric charge can flow freely and *dielectrics*, having a negligible content of free electrons. In conductors the effect of applying an electric field is to redistribute the free charge until the internal field is cancelled. Gauss's law then shows that the net charge inside a conductor is zero under static conditions. When an electric field is applied in a dielectric, each nucleus will tend to displace in the direction of the field and the corresponding electron cloud against the field, as illustrated in Figure 11.5. The relative displacement, given by the position vector $\vec{r}$ of the centre of the negative cloud with respect to the nucleus, allows the restoring force due to the electrons to balance the force on the nucleus due to the applied field. An electric dipole moment, defined as

$$\vec{p} = q\vec{r} \tag{11.22}$$

is produced in both atoms and molecules. As a consequence of the induced dipole moments, layers of negative and positive charges develops on the two surfaces of an element of dielectric, normal to the direction of the applied field $\vec{E}$ (Figure 11.5). The elementary charge induced in a layer of thickness $r = |\vec{r}|$ and area dS can be expressed in terms of the *polarized charge density* $\sigma_P = dq/dS$ as follows

$$\sigma_P dS = \left(\frac{1}{V}\sum q\right) r dS = \left(\frac{1}{V}\sum qr\right) dS = \left(\frac{1}{V}\sum p\right) dS = P dS = \vec{P}\cdot d\vec{S}$$

and this can be interpreted using $\vec{P}\cdot d\vec{S} = \vec{P}\cdot\vec{e}_n dS$ in the sense that the surface charge density σ_P induced on a dielectric equals the component along the outward normal $\vec{P}\cdot\vec{e}_n$ of a vector defined by

$$\vec{P} = \frac{1}{V}\sum \vec{p} \qquad (11.23)$$

and called *polarization* or dipole moment per unit volume of the element of dielectric.

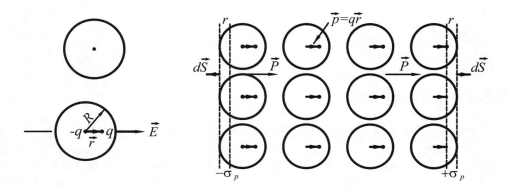

Figure 11.5. Polarization of a molecule and of an element of dielectric

We may integrate the surface charge density over the closed surface of a given volume V to obtain

$$\int_S \sigma_P dS = \int_S \vec{P}\cdot d\vec{S} = \int_V \nabla\cdot\vec{P}\, dV$$

where we have used Gauss's divergence theorem, given in Appendix I, Eq.(I.3). Since each element of dielectric remains neutral under an applied electric field we must introduce a volume polarization charge, of density $\rho_P = dq/dV$, such that the total charge is zero

$$\int_S \sigma_P dS + \int_V \rho_P dV = 0$$

Comparison between the foregoing equations gives

$$\int_V \rho_P dV = -\int_S \sigma_P dS = -\int_V \nabla\cdot\vec{P}\, dV$$

and this is true however small the volume V. It follows that the charge density resulting from polarization of the bulk is

$$\rho_P = -\mathrm{div}\,\vec{P} = -\nabla \cdot \vec{P} \tag{11.24}$$

Therefore there is a volume polarization charge density distributed through the bulk of dielectric wherever the divergence of polarization is not zero, a result similar to that stated by Gauss's law (11.5). However we must bear in mind that both $\sigma_P = \vec{P} \cdot \vec{e}_n$ and $\rho_P = -\nabla \cdot \vec{P}$ are bound charge densities arising from the bound electrons and nuclei in the dielectric. The polarization is determined by the charge separation in the individual atoms or molecules, so that a functional relationship between $\vec{P}$ and the applied field $\vec{E}$ may be assumed, as in the following example.

EXAMPLE 11.3. The electric dipole moment of an atom

For the individual spherical atom of radius R represented in Figure 11.5, we can define a charge density of the electron cloud by

$$\rho = -\frac{q}{4\pi R^3/3}$$

We can draw a sphere of radius r about the centre of the electron cloud after it has been displaced by $\vec{r}$ under an applied field $\vec{E}$ and apply Gauss's law (11.4) which gives the field at the nucleus, $\vec{E}_c$, arising from the negative cloud

$$E_c 4\pi r^2 = \frac{1}{\varepsilon_0}\left(-\frac{3q}{4\pi R^3}\right)\frac{4\pi}{3}r^3 \qquad \text{or} \qquad E_c = -\frac{qr}{4\pi\varepsilon_0 R^3} = -\frac{p}{4\pi\varepsilon_0 R^3}$$

The condition for equilibrium of the nucleus requires that E_c is equal and opposite to the applied field, so that

$$\vec{p} = 4\pi\varepsilon_0 R^3 \vec{E}$$

which means that the dipole moment is directly proportional to $\vec{E}$.

When the polarization $\vec{P}$ is proportional to the electric field $\vec{E}$, the dielectric is called *linear*. If, in addition, its electrical properties are the same in all directions, that is the dielectric is *isotropic*, we can write

$$\vec{P} = \chi_e \varepsilon_0 \vec{E} \tag{11.25}$$

where χ_e is a scalar called the *electric susceptibility*. In anisotropic dielectrics $\vec{P}$ and $\vec{E}$ may have different directions and the electric susceptibility becomes a second rank tensor relating their components.

When using Gauss's law to investigate electric fields in the presence of dielectrics it is necessary to include contributions from the polarization charges of density ρ_P, so that reserving the symbol ρ for free-charge density only, Eq.(11.5) reads

$$\nabla \cdot \vec{E} = \frac{1}{\varepsilon_0}(\rho + \rho_P) = \frac{1}{\varepsilon_0}(\rho - \nabla \cdot \vec{P})$$

where it is convenient to separate the distribution of free charges from that due to polarization

$$\nabla \cdot (\varepsilon_0 \vec{E} + \vec{P}) = \rho$$

Thus we can write Gauss's law for a dielectric using a vector field $\vec{D}$ whose divergence depends only on the free charge density

$$\nabla \cdot \vec{D} = div \vec{D} = \rho \qquad (11.26)$$

The vector $\vec{D}$ is called the *electric displacement* and must always be associated with a particular dielectric

$$\vec{D} = \varepsilon_0 \vec{E} + \vec{P} \qquad (11.27)$$

whereas the electric field $\vec{E}$ is defined irrespective of the environment. In a linear isotropic dielectric we may use Eq.(11.25) and the displacement $\vec{D}$ becomes

$$\vec{D} = \varepsilon_0 \vec{E} + \chi_e \varepsilon_0 \vec{E} = \varepsilon_0 (1 + \chi_e) \vec{E} = \varepsilon_0 \varepsilon_r \vec{E} \qquad (11.28)$$

which is often called the *constitutive* relation of electrostatics, where

$$\varepsilon_r = 1 + \chi_e \qquad (11.29)$$

is known as the *relative permittivity*: ε_r is a dielectric constant except for anisotropic dielectrics where the electric susceptibility (and so ε_r) is represented by a second rank tensor.

11.4. ELECTROSTATIC ENERGY

The potential energy stored in a system of point charges, under mutual electrostatic internal forces, can be expressed by applying Eq.(2.33), which reads

$$U = \frac{1}{2}\sum_i q_i \sum_{j \neq i} V_{ij}$$

showing that each charge experiences the potential due to all other charges. For a continuous charge distribution, the potential energy takes the form

$$U = \frac{1}{2}\int_V \rho(\vec{r})V(\vec{r})dV \qquad (11.30)$$

In the presence of a dielectric, the free-charge density is given by Gauss's law (11.26), so that

$$U_E = \frac{1}{2}\int_V (\nabla \cdot \vec{D})V(\vec{r})dV = \frac{1}{2}\int_V \left[\nabla \cdot (V\vec{D}) - \vec{D} \cdot \nabla V\right]dV$$

where a well known vector identity (see Eq.(I.10), Appendix I) has been used. The first term on the right hand side can be transformed according to Gauss's divergence theorem and vanishes

$$\int_V \nabla \cdot (V\vec{D})dV = \int_S V\vec{D} \cdot d\vec{S} = 0$$

as the surface S is chosen to be an infinite sphere. It follows that

$$U_E = \frac{1}{2}\int_V \vec{E} \cdot \vec{D}\, dV \qquad (11.31)$$

since $\vec{E} = -\nabla V$, as given by Eq.(11.10). Therefore the electrostatic energy density stored in a charge distribution is

$$u_E = \frac{1}{2}\vec{E} \cdot \vec{D} \qquad (11.32)$$

FURTHER READING
1. F. Melia - ELECTRODYNAMICS, University of Chicago Press, Chicago, 2001.
2. W. N. Cottingham, D. A. Greenwood - ELECTRICITY AND MAGNETISM, Cambridge University Press, Cambridge, 1991.
3. G. G. Skitek, S. V. Marshall - ELECTROMAGNETIC CONCEPTS AND APPLICATIONS, Prentice-Hall, Englewood Cliffs, 1990.

PROBLEMS

11.1. Two charges, $+q$ and $-xq$, are held fixed a distance a apart. Where could we place a grounded thin spherical conducting sheet near the charges, without creating any disturbance? Calculate the radius R and the outer surface charge density of the sheet.

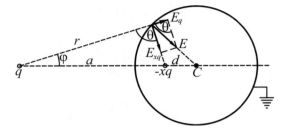

Solution

The conducting sheet is in effect a spherical equipotential surface of radius R, centred on the line passing through the charges, which satisfies the boundary condition that the resultant potential is zero

$$\frac{q}{4\pi\varepsilon_0 r} - \frac{xq}{4\pi\varepsilon_0 r_x} = 0 \quad \text{or} \quad r_x = xr$$

The geometry of the problem is shown in the figure, where

$$\cos\varphi = \frac{r^2 + (a+d)^2 - R^2}{2r(a+d)} = \frac{a^2 + r^2(1-x^2)}{2ra}$$

Rearranging gives $R^2 = r^2\left[x^2 - (1-x^2)d/a\right] + d(a+d)$ and setting the coefficient of r^2 equal to zero, in order for R to be a constant, gives

$$d = a\frac{x^2}{1-x^2} \quad , \quad R = \sqrt{d(a+d)} = a\frac{x}{1-x^2}$$

Thus the conducting sphere can be placed around the charge $-xq$, provided $x < 1$. If $x > 1$ we may use the same geometry, in which q and $-xq$ are interchanged, to show that the centre of a spherical conducting sheet of radius R' must be placed at a distance d' from the charge q, where

$$d' = a\frac{x^2}{x^2-1} \quad , \quad R' = a\frac{x}{x^2-1}$$

Assuming that $x < 1$, the field lines inside the sheet must end on the inner surface of the sheet, and hence, this surface has an induced charge $+xq$. Since the sheet is uncharged, a negative charge $-xq$ is on the outer surface. The surface charge density $\sigma(r)$ at any given point is determined by the resultant electric field $\vec{E}$ of the two point charges, that must everywhere be normal to the surface of the conductor, Eq.(11.10). The components of this radial field, Eq.(11.2), have intensities

$$E_q = \frac{q}{4\pi\varepsilon_0 r^2} \quad , \quad E_{xq} = \frac{xq}{4\pi\varepsilon_0 r_x^2} = \frac{1}{x} E_q$$

and it follows that $E_q/r_x = E_{xq}/r$ are corresponding sides of similar triangles with a congruent angle θ. Thus, we have

$$\frac{E}{a} = \frac{E_q}{r_x} = \frac{E_{xq}}{r} \quad \text{or} \quad E = \frac{aq}{4\pi\varepsilon_0 x} \frac{1}{r^3}$$

The outer surface charge density can be derived from the integral form of Gauss's law, Eq.(11.4), which reads

$$\sigma(r) = \frac{dq}{dS} = \varepsilon_0 E \quad \text{or} \quad \sigma(r) = \frac{-qa}{4\pi x} \frac{1}{r^3}$$

where the sign shows that $\vec{E}$ and $d\vec{S}$ are opposite. The total charge $-xq$ on the outer surface can be obtained by integrating this result.

The point charge $-xq$ can be removed without affecting the external field due to the charge q and the grounded spherical conducting sheet of induced charge $-xq$. The interesting converse is that the electric field produced by a point charge q near a grounded metal sphere of radius R can be calculated by introducing an image charge which, in combination with q, produces a field which satisfies the boundary condition that the resultant field is normal to the conducting surface. The image charge $-xq$ is opposite in sign to the real charge q, and is located on the other side of the spherical surface, a distance d from its centre. If the charge q is a distance $D > R$ from the centre of the metal sphere, we have

$$D = a + d = \frac{a}{1 - x^2} \quad \text{or} \quad dD = R^2$$

It is a straightforward matter to show that $aR = x(D^2 - R^2)$, and hence, the induced charge density on the conducting surface can be rewritten in terms of R and D as

$$\sigma(r) = -\frac{q}{4\pi R}(D^2 - R^2)\frac{1}{r^3}$$

11.2. An infinite line of charge λ per unit length, bent at right angles, lies along the axes of a Cartesian coordinate system as shown in the figure. Calculate the resultant electric field at a distance x from the origin.

Solution

Consider a differential element dl on the horizontal line of charge a distance l from the point where the electric field is calculated. The contribution of the element of charge $dq = \lambda dl$ to the field directed along the x-axis reads

$$dE = \frac{\lambda dl}{4\pi\varepsilon_0 l^2}$$

and hence, the total field produced by a finite line of charge is given by

$$E(x) = \frac{\lambda}{4\pi\varepsilon_0} \int_x^{x+L} \frac{dl}{l^2} = \frac{\lambda}{4\pi\varepsilon_0}\left(-\frac{1}{l}\right)_x^{x+L} = \frac{\lambda}{4\pi\varepsilon_0 x(1+x/L)} \quad \text{or} \quad E(x) = \frac{\lambda}{4\pi\varepsilon_0 x} \quad (x \ll L)$$

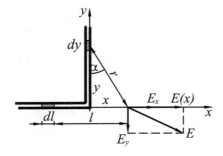

For an element of charge $dq' = \lambda dy$, located on the vertical line at the distance r from the same point, the contribution to the component E_y of the field parallel to the line and to that perpendicular to it, E_x, can be written as

$$dE_y = dE\cos\alpha = \frac{\lambda dy}{4\pi\varepsilon_0 r^2}\frac{y}{r} = \frac{\lambda}{4\pi\varepsilon_0}\frac{y\,dy}{\left(x^2+y^2\right)^{3/2}}$$

$$dE_x = dE\sin\alpha = \frac{\lambda dy}{4\pi\varepsilon_0 r^2}\frac{x}{r} = \frac{\lambda x}{4\pi\varepsilon_0}\frac{dy}{\left(x^2+y^2\right)^{3/2}}$$

Assuming a finite line of charge and integrating along the y-axis gives

$$E_y = \frac{1}{2}\frac{\lambda}{4\pi\varepsilon_0}\int_0^L \frac{d(y^2)}{\left(x^2+y^2\right)^{3/2}} = \frac{\lambda}{4\pi\varepsilon_0}\left(\frac{1}{\sqrt{x^2+y^2}}\right)_0^L = \frac{\lambda}{4\pi\varepsilon_0}\left(\frac{1}{\sqrt{x^2+L^2}}-\frac{1}{x}\right)$$

$$E_x = \frac{\lambda x}{4\pi\varepsilon_0}\int_0^L \frac{dy}{\left(x^2+y^2\right)^{3/2}} = \frac{\lambda}{4\pi\varepsilon_0 x}\left(\frac{y}{\sqrt{x^2+y^2}}\right)_0^L = \frac{\lambda}{4\pi\varepsilon_0 x}\frac{L}{\sqrt{x^2+L^2}}$$

and hence, for a very long line, we obtain

$$E_x = \frac{\lambda}{4\pi\varepsilon_0 x} \quad , \quad E_y = -\frac{\lambda}{4\pi\varepsilon_0 x}$$

It follows that the electric field at a distance x from the origin is given by

$$E(x) + E_x = \frac{2\lambda}{4\pi\varepsilon_0 x} \quad , \quad E_y = -\frac{\lambda}{4\pi\varepsilon_0 x} \quad \text{or} \quad E = \frac{\lambda\sqrt{5}}{4\pi\varepsilon_0 x}$$

11.3. A spherical conducting sheet of radius R and mass M resting on an insulating stand is assembled from two caps which lie above and below a horizontal plane passing a distance z from its centre. How much charge Q can be put on the sheet until the top cap starts to lift off the lower one?

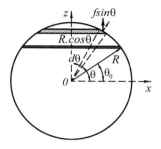

Solution

On a spherical conducting sheet of uniform charge density $\sigma = Q/4\pi R^2$, the electric field is discontinuous at any point by an amount σ/ε_0, since for a spherical surface of radius r Gauss's law gives

$$\int_{r<R} \vec{E} \cdot d\vec{S} = 0 \quad , \quad \int_{r \geq R} \vec{E} \cdot d\vec{S} = \frac{Q}{\varepsilon_0}$$

and hence, the field is zero inside the conductor and is σ/ε_0 just outside. However, if we choose as the Gaussian surface a closed cylinder of cross-sectional area ΔS, the axis of which is perpendicular to the sheet, the field due to the surface charge inside and outside the conductor is found to be $\sigma/2\varepsilon_0$. Hence, there must be an external field $\sigma/2\varepsilon_0$ from charges elsewhere on the conductor, such that, with this additional field the total field outside the conductor is σ/ε_0 and the field inside the conductor is zero. This external field acts on the surface charge $\sigma \Delta S$ with a resulting force per unit area given by

$$f = \frac{1}{\Delta S}(\sigma \Delta S)\frac{\sigma}{2\varepsilon_0} = \frac{\sigma^2}{2\varepsilon_0}$$

This force is always outwards, regardless of whether σ is positive or negative. Thus, the vertical force pushing normally against the upper cap is given by

$$F_y = \int_{\theta_0}^{\pi-\theta_0} f \sin\theta \, dS = f \int_{\theta_0}^{\pi-\theta_0} \sin\theta \, 2\pi R^2 \cos\theta \, d\theta = 2\pi R^2 f \int_{\pi-\theta_0}^{\theta_0} \cos\theta \, d(\cos\theta) = \pi R^2 f \cos^2\theta_0$$

where $dS = 2\pi(R\cos\theta)R d\theta = 2\pi R^2 \cos\theta \, d\theta$ and $R^2 \cos^2\theta_0 = R^2 - z^2$. Substituting f and σ into the last equation yields

$$F_y = \pi f R^2 \left(1 - \frac{z^2}{R^2}\right) = \frac{1}{4\pi\varepsilon_0} \frac{Q^2}{8R^2}\left(1 - \frac{z^2}{R^2}\right)$$

and this should be equal to the weight of the cap, given by

$$mg = \frac{M}{4\pi R^2} 2\pi R(R-z)g = \frac{Mg}{2}\left(1-\frac{z}{R}\right)$$

Thus, one obtains

$$Q^2 = 4\pi\varepsilon_0 R^2 \frac{4Mg}{1+z/R} \quad \text{or} \quad V^2 = \frac{Mg}{\pi\varepsilon_0(1+z/R)}$$

where $V = Q/4\pi\varepsilon_0 R = Q/C$ is the potential above the ground of the spherical sheet, and $C = 4\pi\varepsilon_0 R$ stands for its capacitance.

11.4. A capacitor consists of two air-spaced concentric spheres, of radii R and r, where R is fixed and $r < R$. What is the maximum voltage that can be placed across this capacitor without risking discharge through air breakdown, which occurs at field strengths greater than $E_a = 3 \times 10^6$ V/m? Solve the same problem for a capacitor consisting of two long concentric cylinders of radii R and r.

Solution

Assuming a charge q placed on the inner sphere, the electric field on a spherical Gaussian surface of radius x is given by Eq.(11.4) as $4\pi\varepsilon_0 x^2 E(x) = q$. Since $E(x)$ is radial, the voltage V across the capacitor, Eq.(11.8), reads

$$V = -\int_R^r \frac{q}{4\pi\varepsilon_0 x^2} dx = \frac{q}{4\pi\varepsilon_0}\left(\frac{1}{r} - \frac{1}{R}\right)$$

and hence, the capacitance defined by $C = q/V$ is given by

$$C = 4\pi\varepsilon_0 \frac{rR}{R-r}$$

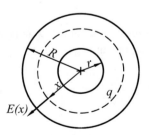

The maximum electric field occurs where x is minimum, that is at $x = r$. If we set the field at $x = r$ equal to the breakdown field E_a, it follows that

$$V = E_a\left(r - \frac{r^2}{R}\right)$$

To find the maximum voltage if R is fixed, we differentiate this expression with respect to r and set $dV/dr = 0$, which gives

$$\frac{dV}{dr} = E_a\left(1 - \frac{2r}{R}\right) = 0 \quad \text{or} \quad r = \frac{R}{2}$$

Since $d^2V/dr^2 = -2E_a/R < 0$, for $r = R/2$, there is a maximum of voltage, given by

$$V_{max} = E_a \frac{R}{4}$$

For a capacitor consisting of two air-spaced cylinders we may assume that the inner cylinder carries a charge per unit length $\lambda = q/L$, so that Gauss's law gives $2\pi\varepsilon_0 x E(x) = \lambda$. The voltage is found by integrating $E(x)$ with respect to x, which gives

$$V = -\int_R^r \frac{\lambda}{2\pi\varepsilon_0 x} dx = \frac{\lambda}{2\pi\varepsilon_0} \ln\left(\frac{R}{r}\right)$$

and the capacitance $C = q/V = \lambda L/V$ follows as

$$C = \frac{2\pi\varepsilon_0 L}{\ln(R/r)}$$

If we set the breakdown field E_a equal to the maximum value of the electric field, which is found at the inner surface, for $x = r$ we similarly obtain

$$V = E_a r \ln\left(\frac{R}{r}\right) \quad \text{i.e.,} \quad \frac{dV}{dr} = E_a \ln\left(\frac{R}{r}\right) - E_a = 0 \quad \text{or} \quad r = R/e$$

where e is the base of natural logarithms. This gives a maximum voltage V_{max} of $E_a R/e$.

11.5. A parallel-plate capacitor with square plates of side l separated by a distance $d \ll l$ is entirely filled with a square slab of dielectric of side l, thickness d, density ρ and relative permittivity ε_r. With a battery maintaining a constant potential difference between the plates, the dielectric slab is withdrawn a distance x in a direction parallel to one side of the plates and then released. Find the period of its oscillatory motion, ignoring friction and edge effects.

Solution

For a parallel-plate capacitor of capacitance C charged to a potential difference V, an electric field $\sigma/2\varepsilon_0$ is due to the uniform surface charges, $+q$ and $-q$. It follows that a uniform field $E = \sigma/\varepsilon_0$ is produced between the plates, and hence, in view of Eq.(11.9), $V = Ed = (\sigma/\varepsilon_0)d$. Thus, the capacitance is $C_0 = Q/V = \varepsilon_0 l^2/d$ for the air-spaced capacitor and $C = \varepsilon_r C_0$ when the gap is entirely filled with dielectric. For a partially inserted dielectric slab, the system consists of two capacitors, connected in parallel across a potential difference V, of capacities

$$C_x = \varepsilon_0 \frac{lx}{d} \quad \text{and} \quad C_{l-x} = \varepsilon_0 \varepsilon_r \frac{l(l-x)}{d}$$

The energy stored in the electric field E, Eq.(11.32), reads

$$U(x) = \frac{\varepsilon_0}{2} E^2 (lxd) + \frac{\varepsilon_0 \varepsilon_r}{2} E^2 l(l-x) d = \frac{1}{2}(C_x + C_{l-x})V^2$$

and hence, it decreases as x increases by $U(x) - U(0) = (C_x + C_{l-x} - C)V^2/2$. The charge stored on the capacitor also decreases by $\Delta q = (C_x + C_{l-x} - C)V$, and hence, the energy of the battery is decreased by ΔqV, which is twice the energy lost by the electric field. The conservation of energy

$$\Delta qV = U(x) - U(0) + \int_0^x F dx \quad \text{or} \quad -\frac{\varepsilon_0(\varepsilon_r - 1)l}{2d} V^2 x = \int_0^x F dx$$

shows that negative work is done against the motion of the dielectric, by a force of constant magnitude

$$F = \frac{\varepsilon_0(\varepsilon_r - 1)l}{2d} V^2$$

which is always directed towards $x = 0$. Since the restoring force F is a constant, the motion with constant acceleration $a = F/m = F/\rho l^2 d$ of the dielectric slab, released at $x > 0$, will be periodic about $x = 0$, with a period T given by

$$x = \frac{a}{2}\left(\frac{T}{4}\right)^2 \quad \text{or} \quad T = 4\sqrt{\frac{2x}{a}} = 8\frac{d}{V}\sqrt{\frac{\rho lx}{\varepsilon_0(\varepsilon_r - 1)}}$$

11.6. Show that in a linear isotropic dielectric each dipole moment $\vec{p}$ experiences an effective field $\vec{E}_{eff} = \vec{E} + \vec{P}/3\varepsilon_0$, where $\vec{E}$ is the applied field and $\vec{P} = N\vec{p}$ is the polarization due to N dipole moments. Assuming that $\vec{p} = \alpha \varepsilon_0 \vec{E}_{eff}$, find an expression for the relative permittivity ε_r in terms of the constant α which stands for the atomic (or molecular) polarizability

Solution

In a condensed phase dielectric (solid or liquid) the dipole moment of each atom or molecule is determined by an effective field, consisting of the sum of the applied field $\vec{E}$ and the total field of all other dipoles. This local field $\vec{E}_{eff}$ can be calculated in a linear isotropic dielectric by considering a spherical region of radius R with the centre at a given molecule. The field determined by a uniform isotropic distribution of molecules inside the sphere averages to zero so that the contribution of all other molecular dipoles to $\vec{E}_{eff}$ reduces to the field associated with the polarization surface charges on the sphere. For the arrangement illustrated in the figure, the surface charge density at an angle θ to the direction of the field is equal to minus the normal component of polarization $\vec{P}$ of the dielectric *outside* the spherical region, that is

$$\sigma = -\vec{P}\cdot\vec{e}_n = -P\cos\theta$$

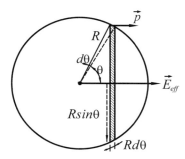

The charge on an annular ring of width $Rd\theta$ and radius $R\sin\theta$ is then given by $dq = (-P\cos\theta)(2\pi R\sin\theta)Rd\theta$ and produces a field at the origin, parallel to the applied field, of magnitude

$$dE_P = \frac{dq\cos\theta}{4\pi\varepsilon_0 R^2} = \frac{2\pi R^2 P\cos^2\theta\sin\theta d\theta}{4\pi\varepsilon_0 R^2} = -\frac{P}{2\varepsilon_0}\cos^2\theta d(\cos\theta)$$

Integration over the sphere is immediate and gives the contribution of the polarization charges to the local field as

$$\vec{E}_P = \frac{\vec{P}}{3\varepsilon_0} \qquad \text{i.e.,} \qquad \vec{E}_{eff} = \vec{E} + \frac{\vec{P}}{3\varepsilon_0}$$

Assuming that the dipole moment of a molecule is proportional to the effective field we may write

$$\vec{p} = \alpha\varepsilon_0 \vec{E}_{eff} = \alpha\varepsilon_0\left(\vec{E} + \frac{\vec{P}}{3\varepsilon_0}\right)$$

where α is called the *polarizability* of the molecule. The polarization $\vec{P}$ for N molecules per unit volume is given by

$$\vec{P} = N\vec{p} = N\alpha\varepsilon_0\left(\vec{E} + \frac{\vec{P}}{3\varepsilon_0}\right) \qquad \text{or} \qquad \vec{P} = \frac{N\alpha}{1-N\alpha/3}\varepsilon_0\vec{E}$$

Comparison with Eqs.(11.25) and (11.29) gives the relative permittivity in terms of the polarizability α and the number of molecules per unit volume as

$$\varepsilon_r = \frac{1+2N\alpha/3}{1-N\alpha/3} \quad \text{or} \quad \frac{\varepsilon_r-1}{\varepsilon_r+2} = \frac{N\alpha}{3}$$

a form which is known as the Clausius-Mossotti equation.

11.7. Two parallel-plate capacitors, one having a charge q_1 and the other q_2, are separated a distance $D \gg d$, where d is the plate spacing. Find the force of attraction between the two capacitors.

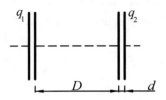

11.8. A dielectric wire of length l and charge q is aligned along a flux line due to an infinite line of charge λ per unit length, with the inner end at a radial distance R from the axis. Assuming that the wire has uniform charge per unit length, find the force felt by the wire.

11.9. A conducting hemisphere of radius R is uniformly charged with a total charge Q. If a metallic ball, uniformly charged with a total charge q, is placed at the centre of the hemisphere, what force is felt by the ball?

11.10. A point charge q is at the centre of a dielectric shell of inner radius R_1, outer radius R_2 and relative permittivity ε_r. Find the electrostatic energy stored in the dielectric shell.

12. Magnetic Induction

12.1. CURRENT FLOW
12.2. MAGNETIC EFFECTS
12.3. MAGNETIC VECTOR POTENTIAL

12.1. Current flow

The assumption of bound charge densities, which is valid in dielectrics, allowed us to describe the electrostatic behaviour of these materials in terms of a vector field $\vec{D}$, proportional to the electric field $\vec{E}$ and obeying the flux and circulation axioms of electrostatics. In conductors, where charges are free to move under an applied electric field $\vec{E}$, an appropriate vector field $\vec{j}$, proportional to $\vec{E}$, will be associated with the steady charge flow. Such a charge flow is called an *electric current*, and its behaviour is described by equations that are formally identical with the axioms described above.

A steady electron flow through a conductor requires that the surface charges, which balance the applied electric field $\vec{E}$, should be continuously removed at one end and replaced at the other, so that the conductor must be part of a complete circuit. The motion of electrons of mass m_e and charge $-e$ under a constant force $-e\vec{E}$ obeys Newton's second law (2.9), which reads

$$m_e \frac{d\vec{v}}{dt} = -e\vec{E} - \alpha\vec{v} \qquad (12.1)$$

where $-\alpha\vec{v}$ represents a damping force, acting against the velocity, due to collisions with the ions. If there are N electrons in a volume V, that is a free charge density $\rho = -Ne/V = -ne$, all moving with the same velocity $\vec{v}$, the magnitude of the electric current is defined as the time rate of charge through the normal cross section of area S, that is

$$I = \frac{dq}{dt} = -neSv = \rho Sv \tag{12.2}$$

and Eq.(12.1) becomes

$$\frac{dI}{dt} = \frac{ne^2 SE}{m_e} - \frac{\alpha}{m_e} I \quad \text{or} \quad \frac{dI}{I - ne^2 SE/\alpha} = -\frac{\alpha}{m_e} dt$$

Integrating gives

$$\ln\left(I - \frac{ne^2 SE}{\alpha}\right) = -\frac{\alpha}{m_e} t + C$$

By applying the condition of initial zero current $I = 0$ at $t = 0$ we obtain

$$I = \frac{ne^2 SE}{\alpha}\left(1 - e^{-\alpha t/m_e}\right) \tag{12.3}$$

which shows an exponential rise, approaching the steady current

$$I = \frac{ne^2 SE}{\alpha} \quad \text{or} \quad I = \frac{ne^2 S}{\alpha l} V \tag{12.4}$$

where the electric field has been expressed in terms of the potential difference V applied across the ends of a conductor of finite length l, according to Eq.(11.8). If we write

$$R = \frac{\alpha}{ne^2} \frac{l}{S} = \frac{l}{\sigma S}$$

the equation (12.4) becomes

$$V = RI \tag{12.5}$$

This equation is known as *Ohm's law*. The constant R is called the *resistance* of the conductor, and the *electrical conductivity* σ is defined as

$$\sigma = \frac{ne^2}{\alpha} \tag{12.6}$$

The charge flow is always tangential to the electric field lines so that it is convenient to define a *current density field* $\vec{j}$, of magnitude given by the current flow per unit normal area, as

$$j = \frac{dI}{dS} = \rho v \quad \text{or} \quad \vec{j} = \rho \vec{v} \tag{12.7}$$

The current I then represents the flux of current density across an arbitrary section of the conductor

$$I = \int_S \vec{j} \cdot d\vec{S} \qquad (12.8)$$

For an elementary volume of the conductor, illustrated in Figure 12.1, substituting Eqs.(12.6) and (12.7) into Eq.(12.4) gives

$$dI = \sigma E dS \quad \text{or} \quad \vec{j} \cdot d\vec{S} = \sigma \vec{E} \cdot d\vec{S} \qquad (12.9)$$

which is true however small the cross-sectional area $d\vec{S}$, irrespective of its orientation. It is therefore clear that $\vec{j}$ is proportional to $\vec{E}$ with σ being the constant of proportionality, that is

$$\vec{j} = \sigma \vec{E} \qquad (12.10)$$

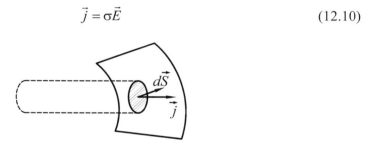

Figure 12.1. A tube of current flow in a conductor

The differential form (12.10) of Ohm's law, which is formally similar to Eq.(11.28) defining the proportionality of $\vec{D}$ and $\vec{E}$ in a dielectric, is often regarded as a constitutive relation for a conductor.

The law of energy conservation in a conductor, under a stationary field $\vec{E}$, is exactly analogous to the form of the circulation law (11.6) for electrostatics. Its alternative form (11.8), upon multiplication of both sides by dq, shows that the decrease $-d(qV)$ of potential energy along the current flow line, in a time interval dt, can be expressed by

$$dP_J = \frac{dq\,\vec{E} \cdot d\vec{r}}{dt} = (\rho dV)\vec{E} \cdot \vec{v} = \vec{j} \cdot \vec{E}\, dV = \sigma E^2 dV \qquad (12.11)$$

which gives the power per unit volume as

$$P_J / V = \sigma E^2 \qquad (12.12)$$

This is *Joule's law*, which can also be written in an integral form by considering a finite volume $V = Sl$ of the conductor. It follows that

$$P_J = (\sigma E)(El)S = (jS)V = IV = RI^2 = \frac{V^2}{R} \qquad (12.13)$$

where V stands for the potential difference applied across the ends of the current flow line. Let us now consider the flux of the current density $\vec{j}$, that is the current crossing a closed surface S, enclosing a volume V (Figure 12.2). The total current flowing out of S must be equal to the time rate of decrease of charge inside S, that is

$$\int_S \vec{j} \cdot d\vec{S} = -\frac{\partial}{\partial t} \int_V \rho dV = -\int_V \frac{\partial \rho}{\partial t} dV \qquad (12.14)$$

The left hand side transforms according to Gauss's divergence theorem and so

$$\int_V \nabla \cdot \vec{j}\, dV = -\int_V \frac{\partial \rho}{\partial t} dV$$

which is equivalent to

$$\nabla \cdot \vec{j} + \frac{\partial \rho}{\partial t} = 0 \qquad (12.15)$$

Equation (12.15) is called the *equation of continuity* for currents, on account of its similarity with Eq.(6.11) for fluid flow.

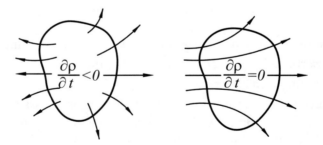

Figure 12.2. Closed surface for the calculation of the continuity equation

For a steady-state flow the charge inside S is stationary, that is $\partial \rho / \partial t = 0$, and Eq.(12.15) reads

$$\nabla \cdot \vec{j} = 0 \qquad (12.16)$$

This equation is analogous to Gauss's law for dielectrics (11.26) and states that the net rate of creation of charge is zero in any element of a conductor.

12. MAGNETIC INDUCTION

> **EXAMPLE 12.1. Kirchhoff's rules**
>
> For a system of conductors connected in an electrical network, the two Kirchhoff's rules represent a specific application of the flux and circulation laws for the current flow. For a closed surface about a junction where several conductors meet, the equation of continuity (12.16) reads
>
> $$\oint_S \vec{j} \cdot d\vec{S} = \sum_k I_k = 0$$
>
> that is the algebraic sum of the currents at a junction must be zero (*Kirchhoff's node rule*). For a path along a closed loop of conductors, the circulation law (11.6) reads
>
> $$\oint_\Gamma \vec{E} \cdot d\vec{r} = \sum_k \Delta V_k = 0$$
>
> which means that the algebraic sum of the potential differences around any closed path in the network must be zero (*Kirchhoff's loop rule*).

It is apparent that considerable similarities have been found between the equations of current flow in conductors, expressed in terms of a current density field $\vec{j}$ and the equations of electrostatic interactions in dielectrics, written in terms of an electric displacement field $\vec{D}$ as follows

(11.28) $\qquad \vec{D} = \varepsilon_0 \varepsilon_r \vec{E} \qquad \vec{j} = \sigma \vec{E}$ (12.10)

(11.26) $\qquad \nabla \cdot \vec{D} = \rho \qquad \nabla \cdot \vec{j} = 0$ (12.16)

(11.7) $\qquad \nabla \times \vec{E} = 0 \qquad \nabla \times \vec{E} = 0$ (11.7)

The proportionality of $\vec{D}$ and $\vec{E}$ given by Eq.(11.28), or $\vec{j}$ and $\vec{E}$ given by Eq.(12.10), is valid for uniform and isotropic materials. Both flux laws (11.26) and (12.16) originate in the equation of continuity for continuous media (6.11), whereas there is the same circulation law (11.7) stating the energy conservation in either dielectrics or conductors.

12.2. MAGNETIC EFFECTS

All magnetic effects, on the macroscopic and microscopic scale, are explained in terms of current flow. Although this assumption does not allow us to ascribe the source of magnetic fields to magnetic poles, which is a concept analogous to electric charge, the theory of magnetostatics can be developed with a formal similarity to that of electrostatics. A vector field $\vec{B}$, called *magnetic induction*, associated with the flow of an elementary charge distribution $dq = \rho dV$ with velocity $\vec{v}$, and given by

$$d\vec{B} = k\frac{dq(\vec{v}\times\vec{r})}{r^3}$$

accounts for the observed deflection of a compass needle by a current. As illustrated in Figure 12.3, the cross product $\vec{v}\times\vec{r}$ shows that the vector $d\vec{B}$ is normal to both $\vec{v}$ and $\vec{r}$.

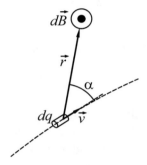

Figure 12.3. Magnetic induction due to an elementary charge flow

The field $\vec{B}$ has an inverse square dependence on position and the similarity to the inverse square law in electrostatics is completed by choosing the constant of proportionality $k = \mu_0/4\pi$ in *SI* units

$$d\vec{B} = \frac{\mu_0}{4\pi}\frac{dq(\vec{v}\times\vec{r})}{r^3} \qquad (12.17)$$

Here v_0 is called the *permeability of free space* and plays a role analogous to that of the reciprocal of ε_0 in electrostatics.

We can now express the magnetic induction in terms of a steady current I flowing in an infinitesimal element $d\vec{l}$ of a conductor in the configuration shown in Figure 12.4. Taking $dq = \rho dV = \rho S dl$, Eq.(12.17) successively transforms, in view of Eq.(12.2), as

$$d\vec{B} = \frac{\mu_0}{4\pi}\frac{\rho S dl(\vec{v}\times\vec{r})}{r^3} = \frac{\mu_0}{4\pi}\frac{S(\rho\vec{v}\times\vec{r})dl}{r^3} = \frac{\mu_0}{4\pi}\frac{S(\vec{j}\times\vec{r})dl}{r^3}$$

Since I and therefore $\vec{j}$ are parallel to $d\vec{l}$, we can equate $\vec{j}dl = jd\vec{l}$ and upon substitution we obtain

$$dB = \frac{\mu_0}{4\pi} \frac{Sj(d\vec{l} \times \vec{r})}{r^3} = \frac{\mu_0 I}{4\pi} \frac{(d\vec{l} \times \vec{r})}{r^3} \quad (12.18)$$

called the *Biot-Savart law*.

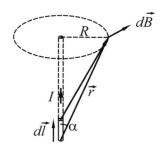

Figure 12.4. Experimental configuration associated with the Biot-Savart law

The magnetic induction is proportional to the electric current, which means that the superposition principle can be applied to magnetic induction vectors due to different current elements, in the same way as to electric fields. Since a steady current flow in an isolated element of conductor causes an accumulation of charge, the Biot-Savart law is not capable of direct experimental check, unlike the inverse square law in electrostatics. However, in its alternative form, which is more convenient for applications

$$dB = \frac{\mu_0 I}{4\pi} \frac{dl \sin\alpha}{r^2} \quad (12.19)$$

the Biot-Savart law always gives the correct magnetic induction for closed current flow circuits (Figure 12.4).

EXAMPLE 12.2. Magnetic induction due to simple circuits

1. Straight-line current. Since each current element makes a parallel contribution to $\vec{B}$, lines of constant $\vec{B}$ are concentric rings around the current I. In the arrangement represented in Figure 12.4 we have

$$r = \frac{R}{\sin\alpha} \quad \text{i.e.,} \quad dl = \frac{rd\alpha}{\sin\alpha} = \frac{Rd\alpha}{\sin^2\alpha}$$

so that Eq.(12.19) becomes

$$dB = \frac{\mu_0}{4\pi} \frac{IRd\alpha \sin\alpha \sin^2\alpha}{R^2 \sin^2\alpha} = \frac{\mu_0}{4\pi} \frac{I}{R} \sin\alpha \, d\alpha$$

For a conducting wire of finite length, the total magnetic induction at a point P making angles α and β with the ends (see Figure 12.5) is obtained upon integration as

$$B = \frac{\mu_0}{4\pi} \frac{I}{R} (\cos\alpha + \cos\beta) \tag{12.20}$$

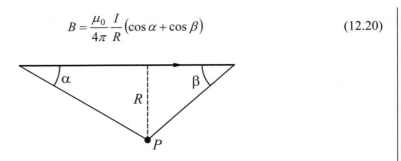

Figure 12.5. Magnetic induction of a finite wire

If the wire is infinite, both α and β become zero and hence

$$B = \frac{\mu_0}{2\pi} \frac{I}{R}$$

2. Current loop. Axial points have position vectors $\vec{r}$ which are always normal to $\vec{dl}$ for all loop elements, so that we must set $\sin\alpha = 1$ in Eq.(12.19). Since the components of $\vec{dB}$ that are perpendicular to the axis cancel around the loop, as illustrated in Figure 12.6, the axial magnetic induction is given by a simple line integral along the circular path of radius R, of the form

$$B_z = \int dB \sin\varphi = \frac{\mu_0 I}{4\pi} \frac{\sin\varphi}{r^2} \int dl = \frac{\mu_0 I R \sin\varphi}{2r^2}$$

It is convenient to set $\varphi = R/r$, which gives

$$B_z = \frac{\mu_0}{4\pi} \frac{2IS}{r^3}$$

where S is the area of the current loop of radius R.

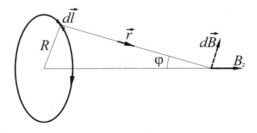

Figure 12.6. Axial magnetic induction due to a circular loop

The magnetic induction $\vec{B}$ is the fundamental magnetic field in terms of which we can formulate the basic equations analogous to those obeyed by the electric field $\vec{E}$. The derivatives of $\vec{E}$ are related to a local charge density, whereas those of $\vec{B}$ will be expressed in terms of the local current density. The Biot-Savart law (12.18) can be written for a current flow line as

$$\vec{B} = \frac{\mu_0 I}{4\pi} \int_\Gamma \frac{d\vec{l} \times \vec{r}}{r^3} \qquad (12.21)$$

where the line integral is carried out over the contributions of the current elements $d\vec{l}$, with coordinates x', y', z', and $\vec{B} = \vec{B}(x, y, z)$. Using Eq.(12.8) we can obtain an equivalent form in terms of volume integral over the coordinates x', y', z' of the current distribution $\vec{j} = \vec{j}(x', y', z')$, which reads

$$\vec{B} = \frac{\mu_0}{4\pi} \int_{V'} \frac{\vec{j} \times \vec{r}}{r^3} dV' \qquad (12.22)$$

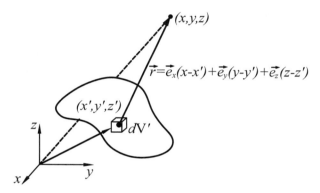

Figure 12.7. Appropriate Cartesian coordinates for the integration of the Biot-Savart law

The distinction is illustrated in Figure 12.7. Equation (12.22) for $\vec{B}$ is the magnetostatic analogue of Eq.(11.3) for an electric field $\vec{E}$ due to a continuous charge distribution. Starting from Eq.(12.22) we obtain the divergence of the magnetic induction, using Eq.(I.14), Appendix I, as

$$\nabla \cdot \vec{B} = \frac{\mu_0}{4\pi} \int_{V'} \frac{\nabla \cdot (\vec{j} \times \vec{r})}{r^3} dV' = \frac{\mu_0}{4\pi} \int_{V'} \left[\frac{\vec{r}}{r^3} \cdot (\nabla \times \vec{j}) - \vec{j} \cdot \left(\nabla \times \frac{\vec{r}}{r^3} \right) \right] dV'$$

where it is obvious that $\nabla \times \vec{j} = 0$ since ∇ operates on x, y, z. Thus, in view of Eq.(I.10), Appendix I, we have

$$\nabla \cdot \vec{B} = \frac{\mu_0}{4\pi} \int_{V'} \vec{j} \cdot \left[\vec{r} \times \nabla \left(\frac{1}{r^3} \right) - \frac{\nabla \times \vec{r}}{r^3} \right] dV'$$

Here $\nabla(1/r^3) = -3\vec{r}/r^5$, as previously shown, thus the two components of the cross product are parallel and reduce to zero. It is also easy to check that $\nabla \times \vec{r} = 0$. It follows that the divergence of $\vec{B}$, which gives the strength of the source field at a point, becomes

$$\nabla \cdot \vec{B} = 0 \tag{12.23}$$

Therefore the magnetic induction $\vec{B}$, which is called a *solenoidal* field has no source, since the field lines have no beginning or end. This result is consistent with our assumption that magnetic poles cannot exist. Using Gauss's divergence theorem, Eq.(12.23) gives

$$\int_S \vec{B} \cdot d\vec{S} = \int_V \nabla \cdot \vec{B} \, dV = 0 \tag{12.24}$$

that is the total flux of magnetic induction through any closed surface is zero. The similarity between Eq.(12.23) and the equation of continuity for the current density (12.16) should be noted, and as a result the vector field $\vec{B}$ is often called the *magnetic flux density*.

The line integral of $\vec{B}$ around an arbitrary closed path Γ encircling an infinite straight line current of value I can be evaluated using Eq.(12.20) in the configuration represented in Figure 12.8, yielding

$$\vec{B} \cdot d\vec{l} = BRd\alpha = \frac{\mu_0}{2\pi} \frac{I}{R} Rd\alpha = \frac{\mu_0 I}{2\pi} d\alpha$$

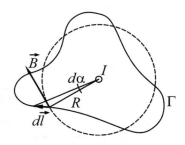

Figure 12.8. Closed path for the calculation of Ampère's law

It follows that

$$\int_\Gamma \vec{B} \cdot d\vec{l} = \frac{\mu_0 I}{2\pi} \int_\Gamma d\alpha = \mu_0 I \tag{12.25}$$

By superposition, Eq.(12.25) which, as derived, only describes currents flowing down the axis of symmetry of the loop, can be extended to a combination of infinite straight-line currents

$$\int_\Gamma \vec{B} \cdot d\vec{l} = \mu_0 \sum_k I_k$$

There are current configurations which cannot be constructed from infinite straight wires, however for an arbitrary distribution of currents, given by the current density $\vec{j}$ at any point, it is correct to assume that Eq.(12.25) reads

$$\int_\Gamma \vec{B} \cdot d\vec{l} = \mu_0 \int_S \vec{j} \cdot d\vec{S} \tag{12.26}$$

This is *Ampère's circuital law* which states that the line integral of $\vec{B}$ around any closed path is equal to μ_0 times the total current I crossing any surface bounded by the path. The left-hand side transforms using Stoke's theorem, which gives

$$\int_S \left(\nabla \times \vec{B} \right) \cdot d\vec{S} = \mu_0 \int_S \vec{j} \cdot d\vec{S}$$

a relation which is valid however small the surface element dS, so that Ampère's law can be rewritten as

$$\nabla \times \vec{B} = \mu_0 \vec{j} \tag{12.27}$$

Equation (12.27) relates the rate of change of the magnetic induction to the current density at a same point. Ampère's law in magnetostatics plays the role assumed by Gauss's law in electrostatics, allowing the calculation of a uniform vector field, given its source distribution.

EXAMPLE 12.3. Uniform magnetic fields

The circuital law (12.26) can be used to calculate a constant magnetic induction $\vec{B}$ along the path of integration, which can be taken out of the line integral. The standard configurations which show the required symmetry are the solenoidal and toroidal coils.

1. *The solenoid*. Let us first consider a *long* close-wound solenoid, represented in Figure 12.9, consisting of N_0 turns per unit length of a cylinder of radius R, carrying a current I. The cylindrical symmetry requires that, both inside and outside the solenoid, $\vec{B}$ should be independent of the coordinates z and φ. There is also no radial component B_r since the integral over any cylindrical surface of radius r and length l must be zero according to Eq.(12.23) which reads

$$\int_S \vec{B} \cdot d\vec{S} = (2\pi r l) B_r = 0$$

This must be true for any r, so that $B_r = 0$. It follows that $\vec{B}$ points parallel to the axis. Now we can apply the circuital law to a rectangular loop which lies entirely outside the solenoid, with sides L at distances $r_1, r_2 > R$ from the axis

$$\int_\Gamma \vec{B} \cdot d\vec{l} = \left[B(r_1) - B(r_2) \right] L = 0 \quad \text{or} \quad B(r_1) = B(r_2)$$

Since the field is independent of the distance from the axis and is zero for large r, it must be zero everywhere outside the solenoid.

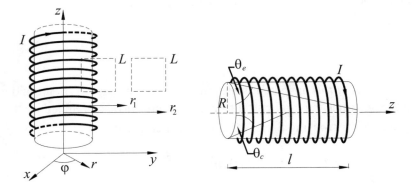

Figure 12.9. Magnetic induction due to an elementary charge flow

For a similar rectangular loop which lies partly inside the solenoid, Eq.(12.26) reads

$$\oint_\Gamma \vec{B} \cdot d\vec{l} = BL = \mu_0 (N_0 I) L \quad \text{or} \quad B = \mu_0 N_0 I$$

since $B = 0$ outside, $B_r = 0$ and there are $N_0 L$ identical currents of value I that cross the loop. In other words the magnetic induction inside a long solenoid is uniform.

For a *short* solenoid, the magnetic field is not uniform and instead of Ampère's law we must use the Biot-Savart law. We obtain the axial magnetic induction by summing the contributions of the individual turns. At an axial point of a solenoid of length l making an angle θ with one end, we have

$$B = \frac{\mu_0 S}{4\pi} \int \frac{N_0 I dz}{r^3} = \frac{\mu_0 N_0 I}{2} \int \frac{R^2 dz}{r^3} = \frac{\mu_0 N_0 I}{2} \int \cos\theta d\theta$$

where $\tan\theta = z/R$ and $\cos\theta = R/r$, so that at the centre of the solenoid we may write, according to Figure 12.9, that

$$B = 2\frac{\mu_0 NI}{2} \int_0^{\theta_c} \cos\theta d\theta = \mu_0 N_0 I \sin\theta_c$$

In the limiting case $\theta_c \to \pi/2$ this formula reduces to that of the magnetic induction inside a long solenoid. At the end of the solenoid we may define the angle θ_e as shown in Figure 12.9 and hence

$$B = \frac{\mu_0 N_0 I}{2} \int_0^{\theta_e} \cos\theta d\theta = \mu_0 N_0 I \frac{\sin\theta_e}{2}$$

indicating that the magnetic induction decreases at both ends of the solenoid.

> **2. *The toroid.*** A close-wound toroidal coil can be regarded as a solenoid with a circumferential axis. The symmetry requires that the magnetic field is circumferential at all points inside the coil and Ampère's circuital law demands that it be zero for points outside the coil. With this assumption we may integrate Eq.(12.26) around the circumferential axis of radius r which gives
>
> $$\oint_\Gamma \vec{B} \cdot d\vec{l} = 2\pi r B = \mu_0 N I$$
>
> where N is the total number of turns. It follows that
>
> $$B = \mu_0 \frac{N}{2\pi r} I$$
>
> For large r, we can set $N \approx 2\pi r N_0$, where N_0 is the number of turns per unit length, so that B is approximately given by the equation valid for a long solenoid, which becomes the limiting case for a toroid.

12.3. Magnetic Vector Potential

Ampère's law (12.27) shows that the vector field $\vec{B}$ is conservative only in regions of space which do not enclose any current. In regions where $\vec{B}$ is conservative it can be derived from a magnetostatic scalar potential ϕ by

$$\vec{B} = -\nabla \phi \qquad (12.28)$$

in an exact analogy to Eq.(11.10) for the conservative electric field. However, in the general case, the magnetic induction $\vec{B}$ is derived from a vector function $\vec{A}$. The flux law (12.23) indicates that a vector field $\vec{A}$, called the *vector potential* can be defined as

$$\vec{B} = \nabla \times \vec{A} \qquad (12.29)$$

because of the vector identity (I.17), Appendix I. The explicit form of the vector potential is obtained in terms of an arbitrary current distribution $\vec{j}$ by using the integral expression (12.21) of the Biot-Savart law. As illustrated in Figure 12.7, we have

$$\vec{r} = \vec{e}_x(x-x') + \vec{e}_y(y-y') + \vec{e}_z(z-z') \quad \text{i.e.,} \quad r = \sqrt{(x-x')^2 + (y-y')^2 + (z-z')^2}$$

and it follows that

$$\frac{\partial}{\partial x}\left(\frac{1}{r}\right) = -\frac{1}{r^2}\frac{\partial r}{\partial x} = -\frac{x-x'}{r^3} \quad \text{i.e.,} \quad \nabla\left(\frac{1}{r}\right) = -\frac{\vec{r}}{r^3}$$

by means of which Eq.(12.21) becomes

$$\vec{B} = -\frac{\mu_0 I}{4\pi}\int_\Gamma \vec{dl} \times \nabla\left(\frac{1}{r}\right) \tag{12.30}$$

The vector identity (I.11), Appendix I, can be written as

$$\vec{dl} \times \nabla\left(\frac{1}{r}\right) = \frac{1}{r}\left(\nabla \times \vec{dl}\right) - \nabla \times \left(\frac{\vec{dl}}{r}\right) = -\nabla \times \left(\frac{\vec{dl}}{r}\right)$$

where $\nabla \times \vec{dl} = 0$, since the differentiation is with respect to x, y, z and $\vec{dl}$ involves the local coordinates x', y', z'. Equation (12.30) then becomes

$$\vec{B} = \frac{\mu_0 I}{4\pi}\int_\Gamma \nabla \times \left(\frac{\vec{dl}}{r}\right) = \frac{\mu_0 I}{4\pi}\nabla \times \int_\Gamma \frac{\vec{dl}}{r} \tag{12.31}$$

because integration with respect to x', y', z' and differentiation with respect to x, y, z can be interchanged. Comparison with Eq.(12.29) shows that $\vec{B}$ is given by the *curl* of the vector potential

$$\vec{A} = \frac{\mu_0 I}{4\pi}\int_\Gamma \frac{\vec{dl}}{r} = \frac{\mu_0}{4\pi}\int_{V'} \frac{\vec{j}}{r}dV' \tag{12.32}$$

which is expressed in terms of either current elements or a current distribution density inside a finite volume V' in space. Equation (12.32) is the magnetostatic analogue of Eq.(11.13) for the electrostatic potential.

Because of the vector identity (I.9), Appendix I, which holds for any scalar function, say φ, $\vec{A} + \nabla\varphi$ will produce the same magnetic induction (12.31) as $\vec{A}$. For the vector function $\vec{A}$ to be defined to within only an additive constant, both its *curl* (12.29) and its divergence $\nabla \cdot \vec{A}$ must be specified. The arbitrary fixing of $\nabla \cdot \vec{A}$ is called the choice of a *gauge* and has no effect on the relation between $\vec{A}$ and $\vec{B}$ which only involves the *curl*. Substituting Eq.(12.29) for $\vec{A}$, into Ampère's law (12.27), we obtain

$$\nabla \times \vec{B} = \nabla \times \left(\nabla \times \vec{A}\right) = \mu_0 \vec{j} \tag{12.33}$$

Using the vector identity (I.19), Appendix I, Eq.(12.33) takes the form

$$\nabla(\nabla \cdot \vec{A}) - \nabla^2 \vec{A} = \mu_0 \vec{j}$$

We may exploit the free choice of the divergence of $\vec{A}$, to eliminate it using the so-called *Coulomb gauge*

$$\nabla \cdot \vec{A} = 0 \qquad (12.34)$$

With this restriction on $\vec{A}$, Ampère's law takes the form

$$\nabla^2 \vec{A} = -\mu_0 \vec{j} \qquad (12.35)$$

which is the magnetostatic equivalent of Poisson's equation (11.14). In regions which contain no current density, Eq.(12.35) reduces to

$$\nabla^2 \vec{A} = 0 \qquad (12.36)$$

a magnetostatic analogue of Laplace's equation (11.15).

The similarities found between the basic equations of magnetic induction $\vec{B}$ and electric field $\vec{E}$ *in free space* can be summarized as follows

(11.3)	$\vec{E} = \dfrac{1}{4\pi\varepsilon_0} \int_{V'} \dfrac{\rho \vec{r}}{r^3} dV'$	$\vec{B} = \dfrac{\mu_0}{4\pi} \int_{V'} \dfrac{\vec{j} \times \vec{r}}{r^3} dV'$	(12.22)
(11.5)	$\nabla \cdot \vec{E} = \dfrac{1}{\varepsilon_0} \rho$	$\nabla \cdot \vec{B} = 0$	(12.23)
(11.7)	$\nabla \times \vec{E} = 0$	$\nabla \times \vec{B} = \mu_0 \vec{j}$	(12.27)
(11.10)	$\vec{E} = -\nabla V$	$\vec{B} = \nabla \times \vec{A}$	(12.29)

The differences are due to the distinct forms of their sources, which are the charge distribution for the electric field and the current distribution for the magnetic induction.

FURTHER READING

1. B. Guru, H. Hiziroglu - ELECTROMAGNETIC FIELD THEORY FUNDAMENTALS, Cambridge University Press, Cambridge, 2004.
2. C. R. Paul, S. A. Nasar - INTRODUCTION TO ELECTROMAGNETIC FIELDS, McGraw-Hill, New York, 1997.
3. E. R. Dobbs - BASIC ELECTROMAGNETISM, Chapman & Hall, London, 1993.

PROBLEMS

12.1. A lossy dielectric shell of relative permittivity ε_r and conductivity σ completely fills the space between two concentric conducting spheres of inner radius R_1 and outer radius R_2. If a charge q_0 is placed on the inner sphere, find the Joule heat dissipated in the dielectric shell.

Solution

The charge flow through a spherical surface of radius r is given by Eq.(12.9) that reads

$$I(r) = 4\pi r^2 \sigma E(r) = 4\pi r^2 \sigma \frac{q}{4\pi\varepsilon r^2} = \frac{\sigma}{\varepsilon} q$$

where use has been made of Gauss's law (11.4) to write down $E(r)$ for $R_1 < r < R_2$ and $\varepsilon = \varepsilon_0 \varepsilon_r$ stands for the dielectric constant. In view of Eq.(12.14) the charge on the inner sphere at time t is given by

$$-\frac{dq}{dt} = \frac{\sigma}{\varepsilon} q \quad \text{or} \quad q = q_0 e^{-\sigma t/\varepsilon}$$

and hence $I = (\sigma q_0/\varepsilon) e^{-\sigma t/\varepsilon}$. The resistance between the two conducting spheres can be obtained by integrating over spherical shells of radius r and thickness dr that carry the same current, and this gives

$$R = \int_{R_1}^{R_2} \frac{dr}{4\pi r^2 \sigma} = \frac{1}{4\pi\sigma}\left(\frac{1}{R_1} - \frac{1}{R_2}\right)$$

We can now integrate Eq.(12.13) to obtain the Joule heat as a function of time in the form

$$Q(t) = \int_0^t RI^2 dt = \frac{\sigma q_0^2}{4\pi\varepsilon^2}\left(\frac{1}{R_1} - \frac{1}{R_2}\right)\int_0^t e^{-2\sigma t/\varepsilon} dt = \frac{q_0^2}{8\pi\varepsilon}\left(\frac{1}{R_1} - \frac{1}{R_2}\right)\left(1 - e^{-2\sigma t/\varepsilon}\right)$$

We can see, for $t \to \infty$, that the total heat dissipated in the dielectric shell, given by

$$Q_0 = \frac{q_0^2}{8\pi\varepsilon}\left(\frac{1}{R_1} - \frac{1}{R_2}\right) = \frac{q_0^2}{2C_1} - \frac{q_0^2}{2C_2}$$

is a result of charge flow from the inner-conducting sphere, of capacitance C_1, to the outer sphere, of capacitance C_2.

12.2. A stable and uniform electron beam is projected onto a homogeneous metallic rod of resistance R_0, so that a total current I_0 is drawn in the circuit. If we set at zero the voltage at the midpoint of the rod, calculate the voltage at each of its terminals.

12. MAGNETIC INDUCTION

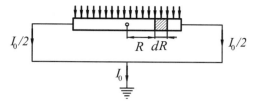

Solution

There will be a linear increase in the electric current I from 0 to $I_0/2$ as we move from the midpoint of the rod toward one of its terminals, because of the steady increase in the carrier concentration. Thus, a conductor element of resistance dR, a distance R from the midpoint, will carry a current $I = R(I_0/2)/(R_0/2)$. The voltage across this element is $dV = IdR$, and hence, as all elements are connected in series, one obtains the voltage at each terminal of the rod in the form

$$V = \int_0^{R_0/2} IdR = \frac{I_0}{R_0} \int_0^{R_0/2} IdR = \frac{1}{8}R_0 I_0$$

12.3. A spherical conducting sheet of radius R and potential V above ground is spinning about a horizontal axis through its centre with angular speed ω. Find the axial magnetic induction at the centre of the sheet.

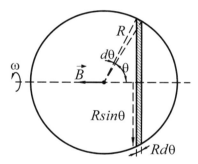

Solution

Consider an annular ring of radius $r = R\sin\theta$, width $Rd\theta$ and area $dS = 2\pi R^2 \sin\theta\, d\theta$. The uniform charge distribution implies that the charge density is given by

$$\frac{dq}{dS} = \frac{q}{4\pi R^2} = \frac{\varepsilon_0 V}{R} \quad \text{or} \quad dq = 2\pi\varepsilon_0 VR \sin\theta d\theta$$

As this charge is spinning at a rate ω, the annular ring turns into a current loop carrying a current

$$dI = \frac{dq}{dt} = \frac{\omega}{2\pi} dq = \varepsilon_0 \omega VR \sin\theta d\theta$$

and producing an axial magnetic induction at the centre of the spherical sheet of the form

$$dB = \frac{\mu_0}{4\pi}\frac{2S}{R^3}dI = \frac{\mu_0}{2R}\sin^2\theta\, dI = \frac{\varepsilon_0\mu_0\omega V}{2}\sin^3\theta\, d\theta$$

Integration over the spherical sheet is immediate and gives the axial magnetic induction at its centre as

$$B = \int_0^\pi dB = \frac{\varepsilon_0\mu_0\omega V}{2}\int_0^\pi \sin^3\theta\, d\theta = \frac{2}{3}\varepsilon_0\mu_0\omega V$$

12.4. A cylindrical conductor of radius R, carrying a current I which is uniformly distributed throughout the cross section, has within it a cylindrical cavity of radius a whose axis is a distance R_0 from that of the conductor. Show that the magnetic induction within the cavity is uniform and determine its magnitude and direction.

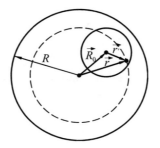

Solution

Let us consider first the magnetic induction inside a cylindrical conductor carrying a uniformly distributed current I. At a radius $r < R$, Ampere's circuital law (12.26) gives the magnitude of B in terms of the current crossing the surface bounded by the path as

$$2\pi r B = \mu_0 I \frac{r^2}{R^2} \quad \text{or} \quad B = \frac{\mu_0 I}{2\pi R^2}r$$

The vector $\vec{B}$ is normal to both the position vector $\vec{r}$ and the current density vector $\vec{j}$, of magnitude $j = I/\pi R^2$, and hence, it can be written as

$$\vec{B} = \frac{\mu_0}{2}\vec{j}\times\vec{r}$$

Note that this is a form suitable to obey the definition constraint (12.27) of the magnetic induction

$$\nabla\times\vec{B} = \nabla\times\left(\frac{\mu_0}{2}\vec{j}\times\vec{r}\right) = \frac{\mu_0}{2}\left[(\nabla\cdot\vec{r})\vec{j} - (\vec{j}\cdot\nabla)\vec{r}\right] = \frac{\mu_0}{2}(3\vec{j}-\vec{j}) = \mu_0\vec{j}$$

Consider now a point within the cavity such that the vector displacement from the axis of the cavity is $\vec{r}' = \vec{r} - \vec{R}_0$, where $\vec{r}$ is the position vector relative to the axis of the conductor. By the principle of superposition we know that the magnetic induction in the cavity is equal to that of the original conductor minus the magnetic induction of the material which has been removed, namely

$$\vec{B}_0 = \frac{\mu_0}{2}\vec{j}\times\vec{r} - \frac{\mu_0}{2}\vec{j}\times(\vec{r}-\vec{R}_0) = \frac{\mu_0}{2}\vec{j}\times\vec{R}_0$$

where $j = I/\pi(R^2 - a^2)$. The magnetic induction within the cavity is thus uniform, of magnitude

$$B_0 = \frac{\mu_0 I R_0}{2\pi(R^2 - a^2)}$$

12.5. Calculate the magnitude of the vector potential inside and outside a cylindrical conductor of radius R carrying a current I which is uniformly distributed throughout the cross section.

Solution

It is convenient to use cylindrical coordinates, as defined in Appendix I, with the z-axis parallel to the axis of the conductor. From Eq.(12.3), it is apparent that the orientation of the vector potential $\vec{A}$ is that of the current density $\vec{j}$, and hence, $A_r = A_\theta = 0$ or $A = A_z$. Under Coulomb gauge (12.34), written down in cylindrical coordinates, Eq.(I.26), it follows that $\partial A_z / \partial z = 0$, thus A is independent of z. At a radial distance r from the axis, the Ampere circuital law yields

$$B_\theta = \frac{\mu_0 I}{2\pi R^2} r \quad (r \le R), \qquad B_\theta = \frac{\mu_0 I}{2\pi}\frac{1}{r} \quad (r \ge R)$$

And, since $B_r = B_z = 0$, the definition (12.29) of the vector potential reduces to

$$0 = \frac{1}{r}\frac{\partial A_z}{\partial \theta} - \frac{\partial A_\theta}{\partial z}, \qquad B_\theta = \frac{\partial A_r}{\partial z} - \frac{\partial A_z}{\partial r}, \qquad 0 = \frac{1}{r}\frac{\partial}{\partial r}(rA_\theta) - \frac{1}{r}\frac{\partial A_r}{\partial \theta}$$

The first equation shows that $\partial A_z / \partial \theta = 0$, and hence, A is also independent of θ. The second equation provides the radial dependence of A as

$$A = -\int B_\theta dr = -\frac{\mu_0 I}{4\pi R^2}r^2 + C \quad (r \le R), \qquad A = -\frac{\mu_0 I}{2\pi}\ln r + C' \quad (r \ge R)$$

The two constants are related by a matching condition at $r = R$, across the surface of the conductor, so that one finally obtains

$$A = \frac{\mu_0 I}{4\pi}\left(1 - \frac{r^2}{R^2}\right) - \frac{\mu_0 I}{2\pi}\ln R + C \qquad r \le R$$

$$A = -\frac{\mu_0 I}{2\pi}\ln r \qquad r \ge R$$

12.6. A parallel-plate capacitor is entirely filled with a slab of lossy dielectric of permittivity ε. If it is found that the voltage across the capacitor decreases to $1/n$ of its original value after a time τ, determine the conductivity σ of the dielectric material.

12.7. A sphere of lossy dielectric of radius R, permittivity ε and conductivity σ is uniformly charged throughout its volume with charge density ρ_0. Obtain an expression for the current density as a function of time and position within the dielectric sphere.

12.8. Use the Biot-Savart law to find the axial magnetic induction at the centre of an n-sided regular polygonal loop carrying a current I.

12.9. A coaxial conductor circuit consists of a metal rod concentrically surrounded by a metallic cylindrical sheath of inner radius R_1 and outer radius R_2, so that the same current I passes in opposite directions through the rod and the sheath. If the current density in each of the two conductors is uniform throughout the cross section, determine the radial distribution of the magnetic induction within the sheath.

12.10. A uniformly charged disc of radius R carrying a total charge q is set rotating at a constant angular speed ω about its axis of symmetry. Find the magnetic induction along the axis at a distance x from the plane of the disc.

13. MAGNETIC FIELDS

13.1. MAGNETIZATION
13.2. FARADAY'S LAW
13.3. ENERGY IN MAGNETIC FIELDS

13.1. MAGNETIZATION

In the presence of neutral matter in the space between current distributions the laws of magnetostatics can be modified in close parallel to those of electrostatics, by introducing a vector field independent of $\vec{B}$ and $\vec{A}$. The macroscopic polarization $\vec{P}$ of a volume of dielectric was defined in electrostatics in terms of the induced electric dipole moments of atoms or molecules in the dielectric. In a similar manner, neutral matter can be regarded in magnetostatics as an assembly of magnetic dipole moments that give rise to a macroscopic magnetization $\vec{M}$. We make the assumption that the magnetic dipole moments are produced by *circulating* atomic charges instead of static charges, without differentiating between the diamagnetic, paramagnetic and ferromagnetic behaviour of materials in external magnetic induction $\vec{B}$. This will be dealt with in Chapter 44.

Let us model a current loop which gives rise to a $\vec{B}$ field at a point P with coordinates x, y and z, as a summation of the contributions of its elements $\vec{dl}$ with local coordinates x', y' and z'. The magnetic induction $\vec{B}$ at P can be expressed in terms of the solid angle Ω of the cone associated with the area of the loop, as represented in Figure 13.1. The same variation $d\Omega$ is obtained whether we move P through $\vec{dr}$ relative to the loop or move each element $\vec{dl}$ by an amount $\vec{dr'} = -\vec{dr}$, keeping P fixed. The second alternative has a simple geometrical interpretation. The element $\vec{dl}$ sweeps out an area $\vec{dr'} \times \vec{dl} = -\vec{dr} \times \vec{dl}$, in moving through the distance $\vec{dr'}$, so that the scalar product $\left(-\vec{dr} \times \vec{dl}\right) \cdot \vec{e}_r$ gives the component of the area vector along the direction of $\vec{r}$, which is the projection of the area itself on a plane normal to $\vec{r}$.

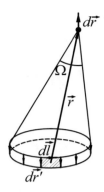

Figure 13.1. Configuration for the calculation of $\vec{B}$ in terms of a solid angle associated with a closed loop

The area projection divided by r^2 defines the solid angle subtended by that area at P. Then $d\Omega$ is represented by the solid angle which the band-like surface, corresponding to the displacement of the whole current loop, intercepts at the point P, given by

$$d\Omega = \int \frac{(-d\vec{r} \times d\vec{l}) \cdot \vec{e}_r}{r^2} = -d\vec{r} \cdot \int_\Gamma \frac{d\vec{l} \times \vec{e}_r}{r^2} = -d\vec{r} \cdot \int_\Gamma \frac{d\vec{l} \times \vec{r}}{r^3} \qquad (13.1)$$

In Cartesian coordinates, the solid angle Ω can be expressed as a function of space coordinates x, y and z, so that we formally have

$$d\Omega = \frac{\partial \Omega}{\partial x}dx + \frac{\partial \Omega}{\partial y}dy + \frac{\partial \Omega}{\partial z}dz = grad\Omega \cdot d\vec{r} = \nabla\Omega \cdot d\vec{r}$$

and Eq.(13.1) becomes

$$\nabla\Omega = -\int_\Gamma \frac{d\vec{l} \times \vec{r}}{r^3} \qquad (13.2)$$

Comparison with the integral form (12.21) shows that the Biot-Savart law reads

$$\vec{B} = -\frac{\mu_0 I}{4\pi}\nabla\Omega = -\nabla\left(\frac{\mu_0 I \Omega}{4\pi}\right) = -\nabla\phi \qquad (13.3)$$

and gives the magnetic induction produced by a current loop at a given point in terms of the space rates of change of the solid angle which the surface of the loop intercepts at that point. It is now convenient to interpret this result in the sense of Eq.(12.28), that is the magnetic induction $\vec{B}$ due to a current loop can be derived from a scalar magnetostatic potential defined as

$$\phi = \frac{\mu_0 I \Omega}{4\pi} \tag{13.4}$$

Assuming a small loop, as given in Figure 13.1, of area S, so that $\sqrt{S} \ll r$, Eq.(13.4) becomes

$$\phi = \frac{\mu_0 I}{4\pi}\left(\frac{S\cos\theta}{r^2}\right) = \frac{\mu_0}{4\pi}\frac{(I\vec{S})\cdot\vec{r}}{r^3}$$

This expression should be compared with Eq.(11.12) for the electric dipole, to which it becomes identical, if $\vec{p}/\varepsilon_0$ is replaced by $\mu_0 I\vec{S}$. Thus a small current loop behaves as a dipole source for the magnetic field, given by a *magnetic dipole moment* $\vec{\mu}$ of the form

$$\vec{\mu} = I\vec{S} \tag{13.5}$$

which is defined as a vector whose magnitude is the product of the current and the area of the loop. Hence we have

$$\phi = \frac{\mu_0}{4\pi}\frac{\vec{\mu}\cdot\vec{r}}{r^3} \tag{13.6}$$

and this is the magnetostatic analogue of Eq.(11.12).

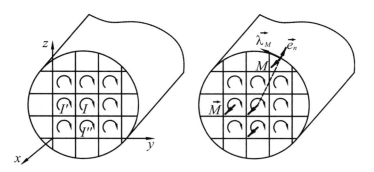

Figure 13.2. Magnetization current density in a volume element and on its surface

The description of neutral matter, introduced in electrostatics, as an aggregate of atoms in fixed average positions, consisting of positively charged nuclei surrounded by negative electron clouds, is now completed by the assumption that circulation of *negative electrons* around closed paths accounts for the magnetic properties. For an assembly of such elementary current loops, each of dipole moment $\vec{\mu}$, located in an elementary volume V which is small compared to macroscopic dimensions but large compared to atomic dimensions, it is convenient to define the magnetic moment per unit volume as

$$\vec{M} = \frac{1}{V}\sum \vec{\mu} \qquad (13.7)$$

This is called the *magnetization* and is the analogue of the electrostatic polarization $\vec{P}$ introduced by Eq.(11.23). A physical picture of a volume V of magnetized material is given in Figure 13.2 in a Cartesian coordinate system. Assuming a rate of change of magnetization, three neighbouring volume elements, each of dimensions $dV = dxdydz$, will have magnetic moments $d\vec{\mu}, d\vec{\mu}', d\vec{\mu}''$ which must be associated with different currents I, I', I''. We can write the *x*-components of these magnetic moments as

$$d\mu_x = Idydz = M_x dV$$

$$d\mu'_x = I'dydz = \left(M_x + \frac{\partial M_x}{\partial y}dy\right)dV$$

$$d\mu''_x = I''dydz = \left(M_x + \frac{\partial M_x}{\partial z}dz\right)dV$$

so that the rate of change of M_x is related to a net current flow in both *z*- and *y*-directions, which reads

$$(I - I')dydz = \left(-\frac{\partial M_x}{\partial y}dy\right)dxdydz \quad \text{or} \quad I_{(x)z} = I - I' = -\frac{\partial M_x}{\partial y}dxdy$$

$$(I'' - I)dydz = \left(\frac{\partial M_x}{\partial z}dz\right)dxdydz \quad \text{or} \quad I_{(x)y} = I'' - I = \frac{\partial M_x}{\partial z}dxdz$$

By similar arguments the rate of change of M_y must be related to a net current flow in both the *x*- and *z*-directions given by

$$I_{(y)z} = -\frac{\partial M_y}{\partial z}dydz \quad , \quad I_{(y)z} = \frac{\partial M_x}{\partial x}dxdy$$

and a net current flow along the *x*- and *y*-axes is associated to the change in M_z as

$$I_{(z)x} = \frac{\partial M_z}{\partial y}dydz \quad , \quad I_{(z)y} = -\frac{\partial M_z}{\partial x}dxdz$$

It follows that we can introduce a magnetization current density $\vec{j}_M$ of components

$$j_{M_x} = \frac{I_{(z)x} + I_{(y)x}}{dydz} = \frac{\partial M_z}{\partial y} - \frac{\partial M_y}{\partial z} \quad , \quad j_{M_y} = \frac{\partial M_x}{\partial z} - \frac{\partial M_z}{\partial x} \quad , \quad j_{M_z} = \frac{\partial M_x}{\partial y} - \frac{\partial M_y}{\partial x}$$

which can be written in a compact form as

$$\vec{j}_M = \nabla \times \vec{M} \tag{13.8}$$

There is a *magnetization current density* $\vec{j}_M$ arising from atomic current loops wherever there is a space rate of change of magnetization, a situation analogous to that described by Eq.(11.24) in a dielectric, where the polarization charge requires that the divergence of polarization should not be zero. Hence, in a uniformly magnetized material, the magnetization current density is zero. However the internal cancellation between currents associated with neighbouring dipoles does not occur at the surface, and the net result is a *surface current density*, defined as

$$\vec{\lambda}_M = \vec{M} \times \vec{e}_n$$

which is illustrated in Figure 13.2. It is obvious that $\vec{j}_M$ and $\vec{\lambda}_M$ are *bound* current densities, arising from bound current loops.

When using Ampère's law to investigate the magnetic induction in the presence of a magnetic material, it is necessary to include contributions from the magnetization currents of density $\vec{j}_M$ so that, reserving the symbol $\vec{j}$ for the free current density only, Eq.(12.27) reads

$$\nabla \times \vec{B} = \mu_0 \left(\vec{j} + \vec{j}_M \right) = \mu_0 \left(\vec{j} + \nabla \times \vec{M} \right) \tag{13.9}$$

For convenience this relation is rearranged so that only the free current density appears on the right-hand side, that is

$$\nabla \times \left(\frac{\vec{B}}{\mu_0} - \vec{M} \right) = \vec{j} \tag{13.10}$$

and Ampère's law in a magnetic material takes the differential form

$$\nabla \times \vec{H} = \vec{j} \tag{13.11}$$

provided we define the *magnetic field* $\vec{H}$ as

$$\vec{H} = \frac{\vec{B}}{\mu_0} - \vec{M} \tag{13.12}$$

This is the analogue of the electric displacement $\vec{D}$ in a dielectric, given by Eq.(11.27).

The field $\vec{H}$ must be associated with a particular magnetic material, whereas $\vec{B}$ is defined irrespective of its presence. The integral form of Ampère's law in terms of the magnetic field $\vec{H}$ follows from Eq.(13.11), using Stokes theorem, as

$$\oint_\Gamma \vec{H} \cdot d\vec{l} = \int_S (\nabla \times \vec{H}) \cdot d\vec{S} = \int_S \vec{j} \cdot d\vec{S} \qquad (13.13)$$

A *linear isotropic* magnetic material is defined by the linear dependence of magnetization $\vec{M}$ on $\vec{H}$ written as

$$\vec{M} = \chi_m \vec{H} \qquad (13.14)$$

where χ_m is called the *magnetic susceptibility*. Then the relationship (13.12) between $\vec{B}$, $\vec{H}$ and $\vec{M}$ becomes

$$\vec{B} = \mu_0 (1 + \chi_m) \vec{H} = \mu_0 \mu_r \vec{H} \qquad (13.15)$$

which is the *constitutive* equation of magnetostatics, analogous to Eq.(11.28). The dimensionless parameter

$$\mu_r = 1 + \chi_m \qquad (13.16)$$

is known as the *relative permeability* of a magnetic material.

13.2. FARADAY'S LAW

A connection between electric field and magnetic induction in free space was experimentally discovered by Faraday when he showed that a potential difference appears across the ends of a conductor or in a closed loop, if moving in a magnetic field. The effect was ascribed to the change in the total magnetic flux linked with the area of the loop, defined as

$$\Phi = \int_S \vec{B} \cdot d\vec{S} \qquad (13.17)$$

If Γ is the closed path along the loop, from the definition (12.29) of the vector potential and using Stoke's theorem, Eq.(13.17) can be rewritten as

$$\Phi = \int_S (\nabla \times \vec{A}) \cdot d\vec{S} = \int_\Gamma \vec{A} \cdot d\vec{l} \qquad (13.18)$$

The potential difference in any closed loop is given in terms of the *electromotive force*, defined as the line integral of the electric field

$$e.m.f. = \int_\Gamma \vec{E} \cdot d\vec{r} \qquad (13.19)$$

as illustrated in Figure 13.3.

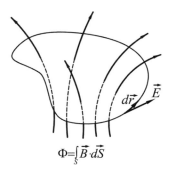

Figure 13.3. Faraday's law

Faraday's law states that the *e.m.f. around any closed path is minus the time rate of change of the magnetic flux over an arbitrary surface bounded by the path*, or

$$\int_\Gamma \vec{E} \cdot d\vec{r} = -\frac{d}{dt} \int_S \vec{B} \cdot d\vec{S} \qquad (13.20)$$

and reduces to the circulation law (11.6) for a conservative electrostatic field, when the magnetic flux linking the path is constant in time. The induced *e.m.f.* causes currents in a closed circuit following the path Γ which tend to oppose the original change of flux, as established by *Lenz's law*.

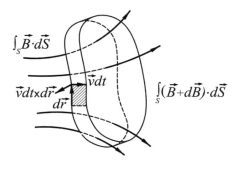

Figure 13.4. Motion of a closed loop of surface area S in a field $\vec{B}$

It is convenient to separate the flux change through a static loop due to a change with time of magnetic induction from that due to the motion of the loop in a stationary magnetic induction, by considering a loop of surface S that moves as represented in Figure 13.4 during a time interval dt. The magnetic induction at a given element $d\vec{r}$ changes from $\vec{B}$, at time t, to $\vec{B}+d\vec{B}$ at $t+dt$. The displacement of each element $d\vec{r}$ with velocity $\vec{v}$ sweeps out an elementary surface $\vec{v}dt \times d\vec{r}$, so that the surface S is augmented by a ribbon surface between the two successive positions of the loop. The change in the magnetic flux through the loop can be written, to a first approximation, as

$$d\Phi = \int_S (\vec{B}+d\vec{B}) \cdot d\vec{S} + \int_\Gamma (\vec{B}+d\vec{B}) \cdot (\vec{v}dt \times d\vec{r}) - \int_S \vec{B} \cdot d\vec{S}$$

$$\cong \int_S d\vec{B} \cdot d\vec{S} + dt \int_\Gamma \vec{B} \cdot (\vec{v} \times d\vec{r}) = \int_S d\vec{B} \cdot d\vec{S} - dt \int_\Gamma (\vec{v} \times \vec{B}) \cdot d\vec{r}$$

$$= dt \left[\int_S \frac{\partial \vec{B}}{\partial t} \cdot d\vec{S} - \int_\Gamma (\vec{v} \times \vec{B}) \cdot d\vec{r} \right]$$

Substituting into Eq.(13.20) gives

$$\int_\Gamma \vec{E} \cdot d\vec{r} = -\int_S \frac{\partial \vec{B}}{\partial t} \cdot d\vec{S} + \int_\Gamma (\vec{v} \times \vec{B}) \cdot d\vec{r} \tag{13.21}$$

This can be turned into a differential law by applying Stoke's theorem

$$\int_S (\nabla \times \vec{E}) \cdot d\vec{S} = -\int_S \frac{\partial \vec{B}}{\partial t} \cdot d\vec{S} + \int_S \nabla \times (\vec{v} \times \vec{B}) \cdot d\vec{S}$$

which is valid for any surface however small the loop, so that

$$\nabla \times \vec{E} = -\frac{\partial \vec{B}}{\partial t} + \nabla \times (\vec{v} \times \vec{B}) \tag{13.22}$$

This is the differential form of *Faraday's law*, which reduces for static circuits to

$$\nabla \times \vec{E} + \frac{\partial \vec{B}}{\partial t} = 0 \tag{13.23}$$

From the definition of $\vec{A}$, Eq.(12.29), we can rewrite this equation in the form

$$\nabla \times \left(\vec{E} + \frac{\partial \vec{A}}{\partial t} \right) = 0 \tag{13.24}$$

which defines a *conservative* vector field, that can be derived from a scalar potential. If this is chosen to be the electrostatic potential V, one obtains

$$\vec{E} + \frac{\partial \vec{A}}{dt} = -\nabla V \qquad (13.25)$$

Eq.(13.25) reduces to Eq.(11.10) in the static case, and therefore, $\vec{E}$ depends on both $\vec{A}$ and V in the general case.

EXAMPLE 13.1. Inductance

Let us consider two *static* circuits Γ_1 and Γ_2 carrying currents I_1 and I_2. An *e.m.f.* can be induced in one of the circuits by a time variation of $\vec{B}$ which is obtained by changing the current in the other circuit, as represented in Figure 13.5. The flux through circuit Γ_1, linked to the varying current I_2, is given by Eq.(13.18), which reads

$$\Phi_1 = \int_{\Gamma_1} \vec{A}_2 \cdot d\vec{l}_1$$

where the vector potential $\vec{A}_2$ can be expressed in terms of the current I_2 by Eq.(12.32), so that

$$\Phi_1 = \int_{\Gamma_1} \left[\frac{\mu_0}{4\pi} \int_{\Gamma_2} \frac{I_2 \cdot d\vec{l}_2}{r} \right] \cdot d\vec{l}_1 = \frac{\mu_0 I_2}{4\pi} \int_{\Gamma_1} \int_{\Gamma_2} \frac{d\vec{l}_1 \cdot d\vec{l}_2}{r} = MI_2$$

where r is the distance between current elements $d\vec{l}_1, d\vec{l}_2$ of the two circuits.

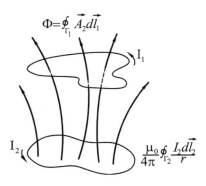

Figure 13.5. Mutual inductance of two static circuits

The symmetrical equation

$$M = \frac{\mu_0}{4\pi} \int_{\Gamma_1} \int_{\Gamma_2} \frac{d\vec{l}_1 \cdot d\vec{l}_2}{r} \qquad (13.26)$$

is known as *Neumann's formula* for the *mutual inductance* M of the two circuits and this, if multiplied by the current in one circuit, gives the magnetic flux threading the other one. Since M is a geometric factor, the induced *e.m.f.* is linked to the varying current in the form

$$(e.m.f.)_1 = -M\frac{dI_2}{dt} \quad , \quad (e.m.f.)_2 = -M\frac{dI_1}{dt}$$

For a single circuit, a *self-inductance* L will be defined by

$$\Phi = LI \tag{13.27}$$

linking the change with time of the current flow to the set up of a *back e.m.f.* which tends to oppose the current change, that is

$$e.m.f. = -L\frac{dI}{dt} \tag{13.28}$$

The currents caused by the induced *e.m.f.* in a closed circuit following a path Γ can also be described in terms of forces acting on single charges in the presence of magnetic $\vec{B}$ and electric fields $\vec{E}$, known as Lorentz forces.

EXAMPLE 13.2. The Lorentz force law

If there is a force $\vec{F}$ acting on a closed circuit Γ, so that a linear displacement $d\vec{r}$ takes place in a time dt, the energy stored by the circuit is equal to minus the work done by the force, according to the energy conservation law (11.8). The potential energy is expressed in terms of the energy supplied by an *e.m.f.*, in a time dt, maintaining its own current

$$\vec{F} \cdot d\vec{r} = -(e.m.f.)I dt$$

Assuming a displacement in a constant magnetic induction, and taking the *e.m.f.* as given by Eq.(13.21), we obtain

$$\vec{F} \cdot d\vec{r} = -Idt \int_\Gamma (\vec{v} \times \vec{B}) \cdot d\vec{r} = Idt \int_\Gamma (d\vec{r} \times \vec{B}) \cdot \vec{v}$$

Setting $d\vec{r}/dt = \vec{v}$, which holds for a translational motion of the circuit, we can write

$$\vec{F} \cdot \vec{v} = \left[I \int_\Gamma d\vec{r} \times \vec{B} \right] \cdot \vec{v}$$

The result is valid for any velocity $\vec{v}$, so that we can express the force as

$$\vec{F} = \int_\Gamma I d\vec{r} \times \vec{B}$$

as a sum over the contributions from forces acting on current elements, each given by the differential law

$$d\vec{F} = Id\vec{r} \times \vec{B} \qquad (13.29)$$

Eq.(13.29) can be generalized as a volume integral over a current distribution of density (12.7), which gives

$$\vec{F} = \int_V (\vec{j} \times \vec{B})dV = \int_V nq(\vec{v} \times \vec{B})dV \to \sum q(\vec{v} \times \vec{B})$$

where $\vec{j}$ is given in terms of the number n of particles per unit volume, each carrying a charge q moving with velocity $\vec{v}$. The total force is now interpreted as a sum of contributions from forces acting on single charges

$$\vec{F}_L = q(\vec{v} \times \vec{B})$$

called *Lorentz forces*. The trajectory of charged particles in fields $\vec{E}$ and $\vec{B}$ is the solution of Newton's equation of motion (2.9) which reads

$$m\frac{d\vec{v}}{dt} = q(\vec{E} + \vec{v} \times \vec{B})$$

where

$$\vec{F} = q(\vec{E} + \vec{v} \times \vec{B}) \qquad (13.30)$$

is called the *Lorentz force law*.

13.3. Energy in Magnetic Fields

The energy stored in a circuit of geometry defined by its self-inductance L, is equal to the work done against the back *e.m.f.* in order to maintain a given current, that is

$$dU = -(e.m.f.)Idt = LIdI \qquad (13.31)$$

Starting with no current and building up to a final value I, the energy becomes

$$U = \frac{1}{2}LI^2 \qquad (13.32)$$

This equation can be generalized by combining Eqs.(13.18) and (13.27) which give

$$LI = \int_\Gamma \vec{A} \cdot d\vec{l}$$

so that Eq.(13.32) transforms as

$$U = \frac{1}{2}I\int_\Gamma \vec{A} \cdot d\vec{l} = \frac{1}{2}\int_\Gamma \vec{I} \cdot \vec{A} dl$$

where we made use of $I d\vec{l} = \vec{I} dl$. For an arbitrary system of currents we then obtain

$$U = \frac{1}{2}\int_V \vec{j} \cdot \vec{A} dV$$

In the presence of magnetic materials, the free-current density $\vec{j}$ is given by Ampère's law (13.11), so that

$$U_H = \frac{1}{2}\int_V (\nabla \times \vec{H}) \cdot \vec{A} dV = \frac{1}{2}\int_V \vec{H} \cdot (\nabla \times \vec{A}) dV - \int_V \nabla \cdot (\vec{A} \times \vec{H}) dV$$

The second term on the right-hand side can be transformed according to Gauss's theorem and vanishes, that is

$$\int_V \nabla \cdot (\vec{A} \times \vec{H}) dV = \int_S (\vec{A} \times \vec{H}) \cdot d\vec{S} = 0$$

since the surface S can be chosen to be an infinite sphere and both the magnetic field and the vector potential fall off as $1/r^2$ at large distances, whereas the surface area increases as r^2. Substituting Eq.(12.29) in the first term on the right-hand side, the energy becomes

$$U_H = \frac{1}{2}\int_V \vec{H} \cdot \vec{B} dV \qquad (13.33)$$

Therefore the energy density associated with a magnetostatic field is

$$u_H = \frac{1}{2}\vec{H} \cdot \vec{B} \qquad (13.34)$$

FURTHER READING
1. D. Dugdale - ESSENTIALS OF ELECTROMAGNETISM, Springer-Verlag, New York, 1998.
2. D. J. Griffiths - INTRODUCTION TO ELECTRODYNAMICS, Prentice Hall, Englewood Cliffs, 1990.
3. W. J. Duffin - ELECTRICITY AND MAGNETISM, McGraw-Hill, New York, 1990.

PROBLEMS

13.1. A homogeneous solid of mass m, uniformly charged throughout its volume with a total charge q, is set rotating at a constant angular speed ω about an arbitrary axis. Find the gyromagnetic ratio of the magnetic dipole moment to the axial angular momentum of the solid.

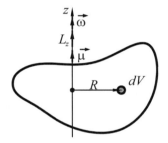

Solution

A volume element dV of mass $dm = (m/V)dV$ and charge $dq = (q/V)dV$, a distance R from the axis, develops a current loop, and hence, a magnetic dipole moment (13.5) given by

$$d\mu = SdI = \pi R^2 \frac{\omega}{2\pi} dq = \frac{\omega}{2} R^2 \frac{q}{V} dV = \frac{\omega}{2} R^2 \frac{q}{m} dm$$

Integrating over the homogeneous solid and taking the z-direction to coincide with that of the axis yields

$$\mu = \frac{\omega}{2} \frac{q}{m} \int R^2 dm = \frac{\omega}{2} \frac{q}{m} I_z$$

where I_z stands for the moment of inertia with respect to the axis. As the axial angular momentum is given by $L_z = I_z \omega$, it follows that the gyromagnetic ratio has the form

$$\frac{\mu}{L_z} = \frac{q}{2m}$$

irrespective of the size and shape of the homogeneous solid.

13.2. Two vertical metal rails, separated by a distance l, run parallel to the z-axis. At $z = 0$ an inductance L is connected between the rails and a closed circuit is formed by a metal rod of mass m which slides along the rails. If a constant magnetic induction B is perpendicular to the plane of the rails, find the equation of vertical motion of the rod. Solve the same problem assuming that a capacitance C is connected between the rails. Neglect the resistance of the rails and the rod.

Solution

The magnetic flux linked by the circuit is Blz, so that the induced e.m.f. is the rate of the change of the flux linked, or $Bldz/dt = Blv$. Let I be the current flowing in the circuit, such that the force acting on the straight rod of length l carrying it perpendicular to the magnetic induction B is $F = BIl$. Kirchhoff's loop rule yields

$$Bl\frac{dz}{dt} - L\frac{dI}{dt} = 0 \quad \text{or} \quad I = \frac{Bl}{L}z$$

and hence, the equation of vertical motion reads

$$m\frac{d^2z}{dt^2} = mg - \frac{B^2l^2}{L}z = -\frac{B^2l^2}{L}\left(z - \frac{mgL}{B^2l^2}\right) = -k(z - z_0)$$

This is the equation of simple harmonic motion about the equilibrium position $z_0 = mgL/B^2l^2$, where

$$k = \frac{B^2l^2}{L}, \quad \omega = \sqrt{\frac{k}{m}} = \frac{Bl}{\sqrt{mL}}$$

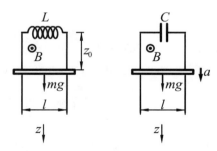

If a capacitance C is connected between the two rails, Kirchhoff's loop rule takes the form

$$Bl\frac{dz}{dt} - \frac{q}{C} = 0 \quad \text{or} \quad I = BlC\frac{d^2z}{dt^2}$$

and the equation of motion becomes

$$m\frac{d^2z}{dt^2} = mg - B^2l^2C\frac{d^2z}{dt^2} \quad \text{or} \quad a = \frac{d^2z}{dt^2} = \frac{mg}{m + B^2l^2C}$$

Thus, the rod merely accelerates in the z-direction.

13.3. A long close-wound solenoid consists of N_0 turns per unit length of a cylinder of radius R. What is the maximum current which passes through this solenoid without exceeding the maximum tension T_0 that its wire can support.

Solution

If the solenoid is long enough for the end effects to be completely neglected, we may consider, to a first approximation, a uniform magnetic induction $B = \mu_0 N_0 I$ inside the solenoid.

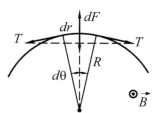

It follows that a centrifugal force (13.29) is acting on a current element $dr = Rd\theta$ of any given turn

$$dF = IBdr = IBRd\theta$$

giving rise to a centripetal reaction $2T \sin(d\theta/2) \cong Td\theta$. The condition of radial equilibrium yields an expression for the tension in the cable

$$T = \mu_0 N_0 I^2 R$$

which should not exceed T_0, and this gives $I \leq \sqrt{T_0 / \mu_0 N_0 R}$.

13.4. An inductance L and a capacitance C are connected in series with two metal balls of radius R. If a charge is put on the capacitor at some instant such that the voltage across the capacitor is V_0, determine the maximum current flowing through the inductance.

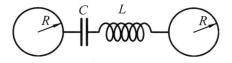

Solution

A short time after charging the capacitor, there is a current I in the circuit, a charge q and $-q$, respectively, on the plates and a charge $q_0 - q$ and $-q_0 + q$, respectively, on the metal balls, where $q_0 = CV_0$. By energy conservation one obtains

$$\frac{1}{2}LI^2 = \frac{q_0^2}{2C} - \frac{q^2}{2C} - \frac{1}{4\pi\varepsilon_0 R}(q_0 - q)^2$$

and hence, the current is a function $f(q)$. The condition that this is a maximum is $df/dq = 0$, so we differentiate f with respect to q to get

$$\frac{df}{dq} = -\frac{q}{C} + \frac{1}{2\pi\varepsilon_0 R}(q_0 - q) \quad , \quad \frac{d^2 f}{dq^2} = -\frac{1}{C} - \frac{1}{2\pi\varepsilon_0 R} < 0$$

Thus we have a maximum for I, when

$$q = q_0 \frac{C}{C + 2\pi\varepsilon_0 R} \quad \text{or} \quad \frac{1}{2} L I_{max}^2 = \frac{\pi\varepsilon_0 R q_0^2}{C(C + 2\pi\varepsilon_0 R)}$$

and this gives

$$I_{max} = q_0 \sqrt{\frac{2\pi\varepsilon_0 R}{LC(C + 2\pi\varepsilon_0 R)}} = V_0 \sqrt{\frac{2\pi\varepsilon_0 RC}{L(C + 2\pi\varepsilon_0 R)}}$$

13.5. Consider a cylindrical sample of magnetic material with a solenoid of length l closely wound over it, consisting of N_0 turns per unit length and carrying a current I. The solenoid provides a uniform magnetic field $H = N_0 I$ parallel to its axis and the magnetic induction B is measured by the magnetic flux Φ through the cross section of the sample. As the current through the solenoid is increased, then decreased and increased again, one obtains a closed B-H diagram called the hysteresis loop. Show that the energy expended in taking the magnetic material around a hysteresis cycle is proportional to the area of the loop.

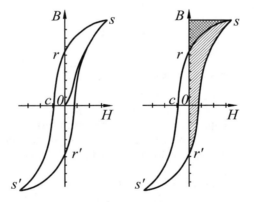

Solution

In a linear magnetic material, the constitutive equation (13.15) shows that the alignment of atomic dipoles is only maintained by an applied magnetic induction. If magnetization is observed even in the absence of external magnetic induction, indicating an alignment of the atomic magnetic dipoles due to some internal interaction, the relationship between $\vec{B}$ and $\vec{H}$ is no longer linear. Such a relationship can be experimentally obtained using a cylindrical sample of magnetic material with a solenoid of length l close-wound over it, consisting of N_0 turns per unit length and carrying a current I. The solenoid provides a uniform magnetic field parallel to its axis $H = N_0 I$. The magnetic induction B is measured in terms of the magnetic flux Φ through the cross sectional area A of the sample. As the current through the solenoid increases we obtain the magnetization curve Os (see the figure) which is characteristic for an unmagnetized material, corresponding to the alignment of magnetic dipoles, until the saturation point s is reached. A differential permeability $dB/\mu_0 dH$ is defined from the curve Os rather than the constant permeability in a linear material. If the current, and therefore the magnetic field H, is reduced to zero, B decreases along the curve sr, whose intercept on the B axis is called the *remanence*. It is a measure of the remaining magnetic flux when the magnetic field vanishes. The remanence can be eliminated by reversing the direction of current and increasing it until the intercept c on the H axis is reached. The corresponding magnitude of H at c is called the *coercivity force* and is a measure of the field required to demagnetize the sample. A further increase of I leads to the alignment of all dipoles in the opposite direction: s' is a saturation point symmetrical to s. If the current is now switched off, the sample is left with a permanent magnetization corresponding to point r'. We may then increase the current again in the positive sense until the saturation point s is reached. The closed curve drawn in the figure is called a hysteresis loop. The energy expended in taking the magnetic material around a hysteresis cycle is proportional to the area of the loop.

If we consider a cycle which starts with the conditions corresponding to point r' shown in the figure, when both the current and the field are increasing from r' to s, the magnetic induction will also be increased, generating a back *e.m.f.* which opposes the increase in current

$$e.m.f. = -\frac{d}{dt}\int B dS = -(N_0 l) S \frac{dB}{dt} = -N_0 V \frac{dB}{dt}$$

where $N_0 l$ is the number of turns of the solenoid, and $V = lS$ is the volume of the sample. The energy spent is measured by the work (13.31) done against the back *e.m.f.*, so that

$$dU = -(e.m.f.)I dt = N_0 V I dB = V H dB$$

We then obtain the energy required in going from r' to s as

$$U_{r's} = V \int_{r'}^{s} H dB$$

which is represented in the figure by the shaded area. When the current is decreasing in the same direction from s to r, the polarity of the back *e.m.f.* is reversed, and the energy

$$U_{sr} = -V \int_{s}^{r} H dB$$

is returned from the sample. The corresponding area must be subtracted from the shaded one and the result is equal to V times half the area contained within the hysteresis loop. In a similar manner the net energy supplied to execute the second half of the cycle is found to be equal to V times the second half of the loop, so that the energy required for one cycle is

$$U = V \int_{\Gamma} H dB$$

where the line integral is evaluated around the hysteresis loop.

For a permanent magnet the energy measured by the area of the loop should be high, requiring high remanence and coercivity. For a material subjected to an alternative current, for instance in a transformer, the power dissipated is proportional to the frequency v, and this gives the number of hysteresis cycles per second as

$$P = \nu V \int_\Gamma H dB$$

Thus, the hysteresis loss is minimized for a narrow hysteresis loop which requires low coercivity and high saturation.

13.6. The center of a horizontal square loop of metallic wire, with sides of length a parallel to the x- and y-axes, respectively, is located at the origin of a Cartesian coordinate system, in a region where the magnetic induction is defined by its components $B_x = -\alpha x$, $B_y = 0$ and $B_z = \alpha z + B_0$. The loop has mass m, self-inductance L, resistance $R = 0$, and there is no current flowing in the circuit. When the loop is released, determine the angular frequency of its vertical harmonic motion.

13.7. A conducting rod of length l rotates about one end, with the other end sliding along a semi-circular metal rail in a plane perpendicular to a uniform magnetic induction B. A resistor R is connected between the rod and the rail. Assuming a constant angular speed ω, what force F should be applied in order to set the rod rotating?

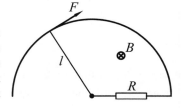

13.8. A beam of electrons of mass m_e, charge $-e$ and velocity v is projected radially inwards through the turns of a long solenoid of radius R, consisting of N_0 turns per unit length which carry a current I. As the beam is deflected in a plane perpendicular to the axial magnetic induction, determine the time of the flight τ of the electrons inside the solenoid.

13.9. Two identical capacitors of capacitance C, connected in series to an e.m.f. E, are removed from the battery and reconnected in an oscillating circuit containing an inductor of inductance L. If one of the capacitors experiences, after a time τ, an electrical break-through, such that the resistance between its plates drops to zero, find the amplitude Q_0 of the oscillation of charge on the other capacitor.

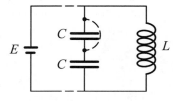

13.10. A capacitor of capacitance C is connected to an oscillating circuit containing two inductors of inductance L_1 and L_2 in parallel. If the initial voltage across the inductors was V_0, determine the amplitudes I_1 and I_2 of the oscillation currents through the inductors.

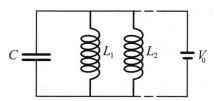

14. Maxwell's Equations

14.1. MAXWELL'S FIELD EQUATIONS
14.2. ELECTROMAGNETIC ENERGY
14.3. POTENTIAL EQUATIONS

14.1. Maxwell's Field Equations

There are four fundamental equations for electromagnetic phenomena that play a role similar to that of Newton's laws for classical mechanics or that of the four principles in thermodynamics, giving a compact and consistent formulation of the laws governing electric and magnetic fields:

1. *Gauss's law* (11.26) in a dielectric

$$\nabla \cdot \vec{D} = \rho$$

2. *Magnetic flux law* (12.23) of the form

$$\nabla \cdot \vec{B} = 0$$

3. *Faraday's law* (13.23) for media at rest

$$\nabla \times \vec{E} = -\frac{\partial \vec{B}}{\partial t}$$

Here $\vec{E}$ is not a conservative field, but depends on both $\vec{A}$ and V, as given by Eq.(13.25), and reduces to the conservative field (11.10) in the static case. It is noteworthy that Eq.(13.23) is consistent with Eq.(12.23) since taking the divergence of each side of Eq.(13.23) gives

$$\nabla \cdot (\nabla \times \vec{E}) = \nabla \cdot \left(-\frac{\partial \vec{B}}{dt} \right) = -\frac{\partial}{dt}(\nabla \cdot \vec{B})$$

where both sides vanish, the left-hand side because divergence of *curl* is always zero and the right-hand side because of Eq.(12.23).

4. *Ampère's law* in a magnetic material *modified by Maxwell* as

$$\nabla \times \vec{H} = \vec{j} + \frac{\partial \vec{D}}{dt} \qquad (14.1)$$

In the case where the fields vary with time, Ampère's circuital law (13.11) is inconsistent with the equation of continuity (12.15), since on applying the divergence to both sides, Eq.(13.11) becomes

$$\nabla \cdot (\nabla \times \vec{H}) = \nabla \cdot \vec{j}$$

where the left-hand side vanishes identically. Maxwell assumed that the right hand side of Ampère's law should be written as a modified current density whose divergence must be zero. Its form was obtained by combining Eq.(11.26) with the equation of continuity (12.15), yielding

$$\nabla \cdot \vec{j} + \frac{\partial}{\partial t}(\nabla \cdot \vec{D}) = 0 \quad \text{or} \quad \nabla \cdot \left(\vec{j} + \frac{\partial \vec{D}}{\partial t} \right) = 0$$

This equation indicates that a current density $\vec{j} + \partial \vec{D}/\partial t$ rather than $\vec{j}$ should be taken on the right-hand side of Ampère's law, leading to Eq.(14.1). This corrected form of Ampère's law is consistent with the conservation of charge and reduces to the original circuital law (13.11) for a stationary charge flow, which obeys Eq.(12.16). The new term is called the *displacement current density*, the source of which is a time-varying electric displacement $\vec{D}$.

EXAMPLE 14.1. The displacement current in a capacitor

The existence of a displacement current is suggested by the case of an alternating current applied to the plates of a capacitor, as illustrated in Figure 14.1. There is no charge transfer between the plates, however a current passes through the capacitor. Drawing two surfaces S and S' bound to the same closed path, Ampère's circuital law (12.26) involving the current through S reads

$$\int_\Gamma \vec{B} \cdot d\vec{l} = \mu_0 \int_S \vec{j} \cdot d\vec{S} = \mu_0 I$$

where I is the current in the circuit. Since there is no charge flow through the capacitor, that is through S', we must take the current density through S' to be the change in the charge density on the plates.

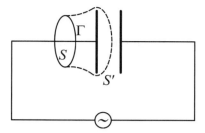

Figure 14.1. A capacitor in an alternating current circuit

For the uniform field $\varepsilon_r E$ between the plates, Gauss's theorem gives

$$\varepsilon_r E S' = \frac{q}{\varepsilon_0} \quad \text{or} \quad \varepsilon \frac{\partial E}{\partial t} = \frac{1}{S'} \frac{dq}{dt} = \frac{I}{S'} = j'$$

where $\varepsilon = \varepsilon_0 \varepsilon_r$ and S' was chosen close to the area of the plates. It follows that

$$\int_\Gamma \vec{B} \cdot d\vec{l} = \mu_0 \int_{S'} \vec{j}' \cdot d\vec{S}' = \mu_0 \int_{S'} \frac{\partial \vec{D}}{dt} \cdot d\vec{S}' = \mu_0 I$$

This displacement current $\vec{j}'$ through S' originates in the time rate of change of the displacement vector, which is associated with the change in the surface charge densities on the plates.

We have seen that taking the divergence of Eqs.(13.23) and (14.1) yields equations that are consistent with Eqs.(12.23) and (11.26), respectively, so that the four Maxwell's field equations form two pairs of differential equations, which can be written in terms of free charges and free currents in materials as

$$\nabla \times \vec{E} = -\frac{\partial \vec{B}}{dt} \qquad \nabla \times \vec{H} = \vec{j} + \frac{\partial \vec{D}}{dt}$$

$$\nabla \cdot \vec{B} = 0 \qquad \nabla \cdot \vec{D} = \rho \tag{14.2}$$

or in integral form

$$\int_\Gamma \vec{E} \cdot d\vec{l} = -\frac{\partial}{\partial t} \int_S \vec{B} \cdot d\vec{S} \qquad \int_\Gamma \vec{H} \cdot d\vec{l} = \int_S \left(\vec{j} + \frac{\partial \vec{D}}{\partial t} \right) \cdot d\vec{S}$$

$$\int_S \vec{B} \cdot d\vec{S} = 0 \qquad \int_S \vec{D} \cdot d\vec{S} = \int_V \rho dV \tag{14.3}$$

Both forms of Eqs.(14.2) and (14.3) are valid for *arbitrary media*, since they contain no parameters other than the free-charge density and free-current density.

The four field vectors $\vec{E}, \vec{D}, \vec{B}, \vec{H}$ are related by the constitutive relations (11.27) and (13.12), whereas the free-current density $\vec{j}$ is given in terms of $\vec{E}$ by Eq.(12.10) as

$$\vec{D} = \varepsilon_0 \vec{E} + \vec{P}$$

$$\vec{H} = \frac{1}{\mu_0}\vec{B} - \vec{M} \qquad (14.4)$$

$$\vec{j} = \sigma \vec{E}$$

Equations (14.2) and (14.4) can be regarded together as the fundamental laws governing electromagnetic phenomena *inside matter*. Assuming that matter behaves in a linear manner, the corresponding constitutive relations (11.28), (13.15) and (12.10) are

$$\vec{D} = \varepsilon_0 \varepsilon_r \vec{E} = \varepsilon \vec{E}$$

$$\vec{B} = \mu_0 \mu_r \vec{H} = \mu \vec{H} \qquad (14.5)$$

$$\vec{j} = \sigma \vec{E}$$

The differential form of Maxwell's equations, restricted to *linear, homogeneous* and *isotropic media* can be obtained by substituting Eqs.(14.5) into Eqs.(14.2) which gives

$$\nabla \times \vec{E} = -\mu \frac{\partial \vec{H}}{dt} \qquad \nabla \times \vec{H} = \sigma \vec{E} + \varepsilon \frac{\partial \vec{E}}{dt}$$

$$\nabla \cdot \vec{H} = 0 \qquad \varepsilon \nabla \cdot \vec{E} = \rho \qquad (14.6)$$

Separate equations for the electric and magnetic fields can be derived in regions inside matter where there is no free charge distribution $\rho = 0$, so that Maxwell's equations (14.6) read

$$\nabla \times \vec{E} = -\mu \frac{\partial \vec{H}}{dt} \qquad \nabla \times \vec{H} = \sigma \vec{E} + \varepsilon \frac{\partial \vec{E}}{dt}$$

$$\nabla \cdot \vec{H} = 0 \qquad \nabla \cdot \vec{E} = 0 \qquad (14.7)$$

By taking the *curl* of the first equation of each pair, we have

$$\nabla \times (\nabla \times \vec{E}) = -\mu \frac{\partial}{\partial t}(\nabla \times \vec{H}) = -\mu \sigma \frac{\partial \vec{E}}{\partial t} - \varepsilon \mu \frac{\partial^2 \vec{E}}{\partial t^2}$$

and, respectively

$$\nabla \times (\nabla \times \vec{H}) = \left(\sigma + \varepsilon \frac{\partial}{\partial t}\right)(\nabla \times \vec{E}) = -\mu\sigma \frac{\partial \vec{H}}{\partial t} - \varepsilon\mu \frac{\partial^2 \vec{H}}{\partial t^2}$$

These equations can be simplified, using Eq.(I.18), Appendix I, to

$$\nabla \times (\nabla \times \vec{E}) = \nabla(\nabla \cdot \vec{E}) - \nabla^2 \vec{E} = -\nabla^2 \vec{E}$$

$$\nabla \times (\nabla \times \vec{H}) = \nabla(\nabla \cdot \vec{H}) - \nabla^2 \vec{H} = -\nabla^2 \vec{H}$$

where we have taken into account the second equation of each pair in Eqs.(14.7). Thus two separate second-order differential equations for $\vec{E}$ and $\vec{H}$ can be written as

$$\nabla^2 \vec{E} - \mu\sigma \frac{\partial \vec{E}}{\partial t} - \varepsilon\mu \frac{\partial^2 \vec{E}}{\partial t^2} = 0$$

$$\nabla^2 \vec{H} - \mu\sigma \frac{\partial \vec{H}}{\partial t} - \varepsilon\mu \frac{\partial^2 \vec{H}}{\partial t^2} = 0$$
(14.8)

Using the Laplacian written in Cartesian coordinates, each component E_x, E_y, E_z and H_x, H_y, H_z becomes a scalar function $\Psi(x,y,z,t)$ with coordinates of both space and time, so that Maxwell's equations can be finally separated into six scalar equations having the form

$$\nabla^2 \Psi - \mu\sigma \frac{\partial \Psi}{\partial t} - \varepsilon\mu \frac{\partial^2 \Psi}{\partial t^2} = 0 \qquad (14.9)$$

In regions inside matter where there is no free-charge or free-current distribution, Maxwell's equations (14.6) exhibit a particular symmetry

$$\nabla \times \vec{E} = -\mu \frac{\partial \vec{H}}{dt} \qquad\qquad \nabla \times \vec{H} = \varepsilon \frac{\partial \vec{E}}{dt}$$

$$\nabla \cdot \vec{H} = 0 \qquad\qquad \nabla \cdot \vec{E} = 0$$
(14.10)

and can be separated into independent scalar equations for the components of $\vec{E}$ and $\vec{H}$, which have the form (4.36) of the *electromagnetic wave equation*, to which Eq.(14.9) reduces by taking $\sigma = 0$.

14.2. ELECTROMAGNETIC ENERGY

From Maxwell's field equations (14.2), an energy relation between time varying fields $\vec{E}$ and $\vec{H}$ can be derived, which is consistent with both the energy densities (11.32) and (13.34) expressed in terms of static fields $\vec{E}$, $\vec{D}$, $\vec{B}$ and $\vec{H}$ and the energy conservation law. A convenient procedure is to take the scalar products of the first two Eqs.(14.2) with $\vec{H}$ and $\vec{E}$, respectively, and this gives

$$\vec{H}\cdot(\nabla\times\vec{E}) = -\vec{H}\cdot\frac{\partial\vec{B}}{\partial t}, \qquad \vec{E}\cdot(\nabla\times\vec{H}) = \vec{E}\cdot\vec{j} + \vec{E}\cdot\frac{\partial\vec{D}}{\partial t}$$

which upon subtraction leads to

$$\vec{E}\cdot(\nabla\times\vec{H}) - \vec{H}\cdot(\nabla\times\vec{E}) = \vec{E}\cdot\vec{j} + \vec{E}\cdot\frac{\partial\vec{D}}{\partial t} + \vec{H}\cdot\frac{\partial\vec{B}}{\partial t}$$

In all the linear isotropic media Eqs.(14.5) are valid, so that the last two terms in the right-hand side become

$$\vec{E}\cdot\frac{\partial\vec{D}}{\partial t} = \varepsilon\vec{E}\cdot\frac{\partial\vec{E}}{\partial t} = \frac{\partial}{\partial t}\left(\frac{1}{2}\varepsilon\vec{E}^2\right) = \frac{\partial}{\partial t}\left(\frac{1}{2}\vec{E}\cdot\vec{D}\right)$$

$$\vec{H}\cdot\frac{\partial\vec{B}}{\partial t} = \mu\vec{H}\cdot\frac{\partial\vec{H}}{\partial t} = \frac{\partial}{\partial t}\left(\frac{1}{2}\mu\vec{H}^2\right) = \frac{\partial}{\partial t}\left(\frac{1}{2}\vec{H}\cdot\vec{B}\right)$$

whereas the left-hand side is equal to minus the divergence of $\vec{E}\times\vec{H}$, according to Eq.(I.14), Appendix I. We thus obtain

$$\nabla\cdot(\vec{E}\times\vec{H}) + \vec{E}\cdot\vec{j} + \frac{\partial}{\partial t}\left(\frac{1}{2}\vec{E}\cdot\vec{D} + \frac{1}{2}\vec{H}\cdot\vec{B}\right) = 0 \qquad (14.11)$$

The scalar product $\vec{E}\cdot\vec{j}$ gives the time rate of change of the mechanical energy density associated with the motion of free charges under a Lorentz force (13.30), as measured by the work performed on an element of charge dq in unit time

$$\frac{d}{dt}(\vec{F}\cdot d\vec{r}) = dq(\vec{E} + \vec{v}\times\vec{B})\cdot\frac{d\vec{r}}{dt} = dq\vec{E}\cdot\vec{v} = \rho dV\vec{E}\cdot\vec{v} = \vec{E}\cdot\vec{j}dV$$

We can then write

14. MAXWELL'S EQUATIONS

$$\vec{E} \cdot \vec{j} = \frac{\partial}{\partial t}\left(\frac{\vec{F} \cdot d\vec{r}}{dV}\right) = \frac{\partial}{\partial t} u_{mec} \qquad (14.12)$$

where u_{mec} stands for the mechanical energy density. Substituting u_{mec} from Eq.(14.12), u_E from Eq.(11.32) and u_H from Eq.(13.34), Eq.(14.11) takes the form

$$\nabla \cdot (\vec{E} \times \vec{H}) + \frac{\partial}{\partial t}(u_{mec} + u_E + u_H) = 0$$

We can define for convenience $u = u_{mec} + u_E + u_H = u_{mec} + u_{em}$ as the total energy density, which is the sum of mechanical and electromagnetic contributions. If the *Poynting vector* $\vec{S}$ is defined by

$$\vec{S} = \vec{E} \times \vec{H} \qquad (14.13)$$

we obtain

$$\nabla \cdot \vec{S} + \frac{\partial u}{\partial t} = 0 \qquad (14.14)$$

This equation is similar to the equation of continuity (12.15), which states, in a differential form, the conservation of free electric charge. Instead of the charge density ρ we have the energy density u and, in addition, the current density $\vec{j}$ is replaced by the Poynting vector. It follows that $\vec{S}$ can be interpreted as an energy flow density similar to the charge flow density $\vec{j}$.

Equation (14.14) represents the differential form of *Poynting's theorem*, describing the conservation of energy in electromagnetic fields. Its integral form over the volume V of a charge and current distribution is obtained from Eq.(14.11), by applying to the left-hand side Gauss's divergence theorem yielding

$$\int_\Sigma (\vec{E} \times \vec{H}) \cdot d\vec{\Sigma} + \int_V \vec{E} \cdot \vec{j}\, dV + \frac{\partial}{\partial t}\int_V \left(\frac{1}{2}\vec{E} \cdot \vec{D} + \frac{1}{2}\vec{H} \cdot \vec{B}\right) dV = 0 \qquad (14.15)$$

The surface integral is the rate at which the electromagnetic energy is flowing outward through the closed surface Σ. Poynting's theorem states that this energy flow is balanced by the rate of change of the mechanical energy of the free charges under the Lorentz force plus the electromagnetic energy stored in the fields within the volume.

> **EXAMPLE 14.2. Momentum and energy of a charge in an electromagnetic field**
>
> The motion of a particle under a conservative force $\vec{F} = -\nabla U$ obeys Eq.(2.9) which reads
>
> $$\frac{d(m\vec{v})}{dt} = \vec{F} = -\nabla U \tag{14.16}$$
>
> Both the momentum and the potential energy of a charge in a conservative electromagnetic field can be expressed in terms of the scalar and vector potentials, which are related to the $\vec{E}$ and $\vec{B}$ fields as defined by Eqs.(12.29) and (13.25). Substituting these relations, the Lorentz force law (13.30) becomes
>
> $$\vec{F} = q\left[-\nabla V - \frac{\partial \vec{A}}{\partial t} + \vec{v} \times (\nabla \times \vec{A})\right] = q\left[-\nabla V - \frac{\partial \vec{A}}{\partial t} + \nabla(\vec{v} \cdot \vec{A}) - (\vec{v} \cdot \nabla)\vec{A}\right]$$
>
> $$= -q\left[\frac{\partial \vec{A}}{\partial t} + (\vec{v} \cdot \nabla)\vec{A} + \nabla(V - \vec{v} \cdot \vec{A})\right]$$
>
> We have used Eq.(I.16), Appendix I, where $\vec{v}$ is not a function of position. The first two terms inside the brackets give the total time derivative of $\vec{A}$, that is the rate of change of $\vec{A}$ with time and space, in the form
>
> $$\frac{d\vec{A}}{dt} = \frac{\partial \vec{A}}{\partial t} + \frac{\partial \vec{A}}{\partial x}\frac{dx}{dt} + \frac{\partial \vec{A}}{\partial y}\frac{dy}{dt} + \frac{\partial \vec{A}}{\partial z}\frac{dz}{dt} = \frac{\partial \vec{A}}{\partial t} + \left(v_x \frac{\partial}{\partial x} + v_y \frac{\partial}{\partial y} + v_z \frac{\partial}{\partial z}\right)\vec{A} = \frac{\partial \vec{A}}{\partial t} + (\vec{v} \cdot \nabla)\vec{A}$$
>
> so that the Lorentz force law can be written as
>
> $$\vec{F} = -q\frac{\partial \vec{A}}{\partial t} - q\nabla(V - \vec{v} \cdot \vec{A})$$
>
> Newton's second law (2.9) then takes the form
>
> $$\frac{d}{dt}(m\vec{v} + q\vec{A}) = -\nabla(qV - q\vec{v} \cdot \vec{A}) \tag{14.17}$$
>
> which is similar to Eq.(14.16) provided the momentum $m\vec{v}$ is replaced by a *generalized momentum* defined as
>
> $$\vec{p} = m\vec{v} + q\vec{A} \quad \text{or} \quad \vec{v} = \frac{1}{m}(\vec{p} - q\vec{A}) \tag{14.18}$$
>
> and the potential energy is taken to be
>
> $$U = qV - q\vec{v} \cdot \vec{A} = qV - \frac{q}{m}(\vec{p} - q\vec{A}) \cdot \vec{A} \tag{14.19}$$
>
> It will be shown in Chapter 24 that, in the canonical formulation of classical mechanics, the two requirements are mutually consistent.

14.3. Potential Equations

Maxwell's equations can be formulated in terms of scalar and vector potentials V and $\vec{A}$, by substituting $\vec{E}$ and $\vec{B}$ as expressed by Eqs.(13.25) and (12.29), which are valid for time-varying fields in the form

$$\vec{E} = -\nabla V - \frac{\partial \vec{A}}{\partial t}, \qquad \vec{B} = \nabla \times \vec{A}$$

The magnetic flux law (12.23) is then automatically fulfilled by the vector potential $\vec{A}$ as given by Eq.(12.29) and so is Faraday's law (13.23) by V and $\vec{A}$, if related by Eq.(13.25). Gauss's law (14.6) for linear isotropic media becomes

$$\nabla \cdot \left(-\nabla V - \frac{\partial \vec{A}}{\partial t} \right) = \frac{\rho}{\varepsilon} \quad \text{or} \quad \nabla^2 V + \frac{\partial}{\partial t}\left(\nabla \cdot \vec{A}\right) = -\frac{\rho}{\varepsilon} \qquad (14.20)$$

which reduces to Poisson's equation (11.14) in the static case. Ampère's corrected law (14.6) becomes

$$\nabla \times \left(\nabla \times \vec{A}\right) = \mu \vec{j} - \mu\varepsilon \nabla\left(\frac{\partial V}{\partial t}\right) - \mu\varepsilon \frac{\partial^2 \vec{A}}{\partial t^2}$$

and, in view of Eq.(I.18), Appendix I, reads

$$\nabla^2 \vec{A} - \nabla\left(\nabla \cdot \vec{A}\right) = -\mu \vec{j} + \mu\varepsilon \nabla\left(\frac{\partial V}{\partial t}\right) + \mu\varepsilon \frac{\partial^2 \vec{A}}{\partial t^2} \qquad (14.21)$$

The potential equations can be further simplified using the choice of a gauge, which means the freedom to adjust the divergence of $\vec{A}$ with no effect on the relation between $\vec{A}$ and the magnetic induction $\vec{B}$, as previously discussed in Chapter 12.

There is a *Coulomb gauge* (12.34) which transforms the potential equations as

$$\nabla^2 V = -\frac{\rho}{\varepsilon}$$

$$\nabla^2 \vec{A} - \varepsilon\mu \frac{\partial^2 \vec{A}}{\partial t^2} = -\mu \vec{j} + \varepsilon\mu \nabla\left(\frac{\partial V}{\partial t}\right) \qquad (14.22)$$

so that V is simply given by Poisson's equation, but it is not simple to find a solution $\vec{A}$ for the second equation. Therefore, in the Coulomb gauge both $\vec{E}$, defined by Eq.(13.25) in terms of V and $\vec{A}$, and $\vec{B}$ are difficult to calculate.

It is convenient to rewrite the potential equation (14.20) in an equivalent form, similar to that of Eq.(14.21), so that they read

$$\nabla^2 V - \varepsilon\mu \frac{\partial^2 V}{\partial t^2} + \frac{\partial}{\partial t}\left(\nabla \cdot \vec{A} + \varepsilon\mu \frac{\partial V}{\partial t}\right) = -\frac{\rho}{\varepsilon}$$

$$\nabla^2 \vec{A} - \varepsilon\mu \frac{\partial^2 \vec{A}}{\partial t^2} - \nabla\left(\nabla \cdot \vec{A} + \varepsilon\mu \frac{\partial V}{\partial t}\right) = -\mu \vec{j}$$

(14.23)

An obvious choice appears now to be the *Lorentz gauge*, defined as

$$\nabla \cdot \vec{A} + \varepsilon\mu \frac{\partial V}{\partial t} = 0 \tag{14.24}$$

which allows us to reduce Eqs.(14.23) to separate equations for V and $\vec{A}$ given by

$$\nabla^2 V - \varepsilon\mu \frac{\partial^2 V}{\partial t^2} = \left(\nabla^2 - \varepsilon\mu \frac{\partial^2}{\partial t^2}\right)V = -\frac{\rho}{\varepsilon}$$

$$\nabla^2 \vec{A} - \varepsilon\mu \frac{\partial^2 \vec{A}}{\partial t^2} = \left(\nabla^2 - \varepsilon\mu \frac{\partial^2}{\partial t^2}\right)\vec{A} = -\mu \vec{j}$$

(14.25)

If we consider the differential operator $\nabla^2 - \varepsilon_0 \mu_0 (\partial^2/\partial t^2) = \nabla^2 - c^{-2}(\partial^2/\partial t^2)$, which is the same as the d'Alembertian (4.37), to be the four-dimensional version of the Laplacian operator, then Eqs.(14.25) can be interpreted as a four-dimensional analogue of Poisson's equation (11.14) and of its magnetostatic equivalent (12.35).

In Cartesian coordinates the differential equation (14.25) for $\vec{A}$ is equivalent to three independent scalar equations for the components of $\vec{A}$. In regions of space where both ρ and $\vec{j}$ are zero, the four scalar equations (14.25) each reduce to the electromagnetic wave equation (4.36), which is the four-dimensional version of Laplace's equations (11.15) or (12.36), written as

$$\left(\nabla^2 - \varepsilon\mu \frac{\partial^2}{\partial t^2}\right)V = 0$$

$$\left(\nabla^2 - \varepsilon\mu \frac{\partial^2}{\partial t^2}\right)\vec{A} = 0$$

(14.26)

We can now define the *Hertz vector* $\vec{Z}$ by

$$V = -\nabla \cdot \vec{Z} \quad \text{and} \quad \vec{A} = \varepsilon\mu \frac{\partial \vec{Z}}{\partial t} \tag{14.27}$$

It is a straightforward matter to show that this definition is consistent with the Lorentz gauge (14.24). Substituting Eq.(14.27), both equations (14.26) reduce to the *Hertz vector equation*, which reads

$$\left(\nabla^2 - \varepsilon\mu\frac{\partial^2}{\partial t^2}\right)\vec{Z} = 0 \qquad (14.28)$$

The potential formulation of electrodynamics is then reduced to a unique vector equation. Both V and $\vec{A}$ can be derived from its solutions, using Eq.(14.27). Therefore $\vec{E}$ follows in terms of $\vec{Z}$ from Eq.(13.25), which gives

$$\vec{E} = \nabla(\nabla \cdot \vec{Z}) - \varepsilon\mu\frac{\partial^2 \vec{Z}}{\partial t^2} = \nabla(\nabla \cdot \vec{Z}) - \nabla^2 \vec{Z} = \nabla \times (\nabla \times \vec{Z}) \qquad (14.29)$$

and $\vec{H}$ is obtained from Eq.(12.29), which becomes

$$\vec{H} = \frac{1}{\mu}(\nabla \times \vec{A}) = \varepsilon \nabla \times \frac{\partial \vec{Z}}{\partial t} = \varepsilon \frac{\partial}{\partial t}(\nabla \times \vec{Z}) \qquad (14.30)$$

A more compact form is obtained by setting

$$\vec{C} = \nabla \times \vec{Z} \qquad (14.31)$$

so that the electric and magnetic fields can be expressed as

$$\vec{E} = \nabla \times \vec{C} \quad \text{and} \quad \vec{H} = \varepsilon\frac{\partial \vec{C}}{\partial t} \qquad (14.32)$$

FURTHER READING
1. G. Pollack, D. Stump - ELECTROMAGNETISM, Addison-Wesley, Reading, 2001.
2. R. K. Wangsness - ELECTROMAGNETIC FIELDS, Wiley, New York, 1999.
3. P. Lorrain and D. R. Corson - ELECTROMAGNETISM: PRINCIPLES AND APPLICATIONS, Freeman, New York, 1990.

Problems

14.1. Show that the equation of continuity is a consequence of Maxwell's equations.

Solution

Maxwell's equations (14.2) relate the sources $\rho, \vec{j}$ and the fields $\vec{D}, \vec{H}$ in the form

$$\nabla \times \vec{H} = \vec{j} + \frac{\partial \vec{D}}{\partial t} \quad , \quad \nabla \cdot \vec{D} = \rho$$

By taking the divergence of the first equation we have

$$\nabla \cdot (\nabla \times \vec{H}) = \nabla \cdot \vec{j} + \frac{\partial}{\partial t}(\nabla \cdot \vec{D}) = 0$$

where use has been made of the vector identity (I.17) from Appendix I. On substitution of $\nabla \cdot \vec{D}$ as given by the second equation, one obtains

$$\nabla \cdot \vec{j} + \frac{\partial \rho}{\partial t} = 0$$

as required.

14.2. A variable voltage $V = V_0 \sin \omega t$ is applied to a circular parallel-plate capacitor filled with a lossy dielectric of permitivity ε and conductivity σ. If the electric field between the plates is homogeneous, due to the small plate separation d, find the magnetic field in the capacitor.

Solution

Neglecting the $\partial \vec{A} / \partial t$ term in Eq.(13.25), which leads to radiation, the homogeneous electric field between the plate reads

$$\vec{E} = -\nabla V \quad \text{or} \quad E = \frac{V_0}{d} \sin \omega t$$

The negative field for a homogeneous medium is given by Eq.(14.3), which can be written as

$$\oint_\Gamma \vec{H} \cdot d\vec{r} = \sigma \int_S \vec{E} \cdot d\vec{S} + \varepsilon_0 \int_S \frac{\partial \vec{E}}{\partial t} \cdot d\vec{S}$$

Due to the symmetry about the $\vec{E}$ direction, this reduces, at a radial distance R from the axis of the capacitor, to

$$2\pi R H(R) = \sigma \int_0^R E 2\pi r dr + \varepsilon_0 \int_0^R \frac{\partial E}{\partial t} 2\pi r dr \quad \text{or} \quad H(R) = \frac{R}{2}\left(\sigma E + \varepsilon \frac{\partial E}{\partial t}\right) = \frac{R}{2}\frac{V_0}{d}(\sigma \sin \omega t + \varepsilon \omega \cos \omega t)$$

If we set $\sigma = A \cos \varphi$ and $\varepsilon \omega = A \sin \varphi$, the required field follows as

$$H(R) = \frac{R}{2}\frac{V_0}{d} A \sin(\omega t + \varphi)$$

where $\tan\varphi = \varepsilon\omega/\sigma$ and $A = (\sigma^2 + \varepsilon^2\omega^2)^{1/2}$.

14.3. An air-filled coaxial cable consists of a metal wire of radius R_1 surrounded by a metal sheat of radius R_2. If a voltage $V = V_0 \cos\omega t$, where V_0 is a real amplitude, is applied at one end, determine the Poynting vector of the electromagnetic field in the air region between the conductors.

Solution

By symmetry, the electric field in the air region between the conductors is directed radially and depends on the distance r from the axis ($R_1 < r < R_2$) only, as given by the Gauss law $E_r = \sigma/2\pi\varepsilon_0 r$, where σ stands for the surface charge density on the inner wire. Now $E_r = -dV/dr$, and this provides the voltage between the conductors as

$$V = -\frac{\sigma}{2\pi\varepsilon_0 r}\int_{R_2}^{R_1}\frac{dr}{r} = \frac{\sigma}{2\pi\varepsilon_0}\ln(R_2/R_1)$$

Combining these expressions to eliminate σ gives

$$E_r = \frac{V}{r\ln(R_2/R_1)}$$

Since the field generated by a time harmonic voltage at the origin $z = 0$, at time t, reaches a point z along the cable at t plus the time z/c it takes to travel along the z-axis at the speed of light c, the electric field should be written as a function of time and position, with cylindrical components

$$E_r = \frac{V_0}{r\ln(R_2/R_1)}\cos\omega\left(t - \frac{z}{c}\right) \quad , \quad E_\theta = 0 \quad , \quad E_z = 0$$

In terms of the cylindrical coordinates given in Appendix I, Eq.(I.27), Faraday's law becomes

$$0 = -\mu_0\frac{\partial H_r}{\partial t} \quad , \quad \frac{\partial E_r}{\partial z} = -\mu_0\frac{\partial H_\theta}{\partial t} \quad , \quad 0 = -\mu_0\frac{\partial H_z}{\partial t}$$

and hence, $H_r = H_z = 0$. Substituting E_r into the second equation yields

$$H_\theta = -\frac{\omega}{\mu_0 c}\frac{V_0}{r\ln(R_2/R_1)}\int\sin\omega\left(t - \frac{z}{c}\right)dt = \frac{1}{\mu_0 c}\frac{V_0}{r\ln(R_2/R_1)}\cos\omega\left(t - \frac{z}{c}\right)$$

The Poynting vector (14.13) follows as

$$S_r = 0 \quad , \quad S_\theta = 0 \quad , \quad S_z = E_r H_\theta = \frac{1}{\mu_0 c}\frac{V_0^2}{r^2\ln^2(R_2/R_1)}\cos^2\omega\left(t - \frac{z}{c}\right)$$

indicating an axial energy flow in the region between the conductors.

14.4. For a dipole antenna at the origin, with complex electric dipole moment $\vec{p}(t) = \vec{e}_z p_0 e^{-i\omega t}$, find the potentials $\vec{A}$ and V under Lorentz gauge.

Solution

The vector potential is given by Eq.(12.32), where $Id\vec{l} = (dq/dt)\vec{d} = \vec{e}_z dp/dt$ (see Figure 11.3), and hence, we have

$$\vec{A}(t) = \vec{e}_z \frac{\mu_0}{4\pi r} \left[\frac{dp}{dt}\right]_{t-r/c}$$

where the potential $\vec{A}$ generated by the harmonic dipole at the origin is retarded at the point P by the time r/c it takes to reach this point. This gives

$$\vec{A}(t) = \vec{e}_z \frac{\mu_0 p_0}{4\pi r} \frac{d}{dt} e^{-i\omega(t-r/c)} = -i\omega \frac{\mu_0 p_0}{4\pi} \frac{e^{i(kr-\omega t)}}{r} \vec{e}_z$$

where k stands for ω/c. It is convenient to write down expressions for the components of $\vec{A}$ in spherical polar coordinates, which read

$$A_r = -i\omega \frac{\mu_0 p_0}{4\pi} \frac{e^{i(kr-\omega t)}}{r} \cos\theta \quad , \quad A_\theta = i\omega \frac{\mu_0 p_0}{4\pi} \frac{e^{i(kr-\omega t)}}{r} \sin\theta \quad , \quad A_\varphi = 0$$

so that, in view of Eq.(I.21) from Appendix I, one obtains

$$\nabla \cdot \vec{A} = i\omega \frac{\mu_0 p_0}{4\pi} \frac{e^{-i\omega t}}{r^2} \left[-\frac{\partial}{\partial r}\left(re^{ikr}\cos\theta\right) + \frac{1}{\sin\theta}\frac{\partial}{\partial \theta}\left(e^{ikr}\sin^2\theta\right)\right]$$

$$= i\omega \frac{\mu_0 p_0}{4\pi} \frac{e^{i(kr-\omega t)}}{r^2}(1-ikr)\cos\theta$$

Under Lorentz gauge, Eq.(14.24), the scalar potential follows as

$$V(t) = -\frac{1}{\varepsilon_0 \mu_0} \int \nabla \cdot \vec{A}\, dt = -\frac{i\omega p_0}{4\pi\varepsilon_0}\frac{e^{ikr}}{r^2}(1-ikr)\cos\theta \int e^{-i\omega t} dt = \frac{p_0}{4\pi\varepsilon_0}\frac{e^{i(kr-\omega t)}}{r^2}(1-ikr)\cos\theta$$

14.5. Show that the potential formulation (14.25) of Maxwell's equations is covariant under Lorentz transformations of special relativity.

Solution

The equation of continuity (12.15), written in the form

$$\nabla \cdot \vec{j} + \frac{\partial \rho}{\partial t} = \nabla \cdot \vec{j} + \frac{1}{ic}\frac{\partial(ic\rho)}{\partial t} = 0$$

can be reformulated, using the *quad* operator (4.37), in terms of the *four-divergence* of a current *density* four-vector $j_\mu = (j_x, j_y, j_z, ic\rho)$ as

$$\sum_\mu \frac{\partial j_\mu}{\partial x_\mu} = 0$$

and this is invariant under Lorentz transformations. In vacuo, the Lorentz gauge (14.24) becomes

$$\nabla \cdot \vec{A} + \varepsilon_0 \mu_0 \frac{\partial V}{\partial t} = \nabla \cdot \vec{A} + \frac{1}{c^2}\frac{\partial V}{\partial t} = \nabla \cdot \vec{A} + \frac{1}{ic}\frac{\partial(iV/c)}{\partial t} = 0$$

where the speed of light c was written as $c = (\varepsilon_0 \mu_0)^{-1/2}$ as will be discussed in more detail in Chapter 15. The equivalent four-divergence form, invariant under Lorentz transformations, is

$$\sum_\mu \frac{\partial A_\mu}{\partial x_\mu} = 0$$

and this defines a *potential* four-vector $A_\mu = (A_x, A_y, A_z, (i/c)V)$. Therefore the potential formulation (14.25) can be expressed in vacuo, using the d'Alembertian operator, the four-current density and the four-potential, in the compact form

$$\sum_\mu \frac{\partial A_\nu}{\partial x_\mu^2} = -\mu_0 j_\nu$$

which represents the covariant formulation of Maxwell's equations under Lorentz transformations.

14.6. Find the charge and current distributions that would lead to the potentials $\vec{A}(\vec{r},t) = \alpha t \vec{r}/r^3$, $V(\vec{r},t) = 0$.

14.7. Show that the conduction and displacement currents cancel each other around a spherically radioactive solid emitting charged particles radially outwards.

14.8. Find the charge and current distributions in a region inside matter of electrostatic polarization $\vec{P}(\vec{r},t)$ and magnetization $\vec{M}(\vec{r},t)$.

14.9. Use the vector and scalar potentials of a radiating electric dipole, derived in Problem 14.4, to determine the nonvanishing components of the electric and magnetic fields and then the Poynting vector at the radiation zone, where one can exclude terms of order $1/r^2$ and higher.

14.10. If $\vec{E}$ and $\vec{B}$ are the fields acting in a reference frame S at rest, write down the Lorentz transformations yielding the components of the fields $\vec{E}'$ and $\vec{B}'$ observed in a reference frame S', which moves with velocity v parallel to the x-axis of S.

15. Wave Equations

15.1. EQUATION OF WAVE MOTION
15.2. ELASTIC WAVES ON A STRING
15.3. SOUND WAVES IN FLUIDS
15.4. ELECTROMAGNETIC WAVES IN ISOTROPIC DIELECTRICS

15.1. Equation of Wave Motion

The concept of a wave is associated with the propagation of a disturbance through a medium at constant velocity, without changing the shape of the disturbance. The wave motion is then carrying energy without disturbing in a permanent way the transmitting medium, which will be regarded as a continuum. For small disturbances a wave can be described in terms of space and time by *linear* differential equations which are exact whatever the nature of the forces or the medium involved.

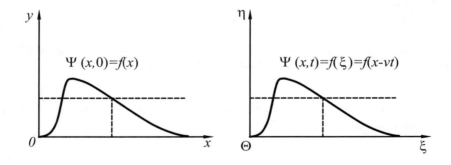

Figure 15.1. Arbitrary disturbance at two successive instants of time

Consider an arbitrary scalar disturbance Ψ which propagates with velocity v along the x-axis, in the laboratory frame S, as illustrated in Figure 15.1. In its rest frame Σ, whose Cartesian system $\Theta\xi\eta\zeta$ coincides with that in S, $Oxyz$, at time $t=0$, we have

$$\xi = x - vt \tag{15.1}$$

Since the disturbance has a constant position and profile with respect to Σ, it must depend on the time t through the coordinate ξ, as expressed by some function

$$\Psi(x,t)= f(\xi)= f(x-vt) \tag{15.2}$$

where $f(\xi)$ defines the profile or the *waveform*. For any profile f, Eq.(15.2) gives

$$\left(\frac{\partial}{\partial x}+\frac{1}{v}\frac{\partial}{\partial t}\right)\Psi = 0 \tag{15.3}$$

which describes a *travelling wave* along the x-axis. If v were negative, the wave would propagate in the opposite direction, obeying the equation

$$\left(\frac{\partial}{\partial x}-\frac{1}{v}\frac{\partial}{\partial t}\right)\Psi = 0 \tag{15.4}$$

and this is satisfied by a function of arbitrary profile g, which is not required to be related with f, that is

$$\Psi(x,t)= g(x+vt) \tag{15.5}$$

By applying the differential operator (15.4), Eq.(15.3) becomes

$$\left(\frac{\partial}{\partial x}-\frac{1}{v}\frac{\partial}{\partial t}\right)\left(\frac{\partial}{\partial x}+\frac{1}{v}\frac{\partial}{\partial t}\right)f(x-vt)=\left(\frac{\partial^2}{\partial x^2}-\frac{1}{v^2}\frac{\partial^2}{\partial t^2}\right)f(x-vt)=0$$

whereas by applying the operator (15.3), Eq.(15.4) reads

$$\left(\frac{\partial}{\partial x}+\frac{1}{v}\frac{\partial}{\partial t}\right)\left(\frac{\partial}{\partial x}-\frac{1}{v}\frac{\partial}{\partial t}\right)g(x+vt)=\left(\frac{\partial^2}{\partial x^2}-\frac{1}{v^2}\frac{\partial^2}{\partial t^2}\right)g(x+vt)=0$$

It follows that the equation

$$\left(\frac{\partial^2}{\partial x^2}-\frac{1}{v^2}\frac{\partial^2}{\partial t^2}\right)\Psi = 0 \quad\text{or}\quad \frac{\partial^2 \Psi}{\partial x^2}=\frac{1}{v^2}\frac{\partial^2 \Psi}{\partial t^2} \tag{15.6}$$

is satisfied by a *linear superposition* of the two wave functions (15.2) and (15.5), of arbitrary form and unrelated to each other. Equation (15.6) is called the *equation of wave motion* in one dimension and permits waves of given profile to progress with velocity v in either direction along the x-axis. Since Eq.(15.6) is valid for functions Ψ of arbitrary profile, which is determined by various sources, it predicts the same wave motion irrespective of nature of the wave. The wave equation is linear, so that each of the wave functions (15.2) and (15.5) can in turn be written as a linear combination of any number of other functions of $x \pm vt$. This is known as the *principle of superposition of waves*, which states that *any solution to the wave equation can be subdivided into simpler partial waves, whose linear combination constitutes the actual wave.*

In two dimensions, a wave travelling along the x-axis can be defined by the wave function

$$\Psi(\vec{r},t) = f(x - vt) + g(x + vt) = f(\vec{r} \cdot \vec{e}_x - vt) + g(\vec{r} \cdot \vec{e}_x + vt)$$

where $\vec{r}$ is the position vector in the xy plane, represented in Figure 15.2. We will assume that on any line perpendicular to the x-axis, Ψ is constant and will maintain the same constant value if the line moves along the x-axis with constant velocity $\pm v$. The moving line is called a *wavefront*. A *line wave* function in the direction of an arbitrary unit vector $\vec{e}_v = \vec{e}_x \cos\varphi + \vec{e}_y \sin\varphi$ has the form

$$\Psi(\vec{r},t) = f(\vec{r} \cdot \vec{e}_x - vt) + g(\vec{r} \cdot \vec{e}_x + vt) = f(x\cos\varphi + y\sin\varphi - vt) + g(x\cos\varphi + y\sin\varphi - vt) \tag{15.7}$$

where the arbitrary functions f and g obey

$$\left(\vec{e}_x \frac{\partial}{\partial x} + \vec{e}_y \frac{\partial}{\partial y} + \vec{e}_v \frac{1}{v}\frac{\partial}{\partial t}\right) f = 0 \tag{15.8}$$

for a wave travelling in the positive $\vec{e}_v$-direction, and

$$\left(\vec{e}_x \frac{\partial}{\partial x} + \vec{e}_y \frac{\partial}{\partial y} - \vec{e}_v \frac{1}{v}\frac{\partial}{\partial t}\right) g = 0 \tag{15.9}$$

if the wave propagates in the opposite direction.

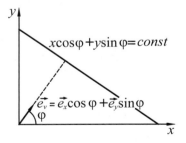

Figure 15.2. Arbitrary line wave in two dimensions

Successive applications of the operators (15.8) and (15.9) leads in a straightforward manner to the wave equation in two dimensions

$$\left(\frac{\partial^2}{\partial x^2} + \frac{\partial^2}{\partial y^2} - \frac{1}{v^2}\frac{\partial^2}{\partial t^2}\right)\Psi = 0 \quad \text{or} \quad \frac{\partial^2 \Psi}{\partial x^2} + \frac{\partial^2 \Psi}{\partial y^2} = \frac{1}{v^2}\frac{\partial^2 \Psi}{\partial t^2} \tag{15.10}$$

In three dimensions, a wave function in the direction of an arbitrary unit vector $\vec{e}_n$ can be defined in a manner similar to Eq.(15.7), that is

$$\Psi(\vec{r},t) = f(\vec{r}\cdot\vec{e}_n - vt) + g(\vec{r}\cdot\vec{e}_n + vt) \qquad (15.11)$$

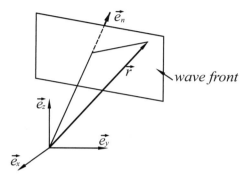

Figure 15.3. Arbitrary plane wave in three dimensions

As illustrated in Figure 15.3, Ψ maintains a constant value on a plane called the *wavefront* which moves along the $\vec{e}_n$-direction with constant velocity $\pm v$, and is said to describe a *plane wave*. The function f satisfies

$$\left(\nabla + \vec{e}_n \frac{1}{v}\frac{\partial}{\partial t}\right)f = 0 \qquad (15.12)$$

if the wave progresses in the positive $\vec{e}_n$-direction, whereas for the opposite direction we have the equation for g which reads

$$\left(\nabla - \vec{e}_n \frac{1}{v}\frac{\partial}{\partial t}\right)g = 0 \qquad (15.13)$$

Thus

$$\left(\nabla + \vec{e}_n \frac{1}{v}\frac{\partial}{\partial t}\right)\left(\nabla - \vec{e}_n \frac{1}{v}\frac{\partial}{\partial t}\right)\Psi = \nabla^2\Psi + \vec{e}_n \cdot \frac{1}{v}\frac{\partial}{\partial t}(\nabla\Psi) - \frac{1}{v}\nabla\cdot\left(\vec{e}_n \frac{\partial \Psi}{\partial t}\right) - \frac{1}{v^2}\frac{\partial^2 \Psi}{\partial t^2}$$

$$= \nabla^2\Psi + \vec{e}_n \cdot \frac{1}{v}\frac{\partial}{\partial t}(\nabla\Psi) - \vec{e}_n \cdot \frac{1}{v}\nabla\left(\frac{\partial \Psi}{\partial t}\right) - \frac{1}{v^2}\frac{\partial^2 \Psi}{\partial t^2} = \left(\nabla^2 - \frac{1}{v^2}\frac{\partial^2}{\partial t^2}\right)\Psi = 0$$

and hence, the *three-dimensional wave equation*, written as

$$\nabla^2 \Psi = \frac{1}{v^2}\frac{\partial^2 \Psi}{\partial t^2} \qquad (15.14)$$

is satisfied by all the plane waves which have the same constant velocity v in both directions, along an axis of arbitrary orientation $\vec{e}_n$, and also by all the waves consisting of linear superposition of such plane waves.

The previous kinematic description of wave motion will be now followed by a discussion of dynamics of several wave conducting systems, using simple models which allow us to consider the forces and fields that generate wave motions obeying the kinematical equations.

15.2. ELASTIC WAVES ON A STRING

Consider the propagation of a small disturbance along a uniform elastic string subject to a tensile stress. If the effect of gravity is neglected, the string is straight in the absence of any disturbance and the x-axis of a Cartesian coordinate system can be taken to lie along the string. We suppose the string to have a density ρ and to be under constant tensile forces $F = S\sigma_x$ applied at its ends, where σ_x is the stress, and S is the cross sectional area of the string.

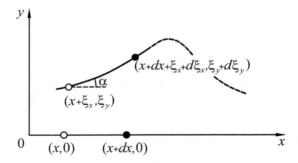

Figure 15.4. Disturbance of a stretched string in the xy-plane

It is usual to separate the longitudinal motion, parallel to the string direction x, from the transverse motion, which can be represented (see Figure 15.4) as a planar motion, confined to the xy-plane. As discussed in Chapter 6, for a length element dx of the string at rest we may either consider a disturbance $d\xi_x$, corresponding to a dilatational strain ε_{xx}, or a disturbance $d\xi_y$, associated with the shear strain ε_{yx}, which are given by

$$d\xi_x = \frac{\partial \xi_x}{\partial x} dx = \varepsilon_{xx} dx \quad , \quad d\xi_y = \frac{\partial \xi_y}{\partial x} dx = \varepsilon_{yx} dx = \alpha dx$$

The equations of motion for both disturbances $\xi_x = \xi_x(x,t)$ and $\xi_y = \xi_y(x,t)$ will be written in terms of the elastic constants of a stretched string.

EXAMPLE 15.1. Elastic constants of a stretched string

If a uniform elastic string is stretched along the x-axis by a tension $F = S\sigma_x$, and all the other stress components are zero, we can choose x, y and z to be the *principal axes* for both the stress and the strain, so that Eqs.(6.29) can be applied in the form

$$\sigma_x = (\lambda + 2\mu)\varepsilon_x + \lambda\varepsilon_y + \lambda\varepsilon_z$$

$$0 = \lambda\varepsilon_x + (\lambda + 2\mu)\varepsilon_y + \lambda\varepsilon_z \tag{15.15}$$

$$0 = \lambda\varepsilon_x + \lambda\varepsilon_y + (\lambda + 2\mu)\varepsilon_z$$

By subtracting the third from the second equation we obtain $\varepsilon_y = \varepsilon_z$, and this, upon substitution into either of them, shows that

$$\varepsilon_y = \varepsilon_z = -\frac{\lambda}{2(\lambda + \mu)}\varepsilon_x = -\upsilon\varepsilon_x \tag{15.16}$$

where υ is called *Poisson's ratio*, given in terms of the Lamé constants by

$$\upsilon = \frac{\lambda}{2(\lambda + \mu)} \tag{15.17}$$

It shows that the elongation ε_x per unit length of the string is accompanied by a decrease of its cross sectional area, measured in terms of the unit transverse contractions ε_y and ε_z. Because λ and μ are always positive, the magnitude of υ never exceeds 1/2. Substituting Eq.(15.16), the first equation in (15.15) becomes

$$\sigma_x = (\lambda + 2\mu - 2\lambda\upsilon)\varepsilon_x = E\varepsilon_x \tag{15.18}$$

where E is called *Young's modulus*, which can be written as

$$E = \lambda + 2\mu - 2\lambda\upsilon = \frac{\mu(3\lambda + 2\mu)}{\lambda + \mu} \tag{15.19}$$

Equations (6.30), relating stress and strain components along *arbitrary axes*, can be then written in terms of E and υ using the property of the diagonal terms

$$\sigma_{xx} + \sigma_{yy} + \sigma_{zz} = (3\lambda + 2\mu)\Delta$$

where Δ is the unit volume expansion defined by Eq.(6.7). The first equation (6.30) reads

$$\sigma_{xx} = \frac{\lambda}{3\lambda + 2\mu}(\sigma_{xx} + \sigma_{yy} + \sigma_{zz}) + 2\mu\varepsilon_{xx}$$

and this can be solved for ε_{xx} which gives

$$\varepsilon_{xx} = \frac{\lambda + \mu}{\mu(3\lambda + 2\mu)}\sigma_{xx} - \frac{\lambda}{2\mu(3\lambda + 2\mu)}(\sigma_{yy} + \sigma_{zz}) = \frac{1}{E}\sigma_{xx} - \frac{\upsilon}{E}(\sigma_{yy} + \sigma_{zz})$$

In a similar manner we obtain the set of symmetric formulae

$$\varepsilon_{xx} = \frac{1}{E}\sigma_{xx} - \frac{\upsilon}{E}(\sigma_{yy} + \sigma_{zz}) \qquad \varepsilon_{xy} = \frac{1+\upsilon}{E}\sigma_{xy}$$

$$\varepsilon_{yy} = \frac{1}{E}\sigma_{yy} - \frac{\upsilon}{E}(\sigma_{zz} + \sigma_{xx}) \qquad \varepsilon_{yz} = \frac{1+\upsilon}{E}\sigma_{yz} \qquad (15.20)$$

$$\varepsilon_{zz} = \frac{1}{E}\sigma_{zz} - \frac{\upsilon}{E}(\sigma_{xx} + \sigma_{yy}) \qquad \varepsilon_{xz} = \frac{1+\upsilon}{E}\sigma_{xz}$$

If all the stresses but σ_{xx} are zero, the properties (15.18) and (15.16) of the stretched string appear to be particular cases of Eq.(15.20).

The *longitudinal motion* of a disturbance $\xi_x(x,t)$ along the string obeys Newton's second law (2.9), applied to the element dx, which reads

$$\rho S dx \frac{\partial^2 \xi_x}{\partial t^2} = F(x+dx) - F(x) = \frac{\partial F}{\partial x} dx \quad \text{or} \quad \rho \frac{\partial^2 \xi_x}{\partial t^2} = \frac{\partial \sigma_{xx}}{\partial x} \qquad (15.21)$$

From Eq.(15.20) it follows that

$$\frac{\partial \sigma_{xx}}{\partial x} = E \frac{\partial \varepsilon_{xx}}{\partial x} = E \frac{\partial^2 \xi_x}{\partial x^2}$$

and Eq.(15.21) becomes

$$\frac{\partial^2 \xi_x}{\partial t^2} = \frac{E}{\rho} \frac{\partial^2 \xi_x}{\partial x^2} \quad \text{or} \quad \frac{\partial^2 \xi_x}{\partial x^2} = \frac{1}{v_0^2} \frac{\partial^2 \xi_x}{\partial t^2} \qquad (15.22)$$

which has the form (15.6) of the wave equation in one dimension. It shows that the longitudinal motion in a stretched string is a *dilatational wave* which propagates with velocity

$$v_0 = \sqrt{\frac{E}{\rho}} \qquad (15.23)$$

The velocity of longitudinal waves v_l in a stretched string which is allowed to deform sideways is larger than that in an elastic body (for instance a piece of rubber filling a cylinder) which is not allowed to expand at right angles to the motion of the disturbance (given for instance by a piston). Taking for simplicity the x-axis as a principal axis for stress and strain and assuming $\xi_y = \xi_z = 0$, that is, $\varepsilon_y = \varepsilon_z = 0$, Eq.(6.29) gives

$$\sigma_x = (\lambda + 2\mu)\varepsilon_x$$

The elastic modulus $\lambda + 2\mu$ is now larger than Young's modulus (15.19) as is the velocity of the longitudinal wave v_l, which coincides with that of longitudinal plane waves, Eq.(6.33).

The *tranverse* motion of a disturbance $\xi_y(x,t)$ along the string is described by Newton's second law (2.9), which in the arrangement shown in Figure 15.5, where we assume $F = \text{const}$, reads

$$\rho S dx \frac{\partial^2 \xi_y}{\partial t^2} = F\sin(\alpha + d\alpha) - F\sin\alpha = F\alpha(x+dx) - F\alpha(x) = F\frac{\partial \alpha}{\partial x}dx$$

and hence

$$\rho S \frac{\partial^2 \xi_y}{\partial t^2} = F\frac{\partial \alpha}{\partial x} \qquad (15.24)$$

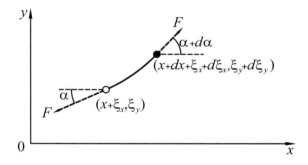

Figure 15.5. The net force in a transverse wave motion

It is usual to set the *linear density* $\mu = \rho S$ for the string, which is not to be confused with the Lamé constant, and to replace

$$\frac{\partial \alpha}{\partial x} = \frac{\partial \varepsilon_{yx}}{\partial x} = E\frac{\partial^2 \xi_y}{\partial x^2}$$

so that Eq.(15.24) reads

$$\frac{\partial^2 \xi_y}{\partial t^2} = \frac{F}{\mu}\frac{\partial^2 \xi_y}{\partial x^2} \qquad \text{or} \qquad \frac{\partial^2 \xi_y}{\partial x^2} = \frac{1}{v^2}\frac{\partial^2 \xi_y}{\partial t^2} \qquad (15.25)$$

It follows that the small transverse motions of a string obey an equation similar to Eq.(15.6), which is called the equation of *shear waves* in one dimension. The transverse waves propagate along the string with velocity

$$v = \sqrt{\frac{F}{\mu}} \qquad (15.26)$$

The wave or phase velocity (15.26) gives the rate at which the disturbance moves along the string, that is $\partial x/\partial t$, whereas the velocity of a length element is $\partial \xi_y/\partial t$. Their ratio

$$\frac{\partial \xi_y/\partial t}{\partial x/\partial t} = \frac{\partial \xi_y}{\partial x} = \alpha$$

gives the gradient of the transverse wave profile, as illustrated in Figure 15.4.

15.3. Sound Waves in Fluids

Sound travels through a fluid as a *compressional wave*. If viscous effects are assumed to be negligible, the hydrostatic pressure p is the only stress component in the fluid. The local pressure $p(\vec{r},t)$ and density $\rho(\vec{r},t)$ are scalar functions of position and time, whereas the related displacement of the fluid, during compressions and rarefactions which compose the wave, is a vector function $\vec{\xi}(\vec{r},t)$. Because the compressions and rarefactions are so widely separated that a negligible flow of heat takes place, the changes of p and ρ may be assumed *adiabatic*, and the relation between pressure and density will be written in terms of the adiabatic compressibility k_S, as defined by Eq.(8.44) which reads

$$k_S = -\frac{1}{V}\left(\frac{\partial V}{\partial p}\right)_S = \frac{1}{\rho}\left(\frac{\partial \rho}{\partial p}\right)_S \qquad (15.27)$$

where use has been made of the conservation of mass ρV, which gives $-dV/V = d\rho/\rho$.

Consider first the one-dimensional problem of the motion of fluid where the displacement $\xi_x = \xi_x(x,t)$ depends on x and t only. We assume the displacement is constant in the y and z directions. This corresponds to a sound wave travelling along a uniform pipe, of constant cross sectional area S. In the absence of wave motion, p and ρ have equilibrium values p_0 and ρ_0. A volume element of the fluid, between the planes x and $x+dx$, represented in Figure 15.6, will increase as a result of a displacement $\xi_x(x,t)$ because we have

$$\xi_x(x+dx,t) = \xi_x(x,t) + \frac{\partial \xi_x}{\partial x}dx = \xi_x(x,t) + d\xi_x$$

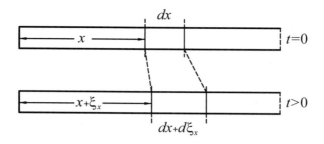

Figure 15.6. The longitudinal displacement in a fluid

The decrease in the local density follows from the conservation of mass as

$$\rho(x,t) = \rho_0 \frac{dx}{dx + d\xi_x} = \rho_0 \left(1 + \frac{\partial \xi_x}{\partial x}\right)^{-1} = \rho_0 \left(1 - \frac{\partial \xi_x}{\partial x}\right) \qquad (15.28)$$

where, for small displacements, we may neglect powers of $\partial \xi_x / \partial x$ higher than the first. Equation (15.28) can alternatively be written in terms of a scalar quantity called the *condensation s*, defined as

$$s = -\frac{\partial \xi_x}{\partial x} \qquad \text{i.e.,} \qquad \rho = \rho_0 (1+s) \qquad (15.29)$$

The change in density causes a pressure gradient which gives the net force on the volume element as

$$dF = -S\,dp = -S \frac{\partial p}{\partial x} dx$$

Newton's second law (2.9) for the longitudinal motion of the disturbance ξ_x, applied to the fluid mass element $\rho_0 S dx$, reads

$$\rho_0 S dx \frac{\partial^2 \xi_x}{\partial t^2} = dF = -S \frac{\partial p}{\partial x} dx \qquad \text{or} \qquad \rho_0 \frac{\partial^2 \xi_x}{\partial t^2} = -\frac{\partial p}{\partial x} = -\frac{\partial p}{\partial \rho} \frac{\partial \rho}{\partial x} \qquad (15.30)$$

From Eq.(15.28) it follows that

$$\frac{\partial^2 \xi_x}{\partial t^2} = \frac{\partial p}{\partial \rho} \frac{\partial^2 \xi_x}{\partial x^2} \qquad (15.31)$$

which has the form of Eq.(15.6), and hence, defines the compressional sound wave motion with velocity

$$v_S = \sqrt{\frac{\partial p}{\partial \rho}} = \sqrt{\frac{1}{k_S \rho}} \qquad (15.32)$$

This expression is valid for the speed of sound in compressible fluids and real gases. It is useful to derive an alternative to Eq.(15.32) which is relevant for a perfect gas. We will assume that compressions and rarefactions are *reversible* processes and that the perfect gas obeys Eq.(9.10) which reads

$$pV^\gamma = \text{const} \quad \text{or} \quad p = \alpha \rho^\gamma$$

where γ is the adiabatic exponent (8.45) and $\alpha = p_0 / p_0^\gamma$. Using Eq.(15.28) yields

$$\frac{\partial p}{\partial \rho} = \gamma \alpha \rho^{\gamma-1} = \gamma \frac{p_0}{\rho_0} \left(1 - \frac{\partial \xi_x}{\partial x}\right)^{\gamma-1}$$

Substitution of the equation of state of a perfect gas (9.5) then results in

$$\frac{\partial p}{\partial \rho} = \gamma \frac{p_0}{\rho_0} = \gamma \frac{RT}{M}$$

where M denotes the molar mass, and the displacements are considered to be small. Therefore the speed of sound in a perfect gas is a function of temperature only

$$v_S = \sqrt{\gamma \frac{p_0}{\rho_0}} = \sqrt{\frac{\gamma R}{M} T} \qquad (15.33)$$

The sound wave equation (15.31) can be rewritten in terms of condensation s defined in Eq.(15.29), by taking the derivative with respect to x which gives

$$\frac{\partial^3 \xi_x}{\partial x \partial t^2} = \frac{\partial p}{\partial \rho} \frac{\partial^3 \xi_x}{\partial x^3} \quad \text{or} \quad \frac{\partial^2}{\partial t^2}\left(\frac{\partial \xi_x}{\partial x}\right) = \frac{\partial p}{\partial \rho} \frac{\partial^2}{\partial x^2}\left(\frac{\partial \xi_x}{\partial x}\right) \quad \text{i.e.,} \quad \frac{\partial^2 s}{\partial t^2} = \frac{1}{v_S^2} \frac{\partial^2 s}{\partial x^2} \qquad (15.34)$$

In view of Eq.(15.29) this equation is equivalent to

$$\frac{\partial^2 \rho}{\partial t^2} = \frac{1}{v_S^2} \frac{\partial^2 \rho}{\partial x^2} \qquad (15.35)$$

The functional relationship $p = p(\rho)$ allows us to write

$$\frac{\partial^2 p}{\partial x^2} = \left(\frac{\partial p}{\partial \rho}\right) \frac{\partial^2 \rho}{\partial x^2} \quad \text{and} \quad \frac{\partial^2 p}{\partial t^2} = \left(\frac{\partial p}{\partial \rho}\right) \frac{\partial^2 \rho}{\partial t^2}$$

so that Eq.(15.34) takes the form

$$\frac{\partial^2 p}{\partial t^2} = \frac{1}{v_S^2} \frac{\partial^2 p}{\partial x^2} \qquad (15.36)$$

From Eqs.(15.31), (15.34), (15.35) and (15.36) it follows that the displacement ξ_x, the condensation s and the local density ρ and pressure p all propagate with the speed of sound (15.32). We are therefore free to choose the appropriate variable to describe the wave motion in any particular case, and it is possible to pass from one to the other, using Eqs.(15.29) and (15.33).

We might also consider the three-dimensional motion of a disturbance in a non-viscous fluid as described by Euler's equation of hydrodynamics (6.41), which reads

$$\rho_0 \frac{d^2 \vec{\xi}}{dt^2} = \rho_0 \vec{f} - \nabla p \qquad (15.37)$$

Using Eq.(6.1) we can assume that $d\vec{\xi}/dt \approx \partial \vec{\xi}/\partial t$ since $|(\vec{v}_S \cdot \nabla)\vec{\xi}| << |\partial \vec{\xi}/\partial t|$, and hence, we have $d^2\vec{\xi}/dt^2 \approx \partial^2 \vec{\xi}/\partial t^2$. If we neglect all the external forces, by taking $\vec{f} = 0$, Eq.(15.37) reduces to

$$\rho_0 \frac{d^2 \vec{\xi}}{dt^2} = -\nabla p = -\frac{\partial p}{\partial \rho} \nabla \rho \qquad (15.38)$$

The local density is given in terms of the vector displacement $\vec{\xi}$ by the three-dimensional analogue of Eq.(15.28) which, for small displacements, can be written as

$$\rho = \rho_0 \left(1 - \frac{\partial \xi_x}{\partial x}\right)\left(1 - \frac{\partial \xi_y}{\partial y}\right)\left(1 - \frac{\partial \xi_z}{\partial z}\right) \approx \rho_0 \left(1 - \nabla \cdot \vec{\xi}\right) \qquad (15.39)$$

Differentiating Eq.(15.39) gives

$$\frac{\partial^2 \rho}{\partial t^2} = -\rho_0 \frac{\partial^2}{\partial t^2}\left(\nabla \cdot \vec{\xi}\right) \quad , \quad \nabla^2 \rho = -\rho_0 \nabla^2 \left(\nabla \cdot \vec{\xi}\right)$$

Thus taking the divergence of both sides of Eq.(15.38) we obtain

$$\rho_0 \frac{\partial^2}{\partial t^2}\left(\nabla \cdot \vec{\xi}\right) = -\frac{\partial p}{\partial \rho} \nabla^2 \rho = \rho_0 \frac{\partial p}{\partial \rho} \nabla^2 \left(\nabla \cdot \vec{\xi}\right)$$

which, in view of Eq.(15.39), is equivalent to the three-dimensional wave equation (15.14), written in terms of local density as

$$\frac{\partial^2 \rho}{\partial t^2} = \frac{\partial p}{\partial \rho} \nabla^2 \rho \qquad (15.40)$$

Using then

$$\nabla^2 p = \left(\frac{\partial p}{\partial \rho}\right)\nabla^2 \rho \quad , \quad \frac{\partial^2 p}{\partial t^2} = \left(\frac{\partial p}{\partial \rho}\right)\frac{\partial^2 \rho}{\partial t^2}$$

Eq.(15.40) becomes

$$\frac{\partial^2 p}{\partial t^2} = \frac{\partial p}{\partial \rho}\nabla^2 p \tag{15.41}$$

From Eqs.(15.40) and (15.41) it follows that the density and pressure disturbances, given by the local ρ and p, propagate *isotropically* in a fluid with the sound velocity. We can write Euler's equation of motion (15.37) in terms of the vector displacement $\vec{\xi}$ by substituting Eq.(15.39) into Eq.(15.38), which gives

$$\frac{\partial^2 \vec{\xi}}{\partial t^2} = \frac{\partial p}{\partial \rho}\nabla(\nabla \cdot \vec{\xi}) \tag{15.42}$$

The components of Eq.(15.42) are different from scalar wave equations, since

$$\frac{\partial^2 \xi_x}{\partial t^2} = \frac{\partial p}{\partial \rho}\left(\frac{\partial^2 \xi_x}{\partial x^2} + \frac{\partial^2 \xi_y}{\partial x \partial y} + \frac{\partial^2 \xi_z}{\partial x \partial z}\right)$$

and this reduces to Eq.(15.31) for $\xi_y = \xi_z = 0$. This means that the longitudinal displacement ξ_x propagates with the speed of sound, while transverse displacements ξ_y, ξ_z show no wave behaviour. Sound waves in three dimensions appear as *longitudinal plane waves*, of vector displacement $\vec{\xi}$ always parallel to the direction of motion that can only be specified by the magnitude ξ. Therefore sound waves can be treated as longitudinal scalar waves.

15.4. Electromagnetic Waves in Isotropic Dielectrics

Electromagnetic waves are characterized by the coexistence of time-varying electric and magnetic fields, as stated by both Faraday's law (13.23) for media at rest

$$\nabla \times \vec{E} = -\frac{\partial \vec{B}}{\partial t} \quad \text{or} \quad \oint_\Gamma \vec{E} \cdot d\vec{r} = -\frac{\partial}{\partial t}\int_S \vec{B} \cdot d\vec{S} \tag{15.43}$$

and the modified version of Ampère's law (14.1), which in the absence of free currents reads

$$\nabla \times \vec{H} = \frac{\partial \vec{D}}{\partial t} \quad \text{or} \quad \oint_\Gamma \vec{H} \cdot \vec{dl} = \frac{\partial}{\partial t} \int_S \vec{D} \cdot \vec{dS} \qquad (15.44)$$

Consider the propagation in one dimension of a small disturbance of electric and magnetic fields $\vec{E}(x,t)$, $\vec{H}(x,t)$ in a linear, homogenous and isotropic medium, defined by the constitutive relations (14.5). Since we let $\vec{E}$ and $\vec{H}$ depend only on t and the coordinate x, the differential form of the preceding laws reduces to the component equations

$$\frac{\partial E_z}{\partial x} = \frac{\partial B_y}{\partial t} \quad , \quad \frac{\partial E_y}{\partial x} = -\frac{\partial B_z}{\partial t} \qquad (15.45)$$

and

$$\frac{\partial H_y}{\partial x} = \frac{\partial D_z}{\partial t} \quad , \quad \frac{\partial H_z}{\partial x} = -\frac{\partial D_y}{\partial t} \qquad (15.46)$$

which show that the electric and magnetic fields are perpendicular to each other and are both perpendicular to the chosen direction x. Taking the z-axis along the magnetic field direction, that is $H \equiv H_z$, and the y-axis along the electric field direction, $E \equiv E_y$, Eq.(15.45) takes the equivalent form

$$\frac{\partial E}{\partial x} = -\mu \frac{\partial H}{\partial t} \quad \text{or} \quad \frac{\partial E}{\partial x} dxdy = -\mu \frac{\partial H}{\partial t} dxdy \quad \text{i.e.,} \quad dEdy = -\mu \frac{\partial H}{\partial t} dxdy \qquad (15.47)$$

This is the integral form (15.43) of Faraday's law applied to the surface element $dxdy$, as represented in Figure 15.7: a disturbance of the magnetic field, described by a time change of the magnetic flux across the surface element, generates an *e.m.f.*, in other words, it gives rise to an electric field gradient along the x-direction, as described by

$$\oint_\Gamma \vec{E} \cdot \vec{dr} = -Edy + (E + dE)dy = dEdy \qquad (15.48)$$

In a similar manner, Eq.(15.46) takes the form

$$\frac{\partial H}{\partial x} = -\varepsilon \frac{\partial E}{\partial t} \quad \text{or} \quad dHdz = -\varepsilon \frac{\partial E}{\partial t} dxdz \qquad (15.49)$$

which states that the flux of the time-rate of change of the electric field, which is the displacement current density, gives rise to a gradient in the magnetic field along the x-direction, as shown in Figure 15.7.

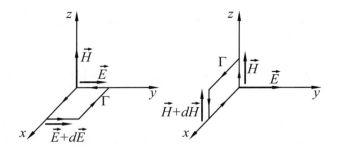

Figure 15.7. The mutual influence between $\vec{E}$ and $\vec{H}$ fields

Equations (15.47) and (15.49) must be simultaneously satisfied. Differentiating Eq.(15.47) with respect to x and Eq.(15.49) with respect to t gives

$$\frac{\partial^2 E}{\partial x^2} = \varepsilon\mu \frac{\partial^2 E}{\partial t^2} \tag{15.50}$$

whereas, if we differentiate Eq.(15.49) with respect to x and Eq.(15.47) with respect to t, one obtains

$$\frac{\partial^2 H}{\partial x^2} = \varepsilon\mu \frac{\partial^2 H}{\partial t^2} \tag{15.51}$$

It follows that varying electric and magnetic fields propagate in a given direction as *transverse waves*, obeying the wave equation of motion (15.6). The corresponding form Eq.(15.14) in three-dimensions is obtained as a particular case of Maxwell's field equations (14.8) which in an isotropic dielectric, where $\sigma = 0$ (and $\rho = 0$), become

$$\nabla^2 \vec{E} = \varepsilon\mu \frac{\partial^2 \vec{E}}{\partial t^2} \quad , \quad \nabla^2 \vec{H} = \varepsilon\mu \frac{\partial^2 \vec{H}}{\partial t^2} \tag{15.52}$$

These equations represent electromagnetic waves travelling with velocity $v = 1/\sqrt{\varepsilon\mu}$. In free space we set

$$c = \frac{1}{\sqrt{\varepsilon_0 \mu_0}} \cong 3 \times 10^8 \, \text{m/s} \tag{15.53}$$

where the SI values for ε_0 and μ_0, given in Chapter 1, have been used. Experiment has shown that the velocity of light in free space has exactly this same value (see Chapter 1) so that light waves are electromagnetic in nature. The ratio given by

$$\frac{c}{v} = \sqrt{\frac{\mu\varepsilon}{\mu_0\varepsilon_0}} = \sqrt{\mu_r \varepsilon_r} = n \tag{15.54}$$

defines the *refractive index n* of a dielectric. For light waves in the visible region of the spectrum μ_r is nearly unity, so that

$$n^2 = \varepsilon_r \qquad (15.55)$$

which is known as Maxwell's relation and holds for many transparent dielectrics.

FURTHER READING

1. D. T. Blackstock – FUNDAMENTALS OF PHYSICAL ACOUSTICS, Wiley, New York, 2000.
2. J. L. Davis – MATHEMATICS OF WAVE PROPAGATION, Princeton University Press, Princeton, 2000.
3. T. G. Main - VIBRATIONS AND WAVES IN PHYSICS, Cambridge University Press, Cambridge, 1993.

PROBLEMS

15.1. Find expressions for the propagation velocities v_l and v_t of the longitudinal and transverse plane waves in an elastic solid of density ρ, Young's modulus E and Poisson's ratio υ. If the elastic body is not allowed to expand sideways, show that velocity v_0 of dilatational waves is greater than v_t and less than v_l.

Solution

The velocities (6.33) of the plane waves propagating in an unbounded elastic medium, given by

$$v_l = \sqrt{\frac{\lambda + 2\mu}{\rho}} \quad , \quad v_t = \sqrt{\frac{\mu}{\rho}}$$

can be written in terms of E and υ by substituting $\mu = E/2(1+\upsilon)$, as derived in Problem 6.1, into Eq.(15.19) to obtain

$$\lambda = \frac{E\upsilon}{(1+\upsilon)(1-2\upsilon)}$$

It follows that

$$v_l = v_0\sqrt{\frac{1-\upsilon}{(1+\upsilon)(1-2\upsilon)}} \quad , \quad v_t = \frac{v_0}{\sqrt{2(1+\upsilon)}}$$

where $v_0 = \sqrt{E/\rho}$ is the phase velocity in a bounded elastic body, given by Eq.(15.23). Since

$$\frac{1}{2(1+\upsilon)} \leq \frac{1-\upsilon}{(1+\upsilon)(1-2\upsilon)}$$

because $0 \leq \upsilon < 1/2$, one obtains $v_t < v_0 \leq v_l$ as required.

15.2. Find the phase velocity for tidal waves in shallow water of depth $h \ll \lambda$.

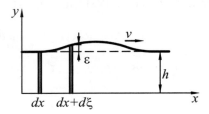

Solution

Consider that a constant mass of water in an element of unit width, height h and length dx moves along the x-axis and assumes a new height $h+\varepsilon$ and length $dx+d\zeta$, but retains unit width. To a first approximation, one obtains

$$hdx = (h+\varepsilon)\left(1+\frac{\partial \xi}{\partial x}\right)dx \quad \text{or} \quad \varepsilon = -h\frac{\partial \xi}{\partial x}$$

Hence, Newton's second law applied to this element reads

$$\rho h dx \frac{\partial^2 \xi}{\partial t^2} = \frac{\partial F}{\partial x}dx$$

where the net force arises from the product of the height and the mean hydrostatic pressure

$$dF = -pdh = -p\frac{\partial \varepsilon}{\partial x}dx \quad \text{or} \quad \frac{\partial F}{\partial x} = -\rho gh\frac{\partial \varepsilon}{\partial x}$$

Substituting into the equation of motion gives

$$\frac{\partial^2 \xi}{\partial t^2} = gh\frac{\partial^2 \xi}{\partial x^2}$$

and hence, tidal waves propagate in shallow water with a velocity $v = \sqrt{gh}$.

15.3. Find the phase velocity of a transverse harmonic wave propagating with angular frequency ω along a linear chain of equal masses m spaced at equal distance a and coupled by light springs of spring constant $k = m\omega_0^2$. Assume that simple harmonic oscillations of the masses, with amplitude much less than a, are allowed in only one plane.

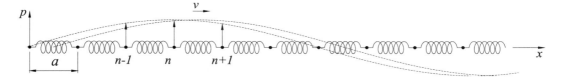

Solution

The equation of motion of the nth particle, which is pulled downwards towards the equilibrium position by restoring forces due to the elastic interactions with nearest neighbours, reads

$$m\frac{d^2 y_n}{dt^2} = -k(y_n - y_{n-1}) - k(y_n - y_{n+1}) = m\omega_0^2 (y_{n+1} + y_{n-1} - 2y_n)$$

where the harmonic displacement due to the transverse wave of angular frequency ω can be expressed as

$$y_n = A_n \cos\omega\left(t - \frac{x}{v}\right) = A_n \cos\omega\left(t - \frac{na}{v}\right)$$

since $x = na$. The equation of motion then becomes

$$-\omega^2 y_n = \omega_0^2 \left\{ \cos\omega\left[t - \frac{(n+1)a}{v}\right] + \cos\omega\left[t - \frac{(n-1)a}{v}\right] - 2\cos\omega\left(t - \frac{na}{v}\right) \right\}$$

$$= -2\omega_0^2 \left(1 - \cos\frac{\omega a}{v}\right) y_n = -4\omega_0^2 \sin^2\left(\frac{\omega a}{2v}\right)$$

Thus, the phase velocity v is given by

$$\omega = 2\omega_0 \sin\left(\frac{\omega a}{2v}\right) \quad \text{or} \quad v = \frac{\omega a}{2\sin^{-1}(\omega/2\omega_0)}$$

15.4. An infinitely long transmission line consists of lumped circuit elements C_0 and L_0, as it can be represented by a pair of parallel wires into one end of which power is fed by an *a.c.* generator. There is a capacitance C_0 per unit length between the lines, which form a condenser, and a self-inductance L_0 per unit length due to the magnetic flux lines generated by the current flow. Find the velocity of propagation for the current waves in the absence of any resistance.

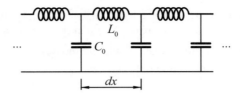

Solution

For a short element of the lossless transmission line of length dx, the self-inductance is $L_0 dx$, and its capacitance is $C_0 dx$. The rate of change of voltage across the element at constant time equals the voltage drop from the self-inductance

$$\frac{\partial V}{\partial x} dx = -(L_0 dx)\frac{dI}{dt} \quad \text{or} \quad \frac{\partial V}{\partial x} = -L_0 \frac{\partial I}{\partial t}$$

There is also a loss of current along the length dx at constant time because the capacitance $C_0 dx$ is charged to a voltage V and this gives

$$-\frac{\partial I}{\partial x} dx = \frac{\partial q}{\partial t} = \frac{\partial}{\partial t}\left[(C_0 dx)V\right] \quad \text{or} \quad -\frac{\partial I}{\partial t} = C_0 \frac{\partial V}{\partial t}$$

Taking $\partial/\partial t$ of the first equation and $\partial/\partial x$ of the second one gives

$$\frac{\partial^2 I}{\partial x^2} = L_0 C_0 \frac{\partial^2 I}{\partial t^2}$$

This is a wave equation for the current, showing that the current waves propagate with velocity

$$v = \frac{1}{\sqrt{L_0 C_0}}$$

A similar equation is obtained for the voltage, with the same velocity of propagation.

15.5. Show that for any two scalar wave functions Ψ and Ψ', which are solutions to the three-dimensional wave equation, the relation $\int_S (\Psi' \nabla \Psi - \Psi \nabla \Psi') \cdot d\vec{S} = 0$ holds true on any closed surface S.

Solution

Consider two arbitrary vector functions $\vec{A}$ and $\vec{B}$ for which Gauss's divergence theorem reads

$$\int_S \vec{A} \cdot d\vec{S} = \int_V \nabla \cdot \vec{A} \, dV \quad , \quad \int_S \vec{B} \cdot d\vec{S} = \int_V \nabla \cdot \vec{B} \, dV$$

If the two vector functions can be written in terms of two scalar wave functions Ψ and Ψ' as

$$\vec{A} = \Psi \nabla \Psi' \quad , \quad \vec{A} = \Psi' \nabla \Psi$$

then on subtraction it is immediate that

$$\int_S (\Psi\nabla\Psi' - \Psi'\nabla\Psi) \cdot d\vec{S} = \int_V \nabla \cdot (\Psi\nabla\Psi' - \Psi'\nabla\Psi) dV$$

The right hand side transforms in view of Eq.(I.10), Appendix I, as

$$\int_V [\nabla \cdot (\Psi\nabla\Psi') - \nabla \cdot (\Psi'\nabla\Psi)] dV = \int_V (\nabla\Psi \cdot \nabla\Psi' + \Psi\nabla^2\Psi' - \nabla\Psi \cdot \nabla\Psi' - \Psi'\nabla^2\Psi) dV$$

$$= \int_V (\Psi\nabla^2\Psi' - \Psi'\nabla^2\Psi) dV$$

and it follows that the two arbitrary scalar wave functions must obey the relation

$$\int_S (\Psi\nabla\Psi' - \Psi'\nabla\Psi) \cdot d\vec{S} = \int_V (\Psi\nabla^2\Psi' - \Psi'\nabla^2\Psi) dV$$

which is known as Green's theorem. Substitution of $\nabla^2\Psi$, $\nabla^2\Psi'$ as given by the scalar wave equation (14.9), provides the required result.

15.6. A ring of rubber of radius R is spinning in its plane with angular speed ω about the axis of symmetry. Find the phase velocity of a transverse wave of small amplitude which is set propagating on the ring.

15.7. Find the upper frequency limit for a transverse wave of small amplitude propagating along a linear chain of equal masses m spaced at equal distance a by a light elastic string of spring constant $k = m\omega_0^2$. What is the upper limit for its phase velocity?

15.8. A longitudinal wave is propagating with velocity v in a fluid of density ρ. Assuming that only small changes of density take place, $\Delta\rho \ll \rho$, show that the increase Δp in the fluid pressure is linearly dependent on the condensation s.

15.9. A sound wave is propagating with velocity v_S in a perfect gas of pressure p and adiabatic exponent γ. Assuming that u is the velocity of the harmonic molecular motion, show that

$$\frac{dp}{p} = \gamma \frac{u}{v_S}$$

15.10. Determine the inductance L_0 and capacitance C_0 per unit length for a transmission line made in the form of coaxial cable, which consists of a cylinder of dielectric material of permittivity ε and permeability μ, having one conductor of radius R_1 along its axis and another, of radius R_2, surrounding its outer surface. Show that the velocity of current waves along such a cable is wholly determined by the characteristics of the dielectric material.

16. Wave Propagation

16.1. HARMONIC WAVES
16.2. WAVE PROPAGATION IN THREE DIMENSIONS
16.3. STATIONARY WAVES
16.4. CONTINUOUS WAVES
16.5. WAVE DISPERSION

16.1. Harmonic Waves

Progressive solutions of the wave equation can always be written as a superposition of two arbitrary functions $f(x-vt)$ and $g(x+vt)$ unrelated to each other. We can specify the wave profile in one dimension by the requirement that f and g satisfy given *initial conditions*, which define the profile of the wave $\Psi_0(x)$ and its time derivative $\chi_0(x)$ at $t=0$, that is,

$$\Psi(x,0) = f(x) + g(x) = \Psi_0(x)$$

$$\frac{\partial \Psi_0(x)}{\partial t} = v\left[-\frac{df(x)}{dx} + \frac{dg(x)}{dx}\right] = \chi_0(x)$$

(16.1)

Using

$$\frac{d\Psi_0(x)}{dx} = \frac{df(x)}{dx} + \frac{dg(x)}{dx}$$

one obtains

$$\frac{df(x)}{dx} = \frac{1}{2}\frac{d\Psi_0(x)}{dx} - \frac{1}{2v}\chi_0(x) \quad , \quad \frac{dg(x)}{dx} = \frac{1}{2}\frac{d\Psi_0(x)}{dx} + \frac{1}{2v}\chi_0(x)$$

Integrating with respect to x gives f and g in terms of initial conditions as

$$f(x) = \frac{1}{2}\Psi_0(x) - \frac{1}{2v}\int_a^x \chi_0(x)dx \quad , \quad g(x) = \frac{1}{2}\Psi_0(x) - \frac{1}{2v}\int_a^x \chi_0(x)dx \quad (16.2)$$

where *a* is an arbitrary constant. The progressive solution $\Psi(x,t)$ then becomes

$$\Psi(x,t) = f(x-vt) + g(x+vt) = \frac{1}{2}\Psi_0(x-vt) + \frac{1}{2}\Psi_0(x+vt) + \frac{1}{2v}\int_{x-vt}^{x+vt} \chi_0(\xi)d\xi \quad (16.3)$$

which is called *d'Alembert's formula*. If the initial velocity $\chi_0(x)$ is zero for all *x*, Eqs.(16.2) show that *f* and *g* are equal and Eq.(16.3) predicts that two waves, having the profile of the original displacement and half the amplitude, travel along the *x*-axis, with one wave in each direction. If the initial displacement $\Psi_0(x)$ is zero for all *x*, *f* and *g* are equal in magnitude but opposite in sign.

A special case of progressive solutions $f(x-vt)$ and $g(x+vt)$, which is of great practical importance, is given by

$$\Psi(x,t) = A\cos[k(x \pm vt) + \varphi] \quad (16.4)$$

Waves with this form are called *harmonic waves* and are generated by the initial conditions

$$\Psi_0(x) = A\cos(kx + \varphi) \quad , \quad \chi_0(x) = \pm kvA\sin(kx + \varphi)$$

which define the wave profile by a cosine curve. The maximum value of the disturbance *A* is called the *amplitude*. The quantity φ is called the *phase constant* of the wave. Two harmonic waves with the same frequency are said to be *in phase* if the difference between their phase constants is an even integral multiple of π, and *out of phase* otherwise. If the difference is an odd integral multiple of π or $\pi/2$ then the two waves are said to be *exactly out of phase* and in *quadrature* respectively. Equation (16.4) can alternatively be written as

$$\Psi(x,t) = A\cos(kx \pm \omega t + \varphi) \quad (16.5)$$

or

$$\Psi(x,t) = A\cos[\omega t \pm (kx + \varphi)] \quad (16.6)$$

where we have set the *angular frequency* ω in a form which is often used in wave theory, given in terms of the so-called *wave number k* and wave velocity *v* as

$$\omega = kv \quad (16.7)$$

If we write the *x* dependent term as

$$\varepsilon(x) = \pm(kx + \varphi) \quad (16.8)$$

$\varepsilon(x)$ is called the *local phase constant* and Eq.(16.6) reads

$$\Psi(x,t) = A\cos\left[\omega t + \varepsilon(x)\right] \qquad (16.9)$$

Equation (16.9) shows that at any given point x, the disturbance is periodic in time, so that the harmonic waves (16.5) consist of simple harmonic motions. All those motions have the same amplitude A, frequency ν and period T given by

$$\nu = \frac{\omega}{2\pi} \quad , \quad T = \frac{2\pi}{\omega}$$

but a different phase constant, since $\varepsilon(x)$ is a linear function of position. Because the difference between the phase constants at any two given points does not vary with time, it is convenient to define the *wavelength* λ as

$$\lambda = \nu T = 2\pi\frac{\nu}{\omega} = \frac{2\pi}{k} \qquad (16.10)$$

and this gives the distance between two successive points whose oscillations are *in phase*. From Eqs.(16.10) it follows that k is the number of waves in a 2π distance, known as the *wave number*.

Expanding the cosine in Eq.(16.4) gives a linear combination of two waves in quadrature

$$\Psi(x,t) = A_1 \cos k(x \pm vt) + A_2 \sin k(x \pm vt) \qquad (16.11)$$

where the information about the amplitude and the phase constant of the wave is expressed by means of two new constants

$$A_1 = A\cos\varphi \qquad\qquad A^2 = A_1^2 + A_2^2$$
$$\text{i.e.,} \qquad\qquad\qquad\qquad (16.12)$$
$$A_2 = -A\sin\varphi \qquad\qquad \tan\varphi = -\frac{A_2}{A_1}$$

Equation (16.11) can be generalized in the form

$$\sum_{j=1}^{n} A_j \cos(kx \pm \omega t + \varphi_j) = A\cos(kx \pm \omega t + \varphi) \qquad (16.13)$$

which states that any linear combination of harmonic waves, having the same frequency ω and wave number k but arbitrary amplitudes A_j and phase constants φ_j reduces to a harmonic wave of definite amplitude A given by

$$A^2 = \sum_{j=1}^{n} A_j^2 + 2\sum_{l>j}^{n} \sum_{j=1}^{n} A_l A_j \cos(\varphi_j - \varphi_l) \tag{16.14}$$

and phase constant φ of the form

$$\tan \varphi = \frac{\sum_{j=1}^{n} A_j \sin \varphi_j}{\sum_{j=1}^{n} A_j \cos \varphi_j} \tag{16.15}$$

The formulation of wave theory is essentially simplified by means of the *complex* linear combination of two waves in quadrature, having the same amplitude and frequency, which reads

$$\Psi(x,t) = A\cos(kx \pm \omega t + \varphi) + iA\sin(kx \pm \omega t + \varphi) = Ae^{i(kx \pm \omega t + \varphi)} = \text{A}e^{i(kx \pm \omega t)} \tag{16.16}$$

We shall call A the *complex amplitude* of the wave

$$\text{A} = Ae^{i\varphi} \tag{16.17}$$

of modulus equal to the amplitude A and argument equal to the phase constant φ. The complex representation (16.16) can be easily differentiated and integrated, hence providing a convenient way of handling harmonic waves. The real part of the complex function (16.16) has the standard form (16.5) of a harmonic wave

$$\text{Re}\left[\text{A}e^{i(kx \pm \omega t)}\right] = \text{Re}\left[Ae^{i\varphi}e^{i(kx \pm \omega t)}\right] = A\,\text{Re}\left[e^{i(kx \pm \omega t + \varphi)}\right] = A\cos(kx \pm \omega t + \varphi)$$

which leads to the following **rule**:

> *A wave may be represented in an equation by a complex function, in which case the real part of the complex function is to be taken, provided the operations are restricted to addition, subtraction, multiplication or division by a real quantity and differentiation or integration with respect to a real variable.*

The real part of the result of any of these operations on a complex function is equal to the result of the same operation on the real part of that function. The rule does not apply in case of multiplication or division by a complex function, where the real part of the product or ratio is different from the product or ratio of the real parts. Under this restriction, the harmonic wave (16.5) can be written in terms of the complex representation (16.16) as

$$\Psi(x,t) = \text{Re}\left[\text{A}e^{i(kx \pm \omega t)}\right] \tag{16.18}$$

The superposition (16.13) of harmonic waves having the same frequency can then be expressed in terms of complex amplitudes of the form (16.17) as

$$\Psi(x,t) = \text{Re}\left[Ae^{i(kx\pm\omega t)}\right] = \sum_{j=1}^{n} A_j \cos(kx \pm \omega t + \varphi_j) = \sum_{j=1}^{n} \text{Re}\left[A_j e^{i(kx\pm\omega t)}\right]$$

$$= \text{Re}\left[\sum_{j=1}^{n} A_j e^{i(kx\pm\omega t)}\right] \quad (16.19)$$

where a known property of the complex quantities has been used. It follows that the harmonic wave superposition can be represented as an addition of complex amplitudes

$$A = \sum_{j=1}^{n} A_j \quad \text{or} \quad Ae^{i\varphi} = \sum_{j=1}^{n} A_j e^{i\varphi_j} \quad (16.20)$$

which reduces to both Eq.(16.14) and Eq.(16.15) by equating the real and the imaginary parts of the two sides.

16.2. Wave Propagation in Three Dimensions

Progressive solutions of the three-dimensional wave equation (15.14) have been expressed as a superposition of two plane waves (15.11) of arbitrary profile travelling with the same constant velocity along an arbitrary direction $\vec{e}_n$. For a profile defined by a cosine curve, we obtain from Eq.(15.11) the *plane harmonic waves* as

$$\Psi(\vec{r},t) = A\cos\left[k(\vec{r}\cdot\vec{e}_n \pm vt) + \varphi\right] = A\cos(\vec{k}\cdot\vec{r} \pm \omega t + \varphi) \quad (16.21)$$

which is the three-dimensional analogue of Eq.(16.4). If α, β, γ are the direction cosines of $\vec{e}_n$, so that

$$\vec{e}_n = \alpha\vec{e}_x + \beta\vec{e}_y + \gamma\vec{e}_z \quad \text{i.e.,} \quad \alpha^2 + \beta^2 + \gamma^2 = 1 \quad (16.22)$$

we may define a *wave vector* $\vec{k}$ along the direction of propagation as

$$\vec{k} = k\vec{e}_n = k\alpha\vec{e}_x + k\beta\vec{e}_y + k\gamma\vec{e}_z = k_x\vec{e}_x + k_y\vec{e}_y + k_z\vec{e}_z \quad (16.23)$$

Substituting Eq.(16.7), Eq.(16.22) becomes

$$k_x^2 + k_y^2 + k_z^2 = \frac{\omega^2}{v^2} \quad (16.24)$$

It is a common practice to write a plane harmonic wave (16.21) in terms of the complex function

$$\Psi(\vec{r},t) = e^{i(\vec{k}\cdot\vec{r}-\omega t)} = Ae^{i(\vec{k}\cdot\vec{r}-\omega t+\varphi)} \qquad (16.25)$$

leaving implicit the rule to take the real part and the restrictions therein.

If we make the assumption that the wave function $\Psi(\vec{r},t)$ has spherical symmetry about the origin, that is $\Psi(\vec{r},t) = \Psi(r,t)$, independent of θ and φ, the Laplacian operator ∇^2 written in spherical coordinates, Eq.(I.23), Appendix I, takes a simpler form, so that the three-dimensional wave equation (15.14) reads

$$\frac{\partial^2 \Psi}{\partial r^2} + \frac{2}{r}\frac{\partial \Psi}{\partial r} = \frac{1}{v^2}\frac{\partial^2 \Psi}{\partial t^2} \qquad (16.26)$$

This may also be written as

$$\frac{\partial^2}{\partial r^2}(r\Psi) = \frac{1}{v^2}\frac{\partial^2}{\partial t^2}(r\Psi) \qquad (16.27)$$

which is a differential equation for the function $r\Psi(r,t)$ having the same form as the equation of wave motion in one dimension (15.6) for the function $\Psi(x,t)$. Thus it has solutions

$$r\Psi = f(r-vt) + g(r+vt)$$

where f and g are again arbitrary functions. The actual wave then takes the form

$$\Psi(r,t) = \frac{1}{r}f(r-vt) + \frac{1}{r}g(r+vt) \qquad (16.28)$$

The first term represents a spherical disturbance that propagates outward from a source located at the origin with constant velocity v. The amplitude of the disturbance falls off as $1/r$. The second term of Eq.(16.28) is interpreted as an incoming spherical disturbance toward a point detector. The radial propagation of a disturbance, which maintains a constant value on spherical wavefronts, defines the *spherical waves*. By letting f be represented in the form (16.4), the outgoing spherical harmonic wave reads

$$\Psi(r,t) = \frac{A}{r}\cos(kr - \omega t + \varphi) \qquad (16.29)$$

and this, in a complex representation, takes the form

$$\Psi(r,t) = \frac{A}{r}e^{i(kr-\omega t)} \qquad (16.30)$$

Waves that propagate in more than one dimension may change direction as a result of changes in the properties of the medium. If these properties change *continuously*, the wave velocity becomes a function of position $v(\vec{r})$ and so does the wave number $k(\vec{r})$, given by Eq.(16.24), and we may assume for the three-dimensional wave equation (15.14) a trial solution of the form

$$\Psi(\vec{r},t) = A(\vec{r})e^{i[k(\vec{r})x(\vec{r}) \pm \omega t]} \tag{16.31}$$

in analogy with (16.16). It is usual to define the properties of the medium in a similar way to that used in Eq.(15.54), by a *generalized refractive index* defined as

$$n(\vec{r}) = \frac{v_0}{v(\vec{r})} \tag{16.32}$$

where v_0 is the wave velocity in free space. The corresponding wave number $k_0 = \omega/v_0$ is a constant. The number of functions to be determined can be reduced by taking

$$k(\vec{r})x(\vec{r}) = k_0 n(\vec{r})x(\vec{r}) = k_0 S(\vec{r}) \tag{16.33}$$

so that the trial solution (16.31) becomes

$$\Psi(\vec{r},t) = A(\vec{r})e^{i[k_0 S(\vec{r}) \pm \omega t]} \tag{16.34}$$

The function $S(\vec{r})$ has the dimensions of length, and Eq.(16.33) shows that it is somewhat similar to the optical path length defined in geometrical optics. Therefore it is called the *eikonal* (image). The quantity $k_0 S(\vec{r})$ may be regarded as the local phase constant of the harmonic motion at a given position $\vec{r}$, so that the equation

$$S(\vec{r}) = \text{const} \tag{16.35}$$

defines the surface of the *wavefront*. The $A(\vec{r})$ function is the envelope of the local maxima and varies independently over the wavefront. Substituting the trial solution (16.34) and equating the real and the imaginary parts, Eq.(15.14) can be reduced to

$$\nabla^2 A + k_0^2 A \left[n^2 - (\nabla S) \cdot (\nabla S) \right] = 0$$
$$2(\nabla A) \cdot (\nabla S) + A \nabla^2 S = 0 \tag{16.36}$$

For a slowly varying amplitude, the second derivative term $\nabla^2 A$ may be neglected, and this gives the equation for the path function as

$$|\nabla S|^2 = \left(\frac{\partial S}{\partial x}\right)^2 + \left(\frac{\partial S}{\partial y}\right)^2 + \left(\frac{\partial S}{\partial z}\right)^2 = n^2 \tag{16.37}$$

known as the *eikonal equation*. For a uniform medium, where n is a constant, a solution of Eq.(16.37) is a plane wave, which propagates in an arbitrary direction $\vec{e}_n$, defined by the direction cosines α, β, γ as given in Eq.(16.22), that is

$$S = n(\alpha x + \beta y + \gamma z) \quad \text{or} \quad \nabla S = n\vec{e}_n \qquad (16.38)$$

An alternative solution, of spherical symmetry $S(\vec{r}) = S(r)$, can be obtained using

$$\frac{\partial S}{\partial x} = \frac{\partial S}{\partial r} \cdot \frac{\partial r}{\partial x} = \frac{\partial S}{\partial r} \cdot \frac{x}{r} \quad \text{or} \quad \left(\frac{\partial S}{\partial x}\right)^2 = \left(\frac{\partial S}{\partial r}\right)^2 \frac{x^2}{r^2}$$

where $r = \sqrt{x^2 + y^2 + z^2}$. Writing in the same way the other derivatives, Eq.(16.37) takes the form

$$\left(\frac{\partial S}{\partial r}\right)^2 \frac{x^2 + y^2 + z^2}{r^2} = n^2 \quad \text{or} \quad \frac{\partial S}{\partial r} = n \qquad (16.39)$$

which is satisfied by

$$S = nr \quad \text{or} \quad \nabla S = n\vec{e}_n \qquad (16.40)$$

Since a well known property of the gradient operator ∇S is that its vector orientation is perpendicular to the surface $S = \text{const}$, Eq.(16.37) can be rewritten as

$$\nabla S(\vec{r}) = n(\vec{r})\vec{e}_s(\vec{r}) \qquad (16.41)$$

where the direction of the unit vector $\vec{e}_s$ is perpendicular to the wavefront. Therefore, continuous curves called *rays* may be constructed, that are everywhere along the gradient of the path function, that is parallel to the local direction of $\vec{e}_s$. For uniform media, as illustrated by Eqs.(16.38) and (16.40) in the particular cases of the plane and spherical waves, the rays are straight lines. Since the ray will in general follow a curve, it is convenient to derive an equation that tells us how $\vec{e}_s(\vec{r})$ changes along this curve. For a general displacement $d\vec{r}$, having length ds along the curve, the change of any function of position $F(\vec{r})$ can be written as

$$dF = \frac{\partial F}{\partial x}dx + \frac{\partial F}{\partial y}dy + \frac{\partial F}{\partial z}dz = \nabla F \cdot d\vec{r} = (\nabla F \cdot \vec{e}_s)ds \qquad (16.42)$$

where $d\vec{r} = \vec{e}_s ds$. We can define the directional derivative of $F(\vec{r})$ along the curve by

$$\frac{dF}{ds} = (\vec{e}_s \cdot \nabla)F$$

so that the scalar function can be regarded as a component of a vector function $F(\vec{r})$, and hence, similar equations hold for the other components. Thus

$$\frac{d\vec{F}}{ds} = (\vec{e}_s \cdot \nabla)\vec{F} \tag{16.43}$$

By taking $\vec{F} = n(\vec{r})\vec{e}_s(\vec{r})$, and using Eqs.(16.43) and (16.41), we can express the rate of change of this quantity along a ray, which successively becomes

$$\frac{d}{ds}(n\vec{e}_s) = \frac{d}{ds}(\nabla S) = (\vec{e}_s \cdot \nabla)(\nabla S) = \left(\frac{\nabla S}{n} \cdot \nabla\right)(\nabla S) = \frac{1}{2n}\nabla(\nabla S)^2 = \frac{1}{2n}\nabla n^2 = \nabla n \tag{16.44}$$

This result is known as the *equation of a ray* and describes how the ray direction changes. If the ray is given by a vector equation $\vec{r} = \vec{r}(s)$, where $\vec{r}$ lies on the ray and s represents the distance along the ray from an arbitrary origin, the unit vector $\vec{e}_s(\vec{r})$ is given by $\vec{e}_s = d\vec{r}/ds$, and the equation of the ray (16.44) reads

$$\frac{d}{ds}\left(n\frac{d\vec{r}}{ds}\right) = \nabla n \tag{16.45}$$

EXAMPLE 16.1. Snell's law

Under the assumption, found in nature, that the refractive index is a function of only one space coordinate, say y, that is $n = v/v(y)$, according to Eq.(15.54), the three component equations of a ray, given by Eq.(16.45), reduce to

$$\frac{d}{ds}\left(n\frac{dx}{ds}\right) = \frac{\partial n}{\partial x} = 0 \quad \text{or} \quad n\frac{dx}{ds} = \text{const}$$

$$\frac{d}{ds}\left(n\frac{dy}{ds}\right) = \frac{\partial n}{\partial y} = \frac{dn}{dy}$$

$$\frac{d}{ds}\left(n\frac{dz}{ds}\right) = \frac{\partial n}{\partial z} = 0 \quad \text{or} \quad n\frac{dz}{ds} = \text{const}$$

It follows that the ray path must be confined to a plane which is perpendicular to the *xz*-plane, and hence, can be chosen to coincide with the *xy*-plane. For a ray direction θ defined as in Figure 16.1, we have $dx/ds = \sin\theta$, which gives

$$n(y)\sin\theta = \text{const} \quad \text{or} \quad n(y)\sin\theta = n(0)\sin\theta_0 \tag{16.46}$$

known as *Snell's law*. The curvature $d\theta/ds$ of the ray path in an inhomogeneous medium can be expressed by substituting $dy/ds = \cos\theta$ into the y component equation for the ray, which becomes

$$\frac{dn}{dy} = \frac{d}{ds}(n\cos\theta) = -n\sin\theta\frac{d\theta}{ds} + \cos\theta\frac{dn}{dy}\frac{dy}{ds} = -n\sin\theta\frac{d\theta}{ds} + \cos^2\theta\frac{dn}{dy}$$

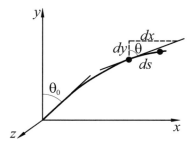

Figure 16.1. Ray path in an inhomogenous medium

This gives

$$\frac{d\theta}{ds} = -\frac{\sin\theta}{n}\frac{dn}{dy} = \frac{\sin\theta}{v}\frac{dv}{dy} = \frac{n\sin\theta}{c}\frac{dv}{dy} = \frac{\text{const}}{c}\frac{dv}{dy}$$

where $dn/n = -dv/v$, since $nv = c$ and $n\sin\theta$ is a constant according to Snell's law. Hence, the curvature of a ray is a measure of the velocity gradient.

16.3. STATIONARY WAVES

The stationary solutions of the wave equations can all be obtained by the method of *separation of variables*. For the equation of wave motion in one dimension (15.6), the desired solutions have the form

$$\Psi(x,t) = X(x)T(t) \tag{16.47}$$

where X and T are functions of x and t only, so that we can write

$$\frac{\partial^2\Psi}{\partial t^2} = X(x)\frac{d^2T(t)}{dt^2} \quad , \quad \frac{\partial^2\Psi}{\partial x^2} = \frac{d^2X(x)}{dx^2}T(t)$$

and Eq.(15.6) takes the form

$$\frac{v^2}{X}\frac{d^2X}{dx^2} = \frac{1}{T}\frac{d^2T}{dt^2} \tag{16.48}$$

where the variables have been separated. Since the left-hand side is independent of t and the right-hand side is independent of x, their equality implies that each side is independent both of x and t, that is both sides are constant. If we set the separation constant equal to $-\omega^2$, Eqs.(16.48) reduce to the harmonic equations

$$\frac{d^2T}{dt^2} + \omega^2 T = 0$$

$$\frac{d^2X}{dx^2} + k^2 X = 0$$
(16.49)

where $k = \omega/v$, as defined in the case of harmonic waves. The stationary solutions (16.47) can take one of the following forms

$$\Psi(x,t) = A\cos(kx)\cos(\omega t + \varphi) \quad , \quad \Psi(x,t) = A\sin(kx)\cos(\omega t + \varphi)$$

$$\Psi(x,t) = A\cos(kx)\sin(\omega t + \varphi) \quad , \quad \Psi(x,t) = A\sin(kx)\sin(\omega t + \varphi)$$
(16.50)

where A, ω and φ are arbitrary. The wave profile is sinusoidal, as in a harmonic travelling wave. The wave displays simple harmonic motion with the same frequency at every point. But all the points now have the same phase constant φ, which means that there is no motion of the wave profile along the wave direction.

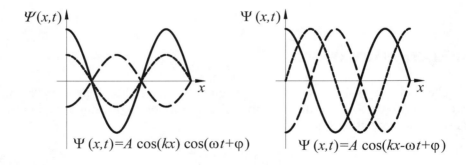

$\Psi(x,t) = A\cos(kx)\cos(\omega t + \varphi)$ $\Psi(x,t) = A\cos(kx - \omega t + \varphi)$

Figure 16.2. Stationary and progressive wave profiles at successive instants of time

The solutions (16.50) describe harmonic motion with *common phase constant* and amplitude that varies sinusoidally from point to point. Such solutions are called *stationary waves*, in contrast to travelling harmonic waves (16.5) which consist of harmonic motion with constant amplitude A and a phase constant φ which varies sinusoidally along the wave direction. From Eq.(16.50) it follows that for stationary waves there are points, called *nodes,* where we have $\Psi = 0$ at all times, given by

$$kx = (2m+1)\frac{\pi}{2} \quad \text{or} \quad kx = 2m\frac{\pi}{2}$$

where m is an integer. The points where the displacement is a maximum, called *antinodes,* are located midway between the nodes, according to

$$kx = 2m\frac{\pi}{2} \quad \text{or} \quad kx = (2m+1)\frac{\pi}{2}$$

Nodes and antinodes are spaced at intervals $\lambda/2$ along the wave direction. Stationary wave profiles at different times are plotted in Figure 16.2 where they are compared to a progressive wave of similar wavelength.

The general solution of the form (16.47) obtained by the principle of superposition from Eqs.(16.50) can be written as

$$\Psi(x,t) = a\cos(kx)\cos(\omega t) + b\sin(kx)\cos(\omega t) + c\cos(kx)\sin(\omega t) + d\sin(kx)\sin(\omega t)$$
(16.51)
$$= A\cos(kx)\cos(\omega t + \alpha) + B\sin(kx)\sin(\omega t + \beta)$$

This reduces to the harmonic solution (16.5) of the equation of wave motion in one dimension for $B = \pm A$. It follows that a travelling harmonic wave can be obtained by a linear superposition of stationary waves of equal amplitudes, whereas the superposition of harmonic waves of equal amplitude, travelling in opposite directions, gives rise to a standing wave.

Stationary waves appear as suitable solutions when the wave motion is subjected to *boundary conditions*. In one dimension these are given by the end constraints. If the motion is restricted to a segment of length l, having nodes at both ends, and the origin is taken at one end, the boundary conditions are, at all times

$$\Psi(0,t) = \Psi(l,t) = 0$$

Then in Eq.(16.51) we must take $a = c = 0$ to ensure that Ψ is zero at the origin at all times. The condition that $\Psi = 0$ at $x = l$ is met by choosing values of k given by

$$\sin k_n l = 0 \quad \text{or} \quad k_n = n\frac{\pi}{l} \quad (n = 1, 2, 3, ...)$$

This also restricts the frequencies to the values $\omega_n = vk_n$, so that Eq.(16.51) reduces to

$$\Psi_n(x,t) = \sin(k_n x)(b_n \cos \omega_n t + d_n \sin \omega_n t) = A_n \sin(k_n x)\cos(\omega_n t + \varphi_n) \quad (16.52)$$

If the ends are both free (antinodes), the boundary conditions take the form

$$\left(\frac{\partial \Psi}{\partial x}\right)_{x=0} = \left(\frac{\partial \Psi}{\partial x}\right)_{x=l} = 0$$

Thus we must take $b = d = 0$ in Eq.(16.51) and the maximum condition on Ψ at $x = l$ gives

$$\cos k_n l = 1 \quad \text{or} \quad k_n = n\frac{\pi}{l} \quad (n=1,2,3,...)$$

which are the same values as before, although the stationary waves have different patterns. Hence, Eq.(16.51) becomes

$$\Psi_n(x,t) = \cos(k_n x)(a_n \cos \omega_n t + c_n \sin \omega_n t) = A_n \cos(k_n x)\cos(\omega_n t + \varphi_n) \quad (16.53)$$

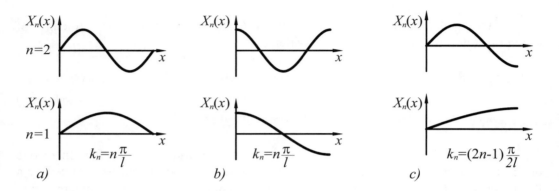

Figure 16.3. Eigenfunctions on a stretched string with end constraints: (a) both ends fixed, (b) both ends free, (c) one end fixed, one end free

If one end of the segment, at $x=0$, is fixed and the other, at $x=l$, is free the solution must obey

$$\Psi(0,t) = \left(\frac{\partial \Psi}{\partial x}\right)_{x=l} = 0$$

which impose that $a = c = 0$, and hence

$$\sin k_n l = 1 \quad \text{or} \quad k_n = (2n-1)\frac{\pi}{2l} \quad (n=1,2,3,...)$$

restricts the solution to the form (16.52). It follows that any stationary wave in one dimension, restricted to a finite length by end constraints, is represented by a function

$$\Psi_n(x,t) = A_n X_n(x)\cos(\omega_n t + \varphi_n) \quad (16.54)$$

where the sinusoidal space functions $X_n(x)$ are known as *eigenfunctions* of the wave motion. The boundary conditions are only satisfied by the eigenfunctions for a given series of wavenumbers k_n called the *eigenvalues* of the problem. The frequencies ω_n of the standing waves with wavevectors k_n are called *eigenfrequencies* of oscillation, since all parts are vibrating in phase or in antiphase. The lowest value $\omega_1 = vk_1$ is known as the *fundamental* frequency and all its multiples are called *harmonics*. A comparative plot of eigenfunctions determined by the end constraints is given in Figure 16.3.

Since the eingenvalues k_n form a series, the most general solution of the equation of wave motion (15.6), given by a linear combination of the eigenfunctions (16.54), appears to consist of an infinite series

$$\Psi(x,t) = \sum_{n=1}^{\infty} A_n X_n(x) \cos(\omega_n t + \varphi_n) \tag{16.55}$$

16.4. Continuous Waves

In view of Eq.(16.11), any harmonic wave (16.4) can be expanded as a linear combination of sine and cosine functions, which reads

$$\Psi(\xi) = A\cos(k\xi + \varphi) = a\cos k\xi + b\sin k\xi = a\cos\frac{2\pi}{\lambda}\xi + b\sin\frac{2\pi}{\lambda}\xi$$

where k is written in terms of a wavelength λ. By a procedure somewhat similar to curve fitting, where an nth degree polynomial is required to fit n points and to determine the expansion coefficients, any periodic function can be represented as the superposition of harmonic terms of frequencies $f_1, f_2 = 2f_1, f_3 = 3f_1, \ldots$, where $f_1 = 1/\lambda$. If we try to fit all points, by letting $n \to \infty$, the representation takes the form

$$\Psi(\xi) = \frac{1}{2}a_0 + \sum_{n=1}^{\infty} a_n \cos\frac{2\pi n\xi}{\lambda} + \sum_{n=1}^{\infty} b_n \sin\frac{2\pi n\xi}{\lambda} \tag{16.56}$$

called the *Fourier series*. The form (16.56) converges to the original function at all points, provided $\Psi(\xi)$ and its first derivative are continuous within each period, conditions which are satisfied for all functions useful in physics. A more convenient alternative to the sine and cosine series is to use the complex notation which transforms Eq.(16.56) into

$$\Psi(\xi) = \frac{1}{2}a_0 + \frac{1}{2}\sum_{n=1}^{\infty}(a_n + ib_n)e^{i2\pi n\xi/\lambda} + \frac{1}{2}\sum_{n=1}^{\infty}(a_n - ib_n)e^{-i2\pi n\xi/\lambda}$$

where the last term can be rewritten in the form

$$\frac{1}{2}\sum_{n=-1}^{-\infty}(a_n - ib_n)e^{i2\pi n\xi/\lambda}$$

If the *complex Fourier coefficients* are now introduced as

$$\Phi_n = \frac{1}{2}(a_n + ib_n), \quad n > 0$$

$$= \frac{1}{2}a_0, \quad n = 0$$

$$= \frac{1}{2}(a_n - ib_n), \quad n < 0$$

Eq.(16.56) takes the compact form

$$\Psi(\xi) = \sum_{n=-\infty}^{\infty} \Phi_n e^{i2\pi n\xi/\lambda} \tag{16.57}$$

Since the exponential functions form an orthogonal set

$$\int_0^\lambda e^{i2\pi m\xi/\lambda} e^{-i2\pi n\xi/\lambda} d\xi = 0 \quad (m \neq n)$$
$$= \lambda \quad (m = n) \tag{16.58}$$

we obtain from Eq.(16.57), by the method of term-by-term integration, the complex coefficients in the form

$$\Phi_n = \frac{1}{\lambda} \int_0^\lambda \Psi(\xi) e^{-i2\pi n\xi/\lambda} d\xi \tag{16.59}$$

The concept of *monochromatic waves*, of one unique frequency, is associated with periodic functions representing disturbances which indefinitely repeat themselves in space and time. In contrast, real waves have a finite extension in space and time and require a representation by nonperiodic functions. If we let the wavelength λ of a periodic wave function $\Psi(\xi)$ to approach infinity, so isolating a *single nonperiodic pulse*, the wave number $k = 2\pi/\lambda$ becomes infinitesimal and the Fourier coefficients take values of the continuous function $\Phi(2\pi/\lambda) = \Phi(k)$. Treating the limit of the sum in Eq.(16.57) as an integral, the wave function of a nonperiodic pulse takes the form

$$\Psi(\xi) = \lim_{\lambda \to \infty} \sum_{n=-\infty}^{\infty} \Phi_n e^{ink\xi} = \int_{-\infty}^{\infty} \Phi(k) e^{ik\xi} dk \tag{16.60}$$

which is called the *Fourier integral formula*. Substituting the complex Fourier coefficients Φ_n, as given by Eq.(16.59), Eq.(16.60) can alternatively be written in the form

$$\Psi(\xi) = \lim_{\lambda \to \infty} \frac{1}{\lambda} \sum_{n=-\infty}^{\infty} \int_0^\lambda \Psi(u) e^{ikn(\xi-u)} du = \frac{1}{2\pi} \int_{-\infty}^{\infty} dk \int_{-\infty}^{\infty} \Psi(u) e^{ik(\xi-u)} du$$

This can be written as

$$\Psi(\xi) = \frac{1}{2\pi} \int_{-\infty}^{\infty} e^{ik\xi} dk \int_{-\infty}^{\infty} \Psi(u) e^{-iku} du$$

and hence, compared to Eq.(16.60), it defines the so-called Fourier spectrum as

$$\Phi(k) = \frac{1}{2\pi} \int_{-\infty}^{\infty} \Psi(u) e^{-iku} du = \frac{1}{2\pi} \int_{-\infty}^{\infty} \Psi(\xi) e^{-ik\xi} d\xi \qquad (16.61)$$

The function $\Phi(k)$ is called the *Fourier transform* of $\Psi(\xi)$. Since the two functions $\Psi(\xi)$ and $\Phi(k)$, as given by Eqs.(16.60) and (16.61), are very nearly symmetrically related, they are said to form a *pair of Fourier transforms*. It is convenient to reformulate Eqs.(16.60) and (16.61) in the symmetrical form

$$\Psi(\xi) = \frac{1}{\sqrt{2\pi}} \int_{-\infty}^{\infty} \Phi(k) e^{ik\xi} dk$$

$$\Phi(k) = \frac{1}{\sqrt{2\pi}} \int_{-\infty}^{\infty} \Psi(\xi) e^{-ik\xi} d\xi \qquad (16.62)$$

The Fourier integrals allow the spectrum analysis of *continuous waves* and therefore the description of waves of arbitrary profile, without any reference to the wave equation.

16.5. Wave Dispersion

Since any solution of the wave equation (15.6) may be expressed as a superposition of harmonic waves by means of Fourier analysis, the medium governed by Eq.(15.6) should transmit harmonic waves of different frequencies at the same velocity v. Consider a harmonic pulse of finite extension in space and time, given by

$$\Psi(x,t) = A(x \pm vt) e^{ik_0(x \pm vt)} = \frac{1}{\sqrt{2\pi}} \int_{-\infty}^{\infty} \Phi(k) e^{ik(x \pm vt)} dk \qquad (16.63)$$

where the envelope $A(x \pm vt)$ is zero outside a specified range of values $\xi = x \pm vt$. This function is said to describe a *wave packet*, as it can always be represented by a continuous superposition of harmonic waves in terms of the Fourier integral (16.62). If v is a constant, and hence, $\Psi(x,t) = \Psi(x \pm vt)$, the wave profile does not change during propagation.

A medium where harmonic waves of different frequencies are transmitted at different velocities is called a *dispersive medium*. This results from a functional relationship

$$\omega = \omega(k) \qquad (16.64)$$

called the *dispersion relation*, representing a generalization of Eq.(16.7), which is valid for constant velocities only. Instead of choosing a particular form of the dispersion relation, it is convenient to expand ω as a function of k in a Taylor series about the wave number k_0 of a wave packet (16.63) as

$$\omega = \omega_0 + v_g(k - k_0) + \zeta(k - k_0)^2 + \ldots \qquad (16.65)$$

where the coefficient of the linear term is

$$v_g = \left(\frac{d\omega}{dk}\right)_{k=k_0} \qquad (16.66)$$

and that of the quadratic term is expressed as

$$\zeta = \frac{1}{2}\left(\frac{d^2\omega}{dk^2}\right)_{k=k_0} \qquad (16.67)$$

If higher order terms are neglected in Eq.(16.65), we obtain an approximation of the dispersion relation in the vicinity of a particular wave number, which becomes exact when ω is a linear or quadratic function of k.

Consider the effect of dispersion on the motion of a wave packet, by taking only the linear term of the dispersion relation (16.65). Substituting into Eq.(16.63) gives

$$\Psi(x,t) = \frac{1}{\sqrt{2\pi}} e^{i(k_0 x \pm \omega_0 t)} \int_{-\infty}^{\infty} \Phi(k) e^{i(k-k_0)(x \pm v_g t)} dk \qquad (16.68)$$

Since v_g is not dependent on k, Eq.(16.68) takes the equivalent form

$$\Psi(x,t) = A(x \pm v_g t) e^{ik_0(x \pm vt)} \qquad (16.69)$$

which describes a harmonic wave of constant frequency ω_0 and phase velocity $v = \omega_0/k_0$, the envelope $A(x \pm v_g t)$ of which is given in terms of the spectral density $\Phi(k)$ of the wave packet. In a frame moving with velocity $\pm v_g$ along the *x*-axis, the envelope appears to be unchanged. It follows that each point of the envelope, for instance its maximum, propagates with velocity v_g, defined by Eq.(16.66) and called the *group velocity*. For a harmonic wave whose wavefront propagates with a *phase velocity* $v = \omega/k$, as defined by Eq.(16.7), substituting ω in Eq.(16.66) gives

$$v_g = v + k\frac{dv}{dk} \qquad (16.70)$$

so that v_g coincides with v in the absence of any dispersion. An alternative form of Eq.(16.70), in terms of wavelength $\lambda = 2\pi/k$, can be derived using

$$\frac{dv}{dk} = \frac{dv}{d\lambda}\frac{d\lambda}{dk} = -\frac{\lambda^2}{2\pi}\frac{dv}{d\lambda}$$

so that

$$v_g = v - \lambda\frac{dv}{d\lambda} \qquad (16.71)$$

Furthermore, it is a straighforward matter to show that

$$v_g = \frac{v}{1-(\omega/v)(dv/d\omega)} \qquad (16.72)$$

Thus the group velocity is always less than the phase velocity in a medium, if the wave or phase velocity decreases with increasing frequency. Note that the quadratic term in the Taylor series expansion of the dispersion relation (16.66) does not contribute directly to the group velocity, causing instead a spatial spreading of the wave packet as it progresses in a dispersive medium.

FURTHER READING
1. H. J. Pain - THE PHYSICS OF VIBRATIONS AND WAVES, Wiley, New York, 1999.
2. F. G. Smith, J. H. Thomson - OPTICS, Wiley, New York, 1990.
3. M. Cartwright - FOURIER METHODS FOR MATHEMATICIANS, SCIENTISTS AND ENGINEERS, Elis Horwood, Chichester, 1990.

PROBLEMS

16.1. A transverse harmonic wave passes over the oscillators in a medium, which are set in motion according to $y = A\cos[\omega t + \varepsilon(x)]$, where $\varepsilon(x)$ is the local phase constant. Find the probability density $P(y)$ of finding one of the oscillators at a displacement between y and $y + dy$.

Solution

The probability of finding the oscillator at a given displacement from equilibrium should be proportional to the amount of time it is located between y and $y + dy$, that is

$$P(y)dy = Cdt \quad \text{or} \quad P(y) = \frac{C}{dy/dt}$$

where

$$\frac{dy}{dt} = -\omega A \sin[\omega t + \varepsilon(x)] = \pm \omega \sqrt{A^2 - y^2}$$

and the constant C is obtained from the normalization condition for probability, which gives

$$1 = \int_{-A}^{A} P(y)dy = \int_{-A}^{A} \frac{Cdy}{\omega\sqrt{A^2 - y^2}} = \frac{C}{\omega}\left[\sin^{-1}\left(\frac{y}{A}\right)\right]_{-A}^{A} = \frac{C\pi}{\omega}$$

Thus one obtains the probability density in the form

$$P(y) = \frac{1}{\pi\sqrt{A^2 - y^2}}$$

which has a sharp increase as $x \to \pm A$.

16.2. Show that ray motion obeys the principle of least time where Snell's law defines the quickest path for a wave to travel between two points, across the boundary between two media, with velocities v_1 and v_2, respectively.

Solution

In the configuration illustrated in Figure (a), let x and x_0 denote the position coordinates, with respect to B_1, of the points O and B_2. The time required for the light ray to travel from P_1 to P_2 is expressed in terms of the variable position x of the point O between B_1 and B_2 in the form

$$t = \frac{P_1O}{v_1} + \frac{P_2O}{v_2} = \frac{n_1}{c}\sqrt{D_1^2 + x^2} + \frac{n_2}{c}\sqrt{D_2^2 + (x_0 - x)^2} = \frac{1}{c}(n_1 l_1 + n_2 l_2)$$

The condition that the time is a minimum reads

$$\frac{dt}{dx} = n_1 \frac{x}{cl_1} - n_2 \frac{x_0 - x}{cl_2} = 0 \quad \text{or} \quad \frac{\sin\theta_1}{\sin\theta_2} = \frac{n_2}{n_1} = \frac{v_1}{v_2}$$

From the configuration illustrated in Figure (b) it is clear that, in a continuous medium, the wave propagation may be regarded as consisting of a series of refractions, which implies that the principle of least time should be formulated as a requirement for the integral

$$t = \frac{1}{c} \int_{P_1}^{P_2} n \, dl$$

to be a minimum. If t is a minimum, any real or virtual variation δt should be equal to zero, which leads to the so-called variational formulation of the principle of least time as

$$\delta \int_{P_1}^{P_2} n \, dl = 0 \quad \text{or} \quad \delta \int_{P_1}^{P_2} \frac{1}{v} dl = 0 \quad \text{or also} \quad \delta \int_{P_1}^{P_2} \frac{1}{\lambda} dl = 0$$

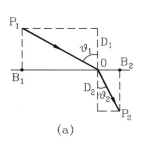
(a)

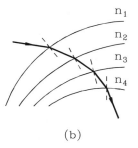
(b)

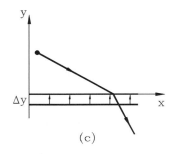
(c)

Thus, a light ray, in going between two points, must transverse an optical path length which is stationary with respect to variations of the path. The similar motion of a particle across the abrupt change in the potential energy at the boundary between two media is illustrated in Figure (c).

16.3. Find the shape of a simple lens of refractive index n and focal length F for minimum spherical aberration to hold if the object is at an infinite distance.

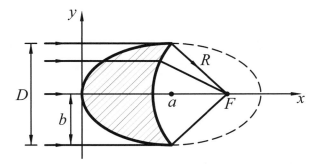

Solution

Consider a set of rays travelling in the direction of the optic axis, x-axis say, of a lens. The position of any ray can be defined by the distance y from the optic axis. Because all the rays are in phase along the y-axis, the optical path from that axis to the focal point should be the same along any ray

$$nF = x + n\sqrt{(F-x)^2 + y^2} \quad \text{or} \quad (nF - y)^2 = n^2 \left[(F-x)^2 + y^2 \right]$$

Rearranging gives

where

$$\frac{(x-a)^2}{a^2} + \frac{y^2}{b^2} = 1$$

$$a = F\frac{n}{n+1} \quad , \quad b = F\sqrt{\frac{n-1}{n+1}}$$

Thus, the front surface of the lens is elliptical with the geometric center at $x = a$. It is apparent that the size of the parallel beam should not exceed the short principal axis

$$D = 2b = 2F\sqrt{\frac{n-1}{n+1}}$$

The second surface of the lens must be spherical with the centre at the focal point and radius R, in order to avoid ray bending by refraction, according to Snell's law.

16.4. A plane membrane of negligible thickness having a mass σ per unit area is stretched under a uniform tension f. Assuming rigid boundaries along the axes $x = 0$, $y = 0$ and the lines $x = a$, $y = b$, find the normal modes of vibration on the membrane.

Solution

A stretched membrane is the two dimensional analogue of a stretched string. The effect of gravity is usually neglected and, in the absence of any disturbance, the membrane is assumed to be flat and lying in the *xy*-plane. There exists a uniform surface tension f per unit length at the ends of any straight line on the membrane. Consider a normal displacement $\xi(x,y,t)$ of a surface element *dxdy*, as represented in the figure.

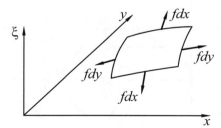

The net force in the *z*-direction is given by the contributions of the two pairs of tensile forces *fdx* and *fdy*. We may regard the second pair as being equivalent to the tension in a string element (of width *dy*) along the *x*-axis, giving rise to the transverse net force which has the form derived in Eq.(15.24) as

$$(fdy)\frac{\partial \alpha}{\partial x}dx = (fdy)\frac{\partial \varepsilon_{zx}}{\partial x}dx = (fdy)\frac{\partial^2 \xi}{\partial x^2}dx$$

In a similar manner the pair of tensile forces *fdx* is equivalent to a normal net force

$$(fdx)\frac{\partial \alpha}{\partial y}dy = (fdx)\frac{\partial \varepsilon_{zy}}{\partial y}dy = (fdx)\frac{\partial^2 \xi}{\partial y^2}dy$$

and Newton's second law (2.9) for the surface element of mass $\sigma dxdy$, where σ denotes the surface density, reads

$$\sigma dxdy \frac{\partial^2 \xi}{\partial t^2} = f dxdy \left(\frac{\partial^2 \xi}{\partial x^2} + \frac{\partial^2 \xi}{\partial y^2} \right) \quad \text{or} \quad \frac{\partial^2 \xi}{\partial x^2} + \frac{\partial^2 \xi}{\partial y^2} = \frac{\sigma}{f} \frac{\partial^2 \xi}{\partial t^2}$$

which is the wave equation in two dimensions (15.10), provided we set the wave velocity on the membrane as $v = \sqrt{f/\sigma}$. In the case of a membrane with *fixed boundaries*, which we take to be the axes $x = 0, y = 0$ and the lines $x = a, y = b$, it is convenient to find the solution by the method of separation of variables. We first put $\xi(x,y,t) = \Phi(x,y)T(t)$ which, upon substitution, gives

$$\frac{1}{T} \frac{\partial^2 T}{\partial t^2} = \frac{v^2}{\Phi} \left(\frac{\partial^2 \Phi}{\partial x^2} + \frac{\partial^2 \Phi}{\partial y^2} \right) = -\omega^2$$

Both sides, which are functions of different independent variables, have been set to be equal to the same constant $-\omega^2$, as before. The time dependent equation reads

$$\frac{\partial^2 T}{\partial t^2} + \omega^2 T = 0 \quad \text{or} \quad T = T_0 e^{\pm i(\omega t + \varphi)}$$

and is now separated from the time-independent wave equation

$$\frac{\partial^2 \Phi}{\partial x^2} + \frac{\partial^2 \Phi}{\partial y^2} + k^2 \Phi = 0$$

where $k = \omega v$. By taking $\Phi(x,y) = X(x)Y(y)$, this equation becomes

$$\frac{1}{X} \frac{\partial^2 X}{\partial x^2} = -k^2 - \frac{1}{Y} \frac{\partial^2 Y}{\partial y^2} = -k_x^2$$

and this splits into the two equations

$$\frac{d^2 X}{dx^2} + k_x^2 X = 0 \quad \text{and} \quad \frac{d^2 Y}{dy^2} + k_y^2 Y = 0$$

where the new constant k_y is related to the separation constants $-k_x^2$ and $-\omega^2$ by

$$k_x^2 + k_y^2 = k^2 = \frac{\omega^2}{v^2}$$

If we choose complex solutions, one obtains the travelling harmonic line waves on the membrane

$$\xi(x,y,t) = A e^{i(k_x x + k_y y - \omega t)}$$

which represent a special case of the general solution (15.7). If we choose real solutions instead, stationary wave functions are obtained in the form

$$\xi(x,y,t) = A\sin(k_x x)\sin(k_y y)e^{-i(\omega t+\varphi)}$$

which satisfy the boundary conditions

$$\xi(0,y,t) = \xi(a,y,t) = \xi(x,0,t) = \xi(x,b,t) = 0$$

In addition we have the restrictions

$$\sin k_x a = 0 \quad \text{or} \quad k_x = m\frac{\pi}{a} \quad (m = 1,2,3,...)$$

$$\sin k_y b = 0 \quad \text{or} \quad k_y = p\frac{\pi}{b} \quad (p = 1,2,3,...)$$

so that the eigenfrequencies are

$$\omega_{mp} = vk = v\sqrt{k_x^2 + k_y^2} = \pi v\sqrt{\left(\frac{m}{a}\right)^2 + \left(\frac{p}{b}\right)^2}$$

For any particular stationary eigenfunction, two sets of straight lines, called the *nodal lines*, which are permanent zeros of the motion, divide the surface into small rectangles, as illustrated in the figure. On crossing a nodal line the motion reverses its phase, so that the differently shaded areas move in opposite phase. If a^2 and b^2 are incommensurable, the eigenfrequencies are all distinct. Otherwise, identical frequencies can be obtained for different pairs (m, p). For instance $\omega_{33} = \omega_{24}$, as illustrated in the figure, and the corresponding eigenfunctions are called *degenerate*.

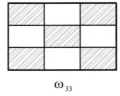
ω_{33}

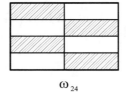
ω_{24}

The general vibration of a rectangular membrane consists of a superposition of wave functions which form the series

$$\xi(x,y,t) = \sum_{m=1}^{\infty}\sum_{p=1}^{\infty} A_{mp}\sin\left(\frac{m\pi}{a}x\right)\sin\left(\frac{p\pi}{b}y\right)e^{-i(\omega_{mp}t+\varphi_{mp})}$$

Note that the amplitude and phase coefficients A_{mp}, φ_{mp} can only be obtained from the initial conditions on both the displacement $\xi(x,y,0) = \xi_0$ and the velocity $(\partial\xi/\partial t)_{t=0} = \chi_0$.

16.5. Two identical sources emitting sound signals of frequency v_0 are moving away and, respectively, toward a receiver at rest, with the same velocity along the same direction. If the observed frequency is $v \ll v_0$ and the velocity of sound is v_S, at what velocity u are the two sources moving?

Solution

For the source moving away from the receiver at a velocity u, the number v_0 of waves emitted per second will be expanded into a distance $v_S + u$ rather than v_S. The observed frequency being the number of waves reaching the receiver per second, it will be given by

$$v_1 = v_0 \frac{v_S}{v_S + u} = \frac{v_0}{1+r}$$

where r stands for u/v_S. For the source moving toward the receiver, the sign of u should be changed in the above formula, so that the observed frequency will be

$$v_2 = \frac{v_0}{1-r}$$

Assuming the receiver is at the origin, the observed signal is obtained by the principle of superposition, which gives

$$\Psi = \Psi_1(0,t) + \Psi_2(0,t) = A\sin(2\pi v_1 t) + A\sin(2\pi v_2 t) = \left[2A\cos 2\pi\left(\frac{v_2 - v_1}{2}\right)t\right]\sin 2\pi \frac{v_1 + v_2}{2}t$$

This is a wave system with a frequency $(v_1 + v_2)/2$, which is very close to the frequency of either source but with a maximum amplitude of $2A$, modulated in time with frequency $(v_2 - v_1)/2$. The maximum sound wave amplitude is observed twice for every period of the modulating frequency, and hence, this occurs at a frequency $v_1 - v_2$. Thus, the observed frequency reads

$$v = \frac{v_0}{1-r} - \frac{v_0}{1+r} = \frac{2rv_0}{1-r^2} \quad \text{or} \quad vr^2 + 2v_0 r - v = 0$$

Since $v \ll v_0$, we have

$$r = \frac{v}{v_0 + \sqrt{v_0^2 + v^2}} \approx \frac{v}{v_0}$$

and it follows that the velocity of each source is given by

$$u = \frac{v_S}{2}\frac{v}{v_0}$$

16.6. A Gaussian harmonic pulse of finite extension in space and time, symmetrically centered at $\xi = 0$, is given by $\Psi(\xi) = A(\xi)e^{ik_0\xi}$, where $A(\xi) = Ae^{-a\xi^2}$ and $\xi = x \pm vt$. The effective width of the pulse is taken to be the interval $\Delta\xi = 2/\sqrt{a}$ over which the amplitude A diminishes by a factor of $1/e$. Find the frequency bandwidth Δk of the Fourier spectrum of the pulse about the spatial frequency k_0.

Solution

Substituting $\Psi(\xi)$ into Eq.(16.62), the spectral density becomes

$$\Phi(k) = \frac{A}{\sqrt{2\pi}} \int_{-\infty}^{\infty} e^{-a\xi^2 + i(k_0-k)\xi} d\xi$$

$$= \frac{A}{\sqrt{2\pi}} e^{-(k_0-k)^2/4a} \int_{-\infty}^{\infty} e^{-a[\xi - i(k_0-k)/2a]^2} d\xi = \frac{A}{\sqrt{2\pi}} e^{-(k_0-k)^2/4a} \int_{-\infty}^{\infty} e^{-au^2} du$$

where we have set $u = \xi - i(k_0 - k)/2a$. The last integral has the Gaussian form (IV.2), Appendix IV, so that we obtain

$$\Phi(k) = \frac{A}{\sqrt{2a}} e^{-(k_0-k)^2/4a}$$

It is apparent that the Fourier transform of the Gaussian pulse is another Gaussian function. The effective width of the spectrum

$$\Delta k = 4\sqrt{a} = \frac{8}{\Delta \xi} \quad \text{i.e.,} \quad \Delta\xi \Delta k = 8$$

is inversely related to the effective width $\Delta\xi$ of the pulse, which means that a narrow pulse has a wide distribution of frequencies. The parameters ξ and k, the product of which is 2π radians, form a pair of *conjugate* parameters. The range in space, defined by the pulse width $\Delta\xi$ and the range in frequency, given by the bandwidth Δk, are inversely related in the form

$$\Delta\xi \Delta k \sim \text{const}$$

where the constant might vary with the shape of pulse and the chosen criterion for the definition of range. It is then reasonable to generalize this as

$$\Delta\xi \Delta k \geq 1$$

which is known as the bandwidth theorem, which states that the product of the allowable range of two conjugated parameters has a lower limit, not equal to zero. If the range in space $\Delta\xi$ is infinite, the range in spatial frequency Δk is zero and one wave number defines the wave, which therefore is monochromatic.

The parameters $\omega = vk$ and $t = \xi/v$ (at a given position x) form another conjugate pair, obeying the bandwidth theorem which reads

$$\Delta t \Delta\omega \geq 1$$

If the range in time Δt, available for the pulse passage through a given point, is small, the range in frequency $\Delta\omega$ is required to be large and vice-versa.

16.7. Two harmonic waves are propagating along the x-axis with velocities v_1 and v_2, respectively. If the planes which are always in phase, and the two waves are spaced at a distance a along the x-axis and progress through the medium with velocity v_0, find the wavelengths λ_1 and λ_2 of the two harmonic waves.

16.8. Use Fermat's principle of least time to derive the focal length f of a thin lens of refractive index n and radii of curvature of the two sides R_1 and R_2, respectively.

16.9. Use Snell's law to find the focal length of a glass marble of refractive index n and radius R in the paraxial approximation of small angles from the optical axis.

16.10. Assuming that Snell's law $n(y)\sin\theta = n_0 \sin\theta_0$ is true in an inhomogeneous medium of refractive index $n(y) = n_0\sqrt{1 - y^2/L^2}$, find the equation of the ray path confined to the xy-plane.

16.11. The amplitude of a monochromatic plane wave of wavelength λ, passing normally through a single slit of width L, may be represented by a rectangular pulse

$$A(\xi) = A, \quad \text{if } |\xi| \leq L/2 \quad \text{and} \quad A(\xi) = 0, \quad \text{otherwise}$$

Show that the Fourier spectral density is shaped by a function $(\sin\alpha)/\alpha$, and find the bandwidth Δk about the spatial frequency k_0.

16.12. An electromagnetic wave, propagating along the x-direction down a waveguide consisting of two parallel conducting planes of separation a in the y-direction, is described by the electric field

$$E_z = E_0 \cos\left(\frac{\pi y}{a}\right) e^{i(k_x x - \omega t)}$$

Find the phase and group velocity of the wave, and show that their product is c^2, where c stands for the speed of light.

17. WAVE ENERGY

17.1. ENERGY DENSITY
17.2. ENERGY FLOW
17.3. WAVE MOMENTUM
17.4. ATTENUATION OF WAVES
17.5. WAVE ENERGY AT AN INTERFACE

17.1. ENERGY DENSITY

If we consider waves travelling in one dimension in a medium with linear density μ, an expression for the total energy of the wave can be obtained by considering a finite section of the one dimensional transmitting medium. D'Alembert's formula (16.3) shows that such waves propagate in both directions with constant velocity and profile and therefore carry energy without loss.

If Ψ is some general function of x, consider stretching a string with this wave form, as in Figure 15.5, to progressively greater amplitudes $\xi_y(x,t) = \alpha \Psi(x,t)$ where $\alpha = 0 \to 1$. The driving force acting on a length element dx, Eq.(15.24), is $F(\partial^2 \Psi / \partial x^2) dx$, so that the work done on this element, in moving it from $\alpha \Psi$ to $(\alpha + d\alpha) \Psi$, is $-F\Psi(\partial^2 \Psi / \partial x^2) \alpha\, dx\, d\alpha$ and the total work can be written as

$$dW = -F\Psi \left(\frac{\partial^2 \Psi}{\partial x^2} \right) dx \int_0^1 \alpha\, d\alpha = -\frac{1}{2} F\Psi \left(\frac{\partial^2 \Psi}{\partial x^2} \right) dx \qquad (17.1)$$

Taking the sum of the contributions from all the elements in a finite section, one obtains

$$W = -\frac{1}{2} \mu v^2 \int_0^x \Psi \left(\frac{\partial^2 \Psi}{\partial x^2} \right) dx \qquad (17.2)$$

where use has been of Eq.(15.26). Integrating by parts Eq.(17.2) and adding the kinetic energy in a finite section of the string gives

$$T = \frac{1}{2}\mu \int_0^x \left(\frac{\partial \Psi}{\partial x}\right)^2 dx \tag{17.3}$$

and this leads to

$$T + W = \frac{1}{2}\mu \int_0^x \left(\frac{\partial \Psi}{\partial x}\right)^2 dx + \frac{1}{2}\mu v^2 \int_0^x \Psi \left(\frac{\partial \Psi}{\partial x}\right)^2 dx - \frac{1}{2}\mu v^2 \left[\Psi \frac{\partial \Psi}{\partial x}\right]_0^x \tag{17.4}$$

The last term on the right-hand side expresses the boundary conditions on the system, given by the end constraints. The second term on the right-hand side defines the potential energy stored in the same section

$$U = \frac{1}{2}\mu v^2 \int_0^x \Psi \left(\frac{\partial \Psi}{\partial x}\right)^2 dx \tag{17.5}$$

so that the total wave energy is given by

$$E(x,t) = \frac{1}{2}\mu \int_0^x \left[\left(\frac{\partial \Psi}{\partial t}\right)^2 + v^2 \left(\frac{\partial \Psi}{\partial x}\right)^2\right] dx \tag{17.6}$$

and the *total energy density* $\varepsilon(x,t)$ can be written as

$$\varepsilon(x,t) = \frac{1}{2}\mu \left[\left(\frac{\partial \Psi}{\partial t}\right)^2 + v^2 \left(\frac{\partial \Psi}{\partial x}\right)^2\right] \tag{17.7}$$

For travelling waves, which obey Eqs.(15.3) or (15.4), the two terms on the right-hand side which give the kinetic and potential energy densities are always equal.

In calculating the energy of *travelling harmonic waves* we must *not* use the complex representation (16.16), since this would involve the product of two complex quantitites, whose real part is not equal to the product of the real parts of the two factors. Applying Eq.(17.7) to calculate the energy density carried by the harmonic wave (16.5), in the positive *x*-direction, we find that

$$\varepsilon(x,t) = \frac{1}{2}\mu \left(\omega^2 + v^2 k^2\right) A^2 \sin^2(kx - \omega t + \varphi) \tag{17.8}$$

The instantaneous energy distribution, which travels along the *x*-axis at the same velocity as the wave profile, is illustrated in Figure 17.1. The kinetic and potential energy densities are equal at any point, the maxima of $\varepsilon(x,t)$ corresponding to those points where the disturbance is zero, whilst the values of both the velocity $\partial \Psi / \partial t$ and the slope $\partial \Psi / \partial x$ are the highest.

For a *stationary wave* (16.52) where the variables *x* and *t* appear in separate factors the kinetic and potential energy densities are no longer equal. Applying Eq.(17.7) gives

$$\varepsilon(x,t) = \frac{1}{2}\mu A_n^2 \left[\omega_n^2 \sin^2(k_n x) \sin^2(\omega_n t + \varphi_n) + v^2 k_n^2 \cos^2(k_n x) \cos^2(\omega_n t + \varphi_n) \right] \quad (17.9)$$

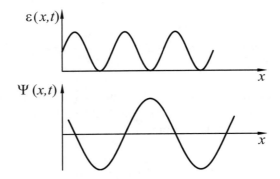

Figure 17.1. Distribution of energy along a travelling harmonic wave

As illustrated in Figure 17.2 for a stretched string fixed at both ends, when the string is straight, the energy (17.9) is all kinetic and located around the antinodes, whereas when the string is stationary, the energy is all potential and located around the nodes. Since energy passes between the nodes and the antinodes, there is no net flow of energy in one direction.

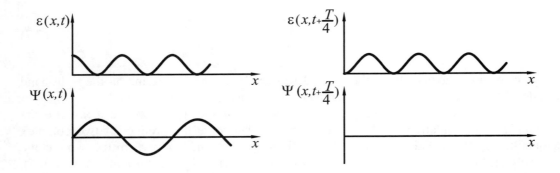

Figure 17.2. Distribution of kinetic and potential energies in a normal mode of a stationary wave

Consider the general case of an arbitrary stationary wave, represented as a superposition of eigenfunctions. The total kinetic and potential energies of a vibrating string of length l, fixed at both ends, are given in terms of the series (16.55) for $\Psi(x,t)$, which reads

$$\Psi(x,t) = \sum_{n=1}^{\infty} A_n \sin(k_n x) \cos(\omega_n t + \varphi_n) \quad (17.10)$$

It follows that

$$\frac{\partial \Psi}{\partial t} = -\sum_{n=1}^{\infty} \omega_n A_n \sin(k_n x) \sin(\omega_n t + \varphi_n) \quad , \quad \frac{\partial \Psi}{\partial x} = \sum_{n=1}^{\infty} k_n A_n \cos(k_n x) \cos(\omega_n t + \varphi_n)$$

The total kinetic energy (17.3) given by integration over the length of the string as

$$T = \frac{1}{2}\mu \int_0^l \left[\sum_{n=1}^{\infty} \omega_n A_n \sin(k_n x) \sin(\omega_n t + \varphi_n)\right]^2 dx = \frac{\mu l}{4} \sum_{n=1}^{\infty} \omega_n^2 A_n^2 \sin^2(\omega_n t + \varphi_n) \quad (17.11)$$

and, in a similar manner, we obtain the total potential energy the form

$$U = \frac{1}{2}\mu v^2 \int_0^l \left[\sum_{n=1}^{\infty} k_n A_n \cos(k_n x) \cos(\omega_n t + \varphi_n)\right]^2 dx = \frac{\mu l}{4} \sum_{n=1}^{\infty} \omega_n^2 A_n^2 \cos^2(\omega_n t + \varphi_n) \quad (17.12)$$

where $\omega_n^2 = v^2 k_n^2$. The total energy of vibration is then given by the addition of Eqs.(17.11) and (17.12) and is independent of time, that is

$$E = \frac{\mu l}{4} \sum_{n=1}^{\infty} \omega_n^2 A_n^2 \quad (17.13)$$

Since there are no cross terms that involve the products of vibrational amplitudes corresponding to different eigenfunctions in the energy equations (17.11), (17.12) and (17.13), the terms of the series (16.55) are called *normal modes*. The energy equations show that either kinetic, potential or the total energy of any vibration of a string is the sum of the energies obtained for each normal mode into which the vibration is resolved.

17.2. Energy Flow

The rate of change of the total wave energy is obtained by differentiating Eq.(17.6), with respect to time, which gives

$$\frac{dE(x,t)}{dt} = \mu \int_0^x \left[\frac{\partial \Psi}{\partial t}\frac{\partial^2 \Psi}{\partial t^2} + v^2 \frac{\partial \Psi}{\partial x}\frac{\partial^2 \Psi}{\partial x \partial t}\right] dx = \mu v^2 \int_0^x \left[\frac{\partial \Psi}{\partial t}\frac{\partial^2 \Psi}{\partial x^2} + \frac{\partial \Psi}{\partial x}\frac{\partial^2 \Psi}{\partial x \partial t}\right] dx$$

$$= \mu v^2 \int_0^x \frac{\partial}{\partial x}\left(\frac{\partial \Psi}{\partial t}\frac{\partial \Psi}{\partial x}\right) dx = \mu v^2 \left(\frac{\partial \Psi}{\partial t}\frac{\partial \Psi}{\partial x}\right)_x - \mu v^2 \frac{\partial}{\partial x}\left(\frac{\partial \Psi}{\partial t}\frac{\partial \Psi}{\partial x}\right)_{x=0} \quad (17.14)$$

Each term of the form

$$P(x,t) = -\mu v^2 \left(\frac{\partial \Psi}{\partial t}\right)\left(\frac{\partial \Psi}{\partial x}\right) \quad (17.15)$$

represents the instantaneous value of *power* passing through a point x at time t in the positive x-direction. This can be seen using Figure 15.5 for the shear motion of a string element, where $\mu v^2 = F$ and $F_d = -F\alpha(x) = -F\varepsilon_{yx} = -F(\partial \xi_y / \partial x) = -F(\partial \Psi / \partial x)$ is the transverse driving force which the part of the string to the left of the point x exerts on the part to the right of that point. Therefore $F_d(\partial \Psi / \partial t) = -F(\partial \Psi / \partial x)(\partial \Psi / \partial t)$, where $\partial \Psi / \partial t = \partial \xi_x / \partial t$ is the transverse velocity of the disturbance at the point x, is the rate at which the left-hand part works on the right-hand part, in other words, it is the rate of energy flow or power in the positive x-direction. The first term on the right-hand side of Eq.(17.14) then gives the energy flux into the segment $(0, x)$ in the negative x-direction, while the second term represents the flow into the segment from the region $x < 0$.

It is a straightforward matter to show that the energy flow (17.15) and the energy density (17.7) are related by the continuity relation

$$\frac{\partial}{\partial x} P(x,t) = -\frac{\partial}{\partial t} \varepsilon(x,t) \qquad (17.16)$$

For convenience, a *characteristic impedance* Z of the transmitting medium can be introduced to be the constant of proportionality between the driving force and the velocity $\partial \Psi / \partial t$. Using Eqs.(15.3) and (15.26), we have for a travelling shear wave that

$$Z = \frac{F_d}{\partial \Psi / \partial t} = \frac{-F(\partial \Psi / \partial x)}{\partial \Psi / \partial t} = \frac{F(\partial \Psi / \partial t)}{v(\partial \Psi / \partial t)} = \frac{F}{v} = \mu v = \sqrt{F\mu} \qquad (17.17)$$

It follows that Z is a constant, independent of both the wave profile $f(x - vt)$ and the frequency ω of the travelling wave. It is common practice to develop the energy formulation in terms of Z, which can be defined for all the media sustaining a wave motion. Substituting $\mu v^2 = Zv$ into Eq.(17.15) the energy flow in the positive x-direction can be written as

$$P(x,t) = -Zv\left(\frac{\partial \Psi}{\partial t}\right)\left(\frac{\partial \Psi}{\partial x}\right) \qquad (17.18)$$

For a travelling wave in the same direction, Eq.(15.3) holds and we obtain the power

$$P(x,t) = Z\left(\frac{\partial \Psi}{\partial t}\right)^2 = v\varepsilon(x,t) \qquad (17.19)$$

where we have used the expression for the total energy density (17.7) carried by the wave. For the wave travelling in the negative x-direction, in view of Eq.(15.4), Eq.(17.19) gives a negative power, so that in general we must write

$$P(x,t) = \pm v\varepsilon(x,t) \qquad (17.20)$$

since the energy travels in the same direction as the wave.

EXAMPLE 17.1. Characteristic impedances

For a *longitudinal travelling wave* in an elastic string, the stress-strain relation (15.18) is

$$\frac{F}{S} = E\frac{\partial \xi_x}{\partial x}$$

Assuming that the dilatational disturbance $\Psi = \xi_x$ obeys Eq.(15.3) which describes a wave travelling along the *x*-direction, we obtain the longitudinal driving force as

$$F_d = -ES\frac{\partial \Psi}{\partial x} = \frac{ES}{v}\frac{\partial \Psi}{\partial t}$$

It follows that the characteristic impedance of the dilatational wave is given by

$$Z = \frac{F_d}{\partial \Psi / \partial t} = \frac{ES}{v} = S\rho v = S\sqrt{\rho E} \tag{17.21}$$

For most purposes, a more useful quantity is the characteristic impedance per unit area, that is

$$Z_0 = \frac{Z}{S} = \sqrt{\rho E}$$

For a *travelling sound wave*, the local pressure was expressed before in terms of condensation (15.29), which becomes proportional to $\partial \Psi / \partial t$ using Eq.(15.3), and this gives

$$s = -\frac{\partial \xi_x}{\partial x} = \frac{1}{v}\frac{\partial \xi_x}{\partial t} = \frac{1}{v}\frac{\partial \Psi}{\partial t}$$

where $\Psi = \xi_x$. The driving force is then obtained as

$$F_d = S\rho_0 v^2 s = S\rho_0 v\frac{\partial \Psi}{\partial t}$$

and the characteristic impedance takes one of the forms

$$Z = \frac{F_d}{\partial \Psi / \partial t} = S\rho_0 v \quad \text{or} \quad Z_0 = \rho_0 v \tag{17.22}$$

For *electromagnetic waves* travelling with velocity c in free space, the appropriate concept of impedance is obtained by comparing the instantaneous values of mechanical power (17.19) per unit area, which reads

$$P(x,t) = Z_0\left(\frac{\partial \Psi}{\partial t}\right)^2$$

and electromagnetic power, that is the Poynting vector (14.13), which for a plane wave travelling in the *x*-direction reduces to

$$S(x,t) = E_y H_z = EH$$

The electric field, of magnitude E, of a travelling electromagnetic wave must obey Eq.(15.3) which takes the form

$$\frac{\partial E}{\partial t} + c\frac{\partial E}{\partial x} = 0$$

so that, substituting Eq.(15.45), we obtain

$$\frac{\partial E}{\partial t} = -c\frac{\partial E}{\partial x} = c\frac{\partial B}{\partial t} = \mu_0 c\frac{\partial H}{\partial t}$$

and this, upon integration, yields

$$E = \mu_0 c H$$

Hence, the magnitude of the Poynting vector becomes

$$S(x,t) = \mu_0 c H^2$$

Comparison with the instantaneous mechanical power given in terms of velocity $\partial \Psi / \partial t$ allows us to introduce the electromagnetic impedance per unit area, in free space, as

$$Z_0 = \mu_0 c = \frac{1}{\varepsilon_0 c} = \sqrt{\frac{\mu_0}{\varepsilon_0}} \qquad (17.23)$$

which has the constant value $Z_0 = 120\pi\ \Omega$ in SI units. Written in terms of field magnitudes as

$$Z_0 = \frac{E}{H} \qquad (17.24)$$

the characteristic impedance becomes similar to the familiar definition from the circuit theory

$$Z_0 = \frac{V}{I} \qquad (17.25)$$

since the electric and magnetic fields are closely related to voltage and current by

$$\int_\Gamma \vec{E} \cdot d\vec{r} = V \quad , \quad \int_\Gamma \vec{H} \cdot d\vec{l} = I$$

For *harmonic* waves (16.5), the power (17.19) carried by the wave is found to be

$$P(x,t) = Z\omega^2 A^2 \sin^2(kx - \omega t + \varphi) \qquad (17.26)$$

and this shows that the power and the proportional energy density, transmitted by a harmonic wave, depend on the product of the square of the disturbance amplitude and the square of frequency. The *average* power carried over a time interval τ is

$$\langle P(\tau)\rangle = \frac{1}{\tau}\int_0^\tau P(x,t)\,dt$$

$$= \frac{Z\omega^2 A^2}{\tau}\int_0^\tau \sin^2(kx-\omega t)\,dt = \frac{Z\omega^2 A^2}{2}\left[1 + \frac{\sin 2(kx-\omega\tau) - \sin 2kx}{2\omega\tau}\right] \quad (17.27)$$

Radiation detectors usually have a slow response time compared to $2\pi/\omega$, the period of the wave oscillation, so that we have $\omega\tau \gg 1$. The *wave intensity* I measured by a detector is defined to be the average power crossing the unit area normal to the direction of propagation, in the limiting case of $\tau \to \infty$, as

$$I = \frac{1}{S}\lim_{\tau\to\infty}\langle P(\tau)\rangle = \frac{1}{2}Z_0\omega^2 A^2 \quad (17.28)$$

If, in one dimension, the harmonic wave is represented by a complex function $\Psi(x,t)$ of the form (16.16), we have $\Psi\Psi^* = |\Psi|^2 = A^2$ so that the intensity (17.28) becomes

$$I = \frac{1}{2}Z_0\omega^2|\Psi|^2 \quad (17.29)$$

Equation (17.29) also holds for plane harmonic waves when using their complex representation $\Psi(\vec{r},t)$ given by Eq.(16.25). A detector is thus able to respond to $|\Psi|^2$ in a given medium Z_0.

EXAMPLE 17.2. Intensity of electromagnetic waves

For an electromagnetic plane wave, travelling in the x-direction in free space, as defined by Maxwell's equations (15.45) and (15.46), where $E \equiv E_y$, $H \equiv H_z$, the instantaneous power per unit area is given by the Poynting vector (14.13) as

$$S(x,t) = EH = Z_0 H^2 = \frac{1}{Z_0}E^2$$

If the travelling wave is sinusoidal, that is

$$E(x,t) = E_0\cos(kx - \omega t + \varphi) = \mathrm{Re}\left[E_0 e^{i(kx-\omega t+\varphi)}\right]$$

$$H(x,t) = H_0\cos(kx - \omega t + \varphi) = \mathrm{Re}\left[H_0 e^{i(kx-\omega t+\varphi)}\right]$$

where E_0 and H_0 are the field amplitudes, the intensity can be written as

$$I = \lim_{\tau\to\infty}\langle S(x,t)\rangle = \frac{1}{2}Z_0 H_0^2 = \frac{1}{2Z_0}E_0^2$$

Since the magnetic field amplitude $H_0 = E_0/Z_0$ is small compared to E_0, effects due to the magnetic field can be usually neglected. In a complex scalar representation of the electric field $E(x,t)$ by a wave function $\Psi(x,t)$, the intensity of electromagnetic waves takes the form

$$I = \frac{1}{2Z_0}|\Psi|^2 \tag{17.30}$$

which is the analogue of Eq.(17.29).

For a *stationary* wave, whose energy density $\varepsilon(x,t)$ was expressed by Eq.(17.9), the instaneous energy flow $P(x,t)$ can be derived using the continuity relation (17.16), where

$$\frac{\partial}{\partial t}\varepsilon(x,t) = -\frac{1}{2}\mu A_n^2 \omega_n^3 \cos(2k_n x)\sin 2(\omega_n t + \varphi_n)$$

for each normal mode, so that

$$P_n(x,t) = \frac{1}{4}Z_0 \omega_n^2 A_n^2 \sin(2k_n x)\sin 2(\omega_n t + \varphi_n)$$

From the sinusoidal time-dependence of the power transported by each normal mode, it follows that the wave intensity is averaged out, as expected, that is

$$I_n = \lim_{\tau \to \infty}\langle P_n(x,t)\rangle = 0$$

17.3. Wave Momentum

There is a close connection between energy transported by a travelling wave and the momentum carried by the wave. We can arrive at an expression for the momentum density by multiplying both sides of the one-dimensional wave equation (15.6) by $\mu \partial\Psi/\partial x$, which gives

$$\mu v^2 \frac{\partial^2\Psi}{\partial x^2}\frac{\partial\Psi}{\partial x} = \mu\frac{\partial^2\Psi}{\partial t^2}\frac{\partial\Psi}{\partial x}$$

We can put this in the form

$$\frac{\partial}{\partial x}\left[\frac{1}{2}\mu v^2\left(\frac{\partial \Psi}{\partial x}\right)^2\right] = \frac{\partial}{\partial t}\left[\mu \frac{\partial \Psi}{\partial x}\frac{\partial \Psi}{\partial t}\right] - \frac{\partial}{\partial x}\left[\frac{1}{2}\mu\left(\frac{\partial \Psi}{\partial t}\right)^2\right]$$

allowing us to express the total energy density (17.7) as

$$\frac{\partial}{\partial x}\varepsilon(x,t) = -\frac{\partial}{\partial t}\left[-\mu\frac{\partial \Psi}{\partial x}\frac{\partial \Psi}{\partial t}\right]$$

Hence, the gradient of the energy density, which has the dimensions of a force density, can be expressed as the time rate of change of the quantity

$$\pi_x(x,t) = -\mu\frac{\partial \Psi}{\partial x}\frac{\partial \Psi}{\partial t} \tag{17.31}$$

which is interpreted, by Newton's second law (2.9), as a localized *momentum density* in the x-direction associated with the wave motion

$$\frac{\partial}{\partial x}\varepsilon(x,t) = -\frac{\partial}{\partial t}\pi_x(x,t) \tag{17.32}$$

A comparison between Eqs.(17.31) and (17.15) shows that the flow of wave energy along the x-direction is given in terms of the momentum density by

$$P(x,t) = v^2 \pi_x(x,t) \tag{17.33}$$

EXAMPLE 17.3. Electromagnetic radiation pressure

Consider an electromagnetic wave, travelling in free space, which is incident on a region containing both charges and currents. We can write the force exerted by the electromagnetic wave on the unit volume of matter to be given by the Lorentz force law (13.30), written as

$$\vec{f}_L = \rho\vec{E} + \mu_0 \vec{j} \times \vec{H}$$

If ρ and $\vec{j}$ are expressed in terms of $\vec{E}$ and $\vec{H}$, using Maxwell's equations (14.2), we obtain

$$\vec{f}_L = \varepsilon_0(\nabla \cdot \vec{E})\vec{E} + \mu_0(\nabla \times \vec{H}) \times \vec{H} - \frac{1}{c^2}\frac{\partial \vec{E}}{\partial t} \times \vec{H}$$

It is convenient to introduce in the last term the Poynting vector $\vec{S}$, Eq.(14.13). Hence, we have

$$\vec{f}_L = \varepsilon_0(\nabla \cdot \vec{E})\vec{E} + \mu_0(\nabla \times \vec{H}) \times \vec{H} - \frac{\partial}{\partial t}\left(\frac{1}{c^2}\vec{S}\right) + \frac{1}{c^2}\vec{E} \times \frac{\partial \vec{H}}{\partial t}$$

that is

$$\vec{f}_L + \frac{\partial}{\partial t}\left(\frac{1}{c^2}\vec{S}\right) = \varepsilon_0(\nabla \cdot \vec{E})\vec{E} + \varepsilon_0(\nabla \times \vec{E}) \times \vec{E} + \mu_0(\nabla \cdot \vec{H})\vec{H} + \mu_0(\nabla \times \vec{H}) \times \vec{H} \tag{17.34}$$

where the term containing $\nabla \cdot \vec{H} = 0$ was introduced for reasons of symmetry. The left-hand side contains the force density $\vec{f}_L$ exerted by the wave on the matter and a second term which has the form of a time derivative, so that it can be associated, by Newton's second law (2.9), with the time rate of change of a *momentum density* $\vec{\pi}$ transported by the electromagnetic wave, given by

$$\vec{\pi} = \frac{1}{c^2}\vec{S} = \frac{1}{c^2}\left(\vec{E} \times \vec{H}\right) \tag{17.35}$$

It is obvious that Eq.(17.35) is the three-dimensional analogue of Eq.(17.33). To explore the significance of Eq.(17.34) we will consider, for simplicity, the special case of a plane electromagnetic wave, described by Eqs.(15.45) and (15.46), where we can put $E \equiv E_y$ and $H \equiv H_z$, thus

$$f_L \vec{e}_x + \frac{\partial}{\partial t}\left(\frac{1}{c^2}EH\right)\vec{e}_x = -\varepsilon_0 E \frac{\partial E}{\partial x}\vec{e}_x - \mu_0 H \frac{\partial H}{\partial x}\vec{e}_x = -\frac{d}{dx}\left\{\frac{1}{2}ED + \frac{1}{2}HB\right\}\vec{e}_x$$

The time average of the second term on the left-hand side is zero for harmonic waves, and we obtain, in view of Eq.(14.11), that

$$\langle f_L(x) \rangle = -\frac{d}{dx}\langle u_{em}(x) \rangle$$

which shows that the force per unit volume exerted by the electromagnetic wave on the matter made up of charges and currents is given by the gradient of the electromagnetic energy density carried by the wave, along the direction in which the wave propagates. The *radiation pressure* exerted on the surface of the absorbing region, describing the net force per unit area, then takes the form

$$p_r = \int_{x_0}^{\infty}\langle f_L(x) \rangle = -\int_{x_0}^{\infty}\frac{d}{dx}\langle u_{em}(x) \rangle dx = \langle u_{em}(x_0) \rangle \tag{17.36}$$

where x_0 specifies the position of the absorbing surface. Therefore the pressure exerted by a plane electromagnetic wave on a perfectly absorbing surface, at normal incidence, is equal to the energy density of the wave.

17.4. ATTENUATION OF WAVES

The propagation of real waves through a medium is described in most cases by a modified form of the wave equation (15.14), written as

$$\nabla^2 \Psi = \frac{1}{v^2}\left(\frac{\partial^2 \Psi}{\partial t^2} + 2\gamma \frac{\partial \Psi}{\partial t}\right) \tag{17.37}$$

which is often known as the *equation of telegraphy*. We derived a similar equation (14.9) for scalar electromagnetic waves, which were propagating in regions of matter containing free-charge and free-current distributions. In one dimension, Eq.(17.37) reads

$$\frac{\partial^2 \Psi}{\partial x^2} = \frac{1}{v^2}\left(\frac{\partial^2 \Psi}{\partial t^2} + 2\gamma \frac{\partial \Psi}{\partial t}\right) \tag{17.38}$$

and a trial solution is

$$\Psi(x,t) = e^{-\gamma t} f(x,t) \tag{17.39}$$

Substituting Eq.(17.39), Eq.(17.38) becomes

$$\frac{\partial^2 f}{\partial x^2} = \frac{1}{v^2}\left(\frac{\partial^2 f}{\partial t^2} - \gamma^2 f\right)$$

and this coincides with the equation of wave motion (15.6), provided γ is so small that we can neglect γ^2. Therefore, under this assumption, we can choose $f(x,t)$ as given by Eq.(15.2), and the solution (17.39) takes the form

$$\Psi(x,t) = e^{-\gamma t} f(x-vt) \tag{17.40}$$

The amplitude of the wave profile f falls off exponentially with time, a property known as *attenuation* which is measured by the reciprocal of the attenuation constant, $1/\gamma$, called the modulus of decay. It is common practice to describe the attenuation *as a function of distance x* rather than time, by this particular choice of the wave profile

$$f(x-vt) = e^{-\gamma(x-vt)/v} h(x-vt)$$

which, by substituting into Eq.(17.40), leads to

$$\Psi(x,t) = e^{-\gamma x/v} h(x-vt) \tag{17.41}$$

In the special case of a harmonic wave, Eq.(17.41) reads

$$\Psi(x,t) = A e^{-\gamma x/v} \cos(kx - \omega t + \varphi) = A(x)\cos(kx - \omega t + \varphi) \tag{17.42}$$

where $A(x)$ denotes a local amplitude.

The exponential decay of the travelling wave amplitude with distance is due to the loss in energy, as the wave propagates through the medium. The power transported by the harmonic wave (17.26) is given at any point x in terms of the local amplitude by

$$P(x,t) = Z\omega^2 A^2(x)\sin^2(kx - \omega t + \varphi) \tag{17.43}$$

so that the decrease in power with distance is given by

$$\frac{\partial}{\partial x}P(x,t) = Z\omega^2 A(x)\left[2\frac{dA(x)}{dx}\sin^2(kx-\omega t+\varphi) + kA(x)\sin 2(kx-\omega t+\varphi)\right]$$

$$= -Z\omega^2 A^2(x)\left[2\frac{\gamma}{v}\sin^2(kx-\omega t+\varphi) - k\sin 2(kx-\omega t+\varphi)\right]$$

The time average of the second term on the right-hand side is zero, so that there is a net power loss of the form

$$\frac{\partial}{\partial x}\langle P(x,t)\rangle = -\frac{Z}{2\gamma v}\left\langle\left(2\gamma\frac{\partial\Psi}{\partial t}\right)^2\right\rangle \tag{17.44}$$

which is associated with the new term introduced in the revised wave equation (17.37), and hence with the attenuation of the wave. The first derivative with respect to time in Eq.(17.37) corresponds to irreversible processes, such as friction or heat transfer, where energy is lost, whereas the wave equation describes reversible and conservative motions.

17.5. WAVE ENERGY AT AN INTERFACE

A wave incident at an interface where the propagation medium changes character abruptly will generally be partly reflected and partly transmitted. We will consider an interface, at $x = x_0$, where the characteristic impedance and wave velocity change from Z_1 and v_1 to Z_2 and v_2, respectively, while the frequency ω remains unchanged. The solution of the wave equation of motion (15.6), written as

$$\frac{\partial^2\Psi}{\partial x^2} = \frac{1}{v_1^2}\frac{\partial^2\Psi}{\partial t^2}, \quad x < x_0$$

$$\frac{\partial^2\Psi}{\partial x^2} = \frac{1}{v_2^2}\frac{\partial^2\Psi}{\partial t^2}, \quad x > x_0 \tag{17.45}$$

must satisfy two *boundary conditions*:

1. Across the interface the wave function Ψ, that is the disturbance, must be continuous as the media do not separate.
2. The derivative $\partial\Psi/\partial x$ must also be continuous, since the interface has no mass (if not, there would be an unbalanced force on the massless interface, giving rise to an infinite acceleration).

The general progressive solution (16.3) takes the form

$$\Psi(x,t) = f_1(x - v_1 t) + g_1(x + v_1 t) \quad , \quad x < x_0$$

$$\Psi(x,t) = f_2(x - v_2 t) + g_2(x + v_2 t) \quad , \quad x > x_0$$
(17.46)

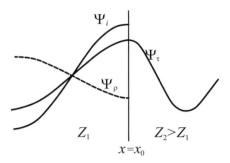

Figure 17.3. Partial reflection and transmission at an interface

The wave incident on the interface from the left $\Psi_i(x,t) = f_1(x - v_1 t)$, in medium 1, is usually known, so that the two boundary conditions allow us to obtain the reflected wave $\Psi_\rho(x,t) = g_1(x + v_1 t)$ and the transmitted wave $\Psi_\tau(x,t) = f_2(x - v_2 t)$ in terms of $\Psi_i(x,t)$, provided that we assume $g_2(x + v_2 t) = 0$ (no wave propagating to the left in medium 2). The continuity of the wave functions at the interface, in the configuration shown in Figure 17.3, reads

$$\Psi_i(x_0,t) + \Psi_\rho(x_0,t) = \Psi_\tau(x_0,t)$$
(17.47)

and that of the first derivatives takes the form

$$\left(\frac{\partial \Psi_i}{\partial x}\right)_{x=x_0} + \left(\frac{\partial \Psi_\rho}{\partial x}\right)_{x=x_0} = \left(\frac{\partial \Psi_\tau}{\partial x}\right)_{x=x_0}$$

or

$$-\frac{1}{v_1}\left(\frac{\partial \Psi_i}{\partial t}\right)_{x=x_0} + \frac{1}{v_1}\left(\frac{\partial \Psi_\rho}{\partial t}\right)_{x=x_0} = -\frac{1}{v_2}\left(\frac{\partial \Psi_\tau}{\partial t}\right)_{x=x_0}$$

where the travelling wave equations (15.3) and (15.4) have been used. Integrating both sides of the last equation with respect to time, we find

$$\Psi_i(x_0,t) - \Psi_\rho(x_0,t) = \frac{v_1}{v_2}\Psi_\tau(x_0,t)$$
(17.48)

Solving Eqs.(17.47) and (17.48) for $\Psi_\rho(x_0,t)$ and $\Psi_\tau(x_0,t)$ one obtains

$$\Psi_\rho(x_0,t) = \frac{v_2 - v_1}{v_2 + v_1} \Psi_i(x_0,t) = \rho \Psi_i(x_0,t)$$

(17.49)

$$\Psi_\tau(x_0,t) = \frac{2v_2}{v_2 + v_1} \Psi_i(x_0,t) = \tau \Psi_i(x_0,t)$$

where ρ and τ are called the *amplitude coefficients* of *reflection* and *transmission*, respectively, given by

$$\rho = \frac{v_2 - v_1}{v_2 + v_1} \quad , \quad \tau = \frac{2v_2}{v_2 + v_1} \quad , \quad \tau = 1 + \rho \qquad (17.50)$$

Assuming a constant force F in both media, and substituting Eq.(17.17), Eqs.(17.50) take the form

$$\rho = \frac{Z_1 - Z_2}{Z_1 + Z_2} \quad , \quad \tau = \frac{2Z_1}{Z_1 + Z_2} \quad , \quad \tau = 1 + \rho \qquad (17.51)$$

The possible values of the reflection coefficient ρ cover the range $-1 \leq \rho \leq 1$. Upon reflection from an interface with a more dense medium, that is $Z_2 > Z_1$ or $v_2 < v_1$, ρ becomes negative, that is the incident wave profile is inverted. If the wave is harmonic, this is represented by a π phase shift. There should be no reflected wave if $Z_2 = Z_1$. The transmission coefficient may have values in the range $0 \leq \tau \leq 2$, so that the transmitted wave is never inverted in keeping with the assumption that $g_2(x + v_2 t) = 0$.

At the interface, the intensity of the incident wave is divided between the reflected and transmitted waves. From Eqs.(17.19), (17.28) and (17.49) we can write the reflected and transmitted intensities as

$$I_\rho = \left\langle Z_1 \left(\frac{\partial \Psi_\rho}{\partial t}\right)^2 \right\rangle = \rho^2 \left\langle Z_1 \left(\frac{\partial \Psi_i}{\partial t}\right)^2 \right\rangle = \rho^2 I_i$$

$$I_\tau = \left\langle Z_2 \left(\frac{\partial \Psi_\tau}{\partial t}\right)^2 \right\rangle = \frac{Z_2}{Z_1} \tau^2 \left\langle Z_1 \left(\frac{\partial \Psi_i}{\partial t}\right)^2 \right\rangle = \frac{Z_2}{Z_1} \tau^2 I_i$$

so that we can define the *reflectivity R* and *transmissivity T* by the ratios

$$R = \frac{I_\rho}{I_i} = \rho^2 = \left(\frac{Z_1 - Z_2}{Z_1 + Z_2}\right)^2 = \left(\frac{v_2 - v_1}{v_2 + v_1}\right)^2 \quad , \quad T = \frac{I_\tau}{I_i} = \frac{Z_2}{Z_1} \tau^2 = \frac{4 Z_1 Z_2}{(Z_1 + Z_2)^2} = \frac{4 v_1 v_2}{(v_2 + v_1)^2}$$

(17.52)

It is a straightforward matter to show that

$$R + T = 1 \qquad (17.53)$$

confirming the conservation of wave energy at the interface.

FURTHER READING
1. S. N. Ghosh - ELECTROMAGNETIC THEORY AND WAVE PROPAGATION, CRC Press, Boca Raton, 2002.
2. W. L. French, Jr. - VIBRATIONS AND WAVES, Wiley, New York, 1999.
3. W. Gough, R. P. Williams, J. P. Richards - VIBRATIONS AND WAVES, Prentice Hall, Englewood Cliffs, 1995.

PROBLEMS

17.1. Find the potential energy distribution in a sound wave by considering the work done on a fixed mass of gas during the adiabatic changes associated with the sound propagation.

Solution

Sound is propagating as a longitudinal wave which stores work during the associated compressions and rarefactions of each volume element V_0 of the wave. The change in volume during the sound propagation is expressed, in one dimension, in terms of condensation s, according to Eq.(15.29) which gives

$$V = V_0(1-s) \qquad \text{or} \qquad dV = -V_0 ds$$

The work done in the compression phase, as given by Eq.(7.17), takes the form

$$-\int_{V_0}^{V_0(1-s)} p\, dV = \int_0^s pV_0\, ds$$

where the pressure can be written, using again Eqs.(15.29) and (15.32), as

$$p = p_0 + \frac{\partial p}{\partial \rho} d\rho = p_0 + \frac{\partial p}{\partial \rho} \rho_0 s = p_0 + \rho_0 v_s^2 s$$

It follows that the contribution of a volume element V_0 to the potential energy can be written as

$$\int_0^s pV_0 ds = \int_0^s (p_0 + \rho_0 v_s^2 s) V_0 ds = V_0 \left(p_0 s + \frac{1}{2} \rho_0 v_s^2 s^2 \right)$$

On integration over the total volume of the fluid, which is left unchanged by the passage of the sound wave, the first term is averaged out, and the potential energy takes the form

$$U = \frac{1}{2} \rho_0 v_s^2 \int \left(\frac{\partial \xi_x}{\partial x} \right)^2 dV$$

which is the same as Eq.(17.5). Assuming adiabatic compressions and rarefactions, the potential energy in a sound wave takes the form of heat, which is associated with the random molecular motion in the medium.

17.2. Show that the average energy flow density for electromagnetic stationary waves is zero in an isotropic dielectric.

Solution

Assuming propagation in one dimension, such waves will be set up between conducting sheets, where the electric field must be zero. In view of Eqs.(16.50), we may set

$$E_z(x,t) = E_0 \sin(kx) \sin(\omega t)$$

and hence, Eq.(15.45) yields

$$B_y(x,t) = \int \frac{\partial E_z}{\partial x} dt = \frac{k}{\omega} E_0 \cos(kx) \cos(\omega t) = \frac{E_0}{c} \cos(kx) \cos(\omega t)$$

The Poynting vector (14.13) is parallel to the x-axis and has the average value

$$\langle S \rangle = \frac{\omega}{2\pi} \int_0^{2\pi/\omega} S(x,t) dt = \frac{\omega E_0^2}{2\pi\mu c} \sin(kx) kx \cos(kx) \int_0^{2\pi/\omega} \sin(\omega t) \cos(\omega t) dt$$

$$= \frac{\omega E_0^2}{4\pi\mu c} \sin(kx) \cos(kx) \left[\sin^2 \omega \right]_0^{2\pi/\omega} = 0$$

as required.

17.3. Find the characteristic impedance for a pair of Lecher wires of radius R and separation a in a medium of permittivity ε and permeability μ.

Solution

Assuming harmonic waves (16.4) moving with velocity $v = 1/\sqrt{L_0 C_0}$ along a transmission line (see Problem 15.4), given by

$$V = V_{max} \cos k(x - vt) \quad , \quad I = I_{max} \cos k(x - vt)$$

we can write

$$\frac{\partial V}{\partial x} = -L_0 \frac{\partial I}{\partial t} \quad \text{or} \quad V_{max} = vL_0 I_{max}$$

so that the characteristic impedance (17.25) of an ideal transmission line reads

$$Z_0 = \frac{V_{max}}{I_{max}} = vL_0 = \sqrt{\frac{L_0}{C_0}}$$

The currents flowing in opposite directions in a pair of parallel wires generate magnetic flux lines, which thread the region between the cables, so that the self-inductance L_0 per unit length is

$$\frac{\Phi}{l} = 2\int_R^{a-R} \frac{\mu}{2\pi} \frac{I}{r} dr = \frac{\mu I}{\pi} \ln\left(\frac{a}{R} - 1\right) \quad \text{or} \quad L_0 = \frac{\mu}{\pi} \ln\left(\frac{a}{R} - 1\right)$$

Between the lines, which form a capacitor, there is a difference of potential, and hence, a capacitance C_0 per unit length given by

$$V_l = -2\int_{a-R}^{R} \frac{\lambda}{2\pi\varepsilon r} dr = \frac{\lambda}{\pi\varepsilon} \ln\left(\frac{a}{R} - 1\right) \quad \text{i.e.} \quad C_0 = \frac{\pi\varepsilon}{\ln\left(\frac{a}{R} - 1\right)}$$

where $C_0 = C/l = \lambda/V_l$. Thus, the characteristic impedance of the Lecher wire system is

$$Z_0 = \frac{1}{\pi}\sqrt{\frac{\mu}{\varepsilon}} \ln\left(\frac{a}{R} - 1\right)$$

17.4. If the temperature T of a metal bar obeys the diffusion equation

$$\frac{\partial T}{\partial t} = D\frac{\partial^2 T}{\partial x^2}$$

where D is the thermal diffusivity, find the temperature distribution $T(x,t)$ due to a point source of heat.

Solution

The diffusion equation results from combining the continuity equation for the heat flow in the bar, expressing the conservation of the amount of heat $c_p TSdx$ in an infinitesimal volume Sdx

$$c_p \frac{\partial T}{\partial t} dtSdx = -S\frac{\partial q}{\partial x} dxdt \quad \text{or} \quad c_p \frac{\partial T}{\partial t} + \frac{\partial q}{\partial x} = 0$$

with Fourier's law of heat conduction, which gives the heat flux in terms of a temperature gradient

$$q = -K\frac{\partial T}{\partial x}$$

where q stands for the flux of heat per unit area per unit time, K is the thermal conductivity, and c_p is the specific heat at constant pressure. It follows that $D = K/c_p$. Assuming that $T(x,t)$ can be separated into a purely time-dependent part and a purely position-dependent part, that is

$$T(x,t) = F(t)X(x)$$

the diffusion equation reduces to

$$\frac{1}{F(t)}\frac{dF(t)}{dt} = \frac{D}{X(x)}\frac{d^2 X(x)}{dx^2}$$

The only way for a function of time to be equal to a function of position is for them both to be constant. It is convenient to choose the constant to be negative and equal to $-1/\tau$, and this gives

$$F(t) = F_0 e^{-t/\tau} \quad , \quad X(x) = X_0 e^{-ikx}$$

where $k^2 = 1/D\tau$. Thus, we have

$$T(x,t) = T_0 e^{-ikx} e^{-k^2 Dt}$$

This is an oscillatory solution having a specific decay time $1/k^2 D$. If we start out with a point source of heat at some position $x = 0$ of the bar, that is

$$T(x,0) = T_0 \delta(x)$$

where $\delta(x)$ is the Dirac δ-function, we may represent $T(x,0)$ by its Fourier spectrum density (16.62), which reads

$$\theta(k,0) = \frac{1}{\sqrt{2\pi}} \int_{-\infty}^{\infty} T(x,0) e^{-ikx} dx = \frac{T_0}{\sqrt{2\pi}} \int_{-\infty}^{\infty} \delta(x) e^{-ikx} dx = \frac{T_0}{\sqrt{2\pi}}$$

because the Dirac δ-function is defined by the property

$$\int_{-\infty}^{\infty} f(x)\delta(x) dx = f(0)$$

The time-dependent density function follows as

$$\theta(k,t) = \theta(k,0) e^{-k^2 Dt} = \frac{T_0}{\sqrt{2\pi}} e^{-k^2 Dt}$$

We now need to transform back to a function of position, using Eqs.(16.62), to get the time-dependent distribution of temperature as

$$T(x,t) = \frac{1}{\sqrt{2\pi}} \int_{-\infty}^{\infty} \frac{T_0}{\sqrt{2\pi}} e^{-k^2 Dt} e^{ikx} dk = \frac{T_0}{2\pi} \int_{-\infty}^{\infty} e^{-\left(Dtk^2 + ixk - x^2/4Dt + x^2/4Dt\right)} dk$$

If we set $u = \sqrt{Dt}\,k + ix/2\sqrt{Dt}$, the last integral takes the Gaussian form (IV.2), Appendix IV, so that we finally obtain

$$T(x,t) = \frac{T_0}{2\pi} e^{-x^2/4Dt} \int_{-\infty}^{\infty} e^{-u^2}\, \frac{du}{\sqrt{Dt}} = \frac{T_0}{\sqrt{4\pi Dt}} e^{-x^2/4Dt}$$

The temperature distribution follows a Gaussian curve that decays in height and broadens with time as the heat spreads to the medium. The total heat Q is conserved at all times, according to

$$c_p S \int_{-\infty}^{\infty} T(x,t)\,dx = Q$$

and hence, the area under the Gaussian curve remains constant.

17.5. Show that the wave energy reflection at the boundary of two media of characteristic impedances Z_1 and Z_2 can be avoided by the smooth insertion at their interface of a particular layer of another medium. Find the width a and the characteristic impedance Z of the coupling medium.

Solution

Upon insertion of a coupling medium of width a, the incident, reflected and transmitted waves at the interfaces $x = 0$ and $x = a$ can be written as

$$\Psi(x,t) = A e^{i(k_1 x - \omega t)} + B e^{i(-k_1 x - \omega t)} \qquad x \leq 0$$

$$\Psi(x,t) = D e^{i(kx - \omega t)} + G e^{i(-kx - \omega t)} \qquad 0 \leq x \leq a$$

$$\Psi(x,t) = C e^{i(k_2 x - \omega t)} \qquad x \geq a$$

The boundary conditions for Ψ and $\partial \Psi / \partial t$ at $x = 0$ take the form

$$A + B = D + G \quad , \quad k_1(A - B) = k(D - G)$$

where $k = \omega/v = \omega Z/F$, from Eq.(17.17). Assuming a constant force F across the interface, it is convenient to put $n_0 = k_1/k = Z_1/Z$ and to eliminate B between the two equations, which yields

$$A = \frac{D(n_0 + 1) + G(n_0 + 1)}{2 n_0}$$

At the second interface $x = a$, the boundary conditions read

$$D e^{ika} + G e^{-ika} = C e^{ik_2 a} \quad , \quad Z\left(D e^{ika} - G e^{-ika}\right) = Z_2 C e^{ik_2 a}$$

If we set $n_a = Z/Z_2$, one obtains

$$D = Ce^{ik_2 a}\frac{e^{-ika}(n_a+1)}{2n_a}, \quad G = Ce^{ik_2 a}\frac{e^{ika}(n_a-1)}{2n_a}$$

Substituting D and G into the expression of A gives

$$A = \frac{C}{2n_a n_0}\left[(n_a n_0 + 1)\cos ka - i(n_a + n_0)\sin ka\right]e^{ik_2 a}$$

$$= \frac{C}{2n_a n_0}\left[(n_a n_0 + 1)^2 \cos^2 ka + (n_a + n_0)^2 \sin^2 ka\right]^{1/2} e^{i(k_2 a + \varphi)}$$

This defines a transmission coefficient of magnitude $\tau = |C/A|$, where C lags A by a phase angle $k_2 a + \varphi$. The transmissivity T is provided by Eqs.(17.52) as

$$T = \frac{Z_2}{Z_1}\tau^2 = \frac{\tau^2}{n_a n_0} = \frac{4 n_a n_0}{(n_a n_0 + 1)^2 \cos^2 ka + (n_a + n_0)^2 \sin^2 ka}$$

The condition $T = 1$ is satisfied if we choose $a = \lambda/4$, as this gives $\cos ka = 0$ and $\sin ka = 1$, and hence, it follows that

$$\frac{4 n_a n_0}{(n_a + n_0)^2} = 1 \quad \text{or} \quad Z = \sqrt{Z_1 Z_2}$$

Thus, the wave energy will be transmitted with zero reflection if the thickness of the coupling medium is $\lambda/4$, and its characteristic impedance is the harmonic mean of the two impedances to be matched.

17.6. Use the value of the inductance and capacitance per unit length for a coaxial cable of permeability μ, permittivity ε and radii R_1 and R_2 of the inner and outer conductors, respectively, to find its characteristic impedance Z_0.

17.7. Determine the inductance L_0 and capacitance C_0 per unit length for a pair of plane parallel conductors of separation a and width b. If μ and ε are the permeability and permittivity of the medium between the conductors, find the characteristic impedance Z_0 of such a transmission line.

17.8. If a sound wave approaches an air-water interface from the water side, with intensity I and speed v_S, find the acoustic radiation pressure on the interface.

17.9. Find the attenuation constant γ of an isotropic sound wave in terms of the values I_1 and I_2 of the intensity measured at distances r_1 and r_2, respectively, from its source.

17.10. If a stationary wave system is obtained in a transmission line where progressive waves are partially reflected from a boundary, find the coefficient ρ of amplitude reflection in terms of the standing wave ratio of the maximum amplitude at an antinode to the minimum amplitude at the adjacent node.

18. INTERFERENCE

18.1. INTERFERENCE OF TWO MONOCHROMATIC WAVES
18.2. INTERFERENCE WITH MULTIPLE BEAMS

18.1. INTERFERENCE OF TWO MONOCHROMATIC WAVES

A suitable description of the phenomena of interference and diffraction, related to the propagation of waves, can be given in terms of complex scalar wave functions $\Psi(\vec{r},t)$, of the form (16.34) in an arbitrary medium. Since the vectorial nature of electromagnetic waves is not taken into account, such a description only represents a *scalar approximation* for optical interference and diffraction. It is rigorously valid for unpolarized light, where the specification of the directions of $\vec{E}$ and $\vec{H}$, is not required. However we have shown in Eqs.(18.29) and (18.30) that the square of the absolute value of a complex scalar wave function Ψ is a measure of the wave *intensity*, which is the quantity of primary interest for the theories of interference and diffraction for either mechanical or electromagnetic waves.

The periodic spatial variation of intensity throughout a region of superposition of waves which originate in the same source is known as *interference*. Its theory is based essentially on the linearity of the wave equation, which allows a linear combination of partial waves to constitute the actual wave at any given point. Consider first the superposition of two monochromatic plane waves, having the complex representation (16.25), that is

$$\Psi_1(\vec{r},t) = A_1 e^{i(\vec{k}_1 \cdot \vec{r} - \omega t + \varphi_1)}$$

$$\Psi_2(\vec{r},t) = A_2 e^{i(\vec{k}_2 \cdot \vec{r} - \omega t + \varphi_2)}$$

(18.1)

Since we are usually interested in the relative variation of the intensity only, in a medium Z_0, and not in its absolute value, it is common practice to adopt simplified wave intensities, compared to those in Eqs.(18.29) or (18.30), which read

$$I_1 = |\Psi_1(\vec{r},t)|^2 \quad , \quad I_2 = |\Psi_2(\vec{r},t)|^2 \tag{18.2}$$

The superposition of monochromatic waves (18.1) results in an intensity

$$I = |\Psi(\vec{r},t)|^2 = |\Psi_1(\vec{r},t) + \Psi_2(\vec{r},t)|^2 = |\Psi_1(\vec{r},t)|^2 + |\Psi_2(\vec{r},t)|^2 + 2\mathrm{Re}\left[\Psi_1(\vec{r},t)\Psi_2^*(\vec{r},t)\right]$$

$$= I_1 + I_2 + 2\sqrt{I_1 I_2}\cos\left(\vec{k}_1 \cdot \vec{r} - \vec{k}_2 \cdot \vec{r} + \varphi_1 - \varphi_2\right) \tag{18.3}$$

where the last term is called the *interference term*. If the *phase difference*, given by

$$\delta = (\vec{k}_1 - \vec{k}_2)\cdot \vec{r} + \varphi_1 - \varphi_2 \tag{18.4}$$

is a constant, the two waves are said to be *mutually coherent*, and the interference term indicates a sinusoidal spatial variation between maxima and minima, of intensity

$$I_{max} = I_1 + I_2 + 2\sqrt{I_1 I_2} \quad , \quad \delta = 2m\pi$$
$$I_{min} = I_1 + I_2 - 2\sqrt{I_1 I_2} \quad , \quad \delta = (2m+1)\pi \tag{18.5}$$

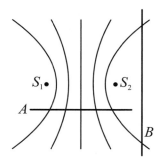

Figure 18.1. Interference pattern produced by two monochromatic waves originating from sources S_1 and S_2

Hence, the stationary interference pattern consists of a succession of surfaces, defined by the condition $\delta = N\pi$ where $N = 2m$ or $2m+1$. It is usually produced by superposition of waves originating in two point sources S_1 and S_2, which can be assumed, for simplicity, to be of equal phase constants $\varphi_1 - \varphi_2 = 0$. If the two sources are far enough from a given point, at distances $r_1, r_2 \gg |\vec{r}_1 - \vec{r}_2|$, the incident waves at the point can be regarded as plane waves, as we shall show in Chapter 19. Under this assumption Eq.(18.4) reduces to

$$\delta = (\vec{k}_1 - \vec{k}_2)\cdot \vec{r} \cong k(r_1 - r_2) = N\pi \quad \text{or} \quad r_1 - r_2 = N\frac{\pi}{k} = N\frac{\lambda}{2} \tag{18.6}$$

Both the *antinodal* $(N=2m)$ and *nodal surfaces* $(N=2m+1)$ are defined by Eq.(18.6) to be hyperboloids of revolution. Their intersection with a plane will produce alternate nodal and antinodal lines called *fringes* of the form of either hyperbolas or circular rings if the plane is parallel (*A*) or normal (*B*) to the line joining the sources (Figure 18.1). The interference fringes are not observed if the phase difference $\varphi_1 - \varphi_2$ has a random variation with time, since the interference term in Eq.(18.3) is averaged out. In this event the waves are said to be *mutually uncoherent* and do not interfere by superposition.

Mutually coherent waves that originate in the same source can be obtained by splitting a single beam, so that on recombining partial waves interference effects occur.

EXAMPLE 18.1. Young's experiment

Young's experiment demonstrates the interference of monochromatic light produced by the division of wavefront. The basic elements are shown in Figure 18.2. Light is first passed through a pinhole which acts as a point source, and then falls on two symmetrical pinholes S_1 and S_2, which divide the wavefront into two beams of equal amplitude. The interference pattern can be observed on a plane normal to the *x*-direction, and we will consider the variation along the *y*-direction in this plane parallel to the line joining the point sources S_1 and S_2.

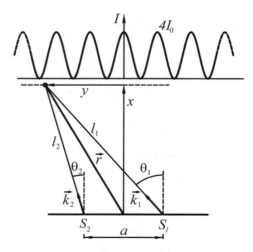

Figure 18.2. Arrangement and fringe pattern for Young's experiment

Provided both the separation *a* of the sources and the distance *y* measured on the screen from the central axis are small compared to the distance *x* between sources and the plane of observation, we can express the path difference between the two beams leaving S_1 and S_2 and arriving at *P* using

$$\vec{k}_1 - \vec{k}_2 = \vec{e}_x(k_{1x} - k_{2x}) + \vec{e}_y(k_{1y} - k_{2y}) \cong \vec{e}_y k(\sin\theta_1 - \sin\theta_2)$$

$$\cong \vec{e}_y k(\theta_1 - \theta_2) \cong \vec{e}_y k \frac{l_1 - l_2}{x} \cong \vec{e}_y \frac{ka}{x}$$

so that one obtains

$$(\vec{k}_1 - \vec{k}_2) \cdot \vec{r} = (\vec{k}_1 - \vec{k}_2) \cdot (\vec{e}_x x + \vec{e}_y y) = \frac{kay}{x}$$

Setting $I_1 = I_2 = I_0$ and $\varphi_1 - \varphi_2 = 0$, Eq.(18.3) becomes

$$I = 2I_0 \left(1 + \cos\frac{kay}{x}\right) = 4I_0 \cos^2\frac{kay}{2x} \qquad (18.7)$$

and hence, the intensity varies between $I_{\max} = 4I_0$ and $I_{\min} = 0$. Bright fringes are located at

$$y = m\frac{\lambda x}{a} \quad , \quad m = 0, \pm 1, \pm 2, \ldots$$

where m is called the *order of interference*, and the angular separation between adjacent fringes is

$$\theta \cong \tan\theta = \frac{\lambda x/a}{x} = \frac{\lambda}{a}$$

Such an interference pattern is observed for any given distance x, that is on any plane normal to the x-direction, in the region of superposition of waves. The fringes are then said to be *nonlocalized* and such fringes are always produced by coherent point sources.

The interference pattern from two identical beams obtained by wavefront division has an average intensity on the plane of superposition equal to the sum of the source intensities, as shown by Eq.(18.7) where $\langle I \rangle = 2I_0$. This shows that the wave energy undergoes by interference a stationary spatial redistribution, whilst obeying the energy conservation law.

An alternative method of demonstrating interference, which can be used with extended sources, is the *division of amplitude*, where a single monochromatic plane wave is divided by partial reflection into a reflected and a transmitted part, which are then brought together to produce interference fringes. Consider first *fringes of equal inclination*, also known as *Haidinger fringes*, which can be observed with an extended monochromatic source and a parallel-sided thin plate, in the configuration shown in Figure 18.3. The fringes result from interference occuring in the focal plane of the lens, which can be replaced by the eye of an observer. The lens brings together at a point P all the beams that leave the plate at a particular angle θ. A pair of beams originates at each point source and since the path difference for all pairs of beams reflected in the front and back surfaces of the plate is the same, interference from an extended source can be observed. The phase difference (18.4) between the two beams emerging from an arbitrary element of the extended source (see Figure 18.3) is given by

$$\delta = AB + BC - AD + \pi = \frac{\omega}{v}\frac{2d}{\cos r} - \frac{\omega}{c}2d\tan(r)\sin i + \pi = \frac{\omega}{c}2nd\cos r + \pi \qquad (18.8)$$

where n is the index of refraction of the plate, and the π phase shift indicates reflection from a more dense medium, as in Eq.(17.50).

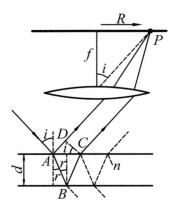

Figure 18.3. Arrangement for observation of fringes of equal inclination

If I_1 and I_2 are the intensities from the front and back surfaces of the plate, Eq.(18.3) reads

$$I_P = I_1 + I_2 - 2\sqrt{I_1 I_2}\cos\left(\frac{\omega}{c}2nd\cos r\right) \qquad (18.9)$$

As the intensity of the fringes are only determined by the angle of incidence i of the primary beam, the interference pattern shows fringes of equal inclination. Because of the rotational symmetry about the axis of the lens, the phase difference will be a constant on a circle in the focal plane, around the lens axis. The interference pattern therefore consists of a series of concentric rings. The conditions for bright and dark fringes are

$$\frac{\omega}{c}2nd\cos r = (2m+1)\pi \qquad \text{or} \qquad 2nd\cos r = \left(m+\frac{1}{2}\right)\lambda_0$$

$$\frac{\omega}{c}2nd\cos r = 2m\pi \qquad \text{or} \qquad 2nd\cos r = m\lambda_0$$

where λ_0 is the wavelength in vacuum. The highest order of interference occurs at the centre, where $\cos r = 1$. If r is small, one may use the approximations $\cos r = 1 - r^2/2$, $i = nr$ and $R = fi$, where f denotes the focal length, as indicated in Figure 18.3. The radius R_m of a bright fringe of order m can be written in terms of plate thickness d as

$$1 - \frac{R_m^2}{2n^2 f^2} = \left(m+\frac{1}{2}\right)\frac{\lambda_0}{2nd}$$

The area between two successive fringes is given by

$$\pi\left(R_m^2 - R_{m-1}^2\right) = \frac{nf^2\lambda_0}{d}$$

and hence, it will diminish, that is the fringes will become closer together, as the plate thickness d increases. With very thick plates, it might be necessary to employ a telescope with suitable magnification to allow visual observation of the interference pattern.

Distorsion of the fringes, due to variations in the plate thickness, becomes significant if the plate is very thin, when variations of $2nd\cos r$ with the angle of incidence are small, at the wavelength λ_0. When d varies across such a thin plate or film, the phase difference will depend upon the local thickness at the point where beam splitting occurs by reflection, regardless of the angle of emergence. The conditions for bright and dark fringes take, respectively, the form

$$2nd = \left(m + \frac{1}{2}\right)\lambda_0 \quad \text{and} \quad 2nd = m\lambda_0$$

Each point element of an extended source provides a pair of coherent beams which produce by interference *fringes of equal thickness*. These so-called *Fizeau fringes* are localized in the film plane.

EXAMPLE 18.2. Newton's rings

If the spherical surface of a long focal length lens is placed on an optical flat, the division of amplitude of a vertical light beam, originating in an extended source, is obtained by partial reflection at the spherical surface and at the surface of the optical flat (see Figure 18.4). The two waves interfere with each other giving rise to a series of concentric bright rings around a central dark spot. The fringes are *localized* near the air wedge between the lens and the flat.

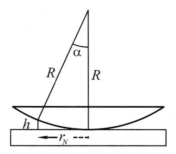

Figure 18.4. Arrangement for observation of Newton's rings

The phase difference results from two sources. Firstly the extra path traversed by the wave reflected by the optical flat, and, secondly, the opposite signs of the reflection coefficients on the two boundaries, which are glass to air and air to glass, respectively. The situation is similar to that described by Eq.(18.8), so that the phase difference can be written as

$$k\left(2\frac{h}{\cos r} - 2h\tan r \sin r\right) + \pi = \frac{2\pi}{\lambda}\left(2h\cos r + \frac{\lambda}{2}\right) = N\pi$$

For normal incidence $(r = 0)$ this gives

$$\frac{4h}{\lambda}\cos r = N-1 \quad \text{or} \quad h = (N-1)\frac{\lambda}{4}$$

where $N = 2m$ for bright fringes and $N = 2m+1$ for dark ones. With h (see Figure 18.4) approximately given by

$$h = R - R\cos\alpha \cong \frac{R\alpha^2}{2} = \frac{r_N^2}{2R}$$

we obtain the radius of the mth *dark* ring as

$$\frac{r_m^2}{R} = m\lambda$$

where $m = (N-1)/2$. Since

$$2r_m \frac{dr_m}{dm} = \lambda R \quad \text{or} \quad \frac{dr_m}{dm} = \frac{\lambda R}{2r_m}$$

it follows that the fringes are more closely spaced for large r_m.

18.2. Interference with Multiple Beams

The fringe pattern resulting from interference is an image of the stationary distribution of the wave energy. A modification of the fringe pattern by, for example, using more than two monochromatic beams provides a method of obtaining intensity distributions of practical interest. Multiple coherent beams can be produced by division of either wavefront or amplitude of a single plane wave.

The *division of the wavefront* of a primary plane wave can be performed with a one dimensional N slit grating, as shown in Figure 18.5. Assuming that the slits are narrow enough for each to be treated as a point source and all the N slits are in phase, the result of superposition of waves leaving the grating in a direction θ can be written as

$$\Psi(\theta) = A\left[e^{i(kr_1-\omega t)} + e^{i(kr_2-\omega t)} + \ldots + e^{i(kr_N-\omega t)}\right] = Ae^{i(kr_1-\omega t)}\left[1 + e^{ik(r_2-r_1)} + \ldots + e^{ik(r_N-r_1)}\right]$$

where the path difference between beams is given in terms of the angle θ and the repeat distance of the grating a by

$$r_2 - r_1 = a\sin\theta, \ldots, r_N - r_1 = (N-1)a\sin\theta$$

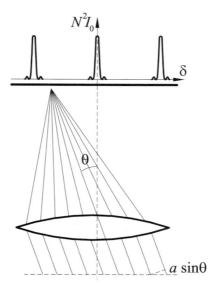

Figure 18.5. Multiple beam interference by division of wavefront

Setting $\delta = ka \sin \theta$, which is the corresponding phase difference, Eq.(18.6), we obtain

$$\Psi(\theta) = Ae^{i(kr_1 - \omega t)}\left[1 + e^{i\delta} + \ldots + e^{i(N-1)\delta}\right] = \Psi_0(r_1, t)\frac{1 - e^{iN\delta}}{1 - e^{i\delta}} \qquad (18.10)$$

Substituting Eq.(18.10) into Eq.(18.2), the intensity distribution takes the form

$$I(\theta) = I_0 \left|\frac{1 - e^{iN\delta}}{1 - e^{i\delta}}\right|^2 = I_0 \frac{\sin^2(N\delta/2)}{\sin^2(\delta/2)} \qquad (18.11)$$

which shows that the main maxima occur when the numerator and denominator both tend to zero, since

$$\lim_{\delta \to 2m\pi} \frac{\sin^2(N\delta/2)}{\sin^2(\delta/2)} = \lim_{\delta \to 2m\pi} \frac{N\sin(N\delta/2)\cos(N\delta/2)}{\sin(\delta/2)\cos(\delta/2)} = \lim_{\delta \to 2m\pi} \frac{N\sin N\delta}{\sin \delta}$$

$$= \lim_{\delta \to 2m\pi} N^2 \frac{\cos N\delta}{\cos \delta} = N^2$$

This happens if $\sin \theta$ is an integral multiple of λ/a, which means that maxima of intensity

$$I_{max} = N^2 I_0 \qquad (18.12)$$

are equally spaced in $\sin \theta$, as represented in Figure 18.6 for $N = 2$ and $N = 4$. Zeros are obtained when the numerator of Eq.(18.11) is zero, whilst the denominator is not, that is, when

$$\frac{N\delta}{2} = m'\pi \quad \text{or} \quad \sin\theta = m'\frac{2\pi}{Na} = m'\frac{\lambda}{Na}$$

unless m'/N is an integer. Between two successive main maxima of intensity (18.11) there are $N-1$ minima, corresponding to the allowed values $m' = 1, 2, \ldots, N-1$. As N increases, the interference pattern consists of narrow fringes of high intensity on a background of almost vanishing intensity. This property is used in the design of radiating aerial arrays for radiowave emission and reception.

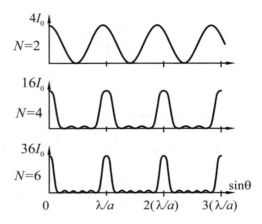

Figure 18.6. Interference pattern for a grating with narrow slits

The *division of amplitude* A_0 of a primary plane wave by multiple reflection between two parallel partially reflecting identical surfaces, in the configuration represented in Figure 18.7, is the most usual way of producing mutually coherent beams. The successive transmitted beams, in the absence of attenuation in the medium between the two surfaces, can be written in terms of amplitude coefficients of reflection $\rho_1 = \rho_2 = \rho$ and transmission $\tau_1 = \tau_2 = \tau$ as

$$A_0\tau^2, \quad A_0\tau^2\rho^2 e^{i\delta}, \quad A_0\tau^2\rho^4 e^{2i\delta}, \ldots$$

where δ denotes the constant phase difference between two successive beams

$$\delta = k\left[2\frac{d}{\cos\theta} - 2d\sin\theta\tan\theta\right] = 2kd\cos\theta = \frac{4\pi}{\lambda}d\cos\theta \qquad (18.13)$$

The transmitted wave in a direction θ is represented by the series

$$\Psi_\tau(\theta) = A_0\tau^2\left(1 + \rho^2 e^{i\delta} + \rho^4 e^{2i\delta} + \ldots\right) = \frac{A_0\tau^2}{1 - \rho^2 e^{i\delta}}$$

and its intensity takes the form

$$I_\tau(\theta) = A_0 \frac{\tau^4}{\left|1-\rho^2 e^{i\delta}\right|^2}$$

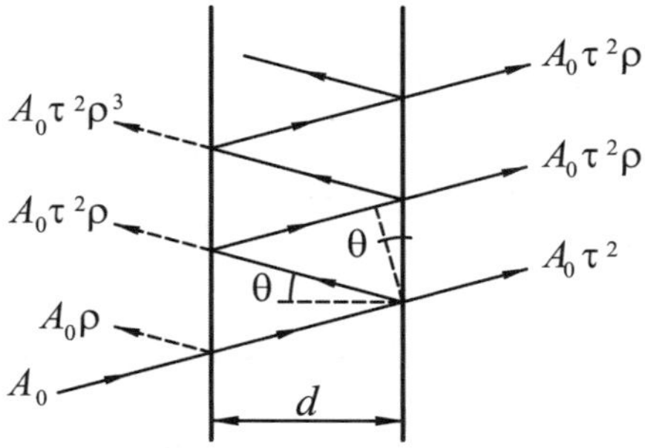

Figure 18.7. Multiple-beam interference by division of amplitude

In terms of reflectivity R and transmissivity T, defined in Eq.(18.52), the transmitted intensity becomes

$$I_\tau(\theta) = I_0 \frac{T^2}{\left|1-Re^{i\delta}\right|^2} = I_0 \frac{T^2}{\left|1-R\cos\delta - iR\sin\delta\right|^2} = I_0 \frac{T^2}{(1-R\cos\delta)^2 + R^2\sin^2\delta}$$

$$= I_0 \frac{T^2}{1+R^2-2R\cos\delta} = I_0 \frac{T^2}{(1-R)^2 + 2R(1-\cos\delta)} = I_0 \frac{T^2}{(1-R)^2 + 4R\sin^2(\delta/2)}$$

and this is usually written in the form

$$I_\tau(\theta) = I_0 \frac{T^2}{(1-R)^2} Y(\delta) = I_0 Y(\delta) \qquad (18.14)$$

where $Y(\delta)$ is called the *Airy function*

$$Y(\delta) = \frac{1}{1+F\sin^2(\delta/2)} \qquad (18.15)$$

Since there is no attenuation, Eq.(18.53) allows us to write $1-R=T$. The parameter F, called the *contrast*, which is defined by

$$F = \frac{4R}{(1-R)^2} \qquad (18.16)$$

increases with reflectivity R and indicates the sharpness of the fringes of equal inclination, represented in Figure 18.8.

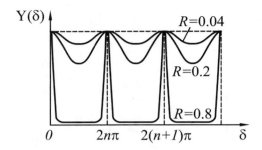

Figure 18.8. Interference pattern shaped by the Airy function

Maximum transmission, when the intensity of the fringes appears to be equal to that of the incident light $I_{\tau\max} = I_0$, occurs for $\delta = 2m\pi$, because $Y(2m\pi) = 1$ for any value of F and it is independent of R. In practice, $I_{\tau\max}$ is always less than I_0 since attenuation can never be neglected. But the intensity of the minima falls as F is increased, so that there is much more contrast when R is close to unity than for small R. The contrast in the intensity distribution for transmitted light is conveniently measured by the ratio

$$\frac{I_{\tau\max}}{I_{\tau\min}} = 1 + F = \left(\frac{1+R}{1-R}\right)^2 \tag{18.17}$$

which can be obtained by setting $\delta = 2m\pi$, or $\sin\theta = m\lambda/a$, and $\delta = (2m+1)\pi$, that is $\sin\theta = (m+1/2)\lambda/a$, respectively, in Eq.(18.14).

FURTHER READING
1. K. D. Möller - OPTICS, Springer-Verlag, New York, 2002.
2. J. R. Meyer-Arendt - INTRODUCTION TO CLASSICAL AND MODERN OPTICS, Simon and Schuster, New York, 1994.
3. F. A. Jenkins, H. E. White - FUNDAMENTALS OF OPTICS, McGraw-Hill, New York, 1990.

18. INTERFERENCE

PROBLEMS

18.1. In an amplitude-splitting interferometer a beam of light is divided into two by a plate M which is half silvered on its rear surface. The two resulting beams are reflected at mirrors M_1 and M_2 and then recombined at the plate. A compensating plate is inserted to ensure the same path in the glass for each beam. Find an expression for the intensity of fringes observed in the focal plane of the lens. Assuming that the whole apparatus is moving in the direction of the initial beam with velocity v, calculate the expected displacement of the fringes as a result of a 90° rotation of the interferometer.

Solution

As shown in the figure, if M'_2 is the virtual image of mirror M_2 in the plate, reflections from M_1 and M'_2 will produce the same fringes as reflections from M_1 and M_2. The interference pattern is similar to that obtained from the two reflecting surfaces of a parallel-sided air plate. Therefore concentric fringes of equal inclination are formed in the focal plane of the lens L. Assuming equal intensities I_0 for the two beams, the intensity distribution (18.9) reads

$$I = 2I_0 \left[1 + \cos\left(\frac{2\pi}{\lambda_0} 2d \cos r\right)\right] = 4I_0 \cos^2\left(\frac{2\pi}{\lambda_0} d \cos r\right)$$

where d is the separation between M_1 and M'_2, and the angle r defines the inclination of the plate. Note that there is no longer a phase shift in this case, as the reflections from M_1 and M'_2 are of the same kind. When the distance between M_1 and M'_2 is small and M_1 is not parallel to M'_2, the circular fringes of equal inclination are replaced by linear fringes of equal thickness across the mirrors.

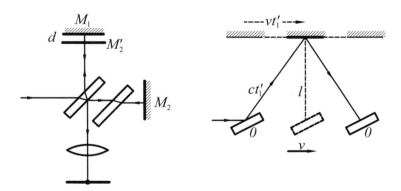

If the arm OM_2 of the interferometer is oriented parallel to the velocity $\vec{v}$ and $OM_1 = OM_2 = l$, the travel time of a light beam towards the mirror M_2 and back is

$$t_2 = \frac{l}{c-v} + \frac{l}{c+v} = \frac{2lc}{c^2 - v^2} = \frac{2l}{c}\gamma^2 \quad \text{or} \quad ct_2 = 2l\gamma^2$$

where $c-v$ and $c+v$ are the relative velocities of light with respect to the moving interferometer. The travel time t_1' towards the mirror M_1 is equal to the time taken to return from M_1 to O, and it can be obtained from the figure as

$$(ct_1')^2 = (vt_1')^2 + l^2 \quad \text{or} \quad t_1' = \frac{l}{c}\gamma$$

so that

$$t_1 = 2t_1' = \frac{2l}{c}\gamma \quad \text{or} \quad ct_1' = 2l\gamma$$

Therefore light traverses paths that differ by

$$c\Delta t = 2l(\gamma^2 - \gamma) \cong 2l\left[1 + \beta^2 - \left(1 + \frac{1}{2}\beta^2\right)\right] = l\beta^2 = l\left(\frac{v}{c}\right)^2$$

where γ and β are the usual abbreviations introduced by Eq.(4.10). A rotation of the interferometer through 90° will interchange the two beams, and since the path lengths are different, the expected displacement of the fringes is given by

$$\Delta N = 2\frac{c\Delta t}{\lambda} = \frac{2l}{\lambda}\left(\frac{v}{c}\right)^2$$

18.2. A laser anemometer issues two equal intensity laser beams, split from a single beam of monochromatic light of wavelength λ, which intersect across a target area at a known angle θ. When the target area is within a flow field, entrained particles passing through the fringes produce a burst of reflected light with flicker frequency v. Find the particle velocity v normal to the fringes.

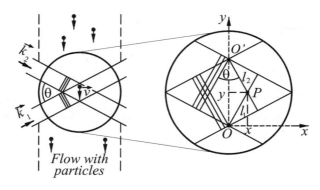

Solution

It is convenient to determine the phase difference between the two laser beams in a Cartesian coordinate system set up in the target area. Let $\delta_1(0,0) = \varphi_1$ and $\delta_2(0,0) = \varphi_2$ be the arbitrary phases of the two beams at O and O', respectively. It is apparent that

$$l_1 = x\cos\frac{\theta}{2} + y\sin\frac{\theta}{2} \quad , \quad l_2 = x\cos\frac{\theta}{2} + (d-y)\sin\frac{\theta}{2}$$

and hence, the phase difference δ at $P(x,y)$ reads

$$\delta = k(2y-d)\sin\frac{\theta}{2}+\varphi_1-\varphi_2$$

Setting $I_1 = I_2 = I_0$, the intensity (19.3) becomes

$$I(y)=2I_0\left[1+\cos\left(2ky\sin\frac{\theta}{2}+\varphi_1-\varphi_2-kd\sin\frac{\theta}{2}\right)\right]$$

and this is a periodic function of y with a fringe spacing ξ given by

$$2k\xi\sin\frac{\theta}{2}=2\pi \quad\text{or}\quad \xi=\pi/\left(k\sin\frac{\theta}{2}\right)=\lambda/\left(2\sin\frac{\theta}{2}\right)$$

The flicker frequency is the velocity of the particle normal to the fringes divided by the fringe spacing, $\nu = v/\xi$, so that

$$v=\nu\xi=\frac{\nu\lambda}{2\sin(\theta/2)}$$

18.3. A two-slit interferometer is illuminated by a star of uniform intensity distribution $I(\beta)=I_s$ and angular diameter $2\beta_0$. Find an expression for the experimental visibility of the fringe pattern, given by

$$V=\frac{I_{max}-I_{min}}{I_{max}+I_{min}}$$

as a function of the variable spacing a of the slits, and show that the apparent diameter of the star can be estimated in terms of the particular spacing a_0 which corresponds to the disappearance of the fringe pattern.

Solution

Since the observed superposition of two light signals Ψ_1 and Ψ_2 depends on the difference in their paths, if τ denotes the extra time taken for the light signal from S_1 compared to that from S_2, one obtains as a result of superposition $\Psi(t)=\Psi_1(t)+\Psi_2(t-\tau)$. Thus, the intensity takes the form

$$I=\left\langle\left[\Psi_1(t)+\Psi_2(t-\tau)\right]\left[\Psi_1^*(t)+\Psi_2^*(t-\tau)\right]\right\rangle$$

$$=\left\langle|\Psi_1(t)|^2\right\rangle+\left\langle|\Psi_2(t-\tau)|^2\right\rangle+2\mathrm{Re}\left\langle\Psi_1(t)\Psi_2^*(t-\tau)\right\rangle$$

The last term is a measure of the degree of correlation between the two light signals at the point of observation, as specified by a correlation function $\Gamma_{12}(\tau)=\left\langle\Psi_1(t)\Psi_2^*(t-\tau)\right\rangle$ which can be written in the normalized form

$$\gamma_{12}(\tau) = \frac{\Gamma_{12}(\tau)}{\sqrt{\Gamma_{11}(0)\Gamma_{22}(0)}} = \frac{\Gamma_{12}(\tau)}{\sqrt{I_1 I_2}}$$

so that we have

$$I = I_1 + I_2 + 2\sqrt{I_1 I_2}\,\text{Re}\,\gamma_{12}(\tau)$$

If we put $\gamma_{12}(\tau) = |\gamma_{12}(\tau)| e^{i\varphi_{12}(\tau)}$, it follows that

$$I = I_1 + I_2 + 2\sqrt{I_1 I_2}\,|\gamma_{12}(\tau)|\cos\varphi_{12}(\tau)$$

and hence, the maximum and minimum values, given by

$$I_{\max} = I_1 + I_2 + 2\sqrt{I_1 I_2}\,|\gamma_{12}(\tau)| \;, \quad I_{\min} = I_1 + I_2 - 2\sqrt{I_1 I_2}\,|\gamma_{12}(\tau)|$$

define the experimental parameter

$$V = \frac{I_{\max} - I_{\min}}{I_{\max} + I_{\min}} = 2\frac{\sqrt{I_1 I_2}}{I_1 + I_2}|\gamma_{12}(\tau)|$$

called *visibility*, which is a measure of the quality of the fringes. For systems which are adjusted such that $I_1 = I_2$, the fringe visibility is identical to the so-called degree of coherence $|\gamma_{12}(\tau)|$. Consider now a star element $d\sigma$, at distances r_1 and r_2 from S_1 and S_2, respectively.

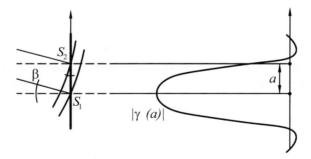

The wave functions at S_1 and S_2 at a time t are given by the superposition of contributions of the form

$$\Psi_1(t) \cong A(t) e^{i[kr_1 - \omega_0 t + \varphi(t)]}, \qquad \Psi_2(t) \cong A(t) e^{i[kr_2 - \omega_0 t + \varphi(t)]}$$

where ω_0 is the mean frequency, and the amplitude of the light signals is assumed to be constant during propagation. If the source is regarded to be divided into many elements, the exponentials with random phase have zero mean value, and we have

$$\gamma_{12}(0) = \frac{1}{I}\int I(\beta) e^{ik(r_1 - r_2)} d\beta$$

The path difference $r_1 - r_2$ can be approximated, far from the source, by the distance between two wavefronts originating from the same source element, and this gives $r_1 - r_2 = \beta a$. For a uniform intensity distribution $I(\beta) = I_s$ and a finite angular diameter $2\beta_0$, that is $I = 2\beta_0 I_s$, the degree of coherence between S_1 and S_2 is obtained as

$$\gamma(a) = \frac{1}{2\beta_0} \int_{-\beta_0}^{\beta_0} e^{ik\beta a} d\beta = \frac{\sin(k\beta_0 a)}{k\beta_0 a}$$

The fringe visibility varies according to the $\gamma(a)$ curve. Thus, by determining the experimental a separation of S_2 from S_1, such that the fringes disappear, we obtain the apparent angular diameter $2\beta_0$ of the star in the form

$$\sin\left(\frac{2\pi}{\lambda}\beta_0 a_0\right) = 0 \quad \text{or} \quad \frac{2\pi}{\lambda}\beta_0 a_0 = \pi \quad \text{i.e.,} \quad 2\beta_0 = \frac{\lambda}{a_0}$$

as required.

18.4. A collimated beam of light with a wavelength range from $\lambda_M = 1000$ nm down to $\lambda_m = 400$ nm falls at normal incidence on an air film of thickness d formed between two parallel sided glass plates of refractive index n. If the interference pattern consists of two bright fringes only, one corresponding to λ_m and another to a wavelength λ less than λ_M, find λ and the thickness d of the air film.

Solution

The condition for a bright Haidinger fringe of order m, observed at normal incidence on an air film of $n = 1$, reads

$$2nd = (2m+1)\frac{\lambda_m}{2} \quad \text{or} \quad \lambda_m = \frac{4d}{2m+1}$$

with the obvious restriction of $\lambda_m < \lambda_{m-1} < \lambda_M < \lambda_{m-2}$ which can be written as

$$\frac{4d}{2(m-1)+1} < \lambda_M < \frac{4d}{2(m-2)+1} \quad \text{or} \quad \frac{2m+1}{2(m-1)+1} < \frac{\lambda_M}{\lambda_m} < \frac{2m+1}{2(m-2)+1}$$

If we let $r = \lambda_M / \lambda_m = 5/2$, rearranging gives

$$\frac{1}{2}\frac{r+1}{r-1} < m < \frac{1}{2}\frac{3r+1}{r-1} \quad \text{or} \quad \frac{7}{6} < m < \frac{17}{6}$$

It follows that $m = 2$, and thus the thickness is $d = (5/4)\lambda_m = 500$ nm. The second bright fringe of order $m - 1 = 1$ corresponds to the wavelength $\lambda = (4/3)d = 666.7$ nm.

18.5. Suppose a spectrum consisting of two closely spaced wavelengths λ and $\lambda + \Delta\lambda$ is to be analyzed with a multiple-beam interferometer, consisting of two plane parallel half silvered glass plates of high reflectivity R separated by an air gap d. Find the resolving power $RP = \lambda/\Delta\lambda$ of the instrument, using Taylor's criterion which states that two equal fringes are considered to be just resolved if the individual curves cross at the half intensity point, so that the total intensity at the saddle point is equal to the maximum intensity of either line alone.

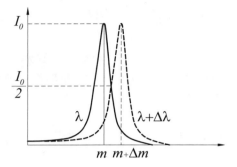

Solution

The bright fringes of each wavelength partially overlap to a greater or lesser extent as $\Delta\lambda$ decreases or increases, as shown in the figure. The combined intensity at the saddle point can be written as

$$I = I_0 \left[\frac{1}{1 + F \sin^2(\delta/2)} + \frac{1}{1 + F \sin^2[(\delta + \Delta\delta)/2]} \right]$$

where the smallest phase increment $\Delta\delta$ can be related with the resolving power using Eq.(18.13), which gives

$$\lambda\delta = 4\pi d \cos\theta \quad \text{or} \quad \lambda\Delta\delta + \delta\Delta\lambda = 0 \quad \text{i.e.,} \quad \frac{\lambda}{\Delta\lambda} = -\frac{\delta}{\Delta\delta} = -\frac{2m\pi}{\Delta\delta}$$

where m is the order of interference. Since the saddle point is midway between the two wavelengths, we may write, in view of Taylor's criterion, that

$$I = 2I_0 \frac{1}{1 + F \sin^2(\Delta\delta/4)} = I_0$$

that is $\sin^2(\Delta\delta/4) = 1/F$. For large values of F, the angle may be equated with its sine, and this gives $\Delta\delta = 4/\sqrt{F}$, so that the resolving power can be written as

$$RP = \frac{\lambda}{|\Delta\lambda|} = m\pi \frac{\sqrt{R}}{1 - R}$$

It follows that the resolving power can be indefinitely increased either by making the reflectivity R closer to unity or by increasing the order of interference $m = 2d\cos\theta/\lambda$, that is by increasing the mirror separation d.

18.6. Young's experiment is performed with monochromatic light λ using a point source placed at a distance h from the line joining two slits S_1 and S_2 of a spacing a. The fringes are measured at a distance x from the double slit, along the axis of the instrument. If the intensity of the central fringe flickers with frequency ν as the point source is set moving uniformly, parallel to the line joining the slits, what is the velocity v of the motion?

18.7. A thin lens of refractive index n is placed on a horizontal optical flat and illuminated by a parallel beam of monochromatic light λ incident normally on it, giving rise to a pattern of bright and dark Newton's rings. If it is found that the radius of the mth dark ring changes from r_m to $2r_m$ when the lens in inserted, find focal length of the thin lens.

18.8. A spectrometer is used to measure the wavelength in a Young's experiment, at distances $x = 10^3 a$ from the line joining the two slits of spacing a, and $y = 10^4 \lambda_M$ from the central axis of the instrument. If the experiment is performed with visible light with a wavelength range from $\lambda_M = 700$ nm down to $\lambda_m = 400$ nm, find the missing wavelengths in the observed spectrum.

18.9. A parallel beam of monochromatic light λ is made normally incident on the upper surface of a thin glass wedge of reactive index n and wedge angle θ. If R is the reflectivity at the air-glass interface, find the visibility V, as defined in Problem 18.3, of the fringes of equal thickness viewed from above.

18.10. Calculate the interference pattern that would be obtained if three identical slits instead of two were used in Young's experiment. Assume equal separation a of the slits.

19. DIFFRACTION

19.1. DIFFRACTION OF SCALAR WAVES
19.2. THE LINEAR APPROXIMATION OF DIFFRACTION
19.3. THE DIFFRACTION GRATING

19.1. DIFFRACTION OF SCALAR WAVES

The deviation of a wave from rectilinear propagation, which can occur whenever a region of the wavefront is obstructed, is called *diffraction*. The elements of the wavefront that are disturbed by the obstruction interfere beyond the obstacle to produce a standing intensity distribution, called the *diffraction pattern*. There is therefore only a formal distinction between interference and diffraction. The superposition of several distinct waves is referred to as interference, whereas the superposition of a continuous distribution of wave elements is treated as diffraction. Therefore diffraction can be described by carrying out a superposition of many individual waves, which originate in a distribution of point sources. This intuitive generalization of the procedure used for interference, is called the *Fresnel theory*. An alternative approach is to express the resultant disturbance at a given point P in terms of wave elements and their derivatives over an arbitrary closed surface surrounding P, without concern about sources. This is the *Kirchhoff formulation* of diffraction, which provides rigorous mathematical expressions based on the wave equation.

EXAMPLE 19.1. Superposition of spherical waves

Consider the superposition of two monochromatic spherical waves, having the complex representation (16.30), which reads

$$\Psi_1(r_1,t) = \frac{A_1}{r_1} e^{i(kr_1 - \omega t)}$$

$$\Psi_2(r_2,t) = \frac{A_2}{r_2} e^{i(kr_2 - \omega t)} \tag{19.1}$$

in the arrangement of Figure 19.1. The complex amplitudes are expected to be proportional to both the incident spherical waves and the area elements of the pinholes S_1 and S_2, that is

$$A_1 = C\frac{A_0}{R_1}e^{ikR_1}\Delta S_1 \quad , \quad A_2 = C\frac{A_0}{R_2}e^{ikR_2}\Delta S_2$$

where R_1 and R_2 are the distances of S_1 and S_2 from a primary source S_0.

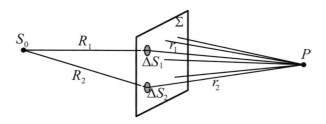

Figure 19.1. Elementary spherical waves at an aperture

We then obtain the resultant wave at P as

$$\Psi_P(r,t) = C\left[\frac{A_0 e^{i(kR_1+kr_1-\omega t)}}{R_1 r_1}\Delta S_1 + \frac{A_0 e^{i(kR_2+kr_2-\omega t)}}{R_2 r_2}\Delta S_2\right] \quad (19.2)$$

For N pinholes with areas ΔS_j, at distances r_j from P and R_j from the primary source, Eq.(19.2) may be generalized in the form

$$\Psi_P(r,t) = CA_0 e^{-i\omega t}\sum_{j=1}^{N}\frac{e^{ik(R_j+r_j)}}{R_j r_j}\Delta S_j \quad (19.3)$$

which can be applied to the total wave due to a large aperture Σ, by dividing the area into small elements dS and summing up the infinitesimal contributions over the surface Σ, which gives

$$\Psi_P(r,t) = CA_0 e^{-i\omega t}\int_{\Sigma}\frac{e^{ik(R+r)}}{Rr}\Delta S \quad (19.4)$$

This basic equation of the Fresnel theory of diffraction was derived by making use of interference concepts only. However, a complete picture of diffraction phenomena requires that $\Psi_P(r,t)$ includes effects such as a phase shift at the aperture or an angular dependence on the inclination to the surface normal. These fundamental properties of wave propagation, which are lacking in the Fresnel treatment, are specified by the Kirchhoff formulation of diffraction phenomena.

Consider two scalar wave functions Ψ and Ψ' which are solutions to the scalar wave equation (15.14), so that, using Green's theorem derived in Problem 15.5, one obtains

$$\int_S (\Psi' \nabla \Psi - \Psi \nabla \Psi') \cdot d\vec{S} = 0 \qquad (19.5)$$

on any closed surface S around a given point P. Let Ψ be an unspecified scalar wave and Ψ' be a spherical wave (16.30) of the form

$$\Psi'(r,t) = \frac{A}{r} e^{i(kr-\omega t)} \qquad (19.6)$$

where r is measured from P, that is Eq.(19.6) represents spherical waves converging to the point P. There is a singularity at P, where $r = 0$, so that this point must be excluded from the region enclosed by S, and this can be achieved by subtracting an integral over a small sphere σ of radius r_0 centred at P.

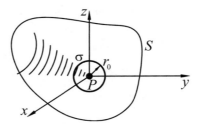

Figure 19.2. Integration of spherical waves converging to a point P

Equation (19.5) takes the form

$$\int_S \left[\frac{e^{ikr}}{r} \nabla \Psi - \Psi \nabla \left(\frac{e^{ikr}}{r} \right) \right] \cdot d\vec{S} + \int_\sigma \left[\frac{e^{ikr}}{r} \nabla \Psi - \Psi \nabla \left(\frac{e^{ikr}}{r} \right) \right]_{r=r_0} \cdot d\vec{\sigma} = 0 \qquad (19.7)$$

where the common factor $Ae^{-i\omega t}$ has been cancelled out. The unit normal $\vec{e}_\sigma$ to the small sphere σ points toward the origin at P, and the gradient is directed radially outward, so that $\nabla = -\vec{e}_\sigma (\partial / \partial r)$. In terms of the solid angle $d\Omega$ measured at P, the element of area on the sphere is $\vec{e}_\sigma \cdot d\vec{\sigma} = d\sigma = r_0^2 d\Omega$ and Eq.(19.7) becomes

$$\int_S \left[\frac{e^{ikr}}{r} \nabla \Psi - \Psi \nabla \left(\frac{e^{ikr}}{r} \right) \right] \cdot d\vec{S} = \int_\sigma \left[\frac{e^{ikr}}{r} \frac{\partial \Psi}{\partial r} - \Psi \frac{\partial}{\partial r} \left(\frac{e^{ikr}}{r} \right) \right]_{r=r_0} r_0^2 d\Omega$$

$$= \int_\sigma \left[\frac{\partial \Psi}{\partial r} - ik\Psi \right]_{r=r_0} e^{ikr_0} r_0 d\Omega + \int_\sigma \Psi e^{ikr_0} d\Omega$$

In the limit as r_0 approaches zero, the first integral on the right-hand side vanishes and $e^{ikr_0} \to 1$, so that the second integral becomes $4\pi \Psi_P$. We then obtain for an arbitrary scalar wave function a relation between its value Ψ_P at any point P inside a closed surface S and its value Ψ at the surface, of the form

$$\Psi_P = \frac{1}{4\pi} \int_S \left[\frac{e^{ikr}}{r} \nabla \Psi - \Psi \nabla \left(\frac{e^{ikr}}{r} \right) \right] \cdot d\vec{S} \qquad (19.8)$$

which is called the *Kirchhoff integral theorem*. The diffraction produced by an aperture Σ of arbitrary shape can now be described by including Σ as part of the closed surface of integration and by assuming that the contributions to the integral from Ψ and $\nabla\Psi$ from elements of S which lie outside the aperture, may be neglected in the geometric configuration illustrated in Figure 19.3.

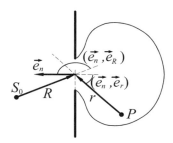

Figure 19.3. Geometry of diffraction by an aperture

For a spherical wave originating at a point source S_0, having a wave function at a point on the aperture given by

$$\Psi = \frac{A_0}{R} e^{i(kR - \omega t)} \qquad (19.9)$$

where R is the distance from S_0, Eq.(19.8) becomes

$$\Psi_P = \frac{1}{4\pi} A_0 e^{-i\omega t} \int_\Sigma \left[\frac{e^{ikr}}{r} \nabla \left(\frac{e^{ikR}}{R} \right) - \frac{e^{ikR}}{R} \nabla \left(\frac{e^{ikr}}{r} \right) \right] \cdot d\vec{S} \qquad (19.10)$$

Taking $d\vec{S} = \vec{e}_n dS$ and denoting the angles between the normal $\vec{e}_n$ to the surface Σ and the directions of wave propagation $\vec{e}_R, \vec{e}_r$ on either side of the aperture by $(\vec{e}_n, \vec{e}_R)$ and $(\vec{e}_n, \vec{e}_r)$, respectively, we obtain

$$\vec{e}_n \cdot \nabla \left(\frac{e^{ikr}}{r} \right) = \cos(\vec{e}_n, \vec{e}_r) \frac{\partial}{\partial r} \left(\frac{e^{ikr}}{r} \right) = \cos(\vec{e}_n, \vec{e}_r) \left(ik - \frac{1}{r} \right) \frac{e^{ikr}}{r}$$

$$\vec{e}_n \cdot \nabla \left(\frac{e^{ikR}}{R} \right) = \cos(\vec{e}_n, \vec{e}_R) \left(ik - \frac{1}{R} \right) \frac{e^{ikR}}{R}$$

When $r, R \gg \lambda$ it follows that $k = 2\pi/\lambda \gg 1/r, 1/R$. Thus, the second term in both parentheses can be neglected, and Eq.(19.10) reduces to

$$\Psi_P = -\frac{ik}{4\pi} A_0 e^{-i\omega t} \int_\Sigma \frac{e^{ik(R+r)}}{Rr} \left[\cos(\vec{e}_n, \vec{e}_r) - \cos(\vec{e}_n, \vec{e}_R)\right] dS \qquad (19.11)$$

which is called the *Fresnel-Kirchhoff diffraction formula*. It predicts two additional effects when compared to the Fresnel formula (19.4):

    1. An angular dependence on both directions of propagation, from the source S_0 and to the point P, defined by the *obliquity factor* which reads

$$K(\vec{e}_r, \vec{e}_R) = \frac{1}{2}\left[\cos(\vec{e}_n, \vec{e}_r) - \cos(\vec{e}_n, \vec{e}_R)\right]$$

In terms of *Huygen's principle*, which states that *each part of the wavefront on the surface Σ of the aperture acts as a source of spherical wavelets which propagate in all directions*, the obliquity factor becomes

$$K(\vec{e}_n, \vec{e}_r) = \cos(\vec{e}_n, \vec{e}_r) \quad \text{in the forward direction}$$

$$= 0 \quad \text{in the backward direction}$$

thus providing an explanation for the absence of the backward wave, which the original formulation could not account for.

    2. A phase shift by $\pi/2$ of the diffracted wave Ψ_P with respect to the primary wave Ψ, at the aperture, as indicated by the phase factor $i = e^{i\pi/2}$.

19.2. The Linear Approximation of Diffraction

A theoretical understanding of the diffraction patterns produced by apertures contained in opaque plane screens (Figure 19.4) usually require several simplifying approximations. If the distances R and r of the points S_0 and P from the aperture Σ are large compared with its linear dimensions, the angular spread of the wave is small enough for the obliquity factor $K(\vec{e}_r, \vec{e}_R)$ to be taken as a constant. The variations of R and r over the aperture will be small compared with R and r, so that the amplitude factor $1/Rr$ can be taken outside the integral. This means that equal area elements dS are assumed to give contributions with equal amplitude or, in other words, the aperture is considered to be *uniform*. The change in $R+r$ affects, however, the phase contribution of different elements, due to the large value of the wave number k. Under these assumptions the diffraction formula (19.11) reduces to

$$\Psi_P = A(\Sigma)e^{-i\omega t}\int_\Sigma e^{ik(R+r)}dS \tag{19.12}$$

where $A(\Sigma)$ includes the phase shift and all slow varying parameters over the aperture.

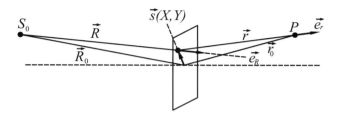

Figure 19.4. Configuration of diffraction experiments

Approximate expressions for R and r may be obtained in terms of the position coordinates $\vec{s}(X,Y)$ of an area element dS in the aperture. From Figure 19.4 we obtain $\vec{r} = \vec{r}_0 - \vec{s}$, that is,

$$r^2 = r_0^2 - 2\vec{s}\cdot\vec{r}_0 + s^2 = r_0^2 - 2r_0\vec{s}\cdot\vec{e}_r + s^2 = r_0^2\left[1 - \frac{2}{r_0}(\alpha X + \beta Y) + \frac{s^2}{r_0^2}\right] = r_0^2(1+\varepsilon)$$

where α and β stand for the direction cosines of $\vec{e}_r$ and $\varepsilon \ll 1$. By expanding r about r_0 according to

$$r = r_0(1+\varepsilon)^{1/2} = r_0\left(1 + \frac{\varepsilon}{2} - \frac{\varepsilon^2}{8}\right)$$

one obtains, to the second order in X and Y, that

$$r = r_0 - (\alpha X + \beta Y) + \frac{1}{2r_0}\left[X^2 + Y^2 - (\alpha X + \beta Y)^2\right] \tag{19.13}$$

In a similar manner we can write $\vec{R} = \vec{R}_0 + \vec{s}$ as

$$R = R_0 + (\alpha_0 X + \beta_0 Y) + \frac{1}{2R_0}\left[X^2 + Y^2 - (\alpha_0 X + \beta_0 Y)^2\right] \tag{19.14}$$

Therefore the diffraction formula (19.12) can be written in the form

$$\Psi_P = A(\Sigma)e^{i[k(R_0+r_0)-\omega t]}\int_\Sigma e^{ik\Delta(X,Y)}dXdY \tag{19.15}$$

where the function $\Delta(X,Y)$, which gives the path difference in terms of aperture coordinates, is derived from Eqs.(19.13) and (19.14) as

$$\Delta(X,Y) = (\alpha_0 - \alpha)X + (\beta_0 - \beta)Y + \frac{R_0 + r_0}{2R_0 r_0}(X^2 + Y^2) - \frac{(\alpha_0 X + \beta_0 Y)^2}{2R_0} - \frac{(\alpha X + \beta Y)^2}{2r} \tag{19.16}$$

Diffraction patterns are determined by the evaluation of the integral (19.15) over the aperture Σ. As long as we are only concerned with the relative amplitude and phase distribution, the phase term in front of the integral can be embedded in a single constant and the diffraction formula reads

$$\Psi_P = \Psi_0 \int_\Sigma e^{ik\Delta(X,Y)} dXdY \tag{19.17}$$

If either the source S_0 or the point of observation P are close to Σ, the curvature of the wavefront, as measured by the quadratic terms in the expansion (19.16), is significant for the diffraction pattern. A recognizable image of the aperture is projected onto the plane of observation, structured with fringes around its periphery. This is known as the *near-field* or *Fresnel diffraction* zone. The calculation becomes simpler if, in the arrangement illustrated in Figure 19.4, we consider that both S_0 and P lie on the same normal to XY-plane and the angular spread of the wave is assumed negligible, so that we can take $\alpha = \alpha_0 = 0$, $\beta = \beta_0 = 0$. Under these assumptions the *Fresnel diffraction formula* reduces to

$$\Psi_P = \Psi_0 \int_\Sigma e^{ik(X^2+Y^2)(R_0+r_0)/2R_0 r_0} dXdY \tag{19.18}$$

When the distances of both the source and the point of observation from the aperture Σ are large enough for the curvature of the wavefront to be neglected, the incident and the diffracted waves can be regarded as plane waves. The stationary intensity distribution which is projected in this case onto the plane of observation is known as the *far-field* or *Fraunhofer diffraction*, which is restricted to *plane waves*. The complex representation of the wave across the aperture $e^{ik\Delta(X,Y)}$, when compared to that of a plane wave (16.25), indicates that the quadratic terms in the expansion (19.16) must be negligibly small in the case of Fraunhofer diffraction, that is

$$k\frac{R_0+r_0}{2R_0 r_0}s^2 \ll 1 \quad \text{or} \quad \pi\frac{R_0+r_0}{R_0 r_0}s^2 \ll \lambda \tag{19.19}$$

where $s = \sqrt{X^2 + Y^2}$ gives the linear size of the aperture. Equation (19.19) provides a criterion for the linear representation of diffraction, which requires that S_0 and P are located at infinity. This is practically achieved by locating the source at the focus of a lens and taking the focal plane of a second lens as the plane of observation (Figure 19.5). In the configuration of Figure 19.5 the point P is the image of the point source S_0 formed by the lens system. If we consider the points of an extended source S_0 and an arbitrary lens system, their images will be given by a Fraunhofer diffraction pattern.

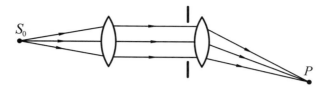

Figure 19.5. Fraunhofer diffraction configuration

If all the quadratic terms in Eq.(19.16) are neglected, we obtain a linear approximation for the dependence of the path difference on the coordinates of a point on the aperture, which in the arrangement of Figure 19.4, if taking $\alpha_0 = \beta_0 = 0$, reads

$$k\Delta(X,Y) = -k(\alpha X + \beta Y) = -2\pi(uX + vY) \tag{19.20}$$

where

$$u = \frac{\alpha}{\lambda} \quad , \quad v = \frac{\beta}{\lambda} \tag{19.21}$$

Both u and v have the dimensions of a reciprocal length and are called *spatial frequencies*. Equation (19.21) shows that for each point P of observation, given by a pair of direction cosines, there is a corresponding pair of spatial frequencies (u,v), such that $\Psi_P \equiv \Psi(u,v)$. Thus, Eq.(19.17) reduces to the *Fraunhofer diffraction formula* given by

$$\Psi(u,v) = \Psi_0 \int_\Sigma e^{-i2\pi(uX+vY)} dXdY \tag{19.22}$$

which is valid for a uniform aperture. If the aperture is not uniform, the wave from each area element $dXdY$ could differ in both amplitude and phase. These variations and the constant Ψ_0 can be combined into a complex *aperture function* written as

$$A(X,Y) = A(X,Y)e^{i\varphi(X,Y)} \tag{19.23}$$

where $A(X,Y)$ describes the amplitude and $\varphi(X,Y)$ the relative phase over the aperture. The diffracted wave originating in an area element $dXdY$ will then be proportional to $A(X,Y)dXdY$. Since the aperture function in nonzero only over the area Σ, Eq.(19.22) can be reformulated, by extending the limits on the integral, as

$$\Psi(u,v) = \int_{-\infty}^{\infty} \int_{-\infty}^{\infty} A(X,Y) e^{-i2\pi(uX+vY)} dXdY \tag{19.24}$$

Therefore, in the linear approximation, the wave distribution $\Psi(u,v)$ in the Fraunhofer diffraction pattern is the Fourier transform of the wave distribution $A(X,Y)$ across the aperture. Similarly to the one-dimensional case, Eq.(17.13), the inverse transform reads

$$A(X,Y) = \frac{1}{(2\pi)^2} \int_{-\infty}^{\infty} \int_{-\infty}^{\infty} \Psi(u,v) e^{i2\pi(uX+vY)} du dv \qquad (19.25)$$

and shows that the wave distribution in the diffraction pattern gives the spatial frequency density of the aperture function. That means that the smaller the diffracting aperture, the larger is the spatial frequency spectrum, and therefore the angular spread of the diffracted beam.

EXAMPLE 19.2. Diffraction by a single-slit

Consider a long narrow slit, of width b in the Y-direction, as represented in Figure 19.6.

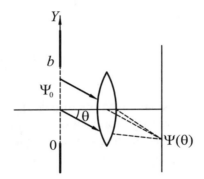

Figure 19.6. Single slit diffraction configuration

Assuming that the slit is illuminated by a monochromatic plane wave, which originates in a point source at infinity, the aperture function (19.23) reduces to

$$A(Y) = \Psi_0 \qquad |Y| \le b$$
$$= 0 \qquad Y > b \qquad (19.26)$$

which is analogous to the slit function (17.15) and shows that there are no amplitude or phase variations across the slit. The problem may be treated as one-dimensional and, in the configuration of Figure 19.6, we can take $\alpha = 0$, $\beta = \sin\theta$, so that Eq.(19.24) reads

$$\Psi(\theta) = \Psi_0 \int_{-b/2}^{b/2} e^{-i2\pi Y \sin\theta / \lambda} dY = \Psi_0 \frac{\sin(\pi b \sin\theta / \lambda)}{\pi \sin\theta / \lambda} = \Psi_0 b \left(\frac{\sin\gamma}{\gamma} \right) \qquad (19.27)$$

where $\gamma = (\pi b / \lambda)\sin\theta$. The amplitude $\Psi(\theta)$ of the wave diffracted in a given direction θ has a form analogous to the spectral density (17.17) and produces in the focal plane of the lens a diffraction pattern having an intensity distribution given by

$$I = |\Psi(\theta)|^2 = I_0 \left(\frac{\sin\gamma}{\gamma} \right)^2 \qquad (19.28)$$

where I_0 is the maximum intensity, which occurs for $\theta = 0$, since $(\sin \gamma)/\gamma \to 1$ as $\theta \to 0$. Zero intensities occur when $\gamma = \pm\pi, \pm 2\pi, \ldots$ and the secondary maxima between them have rapidly diminishing intensities, as shown in Figure 19.7.

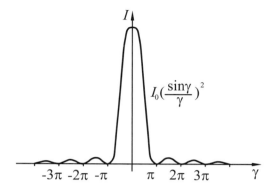

Figure 19.7. Fraunhofer diffraction pattern from a single slit

The image produced by the slit, at the plane of observation, consists of a central bright band of amplitude proportional to the slit area. However, a significant fraction of the incident intensity on the image plane is distributed in the secondary fringes, so that it is convenient to suppress these secondary maxima (the higher spatial frequencies) by a suitable choice of aperture function. Suppose that the slit is covered with a coated flat glass plate, which becomes increasingly opaque off axis, until the transmitted wave is negligible at the edges. If this decrease in amplitude follows a Gaussian law, the aperture function (19.26) takes the form of a Gaussian harmonic pulse centred at $Y = 0$, which can be written as

$$A(Y) = \Psi_0 e^{-aY^2} \tag{19.29}$$

with the effective width $b = 2/\sqrt{a}$.

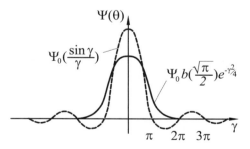

Figure 19.8. Image apodisation by a Gaussian aperture function

Since the transmitted wave is negligible for $|Y| > b$, the amplitude of the diffracted wave may be calculated in a similar way to that used in Problem 16.6 and this gives

$$\Psi(\theta) = \Psi_0 \int_{-\infty}^{\infty} e^{-aY^2 - i(2\pi/\lambda)Y\sin\theta} dY = \Psi_0 \sqrt{\frac{\pi}{a}} e^{-\pi^2 \sin^2\theta/\lambda^2 a} = \Psi_0 \frac{\sqrt{\pi} b}{2} e^{-\gamma^2/4} \tag{19.30}$$

The central band in the image plane is broadened, but the secondary fringes are suppressed, as shown in Figure 19.8. Such a procedure designed to enhance the resolution in the image plane of lens systems, is called *apodisation*. It is seen from both Eqs.(19.27) and (19.30) that the amplitude of the central band is proportional to the width b of the slit, while its angular width (see Figure 19.7) is defined by

$$\gamma = \frac{\pi b}{\lambda}\sin\theta = \pm\pi \quad \text{or} \quad \sin\theta = \pm\frac{\lambda}{b} \qquad (19.31)$$

and it varies inversely with the width b.

19.3. THE DIFFRACTION GRATING

If the aperture consists of a grating, with N identical long slits of width b and repeat distance a, the wave distribution on the image plane can be obtained by applying the superposition principle to the contributions (19.27) arising from each individual slit. If we locate the origins of a series of local coordinate systems $O_j Y_j$, one for each slit, at points O_j such that $Y_j = ja$ $(j = 0, 1, \ldots, N-1)$, as shown in Figure 19.9, we obtain

$$\Psi(\theta) = \Psi_0 \sum_{j=0}^{N-1} \int_{-b/2}^{b/2} e^{-i(2\pi/\lambda)(Y+Y_j)\sin\theta} dY = \left(\Psi_0 \int_{-b/2}^{b/2} e^{-i(2\pi/\lambda)Y\sin\theta} dY \right) \left(\sum_{j=0}^{N-1} e^{-i(2\pi/\lambda)Y_j\sin\theta} \right) \qquad (19.32)$$

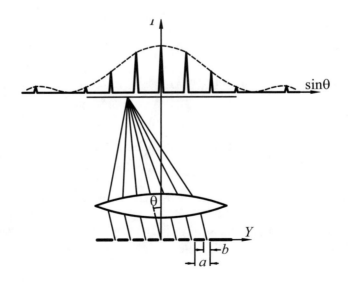

Figure 19.9. Diffraction grating geometry

We have already encountered the integral on the right-hand side, Eq.(19.27), and it is the Fourier transform of the aperture function (19.26) of each slit, that is the wave distribution given by a single slit in the image plane. The sum on the right-hand side represents the interference pattern (18.10) obtained with the same grating, provided the slits are narrow enough for each to be considered as a single wavelet. By combining the results (18.10) and (19.27) we obtain the wave distribution as a product of two independent contributions, of the form

$$\Psi(\theta) = \Psi_0 b e^{i\varphi} \left(\frac{\sin \gamma}{\gamma} \right) \left(\frac{\sin N\delta}{\sin \delta} \right) \quad (19.33)$$

where $\gamma = (\pi b / \lambda) \sin \theta$, $\delta = (\pi a / \lambda) \sin \theta$, and φ denotes the phase factor. The intensity distribution of the Fraunhofer diffraction pattern, given by

$$I = I_0 \left(\frac{\sin \gamma}{\gamma} \right)^2 \left(\frac{\sin N\delta}{\sin \delta} \right)^2 \quad (19.34)$$

is plotted in Figure 19.10, for $N = 2$, $N = 4$ and $N = 6$, and reduces to Eq. (18.11) as b approaches zero or $(\sin \gamma)/\gamma \to 1$. Therefore, the interference pattern is modulated by the single slit diffraction envelope (19.28). Principal maxima within the envelope occur where $\delta = m\pi$, yielding the *grating formula*, which reads

$$a \sin \theta = m\lambda \quad (19.35)$$

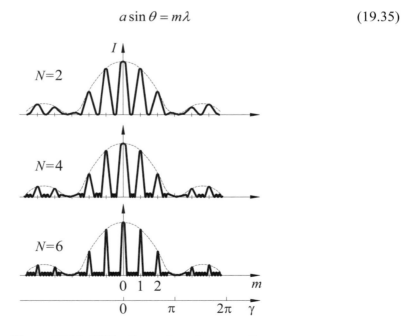

Figure 19.10. Diffraction pattern from a grating

Thus, the angular location of principal maxima is the same, regardless of the number of slits, providing $N \geq 2$. The values of m, called the *order of diffraction*, specify the order of the various principal maxima. Zero intensities occur where $N\delta = m'\pi$, that is where $\delta = m'(\pi/N)$, unless m'/N is an integer. Over a range of π in δ, between successive principal maxima, there are $N-1$ secondary maxima, approximately located at points where $N\delta$ is a maximum, that is near $\delta = (2m'+1)\pi/2N$. For large N, the grating spectrum reduces to the principal maxima, corresponding to different orders of diffraction m. The linewidth of a principal fringe is defined as the angular separation between the zeros on either side of it, that is $\Delta\delta = 2\pi/N$, and the sharpness increases with the overall width of the grating Na as

$$\Delta\left(\frac{\pi a}{\lambda}\sin\theta\right) = \frac{2\pi}{N} \quad \text{or} \quad \frac{\pi a}{\lambda}\cos(\theta)\Delta\theta = \frac{2\pi}{N} \quad \text{i.e.,} \quad \Delta\theta = \frac{2\lambda}{Na\cos\theta} \quad (19.36)$$

The *angular dispersion* $\Delta\theta/\Delta\lambda$, which corresponds to a difference $\Delta\lambda$ in wavelength of two spectral lines, is obtained by differentiating the grating formula (19.35) for a given order m, which gives

$$\frac{\Delta\theta}{\Delta\lambda} = \frac{m}{a\cos\theta} \quad (19.37)$$

that is the angular separation will increase as the order m increases. At the limit of resolution between two lines which partially overlap, the angular separation is half of the linewidth (19.36), and we can equate

$$\frac{\lambda}{Na\cos\theta} = \frac{m\Delta\lambda}{a\cos\theta} \quad \text{or} \quad \frac{\lambda}{\Delta\lambda} = Nm$$

Thus the *resolving power* of a grating spectrometer is obtained as

$$RP = Nm \quad (19.38)$$

A comparison with the resolving power of a multiple-beam interferometer, derived in Problem 18.5, shows that N plays, for the diffraction grating, the role of the contrast parameter F defined by Eq.(18.16).

FURTHER READING
1. M. Mansuripur - CLASSICAL OPTICS AND ITS APPLICATIONS, Cambridge University Press, Cambridge, 2001.
2. J. M. Cowley - DIFFRACTION PHYSICS, Elsevier, Amsterdam, 1995.
3. M. Nielo-Vesperinas - SCATTERING AND DIFFRACTION IN PHYSICAL OPTICS, Wiley, New York, 1991.

19. Diffraction

Problems

19.1. Young's experiment is performed with monochromatic light λ falling normally on two slits of variable width b at a fixed spacing a. What is the maximum value of b which preserves a good visibility of the fringes, if this imposes a limit of $I_{max} \geq 5 I_{min}$ in the fringe pattern?

Solution

The interference pattern arising from two narrow slits, each seen as the source of a single wave, is given by Eq.(18.7), where the intensity varies between $I_{max} = 4 I_0$ and $I_{min} = 0$, so that a maximum visibility $V = (I_{max} - I_{min})/(I_{max} + I_{min}) = 1$ occurs. In the configuration illustrated in Figure 18.2, for an aperture b of the slits, finding the fringe intensity on the screen, at a distance y from the central axis of the instrument, is a simple problem of superposing the small contributions from single waves in the plane of the slits, taking into account the phase differences which arise from the variation in path length for these different waves. The contribution to intensity arising from two single waves situated at distance $\xi, \xi + d\xi$ from the axis of each slit can be written as

$$I(\xi) d\xi = \frac{I_0}{b}\left[1 + \cos\frac{ka(y-\xi)}{x}\right] d\xi$$

so that the total intensity at same position y on the screen is obtained by integration over the slit area as

$$I = \frac{I_0}{b} \int_{-b/2}^{b/2} \left[1 + \cos\frac{ka(y-\xi)}{x}\right] d\xi = I_0 \left[1 + \frac{2x}{kab}\sin\left(\frac{kab}{2x}\right)\cos\left(\frac{kay}{x}\right)\right]$$

It follows that, for appropriate y values, we obtain the intensities at the maxima and minima of the fringe pattern as

$$I_{max} = I_0\left[1 + \frac{2x}{kab}\left|\sin\left(\frac{kab}{2x}\right)\right|\right] \quad , \quad I_{min} = I_0\left[1 - \frac{2x}{kab}\left|\sin\left(\frac{kab}{2x}\right)\right|\right]$$

Thus, the visibility of the fringes is given by

$$V = \frac{I_{max} - I_{min}}{I_{max} + I_{min}} = \left|\frac{\sin(kab/2x)}{kab/2x}\right|$$

and we have $V = 1$ when $b = 0$, as expected. The restriction $I_{max} \geq 5 I_{min}$ implies that $V = 2/3 > 2/\pi$, and this gives the condition for a good visibility as

$$\frac{kab}{2x} < \frac{\pi}{2} \quad \text{or} \quad b\theta < \frac{\lambda}{2}$$

where $\theta \cong a/x$. Thus, the aperture of the slits should not exceed a value of $\lambda/2\theta$.

19.2. Find the far-field diffraction pattern of a monochromatic point source of wavelength λ, formed by a rectangular aperture of length h along the X-direction and width b along the Y-direction (see Figure 19.4).

Solution

When evaluated over a rectangular aperture, the diffraction formula factorizes, and this corresponds to integrating over strips. The Fraunhofer diffraction formula (19.22) takes the form

$$\Psi(u,v) = \Psi_0 \int_0^h \int_0^b e^{-i2\pi(uX+vY)} dX dY = \Psi_0 \left(\int_0^h e^{-i2\pi uX} dX \right)\left(\int_0^b e^{-i2\pi vY} dY \right) = \Psi_0 J_{fX} J_{fY}$$

and hence

$$I_f = I_0 |J_{fX}|^2 |J_{fY}|^2$$

If we regard the *far field* integral as the limit of a sum of infinitesimal vectors of length ΔX and phase angle $\varphi(X) = 2\pi u X$, which can be represented in terms of the coordinates $c(X)$ and $s(X)$ that denote its real and imaginary parts in the complex plane, we have

$$J_{fX} = \lim_{\Delta X \to 0} \sum e^{-i\varphi(X)} \Delta X = \int_0^h e^{-i\varphi(X)} dX = c(h) - i s(h)$$

Integrating gives

$$J_{fX} = \int_0^h e^{-i\varphi(X)} dX = \frac{1}{2\pi u} \int_0^{\varphi(h)} e^{-i\varphi} d\varphi = \frac{1}{2\pi u} \left[\int_0^{\varphi(h)} \cos\varphi\, d\varphi - i \int_0^{\varphi(h)} \sin\varphi\, d\varphi \right]$$

$$= \frac{1}{2\pi u} \{\sin\varphi(h) + i[\cos\varphi(h) - 1]\}$$

It is convenient to take the direction cosine α to be equal to $\sin\rho$, where ρ is the angle between the diffracted ray and the normal to the aperture, so that $u = (\sin\rho)/\rho$, and the phase corresponding to the arc length h becomes $\varphi(h) = (2\pi h/\lambda)\sin\rho$. This yields

$$|J_{fX}| = \frac{\lambda}{2\pi \sin\rho} \left\{ 2\left[1 - \cos\left(\frac{2\pi h}{\lambda}\sin\rho\right)\right] \right\}^{1/2} = \frac{1}{(\pi/\lambda)\sin\rho} \sin\left(\frac{\pi h}{\lambda}\sin\rho\right) = h\frac{\sin\beta}{\beta}$$

where $\beta = (\pi h/\lambda)\sin\rho = \varphi(h)/2$. The angular dependence is the same as that described by Eq.(19.27) for the far-field diffraction by a single slit, and hence, the intensity distribution along the X-direction is given by Eq.(19.28) with γ being replaced by β. From the factorization of the Fraunhofer diffraction formula, it follows that the intensity distribution in the image plane is given by the product of two single-slit distribution functions

$$I_f = |\Psi(u,v)|^2 = I_0 \left(\frac{\sin\beta}{\beta}\right)^2 \left(\frac{\sin\gamma}{\gamma}\right)^2$$

where $\beta = (\pi h/\lambda)\sin\rho$, $\gamma = (\pi b/\lambda)\sin\theta$, and the angles ρ and θ define the direction of the diffracted wave. The pattern is centred at $\rho = 0$ and $\theta = 0$.

19.3. Find the image of a distant monochromatic point source of wavelength λ formed by the uniform circular aperture of diameter D of an optical instrument.

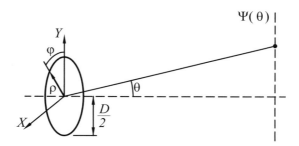

Solution

For a uniform circular aperture of diameter D, the Fraunhofer diffraction formula (19.22) is conveniently expressed in terms of polar coordinates, defined in the configuration shown in the figure by

$$X = \rho\sin\varphi \quad , \quad Y = \rho\cos\varphi$$

so that $dS = \rho d\rho d\varphi$. Assuming that the aperture is normal to the incident wave, we can take $u = 0$, $v = (\sin\theta)/\lambda$ and Eq.(19.22) becomes

$$\Psi(\theta) = \Psi_0 \int_0^{D/2}\int_0^{2\pi} e^{-i2\pi v\rho\cos\varphi}\rho d\rho d\varphi = \Psi_0 \int_0^{D/2}\rho d\rho \int_0^{2\pi} e^{-i(k\rho\sin\theta)\cos\varphi}d\varphi$$

The problem becomes one-dimensional, since the axial symmetry requires a solution which is independent of the polar angle in the plane of observation. However, this may be solved in terms of two standard integrals only, defined as

$$\int_0^{2\pi} e^{i\mu\cos\varphi}d\varphi = 2\pi J_0(\mu) \quad , \quad \int_0^{D/2}\mu J_0(\mu)d\mu = \frac{D}{2}J_1(D/2)$$

where J_0 and J_1 stand for the zero order and first order *Bessel functions*. Their numerical values can be found in tables. It follows that

$$\Psi(\theta) = \Psi_0 \int_0^{D/2} 2\pi J_0(-k\rho\sin\theta)\rho d\rho$$

$$= \Psi_0 \frac{2\pi}{k^2\sin^2\theta}\int_0^{(\pi D\sin\theta)/\lambda}\mu J_0(\mu)d\mu = \Psi_0 \frac{\pi D^2}{4}\cdot 2\frac{J_1\left(\frac{\pi D}{\lambda}\sin\theta\right)}{(\pi D\sin\theta)/\lambda}$$

Thus, the amplitude of the diffracted wave, in a given direction θ, is proportional to the area of the circular aperture and has a form analogous to that found before in Eq.(19.27) for diffraction by a long narrow slit, where the modulation factor has the same behaviour as $(\sin \gamma)/\gamma$ but has a weaker secondary maxima. The intensity distribution of the diffraction pattern is

$$I = |\Psi(\theta)|^2 = 4I_0 \frac{J_1^2\left(\frac{\pi D}{\lambda}\sin\theta\right)}{(\pi D \sin\theta)^2/\lambda^2}$$

and exhibits a central maximum similar to that illustrated in Figure 19.7. However, because of the axial symmetry, it corresponds to a circular pattern called the *Airy disk*. The first zero of the intensity, obtained from standard tables, gives the angular radius of the disk as

$$\frac{\pi D}{\lambda}\sin\theta = 1.22\pi \quad \text{or} \quad \sin\theta = 1.22\frac{\lambda}{D}$$

Since 84 percent of the total intensity in the diffraction pattern is in the first maximum and the intensity drops off rapidly beyond the first zero, circular apertures are preferable to slits or rectangular apertures.

19.4. Find the Fresnel diffraction pattern produced by a wide slit.

Solution

The Fresnel diffraction formula (19.18) can be written as

$$\Psi_P = \Psi_0 \int_0^h \int_0^b e^{ik(X^2+Y^2)(R_0+r_0)/2R_0 r_0} dX dY$$

$$= \Psi_0 \left(\int_0^h e^{ikX^2(R_0+r_0)/2R_0 r_0} dX \right) \left(\int_0^b e^{ikY^2(R_0+r_0)/2R_0 r_0} dY \right) = \Psi_0 J_{nX} J_{nY}$$

so that the wave distribution along a given direction, X say, is described by

$$J_{nX} = \int_0^h e^{ikX^2(R_0+r_0)/2R_0 r_0} dX \quad \text{and} \quad I_n = I_0 |J_{nX}|^2 |J_{nY}|^2$$

The *near-field* diffraction integral is usually rewritten in terms of a dimensionless variable w, defined by

$$w = X\sqrt{\frac{2(R_0+r_0)}{\lambda R_0 r_0}} = X\sqrt{\frac{2}{\lambda L}} \quad \text{with} \quad \frac{1}{L} = \frac{1}{R_0} + \frac{1}{r_0}$$

and this gives

$$J_{nX} = \sqrt{\lambda L/2} \int_0^{h/\sqrt{2/\lambda L}} e^{i\pi w^2/2} dw$$

Since we are only concerned with relative intensities, it is convenient to include the factor

$\sqrt{\lambda L/2}$ in the diffracted wave amplitude Ψ_0 and to adopt the simplified integral

$$J_{nX} = \int_0^l e^{i\pi w^2/2} dw = \int_0^l \cos\left(\frac{\pi w^2}{2}\right) dw + i\int_0^l \sin\left(\frac{\pi w^2}{2}\right) dw$$

where $l = h\sqrt{2/\lambda L}$. The integrals on the right-hand side can be evaluated in terms of the Fresnel integrals, defined as

$$C(l) = \int_0^l \cos\left(\frac{\pi w^2}{2}\right) dw \quad , \quad S(l) = \int_0^l \sin\left(\frac{\pi w^2}{2}\right) dw$$

whose numerical values are tabulated. We may represent the complex Fresnel integral as

$$J(l) = \int_0^l e^{i\pi w^2/2} dw = C(l) + iS(l)$$

so that, for a large slit specified by two arbitrary limit points l_1 and l_2 as in the figure, we can write

$$J_{nX} = J(l_2) - J(l_1) = [C(l) + iS(l)]_{l_1}^{l_2}$$

It can be shown that $C(l)$ and $S(l)$ have the asymptotic forms

$$C(l) = \frac{1}{2} + \frac{1}{\pi l}\sin\left(\frac{\pi l^2}{2}\right) \quad \text{and} \quad S(l) = \frac{1}{2} - \frac{1}{\pi l}\cos\left(\frac{\pi l^2}{2}\right)$$

which gives

$$C(\infty) = S(\infty) = \frac{1}{2} \quad \text{or} \quad J(\infty) = \frac{1}{2}(1+i)$$

and similarly $J(-\infty) = -(1+i)/2$. A normalized near-field integral is then written as

$$J_{nX} = \frac{J(l_2) - J(l_1)}{J(\infty) - J(-\infty)} = \frac{1}{1+i}[C(l) + iS(l)]_{l_1}^{l_2}$$

so that we obtain the intensity distribution in the X-direction in the form

$$|J_{nX}|^2 = \frac{1}{2}[C(l_2) - C(l_1)]^2 + \frac{1}{2}[S(l_2) - S(l_1)]^2$$

for various positions of l_1 and l_2 restricted by $l_2 - l_1 = l$ only. For large values l, the Fresnel diffraction pattern can be understood as a superposition of two independent straight-edge patterns. Taking as a starting point one end point of l, located at $l_2 = \infty$, the intensity becomes a function of the position l_1 of a single diffracting edge, given by

$$|J_{nX}|^2 = \frac{1}{2}\left[\frac{1}{2}-C(l_1)\right]^2 + \frac{1}{2}\left[\frac{1}{2}-S(l_1)\right]^2$$

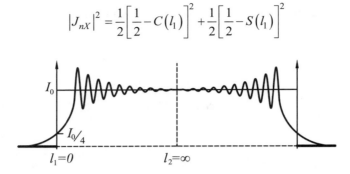

As the length l moves away from $l_2 = \infty$ the intensity increases from zero in the shadow region. The intensity at the limit of the geometrical shadow, which corresponds to l_1 at the origin, is then immediately obtained as $I_n = I_0/4$. In the illuminated region, the intensity rises to a maximum and then oscillates about I_0 with diminishing amplitude.

19.5. Show that a bright spot is observed at the centre of the near-field diffraction pattern of a circular obstacle, and find its intensity.

Solution

In terms of the polar coordinates in the plane of the circular aperture, introduced in Problem 19.3, the near-field diffraction formula (19.18) reads

$$\Psi_P = \Psi_0 \int_0^{D/2}\int_0^{2\pi} e^{ik\rho^2(R_0+r_0)/2R_0 r_0} K(\vec{e}_n,\vec{e}_r)\rho\, d\rho\, d\varphi$$

where $K(\vec{e}_n,\vec{e}_r)$ stands for the obliquity factor. An appropriate dimensionless variable may now be defined as

$$\Phi = k\frac{R_0+r_0}{2R_0 r_0}\rho^2 = \frac{\pi}{\lambda}\frac{R_0+r_0}{R_0 r_0}\rho^2 = \frac{\pi}{\lambda L}\rho^2$$

where L has been defined as in Problem 19.4. Because of the axial symmetry, the polar angle φ ranges from 0 to 2π independently of ρ, so that

$$\rho d\rho \int_0^{2\pi} d\varphi = 2\pi\rho d\rho = \lambda L d\Phi$$

and the obliquity factor can be taken as a function of ρ^2, that is of Φ. This yields

$$\Psi_P = \Psi_0 \lambda L \int_0^{\Phi_{max}} K(\Phi)e^{i\Phi}d\Phi = c(\Phi_{max}) + is(\Phi_{max})$$

Since $K(\Phi)$ is a slowly decreasing function of Φ, the representation in the complex plane consists of a spiral where, for each half turn, the phase Φ increases by π, and this corresponds to a change in the path difference of $\lambda/2$. This suggests the common procedure of imagining the area of the circular aperture as being divided into annular regions called Fresnel zones, defined such that the phase difference changes by π across each zone. If $\Phi = n\pi$ at the boundary of the nth zone, its radius is given by $\rho_n = \sqrt{n\lambda L}$, and hence, all the Fresnel zones have equal areas

$$\Delta S = \pi(\rho_n^2 - \rho_{n-1}^2) = \pi\lambda L = \pi\rho_1^2$$

If K_n is the average value of the obliquity factor over the nth zone, we can approximate the contribution of that zone to the diffracted amplitude as

$$\Delta\Psi_n = \Psi_0 \lambda L \int_{(n-1)\pi}^{n\pi} K_n(\Phi) e^{i\Phi} d\Phi = \Psi_0 \lambda L K_n \left[\int_{(n-1)\pi}^{n\pi} \cos(\Phi) d\Phi + i \int_{(n-1)\pi}^{n\pi} \sin(\Phi) d\Phi \right]$$

$$= \Psi_0 \lambda L 2 i K_n \sin\frac{(2n-1)\pi}{2} = \Psi_0 \lambda L (-1)^{n-1} 2 i K_n$$

and this gives

$$\Psi_P = \sum_{n=1}^{N} \Delta\Psi_n = 2i\Psi_0 \lambda L \sum_{n=1}^{N} (-1)^{n-1} K_n = i\Psi_0 \lambda L \left[K_1 + (K_1 - 2K_2 + K_3) + \ldots + (-1)^{N-1} K_N \right]$$

where the variations in K_n between adjacent zones can be neglected, so that all terms inside parentheses are zero, giving

$$\Psi_P = i\Psi_0 \lambda L \left[K_1 + (-1)^{N-1} K_N \right]$$

For an entire unobstructed wavefront, where $N \to \infty$, the obliquity factor $K_N = \cos(\vec{e}_n, \vec{e}_r) \to 0$ and the disturbance generated at the point of observation P becomes $\Psi_P(\infty) = i\Psi_0 \lambda L$, since $K_1 \approx 1$. Assuming now a circular obstacle that is obstructing the first m zones, one obtains

$$\Psi_P = 2\Psi_P(\infty) \sum_{n=m}^{\infty} (-1)^{n-1} K_n = \Psi_P(\infty) K_m \approx \Psi_P(\infty) \quad \text{or} \quad I_P \approx I_0$$

and this describes the so-called Poisson bright spot appearing in the centre of the near-field diffraction pattern of a circular obstacle, as required. The intensity of this spot roughly equals the axial intensity observed when there is no obstacle.

19.6. Show that the secondary maxima of a Fraunhofer diffraction pattern from a single slit of width b occur at the points for which $\gamma = \tan\gamma$. What is the total number of maxima one could get from such a slit using incident light of wavelength λ?

19.7. In the Fraunhofer diffraction pattern from a double slit, find an expression for the missing orders of interference n, superposed by the diffraction minima of order m, in terms of the ratio of the slit spacing a to the slit width b.

19.8. A parallel beam of monochromatic light of wavelength λ is normally incident on a grating. If the first secondary maximum is observed at an angle $\theta = 9°$, find the highest order of diffraction one could get from such a grating.

19.9. Use the Rayleigh criterion, which states that two equal fringes are resolved when the maximum of one falls upon the first minimum of the other, to find the spatial resolving power of an optical instrument having the objective lens with a circular aperture of diameter D when illuminated normally with light of wavelength λ.

19.10. A circular aperture is filled by a plate consisting of a series of clear and dark concentric rings that coincide with the Fresnel zones defined in Problem 19.5. Assuming a radius ρ_1 for the first transparent Fresnel zone, find the focal length f of such a plate for monochromatic light of wavelength λ.

20. POLARIZATION

20.1. THE TRANSVERSE NATURE OF ELECTROMAGNETIC WAVES
20.2. INTENSITY OF ELECTROMAGNETIC WAVES
20.3. THE STATE OF POLARIZATION
20.4. ALTERNATIVE DESCRIPTIONS OF POLARIZATION

20.1. THE TRANSVERSE NATURE OF ELECTROMAGNETIC WAVES

Many optical phenomena can only be understood in terms of the vectorial nature of electromagnetic waves. While each component of $\vec{E}(\vec{r},t)$ and $\vec{H}(\vec{r},t)$ has a scalar wave-like behaviour, as described by Eq.(14.9), Maxwell's equations (14.7) restrict the relative orientation of the field vectors with respect to the propagation direction. Therefore a complete representation of electromagnetic waves will require the specification of both the magnitude and direction of the field vectors, as a function of position and time.

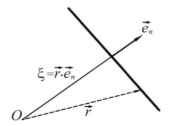

Figure 20.1. The ξ-dependence of the plane wave function

As a result of the additional degree of freedom arising from the transverse orientation of the field vectors, electromagnetic waves exhibit polarization properties, which are not manifested by scalar waves. Consider the plane wave solutions to the vector electromagnetic wave equations (14.7) in linear, homogenous and isotropic media which contain no free charge distribution. As illustrated in Figure 20.1, a plane wave function is conveniently written in terms of the distance $\xi = \vec{r} \cdot \vec{e}_n$ travelled along the direction of propagation $\vec{e}_n$, rather than in terms of the position vector $\vec{r}$ of an arbitrary point on the wavefront. We have

$$\nabla = \vec{e}_x \frac{\partial}{\partial x} + \vec{e}_y \frac{\partial}{\partial y} + \vec{e}_z \frac{\partial}{\partial z} = \left(\vec{e}_x \frac{\partial \xi}{\partial x} + \vec{e}_y \frac{\partial \xi}{\partial y} + \vec{e}_z \frac{\partial \xi}{\partial z} \right) \frac{\partial}{\partial \xi}$$

$$= \left[\vec{e}_x \frac{\partial}{\partial x}(\vec{e}_n \cdot \vec{r}) + \vec{e}_y \frac{\partial}{\partial y}(\vec{e}_n \cdot \vec{r}) + \vec{e}_z \frac{\partial}{\partial z}(\vec{e}_n \cdot \vec{r}) \right] \frac{\partial}{\partial \xi}$$

$$= \left[\vec{e}_x (\vec{e}_n \cdot \vec{e}_x) + \vec{e}_y (\vec{e}_n \cdot \vec{e}_y) + \vec{e}_z (\vec{e}_n \cdot \vec{e}_z) \right] \frac{\partial}{\partial \xi} = \vec{e}_n \frac{\partial}{\partial \xi} \quad (20.1)$$

so that Eqs.(14.7) can be rewritten in terms of ξ as

$$\vec{e}_n \times \frac{\partial \vec{E}}{\partial \xi} = -\mu \frac{\partial \vec{H}}{\partial t} \quad (20.2)$$

$$\vec{e}_n \times \frac{\partial \vec{H}}{\partial \xi} = \sigma \vec{E} + \varepsilon \frac{\partial \vec{E}}{\partial t} \quad (20.3)$$

$$\vec{e}_n \cdot \frac{\partial \vec{H}}{\partial \xi} = 0 \quad (20.4)$$

$$\vec{e}_n \cdot \frac{\partial \vec{E}}{\partial \xi} = 0 \quad (20.5)$$

where $\vec{E} = \vec{E}(\xi,t)$ and $\vec{H} = \vec{H}(\xi,t)$. Equation (20.5) also reads $\partial(\vec{e}_n \cdot \vec{E})/\partial \xi = \partial E_n / \partial \xi = 0$, and this implies that the longitudinal component of the electric field might be a function of time only, which can be obtained from Eq.(20.3) giving

$$\vec{e}_n \cdot \left(\vec{e}_n \times \frac{\partial \vec{H}}{\partial \xi} \right) = 0 = \left(\sigma + \varepsilon \frac{\partial}{\partial t} \right)(\vec{e}_n \cdot \vec{E}) = \left(\sigma + \varepsilon \frac{\partial}{\partial t} \right) E_n \quad (20.6)$$

The solution

$$E_n(t) = E_{n0} e^{-\sigma t / \varepsilon} = E_{n0} e^{-t/\tau} \quad (20.7)$$

shows an exponential decay with relaxation time $\tau = \varepsilon / \sigma$ and this allows us to set $E_n = 0$ in conductors, as τ is very small. In nonconducting media, where $\sigma = 0$, Eq.(20.6) reduces to $\partial E_n / \partial t = 0$, thus the longitudinal component of the electric field is independent of both time and position. A similar result holds for the longitudinal component of the magnetic field, since Eq.(20.4) and Eq.(20.2) can be reduced to

$\partial H_n/\partial\xi = 0$, and $\partial H_n/\partial t = 0$, respectively. Therefore the field vectors of plane electromagnetic waves, defined as functions of both position and time, are not allowed to have components along the propagation direction $\vec{e}_n$. This type of plane wave, where both $\vec{E}$ and $\vec{H}$ are contained in the wavefront plane, is called a *transverse electromagnetic wave*.

If we assume that the field variation with both position and time is *harmonic* and has the form (16.25)

$$\vec{E} = \vec{E}_0 e^{i(\vec{k}\cdot\vec{r}-\omega t)} \quad , \quad \vec{H} = \vec{H}_0 e^{i(\vec{k}\cdot\vec{r}-\omega t)} \tag{20.8}$$

where $\vec{k} = k\vec{e}_n$ is the wave vector, as defined by Eq.(16.23), it is a straightforward matter to derive, by direct substitution of Eq.(20.8), the following properties

$$\nabla\cdot\vec{E} = i\vec{k}\cdot\vec{E} \qquad \nabla\cdot\vec{H} = i\vec{k}\cdot\vec{H}$$

$$\nabla\times\vec{E} = i\vec{k}\times\vec{E} \qquad \nabla\times\vec{H} = i\vec{k}\times\vec{H} \tag{20.9}$$

$$\frac{\partial\vec{E}}{\partial t} = -i\omega\vec{E} \qquad \frac{\partial\vec{H}}{\partial t} = -i\omega\vec{H}$$

Thus, for a plane harmonic wave representation (20.8) of the field vectors, two operator relations hold as

$$\nabla \to i\vec{k} \quad , \quad \frac{\partial}{\partial t} \to -i\omega \tag{20.10}$$

and this allow Maxwell's equations (14.7), in nonconducting media $(\sigma = 0)$, to be reduced to the form

$$\vec{k}\times\vec{E} = \mu\omega\vec{H} \qquad \vec{k}\times\vec{H} = -\varepsilon\omega\vec{E}$$

$$\vec{k}\cdot\vec{H} = 0 \qquad \vec{k}\cdot\vec{E} = 0 \tag{20.11}$$

Equations (20.11) show that not only both $\vec{E}$ and $\vec{H}$ are perpendicular to the direction of propagation $\vec{k}$, but also

$$k\vec{e}_n\times\vec{E} = \mu\omega\vec{H} \quad \text{or} \quad \vec{H} = \frac{k}{\mu\omega}(\vec{e}_n\times\vec{E}) = \sqrt{\frac{\varepsilon}{\mu}}(\vec{e}_n\times\vec{E}) \tag{20.12}$$

where we have used the definitions (16.24) and (15.54) for the wave number k and velocity v. It follows that the three vectors $\vec{E}, \vec{H}$ and $\vec{k}$ constitute a right-handed orthogonal set, as illustrated in Figure 20.2. Since the cross product $\vec{E}\times\vec{H}$, which defines the Poynting vector (14.13), has the same direction as the wave vector $\vec{k}$, the direction of energy flow in isotropic media coincides with the propagation direction.

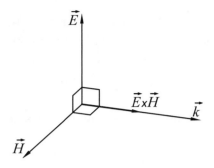

Figure 20.2. The transverse orientation of the field vectors

Equation (20.12) shows that the field vectors $\vec{E}$ and $\vec{H}$ are always *in phase*, as ε and μ are real quantities. The ratio of their magnitudes may be rewritten in terms of the impedance Z_0 per unit area of free space and the index of refraction n of a given medium, expressed by Eqs.(17.23) and (15.55), respectively, as

$$\frac{|\vec{H}|}{|\vec{E}|} = \sqrt{\frac{\varepsilon}{\mu}} = \frac{n}{Z_0} = \frac{1}{Z_n} \tag{20.13}$$

where

$$Z_n = \frac{Z_0}{n} \tag{20.14}$$

is referred to as the *intrinsic impedance* per unit area of the medium.

20.2. INTENSITY OF ELECTROMAGNETIC WAVES

The intensity of electromagnetic waves, as obtained in the form (17.30) in the scalar approximation, must be re-examined in the case of the field vector representation of transverse waves. The appropriate evaluation of the time averaged power transported across the unit area normal to the propagation direction, that is

$$\langle \vec{S} \rangle = \langle \vec{E} \times \vec{H} \rangle$$

must take into account the restrictions introduced by the rule of complex representation (16.18) of any harmonic wave. As we are only concerned with time averaged values of the energy flux, it is convenient to separate, in the representation (20.8), the harmonic time variation of the field, in the form

$$\vec{E} = \left(\vec{E}_0 e^{i\vec{k}\cdot\vec{r}}\right)e^{-i\omega t} = \vec{E}(\vec{r})e^{-i\omega t}$$
$$\vec{H} = \left(\vec{H}_0 e^{i\vec{k}\cdot\vec{r}}\right)e^{-i\omega t} = \vec{H}(\vec{r})e^{-i\omega t} \tag{20.15}$$

so that the actual fields that specify the wave response are

$$\vec{E}(\vec{r},t) = \text{Re}\left[\vec{E}(\vec{r})e^{-i\omega t}\right] = \frac{1}{2}\left[\vec{E}(\vec{r})e^{-i\omega t} + \vec{E}^*(\vec{r})e^{i\omega t}\right]$$
$$\vec{H}(\vec{r},t) = \text{Re}\left[\vec{H}(\vec{r})e^{-i\omega t}\right] = \frac{1}{2}\left[\vec{H}(\vec{r})e^{-i\omega t} + \vec{H}^*(\vec{r})e^{i\omega t}\right] \tag{20.16}$$

where

$$\vec{E}^*(\vec{r}) = \vec{E}_0 e^{-i\vec{k}\cdot\vec{r}} \quad\text{and}\quad \vec{H}^*(\vec{r}) = \vec{H}_0 e^{-i\vec{k}\cdot\vec{r}}$$

It follows that

$$\vec{S} = \vec{E}\times\vec{H} = \frac{1}{2}\text{Re}\left[\vec{E}(\vec{r})\times\vec{H}^*(\vec{r})\right] + \frac{1}{2}\text{Re}\left[\vec{E}(\vec{r})\times\vec{H}(\vec{r})\right]\cos 2\omega t \tag{20.17}$$

Since $\langle\cos 2\omega t\rangle = 0$, we obtain

$$\langle\vec{S}\rangle = \frac{1}{2}\text{Re}\left[\vec{E}(\vec{r})\times\vec{H}^*(\vec{r})\right] = \frac{1}{2}\text{Re}\left(\vec{E}\times\vec{H}^*\right) \tag{20.18}$$

In a similar manner, the time-averaged energy densities of the electric and magnetic fields $\langle u_E\rangle$ and $\langle u_H\rangle$ are obtained by substituting Eq.(20.16) into Eq.(11.32), that is

$$\langle u_E\rangle = \frac{1}{2}\varepsilon\langle\vec{E}(\vec{r},t)\cdot\vec{E}(\vec{r},t)\rangle = \frac{1}{4}\varepsilon\vec{E}(\vec{r})\cdot\vec{E}^*(\vec{r}) + \frac{1}{4}\varepsilon\text{Re}\left[\vec{E}^2(\vec{r})\right]\langle\cos 2\omega t\rangle$$
$$= \frac{1}{4}\varepsilon\vec{E}(\vec{r})\cdot\vec{E}^*(\vec{r}) = \frac{1}{4}\varepsilon\vec{E}\cdot\vec{E}^* \tag{20.19}$$

and, respectively, into Eq.(13.34), which gives

$$\langle u_H\rangle = \frac{1}{2}\mu\langle\vec{H}(\vec{r},t)\cdot\vec{H}(\vec{r},t)\rangle = \frac{1}{4}\mu\vec{H}(\vec{r})\cdot\vec{H}^*(\vec{r}) = \frac{1}{4}\mu\vec{H}\cdot\vec{H}^* \tag{20.20}$$

Therefore the time-averaged energy density of the electromagnetic wave is

$$\langle u_{em}\rangle = \frac{1}{4}\left(\varepsilon\vec{E}\cdot\vec{E}^* + \mu\vec{H}\cdot\vec{H}^*\right) = \frac{1}{4}\left(\vec{E}\cdot\vec{D}^* + \vec{B}\cdot\vec{H}^*\right) \tag{20.21}$$

Substituting the ratio (20.13) of the field vector magnitudes, it becomes apparent that the time-averaged energy densities of the two fields are equal, since

$$\langle u_E \rangle = \frac{1}{4}\varepsilon \vec{E}\cdot\vec{E}^* = \frac{1}{4}\varepsilon |\vec{E}_0|^2 = \frac{1}{4}\mu |\vec{H}_0|^2 = \langle u_H \rangle \qquad (20.22)$$

and that the total energy density (20.21) has the time average

$$\langle u_{em} \rangle = \frac{1}{2}\varepsilon |\vec{E}_0|^2 = \frac{1}{2}\mu |\vec{H}_0|^2 \qquad (20.23)$$

In view of the relative orientation (20.12) of the field vectors, Eq.(20.18) takes the form

$$\langle \vec{S} \rangle = \frac{1}{2}\sqrt{\frac{\varepsilon}{\mu}} \operatorname{Re}\left[\vec{E}\times(\vec{e}_n \times \vec{E}^*)\right] = \frac{1}{2}\sqrt{\frac{\varepsilon}{\mu}}(\vec{E}\times\vec{E}^*)\vec{e}_n \qquad (20.24)$$

where we made use of the vector identity

$$\vec{E}\times(\vec{e}_n \times \vec{E}^*) = \vec{e}_n(\vec{E}\cdot\vec{E}^*) - \vec{E}^*(\vec{E}\cdot\vec{e}_n) = \vec{e}_n(\vec{E}\cdot\vec{E}^*)$$

which is valid for a transverse electromagnetic wave, since $\vec{E}\cdot\vec{e}_n = 0$. Equation (20.24) shows that in an isotropic medium, the energy flux is in the direction of wave propagation, and can be written as

$$\langle \vec{S} \rangle = I\,\vec{e}_n \qquad (20.25)$$

where I is the intensity of plane electromagnetic waves, given by

$$I = \frac{1}{2}\sqrt{\frac{\varepsilon}{\mu}}\vec{E}\cdot\vec{E}^* = \frac{1}{2Z_n}|\vec{E}_0|^2 = \frac{1}{2}Z_n|\vec{H}_0|^2 \qquad (20.26)$$

Substituting Eq.(20.23), Eq.(20.24) can also be written as

$$\langle \vec{S} \rangle = \frac{1}{\sqrt{\varepsilon\mu}}\langle u_{em} \rangle \vec{e}_n \qquad (20.27)$$

so that the intensity (20.25) takes the form

$$I = \langle u_{em} \rangle v \qquad (20.28)$$

where v is the speed (15.54) of electromagnetic waves. Equation (20.28) is the analogue of Eq.(17.19), showing that the energy flux is given by the product of the time averaged energy density and the wave velocity. Note that Eqs.(20.26) and (20.28) are only valid for strictly monochromatic waves that travel with phase velocity $v = \omega/c$ and have the amplitudes and phase factors of the field vectors as defined by Eq.(20.15), which are not time-dependent. For *quasi-monochromatic waves*, the intensity (20.26) must be written as a time average over the field vectors, that is,

$$I = \frac{1}{2Z_n}\left\langle \vec{E}\cdot\vec{E}^*\right\rangle \tag{20.29}$$

and represents an energy flow of the form (20.28), where the phase velocity is replaced by the group velocity (16.66).

20.3. THE STATE OF POLARIZATION

The transverse nature of electromagnetic waves requires that $\vec{E}$ points along an arbitrary direction in the plane perpendicular to $\vec{k}$ and that $\vec{H}$ should lie along the vector $\vec{k}\times\vec{E}$. Consider a monochromatic wave with $\vec{k}$ along the z-axis and the electric vector (20.15) represented by its components $\vec{E}_x, \vec{E}_y$ in two mutually orthogonal directions

$$\vec{E} = \vec{e}_x \vec{E}_x + \vec{e}_y \vec{E}_y = \vec{e}_x A_x e^{i(\varphi_x - \omega t)} + \vec{e}_y A_y e^{i(\varphi_y - \omega t)} \tag{20.30}$$

where $A_x, A_y, \varphi_x, \varphi_y$ are the constant amplitudes and phase factors of the components at a point on the wavefront. The corresponding intensity (20.26) has a constant value which is usually taken as

$$I \sim \vec{E}\cdot\vec{E}^* = A_x^2 + A_y^2 \tag{20.31}$$

In an arbitrary direction α within the wavefront plane (the xy-plane), represented in Figure 20.3, the component of the electric field vector is

$$E(\alpha) = \vec{E}\cdot\vec{e}_\alpha = \left(A_x \cos\alpha\, e^{i\varphi_x} + A_y \sin\alpha\, e^{i\varphi_y}\right)e^{-i\omega t} \tag{20.32}$$

so that the intensity is derived from (20.26) in the form

$$I(\alpha) \sim E(\alpha)E^*(\alpha) = A_x^2 \cos^2\alpha + A_y^2 \sin^2\alpha + \sin\alpha\cos\alpha\left(A_x A_y e^{i\delta} + A_x A_y e^{-i\delta}\right) \tag{20.33}$$

where $\delta = \varphi_x - \varphi_y$ is the phase difference between the orthogonal components. Thus the intensity is a function of the orientation α, with coefficients written in terms of three independent real parameters A_x, A_y and δ, which can be considered for convenience to be elements of a correlation matrix given by

$$\hat{M} = \begin{pmatrix} M_{xx} & M_{xy} \\ M_{yx} & M_{yy} \end{pmatrix} = \begin{pmatrix} A_x^2 & A_x A_y e^{i\delta} \\ A_x A_y e^{-i\delta} & A_y^2 \end{pmatrix} \qquad (20.34)$$

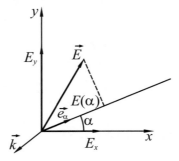

Figure 20.3. Orientation α of an arbitrary direction in the wavefront plane

The matrix (20.34) is Hermitian, as $M_{xx} = M_{xx}^*$, $M_{yy} = M_{yy}^*$, $M_{xy} = M_{yx}^*$ (see Appendix II). The real diagonal elements represent the intensities of the E_x and E_y components of the field, so that their sum, called the *trace* $\text{Tr}(\hat{M})$ of the matrix, gives the constant intensity (20.31) according to

$$\text{Tr}(\hat{M}) = M_{xx} + M_{yy} = I \qquad (20.35)$$

The nondiagonal elements have the form

$$M_{xy} = \sqrt{M_{xx}} \sqrt{M_{yy}} e^{i\delta}$$

so that Eq.(20.33) can be written as

$$I(\alpha) = M_{xx} \cos^2 \alpha + M_{yy} \sin^2 \alpha + \sqrt{M_{xx}} \sqrt{M_{yy}} \sin \alpha \cos \alpha \cos \delta \qquad (20.36)$$

which has the form of the interference law (18.3) written as

$$I = I_1 + I_2 + 2\sqrt{I_1 I_2} \, \text{Re}\, \gamma_{xy} \quad , \quad \gamma_{xy} = \frac{M_{xy}}{\sqrt{M_{xx}} \sqrt{M_{yy}}} = e^{i\delta} \qquad (20.37)$$

The parameter γ_{xy} is a measure of the correlation between the components E_x, E_y of the electric vector, its magnitude $|\gamma_{xy}| = 1$ indicating a complete correlation of the two components, which is referred to as a *complete polarization* of strictly monochromatic waves. When $|\gamma_{xy}| = 0$, the two components are said to be uncorrelated and, in general, a partial polarization described by $0 < |\gamma_{xy}| < 1$ is expected to occur. For all the states of complete polarization the determinant of the correlation matrix is zero

$$\det(\hat{M}) = M_{xx}M_{yy} - M_{xy}M_{yx} = 0 \tag{20.38}$$

The matrix (20.34) defines the state of complete polarization in terms of four real parameters $M_{xx}, M_{yy}, |M_{xy}|$ and δ, related by the identity (20.38), which are thus equivalent to the three independent parameters A_x, A_y and δ. It is common practice to discuss the type of polarization in terms of parameters suitable for graphical representation such as the overall intensity I, given by Eq.(20.31), and the ratio of component amplitudes, written as

$$\tan\theta = \frac{A_y}{A_x} \tag{20.39}$$

Substituting Eq.(20.39), Eq.(20.31) gives

$$I = A_x^2(1 + \tan^2\theta) \quad \text{i.e.,} \quad A_x^2 = I\cos^2\theta \quad \text{and} \quad A_y^2 = I\sin^2\theta$$

so that the correlation matrix takes the form

$$\hat{M} = I\begin{pmatrix} \cos^2\theta & \sin\theta\cos\theta\, e^{i\delta} \\ \sin\theta\cos\theta\, e^{-i\delta} & \sin^2\theta \end{pmatrix} \tag{20.40}$$

which defines the state of polarization in terms of I, θ and δ, where $0 \leq \theta \leq \pi/2$ and $0 \leq \delta < 2\pi$.

In the special cases where $\theta = 0$ or $\theta = \pi/2$, the electromagnetic wave is *linearly polarized* with the electric vector along the x-direction $(A_y = 0)$ and the y-direction $(A_x = 0)$, respectively. The matrices

$$\hat{M}_x = I\begin{pmatrix} 1 & 0 \\ 0 & 0 \end{pmatrix}, \quad \hat{M}_y = I\begin{pmatrix} 0 & 0 \\ 0 & 1 \end{pmatrix} \tag{20.41}$$

form a linearly polarized basis represented in Figure 20.4.

If $0 < \theta < \pi/2$ the state of polarization is determined by the phase difference $\delta = \varphi_x - \varphi_y$. Linear polarization then corresponds to either $\delta = 0$ or $\delta = \pi$, when the matrix (20.40) reduces to

$$\hat{M} = I\begin{pmatrix} \cos^2\theta & \pm\sin\theta\cos\theta \\ \pm\sin\theta\cos\theta & \sin^2\theta \end{pmatrix} \tag{20.42}$$

The matrix (20.42) describes an electric vector that vibrates in either the θ or $-\theta$

directions (illustrated in Figure 20.4). The electric field components can be derived from Eqs.(20.30) and (20.39) in the form

$$E_y = \left(e^{i\delta}\frac{A_y}{A_x}\right)E_x = \left(\pm\frac{A_y}{A_x}\right)E_x = (\pm\tan\theta)E_x \qquad (20.43)$$

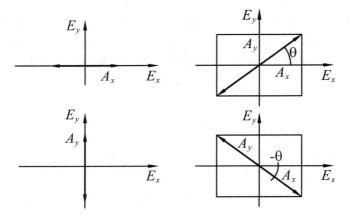

Figure 20.4. States of linear polarization

For all the other values of δ the polarization is *elliptical*, since the matrix (20.40) describes an electric vector that traces out an *ellipse* in the wavefront plane. Its equation can be derived by writing the components of the fields from Eq.(20.30) as

$$E_x(\vec{r},t) = \operatorname{Re} E_x = A_x \cos(\varphi_x - \omega t) = A_x \cos(\varphi_y + \delta - \omega t)$$

$$= A_x\left[\cos(\varphi_y - \omega t)\cos\delta - \sin(\varphi_y - \omega t)\sin\delta\right] \qquad (20.44)$$

$$E_y(\vec{r},t) = \operatorname{Re} E_y = A_y \cos(\varphi_y - \omega t)$$

If we rewrite Eqs.(20.44) in the form

$$\frac{E_y(\vec{r},t)}{A_y}\cos\delta - \frac{E_x(\vec{r},t)}{A_x} = \sin(\varphi_y - \omega t)\sin\delta \quad , \quad \frac{E_y(\vec{r},t)}{A_y}\sin\delta = \cos(\varphi_y - \omega t)\sin\delta$$

$$(23.45)$$

we obtain, by squaring and adding, the general equation of an ellipse inscribed within a rectangle of sides $2A_x$ and $2A_y$ in the form

$$\frac{E_x^2(\vec{r},t)}{A_x^2} - 2\frac{E_x(\vec{r},t)E_y(\vec{r},t)}{A_x A_y}\cos\delta + \frac{E_y^2(\vec{r},t)}{A_y^2} = \sin^2\delta \qquad (20.46)$$

The principal axes of the ellipse are turned by an angle ψ with respect to the axes of the Cartesian system, as in Figure 20.5, and this allows us to define

$$\tan\psi = \left[\frac{E_y(\vec{r},t)}{E_x(\vec{r},t)}\right]_{E=\max} \quad (20.47)$$

The set of three independent quantities needed to determine the polarization ellipse of a monochromatic wave consists either of the amplitudes A_x, A_y and the phase difference δ or the semimajor and semiminor axes of the ellipse, denoted by a and b, and the orientation angle ψ. The two sets are related, such that $a, b,$ and ψ can be found from A_x, A_y and δ.

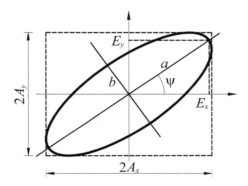

Figure 20.5. Polarization ellipse having orientation angle ψ

Since the field amplitude $E_x^2(\vec{r},t) + E_y^2(\vec{r},t)$ is maximum along the major axis, we have

$$E_x(\vec{r},t)dE_x + E_y(\vec{r},t)dE_y = 0 \quad \text{or} \quad \frac{dE_x}{dE_y} = -\left[\frac{E_y(\vec{r},t)}{E_x(\vec{r},t)}\right]_{E=\max} = -\tan\psi \quad (20.48)$$

so that, by differentiating Eq.(20.46), we obtain

$$\frac{2}{A_x^2}E_x(\vec{r},t)dE_x - \frac{2\cos\delta}{A_x A_y}\left[E_x(\vec{r},t)dE_y + E_y(\vec{r},t)dE_x\right] + \frac{2}{A_y^2}E_y(\vec{r},t)dE_y = 0$$

or, in view of Eqs.(20.47) and (20.48), we have

$$\frac{1}{A_y^2} - \frac{1}{A_x^2} = \frac{\cos\delta}{A_x A_y}(\cot\psi - \tan\psi) = \frac{2\cos\delta}{A_x A_y}\cot 2\psi$$

This gives the orientation angle ψ as

$$\tan 2\psi = \frac{2A_x A_y}{A_x^2 - A_y^2} \cos\delta = \tan(2\theta)\cos\delta \qquad (20.49)$$

where the ratio of the component amplitudes (20.39) was inserted. It is obvious that

$$a^2 + b^2 = A_x^2 + A_y^2 \qquad (20.50)$$

and the axial ratio of the ellipse

$$\tan \chi = \frac{b}{a} \qquad (20.51)$$

is usually defined in terms of the angle χ called the *ellipticity angle*. It can be shown that χ is given by

$$\sin 2\chi = \frac{2A_x A_y}{A_x^2 + A_y^2} \sin\delta = \sin(2\theta)\sin\delta \qquad (20.52)$$

The sign of χ makes the difference between the two senses of polarization, which are called *right-handed*, if the electric vector is rotating clockwise, for $\chi > 0$, and *left-handed*, if it is rotating counterclockwise, for $\chi < 0$.

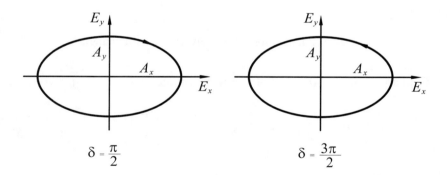

Figure 20.6. States of elliptic polarization

For $\delta = \pi/2$ or $\delta = 3\pi/2$, Eq.(20.46) reduces to the familiar equation of an ellipse with semiaxes $a = A_x$ and $b = A_y$, oriented along the x- and y-axes, which reads

$$\frac{E_x^2(\vec{r},t)}{A_x^2} + \frac{E_y^2(\vec{r},t)}{A_y^2} = 1 \qquad (20.53)$$

and is represented in Figure 20.6. For $\delta = \pi/2$ the electric field traverses the elliptical path *clockwise*, while for $\delta = 3\pi/2$ the sense of traversal is *counterclockwise*. The matrix representation of this polarization state is obtained from Eq.(20.40) as

$$\hat{M} = I \begin{pmatrix} \cos^2\theta & \pm i\cos\theta\sin\theta \\ \mp i\cos\theta\sin\theta & \sin^2\theta \end{pmatrix} \qquad (20.54)$$

For $\theta = \pi/4$, the matrices (20.54) reduce to

$$\hat{M}_r = \frac{I}{2}\begin{pmatrix} 1 & i \\ -i & 1 \end{pmatrix} \quad , \quad \hat{M}_l = \frac{I}{2}\begin{pmatrix} 1 & -i \\ i & 1 \end{pmatrix} \qquad (20.55)$$

and define the states of *right circular polarization*, $\delta = \pi/2$, with the electric vector describing clockwise a circle in the *xy*-plane and *left circular polarization*, where the electric vector traverses the circle counterclockwise $\delta = 3\pi/2$.

EXAMPLE 20.1. Jones vectors

The correlation matrix (20.34) can be formally written as a direct product of a column vector

$$J = \begin{pmatrix} E_x \\ E_y \end{pmatrix} = \begin{pmatrix} A_x e^{i\varphi_x} \\ A_y e^{i\varphi_y} \end{pmatrix} = \sqrt{I}\begin{pmatrix} e^{i\varphi_x}\cos\theta \\ e^{i\varphi_y}\sin\theta \end{pmatrix} \qquad (20.56)$$

called the *Jones vector* of the polarization state and its Hermitian conjugate J^+, which is a row vector (see Appendix II) of the form

$$J^+ = (E_x^*, E_y^*) = (A_x e^{-i\varphi_x}, A_y e^{-i\varphi_y}) = \sqrt{I}(e^{-i\varphi_x}\cos\theta, e^{-i\varphi_y}\sin\theta)$$

so that

$$JJ^+ = \begin{pmatrix} A_x e^{i\varphi_x} \\ A_y e^{i\varphi_y} \end{pmatrix}(A_x e^{-i\varphi_x}, A_y e^{-i\varphi_y}) = \hat{M} \qquad (20.57)$$

The Jones vectors corresponding to the linearly polarized basis (20.41) are obtained in a normalized form using Eq.(20.57) and dividing the result by $\sqrt{I}$, which gives

$$J_x = \begin{pmatrix} 1 \\ 0 \end{pmatrix} \quad , \quad J_y = \begin{pmatrix} 0 \\ 1 \end{pmatrix} \qquad (20.58)$$

In a similar manner, the Jones vectors associated with the right and left circularly polarized waves (20.55) result from Eq.(20.57) in the normalized form

$$J_r = \frac{1}{\sqrt{2}}\begin{pmatrix} 1 \\ -i \end{pmatrix} \quad , \quad J_l = \frac{1}{\sqrt{2}}\begin{pmatrix} 1 \\ i \end{pmatrix} \qquad (20.59)$$

whereas the state of elliptic polarization (20.54) is described by

$$J = \begin{pmatrix} \cos\theta \\ \pm i \sin\theta \end{pmatrix} \quad (20.60)$$

Figure 20.7 illustrates the state of polarization represented by the Jones vectors (20.60) for various values of θ in the range from 0 to $\pi/2$.

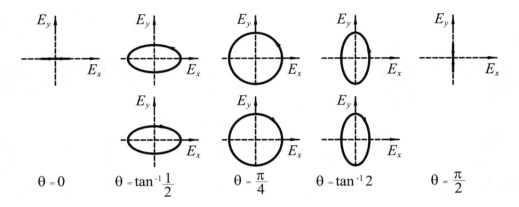

Figure 20.7. States of polarization represented by Jones vectors

The superposition of two or more waves of given polarization can be represented by the addition of the Jones vectors. As an example, a state of elliptic polarization (20.52) can always be resolved into two orthogonal components, either linear

$$J = \cos\theta \begin{pmatrix} 1 \\ 0 \end{pmatrix} \pm i \sin\theta \begin{pmatrix} 0 \\ 1 \end{pmatrix} = J_x \cos\theta \pm J_y \sin\theta \quad (20.61)$$

or circular

$$J = \frac{1}{\sqrt{2}}(\cos\theta \pm \sin\theta)J_r + \frac{1}{\sqrt{2}}(\cos\theta \mp \sin\theta)J_l \quad (20.62)$$

The effect of linear optical elements that change the state of polarization J_i of an incident wave into an emergent state J_f may be represented by a matrix product

$$J_f = \hat{M} J_i \quad (20.63)$$

where $\hat{M}$ are 2×2 matrices of optical elements, called *Jones matrices*, which have the form of the correlation matrices (20.34) or (20.40). In the particular case where J_i is a linearly polarized state J_x or J_y described by Eq.(20.58), substituting Eq.(20.34) into Eq.(20.63) gives

$$\begin{pmatrix} A_x^2 & A_x A_y e^{i\delta} \\ A_x A_y e^{-i\delta} & A_y^2 \end{pmatrix} \begin{pmatrix} 1 \\ 0 \end{pmatrix} = A_x \begin{pmatrix} A_x \\ A_y e^{-i\delta} \end{pmatrix}$$

and also

$$\begin{pmatrix} A_x^2 & A_x A_y e^{i\delta} \\ A_x A_y e^{-i\delta} & A_y^2 \end{pmatrix} \begin{pmatrix} 0 \\ 1 \end{pmatrix} = A_y \begin{pmatrix} A_x e^{i\delta} \\ A_y \end{pmatrix}$$

It is seen that the emergent state of polarization is determined by the polarization parameters A_x, A_y and δ of the optical element. Therefore the matrices $\hat{M}_x$ and $\hat{M}_y$, given by Eq.(20.41), represent linear polarizers having respectively horizontal and vertical transmission axes, while matrices $\hat{M}_r$ and $\hat{M}_l$ from Eq.(20.54) define right and left circular polarizers. The change of an incident polarization state J_i as a result of inserting a succession of optical elements, represented by the matrices $\hat{M}_1,\ldots,\hat{M}_n$, can be evaluated by matrix multiplication as follows

$$\begin{aligned} J_1 &= \hat{M}_1 J_i \\ J_2 &= \hat{M}_2 J_1 = \hat{M}_2 \hat{M}_1 J_i \\ &\vdots \\ J_f &= \hat{M}_n \hat{M}_{n-1} \ldots \hat{M}_1 J_i \end{aligned} \qquad (20.64)$$

where the products are taken in the order 1 to n.

20.4. ALTERNATIVE DESCRIPTIONS OF POLARIZATION

The traditional parameters for the complete description of polarized waves, called the *Stokes parameters*, are defined as

$$I = A_x^2 + A_y^2 \quad , \quad Q = A_x^2 - A_y^2 \quad , \quad U = 2A_x A_y \cos\delta \quad , \quad V = 2A_x A_y \sin\delta \quad (20.65)$$

Only three of the four parameters are independent, because there is an obvious identity given by

$$I^2 = Q^2 + U^2 + V^2 \qquad (20.66)$$

Since the parameter I gives the intensity, as defined by Eq.(20.31), it is usual to express Q, U and V in terms of I, the orientation angle ψ and the ellipticity angle χ. Equation (20.52) immediately gives

$$V = I \sin 2\chi \qquad (20.67)$$

so that from Eqs.(20.66), (20.65) and (20.49) we obtain

$$Q^2 + U^2 = I^2 \cos^2 2\chi \quad , \quad \frac{U}{Q} = \tan 2\psi \qquad (20.68)$$

It follows that

$$Q = I \cos 2\chi \cos 2\psi \quad , \quad U = I \cos 2\chi \sin 2\psi \qquad (20.69)$$

If Q, U and V are interpreted as the Cartesian coordinates of a point on a sphere of radius I, called the *Poincaré sphere*, 2ψ and 2χ are its longitude and latitude, respectively, as shown in Figure 20.8. This way, each point on the sphere represents a state of complete polarization of the wave. The linearly polarized waves, with $\delta = 0$ or $\delta = \pi$ must have $\chi = 0$ according to Eq.(20.52), so that they are represented by points on the equator of the Poincaré sphere. The circularly polarized waves, with $\delta = \pi/2$ or $\delta = 3\pi/2$ and $\theta = \pi/4$, correspond to $2\chi = \pi/2$ and $2\chi = -\pi/2$, respectively, according to Eq.(20.52), that is to the north and south poles. The remaining surface of the sphere represents elliptic polarization, right-handed on the northern hemisphere and left-handed on the southern hemisphere.

By substituting the Stokes parameters (20.65), the correlation matrix (20.34) can be rewritten as

$$\hat{M} = \frac{1}{2}\begin{pmatrix} I+Q & U+iV \\ U-iV & I-Q \end{pmatrix} \qquad (20.70)$$

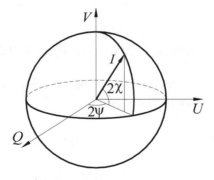

Figure 20.8. The Poincaré sphere

From Eqs.(20.67) and (20.69) we then obtain the form

$$\hat{M} = \frac{I}{2}\begin{pmatrix} 1 + \cos 2\chi \cos 2\psi & \cos 2\chi \sin 2\psi + i \sin 2\chi \\ \cos 2\chi \sin 2\psi - i \sin 2\chi & 1 - \cos 2\chi \cos 2\psi \end{pmatrix} \qquad (20.71)$$

and this defines the state of polarization in terms of I, ψ and χ. If we recall that $\chi = 0$ for linear polarization, from Eq.(20.71) we find that

$$\hat{M} = \frac{I}{2}\begin{pmatrix} 1+\cos 2\psi & \sin 2\psi \\ \sin 2\psi & 1-\cos 2\psi \end{pmatrix} \qquad (20.72)$$

is the correlation matrix of a linearly polarized wave making an angle ψ with the x-axis. It becomes identical to Eq.(20.42) by taking the orientation angle $\psi = \theta$, that is $0 \le \psi \le \pi$. For $\psi = \pi/4$ and $\psi = 3\pi/4$, Eq.(20.72) gives

$$\hat{M}_+ = \frac{I}{2}\begin{pmatrix} 1 & 1 \\ 1 & 1 \end{pmatrix} \quad , \quad \hat{M}_- = \frac{I}{2}\begin{pmatrix} 1 & -1 \\ -1 & 1 \end{pmatrix} \qquad (20.73)$$

Note that the correlation matrix (20.70) may also be expanded in terms of Stokes parameters in the form

$$\hat{M} = \frac{1}{2}\left(I\hat{\sigma}_0 + U\hat{\sigma}_x - V\hat{\sigma}_y + Q\hat{\sigma}_z\right) \qquad (20.74)$$

where $\hat{\sigma}_0$ is the 2×2 unit matrix and the elementary matrices defined as

$$\hat{\sigma}_x = \begin{pmatrix} 0 & 1 \\ 1 & 0 \end{pmatrix} = \frac{1}{I}\left(\hat{M}_+ - \hat{M}_-\right)$$

$$\hat{\sigma}_y = \begin{pmatrix} 0 & -i \\ i & 0 \end{pmatrix} = \frac{1}{I}\left(\hat{M}_r - \hat{M}_l\right) \qquad (20.75)$$

$$\hat{\sigma}_z = \begin{pmatrix} 1 & 0 \\ 0 & 1 \end{pmatrix} = \frac{1}{I}\left(\hat{M}_x - \hat{M}_y\right)$$

are the so-called *Pauli spin matrices* which are widely used in quantum mechanics. Equation (20.75) shows that each of the Pauli spin matrices $\hat{\sigma}_x$, $\hat{\sigma}_y$ and $\hat{\sigma}_z$ can be interpreted as a measure of the excess of an elementary linear or circular state of polarization over the corresponding state of orthogonal polarization.

FURTHER READING
1. W. Greiner, S. Soff - CLASSICAL ELECTRODYNAMICS, Springer-Verlag, New York, 1998.
2. J. Marion, M. Heald - CLASSICAL ELECTROMAGNETIC RADIATION, Brooks/Cole, Pacific Grove, 1994.
3. D. S. Jones - METHODS IN ELECTROMAGNETIC WAVE PROPAGATION, Wiley, New York, 1994.

Problems

20.1. Find the total radiated power by a point charge q moving with constant acceleration a in the nonrelativistic limit.

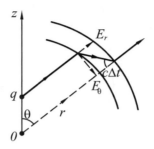

Solution

Consider a charge q at rest at the origin, until time $t=0$, so that electric field lines point away from the origin in all directions. If we briefly apply a uniform acceleration a which brings the velocity of the charge up to Δv in time Δt, at some time r/c much later than Δt, the charge will have moved a distance $r\Delta v/c$. Outside the circle of radius r the field generated at the origin is retarded by the time r/c it takes to reach the circle, so that the field lines point radially away from the origin. In the nonrelativistic limit, where $\Delta v \ll c$, there is a shell of width $c\Delta t$ inside which the field lines point away from the charge within the shell. As the field lines must join up inside the shell, it follows that a sharp kink, which is a pulse of radiation, is propagating outward at speed c. Due to the transversal nature of electromagnetic waves, it is only the θ-component of $\vec{E}$, perpendicular to $\vec{k}$, which contributes to the flux. It is a matter of simple geometry to find E_θ in terms of the radial component E_r of the field as

$$\frac{E_\theta}{E_r} = \frac{(r\Delta v/c)\sin\theta}{c\Delta t} = \frac{r\sin\theta}{c^2}a \quad \text{or} \quad E_\theta = \frac{qa\sin\theta}{4\pi\varepsilon c^2 r}$$

where a stands for the acceleration at the retarded time $t - r/c$, and E_r is just the Coulomb field in the dipole approximation

$$E_r = \frac{q}{4\pi\varepsilon(r+c\Delta t)^2} \simeq \frac{q}{4\pi\varepsilon r^2}$$

In view of Eq.(20.12), one obtains

$$\vec{H} = \sqrt{\frac{\varepsilon}{\mu}}\left(\vec{e}_n \times \vec{E}\right) = \sqrt{\frac{\varepsilon}{\mu}}E_\theta\left(\vec{e}_n \times \vec{e}_\theta\right) = \sqrt{\frac{\varepsilon}{\mu}}E_\theta \vec{e}_\varphi$$

and hence, the Poynting vector (14.13), giving the power radiated per unit area in a given direction, reads

$$\vec{S} = \sqrt{\frac{\varepsilon}{\mu}}E_\theta^2\left(\vec{e}_\theta \times \vec{e}_\varphi\right) = \frac{\mu q^2 a^2}{16\pi^2 c}\frac{\sin^2\theta}{r^2}\vec{e}_n$$

Integrating over all the area elements $r^2 d\Omega$ on the surface of the shell gives

$$P = \frac{\mu q^2 a^2}{16\pi^2 c} \int_0^\pi \sin^3\theta\, d\theta \int_0^{2\pi} d\varphi = \frac{1}{4\pi\varepsilon} \frac{2q^2 a^2}{3c^3}$$

The same expression for the radiated power, known as Larmor's formula, can be derived from the result of Problem 14.9, which gives the time-averaged power transported across unit area, normal to the propagation direction, as

$$\langle S \rangle = \frac{ck^4 p_0^2}{16\pi^2 \varepsilon} \frac{\sin^2\theta}{r^2} \langle \cos^2(kr - \omega t) \rangle = \frac{1}{2} \frac{\mu\omega^4 p_0^2}{16\pi^2 c} \frac{\sin^2\theta}{r^2}$$

The acceleration of the harmonic motion of the charge q is

$$a = -\omega^2 \frac{p_0}{q} \cos\omega t \quad \text{or} \quad \langle a^2 \rangle = \frac{\omega^4 p_0^2}{q^2} \langle \cos^2\omega t \rangle = \frac{1}{2} \frac{\omega^4 p_0^2}{q^2}$$

and hence, the radiated power is proportional to the average acceleration of the harmonic motion, that is

$$\langle S \rangle = \frac{\mu q^2 \langle a^2 \rangle}{16\pi^2 c} \frac{\sin^2\theta}{r^2}$$

as expected.

20.2. If a bound electron of mass m_e, charge $-e$ and natural frequency ω_0 is set oscillating in the harmonic field of amplitude E_0 and angular frequency ω of a linearly polarized wave, find the ratio of the time-averaged power radiated by the electron to the intensity of the incident wave.

Solution

Assuming a linearly polarized wave with the electric vector in the x-direction, the equation of motion of the bound electron under both the restoring and electric forces reads

$$\frac{d^2 x}{dt^2} + \omega_0^2 x = -\frac{e}{m_e} E_0 \cos\omega t$$

Substituting a trial solution $x = x_0 \cos\omega t$ yields

$$x = \frac{-eE_0}{m_e(\omega_0^2 - \omega^2)} \cos\omega t \quad \text{or} \quad a = \frac{eE_0 \omega^2}{m_e(\omega_0^2 - \omega^2)} \cos\omega t$$

Using Larmor's formula, the average radiated power can be written as

$$\langle P \rangle = \frac{1}{4\pi\varepsilon_0} \frac{2e^2 \langle a^2 \rangle}{3c^3} = \frac{1}{4\pi\varepsilon_0} \frac{e^4 \omega^4 E_0^2}{3c^3 m_e^2 (\omega_0^2 - \omega^2)^2}$$

In view of Eq.(20.26) for the wave intensity, it follows that

$$\frac{\langle P \rangle}{I} = \frac{\mu^2}{6\pi} \frac{e^4 \omega^4}{m_e^2 (\omega_0^2 - \omega^2)^2}$$

20.3. Find the Jones eigenvectors associated with the states of polarization of a wave which, upon passing through an optical element described by the Jones matrix (20.40), emerges with the same polarization as when it entered.

Solution

In view of Eq.(20.63) for the emergent state of polarization, we must solve the eigenvalue equation

$$I \begin{pmatrix} \cos^2 \theta & \sin \theta \cos \theta e^{i\delta} \\ \sin \theta \cos \theta e^{-i\delta} & \sin^2 \theta \end{pmatrix} \begin{pmatrix} E_x \\ E_y \end{pmatrix} = \lambda \begin{pmatrix} E_x \\ E_y \end{pmatrix}$$

which is equivalent to the algebraic equations

$$\left(I \cos^2 \theta - \lambda \right) E_x + I \sin \theta \cos \theta e^{i\delta} E_y = 0$$

$$I \sin \theta \cos \theta e^{-i\delta} E_x + \left(I \sin^2 \theta - \lambda \right) E_y = 0$$

In order that a nontrivial solution exists, the determinant of the matrix must vanish

$$\begin{vmatrix} I \cos^2 \theta - \lambda & I \sin \theta \cos \theta e^{i\delta} \\ I \sin \theta \cos \theta e^{-i\delta} & I \sin^2 \theta - \lambda \end{vmatrix} = 0$$

and this provides the characteristic equation

$$\lambda(\lambda - I) = 0$$

Substituting the eigenvalue $\lambda = I$ into either equation, we find the ratio of E_x to E_y as

$$\frac{E_x}{E_y} = \frac{\cos \theta}{\sin \theta} e^{i\delta} \quad \text{or} \quad J = \begin{pmatrix} \cos \theta \\ \sin \theta e^{-i\delta} \end{pmatrix}$$

Thus, light in this particular state of polarization, specified in terms of θ and δ, can be transmitted without change of polarization. There will be no change in intensity either, since the eigenvalue $\lambda = I$ stands for the intensity of the transmitted wave in the state of polarization defined by the corresponding Jones eigenvector.

20.4. A monochromatic wave propagates along the z-axis in a state of elliptic polarization described by the Stokes parameters $I = 100 \text{ V}^2/\text{m}^2$, $Q = 18 \text{ V}^2/\text{m}^2$ and $U = 48 \text{ V}^2/\text{m}^2$. Find the energy flux $\langle S_z \rangle$ in the direction of wave propagation and the parameters a, b, ψ of the polarization ellipse.

Solution

In view of Eqs.(20.6), one obtains

$$A_x = \sqrt{\frac{I+Q}{2}} = 8 \text{ V/m} \quad , \quad A_y = \sqrt{\frac{I-Q}{2}} = 6 \text{ V/m} \quad , \quad \cos\delta = \frac{U}{2A_x A_y} = \frac{1}{2}$$

and hence, $\delta = \pi/3$. It follows that

$$\vec{E}(z,t) = A_x \cos(kz - \omega t)\vec{e}_x + A_y \cos\left(kz - \omega t + \frac{\pi}{3}\right)\vec{e}_y$$

and the magnetic field is given by Eq.(14.6), which reads

$$-\mu_0 \frac{\partial \vec{H}}{\partial t} = \begin{vmatrix} \vec{e}_x & \vec{e}_y & \vec{e}_z \\ \frac{\partial}{\partial x} & \frac{\partial}{\partial y} & \frac{\partial}{\partial z} \\ E_x & E_y & 0 \end{vmatrix} = -\frac{\partial E_y}{\partial z}\vec{e}_x + \frac{\partial E_x}{\partial z}\vec{e}_y = kA_y \sin\left(kz - \omega t + \frac{\pi}{3}\right)\vec{e}_x - kA_x \sin(kz - \omega t)\vec{e}_y$$

and one obtains by integration that

$$\vec{H}(z,t) = -\frac{A_y}{Z_0}\cos\left(kz - \omega t + \frac{\pi}{3}\right)\vec{e}_x + \frac{A_x}{Z_0}\cos(kz - \omega t)\vec{e}_y$$

where $Z_0 = \sqrt{\mu_0/\varepsilon_0}$ stands for the impedance per unit area of the free space, Eq.(17.23). The Poynting vector results as

$$\vec{S} = \begin{vmatrix} \vec{e}_x & \vec{e}_y & \vec{e}_z \\ E_x & E_y & 0 \\ H_x & H_y & 0 \end{vmatrix} = (E_x H_y - E_y H_x)\vec{e}_z = \frac{1}{Z_0}\left[A_x^2 \cos^2(kz - \omega t) + A_y^2 \cos^2\left(kz - \omega t + \frac{\pi}{3}\right)\right]\vec{e}_z$$

The time average $\langle \cos^2(kz - \omega t) \rangle = \langle \cos^2(kz - \omega t + \pi/3) \rangle = 1/2$ implies that the energy flux is given by

$$\langle S_z \rangle = \frac{1}{2Z_0}\left(A_x^2 + A_y^2\right) = \frac{5}{12\pi} \text{ W/m}^2$$

where $Z_0 = 120\pi \, \Omega$. The semiaxes of the polarization ellipse can be obtained by combining Eqs.(20.50), (20.51) and (20.52), which gives

$$a^2 + b^2 = A_x^2 + A_y^2 \quad , \quad ab = A_x A_y \sin\delta$$

and this yields $a = 8.8$ and $b = 4.7$. Finally, from Eq.(20.68) one obtains $\tan 2\psi = 8/3$.

20.5. The polarization state of a quasi-monochromatic wave propagating along the z-axis is represented by a coherency matrix

$$\hat{C} = \begin{pmatrix} C_{xx} & C_{xy} \\ C_{yx} & C_{yy} \end{pmatrix} = \begin{pmatrix} \langle A_x^2 \rangle & \langle A_x A_y e^{i\delta} \rangle \\ \langle A_x A_y e^{-i\delta} \rangle & \langle A_y^2 \rangle \end{pmatrix}$$

which reduces to the correlation matrix $\hat{M}$, Eq.(20.34), in the special case of a monochromatic wave. Show that the wave can be described as a superposition of an unpolarized beam and a completely polarized beam, and derive an expression for the degree of polarization P, defined as the ratio of the intensity of the polarized part to that of the total beam.

Solution

If the components E_x and E_y of the electric vector (20.15) are written in terms of time-dependent amplitudes A_x, A_y and phase factors φ_x, φ_y in the form

$$\vec{E}(t) = \vec{e}_x A_x(t) e^{i[\varphi_x(t) - \omega t]} + \vec{e}_y A_y(t) e^{i[\varphi_y(t) - \omega t]}$$

the intensity along an arbitrary direction α within the wavefront plane can be derived from Eqs.(20.29) and (20.32) in the form

$$I(\alpha) \sim \langle \vec{E}(t,\alpha) \cdot \vec{E}^*(t,\alpha) \rangle = \left[\langle A_x^2 \rangle \cos^2\alpha + \langle A_y^2 \rangle \sin^2\alpha + \sin\alpha\cos\alpha \left(\langle A_x A_y e^{i\delta} \rangle + \langle A_x A_y e^{-i\delta} \rangle \right) \right]$$

This can be written in terms of the coherency matrix elements as

$$I(\alpha) = C_{xx} \cos^2\alpha + C_{yy} \sin^2\alpha + 2\sqrt{C_{xx}}\sqrt{C_{yy}} |\gamma_{xy}| \cos\delta_{xy}$$

where $\gamma_{xy} = C_{xy}/(\sqrt{C_{xx}}\sqrt{C_{yy}}) = |\gamma_{xy}| e^{i\delta_{xy}}$ is a complex factor, and δ_{xy} constitutes the time-averaged phase difference between the two components of the electric vector. A constant intensity $I(\alpha) = $ const, independent of both α and δ_{xy}, defines the completely unpolarized waves, referred to as natural light in the range of the visible spectrum. Such an intensity is obtained if $\gamma_{xy} = 0$ and $C_{xx} = C_{yy}$, and this requires the components E_x and E_y to be mutually incoherent and of equal amplitudes. The coherency matrix reduces to

$$\hat{C}_n = \frac{I}{2}\begin{pmatrix} 1 & 0 \\ 0 & 1 \end{pmatrix}$$

It is convenient to introduce the trace and the determinant of the matrix, given by

$$\text{Tr}(\hat{C}) = \langle A_x^2 \rangle + \langle A_y^2 \rangle = I \quad , \quad \det(\hat{C}) = C_{xx}C_{yy} - C_{xy}C_{yx} \geq 0$$

and to write the coherency matrix of a partially polarized wave as a superposition of an unpolarized beam and a completely polarized beam

$$\begin{pmatrix} C_{xx} & C_{xy} \\ C_{xy}^* & C_{yy} \end{pmatrix} = (1-P)\frac{I}{2}\begin{pmatrix} 1 & 0 \\ 0 & 1 \end{pmatrix} + PI\begin{pmatrix} M_{xx} & M_{xy} \\ M_{xy}^* & M_{yy} \end{pmatrix}$$

where the degree of polarization P gives the ratio of the intensity of the polarized part to that of the total beam $I = \text{Tr}(\hat{C})$. For the state of complete polarization we have $\det(\hat{M}) = 0$, as required by Eq.(20.38), and $\text{Tr}(\hat{M}) = 1$. The matrix equation reduces to

$$C_{xx} = I\left(\frac{1-P}{2} + PM_{xx}\right), \quad C_{yy} = I\left(\frac{1-P}{2} + PM_{yy}\right), \quad C_{xy} = IPM_{xy}$$

and these can be solved for M_{xx}, M_{yy} and M_{xy}. Upon substitution into Eq.(20.38) one obtains

$$(1-P)^2 - 2(1-P) + \frac{4\det(\hat{C})}{I^2} = 0$$

which has the physical solution

$$P = \sqrt{1 - \frac{4\det(\hat{C})}{\left[\text{Tr}(\hat{C})\right]^2}}$$

since we must have $0 \leq P \leq 1$, where $P = 1$ for completely polarized waves, $P = 0$ for unpolarized waves and $0 < P < 1$ for partially polarized waves.

20.6. A free electron of mass m_e and charge $-e$ is set oscillating in the harmonic field of a linearly polarized wave $E_x = E_0(z_0)\cos\omega t$. Find the logarithmic decrement of its damped harmonic motion due to the power that is radiated.

20.7. Find the time-averaged power radiated by a bound electron of mass m_e, charge $-e$ and natural frequency ω_0, which is set oscillating in the harmonic field of amplitude E_0 and angular frequency ω of a circularly polarized wave.

20.8. Find the normalized Jones eigenvectors of a quarter-wave plate described by the matrix

$$\hat{M} = \begin{pmatrix} 1 & 0 \\ 0 & i \end{pmatrix}$$

20.9. A linear polarizer of matrix $\hat{M}$, Eq.(23.42), is placed in a beam described by the Jones vector

$$J = \begin{pmatrix} 2-i \\ 1-2i \end{pmatrix}$$

Find the orientation θ for which the largest possible fraction of the original intensity I_i is transmitted and the value I_f / I_i of that fraction.

20.10. A plane electromagnetic wave is described by

$$\vec{E}(z,t) = 2\sqrt{5}\cos(kz-\omega t)\vec{e}_x + 4\cos\left(kz-\omega t+\frac{\pi}{6}\right)\vec{e}_y$$

Find the Stokes parameters of the polarization state and the parameters a, b and ψ of the polarization ellipse.

20.11. Assuming that the coherency matrix introduced in Problem 20.5 to describe the polarization state of a quasi-monochromatic wave has two real eigenvalues λ_1 and λ_2, show that the degree of polarization P is given by

$$P = \frac{\lambda_1 - \lambda_2}{\lambda_1 + \lambda_2}$$

21. REFLECTION AND REFRACTION

21.1. ELECTROMAGNETIC WAVES AT AN INTERFACE
21.2. FRESNEL'S EQUATIONS

21.1. ELECTROMAGNETIC WAVES AT AN INTERFACE

Due to the vector nature of electromagnetic waves, their propagation at an interface between two different media is determined by the requirements that must be met by the field vector components on both sides of the interface called the *boundary conditions*. These are obtained by applying the integral form of Maxwell's equations to suitable regions containing the interface. We are concerned with linear, homogenous and isotropic media and assume that there is an abrupt change in both permittivity and permeability from ε_1 and μ_1 to ε_2 and μ_2 at the interface.

Consider a contour Γ as in Figure 21.1 and let h be so small that the enclosed surface hl becomes negligible. In the limit of a vanishing h, Faraday's law (13.20) reads

$$\oint_\Gamma \vec{E} \cdot d\vec{r} = -\mu_0 \mu_r \int_S \frac{\partial \vec{H}}{\partial t} \cdot d\vec{S} \to 0 \quad \text{or} \quad \left(\vec{E}_1 - \vec{E}_2 \right) \cdot \vec{l} = 0$$

as long as the quantity $\partial \vec{H} / \partial t$, which for an harmonic wave is given by $i\omega \vec{H}$, remains finite at the interface. Thus we obtain

$$E_{1\tau} = E_{2\tau} \quad \text{or} \quad \vec{e}_n \times \vec{E}_1 = \vec{e}_n \times \vec{E}_2 \qquad (21.1)$$

which states that the tangential component of $\vec{E}$ is continuous across the interface. An application of the modified Ampère law (14.3) to a similar contour, under the assumption that the current densities $\vec{j}$ and $\partial \vec{D}/\partial t$ remain finite at the interface, where there are no free charges, gives

$$\oint_\Gamma \vec{H}\cdot d\vec{l} = \int_S \left(\vec{j} + \frac{\partial \vec{D}}{\partial t}\right)\cdot d\vec{S} \to 0 \quad \text{or} \quad \left(\vec{H}_1 - \vec{H}_2\right)\cdot \vec{l} = 0$$

and hence, the tangential component of $\vec{H}$ is also continuous across the interface, that is

$$\vec{H}_{1\tau} = \vec{H}_{2\tau} \quad \text{or} \quad \vec{e}_n \times \vec{H}_1 = \vec{e}_n \times \vec{H}_2 \qquad (21.2)$$

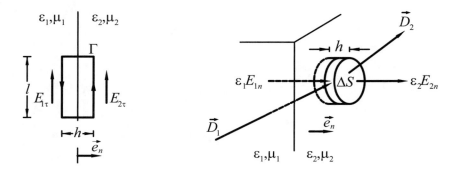

Figure 21.1. Interface between two different media

There is, however, a discontinuity in the normal component of both $\vec{E}$ and $\vec{H}$ across the boundary, and this can be derived from the integral form (14.3) of Gauss's law. Consider a cylindrical volume of height h with faces on the two sides of the interface, as illustrated in Figure 21.1. As h is made negligibly small, so that the flux of $\vec{D}$ through the sides of the cylinder is negligible, we obtain

$$\oint_S \vec{D}\cdot d\vec{S} = \int_V \rho\, dV \to 0 \quad \text{or} \quad \left(\vec{D}_1 - \vec{D}_2\right)\cdot \vec{e}_n \Delta S = 0$$

as long as the charge density ρ remains finite. Hence

$$\varepsilon_1 E_{1n} = \varepsilon_2 E_{2n} \quad \text{or} \quad \varepsilon_1 \vec{E}_1 \cdot \vec{e}_n = \varepsilon_2 \vec{E}_2 \cdot \vec{e}_n \qquad (21.3)$$

The same derivation applies to the magnetic induction $\vec{B}$ and results in

$$\mu_1 H_{1n} = \mu_2 H_{2n} \quad \text{or} \quad \mu_1 \vec{H}_1 \cdot \vec{e}_n = \mu_2 \vec{H}_2 \cdot \vec{e}_n \qquad (21.4)$$

The laws of reflection and transmission of electromagnetic waves follow from the continuity requirements for the tangential components of both $\vec{E}$ and $\vec{H}$. If a plane harmonic wave is incident upon an interface, there will generally be a reflected and a transmitted wave, as previously shown for scalar plane waves. The transverse fields have the form

$$\vec{E}_i(\vec{r},t) = \vec{E}_i e^{i(\vec{k}_i \cdot \vec{r} - \omega t)} \quad , \quad \vec{H}_i(\vec{r},t) = \vec{H}_i e^{i(\vec{k}_i \cdot \vec{r} - \omega t)}$$

$$\vec{E}_\rho(\vec{r},t) = \vec{E}_\rho e^{i(\vec{k}_\rho \cdot \vec{r} - \omega t)} \quad , \quad \vec{H}_\rho(\vec{r},t) = \vec{H}_\rho e^{i(\vec{k}_\rho \cdot \vec{r} - \omega t)} \qquad (21.5)$$

$$\vec{E}_\tau(\vec{r},t) = \vec{E}_\tau e^{i(\vec{k}_\tau \cdot \vec{r} - \omega t)} \quad , \quad \vec{H}_\tau(\vec{r},t) = \vec{H}_\tau e^{i(\vec{k}_\tau \cdot \vec{r} - \omega t)}$$

where the subscripts i, ρ, τ refer to the incident, reflected and transmitted waves, respectively. The amplitudes $\vec{A}_i, \vec{A}_\rho$ and $\vec{A}_\tau$ (where $\vec{A}$ stands for either $\vec{E}$ or $\vec{H}$) may be resolved into their components parallel and perpendicular to the plane of incidence. A field vector of arbitrary orientation in the wavefront can be written as a linear combination of two independent states of polarization: *transverse electric* (TE) *polarization* with $\vec{H}$ in the plane of incidence and $\vec{E}$ perpendicular to it, and *transverse magnetic* (TM) *polarization* where $\vec{E}$ is in the plane of incidence and $\vec{H}$ is perpendicular to it. On substituting Eq.(21.5), Eqs.(21.1) and (21.2) allow us to write the continuity of the tangential component across the interface, for all values of time, as

$$\vec{e}_n \times \vec{A}_i e^{i\vec{k}_i \cdot \vec{r}} + \vec{e}_n \times \vec{A}_\rho e^{i\vec{k}_\rho \cdot \vec{r}} = \vec{e}_n \times \vec{A}_\tau e^{i\vec{k}_\tau \cdot \vec{r}} \qquad (21.6)$$

which may be separated into a *phase-matching condition* at the interface, which reads

$$\vec{k}_i \cdot \vec{r} = \vec{k}_\rho \cdot \vec{r} = \vec{k}_\tau \cdot \vec{r} \qquad (21.7)$$

and an *amplitude-matching condition*, given by

$$\vec{e}_n \times \vec{A}_i + \vec{e}_n \times \vec{A}_\rho = \vec{e}_n \times \vec{A}_\tau \qquad (21.8)$$

In the configuration represented in Figure 21.2, where the interface lies in the *xz*-plane and the plane of incidence is the *xy*-plane, Eq.(21.7) reduces to

$$k_{ix} x = k_{\rho x} x + k_{\rho z} z = k_{\tau x} x + k_{\tau z} z \qquad (21.9)$$

since $y = 0$ at the interface. This equation is valid for all x and z provided that $\vec{k}_\rho$ and $\vec{k}_\tau$ are coplanar with $\vec{k}_i$, that is,

$$k_{\rho z} = k_{\tau z} = 0 \qquad (21.10)$$

Substituting Eq.(21.10), Eq.(21.9) becomes

$$k_i \sin\theta_i = k_\rho \sin\theta_\rho = k_\tau \sin\varphi \qquad (21.11)$$

where θ_i, θ_ρ and φ are the angles between the wave vectors and the *y*-axis.

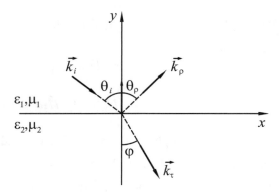

Figure 21.2. Geometry of reflection and refraction at an interface

In view of Eq.(16.24), we obtain

$$\frac{1}{v_1}\sin\theta_i = \frac{1}{v_1}\sin\theta_\rho = \frac{1}{v_2}\sin\varphi \tag{21.12}$$

which states that the directions of propagation follow both the law of reflection, that is

$$\theta_i = \theta_\rho = \theta \tag{21.13}$$

and *Snell's law* of refraction, given by

$$n_1 \sin\theta = n_2 \sin\varphi \tag{21.14}$$

which has been previously derived in Eq.(16.46) from the ray theory of light.

21.2. Fresnel's Equations

The amplitude-matching condition (21.8) allows us to derive the intensities and states of polarization of the reflected and transmitted waves. Substituting Eq.(20.11), Eq.(21.8) can be written in terms of electric fields only as

$$\vec{e}_n \times \vec{E}_i + \vec{e}_n \times \vec{E}_\rho = \vec{e}_n \times \vec{E}_\tau$$

$$n_1 \vec{e}_n \times (\vec{e}_i \times \vec{E}_i) + n_1 \vec{e}_n \times (\vec{e}_\rho \times \vec{E}_\rho) = n_2 \vec{e}_n \times (\vec{e}_\tau \times \vec{E}_\tau) \tag{21.15}$$

where $\vec{e}_i$, $\vec{e}_\rho$, $\vec{e}_\tau$ are the unit vectors of the directions of propagation, such that $\vec{k}_i = k_1 \vec{e}_i$,

$\vec{k}_\rho = k_1 \vec{e}_\rho$ and $\vec{k}_\tau = k_1 \vec{e}_\tau$, and we have used Eq.(20.13) to set

$$\frac{k_1}{\mu_1 \omega} = \sqrt{\frac{\varepsilon_1}{\mu_1}} = \frac{n_1}{Z_0} \quad , \quad \frac{k_2}{\mu_2 \omega} = \frac{n_2}{Z_0}$$

Figure 21.3 shows the configuration for both the TM case, with coplanar directions for $\vec{E}_i, \vec{E}_\rho$ and $\vec{E}_\tau$ and the TE case, where the electric vector has only a z-component $\vec{E} = \vec{e}_z E$.

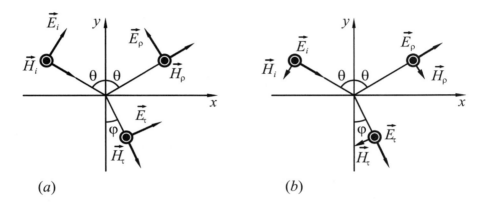

Figure 21.3. Field orientations in (a) TM case, (b) TE case

For a TM *polarization*, Figure 21.3 (a) gives

$$\vec{e}_y \times \vec{E}_i = -\vec{e}_z E_i \cos\theta \quad , \quad \vec{e}_y \times \vec{E}_\rho = \vec{e}_z E_\rho \cos\theta \quad , \quad \vec{e}_y \times \vec{E}_\tau = -\vec{e}_z E_\tau \cos\theta$$

and also $\vec{e}_y \times (\vec{e}_i \times \vec{E}_i) = (\vec{e}_y \times \vec{e}_z) E_i = \vec{e}_x E_i$ and two other similar identities for $\vec{E}_\rho$ and $\vec{E}_\tau$, so that Eq.(21.15) reads

$$\vec{e}_z \left[(E_i - E_\rho) \cos\theta - E_\tau \cos\varphi \right] = 0$$

$$\vec{e}_x \left[n_1 (E_i + E_\rho) - n_2 E_\tau \right] = 0$$

(21.16)

The field amplitudes are then solutions to the equations

$$(E_i - E_\rho) \cos\theta = E_\tau \cos\varphi$$

$$E_i + E_\rho = n E_\tau$$

(21.17)

where $n = n_2 / n_1$ is the relative refractive index of the two media. From Eqs.(21.17) the amplitude coefficients of reflection ρ_{TM} and transmission τ_{TM} are

$$E_\rho = E_i \frac{n\cos\theta - \cos\varphi}{n\cos\theta + \cos\varphi} = E_i \rho_{TM}$$

$$E_\tau = E_i \frac{2\cos\theta}{n\cos\theta + \cos\varphi} = E_i \tau_{TM}$$
(21.18)

Upon substitution of Eq.(21.14), we obtain *Fresnel's equations* for the TM case as

$$\rho_{TM} = \frac{\tan(\theta - \varphi)}{\tan(\theta + \varphi)}$$

$$\tau_{TM} = \frac{2\cos\theta \sin\varphi}{\sin(\theta + \varphi)\cos(\theta - \varphi)}$$
(21.19)

EXAMPLE 21.1. Polarization by reflection

Let us examine the form of the amplitude coefficient of reflection ρ_{TM} over the entire range of angles of incidence. Eliminating φ in Eq.(21.18) by use of Snell's law, we obtain

$$\rho_{TM} = \frac{n^2\cos\theta - \sqrt{n^2 - \sin^2\theta}}{n^2\cos\theta + \sqrt{n^2 - \sin^2\theta}}$$
(21.20)

which is plotted in Figure 21.4 for an air/glass interface, with $n_1 = 1$ and $n_2 = 1.5$, in two cases: *external reflection*, where the incident wave approaches the interface from the side with the smaller refractive index, so that $n = n_2/n_1 = 3/2 > 1$, and *internal reflection*, in which the incident wave is in the medium with a larger refractive index, that is $n = n_2/n_1 = 2/3 < 1$. Equation (21.20) shows that reflection is zero for a particular angle of incidence θ_p given by

$$\tan\theta_p = n$$
(21.21)

which is referred to as the *polarization angle* or *Brewster angle*. At the air-glass interface, the polarization angle is $\theta_{pe} = 56.3°$ when the incident wave is in air (external reflection). If the roles of incident and transmitted waves are interchanged (internal reflection), ρ_{TM} passes through zero at a different polarization angle $\theta_{pi} = 33.7°$ which is the complement of θ_{pe}, since

$$\tan(\theta_{pe} + \theta_{pi}) = \frac{(n_2/n_1)(n_1/n_2)}{1 - (n_2/n_1)(n_1/n_2)} \to \infty \quad \text{or} \quad \theta_{pe} + \theta_{pi} = \frac{\pi}{2}$$

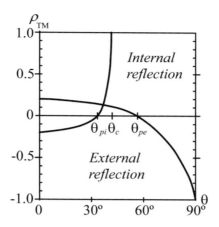

Figure 21.4. ρ_{TM} as a function of θ at an air-glass interface

If an unpolarized wave, which can be regarded as a superposition of two mutually incoherent TE and TM polarized beams of equal intensities, is incident on the interface at θ_p, the reflected wave is completely TE polarized (with the electric field perpendicular to the plane of incidence), whereas the transmitted wave is partially polarized. This provides a procedure to obtain linearly polarized beams by either external or internal reflection at the polarization angle.

If the incoming light at θ_{pe} on a glass parallel plate is completely TM polarized, there is no reflected light from the first face, as $\rho_{TM}(\theta_{pe}) = 0$. The light is transmitted entirely through the face and leaves at an angle $\varphi = \theta_{pi}$, since

$$\tan \theta_{pe} = n \quad \text{or} \quad \sin \theta_{pe} = n\sin(\pi/2 - \theta_{pe}) = n\sin\theta_{pi}$$

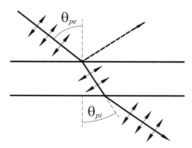

Figure 21.5. Brewster window

Therefore there will be no reflected light from the second face either, as $\rho_{TM}(\theta_{pi}) = 0$ for internal reflection. The result is a total transparency that makes the configuration represented in Figure 21.5, called a *Brewster window*, suitable for use as a perfect window for TM linearly polarized beams.

For a TE *polarization*, as $\vec{e}_n = \vec{e}_y$, Eqs.(21.15) reduce to

$$(\vec{e}_y \times \vec{e}_z)(E_i + E_\rho - E_\tau) = 0$$

$$\vec{e}_z\left[n_1(\vec{e}_y \cdot \vec{e}_i)E_i + n_1(\vec{e}_y \cdot \vec{e}_\rho)E_\rho - n_2(\vec{e}_y \cdot \vec{e}_\tau)E_\tau\right] = 0$$

(21.22)

where we made use of the identity

$$\vec{e}_y \times (\vec{e}_i \times \vec{e}_z) = \vec{e}_i(\vec{e}_y \cdot \vec{e}_z) - \vec{e}_z(\vec{e}_y \cdot \vec{e}_i) = -\vec{e}_z(\vec{e}_y \cdot \vec{e}_i)$$

which also holds for $\vec{e}_\rho$ and $\vec{e}_\tau$. It follows that, in the TE case, the field amplitudes result from the equations

$$E_i + E_\rho = E_\tau$$

$$n_1(E_i - E_\rho)\cos\theta = n_2 E_\tau \cos\varphi$$

(21.23)

which are readily solved for E_ρ and E_τ, to give

$$E_\rho = E_i \frac{\cos\theta - n\cos\varphi}{\cos\theta + n\cos\varphi} = E_i \rho_{\text{TE}}$$

$$E_\tau = E_i \frac{2\cos\theta}{\cos\theta + n\cos\varphi} = E_i \tau_{\text{TE}}$$

(21.24)

Substituting Eq.(21.14), it is a straightforward matter to derive the amplitude coefficients of reflection and transmission in the form

$$\rho_{\text{TE}} = -\frac{\sin(\theta - \varphi)}{\sin(\theta + \varphi)} \quad , \quad \tau_{\text{TE}} = \frac{2\cos\theta\sin\varphi}{\sin(\theta + \varphi)}$$

(21.25)

known as *Fresnel's equations* for the TE *case*. As $e^{i\pi} = -1$, the sign of ρ_{TE} is associated with the relative directions of $\vec{E}_i$ and $\vec{E}_\rho$, which are in phase, if parallel, and π out of phase, if antiparallel. It follows that the electric field normal to the plane of incidence undergoes a phase shift of π upon external reflexion, as $n_2 > n_1$ or $\theta > \varphi$ while no phase shift is introduced by internal reflection, where $n_2 < n_1$ or $\theta < \varphi$.

Suppose I_i, I_ρ and I_τ are the incident, reflected and transmitted intensities, as defined by Eq.(20.29), for quasi-monochromatic beams. The average power crossing the unit area of the interface *in the normal direction* is obtained by taking the *y*-components of the time-averaged Poynting's vector, which are $I_i\cos\theta$, $I_\rho\cos\theta$ and $I_\tau\cos\varphi$, respectively. In view of Eq.(20.29) for intensity, the *reflectivity R* and *transmissivity T*, introduced by Eq.(17.52), become

$$R = \frac{I_\rho \cos\theta}{I_i \cos\theta} = \frac{I_\rho}{I_i} = \frac{\langle \vec{E}_\rho \cdot \vec{E}_\rho^* \rangle}{\langle \vec{E}_i \cdot \vec{E}_i^* \rangle} = |\rho|^2$$

(21.26)

$$T = \frac{I_\tau \cos\varphi}{I_i \cos\theta} = \frac{\cos\varphi}{\cos\theta}\frac{Z_{n_1}}{Z_{n_2}}\frac{\langle \vec{E}_\tau \cdot \vec{E}_\tau^* \rangle}{\langle \vec{E}_i \cdot \vec{E}_i^* \rangle} = \frac{n_2 \cos\varphi}{n_1 \cos\theta}\frac{\langle \vec{E}_\tau \cdot \vec{E}_\tau^* \rangle}{\langle \vec{E}_i \cdot \vec{E}_i^* \rangle} = \frac{\tan\theta}{\tan\varphi}|\tau|^2$$

where the intrinsic impedances per unit area Z_{n_1}, Z_{n_2} have been inserted as defined by Eq.(20.14). Since the average power per unit area of the interface, in the incident beam, equals the sum of average powers in the reflected and transmitted beams, the conservation of energy flow through any area S of the interface requires that

$$I_i S \cos\theta = I_\rho S \cos\theta + I_\tau S \cos\varphi \quad \text{or} \quad 1 = \frac{I_\rho}{I_i} + \frac{I_\tau \cos\varphi}{I_i \cos\theta} = R + T \quad (21.27)$$

which has the same form as Eq.(17.53), written for scalar waves. Equations (21.26) can be applied in either the TM polarization case, where squaring Fresnel's equations (21.19) yields

$$R_{TM} = \frac{\tan^2(\theta - \varphi)}{\tan^2(\theta + \varphi)} \quad , \quad T_{TM} = \frac{\sin 2\theta \sin 2\varphi}{\sin^2(\theta + \varphi)\cos^2(\theta - \varphi)} \quad (21.28)$$

or the TE polarization case, where we obtain

$$R_{TE} = \frac{\sin^2(\theta - \varphi)}{\sin^2(\theta + \varphi)} \quad , \quad T_{TE} = \frac{\sin 2\theta \sin 2\varphi}{\sin^2(\theta + \varphi)} \quad (21.29)$$

It is a straightforward matter to show that

$$R_{TM} + T_{TM} = 1 \quad , \quad R_{TE} + T_{TE} = 1 \quad (21.30)$$

EXAMPLE 21.2. Total internal reflection

Reflectivity and transmissivity are plotted in Figure 21.6 for radiation incident at a glass-air interface where the refractive index of glass is taken to be 1.5. For light incident at an angle greater then the *critical angle* θ_c, both the TE and TM geometries show the same behaviour. For the glass-air interface, $\theta_c = \arcsin(2/3) = 42°$. We can understand such a behaviour if we consider the amplitude coefficients of reflection ρ defined over the entire range of θ values. For the TM case ρ_{TM} is given by Eq.(21.20), while for the TE case we can use Eq.(21.24) and Snell's law to obtain

$$\rho_{TE} = \frac{\cos\theta - \sqrt{n^2 - \sin^2\theta}}{\cos\theta + \sqrt{n^2 - \sin^2\theta}} \quad (21.31)$$

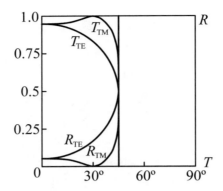

Figure 21.6. Reflectivity and transmissivity in internal reflection

Since $n^2 - \sin^2\theta$ is negative for internal angles of incidence exceeding the critical value $\theta_c = \arcsin n$, it is convenient to write the amplitude coefficients of reflection for this range of values of θ in the form

$$\rho_{\text{TE}} = \frac{\cos\theta - i\sqrt{\sin^2\theta - n^2}}{\cos\theta + i\sqrt{\sin^2\theta - n^2}} = \frac{1 - i\gamma}{1 + i\gamma}$$

$$\rho_{\text{TM}} = \frac{n^2\cos\theta - i\sqrt{\sin^2\theta - n^2}}{n^2\cos\theta + i\sqrt{\sin^2\theta - n^2}} = \frac{1 - i\gamma/n^2}{1 + i\gamma/n^2}$$

(21.32)

where γ is a real parameter defined as

$$\gamma = \frac{n}{\cos\theta}\sqrt{\frac{\sin^2\theta}{n^2} - 1} \qquad (21.33)$$

Substituting Eq.(21.32), Eq.(21.26) shows that reflectivities are unity in either the TE or TM case for $\theta > \theta_c$, because

$$R_{\text{TE}} = \rho_{\text{TE}}\rho_{\text{TE}}^* = 1 \quad , \quad R_{\text{TM}} = \rho_{\text{TM}}\rho_{\text{TM}}^* = 1$$

This is known as *total internal reflection*: light is totally reflected when incident on the interface from the more dense medium, at an angle θ equal or greater than the critical angle. There must be a transmitted wave which cannot, however, carry energy across the interface, as the transmissivities T_{TE} and T_{TM} are zero, according to Eq.(21.30). In the configuration illustrated in Figure 21.2 the electric field (21.5) of the *transmitted* wave reduces to

$$\vec{E}_\tau(\vec{r},t) = \vec{E}_\tau e^{i(k_{\tau x}x + k_{\tau y}y - \omega t)} = \vec{E}_\tau e^{i(k_\tau x \sin\varphi - k_\tau y \cos\varphi - \omega t)} \qquad (21.34)$$

If the angle of internal incidence exceeds the critical angle, $\cos\varphi$ can be written in terms of the attenuation parameter (21.33) as

$$\cos\varphi = \sqrt{1 - \frac{\sin^2\theta}{n^2}} = i\sqrt{\frac{\sin^2\theta}{n^2} - 1} = i\frac{\gamma\cos\theta}{n}$$

so that the field amplitude (21.34) drops off exponentially, if the wave penetrates the less dense medium, as given by

$$\vec{E}_\tau(\vec{r},t) = \left(\vec{E}_\tau e^{-\gamma(k_\tau \cos\theta)|y|/n}\right) e^{i\left[(k_\tau \sin\theta)x/n - \omega t\right]} \tag{21.35}$$

The result is a boundary wave, with the wavefront of constant phase moving along the interface and the amplitude decaying rapidly in the y-direction (Figure 21.2).

The electric field of the *reflected wave* $\vec{E}_\rho$, as given by Eq.(21.18) or Eq.(21.24), remains unchanged in amplitude, since the absolute values (21.32) are unity for both ρ_{TE} and ρ_{TM}, but $\vec{E}_\rho$ undergoes a change of phase, which is a function of γ, and therefore of the angle of incidence. It is usual to set

$$\gamma = \tan\frac{\varphi_{TE}}{2} = n^2 \tan\frac{\varphi_{TM}}{2} \tag{21.36}$$

so that Eq.(21.32) gives the phase shifts occurring in total internal reflection as

$$\rho_{TE} = \frac{\sqrt{1+\gamma^2}\, e^{-i\varphi_{TE}/2}}{\sqrt{1+\gamma^2}\, e^{i\varphi_{TE}/2}} = e^{-i\varphi_{TE}} \quad , \quad \rho_{TM} = e^{-i\varphi_{TM}}$$

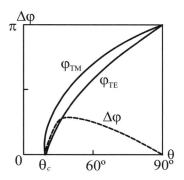

Figure 21.7. Phase difference $\Delta\varphi$ in total internal reflection

As a result of the phase difference $\Delta\varphi = \varphi_{TM} - \varphi_{TE}$, which is readily evaluated to be

$$\tan\frac{\Delta\varphi}{2} = \gamma \frac{\cos^2\theta}{\sin^2\theta} \tag{21.37}$$

and is illustrated in Figure 21.7, a change in the state of polarization of the incident wave is produced by total internal reflection.

At *normal incidence* the distinction between TM and TE cases is lost, as the incident plane becomes undefined for $\theta = 0$. Equations (21.18) and (21.24) reduce to

$$\rho = \rho_{TE} = \frac{1-n}{1+n} = -\rho_{TM}$$

$$\tau = \tau_{TE} = \frac{2}{1+n} = \tau_{TM}$$

(21.38)

where the change in sign in the first equation has no physical significance, being simply a consequence of the sign convention adopted in Figure 21.3 for the TM case. The vector nature of electromagnetic waves is not apparent at normal incidence, since Eqs.(21.38) can be derived from the relations (17.50) obtained for scalar waves, by substituting $v_1 = c/n_1$ and $v_2 = c/n_2$. The reflectivity and transmissivity at normal incidence, as given by Eqs.(21.26) and (21.38) as

$$R = R_{TE} = R_{TM} = |\rho|^2 = \left(\frac{1-n}{1+n}\right)^2$$

$$T = T_{TE} = T_{TM} = n|\tau|^2 = \frac{4n}{(1+n)^2}$$

(21.39)

are the same as those defined by Eq.(17.52) for scalar waves.

FURTHER READING

1. M. Katz - INTRODUCTION TO GEOMETRICAL OPTICS, World Scientific, Singapore, 2003.
2. G. Waldman - INTRODUCTION TO LIGHT, Dover Publications, New York, 2002.
3. R. Ditteon - OPTICAL MODERN GEOMETRICAL OPTICS, Wiley, New York, 1997.

PROBLEMS

21.1. Find an expression for the polarization angle θ_p in external reflection at the air/dielectric interface in terms of the relative permittivity ε_r and permeability μ_r of the dielectric material.

Solution

It is convenient to express the amplitude coefficient of reflection ρ_{TM} in terms of ε_r and μ_r, using Eq.(15.54) for the refractive index instead of Maxwell's relation (15.55). It is straightforward, in view of Eq.(20.13), that Eqs.(21.18) take the form

$$\rho_{TM} = \frac{\frac{1}{Z_2}\cos\theta - \frac{1}{Z_1}\cos\varphi}{\frac{1}{Z_2}\cos\theta + \frac{1}{Z_1}\cos\varphi}, \quad \tau_{TM} = \frac{\frac{2}{Z_1}\cos\theta}{\frac{1}{Z_2}\cos\theta + \frac{1}{Z_1}\cos\varphi}$$

so that the external reflection is zero when

$$\sqrt{\frac{\varepsilon_r}{\mu_r}}\cos\theta_p = \cos\varphi$$

Eliminating φ by use of Snell's law (21.14), which reads $\sin\theta_p = \sqrt{\varepsilon_r \mu_r}\sin\varphi$, one obtains

$$\tan\theta_p = \sqrt{\frac{\varepsilon_r(\varepsilon_r - \mu_r)}{\varepsilon_r \mu_r - 1}}$$

and this reduces to Eq.(21.21) for $\mu_r = 1$.

21.2. If an unpolarized beam of light is incident at the Brewster angle θ_p on the air/glass interface $(n_1 = 1, n_2 = 1.5)$, find the degree of polarization P of the transmitted light.

Solution

The degree of polarization can be defined, according to Problem 20.11, in terms of the intensities of two orthogonal states of polarization TM and TE of the transmitted light, in the form

$$P = \frac{I_{TM} - I_{TE}}{I_{TM} - I_{TE}} = \frac{T_{TM} - T_{TE}}{T_{TM} + T_{TE}}$$

where T_{TM} and T_{TE} are given by Eqs.(21.28) and (21.29), respectively, since the incident unpolarized light can be represented as a superposition of two orthogonally polarized beams of equal intensities. Thus, we have

$$P = \frac{1 - \cos^2(\theta_p - \varphi)}{1 + \cos^2(\theta_p - \varphi)}$$

where the Brewster angle is given by $\tan\theta_p = n$. In view of Snell's law, we have

$$\sin\theta_p = n\sin\varphi = n\cos\theta_p \quad \text{or} \quad \theta_p + \varphi = \frac{\pi}{2}$$

Eliminating φ gives

$$P = \frac{1-\sin^2 2\theta_p}{1+\sin^2 2\theta_p} = \frac{1-\left(\frac{2n}{n^2+1}\right)^2}{1+\left(\frac{2n}{n^2+1}\right)^2} = \frac{\left(n^2-1\right)^2}{\left(n^2+1\right)^2+4n^2}$$

A fairly low degree of polarization of about 8 percent can be achieved for an air/glass interface.

21.3. If a linearly polarized beam of light is totally reflected at the glass/air interface $(n_1 = 1.5,\ n_2 = 1)$, find the maximum phase difference $\Delta\varphi = \varphi_{TM} - \varphi_{TE}$ achievable in internal reflection and the angle of incidence θ at which this is expected to occur.

Solution

By combining Eqs.(21.37) and (21.33), where $n = n_2/n_1 = 2/3$, one obtains

$$\tan\frac{\Delta\varphi}{2} = \frac{\cos\theta\sqrt{\sin^2\theta - n^2}}{\sin^2\theta}$$

To find the extremum, we set

$$\frac{d}{d\theta}\left(\tan\frac{\Delta\varphi}{2}\right) = 0 \quad \text{or} \quad \frac{d}{du}\frac{\sqrt{(1-u)(u-n^2)}}{u} = 0$$

where $u = \sin^2\theta$, and this yields

$$\sin^2\theta = \frac{2n^2}{n^2+1} \quad \text{or} \quad \theta \approx 51° > \theta_c$$

as illustrated in Figure 21.7. Substituting this value for θ, the maximum $\Delta\varphi$ results as

$$\tan\frac{\Delta\varphi}{2} = \frac{1-n^2}{2n} \quad \text{or} \quad \Delta\varphi \approx 46°$$

21.4. Use the Jones calculus to show that total internal reflection is producing, in general, elliptically polarized light and find an expression for the orientation angle ψ of the polarization ellipse in terms of the internal angle of incidence θ.

Solution

The incident state of polarization may be described by a Jones vector (20.56) in terms of a linearly polarized basis consisting of the orthogonal TM and TE components

$$J = \begin{pmatrix} E_x \\ E_y \end{pmatrix} = \begin{pmatrix} E_{TM} \\ E_{TE} \end{pmatrix}$$

Since total reflection produces fields of amplitudes (21.18) and (21.24), which can be written as

$$E_{\tau x} = e^{-i\varphi_{TM}} E_{TM} \quad , \quad E_{\tau y} = e^{-i\varphi_{TE}} E_{TE}$$

it may be suitably represented by a 2×2 matrix

$$\begin{pmatrix} e^{-i\varphi_{TM}} & 0 \\ 0 & e^{-i\varphi_{TE}} \end{pmatrix} = e^{-i\varphi_{TM}} \begin{pmatrix} 1 & 0 \\ 0 & e^{i\Delta\varphi} \end{pmatrix}$$

where $\Delta\varphi$ is given by Eq.(21.37). It follows from Eq.(20.63) that the reflected wave, in general, has an elliptic state of polarization

$$J_f = \begin{pmatrix} 1 & 0 \\ 0 & e^{i\Delta\varphi} \end{pmatrix} \begin{pmatrix} E_{TM} \\ E_{TE} \end{pmatrix} = \begin{pmatrix} E_{TM} \\ E_{TE} e^{i\Delta\varphi} \end{pmatrix}$$

In view of Eq.(20.49), the orientation angle of the polarization ellipse upon total internal reflection results, after a little algebra, as

$$\tan 2\psi = \frac{2 E_{TM} E_{TE}}{E_{TM}^2 - E_{TE}^2} \cos \Delta\varphi = \frac{2 E_{TM} E_{TE}}{E_{TM}^2 - E_{TE}^2} \frac{n^2 \cos^2\theta - \sin^2\theta \cos 2\theta}{\sin^2\theta - n^2 \cos^2\theta}$$

21.5. Calculate the reflectivity R at normal incidence of a dielectric layer of refractive index n and thickness d, which is coated on an optical glass surface of refractive index n_τ.

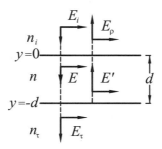

Solution

Consider a dielectric layer sandwiched between an interface located in the plane $y = 0$ and a second interface in the plane $y = -d$. The continuity requirements for the tangential component of the field vectors at the interface $y = 0$ are simply the generalizations of Eqs.(21.23) in order to include backward travelling waves. At normal incidence, we set $\theta = \varphi = 0$. In the configuration shown in the figure, Eqs.(21.23) become

$$E_i + E_\rho = E + E' \quad , \quad n_i(E_i - E_\rho) = n(E - E')$$

The propagation across the layer results in a phase shift e^{iky} along the y-direction so that at the second interface $y = -d$, we obtain

$$E e^{-ikd} + E' e^{ikd} = E_\tau \quad , \quad n(E e^{-ikd} - E' e^{ikd}) = n_\tau E_\tau$$

where the phase of the transmitted wave was chosen to be $E_\tau(\vec{r},t) = E_\tau e^{i[k(y+d)-\omega t]}$. Expressing E and E' in terms of E_τ as

$$E = \frac{E_\tau}{2}\left(1 + \frac{n_\tau}{n}\right)e^{ikd} \quad , \quad E' = \frac{E_\tau}{2}\left(1 - \frac{n_\tau}{n}\right)e^{-ikd}$$

one obtains

$$E_i + E_\rho = \left(\cos kd + i\frac{n_\tau}{n}\sin kd\right)E_\tau \quad , \quad n_i E_i - n_i E_\rho = (in\sin kd + n_\tau \cos kd)E_\tau$$

and the amplitude coefficient of reflection takes the form

$$\rho = \frac{E_\rho}{E_i} = \frac{n(n_i - n_\tau)\cos kd + i(n_i n_\tau - n^2)\sin kd}{n(n_i + n_\tau)\cos kd + i(n_i n_\tau + n^2)\sin kd}$$

A layer of optical thickness $d = \lambda/4$ produces a phase shift $kd = \pi/2$ and its reflectivity is

$$\rho = \frac{n_i n_\tau - n^2}{n_i n_\tau + n^2} \quad \text{or} \quad R = |\rho|^2 = \left(\frac{n_i n_\tau - n^2}{n_i n_\tau + n^2}\right)^2$$

As the reflectivity can be made zero, provided $n = \sqrt{n_i n_\tau}$, a single quarter-wave layer of suitable refractive index constitutes a perfect antireflection film for a given wavelength. This result is consistent with that derived in Problem 17.5.

21.6. Find an expression for the transfer matrix connecting the incident and emergent field components under the restriction of normal incidence of light on a dielectric film consisting of N different layers.

Solution

For a film consisting of several layers, it is convenient to introduce a transfer matrix for each layer that relates the incident and emergent field components at its interfaces. For a single layer of thickness d and refractive index n we can set

$$\hat{M}(d) = \begin{pmatrix} \cos kd & \dfrac{i}{n}\sin kd \\ in\sin kd & \cos kd \end{pmatrix}$$

so that the field components are related by

$$\begin{pmatrix} E_i + E_\rho \\ n_i E_i - n_i E_\rho \end{pmatrix} = \hat{M}(d)\begin{pmatrix} E_\tau \\ n_\tau E_\tau \end{pmatrix}$$

For a bilayer we may successively write

$$\begin{pmatrix} E_i + E_\rho \\ n_i E_i - n_i E_\rho \end{pmatrix} = \hat{M}_1(d_1)\begin{pmatrix} E_2 + E_2' \\ n_2 E_2 - n_2 E_2' \end{pmatrix} \quad \text{and} \quad \begin{pmatrix} E_1 + E_1' \\ n_1 E_1 - n_1 E_1' \end{pmatrix} = \hat{M}_2(d_2)\begin{pmatrix} E_\tau \\ n_2 E_\tau \end{pmatrix}$$

and, since the field is continuous across the interface between the two layers, it follows that

$$\begin{pmatrix} E_i + E_\rho \\ n_i E_i - n_i E_\rho \end{pmatrix} = \hat{M}_1(d_1)\hat{M}_2(d_2)\begin{pmatrix} E_\tau \\ n_\tau E_\tau \end{pmatrix}$$

For instance, a bilayer that consists of two quarter-wave films of different refractive indices n and n', has a transfer matrix

$$\hat{M}\left(\frac{\lambda}{4}\right)\hat{M}'\left(\frac{\lambda}{4}\right) = \begin{pmatrix} 0 & \dfrac{i}{n} \\ in & 0 \end{pmatrix}\begin{pmatrix} 0 & \dfrac{i}{n'} \\ in' & 0 \end{pmatrix} = \begin{pmatrix} -\dfrac{n'}{n} & 0 \\ 0 & -\dfrac{n}{n'} \end{pmatrix}$$

The result may be generalized for a multilayer film, consisting of N layers, in the form

$$\begin{pmatrix} E_i + E_\rho \\ n_i E_i - n_i E_\rho \end{pmatrix} = \hat{M}_1(d_1)\hat{M}_2(d_2)\ldots\hat{M}_N(d_N)\begin{pmatrix} E_\tau \\ n_\tau E_\tau \end{pmatrix} = \begin{pmatrix} M_{11} & M_{12} \\ M_{21} & M_{22} \end{pmatrix}\begin{pmatrix} E_\tau \\ n_\tau E_\tau \end{pmatrix}$$

21.7. A beam of light falls at oblique incidence on a plate of glass of refractive index n. Use Fresnel's equations to find a relation between the amplitude coefficients of reflection and transmission ρ_1, τ_1 at the air/glass interface and ρ_2, τ_2 at the glass/air interface.

21.8. A linearly polarized beam of light undergoes two total internal reflections in a Fresnel prism, as shown in the figure. If the emergent beam is circularly polarized, find the refractive index n of the prism.

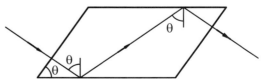

21.9. Show that, in the case of oblique incidence of TE polarized light, the tangential fields at both ends of a single layer are related by

$$\begin{pmatrix} E_i + E_\rho \\ \gamma_i E_i - \gamma_i E_\rho \end{pmatrix} = \begin{pmatrix} \cos\beta & \dfrac{i}{\gamma}\sin\beta \\ i\gamma\sin\beta & \cos\beta \end{pmatrix} \begin{pmatrix} E_\tau \\ \gamma_\tau E_\tau \end{pmatrix}$$

Find expressions for the coefficients β and γ, γ_i, γ_τ in terms of the refraction angles θ, θ_i, θ_τ inside and, respectively, outside the layer.

21.10. Find the reflectivity R of a multilayer film consisting on an even number N of quarter-wave layers whose refractive indices alternate between n and n'. Assume that $n_i = n_\tau = 1$.

21.11. A trilayer structure $\lambda/4 - \lambda/4 - \lambda/4$ consisting of quarter-wave films of different refractive indices n_1, n_2 and n_3 is coated on a glass lens of refractive index n_τ. Find the condition for antireflective coating.

21.12. Find the reflectivity of a trilayer structure $\lambda/4 - \lambda/2 - \lambda/4$ consisting of films of different refractive indices n_1, n_2 and n_3, which is coated on an optical glass surface of refractive index n_τ.

22. WAVES IN ANISOTROPIC MEDIA

22.1. THE PLANE WAVE EQUATION
22.2. OPTICAL ANISOTROPIC MEDIA
22.3. RAY AND PHASE VELOCITY

22.1. THE PLANE WAVE EQUATION

The propagation of electromagnetic waves in a medium is described using wave equations that can be derived by combining Maxwell's field equations (14.2), valid for arbitrary media, with appropriate constitutive relations in the form of either Eq.(14.4) or Eq.(14.5). We will concentrate on nonmagnetic media ($\vec{M} = 0$ or $\mu_r = 1$) which contain no volume charge ($\rho = 0$) and no conduction current ($\vec{j} = 0$) so that Eqs.(14.2) become

$$\nabla \times \vec{E} = -\mu_0 \frac{\partial \vec{H}}{\partial t} \qquad \nabla \times \vec{H} = \frac{\partial \vec{D}}{\partial t}$$

$$\nabla \cdot \vec{H} = 0 \qquad \nabla \cdot \vec{D} = 0$$

(22.1)

Assuming that these media can sustain monochromatic plane waves, propagating in any direction $\vec{k} = k\vec{e}_n$, the field vectors $\vec{E}$, $\vec{H}$ and $\vec{D}$ may be given the harmonic representation (20.8) that reads

$$\vec{E} = \vec{E}_0 e^{i(\vec{k}\cdot\vec{r}-\omega t)} \;,\; \vec{H} = \vec{H}_0 e^{i(\vec{k}\cdot\vec{r}-\omega t)} \;,\; \vec{D} = \vec{D}_0 e^{i(\vec{k}\cdot\vec{r}-\omega t)} \qquad (22.2)$$

Substituting the operator relations (20.10), Eqs.(22.1) take the form

$$\vec{k} \times \vec{E} = \mu_0 \omega \vec{H} \qquad \vec{k} \times \vec{H} = -\omega \vec{D}$$

$$\vec{k} \cdot \vec{H} = 0 \qquad \vec{k} \cdot \vec{D} = 0$$

(22.3)

which shows that $\vec{H}$ and $\vec{D}$ are always perpendicular to $\vec{k}$ and to each other. Thus both lie in the wavefront, whereas, in general, $\vec{E}$ is not perpendicular to the wave vector, as highlighted by Eqs.(20.11) for an isotropic medium. As Faraday's law (22.3) shows that $\vec{H}$ is normal to $\vec{E}$, the electric vector must be coplanar with $\vec{D}$ and $\vec{k}$, and can be resolved into components parallel and perpendicular to $\vec{e}_n$

$$\vec{E}_n = (\vec{E} \cdot \vec{e}_n)\vec{e}_n \quad , \quad \vec{E}_\pi = \vec{E} - (\vec{E} \cdot \vec{e}_n)\vec{e}_n$$

The transverse nature of electromagnetic waves is preserved with respect to $\vec{H}$ and $\vec{D}$, as illustrated in Figure 22.1.

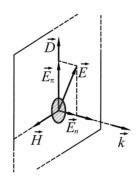

Figure 22.1. Orientation of field vectors in anisotropic media

By eliminating $\vec{H}$ from Eqs.(22.3) we obtain the plane wave equation as

$$\vec{k} \times (\vec{k} \times \vec{E}) = \mu_0 \omega (\vec{k} \times \vec{H}) = -\mu_0 \omega^2 \vec{D} \tag{22.4}$$

With $\vec{k} = k\vec{e}_n$, we then have

$$\vec{e}_n \times (\vec{e}_n \times \vec{E}) = \vec{e}_n (\vec{e}_n \cdot \vec{E}) - (\vec{e}_n \cdot \vec{e}_n)\vec{E} = -\mu_0 \frac{\omega^2}{k^2} \vec{D} \tag{22.5}$$

which gives an expression for $\vec{D}$ in terms of $\vec{E}$ in the form

$$\vec{D} = \frac{k^2}{\mu_0 \omega^2}\left[\vec{E} - (\vec{E} \cdot \vec{e}_n)\vec{e}_n\right] = \frac{k^2}{\mu_0 \omega^2} \vec{E}_\pi \tag{22.6}$$

Therefore the relation between $\vec{D}$ and $\vec{E}$ for a linear medium, as formulated by the constitutive relation (14.5), cannot always be expressed in terms of just a real scalar dielectric constant ε only, as we have seen in the case of isotropic media, but must assume the general form of a linear relation between the field components with respect to a set of Cartesian axes fixed in the medium, given by

$$D_i = \varepsilon_0 \sum_{j=1}^{3} \varepsilon_{ij} E_j \tag{22.7}$$

where i, j stand for x, y and z. There are nine coefficients ε_{ij} which form the elements of the *dielectric tensor* $\hat{\varepsilon}_r$, so that Eq.(22.7) can also be written as

$$\begin{pmatrix} D_x \\ D_y \\ D_z \end{pmatrix} = \varepsilon_0 \begin{pmatrix} \varepsilon_{xx} & \varepsilon_{xy} & \varepsilon_{xz} \\ \varepsilon_{yx} & \varepsilon_{yy} & \varepsilon_{yz} \\ \varepsilon_{zx} & \varepsilon_{zy} & \varepsilon_{zz} \end{pmatrix} \begin{pmatrix} E_x \\ E_y \\ E_z \end{pmatrix} \tag{22.8}$$

The propagation properties of electromagnetic waves in anisotropic media can be understood by assuming that the dielectric matrix is always Hermitian, with real diagonal elements ε_{xx}, ε_{yy}, ε_{zz}, and complex conjugated off-diagonal elements, that is

$$\varepsilon_{ij} = \varepsilon_{ji}^* \tag{22.9}$$

If all elements ε_{ij} are real, the matrix is symmetric about the diagonal and describes a *nonactive medium*, which exhibits no optical activity, whereas in any optically active medium, some elements ε_{ij} must be assumed complex. For real symmetric matrices there always exists a set of Cartesian axes, called *principal axes*, such that the dielectric matrix takes the diagonal form

$$\hat{\varepsilon}_r = \begin{pmatrix} \varepsilon_x & 0 & 0 \\ 0 & \varepsilon_y & 0 \\ 0 & 0 & \varepsilon_z \end{pmatrix} = \begin{pmatrix} n_x^2 & 0 & 0 \\ 0 & n_y^2 & 0 \\ 0 & 0 & n_z^2 \end{pmatrix} \tag{22.10}$$

where $\varepsilon_x, \varepsilon_y$ and ε_z are called *principal dielectric constants* and the *principal refractive indices* n_x, n_y, n_z are defined according to Eq.(15.55).

Substituting Eq.(22.8), Eq.(22.4) takes the form of the *general plane wave equation* in anisotropic media

$$\vec{k} \times (\vec{k} \times \vec{E}) + \frac{\omega^2}{c^2} \hat{\varepsilon}_r \vec{E} = 0 \tag{22.11}$$

which gives the allowed values of k, and therefore the distribution of the phase velocity ω/k for the wave. Using a vector identity similar to that used before in Eq.(22.5), Eq.(22.11) becomes

$$\vec{k}(\vec{k} \cdot \vec{E}) - k^2 \vec{E} + \frac{\omega^2}{c^2} \hat{\varepsilon}_r \vec{E} = 0 \tag{22.12}$$

If $\hat{\varepsilon}_r$ is taken in the principal-axis representation (22.10), the plane wave solutions result from three equivalent scalar equations which are readily obtained as

$$\left(\frac{\omega^2}{c^2}n_x^2 - k_y^2 - k_z^2\right)E_x + k_x k_y E_y + k_x k_z E_z = 0$$

$$k_y k_x E_x + \left(\frac{\omega^2}{c^2}n_y^2 - k_x^2 - k_z^2\right)E_y + k_y k_z E_z = 0 \quad (22.13)$$

$$k_z k_x E_x + k_z k_y E_y + \left(\frac{\omega^2}{c^2}n_z^2 - k_x^2 - k_y^2\right)E_z = 0$$

For a given wave vector $\vec{k} = (k_x, k_y, k_z)$, Eqs.(22.13) have a nontrivial solution for E_x, E_y and E_z if and only if the characteristic equation, given by

$$\begin{vmatrix} (\omega n_x/c)^2 - k_y^2 - k_z^2 & k_x k_y & k_x k_z \\ k_y k_x & (\omega n_y/c)^2 - k_x^2 - k_z^2 & k_y k_z \\ k_z k_x & k_z k_y & (\omega n_z/c)^2 - k_x^2 - k_y^2 \end{vmatrix} = 0 \quad (22.14)$$

is satisfied. Equation (22.14) can be represented by a *wave vector surface* in $\vec{k}$-space, illustrated in Figure 22.2. The algebraic complications related to the physical interpretation of this two sheet surface can be avoided if we consider in Eq.(22.12) only directions of $\vec{k}$ that lie normal to a principal axis of the dielectric tensor rather than the general characteristic equation (22.14).

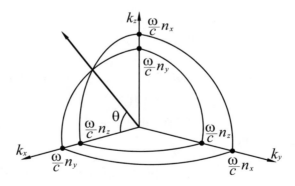

Figure 22.2. The wave vector surface in $\vec{k}$-space

22.2. OPTICAL ANISOTROPIC MEDIA

Suppose first that $\vec{k}$ lies in the xz-plane, that is $k_y = 0$ along the principal axis y of the dielectric tensor and $k^2 = k_x^2 + k_z^2$. The component scalar Eqs.(22.13) reduce to

$$\left(\frac{\omega^2}{c^2}n_x^2 - k_z^2\right)E_x + k_x k_z E_z = 0$$

$$\left(\frac{\omega^2}{c^2}n_y^2 - k_x^2 - k_z^2\right)E_y = 0 \qquad (22.15)$$

$$k_z k_x E_x + \left(\frac{\omega^2}{c^2}n_z^2 - k_x^2\right)E_z = 0$$

Nontrivial solutions can be obtained provided that the determinant of the coefficients vanishes, that is,

$$\begin{vmatrix} (\omega n_x/c)^2 - k_z^2 & 0 & k_x k_z \\ 0 & (\omega n_y/c)^2 - k_x^2 - k_z^2 & 0 \\ k_z k_x & 0 & (\omega n_z/c)^2 - k_x^2 \end{vmatrix} = 0 \qquad (22.16)$$

which factorizes as

$$\left(\frac{\omega^2}{c^2}n_y^2 - k_x^2 - k_z^2\right)\left[\left(\frac{\omega^2}{c^2}n_x^2 - k_z^2\right)\left(\frac{\omega^2}{c^2}n_z^2 - k_x^2\right) - k_x^2 k_z^2\right] = 0 \qquad (22.17)$$

Thus, the characteristic equation reduces to the independent equations of a circle

$$k_x^2 + k_z^2 = \frac{\omega^2}{c^2}n_y^2 \qquad (22.18)$$

and of an ellipse

$$\frac{k_x^2}{(\omega n_z/c)^2} + \frac{k_z^2}{(\omega n_x/c)^2} = 1 \qquad (22.19)$$

which are represented in Figure 22.2 by the intercept of the wave vector surface (22.14) with the wave vector plane xz. Each of Eqs.(22.18) and (22.19) provides a value for the wave number k. Thus, for a given direction of propagation $\vec{k}$ there are two possible solutions k_1 and k_2, that is, there exist two waves which propagate with phase velocities ω/k_1 and ω/k_2. The restrictions on their state of polarization, as given by the orientation of the electric field components $\vec{E}_{1\pi}$ and $\vec{E}_{2\pi}$ that lie in the wavefront plane, can be obtained using Eq.(22.6) which reads

$$\vec{D}_1 = \frac{k_1^2}{\mu_0 \omega^2} \vec{E}_{1\pi} \quad , \quad \vec{D}_2 = \frac{k_2^2}{\mu_0 \omega^2} \vec{E}_{2\pi} \tag{22.20}$$

The symmetry of the dielectric tensor (22.7) implies the symmetry of the scalar product

$$\vec{D}_1 \cdot \vec{E}_2 = \sum_{i=1}^{3} D_{1i} E_{2i} = \varepsilon_0 \sum_{i=1}^{3} \left(\sum_{j=1}^{3} \varepsilon_{ij} E_{1j} \right) E_{2i} = \varepsilon_0 \sum_{j=1}^{3} \left(\sum_{i=1}^{3} \varepsilon_{ji} E_{2i} \right) E_{1j}$$

$$= \sum_{j=1}^{3} D_{2j} E_{1j} = \vec{D}_2 \cdot \vec{E}_1 \tag{22.21}$$

Substituting Eq.(22.20), Eq.(22.21) becomes

$$k_1^2 \vec{E}_{1\pi} \cdot \vec{E}_2 = k_2^2 \vec{E}_{2\pi} \cdot \vec{E}_1$$

Since $\vec{E}_{1\pi} \cdot \vec{E}_{2n} = \vec{E}_{2\pi} \cdot \vec{E}_{1n} = 0$, it follows that

$$\left(k_1^2 - k_2^2 \right) \left(\vec{E}_{1\pi} \cdot \vec{E}_{2\pi} \right) = 0 \tag{22.22}$$

which shows that when $k_1 \neq k_2$ the two waves travelling with different phase velocities in a particular direction $\vec{k}$ must be polarized orthogonally with respect to each other.

In the general case, we must assume that the three principal refractive indices are different and, for convenience, we can order them as $n_x > n_y > n_z$. The end points of possible wave vectors in the xz-plane are distributed on the circle (22.18) and the ellipse (22.19), and are plotted in Figure 22.3.

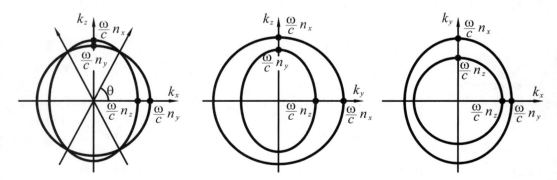

Figure 22.3. Intersections of the wave vector surface with the xz, yz and xy-planes

Similar results are obtained if $\vec{k}$ is perpendicular to the x- and z-axes, respectively. The former case, described by the equations

$$k_y^2 + k_z^2 = \frac{\omega^2}{c^2} n_x^2 \quad , \quad \frac{k_y^2}{(\omega n_z/c)^2} + \frac{k_z^2}{(\omega n_y/c)^2} = 1 \tag{22.23}$$

involves an outer circle of radius $(\omega/c)n_x$ and an inner ellipse in the yz-plane, whereas in the latter case, we have

$$k_x^2 + k_y^2 = \frac{\omega^2}{c^2} n_z^2 \quad , \quad \frac{k_x^2}{(\omega n_y/c)^2} + \frac{k_y^2}{(\omega n_x/c)^2} = 1 \tag{22.24}$$

with an inner circle of radius $(\omega/c)n_z$. Both cases are illustrated in Figure 22.3. Since in the case of the xz-plane, where the ellipse intersects the circle of radius $(\omega/c)n_y$, the crossing points define two directions called the *optic axes*, the medium is referred to as *biaxial*. Along the optic axes the two values of $\vec{k}$ are equal, which means that the two orthogonally polarized waves propagate in the direction of an optic axis with the same phase velocity. The orientation θ of the two optic axes is used to separate *positive biaxial* crystals with $\theta < 45°$ from *negative biaxial* ones, where $\theta > 45°$. For $\theta = 0$ and $\theta = 90°$ the two optic axes coincide and the medium having a single optic axis is said to be *uniaxial*. Substituting $k_x = k\cos\theta$ and $k_z = k\sin\theta$ in both Eqs.(22.18) and (22.19), it is a straightforward matter to derive, by eliminating k between these equations, that

$$\tan^2\theta = \frac{1/n_z^2 - 1/n_y^2}{1/n_y^2 - 1/n_x^2} \tag{22.25}$$

If $\theta = 0$ we obtain $n_x > n_y = n_z$ and, if $\theta = 90°$, we have $n_x = n_y > n_z$. Either case is characterized by the equality of two of the principal refractive indices.

In a uniaxial medium, the dielectric tensor has two equal elements and assumes one of the forms

$$\hat{\varepsilon}_r^+ = \begin{pmatrix} n_e^2 & 0 & 0 \\ 0 & n_o^2 & 0 \\ 0 & 0 & n_o^2 \end{pmatrix} \quad , \quad \hat{\varepsilon}_r^- = \begin{pmatrix} n_o^2 & 0 & 0 \\ 0 & n_o^2 & 0 \\ 0 & 0 & n_e^2 \end{pmatrix} \tag{22.26}$$

corresponding to *positive* crystals, where $n_o < n_e$ ($\theta = 0$) and to *negative* crystals, if $n_o > n_e$ ($\theta = 90°$). The magnitude of the wave vector, whose end point lies on the circle (22.18), which reads

$$k^2 = k_x^2 + k_z^2 = \frac{\omega^2}{c^2} n_o^2 \tag{22.27}$$

is independent of direction. Substituting Eq.(22.27), Eq.(22.15) has the solution

$$E_y \neq 0 \quad , \quad E_x = E_z = 0$$

which describes a linearly polarized wave, with the electric vector along the y direction, that is perpendicular to both $\vec{k}$ and the optic axis, which therefore means that $\vec{E} = \vec{E}_\pi$. It follows from Eq.(22.6) that $\vec{D}$ is collinear with $\vec{E}$. As the properties of this solution are similar to those of a plane wave in an isotropic medium, it is called an *ordinary wave*. The second solution to Eq.(22.15) is obtained by substituting the k values given by the ellipse (22.19) which can be written either as

$$\frac{k_x^2}{(\omega n_o /c)^2} + \frac{k_z^2}{(\omega n_e /c)^2} = 1 \tag{22.28}$$

for *positive* uniaxial crystals $(n_o = n_y = n_z < n_e = n_x)$ or

$$\frac{k_x^2}{(\omega n_e /c)^2} + \frac{k_z^2}{(\omega n_o /c)^2} = 1 \tag{22.29}$$

for *negative* uniaxial media $(n_o = n_x = n_y > n_e = n_z)$. Equations (22.15) give $E_y = 0$ and $E_x \neq 0, E_z \neq 0$. The result is a wave linearly polarized in the plane defined by $\vec{k}$ and the optical axis, which is called the *extraordinary wave*.

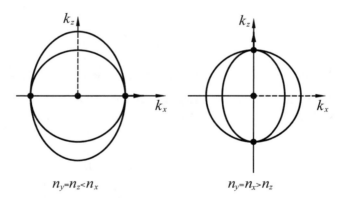

Figure 22.4. Wave vector surfaces for uniaxial crystals

The orientation of the electric vector, given by the ratio of the electric field components, is obtained from the first Eq.(22.15) as

$$\frac{E_z}{E_x} = -\frac{k_x}{k_z}\frac{n_e^2}{n_o^2} \quad , \quad \frac{E_z}{E_x} = -\frac{k_x}{k_z}\frac{n_o^2}{n_e^2} \tag{22.30}$$

for positive and negative uniaxial crystals, respectively. The plot given in Figure 22.4 for Eqs.(22.27), (22.28) and (22.29) shows that the wavevector surface of an uniaxial crystal consists of a sphere and an ellipsoid of revolution, which are tangents to one another along the optic axis. If all the three indices are equal, the wavevector surface reduces to a single sphere indicating an optically isotropic medium.

22.3. Ray and Phase Velocity

In an anisotropic medium the direction of the energy flow, as defined by the Poynting vector (14.13), is different from the direction $\vec{k}$ normal to the wavefront. The Poynting vector $\vec{S} = S\vec{e}_S$ is normal to the plane of $\vec{E}$ and $\vec{H}$, and since $\vec{H}$ is normal to the plane containing the three vectors $\vec{D}$, $\vec{E}$ and $\vec{e}_n$, $\vec{S}$ will lie in that plane, as illustrated in Figure 22.5. $\vec{S}$ makes the same angle α with $\vec{e}_n$ as does $\vec{E}$ with $\vec{D}$. The direction of the Poynting vector defines the *ray direction*.

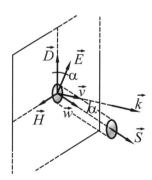

Figure 22.5. Orientation of the wave normal and ray

Therefore there is a distinction between the propagation of the wavefront of constant phase with *phase velocity* $\vec{v}$ in the direction $\vec{e}_n$ of the wavevector $\vec{k}$, given by

$$\vec{v} = \frac{\omega}{k}\vec{e}_n = \frac{c}{n}\vec{e}_n \qquad (22.31)$$

and the energy transport with *ray velocity* $\vec{w}$ in the direction $\vec{e}_S$ of the Poynting vector $\vec{S}$, which is defined as

$$\vec{w} = \frac{\langle S \rangle}{\langle u_{em} \rangle}\vec{e}_S \qquad (22.32)$$

In the case of an isotropic medium, Eq.(22.32) reduces to Eq.(20.27), as the difference between $\vec{e}_S$ and $\vec{e}_n$ is lost and w coincides with the wave velocity (22.31). The wave energy always propagates in the direction $\vec{e}_s$, so that the phase velocity can be regarded in an anisotropic medium as the component of the ray velocity along the direction $\vec{e}_n$ of the wave vector, given by

$$v = w\cos\alpha \tag{22.33}$$

The two velocities are equal for waves propagating along one of the principal axes of the dielectric tensor only, where we have seen that $\vec{S}$ and $\vec{k}$ have the same direction.

The general plane wave equation (22.12) can be rewritten in terms of $\vec{D}$ and $\vec{w}$ in order to obtain the allowed values of the ray velocity in anisotropic media. Taking the scalar product with $\vec{D}$, Eq.(22.12) reduces to

$$k^2 \vec{E}\cdot\vec{D} = \frac{\omega^2}{\varepsilon_0 c^2}\vec{D}\cdot\vec{D} \tag{22.34}$$

since $\vec{k}\cdot\vec{D}=0$, from Eq.(22.3) and $\varepsilon_0\hat{\varepsilon}_r\vec{E}=\vec{D}$. Since $\vec{E}$ makes an angle α with $\vec{D}$, substituting Eqs.(22.31) and (22.33), Eqs.(22.34) takes the form

$$\frac{D}{E}\cos\alpha = \varepsilon_0 \frac{c^2}{w^2} \tag{22.35}$$

The vector $\vec{D}$ can be resolved into components along the orthogonal directions of $\vec{E}$ and $\vec{w}$ as

$$\vec{D} = \vec{E}\frac{(\vec{E}\cdot\vec{D})}{E^2} + \vec{w}\frac{(\vec{w}\cdot\vec{D})}{w^2} = \vec{E}\frac{D}{E}\cos\alpha + \vec{w}\frac{(\vec{w}\cdot\vec{D})}{w^2} \tag{22.36}$$

so that, upon substitution of Eq.(22.35), we obtain

$$\vec{D} = \vec{E}\varepsilon_0\frac{c^2}{w^2} + \vec{w}\frac{(\vec{w}\cdot\vec{D})}{w^2} \tag{22.37}$$

If the relation of $\vec{D}$ to $\vec{E}$ is taken in the principal-axis representation (22.10), the components of $\vec{E}$ in terms of those of $\vec{D}$, which result by inversion of Eq.(22.7), are simply given by

$$E_i = \frac{1}{\varepsilon_0}\varepsilon_i^{-1}D_i = \frac{1}{\varepsilon_0 n_i^2}D_i \tag{22.38}$$

where i stands for x, y and z. Thus Eq.(22.37) takes a form similar to that of Eq.(22.12), that is

$$\vec{w}(\vec{w}\cdot\vec{D}) - w^2\vec{D} + c^2\hat{\varepsilon}_r^{-1}\vec{D} = 0 \tag{22.39}$$

where $\hat{\varepsilon}_r^{-1}$, the inverse matrix of $\hat{\varepsilon}_r$ defined by (22.10), is reduced to its diagonal elements $\hat{\varepsilon}_i^{-1} = 1/n_i^2$. The scalar component equations equivalent to Eq.(22.39) can then be formally obtained from the analogous Eqs.(22.13) by replacing

$$E_i \to D_i \quad , \quad k_i \to w_i \quad , \quad \frac{w}{c} n_i \to \frac{c}{n_i} = v_i \quad (22.40)$$

as follows

$$\left(v_x^2 - w_y^2 - w_z^2\right) D_x + w_x w_y D_y + w_x w_z D_z = 0$$

$$w_y w_x D_x + \left(v_y^2 - w_x^2 - w_z^2\right) D_y + w_y w_z D_z = 0 \quad (22.41)$$

$$w_z w_x D_x + w_z w_y D_y + \left(v_z^2 - w_x^2 - w_y^2\right) D_z = 0$$

where $v_i = c/n_i$ are the phase velocity components along the principal axes. A nontrivial solution for D_x, D_y and D_z can be obtained, given the ray velocity $\vec{w} = (w_x, w_y, w_z)$, if and only if the determinant of the coefficients vanishes, that is

$$\begin{vmatrix} v_x^2 - w_y^2 - w_z^2 & w_x w_y & w_x w_z \\ w_y w_x & v_y^2 - w_x^2 - w_z^2 & w_y w_z \\ w_z w_x & w_z w_y & v_z^2 - w_x^2 - w_y^2 \end{vmatrix} = 0 \quad (22.42)$$

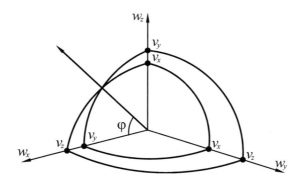

Figure 22.6. The ray-velocity surface in $\vec{w}$-space

The graphical representation of Eq.(22.42) in $\vec{w}$-space, plotted in Figure 22.6, is referred to as the *ray-velocity surface* and can be interpreted along the lines used before to discuss the wave vector surface. The intercept of the ray velocity surface (22.42) with the *xz*-plane is given by a circle of equation

$$w_x^2 + w_z^2 = v_y^2 \quad (22.43)$$

and an ellipse represented by

$$\frac{w_x^2}{v_z^2} + \frac{w_z^2}{v_x^2} = 1 \tag{22.44}$$

which result from Eqs.(22.18) and (22.19) according to the rule (22.40). Thus there are two allowed ray velocities w_1 and w_2 for a given ray direction $\vec{e}_s$. With $n_x > n_y > n_z$ as before, the ellipse (22.44) must intercept the circle of radius v_y and the crossing points define, in general, two directions called the *ray axes* of the medium, along which the ray velocity is single valued.

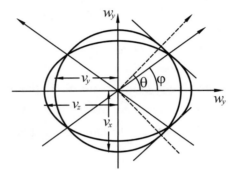

Figure 22.7. Orientation of the ray and optic axes

The orientation φ of the ray axes, in the configuration of Figure 22.7, is obtained by eliminating w from Eqs.(22.43) and (22.44), and is written in terms of $w_x = w\cos\varphi$ and $w_z = w\sin\varphi$ as

$$\tan^2 \varphi = \frac{n_y^2 - n_z^2}{n_x^2 - n_y^2} \tag{22.45}$$

If compared to Eq.(22.25), Eq.(22.45) gives

$$\tan \varphi = \frac{n_z}{n_x} \tan \theta \tag{22.46}$$

and hence, ray axes are distinct from the optic axes of biaxial media defined before, as in the xz-plane we always have $n_z \neq n_x$. It is straightforward to show that the orientation θ of the optic axis with respect to the x-axis, as given by Eq.(22.25), which reads

$$\tan^2 \theta = \frac{v_z^2 - v_y^2}{v_y^2 - v_x^2} \tag{22.47}$$

defines a direction perpendicular to the tangent common to the circle and the ellipse, as drawn in Figure 22.7. As the optic axis gives the direction of the wave normal, this tangent must be parallel to the wavefront of a plane wave which propagates with phase velocity v_y along the optic axis. In the case of uniaxial media, where $\theta = 0$ or $\theta = \pi/2$, Eq.(22.46) shows that the ray axis coincides with the optic axis.

FURTHER READING
1. M. Born, E. Wolf - PRINCIPLES OF OPTICS, Cambridge University Press, Cambridge, 2000.
2. M. H. H. Freeman , W. H. Fincham - OPTICS, Elsevier, Amsterdam, 1999.
3. D. S. Kliger, J. W. Lewis, C. E. Randall - POLARIZED LIGHT IN OPTICS AND SPECTROSCOPY, Elsevier, Amsterdam, 1990.

PROBLEMS

22.1. A plane wave is incident at an angle θ of the surface on a uniaxial crystal. Show that, owing to the double nature of the wave vector surface, there are two possible refracted paths corresponding to mutually orthogonal polarizations.

Solution

In the case of wave propagation along one of the principal axes of the dielectric tensor, say the y-axis, such that $k_x = k_z = 0$ and $k_y = k$, Eqs.(22.13) take the form

$$\left(\frac{\omega^2}{c^2}n_x^2 - k^2\right)E_x = 0, \qquad \frac{\omega^2}{c^2}n_y^2 E_y = 0, \qquad \left(\frac{\omega^2}{c^2}n_z^2 - k^2\right)E_z = 0$$

The second equation demonstrates that the electric vector has no longitudinal component, since $E_y = 0$, while the other two equations show that the transverse components $E_x \neq 0$ and $E_z \neq 0$ correspond to different k values

$$k_1 = \frac{\omega}{c} n_x \quad , \quad k_2 = \frac{\omega}{c} n_z$$

If we take $n_x > n_z$ in the configuration shown in Figure 22.4, we can obtain the intercepts of the outer and inner sheets of the wave vector surface with the principal axis to be at k_1 and k_2, respectively. The two k values correspond to mutually orthogonal states of linear polarization, in the x- and z-directions, travelling with different phase velocities ω/k_1 and ω/k_2. Consider now the surface of a uniaxial crystal, cut so that the two waves travelling along the principal axis

y are incident on the interface at an angle θ. The direction of the transmitted beam is derived by applying the phase matching condition (21.7) at the interface, which gives

$$(\vec{k}_\tau - k\vec{e}_i) \cdot \vec{r} = 0$$

where the field vectors have been assumed to be of the form (21.5), with $\vec{k}_i = k\vec{e}_i$. Since the end point of $\vec{k}_\tau$ lies on either the circle or the ellipse which constitutes the intercepts of the wave vector surface with the plane of incidence, its possible directions will be determined by the crossing points between the normal to the interface (through the end point of $\vec{k}_i$) and the inner and outer curves.

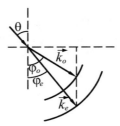

In general there will be two crossing points, which define two different possible wave vectors $\vec{k}_o$ and $\vec{k}_e$, so that there are two refracted paths, a phenomenon known as birefringence or double refraction. Thus, there are two scalar equations

$$(\vec{k}_o - k\vec{e}_i) \cdot \vec{r} = 0 \quad \text{or} \quad k_o \sin \varphi_o = k \sin \theta$$

$$(\vec{k}_e - k\vec{e}_i) \cdot \vec{r} = 0 \quad \text{or} \quad k_e \sin \varphi_e = k \sin \theta$$

which reduce to

$$n_o \sin \varphi_o = n_e \sin \varphi_e = n \sin \theta$$

The refractive index n_o is constant for all directions, and hence, Snell's law (21.14) always holds for the ordinary wave. However, the requirement of a constant n_e is only satisfied if the plane of incidence is perpendicular to the optic axis, in the special case of a crystal cut with its surface parallel to the optic axis. This corresponds to an ordinary wave with the $\vec{E}$ vector perpendicular to the optic axis and an extraordinary wave which is linearly polarized parallel to the optic axis.

22.2. Optical activity of certain uniaxial crystals which have the ability to rotate the plane of polarization of light passing through them can be assigned to a different speed of propagation for the right- and left-circularly polarized light. Find an expression for the dielectric tensor of such a crystal.

Solution

Linearly polarized waves can be regarded as a superposition of circularly polarized beams of equal intensity and opposite handedness, derived in the Jones representation from Eqs.(20.58) and (20.59) as

$$J_x = \frac{1}{\sqrt{2}}(J_r + J_l) \quad , \quad J_y = \frac{i}{\sqrt{2}}(J_r - J_l)$$

Optical activity, assigned to the existence of different refractive indices for the right- and left-circularly polarized waves, will be properly described by taking $\vec{e}_r, \vec{e}_l$ and $\vec{e}_z$ as the principal axes of the dielectric tensor rather than $\vec{e}_x, \vec{e}_y$ and $\vec{e}_z$, which are suitable for nonactive crystals. Solving for J_r and J_l gives

$$J_r = \frac{1}{\sqrt{2}}(J_x - iJ_y) \quad , \quad J_l = \frac{1}{\sqrt{2}}(J_x + iJ_y)$$

so that the unit orthogonal vectors of the principal axes are actually given by

$$\vec{e}_r = \frac{1}{\sqrt{2}}(\vec{e}_x - i\vec{e}_y) \quad , \quad \vec{e}_l = \frac{1}{\sqrt{2}}(\vec{e}_x + i\vec{e}_y) \quad , \quad \vec{e}_z$$

and the field vector components along the new principal axes become

$$E_r = \vec{E} \cdot (\vec{e}_x - i\vec{e}_y) = (\vec{e}_x E_x + \vec{e}_y E_y + \vec{e}_z E_z) \cdot (\vec{e}_x - i\vec{e}_y) = E_x - iE_y$$

$$E_l = \vec{E} \cdot (\vec{e}_x + i\vec{e}_y) = E_x + iE_y$$

The relation between $\vec{D}$ and $\vec{E}$ takes the matrix form

$$\begin{pmatrix} D_x - iD_y \\ D_x + iD_y \\ D_z \end{pmatrix} = \varepsilon_0 \begin{pmatrix} \varepsilon_{rr} & 0 & 0 \\ 0 & \varepsilon_{ll} & 0 \\ 0 & 0 & \varepsilon_{zz} \end{pmatrix} \begin{pmatrix} E_x - iE_y \\ E_x + iE_y \\ E_z \end{pmatrix}$$

and this can be reduced to a Cartesian representation (22.8), with respect to the x, y and z-axes of the dielectric tensor, by taking $\varepsilon_{rr} = \varepsilon - \varepsilon_i$, and $\varepsilon_{ll} = \varepsilon + \varepsilon_i$, so that

$$D_x - iD_y = \varepsilon_0(\varepsilon - \varepsilon_i)(E_x - iE_y) \quad , \quad D_x + iD_y = \varepsilon_0(\varepsilon + \varepsilon_i)(E_x + iE_y)$$

which gives $D_x = \varepsilon_0 \varepsilon E_x + i\varepsilon_0 \varepsilon_i E_y$ and $D_y = -i\varepsilon_0 \varepsilon_i E_x + \varepsilon_0 \varepsilon E_y$. Since $D_z = \varepsilon_0 \varepsilon_{zz} E_z$, it follows that

$$\begin{pmatrix} D_x \\ D_y \\ D_z \end{pmatrix} = \varepsilon_0 \begin{pmatrix} \varepsilon & i\varepsilon_i & 0 \\ -i\varepsilon_i & \varepsilon & 0 \\ 0 & 0 & \varepsilon_{zz} \end{pmatrix} \begin{pmatrix} E_x \\ E_y \\ E_z \end{pmatrix}$$

Thus, in an optically active uniaxial crystal, the dielectric tensor differs from that given in Eq.(22.26) by the presence of two conjugate imaginary off diagonal elements, that is

$$\hat{\varepsilon}_r = \begin{pmatrix} \varepsilon & i\varepsilon_i & 0 \\ -i\varepsilon_i & \varepsilon & 0 \\ 0 & 0 & \varepsilon_{zz} \end{pmatrix} = \begin{pmatrix} n_o^2 & i\varepsilon_i & 0 \\ -i\varepsilon_i & n_o^2 & 0 \\ 0 & 0 & n_e^2 \end{pmatrix}$$

Substituting into Eq.(22.12) provides three scalar equations for propagation along the z-axis $\left(\vec{k} = k\vec{e}_z\right)$ given by

$$\left(\frac{\omega^2}{c^2}n_o^2 - k^2\right)E_x + i\frac{\omega^2}{c^2}\varepsilon_i E_y = 0 \quad , \quad -i\frac{\omega^2}{c^2}\varepsilon_i E_y + \left(\frac{\omega^2}{c^2}n_o^2 - k^2\right)E_y = 0 \quad , \quad \frac{\omega^2}{c^2}n_e^2 E_z = 0$$

The last equation gives $E_z = 0$, which shows that $\vec{E}$ lies in the plane of the wavefront. We have a solution for the transverse components E_x and E_y only if the determinant of their coefficients vanishes, that is

$$\begin{vmatrix} (\omega n_o/c)^2 - k^2 & i\omega^2 \varepsilon_i/c^2 \\ -i\omega^2 \varepsilon_i/c^2 & (\omega n_o/c)^2 - k^2 \end{vmatrix} = 0$$

This condition factorizes into

$$\left[k^2 - \frac{\omega^2}{c^2}\left(n_o^2 + \varepsilon_i\right)\right]\left[k^2 - \frac{\omega^2}{c^2}\left(n_o^2 - \varepsilon_i\right)\right] = 0$$

and gives two positive solutions for k of the form

$$k_r = \frac{\omega}{c}n_o\sqrt{1 + \frac{\varepsilon_i}{n_0^2}} \cong \frac{\omega}{c}n_o\left(1 + \frac{\varepsilon_i}{2n_0^2}\right) = \frac{\omega}{c}n_r \quad , \quad k_l \cong \frac{\omega}{c}n_o\left(1 - \frac{\varepsilon_i}{2n_0^2}\right) = \frac{\omega}{c}n_l$$

and this defines the refractive indices n_r and n_l. Either of the first two equations for E_x and E_y yields $E_x = iE_y$ when substituting k_r, and $E_x = -iE_y$, when $k = k_l$. This means that different indices n_r and n_l must be associated with the right- and left-circularly polarized wave, respectively, represented by Jones vectors J_r and J_l and propagating through an active medium with phase velocities c/n_r and c/n_l. Therefore in an active uniaxial crystal only circularly polarized waves can propagate in the direction of the optic axis and the phase velocity is different for the two orthogonal states, unlike the case of nonactive media where a single phase velocity $(\omega/c)n_o$ is observed along the optic axis.

22.3. A beam of linearly polarized light of wavelength λ_0 in vacuum travels through an optically active crystal. Use the Jones calculus to find the specific rotatory power $\delta(\lambda_0) = \partial\varphi/\partial z$ of the active medium in terms of the imaginary component ε_i of the dielectric tensor.

Solution

The superposition of the two orthogonal waves of equal amplitude propagating along the optic axis z reads

$$\vec{E}(z,t) = \vec{e}_r E_0 e^{i(k_r z - \omega t)} + \vec{e}_l E_0 e^{i(k_l z - \omega t)} = \frac{1}{\sqrt{2}}\left(\vec{e}_x - i\vec{e}_y\right)E_0 e^{i(k_r z - \omega t)} + \frac{1}{\sqrt{2}}\left(\vec{e}_x + i\vec{e}_y\right)E_0 e^{i(k_l z - \omega t)}$$

and can be represented by a Jones vector dependent on both the position and time as

$$J(z,t)=\begin{pmatrix}E_x\\E_y\end{pmatrix}=\frac{1}{\sqrt{2}}\begin{pmatrix}1\\-i\end{pmatrix}E_0 e^{i(k_r z-\omega t)}+\frac{1}{\sqrt{2}}\begin{pmatrix}1\\i\end{pmatrix}E_0 e^{i(k_l z-\omega t)}$$

This depends on the distance z travelled through the crystal in the form

$$J(z,t)=\frac{1}{\sqrt{2}}E_0 e^{-i\omega t}\left[\begin{pmatrix}1\\-i\end{pmatrix}e^{ik_r z}+\begin{pmatrix}1\\i\end{pmatrix}e^{ik_l z}\right]=\frac{1}{\sqrt{2}}E_0 e^{i(k_o z-\omega t)}\left[\begin{pmatrix}1\\-i\end{pmatrix}e^{i\varphi(z)}+\begin{pmatrix}1\\i\end{pmatrix}e^{-i\varphi(z)}\right]$$

where we have set

$$k_o=\frac{1}{2}(k_r+k_l)\cong\frac{\omega}{c}n_o\quad\text{and}\quad\varphi(z)=\frac{1}{2}(k_r-k_l)z=\frac{\omega}{2c}(n_r-n_l)z=\frac{\omega}{2c}\frac{\varepsilon_i}{n_o}z$$

Thus, the Jones vector can be written as

$$J(z,t)=\sqrt{2}\begin{pmatrix}\cos\varphi(z)\\\sin\varphi(z)\end{pmatrix}E_0 e^{i(k_o z-\omega t)}$$

and this represents a linearly polarized wave for any distance z travelled along the z-direction, with the $\vec{E}$ field making an angle $\varphi(z)$ with respect to the x-axis. As φ has a linear dependence on z, we can derive the law for optical rotation by assuming that the wave is incident at time zero on the boundary of the active crystal, located at $z = 0$, where

$$J(0,0)=\sqrt{2}\begin{pmatrix}1\\0\end{pmatrix}E_0\quad\text{or}\quad\vec{E}(0,0)=\vec{e}_x\sqrt{2}E_0$$

The incident wave is represented as being in a state of linear polarization, with $\vec{E}$ along the x-axis, so that its plane of polarization is rotated through an angle $\varphi(z)$ during propagation over a distance z. The amount of rotation per unit length, called the *specific rotatory power*, of the active medium is

$$\delta(\lambda_0)=\frac{\partial\varphi}{\partial z}=\frac{1}{2}(k_r-k_l)=(n_r-n_l)\frac{\pi}{\lambda_0}=\frac{\varepsilon_i}{n_o}\frac{\pi}{\lambda_0}$$

for a given wavelength λ_0 in vacuum. It is apparent that optical activity can be regarded as arising from the existence of the imaginary component ε_i of the dielectric tensor, as required. The medium is said to be dextrorotatory if $\delta(\lambda_0)>0$ and levorotatory if $\delta(\lambda_0)<0$.

22.4. Show that an isotropic dielectric placed in a magnetic field $\vec{B}$ becomes electrically anisotropic for electromagnetic waves and derive an expression for its dielectric tensor by considering the motion of bound electrons of natural frequency ω_0 in the presence of both a harmonic field $\vec{E}(\vec{r},t)=\vec{E}e^{-i\omega t}$ and a static field $\vec{B}$.

Solution

In view of the Lorentz force law (13.30), the equation of motion reads

$$m_e \ddot{\vec{r}} + k_0 \vec{r} = -e\vec{E}e^{-i\omega t} - e\dot{\vec{r}} \times \vec{B}$$

where $\vec{r}$ is the electron displacement from the equilibrium position, $k_0 = m_e \omega_0^2$ is the elastic binding force constant, and ω_0 is the natural frequency of the bound electron. Assuming that the displacement $\vec{r}$ acquires the same harmonic time dependence as the wave, we obtain the time-independent relation

$$\left(-\omega^2 + \omega_0^2\right)\vec{r} - i\omega \frac{e}{m_e}\left(\vec{r} \times \vec{B}\right) = -\frac{e}{m_e}\vec{E}$$

Rearranging in terms of polarization $\vec{P} = -Ne\vec{r}$, Eq.(11.23), gives

$$\left(\omega_0^2 - \omega^2\right)\vec{P} - i\omega \frac{e}{m_e}\left(\vec{P} \times \vec{B}\right) = \frac{Ne^2}{m_e}\vec{E}$$

It is usual to choose a Cartesian system of coordinates whose z-axis is parallel to $\vec{B}$, that is $\vec{B} = \vec{e}_z B$, such that

$$\vec{P} \times \vec{B} = \begin{vmatrix} \vec{e}_x & \vec{e}_y & \vec{e}_z \\ P_x & P_y & P_z \\ 0 & 0 & B \end{vmatrix} = \vec{e}_x P_y B - \vec{e}_y P_x B$$

If we set the cyclotron frequency as $\omega_c = eB/m_e$, the components of the polarization follow from the algebraic equations

$$\left(\omega_0^2 - \omega^2\right)P_x - i\omega\omega_c P_y = \frac{Ne^2}{m_e}E_x, \quad i\omega\omega_c P_x + \left(\omega_0^2 - \omega^2\right)P_y = \frac{Ne^2}{m_e}E_y, \quad \left(\omega_0^2 - \omega^2\right)P_z = \frac{Ne^2}{m_e}E_z$$

in the form

$$P_x = \frac{Ne^2/m_e}{\left(\omega_0^2 - \omega^2\right)^2 - \omega^2\omega_c^2}\left[\left(\omega_0^2 - \omega^2\right)E_x + i\omega\omega_c E_y\right]$$

$$P_y = \frac{Ne^2/m_e}{\left(\omega_0^2 - \omega^2\right)^2 - \omega^2\omega_c^2}\left[-i\omega\omega_c E_x + \left(\omega_0^2 - \omega^2\right)E_x\right] \quad , \quad P_z = \frac{Ne^2/m_e}{\omega_0^2 - \omega^2}E_z$$

In view of Eq.(11.25), there is a tensor relation between the polarization and electric field vectors, with components χ_{ik} of the electric susceptibility, where $i, k = x, y, z$, which can be readily identified. Using Eq.(11.29) the dielectric tensor then takes a form similar to that of an optically active crystal

$$\hat{\varepsilon}_r = \begin{pmatrix} 1 & 0 & 0 \\ 0 & 1 & 0 \\ 0 & 0 & 1 \end{pmatrix} + \begin{pmatrix} \chi_{xx} & \chi_{xy} & 0 \\ \chi_{yx} & \chi_{yy} & 0 \\ 0 & 0 & \chi_{zz} \end{pmatrix} = \begin{pmatrix} \varepsilon & i\varepsilon_i & 0 \\ -i\varepsilon_i & \varepsilon & 0 \\ 0 & 0 & \varepsilon_{zz} \end{pmatrix}$$

where it is apparent that

$$\varepsilon = 1 + \frac{\omega_p^2 (\omega_0^2 - \omega^2)}{(\omega_0^2 - \omega^2)^2 - \omega^2 \omega_c^2}, \quad \varepsilon_i = +\frac{\omega_p^2 \omega \omega_c}{(\omega_0^2 - \omega^2)^2 - \omega^2 \omega_c^2}, \quad \varepsilon_{zz} = 1 + \frac{\omega_p^2}{\omega_0^2 - \omega^2}$$

The elements of the dielectric tensor are given in terms of the resonance frequency ω_0, the cyclotron frequency ω_c and the plasma frequency defined as $\omega_p = \sqrt{Ne^2/m_e\varepsilon_0}$.

22.5. Show that an isotropic dielectric placed in a magnetic field $\vec{B}$ becomes double refracting for light propagating at right angles to the field, and prove that the effect is proportional to the square of the magnetic field strength.

Solution

Suppose that light propagates along a direction $\vec{k}$, contained in the yz-plane of the Cartesian system having $\vec{B}$ parallel to the z-axis, and let θ be the angle between $\vec{k}$ and $\vec{B}$. Substituting the dielectric tensor derived in Problem 22.4 and inserting $k_x = 0$, $k_y = k\sin\theta$ and $k_z = k\cos\theta$, the general plane wave equation (22.12) reduces to

$$\left(\frac{\omega^2}{c^2} n_o^2 - k^2\right) E_x + i\frac{\omega^2}{c^2} \varepsilon_i E_y = 0$$

$$-i\frac{\omega^2}{c^2} \varepsilon_i E_x + \left(\frac{\omega^2}{c^2} n_o^2 - k^2 \cos^2\theta\right) E_y + k^2 \sin\theta \cos\theta\, E_z = 0$$

$$k^2 \sin\theta \cos\theta\, E_y + \left(\frac{\omega^2}{c^2} n_e^2 - k^2 \sin^2\theta\right) E_z = 0$$

If the wave is propagating perpendicular to $\vec{B}$, for instance along the y-axis, we have $\theta = \pi/2$ and this reduce to the algebraic equations

$$\left(\frac{\omega^2}{c^2} n_o^2 - k^2\right) E_x + i\frac{\omega^2}{c^2} \varepsilon_i E_y = 0, \quad \frac{\omega^2}{c^2}\left(-i\varepsilon_i E_x + n_o^2 E_y\right) = 0, \quad \left(\frac{\omega^2}{c^2} n_e^2 - k^2\right) E_z = 0$$

Eliminating E_y as given by the second equation, and setting $D_y = -i\varepsilon_0 \varepsilon_i E_x + \varepsilon_0 \varepsilon E_y$ as in Problem 2.2, one obtains the wave equations for a double refracting medium as

$$\left[\frac{\omega^2}{c^2}\left(n_o^2 - \frac{\varepsilon_i^2}{n_o^2}\right) - k^2\right] E_x = 0, \quad \frac{\omega^2}{c^2 \varepsilon_0} D_y = 0, \quad \left(\frac{\omega^2}{c^2} n_e^2 - k^2\right) E_z = 0$$

Thus a medium placed in a magnetic field behaves like a biaxial crystal for light propagating at right angles to the field. There are two linearly polarized waves with mutually perpendicular vibrations, propagating with different phase velocities. One solution has $E_x = 0$, when the electric vector is parallel to $\vec{B}$, while the second solution has the electric vector perpendicular to $\vec{B}$ ($E_z = 0$). The wave numbers $k_\parallel$ and $k_\perp$ for the two cases are given by

$$k_\parallel = \frac{\omega}{c} n_e = \frac{\omega}{c} \varepsilon_{zz}^{1/2} \quad , \quad k_\perp = \frac{\omega}{c}\left(n_o^2 - \frac{\varepsilon_i^2}{n_o^2}\right)^{1/2} = \frac{\omega}{c} \varepsilon^{1/2}\left(1 - \frac{\varepsilon_i^2}{\varepsilon^2}\right)^{1/2}$$

For frequencies away from resonance, where $\omega\omega_c \ll |\omega_0^2 - \omega^2|$, we can drop the dependence on ε_i, to obtain

$$k_\perp = \frac{\omega}{c}\varepsilon^{1/2} = \frac{\omega}{c}\left\{1 + \frac{\omega_p^2}{\omega_0^2 - \omega^2}\left[1 - \frac{\omega^2\omega_c^2}{(\omega_0^2 - \omega^2)^2}\right]^{-1}\right\}^{1/2}$$

$$\cong \frac{\omega}{c}\left[1 + \frac{\omega_p^2}{2(\omega_0^2 - \omega^2)} + \frac{\omega_p^2\omega^2\omega_c^2}{2(\omega_0^2 - \omega^2)^3}\right] = k_\parallel + \frac{\omega_p^2\omega^3\omega_c^2}{2c(\omega_0^2 - \omega^2)^3}$$

This indicates an effect, called the Cotton-Mouton effect, that is proportional to the square of the magnetic induction

$$k_\perp - k_\parallel = \frac{e^2\omega_p^2\omega^3}{2m_e^2 c(\omega_0^2 - \omega^2)^3} B^2$$

as required.

22.6. Find the range of the internal angle of incidence on the plane boundary of a negative uniaxial crystal $(n_o > n_e)$, cut so that the optic axis is perpendicular to the plane of incidence, in order to obtain a linearly polarized transmitted beam.

22.7. When light is striking at right angles the surface of a Glan prism, made up of two prisms of negative uniaxial material (calcite), the ordinary ray is removed by total internal reflection, and the extraordinary ray continues straight through the device. If the intensity of the emergent beam is reduced to a half of its maximum value when the device is rotated by $\pi/3$ about the direction of propagation, find the degree of polarization P of the incident light.

22.8. A beam of light falls at oblique incidence on a negative uniaxial crystal $(n_o > n_e)$ cut so that the optic axis is parallel to the boundary. Find the maximum possible angle α between the ray and the phase velocities inside the crystal.

22.9. A parallel-sided plate of thickness d, made with positive uniaxial material $(n_e > n_o)$, is cut so that the optic axis is parallel to the corner edges. If linearly polarized light of wavelength λ_0 in vacuum is normally incident on a boundary, show that the polarization of the emergent beam depends on the orientation α of the incident plane of polarization with respect to the optic axis, and determine α and d when circularly polarized light is produced.

22.10. Show that an isotropic dielectric rotates the plane of a linearly polarized wave of wavelength λ_0 in vacuum when a magnetic field $\vec{B}$ is applied along the direction of propagation (Faraday rotation), and find the specific rotatory power $\delta(\lambda_0)$ of such an active medium.

23. ABSORPTION AND DISPERSION

23.1. OPTICAL PROPERTIES OF CONDUCTING MEDIA
23.2. ORIGIN OF COMPLEX CONSTITUTIVE PARAMETERS

23.1. OPTICAL PROPERTIES OF CONDUCTING MEDIA

The electromagnetic waves passing through an electrically neutral medium $(\rho = 0)$ will interact with both the conduction currents and polarization charges, if present. For harmonic waves of angular frequency ω, represented in the form (20.15), Maxwell's equations (14.6) for linear, homogenous and isotropic media take the form

$$\nabla \times \vec{E} = i\omega\mu\vec{H} \qquad \nabla \times \vec{H} = \sigma\vec{E} - i\omega\varepsilon\vec{E}$$
$$\nabla \cdot \vec{H} = 0 \qquad \nabla \cdot \vec{E} = 0$$
(23.1)

where the presence of imaginary terms is related to the first order time derivatives of the field vectors, which in turn results in the presence of damping. Such media can be suitably described in terms of *complex constitutive parameters*. In view of Eq.(11.28), Ampère's law (23.1) takes the form

$$\nabla \times \vec{H} = (\sigma - i\omega\varepsilon_0\chi)\vec{E} - i\omega\varepsilon_0\vec{E} = \sigma_c\vec{E} - i\omega\varepsilon_0\vec{E}$$
(23.2)

which suggests that we can think of a *complex conductivity* σ_c given by

$$\sigma_c = \sigma - i\omega\varepsilon_0\chi = \sigma - i\omega(\varepsilon - \varepsilon_0)$$
(23.3)

Thus a possible approach is to describe the medium as a conductor of permittivity ε_0, permeability $\mu = \mu_0\mu_r$ and complex conductivity σ_c. Alternatively, we can eliminate either $\vec{H}$, by taking the *curl* of Faraday's law, or $\vec{E}$, by taking the *curl* of Ampère's law, to obtain wave equations similar to Eq.(14.8), which read

$$\nabla^2 \vec{E} + i\mu\sigma\omega\vec{E} + \varepsilon\mu\omega^2 \vec{E} = 0$$
$$\nabla^2 \vec{H} + i\mu\sigma\omega\vec{H} + \varepsilon\mu\omega^2 \vec{H} = 0 \tag{23.4}$$

We may now consider the medium as a lossy dielectric by writing Eqs.(23.4) in the usual Helmholtz form

$$\nabla^2 \vec{E} + \varepsilon_c \mu\omega^2 \vec{E} = 0$$
$$\nabla^2 \vec{H} + \varepsilon_c \mu\omega^2 \vec{H} = 0 \tag{23.5}$$

where the real permittivity is replaced by a *complex permittivity* ε_c written as

$$\varepsilon_c = \varepsilon + i\frac{\sigma}{\omega} = \varepsilon_0(\varepsilon_r + i\varepsilon_i) = \varepsilon_0\left(\varepsilon_r + i\frac{\sigma}{\varepsilon_0\omega}\right) = \varepsilon_0\left(1 + \chi_e + i\frac{\sigma}{\varepsilon_0\omega}\right) \tag{23.6}$$

The constitutive parameters of a conducting medium, if regarded to be a lossy dielectric, are the conductivity σ, the permeability μ and the complex permittivity ε_c. In this case, the optical properties become formally identical with those of transparent media, if the quantity $\varepsilon = \varepsilon_0\varepsilon_r = \varepsilon_0 n^2$ is replaced by the complex quantity

$$\varepsilon_c = \varepsilon_0 n_c^2 = \varepsilon_0(n + in_i)^2 \tag{23.7}$$

which can be regarded as a definition of the *complex refractive index* n_c. Substituting Eq.(23.7) and then equating the real and imaginary parts, Eq.(23.6) becomes

$$n^2 - n_i^2 = \varepsilon_r \quad , \quad 2nn_i = \frac{\sigma}{\varepsilon_0\omega} = \varepsilon_i \tag{23.8}$$

On solving these equations we obtain expressions for both n, called the *refractive index* and n_i, called the *extinction index*, in the form

$$n^2 = \frac{\varepsilon_r}{2}\left[\left(1 + \frac{\sigma^2}{\varepsilon^2\omega^2}\right)^{1/2} + 1\right] \quad , \quad n_i^2 = \frac{\varepsilon_r}{2}\left[\left(1 + \frac{\sigma^2}{\varepsilon^2\omega^2}\right)^{1/2} - 1\right] \tag{23.9}$$

One also may leave the constitutive parameters σ and ε unspecified, so that the plane wave solutions of Eqs.(23.4) must be written in all but special cases in terms of *complex wave vectors* $\vec{k}_c$ as

$$\vec{E}(\vec{r},t) = \vec{E}(\vec{r})e^{-i\omega t} = \vec{E}e^{i(\vec{k}_c \cdot \vec{r} - \omega t)}$$
$$\vec{H}(\vec{r},t) = \vec{H}(\vec{r})e^{-i\omega t} = \vec{H}e^{i(\vec{k}_c \cdot \vec{r} - \omega t)} \tag{23.10}$$

Since the operator relations (20.10) are valid under the assumptions of Eq.(23.10), that is when $\vec{k}_c$ are complex quantities, Eqs.(23.4) reduce to the following algebraic *plane wave equation* for conducting media

$$k_c^2 \vec{E} = \varepsilon\mu\omega^2 \vec{E} + i\mu\sigma\omega\vec{E}$$
$$k_c^2 \vec{H} = \varepsilon\mu\omega^2 \vec{H} + i\mu\sigma\omega\vec{H}$$
(23.11)

Each of Eqs.(23.11) requires that

$$k_c^2 = \varepsilon\mu\omega^2\left(1 + i\frac{\sigma}{\varepsilon\omega}\right) \tag{23.12}$$

and hence, the wave number k_c may be written in complex representation as

$$k_c = k + ik_i \tag{23.13}$$

Specific expressions for k and k_i follow by squaring Eq.(23.13) and then equating the real and imaginary parts to the real and imaginary parts of Eq.(23.12), which gives

$$k^2 - k_i^2 = \varepsilon\mu\omega^2 \quad , \quad 2kk_i = \mu\sigma\omega \tag{23.14}$$

The resulting expressions are

$$k = \omega\sqrt{\frac{\varepsilon\mu}{2}}\left[\left(1 + \frac{\sigma^2}{\varepsilon^2\omega^2}\right)^{1/2} + 1\right]^{1/2} \quad , \quad k_i = \omega\sqrt{\frac{\varepsilon\mu}{2}}\left[\left(1 + \frac{\sigma^2}{\varepsilon^2\omega^2}\right)^{1/2} - 1\right]^{1/2} \tag{23.15}$$

and their similarity with Eqs.(23.9) allows us to define the complex wave number in terms of the complex refractive index (23.7) as

$$k_c = k + ik_i = \sqrt{\mu_r}\frac{\omega}{c}(n + in_i) = \sqrt{\mu_r}\frac{\omega}{c}n_c \tag{23.16}$$

Using the configuration illustrated before in Figure 20.1 and Eq.(20.12), the transverse fields (23.10) can be written as

$$\vec{E}(\vec{r},t) = \vec{E}e^{i(k_c\xi - \omega t)} = \vec{E}e^{-k_i\xi}e^{i(k\xi - \omega t)}$$

$$\vec{H}(\vec{r},t) = \frac{k + ik_i}{\mu\omega}(\vec{e}_n \times \vec{E})e^{-k_i\xi}e^{i(k\xi - \omega t)}$$
(23.17)

where $\xi = \vec{r}\cdot\vec{e}_n$. Equations (23.17) show that the imaginary part k_i of the complex wave number, called the *absorption coefficient*, is a measure of attenuation of a plane wave during propagation in metals, whereas its real part k must preserve the same relationship with the wave vector $\vec{k} = k\vec{e}_n$ as in the absence of attenuation. The planes of constant phase are propagated with a velocity defined in terms of k and derived from Eqs.(23.15) in the form

$$v = \frac{\omega}{k} = \frac{c}{\left\{(\varepsilon_r \mu_r / 2)\left[\left(1 + \sigma^2 / \varepsilon^2 \omega^2\right)^{1/2} + 1\right]\right\}^{1/2}} \quad (23.18)$$

where use has been made of Eq.(15.53) for c. Hence, conducting media must be referred to not only as absorbing but also as *dispersive* media, since waves of different frequencies are transmitted at different velocities. It is apparent that phase velocities decrease with increasing conductivity σ and increase with increasing frequency ω as long as σ and ε are independent of ω. If k_c is written in the form

$$k_c = \sqrt{k^2 + k_i^2}\, e^{-i\varphi} \quad (23.19)$$

it follows from Eq.(23.19) that the fields inside a conducting medium are out of phase by an angle φ, given by

$$\tan \varphi = \frac{k_i}{k} = \frac{n_i}{n} \quad (23.20)$$

The first term on the right hand side of the plane wave equation (23.11) arises from the displacement current density $\partial \vec{D} / \partial t = \varepsilon \partial \vec{E} / \partial t = -i\omega\varepsilon\vec{E}$ whereas the second term arises from the conduction current density $\vec{j} = \sigma \vec{E}$. The ratio of their magnitudes $\sigma / \varepsilon \omega$ appears to be a macroscopic criterion for the classification of conducting media. In the case of *poor conductors* the displacement current is of importance, as $\sigma / \varepsilon \omega \ll 1$, so that the components (23.15) of the complex wave number can be approximated by

$$k \cong \omega \sqrt{\frac{\varepsilon \mu}{2}} \left(2 + \frac{\sigma^2}{2\varepsilon^2 \omega^2}\right)^{1/2} \cong \omega \sqrt{\varepsilon \mu}\left(1 + \frac{\sigma^2}{8\varepsilon^2 \omega^2}\right)$$

$$k_i \cong \omega \sqrt{\frac{\varepsilon \mu}{2}} \left(\frac{\sigma^2}{2\varepsilon^2 \omega^2}\right)^{1/2} = \frac{\sigma}{2}\sqrt{\frac{\mu}{\varepsilon}} \quad (23.21)$$

For *good conductors* it is the conduction current that is important, $\sigma / \varepsilon \omega \gg 1$, so that the real and imaginary parts of k_c are equal, as given by

$$k = k_i = \sqrt{\frac{\omega \mu \sigma}{2}} \quad (23.22)$$

EXAMPLE 23.1. The skin effect

It is usual to define the so-called *skin depth* δ of a good conductor to be the distance the electromagnetic wave penetrates the surface before its amplitude drops to $1/e$ times its surface value. Using Eqs.(23.17) and (23.22) this may be expressed in terms of k_i as

$$\delta = \frac{1}{k_i} = \sqrt{\frac{2}{\omega\mu\sigma}} \qquad (23.23)$$

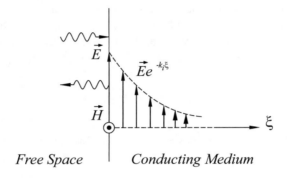

Figure 23.1. Field vector attenuation in a conducting medium

High values for either the conductivity σ or frequency ω will lead to a small skin depth. The fields inside a good conductor, as represented in Figure 23.1, are out of phase by an angle $\varphi = \pi/4$ and result from Eqs.(23.17) as

$$\vec{E}(\vec{r},t) = \vec{E}e^{-k_i\xi}e^{i(k\xi-\omega t)}$$
$$\vec{H}(\vec{r},t) = \sqrt{\frac{\sigma}{\mu\omega}}(\vec{e}_n \times \vec{E})e^{-k_i\xi}e^{i(k\xi-\omega t+\pi/4)} \qquad (23.24)$$

where we have used $k + ik_i = k(1+i) = \sqrt{\omega\mu\sigma}e^{i\pi/4}$. The time-averaged power transported across the unit area inside the conductor is given by Eq.(20.18), which reads

$$\langle \vec{S} \rangle = \frac{1}{2}\text{Re}\left[\vec{E}(\vec{r}) \times \vec{H}^*(\vec{r})\right] = \vec{e}_n \frac{1}{2}\sqrt{\frac{\sigma}{\mu\omega}}|\vec{E}|^2 e^{-2k_i\xi}\text{Re}\left(e^{-i\pi/4}\right)$$

and gives the plane wave intensity (20.25), as a function of the penetration ξ into the conductor, in the form

$$I(\xi) = \frac{1}{2\mu\omega\delta}|\vec{E}|^2 e^{-2\xi/\delta} \qquad (23.25)$$

The complex representation (23.16) of the wave number, and therefore, the complex refractive index, are associated with the fact that waves are attenuated in conducting media. However, only a small part of the incident wave intensity on the surface of a conducting medium undergoes a true absorption (i.e., conversion into Joule heat). A large fraction is propagated back into the external medium as a high intensity reflected wave. Consider the reflection of a plane wave which falls normally on the surface of a conducting medium, lying in the *xz*-plane. The configuration illustrated in Figure 23.2 is a special case of that given before in Figure 21.3, where $\theta = \varphi = 0$, thus the distinction between the TM and TE polarizations is lost.

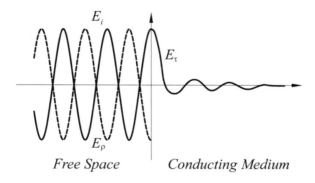

Figure 23.2. Configuration of normal incidence on the surface of a conducting medium

In view of Eqs.(23.17), the amplitude matching conditions (21.23) read

$$E_i + E_\rho = E_\tau$$

$$\frac{k_1}{\mu_1 \omega}(E_i - E_\rho) = \frac{k_2}{\mu_2 \omega} E_\tau$$

(23.26)

where, in vacuum, $k_1 = \omega/c$ and $\mu_1 = \mu_0$, whereas in the conducting medium we have $k_2 = k_c = \sqrt{\mu_r}\,\frac{\omega}{c}(n + in_i)$, as given by Eq.(23.16), and $\mu_2 = \mu_0 \mu_r$, so that

$$E_\rho = E_i \frac{n - \sqrt{\mu_r} + in_i}{n + \sqrt{\mu_r} + in_i} = \rho E_i$$

$$E_\tau = E_i \frac{2\sqrt{\mu_r}}{n + \sqrt{\mu_r} + in_i} = \tau E_i$$

(23.27)

The reflectivity (21.26) is obtained as

$$R = |\rho|^2 = \frac{\left(n - \sqrt{\mu_r}\right)^2 + n_i^2}{\left(n + \sqrt{\mu_r}\right)^2 + n_i^2} \tag{23.28}$$

and, in the visible and ultraviolet regions, becomes

$$R = \frac{(n-1)^2 + n_i^2}{(n+1)^2 + n_i^2} \tag{23.29}$$

since at these frequencies the relative permeability is practically equal to unity. Equation (23.29) reduces to the simple form of reflectivity in transparent media (21.39), by taking $n_i = 0$. For a good conductor, Eqs.(23.9) give

$$n = n_i = \sqrt{\frac{\sigma}{2\varepsilon_0 \omega}}$$

so that Eq.(23.29) can be approximated by

$$R = 1 - \frac{2}{n} = 1 - 2\sqrt{\frac{2\varepsilon_0 \omega}{\sigma}} \tag{23.30}$$

If a fraction a of the wave energy is absorbed in a medium, the energy conservation law (17.53) should be written in the form

$$R + T + a = 1$$

and, because T is negligible for good conductors, Eq.(23.30) provides the fraction of the absorbed energy as

$$a = 2\sqrt{\frac{2\varepsilon_0 \omega}{\sigma}} \tag{23.31}$$

The amount of energy is much smaller than R and decreases with increasing conductivity, while R increases with increasing σ. It follows that the absorption coefficient k_i or the extinction index n_i are measures of the damping of waves in conducting media as a result of both the skin effect and reflection.

23.2. ORIGIN OF COMPLEX CONSTITUTIVE PARAMETERS

The conditions that lead to complex constitutive parameters cannot be explained in purely phenomenological terms and require a model for the interaction of electromagnetic waves with matter. It is sufficient to assume that the medium consists of a large number of electrons moving in the harmonic field of the incident wave. If the electrons are bound to fixed sites by quasi-elastic forces, and therefore, have a resonance frequency of natural oscillation ω_0, their motion gives rise to a polarization current density, which is of importance for poor conductors and dielectrics. Since free electrons, of importance for good conductors, can be regarded to be oscillators with resonance frequencies equal to zero, the free electron model will be derived as a special case of the *Lorentz model for bound electrons*.

The equation of motion for a bound electron in the harmonic field of amplitude $\vec{E}_0$ of a linearly polarized wave, involving a restoring force and a phenomenological representation of the electron momentum loss as a result of collisions, has the form

$$m_e \ddot{\vec{r}} + 2m_e \beta \dot{\vec{r}} + k_0 \vec{r} = -e\vec{E}_0 e^{-i\omega t} \tag{23.32}$$

where $k_0 = m_e \omega_0^2$ is the elastic binding force constant defined in terms of the natural frequency ω_0 of the bound electron, and 2β stands for the damping constant. Equation (23.32) represents the three-dimensional forced damped harmonic motion and can be written as

$$\ddot{\vec{r}} + 2\beta \dot{\vec{r}} + \omega_0^2 \vec{r} = -\frac{e}{m_e} \vec{E}_0 e^{-i\omega t} \tag{23.33}$$

It is satisfied by substituting $\vec{r} = \vec{r}_0 e^{-i\omega t}$, which gives

$$\vec{r} = \frac{-e/m_e}{\omega_0^2 - \omega^2 - 2i\beta\omega} \vec{E}$$

This solution describes the *steady-state* oscillation of electrons at the forcing frequency of the harmonic field, as outlined in Chapter 2. If N is the number of bound electrons per unit volume, each of them displaced a distance $\vec{r}$ from their equilibrium position, it is convenient to define the *plasma frequency* ω_p as

$$\omega_p = \sqrt{\frac{Ne^2}{m_e \varepsilon_0}} \tag{23.34}$$

and to introduce the macroscopic polarization of the medium $\vec{P} = -Ne\vec{r}$ in the form

$$\vec{P} = \frac{Ne^2/m_e}{\omega_0^2 - \omega^2 - 2i\beta\omega} \vec{E} = \frac{\varepsilon_0 \omega_p^2}{\omega_0^2 - \omega^2 - 2i\beta\omega} \vec{E} \tag{23.35}$$

If compared with Eq.(11.25), Eq.(23.35) defines a complex electric susceptibility

$$\chi_e = \frac{\omega_p^2}{\omega_0^2 - \omega^2 - 2i\beta\omega} \tag{23.36}$$

which describes a lossy dielectric, even if the conduction charges are absent $(\sigma = 0)$. The complex permittivity can be obtained by substituting Eq.(23.36) into Eq.(23.6) which, in this case, takes the form

$$\varepsilon_c = \varepsilon_0 \left(1 + \frac{\omega_p^2}{\omega_0^2 - \omega^2 - 2i\beta\omega} \right) \tag{23.37}$$

In view of Eq.(23.7), if ω_p is small enough for the absolute value of the complex number on the right-hand side to be small when compared with unity for all frequencies, we may approximate

$$n + in_i = \left(1 + \frac{\omega_p^2}{\omega_0^2 - \omega^2 - 2i\beta\omega} \right)^{1/2} \cong 1 + \frac{1}{2} \frac{\omega_p^2}{\omega_0^2 - \omega^2 - 2i\beta\omega} \tag{23.38}$$

and this gives the optical constants n and n_1 as

$$n \cong 1 + \frac{1}{2} \frac{\omega_p^2 (\omega_0^2 - \omega^2)}{(\omega_0^2 - \omega^2)^2 + 4\beta^2 \omega^2} \quad , \quad n_i \cong \frac{\omega_p^2 \beta \omega}{(\omega_0^2 - \omega^2)^2 + 4\beta^2 \omega^2} \tag{23.39}$$

In the so-called transparent region, which occurs for frequencies well below the natural frequency ω_0, where $(\omega_0^2 - \omega^2)^2 \gg 4\beta^2 \omega^2$, Eqs.(23.39) show that $n_i \cong 0$, and we obtain the *dispersion relation* which gives the frequency dependence of n as

$$n \cong 1 + \frac{1}{2} \frac{\omega_p^2}{\omega_0^2 - \omega^2} \tag{23.40}$$

Equation (23.40) indicates a refractive index which is greater than unity and increases with increasing frequency, as $dn/d\omega > 0$. Such a behaviour, characteristic for most of the ionic and molecular crystals in the visible region of the spectrum, is called *normal dispersion*.

In the vicinity of the resonance frequency ω_0 we may set $\omega \cong \omega_0$ so that $\omega_0^2 - \omega^2 \cong 2\omega_0(\omega_0 - \omega)$ and Eqs.(23.39) reduce to

$$n \cong 1 + \frac{\omega_p^2}{4\omega_0} \frac{(\omega_0 - \omega)}{(\omega_0 - \omega)^2 + \beta^2} \quad , \quad n_i \cong \frac{\omega_p^2}{4\omega_0} \frac{\beta}{(\omega_0 - \omega)^2 + \beta^2} \tag{23.41}$$

Both n and n_i depend on frequency, as plotted in Figure 23.3. The frequency range $\omega = \omega_0 \pm \beta$ over which n_i is significantly different from zero is called the *absorption region*. In this region n reaches a maximum for $\omega = \omega_0 - \beta$ and then falls to a minimum at $\omega = \omega_0 + \beta$. This behaviour is referred to as *anomalous dispersion*, with $dn/d\omega < 0$.

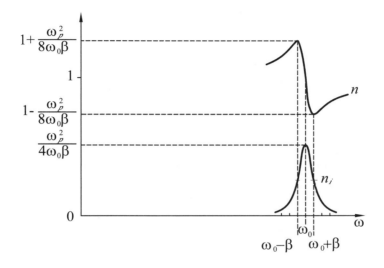

Figure 23.3. Dependence of n and n_i on frequency in the vicinity of the resonance frequency ω_0

In the condensed state we must consider the steady-state motion of bound electrons in a local field $\vec{E}_{\mathit{eff}}$ which is different from the macroscopic field, due to the contribution of polarization charges, as shown in Problem 11.6. Instead of Eq.(23.35), the appropriate expression is

$$\vec{P} = \frac{\varepsilon_0 \omega_p^2}{\left(k_0/m_e - \omega^2\right) - 2i\beta\omega}\left(\vec{E} + \frac{\vec{P}}{3\varepsilon_0}\right) \tag{23.42}$$

which can be solved for $\vec{P}$ to give

$$\vec{P} = \frac{\varepsilon_0 \omega_p^2}{\left(k_0/m_e - \omega_p^2/3\right) - \omega^2 - 2i\beta\omega}\vec{E} \tag{23.43}$$

This result can be regarded to be identical with Eq.(23.35), providing the resonance frequency in the condensed state is taken as

$$\omega_0 = \sqrt{\frac{k_0}{m_e} - \frac{\omega_p^2}{3}} \tag{23.44}$$

Thus, the resonance frequency of media consisting of isolated atoms or molecules, such as gases, shifts downward when the molecules are assembled into a condensed (liquid or solid) structure.

If there are several kinds of elastically bound electrons, we may think of certain fractions f_j with resonance frequencies ω_j and the dispersion relation (23.38) becomes

$$n_c \cong 1 + \frac{\omega_p^2}{2} \sum_j \frac{f_j}{\omega_j^2 - \omega^2 - 2i\beta\omega} \qquad (23.45)$$

In the transparent region Eq.(23.45) reduces to

$$n \cong 1 + \frac{\omega_p^2}{2} \sum_j \frac{f_j}{\omega_j^2 - \omega^2} \qquad (23.46)$$

which, when given in terms of wavelength instead of frequency, is known as *Sellmeyer's equation* and is commonly used for experimental fitting of the measured values of n. At very low frequencies, Eq.(23.46) takes the form of *Maxwell's relation* for the static refractive index of transparent media, given by

$$n = 1 + \frac{\omega_p^2}{2} \sum_j \frac{f_j}{\omega_j^2} \qquad (23.47)$$

The approach used above can be applied to the classical *free-electron model*, which is appropriate for good conductors, by setting $\omega_0 = 0$, that is by neglecting the binding force. However, the constant 2β now has a different meaning from that in Eq.(23.32) and represents the damping due to the resistance of the conductor. It is common practice to introduce the relaxation time $\tau = 1/2\beta$, which is a measure of the mean time between collisions of the conduction electrons with the ionic lattice of the conductor. By taking $\omega_0 = 0$, the permittivity (23.37) is obtained in terms of τ as

$$\varepsilon_c = \varepsilon_0 \left(1 - \frac{\omega_p^2}{\omega^2 + i\omega/\tau}\right) \qquad (23.48)$$

Since, in a free-electron gas there is no polarization charge, Eq.(23.48) must assume the form (23.6) with $\chi_e = 0$, so that a frequency-dependent conductivity σ is obtained as

$$\sigma = \frac{\omega_p^2 \varepsilon_0 \tau}{1 - i\omega\tau} = \frac{\sigma_0}{1 - i\omega\tau} \qquad (23.49)$$

where σ_0 is the low frequency limit, called the *static conductivity*, given by

$$\sigma_0 = \frac{Ne^2}{m_e} \tau \qquad (23.50)$$

In view of Eqs.(23.8), when separated into real and imaginary parts, Eq.(23.48) gives

$$\varepsilon_r = n^2 - n_i^2 = 1 - \frac{\omega_p^2}{\omega^2 + 1/\tau^2} \quad , \quad \varepsilon_i = 2nn_i = \frac{\omega_p^2}{\omega^2 + 1/\tau^2}\left(\frac{1}{\omega\tau}\right) \qquad (23.51)$$

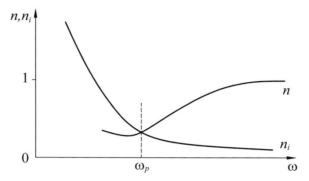

Figure 23.4. Dependence of n and n_i on frequency for a good conductor

A plot of n and n_i as functions of frequency is given in Figure 23.4. It shows that the extinction index n_i assumes large values at low frequencies, where the refractive index is less than unity. Neglecting n in Eq.(23.29) shows that there is a region of high reflectivity for $\omega < \omega_p$. Plasma frequencies for metals correspond to the visible region of the spectrum.

At high frequencies, good conductors are transparent, since n_i becomes negligible for $\omega > \omega_p$. The onset of transparency is defined, using Eq.(23.51), by the condition that ε_r must be zero, and hence

$$\omega^2 = \omega_p^2 - 1/\tau^2 \qquad (23.52)$$

EXAMPLE 23.2. Optical properties of a plasma

A plasma may be assumed to be a collisionless electron gas. This assumption is equivalent to that of a perfect metal, where the relaxation time τ is infinite. Eq.(23.48) then gives

$$\varepsilon = \varepsilon_0\left(1 - \frac{\omega_p^2}{\omega^2}\right) \qquad (23.53)$$

It follows from Eq.(23.7) that the complex refractive index is purely imaginary $n_c = in_i$ for $\omega < \omega_p$, where

$$n_i = \sqrt{\frac{\omega_p^2}{\omega^2} - 1}$$

so that, in this frequency region, a plasma cannot propagate electromagnetic waves. The equation

(23.29) shows that $R=1$, and the waves are completely reflected at the free space/plasma interface. There is a skin depth, given by Eq.(23.23) as

$$\delta = \frac{1}{k_i} = \frac{c}{\omega n_i} = \frac{c}{\sqrt{\omega_p^2 - \omega^2}}$$

where for a plasma it is convenient to set $\mu_r = 1$.

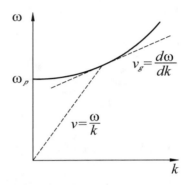

Figure 23.5. Dispersion curve for a plasma

For $\omega > \omega_p$ it follows that $\varepsilon > 0$, and one may substitute Eq.(23.53) into Eq.(23.14), which reads

$$k^2 = \varepsilon \mu \omega^2 = \varepsilon_0 \mu_0 \omega^2 \left(1 - \frac{\omega_p^2}{\omega^2}\right)$$

This constitutes the dispersion relation, usually written as

$$\omega^2 = \omega_p^2 + k^2 c^2 \tag{23.54}$$

and plotted in Figure 23.5. The phase velocity

$$v = \frac{\omega}{k} = \frac{c\omega}{\sqrt{\omega^2 - \omega_p^2}} \tag{23.55}$$

is always greater than c and depends on frequency, so that a plasma is dispersive. The group velocity for the propagation of a wave packet is

$$v_g = \frac{d\omega}{dk} = \frac{c^2 k}{\omega} = c\sqrt{1 - \frac{\omega_p^2}{\omega^2}} < c \tag{23.56}$$

FURTHER READING

1. N. Ida - ENGINEERING ELECTROMAGNETICS, Springer-Verlag, New York, 2004.
2. W. H. Hayt, J. A. Buck - ENGINEERING ELECTROMAGNETICS, McGraw-Hill, New York, 2001.
3. K. R. Demarest - ENGINEERING ELECTROMAGNETICS, Pearson Education, Boston, 1997.

PROBLEMS

23.1. Find an expression for the complex impedance Z_c of a conducting medium of properties μ, ε and σ to electromagnetic waves.

Solution

If we set $\xi \equiv z$ in the configuration of Figure 23.1, the fields inside the conducting medium are given by Eqs.(23.24) as

$$E_x = E_0 e^{-k_i z} e^{i(kz-\omega t)} \quad , \quad H_y = \sqrt{\frac{\sigma}{\mu\omega}} E_0 e^{-k_i z} e^{i(kz-\omega t + \pi/4)}$$

so that the impedance reads

$$Z_c = \frac{E_x}{H_y} = \sqrt{\frac{\mu\omega}{\sigma}} e^{i\pi/4} = (1+i)\sqrt{\frac{\mu\omega}{2\sigma}}$$

Thus, H_y lags E_x by $\pi/4$, and the magnitude of the impedance is

$$Z = \left|\frac{E_x}{H_y}\right| = \sqrt{\frac{\mu\omega}{\sigma}}$$

This also may be expressed in terms of the characteristic impedance in free space Z_0 in the form

$$Z = \sqrt{\frac{\mu}{\varepsilon}\frac{\varepsilon\omega}{\sigma}} = Z_0\sqrt{\frac{\mu_r}{\varepsilon_r}}\sqrt{\frac{\varepsilon\omega}{\sigma}}$$

23.2. Show that the refractive index and the reflectivity of a conducting medium of properties μ, ε and σ can be interpreted in terms of absorption because of a complex impedance Z_c, and find expressions for n_c and R in a good conductor.

Solution

At the air/conductor interface, where Z_0 is the impedance in free space, given by Eq.(17.23), and $Z_c = (1+i)X$ where $X = \sqrt{\mu\omega/2\sigma}$, there is a complex refractive index of the form

$$n_c = \frac{Z_0}{Z_c} = \frac{Z_0}{(1+i)X} = \frac{1}{2}(1-i)\frac{Z_0}{X} = (1-i)\sqrt{\frac{\sigma}{2\mu_r\varepsilon_0\omega}}$$

as expected from Eq.(23.16). Since the amplitude coefficient of reflection (17.51) has the complex form

$$\rho = \frac{Z_0 - Z_c}{Z_0 + Z_c} = \frac{(Z_0 - X) - iX}{(Z_0 + X) + iX}$$

the reflectivity (17.52) should be written in the real form

$$R = \frac{|Z_0 - Z_c|^2}{|Z_0 + Z_c|^2} = \frac{(Z_0/X - 1)^2 + 1}{(Z_0/X + 1)^2 + 1} = 1 - \frac{4Z_0/X}{(Z_0/X)^2 + 2Z_0/X + 2}$$

For a good conductor $\sigma/\mu_r\varepsilon_0\omega \gg 1$, and hence, $Z_0/X \gg 1$, so that the reflectivity reduces to

$$R = 1 - 4\frac{X}{Z_0} = 1 - 2\sqrt{\frac{2\mu_r\varepsilon_0\omega}{\sigma}}$$

which is the same result as that given by Eq.(23.30) for the visible and ultraviolet regions, assuming that $\mu_r \approx 1$.

23.3. The refractive index of a crystal in the transparent region is given by

$$n = a + \frac{b}{\lambda^2}$$

where $a = 2.971$, $b = 0.035$ and λ is given in microns. Use the Sellmeyer coefficients a and b to determine the resonance wavelength λ_0 and then the ratio of the phase to the group velocities of light in such a crystal.

Solution

Sellmeyer's equation can be derived from Eq.(23.46) as

$$n \approx 1 + \frac{\omega_p^2}{2(\omega_0^2 - \omega^2)} \approx 1 + \frac{\omega_p^2}{2\omega_0^2}\left(1 + \frac{\omega^2}{\omega_0^2}\right) = 1 + \frac{\omega_p^2\lambda_0^2}{8\pi^2 c^2}\left(1 + \frac{\lambda_0^2}{\lambda^2}\right)$$

It follows that $a = 1 + \omega_p^2\lambda_0^2/8\pi^2 c^2$ and $b = (a-1)\lambda_0^2$, thus $\lambda_0 = \sqrt{b/(a-1)} = 0.133$ microns. The phase velocity is $v = c/n$ and the group velocity (16.71) can be expressed in terms of n in the form

$$\frac{1}{v_g} \approx \frac{1}{v}\left(1+\frac{\lambda}{v}\frac{dv}{d\lambda}\right) = \frac{1}{v}+\frac{\lambda}{v^2}\frac{dv}{d\lambda} = \frac{1}{v}-\frac{\lambda}{c}\frac{dn}{d\lambda}$$

where

$$\frac{dn}{d\lambda} = \frac{d}{d\lambda}\left(\frac{c}{v}\right) = -\frac{c}{v^2}\frac{dv}{d\lambda} \quad \text{or} \quad \frac{dv}{d\lambda} = -\frac{v^2}{c}\frac{dn}{d\lambda}$$

Substituting $dn/d\lambda$ one obtains the required ratio as

$$\frac{v}{v_g} = 1 - \frac{\lambda}{n}\frac{dn}{d\lambda} = 1 + \frac{2b/\lambda^2}{a+b/\lambda^2} = 1 + \frac{2}{1+1.5\lambda^2/\lambda_0^2}$$

23.4. A linearly polarized electromagnetic wave of frequency $\omega = 2\omega_p$ is propagating through a plasma layer of frequency ω_p in the ionosphere, along the direction $\vec{B}_0$ of Earth's magnetic field. Show that the specific rotatory power of the plane of polarization of the wave is proportional to the magnetic induction.

Solution

The equation of motion (13.30) for an electron of velocity $\vec{v}$ reads

$$m_e \frac{d\vec{v}}{dt} = -e\left(\vec{E} + \vec{v} \times \vec{B}_0\right)$$

where the electric field can be written as a superposition of right and left circularly polarized components (see Problem 22.2) in the form

$$\vec{E} = E_0\left(\vec{e}_x \pm i\vec{e}_y\right)e^{-i\omega t}$$

and $\vec{B}_0 = B_0 \vec{e}_z$. It is convenient to set

$$\vec{v} = v_0\left(\vec{e}_x \pm i\vec{e}_y\right)e^{-i\omega t} \quad \text{and} \quad \vec{j} = -Nev_0\left(\vec{e}_x \pm i\vec{e}_y\right)e^{-i\omega t}$$

where $\vec{j} = -Ne\vec{v}$ is the current density in the plasma, and this gives

$$\vec{v} \times \vec{B}_0 = \begin{vmatrix} \vec{e}_x & \vec{e}_y & \vec{e}_z \\ v_0 e^{-i\omega t} & \pm i v_0 e^{-i\omega t} & 0 \\ 0 & 0 & B_0 \end{vmatrix} = \pm i\left(\vec{e}_x \pm i\vec{e}_y\right) v_0 B_0$$

Substituting into the equation of motion yields

$$v_0 = \frac{-ieE_0}{m_e(\omega \pm \omega_c)} \quad \text{and} \quad \vec{j} = \frac{iNe^2}{m_e(\omega \pm \omega_c)} \vec{E}$$

where $\omega_c = eB_0/m_e$ is the cyclotron frequency. The current density due to the electron motion will modify Ampère's law (14.1) which becomes

$$\nabla \times \vec{H} = \frac{\partial}{\partial t}(\varepsilon_0 \vec{E}) + \vec{j} = -i\omega\varepsilon_0 \vec{E} + \frac{iNe^2}{m_e(\omega \pm \omega_c)}\vec{E}$$

and can be written as

$$\nabla \times \vec{H} = -i\omega\varepsilon_0\left[1 - \frac{\omega_p^2}{\omega(\omega \pm \omega_c)}\right]\vec{E} = -i\omega\varepsilon\vec{E}$$

giving

$$\varepsilon_r = 1 - \frac{\omega_p^2}{\omega(\omega \pm \omega_c)} \quad \text{or} \quad n_\pm \approx 1 - \frac{\omega_p^2}{2\omega(\omega \pm \omega_c)}$$

There are two different values $n_\pm$ for the right- and left-circularly polarized waves, which propagate at different velocities. The specific rotatory power, as derived in Problem 22.3, becomes

$$\delta(\omega) = \frac{\omega}{2c}(n_+ - n_-) = \frac{\omega_c \omega_p^2}{2c(\omega^2 - \omega_c^2)} = \frac{e\omega_p^2}{2m_e c(\omega^2 - \omega_c^2)} B_0$$

as required.

23.5. The electron number density N in an ionospheric layer is found to increase with height until it reaches the maximum. Assuming a constant gradient $\delta = dN/dz$, find the curvature of the ray path for an electromagnetic wave striking the layer surface at oblique incidence θ_0.

Solution

In view of Eq.(23.53), the relative permittivity, and hence, the refractive index, is a function of z only, given by

$$\varepsilon_r = 1 - \frac{Ne^2}{m_e \varepsilon_0 \omega^2} \quad \text{or} \quad \frac{d\varepsilon_r}{dz} = -\frac{e^2}{m_e \varepsilon_0 \omega^2}\frac{dN}{dz} = -\frac{\omega_p^2}{N\omega^2}\delta$$

It follows from Eq.(16.46) that the ray path will be confined to a vertical plane where

$$n(z)\sin\theta = n_0 \sin\theta_0$$

The curvature of the ray path is given by

$$\frac{1}{R} = \frac{d\theta}{ds} = -\frac{\sin\theta}{n}\frac{dn}{dz} = -\frac{n_0 \sin\theta_0}{n^2}\frac{dn}{dz} \approx -\frac{\sin\theta_0}{2\varepsilon_r}\frac{d\varepsilon_r}{dz}$$

Substituting $d\varepsilon_r / dz$ immediately yields

$$\frac{1}{R} = \frac{\omega_p^2 \sin\theta_0}{2N(\omega^0 - \omega_p^2)}\delta$$

23.6. Show that in a conductor of properties σ, ε the ratio of the magnetic to electric energy density in an electromagnetic wave of frequency ω is the same as that of the conduction to displacement current and find its expression.

23.7. A perfect conductor develops a charge and current distribution on the surface so that the external fields do not penetrate it. Find the charge density σ_f and current density j_f at an air/conductor flat interface due to the external reflection at oblique incidence θ of a TM polarized electromagnetic wave of electric field amplitude E_0.

23.8. The Sellmeyer formula for a monovalent ionic crystal is given by

$$n^2 = A + \frac{B}{\lambda^2 - \lambda_1^2} + \frac{C}{\lambda^2 - \lambda_2^2}$$

where $A = 1.23$, $B = 0.00324$ and $C = 9.03$, the ultraviolet resonance wavelength $\lambda_1 = 0.134$ microns is due to the bound electrons and the infrared wavelength $\lambda_2 = 13.4$ microns is assigned to an ionic resonance. If the ratio of the proton to electron masses is $m_p / m_e \approx 1840$, estimate the reduced ionic mass μ in such a crystal.

23.9. For an optical prism with apex angle A and refractive index $n(\lambda)$, the dispersion $dn/d\lambda$ of the glass used in the prism can be evaluated by measuring the angular dispersion $d\delta(\lambda)/d\lambda$ of the deviation $\delta(\lambda)$ of an incident ray, at the position of minimum deviation, where the following relation holds

$$n(\lambda)\sin\frac{A}{2} = \sin\frac{A + \delta(\lambda)}{2}$$

Find the relation between the angular dispersion $d\delta(\lambda)/d\lambda$ and the glass dispersion $dn/d\lambda$.

23.10. Find an expression for the dispersion relation $n^2 = n^2(\omega)$ of a medium where the phase and group velocities of an electromagnetic wave are related by $vv_g = c^2$.

24. GENERALIZED MECHANICS

24.1. D'ALEMBERT'S PRINCIPLE
24.2. HAMILTON'S PRINCIPLE
24.3. LAGRANGE'S EQUATIONS
24.4. THE CANONICAL EQUATIONS
24.5. THE POISSON BRACKETS
24.6. THE HAMILTON-JACOBI EQUATION

24.1. D'ALEMBERT'S PRINCIPLE

The generalized formulation of Newtonian mechanics in an arbitrary coordinate system, which is concerned with the description of many-particle systems, provides the transition to the statistical interpretation of physical phenomena.

The motion of a system of N particles, of masses m_i, is described in a three-dimensional Cartesian frame by a set of $3N$ differential equations (2.15). It is usual to assume that the positions $\vec{r}_i$ of the particles are not independent of one another, but are interconnected through a certain number of independent equations

$$G_k(\vec{r}_1,\ldots,\vec{r}_i,\ldots,\vec{r}_n,t) = 0 \qquad k = 1,\ldots,l \qquad (24.1)$$

which define the *constraints* imposed upon the system.

EXAMPLE 24.1. Constraints

For the motion on a given surface the equation of this surface will define a constraint. For instance, along the surface of a sphere of constant radius R, the constraining equations are

$$|\vec{r}_i| = a \quad \text{or} \quad (x_i - x_0)^2 + (y_i - y_0)^2 + (z_i - z_0)^2 - a^2 = 0$$

where $\vec{r}_i$ are the position vectors, and x_0, y_0, z_0 are the coordinates of the centre of the sphere. In a rigid body the distances between the N component particles are all constant, and the constraints are given by the equations

$$r_{ij} = a_{ij} \quad \text{or} \quad (\vec{r}_i - \vec{r}_j)^2 - a_{ij}^2 = 0$$

Such constraints, which are independent of time, are said to be *scleronomous*. If the given surface or body is moving, the time will enter the previous equations as an explicit parameter, and the constraints become *rheonomous*.

The independent coordinates x_i, y_i, z_i of a system of N particles, free from constraints, represent $3N$ *degrees of freedom* of their motion. The l independent equations (24.1) enable us to express l of these coordinates in term of the remaining $3N - l$ such that the system is left with a reduced number of $f = 3N - l$ degrees of freedom. However the constraints necessarily imply unknown reaction forces or *forces of constraint* $\vec{R}_i$ which are only specified in terms of their effect on the motion. Their components must be considered as new degrees of freedom in place of the eliminated coordinates, so that the set of equations of motion (2.15) must be written in the form

$$m_i \ddot{\vec{r}}_i = \vec{F}_i + \vec{R}_i \qquad (24.2)$$

where the vector sum $\vec{R}_i$ of forces of constraint experienced by the ith particle is

$$\vec{R}_i = \sum_k \vec{R}_{ik}$$

In Eqs.(24.2) we know the applied forces $\vec{F}_i$ only, given in terms of $\vec{F}_{ei}$ and $\vec{F}_{ij}$ by Eq.(2.15), and it is appropriate to reformulate these equations so that the unknown reaction forces $\vec{R}_i$ disappear. The motion of particles with given positions $\vec{r}_i$ consistent with the imposed constraints, at a given instant t, must obey the equations obtained by differentiation with respect to time of the conditions of constraint (24.1), which gives

$$\sum_i \frac{\partial G_k}{\partial \vec{r}_i} \cdot \dot{\vec{r}}_i + \frac{\partial G_k}{\partial t} = 0 \quad \text{or} \quad \sum_i (\nabla_i G_k) \cdot \dot{\vec{r}}_i + \frac{\partial G_k}{\partial t} = 0, \qquad k = 1, \ldots, l$$

where $\dot{\vec{r}}_i$ are the velocities of the various particles. The actual displacements in a time interval dt are $d\vec{r}_i = \dot{\vec{r}}_i dt$, and we have

$$\sum_i (\nabla_i G_k) \cdot d\vec{r}_i + \frac{\partial G_k}{\partial t} dt = 0$$

During the time interval dt the constraints might change, so that we need to consider what that equation would reduce to if the explicit time dependence is eliminated. This can be achieved by equating the difference between possible displacements, in a given time interval, with a *virtual displacement* $\delta \vec{r}_i$ defined as

$$\delta \vec{r}_i = d\vec{r}_i - d'\vec{r}_i$$

where $d\vec{r}_i = \dot{\vec{r}}_i dt$ and $d'\vec{r}_i = \dot{\vec{r}}'_i dt$. It follows that the displacements $\delta \vec{r}_i$ are consistent with the conditions of constraint, as given by the time independent equation

$$\sum_i (\nabla_i G_k) \cdot \delta \vec{r}_i = 0 \quad , \quad k = 1, \ldots, l$$

The virtual displacements $\delta \vec{r}_i$ express all possible motions of the system allowed under given constraints, having a geometrical rather than a physical significance.

EXAMPLE 24.2. Virtual displacements

A virtual or geometric differential $\delta \vec{r}$ must not be confused with a displacement $d\vec{r}$. The difference between the two can be illustrated as follows. For particles constrained to stay on a moving surface, the displacement $d\vec{r}$ gives the actual direction of their motion at a given instant, while the virtual displacement $\delta \vec{r}$ gives a direction which is only allowed to be tangential to the moving surface, at the same instant. It is obvious that the direction of motion of the particles will be not tangential, in general, to the moving surface, since it must have a component along the direction of motion of this surface.

The reaction force $\vec{R}$ that a frictionless surface exerts on a particle is always in the direction normal to the constraining surface and defined by a condition of constraint similar to Eq.(24.1). Since the vector ∇G is always normal to the surface and two parallel vectors can differ only in magnitude, we can express the constraint of a surface by a reaction force $\vec{R} = \alpha \nabla G$. For a particle confined to a motion along a certain path there are two constraining conditions and the corresponding reaction forces are expressible as $\vec{R}_1 = \alpha_1 \nabla G_1$, $\vec{R}_2 = \alpha_2 \nabla G_2$. It follows that we can use the conditions (24.1) to define the forces of constraint as

$$\vec{R}_{ik} = \alpha_k (\nabla_i G_k) \quad \text{or} \quad \vec{R}_i = \sum_k \alpha_k (\nabla_i G_k)$$

If we define by $\vec{R}_i \delta \vec{r}_i$ the *virtual work* of the reaction force $\vec{R}_i$ in a virtual displacement, one obtains

$$\sum_i \vec{R}_i \cdot \delta \vec{r}_i = \sum_i \sum_k \alpha_k (\nabla_i G_k) \cdot \delta \vec{r}_i = \sum_k \alpha_k \sum_i (\nabla_i G_k) \cdot \delta \vec{r}_i = 0 \qquad (24.3)$$

that is *the virtual work of the forces of constraint is zero*, a result which is only true if friction forces are excluded. It is noteworthy that since friction is a macroscopic phenomenon, this condition will not restrict the applicability of our results to systems of N microscopic particles. Inserting Eq.(24.3) into Eq.(24.2) we obtain

$$\sum_i \left(m_i \ddot{\vec{r}}_i - \vec{F}_i \right) \cdot \delta \vec{r}_i = 0 \tag{24.4}$$

called *d'Alembert's principle*, where the reaction forces no longer appear.

24.2. HAMILTON'S PRINCIPLE

D'Alembert's principle, which is concerned with the instantaneous state of a many particle system and with small virtual displacements about this state, is a differential principle. We can obtain an equivalent integral principle by considering the change of state of the system during its motion between times t_1 and t_2.

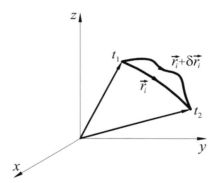

Figure 24.1. Actual and virtual trajectories in Cartesian coordinates

The instantaneous state, which is described by the values of the coordinates $\vec{r}_i(t)$ of the various particles, can be represented by a point in a $3N$-dimensional space. The entire motion of the system is then described by a curve in this space, illustrated for convenience in three dimensions in Figure 24.1. A small virtual displacement $\delta \vec{r}_i(t)$ from the instantaneous state $\vec{r}_i(t)$ will result in two different paths which only coincide at t_1 and t_2, as represented in Figure 24.1, one described by the coordinates $\vec{r}_i(t)$ of the actual motion and the other described by $\vec{r}_i(t) + \delta \vec{r}_i(t)$ for a virtual motion of the system. We have

$$\delta \dot{\vec{r}}_i = \frac{d}{dt}\left(\vec{r}_i + \delta \vec{r}_i \right) - \frac{d\vec{r}_i}{dt} = \frac{d}{dt} \delta \vec{r}_i \tag{24.5}$$

Integrating Eq.(24.4) along the two paths, and then subtracting, we obtain

$$\int_{t_1}^{t_2}\sum_i\left(m_i\ddot{\vec{r}}_i - \vec{F}_i\right)\cdot\delta\vec{r}_i\,dt = 0 \tag{24.6}$$

Integrating by parts, the integral of the first sum becomes

$$\int_{t_1}^{t_2}\sum_i m_i\ddot{\vec{r}}_i\cdot\delta\vec{r}_i\,dt = \sum_i m_i\dot{\vec{r}}_i\cdot\delta\vec{r}_i\Big|_{t_1}^{t_2} - \sum_i\int_{t_1}^{t_2} m_i\dot{\vec{r}}_i\cdot\left(\frac{d}{dt}\delta\vec{r}_i\right)dt = -\int_{t_1}^{t_2}\sum_i m_i\dot{\vec{r}}_i\cdot\delta\dot{\vec{r}}_i\,dt = -\int_{t_1}^{t_2}\delta T\,dt \tag{24.7}$$

where the virtual displacements vanish, that is $\delta\vec{r}_i = 0$ at $t = t_1$ and $t = t_2$, and the kinetic energy (2.29) of the system, written in the form

$$T = \frac{1}{2}\sum_i m_i\dot{\vec{r}}_i^{\,2} \tag{24.8}$$

can be recognized. Therefore, Eq.(24.6) reads

$$0 = -\int_{t_1}^{t_2}\left[\delta T + \sum_i \vec{F}_i\cdot\delta\vec{r}_i\right]dt \tag{24.9}$$

For a conservative force field, as defined in Eq.(2.26), the virtual work of the applied forces $\vec{F}_i$ takes the form

$$\sum_i \vec{F}_i\cdot\delta\vec{r}_i = -\sum_i(\nabla_i U)\cdot\delta\vec{r}_i = -\delta U$$

which defines the *variation* of the potential energy U. Therefore Eq.(24.9) becomes

$$\int_{t_1}^{t_2}(\delta T - \delta U)dt = \delta\int_{t_1}^{t_2}(T - U)dt = \delta\int_{t_1}^{t_2} L\,dt = 0 \tag{24.10}$$

where the function L, called the *Lagrangian*, is defined as

$$L = T - U \tag{24.11}$$

Equation (24.10) is an expression of **Hamilton's principle** for conservative systems: *the path of motion of the system from time t_1 to time t_2 is such that the line integral of the Lagrangian is an extremum.* Hamilton's integral formulation of the law of motion defines a path, which must obviously obey both d'Alembert's principle (24.4), from which Eq.(24.10) was derived and the set of equations of motion (24.2). However, Hamilton's principle involves only the kinetic and potential energies, which can be defined independently of the choice of coordinates, so that the formulation (24.10) is true in an arbitrary coordinate system.

24.3. LAGRANGE'S EQUATIONS

A symmetrical and convenient way to define an arbitrary coordinate system is to express the $3N$ coordinates of a system of N particles in terms of $f = 3N - l$ independent variables $q_1, q_2, \ldots, q_f$ called *generalized coordinates*. For scleronomous constraints, independent of time, the transformation equations are

$$\vec{r}_i = \vec{r}_i(q_1, \ldots, q_k, \ldots, q_f) \quad , \quad i = 1, 2, \ldots, N \tag{24.12}$$

Note that if constraints are rheonomous, the time will explicitly occur in the expressions giving the position vectors of the particles. The variables q_k, which are expressed by the reciprocal equations

$$q_k = q_k(\vec{r}_1, \ldots, \vec{r}_i, \ldots, \vec{r}_N) \quad , \quad k = 1, 2, \ldots, f \tag{24.13}$$

need not have the dimensions of a length. On differentiation with respect to time, we then derive from Eqs.(24.12) the velocities of the various particles as

$$\dot{\vec{r}}_i = \sum_k \frac{\partial \vec{r}_i}{\partial q_k} \dot{q}_k \tag{24.14}$$

in terms of *generalized velocities* $\dot{q}_k$. Inserting this result into Eq.(24.8) we obtain the kinetic energy of scleronomous systems as a homogenous quadratic form in the generalized velocities, which reads

$$T = \frac{1}{2} \sum_{k,l} a_{kl} \dot{q}_k \dot{q}_l \tag{24.15}$$

where the functions a_{kl} are given by

$$a_{kl} = a_{lk} = \sum_i m_i \frac{\partial \vec{r}_i}{\partial q_k} \frac{\partial \vec{r}_i}{\partial q_l} \tag{24.16}$$

Provided the potential energy of the system is a function of position only (the velocity-dependent potential corresponds to the electromagnetic forces of moving charges), its Lagrangian (24.11) takes the form

$$L(q_k, \dot{q}_k, t) = T(q_k, \dot{q}_k) - U(q_k, t) \tag{24.17}$$

which is a function of the independent variables q_k, $\dot{q}_k$ at a time t. The variation of the Lagrangian at a given point on the path of motion is expressed in terms of the variations δq_k, $\delta \dot{q}_k$ of its independent variables, at a given instant t, as

$$\delta L = \sum_k \frac{\partial L}{\partial \dot{q}_k} \delta \dot{q}_k + \sum_k \frac{\partial L}{\partial q_k} \delta q_k \qquad (24.18)$$

where the variations δq_k are zero at the two limits of the independent variable t and are arbitrary between the two limits. This may be combined with Hamilton's principle (24.10) which successively becomes

$$0 = \int_{t_1}^{t_2} \delta L \, dt = \int_{t_1}^{t_2} \sum_k \frac{\partial L}{\partial \dot{q}_k} \delta \dot{q}_k \, dt + \int_{t_1}^{t_2} \sum_k \frac{\partial L}{\partial q_k} \delta q_k \, dt$$

$$= \sum_k \frac{\partial L}{\partial \dot{q}_k} \delta q_k \Big|_{t_1}^{t_2} - \int_{t_1}^{t_2} \sum_k \frac{d}{dt}\left(\frac{\partial L}{\partial \dot{q}_k}\right) \delta q_k \, dt + \int_{t_1}^{t_2} \sum_k \frac{\partial L}{\partial q_k} \delta q_k \, dt$$

$$= \int_{t_1}^{t_2} \sum_k \left[\frac{\partial L}{\partial q_k} - \frac{d}{dt}\left(\frac{\partial L}{\partial \dot{q}_k}\right)\right] \delta q_k \, dt$$

This equation requires that the coefficients of δq_k separately vanish, and this yields a set of f equations

$$\frac{d}{dt}\left(\frac{\partial L}{\partial \dot{q}_k}\right) - \frac{\partial L}{\partial q_k} = 0 \qquad (24.19)$$

which are known as *Lagrange's equations* and follow from Hamilton's principle for conservative systems. Defining the *generalized momenta* as

$$p_k = \frac{\partial L}{\partial \dot{q}_k} \qquad (24.20)$$

the set of Lagrange's equations (24.19) takes the form

$$\dot{p}_k = \frac{\partial L}{\partial q_k} \qquad (24.21)$$

and this is analogous to that of Newton's equations of motion, expressed in terms of generalized coordinates. Using the Lagrangian (24.17) of a conservative system, we can identify the *generalized forces* as

$$Q_k = \frac{\partial L}{\partial q_k} = -\frac{\partial U}{\partial q_k}$$

Since any independent variables that determine a definite configuration of a system can be chosen to be generalized coordinates, the Lagrange formulation of the law of motion can be extended to describe the evolution of systems beyond classical dynamics, such as electromagnetic and quantum systems.

24.4. THE CANONICAL EQUATIONS

Using the transformation of coordinates

$$q_1,\ldots,q_k,\ldots,q_f,\dot{q}_1,\ldots,\dot{q}_k,\ldots,\dot{q}_f,t \quad \rightarrow \quad q_1,\ldots,q_k,\ldots,q_f,p_1,\ldots,p_k,\ldots,p_f,t$$

where the generalized momenta are defined by Eq.(24.20), the f differential equations of the *second order* (24.19) for the coordinates q as functions of the time t can be replaced by $2f$ differential equations of the *first order* for the coordinates q and p as functions of t. These equations may be given a symmetrical form if we introduce the Hamiltonian function defined by

$$H(q_k,p_k,t) = \sum_k p_k \dot{q}_k - L(q_k,\dot{q}_k,t) \tag{24.22}$$

The complete differential of the Hamiltonian follows from Eq.(24.22) as

$$dH = \sum_k p_k d\dot{q}_k + \sum_k \dot{q}_k dp_k - dL \tag{24.23}$$

while for the Lagrangian we have

$$dL = \sum_k \frac{\partial L}{\partial q_k} dq_k + \sum_k \frac{\partial L}{\partial \dot{q}_k} d\dot{q}_k + \frac{\partial L}{\partial t} dt$$

and this can be written, in view of Eqs.(24.20) and (24.21), in the form

$$dL = \sum_k \dot{p}_k dq_k + \sum_k p_k d\dot{q}_k + \frac{\partial L}{\partial t} dt \tag{24.24}$$

Hence, substituting for dL in Eq.(24.23), we have

$$dH = \sum_k \dot{q}_k dp_k - \sum_k \dot{p}_k dq_k - \frac{\partial L}{\partial t} dt \tag{24.25}$$

Equation (24.25) is equivalent to

$$dH = \sum_k \frac{\partial H}{\partial p_k} dp_k + \sum_k \frac{\partial H}{\partial q_k} dq_k + \frac{\partial H}{\partial t} dt \tag{24.26}$$

provided a system of $2f$ dynamic equations of the first order, called the *canonical equations*, hold in the form

$$\dot{q}_k = \frac{\partial H}{\partial p_k} \quad , \quad \dot{p}_k = -\frac{\partial H}{\partial q_k} \quad (k=1,2,\ldots,f) \tag{24.27}$$

and also, provided that

$$\frac{\partial H}{\partial t} + \frac{\partial L}{\partial t} = 0$$

The Hamiltonian takes a convenient form for conservative systems, under constraints independent of time, as expressed by the transformation equations (24.12). By applying Euler's theorem for homogenous functions, Eq.(III.1) in Appendix III, the first term in Eq.(24.22) can be reduced to the quadratic form (24.15) of the kinetic energy T, as we have

$$\sum_k p_k \dot{q}_k = \sum_k \frac{\partial L}{\partial \dot{q}_k} \dot{q}_k = \sum_k \frac{\partial T}{\partial \dot{q}_k} \dot{q}_k = 2T \qquad (24.28)$$

so that the Hamiltonian gives the total energy of the system, in terms of q_k, p_k, as

$$H = \sum_k p_k \dot{q}_k - L = 2T - (T - U) = T + U \qquad (24.29)$$

If the potential energy term in the Hamiltonian does not involve time explicitly, in other words if H is completely determined in terms of the solutions q_k, p_k of the canonical equations (24.27), then the rate of change of H with respect to the independent variable t vanishes as the system moves, and hence

$$\frac{dH}{dt} = \sum_k \frac{\partial H}{\partial p_k} \dot{p}_k + \sum_k \frac{\partial H}{\partial q_k} \dot{q}_k + \frac{\partial H}{\partial t} = \frac{\partial H}{\partial t} = 0 \qquad (24.30)$$

The Hamiltonian is therefore a constant of motion for a conservative system under scleronomous constraints.

The solutions to Eqs.(24.27) consist of a set of $2f$ equations

$$q_k = q_k(q_0, p_0, t) \quad , \quad p_k = p_k(q_0, p_0, t)$$

where $q_0 = (q_{01}, \ldots, q_{0f})$, $p_0 = (p_{01}, \ldots, p_{0f})$ are the initial values of q_k and p_k at $t = 0$. Thus, the motion of a system can be described in terms of its generalized coordinates and momenta, which define a *phase* of the system at any given instant t. For a conservative system, H is completely determined by the phase, and it is not necessary to know the time at which this phase was assumed. It is convenient to represent the phase of the system by a point in a $2f$-dimensional *phase space* with coordinates q_k and p_k. As the system moves, and hence it takes different phases, the representative point moves along a curve called the *trajectory* of the system in phase space, with velocities $\dot{q}_k, \dot{p}_k$ given by the canonical equations (24.27).

EXAMPLE 24.3. The action integral of a linear harmonic oscillator

Integrating the equation of oscillatory motion in one dimension, one obtains

$$\frac{d}{dt}\left(\frac{m\dot{x}^2}{2} + \frac{kx^2}{2}\right) = 0 \quad \text{or} \quad E = \frac{m\dot{x}^2}{2} + \frac{kx^2}{2} = \text{const}$$

where E is the total energy.

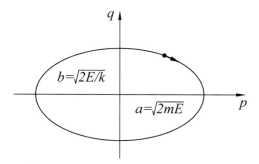

Figure 24.2. Elliptical representation of an oscillatory motion in phase space

With the obvious choice of $q = x$, $p = \partial T / \partial \dot{x} = m\dot{x}$, we may express the Hamiltonian as

$$H(q,p) = \frac{p^2}{2m} + \frac{kq^2}{2} = E \quad \text{or} \quad \frac{p^2}{2mE} + \frac{kq^2}{2E/k} = 1$$

It follows that the representative point (p,q) of an oscillatory motion in one dimension describes an ellipse in phase space, with semimajor axis $a = \sqrt{2mE}$ and semiminor axis $b = \sqrt{2E/k}$, as illustrated in Figure 24.2. The area πab of the ellipse may alternatively be evaluated by a line integral along the closed trajectory Γ, which is called the *action integral*, defined as

$$J = \oint_\Gamma p\,dq$$

Using $\pi ab = \pi\sqrt{2mE}\sqrt{2E/k} = 2\pi\sqrt{m/k}\,E = (2\pi/\omega_0)E$, where ω_0 is the angular frequency, the integral J is expressed in terms of the energy and angular frequency of the oscillatory motion as

$$J = \oint_\Gamma p\,dq = \frac{2\pi}{\omega_0} E$$

and this shows that the action integral is a constant of motion.

Note that for a complete description of a state of the system we need to know not only the phase but also the particular instant t to which this phase corresponds. If it is necessary to be concerned about the time at which the various phases are assumed, a phase-time representation in a state space of $2f+1$ dimensions (q,p,t) must be given. The trajectory of a representative point in such a space may be regarded to be the phase history of the system.

24.5. THE POISSON BRACKETS

The time evolution for an arbitrary function $F(q_k, p_k, t)$ is expressed by the rate of change of F with respect to time as the system moves. Substituting the canonical equations (24.27) into the total time derivative dF/dt, defined in terms of q_k, p_k and t, we obtain

$$\frac{dF}{dt} = \sum_k \left(\frac{\partial F}{\partial q_k} \dot{q}_k + \frac{\partial F}{\partial p_k} \dot{p}_k \right) + \frac{\partial F}{\partial t} = \sum_k \left(\frac{\partial F}{\partial q_k} \frac{\partial H}{\partial p_k} - \frac{\partial F}{\partial p_k} \frac{\partial H}{\partial q_k} \right) + \frac{\partial F}{\partial t} \quad (24.31)$$

For any two arbitrary functions f and g, the *Poisson bracket* with respect to q_k, p_k is introduced as

$$\{f, g\} = \sum_k \left(\frac{\partial f}{\partial q_k} \frac{\partial g}{\partial p_k} - \frac{\partial f}{\partial p_k} \frac{\partial g}{\partial q_k} \right) = -\{g, f\} \quad (24.32)$$

so that we can identify the Poisson bracket of F and H in Eq.(24.31), which can be put in the form

$$\frac{dF}{dt} = \{F, H\} + \frac{\partial F}{\partial t} \quad (24.33)$$

Equation (24.33) is called the *evolution equation* of $F(q_k, p_k, t)$. If t does not occur explicitly, that is if $\partial F/\partial t = 0$, the equation reduces to

$$\frac{dF}{dt} = \{F, H\} \quad (24.34)$$

A function $F(q_k, p_k, t)$ is called a *constant of motion* if

$$\frac{dF}{dt} = 0 \quad \text{or} \quad \{F, H\} + \frac{\partial F}{\partial t} = 0 \quad (24.35)$$

and, respectively, if

$$\{F, H\} = 0 \quad (24.36)$$

In other words, any function which does not depend explicitly on time is a constant of motion, provided its Poisson bracket with the Hamiltonian vanishes. If two constants of motion are known, the properties of the Poisson brackets allow us to find other new constants.

The basic properties of the Poisson brackets, that is, their linearity

$$\{f_1 + f_2, g\} = \{f_1, g\} + \{f_2, g\}$$

the product rule

$$\{f_1 f_2, g\} = f_1 \{f_2, g\} + f_2 \{f_1, g\}$$

and the time derivative

$$\frac{\partial}{\partial t}\{f, g\} = \left\{\frac{\partial f}{\partial t}, g\right\} + \left\{f, \frac{\partial g}{\partial t}\right\}$$

derive in a straightforward manner from those of the partial derivatives. If we consider the special cases where $g = q_k$ or $g = p_k$ the corresponding Poisson brackets reduce to

$$\{f, q_k\} = -\frac{\partial f}{\partial p_k} \quad , \quad \{f, p_k\} = -\frac{\partial f}{\partial q_k} \qquad (24.37)$$

and further, if f is chosen to be the Hamiltonian, we obtain the canonical equations (24.27) in terms of Poisson brackets as

$$\{H, q_k\} = -\frac{\partial H}{\partial p_k} = -\dot{q}_k \qquad \text{i.e.} \qquad \dot{q}_k = \{q_k, H\}$$

$$\{H, p_k\} = \frac{\partial H}{\partial q_k} = -\dot{p}_k \qquad \text{i.e.} \qquad \dot{p}_k = \{p_k, H\}$$

Substituting the function f in Eqs.(24.37) by q_i and p_i, respectively, we obtain

$$\{q_i, q_k\} = 0 \quad , \quad \{q_i, p_k\} = \delta_{ik} \quad , \quad \{p_i, p_k\} = 0 \qquad (24.38)$$

where δ_{ik} is the Kronecker symbol, which is defined as being 1 for $i = k$ and 0 for $i \neq k$. Equations (24.38) are called the *fundamental Poisson bracket* relations, which hold for any particular set of canonical variables. The symmetry of the Poisson bracket expression leads to *Jacobi's identity*, whose terms are homogenous linear functions of the second order derivatives of three arbitrary functions f, g and h, respectively, so that their sum must vanish identically, according to

$$\{f, \{g, h\}\} + \{g, \{h, f\}\} + \{h, \{f, g\}\} = 0 \qquad (24.39)$$

Based on the previous properties of the Poisson brackets, the approach to find new constants of motion is stated by

Poisson's theorem: *If f and g are two arbitrary constants of motion, their Poisson bracket $\{f,g\}$ is also a constant of motion.*

Proof. The total time derivative of the Poisson bracket $\{f,g\}$ successively transforms as

$$\frac{d}{dt}\{f,g\} = \frac{\partial}{\partial t}\{f,g\} + \{\{f,g\},H\} = \left\{\frac{\partial f}{\partial t},g\right\} + \left\{f,\frac{\partial g}{\partial t}\right\} - \{\{H,f\},g\} - \{\{g,H\},f\}$$

$$= \left\{\frac{\partial f}{\partial t} + \{f,H\},g\right\} + \left\{f,\frac{\partial g}{\partial t} + \{g,H\}\right\} = \left\{\frac{\partial f}{\partial t},g\right\} + \left\{f,\frac{\partial g}{\partial t}\right\}$$

so that, assuming that $df/dt = 0$ and $dg/dt = 0$, it follows that $d\{f,g\}/dt$ also vanishes, and hence, $\{f,g\}$ is a constant of motion.

24.6. THE HAMILTON-JACOBI EQUATION

The function obtained by integrating the Lagrangian $L(q_k, p_k, t)$ along the trajectory of motion of the system, between an initial instant $t_1 = t_0$ and a final instant $t_2 = t$, has the form

$$S = \int_{t_0}^{t} L \, dt \qquad (24.40)$$

and this is called the *principal function of Hamilton*. We obtain from this definition the variation δS using Eq.(24.18) for the Lagrangian, so that

$$\delta S = \int_{t_0}^{t} \delta L \, dt = \int_{t_0}^{t} \sum_k \frac{\partial L}{\partial \dot{q}_k} \delta \dot{q}_k \, dt + \int_{t_0}^{t} \sum_k \frac{\partial L}{\partial q_k} \delta q_k \, dt$$

$$= \sum_k \frac{\partial L}{\partial \dot{q}_k} \delta q_k \bigg|_{t_0}^{t} + \int_{t_0}^{t} \sum_k \left[\frac{\partial L}{\partial q_k} - \frac{d}{dt}\left(\frac{\partial L}{\partial \dot{q}_k}\right)\right] \delta q_k \, dt$$

In view of Lagrange's equations (24.19), the integral vanishes and, making the choice

$\delta q_k(t_0) = 0$, δS reduces to

$$\delta S = \sum_k \frac{\partial L}{\partial \dot{q}_k} \delta q_k = \sum_k p_k \delta q_k \qquad (24.41)$$

The principal function S defined by (24.40) is a function of $4f+1$ variables $q_k, p_k, q_{k0}, p_{k0}, t$ but it is more convenient to express it as a function of $2f+1$ variables q_k, q_{k0}, t by means of the $2f$ equations $q_k = q_k(q_0, p_0, t)$ and $p_k = p_k(q_0, p_0, t)$ which give the solution to the canonical equations. Therefore Eq.(24.41) can be written as an expansion in terms of coordinates

$$\delta S = \sum_k \frac{\partial S}{\partial q_k} \delta q_k$$

and it follows that

$$\frac{\partial S}{\partial q_k} = p_k \qquad (24.42)$$

From $S = S(q_k, q_{k0}, t)$ and Eq.(24.42) we obtain

$$\frac{dS}{dt} = \sum_k \frac{\partial S}{\partial q_k} \dot{q}_k + \frac{\partial S}{\partial t} = \sum_k p_k \dot{q}_k + \frac{\partial S}{\partial t} \qquad (24.43)$$

Substituting the expression (24.40) for the principal function and using Eq.(24.22) for the Hamiltonian function we have

$$\frac{dS}{dt} = L - \sum_k p_k \dot{q}_k = -H \qquad (24.44)$$

which is equivalent to

$$\frac{\partial S}{\partial t} + H(q_k, p_k, t) = 0 \qquad (24.45)$$

where the generalized momenta are given by Eq.(24.42), so that

$$\frac{\partial S}{\partial t} + H\left(q_k, \frac{\partial S}{\partial q_k}, t\right) = 0 \qquad (24.46)$$

This equation for $S(q_k, q_{k0}, t)$ is called the *Hamilton-Jacobi equation*.

EXAMPLE 24.4. The Hamilton-Jacobi equation for a system of N particles

In Cartesian coordinates, substituting the kinetic energy (2.29), the Hamiltonian (24.29) for a system of N particles takes the form

$$H = \sum_{i=1}^{N} \frac{1}{2m_i} p_i^2 + U(\vec{r}_i,t) = \sum_{i=1}^{N} \frac{1}{2m_i}(\nabla_i S)\cdot(\nabla_i S) + U(\vec{r}_i,t)$$

where the momenta $\vec{p}_i$ of an arbitrary particle, in view of Eq.(24.42), are expressed in terms of the principal function $S(\vec{r}_i,\vec{r}_{i0},t)$ as

$$\vec{p}_i = p_{xi}\vec{e}_{xi} + p_{yi}\vec{e}_{yi} + p_{zi}\vec{e}_{zi} = \frac{\partial S}{\partial x_i}\vec{e}_{xi} + \frac{\partial S}{\partial y_i}\vec{e}_{yi} + \frac{\partial S}{\partial z_i}\vec{e}_{zi} = \nabla_i S$$

It follows from Eq.(24.46) that the Hamilton-Jacobi equation for a system of N particles, can be written in Cartesian coordinates as

$$\sum_{i=1}^{N} \frac{1}{2m_i}(\nabla_i S)\cdot(\nabla_i S) + U(\vec{r}_i,t) + \frac{\partial S(\vec{r}_i,\vec{r}_{i0},t)}{\partial t} = 0 \tag{24.47}$$

Substituting Eq.(24.44), Eq.(24.43) becomes

$$\frac{dS}{dt} = \sum_k p_k \dot{q}_k - H \tag{24.48}$$

and this allows us to simplify the Hamilton-Jacobi equation in the case of conservative systems, where the Hamiltonian is not an explicit function of time, that is $H = E = \text{const}$. It is convenient to denote by S_0 the integral

$$S_0 = \int_{t_0}^{t} \sum_k p_k \dot{q}_k \, dt = \int_{t_0}^{t} 2T \, dt \tag{24.49}$$

where use has been made of Eq.(24.28). Integrating Eq.(24.48), we obtain

$$S = S_0 - E(t - t_0) \tag{24.50}$$

so that, substituting Eq.(24.50), Eq.(24.43) becomes

$$L = \frac{dS}{dt} = \sum_k \frac{\partial S_0}{\partial q_k}\dot{q}_k + \frac{\partial S_0}{\partial t} - E = 2T + \frac{\partial S_0}{\partial t} - E$$

On comparison with Eq.(24.11), which reads $L = 2T - E$, it follows that

$$\frac{\partial S_0}{\partial t} = 0 \tag{24.51}$$

In other words, $\frac{\partial S}{\partial t} = -E$ and $\frac{\partial S}{\partial q_k} = \frac{\partial S_0}{\partial q_k}$, so that Eq.(24.46) reduces to

$$H\left(q_k, \frac{\partial S_0}{\partial q_k}\right) = E \qquad (24.52)$$

which is the desired form of the Hamilton-Jacobi equation for conservative systems.

FURTHER READING
1. M. G. Calkin - LAGRANGIAN AND HAMILTONIAN MECHANICS, World Scientific, Singapore, 2000.
2. L. N. Hand, J. D. Finch - ANALYTICAL MECHANICS, Cambridge University Press, Cambridge, 1998.
3. G. R. Fowles, G. L. Cassiday - ANALYTICAL MECHANICS, Brooks/Cole, Pacific Grove, 1998

PROBLEMS

24.1. An elliptic pendulum consists of a pivot of mass m_1 which may slide freely on a smooth horizontal table, being connected by a light stick of length l to a bob of mass m_2. If the pendulum is released from an angle θ_0, apply Lagrange's equations of motion to find the trajectory and the period of the pendulum bob.

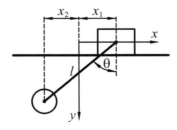

Solution

It is convenient to choose a vertical axis passing through the centre of the mass of the system and to take $q_1 = x_1$ and $q_2 = \theta$. It follows that $x_2 = x_1 - l\sin\theta$ and $y_2 = l\cos\theta$ which gives

$$T = T_1 + T_2 = \frac{m_1 + m_2}{2}\dot{x}_1^2 + \frac{m_2}{2}l^2\dot{\theta}^2 - m_2 l\dot{x}_1\dot{\theta}\cos\theta \quad , \quad U = U_1 + U_2 = -m_2 g l\cos\theta$$

It is apparent that the Lagrangian (24.11), that is $T-U$, does not depend explicitly on x_1, and this is called a cyclic or ignorable coordinate. The first equation of motion (24.19) reduces to

$$\frac{d}{dt}\left(\frac{\partial L}{\partial \dot{x}_1}\right) - \frac{\partial L}{\partial x_1} = 0 \quad \text{or} \quad \frac{\partial L}{\partial \dot{x}_1} = A$$

where A is a constant. This can be written as

$$(m_1 + m_2)\dot{x}_1 - m_2 l \dot{\theta} \cos\theta = A \quad \text{or} \quad (m_1 + m_2)x_1 - m_2 l \sin\theta = At + B$$

and we have $\dot{x}_1 = 0$ and $\dot{\theta} = 0$, when $t = 0$, , thus $A = 0$. We also have

$$B = (m_1 + m_2)x_{10} - m_2 l \sin\theta_0 = m_1 x_{10} + m_2(x_{10} - l\sin\theta_0) = 0$$

because the centre of the mass is initially at $x = 0$ and, in view of Eq.(2.17), this gives

$$\frac{m_1 x_{10} + m_2 x_{20}}{m_1 + m_2} = \frac{m_1 x_{10} + m_2(x_{10} - l\sin\theta_0)}{m_1 + m_2} = 0$$

Substituting $A = B = 0$, one obtains the solution

$$x_1 = \frac{m_2 l}{m_1 + m_2}\sin\theta$$

which implies that the abscissa of the centre of mass will always be zero. The trajectory of the pendulum bob is defined by

$$x_2 = \frac{m_2 l}{m_1 + m_2}\sin\theta - l\sin\theta = -\frac{m_1}{m_1 + m_2}l\sin\theta \quad , \quad y_2 = l\sin\theta$$

and eliminating θ gives, as expected, an ellipse of equation

$$\frac{x_2^2}{\left(\dfrac{m_1 l}{m_1 + m_2}\right)^2} + \frac{y_2^2}{l^2} = 1$$

The second Lagrange's equation should be written in terms of the angular variable θ, using

$$\frac{\partial L}{\partial \theta} = m_2 l(\dot{x}_1 \dot{\theta} - g)\sin\theta = m_2 l\left(\frac{m_2 l}{m_1 + m_2}\dot{\theta}^2 \cos\theta - g\right)\sin\theta$$

$$\frac{\partial L}{\partial \dot{\theta}} = m_2 l^2 \dot{\theta} - m_2 l\dot{x}_1 \cos\theta = m_2 l^2 \dot{\theta}\left(1 - \frac{m_2 \cos^2\theta}{m_1 + m_2}\right)$$

where for small oscillations we may set

$$\sin\theta = \theta - \frac{\theta^3}{3!} + \cdots \approx \theta \quad , \quad \cos\theta = 1 - \frac{\theta^2}{2} + \cdots \approx 1$$

Thus, the equation of motion (24.19) simplifies to

$$\frac{m_1 m_2}{m_1 + m_2} l^2 \ddot{\theta} - m_2 l \left(\frac{m_2 l}{m_1 + m_2} \dot{\theta}^2 - g \right) \theta = 0$$

Excluding terms of the second order, it reduces to

$$\ddot{\theta} + \left(1 + \frac{m_2}{m_1}\right) \frac{g}{l} \theta = 0 \quad \text{or} \quad \ddot{\theta} + \omega^2 \theta = 0$$

and this describes angular harmonic motion $\theta = \theta_0 \cos \omega t$, where $\omega = \left(1 + \frac{m_2}{m_1}\right) \frac{g}{l}$.

24.2. A ball of radius r and moment of inertia I rolls without sliding on a circular track of radius R. Use Lagrange's equations to find the period of small oscillations of the ball.

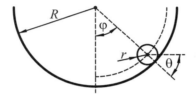

Solution

In terms of the angular coordinates $q_1 = \theta$ and $q_2 = \varphi$ we have

$$T = \frac{1}{2} I \dot{\theta}^2 + \frac{1}{2} m(R-r)^2 \dot{\varphi}^2 \quad , \quad U = mgR - mg(R-r)\cos\varphi$$

and $L = T - U$. The two coordinates are related by the constraint condition $(R-r)d\varphi = rd\theta$ which implies that

$$G(\theta, \varphi) = (R-r)\varphi - r\theta = 0$$

It follows that the Lagrangian can be written as

$$L_\mu(\theta, \varphi, \dot{\theta}, \dot{\varphi}, \mu, t) = L(\theta, \varphi, \dot{\theta}, \dot{\varphi}, t) + \mu G(\theta, \varphi)$$

where $\mu = \mu(t)$ is an additional coordinate function called a Lagrange multiplier, to be determined from two Lagrange's equations (24.19), which take the form

$$\frac{d}{dt}\left(\frac{\partial L_\mu}{\partial \dot{\theta}}\right) - \frac{\partial L_\mu}{\partial \theta} = 0 \quad \text{or} \quad \frac{d}{dt}\left(\frac{\partial L}{\partial \dot{\theta}}\right) - \frac{\partial L}{\partial \theta} - \mu \frac{\partial G}{\partial \theta} = 0$$

$$\frac{d}{dt}\left(\frac{\partial L_\mu}{\partial \dot{\varphi}}\right) - \frac{\partial L_\mu}{\partial \varphi} = 0 \quad \text{or} \quad \frac{d}{dt}\left(\frac{\partial L}{\partial \dot{\varphi}}\right) - \frac{\partial L}{\partial \varphi} - \mu \frac{\partial G}{\partial \varphi} = 0$$

Thus, one obtains the equations

$$I\ddot{\theta} + \mu r = 0 \quad , \quad m(R-r)^2 \ddot{\varphi} + mg(R-r)\sin\varphi - \mu(R-r) = 0$$

and hence, we obtain $\mu = -I\ddot{\theta}/r = -I(R-r)\ddot{\varphi}/r^2$, so that

$$\ddot{\varphi}(R-r)\left(1 + \frac{I}{mr^2}\right) + g\sin\varphi = 0$$

For small oscillations, where $\sin\varphi \approx \varphi$, this describes a simple harmonic motion $\ddot{\varphi} + \omega_0^2 \varphi = 0$, of period given by

$$T = \frac{2\pi}{\omega_0} = 2\pi\sqrt{\frac{R-r}{g}\left(1 + \frac{I}{mr^2}\right)}$$

24.3. Find Lagrange's formulation of the laws of motion in the presence of dissipative frictional forces, and determine the rate of energy dissipation.

Solution

In the presence of some dissipative forces $\vec{f}_i$, it is convenient to separate their contribution from that of the conservative forces

$$\sum_i \vec{F}_i \cdot \delta\vec{r}_i = \sum_i \left(-\nabla_i U + \vec{f}_i\right)_p \cdot \delta\vec{r}_i = -\delta U + \sum_i \vec{f}_i \cdot \delta\vec{r}_i$$

so that Eq.(24.10) becomes

$$\int_{t_1}^{t_2}\left(\delta L + \sum_i \vec{f}_i \cdot \delta\vec{r}_i\right) dt = 0$$

We can write

$$\sum_i \vec{f}_i \cdot \delta\vec{r}_i = \sum_k \sum_i \vec{f}_i \cdot \frac{\partial \vec{r}_i}{\partial q_k} \delta q_k = \sum_k Q_k \delta q_k$$

where Q_k stands for generalized dissipative forces. Substituting δL from Eq.(24.18) and following the approach in the main text gives

$$0 = \int_{t_1}^{t_2} \sum_k \left[\frac{\partial L}{\partial q_k} - \frac{d}{dt}\left(\frac{\partial L}{\partial \dot{q}_k}\right) + Q_k\right] \delta q_k \, dt$$

so that Lagrange's equations read

$$\frac{d}{dt}\left(\frac{\partial L}{\partial \dot{q}_k}\right) - \frac{\partial L}{\partial q_k} = Q_k = \sum_i \vec{f}_i \cdot \frac{\partial \vec{r}_i}{\partial q_k}$$

A common type of frictional force $\vec{f}_i = -\alpha_i \dot{\vec{r}}_i$ is derived from the Rayleigh dissipation function

$$\Phi = \frac{1}{2}\sum \alpha_i \dot{\vec{r}}_i^{\,2}$$

24. GENERALIZED MECHANICS

so that we can write

$$Q_k = -\sum_i \alpha_i \dot{r}_i^2 \cdot \frac{\partial \vec{r}_i}{\partial q_k} = -\sum_i \alpha_i \dot{\vec{r}}_i \cdot \frac{\partial \dot{\vec{r}}_i}{\partial \dot{q}_k} = -\frac{\partial}{\partial \dot{q}_k}\left(\frac{1}{2}\sum_i \alpha_i \dot{r}_i^2\right) = -\frac{\partial \Phi}{\partial \dot{q}_k}$$

where it is left as an exercise for the reader (Problem 24.6) to show that $\partial \vec{r}_i / \partial q_k = \partial \dot{\vec{r}}_i / \partial \dot{q}_k$. It follows that Lagrange's equations have the form

$$\frac{d}{dt}\left(\frac{\partial L}{\partial \dot{q}_k}\right) - \frac{\partial L}{\partial q_k} + \frac{\partial \Phi}{\partial \dot{q}_k} = 0$$

The energy loss is given by the work of dissipative forces, written as

$$dE = \sum_k Q_k dq_k = -\sum_k \frac{\partial \Phi}{\partial \dot{q}_k} dq_k = -\sum_k \frac{\partial \Phi}{\partial \dot{q}_k} \dot{q}_k dt$$

and hence, using Euler's theorem for homogeneous functions (III.1), given in Appendix III, the rate of energy dissipation is given by

$$\frac{dE}{dt} = -\sum_k \frac{\partial \Phi}{\partial \dot{q}_k} \dot{q}_k = -2R$$

24.4. Find an expression for the Lagrangian of a relativistic particle of charge q and rest mass m_0 moving with velocity $\vec{v}$ in an electromagnetic field specified by the scalar and vector potentials V and $\vec{A}$.

Solution

The equation of motion of a charged particle under the Lorentz force, Eq.(14.17), has the form

$$\frac{d}{dt}\left(\frac{m_0 \vec{v}}{\sqrt{1-v^2/c^2}} + q\vec{A}\right) + \nabla(qV - q\vec{V}\cdot\vec{A}) = 0$$

and this has already the form of Lagrange's equations (24.19), provided we set

$$\frac{\partial L}{\partial \dot{x}_k} = \frac{\partial L}{\partial v_k} = \frac{m_0 v_k}{\sqrt{1-v^2/c^2}} + qA_k \quad , \quad \frac{\partial L}{\partial x_k} = -q\frac{\partial V}{\partial x_k} + q\sum_{i=1}^{3} v_i \frac{\partial A_i}{\partial x_k}$$

It follows that the Lagrangian should have the form

$$L = m_0 c^2\left(1 - \sqrt{1-v^2/c^2}\right) - qV + q\vec{v}\cdot\vec{A}$$

where we may drop the constant to arrive at

$$L = -m_0 c^2 \sqrt{1-v^2/c^2} - qV + q\vec{v}\cdot\vec{A}$$

24.5. Solve the Hamiltonian-Jacobi equation for simple harmonic motion.

Solution

The Hamiltonian of simple harmonic motion, given by

$$H = \frac{1}{2m}p^2 + \frac{k}{2}q^2$$

where $q = x$ and $p = m\dot{x}$, should be written in terms of the principal function S using Eq.(24.42), and this gives

$$H = \frac{1}{2m}\left(\frac{\partial S}{\partial q}\right)^2 + \frac{k}{2}q^2$$

Since the Hamiltonian is not an explicit function of time, it is convenient to take S to be given by Eq.(24.50), and hence, the Hamiltonian-Jacobi equation reduces to

$$\frac{1}{2m}\left(\frac{\partial S_0}{\partial q}\right)^2 + \frac{k}{2}q^2 = E$$

It follows immediately that

$$p = \frac{\partial S_0}{\partial q} = \sqrt{2mE - mkq^2} \quad \text{or} \quad S_0 = \sqrt{mk}\int\sqrt{2E/k - q^2}\,dq$$

and substituting in Eq.(24.50) results in

$$S = \sqrt{mk}\int\sqrt{2E/k - q^2}\,dq - E(t - t_0)$$

Differentiating with respect to E gives

$$\frac{\partial S}{\partial E} = -\sqrt{\frac{m}{k}}\int\frac{dq}{k\sqrt{2E/k - q^2}} - t + t_0 = -\sqrt{\frac{m}{k}}\cos^{-1}\left(q\sqrt{\frac{k}{2E}}\right) - t + t_0$$

Assuming that $q = 0$ and $p = \sqrt{2mE}$, when $t = t_0$, it follows that

$$\frac{\partial S}{\partial E} = -\frac{\pi}{2}\sqrt{\frac{m}{k}} \quad \text{or} \quad \frac{\pi}{2}\sqrt{\frac{m}{k}} = \sqrt{\frac{m}{k}}\cos^{-1}\left(q\sqrt{\frac{k}{2E}}\right) + t - t_0$$

One obtains the required solution as

$$q = \sqrt{\frac{2E}{k}}\cos\left[\frac{\pi}{2} - \sqrt{\frac{k}{m}}(t - t_0)\right] \quad \text{or} \quad q = \sqrt{\frac{2E}{k}}\sin\sqrt{\frac{k}{m}}(t - t_0)$$

and the momentum is given by

$$p = \sqrt{2mE}\cos\sqrt{\frac{k}{m}}(t - t_0)$$

24.6. Prove the following equations:

$$\frac{\partial \vec{v}_i}{\partial \dot{q}_k} = \frac{\partial \vec{r}_i}{\partial q_k} \quad \text{and} \quad \frac{\partial \vec{v}_i}{\partial q_k} = \frac{d}{dt}\left(\frac{\partial \vec{r}_i}{\partial q_k}\right)$$

24.7. Use Lagrange's equations to describe the motion of a pendulum bob of mass m suspended by a light cord of length l, if the pivot is set in horizontal harmonic motion given by $x = A\sin\omega t$.

24.8. Find the Hamiltonian of a pendulum consisting of a bob of mass m suspended by a light cord of length l, assuming that the pivot is set rotating with constant angular velocity ω on a vertical circular loop of radius R.

24.9. Solve Lagrange's equations for a hoop of mass M and moment of inertia $I = Mr^2/2$ rolling down a plane of inclination α with nonsliding contact.

24.10. Find an expression for the Hamiltonian of a relativistic particle of charge q and rest mass m_0 moving with velocity $\vec{v}$ in an electromagnetic field specified by V and $\vec{A}$.

24.11. Find the relativistic Hamilton-Jacobi equation for a particle of charge q, rest mass m_0 and velocity $\vec{v}$ in an electromagnetic field V, $\vec{A}$.

25. PRINCIPLES OF STATISTICAL MECHANICS

25.1. LIOUVILLE'S THEOREM
25.2. BASIC STATISTICAL ASSUMPTIONS
25.3. BOLTZMANN'S PRINCIPLE
25.4. MICROCANONICAL ENSEMBLE
25.5. CANONICAL ENSEMBLE
25.6. GRAND CANONICAL ENSEMBLE

25.1. LIOUVILLE'S THEOREM

The behaviour of solids and fluids consisting of many particles can be described in statistical mechanics by assuming them to be dynamical systems with N particles, each having f degrees of freedom, whose motion satisfies the canonical equations (24.27). We have seen that motion with f degrees of freedom is suitably described in a $2f$ dimensional *phase space*, where the instantaneous state of the system at any time is determined by $2f$ generalized coordinates q_k and momenta p_k $(k=1,2,...,f)$, and it is graphically specified by the position of a *representative* or *phase point*. As q_k and p_k vary with the time, the evolution of the system is represented by a single trajectory of its phase point, which either describes a closed curve or goes to infinity but does not cross other trajectories, since motion through each point in phase space is uniquely determined by Eqs.(24.27). The macroscopic value of any physical observable of this system can be understood as a *time average* over a segment of the trajectory in phase space, corresponding to the time interval during which its measurement is carried out.

The procedure of taking time averages for a given system becomes highly complicated if the system consists of a large number of particles. Following the suggestion of Gibbs, it is more appropriate to adopt an alternative procedure of taking *ensemble averages* at *a fixed time* over a collection of dynamical systems each having the same construction as the actual system. The collection of systems is called an *ensemble*. All the copies of the given system, called *elements* of the ensemble, consist of N particles with f degrees of freedom and have the same Hamiltonian. Their representative points in phase space are assumed to be suitably random so that every microstate accessible to the actual system during its motion is represented at any given instant by at least one system. Each of these microstates must be consistent with the same macrostate associated with the ensemble. In any volume element $d\Omega$ of phase space there will be at time t a certain number of representative points dn. If we consider a sufficiently large number of elements of the ensemble, it is convenient to define a *phase density* $D(q_k, p_k, t)$ of the representative points in phase space in the form

$$D(q_k, p_k, t) = \frac{dn}{d\Omega} \tag{25.1}$$

where $d\Omega$ is the volume element in phase space, given by

$$d\Omega = dq_1 dq_2 \ldots dq_f dp_1 dp_2 \ldots dp_f \tag{25.2}$$

In the course of time, every element of the ensemble undergoes a continual change of microstates, so that the assembly of representative points moves with time through phase space like a fluid of density D. The number dn of phase points in the volume element $d\Omega$ will change with time at a rate given by

$$\frac{\partial n}{\partial t} = \frac{\partial D}{\partial t} d\Omega \tag{25.3}$$

where $\partial / \partial t$ gives the rate of change with respect to a fixed point q_k, p_k in phase space, and this can be estimated by taking the difference between the number of points entering and leaving $d\Omega$, as suggested in Figure 25.1.

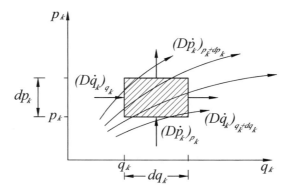

Figure 25.1. The flux of representative points in phase space

For each given coordinate q_k there is a velocity component $\dot{q}_k$ perpendicular to the surface $d\Omega / dq_k$, so that the number of points entering $d\Omega$ per unit time at q_k and leaving $d\Omega$ at $q_k + dq_k$, respectively, is given by

$$(D\dot{q}_k) \frac{d\Omega}{dq_k}, \qquad \left(D + \frac{\partial D}{\partial q_k} dq_k \right) \left(\dot{q}_k + \frac{\partial \dot{q}_k}{\partial q_k} dq_k \right) \frac{d\Omega}{dq_k}$$

As a result, neglecting second-order terms, the number of representative points in the volume element diminishes by an amount of

$$-\left(\frac{\partial D}{\partial q_k}\dot{q}_k dq_k + D\frac{\partial \dot{q}_k}{\partial q_k}dq_k\right)\frac{d\Omega}{dq_k} = -\frac{\partial}{\partial q_k}(D\dot{q}_k)d\Omega$$

By considering in addition similar expressions for each coordinate p_k, the rate of change with time of the number of phase points in a volume element is

$$\frac{\partial n}{\partial t} = -\sum_{k=1}^{f}\left[\frac{\partial D}{\partial q_k}\dot{q}_k + \frac{\partial D}{\partial p_k}\dot{p}_k + D\left(\frac{\partial \dot{q}_k}{\partial q_k} + \frac{\partial \dot{p}_k}{\partial p_k}\right)\right]d\Omega = -\sum_{k=1}^{f}\left(\frac{\partial D}{\partial q_k}\dot{q}_k + \frac{\partial D}{\partial p_k}\dot{p}_k\right)d\Omega \quad (25.4)$$

where a group of terms vanishes identically, according to the canonical equations (24.27) which lead, for all k, to

$$\frac{\partial \dot{q}_k}{\partial q_k} = \frac{\partial^2 H(q_k,p_k)}{\partial q_k \partial p_k} = -\frac{\partial \dot{p}_k}{\partial p_k} \quad (25.5)$$

From Eqs.(25.3) and (25.4) it follows that

$$\frac{\partial D}{\partial t} + \sum_{k=1}^{f}\left(\frac{\partial D}{\partial q_k}\dot{q}_k + \frac{\partial D}{\partial p_k}\dot{p}_k\right) = \frac{dD}{dt} = 0 \quad (25.6)$$

and this, in terms of Poisson's brackets (24.31), reads

$$\frac{dD}{dt} = \frac{\partial D}{\partial t} + \{D,H\} = 0 \quad (25.7)$$

It follows from either Eq.(25.6) or Eq.(25.7) that the total time derivative dD/dt, which gives the rate of change of the phase density in a coordinate system moving with a representative point through phase space, is zero. This result is known as *Liouville's theorem* and states that the phase density remains constant in the neighbourhood of a point that is moving in phase space. In other words, the number of representative points contained in a volume $\Delta\Omega$, as we follow its motion along trajectories which are analogous to the streamlines in a fluid, remains constant, like the density in an *incompressible* fluid. As a result, although the shape of a given volume distorts in the course of time, its magnitude will remain constant with time, since the number of representative points remains constant, and D is uniform throughout phase space. This geometrical interpretation is obtained by integration of Eq.(25.1), which gives

$$n = D\int_{\Delta\Omega}d\Omega = D\Delta\Omega \quad \text{or} \quad \Delta\Omega = \text{const} \quad (25.8)$$

and it is illustrated in Figure 25.2.

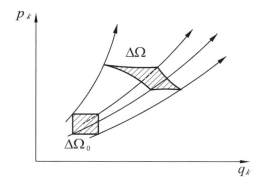

Figure 25.2. Geometrical interpretation of Liouville's theorem where $\Delta\Omega_0 = \Delta\Omega$

EXAMPLE 25.1. Motion in phase space under a constant force

Consider the motion of a particle of mass m in a constant force field F directed along the positive q-axis, so that the Hamiltonian is given by $H = \left(p^2/2m\right) + Fq$. The motion with constant energy $H = E$ will be described by an ensemble containing elements represented by phase points located on a parabola in the qp-plane, as in Figure 25.3.

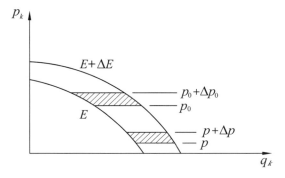

Figure 25.3. Conservation of phase density for a particle motion under a constant force

If the total energy lies between E and $E + \Delta E$, the representative points of the ensemble will be initially located in a phase volume $\Delta\Omega$ which can be approximated by the area of the parallelogram which is contained between the parabolas with energies E and $E + \Delta E$ and the lines parallel to the q-axis at p_0 and $p_0 + \Delta p_0$, and this gives

$$\Delta\Omega_0 = \Delta q \Delta p = \frac{\Delta E}{F} \Delta p_0$$

After a time t, the representative points move to a new phase region of equal volume

$$\Delta\Omega = \Delta q \Delta p = \frac{\Delta E}{F} \Delta p = \Delta\Omega_0$$

since the canonical equations (24.27) give $\dot{p} = -\partial H / \partial q = -F$ and this yields

$$p = p_0 - Ft \quad , \quad p + \Delta p = p_0 + \Delta p_0 - Ft$$

that is $\Delta p = \Delta p_0$. It follows that the phase density is conserved, as required by Liouville's theorem, although the shape of a given phase region changes with time.

In the special case of an ensemble of phase density D which is a function of energy E only, in view of Eqs.(24.27), we can write

$$\frac{\partial D}{\partial q_k} = \frac{\partial D}{\partial E}\frac{\partial E}{\partial q_k} = \frac{\partial D}{\partial E}\frac{\partial H}{\partial q_k} = -\frac{\partial D}{\partial E}\dot{p}_k \quad , \quad \frac{\partial D}{\partial p_k} = \frac{\partial D}{\partial E}\frac{\partial E}{\partial p_k} = \frac{\partial D}{\partial E}\frac{\partial H}{\partial p_k} = \frac{\partial D}{\partial E}\dot{q}_k \quad (25.9)$$

which gives

$$\frac{\partial D}{\partial q_k}\dot{q}_k + \frac{\partial D}{\partial p_k}\dot{p}_k = \frac{\partial D}{\partial E}(-\dot{q}_k\dot{p}_k + \dot{q}_k\dot{p}_k) = 0 \qquad (25.10)$$

Substituting Eq.(25.10), Liouville's theorem (25.6) becomes

$$\frac{\partial D}{\partial t} = 0 \qquad (25.11)$$

It follows that the composition of the ensemble is independent of time, and so are the ensemble averages, such that the macroscopic values of all physical observables will not change with time. Therefore the system represented by such an ensemble, defined by the condition (25.11), is called a *stationary ensemble* and is in thermodynamic equilibrium. Liouville's theorem indicates that any ensemble for which the phase density is a function of energy only must be stationary.

25.2. Basic Statistical Assumptions

The statistical procedure of averaging over a very large number of microstates of a mechanical model, each microstate being compatible with the same equilibrium macrostate of a given system, dispenses with a detailed enumeration of the parameters of a particular state and concentrates on the assignment of probabilities and the formation of averages. Since time averages over a single system are replaced by ensemble averages over the ensemble at a fixed time, a connection must be provided between the dynamical system and the abstract collection of copies which are meant to describe it, by means of a few basic statistical assumptions.

Consider an ensemble with n elements and phase density $D(q_k, p_k, t)$. It is convenient to introduce the *distribution function* $f(q_k, p_k, t)$ by choosing at random an element of representative point which is contained in a phase volume $d\Omega$ about a given phase point (q_k, p_k) as

$$f(q_k, p_k, t)d\Omega = \frac{dn}{n} = \frac{D(q_k, p_k, t)d\Omega}{n} \qquad (25.12)$$

This provides $f(q_k, p_k, t)$, in view of Eq.(25.1), to be a normalized density of representative points, given by

$$f(q_k, p_k, t) = \frac{D(q_k, p_k, t)}{n} = \frac{D(q_k, p_k, t)}{\int_{\Delta\Omega} D(q_k, p_k, t) d\Omega} \qquad (25.13)$$

which satisfies the condition

$$\int_{\Delta\Omega} f(q_k, p_k, t) d\Omega = 1 \qquad (25.14)$$

The ensemble averages of any observable of an actual system, defined in terms of the distribution function $f(q_k, p_k, t)$, are given a physical meaning through the following assumption:

Statistical hypothesis 1. *The probability $f(q_k, p_k, t)d\Omega$ of choosing at random from an ensemble an element contained in the phase volume $d\Omega$ is the same as the probability that the actual system described by the ensemble (which assumes its possible values one after another in the course of time) has its representative point in the phase region $d\Omega$, at a given instant in time.*

Since $D(q_k, p_k, t)$ remains constant in Eq.(25.13), so does $f(q_k, p_k, t)$ according to Liouville's theorem (25.6). This means that the probability $f(q_k, p_k, t)d\Omega$ is a function of the phase volume element $d\Omega$ only, and hence, we can adopt the following hypothesis:

Statistical hypothesis 2. *The probability $f(q_k, p_k, t)d\Omega$ that the actual system has its phase included in the volume element $d\Omega$, at a given instant of time, is proportional to the magnitude of the phase region $d\Omega$.*

Consider now the evolution of a system in a time interval τ during which a macroscopic measurement is carried out, as represented by the motion of its phase point along a trajectory in phase space. Let π_0, π_0' and π, π' be the ends of two segments of the trajectory, over which the time averages must be performed, between t_0 and $t_0 + \tau$ and t and $t + \tau$, respectively, as in Figure 25.4.

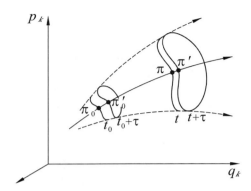

Figure 25.4. Phase volumes traversed by the representative point in the same time interval

The phase volumes $d\Omega_0$ and $d\Omega$ include the elements of the ensemble distributed in all the different sequences of microstates (segments of different trajectories) compatible with the given sequence of macroscopic states, traversed by the system between t_0 and $t_0 + \tau$ and t and $t + \tau$, respectively. Since the volume element $d\Omega_0$, at time t_0, moves along the trajectory as time progresses, changing its shape but not its magnitude, according to Eq.(25.8), it follows that, at any time t, equal phase volumes are traversed in the same time interval τ. In other words, any time average to be compared with the macroscopic value of a physical observable of an actual system is performed over a time interval proportional to the magnitude of a corresponding phase volume $d\Omega$ that is proportional to the probability that the actual system has its representative point included in $d\Omega$ (see statistical hypothesis 2). On this grounds we can formulate the following basic assumption:

Statistical hypothesis 3. *The time average of any physical observable over a segment of trajectory in phase space of an actual system is equal to the ensemble average of it, at a fixed time, over the ensemble which represents the system.*

Thus, if $F(q_k, p_k)$ is an arbitrary physical observable which is a function of the conjugate coordinates of the actual system, its time average may be replaced by an ensemble average over a volume $\Delta\Omega$ in phase space. The ensemble average is computed by means of the probability that the actual system has its phase included in a volume element $d\Omega$ adopted by statistical hypothesis 2, that is

$$\langle F(q_k, p_k) \rangle = \int_{\Delta\Omega} F(q_k, p_k) f(q_k, p_k, t) d\Omega \qquad (25.15)$$

Since the ensemble average $\langle F(q_k, p_k) \rangle$ is taken to be the actual physical value of $F(q_k, p_k)$, which could be obtained by experimental measurement on the system, it must be *independent of time*, otherwise the system represented by the ensemble is not in a state of thermodynamic equilibrium, which is the only macroscopic state compatible with a measurement procedure. This requirement can be met only if the distribution function does not depend explicitly on time, that is,

$$\frac{\partial f}{\partial t} = 0 \quad \text{or} \quad \frac{\partial D}{\partial t} = 0$$

In other words, for a stationary ensemble, from Eqs.(25.7) and (25.13), the thermodynamic equilibrium is defined by

$$\{f, H\} = \sum_k \left(\frac{\partial f}{\partial q_k} \dot{q}_k + \frac{\partial f}{\partial p_k} \dot{p}_k \right) = 0 \qquad (25.16)$$

As compared to Eq.(25.10), Eq.(25.16) shows that the distribution function f must depend on the conjugate coordinates through the Hamiltonian

$$f(q_k, p_k) = f'[H(q_k, p_k)] \qquad (25.17)$$

This requirement may be given a geometrical interpretation in phase space. Consider the motion of a representative point which is confined to a surface of constant energy in phase space

$$H(q_k, p_k) = E_0 \qquad (25.18)$$

such that the distribution function (25.17) is a constant.

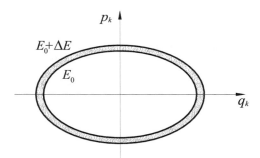

Figure 25.5. Energy shell of a stationary ensemble

The ensemble average (25.15) must be performed over all the phase points compatible with the energy E_0 of the system, in other words, the trajectory of a representative point, which always stays on the same energy surface (25.18), must pass in succession through every point of this surface before returning to its original position. This assumption was introduced by Boltzmann, who called it the *ergodic hypothesis*. Since it was shown that this strict assumption leads to a logical contradiction, the ergodic hypothesis was replaced by the less stringent *quasi-ergodic hypothesis*, which states that *the trajectory of a representative point passes arbitrarily close to any point in phase space whose energy is compatible with that of the system*. Under this assumption, the surface (25.18) can be imagined to be an *energy shell* (Figure 25.5) enclosing all the phase points with energies in the interval from E_0 to $E_0 + \Delta E$, for which we have

$$E_0 \leq H(q_k, p_k) \leq E_0 + \Delta E \tag{25.19}$$

We shall assume that the representative points of the ensemble fill this shell densely everywhere, in other words we adopt the following assumption:

Statistical hypothesis 4. *For a system in thermodynamic equilibrium all the phase points of equal energy have equal* a priori *probability.*

This statement is also known as the *assumption of equal* a priori *probability*, and it implies that every allowed microstate of a system is equally probable, that is it is represented by the same number of elements in the ensemble. Statistical hypothesis 4 seems less arbitrary than the previous ones, as the canonical equations show no tendency for the phase points (q_k, p_k) to crowd anywhere in phase space. On the basis of either the quasi-ergodic hypothesis or the assumption of equal *a priori* probabilities the distribution function of an actual system in thermodynamic equilibrium is expressible in the form (25.17) as a function of the Hamiltonian H alone.

The set of hypotheses that we have obtained by inference will only be justified if we will find complete agreement between the application of such hypotheses and the *macroscopic* measurements.

25.3. Boltzmann's Principle

The number of microstates compatible with a macroscopic state of an actual system is called the *thermodynamic probability W*. Consider a system which consists of N identical noninteracting particles in a space of volume V, given by

$$N = \sum_i N_i \tag{25.20}$$

where N_i represents the number of particles with energy ε_i, such that

$$E = \sum_i N_i \varepsilon_i \tag{25.21}$$

A given macroscopic state, defined by the parameters E, V and N will be represented by phase points in a volume $\Delta\Omega$ in phase space. The phase volume is an increasing function of E and will also depend on both N, as the number of conjugate coordinates is proportional to the number of particles, and V, since by increasing the volume of the system the possible range of the position coordinates q_k is increased. According to statistical hypothesis 2, the thermodynamic probability is proportional to the phase volume $\Delta\Omega(E, V, N)$, so that its magnitude $W(E, V, N)$ and the nature of its dependence on the parameters E, V and N will provide us with a bridge between statistical mechanics and thermodynamics. As N is typically a very large number, the analysis will be carried out in the thermodynamical limit, where $N \to \infty$ and $V \to \infty$ (while N/V and E/N remain finite), such that a continuous distribution of the single particle energies ε_i is obtained. We recall that the extensive variables, such as energy and entropy, which refer to the actual system as a whole, become in this limit proportional to both V and N, while the intensive variables, such as temperature and pressure, remain independent of these parameters.

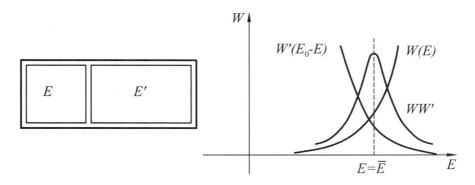

Figure 25.6. Thermal contact between two subsystems

A direct relation between the thermodynamic probability $W(E, V, N)$ and an extensive thermodynamic function can be derived by dividing the actual system into two subsystems which are separately in states of thermodynamic equilibrium specified by E, V, N and $W(E, V, N)$, and E', V', N' and $W'(E', V', N')$, respectively. In the configuration of Figure 25.6, we consider the case of rigid walls so that V, V', N and N' remain unchanged and the exchange of energy is only allowed under the restriction

$$E_0 = E + E' = \text{const} \qquad (25.22)$$

Because any one microstate of $W(E)$ can combine with any microstate of $W'(E')$ to yield a microstate of the system $W_0(E, E')$, the rule for combining microstates reads

$$W_0(E, E') = W(E) W'(E') = W(E) W'(E_0 - E) = W_0(E, E_0 - E) \qquad (25.23)$$

The thermodynamic equilibrium of the system will be reached at that value of E which maximizes the thermodynamic probability $W_0(E, E_0 - E)$, since this leads, according to statistical hypothesis 3, to a maximum time interval spent by the system in the corresponding macrostate, which thus becomes its state of macroscopic equilibrium. In other words, the most probable state of the system is that of thermodynamic equilibrium. When $W(E)$ and $W'(E')$ are both rapidly increasing functions of energy, as in Figure 25.6, then $W_0(E, E_0 - E)$ will have a very sharp maximum, which gives the most probable value of energy. Let $\bar{E}$ and $\bar{E}' = E_0 - \bar{E}$ be the definite values of E and E' that maximize the function (25.23) under the restriction (25.22). That is

$$d[W(E)W'(E')] = 0 \quad , \quad dE + dE' = 0$$

which gives

$$\left[\frac{\partial}{\partial E} \ln W(E)\right]_{E=\bar{E}} = \left[\frac{\partial}{\partial E'} \ln W'(E')\right]_{E'=\bar{E}'} \tag{25.24}$$

It is convenient to define a parameter β, governing the thermal equilibrium between one part of the system and another, by

$$\beta = \left[\frac{\partial}{\partial E} \ln W(E, V, N)\right]_{V,N} \tag{25.25}$$

so that the condition for equilibrium (25.24) reads

$$\beta = \beta' \tag{25.26}$$

The physical significance of β is clear when taking into account *Boltzmann's principle* that *the thermodynamic entropy $S(E, V, N)$ is the extensive function directly related to the thermodynamic probability $W(E, V, N)$*. We recall Eq.(8.21), written as

$$\left[\frac{\partial S(E, V, N)}{\partial E}\right]_{V,N} = \frac{1}{T} \tag{25.27}$$

which, upon substitution into the zeroth law of thermodynamics (7.5), gives the condition for equilibrium as

$$T = T' \quad \text{or} \quad \left[\frac{\partial S(E)}{\partial E}\right]_{E=\bar{E}} = \left[\frac{\partial S'(E')}{\partial E'}\right]_{E'=\bar{E}'} \tag{25.28}$$

The correspondence between the result (25.24) of the statistical description and that of the thermodynamic approach, Eq.(25.28), can be obtained by combining Eqs.(25.25) and (25.27) in the form

$$\frac{1}{\beta T} = \left[\frac{\partial S(E)/\partial E}{\partial \ln W(E)/\partial E}\right]_{V,N} = \frac{\Delta S}{\Delta(\ln W)} = \frac{S - S_0}{\ln W - \ln W_0} = k_B \tag{25.29}$$

where the additive constant S_0, in view of the law of increase of entropy (8.10), must give the minimum entropy associated with the state of perfect order. This is the macrostate realized through a single microstate, that is $W_0 = 1$. Equation (25.29) can then be rewritten in the form

25. PRINCIPLES OF STATISTICAL MECHANICS

$$S = k_B \ln W + S_0$$

At absolute zero the entropy must vanish if the third law of thermodynamics is to hold, as discussed in Chapter 10. If we assume that the thermodynamic probability is one at absolute zero, then the third law allows us to set S_0 equal to zero, which gives

$$S = k_B \ln W \qquad (25.30)$$

This equation is called the *Boltzmann relation*. Written in the form

$$W = e^{S/k_B} \qquad (25.31)$$

it shows that an increase in entropy of a system which is in thermodynamic equilibrium is associated with an increase in the number of accessible microstates in phase space. Any sequence of microstates presents an ever changing appearance, which means that disorder increases with entropy. The absolute value of entropy in terms of the number W of accessible microstates, as given by Eq.(25.30), can thus be taken as a measure of disorder in the system. The constant k_B, known as the *Boltzmann constant* allows us to express the statistical parameter β from Eq.(25.29) as

$$\beta = \frac{1}{k_B T} \qquad (25.32)$$

If the wall between the two subsystems described in Figure 25.6 is allowed to move, in the configuration of Figure 25.7, so that only N and N' remain unchanged during the thermal contact, the thermodynamic probabilities become functions of both E and V. In addition to the restriction (25.22) on the total energy E_0 there is a restriction

$$V_0 = V + V' = \text{const} \qquad (25.33)$$

on the volume V_0 of the system. The thermodynamic probabilities are functions of both the energy and volume, so that the most probable state of the system is defined by

$$d[W(E,V)W'(E',V')] = 0 \quad , \quad dE + dE' = 0 \quad , \quad dV + dV' = 0$$

The equilibrium condition for a variation of energy was derived in Eq.(25.24). The condition for a maximum in WW' for a variation of V is obtained in a similar form as

$$\left[\frac{\partial}{\partial V} \ln W(E,V)\right]_{V=\overline{V}} = \left[\frac{\partial}{\partial V'} \ln W'(E',V')\right]_{V'=\overline{V}'} \qquad (25.34)$$

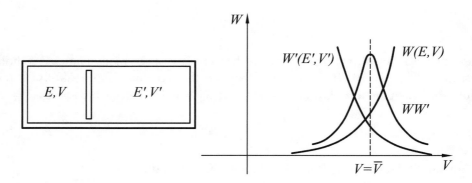

Figure 25.7. Thermal contact between subsystems with a variable volume

If a new statistical parameter γ is defined as

$$\gamma = \left[\frac{\partial}{\partial V} \ln W(E, V, N)\right]_{E,N} \quad (25.35)$$

the appropriate equilibrium conditions can be written in terms of β, defined by Eq.(25.25), and γ as

$$\beta = \beta' \quad , \quad \gamma = \gamma' \quad (25.36)$$

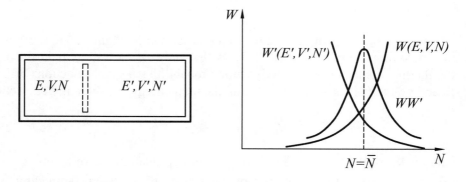

Figure 25.8. Thermal contact between two subsystems with a variable number of particles

If the exchange of particles between subsystems is also allowed, in the arrangement illustrated in Figure 25.8, and we treat N and N' as continuous variables, we may write

$$d\left[W(E, V, N)W'(E', V', N')\right] = 0 \quad , \quad dE + dE' = 0 \quad , \quad dV + dV' = 0 \quad , \quad dN + dN' = 0$$

The previous equilibrium conditions for energy (25.24) and volume (25.34) must now be augmented by the equation

$$\left[\frac{\partial}{\partial N} \ln W(E, V, N)\right]_{N=\overline{N}} = \left[\frac{\partial}{\partial N'} \ln W'(E', V', N')\right]_{N'=\overline{N}'} \quad (25.37)$$

and will be expressed in terms of appropriate statistical parameters in the form

$$\beta = \beta' \quad , \quad \gamma = \gamma' \quad , \quad \delta = \delta' \tag{25.38}$$

where

$$\delta = \left[\frac{\partial}{\partial N} \ln W(E, V, N)\right]_{E,V} \tag{25.39}$$

The statistical conditions for equilibrium may be given a physical interpretation using the Boltzmann relation (25.30), where it is understood that both S and W are functions of E, V and N, so that

$$dS = k_B d(\ln W) = k_B \left[\frac{\partial}{\partial E} \ln W\right]_{V,N} dE + k_B \left[\frac{\partial}{\partial V} \ln W\right]_{E,N} dV + k_B \left[\frac{\partial}{\partial N} \ln W\right]_{E,V} dN$$

and hence

$$dE = TdS - k_B T \gamma dV - k_B T \delta dN \tag{25.40}$$

where use has been made of Eqs.(25.25), (25.35) and (25.39). It is convenient to compare this equation with the differential energy dE given by Eqs.(8.17) and (10.23) as

$$dE = dG + d(TS) - d(pV) = (-SdT + Vdp + \mu_0 dN) + TdS + SdT - pdV - Vdp$$

$$= TdS - pdV + \mu_0 dN \tag{25.41}$$

where the Gibbs function (10.23) was written for a single phase of a system of one component, in terms of the *chemical potential per particle* μ_0, which is related to the molar chemical potential μ, defined by Eq.(10.24), and the chemical potential g by

$$\mu_0 = \frac{\mu}{N_A} = \frac{M}{N_A} g = \frac{m}{N_A \nu} g = \frac{m}{N} g \tag{25.42}$$

so that $gdm = \mu_0 dN$. From Eqs.(24.40) and (24.41) we can write

$$\gamma = \frac{p}{k_B T} \quad , \quad \delta = \frac{\mu_0}{k_B T} \tag{25.43}$$

and this shows that the statistical conditions for equilibrium (25.36) reduce to the thermodynamic conditions (10.8) providing that the temperatures and pressures of the two subsystems in the configuration of Figure 25.7 are the same, that is

$$T = T' \quad , \quad p = p'$$

Substituting Eq.(25.43), the conditions (25.38) can be rewritten as

$$T = T' \quad , \quad p = p' \quad , \quad \mu_0 = \mu_0'$$

which are similar to those given before in Eqs.(10.8) and (10.25). In other words, in the arrangement of Figure 25.8, the state of equilibrium requires that the temperatures, pressures and chemical potentials of the two subsystems should be the same.

25.4. MICROCANONICAL ENSEMBLE

We cannot apply the basic assumptions of statistical mechanics for a state of equilibrium unless we have a definite distribution for the elements of the representative ensemble which are compatible with the given state. Special types of ensembles can be obtained by choosing the set of macroscopic parameters that uniquely determines the state of equilibrium. There are three forms of the distribution function $f(q_k, p_k)$ which play an important part in statistical thermodynamics, corresponding in turn to a *microcanonical ensemble*, whose elements have the same energy E, volume V and number of particles N, a *canonical ensemble*, with elements of the same temperature T, volume V and number of particles N and a *grand canonical ensemble*, a collection of elements having equal temperatures, volumes and chemical potentials. Each distribution function will lead to a different basic equation, which is appropriate to the given set of macroscopic constraints. However, the equilibrium properties must be the same, regardless of constraints, so that it is expected that the basic statistical equations should be equivalent.

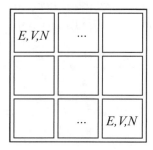

Figure 25.9. Microcanonical ensemble

The microcanonical ensemble represents an isolated system of known energy E which consists of N particles, constrained to move within a given volume V. The individual elements of this ensemble, which are macroscopically identical, may be imagined to be separated by adiabatic, rigid and impermeable walls, as in Figure 25.9.

In view of the quasi-ergodic hypothesis, the macroscopic energy E cannot be specified exactly, but can be defined to lie within some narrow range (25.19), from E_0 to $E_0 + \Delta E$, as in Figure 25.5. The microcanonical thermodynamic probability is given by the number of representative points of the ensemble lying in a phase volume

$$W = \Delta\Omega = \int_{E_0 \leq H \leq E_0 + \Delta E} d\Omega = \int_{0 \leq H \leq E_0 + \Delta E} d\Omega - \int_{0 \leq H \leq E_0} d\Omega = \Omega(E_0 + \Delta E) - \Omega(E_0) = \left(\frac{\partial \Omega}{\partial E}\right)_{E_0} \Delta E$$
(25.44)

Each microstate in this range is assumed to have the same probability, so that the distribution function must be written as

$$f(q_k, p_k) = \text{const} \quad , \quad E_0 \leq H \leq E_0 + \Delta E$$

$$f(q_k, p_k) = 0 \quad , \quad \text{otherwise}$$
(25.45)

and this is illustrated in Figure 25.10 (a).

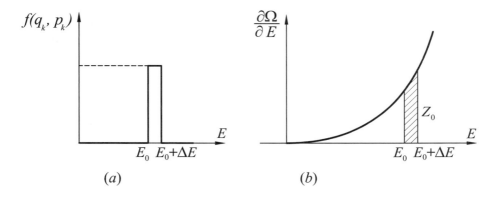

Figure 25.10. Microcanonical distribution function f (a) and the partition function Z_0 (b)

Substituting Eq.(25.45) into the normalization condition (25.14) gives

$$\text{const} \cdot \int_{E_0 \leq H \leq E_0 + \Delta E} d\Omega = \text{const} \cdot \Delta\Omega = 1$$

so that the *microcanonical distribution function* may be written as

$$f(q_k, p_k) = \frac{1}{(\partial\Omega/\partial E)_{E_0} \Delta E} = \frac{1}{Z_0} \quad , \quad E_0 \leq H \leq E_0 + \Delta E$$

$$f(q_k, p_k) = 0 \quad , \quad \text{otherwise}$$
(25.46)

where Z_0, called the *microcanonical partition function* or *the sum over the states*, is a measure of the number of microstates compatible with the macrostate E, V, N and therefore of the thermodynamic probability W. Z_0 is illustrated in Figure 25.10(b) for a system whose thermodynamic probability is a rapidly increasing function of energy. The ensemble average (25.15) over the microcanonical ensemble becomes

$$\langle F(q_k,p_k)\rangle = \int_{E_0\leq H\leq E_0+\Delta E} F(q_k,p_k)f(q_k,p_k)d\Omega = \frac{1}{Z_0}\int_{E_0\leq H\leq E_0+\Delta E} F(q_k,p_k)d\Omega \quad (25.47)$$

The integral on the right-hand side can be written, to a first approximation, in the form

$$\int_{0\leq H\leq E_0+\Delta E} F(q_k,p_k)d\Omega - \int_{0\leq H\leq E_0} F(q_k,p_k)d\Omega \cong \Delta E\frac{\partial}{\partial E}\int_{0\leq H\leq E_0} F(q_k,p_k)d\Omega \quad (25.48)$$

and this allows us to eliminate ΔE in Eq.(25.47), in view of Eq.(25.46), that is

$$\langle F(q_k,p_k)\rangle = \frac{\Delta E}{Z_0}\frac{\partial}{\partial E}\int_{0\leq H\leq E_0} F(q_k,p_k)d\Omega = \left(\frac{1}{\partial\Omega/\partial E}\right)_{E_0}\frac{\partial}{\partial E}\int_{0\leq H\leq E_0} F(q_k,p_k)d\Omega \quad (25.49)$$

This is the *microcanonical ensemble average* which can be applied to actual isolated systems of known energy. All the equilibrium properties of the ensemble can be calculated if Z_0 is known as a function of E, V and N. The substitution of Z_0 from Eq.(25.46) for the thermodynamic probability W, Eq.(25.44), allows Eq.(25.29) to be written as

$$k_B = \frac{\Delta S}{\Delta\ln\left[(\partial\Omega/\partial E)_{E_0}\Delta E\right]} = \frac{\Delta S}{\Delta\ln(\partial\Omega/\partial E)_{E_0} + \Delta\ln(\Delta E)} = \frac{\Delta S}{\Delta\ln(\partial\Omega/\partial E)_{E_0}} \quad (25.50)$$

where

$$\Delta\ln(\Delta E) = \frac{\partial\ln(\Delta E)}{\partial E}\Delta E = \frac{1}{\Delta E}\frac{\partial(\Delta E)}{\partial E}\Delta E = \Delta\left(\frac{\partial E}{\partial E}\right) = 0$$

The *microcanonical entropy* can therefore be written as

$$S(E,V,N) = k_B \ln\left(\frac{\partial\Omega}{\partial E}\right)_{E_0} \quad (25.51)$$

and this represents the basic equation providing a link between the microcanonical ensemble and thermodynamics. If the Hamiltonian of an isolated system of volume V is presumed known, the thermodynamic description of the system begins by calculating the *density of states* $\partial\Omega/\partial E$ in phase space.

25.2. Canonical Ensemble

Since it is difficult to measure and keep under physical control the total energy E of an actual system, an alternative approach is to define the macroscopic state of equilibrium in terms of a fixed temperature T, which is a parameter that can be both measured by a thermometer and controlled by keeping the system in contact with a larger system of a very large heat capacity, called a *heat reservoir*. The temperature of the reservoir T is considered to be unchanged even if some energy is given or taken by a system in contact with it. The heat reservoir therefore serves as a thermostat.

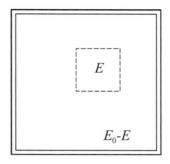

Figure 25.11. Subsystem with a variable energy E

A new type of ensemble, in which the energy is allowed to vary continuously, instead of being restricted to lie in a shell, can be obtained by considering an isolated system which is made up of two subsystems, of Hamiltonians $H(q_k, p_k)$ and $H'(q'_k, p'_k)$, and number of particles N and N', respectively, as in Figure 25.11. The larger subsystem is assumed to be a heat reservoir, with $N' \gg N$. Since the composite system is isolated, its energy E_0 is a constant, that is

$$E_0 = E + E' = \text{const} \tag{25.52}$$

where it is reasonable to assume that $E \ll E'$, or

$$\frac{E}{E_0} = 1 - \frac{E'}{E_0} \ll 1 \tag{25.53}$$

As the composite system is part of a microcanonical ensemble, the probability to find its representative point in a volume $d\Omega_0$ of phase space is given by

$$f(q_k, p_k) d\Omega_0 = \frac{1}{Z_0} d\Omega_0 = \frac{1}{Z_0} d\Omega_0 d\Omega' \tag{25.54}$$

for $E_0 \leq E + E' \leq E_0 + \Delta E$, and it is zero otherwise. In Eq.(25.54), $d\Omega$ contains only the coordinates and momenta belonging to the subsystem and $d\Omega'$ contains only the coordinates and momenta of the reservoir. The probability that the representative point of the subsystem is in the volume element $d\Omega$, without specifying the condition of the reservoir, but still requiring that E_0 is a constant, is obtained integrating Eq.(25.54) over the volume of phase space accessible to the phase point of the reservoir, which gives

$$f(q_k,p_k)d\Omega = \frac{d\Omega}{(\partial\Omega/\partial E)_{E_0}\Delta E}\int_{E_0-E\leq H\leq E_0-E+\Delta E}d\Omega' = \frac{d\Omega}{(\partial\Omega/\partial E)_{E_0}\Delta E}\frac{\partial\Omega'(E_0-E)}{\partial E'}\Delta E$$

or

$$f(q_k,p_k) = \text{const}\cdot\frac{\partial\Omega'(E_0-E)}{\partial E'} \qquad (25.55)$$

Since only the total energy $E + E'$ is restricted to a shell from E_0 to $E_0 + \Delta E$, while both E and E' are allowed to vary continuously from zero within the limitation of Eq.(25.52), the thermodynamic probabilities will be given by

$$W = \Omega(E) = \int_{0\leq H\leq E}d\Omega\;,\quad W' = \Omega'(E') = \int_{0\leq H'\leq E'}d\Omega' \qquad (25.56)$$

Substituting Eq.(25.56), Eq.(25.25) reads, if applied to the reservoir, as

$$\beta = \left[\frac{\partial}{\partial E'}\ln W'(E')\right]_{E'=\overline{E}'} = \left[\frac{\partial}{\partial E'}\ln\Omega'(E')\right]_{E'=\overline{E}'} = \frac{1}{\Omega'(E')}\frac{\partial\Omega'(E')}{\partial E'} \qquad (25.57)$$

where $\beta = k_B T = \text{const}$, for the reservoir, so that

$$\frac{\partial\Omega'(E')}{\partial E'} = \frac{\partial\Omega'(E_0-E)}{\partial E'} = \text{const}\cdot\Omega'(E') = \text{const}\cdot\Omega'(E_0-E)$$

and the distribution function (25.55) becomes

$$f(q_k,p_k) = \text{const}\cdot\Omega'(E_0-E) \qquad (25.58)$$

Under the assumption (25.53) for a subsystem which is small in comparison with the reservoir, we may use Taylor's expansion, which gives

$$\ln\Omega'(E_0-E) = \ln\Omega'(E_0) - E\left(\frac{\partial}{\partial E'}\ln\Omega'\right)_{E'=E_0} + \ldots \cong \ln\Omega'(E_0) - \beta E \qquad (25.59)$$

Substituting Eq.(25.59) into Eq.(25.58) yields a distribution function which is independent of the reservoir, apart from the information on its temperature T, that is

$$f(q_k,p_k) = \text{const}\cdot e^{-\beta E} = \text{const}\cdot e^{-E/k_B T} \qquad (25.60)$$

Owing to the fact that $E = H(q_k,p_k)$ we may take the distribution function of the subsystem to be

$$f(q_k,p_k) = \text{const}\cdot e^{-H(q_k,p_k)/k_B T} \qquad (25.61)$$

where the constant is determined by the normalization condition

$$\text{const} \cdot \int e^{-H(q_k,p_k)/k_B T} d\Omega = 1 \tag{25.62}$$

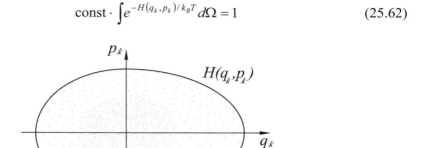

Figure 25.12. Distribution of phase points for a canonical ensemble

The sum over all the representative points accessible to a macroscopic state of the subsystem is called the *canonical partition function*, defined as

$$Z = \int e^{-H(q_k,p_k)/k_B T} d\Omega \tag{25.63}$$

so that the distribution function (25.61) becomes

$$f(q_k, p_k) = \frac{1}{Z} e^{-H(q_k,p_k)/k_B T} \tag{25.64}$$

This distribution of representative points plotted in Figure 25.12 is called the *canonical distribution*. The density of points falls off exponentially as we go away from the origin of phase space.

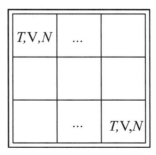

Figure 25.13. Canonical ensemble

The canonical ensemble, which represents a system in thermal contact with a reservoir, can be imagined to be a collection of elements having the same T, V and N. Each element is in thermal contact with a reservoir at temperature T or, alternatively, all the elements are in thermal contact with one another. The elements are shown in Figure 25.13 to be separated by rigid, impermeable but diathermic walls. They all arrive at a common temperature, as energy can be exchanged among the elements. The ensemble average over a canonical ensemble is given by Eq.(25.58) and Eq.(25.64) which yields

$$\langle F(q_k, p_k) \rangle = \frac{\int F(q_k, p_k) e^{-H/k_B T} d\Omega}{\int e^{-H/k_B T} d\Omega} = \frac{1}{Z} \int F(q_k, p_k) e^{-H/k_B T} d\Omega \quad (25.65)$$

The link with thermodynamics is provided by the change in the partition function (25.63) in an elementary reversible process for a closed system $(N = \text{const})$. Since $\ln Z$ is additive, it is common practice to give preference to $\ln Z$ over Z itself, and this can be expressed as

$$d \ln Z(T, V) = \frac{\partial \ln Z}{\partial T} dT + \frac{\partial \ln Z}{\partial V} dV = \frac{1}{Z} \frac{\partial Z}{\partial T} dT + \frac{1}{Z} \frac{\partial Z}{\partial V} dV$$

$$= \frac{dT}{Z} \frac{1}{k_B T^2} \int H e^{-H/k_B T} d\Omega - \frac{dV}{Z} \frac{1}{k_B T} \int \frac{\partial H}{\partial V} e^{-H/k_B T} d\Omega \quad (25.66)$$

In view of Eq.(25.65) we have

$$\frac{1}{Z} \int H e^{-H/k_B T} d\Omega = \langle H \rangle = \overline{E}$$

$$\frac{1}{Z} \int \frac{\partial H}{\partial V} e^{-H/k_B T} d\Omega = \left\langle \frac{\partial H}{\partial V} \right\rangle = \frac{\partial \overline{E}}{\partial V} = -p \quad (25.67)$$

where the link with pressure p was obtained from Eq.(8.21). Substituting Eq.(25.67), Eq.(25.66) takes the form

$$d \ln Z = \frac{\overline{E}}{k_B T^2} dT + \frac{p}{k_B T} dV = -d\left(\frac{\overline{E}}{k_B T}\right) + \frac{1}{k_B T}(d\overline{E} + p dV) \quad (25.68)$$

where

$$\frac{\overline{E}}{T^2} = -\overline{E} d\left(\frac{1}{T}\right) = -d\left(\frac{\overline{E}}{T}\right) + \frac{1}{T} d\overline{E}$$

Taking Eq.(8.13) into account, Eq.(25.68) can be written as

$$d \ln Z + d\left(\frac{\overline{E}}{k_B T}\right) = \frac{dS}{k_B}$$

which gives the entropy of the system in the form

$$S = k_B \ln Z + \frac{\overline{E}}{T} \quad (25.69)$$

From Eq.(8.16) the Helmholtz function can now be written as

$$F = -k_B T \ln Z \tag{25.70}$$

Thus, the canonical partition function is given in terms of the Helmholtz function F by

$$Z = \int e^{-H/k_B T} d\Omega = e^{-F/k_B T} \tag{25.71}$$

Equation (25.70) plays the same role for the canonical ensemble as Eq.(25.51) does for the microcanonical ensemble. It is a starting point for its complete thermodynamic description. This is made easier since Eq.(25.68) also gives the internal energy $U = \overline{E}$ and the pressure p in terms of the canonical partition function Z as

$$U = k_B T^2 \frac{\partial \ln Z}{\partial T} \tag{25.72}$$

and

$$p = k_B T \frac{\partial \ln Z}{\partial V} \tag{25.73}$$

Therefore all the calculations for canonical ensembles begin with the determination of the partition function (25.63).

25.3. GRAND CANONICAL ENSEMBLE

The canonical ensemble corresponds more closely to physical situations than the microcanonical ensemble, because we never deal with completely isolated systems but with systems of known temperature. However, its usefulness for a number of problems is severely limited, since not only do we not know the energy of the system but also the number of particles is never precisely known. We must regard both E and N as variables of the system, with average values determined by conditions which are external to the system. This is the motivation for introducing the *grand canonical ensemble*, where the elements can have any number of particles, with the average number controlled by keeping the system in contact with a reservoir of particles.

If both E and N are allowed to vary continuously, we may consider an isolated system consisting of a subsystem $H(q_k, p_k, N)$ in contact with a large reservoir $H'(q'_k, p'_k, N')$, with which it can exchange both energy and particles, as in Figure 25.14. The distribution function of the phase points representing a state of energy E and number of particles N of the subsystem must be calculated under the conditions

$$E_0 = E + E' = \text{const} \quad , \quad N_0 = N + N' = \text{const} \tag{25.74}$$

where

$$\frac{E}{E_0} = 1 - \frac{E'}{E_0} \ll 1 \quad , \quad \frac{N}{N_0} = 1 - \frac{N'}{N_0} \ll 1 \tag{25.75}$$

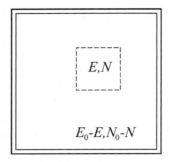

Figure 25.14. Subsystem with variable energy and number of particles

These conditions are analogous with those found for the canonical ensemble, so that an approach similar to that used before leads to a distribution function

$$f_G(q_k, p_k) = \text{const} \cdot \Omega'(E_0 - E, N_0 - N) \tag{25.76}$$

which is an extension of the canonical distribution function (25.58). In view of the assumption (25.75), we may carry out a Taylor's expansion of the logarithm of Eq.(25.76) which, to a first approximation, gives

$$\ln \Omega'(E_0 - E, N_0 - N) = \ln \Omega'(E_0, N_0) - E\left(\frac{\partial}{\partial E'} \ln \Omega'\right)_{E'=E_0} - N\left(\frac{\partial}{\partial N'} \ln \Omega'\right)_{N'=N_0} + \ldots$$

$$\cong \ln \Omega'(E_0, N_0) - \beta E - \delta N \tag{25.77}$$

where use has been made of the definitions (25.68) and (25.82) for β and δ. Recalling that β and δ are directly related to the temperature T and the chemical potential μ_0 of the reservoir through the definitions (25.32) and (25.43), Eq.(25.76) becomes

$$f_G(q_k, p_k) = \text{const} \cdot e^{-H/k_BT + \mu_0 N/k_BT} \tag{25.78}$$

On normalization over all the microstates accessible to the subsystem, we obtain

$$\text{const} \cdot \sum_{N=0}^{\infty} \int e^{-H/k_BT + \mu_0 N/k_BT} d\Omega_N = 1 \tag{25.79}$$

where the sum over the states for a grand canonical ensemble is called the *grand partition function*, given by

$$Z_G = \sum_{N=0}^{\infty} e^{\mu_0 N/k_B T} \int e^{-H/k_B T} d\Omega_N = \sum_{N=0}^{\infty} e^{\mu_0 N/k_B T} Z(N) \qquad (25.80)$$

Here $d\Omega_N$ and $Z(N)$ stand for the phase volume element and the canonical partition function for a fixed number of particles N. Substituting Eqs.(25.79) and (25.80), Eq.(25.78) gives the *grand distribution function* as

$$f_G(q_k, p_k) = \frac{1}{Z_G} e^{-H/k_B T + \mu_0 N/k_B T} \qquad (25.81)$$

The distribution of representative points according to Eq.(25.81) corresponds to the *grand canonical ensemble*, which is a collection of elements having the same temperature T, volume V and chemical potential μ_0. Individual elements are illustrated in Figure 25.15, each occupying a volume V separated by rigid but diathermic and permeable walls. Energy and particles can be exchanged through the walls, all the elements arriving at a common temperature and a common chemical potential.

The ensemble average over the grand canonical ensemble is obtained from Eq.(25.58) in the form

$$\langle F(q_k, p_k) \rangle = \frac{1}{Z_G} \sum_{N=0}^{\infty} e^{\mu_0 N/k_B T} \int F(q_k, p_k) e^{-H/k_B T} d\Omega_N \qquad (25.82)$$

The link between the thermodynamics of an actual system and the statistics of the representative grand canonical ensemble can be developed in the same manner as before, by considering a reversible process during which the logarithm of the partition function (25.80) changes by $d \ln Z_G$ as a result of changes in the independent variables T, V and $\mu_0/k_B T$, and this gives

$$\begin{aligned} d \ln Z_G &= \frac{\partial \ln Z_G}{\partial T} dT + \frac{\partial \ln Z_G}{\partial V} dV + \frac{\partial \ln Z_G}{\partial (\mu_0/k_B T)} d\left(\frac{\mu_0}{k_B T}\right) \\ &= \frac{\overline{E}}{k_B T^2} dT + \frac{p}{k_B T} dV + \overline{N} d\left(\frac{\mu_0}{k_B T}\right) \end{aligned} \qquad (25.83)$$

where the ensemble averages have been replaced as given by Eq.(25.82), that is,

$$\overline{E} = \langle H \rangle = \frac{1}{Z_G} \sum_{N=0}^{\infty} e^{\mu_0 N/k_B T} \int H e^{-H/k_B T} d\Omega_N$$

$$p = \left\langle \frac{\partial H}{\partial V} \right\rangle = \frac{1}{Z_G} \sum_{N=0}^{\infty} e^{\mu_0 N/k_B T} \int \frac{\partial H}{\partial V} e^{-H/k_B T} d\Omega_N \qquad (25.84)$$

$$\overline{N} = \langle N \rangle = \frac{1}{Z_G} \sum_{N=0}^{\infty} N e^{\mu_0 N/k_B T} \int e^{-H/k_B T} d\Omega_N$$

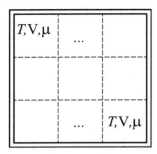

Figure 25.15. Grand canonical ensemble

If we take into account the properties of the differentials expressed as

$$\frac{\overline{E}}{k_B T^2} dT = -d\left(\frac{\overline{E}}{k_B T}\right) + \frac{d\overline{E}}{k_B T} \quad , \quad \overline{N} d\left(\frac{\mu_0}{k_B T}\right) = d\left(\frac{\mu_0 \overline{N}}{k_B T}\right) - \frac{\mu_0}{k_B T} d\overline{N}$$

it follows from Eq.(25.83) that

$$d\left(\ln Z_G + \frac{\overline{E}}{k_B T} - \frac{\mu_0 \overline{N}}{k_B T}\right) = \frac{1}{k_B T}\left(d\overline{E} + pdV - \mu_0 d\overline{N}\right) \tag{25.85}$$

and hence, in view of Eq.(25.84), Eq.(25.85) can be written as

$$S = k_B \ln Z_G + \frac{\overline{E}}{T} - \frac{\mu_0 \overline{N}}{T} \tag{25.86}$$

Since there is equilibrium in both temperature and pressure between the elements of a grand canonical ensemble, $\mu_0 \overline{N}$ is identically equal to the Gibbs function G, written for one phase and one component, as in Eq.(25.41), so that Eq.(25.86) becomes

$$k_B T \ln Z_G = TS - \overline{E} + G = TS - \overline{E} + \left(\overline{E} - TS + pV\right) = pV = -q \tag{25.87}$$

where the so-called *q-potential* plays for the grand canonical distribution the same role the Helmholtz function F plays for the canonical distribution, because

$$Z_G = e^{-q/k_B T} \tag{25.88}$$

The thermodynamic description of the grand canonical ensemble is obtained in terms of the grand partition function Z_G making use of Eq.(25.88) and of the equations for $U = \overline{E}$, p and $\overline{N}$ derived from Eq.(25.83) in the form

$$U = k_B T^2 \frac{\partial \ln Z_G}{\partial T} \tag{25.89}$$

$$p = k_B T \frac{\partial \ln Z_G}{\partial V} \qquad (25.90)$$

$$\overline{N} = \frac{\partial \ln Z_G}{\partial(\mu_0 / k_B T)} \qquad (25.91)$$

As far as the thermodynamic behaviour of actual systems is concerned, the principal problem in statistical mechanics has been reduced to that of computing their partition functions, on the grounds of the particular mechanical models that we assume.

FURTHER READING
1. N. Davidson - STATISTICAL MECHANICS, Dover Publications, New York, 2003.
2. D. A. A. McQuarrie - STATISTICAL MECHANICS, University Science Books, Sausalito, 2000.
3. R. Bowley, M. Sanchez - INTRODUCTORY STATISTICAL MECHANICS, Oxford University Press, Oxford, 1999.

PROBLEMS

25.1. If Δ is the work necessary to create a single vacancy, at constant temperature and pressure, in a crystal of N atomic sites, find the fraction $f = n/N$ of vacant sites as a function of Δ and temperature T.

Solution

The work done in creating n vacant lattice points, assumed to be statistically independent, is $n\Delta$, and hence, their equilibrium state under constant temperature and pressure, as determined by the minimum of the Gibbs function, Eq.(8.28), can be derived from the external condition

$$\frac{\partial G}{\partial n} = \frac{\partial}{\partial n}(U + n\Delta - TdS) = 0$$

Because U and Δ are independent of n, this reduces to $\Delta = T\partial S/\partial n$, where S stands for the configurational entropy given by the Boltzmann relation, Eq.(25.30). The number of ways of arranging n vacancies follows the binomial distribution

$$W = \frac{N!}{n!(N-n)!} \quad \text{or} \quad S = k_B \ln \frac{N!}{n!(N-n)!}$$

Using Stirling's approximation derived in Appendix IV, Eq.(IV.16), one obtains

$$S = k_B\left[N \ln N - n \ln n - (N-n)\ln(N-n)\right] \quad \text{or} \quad \frac{\partial S}{\partial n} = k_B \ln\frac{N-n}{n} = k_B \ln\frac{1-f}{f}$$

If the vacancy concentration is small, $f \ll 1$, one obtains

$$\Delta = k_B T \ln\frac{1-f}{f} = -k_B T \ln f \quad \text{or} \quad f = e^{-\Delta/k_B T}$$

25.2. Derive an expression for the statistical entropy of a system in contact with a heat bath in terms of the probabilities f_k of finding it at temperature T in particular microstates of energy E_k.

Solution

The probability that the system is in a microstate of energy E_k, Eq.(25.17), can be written as

$$f_k = \frac{1}{Z}e^{-E_k/k_B T} \quad , \quad Z = \sum_k e^{-E_k/k_B T}$$

where the partition function Z is found from the normalization condition $\sum_k f_k = 1$ over the accessible microstates. The ensemble average over the canonical ensemble, and hence, Eqs.(25.24) become

$$\overline{E} = \sum f_k E_k \quad , \quad \frac{\partial \overline{E}}{\partial V} = \sum_k f_k \frac{\partial E_k}{\partial V} = -p$$

We can thus write the total change in energy as

$$d\overline{E} = \sum_k E_k df_k + \sum_k f_k \frac{\partial E_k}{\partial V} dV = \sum_k E_k df_k - pdV$$

and, in view of Eq.(8.13), it follows that

$$TdS = \sum_k E_k df_k$$

Rewriting the expression for the probability f_k as $E_k = -k_B T(\ln Z + \ln f_k)$ and substituting for E_k into the last equation gives

$$TdS = -k_B T \ln Z \sum_k f_k - k_B T \sum_k \ln f_k df_k = -k_B T \sum_k \ln f_k df_k$$

The total change in probability must sum to zero, i.e., $\sum_k df_k = d\left(\sum_k f_k\right) = 0$. Thus we have

$$dS = -k_B \sum_k \ln f_k df_k - k_B \sum_k df_k = -k_B \sum_k d(f_k \ln f_k)$$

and integrating both sides one obtains the required expression for the statistical entropy as

$$S = -k_B \sum_k f_k \ln f_k$$

We may define the probabilities f_k by considering a canonical ensemble having n_k elements in the nth microstate, so that the number of ways of arranging this is

$$W = \frac{n!}{n_1! n_2! n_3! \ldots}$$

where $n = n_1 + n_2 + n_3 + \ldots$ is the number of replications of the system. Using Stirling's approximation derived in Appendix IV, Eq.(IV.16), gives

$$\ln W = n \ln n - n - \sum_k (n_k \ln n_k - n_k) = \sum_k n_k (\ln n - \ln n_k) = -\sum_k n_k \ln\left(\frac{n_k}{n}\right) = -n \sum f_k \ln f_k$$

where use has been made of $n = \sum_k n_k$. Since the entropy of the system is defined as the entropy of the ensemble divided by the number n of elements in it, one obtains

$$S = k_B \ln W$$

as expected from Eq.(25.30).

25.3. If a system consisting of two noninteracting distinguishable particles, each having two states with single particle energies 0 and ε, is in equilibrium with a thermal bath at temperature T, work out the internal energy and the entropy of the system.

Solution

The canonical partition function for one particle reads

$$z = \sum_{k=1}^{2} e^{-\varepsilon_k / k_B T} = 1 + e^{-\varepsilon / k_B T}$$

For the two-particle system, the available microstates are $(0,0), (0,\Delta), (\Delta,0)$ and (Δ,Δ), so that the partition function is

$$Z = \sum_{k=1}^{4} e^{-\varepsilon / k_B T} = 1 + 2e^{-\varepsilon / k_B T} + e^{-2\varepsilon / k_B T} = \left(1 + e^{-\varepsilon / k_B T}\right)^2 = z^2$$

Note that if the two particles were indistinguishable this relation would not hold, as we will prove later on. However, the relation holds for a system of more than two particles, whose partition function can be factored in general as $Z = z^N$, provided the particles are distinguishable. It follows that our problem reduces to the statistical properties of a single particle. The probability that a particle is in each of its two possible states is given by

$$p_1 = \frac{1}{z} = \frac{1}{1 + e^{-\varepsilon / k_B T}} \quad , \quad p_2 = \frac{e^{-\varepsilon / k_B T}}{z} = \frac{e^{-\varepsilon / k_B T}}{1 + e^{-\varepsilon / k_B T}}$$

so that the average energy per particle is

$$\bar{\varepsilon} = \sum_{k=1}^{2} p_k \varepsilon_k = \frac{\varepsilon e^{-\varepsilon / k_B T}}{1 + e^{-\varepsilon / k_B T}} = k_B T \frac{\partial \ln z}{\partial T}$$

The average energy of two distinguishable particles of the same type is obtained, in view of Eq.(25.29), as

$$\overline{E} = k_B T^2 \frac{\partial \ln Z}{\partial T} = 2\overline{\varepsilon}$$

The free energy per particle, Eq.(25.27), is given by

$$f = -k_B T \ln z = -k_B T \ln\left(1 + e^{-\varepsilon/k_B T}\right)$$

and the entropy per particle, Eq.(8.25), results as

$$s = -\left(\frac{\partial f}{\partial T}\right)_V = k_B \ln\left(1 + e^{-\varepsilon/k_B T}\right) + \frac{\varepsilon/T}{1 + e^{-\varepsilon/k_B T}}$$

25.4. Find the probability for the fluctuation of a macroscopic variable about its mean value in an isolated system with temperature T_0 and pressure p_0.

Solution

A deviation of some macroscopic variable ξ from its mean value can be described in terms of a nonisolated body at temperature T and pressure p, surrounded by an environment with temperature T_0 and pressure p_0. For the microcanonical ensemble associated with a body in an environment, this implies a change in the density of states $\partial \Omega / \partial E$ in phase space, and hence, in the entropy, Eq.(25.8), so that the probability that a fluctuation of ξ may occur is given by

$$P(\xi) = \text{const} \cdot e^{(\Delta S_0 + \Delta S)/k_B}$$

The change in the entropy of the environment ΔS_0 is given by Eq.(8.13) as

$$\Delta S_0 = \frac{\Delta E_0 + p_0 \Delta V_0}{T_0}$$

but the body entropy change ΔS should also include the work $\Delta W(\xi)$ from an external source as

$$\Delta S = \frac{\Delta E + p\Delta V - \Delta W(\xi)}{T} \approx \frac{\Delta E + p_0 \Delta V - \Delta W(\xi)}{T_0}$$

where it was assumed, for small fluctuations, that $T = T_0 + \Delta T \approx T_0$ and $p \approx p_0$. Since $V + V_0 = \text{const}$, or $\Delta V = -\Delta V_0$ and $E + E_0 = \text{const}$, so that $\Delta E = -\Delta E_0$, we have

$$\Delta S_0 + \Delta S = -\frac{\Delta W(\xi)}{T_0} \quad \text{or} \quad P(\xi) = \text{const} \cdot e^{-\Delta W(\xi)/k_B T_0}$$

For irreversible processes, the total entropy increases as $\Delta S_0 + \Delta S \geq 0$, and hence, $\Delta W(\xi) \leq 0$. A minimal amount of work will be needed to create the fluctuation, and this can be written in terms of the change in the entropy of the body as

$$\Delta W(\xi) = \Delta E - T_0 \Delta S + p_0 \Delta V = \Delta(E - T_0 S + p_0 V)$$

In view of Eq.(8.17), the probability for a fluctuation at $T = T_0$ and $p = p_0$ (a nonisolated container under flexible stopper) reads

$$P(\xi) = \text{const} \cdot e^{-\Delta G / k_B T}$$

In the particular case of fluctuations at $T = T_0$ and $\Delta V = 0$, whenre $\Delta W(\xi) = \Delta(E - T_0 S) = \Delta F$, see Eq.(8.17), we should take

$$P(\xi) = \text{const} \cdot e^{-\Delta F / k_B T}$$

whereas for fluctuations at $p = p_0$ and $\Delta S = 0$ (an isolated container), where $\Delta W(\xi) = \Delta(E + p_0 V) = \Delta H$, the required probability is

$$P(\xi) = \text{const} \cdot e^{-\Delta H / k_B T}$$

25.5. Show that the probability for a fluctuation of an extensive variable x is given by a Gaussian distribution and find expressions for the expectation of x and its variance σ^2.

Solution

The probability has the form derived in Problem 25.4, where it is convenient to set $G(x) = F(x) + Xx$ in terms of the free energy for a given x and the conjugated intensive variable X. Assuming that $G(x)$ has a minimum at $x = x_0$, at constant temperature and pressure, we may use the expansion

$$\Delta G = \Delta F + X \Delta x = \left(\frac{\partial F}{\partial x}\right)_T \Delta x + \frac{1}{2}\left(\frac{\partial^2 F}{\partial x^2}\right)_T (\Delta x)^2 + X \Delta x + \cdots$$

where $\Delta x = x - x_0$. Since $(\partial F / \partial x)_T = -X$, in view of Eq.(8.25), it is convenient to set $\gamma = -\partial X / \partial x$, and we are left with

$$P(x) = \text{const} \cdot e^{-\gamma (\Delta x)^2 / 2 k_B T}$$

Normalization according to Eq.(IV.2), derived in Appendix IV, gives

$$1 = \text{const} \cdot \int_{-\infty}^{\infty} e^{-\gamma (\Delta x)^2 / 2 k_B T} dx = \text{const} \cdot \sqrt{\frac{2 \pi k_B T}{\gamma}} \quad \text{or} \quad P(x) = \sqrt{\frac{\gamma}{2 \pi k_B T}} e^{-\gamma (\Delta x)^2 / 2 k_B T}$$

It immediately follows that the expectation value is

$$\langle x - x_0 \rangle = \int_{-\infty}^{\infty} (x - x_0) P(x) dx = 0 \quad \text{or} \quad \langle x \rangle = x_0$$

and the variance can be written as

$$\sigma^2 = \langle (x-x_0)^2 \rangle = \int_{-\infty}^{\infty} (x-x_0)^2 P(x)\,dx = \frac{k_B T}{\gamma}$$

where use has been made of Eqs.(IV.10), derived in Appendix IV. The probability follows as

$$P(x) = \frac{1}{\sqrt{2\pi\sigma^2}} e^{-(x-x_0)^2/2\sigma^2}$$

Sometimes $1/\gamma$ is called generalized susceptibility. In the case of density fluctuations at $T = \text{const}$, $p = \text{const}$, where x is a volume, from Eqs.(8.25) and (8.42) one obtains

$$\gamma = -\left(\frac{\partial p}{\partial V}\right)_T = \frac{1}{Vk_T} \quad , \quad \sigma^2 = \langle (V-V_0)^2 \rangle = k_B T V k_T$$

and hence, V/γ stands for the coefficient of isothermal compressibility. In the case of entropy fluctuations at constant pressure, we have

$$\Delta G = \Delta(E + p_0 V) - T\Delta S = \Delta H - T\Delta S$$

and hence, x is an entropy. It follows, in view of Eq.(8.24), that

$$\gamma = \left(\frac{\partial T}{\partial S}\right)_p = \frac{T}{C_p} \quad , \quad \sigma^2 = \langle (S-S_0)^2 \rangle = k_B C_p$$

and T/γ stands for the heat capacity C_p.

25.6. If the mean energy of a system in thermal equilibrium at temperature T is $\overline{E}$, find an expression for the variance of energy, and show that the relative fluctuations in energy in the canonical ensemble are vanishingly small in the limit of large N.

Solution

It is convenient to use Eq.(25.32), which gives

$$\frac{\partial}{\partial \beta} = -k_B T^2 \frac{\partial}{\partial T}$$

so that the mean energy (25.72) becomes

$$\overline{E} = -\frac{\partial}{\partial \beta} \ln Z = -\frac{1}{Z}\frac{\partial Z}{\partial \beta}$$

and it follows that

$$\frac{\partial \overline{E}}{\partial \beta} = \frac{1}{Z^2}\left(\frac{\partial Z}{\partial \beta}\right)^2 - \frac{1}{Z}\frac{\partial^2 Z}{\partial \beta^2}$$

From Eqs.(25.65) and (25.63) it is apparent that

$$\overline{E}^2 = \frac{1}{Z}\int H^2 e^{-\beta H}d\Omega = \frac{1}{Z}\frac{\partial^2 Z}{\partial \beta^2}$$

and hence, the variance of energy results as

$$\overline{E^2} - \overline{E}^2 = -\frac{\partial \overline{E}}{\partial \beta} = k_B T^2 \frac{\partial \overline{E}}{\partial T} = k_B T^2 C_V$$

where C_V stands for the heat capacity, Eq.(7.19). Because both E and C_V are extensive quantities, they are proportional to N, and hence the relative fluctuations of energy are given by

$$\frac{\sqrt{\overline{E^2} - \overline{E}^2}}{\overline{E}} = \frac{\sqrt{k_B T^2 C_V}}{\overline{E}} \sim \frac{N^{1/2}}{N} \sim N^{-1/2}$$

Thus, for a macroscopic system, the relative fluctuations in the values of E are very small, so that the mean energy in the canonical ensemble may be considered to be constant at thermal equilibrium.

25.7. A system consisting of three distinguishable particles, each having three possible states of energy ε_1, ε_2 and ε_3, is in equilibrium with a thermal bath at temperature T. Find the free energy F for such a system.

25.8. A crystalline solid in thermal equilibrium at temperature T contains N noninteracting defects of the same type. Assuming that each defect can be in one of the five states of energy $\varepsilon_1 = \varepsilon_2 = 0$ and $\varepsilon_3 = \varepsilon_4 = \varepsilon_5 = \varepsilon$, work out the defect contribution to the internal energy of the solid.

25.9. Consider a solid where electrical conduction is due to ions hopping between point defect sites in the crystal structure. If Δ is the work needed to create a single point defect at constant temperature and pressure, find an expression for the resistance ratio $R(T)/R(T_0)$ at two different temperatures in such a crystal.

25.10. If the intensive variable X is a conjugated field to the extensive variable x, that is $x_k = -\partial E_k / \partial X$, prove the following relations

$$\frac{\partial F}{\partial X} = -\langle x \rangle \quad , \quad \frac{\partial^2 F}{\partial X^2} = \langle x \rangle^2 - \langle x^2 \rangle$$

where F stands for the free energy of a canonical ensemble.

25.11. Find the pressure variance at constant entropy for an isolated system with temperature T_0 and pressure p_0, and write down the normalized Gaussian probability distribution function that would have such a variance.

26. STATISTICAL THERMODYNAMICS

26.1. EQUIPARTITION THEOREM
26.2. STATISTICS OF THE PERFECT GAS
26.3. THE MAXWELL-BOLTZMANN DISTRIBUTION LAW
26.4. THE GIBBS PARADOX

26.1. EQUIPARTITION THEOREM

The purpose of statistical thermodynamics is to apply the general methods of ensemble averaging over the distribution of phase points in the appropriate statistical ensemble, to derive thermal properties from the known Hamiltonians of various actual systems.

We first calculate the ensemble average of $x_i(\partial H / \partial x_j)$, where H is the Hamiltonian of a system which consists of N particles and x_i is either q_i or p_i ($i=1,\ldots,3N$). By averaging over a canonical ensemble, according to Eq.(25.65), we obtain

$$\left\langle x_i \frac{\partial H}{\partial x_j} \right\rangle = \frac{1}{Z} \int \left(x_i \frac{\partial H}{\partial x_j} \right) e^{-H/k_B T} d\Omega \tag{26.1}$$

Integrating by parts gives

$$\int \left(x_i \frac{\partial H}{\partial x_j} \right) e^{-H/k_B T} d\Omega = -k_B T \int \frac{\partial}{\partial x_j}\left(x_i e^{-H/k_B T} \right) d\Omega + k_B T \int \frac{\partial x_i}{\partial x_j} e^{-H/k_B T} d\Omega$$

$$= -k_B T \left[x_i e^{-H/k_B T} \right]_{x_i=-\infty}^{x_i=\infty} + k_B T \delta_{ij} \int e^{-H/k_B T} d\Omega = k_B T Z \delta_{ij} \tag{26.2}$$

where the first term on the right-hand side is zero, since for $x_i = q_i = \pm\infty$ the potential energy in infinite, whilst for $x_i = p_i = \pm\infty$ the kinetic energy is infinite, that is in either case H becomes infinite. It follows that Eq.(26.1) reduces to

$$\left\langle x_i \frac{\partial H}{\partial x_j} \right\rangle = \delta_{ij} k_B T \tag{26.3}$$

which is known as the *generalized equipartition theorem*. The same result can be derived by averaging over a microcanonical ensemble.

For systems described by Hamiltonians given by homogeneous quadratic functions of the conjugate coordinates and reducible to the form

$$H = \sum_i a_i p_i^2 + \sum_i b_i q_i^2 \tag{26.4}$$

we have, according to Eq.(III.1), Appendix III, the relation

$$\sum_i \left(q_i \frac{\partial H}{\partial q_i} + p_i \frac{\partial H}{\partial p_i} \right) = 2H \tag{26.5}$$

Assuming that f of the coefficients a_i and b_i are nonvanishing and substituting Eq.(26.3), Eq.(26.5) gives

$$\langle H \rangle = \frac{1}{2} \left\langle \sum_{i=1}^{f} x_i \frac{\partial H}{\partial x_i} \right\rangle = \frac{1}{2} f k_B T \tag{26.6}$$

This is called the *theorem of equipartition of energy* which states that each harmonic term in the Hamiltonian contributes $k_B T / 2$ the average energy of the system. Therefore, we can say that energy is equally distributed among the degrees of freedom contributing the kinetic energy and those contributing a quadratic term the potential energy of the system.

EXAMPLE 26.1. Heat capacities

For a perfect monatomic gas consisting of N noninteracting molecules we have $f = 3N$ degrees of freedom of translational motion, that is an average energy given by

$$\overline{E} = \langle H \rangle = \frac{3}{2} N k_B T$$

The heat capacity at constant volume follows from Eq.(7.19) as

$$C_V = \left(\frac{\partial \overline{E}}{\partial T} \right)_V = \frac{3}{2} N k_B \tag{26.7}$$

or $C_V = 3R/2$ per mole. In the classical theory of solids, as it will be shown in Chapter 38, a crystal is considered to be a collection of N harmonic oscillators in three dimensions each having a potential energy which is a quadratic function of the position coordinates. Thus $f = 6N$ and Eqs.(26.6) and (7.19) give

$$C_V = 3Nk_B \qquad (26.8)$$

or $C_V = 3R$ per mole. Although it is in rather good agreement with the experimental value for monatomic crystals at room temperature, this expression, called the *Dulong-Petit law*, fails to account for the variation of the molar heat capacity with the temperature, contradicting the predictions (10.29) of the third law of thermodynamics.

For the special case where $i = j$, $x_i = q_i$, Eq.(26.3) reduces to

$$\left\langle q_i \frac{\partial H}{\partial q_i} \right\rangle = -\langle q_i \dot{p}_i \rangle = k_B T \qquad (26.9)$$

while for $i = j$ and $x_i = p_i$ we obtain

$$\left\langle p_i \frac{\partial H}{\partial p_i} \right\rangle = \langle p_i \dot{q}_i \rangle = k_B T \qquad (26.10)$$

where use has been made of the canonical equations (24.27). By applying Eq.(26.9) to a system of N particles, whose state is specified by $f = 3N$ generalized coordinates and $f = 3N$ generalized momenta, we obtain

$$\left\langle \sum_{i=1}^{3N} q_i \dot{p}_i \right\rangle = -3Nk_B T \qquad (26.11)$$

where the sum over the *i*th coordinates times the *i*th components of the generalized force is called the *virial* of the system. If Eq.(26.6) is limited to the degrees of freedom contributing the kinetic energy only (which will be denoted for convenience by T rather than *T*), it reduces to

$$\langle T \rangle = \frac{1}{2} \left\langle \sum_{i=1}^{3N} p_i \frac{\partial H}{\partial p_i} \right\rangle = \frac{3}{2} Nk_B T \qquad (26.12)$$

Substituting Eq.(26.12), Eq.(26.11) becomes

$$\langle T \rangle = -\frac{1}{2} \left\langle \sum_{i=1}^{3N} q_i \dot{p}_i \right\rangle \qquad (26.13)$$

This relationship between the average total kinetic energy T and the virial of the system is known as the *virial theorem*. In Cartesian coordinates Eq.(26.13) takes the form

$$\langle T \rangle = -\frac{1}{2}\left\langle \sum_{i=1}^{N}(X_i x_i + Y_i y_i + Z_i z_i) \right\rangle \tag{26.14}$$

where for the ith particle X_i, Y_i and Z_i stand for the components of force in the x_i-, y_i- and z_i-directions, respectively. In the case of a collection of N harmonic oscillators, for instance, which are bound by harmonic forces of the form

$$X_i = -\frac{\partial U}{\partial x_i} = -k_i x_i \quad , \quad Y_i = -\frac{\partial U}{\partial y_i} = -k_i y_i \quad , \quad Z_i = -\frac{\partial U}{\partial z_i} = -k_i z_i$$

where U is the potential energy of the system, Eq.(2.27), we have

$$\sum_{i=1}^{N}(X_i x_i + Y_i y_i + Z_i z_i) = -\sum_{i=1}^{N} k(x_i^2 + y_i^2 + z_i^2) = -2U$$

Hence, the virial theorem (26.14) simply states that the average values of the kinetic and potential energies of a system of harmonic oscillators are equal.

26.2. Statistics of the Perfect Gas

A perfect gas, consisting of N identical molecules, is a suitable thermodynamic system for illustrating the appropriate method of calculation for each of the three types of ensembles. The Hamiltonian for the perfect gas can be written as

$$H = \frac{1}{2m}\sum_{i=1}^{3N} p_i^2$$

Our task is to derive the equation of state (9.5) and the internal energy function (9.25) for a perfect gas, from which all the other potential functions, that is the macroscopic equilibrium behaviour of the system, follow from the formalism given in Chapter 9.

Microcanonical ensemble

Assuming an isolated perfect gas, of known energy E_0, we must derive the density of states $\partial \Omega / \partial E$ of the system from the Hamiltonian and then the entropy (25.51) can be expressed. We first calculate the phase volume

$$\Omega(E_0) = \int_{0 \leq H \leq E_0} dq_1 dq_2 \ldots dq_{3N} dp_1 dp_2 \ldots dp_{3N} = \Omega_q \Omega_p(E_0)$$

Since the Hamiltonian of the perfect gas is independent of q_k, the integration over

$dq_1 dq_2 dq_3$, which specifies the position of a first molecule, gives a factor of V and so does the integration over the position coordinates of each of the other molecules, so that

$$\Omega_q = V^N$$

where V is the volume of space occupied by the perfect gas. Thus, $\Omega_p(E_0)$ can be interpreted as the volume of a 3N-dimensional sphere, of surface given by the equation

$$H = \frac{1}{2m}\sum p_i^2 = E_0$$

and radius defined by the condition

$$\sum p_i^2 = R^2 \quad \text{or} \quad R = \sqrt{2mE_0}$$

In a 3N-dimensional space, the volume of a sphere must be proportional to the 3Nth power of the radius, that is,

$$\Omega_p(E_0) = \text{const} \cdot R^{3N} = \text{const} \cdot E_0^{3N/2} \quad \text{or} \quad \Omega(E_0) = \text{const} \cdot V^N E_0^{3N/2}$$

The density of states follows as

$$\left(\frac{\partial \Omega}{\partial E}\right)_{E_0} = \text{const} \cdot V^N E_0^{3N/2 - 1}$$

which, upon substitution into Eq.(25.51), gives

$$S = \text{const} + Nk_B \ln V + k_B\left(\frac{3N}{2} - 1\right)\ln E_0 \cong \text{const} + Nk_B \ln V + \frac{3Nk_B}{2}\ln E_0 \quad (26.15)$$

where, for very large N, the unity is negligible. For an isolated system $(dU = 0)$, Eq.(8.13) reads

$$\left(\frac{\partial S}{\partial V}\right)_{E_0} = \frac{p}{T}$$

Substituting Eq.(26.15) gives the equation of state as

$$\frac{Nk_B}{V} = \frac{p}{T} \quad \text{or} \quad pV = Nk_BT$$

From Eq.(25.27), upon substitution of Eq.(26.15), we obtain the internal energy as

$$\frac{3Nk_B}{2E_0} = \frac{1}{T} \quad \text{or} \quad U = E_0 = \frac{3}{2}Nk_BT$$

It is then possible to express the entropy (26.15) in terms of V and T by

$$S(V,T) = S_0 + Nk_B \ln V + \frac{3}{2} Nk_B \ln T \qquad (26.16)$$

As compared to the thermodynamic relation (9.13), Eq.(26.16) allows us to express Boltzmann's constant as

$$k_B = \frac{R}{N_A} \qquad (26.17)$$

Canonical ensemble

For a perfect gas of known temperature T, the calculation begins with that of the canonical partition function (26.63), which reads

$$Z = \int e^{-\left(\sum_{i=1}^{3N} p_i^2\right)/2mk_B T} dq_1 dq_2 \ldots dq_{3N} dp_1 dp_2 \ldots dp_{3N} = V^N \prod_{i=1}^{3N} \int e^{-p_i^2/2mk_B T} dp_i$$

where use has been made of the Hamiltonian H and $\Omega_q = V^N$ as before. Each integral on the right-hand side is a Gaussian integral (IV.2), Appendix IV, so that we have

$$\int e^{-p_i^2/2mk_B T} dp_i = (2\pi m k_B T)^{1/2}$$

It follows that the canonical partition function can be written as

$$Z = V^N (2\pi m k_B T)^{3N/2} \qquad (26.18)$$

which gives

$$\ln Z = N \ln V + \frac{3N}{2} \ln(2\pi m k_B T)$$

Substituting the last equation, Eqs.(25.72) and (25.73) read

$$U = k_B T^2 \frac{3N}{2T} = \frac{3}{2} Nk_B T \quad , \quad p = k_B T \frac{N}{V} = \frac{Nk_B T}{V} \qquad (26.19)$$

and the entropy (25.69) reduces to the form (26.16) derived before, that is

$$S = Nk_B \ln V + \frac{3}{2} Nk_B \ln(2\pi m k_B T) + \frac{3}{2} Nk_B = Ns_0 + Nk_B \ln V + \frac{3}{2} Nk_B \ln T \quad (26.20)$$

where

$$s_0 = \frac{3}{2} k_B \left[1 + \ln(2\pi m k_B) \right]$$

By Eq.(26.20), where the additive constant is $S_0 = Ns_0$, the entropy becomes a truly extensive variable of the system. It can be seen that the canonical ensemble, which gives results which are equivalent to those of the microcanonical ensemble, is more convenient for practical calculations.

Grand canonical ensemble

In the case of a perfect gas of known temperature T and average energy $\overline{E}$ and number of molecules $\overline{N}$ determined by external conditions, we must first calculate the grand partition function (25.80), which becomes

$$Z_G = \sum_{N=0}^{\infty} e^{\mu_0 N/k_B T} \int e^{-\left(\sum_{i=1}^{3N} p_i^2\right)/2mk_B T} dq_1 dq_2 \ldots dq_{3N} dp_1 dp_2 \ldots dp_{3N}$$

$$= \sum_{N=0}^{\infty} e^{\mu_0 N/k_B T} V^N (2\pi m k_B T)^{3N/2} = \sum_{N=0}^{\infty} A^N = \frac{1}{1-A} \qquad (26.21)$$

where

$$A = e^{\mu_0/k_B T} V (2\pi m k_B T)^{3/2}$$

The average number of molecules (25.91) is then given by

$$\overline{N} = -\frac{\partial \ln(1-A)}{\partial(\mu_0/k_B T)} = -\frac{\partial \ln(1-A)}{\partial A} \frac{\partial A}{\partial(\mu_0/k_B T)} = \frac{A}{A-1} \qquad (26.22)$$

The equation of state results from Eq.(25.90) which reads

$$p = -k_B T \frac{\partial \ln(1-A)}{\partial V} = -k_B T \frac{\partial \ln(1-A)}{\partial A} \frac{\partial A}{\partial V} = k_B T \frac{1}{A-1} \frac{A}{V} = \frac{\overline{N} k_B T}{V}$$

where $\partial A / \partial V = A/V$ and $\overline{N}$ were obtained from Eq.(26.22). Finally, from Eq.(25.89) we have

$$U = -k_B T^2 \frac{\partial \ln(1-A)}{\partial T} = -k_B T^2 \frac{\partial \ln(1-A)}{\partial A} \frac{\partial A}{\partial T} = k_B T^2 \frac{1}{A-1} \frac{3A}{2T} = \frac{3}{2} \overline{N} k_B T$$

where $\partial A/\partial T = 3A/2T$, as T and $\mu_0/k_B T$ are independent variables.

Therefore an approach based on the grand canonical ensemble always agrees with the result obtained using the canonical ensemble, providing the physical situation allows us to replace the number of particles N of the system by an average number $\overline{N}$.

26.3. THE MAXWELL-BOLTZMANN DISTRIBUTION LAW

A system consisting of N independent molecules can be described by a canonical ensemble at a known temperature T in a volume V, if each molecule is regarded to be an element of the ensemble. We are not restricted to ideal monatomic gases only, as we can introduce a potential energy function in the Hamiltonian. The Maxwell-Boltzmann distribution law of the kinetic theory of gases can be obtained by applying the canonical ensemble distribution to each of the individual molecules of the macroscopic system, by taking the Hamiltonian in the form

$$H = \frac{1}{2m}\left(p_x^2 + p_y^2 + p_z^2\right) + U(x,y,z) + H'(q_i, p_i) \qquad (26.23)$$

where $U(x,y,z)$ is the potential energy that depends on the location (x,y,z) of a given molecule in the volume V and $q_i = (q_4,\ldots,q_{3N})$, $p_i = (p_4,\ldots,p_{3N})$ denote the remaining degrees of freedom of all the other molecules. If dN is the number of molecules whose position and momenta are confined within the interval

$$x, x+dx \qquad y, y+dy \qquad z, z+dz$$

$$p_x, p_x+dp_x \qquad p_y, p_y+dp_y \qquad p_z, p_z+dp_z$$

the canonical distribution function (25.64) for one molecule can be inserted into Eq.(25.12) to give

$$dP = \frac{dN}{N} = fd\Omega = \frac{e^{-\left(p_x^2+p_y^2+p_z^2\right)/2mk_BT - U/k_BT - H'/k_BT} dxdydzdp_xdp_ydp_zdq_idp_i}{\int e^{-\left(p_x^2+p_y^2+p_z^2\right)/2mk_BT - U/k_BT} dxdydzdp_xdp_ydp_z \int e^{-H'/k_BT} dq_idp_i}$$

The probability that any one molecule is in the location $dxdydz$ and in the momentum state $dp_xdp_ydp_z$, irrespective of other conditions, is obtained by integration over q_i, p_i, which eliminates the dependence on $H'(q_i, p_i)$ and gives

$$dN = \text{const} \cdot e^{-m\left(v_x^2+v_y^2+v_z^2\right)/2k_BT - U/k_BT} dxdydzdv_xdv_ydv_z \qquad (26.24)$$

where the common practice is to express translational momenta in terms of molecular velocities. Equation (26.24) is called the *Maxwell-Boltzmann distribution law*, where the value of the constant depends on the potential energy function $U(x,y,z)$ and can be specified by suitable normalizing conditions.

The distribution of molecules as a result of the forces acting on the system only is obtained from Eq.(26.24), by integrating over the velocity components, which gives

$$dN = \text{const} \cdot e^{-U(x,y,z)/k_BT} dxdydz$$

It is convenient to introduce the number of molecules per unit volume $n = dN/dV = dN/dxdydz$ and to set

$$n_0 = \left(\frac{dN}{dV}\right)_{U=0} = \text{const}$$

so that the last equation can be rewritten as

$$n = n_0 e^{-U(x,y,z)/k_B T} \qquad (26.25)$$

and this is called the *Boltzmann distribution law*.

EXAMPLE 26.2. The barometric equation

Consider the effect of an external field, for instance gravitation, on a perfect gas. Substituting $U(x,y,z) = mgz$, Eq.(26.25) becomes

$$n = n_0 e^{-mgz/k_B T}$$

Using the equation of state (26.19) for a perfect gas, we obtain

$$p = p_0 e^{-mgz/k_B T} = p_0 e^{-Mgz/RT}$$

which is known as the *barometric pressure equation*. For an isothermal atmosphere of a single chemical species, it states that the pressure decreases exponentially with height.

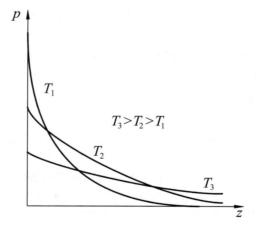

Figure 26.1. Pressure dependence on altitude for an isothermal atmosphere

The height dependence on the isothermal pressure is plotted in Figure 26.1 for different temperatures. The effect of the gravitational field is to increase the pressure at low altitudes whereas increasing temperature makes the distribution of pressure more uniform with altitude.

As we have seen in the preceding section, the perfect gas laws result from consideration of the translational motion alone, and hence, it is instructive to obtain a representation of the velocity distribution which exists in a perfect gas in equilibrium. Assuming that $U = 0$ and integrating over the position coordinates in Eq.(26.24) yields

$$dn = \text{const} \cdot e^{-m(v_x^2 + v_y^2 + v_z^2)/2k_B T} dv_x dv_y dv_z \tag{26.26}$$

The normalizing condition reads

$$n = \text{const} \cdot \int_{-\infty}^{\infty} e^{-mv_x^2/2k_B T} dv_x \int_{-\infty}^{\infty} e^{-mv_y^2/2k_B T} dv_y \int_{-\infty}^{\infty} e^{-mv_z^2/2k_B T} dv_z = \text{const} \cdot \left(\frac{2\pi k_B T}{m} \right)^{3/2}$$

where the integrals over v_x, v_y and v_z have the Gaussian form (IV.2), Appendix IV, and this yields the result shown on the right-hand side. From Eq.(26.26) it follows that

$$\frac{dn}{n} = \left(\frac{m}{2\pi k_B T} \right)^{3/2} e^{-m(v_x^2 + v_y^2 + v_z^2)/2k_B T} dv_x dv_y dv_z \tag{26.27}$$

The result (26.27) represents a *Gaussian probability distribution* curve as plotted in Figure 26.2, when applied to one velocity component

$$f(v_k) = \left(\frac{m}{2\pi k_B T} \right)^{1/2} e^{-mv_k^2/2k_B T} \tag{26.28}$$

where v_k stands for v_x, v_y or v_z. The factorization of Eq.(26.27) is allowed because the three velocity components are statistically independent.

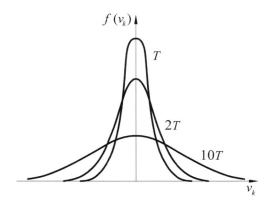

Figure 26.2. One velocity component distribution function

A more important representation is given by the probability distribution function $f(v)$ for molecular speeds. If we change from Cartesian velocity components v_x, v_y, v_z

to the corresponding spherical coordinates v, θ, φ we obtain the volume element in the velocity space as $dv_x dv_y dv_z = v^2 dv \sin\theta d\theta d\varphi$, so that the fraction of molecules in the velocity state $dv d\theta d\varphi$ is

$$\frac{dn}{n} = \left(\frac{m}{2\pi k_B T}\right)^{3/2} e^{-mv^2/2k_B T} v^2 dv \sin\theta d\theta d\varphi \tag{26.29}$$

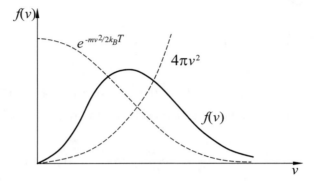

Figure 26.3. Maxwellian distribution of speeds

On integrating over θ and φ we have

$$\frac{dn}{n} = f(v)dv = \left(\frac{m}{2\pi k_B T}\right)^{3/2} e^{-mv^2/2k_B T} 4\pi v^2 dv \tag{26.30}$$

This gives the *Maxwell distribution law* in the form

$$f(v) = 4\pi v^2 \left(\frac{m}{2\pi k_B T}\right)^{3/2} e^{-mv^2/2k_B T} \tag{26.31}$$

As illustrated in Figure 26.3, the Maxwellian distribution is a product of a Boltzmann factor $e^{-mv^2/2k_B T}$ that follows the Gaussian distribution and a density of states factor $4\pi v^2$ which increases with energy, so that $f(v)$ has a maximum. At a given temperature, the function $f(v)$ indicates the distribution of translational kinetic energy for one molecule of a perfect gas. The change in that distribution at various temperatures is plotted in Figure 26.4, which shows that the maximum is shifted towards higher speeds as temperature increases and, because more and more particles move at speeds in this range, the probability tends to a more symmetrical distribution.

26. STATISTICAL THERMODYNAMICS

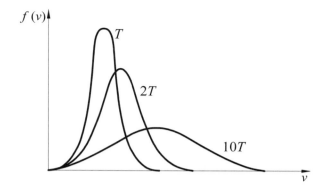

Figure 26.4. Maxwellian speed distribution at various temperatures

---

EXAMPLE 26.3. Most probable, average and root-mean-square speeds

The *most probable molecular speed* v_p is obtained from Eq.(26.31), which gives

$$\frac{d}{dv}\left(v^2 e^{-mv^2/2k_BT}\right) = e^{-mv^2/2k_BT}\left(2v - \frac{mv}{k_BT}v^2\right) = 0 \quad \text{or} \quad v_p = \left(\frac{2k_BT}{m}\right)^{1/2} = \left(\frac{2RT}{M}\right)^{1/2}$$

where $M = mN_A$. Figure 26.4 indicates that v_p increases as the temperature increases. The *average molecular speed* $\bar{v}$ is obtained by using the distribution function (26.31), which gives

$$\bar{v} = \langle v \rangle = \int_0^\infty vf(v)dv = 4\pi\left(\frac{m}{2\pi k_BT}\right)^{3/2}\int_0^\infty v^3 e^{-mv^2/2k_BT}dv$$

and this can be evaluated (see Eq.(IV.8), Appendix IV) as

$$\bar{v} = 4\pi\left(\frac{m}{2\pi k_BT}\right)^{3/2}\frac{1}{2}\left(\frac{2k_BT}{m}\right)^2 = \left(\frac{8k_BT}{\pi m}\right)^{1/2} = \left(\frac{8RT}{\pi M}\right)^{1/2}$$

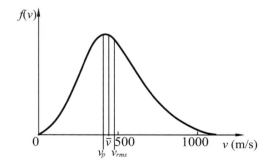

Figure 26.5. Relative positions of $v_p, \bar{v}$ and v_{rms} on the Maxwellian distribution curve

In a similar manner we obtain the average of the speed squared as

$$\langle v^2 \rangle = \int_0^\infty v^2 f(v)dv = 4\pi \left(\frac{m}{2\pi k_B T}\right)^{3/2} \int_0^\infty v^4 e^{-mv^2/2k_B T} dv$$

so that, in view of Eq.(IV.5), Appendix IV, we obtain

$$\langle v^2 \rangle = 4\pi \left(\frac{m}{2\pi k_B T}\right)^{3/2} \frac{3\sqrt{\pi}}{8} \left(\frac{2k_B T}{m}\right)^{5/2} = \frac{3k_B T}{m}$$

The *root-mean-square speed* is then defined by

$$v_{rms} = \sqrt{\langle v^2 \rangle} = \left(\frac{3k_B T}{m}\right)^{1/2} = \left(\frac{3RT}{m}\right)^{1/2}$$

The relationship between the three characteristic molecular speeds in the Maxwellian distribution for a perfect gas is illustrated in Figure 26.5. They are in a constant ratio to one another, that is

$$v_p : \bar{v} : v_{rms} = \sqrt{2} : \sqrt{\frac{8}{\pi}} : \sqrt{3} = 1.41 : 1.59 : 1.73$$

Their order of magnitude is indicated in Figure 26.5 for nitrogen molecules at room temperature.

26.4. THE GIBBS PARADOX

Consider two perfect gases of N_1 and N_2 molecules, contained in two separate volumes V_1 and V_2, respectively, at the same temperature T. The removal of the wall separating the gases permits them to mix by molecular diffusion as in Figure 26.6. After sufficient time, a uniform mixture of $N_1 + N_2$ molecules will exist throughout the whole volume $V_1 + V_2$. If the system is isolated during the mixing process, so that no heat enters or leaves, the internal energy is conserved, following the first law of thermodynamics. Since the internal energy (26.19) depends on temperature and the number of molecules only, it follows that the temperature remains unchanged, as N_1 and N_2 are constant.

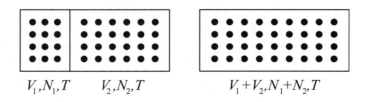

V_1, N_1, T V_2, N_2, T V_1+V_2, N_1+N_2, T

Figure 26.6. Configuration for the calculation of the entropy of mixing

The entropies of the separate gases, as given by Eq.(26.20), are

$$S_1 = N_1 s_0 + N_1 k_B \ln V_1 + \frac{3}{2} N_1 k_B \ln T$$

$$S_2 = N_2 s_0 + N_2 k_B \ln V_2 + \frac{3}{2} N_2 k_B \ln T$$

(26.32)

and their sum is the initial entropy of the system

$$S_i = S_1 + S_2 \tag{26.33}$$

The final entropy is that of the mixture, which also has the form (26.20), that is

$$S_f = (N_1 + N_2)s_0 + (N_1 + N_2)k_B \ln(V_1 + V_2) + \frac{3}{2}(N_1 + N_2)k_B \ln T \tag{26.34}$$

The change in entropy, resulting from Eqs.(26.32) to (26.34), becomes

$$\Delta S = S_f - S_i = N_1 k_B \ln\left(1 + \frac{V_2}{V_1}\right) + N_2 k_B \ln\left(1 + \frac{V_1}{V_2}\right) > 0 \tag{26.35}$$

which is always positive and gives the entropy of mixing. The result is experimentally correct if the two gases are different. It is due to the volume increase for each component, which can be thought of as the increased range of the position coordinates of the gas molecules. However, since the derivation of Eq.(26.35) does not depend on the identity of the gases, in the special case where the two mixing gases are identical we would obtain the same increase in entropy. In other words, removing the wall in Figure 26.6 would always result in an entropy increase even if the initial pressures on either side were equal. This conclusion is known as the *Gibbs paradox* and not only violates our intuition, but also implies that the entropy depends on the history of the gas and thus cannot be a function of the thermodynamic state alone.

The explanation of the Gibbs paradox is contained in the expression (26.18) of the canonical partition function Z for a perfect gas, which may be expressed, since there are no interactions, in terms of the partition function z for a single molecule

$$Z = V^N (2\pi m k_B T)^{3N/2} = \left(\int dq_i \int e^{-p_i^2/2mk_B T} dp_i\right)^{3N} = z^N \tag{26.36}$$

Consider a system of two identical molecules, labelled as 1 and 2, with the partition function

$$Z = z_1 z_2 = \int dq_{1x} dq_{1y} dq_{1z} \int e^{-p_1^2/2mk_B T} dp_{1x} dp_{1y} dp_{1z} \int dq_{2x} dq_{2y} dq_{2z} \int e^{-p_2^2/2mk_B T} dp_{2x} dp_{2y} dp_{2z}$$

The symmetrical representative points $(p_1, p_2) = (a,b)$ and (b,a) in the plot given in Figure 26.7, where the molecules are interchanged, contribute separately to the partition function (26.36), in which the integration is carried out over p_1, at fixed p_2, then over p_2 at fixed p_1. Thus, when the molecules 1 and 2 are different, the corresponding microstate is duplicated in Figure 26.7 as well as in phase space. But when the two molecules are *indistinguishable*, it is unphysical to label them as 1 and 2, because the phase points (a,b) and (b,a) correspond to the same microstate. As a result, we must divide $z_1 z_2$ by 2! to obtain the correct counting of microstates.

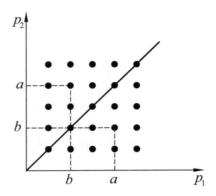

Figure 26.7. Phase points corresponding to a permutation of two molecules

Similarly, for an N-molecule system, the number of ways of interchanging the molecules is the number of permutations $N!$. Thus, for a perfect gas consisting of N indistinguishable molecules, the total number of microstates (26.36) compatible with a given macrostate (T, V, N) must be divided by $N!$. This rule of counting is known as the *correct Boltzmann counting* and assumes that the partition function (26.36) must be written as

$$Z = \frac{1}{N!} z^N = \frac{1}{N!} V^N (2\pi m k_B T)^{3N/2} \qquad (26.37)$$

It follows that the canonical partition function (25.63), if applied to systems of *identical particles*, must be modified according to the same rule

$$Z = \frac{1}{N!} \int e^{-H/k_B T} d\Omega \qquad (26.38)$$

This implies that the grand partition function (25.80) is also similarly modified. By the assumption of Eq.(26.37), we obtain

$$\ln Z = N \ln V + \frac{3}{2} N \ln(2\pi m k_B T) - \ln N! = N \ln \frac{V}{N} + \frac{3}{2} N \ln(2\pi m k_B T) + N \qquad (26.39)$$

where use has been made of Stirling's approximation for $\ln N!$ derived in Appendix IV, Eq.(IV.16). Substituting Eq.(26.39), Eq.(25.69) gives the entropy as

$$S = N s_0 + N k_B \ln \frac{V}{N} + \frac{3}{2} N k_B \ln T \qquad (26.40)$$

where
$$s_0 = \frac{3}{2} k_B \left[\frac{5}{3} + \ln(2\pi m k_B) \right] \qquad (26.41)$$

Equation (26.40), which represents the classical limit of the so-called *Sackur-Tetrode equation*, leads to the same equation of state and internal energy (26.19) of the perfect gas as the entropy (26.20) derived earlier, because the subtracted term $\ln N!$ is independent of T and V. As compared to Eq.(26.20), it shows, however, that the entropy depends on the specific volume V/N instead of the volume V of the system. For the mixing of two different gases, Eq.(26.40) predicts that

$$\Delta S = (N_1 + N_2) k_B \ln \frac{V_1 + V_2}{N_1 + N_2} - N_1 k_B \ln \frac{V_1}{N_1} - N_2 k_B \ln \frac{V_2}{N_2}$$

$$= N_1 k_B \ln \frac{(V_1 + V_2) N_1}{(N_1 + N_2) V_1} + N_2 k_B \ln \frac{(V_1 + V_2) N_2}{(N_1 + N_2) V_2} \qquad (26.42)$$

When V_2 is a vacuum, N_2 is zero, and hence, the entropy increase is $N_1 k_B \ln(1 + V_2/V_1)$ in agreement with Eq.(26.35). If the two gases are of the same kind, and the pressures in V_1 and V_2 are initially equal, the equation of state (26.19) requires that the specific volume is the same before and after mixing, that is

$$\frac{V_1}{N_1} = \frac{V_2}{N_2} = \frac{V_1 + V_2}{N_1 + N_2}$$

so that Eq.(26.42) gives no entropy of mixing.

Note that the rule of the correct Boltzmann counting, assumed by Gibbs in order to resolve the paradox, is not consistent with the principles of Newtonian mechanics, where the identical molecules in a gas must be regarded as distinguishable. As we shall show in Chapter 35, the reason why identical particles are indistinguishable is inherently quantum mechanical.

FURTHER READING
1. R. L. Liboff - KINETIC THEORY, Springer-Verlag, New York, 2003.
2. A. Maczek - STATISTICAL THERMODYNAMICS, Oxford University Press, Oxford, 1998.
3. W. Greiner, H. Stocker, L. Neise - THERMODYNAMICS AND STATISTICAL MECHANICS, Springer-Verlag, New York, 1994.

Problems

26.1. A perfect gas consisting of N molecules is in thermal equilibrium at a temperature T. If the molecules move in an external potential energy $U(x,y,z) = \alpha\sqrt{x^2+y^2+z^2}$, find the contribution of that potential field to the internal energy of the gas.

Solution

For systems of identical particles the partition function has the form (26.38), which reads

$$Z = \frac{1}{N!}\left(\int e^{-H/k_BT}d\Omega\right)^N = \frac{1}{N!}z^N$$

where H is the single particle Hamiltonian, given by Eq.(26.23) as

$$H = \frac{1}{2m}\left(p_x^2 + p_y^2 + p_z^2\right) + \alpha\sqrt{x^2+y^2+z^2}$$

The canonical partition function z for one molecule is

$$z = \int e^{-\left(p_x^2+p_y^2+p_z^2\right)/2mk_BT}dp_xdp_ydp_z \int e^{-\alpha\sqrt{x^2+y^2+z^2}}dxdydz$$

$$= \left(\int_{-\infty}^{\infty} e^{-p_i^2/2mk_BT}dp_i\right)^3 4\pi\int_0^{\infty} e^{-\alpha r/k_BT}r^2 dr = (2\pi mk_BT)^{3/2} 4\pi\left(\frac{k_BT}{\alpha}\right)^3 \int_0^{\infty} u^2 e^{-u}du = \text{const}\cdot T^{9/2}$$

and that of the gas reads

$$Z = \frac{\text{const}}{N!}T^{9N/2}$$

It follows immediately that

$$\ln Z = \frac{9N}{2}\ln T - N\ln N + N + \text{const}$$

and hence, Eq.(25.72) gives

$$U = k_BT\frac{9N}{2T} = \frac{9}{2}Nk_BT$$

In view of Eq.(26.19), it follows that the external field contributes $3Nk_BT$ the internal energy of the gas.

26.2. A perfect gas is in thermal equilibrium at temperature T inside a cylinder of radius R and height h. If the cylinder is set rotating with constant angular velocity ω about its vertical axis, find the change in the internal energy of the gas.

26. STATISTICAL THERMODYNAMICS

Solution

It is convenient to use cylindrical coordinates, so that the distribution of molecules, as a result of external forces only, becomes

$$dN(r,\theta,z) = \text{const} \cdot e^{-U(r,\theta,z)/k_BT} r\,dr\,d\theta\,dz$$

where the potential energy of a single molecule reads

$$U = -mgz - \int_0^r m\omega^2 r\,dr = mgz - \frac{1}{2}m\omega^2 r^2$$

and the normalizing condition gives

$$N = \text{const} \cdot \int_0^R e^{m\omega^2 r^2/2k_BT} r\,dr \int_0^h e^{-mgz/k_BT} dz = \text{const} \cdot \frac{k_BT}{m\omega^2}\left(e^{m\omega^2 R^2/2k_BT} - 1\right)\frac{k_BT}{mg}\left(1 - e^{-mgh/k_BT}\right)$$

As there is symmetry about the z-axis, integrating $dN(r,\theta,z)$ over θ gives

$$dN(r,z) = Ng\left(\frac{m\omega}{k_BT}\right)^2 \frac{e^{m\omega^2 r^2/2k_BT}}{e^{m\omega^2 R^2/2k_BT} - 1}\frac{e^{-mgz/k_BT}}{1 - e^{-mgh/k_BT}} r\,dr\,dz$$

We may further separate the radial from the vertical part of the distribution in the form

$$dN(r) = \int_0^h dN(r,z) = N\frac{m\omega^2}{k_BT}\frac{e^{m\omega^2 r^2/2k_BT}}{e^{m\omega^2 R^2/2k_BT} - 1} r\,dr$$

$$dN(z) = \int_0^R dN(r,z) = N\frac{mg}{k_BT}\frac{e^{-mgz/k_BT}}{1 - e^{-mgh/k_BT}} dz$$

The additional internal energy due to the external forces is given by

$$\Delta U = \int U(r,z)\,dN(r,z) = \int_0^R\left(-\frac{1}{2}m\omega^2 r^2\right)dN(r) + \int_0^z mgz\,dN(z)$$

$$= N\frac{m\omega^2/k_BT}{e^{m\omega^2 R^2/2k_BT} - 1}\int_0^R e^{m\omega^2 r^2/2k_BT}\left(-m\omega^2 r^2\right)d(r^2) + N\frac{mg/k_BT}{1 - e^{-mgh/k_BT}}\int_0^h e^{-mgz/k_BT} mgz\,dz$$

We may set $u = m\omega^2 r^2/2k_BT$ and $v = mgz/k_BT$, and this gives

$$\Delta U = \frac{Nk_BT}{1 - e^{u_0}}\int_0^{u_0} e^u u\,du + \frac{Nk_BT}{1 - e^{-v_0}}\int_0^{v_0} e^{-v} v\,dv = Nk_BT\left(1 - \frac{u_0}{1 - e^{-u_0}}\right) + Nk_BT\left(1 - \frac{v_0}{e^{v_0} - 1}\right)$$

Using the expansions

$$\frac{u_0}{1 - e^{-u_0}} = 1 + \frac{u_0}{2} + \frac{u_0^2}{12} + \cdots \text{ if } u_0 = \frac{m\omega^2 R^2}{2k_BT} \ll 1 \quad , \quad \frac{v_0}{e^{v_0} - 1} = 1 - \frac{v_0}{2} + \frac{v_0^2}{12} + \cdots \text{ if } v_0 = \frac{mgh}{k_BT} \ll 1$$

one obtains the change in the internal energy of the gas as

$$\Delta U \cong Nk_BT\left(-\frac{u_0}{2}-\frac{u_0^2}{12}+\frac{v_0}{2}-\frac{v_0^2}{12}\right)=\frac{N}{2}\left(mgh-\frac{m\omega^2R^2}{2}\right)-\frac{N}{12k_BT}\left[(mgh)^2+\left(\frac{m\omega^2R^2}{2}\right)^2\right]$$

26.3. If a particle is in Brownian motion at a finite temperature caused by collisions of neighbouring gas particle, which can be represented by a random force $f(t)$, find its mean square displacement from the origin after a time t.

Solution

The Brownian particle experiences a viscous drag $-v_x/B$, where B stands for its mobility, and a rapidly fluctuating force $f(t)$ which averages out to zero over long intervals of time, so that we may write

$$\frac{dv_x}{dt}=-\frac{v_x}{mB}-\frac{f(t)}{m} \quad \text{or} \quad \frac{d}{dt}\langle v_x\rangle=-\frac{1}{mB}\langle v_x\rangle$$

because $\langle f(t)\rangle=0$. It follows that the drift velocity decays exponentially at a rate determined by the relaxation time $\tau=mB$. It is convenient to calculate, at some instantaneous position x, the average product

$$\left\langle x\frac{dv_x}{dt}\right\rangle=-\frac{1}{\tau}\langle xv_x\rangle+\frac{1}{m}\langle xf(t)\rangle$$

taking into account that $2x(dv_x/dt)=d^2\langle x^2\rangle/dt^2-2v_x^2$ and $2xv_x=d\langle x^2\rangle/dt$. Thus, we obtain

$$\frac{d^2}{dt^2}\langle x^2\rangle+\frac{1}{\tau}\frac{d}{dt}\langle x^2\rangle=2\langle v_x^2\rangle$$

where $\langle v_x^2\rangle$ is given, in view of Eq.(26.28), by

$$\langle v_x^2\rangle=\int_{-\infty}^{\infty}v_x^2 f(v_x)dv_x=\left(\frac{m}{2\pi k_BT}\right)^{1/2}\int_{-\infty}^{\infty}v_x^2 e^{-mv_x^2/2k_BT}dv_x=\frac{k_BT}{m}$$

The equation for $\langle x^2\rangle$ is readily integrated with the result

$$\langle x^2\rangle=\frac{2k_BT\tau^2}{m}\left[\frac{t}{\tau}-\left(1-e^{-t/\tau}\right)\right] \quad \text{or} \quad \langle x^2\rangle=2Bk_BTt \quad (t\gg\tau)$$

The irreversible character of this result arises essentially from the viscosity of the system.

26.4. A constant volume gas thermometer contains a perfect gas of N particles at thermal equilibrium. Find an expression for the variance of temperature and use it to estimate the accuracy with which the instrument reads the temperature of the thermal bath.

26. Statistical Thermodynamics

Solution

The probability for a fluctuation at constant volume has the form derived in Problem 25.4, where ΔF is given by the expansion

$$\Delta F = \Delta E - T\Delta S = \left(\frac{\partial E}{\partial S}\right)_V \Delta S + \frac{1}{2}\left(\frac{\partial^2 E}{\partial S^2}\right)_V (\Delta S)^2 - T\Delta S + \cdots$$

In view of Eqs.(8.21) and (8.22), one obtains

$$\Delta F = \frac{1}{2}\left(\frac{\partial T}{\partial S}\right)_V (\Delta S)^2 = \frac{1}{2}\left(\frac{\partial S}{\partial T}\right)_V (\Delta T)^2 = \frac{C_V}{2T}(\Delta T)^2$$

since we have $\Delta S = (\partial S/\partial T)\Delta T$. If we put $\gamma = C_p/T$ it follows, as in Problem 25.5, that

$$P(T) = \sqrt{\frac{C_V}{2\pi k_B T^2}} e^{-C_V(\Delta T)^2/2k_B T^2}$$

and the variance of temperature is given by

$$\left\langle (\Delta T)^2 \right\rangle = \frac{k_B T^2}{C_V}$$

Substituting C_V for a perfect gas from Eq.(26.7), we can estimate the accuracy of the thermometer in terms of the relative fluctuation

$$\frac{\sqrt{\left\langle (\Delta T)^2 \right\rangle}}{T} = \sqrt{\frac{2}{3N}} \sim N^{-1/2}$$

which is a vanishingly small value for a real thermometer.

26.5. Consider a gas of particles interacting through a two-particle potential

$$u(r) = \infty \quad , \quad r < \sigma$$
$$= -u_0 \quad , \quad \tau < r < 2\tau$$
$$= 0 \quad , \quad r > \tau$$

where r denotes the distance between two particles, σ is the molecular diameter, and u_0 is some positive constant. Derive the virial expansion of the equation of state for such a gas.

Solution

The virial expansion (9.1) can be obtained from Eq.(25.73), where the partition function should include the contribution from particle interactions. Thus, Eq.(26.18) becomes

$$Z = (2\pi m k_B T)^{3N/2} \int e^{-\left[\sum_{i>j} u(r_{ij})\right]/k_B T} dq_1 dq_2 \ldots dq_{3N}$$

This can be written as

$$Z = (2\pi m k_B T)^{3N/2} \int \prod_{i>j} e^{-u(r_{ij})/k_B T} dq_1 dq_2 \ldots dq_{3N}$$

$$= (2\pi m k_B T)^{3N/2} \int \prod_{i>j} \left[1 + f(r_{ij})\right] dq_1 dq_2 \ldots dq_{3N}$$

where, for convenience, the interaction is represented in terms of the Mayer function as

$$f(r) = e^{u(r)/k_B T} - 1$$

It is apparent that $f(r)$ is small, as $u(r)$ represents a weak interaction and $f(r) \to -1$ when $u(r) \to \infty$, which corresponds to a collision between hard spheres. Assuming a low density gas, where collisions are rare, we may neglect products of Mayer functions and Z reduces to

$$Z = (2\pi m k_B T)^{3N/2} \int \left[1 + \sum_{i>j} f(r_{ij}) + \cdots\right] dq_1 dq_2 \ldots dq_{3N}$$

$$= (2\pi m k_B T)^{3N/2} \left[V^N + \sum_{i>j} \int f(r_{ij}) dq_1 dq_2 \ldots dq_{3N}\right]$$

The last sum has $N(N-1)/2 \approx N^2/2$ terms of the form

$$\int f(r_{ij}) dq_1 dq_2 \ldots dq_{3N} = V^{N-1} \int f(\vec{r}) d\vec{r} = -2V^{N-1} B$$

where B stands for the second virial coefficient defined as $B = -\frac{1}{2} \int f(\vec{r}) d\vec{r}$. Thus we obtain

$$Z = (2\pi m k_B T)^{3N/2} V^N \left(1 - \frac{BN^2}{V} + \cdots\right)$$

and hence

$$\ln Z = N \ln V + \ln\left(1 - \frac{BN^2}{V} + \cdots\right) + \frac{3N}{2} \ln(2\pi m k_B T)$$

$$\approx N \ln V + \left(1 - \frac{BN^2}{V} + \cdots\right) + \frac{3N}{2} \ln(2\pi m k_B T)$$

because $\ln(1-x) \approx 1-x$. From Eq.(25.73) we obtain the virial expansion as

$$p = \frac{N k_B T}{V}\left(1 + \frac{NB}{V} + \cdots\right)$$

The second virial coefficient has the explicit form

$$B = -\frac{1}{2}\int_0^\sigma \left[e^{-u(r)/k_BT} - 1\right]4\pi r^2 dr - \frac{1}{2}\int_\sigma^\infty \left[e^{-u(r)/k_BT} - 1\right]4\pi r^2 dr$$

$$= \frac{2\pi}{3}\sigma^3 + \frac{2\pi}{k_BT}\int_\sigma^\infty u(r)r^2 dr = b - \frac{a}{k_BT}$$

where b, which is four times the volume of one molecule, describes the steric repulsion, and a is a coefficient associated with long-range interactions given by

$$a = -2\pi \int_\sigma^\infty u(r)r^2 dr$$

which is positive provided the potential $u(r) \ll k_BT$ is attractive for $r > \sigma$. Assuming a finite-range attraction of the form $u(r) = -u_0$, we have

$$a = 2\pi u_0 \int_\sigma^{2\sigma} r^2 dr = 7u_0 b \quad \text{or} \quad B = b\left(1 - \frac{7u_0}{k_BT}\right)$$

and the equation of state can be approximated by

$$\frac{pV}{Nk_BT} = 1 + \frac{Nb}{V}\left(1 - \frac{7u_0}{k_BT}\right)$$

26.6. Use the theorem of equipartition of energy to find the energy relative fluctuation for a perfect gas of f degrees of freedom at thermal equilibrium.

26.7. Find the distribution of the molecular kinetic energy in a perfect gas at thermal equilibrium, and show that the most probable kinetic energy is different from that corresponding to the most probable molecular speed.

26.8. Find the ratio of the variance of the molecular speed $\sigma^2(v)$ to the variance of the vector velocity $\sigma^2(\vec{v})$ for a perfect gas at thermal equilibrium.

26.9. Find the average speed $\bar{v}_R$ of the relative molecular motion in a perfect gas that is in thermal equilibrium at a temperature T.

26.10. Put the van der Waals equation into virial form, by expanding it into a power series, and find the second and the third virial coefficients as functions of the constants a and b.

27. Thermal Radiation

27.1. THERMODYNAMICS OF RADIATION
27.2. STATISTICS OF RADIATION
27.3. PLANCK'S RADIATION FORMULA

27.1. Thermodynamics of Radiation

When electromagnetic radiation hits the surface of a body, it is partially reflected and partially penetrates the body, either passing through it or being absorbed by it. At the same time, a body at a temperature above absolute zero will radiate heat in the form of electromagnetic waves. The laws governing the emitted radiation are especially simple for an opaque body (which does not transmit radiation) if it absorbs all incident waves and reflects none from its surface. Such a body is called a *black body*. In a cavity where the walls are kept at constant temperature T, which is an approximation to a black body, the state of thermodynamic equilibrium can be experimentally obtained with respect to the exchange of electromagnetic radiation. For this reason it is commonly known as *thermal radiation*. The condition for equilibrium is that a balance should be reached between the rates of emission and absorption by unit area of the inner cavity surface. Any incident radiation entering by a small aperture in such a cavity is reflected repeatedly inside before leaving again, so that most of it is absorbed. Thus, the cavity aperture acts as a black body and, if heated, it emits the so called *black body radiation*.

The energy density u of the radiation in the cavity is dependent on the temperature only, and not on the shape of the cavity or on the material of its walls. In order to prove this statement let us examine the hypothetical contact of two cavities 1 and 2, different in shape and wall material but at the same temperature T, as illustrated in Figure 27.1. According to the second law of thermodynamics, the energy density will be the same in both cavities, otherwise an energy transfer would hold from one cavity to the other, at constant temperature. It follows that

$$u_1 = u_2 = u(T)$$

Therefore the radiation in the cavity can be treated as a hydrostatic system, in the sense given in Chapter 7, using the temperature T of the walls and the volume V of the cavity.

Hence, all the extensive thermodynamic variables of the radiation are proportional to the volume. For instance $U = Vu$, $S = Vs$ where the energy and entropy densities are functions of temperature only.

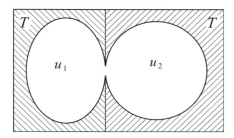

Figure 27.1. Hypothetical contact between two different cavities

The pressure of the thermal radiation can be expressed using Eq.(17.36) under the assumption of an isotropic incidence on the inner cavity surface. An electromagnetic radiation incident on an element of surface dA, at an angle θ from the normal, as illustrated in Figure 27.2, transfers to the surface the component $|\vec{\pi}|\cos\theta$ of the momentum density $\vec{\pi}$ transported by the wave. It follows that the component of the transferred momentum normal to the surface element is $(|\vec{\pi}|\cos\theta)2\cos\theta = 2|\vec{\pi}|\cos^2\theta$, where the factor 2 is caused by the recoil of the radiation emitted from the surface element dA, which transfers the same momentum as the incident radiation.

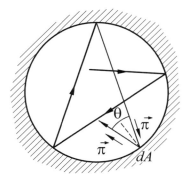

Figure 27.2. Radiation pressure on the cavity inner surface

Equation (17.36) must then be integrated over the solid angle about the surface element

$$p = \frac{1}{4\pi}\int \langle u_{em}\rangle 2\cos^2\theta d\Omega = \frac{u}{4\pi}\int_0^{2\pi}\int_0^{\pi/2} 2\cos^2\theta \sin\theta d\theta d\varphi = \frac{1}{3}u \quad (27.1)$$

where $u \equiv \langle u_{em}\rangle$. Comparison between Eqs.(27.1) and (17.36) shows that the thermal radiation pressure on the inner surface of the cavity obeys the theorem of equipartition of energy (26.6), since its energy density along a given direction of propagation is a third of the total energy density.

We may now apply the internal energy equation (8.40), where the radiation energy is expressed by

$$U(T) = Vu(T) \tag{27.2}$$

and the radiation pressure is given by Eq.(27.1), so that we obtain

$$u = \frac{T}{3}\frac{du}{dT} - \frac{u}{3} \quad \text{or} \quad \frac{du}{u} = 4\frac{dT}{T} \quad \text{i.e.,} \quad \ln u = \ln T^4 + \ln a$$

where a is a constant, so that

$$u = aT^4 \tag{27.3}$$

This is the *Stefan-Boltzmann law* which states that the total amount of radiation energy per unit volume is proportional to T^4. It has been found experimentally that $a = 7.6 \times 10^{-16}$ J/m³K⁴. Hence, the total radiation energy is given by

$$U = aVT^4 \tag{27.4}$$

and Eq.(27.1) reads

$$p = \frac{1}{3}aT^4 \tag{27.5}$$

The entropy change in an elementary reversible process, given by Eq.(8.13), becomes

$$dS = \frac{1}{T}dU + \frac{p}{T}dV = 4aVT^2 dT + \frac{4}{3}aT^3 dV$$

so that the radiation entropy has a temperature dependence given by

$$S = \frac{4}{3}aVT^3 \tag{27.6}$$

The condition for entropy to be a constant gives the adiabatic equation as

$$VT^3 = \text{const} \tag{27.7}$$

Since the thermal radiation has a continuous spectrum of frequencies extending over a wide range, it is convenient to characterize this mixture of monochromatic waves by a function $u_\omega = u(\omega, T)$ called the *spectral energy density*, which defines the radiation energy per unit volume and unit frequency interval as

$$u(T) = \int u(\omega, T) d\omega \quad \text{or} \quad u(T) = \int u_\omega d\omega \tag{27.8}$$

Experiments have confirmed that u_ω only depends on the frequency ω of the radiation and the temperature T of the cavity, as illustrated in Figure 27.3. From Eqs.(27.8) and (27.3) it follows that the area under the energy distribution curve is proportional to T^4. The maximum spectral energy density and its position on the distribution curve can also be derived from thermodynamic arguments only.

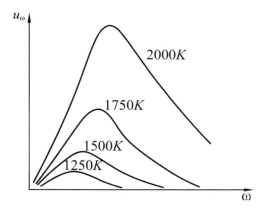

Figure 27.3. Curves of spectral energy density u_ω versus frequency ω

Consider a hypothetical adiabatic expansion of the isothermal cavity, where all the linear dimensions, including the wavelength of every thermal radiation mode inside, are increasing at a constant ratio α. Hence, we have

$$\frac{l'}{l} = \alpha \quad \text{or} \quad \frac{V'}{V} = \alpha^3 \quad , \quad \frac{\omega'}{\omega} = \frac{\lambda}{\lambda'} = \frac{1}{\alpha} \quad \text{or} \quad \frac{d\omega'}{d\omega} = \frac{1}{\alpha} \tag{27.9}$$

where the frequency $\omega = 2\pi c/\lambda$ transforms itself as the reciprocal of the wavelength. From the adiabatic equation (27.7) it follows that

$$V'T'^3 = VT^3 \quad \text{or} \quad \frac{T'}{T} = \left(\frac{V}{V'}\right)^{1/3} = \frac{1}{\alpha}$$

Therefore,

$$\frac{\omega'}{T'} = \frac{\omega}{T} \tag{27.10}$$

Using Eq.(27.3) we obtain

$$\frac{U'}{U} = \left(\frac{T'}{T}\right)^4 = \frac{1}{\alpha^4} \tag{27.11}$$

so that, in view of Eq.(27.9), we can write

$$\frac{u'(\omega',T')}{u(\omega,T)} = \frac{dU'/d\omega'}{dU/d\omega} = \frac{1}{\alpha^3}$$

It follows that the expression

$$\frac{u'(\omega',T')}{\omega'^3} = \frac{u(\omega,T)}{\omega^3}$$

is a constant during the adiabatic expansion, and therefore, being a function of both ω and T, it must depend on a constant combination of these two variables only. Such a combination has been previously derived in Eq.(27.10), so that we must choose

$$u(\omega,T) = \omega^3 F\left(\frac{\omega}{T}\right) \tag{27.12}$$

and this is known as the *Wien law*. Although the thermodynamical arguments alone cannot determine the form of the function F, they allow us to eliminate any result which does not obey the Wien law.

Information on the temperature dependence of the spectral energy density can be derived from the position of its maximum, given by Eq.(27.12) as

$$\frac{\partial}{\partial \omega}\left[\omega^3 F\left(\frac{\omega}{T}\right)\right]_{\omega=\omega_0} = T^4 \frac{\partial}{\partial x}\left[x^3 F(x)\right]_{x=x_0} = T^4\left[x_0^3\left(\frac{\partial F}{\partial x}\right)_{x=x_0} - 3x_0^2 F(x_0)\right] = 0$$

where the equation $x_0 (\partial F/\partial x)_{x=x_0} - 3F(x_0) = 0$ may be solved for x_0 to obtain

$$x_0 = \frac{\omega_0}{T} = \text{const} \quad \text{or} \quad \omega_0 = \text{const} \cdot T \tag{27.13}$$

This shows that the frequency corresponding to the maximum spectral energy density is linearly increasing as temperature increases, corresponding to the data plotted in Figure 27.3. Inserting this result into Eq.(27.12), it follows that the maximum spectral energy density itself is increasing with increasing temperature, according to

$$u_{\omega \max} = u(\omega_0, T) = \text{const} \cdot T^3 \tag{27.14}$$

Both relations (27.13) and (27.14) express *Wien's displacement law* which has been found to agree with experimental data. The magnitude of the two constants is related to the function F, whose form can be derived from statistical arguments only. Such arguments must account for both the parabolic increase of u_ω, which is proportional to ω^2, at low frequencies, and the exponential decrease of u_ω at sufficiently high frequencies.

27.2. STATISTICS OF RADIATION

The energy of radiation in the cavity can be interpreted in terms of the electromagnetic field energy. As described in Chapter 14, the electromagnetic radiation is represented by a vector potential $\vec{A}(\vec{r},t)$ which satisfies Eq.(14.26). Since thermal radiation is independent of either the shape or the material of the cavity let us consider, without loss of generality, a cubic metallic cavity where the solutions $\vec{A}(\vec{r},t)$ obey the continuity condition at boundaries and have vanishing tangential components at the inner surfaces. The problem is analogous to that solved in Chapter 16 for the vibrating string in one dimension. In the case of a wave motion restricted inside a segment of length l the solutions to the wave equation are standing waves given by Eq.(16.52) as

$$\Psi_n(x,t) = \sum_{n=1}^{\infty} A_n \sin(k_n x) \cos(\omega_n t + \varphi_n)$$

so that only a set of *normal* frequencies $\omega_n = k_n c = nc\pi/l$ are allowed. Each term of the expansion is called a *normal mode* of vibration.

For thermal radiation in a cubic metallic cavity of edge l, the normal modes can be derived from the periodicity condition on the plane wave solutions to Eq.(14.26), which have the form $Ae^{i(\vec{k}\cdot\vec{r}-\omega t)}$ according to Eq.(16.25), so that

$$Ae^{ik_x(x+l)+ik_y(y+l)+ik_z(z+l)-i\omega t} = Ae^{ik_x x+ik_y y+ik_z z-i\omega t} \tag{27.15}$$

For an electromagnetic wave the condition (16.24) also holds as $k_x^2 + k_y^2 + k_z^2 = \omega^2/c^2$, and hence, the normal frequencies of thermal radiation are given by the allowed values $k_i = n_i \pi / l$ (where n_i are restricted to positive integer values only, and i stands for x, y and z) in the form

$$\omega = \frac{c\pi}{l}\left(n_x^2 + n_y^2 + n_z^2\right)^{1/2} \tag{27.16}$$

The problem of finding the spectral energy density distribution over various frequencies is one of determining the number of allowed independent electromagnetic modes $dN(\omega) = (dN/d\omega)d\omega$ in any range of frequency. Let $\bar{\varepsilon} = \langle \varepsilon(\omega,T) \rangle$ be the average energy per electromagnetic mode of frequency ω at temperature T. The energy in the cavity corresponding to a frequency interval $\omega, \omega + d\omega$ is

$$Vu(\omega,T)d\omega = \langle \varepsilon(\omega,T) \rangle \frac{dN(\omega)}{d\omega} d\omega$$

and hence, one obtains

$$u(\omega,T) = \frac{1}{V}\frac{dN(\omega)}{d\omega}\bar{\varepsilon} \tag{27.17}$$

where expressions for both $dN(\omega)/d\omega$ and $\bar{\varepsilon} = \langle \varepsilon(\omega,T) \rangle$ can be derived only by using statistical thermodynamic arguments.

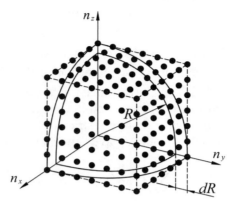

Figure 27.4. Representative points of the electromagnetic normal modes inside a cavity

Different electromagnetic modes of the same frequency ω are normally *degenerate,* in the sense that each frequency has an intrinsic multiplicity equal to the number of arrangements of a given set of integers n_x, n_y, n_z leading to the same magnitude of ω in Eq.(27.16). It is common practice to calculate the sum over all the allowed normal frequencies from a plot of their expression (27.16) in terms of the independent coordinates n_x, n_y, n_z illustrated in Figure 27.4. Each allowed mode is represented by a point at one corner of a unit cube, whose distance R from the origin gives the corresponding normal mode frequency, Eq.(27.16), as

$$\omega = \frac{c\pi R}{l} \quad , \quad R = \left(n_x^2 + n_y^2 + n_z^2\right)^{1/2} \qquad (27.18)$$

In view of the frequency range of thermal radiation, which is concentrated about the infrared region of the electromagnetic spectrum, the representative points are situated far from the origin and close to each other, so that it is reasonable to look for the number of modes in the frequency range between ω and $\omega + d\omega$. The number $dN(\omega)$ of linearly independent functions $\vec{A}(\vec{r},t)$ in that frequency range is obtained by taking into account the two independent directions of polarization for every normal mode of the plane electromagnetic waves. Therefore, since each unit cube in Figure 27.4 is associated with one mode, the number $dN(\omega)$ is twice the volume between the first octants of the spheres of radii R and $R + dR$, that is

$$dN(\omega) = 2\frac{1}{8}4\pi R^2 dR = \pi R^2 dR \qquad (27.19)$$

We may use Eq.(27.18) to make explicit the dependence on frequency, which is

$$R^2 dR = \left(\frac{1}{c\pi}\right)^3 \omega^2 d\omega = \frac{V}{c^3 \pi^3} \omega^2 d\omega$$

so that

$$dN(\omega) = \frac{V}{c^3 \pi^2} \omega^2 d\omega \qquad (27.20)$$

The spectral energy density may now be expressed to be the energy per volume and frequency interval $d\omega$ in the form

$$u_\omega = \frac{dU}{Vd\omega} = \frac{1}{V}\frac{dN(\omega)}{d\omega}\bar{\varepsilon} = \frac{\omega^2}{c^3 \pi^2}\bar{\varepsilon} \qquad (27.21)$$

Since an electromagnetic mode inside a cavity with conducting walls behaves like a harmonic oscillator, passing energy back and forth between electric and magnetic forms, it has been assumed that an ensemble of linear harmonic oscillators with the same frequency ω and energy given by

$$\varepsilon = \frac{p_x^2}{2m} + \frac{1}{2}m\omega^2 x^2$$

can be used to derive the average energy $\bar{\varepsilon}$ of a normal mode. If such an ensemble is in equilibrium state at a temperature T, it has the characteristics of a canonical ensemble, and hence, the average energy may be calculated using Eq.(25.65), which reads

$$\bar{\varepsilon} = \langle \varepsilon(\omega,T) \rangle = \frac{\int_{-\infty}^{\infty}\int_{-\infty}^{\infty} \varepsilon e^{-\varepsilon/k_B T} dx dp_x}{\int_{-\infty}^{\infty}\int_{-\infty}^{\infty} e^{-\varepsilon/k_B T} dx dp_x} = k_B T \qquad (27.22)$$

This result obeys the theorem of equipartition of energy (26.6), that is, the average energy of an ensemble of oscillators, at a given temperature T, is independent of frequency ω. In view of further developments, it is convenient to introduce the average entropy $\bar{s}$ of a normal oscillator, related to its average energy $\bar{\varepsilon}$, as required by Eq.(8.21), in the form

$$\left(\frac{\partial \bar{s}}{\partial \bar{\varepsilon}}\right)_V = \frac{1}{T} = \frac{\text{const}}{\bar{\varepsilon}}$$

so that

$$\left(\frac{\partial^2 \bar{s}}{\partial \bar{\varepsilon}^2}\right)_V = \frac{1}{T} = \frac{\text{const}}{\bar{\varepsilon}^2} \qquad (27.23)$$

Upon inserting the average energy (27.22) into Eq.(27.21), an expression for the spectral energy density, called the *Rayleigh-Jeans law*, follows as

$$u(\omega,T)_{RJ} = \frac{\omega^2}{c^3 \pi^2} k_B T \qquad (27.24)$$

Comparison with the Wien law (27.12) indicates that

$$F_{RJ}\left(\frac{\omega}{T}\right) = \frac{k_B}{c^3 \pi^2}\left(\frac{\omega}{T}\right)^{-1} \qquad (27.25)$$

The Rayleigh-Jeans law, plotted in Figure 27.5, agrees with experimental data for low frequencies only. The frequency-squared dependence predicted by Eq.(27.24) for the spectral energy density shows no maximum, in variance with Wien's displacement law (27.13). Moreover, in contradiction to the Stefan-Boltzmann law (27.3), the energy density u_{RJ} becomes infinite upon inserting Eq.(27.24) into Eq.(27.8).

The experimental decrease in the spectral energy density towards higher frequencies means, from the classical point of view, that the normal modes of electromagnetic radiation with higher frequencies contain less average energy $\bar{\varepsilon}$ than it is expected from the theorem of equipartition of energy. As an approximation to the measured spectral energy distribution, represented in Figure 27.5, Wien has proposed the formula

$$u(\omega,T)_W = c_1 \omega^3 e^{-c_2 \omega / T} \qquad (27.26)$$

which obeys the Wien law (27.12), provided we choose

$$F_W\left(\frac{\omega}{T}\right) = c_1 e^{-c_2 \omega / T} \qquad (27.27)$$

The constants c_1 and c_2 can be derived by fitting the exponential decrease of the spectral energy density found experimentally at high frequencies.

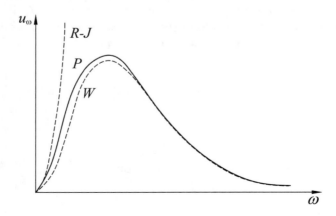

Figure 27.5. Rayleigh-Jeans law (*RJ*) and Wien formula (*W*) referred to the experimental data

Comparison of the Wien formula (27.26) with the Rayleigh-Jeans law (27.24) allows us to assign to a normal oscillator at high frequencies the average energy

$$\bar{\varepsilon} = \text{const} \cdot \omega e^{-c_2 \omega / T} \tag{27.28}$$

which gives an underestimate of the real energy of normal electromagnetic modes. This arises because the Wien formula (27.26) does not account for the low frequency behaviour of u_ω. Equation (27.28) may also be written as

$$T = \frac{\text{const} \cdot \omega}{\ln(\bar{\varepsilon}/\omega)}$$

so that, at a given high frequency ω, we obtain

$$\left(\frac{\partial \bar{s}}{\partial \bar{\varepsilon}}\right)_V = \frac{1}{T} = \text{const} \cdot \ln \bar{\varepsilon} \quad \text{or} \quad \left(\frac{\partial^2 \bar{s}}{\partial \bar{\varepsilon}^2}\right)_V = \frac{\text{const}}{\bar{\varepsilon}} \tag{27.29}$$

27.3. PLANCK'S RADIATION FORMULA

Planck has removed the discrepancy between theory and experiment by using an interpolation between the low and high frequency properties of the normal electromagnetic modes in the cavity, as expressed by Eqs.(27.23) and (27.29). He made the assumption that

$$\left(\frac{\partial^2 \bar{s}}{\partial \bar{\varepsilon}^2}\right)_V = \frac{\alpha}{\bar{\varepsilon}^2 + \beta \bar{\varepsilon}} \tag{27.30}$$

where α, β are constant at a given frequency. Integration of this relation gives

$$\left(\frac{\partial \bar{s}}{\partial \bar{\varepsilon}}\right)_V = \ln\left(1 + \frac{\beta}{\bar{\varepsilon}}\right)^{-\alpha/\beta}$$

and this result should be set equal to $1/T$, according to Eq.(8.21). It follows that

$$\bar{\varepsilon} = \frac{\beta}{e^{-\beta/\alpha T} - 1}$$

Therefore, the spectral energy density (27.21) becomes

$$u_\omega = \frac{\omega^2}{c^3 \pi^2} \frac{\beta}{e^{-\beta/\alpha T} - 1}$$

where the requirements of the Wien law (27.12) impose that $\beta \sim \omega$, so that finally we obtain the formula proposed by Planck in the form

$$u_\omega = \omega^2 \frac{\gamma\omega}{e^{\delta\omega/T} - 1} \qquad (27.31)$$

This expression was found to be in agreement with experimental data for the whole range of frequencies. This empirical formula is readily derived, provided we accept **Planck's hypothesis**:

The energy of a harmonic oscillator is an integer multiple of an energy quantum and proportional to its frequency as

$$\varepsilon_n = nh\nu = n\hbar\omega, \qquad n = 0,1,2,\ldots \qquad (27.32)$$

where h is called the *Planck constant* and $\hbar = h/2\pi$. For a system of N oscillators, each of average energy $\bar{\varepsilon}$, this means that the total energy $N\bar{\varepsilon}$ is distributed not in a continuously divisible manner, as it is assumed in classical physics, but in discrete amounts, which are multiples of $\hbar\omega$. According to the basic assumption (27.32), the canonical distribution formula (27.22) for the average energy should be written as

$$\bar{\varepsilon} = \langle \varepsilon(\omega, T) \rangle = \frac{\sum_{n=0}^{\infty} \varepsilon_n e^{-\varepsilon_n/k_B T}}{\sum_{n=0}^{\infty} e^{-\varepsilon_n/k_B T}} \qquad (27.33)$$

Setting $y = 1/k_B T$, this relation reduces to

$$\bar{\varepsilon} = -\frac{\partial}{\partial y}\left(\ln \sum_{n=0}^{\infty} e^{-y\varepsilon_n} \right) \qquad (27.34)$$

and using Eq.(27.32) to express ε_n the series successively becomes

$$\sum_{n=0}^{\infty} e^{-ny\hbar\omega} = \sum_{n=0}^{\infty} \left(e^{-y\hbar\omega}\right)^n = \sum_{n=0}^{\infty} z^n = \frac{1}{1-z} = \frac{1}{1 - e^{-y\hbar\omega}}$$

Thus, the average energy (27.34) is given by

$$\bar{\varepsilon} = \bar{\varepsilon}(\omega, T) = \frac{\hbar\omega}{e^{\hbar\omega/k_B T} - 1} \qquad (27.35)$$

which, in variance with Eq.(27.22), is a function of both T and ω. In other words, the quantization condition (27.32) invalidates the theorem of equipartition of energy. Inserting this expression for the average energy into Eq.(27.21) we obtain the spectral energy density as

$$u(\omega,T)_P = \frac{\omega^2}{c^3\pi^2}\frac{\hbar\omega}{e^{\hbar\omega/k_BT}-1} \qquad (27.36)$$

which is the same as Eq.(27.31), but now a physical significance is assigned to the empirical constants γ and δ. Equation (27.36) is known as *Planck's radiation formula* and obeys the Wien law (27.12) provided that we choose

$$F_P\left(\frac{\omega}{T}\right) = \frac{\hbar}{c^3\pi^2}\frac{1}{e^{\hbar\omega/k_BT}-1} \qquad (27.37)$$

In the low frequency range, where $\hbar\omega/k_BT \ll 1$, we can use $e^{\hbar\omega/k_BT} \cong 1+\hbar\omega/k_BT$ and the Planck radiation formula (27.36) reduces to the Rayleigh-Jeans law (27.24), whereas in the high frequency range, where $\hbar\omega/k_BT \gg 1$, the unity is negligibly small in the denominator of Eq.(27.36), which then takes the form of Wien formula (27.26)

$$u(\omega,T)_W = \frac{\hbar}{c^3\pi^2}\omega^3 e^{-\hbar\omega/k_BT} \qquad (27.38)$$

All the thermodynamic properties of thermal radiation, described earlier, can be derived using Planck's radiation formula. In this respect it is convenient to set $x = \hbar\omega/k_BT$ in Eq.(27.36), which then becomes

$$u_\omega(x) = \frac{k_B^3 T^3}{c^3\pi^2\hbar^2}\frac{x^3}{e^x-1} \qquad (27.39)$$

The maximum condition on the spectral energy density gives

$$\left[\frac{du_\omega(x)}{dx}\right]_{x=x_0} = \frac{k_B^3 T^3}{c^3\pi^2\hbar^2}\frac{x^3}{(e^{x_0}-1)^2}\left[3x_0^2(e^{x_0}-1)-x_0^3 e^{x_0}\right] = 0$$

and this is equivalent to an equation which can be solved numerically, that is

$$e^{x_0} = \frac{3}{3-x_0} \quad \text{or} \quad x_0 = 2.822 \quad \text{i.e.,} \quad \omega_0 = 2.822\frac{k_B T}{\hbar}$$

allowing us to prove Wien's displacement law (27.13). The total energy density (27.8) can be written, using Eq.(27.39), in the form

$$u = \int_0^\infty u(\omega,T)d\omega = \frac{k_B T}{\hbar}\int_0^\infty u_\omega(x)dx = \frac{k_B^4 T^4}{c^3\pi^2\hbar^3}\int_0^\infty \frac{x^3}{e^x-1}dx = aT^4$$

which is the Stefan-Boltzmann law (27.3). Since the integral is equal to $\pi^4/15$, as shown in Appendix IV, Eq.(IV.14), the Stefan constant a is related to the Planck constant h as given by

$$h = \left(\frac{8}{15} \frac{k_B^4 \pi^5}{c^3 a} \right)^{1/3} \tag{27.40}$$

The SI value for h that appears throughout the modern physics is given in Chapter 1.

FURTHER READING
1. P. W. Atkins - MOLECULAR QUANTUM MECHANICS, Oxford University Press, Oxford, 2005.
2. K. G. T. Hollands - THERMAL RADIATION FUNDAMENTALS, Begell House, New York, 2004.
3. D. V. Schröder - AN INTRODUCTION TO THERMAL PHYSICS, Addison-Wesley, San Francisco, 1999.

PROBLEMS

27.1. If a cavity enclosure at some temperature is filled with radiation of energy density u, find the intensity I of the radiation emanating from a small hole pierced in the side of the cavity. The propagation of radiation is presumed to be isotropic.

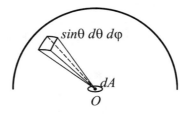

Solution

There is hemispherical symmetry about the hole, so that we may consider for simplicity the centre O of the base of a hemisphere being hit by radiation coming from all directions. A fraction $\sin\theta d\theta d\varphi / 4\pi$ is coming from directions within the elemental solid angle θ, $\theta + d\theta$ and φ, $\varphi + d\varphi$ so that the rate at which the energy, of energy density u, is flowing at speed c through an elemental area dA at O is

$$\frac{\sin\theta d\theta d\varphi}{4\pi} \times ncdA\cos\theta$$

because the elemental area presents a projected area $dA\cos\theta$ to the radiation arriving from that particular direction. The rate at which energy is flowing outward, per unit area, from the entire hemisphere is obtained by dividing by dA and integrating over θ and φ. This gives the radiation intensity as

$$I = \frac{uc}{4\pi}\int_0^{2\pi}\int_0^{\pi/2}\sin\theta\cos\theta d\theta d\varphi = \frac{uc}{4}$$

27.2. Find a rule for counting the microstates of a classical harmonic oscillator which is consistent with Planck's hypothesis, and derive an expression for the partition function of such an oscillator.

Solution

Since the energy ε of a harmonic oscillator is a constant of motion and the energy values should be given by Eq.(27.32), the number of available microstates in phase space is

$$\Omega_P(\varepsilon) = n = \frac{\varepsilon}{\hbar\omega}$$

The harmonic oscillator in one dimension can be represented in phase space by an ellipse (see Figure 24.2) given by

$$\frac{p^2}{2m\varepsilon} + \frac{q^2}{2\varepsilon/m\omega^2} = 1 \quad \text{or} \quad \frac{p^2}{a^2} + \frac{q^2}{b^2} = 1$$

where $a^2 = 2m\varepsilon$, $b^2 = 2\varepsilon/(m\omega^2)$. Since the possible values of p and q are continuous, in order to count the microstates within the ellipse we must divide its area $\pi ab = (2\pi/\omega)\varepsilon$ into cells of area $\Delta p\Delta q$. The number of states with energy less or equal to ε is thus given by

$$\Omega(\varepsilon) = \frac{\pi ab}{\Delta p\Delta q} = \frac{2\pi\varepsilon}{\omega\Delta p\Delta q}$$

The counting of microstates of the classical oscillator becomes consistent with that of Planck's oscillator provided that $\Omega(\varepsilon) = \Omega_P(\varepsilon)$ or $h = \Delta p\Delta q$. In other words, Planck's hypothesis implies that we cannot specify a microstate of the classical oscillator more precisely than by assigning it to a cell of area h in phase space. Thus, Planck's constant h can be interpreted to be the area (volume) of the unit cell in phase space. As a result, the partition function of a single classical oscillator should be written as

$$z = \frac{1}{h}\int_{-\infty}^{\infty}\int_{-\infty}^{\infty}e^{-(p^2/2m+m\omega^2q^2/2)/k_BT}dqdp = \frac{1}{h}(2\pi mk_BT)^{1/2}\left(\frac{2\pi k_BT}{m\omega^2}\right)^{1/2} = \frac{k_BT}{\hbar\omega}$$

In view of Eq.(25.72), this provides the mean energy as

$$\bar{\varepsilon} = k_B T^2 \frac{\partial \ln z}{\partial T} = k_B T$$

and this is in agreement with the equipartition theorem, Eq.(26.6), because we have two independent quadratic terms in the Hamiltonian of the single oscillator, that is, $f = 2$.

27.3. Determine the thermodynamic properties of Planck's harmonic oscillator in equilibrium with a thermal bath at temperature T.

Solution

The partition function corresponding to the energy states (27.32) is given by

$$z = \sum_{n=0}^{\infty} e^{-n\hbar\omega/k_B T} = \frac{1}{1 - e^{-\hbar\omega/k_B T}}$$

From Eq.(25.70) the free energy results as $f_\omega = k_B T \ln\left(1 - e^{-\hbar\omega/k_B T}\right)$ and then, in view of Eq.(8.25), one obtains

$$s_\omega = -\left(\frac{\partial f_\omega}{\partial T}\right)_V = k_B \left[\frac{\hbar\omega/k_B T}{e^{\hbar\omega/k_B T} - 1} - \ln\left(1 - e^{-\hbar\omega/k_B T}\right)\right]$$

Substituting into Eq.(25.69) yields the mean energy

$$\bar{\varepsilon} = \frac{\hbar\omega}{e^{\hbar\omega/k_B T} - 1}$$

which is the same as that given by Eq.(27.35). If $\hbar\omega \ll k_B T$, we may approximate the denominator as $\hbar\omega/k_B T$ and the mean energy reduces to that of the classical oscillator

$$\bar{\varepsilon} \to k_B T \quad \text{or} \quad \frac{\bar{\varepsilon}}{\hbar\omega} \to \frac{k_B T}{\hbar\omega} \gg 1$$

Thus, the classical treatment appears to be a limit corresponding to a large occupancy n (number of microstates) of a given mode. At the other extreme, if $\hbar\omega \gg k_B T$, then the exponential in the denominator is large compared to unity and the mean energy reads

$$\bar{\varepsilon} \to \hbar\omega e^{-\hbar\omega/k_B T} \quad \text{or} \quad \frac{\bar{\varepsilon}}{\hbar\omega} \to e^{-\hbar\omega/k_B T} \ll 1$$

The occupancy is low, because it is difficult to get energy much larger than $k_B T$, which is typical for a heat bath, as it is required to excite a high energy mode.

27.4. Find an expression for the entropy of a single mode of thermal radiation in terms of the average number of photons in the mode.

Solution

Using the partition function z for a single mode ω, derived in Problem 27.3, the occupancy n_ω in the mode reads

$$n_\omega = \langle n \rangle = \frac{1}{z} \sum_{n=0}^{\infty} n e^{-n\hbar\omega/k_B T} = \frac{e^{-\hbar\omega/k_B T}}{z} \sum_{n=0}^{\infty} n \left(e^{-\hbar\omega/k_B T} \right)^{n-1}$$

$$= \frac{e^{-\hbar\omega/k_B T}}{z} \frac{\partial}{\partial x} \left(\sum_{n=0}^{\infty} x^n \right) = \frac{e^{-\hbar\omega/k_B T}}{z} \frac{\partial z}{\partial x}$$

where $x = e^{-\hbar\omega/k_B T}$, and hence

$$\sum_{n=0}^{\infty} x^n = \frac{1}{1-x} = z \quad , \quad \frac{\partial z}{\partial x} = \frac{\partial}{\partial x}\left(\frac{1}{1-x}\right) = \frac{1}{(1-x)^2}$$

It follows that

$$n_\omega = \frac{e^{-\hbar\omega/k_B T}}{1 - e^{-\hbar\omega/k_B T}} = \frac{1}{e^{\hbar\omega/k_B T} - 1}$$

from which we find

$$e^{\hbar\omega/k_B T} = 1 + \frac{1}{n_\omega} \quad \text{or} \quad e^{-\hbar\omega/k_B T} = \frac{n_\omega}{n_\omega + 1} \quad \text{i.e.,} \quad \frac{\hbar\omega}{k_B T} = \ln \frac{n_\omega + 1}{n_\omega}$$

Substituting into the expression for entropy derived in Problem 27.3 yields

$$s_\omega = k_B \left(n_\omega \ln \frac{n_\omega + 1}{n_\omega} - \ln \frac{1}{n_\omega + 1} \right) = (n_\omega + 1) k_B \ln(n_\omega + 1) - n_\omega k_B \ln n_\omega$$

as required.

27.5. Find the average value of the ratio $\hbar\omega/k_B T$ for thermal radiation in the volume V of a cavity enclosure at temperature T.

Solution

The mean energy per mode in a volume V is obtained if we divide the total energy U, Eq.(27.4), by the total number N of modes in the cavity. In the frequency range between ω and $\omega + d\omega$ the energy is given by the number of modes times the energy $\hbar\omega$ of each

$$V u_\omega d\omega = \hbar\omega N(\omega) d\omega$$

where u_ω is obtained from Eq.(27.36), that is,

$$N(\omega) d\omega = \frac{V}{c^3 \pi^2} \frac{\omega^2 d\omega}{e^{\hbar\omega/k_B T} - 1}$$

The total number of modes is determined by integrating this expression over the infinite range of frequencies, and this gives

$$N = \frac{V}{c^3\pi^2}\int_0^\infty \frac{\omega^2 d\omega}{e^{\hbar\omega/k_BT}-1} = \frac{V}{c^3\pi^2}\left(\frac{k_BT}{\hbar}\right)^3\int_0^\infty \frac{x^2 dx}{e^x-1}$$

where the substitution $x = \hbar\omega/k_BT$ has been made. The integral can be evaluated using the Riemann zeta function, as shown in Appendix IV, and has the numerical value

$$\int_0^\infty \frac{x^2 dx}{e^x-1} = 2.404$$

It follows that $N = 2.02\times 10^7\, VT^3$ and, in view of Eq.(27.4), one obtains

$$\hbar\omega = \frac{7.6\times 10^{-16}\, VT^4}{2.02\times 10^7\, VT^3} = 3.74\times 10^{-23}T \approx 2.7\, k_BT$$

Thus, the average value of $x = \hbar\omega/k_BT$ is of the order of unity.

27.6. Show that the pressure of thermal radiation on the walls of a cavity of volume V and temperature T is different from that obtained for a perfect monatomic gas having the same internal energy U. What pressure rises faster for a given amount of adiabatic compression?

27.7. Find an expression for the spectral energy density u_λ per unit wavelength, and use it to derive Wien's displacement law in λ space.

27.8. Find the thermodynamic potentials of a system of N independent classical harmonic oscillators in equilibrium at a temperature T.

27.9. Find the thermodynamic potentials of a system of N Planck's oscillators in equilibrium with a heat bath at temperature T.

27.10. If the mean value of the spectral energy $U_\omega = Vu(\omega,T)$ of thermal radiation in a volume V at a temperature T is given by Planck's radiation formula, find an expression for the variance of this energy.

28. Wave Mechanics

28.1. THE CORPUSCULAR NATURE OF RADIATION
28.2. THE OLD QUANTUM THEORY
28.3. THE WAVE NATURE OF PARTICLES
28.4. THE UNCERTAINTY PRINCIPLE

28.1. The Corpuscular Nature of Radiation

It has been proposed that electromagnetic radiation exists in discrete quanta either when propagating in space or during the emission and absorption processes. The thermal radiation inside a cavity might thus be regarded, from the point of view of entropy, as obeying the canonical distribution of n molecules of an ideal gas in a volume V, that is, Eq.(26.20), which can be written in the form

$$S = ns_0 + nk_B \ln V + \frac{3}{2} nk_B \ln T \tag{28.1}$$

The radiation entropy S can be expanded in terms of the spectral entropy density s_ω, which is the entropy per unit volume in the frequency range between ω and $\omega + d\omega$, as

$$S = V \int_0^\infty s_\omega \, d\omega \tag{28.2}$$

where $s_\omega = s(u_\omega, \omega)$ must depend on the spectral energy density u_ω, defined in Eq.(27.8) which also reads

$$U = V \int_0^\infty u_\omega \, d\omega \tag{28.3}$$

Since the conditions for equilibrium, $\delta S = 0$, and $\delta U = 0$ must hold simultaneously, their integrands should be proportional, that is,

$$\delta s_\omega = C\delta u_\omega \quad \text{or} \quad \frac{\partial s_\omega}{\partial u_\omega}\delta u_\omega = C\delta u_\omega \quad \text{i.e.,} \quad \frac{\partial s_\omega}{\partial u_\omega} = C \tag{28.4}$$

Upon differentiation of Eqs.(28.2) and (28.3) we obtain the relation

$$dS = V\int_0^\infty \frac{\partial s_\omega}{\partial u_\omega} du_\omega d\omega = CV\int_0^\infty du_\omega d\omega = CdU \tag{28.5}$$

which might be compared with the definition (8.7) of the entropy change in an infinitesimal process. The interpretation of dU as reversible added heat gives

$$C = \frac{1}{T} = \frac{\partial s_\omega}{\partial u_\omega} \tag{28.6}$$

where we made use of Eq.(28.4). The temperature can be eliminated between Eq.(28.6) and the Wien formula (27.38) by writing

$$\frac{1}{T} = -\frac{k_B}{\hbar\omega}\ln\frac{c^3\pi^2 u_\omega}{\hbar\omega^3} \quad \text{or} \quad \frac{\partial s_\omega}{\partial u_\omega} = -\frac{k_B}{\hbar\omega}\ln\frac{c^3\pi^2 u_\omega}{\hbar\omega^3}$$

so that, on integrating, we obtain the spectral entropy density as

$$s_\omega = -\frac{k_B u_\omega}{\hbar\omega}\left(\ln\frac{c^3\pi^2 u_\omega}{\hbar\omega^3} - 1\right) \tag{28.7}$$

A similar relation should be true between $S_\omega = Vs_\omega$ and $U_\omega = Vu_\omega$ as

$$S_\omega = -\frac{k_B U_\omega}{\hbar\omega}\left(\ln\frac{c^3\pi^2 U_\omega}{\hbar\omega^3 V} - 1\right) \tag{28.8}$$

so that the entropy change in the spectral range between ω and $\omega + d\omega$, associated with a reversible change of volume from V to V_0, can be written as

$$\Delta S_\omega d\omega = \frac{k_B U_\omega d\omega}{\hbar\omega}\ln\frac{V}{V_0}$$

It has been emphasized by Einstein that this relation takes the form

$$\Delta S = nk_B \ln\frac{V}{V_0}$$

used to describe the n-particle gas behaviour, according to Eq.(28.1), provided that

$$\frac{U_\omega d\omega}{\hbar\omega} = n \quad \text{or} \quad U_\omega d\omega = n\hbar\omega \tag{28.9}$$

The similarity of this relation to Eq.(27.32), assumed in Planck's theory, has been considered to be an argument for the discrete nature of electromagnetic radiation. In other words, an energy

$$\varepsilon = \hbar\omega \qquad (28.10)$$

can be associated with each radiation quantum, later called the *photon*.

28.2. THE OLD QUANTUM THEORY

An analysis of the radiation spectrum emitted by hydrogen atoms shows intense lines, of definite frequencies v, which can be clustered into several series that fit the empirical formula

$$v = R\left(\frac{1}{n^2} - \frac{1}{m^2}\right) \quad , \qquad m > n \geq 1 \qquad (28.11)$$

where m and n are integers, and $R = 3.29 \times 10^{15}$ Hz is the Rydberg constant. The significant fact is that the frequency of the observed spectral lines of each series can be represented by the difference between two quadratic terms, one of which, T_n, is fixed and the other, T_m, is a variable. Similar regularities have been found for the emission spectra of alkali metal atoms. Furthermore, in addition to the lines represented in the various series (28.11), there are other lines not included in this series, but it was observed that the frequencies of all spectral lines also correspond to the differences between two quadratic terms

$$v_{mn} = T_n - T_m \qquad (28.12)$$

and this is called the *Ritz combination principle*. These results are conflicting with the classical picture of an electron orbiting around a nucleus, which radiates electromagnetic waves, and hence, leads to an unstable atom: since the electron frequency of revolution would change smoothly, the emission spectrum should be expected to be continuous.

A successful interpretation of the discrete spectrum of one-electron atoms has been proposed on the grounds of two assumptions known as *Bohr's postulates*:

1. **Postulate of stationary states**: *an electron moves only in certain permissible orbits which are stationary states, in the sense that no radiation is emitted. The condition for such states is that the orbital angular momentum of the electron equals an integer times $\hbar = h/2\pi$.*

This postulate states that an electron in a stable orbit is exempt from the requirement that an accelerated charge must radiate energy, in spite of its macroscopic validity. However, the reality of stationary states has received direct experimental evidence.

EXAMPLE 28.1. The Franck-Hertz experiment

In the Franck-Hertz experiment electrons were accelerated through mercury vapour from a cathode to a positive plate acting as an anode. In collisions of the mercury vapour atoms with accelerated electrons of known kinetic energy, a decrease in the current produced by electrons reaching the plate was found to occur at certain plate voltages, as illustrated in Figure 28.1.

The first plate current drop occurs at 4.9 V, and a spectral line appears simultaneously in the emission spectrum of the vapour at 253.6 nm, a value which corresponds to a photon energy of 4.9 eV, according to Eq.(28.10). A similar behaviour is found at multiples of 4.9 V. It follows that the energy is transferred to the target atoms in discrete quantities, which can be simply explained by the existence of stationary states in the atoms.

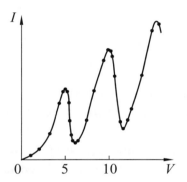

Figure 28.1 Plot of plate current versus voltage in Franck-Hertz experiment

---

In the case of an electron in a circular orbit of radius r, the Bohr condition reads

$$m_e v r = n\hbar \tag{28.13}$$

where m_e is the mass of the electron, v its velocity, and n is a positive integer. The dynamical stability of the circular orbit requires that

$$\frac{Ze^2}{4\pi\varepsilon_0 r^2} = \frac{m_e v^2}{r} \quad \text{or} \quad \frac{Ze_0^2}{r^2} = \frac{m_e v^2}{r} \tag{28.14}$$

where Ze is the nuclear charge of the one electron atom and $e_0 = e/\sqrt{4\pi\varepsilon_0}$. Eliminating r from Eqs.(28.13) and (28.14) we obtain the electron velocity as

$$v_n = \frac{Z e_0^2}{n \hbar} \qquad (28.15)$$

and this gives, on substitution into Eq.(28.13), the orbit radius as

$$r_n = \frac{n^2}{Z} \frac{\hbar^2}{m_e e_0^2} = \frac{n^2}{Z} a_0 \qquad (28.16)$$

The lowest radius a_0, given by

$$a_0 = \frac{\hbar^2}{m_e e_0^2} = 0.053 \quad (\text{nm}) \qquad (28.17)$$

is called the *Bohr radius*. Finally the orbital energy of state n may be written as

$$E_n = \frac{m_e v^2}{2} - \frac{Z e_0^2}{r_n} = -\frac{Z^2}{n^2} \frac{m_e e_0^4}{2\hbar^2} = -13.6 \frac{Z^2}{n^2} \quad (\text{eV}) \qquad (28.18)$$

2. **Postulate of discrete transitions**: *emission and absorption of radiation occurs only when an electron makes a transition from one stationary state to another. The radiation has a definite frequency v_{mn} given by the condition*

$$h v_{mn} = \varepsilon_m - \varepsilon_n \qquad (28.19)$$

An analogy between the frequency condition (28.19) and the empirical combination principle (28.12) follows immediately. Since the frequency is expressed as the energy difference between two stationary states divided by h, this postulate may also be linked with the Planck assumption (27.32) concerning the discrete emission and absorption of radiation by oscillators.

The stationary state for which $n=1$ is called the *ground state*, while states for which $n>1$ are said to be *excited states*. The condition (28.19) states that radiation of frequency given by

$$v_{mn} = Z^2 \frac{m_e e_0^4}{4\pi \hbar^3} \left(\frac{1}{n^2} - \frac{1}{m^2} \right) \qquad (28.20)$$

is emitted when the electron drops from the mth to the nth state, where $m > n \geq 1$. Comparing Eqs.(28.20) and (28.11) we see that Bohr's postulates explain the emission spectra of one electron atoms, provided the Rydberg constant is expressed as

$$R = \frac{m_e e_0^4}{4\pi \hbar^3} \qquad (28.21)$$

> **EXAMPLE 28.2. The correspondence principle**
>
> The *correspondence principle*, formulated by Bohr, states that quantum mechanics must give the same results as classical mechanics in the limit of large quantum numbers $(n \to \infty)$ and low quantum jumps $(h \to 0)$. A simple illustration is given by a comparison between the classical and Bohr model predictions for the line spectra of one electron atoms.
>
> During the periodic motion on a stationary orbit, the classical electron should emit radiation with frequencies equal to those of its rotation, which can be obtained by combining Eqs.(28.15) and (28.16) as
>
> $$v_{rot} = \frac{v_n}{2\pi r_n} = Z^2 \frac{m_e e_0^4}{2\pi \hbar^3} \frac{1}{n^3}$$
>
> Assuming that a Bohr transition $n+1 \to n$ occurs, Eq.(28.20) gives the frequency of the emitted radiation as
>
> $$v_B = Z^2 \frac{m_e e_0^4}{4\pi \hbar^3} \left[\frac{1}{n^2} - \frac{1}{(n+1)^2} \right] = \left(Z^2 \frac{m_e e_0^4}{2\pi \hbar^3} \right) \frac{2n+1}{2n^2(n+1)^2}$$
>
> and it is clear that $v_B \to v_{rot}$ if $n \to \infty$, as required by the correspondence principle.

The definition of stationary states by quantum conditions in terms of quantum numbers is a basic procedure, followed by Sommerfeld in an extension of the Bohr model, which takes into account the classical solution for the motion of a particle in a central force field, described in Chapter 3. Since the orbital angular momentum $L = mvr$ is a constant of motion in a circular orbit, the quantization condition (28.13) may be written as

$$2\pi L = nh \quad \text{or} \quad \int_0^{2\pi} L d\varphi = nh$$

This condition was generalized in the form of a *quantization rule*, given in terms of the action integral introduced in Chapter 24 by

$$\oint_\Gamma p_i dq_i = n_i h \qquad (28.22)$$

which reads: *the action integral of any periodic motion equals an integer times h.* The integral is over the range of variation of the position coordinate q_i during a period of motion. Applying this quantization rule to the coordinates r and α of an elliptical electron motion around the nucleus, described in Chapter 3 and illustrated in Figure 28.2, we have

$$\oint_\Gamma p_\alpha d\alpha = \int_0^{2\pi} L d\alpha = 2\pi L = lh \quad , \quad \oint_\Gamma p_r dr = n_r h \qquad (28.23)$$

where the azimuthal quantum number $n_\alpha = l$ and the radial quantum number n_r are integers, and $p_\alpha = L$ is the angular momentum.

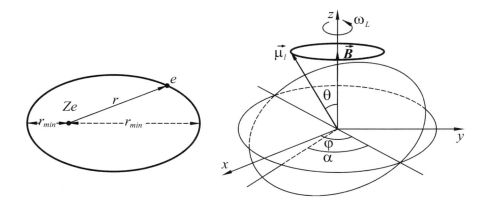

Figure 28.2. Elliptical electron motion

The radial momentum p_r can be derived from the energy equation (3.11) which, for the electron orbit, reads

$$\varepsilon = \frac{m_e \dot{r}^2}{2} + \frac{L^2}{2m_e r^2} - \frac{Ze_0^2}{r} = \frac{1}{2m_e}\left(p_r^2 + \frac{L^2}{r^2}\right) - \frac{Ze_0^2}{r}$$

where $\varepsilon < 0$, as required for bound elliptical orbits. Hence, we have

$$p_r = m_e \dot{r} = \pm\sqrt{2m_e\left(\varepsilon + \frac{Ze_0^2}{r} - \frac{L^2}{2m_e r^2}\right)}$$

and it is obvious that p_r should be positive as r increases from r_{min} to r_{max} and negative along the other half of the elliptical path. The action integral can be evaluated as

$$\oint_\Gamma p_r dr = 2\int_{r_{min}}^{r_{max}} \sqrt{2m_e\left(\varepsilon + \frac{Ze_0^2}{r} - \frac{L^2}{2m_e r^2}\right)}\, dr = -2\pi L + 2\pi Ze_0^2\sqrt{\frac{m_e}{-2\varepsilon}}$$

and inserting this result into Eq.(28.23) yields

$$2\pi Ze_0^2\sqrt{\frac{m_e}{-2\varepsilon}} = (n_r + l)h = nh \tag{28.24}$$

where the principal quantum number is set as $n = n_r + l$. Solving for ε, we obtain the same energy levels as in Eq.(28.18) but without restriction to circular orbits. If $p_r = m_e \dot{r} = 0$, which happens at $r = r_{min}$ and $r = r_{max}$, the energy equation reduces to

$$\varepsilon = \frac{L^2}{2m_e r^2} - \frac{Ze_0^2}{r} \quad \text{or} \quad r^2 + \frac{Ze_0^2}{\varepsilon} r - \frac{L^2}{2m_e \varepsilon} = 0$$

and the sum of the roots $r_{\min}$ and $r_{\max}$, i.e., the major axis $2a$ of the ellipse, is given by

$$r_{\min} + r_{\max} = -\frac{Ze_0^2}{\varepsilon} \quad \text{or} \quad a = -\frac{Ze_0^2}{2\varepsilon}$$

Thus, the semiaxis a of the nth stable orbit is related to the principal quantum number by

$$a_n = n^2 a_0 \tag{28.25}$$

where a_0 is the Bohr radius (28.17) of the circular ground state orbit.

Any elliptic orbit contains the nucleus in its plane, whose orientation is defined by a polar angle measured from the z-axis. The projection $L_z = L\cos\theta$ of the orbital angular momentum is also a constant, for a given orbit, and we may set $p_\varphi = L_z$ as φ is the azimuthal angle about the z-axis. The quantization rule (28.22) for p_φ reads

$$\oint_\Gamma p_\varphi d\varphi = 2\pi p_\varphi = mh \tag{28.26}$$

where m is an integer. A discrete set of orientations θ is allowed according to

$$\cos\theta = \frac{p_\varphi}{p_\alpha} = \frac{L_z}{L} = \frac{m}{l} \quad , \quad -l < m < l$$

so that the Sommerfeld model accounts for a *spatial quantization* of the orbital angular momentum. Since the spatial orientation of orbits is apparent when a magnetic field is present, m is called the *magnetic quantum number*. This effect is classically assigned to an interaction of an applied field $\vec{B}$ with a magnetic dipole $\vec{\mu}_l$ defined in Eq.(13.5), which is associated with the circular electron path in the form

$$\mu_l = IS = -ev\pi r^2 = -\frac{e\omega}{2} r^2 = -\frac{e}{2m_e} L \quad \text{or} \quad \vec{\mu}_l = -\frac{e}{2m_e}\vec{L} \tag{28.27}$$

The action of a magnetic field $\vec{B}$ gives a torque on $\vec{\mu}_l$ which is equal to the rate of change of the angular momentum (see Chapter 3), that is,

$$\frac{d\vec{L}}{dt} = \vec{\mu}_l \times \vec{B} = -\frac{e}{2m_e}\vec{L} \times \vec{B}$$

Taking $|d\vec{L}| = L\sin\theta d\varphi$, as represented in Figure 28.4 (b), one obtains

$$L\sin\theta \frac{d\varphi}{dt} = \frac{e}{2m_e} LB \sin\theta \quad \text{or} \quad \omega_L = \frac{e}{2m_e} B$$

The angular velocity $\omega_L = d\varphi/dt$ is known as *Larmor frequency*. It is a measure of the precession motion of the angular momentum vector about the z-axis, at constant polar orientation θ. An energy $\hbar\omega_L$ can be formally associated with the Larmor precession, on the grounds of the Planck hypothesis (27.32), although its significance only becomes apparent in quantum mechanics, as will be shown in Chapter 32.

28.3. THE WAVE NATURE OF PARTICLES

The stable motion of electrons in the atom introduces integers, which have been classically involved only in wave phenomena such as the normal modes of standing waves, described in Chapter 16. A similar periodicity has been assigned to electrons by de Broglie, who derived the quantization condition (28.13) for Bohr orbits by fitting a standing wave around each circumference. In other words, he assumed that the electrons were accompanied by matter waves, which were considered to be localized with the particle, in contrast to the situation for classical waves.

Highly localized wave configurations can be described by means of the wave packets, as discussed in Chapter 16, and hence, the representation of a particle in space and time in terms of a complex wave packet is a basic assumption of wave mechanics. The behaviour of the wave packet as a whole accounts for experiments where particle-like properties are observed, whereas individual wavelength components of the packet appear in experiments which depend upon interference and diffraction effects. A convenient representation of the wave packet by a Fourier integral, in the coordinate-wave vector domain, was introduced in Eq.(16.63) which, for electromagnetic waves with phase velocity $c = \omega/k$, reads

$$\Psi(x,t) = \frac{1}{\sqrt{2\pi}} \int_{-\infty}^{\infty} \Phi(k) e^{ik(x-ct)} dk = \frac{1}{\sqrt{2\pi}} \int_{-\infty}^{\infty} \Phi(k) e^{i(kx-\omega t)} dk \qquad (28.28)$$

and represents the superposition of a continuum of plane wave states. The $\Phi(k)$ values, which can be regarded to be weighting factors for the component plane waves, give the spectral distribution of states in *k*-space. The matter wave packet is highly localized in the neighbourhood of the particle position, since the component waves interfere destructively except in a limited region.

Consider a wave packet with a Gaussian spectrum of *k*-values given by

$$\Phi(k) = e^{-\alpha(k-k_0)^2} = e^{-\alpha(\Delta k)^2} \qquad (28.29)$$

The dispersion of waves of different frequencies is given by the Taylor expansion (16.65) about the wave vector value k_0 as

$$\omega(k) = \omega_0 + v_g \Delta k + \zeta (\Delta k)^2 + \ldots \qquad (28.30)$$

where the *group velocity* v_g and the coefficient ζ are defined as in Eqs.(16.66) and (16.67). The function (28.28) becomes

$$\Psi(x,t) = \frac{1}{\sqrt{2\pi}} e^{i(k_0 x - \omega_0 t)} \int_{-\infty}^{\infty} e^{i\Delta k(x - v_g t)} e^{-(\alpha + i\zeta t)(\Delta k)^2} dk$$

$$= \frac{1}{\sqrt{2\pi}} e^{i(k_0 x - \omega_0 t)} \int_{-\infty}^{\infty} e^{-(\alpha + i\zeta t)\left[\Delta k - i(x - v_g t)/2(\alpha + i\zeta t)\right]^2} e^{-(x - v_g t)^2 / 4(\alpha + i\zeta t)} dk$$

$$= \frac{1}{\sqrt{2\pi}} e^{i(k_0 x - \omega_0 t)} e^{-(x - v_g t)^2 / 4(\alpha + i\zeta t)} \int_{-\infty}^{\infty} e^{-(\alpha + i\zeta t) u^2} du$$

The last integral has the Gaussian form (IV.2), Appendix IV, so that we finally obtain

$$\Psi(x,t) = \frac{1}{\sqrt{2(\alpha + i\zeta)}} e^{-(x - v_g t)^2 / 4(\alpha + i\zeta t)} e^{i(k_0 x - \omega_0 t)} \qquad (28.31)$$

The function $\Psi(x,t)$ gives the coordinate representation of a travelling wave packet of phase velocity $c = \omega_0 / k_0$ and amplitude

$$|\Psi(x,t)|^2 = \frac{1}{2\sqrt{\alpha^2 + \zeta^2 t^2}} e^{-\alpha(x - v_g t)^2 / 2(\alpha^2 + \zeta^2 t^2)} \qquad (28.32)$$

A spreading of the Gaussian wave packet occurs if dispersion is present, as illustrated in Figure 28.3, and this occurs if the amplitude (or group) velocity v_g is different from the phase velocity c. This seems incompatible with our expectations that the matter waves associated with particles should remain highly localized.

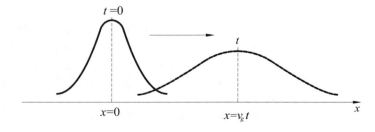

Figure 28.3. The spreading of the Gaussian matter wave packet

The de Broglie assumption was to assign the group velocity v_g to a classical free particle of momentum p in the form

$$v_g = \frac{d\omega}{dk} = \frac{p}{m} \tag{28.33}$$

If the free-particle energy ε obeys the Einstein relation (28.10) for electromagnetic radiation, that is

$$\varepsilon = \hbar\omega = \frac{p^2}{2m} \quad \text{or} \quad \omega = \frac{p^2}{2m\hbar} \tag{28.34}$$

then Eq.(28.33) only holds provided that

$$p = \hbar k \tag{28.35}$$

and this is known to be the *de Broglie hypothesis*. The matter wavelength given by

$$\lambda = \frac{2\pi}{k} = \frac{h}{p} \tag{28.36}$$

has been experimentally measured by Davisson and Germer in the case of electrons, providing a strong reason to retain the de Broglie concept of matter waves and to modify classical concepts concerning the meaning of position and velocity of a particle. According to classical ideas, we should expect that the matter waves should travel without dispersion.

EXAMPLE 28.3. The Davisson-Germer experiment

The matter wavelengths associated with low energy electrons correspond to the atomic spacing of most crystalline solids. If Eq.(28.36) were correct, such electrons should exhibit diffraction effects. A diffraction pattern for low energy electrons scattered from a nickel crystal was observed by Davisson and Germer. An angular plot of the intensity of the scattered electron current is given in Figure 28.4(b), showing a distinct peak at $\theta = 50°$, for electrons of 54 eV, having an associated wavelength $\lambda = 0.167$ nm according to Eq.(28.36) which reads

$$\lambda = \frac{h}{\sqrt{2m_e\varepsilon}} = \frac{h}{\sqrt{2m_e eV}} = \sqrt{\frac{1.5}{V}} \quad \text{(nm)}$$

The condition of constructive interference (18.6), when applied to beams reflected from adjacent crystal planes, as illustrated in Figure 28.4(a), yields

$$m\lambda = 2\frac{d}{\cos(\theta/2)} - 2d\tan\left(\frac{\theta}{2}\right)\sin\left(\frac{\theta}{2}\right) = 2d\cos\left(\frac{\theta}{2}\right) = a\sin\theta$$

where $d = a\sin(\theta/2)$ is the distance between the reflecting planes, and $a = 0.215$ (nm) is the lattice constant of the Ni crystal. The calculated value of $\theta = 51°$ for $m=1$ agrees very well with the results described above.

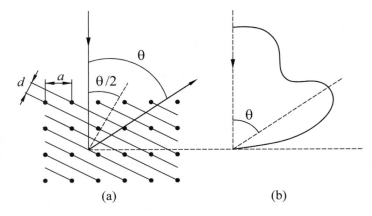

Figure 28.4. Reflection of electron waves on a crystal (a) and the corresponding intensity of electron current (b)

The matter wave function (28.28) may now be expressed in terms of the corpuscular properties ε and p, given by Eqs.(28.34) and (28.35), as

$$\Psi(x,t) = \frac{1}{\sqrt{2\pi\hbar}} \int_{-\infty}^{\infty} \Phi(p_x) e^{i(p_x x - \varepsilon t)/\hbar} dp_x \qquad (28.37)$$

where the constant is obtained by normalizing $\Psi(x,t)$. The function $\Psi(x,t)$ is a solution to the equation

$$i\hbar \frac{\partial \Psi(x,t)}{\partial t} = \frac{1}{\sqrt{2\pi\hbar}} \int_{-\infty}^{\infty} \Phi(p_x) \varepsilon e^{i(p_x x - \varepsilon t)/\hbar} dp_x$$

$$= \frac{1}{\sqrt{2\pi\hbar}} \int_{-\infty}^{\infty} \Phi(p_x) \frac{p_x^2}{2m} e^{i(p_x x - \varepsilon t)/\hbar} dp_x = -\frac{\hbar^2}{2m} \frac{\partial^2 \Psi(x,t)}{\partial x^2} \qquad (28.38)$$

which will be associated with the free-particle motion, in the so-called *coordinate representation*.

Since the functions $\Psi(x,t)$, $\Phi(p_x)$ are Fourier transforms of each other, in view of Eq.(16.62), we can also define the normalized matter wave function as

$$\Phi(p_x,t) = \frac{1}{\sqrt{2\pi\hbar}} \int_{-\infty}^{\infty} \Psi(x) e^{-i(p_x x - \varepsilon t)/\hbar} dx \qquad (28.39)$$

This is a solution to the equation

$$i\hbar \frac{\partial \Phi(p_x,t)}{\partial t} = -\frac{1}{\sqrt{2\pi\hbar}} \int_{-\infty}^{\infty} \Psi(x) \varepsilon e^{-i(p_x x - \varepsilon t)/\hbar} dx = -\varepsilon \Phi(p_x,t) = -\frac{p_x^2}{2m} \Phi(p_x,t) \quad (28.40)$$

which is associated with the free-particle motion in the *momentum representation*.

28.4. THE UNCERTAINTY PRINCIPLE

The experimental support for the wave-particle relations written as

$$\varepsilon = \hbar\omega \quad , \quad p = \hbar k$$

ascribes to the same entity attributes considered mutually exclusive in classical physics. This has required that we reconsider the classical concepts concerning the meaning of simultaneous position x and its conjugate momentum p_x. The basic difference between the classical and quantum interpretation of these concepts was demonstrated by Heisenberg in a statistical examination of matter wave spreading. At a given instant of time, which is conveniently chosen to be $t = 0$, the Gaussian representation (28.32) associated with a particle as shown in Figure 28.3, becomes

$$|\Psi(x)|^2 = \frac{C}{2\alpha} e^{-x^2/2\alpha} \quad \text{or} \quad |\Psi(x)|^2 = \frac{1}{\sqrt{2\pi\alpha}} e^{-x^2/2\alpha} \quad (28.41)$$

where the normalization constant C is obtained from the condition

$$\int_{-\infty}^{\infty} |\Psi(x)|^2 dx = \frac{C}{2\alpha} \int_{-\infty}^{\infty} e^{-x^2/2\alpha} dx = \frac{C}{2\alpha} \sqrt{2\pi\alpha} = 1$$

using Eq.(IV.2), Appendix IV. The degree of spreading of the wave about the point $x = 0$ is given by the *standard deviation* σ_x defined as

$$\Delta x = \sigma_x = \sqrt{\langle x^2 \rangle - \langle x \rangle^2} = \sqrt{\langle x^2 \rangle} \quad (28.42)$$

where $\langle x \rangle = 0$. The mean square value can be evaluated, using Eq.(IV.4), Appendix IV, as follows

$$\langle x^2 \rangle = \frac{1}{\sqrt{2\pi\alpha}} \int_{-\infty}^{\infty} x^2 e^{-x^2/2\alpha} dx = \frac{1}{\sqrt{2\pi\alpha}} \frac{1}{2} \sqrt{\pi(2\alpha)^3} = \alpha$$

so that the spreading of the matter wave packet can be written as

$$\Delta x = \sigma_x = \sqrt{\alpha} \tag{28.43}$$

In view of Eqs.(28.32) and (28.29), the wave number representation of the same wave packet is also given by a Gaussian distribution associated with the spectral density function

$$|\Phi(k)|^2 = e^{-2\alpha(k-k_0)^2} \tag{28.44}$$

The k-values spreading about the mean value k_0 is measured by a standard deviation σ_k and it is straightforward to obtain, from Eq.(28.44), that

$$\Delta k = \sigma_k = \frac{1}{2\sqrt{\alpha}} \tag{28.45}$$

Thus

$$\Delta x \Delta k \geq \frac{1}{2} \tag{28.46}$$

The equality defines the Gaussian wave packets of minimum uncertainty. In view of the de Broglie hypothesis (28.35), the relation (28.46) also reads

$$\Delta x \Delta p_x \geq \frac{\hbar}{2} \tag{28.47}$$

and similar constraints apply to any pair of coordinate and conjugate momentum. It is for instance a straightforward matter to show that

$$\frac{m\Delta x}{p}\frac{p\Delta p}{m} \geq \frac{\hbar}{2} \quad \text{or} \quad \Delta t \Delta \varepsilon \geq \frac{\hbar}{2} \tag{28.48}$$

The expressions (28.47) and (28.48) illustrate the **Heisenberg uncertainty principle**, which states that *the product of uncertainties in the measurement of two canonically conjugate variables must be greater than* $\hbar/2$.

The uncertainty principle can also be interpreted in the sense that it is not possible to define simultaneously the conjugate variables q and p in the classical sense. Therefore the state of a system cannot be specified by both its coordinate and conjugate momentum at a given instant of time, so that the classical equations of motion do not have an exact meaning. It follows that we must define either the coordinates of a state or its momenta. In other words, we must choose either a *coordinate representation* or a *momentum representation* for describing the time evolution of a system.

FURTHER READING

1. R. Gilmore - ELEMENTARY QUANTUM MECHANICS IN ONE DIMENSION, Johns Hopkins University Press, Baltimore, 2005.
2. F. L. Pilar - ELEMENTARY QUANTUM CHEMISTRY, Dover Publications, New York, 2001.
3. D. F. Jackson - ATOMS AND QUANTA, Elsevier, Amsterdam, 1990.

PROBLEMS

28.1. When a beam of light of frequency ω strikes a metallic surface liberating electrons from the surface atoms, it is found that a retarding potential we need to apply to just stop all the photoelectrons is linearly dependent on ω. Find the threshold frequency ω_0, below which no electrons are emitted, in terms of the work W required to free an electron from the metallic surface.

Solution

When light is incident upon certain metallic surfaces, a radiative energy ε is transferred to bound electrons, liberating them from the surface atoms. The kinetic energy of an electron produced by photoemission is simply given by

$$\frac{mv^2}{2} = \varepsilon - W$$

where W is the work required to free an electron from the metallic surface. If a retarding potential V_0 is experimentally applied to just stop all the photoelectrons, the conservation of energy reads $eV_0 = \varepsilon - W$.

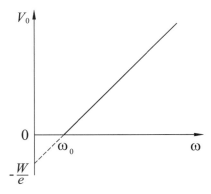

Assuming that the exchange of energy occurs between a single electron and a single photon, we may set the photon energy as given by Eq.(28.10), so that the conservation of energy becomes

$$eV_0 = \hbar\omega - W = \hbar(\omega - \omega_0)$$

where the threshold frequency ω_0 is expressed by $\omega_0 = W/\hbar$ and the slope of the straight line, represented in the figure, is $\hbar/e$. This relation gives a direct method for determining $\hbar$, provided e is known. Note that the wave theory of light provides no explanation for either the measured linear dependence of V_0 on the frequency ω of the incident light or the threshold frequency ω_0 below which no electrons are emitted.

28.2. Find the wavelength shift $\Delta\lambda$ of a beam of X-rays that is scattered through an angle θ by collisions with free electrons of mass m_e.

Solution

Treating the X-ray as a photon which undergoes a collision with an electron of rest mass m_e, as in the figure, the conservation of energy and momentum reads

$$\varepsilon + m_e c^2 = \varepsilon' + \left(p_e^2 c^2 + m_e^2 c^4\right)^{1/2} \quad , \quad \vec{p} = \vec{p}' + \vec{p}_e$$

where $\vec{p}, \vec{p}'$ are photon momenta before and after the collision.

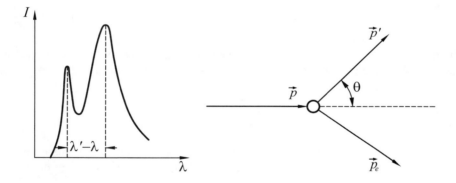

The momentum $\vec{p}_e$ of the recoil electron, whose energy is written according to the relativistic energy-momentum relation (5.10), can be independently expressed from the previous equations as

$$p_e^2 = p^2 + p'^2 - 2\vec{p} \cdot \vec{p}' = \frac{1}{c^2}\left(\varepsilon^2 + \varepsilon'^2 - 2\varepsilon\varepsilon' \cos\theta\right)$$

$$p_e^2 = \frac{1}{c^2}\left(\varepsilon - m_e c^2 - \varepsilon'\right)^2 - m_e c^2 = \frac{1}{c^2}\left[\varepsilon^2 + \varepsilon'^2 + 2m_e c^2\left(\varepsilon - \varepsilon'\right) - 2\varepsilon\varepsilon'\right]$$

so that we have

$$\varepsilon\varepsilon'\left(1 - \cos\theta\right) = m_e c^2\left(\varepsilon - \varepsilon'\right) \quad \text{or} \quad \frac{1}{\varepsilon'} - \frac{1}{\varepsilon} = \frac{1}{m_e c^2}\left(1 - \cos\theta\right)$$

This leads to the required $\Delta\lambda$ provided $\varepsilon, \varepsilon'$ are given by Eq.(28.10), that is,

$$\lambda' - \lambda = \Lambda_0 \left(1 - \cos\theta\right)$$

where the so-called Compton wavelength is given by $\Lambda_0 = h/m_e c$.

28.3. Use the Sommerfeld quantization rule to find the energy levels of a ball of mass m that is elastically bouncing on a flat surface under a gravitational field.

Solution

The total energy of the ball, which is a constant of motion, is given in terms of the height z from the surface as $E = p_z^2/2m + mgz$, where g is the acceleration due to gravity. The linear momentum results as

$$p_z = \sqrt{2m(E - mgz)}$$

and it is obvious that the admissible range of heights for the ball is $z \le z_{max} = E/mg$, where z_{max} corresponds to the turning point where $p_z = 0$. The quantization rule, Eq.(28.22), written for a cycle of motion, reads

$$2 \int_0^{E/mg} \sqrt{2m(E - mgz)}\, dz = nh$$

Using the integral formula $\int (a+bz)^{1/2} dz = 2(a+bz)^{3/2}/3b$, which can be proved by simple differentiation, it follows that

$$\frac{3}{4}\sqrt{2m}\frac{E^{3/2}}{mg} = nh \quad \text{or} \quad E_n = \left(\frac{3n}{4}gh\sqrt{\frac{m}{2}}\right)^{2/3}$$

28.4. Use the Maxwell distribution law for molecular speeds in a perfect gas in equilibrium at a temperature T to derive the distribution function for the de Broglie wavelengths associated with the same molecules. Find the most probable de Broglie wavelength for such a gas.

Solution

In view of Eq.(28.36) we may set $v = h/m\lambda$ in Eq.(26.31), where the distribution function within a range of speeds should be the same as the distribution function within the corresponding range of wavelengths $f(v)|dv| = f(\lambda)|d\lambda|$ which gives

$$f(\lambda) = f(v)\left|\frac{dv}{d\lambda}\right| = \frac{h}{m\lambda^2} f(v) = \frac{4\pi}{\lambda^4}\left(\frac{h^2}{2\pi m k_B T}\right)^{3/2} e^{-h^2/2mk_B T\lambda^2}$$

It is convenient to introduce the so-called thermal de Broglie wavelength, defined as

$$\lambda_B = \left(\frac{h^2}{2\pi m k_B T}\right)^{1/2}$$

so that the required distribution becomes

$$f(\lambda) = \frac{4\pi\lambda_B^2}{\lambda^4} e^{-\pi\lambda_B^2/\lambda^2}$$

The most probable wavelength λ_p is given by the maximum condition on $f(x)$, which is

$$\frac{d}{d\lambda}\left(\frac{1}{\lambda^4} e^{-\pi\lambda_B^2/\lambda^2}\right) = \frac{1}{\lambda^4} e^{-\pi\lambda_B^2/\lambda^2}\left(-\frac{4}{\lambda} + \frac{2\pi\lambda_B^2}{\lambda^3}\right) = 0$$

so that

$$\lambda_p = \sqrt{\frac{\pi}{2}}\lambda_B = \left(\frac{h^2}{4mk_B T}\right)^{1/2}$$

28.5. Find an expression for the dispersion relation of the matter waves associated with a particle of mass m and velocity v.

Solution

In the relativistic case, the relations between the corpuscular and wave parameters, Eqs.(28.10) and (28.36), take the form

$$E = mc^2 = \frac{m_0 c^2}{\sqrt{1 - v^2/c^2}} = \hbar\omega \quad , \quad \vec{p} = m\vec{v} = \frac{m_0 \vec{v}}{\sqrt{1 - v^2/c^2}} = \hbar\vec{k}$$

Substituting these results into the relativistic momentum-energy relation, where the norm $\sum_\mu p_\mu^2$ of the four-momentum p_μ, which is defined similarly to that of the three-dimensional vector $\vec{p}$, is invariant under Lorentz transformation, namely $p^2 - E^2/c^2 = -m_0^2 c^2$, one obtains the relativistic dispersion relation for the matter waves

$$\frac{\omega^2}{c^2} = k^2 + \frac{m_0^2 c^2}{\hbar^2}$$

The phase velocity is readily derived as

$$u = \frac{\omega}{k} = c\left(1 + \frac{m_0^2 c^2}{h^2}\lambda^2\right)^{1/2}$$

which obviously exceeds c for all material particles with $m_0 \neq 0$. However, there is no violation of the principle of special relativity which states that the velocity of material particles may not exceed c, because the velocity of propagation u of the phase waves associated with a particle is not to be confused with the velocity v of this particle. The last result may be reformulated as

$$u = c\left(1 + \frac{m_0^2 c^2}{m^2 v^2}\right)^{1/2} = c\left[1 + (1 - v^2/c^2)\frac{c^2}{v^2}\right]^{1/2} = \frac{c^2}{v}$$

In the nonrelativistic case, we have

$$E = \frac{p^2}{2m} + U(\vec{r}) = \frac{1}{2m}(p_x^2 + p_y^2 + p_z^2) + U(\vec{r})$$

and the dispersion relation for the de Broglie waves reads

$$\hbar\omega = \frac{\hbar^2 k^2}{2m} + U(\vec{r}) \quad \text{or} \quad \hbar\omega = \frac{mv^2}{2} + U(\vec{r})$$

which is the same as Eq.(28.10) written for a nonrelativistic material particle. Thus, the phase velocity u of the wave associated with a nonrelativistic particle may be expressed in terms of the velocity v of this particle by

$$u = \frac{\omega}{k} = \frac{\hbar k}{2m} + \frac{U(\vec{r})}{\hbar k} = \frac{v}{2} + \frac{U(\vec{r})}{mv}$$

For a free particle, for instance, the phase velocity is one-half of the particle velocity, and therefore it is of no physical significance.

28.6. Use the Heisenberg uncertainty principle to estimate the size of the ground state electron orbit in the Bohr model.

Solution

For an electron moving in a circular orbit of radius r, about a nucleus of nuclear charge Ze, the total energy is given by

$$E = \frac{p^2}{2m_e} - \frac{Ze_0^2}{r} = \frac{1}{2m_e}\langle p^2 \rangle - Ze_0^2 \left\langle \frac{1}{r} \right\rangle$$

and the uncertainty relation reads $\langle p^2 \rangle \langle r^2 \rangle \sim \hbar^2$. Using the approximations $\langle r^2 \rangle \cong \langle r \rangle^2$ and $\langle 1/r \rangle = 1/\langle r \rangle$ which, however, are valid to an order of magnitude only, the total energy becomes

$$E \geq \frac{\hbar^2}{2m_e \langle r \rangle^2} - \frac{Ze_0^2}{\langle r \rangle}$$

The right-hand side is then minimized by taking

$$\langle r \rangle = \frac{\hbar^2}{Zm_e e_0^2}$$

This value is the same as the Bohr radius for the ground state of one-electron atoms, Eq.(28.16), and hence, the same energy as that determined from the Bohr model, Eq.(28.18), is obtained.

28.7. A metallic ball of radius R produces photoelectrons when irradiated with light of wavelength λ. Giving the threshold value λ_0 of the photoelectric effect, find the maximum number N of photoelectrons which might be emitted.

28.8. If a photon of wavelength λ undergoes an elastic collision with an electron of mass m_e at rest and is scattered through an angle θ, find the orientation φ of the recoil electron.

28.9. Use the Sommerfeld quantization rule to find the energy levels of a rigid plane rotator of moment of inertia I at constant angular velocity.

28.10. If two hydrogen atoms of mass M, assumed to be in their ground state ε_1, collide with each other, find the limiting value of their relative velocity v_0 in order for the collision to be elastic.

28.11. Consider the reflection of electron waves on a crystal of lattice constant a, as illustrated in Figure 28.6. If the order of interference $m = 1$ is under observation at an angle θ, find the velocity of the relative motion of scattered electrons.

28.12. Use the Heisenberg uncertainty principle to estimate the ground state energy of a He atom.

29. POSTULATES OF QUANTUM MECHANICS

29.1. POSTULATE 1: WAVE FUNCTIONS
29.2. POSTULATE 2: OPERATORS
29.3. POSTULATE 3: EIGENVALUES
29.4. POSTULATE 4: COMMUTATION RELATIONS

29.1. POSTULATE 1: WAVE FUNCTIONS

Let us assume that the state of a free particle, of energy ε and momentum p, may be described by the matter wave function $\Psi(x,t)$ defined in Eq.(28.37). A physical significance must now be assigned to this wave function bearing in mind that it obeys Eq.(28.38), which reads

$$i\hbar \frac{\partial \Psi}{\partial t} = -\frac{\hbar^2}{2m}\frac{\partial^2 \Psi}{\partial x^2}$$

The solution $\Psi(x,t)$ is essentially a complex function which accounts for the matter wave spreading illustrated in Figure 28.3. As the wave function itself cannot be identified with a physical property of the free particle, an indirect interpretation was suggested by Born. He pointed out that its modulus $|\Psi(x,t)|$ is large where the particle becomes localized and small elsewhere, and hence

$$P(x,t)dx = |\Psi(x,t)|^2 dx \qquad (29.1)$$

is the *probability* of finding the particle described by $\Psi(x,t)$ at a point between x and $x+dx$ at a time t. Therefore the matter wave spreading is associated with a decrease in the *probability density* $P(x,t)$. As the total probability of finding the particle somewhere on the x-axis must be unity, the wave function must be *normalized* to unity, that is

$$\int_{-\infty}^{\infty} P(x,t)dx = \int_{-\infty}^{\infty} \Psi^*(x,t)\Psi(x,t)dx = 1 \qquad (29.2)$$

The behaviour of the wave function $\Psi(x,t)$ is determined by Eq.(28.38), which may be rewritten as

$$\Psi(x,t+\Delta t) = \Psi(x,t) + \frac{i\hbar}{2m}\frac{\partial^2 \Psi(x,t)}{\partial x^2}\Delta t \qquad (29.3)$$

where the expression $\partial\Psi/\partial t \cong [\Psi(x,t+\Delta t) - \Psi(x,t)]/\Delta t$, which holds for a small Δt, has been used. In view of Eq.(29.3) the values of the wave function at any time can be derived from the initial value $\Psi(x,0)$ that describes a reference state of the particle.

Since Eq.(28.38) is linear, if the wave functions $\Psi_1(x,t)$ and $\Psi_2(x,t)$ of two particular states of the particle, at a given position and time, are both solutions, so is the function

$$\Psi(x,t) = a_1\Psi_1(x,t) + a_2\Psi_2(x,t)$$

where a_1 and a_2 are arbitrary complex numbers. This result is generalized as the **principle of superposition of states**: *if n particular states of an arbitrary system are described by wave functions* $\Psi_1(x,t), \Psi_2(x,t),\ldots, \Psi_n(x,t)$, *their linear combination*

$$\Psi(x,t) = \sum_i a_i \Psi_i(x,t) \qquad (29.4)$$

describes a new particular state of the system. It follows that, although $|\Psi(x,t)|$ is the physically significant quantity, the phase factor of the solution cannot be ignored, since the modulus of $\Psi(x,t)$ depends on the relative phases of $\Psi_i(x,t)$. An optical analogue of this behaviour was described in Chapter 18, where the interference pattern, produced as a result of superposition of coherent beams obtained by the division of a wavefront, is determined by the phase relation between the parts of the wave function, each associated with one of the slits, in a multiple beam experiment.

Provided the wave functions $\Psi_1(x,t), \Psi_2(x,t),\ldots, \Psi_n(x,t)$ are linearly independent, that is none of these functions can be expressed in terms of the others, their linear combination Eq.(29.4) defines an *n*-dimensional function space. A *complete* set of wave functions is the minimum number such that any other wave function can be expanded as a member of this set. The function space is similar to the three-dimensional space which includes all the vectors of the form $\vec{a} = a_x\vec{e}_x + a_y\vec{e}_y + a_z\vec{e}_z$. As the unit vectors $\vec{e}_i$ are mutually orthogonal, that is $\vec{e}_i \cdot \vec{e}_j = \delta_{ij}$, the wave functions $\Psi_i(x,t)$, normalized to unity according to Eq.(29.2), are also required to obey an orthogonality relation which has the form

$$\int_{-\infty}^{\infty} \Psi_i^*(x,t)\Psi_j(x,t)dx = \delta_{ij} \qquad (29.5)$$

In such a case the wave function defined by Eq.(29.4) may also be called a *state vector*. When its explicit dependence on the position coordinates and time is not given, the principle of superposition of states can be interpreted as an *expansion theorem* which states that: *any state vector* Ψ *can be expanded in terms of a complete set of orthonormal vectors* Ψ_i *in an n-dimensional function space.*

EXAMPLE 29.1. The Dirac notation

Following Dirac, a convenient notation for a state vector Ψ is the *ket* vector $|\Psi\rangle$ and for its complex conjugate Ψ^* is the *bra* vector $\langle\Psi|$, so that the expansion (29.4) and its complex conjugate form may be written as

$$|\Psi\rangle = \sum_i a_i |\Psi_i\rangle \quad , \quad \langle\Psi| = \sum_i a_i^* \langle\Psi_i| \qquad (29.6)$$

Given a second state vector $|\Phi\rangle$ in the same function space, expressed by

$$|\Phi\rangle = \sum_i b_i |\Psi_i\rangle \quad , \quad \langle\Phi| = \sum_i b_i^* \langle\Psi_i|$$

the *scalar product* of the two vectors can be defined as follows

$$\langle\Psi|\Phi\rangle = \int_{-\infty}^{\infty} \Psi^*(x,t)\Phi(x,t)\,dx = \sum_i a_i^* b_i \qquad (29.7)$$

and this is an obvious analogue of the usual scalar product of two vectors in the Cartesian space, given by $\vec{a}\cdot\vec{b} = a_x b_x + a_y b_y + a_z b_z$.

The first postulate of quantum mechanics states the existence of a wave function that depends upon the position coordinates and time and describes as completely as possible the physical properties of a given state in the coordinate representation.

Postulate 1: *Every state of a dynamical system is described by a normalized wave function with state vector properties.*

In view of the probability density (29.1), the average value of an arbitrary function of position, in the state $\Psi(x,t)$ of a given system, called the *expectation value*, can be written as

$$\langle f(x)\rangle = \int_{-\infty}^{\infty} f(x)P(x,t)\,dx = \int_{-\infty}^{\infty} \Psi^*(x,t)f(x)\Psi(x,t)\,dx = \langle\Psi|f|\Psi\rangle \qquad (29.8)$$

The advantage of the Dirac notation for integrals in Eq.(29.8) is that we do not have to write out all the integration variables. Since the square modulus of the wave function represents a probability density, there are certain restrictions to be imposed on the state function, usually formulated as the following *rule*:

A state function and its derivative describing a dynamical system must be finite and continuous everywhere.

29.2. POSTULATE 2: OPERATORS

One may use Eq.(29.8), given in the coordinate representation, to calculate the expectation value of physical observables, provided there is a way of representing these variables as functions of position. Consider first the momentum, whose component p_x is classically expressed in terms of the position coordinate x, according to Eq.(2.8), as

$$p_x = mv_x = m\frac{dx}{dt}$$

so that its expectation value (29.8) becomes

$$\langle p_x \rangle = m\frac{d}{dt}\langle x \rangle = m\frac{d}{dt}\int_{-\infty}^{\infty} \Psi^*(x,t)x\Psi(x,t)dx = m\int_{-\infty}^{\infty}\left(\frac{\partial \Psi^*}{\partial t}x\Psi + \Psi^* x\frac{\partial \Psi}{\partial t}\right)dx \quad (29.9)$$

The partial derivative $\partial x/\partial t$ has been ignored under the integral, since both the position coordinate and time are independent variables and the state vector $\Psi(x,t)$ itself only varies with time. Inserting Eq.(28.38) and its complex conjugate into (29.9), one obtains

$$\langle p_x \rangle = \frac{\hbar}{2i}\int_{-\infty}^{\infty}\left(\frac{\partial^2 \Psi^*}{\partial x^2}x\Psi - \Psi^* x\frac{\partial^2 \Psi}{\partial x^2}\right)dx$$

As it is straightforward to show that

$$\frac{\partial^2 \Psi^*}{\partial x^2}x\Psi - \Psi^* x\frac{\partial^2 \Psi}{\partial x^2} = \frac{\partial}{\partial x}\left(\frac{\partial \Psi^*}{\partial x}x\Psi\right) - \frac{\partial}{\partial x}(\Psi^*\Psi) + \Psi^*\frac{\partial \Psi}{\partial x} - \frac{\partial}{\partial x}\left(\Psi^* x\frac{\partial \Psi}{\partial x}\right) + \Psi^*\frac{\partial \Psi}{\partial x}$$

the result of integration is

$$\langle p_x \rangle = \frac{\hbar}{2i}\left[\left(\frac{\partial \Psi^*}{\partial x}x\Psi - \Psi^*\Psi - \Psi^* x\frac{\partial \Psi}{\partial x}\right)_{-\infty}^{\infty} + \int_{-\infty}^{\infty} 2\Psi^*\frac{\partial \Psi}{\partial x}dx\right]$$

Since the wave functions are zero at infinity, where the probability density is considered to be vanishingly small, the expectation value of the x component of the momentum can be written as

$$\langle p_x \rangle = \int_{-\infty}^{\infty} \Psi^*(x,t) \frac{\hbar}{i} \frac{\partial}{\partial x} \Psi(x,t) dx = \int_{-\infty}^{\infty} \Psi^*(x,t) \hat{p}_x \Psi(x,t) dx \qquad (29.10)$$

The quantity denoted by

$$\hat{p} = \frac{\hbar}{i} \frac{\partial}{\partial x} \qquad (29.11)$$

is an an *operator* representing the x-component of the momentum in terms of the position coordinate x. Any classical observable, for instance the kinetic energy T, which is a function $f(p_x)$, can be represented by an operator $f(-i\hbar \partial / \partial x)$, such that

$$\langle f(p_x) \rangle = \int_{-\infty}^{\infty} \Psi^*(x,t) f\left(\frac{\hbar}{i} \frac{\partial}{\partial x} \right) \Psi(x,t) dx \qquad (29.12)$$

A similar interpretation given to Eq.(29.8) shows that the operator representing the position coordinate x is just the algebraic variable x and that any classical observable, the potential energy U say, which is a function $f(x)$ of the position coordinate, can be represented by the algebraic function $f(x)$ as

$$\hat{x} \equiv x \quad , \quad \hat{f}(x) \equiv f(x)$$

It is clear that the expectation value $\langle x \rangle$ given by Eq.(29.8) is real and it can be shown that $\langle p_x \rangle$ expressed by Eq.(29.10) is also real, because

$$\langle p_x \rangle = \int_{-\infty}^{\infty} \Psi^* \frac{\hbar}{i} \frac{\partial \Psi}{\partial x} dx = \frac{\hbar}{i} \int_{-\infty}^{\infty} \left(\Psi^* \frac{\partial \Psi}{\partial x} + \frac{\partial \Psi^*}{\partial x} \Psi \right) dx + \int_{-\infty}^{\infty} \Psi \left(-\frac{\hbar}{i} \frac{\partial}{\partial x} \right) \Psi^* dx$$

$$= \frac{\hbar}{i} \int_{-\infty}^{\infty} \frac{\partial}{\partial x} (\Psi^* \Psi) dx + \langle p_x \rangle^* = \frac{\hbar}{i} (\Psi^* \Psi) \Big|_{-\infty}^{\infty} + \langle p_x \rangle^* = \langle p_x \rangle^*$$

where we have used the property that the state functions are vanishing at infinity.

One class of operators that always have real expectation values are the so-called Hermitian operators, as we shall prove in the next section. The operators representing position and momentum in one dimension, usually denoted by q_i and $-i\hbar \partial / \partial q_i$, where q_i stands for x, y or z are Hermitian and linear. The second postulate states how the physical observables are represented in quantum mechanics.

> **Postulate 2.** *Every physical observable may be represented by a Hermitian linear operator. The operators representing the position and momentum of a particle in one dimension are q_i and $-i\hbar \partial / \partial q_i$, respectively. Operators representing other physical observables bear the same functional relation to these as do the corresponding classical*

quantities to the classical position and momentum variables.

As a consequence, the usual operators representing the physical observables, described in classical physics by functions $f(q_i)$ and $f(p_i)$, take the following forms

1. *Position operator* $\hat{\vec{r}}$, according to Eq.(2.1), can be written as

$$\vec{r} = x\vec{e}_x + y\vec{e}_y + z\vec{e}_z \quad \text{or} \quad \hat{\vec{r}} = \vec{r}$$

2. *Momentum operator* $\hat{\vec{p}}$, starting from Eq.(2.8), reads

$$\vec{p} = p_x \vec{e}_x + p_y \vec{e}_y + p_z \vec{e}_z \quad \text{or} \quad \hat{\vec{p}} = \frac{\hbar}{i}\left(\vec{e}_x \frac{\partial}{\partial x} + \vec{e}_y \frac{\partial}{\partial y} + \vec{e}_z \frac{\partial}{\partial z}\right) = \frac{\hbar}{i}\nabla \quad (29.13)$$

3. *Kinetic energy operator* $\hat{T}$, using Eq.(2.22), is

$$T = \frac{p^2}{2m} = \frac{1}{2m}\left(p_x^2 + p_y^2 + p_z^2\right) \quad \text{or} \quad \hat{T} = -\frac{\hbar^2}{2m}\left(\frac{\partial^2}{\partial x^2} + \frac{\partial^2}{\partial y^2} + \frac{\partial^2}{\partial z^2}\right) = -\frac{\hbar^2}{2m}\nabla^2 \quad (29.14)$$

4. *Hamiltonian operator* $\hat{H}$, in view of Eq.(24.29), has the form

$$H = \frac{p^2}{2m} + U(x,y,z) \quad \text{or} \quad \hat{H} = -\frac{\hbar^2}{2m}\nabla^2 + U(x,y,z) \quad (29.15)$$

and it is a Hermitian operator, provided the potential energy $U(r)$ is real.

5. *Angular momentum operator*, according to Eq.(3.1), becomes

$$\vec{L} = \vec{r} \times \vec{p} = \begin{vmatrix} \vec{e}_x & \vec{e}_y & \vec{e}_z \\ x & y & z \\ p_x & p_y & p_z \end{vmatrix}$$

so that the operators representing its components are

$$L_x = yp_z - zp_y \quad \text{or} \quad \hat{L}_x = \frac{\hbar}{i}\left(y\frac{\partial}{\partial z} - z\frac{\partial}{\partial y}\right)$$

$$L_y = zp_x - xp_z \quad \text{or} \quad \hat{L}_y = \frac{\hbar}{i}\left(z\frac{\partial}{\partial x} - x\frac{\partial}{\partial z}\right) \quad (29.16)$$

$$L_z = xp_y - yp_x \quad \text{or} \quad \hat{L}_x = \frac{\hbar}{i}\left(x\frac{\partial}{\partial y} - y\frac{\partial}{\partial x}\right)$$

The form of the operators given by Postulate 2 derives from the basic definitions of the position and momentum operators in the coordinate representation. In the momentum representation, where the matter waves are described by a function $\Phi(p_x,t)$ given by Eq.(28.39), which is a solution to Eq.(28.40), the representative operators of the position and momentum take the different form

$$\hat{q}_i = i\hbar \frac{\partial}{\partial q_i} \quad , \quad \hat{p}_i = p_i$$

and hence, the Hamiltonian (29.15) becomes

$$\hat{H} = \frac{1}{2m}\left(p_x^2 + p_y^2 + p_z^2\right) + \hat{U}\left(i\hbar\frac{\partial}{\partial x}, i\hbar\frac{\partial}{\partial y}, i\hbar\frac{\partial}{\partial z}\right)$$

Since the form of the potential energy function makes this representation difficult, the coordinate representation of quantum operators is usually preferred.

29.3. POSTULATE 3: EIGENVALUES

Let us consider two arbitrary state functions denoted by Φ and Ψ. The operator $\hat{f}$ is said to be *Hermitian* if and only if we have

$$\int_{-\infty}^{\infty}\Phi^*\hat{f}\,\Psi dx = \int_{-\infty}^{\infty}\left(\hat{f}\,\Phi\right)^*\Psi dx \qquad (29.17)$$

We can extend the Dirac notation, in order to include operators, in the following way

$$\int_{-\infty}^{\infty}\Phi^*\hat{f}\,\Psi dx = \left\langle\Phi\middle|\hat{f}\,\Psi\right\rangle = \left\langle\Phi\middle|\hat{f}\middle|\Psi\right\rangle$$

$$\int_{-\infty}^{\infty}\left(\hat{f}\,\Phi\right)^*\Psi dx = \left\langle\hat{f}\,\Phi\middle|\Psi\right\rangle = \left\langle\Phi\middle|\hat{f}^\dagger\middle|\Psi\right\rangle$$

(29.18)

so that we find that the operator $\hat{f}$ is Hermitian if

$$\left\langle\Phi\middle|\hat{f}\middle|\Psi\right\rangle = \left\langle\Phi\middle|\hat{f}^\dagger\middle|\Psi\right\rangle = \left\langle\Psi\middle|\hat{f}\middle|\Phi\right\rangle^* \qquad (29.19)$$

where the dagger denotes an adjoint operator. It is seen that a Hermitian operator is self-adjoint, since Eq.(29.19) yields $\hat{f}^\dagger = \hat{f}$. Provided the equation

$$\hat{f}\Psi_i = \lambda_i \Psi_i \qquad (29.20)$$

where λ_i is a constant, holds for the quantum operator $\hat{f}$, the state function Ψ_i is called *eigenfunction* of $\hat{f}$ and λ_i is the corresponding *eigenvalue*. The number of eigenvalues of a quantum operator is, in general, infinite and their spectrum may be discrete or continuous. The requirement that physical observables are represented by Hermitian operators results from the properties of their eigenvalues and eigenfunctions, as expressed by the following theorems

Theorem 1. *The eigenvalues of Hermitian operators are real.*

Proof. Since any eigenvalue λ_i in Eq.(29.20) is a constant, we can multiply both sides of this relation by Ψ_i^* and then integrate to obtain

$$\int_{-\infty}^{\infty}\Psi_i^* \hat{f}\Psi_i dx = \int_{-\infty}^{\infty}\Psi_i^* \lambda_i \Psi_i dx = \lambda_i \int_{-\infty}^{\infty}|\Psi_i|^2 dx$$

Using Eq.(29.17) where $\Phi = \Psi = \Psi_i$, this result may also be expressed as

$$\int_{-\infty}^{\infty}\Psi_i^* \hat{f}\Psi_i dx = \int_{-\infty}^{\infty}(\hat{f}\Psi_i)^* \Psi_i dx = \lambda_i^* \int_{-\infty}^{\infty}|\Psi_i|^2 dx$$

so that it follows that any eigenvalue λ_i is real, since

$$\lambda_i = \lambda_i^* \qquad (29.21)$$

Theorem 2. *Any two eigenfunctions of a Hermitian operator that belong to different eigenvalues are orthogonal.*

Proof. Let λ_i, λ_k be different eigenvalues of the operator $\hat{f}$ so that we have

$$\hat{f}\Psi_i = \lambda_i \Psi_i$$
$$\hat{f}\Psi_k = \lambda_k \Psi_k \quad \text{or} \quad (\hat{f}\Psi_k)^* = \lambda_k^* \Psi_k^* = \lambda_k \Psi_k^* \qquad (29.22)$$

where Eq.(29.21) has been used. It follows that

$$\int_{-\infty}^{\infty}\Psi_k^* \hat{f}\Psi_i dx = \lambda_i \int_{-\infty}^{\infty}\Psi_k^* \Psi_i dx$$

Since $\hat{f}$ is Hermitian, according to Eq.(29.17) we may write

$$\int_{-\infty}^{\infty} \Psi_k^* \hat{f} \Psi_i \, dx = \int_{-\infty}^{\infty} \left(\hat{f} \Psi_k \right)^* \Psi_i \, dx = \lambda_k \int_{-\infty}^{\infty} \Psi_k^* \Psi_i \, dx$$

Provided $\lambda_i \neq \lambda_k$, the last two equations lead to

$$\int_{-\infty}^{\infty} \Psi_k^* \Psi_i \, dx = 0 \tag{29.23}$$

It is a reasonable extension of Eqs.(29.8) and (29.12) to consider that, if a physical observable is represented by a Hermitian operator $\hat{f}$, then the expectation value of this operator, in a state Ψ, which is given by

$$\langle \hat{f} \rangle = \int_{-\infty}^{\infty} \Psi^* \hat{f} \Psi \, dx = \langle \Psi | \hat{f} | \Psi \rangle \tag{29.24}$$

corresponds to the average value that will be obtained from a large number of measurements of the same observable f, carried out on systems with state functions which are all identical with Ψ. The third postulate states the significance of quantum operator eigenvalues in relation to the physical measurements.

Postulate 3. *The eigenvalues of a quantum operator represent the possible results of carrying out a measurement of the associated physical observable. If the system is in a state described by an eigenfunction of a quantum operator, the result of the measurement is given by the corresponding eigenvalue.*

Taking the state function Ψ of the system identical to an eigenfunction Ψ_i of the operator $\hat{f}$, which obeys Eq.(29.20), the expectation value (29.24) becomes

$$\langle \Psi | \hat{f} | \Psi \rangle = \langle \Psi_i | \hat{f} | \Psi_i \rangle = \lambda_i \tag{29.25}$$

The *deviation* operator, defined as

$$\Delta \hat{f} = \hat{f} - \langle \Psi | \hat{f} | \Psi \rangle \tag{29.26}$$

may be then used to express the standard deviation associated with any measurement as

$$\langle \Psi | (\Delta \hat{f})^2 | \Psi \rangle = \langle \Psi | \hat{f}^2 | \Psi \rangle - \langle \Psi | \hat{f} | \Psi \rangle^2 \tag{29.27}$$

From Eq.(29.24), taking $\Psi = \Psi_i$, we obtain

$$\langle \Psi | \hat{f}^2 | \Psi \rangle = \lambda_i^2$$

so that the standard deviation vanishes. Therefore the measurement of a physical observable represented by $\hat{f}$ gives precisely a result equal to the real eigenvalue λ_i, in a state described by the corresponding eigenfunction Ψ_i.

Taking now the state function of a system to be a linear combination (29.4) of eigenfunctions Ψ_i we may derive

$$a_i = \langle \Psi_i | \Psi \rangle = \int_{-\infty}^{\infty} \Psi_i^* \Psi \, dx$$

so that Eq.(29.4) becomes

$$\Psi = \sum_i \langle \Psi_i | \Psi \rangle \Psi_i \quad (29.28)$$

The expectation value (29.24) of the operator $\hat{f}$ in this state is then given by

$$\langle \Psi | \hat{f} | \Psi \rangle = \sum_{i,k} \langle \Psi_k | \Psi \rangle^* \langle \Psi_i | \Psi \rangle \int_{-\infty}^{\infty} \Psi_k^* \hat{f} \Psi_i \, dx = \sum_i |\langle \Psi_i | \Psi \rangle|^2 \lambda_i \quad (29.29)$$

This expression shows that $|\langle \Psi_i | \Psi \rangle|^2$ gives the probability of obtaining the eigenvalue λ_i as a result of the measurement. Since a measurement can only give a result λ_i if the system is in a corresponding eigenstate Ψ_i, each weighting factor $|\langle \Psi_i | \Psi \rangle|^2$ may also be interpreted as the probability for the state function Ψ of the system be identical to the eigenfunction Ψ_i of the quantum operator $\hat{f}$, immediately after such a measurement.

EXAMPLE 29.2. The eigenfunction of the position operator in one dimension

For any eigenvalue x_0 of the position operator $\hat{x}$ the eigenfunction must be very large at the point $x = x_0$ and vanishingly small everywhere else, in order to describe the state of the system immediately after a position measurement which yields the result x_0. A function with such properties is called Dirac's delta function $\delta(x - x_0)$, and Eq.(29.20) can be written as

$$\hat{x}\delta(x - x_0) = x_0 \delta(x - x_0)$$

Therefore the eigenfunction $\delta(x - x_0)$ describes a particle precisely localized at x_0. The probability of finding the particle close to x_0, in a state $\Psi(x,t)$, is given by Eq.(29.29) as

$$|\langle \delta(x - x_0) | \Psi \rangle|^2 = \left| \int_{-\infty}^{\infty} \delta(x - x_0) \Psi(x,t) \, dx \right|^2 = |\Psi(x_0, t)|^2$$

Thus, $|\Psi(x_0, t)|^2$ may indeed be interpreted as the probability density of finding the particle at x_0 at a time t.

29.4. POSTULATE 4: COMMUTATION RELATIONS

Let us consider two quantum operators $\hat{f}$ and $\hat{g}$ representing physical observables. The effect of their product on an arbitrary state function is, in general, dependent on the order in which the operators are applied, that is, unlike algebraic variables, quantum operators do not, in general, commute. Hence, a commutator of the two operators is defined as

$$[\hat{f},\hat{g}] = \hat{f}\hat{g} - \hat{g}\hat{f} \qquad (29.30)$$

EXAMPLE 29.3. The commutator $[\hat{x},\hat{p}_x]$

Suppose the commutator of the operators $\hat{x}$ and $\hat{p}_x$ representing the position and momentum of a particle in one dimension is applied to an arbitrary state function $\Psi = \Psi(x)$, thus

$$[\hat{x},\hat{p}_x]\Psi = \frac{\hbar}{i}\left[x,\frac{\partial}{\partial x}\right]\Psi = \frac{\hbar}{i}x\frac{\partial\Psi}{\partial x} - \frac{\hbar}{i}\frac{\partial}{\partial x}(x\Psi) = \frac{\hbar}{i}\left(x\frac{\partial\Psi}{\partial x} - \Psi - x\frac{\partial\Psi}{\partial x}\right) = i\hbar\Psi$$

Since the result should be independent of the particular form of the function Ψ, we may write this in operatorial form as

$$[\hat{x},\hat{p}_x] = i\hbar$$

which shows that the operators $\hat{x}$ and $\hat{p}_x$ do not commute. Similar arguments prove that $\hat{x}$ commutes with $\hat{p}_y$ and $\hat{p}_z$ and also with $\hat{y}$ and $\hat{z}$.

The different components of the position and momentum operators, expressed according to Postulate 2 as

$$\hat{q}_i = q_i \quad , \quad \hat{p}_i = \frac{\hbar}{i}\frac{\partial}{\partial q_i} \qquad (29.31)$$

obey commutation relations similar to those proved in the previous example, that is

$$[\hat{q}_i,\hat{q}_k] = 0 \quad , \quad [\hat{q}_i,\hat{p}_k] = i\hbar\delta_{ik} \quad , \quad [\hat{p}_i,\hat{p}_k] = 0 \qquad (29.32)$$

Therefore the quantum operators $\hat{q}_i$ and $\hat{p}_i$ are such that their commutators are proportional to the corresponding Poisson brackets (24.38) of the position and momentum components $\hat{q}_i$ and $\hat{p}_i$ with $i\hbar$ as the constant of proportionality. The fourth postulate introduces a quantum analogue of the generalized mechanics formalism, which relies on the analogy between Eq.(29.32) and Eq.(24.38), extended to any pair of operators representing physical observables.

Postulate 4. *The commutation relation between any pair of quantum operators is derived from the Poisson bracket of the corresponding pair of classical variables, according to the correspondence rule*

$$\{f,g\}=k \quad \rightarrow \quad [\hat{f},\hat{g}]=i\hbar\hat{k} \tag{29.33}$$

The concept of commutator provides a direct way to discuss the compatibility of physical observables, the uncertainty relations and the evolution of quantum operators.

Compatibility

According to Postulate 3 if two observables are simultaneously measurable in a particular state of a given system, and a unique result is obtained if either one is measured, then the state function is an eigenfunction of each representative operator. Two observables are said to be *compatible* if the operators representing them have a common set of eigenfunctions. The following two theorems show the connection between compatible observables and commuting operators.

Theorem 3. *If the operators $\hat{f}$ and $\hat{g}$ commute, they have a common set of eigenfunctions, and hence, the corresponding observables are compatible.*

Proof. Let $\Psi_i, \Psi_k,\ldots$ be the eigenfunctions of the operator $\hat{f}$ with eigenvalues $\lambda_i, \lambda_k,\ldots$, respectively. Assuming that Ψ_i is not necessarily an eigenfunction of the operator $\hat{g}$, the function $\hat{g}\Psi_i$ may be expressed in terms of the eigenfunctions Ψ_k of $\hat{f}$, according to Eq.(29.28), as

$$\hat{g}\Psi_i = \sum_k \langle \Psi_k|\hat{g}|\Psi_i\rangle \Psi_k \tag{29.34}$$

On the other hand, the effect of the commutator $[\hat{f},\hat{g}]$ on Ψ_i can be written as

$$[\hat{f},\hat{g}]\Psi_i = \hat{f}\hat{g}\Psi_i - \hat{g}\hat{f}\Psi_i = \sum_k \hat{f}\Psi_k\langle \Psi_k|\hat{g}|\Psi_i\rangle - \hat{g}(\lambda_i\Psi_i)$$

$$= \sum_k \lambda_k \Psi_k\langle \Psi_k|\hat{g}|\Psi_i\rangle - \sum_k \lambda_i \langle \Psi_k|\hat{g}|\Psi_i\rangle \Psi_k = \sum_k \Psi_k\langle \Psi_k|\hat{g}|\Psi_i\rangle(\lambda_k - \lambda_i) \tag{29.35}$$

Since $\hat{f}$ and $\hat{g}$ commute, the result must be set equal to zero, and hence, each coefficient of the eigenfunctions Ψ_k will vanish, that is

$$\langle \Psi_k|\hat{g}|\Psi_i\rangle(\lambda_k - \lambda_i) = 0 \tag{29.36}$$

Assuming distinct or *nondegenerate* eigenvalues, it follows that every $\langle \Psi_k|\hat{g}|\Psi_i\rangle$ vanishes unless $k=i$, when its value may be denoted by μ_i, that is,

$$\langle \Psi_k | \hat{g} | \Psi_i \rangle = \mu_i \delta_{ik} \tag{29.37}$$

If this result is inserted into Eq.(29.34), the eigenvalue equation of the operator $\hat{g}$ reads

$$\hat{g} \Psi_i = \mu_i \Psi_i$$

and this shows that the eigenfunctions Ψ_i of $\hat{f}$ are also eigenfunctions of $\hat{g}$.

Theorem 4. *If the operators $\hat{f}$ and $\hat{g}$ have a common set of eigenfunctions, then the operators commute.*

Proof. Let Ψ_i denote the common set of eigenfunctions of the operators $\hat{f}$ and $\hat{g}$, with respective eigenvalues λ_i and μ_i, which obey Eq.(29.20). Consider the effect of their commutator on an arbitrary state vector Ψ given by Eq.(29.28), which reads

$$[\hat{f}, \hat{g}] \Psi = \sum_i \langle \Psi_i | \Psi \rangle (\hat{f} \hat{g} \Psi_i - \hat{g} \hat{f} \Psi_i) = \sum_i \langle \Psi_i | \Psi \rangle (\hat{f} \mu_i \Psi_i - \hat{g} \lambda_i \Psi_i)$$

$$= \sum_i \langle \Psi_i | \Psi \rangle (\lambda_i \mu_i \Psi_i - \mu_i \lambda_i \Psi_i) = 0$$

Since Ψ is arbitrary, it follows that $[\hat{f}, \hat{g}] = 0$.

Therefore the compatibility test for any pair of physical observables is to check whether or not the commutator of their representative operators is zero. In view of the commutation relations (29.32), the three position coordinates and the three components of momentum are compatible.

Heisenberg inequality

The uncertainty principle follows directly from the basic postulates of quantum mechanics, as the next theorem shows.

Theorem 5. *If the operators $\hat{f}$, $\hat{g}$ and $\hat{k}$ representing physical observables are related by*

$$[\hat{f}, \hat{g}] = i\hat{k} \tag{29.38}$$

then the deviations of the result of measurements of the corresponding observables, in an arbitrary state, obey the inequality

$$\langle (\Delta \hat{f})^2 \rangle \langle (\Delta \hat{g})^2 \rangle \geq \frac{1}{4} \langle \hat{k} \rangle^2 \tag{29.39}$$

Proof. If the definition (29.26) of the deviation operator $\Delta \hat{f}$ is inserted into Eq.(29.38), it is straightforward to obtain that

$$[\Delta\hat{f}, \Delta\hat{g}] = i\hat{k} \tag{29.40}$$

Denoting by α an arbitrary real number, it is convenient to define a nonnegative expression, which successively transforms, in view of Eq.(29.40), as follows

$$\int_{-\infty}^{\infty} |(\alpha\Delta\hat{f} - i\Delta\hat{g})\Psi|^2 \, dx = \int_{-\infty}^{\infty} (\alpha\Delta\hat{f} + i\Delta\hat{g})^* \Psi^* (\alpha\Delta\hat{f} - i\Delta\hat{g})\Psi \, dx$$

$$= \int_{-\infty}^{\infty} \Psi^* (\alpha\Delta\hat{f} + i\Delta\hat{g})^\dagger (\alpha\Delta\hat{f} - i\Delta\hat{g})\Psi \, dx$$

$$= \langle (\alpha\Delta\hat{f} + i\Delta\hat{g})(\alpha\Delta\hat{f} - i\Delta\hat{g}) \rangle = \alpha^2 \langle (\Delta\hat{f})^2 \rangle - \alpha \langle \hat{k} \rangle + \langle (\Delta\hat{g})^2 \rangle \geq 0$$

where the minimum is attained where the derivative with respect to α is zero, that is, for

$$\alpha = \frac{\langle \hat{k} \rangle}{2\langle (\Delta\hat{f})^2 \rangle} = \frac{\langle \hat{k} \rangle}{2(\Delta f)^2}$$

Inserting this into the inequality, one obtains

$$4\langle (\Delta\hat{f})^2 \rangle \langle (\Delta\hat{g})^2 \rangle - \langle \hat{k} \rangle^2 \geq 0$$

which is the same as Eq.(29.39), as required.

In the particular case where $\hat{f} = \hat{x}$ and $\hat{g} = \hat{p}_x$ are the operators representing a position coordinate and the corresponding component of momentum, respectively, the commutation relation (29.38) gives $\hat{k} = \hbar$, and the inequality takes the familiar form

$$\langle (\Delta\hat{x})^2 \rangle \langle (\Delta\hat{p}_x)^2 \rangle \geq \frac{1}{4}\hbar^2 \tag{29.41}$$

already introduced in (28.47) as the Heisenberg uncertainty relation.

Evolution of quantum operators

In view of the correspondence rule (29.33), Eq.(24.33) for an observable f leads to the *evolution equation* of the representative quantum operator $\hat{f}$, which reads

$$i\hbar \frac{d\hat{f}}{dt} = [\hat{f}, \hat{H}] + i\hbar \frac{\partial \hat{f}}{\partial t} \tag{29.42}$$

where $\hat{H}$ is the Hamiltonian, as introduced by Eq.(29.15). Provided the operator $\hat{f}$ does not have an explicit dependence on time, the evolution equation reduces to

$$i\hbar \frac{d\hat{f}}{dt} = \left[\hat{f}, \hat{H}\right] \qquad (29.43)$$

If the operator $\hat{f}$ also commutes with the Hamiltonian, we obtain

$$\left[\hat{f}, \hat{H}\right] = 0 \quad \text{or} \quad \frac{d\hat{f}}{dt} = 0 \qquad (29.44)$$

In view of Eq.(29.24), Eq.(29.44) also holds for expectation values, so that the observable $\hat{f}$ is then said to be conserved or to be a constant of motion. The evolution of quantum systems is described in Eqs.(29.42) and (29.43) by time-dependent operators and time-independent state vectors, since both equations reflect only the operator evolution, independent of how the state to which they are applied changes with time. Such a description of the dynamical systems is called the *Heisenberg picture*.

FURTHER READING
1. A. I. M. Rae - QUANTUM MECHANICS, Institute of Physics Publishing, Bristol, 2002.
2. P. C. W. Davies, D. S. Betts - QUANTUM MECHANICS, Nelson Thornes, Cheltenham, 1994.
3. F.Mandl - QUANTUM MECHANICS, Wiley, New York, 1992.

PROBLEMS

29.1. A one-dimensional system has a state function $\Psi(x) = Nxe^{-\alpha x^2}$. Show that $\Psi(x)$ is an eigenfunction of the operator

$$\hat{A} = -\frac{d^2}{dx^2} + x^2$$

for a given value of α, and find the corresponding eigenvalue. Determine the expectation value of the position in the state $\Psi(x)$ of the system.

It is straightforward that

$$\hat{A}\Psi(x) = -\frac{d^2\Psi}{dx^2} + x^2\Psi = -\frac{d^2}{dx^2}\left(Nxe^{-\alpha x^2}\right) + Nx^3 e^{-\alpha x^2} = -N\frac{d}{dx}\left(e^{-\alpha x^2} - 2\alpha x^2 e^{-\alpha x^2}\right) + Nx^3 e^{-\alpha x^2}$$

$$= -N\frac{d}{dx}\left[e^{-\alpha x^2}\left(1-2\alpha x^2\right)\right] + Nx^3 e^{-\alpha x^2} = Ne^{-\alpha x^2}\left[4\alpha x + 2\alpha x\left(1-2\alpha x^2\right) + x^3\right]$$

and hence, we can write

$$\hat{A}\Psi(x) = 6\alpha \Psi(x) + \left(1-4\alpha^2\right)Nx^3 e^{-\alpha x^2}$$

The eigenvalue equation (29.20) is satisfied provided that

$$1-4\alpha^2 = 0 \quad \text{or} \quad \alpha = \frac{1}{2}$$

and it follows that the corresponding eigenvalue is $6\alpha = 3$. The normalized state function is obtained from Eq.(29.2), which reads

$$|N|^2 \int_{-\infty}^{\infty} x^2 e^{-x^2} dx = 1 \quad \text{or} \quad |N|^2 = \frac{2}{\sqrt{\pi}}$$

where use has been made of Eq.(IV.10) given in Appendix IV. Thus, in view of Eqs.(29.8) and (IV.10), one obtains

$$\langle x \rangle = |N|^2 \int_{-\infty}^{\infty} x^3 e^{-x^2} dx = 0$$

29.2. A particle of mass m moving in one dimension has a state function $\Psi(x) = Ne^{-\alpha x^2}$. Find an expression for the potential energy $U(x)$, and determine the energy eigenvalue ε.

Solution

The energy eigenvalue equation (29.20) reads

$$-\frac{\hbar^2}{2m}\frac{d^2\Psi}{dx^2} + U(x)\Psi = \varepsilon \Psi$$

where

$$\frac{d\Psi}{dx} = -2\alpha x e^{-\alpha x^2} \quad \text{and} \quad \frac{d^2\Psi}{dx^2} = -2\alpha\left(1-2\alpha x^2\right)e^{-\alpha x^2}$$

Substituting into the eigenvalue equation gives

$$\frac{\hbar^2}{2m}2\alpha\left(1-2\alpha x^2\right) + U(x) = \varepsilon \quad \text{or} \quad \left[U(x) - \frac{2\alpha^2 \hbar^2}{m}x^2\right] + \left(\frac{\alpha\hbar^2}{m} - \varepsilon\right) = 0$$

and hence, one obtains

$$U(x) = \frac{2\alpha^2\hbar^2}{m}x^2 \quad , \quad \varepsilon = \frac{\alpha\hbar^2}{m}$$

The potential energy is that of a simple harmonic oscillator, given by $kx^2/2 = m\omega^2 x^2/2$. The angular frequency ω has a value of $2\alpha\hbar/m$. It follows that $\varepsilon = \hbar\omega/2$, which is known to be the zero point energy of the one-dimensional harmonic oscillator, as we will show in Chapter 30.

29.3. Show that the parity operator $\hat{\Pi}$, defined by $\hat{\Pi}\Psi(x) = \Psi(-x)$, is Hermitean.

Solution

We may use, for simplicity, the definition (29.19) of a Hermitian operator, written as

$$\langle\Psi|\hat{f}|\Psi\rangle = \langle\Psi|\hat{f}^\dagger|\Psi\rangle$$

In the case of the parity operator, we have

$$\langle\Psi|\hat{\Pi}|\Psi\rangle = \int_{-\infty}^{\infty}\Psi^*(x,t)\hat{\Pi}\Psi(x,t)dx = \int_{-\infty}^{\infty}\Psi^*(x,t)\Psi(-x,t)dx = \left[\int_{-\infty}^{\infty}\Psi^*(-x,t)\hat{\Pi}\Psi(-x,t)dx\right]^*$$

$$= \int_{-\infty}^{\infty}\left[\hat{\Pi}\Psi(x,t)\right]^*\Psi(x,t)dx = \langle\hat{\Pi}\Psi|\Psi\rangle = \langle\Psi|\hat{\Pi}^\dagger|\Psi\rangle$$

and hence, $\hat{\Pi}$ is Hermitian as required.

29.4. Use the uncertainty relations to find an expression for the function of position $\Psi(x)$ representing the minimum uncertainty state of a classical particle, which ensures the highest precision for the simultaneous determination of its position x and momentum p_x.

Solution

Consider the minimum condition on the non-negative expression

$$\langle\Psi|\left(\alpha\Delta\hat{f}+i\Delta\hat{g}\right)^\dagger\left(\alpha\Delta\hat{f}-i\Delta\hat{g}\right)|\Psi\rangle \geq 0 \quad \text{or} \quad \langle\Psi|\left(\frac{\langle\hat{k}\rangle}{2(\Delta f)^2}\Delta\hat{f}+i\Delta\hat{g}\right)^\dagger\left(\frac{\langle\hat{k}\rangle}{2(\Delta f)^2}\Delta\hat{f}-i\Delta\hat{g}\right)|\Psi\rangle = 0$$

where the α value for which the derivative of the left-hand side is zero has been inserted as given in the proof of Theorem 5. This implies, in view of Eq.(29.38), that

$$\left(\frac{\langle\hat{k}\rangle}{2(\Delta f)^2}\Delta\hat{f}-i\Delta\hat{g}\right)\Psi = 0 \quad \text{or} \quad \left[\frac{1}{2}\frac{\langle[\hat{f},\hat{g}]\rangle}{(\Delta f)^2}(\hat{f}-\langle f\rangle)-(\hat{g}-\langle g\rangle)\right]\Psi = 0$$

and allows us to determine the minimum uncertainty states. Substituting $\hat{f} = x$ and $\hat{g} = \hat{p}_x = -i\hbar(d/dx)$ leads to the differential equation

$$\left(i\hbar \frac{d}{dx} + \langle p_x \rangle + \frac{i\hbar}{2} \frac{x - \langle x \rangle}{(\Delta x)^2} \right) \Psi(x) = 0$$

and this is solved by the normalized wave function

$$\Psi(x) = \frac{1}{\left[2\pi(\Delta x)^2 \right]^{1/4}} e^{-(x-\langle x \rangle)^2/4(\Delta x)^2} e^{i\langle \hat{p}_x \rangle x/\hbar} = A(x) e^{i\langle \hat{p}_x \rangle x/\hbar}$$

The amplitude $A(x)$ of the wave function corresponds to a Gaussian distribution of the position values about $\langle x \rangle$. If $\langle x \rangle = 0$ we obtain the wave packet introduced by Eq.(28.41) as

$$|\Psi(x)|^2 = \frac{1}{\sqrt{2\pi(\Delta x)^2}} e^{-x^2/2(\Delta x)^2}$$

This distribution has a dispersion of $\Delta x = \hbar/2\Delta p_x$, as determined by the equality $\Delta x \Delta p_x = \hbar/2$. It follows that the Gaussian wave packet provides the best approximation, compatible with the uncertainty relations, for the description of a classical particle.

29.5. Derive the evolution equations for the expectation values of position $\langle \hat{x} \rangle$ and momentum $\langle p_x \rangle$, and discuss their physical significance in terms of the correspondence principle.

Solution

Since the operators $\hat{q}_i$ and $\hat{p}_i$ do not depend explicitly on time, their appropriate evolution equation is (29.43), which reads

$$\frac{d\hat{q}_i}{dt} = \frac{i}{\hbar} \left[\hat{H}, \hat{q}_i \right] \quad , \quad \frac{d\hat{p}_i}{dt} = \frac{i}{\hbar} \left[\hat{H}, \hat{p}_i \right]$$

By taking $\hat{q}_i = \hat{x}$ we can then write the evolution equation for position as

$$\frac{d\hat{x}}{dt} = \frac{i}{\hbar} \left[\hat{H}, \hat{x} \right] = \frac{i}{\hbar} \left(\hat{H}\hat{x} - \hat{x}\hat{H} \right)$$

where $\hat{H}$ is the Hamiltonian operator in one dimension, given by Eq.(29.15). For an arbitrary time-independent state function $\Psi(x)$, we may write

$$\left(\hat{H}\hat{x} - \hat{x}\hat{H} \right)\Psi = -\frac{\hbar^2}{2m} \frac{d^2}{dx^2}(x\Psi) + Ux\Psi + x \frac{\hbar^2}{2m} \frac{d^2\Psi}{dx^2} - xU\Psi = \left(-\frac{\hbar^2}{m} \frac{d}{dx} \right) \Psi$$

so that, independent of the particular state $\Psi(x)$, the evolution of the position operator reads

$$\frac{d\hat{x}}{dt} = \frac{i}{\hbar} \left[\hat{H}, \hat{x} \right] = -\frac{i\hbar}{m} \frac{d}{dx} = \frac{\hat{p}_x}{m}$$

In a similar way, by taking $\hat{p}_i = \hat{p}_x$, the evolution equation for momentum becomes

$$\frac{d\hat{p}_x}{dt} = \frac{i}{\hbar}\left[\hat{H}, \hat{p}_x\right] = \frac{i}{\hbar}\left(\hat{H}\hat{p}_x - \hat{p}_x\hat{H}\right)$$

The effect of the commutator $\left[\hat{H}, \hat{p}_x\right]$ on an arbitrary state function $\Psi(x)$ is

$$\left(\hat{H}\hat{p}_x - \hat{p}_x\hat{H}\right)\Psi = \left[\left(-\frac{\hbar^2}{2m}\frac{d^2}{dx^2} + U\right)\left(-i\hbar\frac{d}{dx}\right) - \left(-i\hbar\frac{d}{dx}\right)\left(-\frac{\hbar^2}{2m}\frac{d^2}{dx^2} + U\right)\right]\Psi$$

$$= -i\hbar\left(U\frac{d}{dx} - \frac{d}{dx}U\right)\Psi = -i\hbar\left(U\frac{d\Psi}{dx} - \frac{dU}{dx}\Psi - U\frac{d\Psi}{dx}\right) = i\hbar\frac{dU}{dx}\Psi$$

Therefore, we have the operator relation

$$\frac{d\hat{p}_x}{dt} = -\frac{dU}{dx} = F(\hat{x})$$

where the operator $F(\hat{x})$ must be associated with a conservative force, according to Eq.(2.27). It follows that the expectation values of position, momentum and force are related by

$$\left\langle\frac{d\hat{x}}{dt}\right\rangle = \frac{1}{m}\langle\hat{p}_x\rangle \quad , \quad \left\langle\frac{d\hat{p}_x}{dt}\right\rangle = \langle F(\hat{x})\rangle$$

The first equation is almost identical to the relation between the classical momentum p_x and velocity dx/dt. It would be exactly identical, if and only if we have

$$\left\langle\frac{d\hat{x}}{dt}\right\rangle = \frac{d}{dt}\langle\hat{x}\rangle \quad \text{i.e.,} \quad \frac{d}{dt}\langle\hat{x}\rangle = \frac{1}{m}\langle\hat{p}_x\rangle$$

which is the same as the classical relation (2.8), in one dimension, provided we identify $\langle\hat{x}\rangle$ with the position coordinate x. Assuming also that

$$\left\langle\frac{d\hat{p}_x}{dt}\right\rangle = \frac{d}{dt}\langle\hat{p}_x\rangle \quad \text{i.e.,} \quad \frac{d}{dt}\langle\hat{p}_x\rangle = \langle F(\hat{x})\rangle$$

it follows that

$$\frac{d}{dt}\left(m\frac{d}{dt}\langle\hat{x}\rangle\right) = \langle F(\hat{x})\rangle \quad \text{or} \quad \frac{d^2}{dt^2}\langle\hat{x}\rangle = \frac{1}{m}\langle F(\hat{x})\rangle$$

This equation would read

$$\frac{d^2}{dt^2}\langle\hat{x}\rangle = \frac{1}{m}F(\langle\hat{x}\rangle)$$

and hence, it would be identical to Newton's second law (2.11), provided that

$$\langle F(\hat{x})\rangle = F(\langle\hat{x}\rangle)$$

Thus, the evolution of $\langle\hat{x}\rangle$ and $\langle\hat{p}_x\rangle$ coincides with that of x and $\hat{p}_x$, respectively, and hence, quantum mechanics corresponds to classical dynamics, provided that the evolution of any expectation value $\langle\hat{f}\rangle$ is analogous to that of the corresponding classical function f, according to

$$\left\langle\frac{d\hat{f}}{dt}\right\rangle = \frac{d}{dt}\langle\hat{f}\rangle$$

a statement which is known as the Ehrenfest theorem.

29.6. Find the eigenvalues and eigenfunctions of the differential operator $\hat{A} = \dfrac{1}{x^2}\dfrac{d}{dx}\left(x^2\dfrac{d}{dx}\right)$.

29.7. Find the eigenvalues and eigenfunctions of the parity operator, defined in Problem 29.3.

29.8. If the potential energy is a periodic function with period a, such that $U(x) = U(x-a)$, show that the translation operator given by $\hat{S}(a)\Psi(x) = \Psi(x-a)$ commutes with the Hamiltonian.

29.9. Prove that the commutation relation

$$\left[f(\hat{A}),\hat{B}\right] = \frac{d}{d\hat{A}}f(\hat{A})$$

holds for any function $f(\hat{A})$, provided that $\left[\hat{A},\hat{B}\right] = \hat{I}$ is the identity operator.

29.10. Use the evolution equation to derive expressions for the position and momentum operators $\hat{x}$ and $\hat{p}_x$ of a linear harmonic oscillator of mass m and angular frequency ω.

30. THE SCHRÖDINGER PICTURE

30.1. POSTULATE 5: THE TIME-DEPENDENT SCHRÖDINGER EQUATION
30.2. THE TIME-INDEPENDENT SCHRÖDINGER EQUATION
30.3. UNBOUND STATES. PROBABILITY CURRENT DENSITY
30.4. BOUND STATES. THE HARMONIC OSCILLATOR

30.1. POSTULATE 5: THE TIME-DEPENDENT SCHRÖDINGER EQUATION

Let us examine the evolution of the expectation values of the quantum operators, given by Eq.(29.24), which are related to the measured values of the corresponding observables. The expectation value of an operator $\hat{f}$ may change with time because the operator has an explicit time dependence or because it is taken with respect to a state function that itself changes with time. By the correspondence principle, the dynamics we introduce via operators must ensure that we recover Newtonian behaviour for localized wave packets, and hence, the evolution of the measured value of any classical observable must be identical to that of the expectation value of the representative operator, as required by the Ehrenfest theorem. Therefore, we may assume for any quantum operator $\hat{f}$ that

$$\left\langle \frac{d\hat{f}}{dt} \right\rangle = \frac{d}{dt}\left\langle \hat{f} \right\rangle \qquad (30.1)$$

In the Heisenberg picture this result can be inserted into Eq.(29.42) to show that the evolution equation of a quantum operator is also obeyed by its expectation values

$$i\hbar \frac{d}{dt}\left\langle \hat{f} \right\rangle = \left\langle \left[\hat{f}, \hat{H} \right] \right\rangle + i\hbar \left\langle \frac{\partial \hat{f}}{\partial t} \right\rangle \qquad (30.2)$$

We now may bring the time-dependent state functions $\Psi(x,t)$ into the picture, taking into account, as we already did in Eq.(29.9), that only the state function but not its position coordinate x varies with time, so that

$$\frac{d\Psi(x,t)}{dt} \equiv \frac{\partial \Psi(x,t)}{\partial t} \tag{30.3}$$

Upon insertion of the expression (29.24) for the expectation value, Eq.(30.2) becomes

$$i\hbar \frac{d}{dt}\langle \hat{f} \rangle = i\hbar \frac{d}{dt}\int_{-\infty}^{\infty} \Psi^* \hat{f} \Psi dx = i\hbar \int_{-\infty}^{\infty}\left(\frac{\partial \Psi^*}{\partial t}\hat{f}\Psi + \Psi^* \hat{f}\frac{\partial \Psi}{\partial t}\right)dx + i\hbar \int_{-\infty}^{\infty}\Psi^* \frac{\partial \hat{f}}{\partial t}\Psi dx \tag{30.4}$$

On the other hand, as the Hamiltonian $\hat{H}$ is a Hermitian operator, we have

$$\langle [\hat{f},\hat{H}] \rangle + i\hbar \left\langle \frac{\partial \hat{f}}{\partial t} \right\rangle = \int_{-\infty}^{\infty}\left(-\Psi^*\hat{H}\hat{f}\Psi + \Psi^*\hat{f}\hat{H}\Psi\right)dx + i\hbar \int_{-\infty}^{\infty}\Psi^*\frac{\partial \hat{f}}{\partial t}\Psi dx$$

$$= \int_{-\infty}^{\infty}\left[\left(-\hat{H}\Psi\right)^* \hat{f}\Psi + \Psi^*\hat{f}\hat{H}\Psi\right]dx + i\hbar \int_{-\infty}^{\infty}\Psi^*\frac{\partial \hat{f}}{\partial t}\Psi dx \tag{30.5}$$

From Eqs.(30.4) and (30.5) it follows that the evolution equation (30.2) is true only if

$$i\hbar \frac{\partial \Psi}{\partial t} = \hat{H}\Psi \tag{30.6}$$

and this is a basic result called the *time-dependent Schrödinger equation*. This equation describes the evolution of quantum systems in terms of time-dependent state functions that obey Eq.(30.6), but neglect the time dependence of the quantum operators. Such a description is known as the *Schrödinger picture*. The fifth postulate states that the time-dependent Schrödinger equation must always be satisfied.

Postulate 5. *The evolution of the state function of a quantum system is given by the time-dependent Schrödinger equation (30.6), where $\hat{H}$ is the Hamiltonian of the system.*

The physical significance of Eq.(30.6) may be emphasized by considering a three-dimensional solution of the form

$$\Psi(\vec{r},t) = e^{iS(\vec{r},t)/\hbar} \tag{30.7}$$

where the function $S(\vec{r},t)$ has the dimensions of action (energy × time). It is a straightforward matter to show that

$$\frac{\partial \Psi}{\partial t} = \frac{i}{\hbar}\frac{\partial S}{\partial t}\Psi \quad , \quad \frac{\partial \Psi}{\partial x} = \frac{i}{\hbar}\frac{\partial S}{\partial x}\Psi \quad , \quad \frac{\partial^2 \Psi}{\partial x^2} = \left[-\frac{i}{\hbar^2}\left(\frac{\partial S}{\partial x}\right)^2 + \frac{i}{\hbar}\frac{\partial^2 S}{\partial x^2}\right]\Psi$$

and similarly for *y*- and *z*-derivatives, so that Eq.(30.6) takes the form

$$i\hbar\left(\frac{i}{\hbar}\frac{\partial S}{\partial t}\Psi\right) = -\frac{\hbar^2}{2m}\left[-\frac{1}{\hbar^2}(\nabla S)^2 + \frac{i}{\hbar}\nabla^2 S\right]\Psi + U\Psi$$

which may also be written as

$$\frac{1}{2m}(\nabla S)\cdot(\nabla S) + U + \frac{\partial S}{\partial t} = \frac{i\hbar}{2m}\nabla^2 S \tag{30.8}$$

In the limit where the terms in $\hbar$ may be ignored, this form of the time-dependent Schrödinger equation reduces to the Hamilton-Jacobi equation (24.47), written for a single particle as

$$\frac{1}{2m}(\nabla S)\cdot(\nabla S) + U + \frac{\partial S}{\partial t} = 0 \tag{30.9}$$

Therefore, if $\hbar$ can be replaced by zero, the time-dependent equation (in the Schrödinger picture) reduces to the basic equation of classical dynamics (in the Hamilton-Jacobi form). From Eq.(29.33) it also follows that, if $\hbar \to 0$, all the commutators are zero, which means that the quantum operators may be replaced by classical variables. Hence, classical mechanics is included as a limiting case of the quantum mechanics.

Provided the Hamiltonian does not depend explicitly on time, the solution to Eq.(30.6) can be formally written as

$$\Psi(\vec{r},t) = e^{-i\hat{H}t/\hbar}\Psi(\vec{r},0) \tag{30.10}$$

where $\Psi(\vec{r},0)$ describes the initial state of the system. By differentiating Eq.(30.10) one obtains

$$\frac{\partial \Psi(\vec{r},t)}{\partial t} = -\frac{i}{\hbar}\hat{H}e^{-i\hat{H}t/\hbar}\Psi(\vec{r},0) = -\frac{i}{\hbar}\hat{H}\Psi(\vec{r},t) \tag{30.11}$$

and this shows that the state function (30.10) is indeed a solution to the time-dependent Schrödinger equation. The *evolution operator* $\hat{U}(t)$ defined as

$$\hat{U}(t) = e^{-i\hat{H}t/\hbar} \tag{30.12}$$

gives the time dependence of the state functions, in the Schrödinger picture, according to

$$\Psi(\vec{r},t) = \hat{U}(t)\Psi(\vec{r},0) \tag{30.13}$$

for systems described by Hamiltonians with no explicit dependence on time.

30.2. The Time-Independent Schrödinger Equation

If the initial state of a system is the eigenfunction of a Hamiltonian $\hat{H}$ which has no explicit dependence on time, its energy is given by the corresponding eigenvalue ε, according to

$$\hat{H}\Psi(x) = \varepsilon \Psi(x) \qquad (30.14)$$

From the series expansion of the operator $\hat{U}(t)$ given by Eq.(30.12), which reads

$$e^{-i\hat{H}t/\hbar} = \sum_n \frac{1}{n!}\left(-\frac{i}{\hbar}\hat{H}t\right)^n \qquad (30.15)$$

it follows that $\hat{U}(t)$ and $\hat{H}$ commute, thus the eigenfunction $\Psi(x) = \Psi(x,0)$ of $\hat{H}$ is also an eigenfunction of $\hat{U}(t)$ with the eigenvalue $e^{-i\varepsilon t/\hbar}$. The evolution (30.12) of the state function thus takes the form

$$\Psi(x,t) = e^{-i\varepsilon t/\hbar}\Psi(x,0) \qquad (30.16)$$

Hence

$$|\Psi(x,t)|^2 = \Psi^*(x,0)e^{i\varepsilon t/\hbar}e^{-i\varepsilon t/\hbar}\Psi(x,0) = |\Psi(x,0)|^2$$

This shows that the state described by $\Psi(x,t)$ is physically equivalent to the initial state $\Psi(x,0)$, since the probability density does not vary with time and no evolution occurs. This particular result may be generalized as a *rule*:

> *If a system is initially in an eigenstate of the Hamiltonian, it remains in that state indefinitely and all the measured values of its physical observables are stationary.*

This rule substantiates Bohr's postulate of stationary states for one-electron atoms. It emphasizes that the Hamiltonian of a quantum system, which is governing its evolution according to Postulate 5, has stationary eigenstates of energies that are definite eigenvalues of Eq.(30.14). This equation can be written in one dimension in the form

$$-\frac{\hbar^2}{2m}\frac{d^2\Psi(x)}{dx^2} + U(x)\Psi(x) = \varepsilon \Psi(x) \qquad (30.17)$$

where the one-dimensional Hamiltonian, given by Eq.(29.15), was inserted. It can be generalized to three dimensions, using Eq.(29.15), in the form

$$\hat{H}\Psi(\vec{r}) = -\frac{\hbar^2}{2m}\nabla^2\Psi(\vec{r}) + U(\vec{r})\Psi(\vec{r}) = \varepsilon\Psi(\vec{r}) \qquad (30.18)$$

and this is called the *time-independent Schrödinger equation*.

The evolution of stationary states obeys Eq.(30.16), as they are eigenfunctions of the Hamiltonian according to Eq.(30.14). Hence, from Eqs.(29.4) and (30.16), a general solution to the time-dependent Schrödinger equation (30.6) in one dimension is given by the linear combination

$$\Psi(x,t) = \sum_k a_k e^{-i\varepsilon_k t/\hbar} \Psi_k(x) \qquad (30.19)$$

The general three-dimensional solution has a similar form

$$\Psi(\vec{r},t) = \sum_k a_k e^{-i\varepsilon_k t/\hbar} \Psi_k(\vec{r}) \qquad (30.20)$$

provided $\Psi_k(\vec{r})$ are solutions to Eq.(30.18).

A stationary state is classified as bound or unbound according to whether the corresponding state function $\Psi(\vec{r})$, and hence, the probability density $|\Psi(\vec{r})|^2$, vanish at infinity. In other words a state is *bound* if

$$\lim_{r\to\infty} \Psi(\vec{r}) = 0 \qquad (30.21)$$

and is unbound otherwise. The following *rule* determines the type of state associated with a given Hamiltonian:

> *The Hamiltonian operator has bound eigenstates that are part of a discrete spectrum if the corresponding classical Hamiltonian supports bound orbits. Otherwise the eigenstates of the Hamiltonian operator are unbound and are part of a continuous spectrum.*

The best example is the motion of a particle in a central field $U(r) = -C/r$, as defined in Eq.(3.13). In view of Eq.(3.25), the particle motion follows a conic section and might be either a bound periodical elliptical motion (for negative energies i.e. for an attractive interaction) or an unbound non-periodical motion, which is hyperbolic (for positive energies, i.e. for a repulsive interaction) or parabolic (in the absence of any interaction). According to the given rule we must expect a discrete spectrum of the Hamiltonian for negative eigenvalues $\varepsilon < 0$ and a continuous spectrum for $\varepsilon \geq 0$.

The energy eigenvalue problem, i.e., the time-independent Schrödinger equations (30.17) and (30.18), has to be solved differently for bound and unbound states:

- for *bound states* the discrete spectrum is determined by the condition that the eigenfunctions must vanish at infinity. Then the eigenfunctions corresponding to the allowed eigenvalues must be found in order to evaluate the expectation values of various operators.
- for *unbound states* the energy eigenvalues are not quantized and the form of the eigenfunctions is determined by asymptotic conditions which impose their nature as we approach infinity. For this reason the interpretation of an unbound state does

not rely on the form of its eigenfunctions but on a related quantity, called the probability current density $\vec{j}$ that can be used to find the probability for a particle in an unbound state to cross a given surface in unit time.

30.3. Unbound States. Probability Current Density

The probability of finding a system, described by a state function $\Psi = \Psi(\vec{r},t)$, in a finite volume V in space, has a three-dimensional form analogous to Eq.(29.1), that is,

$$\int_V \Psi^* \Psi dV$$

and changes as the state function evolves in time. In other words, the evolution of the state function determines a *probability current* from one region to another. This current can be evaluated by using Eq.(30.6) and its complex conjugate form

$$i\hbar \frac{\partial \Psi}{\partial t} = \left[-\frac{\hbar^2}{2m}\nabla^2 + U(\vec{r}) \right]\Psi \quad , \quad -i\hbar \frac{\partial \Psi^*}{\partial t} = \left[-\frac{\hbar^2}{2m}\nabla^2 + U(\vec{r}) \right]\Psi^*$$

Multiplication of the first equation by Ψ^* and of the second by Ψ and subtraction gives

$$i\hbar \left(\Psi^* \frac{\partial \Psi}{\partial t} + \Psi \frac{\partial \Psi^*}{\partial t} \right) = -\frac{\hbar^2}{2m} \left(\Psi^* \nabla^2 \Psi - \Psi \nabla^2 \Psi^* \right)$$

It is straightforward to show that an equivalent form can be written in the form

$$\frac{\partial}{\partial t}(\Psi^* \Psi) = \frac{i\hbar}{2m} \nabla \cdot \left(\Psi^* \nabla \Psi - \Psi \nabla \Psi^* \right)$$

and hence, a vector function called *probability current density* may be defined as

$$\vec{j}(\vec{r},t) = \frac{i\hbar}{2m}\left(\Psi \nabla \Psi^* - \Psi^* \nabla \Psi \right) \tag{30.22}$$

with the property that if Ψ is a real state function the vector $\vec{j}$ vanishes. As a result an equation of continuity for probability is obtained as

$$\frac{\partial}{\partial t}(\Psi^* \Psi) + \nabla \cdot \vec{j} = 0 \tag{30.23}$$

This result is an analogue of the already introduced equations of continuity for hydrodynamics, Eq.(6.11), and electrodynamics, Eq.(12.15), and shows that probability is locally conserved. If a system is in a stationary state, that is if $\partial(\Psi^*\Psi)/\partial t = 0$, it follows that $\nabla \cdot \vec{j} = 0$, which means that the probability current is solenoidal. The integral form of Eq.(30.23) in a finite volume V enclosed by a surface S is

$$\frac{\partial}{\partial t}\int_V \Psi^*\Psi dV = -\int_V (\nabla \cdot \vec{j})dV = -\int_S \vec{j} \cdot d\vec{S}$$

where Gauss's theorem (I.3) from Appendix I was used. This implies that any decrease of probability within a region in space is accompanied by an outflow of probability across its surface. The concept of probability current density simplifies the description of the evolution of bound states, as illustrated by the following example.

EXAMPLE 30.1. The tunnel effect

A continuous energy spectrum occurs in the motion of particles past a potential barrier $U(x)$, illustrated in Figure 30.1. We assume that particles with kinetic energy ε approach a potential barrier of height $U_0 > \varepsilon$ so that classical particles would be reflected. The barrier may be qualitatively characterized by its height U_0 and width 2α. For simplicity we assume discontinuous transitions in the barrier from $U = 0$ to $U = U_0$ at $x = -\alpha$ and $x = \alpha$. The energy eigenfunctions associated to this potential are found by solving Eq.(30.17) which reads

$$-\frac{\hbar^2}{2m}\frac{d^2\Psi}{dx^2} = \varepsilon\Psi \qquad x < -\alpha \quad , \quad x > \alpha$$

$$\frac{\hbar^2}{2m}\frac{d^2\Psi}{dx^2} = (V_0 - \varepsilon)\Psi \qquad -\alpha < x < \alpha$$

The first equation has e^{ikx} and e^{-ikx}, where $k = \sqrt{2m\varepsilon/\hbar^2}$, as linearly independent solutions which represent incident and reflected particles to the left of the barrier. If we impose the asymptotic condition that, far to the right of the barrier, we must have only transmitted particles, the acceptable solutions are

$$\Psi(x) = Ae^{ikx} + Be^{-ikx} \qquad , \qquad x < -\alpha$$

$$\Psi(x) = Ce^{ikx} \qquad , \qquad x > \alpha$$

The probability flow through the barrier is related to both the incident probability current, given by Eq.(30.22) as

$$j_i = \frac{i\hbar}{2m}\left[Ae^{ikx}\frac{d}{dx}(A^*e^{-ikx}) - A^*e^{-ikx}\frac{d}{dx}(Ae^{ikx})\right] = \frac{\hbar k}{m}|A|^2$$

and the transmitted one, which has the similar expression $j_t = \frac{\hbar k}{m}|C|^2$.

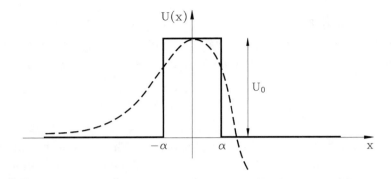

Figure 30.1. A potential barrier of arbitrary shape and its rectangular approximation of height U_0 and width 2α

Hence, the transmitted fraction of the incident beam is given by the *transmission coefficient*

$$T = \frac{j_t}{j_i} = \left|\frac{C}{A}\right|^2$$

For $-\alpha < x < \alpha$ the general solution of time-independent Schrödinger equation is

$$\Psi = De^{\gamma x} + Fe^{-\gamma x} \quad , \quad \gamma = \sqrt{2m(U_0 - \varepsilon)/\hbar^2} \qquad (30.24)$$

Since Ψ must be finite far to the right of the barrier, the required condition $D = 0$, although it is not true for negative values of x, allows a simple outline of the particle behaviour. The continuity of Ψ and its derivative at $x = \pm\alpha$ gives

$$Ae^{-ik\alpha} + Be^{ik\alpha} = De^{-\gamma\alpha} + Fe^{\gamma\alpha} \quad , \quad Ae^{-ik\alpha} - Be^{ik\alpha} = \frac{\gamma}{ik}\left(De^{-\gamma\alpha} - Fe^{\gamma\alpha}\right)$$

$$De^{\gamma\alpha} + Fe^{-\gamma\alpha} = Ce^{ik\alpha} \quad , \quad De^{\gamma\alpha} - Fe^{-\gamma\alpha} = \frac{ik}{\gamma}Ce^{ik\alpha}$$

The last two equations yield

$$D = \frac{C}{2}e^{-\gamma\alpha}\left(1 + \frac{ik}{\gamma}\right)e^{ik\alpha} \quad , \quad F = \frac{C}{2}e^{\gamma\alpha}\left(1 - \frac{ik}{\gamma}\right)e^{ik\alpha}$$

and then, substituting into the other two equations, and solving for A and B, we obtain

$$A = Ce^{2k\alpha}\left[\cosh(2\gamma\alpha) + \frac{i}{2}\left(\frac{\gamma}{k} - \frac{k}{\gamma}\right)\sinh(2\gamma\alpha)\right] \quad , \quad B = \frac{C}{2i}\left(\frac{\gamma}{k} + \frac{k}{\gamma}\right)\sinh(2\gamma\alpha)$$

As a result, the transmission coefficient is given by

$$T = \left|\frac{C}{A}\right|^2 = \frac{1}{1+\left(\frac{\gamma^2+k^2}{2\gamma k}\right)^2 \sinh^2(2\gamma\alpha)} \cong \frac{1}{1+\left(\frac{\gamma^2+k^2}{2\gamma k}\right)^2 e^{4\gamma\alpha}}$$

where $2\gamma\alpha \gg 1$, so that T falls off exponentially as

$$T \cong \frac{4\gamma^2 k^2}{\left(k^2+\gamma^2\right)^2} e^{-4\gamma\alpha} = \frac{4\gamma^2 k^2}{\left(k^2+\gamma^2\right)^2} e^{-4\alpha\sqrt{2m(U_0-\varepsilon)}/\hbar}$$

It appears that a flux of particles can penetrate a potential barrier of finite width 2α even though the barrier is too high $(U_0 > \varepsilon)$ for the particles to pass over it. This effect is known as *tunnelling* and occurs for instance in a radioactive decay, as we shall see later on in Chapter 47. It can be seen from the expression for γ that the classical limit of no transmission through the barrier is obtained in the case where $\hbar \to 0$.

30.4. Bound States. The Harmonic Oscillator

For bound states, the eigenfunctions of the time-independent Schrödinger equation (30.17) must be subjected to the boundary condition (30.21), and hence, must tend to zero as x tends to infinity. Acceptable solutions $\Psi(x)$ will be not available for any ε but only for certain values $\varepsilon = \varepsilon_n$ which belong to a discrete spectrum.

The harmonic oscillator in one dimension is an example of bound motion governed by an elastic potential $U(x) = kx^2/2$. The Schrödinger time-independent equation (30.17) of a quantum oscillator takes the form

$$\frac{d^2\Psi(x)}{dx^2} + \frac{2m}{\hbar^2}\left(\varepsilon - \frac{m\omega^2}{2}x^2\right)\Psi(x) = 0 \qquad (30.25)$$

where the oscillation frequency $\omega = \sqrt{k/m}$. It is convenient to use

$$\lambda = \frac{2\varepsilon}{\hbar\omega} \quad , \quad \xi = \sqrt{\frac{m\omega}{\hbar}}x \qquad (30.26)$$

where λ and ξ are dimensionless parameters, and hence, we obtain

$$\frac{d^2\Psi(\xi)}{d\xi^2} + \left(\lambda - \xi^2\right)\Psi(\xi) = 0 \qquad (30.27)$$

When $\xi^2 \to \infty$ the term involving the constant λ becomes negligible, for any eigenvalue ε, so that $\Psi(\xi)$ must have an asymptotic form $\Psi_0(\xi)$, which is a solution to

$$\frac{d^2\Psi_0(\xi)}{d\xi^2} - \xi^2 \Psi_0(\xi) = 0$$

On multiplication by $2(d\Psi_0/d\xi)$ this becomes

$$\frac{d}{d\xi}\left(\frac{d\Psi_0}{d\xi}\right)^2 - \xi^2 \frac{d}{d\xi}(\Psi_0^2) = 0 \quad \text{or} \quad \frac{d}{d\xi}\left[\left(\frac{d\Psi_0}{d\xi}\right)^2 - \xi^2 \Psi_0^2\right] = -2\xi\Psi_0^2$$

We may drop the right-hand side, which is negligible in the asymptotic region, as $2\xi\Psi_0^2 \ll d(\xi^2\Psi_0^2)/d\xi$, and the equation reduces to

$$\left(\frac{d\Psi_0}{d\xi}\right)^2 = \left(\text{const} + \xi^2\Psi_0^2\right) \quad \text{or} \quad \frac{d\Psi_0}{d\xi} = \pm\xi\Psi_0$$

since the constant of integration must be set to zero in order for $\Psi_0(\xi)$ and its derivative to vanish at infinity. It follows that

$$\Psi_0(\xi) = e^{-\xi^2/2} \tag{30.28}$$

It is now straightforward, using Eq.(30.28), to show that the asymptotic inequality stated above is satisfied. We must therefore look for solutions to Eq.(30.27) having the form

$$\Psi(\xi) = u(\xi)e^{-\xi^2/2} \tag{30.29}$$

so that a differential equation for the function $u(\xi)$ is now obtained as

$$\frac{d^2u(\xi)}{d\xi^2} - 2\xi\frac{du(\xi)}{d\xi} + (\lambda - 1)u(\xi) = 0 \tag{30.30}$$

Equation (30.30) is known as the *Hermite equation*. We look for $u(\xi)$ in the form of a power series

$$u(\xi) = \sum_{k=0}^{\infty} a_k \xi^k \tag{30.31}$$

Thus

$$\frac{du(\xi)}{d\xi} = \sum_{k=0}^{\infty} a_k k \xi^{k-1} \quad , \quad \frac{d^2u(\xi)}{d\xi^2} = \sum_{k=0}^{\infty} a_k k(k-1)\xi^{k-2} = \sum_{k=0}^{\infty} a_{k+2}(k+2)(k+1)\xi^k$$

where we have redefined every index k to be equal to its previous value plus two. Substituting into Eq.(30.30) we obtain

$$\sum_{k=0}^{\infty}\left[(k+1)(k+2)a_{k+2}-(2k+1-\lambda)a_k\right]\xi^k = 0$$

This can be true only if the coefficient of each power of ξ vanishes, so that we obtain a recurrence relation of the form

$$a_{k+2} = \frac{2k+1-\lambda}{(k+1)(k+2)}a_k \quad \text{or} \quad a_{k+2} \cong \frac{2}{k}a_k \tag{30.32}$$

The last expression only holds as $k \to \infty$, and it is identical to the recurrence relation between the successive terms of the power series

$$e^{\xi^2} = \sum_{k=0,2,4,\ldots} \frac{1}{(k/2)!}\xi^k$$

Therefore $u(\xi)$ will tend to infinity with ξ like e^{ξ^2}, and hence, $\Psi(\xi)$ given by Eq.(30.29) will be divergent like $e^{\xi^2/2}$. This unrealistic solution can be avoided only if the power series (30.31) terminates after a finite number of terms. This is accomplished by choosing λ such that the numerator of Eq.(30.32) vanishes for some finite value $k = n$, that is, by choosing $\lambda = 2n+1$. If this condition is combined with the definition of λ, Eq.(30.26), it follows that the total energy of the system is quantized according to

$$\varepsilon_n = \hbar\omega\left(n+\frac{1}{2}\right) \qquad n = 0,1,2,\ldots \tag{30.33}$$

The term $n\hbar\omega$ gives Planck's series of energy levels, as postulated by Eq.(27.32) for the normal modes of the thermal radiation. However, the minimum energy of the quantum oscillator is not zero but takes the value $\hbar\omega/2$, known as the *zero point energy*, which is a manifestation of the uncertainty principle. Since a classical harmonic oscillator, bound to the origin, has average position and momentum equal to zero, this must also be true for the expectation values $\langle \hat{x} \rangle_n = 0$ and $\langle \hat{p}_x \rangle_n = 0$ of any oscillator eigenstate, as stated by Ehrenfest's theorem. It has been shown before, using the virial theorem, Eq.(26.14), that the average values of the kinetic and potential energies of an harmonic oscillator are equal. Assuming that this also holds for the expectation values of the quantum oscillator, we must have

$$\frac{1}{2m}\langle \hat{p}_x^2 \rangle_n = \frac{1}{2}m\omega^2\langle \hat{x}^2 \rangle_n = \frac{1}{2}\varepsilon_n = \frac{1}{2}\hbar\omega\left(n+\frac{1}{2}\right)$$

It follows that the uncertainties of position and momentum can be expressed as

$$(\Delta \hat{p}_x)_n^2 = \langle \hat{p}_x^2 \rangle_n - \langle \hat{p}_x \rangle_n^2 = m\hbar\omega\left(n+\frac{1}{2}\right) \quad , \quad (\Delta \hat{x})_n^2 = \langle \hat{x}^2 \rangle_n - \langle \hat{x} \rangle_n^2 = \frac{\hbar}{m\omega}\left(n+\frac{1}{2}\right)$$

so that

$$(\Delta \hat{x})_n (\Delta \hat{p}_x)_n = \left(n+\frac{1}{2}\right)\hbar$$

As stated by the uncertainty principle (29.41), this product must never be smaller than $\hbar/2$, even in the lowest energy state $n=0$, where a zero point energy $\hbar\omega/2$ is thus required to exist.

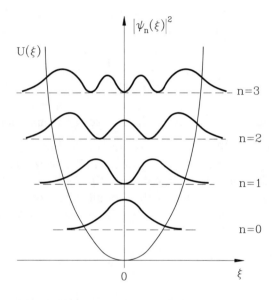

Figure 30.2. Probability densities of harmonic oscillator for the lowest quantum numbers n

If we let $\lambda = 2n+1$, the eigenfunctions (30.29) can be expressed as

$$\Psi(\xi) = N_n e^{-\xi^2/2} H_n(\xi) \qquad (30.34)$$

where N_n is a normalization constant, and $H_n(\xi)$, called the *Hermite polynomials*, are solutions to Eq.(30.30) which becomes

$$\frac{d^2 H_n(\xi)}{d\xi^2} - 2\xi \frac{dH_n(\xi)}{d\xi} + 2n H_n(\xi) = 0 \qquad (30.35)$$

The first four Hermite polynomials have the form

$$H_0(\xi) = 1 \quad , \quad H_1(\xi) = 2\xi \quad , \quad H_2(\xi) = 4\xi^2 - 2 \quad , \quad H_3(\xi) = 8\xi^3 - 12\xi$$

The corresponding eigenfunctions $\Psi_n(\xi)$ have n nodes and are either odd or even functions. The probability densities $|\Psi_n(\xi)|^2$, which describe the probability of finding the quantum oscillator at various distances x from the origin, are plotted in Figure 30.2, showing that the quantum oscillator can also be found outside the parabolic potential barrier $U(x)$, in a region that is forbidden for the classical oscillator.

FURTHER READING
1. D. J. Griffiths - INTRODUCTION TO QUANTUM MECHANICS, Prentice Hall, Upper Saddle River, 2004.
2. D. R. Bes - QUANTUM MECHANICS, Springer Publishing Company, New York, 2004.
3. J. P. Lowe - QUANTUM CHEMISTRY, Elsevier, Amsterdam, 1993.

PROBLEMS

30.1. If the solution to the time-dependent Schrödinger equation is written in terms of a real amplitude $A(\vec{r},t)$ and a phase function $S(\vec{r},t)$, having the dimensions of action, as $\Psi(\vec{r},t) = A(\vec{r},t) e^{iS(\vec{r},t)/\hbar}$, derive the equations satisfied by $S(\vec{r},t)$ and $A(\vec{r},t)$ in the quasi-classical approximation, where one can exclude terms of order $\hbar^2$ and higher.

Solution

It is a straightforward matter to express the derivatives in the form

$$\frac{\partial \Psi}{\partial t} = \frac{\partial A}{\partial t} e^{iS/\hbar} + \frac{i}{\hbar} \frac{\partial S}{\partial t} A e^{iS/\hbar} \quad , \quad \frac{\partial \Psi}{\partial x_i} = \frac{\partial A}{\partial x_i} e^{iS/\hbar} + \frac{i}{\hbar} \frac{\partial S}{\partial x_i} A e^{iS/\hbar}$$

$$\frac{\partial^2 \Psi}{\partial x_i^2} = \frac{\partial^2 A}{\partial x_i^2} e^{iS/\hbar} + 2\frac{i}{\hbar} \frac{\partial S}{\partial x_i} \frac{\partial A}{\partial x_i} e^{iS/\hbar} + \frac{i}{\hbar} \frac{\partial^2 S}{\partial x_i^2} A e^{iS/\hbar} - \frac{1}{\hbar^2} \left(\frac{\partial S}{\partial x_i} \right)^2 A e^{iS/\hbar}$$

It follows that Eq.(30.6) takes the form

$$i\hbar \left[\frac{\partial A}{\partial t} + \frac{i}{\hbar} \frac{\partial S}{\partial t} A \right] = -\frac{\hbar^2}{2m} \sum_{i=x,y,z} \left[\frac{\partial^2 A}{\partial x_i^2} + 2\frac{i}{\hbar} \frac{\partial S}{\partial x_i} \frac{\partial A}{\partial x_i} + \frac{i}{\hbar} \frac{\partial^2 S}{\partial x_i^2} A - \frac{1}{\hbar^2} \left(\frac{\partial S}{\partial x_i} \right)^2 A \right] + UA$$

where the exponential common factor has been simplified. If we limit ourselves to the first-order terms in $\hbar$, the first term on the right-hand side should be neglected, and this equation reduces to

$$i\hbar\left[\frac{\partial A}{\partial t}+\frac{1}{2m}\sum_{i=x,y,z}\left(2\frac{\partial A}{\partial x_i}\frac{\partial S}{\partial x_i}+A\frac{\partial^2 S}{\partial x_i^2}\right)\right]=\left[\frac{1}{2m}\sum_{i=x,y,z}\left(\frac{\partial S}{\partial x_i}\right)^2+U+\frac{\partial S}{\partial t}\right]A$$

The real and imaginary parts should each be equal to zero, and this provides the required equations for the phase and amplitude functions as

$$\frac{1}{2m}\sum_{i=x,y,z}\left(\frac{\partial S}{\partial x_i}\right)^2+U+\frac{\partial S}{\partial t}=0 \quad\text{and}\quad \frac{\partial A}{\partial t}+\frac{1}{2m}\sum_{i=x,y,z}\left(2\frac{\partial A}{\partial x}\frac{\partial S}{\partial x_i}+A\frac{\partial^2 S}{\partial x_i^2}\right)=0$$

The equation for $S(\vec{r},t)$ reduces to the Hamilton-Jacobi equation (24.45) by making the choice $p_i = m\dot{x}_i = \partial S/\partial x_i$. The amplitude equation can be rewritten in terms of the probability amplitude A^2, in a form reducible to the continuity equation for probabilities, through multiplication by $2A$, which gives

$$\frac{\partial A^2}{\partial t}+\sum_{i=x,y,z}\frac{\partial}{\partial x_i}\left(A^2\frac{1}{m}\frac{\partial S}{\partial x_i}\right)=0$$

30.2. Prove the equivalence between the Schrödinger description of the evolution of a system, where the time dependence of its state function is assumed, and its Heisenberg description, where the evolution with time is introduced through the time dependence of quantum operators.

Solution

In the Schrödinger picture, the evolution of a system whose Hamiltonian does not depend explicitly on time is described by the time development of its state function, Eq.(30.10), as

$$\Psi(x,t) = e^{-i\hat{H}t/\hbar}\Psi(x,0)$$

and hence, the expectation value of a time-dependent quantum operator $\hat{f}_S = \hat{f}(t)$ is

$$\langle\hat{f}(t)\rangle = \langle\Psi(x,t)|\hat{f}(t)|\Psi(x,t)\rangle = \langle e^{i\hat{H}t/\hbar}\Psi(x,0)|\hat{f}(t)|e^{-i\hat{H}t/\hbar}\Psi(x,0)\rangle$$

$$= \langle\Psi(x,0)|e^{i\hat{H}t/\hbar}\hat{f}(t)e^{-i\hat{H}t/\hbar}|\Psi(x,0)\rangle$$

It is obvious that the relation can be obtained as a result of the evolution of the operator $\hat{f}_S$ according to

$$\hat{f}_H = e^{i\hat{H}t/\hbar}\hat{f}_S e^{-i\hat{H}t/\hbar}$$

whereas the state vector preserves its initial value $\Psi(x,0)$. This evolution of a quantum operator in the Heisenberg picture plays the same role as the evolution given by Eq.(30.10) for the state function in the Schrödinger picture. The result is the same, whatever picture is used. The choice between the two pictures is one of convenience, as it is for instance the option of describing a body at rest in a rotating set of axes or the body rotating with respect to a fixed coordinate system. By taking the total differential of the operator $\hat{f}_H$ one obtains

$$\frac{d}{dt}\hat{f}_H = \frac{i}{\hbar}\hat{H}e^{i\hat{H}t/\hbar}\hat{f}_S e^{-i\hat{H}t/\hbar} - \frac{i}{\hbar}e^{i\hat{H}t/\hbar}\hat{f}_S\hat{H}e^{-i\hat{H}t/\hbar} + e^{i\hat{H}t/\hbar}\frac{\partial \hat{f}_S}{\partial t}e^{-i\hat{H}t/\hbar}$$

where the last term can be set equal to the partial time derivative of $\hat{f}_H$, that is,

$$\frac{d\hat{f}_H}{dt} = e^{i\hat{H}t/\hbar}\frac{\partial \hat{f}_S}{\partial t}e^{-i\hat{H}t/\hbar}$$

It follows that the relation reduces to the evolution equation (29.42) for the Heisenberg operator $\hat{f}_H$, which reads

$$i\hbar\frac{d\hat{f}_H}{dt} = \left[\hat{f}_H, \hat{H}\right] + i\hbar\frac{\partial \hat{f}_H}{\partial t}$$

This equation plays in the Heisenberg picture the role of the time-dependent Schrödinger equation (30.6) in the Schrödinger picture.

30.3. Find the probability density $P(x)$ and the probability current density $j(x)$ for a particle described by the wave function $\Psi(x) = Ae^{ip_x x/\hbar - x^2/2a^2}$, where the amplitude A is a constant.

Solution

The normalizing condition reads $\int_{-\infty}^{\infty}|\Psi(x)|^2 dx = |A|^2 \int_{-\infty}^{\infty} e^{-x^2/a^2} dx = |A|^2 a\sqrt{\pi} = 1$. Thus we have

$$P(x) = |\Psi(x)|^2 = |A|^2 e^{-x^2/a^2} = \frac{1}{a\sqrt{\pi}}e^{-x^2/a^2}$$

and the probability current density follows from Eq.(30.22) as

$$j(x) = \frac{i\hbar}{2m}\left(\Psi\frac{\partial \Psi^*}{\partial x} - \Psi^*\frac{\partial \Psi}{\partial x}\right) = |A|^2 \frac{p_x}{m}e^{-x^2/a^2} = \frac{p_x}{m}P(x)$$

30.4. Find the transmission coefficient T for an electron of mass m_e and energy $\varepsilon < U_0$ through a potential barrier given by

$$U(x) = 0 \quad \text{if} \quad x < 0 \quad , \quad U(x) = V_0\left(1 - \frac{x}{\alpha}\right) \quad \text{if} \quad x \geq 0$$

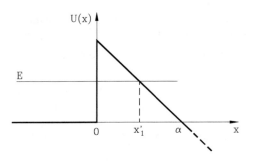

Solution

Since the transmission coefficient through a rectangular barrier of height V_0 and width 2α has the form

$$T = T_0 e^{-4\alpha\sqrt{2m(U_0-\varepsilon)}/\hbar}$$

it is convenient to divide the actual barrier into a series of small rectangular regions Δx_i of heights $U(x_i)$, so that the probability for tunneling will be given by

$$T = \prod_{i=1}^{n} T_i = T_0 e^{-2\sum \Delta x_i \sqrt{2m[U(x_i)-\varepsilon]}/\hbar}$$

As $\Delta x_i \to 0$, we have $\dfrac{2}{\hbar}\sum \Delta x_i \sqrt{2m[U(x_i)-\varepsilon]} \to \dfrac{2}{\hbar}\int_0^{x_1'}\sqrt{2m[U(x)-\varepsilon]}\,dx$, where x_1' is given by

$$U_0\left(1 - \frac{x_1'}{\alpha}\right) = \varepsilon \qquad \text{or} \qquad x_1' = \alpha\sqrt{1 - \frac{\varepsilon}{U_0}}$$

Thus, one obtains

$$\int_0^{x_1'}\sqrt{U(x)-\varepsilon}\,dx = \int_0^{x_1'}\sqrt{(U_0-\varepsilon)-U_0 x/\alpha}\,dx = \frac{2}{3}\frac{\alpha}{U_0}(U_0-\varepsilon)^{3/2}$$

where use has been made of the standard integral $\int_0^x \sqrt{a+bx}\,dx = \dfrac{2}{3b}(a+bx)^{3/2}$. This finally yields

$$T = T_0 e^{-4\alpha\sqrt{2m}(U_0-\varepsilon)^{3/2}/3\hbar U_0}$$

30.5. Derive the eigenvalues and eigenfunctions of a quantum oscillator using the operator formalism of the Heisenberg description, where the Hamiltonian $\hat{H} = \omega \hat{a}^\dagger \hat{a} + \hbar\omega/2$ is written in terms of two operators defined as

$$\hat{a} = \sqrt{\frac{m\omega}{2}}\hat{x} + i\frac{\hat{p}_x}{\sqrt{2m\omega}} \qquad \text{and} \qquad \hat{a}^\dagger = \sqrt{\frac{m\omega}{2}}\hat{x} - i\frac{\hat{p}_x}{\sqrt{2m\omega}}$$

Solution

Since the Hamiltonian of a classical oscillator may be factorized into

$$H = \frac{p_x^2}{2m} + \frac{m\omega^2}{2}x^2 = \omega\left(\sqrt{\frac{m\omega}{2}}x - i\frac{p_x}{\sqrt{2m\omega}}\right)\left(\sqrt{\frac{m\omega}{2}}x + i\frac{p_x}{\sqrt{2m\omega}}\right)$$

it follows that the operators $\hat{x}$ and $\hat{p}_x$ must be related by

$$\omega\left(\sqrt{\frac{m\omega}{2}}\hat{x} - i\frac{\hat{p}_x}{\sqrt{2m\omega}}\right)\left(\sqrt{\frac{m\omega}{2}}\hat{x} + i\frac{\hat{p}_x}{\sqrt{2m\omega}}\right) = \frac{\hat{p}_x^2}{2m} + \frac{m\omega^2}{2}\hat{x}^2 - \frac{i\omega}{2}(\hat{p}_x\hat{x} - \hat{x}\hat{p}_x) = \hat{H} - \frac{1}{2}\hbar\omega$$

in view of Eq.(29.32). Thus the Hamiltonian takes the required form $\hat{H} = \omega\hat{a}^\dagger\hat{a} + \frac{1}{2}\hbar\omega$. We have

$$\left[\hat{a}, \hat{a}^\dagger\right] = \left[\sqrt{\frac{m\omega}{2}}\hat{x}, -i\frac{\hat{p}_x}{\sqrt{2m\omega}}\right] + \left[i\frac{\hat{p}_x}{\sqrt{2m\omega}}, \sqrt{\frac{m\omega}{2}}\hat{x}\right] = \hbar$$

$$\left[\hat{H}, \hat{a}\right] = \left[\omega\hat{a}^\dagger\hat{a}, \hat{a}\right] = \omega\left[\hat{a}^\dagger, \hat{a}\right]\hat{a} = -\hbar\omega\hat{a} \quad , \quad \left[\hat{H}, \hat{a}^\dagger\right] = \left[\omega\hat{a}^\dagger\hat{a}, \hat{a}^\dagger\right] = \omega\hat{a}^\dagger\left[\hat{a}, \hat{a}^\dagger\right] = \hbar\omega\hat{a}^\dagger$$

If we let Ψ_n be an eigenfunction of $\hat{H}$ corresponding to the eigenvalue ε_n, that is $\hat{H}\Psi_n = \varepsilon_n\Psi_n$, and apply the operator $\hat{a}$ to this eigenvalue equation, one obtains

$$\hat{a}\hat{H}\Psi_n = (\hat{H}\hat{a} + \hbar\omega\hat{a})\Psi_n = (\hat{H} + \hbar\omega)\hat{a}\Psi_n = \varepsilon_n\hat{a}\Psi_n \quad \text{or} \quad \hat{H}(\hat{a}\Psi_n) = (\varepsilon_n - \hbar\omega)(\hat{a}\Psi_n)$$

Thus, if Ψ_n is an eigenfunction of $\hat{H}$, then $\hat{a}\Psi_n$ is also eigenfunction of $\hat{H}$, but corresponding to an eigenvalue lowered by $\hbar\omega$. In the same way we can show that

$$\hat{H}(\hat{a}^\dagger\Psi_n) = (\varepsilon_n + \hbar\omega)(\hat{a}^\dagger\Psi_n)$$

that is $\hat{a}^\dagger\Psi_n$ is also an eigenfunction of $\hat{H}$ with the eigenvalue raised by $\hbar\omega$. The operators $\hat{a}$ and $\hat{a}^\dagger$ are called the lowering and, respectively, raising operators. Denoting by Ψ_0 the ground state function corresponding to the minimum eigenvalue ε_0, the condition $\hat{a}\Psi_0 = 0$ must be true. If this condition is substituted into the eigenvalue equation, the energy of the ground state ε_0 is obtained as

$$\hat{H}\Psi_0 = \left(\omega\hat{a}^\dagger\hat{a} + \frac{1}{2}\hbar\omega\right)\Psi_0 = \frac{1}{2}\hbar\omega\Psi_0 \quad \text{or} \quad \varepsilon_0 = \frac{1}{2}\hbar\omega$$

In view of the representation (29.11) of the momentum operator, the same condition gives the ground state function in the form

$$\left(m\omega x + \hbar\frac{\partial}{\partial x}\right)\Psi_0 = 0 \quad \text{or} \quad \Psi_0 = A_0 e^{-m\omega x^2/2\hbar}$$

where A_0 is a normalization constant. The whole set of eigenfunctions can be generated by successive raising operations on Ψ_0 as

$$\hat{H}(\hat{a}^\dagger)^n \Psi_0 = (\varepsilon_0 + n\hbar\omega)(\hat{a}^\dagger)^n \Psi_0$$

It follows that the energy eigenvalues are given by

$$\varepsilon_n = \varepsilon_0 + n\hbar\omega = \hbar\omega\left(n + \frac{1}{2}\right), \quad n = 0, 1, 2, \ldots$$

and the eigenfunctions have the form

$$\Psi_n = A_n(\hat{a}^\dagger)^n \Psi_0 = A'_n\left(m\omega x - \hbar\frac{\partial}{\partial x}\right)^n e^{-m\omega x^2/2\hbar}$$

where A_n, A'_n denote normalization constants. A comparison of Eqs.(30.40) and (30.41) with Eqs.(30.33) and (30.34) shows that this result is identical to those derived in the Schrödinger picture.

30.6. Show that the probability current density can be written as $\vec{j} = P\vec{v}$, where P is the probability density, and find an expression for the vector function $\vec{v}$.

30.7. Find the potential energy $U(x)$ and the energy eigenvalue ε if the corresponding eigenfunction is given by

$$\psi(x) = Ax^n e^{-\gamma x} \quad \text{if} \quad x \geq 0, \quad \psi(x) = 0 \quad \text{if} \quad x < 0$$

30.8. If the solution of the time-independent Schrödinger equation is $\psi(\vec{r}) = A_0(\vec{r}) e^{iS_0(\vec{r})/\hbar}$, where $A_0(\vec{r})$ is a real amplitude and the phase function $S_0(\vec{r})$ has the dimensions of action, find the equations satisfied by $S_0(\vec{r})$ and $A_0(\vec{r})$ in the quasi-classical approximation.

30.9. Find expressions for $\langle(\Delta x)^2\rangle$ and $\langle(\Delta p_x)^2\rangle$ and show that the uncertainty relations are satisfied in the case of a particle described by the wave function of constant amplitude

$$\Psi(x) = A e^{i p_x x/\hbar - x^2/2a^2}$$

30.10. Find the energy eigenvalues ε_n for a harmonic oscillator described by the Hamiltonian

$$\hat{H} = -\frac{\hbar}{2m}\nabla^2 + \frac{m}{2}\left(\omega_1^2 x^2 + \omega_2^2 y^2 + \omega_3^2 z^2\right)$$

and show that there is degeneracy when the oscillator is isotropic ($\omega_1 = \omega_2 = \omega_3$). What is the degeneracy of the level ε_n?

31. ORBITAL ANGULAR MOMENTUM

31.1. ORBITAL ANGULAR MOMENTUM OPERATORS
31.2. EIGENVALUE EQUATIONS IN A CENTRAL FIELD
31.3. QUANTIZATION OF ORBITAL ANGULAR MOMENTUM

31.1. ORBITAL ANGULAR MOMENTUM OPERATORS

If we consider the motion of a particle in a spherically symmetric potential energy $U(r)$ that is a function solely of the distance r from the origin and independent of the angular variables, then the Hamiltonian has the form

$$\hat{H} = \frac{\hat{p}^2}{2m} + U(r) = -\frac{\hbar^2}{2m}\nabla^2 + U(r) \tag{31.1}$$

and is invariant under rotations. We expect the angular momentum to be a constant of motion, in order to parallel the classical result referred to in Chapter 3.

Explicit expressions for the operators $\hat{L}_x, \hat{L}_y$ and $\hat{L}_z$ representing the angular momentum components were already introduced in Eq.(29.16), according to Postulate 2. It is easy to show that the observables $\hat{L}_x, \hat{L}_y$ and $\hat{L}_z$ are not compatible, since the representative operators do not commute with one another. For instance,

$$\left[\hat{L}_x, \hat{L}_y\right] = \left[y\hat{p}_z - z\hat{p}_y, z\hat{p}_x - x\hat{p}_z\right] = \left[y\hat{p}_z, z\hat{p}_x\right] - \left[z\hat{p}_y, z\hat{p}_x\right] - \left[y\hat{p}_z, x\hat{p}_z\right] + \left[z\hat{p}_y, x\hat{p}_z\right]$$

$$= y[\hat{p}_z, z]\hat{p}_x + x[z, \hat{p}_z]\hat{p}_y = \frac{\hbar}{i}\left(y\hat{p}_x - x\hat{p}_y\right) = i\hbar\hat{L}_z \tag{31.2}$$

and similarly

$$\left[\hat{L}_y, \hat{L}_z\right] = i\hbar\hat{L}_x \quad , \quad \left[\hat{L}_z, \hat{L}_x\right] = i\hbar\hat{L}_y \tag{31.3}$$

However, each component of the angular momentum is compatible with the Hamiltonian as we shall prove for $\hat{L}_z$. It is a straightforward matter to show that $\left[\hat{L}_z, \nabla^2\right] = 0$ since all the derivatives commute, and furthermore we have

$$\left[\hat{L}_z U(r) - U(r)\hat{L}_z\right]\Psi = \frac{\hbar}{i}\left(x\frac{\partial}{\partial y} - y\frac{\partial}{\partial x}\right)(U(r)\Psi) - U(r)\frac{\hbar}{i}\left(x\frac{\partial}{\partial y} - y\frac{\partial}{\partial x}\right)\Psi$$

$$= \frac{\hbar}{i}\left[x\frac{\partial U(r)}{\partial y} - y\frac{\partial U(r)}{\partial x}\right]\Psi = \frac{\hbar}{i}\left[x\frac{dU(r)}{dr}\frac{y}{r} - y\frac{dU(r)}{dr}\frac{x}{r}\right]\Psi = 0$$

In view of Eq.(31.1) it follows that

$$\left[\hat{H}, \hat{L}_z\right] = 0 \tag{31.4}$$

and similarly

$$\left[\hat{H}, \hat{L}_x\right] = 0 \quad , \quad \left[\hat{H}, \hat{L}_y\right] = 0 \tag{31.5}$$

The magnitude of orbital angular momentum is represented by the operator

$$\hat{L}^2 = \hat{L}_x^2 + \hat{L}_y^2 + \hat{L}_z^2 \tag{31.6}$$

which commutes with each of its components, as for instance

$$\hat{L}_z, \hat{L}^2 = \left[\hat{L}_z, \hat{L}_x^2 + \hat{L}_y^2 + \hat{L}_z^2\right] = \left[\hat{L}_z, \hat{L}_x^2\right] + \left[\hat{L}_z, \hat{L}_y^2\right]$$

$$= \hat{L}_x\left[\hat{L}_z, \hat{L}_x\right] + \left[\hat{L}_z, \hat{L}_x\right]\hat{L}_x + \hat{L}_y\left[\hat{L}_z, \hat{L}_y\right] + \left[\hat{L}_z, \hat{L}_y\right]\hat{L}_y$$

$$= i\hbar\hat{L}_x\hat{L}_y + i\hbar\hat{L}_y\hat{L}_x - i\hbar\hat{L}_y\hat{L}_x - i\hbar\hat{L}_x\hat{L}_y = 0 \tag{31.7}$$

As each of $\hat{L}_x, \hat{L}_y, \hat{L}_z$ commutes with the Hamiltonian, this must be true for $\hat{L}^2$, that is,

$$\left[\hat{H}, \hat{L}^2\right] = 0 \tag{31.8}$$

Therefore we can choose among the three sets of compatible observables, given by their representative operators as

$$\{\hat{H}, \hat{L}^2, \hat{L}_x\} \quad , \quad \{\hat{H}, \hat{L}^2, \hat{L}_y\} \quad , \quad \{\hat{H}, \hat{L}^2, \hat{L}_z\} \tag{31.9}$$

The operator $\hat{L}^2$ is contained in the expression (31.1) of the central field Hamiltonian, and this can be shown using Eq.(31.6), where the component operators are replaced by their forms (29.16) written in Cartesian coordinates, and this gives

$$\hat{L}^2 = -\hbar^2\left(y\frac{\partial}{\partial z}-z\frac{\partial}{\partial y}\right)\left(y\frac{\partial}{\partial z}-z\frac{\partial}{\partial y}\right) - \hbar^2\left(z\frac{\partial}{\partial x}-x\frac{\partial}{\partial z}\right)\left(z\frac{\partial}{\partial x}-x\frac{\partial}{\partial z}\right)$$

$$-\hbar^2\left(x\frac{\partial}{\partial y}-y\frac{\partial}{\partial x}\right)\left(x\frac{\partial}{\partial y}-y\frac{\partial}{\partial x}\right)$$

It is a straightforward matter to show that this relation may also be written as

$$\hat{L}^2 = -\hbar^2\left[x^2\left(\frac{\partial^2}{\partial y^2}+\frac{\partial^2}{\partial z^2}\right)+y^2\left(\frac{\partial^2}{\partial z^2}+\frac{\partial^2}{\partial x^2}\right)+z^2\left(\frac{\partial^2}{\partial x^2}+\frac{\partial^2}{\partial y^2}\right)\right]$$

$$+\hbar^2\left[2xy\frac{\partial^2}{\partial x\partial y}+2yz\frac{\partial^2}{\partial y\partial z}+2zx\frac{\partial^2}{\partial z\partial x}+2x\frac{\partial}{\partial x}+2y\frac{\partial}{\partial y}+2z\frac{\partial}{\partial z}\right]$$

We now use the expansion

$$\left(\hat{\vec{r}}\cdot\hat{\vec{p}}\right)^2 = -\hbar^2\left(x\frac{\partial}{\partial x}+y\frac{\partial}{\partial y}+z\frac{\partial}{\partial z}\right)\left(x\frac{\partial}{\partial x}+y\frac{\partial}{\partial y}+z\frac{\partial}{\partial z}\right)$$

$$= -\hbar^2\left[x^2\frac{\partial^2}{\partial x^2}+y^2\frac{\partial^2}{\partial y^2}+z^2\frac{\partial^2}{\partial z^2}+2xy\frac{\partial^2}{\partial x\partial y}+2yz\frac{\partial^2}{\partial y\partial z}+2zx\frac{\partial^2}{\partial z\partial x}+x\frac{\partial}{\partial x}+y\frac{\partial}{\partial y}+z\frac{\partial}{\partial z}\right]$$

so that the addition of the last two relations yields

$$\hat{L}^2+\left(\hat{\vec{r}}\cdot\hat{\vec{p}}\right)^2 = -\hbar^2\left(x^2+y^2+z^2\right)\left(\frac{\partial^2}{\partial x^2}+\frac{\partial^2}{\partial y^2}+\frac{\partial^2}{\partial z^2}\right)+\hbar^2\left(x\frac{\partial}{\partial x}+y\frac{\partial}{\partial y}+z\frac{\partial}{\partial z}\right)$$

$$= r^2\hat{p}^2+i\hbar\left(\hat{\vec{r}}\cdot\hat{\vec{p}}\right) \tag{31.10}$$

It follows that the momentum can be expressed as

$$\left(\hat{p}\right)^2 = \frac{1}{r^2}\left[\hat{L}^2+\left(\hat{\vec{r}}\cdot\hat{\vec{p}}\right)^2-i\hbar\left(\hat{\vec{r}}\cdot\hat{\vec{p}}\right)\right] = \frac{1}{r^2}\hat{L}^2-\frac{\hbar^2}{r^2}\left(r\frac{\partial}{\partial r}\right)^2-\frac{\hbar^2}{r}\frac{\partial}{\partial r} \tag{31.11}$$

where the radial part results from the substitution of Eq. (I.20), Appendix I, which gives

$$\hat{p} = -i\hbar\nabla = -i\hbar\left(\vec{e}_r\frac{\partial}{\partial r}+\vec{e}_\theta\frac{1}{r}\frac{\partial}{\partial\theta}+\vec{e}_\varphi\frac{1}{r\sin\theta}\frac{\partial}{\partial\varphi}\right)$$

and hence

$$\hat{\vec{r}}\cdot\hat{\vec{p}} = -i\hbar r\frac{\partial}{\partial r}$$

Therefore the Hamiltonian (31.1) becomes

$$\hat{H} = -\frac{\hbar^2}{2m}\left[\frac{1}{r^2}\left(r\frac{\partial}{\partial r}\right)^2 + \frac{1}{r}\frac{\partial}{\partial r} - \frac{1}{\hbar^2 r^2}\hat{L}^2\right] + U(r)$$

and reduces to the equivalent form

$$\hat{H} = -\frac{\hbar^2}{2m}\left[\frac{1}{r^2}\frac{\partial}{\partial r}\left(r^2\frac{\partial}{\partial r}\right) - \frac{1}{\hbar^2 r^2}\hat{L}^2\right] + U(r) \qquad (31.12)$$

used for a particle bound by a central force field.

31.2. Eigenvalue Equations in a Central Field

Spherically symmetric systems, where the potential energy $U(r)$ is independent of the direction of $\vec{r}$, are usually treated using the spherical coordinates (r,θ,φ) related to the Cartesian coordinates (x,y,z) as referred to in Appendix I, Eq.(I.19). As the angle θ is defined by the inclination to the z-axis, it is mathematically convenient to take the direction of the component of $\vec{L}$ that is compatible with $\hat{L}^2$ (in the sense introduced in Chapter 34) to be the z-axis. Therefore we consider the eigenvalue equations of the operators $\hat{H}, \hat{L}_z$ and $\hat{L}^2$ given by

$$\hat{H}\Psi(r,\theta,\varphi) = \varepsilon\Psi(r,\theta,\varphi)$$

$$\hat{L}_z\Psi(r,\theta,\varphi) = L_z\Psi(r,\theta,\varphi) \qquad (31.13)$$

$$\hat{L}^2\Psi(r,\theta,\varphi) = L^2\Psi(r,\theta,\varphi)$$

The appropriate form of the Hamiltonian $\hat{H}$ is conveniently written using the standard expression (I.23) of the Laplace operator ∇^2 in spherical coordinates as

$$\hat{H} = -\frac{\hbar^2}{2m}\left[\frac{1}{r^2}\frac{\partial}{\partial r}\left(r^2\frac{\partial}{\partial r}\right) + \frac{1}{r^2\sin\theta}\frac{\partial}{\partial\theta}\left(\sin\theta\frac{\partial}{\partial\theta}\right) + \frac{1}{r^2\sin^2\theta}\frac{\partial^2}{\partial\varphi^2}\right] + U(r) \quad (31.14)$$

Using the transformation (I.19) from Cartesian to spherical coordinates and the chain rule for differentiation, we find

$$\frac{\partial}{\partial\varphi} = \frac{\partial x}{\partial\varphi}\frac{\partial}{\partial x} + \frac{\partial y}{\partial\varphi}\frac{\partial}{\partial y} + \frac{\partial z}{\partial\varphi}\frac{\partial}{\partial z} = -r\sin\theta\sin\varphi\frac{\partial}{\partial x} + r\sin\theta\cos\varphi\frac{\partial}{\partial y} = x\frac{\partial}{\partial y} - y\frac{\partial}{\partial x}$$

so that the operator $\hat{L}_z$ takes the form

$$\hat{L}_z = \frac{\hbar}{i}\left(x\frac{\partial}{\partial y} - y\frac{\partial}{\partial x}\right) = \frac{\hbar}{i}\frac{\partial}{\partial \varphi} \qquad (31.15)$$

Furthermore, from Eqs.(31.12) and (31.14) it follows that $\hat{L}^2$ is in fact expressed by that part of Hamiltonian which operates on the angular variables θ, φ only, that is,

$$\hat{L}^2 = -\hbar^2\left[\frac{1}{\sin\theta}\frac{\partial}{\partial \theta}\left(\sin\theta\frac{\partial}{\partial \theta}\right) + \frac{1}{\sin^2\theta}\frac{\partial^2}{\partial \varphi^2}\right] \qquad (31.16)$$

The energy eigenfunctions in central field, which are simultaneously eigenfunctions of $\hat{L}_z$ and $\hat{L}^2$, are subjected, in addition to the usual bound state requirement (30.21), to the *single-valueness* condition

$$\Psi(r,\theta,\varphi) = \Psi(r,\theta,\varphi + 2\pi)$$

since the points φ and $\varphi + 2\pi$ are in fact the same physical point.

The solution to the time-independent Schrödinger equation (31.13) in central field can be found by a process of separation of variables, yielding three ordinary differential equations in the variables r, θ and φ, respectively, to which a physical significance will be assigned. It is common practice to separate the variables in two stages, first substituting the trial solution

$$\Psi(r,\theta,\varphi) = R(r)Y(\theta,\varphi) \qquad (31.17)$$

into the energy eigenvalue equation. If the Hamiltonian is expanded in spherical coordinates, as given by Eq.(31.14), Eq.(31.13) becomes

$$\frac{\hbar^2}{2m}\left[\frac{1}{R}\frac{\partial}{\partial r}\left(r^2\frac{\partial R}{\partial r}\right)\right] + \left[\varepsilon - U(r)\right]r^2 = -\frac{\hbar^2}{2m}\frac{1}{Y}\left[\frac{1}{\sin\theta}\frac{\partial}{\partial \theta}\left(\sin\theta\frac{\partial Y}{\partial \theta}\right) + \frac{1}{\sin^2\theta}\frac{\partial^2 Y}{\partial \varphi^2}\right]$$
(31.18)

One side is independent of θ and φ, and the other side is independent of r, so they must be separately equal to some constant, which is conveniently chosen to be equal to $L^2/2m$. In this way we obtain a *radial equation*, given by

$$\left[-\frac{\hbar^2}{2m}\frac{1}{r^2}\frac{d}{dr}\left(r^2\frac{d}{dr}\right) + U(r) + \frac{L^2}{2mr^2}\right]R(r) = \varepsilon R(r) \qquad (31.19)$$

where $\partial/\partial r$ was replaced by d/dr for obvious reasons, and an *angular equation*, having the so-called *spherical harmonics* $Y(\theta,\varphi)$ as solutions, which reads

$$-\hbar^2\left[\frac{1}{\sin\theta}\frac{\partial}{\partial\theta}\left(\sin\theta\frac{\partial}{\partial\theta}\right)+\frac{1}{\sin^2\theta}\frac{\partial^2}{\partial\varphi^2}\right]Y = L^2 Y \quad \text{or} \quad \hat{L}^2 Y = L^2 Y \quad (31.20)$$

The form of the spherical harmonics is obtained as a result of the second stage of the separation process, where we assume that

$$Y(\theta,\varphi) = P(\theta)\Phi(\varphi)$$

so that Eq.(31.20) becomes

$$\hbar^2\left[\frac{\sin\theta}{P}\frac{\partial}{\partial\theta}\left(\sin\theta\frac{\partial P}{\partial\theta}\right)+\frac{L^2}{\hbar^2}\sin^2\theta\right] = -\hbar^2\frac{1}{\Phi}\frac{\partial^2\Phi}{\partial\varphi^2} \quad (31.21)$$

Since one side is independent of φ while the other is independent of θ, they must each be equal to a constant, which we call L_z^2. We thus obtain the differential equations

$$-\hbar^2\frac{d^2\Phi}{d\varphi^2} = L_z^2 \Phi \quad (31.22)$$

where again $\partial/\partial\varphi$ has been replaced by $d/d\varphi$, and

$$\hbar^2\left[\frac{1}{\sin\theta}\frac{d}{d\theta}\left(\sin\theta\frac{dP}{d\theta}\right)+\left(\frac{L^2}{\hbar^2}-\frac{L_z^2}{\hbar^2\sin^2\theta}\right)P\right] = 0 \quad (31.23)$$

where $\partial/\partial\theta$ has been replaced by $d/d\theta$.

Therefore the energy equation (31.13) completely separates, in central field, into a radial equation (31.19) and two angular equations (31.22) and (31.23). The radial equation, giving the energy levels ε of the system, takes the form

$$-\frac{\hbar^2}{2m}\left(\frac{d^2}{dr^2}+\frac{2}{r}\frac{d}{dr}\right)R(r)+\left[U(r)+\frac{L^2}{2mr^2}\right]R(r) = \varepsilon R(r) \quad (31.24)$$

which can be simplified by making the substitution

$$\xi(r) = rR(r) \quad (31.25)$$

and using the identity

$$\left(\frac{d^2}{dr^2}+\frac{2}{r}\frac{d}{dr}\right)\frac{\xi(r)}{r} = \frac{1}{r}\frac{d^2\xi(r)}{dr^2}$$

Thus Eq.(31.24) reduces to

$$-\frac{\hbar^2}{2m}\frac{d^2\xi(r)}{dr^2}+\left[U(r)+\frac{L^2}{2mr^2}\right]\xi(r) = \varepsilon\xi(r) \quad (31.26)$$

which is formally identical to the time-independent Schrödinger equation in one dimension (30.17), provided the central force potential $U(r)$ is replaced by an *effective potential energy* of the form

$$U_{eff}(r) = U(r) + \frac{L^2}{2mr^2}$$

The repulsive potential, associated with the angular momentum, modifies the central potential as illustrated before, see Figure 3.3, for $U(r) = -C/r$. Further progress with the solution of the radial equation depends on the particular form of the potential function $U(r)$. We shall consider the significant example of one-electron atoms in the next chapter.

31.3. Quantization of Orbital Angular Momentum

The solutions to the eigenvalue equation (31.13) for $\hat{L}_z$, which reads

$$\hat{L}_z \Phi = \frac{\hbar}{i} \frac{d\Phi}{d\varphi} = L_z \Phi \qquad (31.27)$$

have the form $\Phi = e^{iL_z \varphi / \hbar}$ which also satisfies Eq.(31.22). Applying the single-valueness condition, we obtain

$$\Phi = e^{iL_z(\varphi + 2\pi)/\hbar} = e^{iL_z \varphi / \hbar} \qquad \text{or} \qquad e^{iL_z 2\pi / \hbar} = 1$$

which means that the eigenvalues of the operator $\hat{L}_z$ are

$$L_z = \pm m\hbar \quad , \quad m = 0,1,2,\ldots \qquad (31.28)$$

The integer *m*, which can either be positive, negative or zero, is called the *magnetic quantum number*. The corresponding eigenfunctions of Eq.(31.27) should be written as

$$\Phi_m(\varphi) = N e^{\pm im\varphi} \qquad (31.29)$$

where N is a normalization constant. In this way, one angular differential equation (31.22) has been associated with the eigenvalue equation of $\hat{L}_z$ and has been solved such that the interpretation of its solutions as single valued state functions yields the quantum condition (31.28).

The other angular differential equation (31.23), where we now substitute the allowed values of L_z given by Eq.(31.28), becomes

$$\frac{1}{\sin\theta}\frac{d}{d\theta}\left(\sin\theta\frac{dP}{d\theta}\right)+\left(\frac{L^2}{\hbar^2}-\frac{m^2}{\sin^2\theta}\right)P=0 \tag{31.30}$$

and will be associated with the eigenvalue equation (31.20) of $\hat{L}^2$. The procedure is made simpler if we set $u=\cos\theta$, so that

$$\frac{d}{d\theta}=-\sin\theta\frac{d}{du}$$

and Eq.(31.30) transforms into

$$-\frac{d}{du}\left(-\sin^2\theta\frac{dP}{du}\right)+\left(\frac{L^2}{\hbar^2}-\frac{m^2}{\sin^2\theta}\right)P=0 \quad\text{or}\quad \frac{d}{du}\left[(1-u^2)\frac{dP}{du}\right]+\left(\frac{L^2}{\hbar^2}-\frac{m^2}{1-u^2}\right)P=0 \tag{31.31}$$

It is convenient to find first a particular solution to this equation for $m=0$, as in this case it reduces to

$$\frac{d}{du}\left[(1-u^2)\frac{dP}{du}\right]+\frac{L^2}{\hbar^2}P=0 \tag{31.32}$$

Appropriate finite solutions can be obtained by the method used earlier for the case of the Hermite equation (30.30), that is, by writing $P=P(u)$ as a power series

$$P(u)=\sum_{k=0}^{\infty}a_k u^k \quad\text{or}\quad \frac{dP(u)}{du}=\sum_{k=1}^{\infty}ka_k u^{k-1} \tag{31.33}$$

and substituting into Eq.(31.32) to obtain

$$\sum_{k=0}^{\infty}(k+1)(k+2)a_{k+2}u^k-\sum_{k=0}^{\infty}k(k+1)a_k u^k+\frac{L^2}{\hbar^2}\sum_{k=0}^{\infty}a_k u^k=0$$

The condition that the coefficient of each power must be zero yields

$$a_{k+2}=\frac{k(k+1)-L^2/\hbar^2}{(k+1)(k+2)}a_k \quad\text{or}\quad \lim_{k\to\infty}\frac{a_{k+2}}{a_k}=1 \tag{31.34}$$

and this becomes, for large k, identical to the recurrence relation between the successive terms of the expansion of $(1-u^2)^{-1}$. Since this function is divergent for $u=1$, we must impose the condition that the series terminates for a finite value $k=l$. By taking $a_l\neq 0$ and $a_{l+2}=0$, Eq.(31.34) gives

$$L^2=l(l+1)\hbar^2 \tag{31.35}$$

Comparison with Eq.(31.20) shows that Eq.(31.35) is a second quantum condition, referring to the eigenvalues of $\hat{L}^2$. The integer l, which must obviously be positive or zero, is called the *orbital quantum number*. The series (31.33) reduces to the *Legendre polynomials* $P_l(u)$ that are solutions to the *Legendre equation*, which reads

$$\frac{d}{du}\left[(1-u^2)\frac{dP_l(u)}{du}\right] + l(l+1)P_l(u) = 0 \tag{31.36}$$

In view of further applications to atomic systems, we must also solve Eq.(31.31) and find the eigenfunctions for nonzero values of m. Upon insertion of quantum condition (31.35), Eq.(31.31) becomes

$$\frac{d}{du}\left[(1-u^2)\frac{dP(u)}{du}\right] + \left[l(l+1) - \frac{m^2}{1-u^2}\right]P(u) = 0 \tag{31.37}$$

Equation (31.37) is known as the *associated Legendre equation*. A trial solution is

$$P(u) = (u^2 - 1)^{m/2} F(u) \tag{31.38}$$

where the function $F(u)$ is related to the Legendre polynomials as we shall show below.

Substituting $P(u)$ and its derivatives $dP(u)/du$, $d^2P(u)/du^2$ into Eq.(31.37), the equation obeyed by $F(u)$ is obtained as

$$(u^2-1)\frac{d^2F(u)}{du^2} + 2u(m+1)\frac{dF(u)}{du} - [l(l+1) - m(m+1)]F(u) = 0 \tag{31.39}$$

An analogous equation is obtained through m-fold differentiation of the Legendre equation (31.36), which gives

$$-m(m+1)\frac{d^m P_l}{du^m} - 2u(m+1)\frac{d^{m+1} P_l}{du^{m+1}} + (1-u^2)\frac{d^{m+2} P_l}{du^{m+2}} + l(l+1)\frac{d^m P_l}{du^m} = 0 \tag{31.40}$$

and takes the form

$$(u^2-1)\frac{d^2}{du^2}\left(\frac{d^m P_l}{du^m}\right) + 2u(m+1)\frac{d}{du}\left(\frac{d^m P_l}{du^m}\right) - [l(l+1) - m(m+1)]\frac{d^m P_l}{du^m} = 0 \tag{31.41}$$

Equation (31.41) becomes identical to Eq.(31.39) if and only if

$$F(u) = \frac{d^m P_l(u)}{du^m}$$

It follows that the solution (31.38) to the associated Legendre equation (31.37) is

$$P_l^{|m|}(u) = (u^2 - 1)^{|m|/2} \frac{d^{|m|} P_l(u)}{du^{|m|}} \tag{31.42}$$

These are known to be the *associated Legendre functions* and are characterized by $|m|$, because Eq.(31.37) is independent of the sign of *m*. Since $P_l(u)$ are polynomials of degree *l*, their $|m|$ th derivative will be zero if $|m|$ is greater than *l*, that is, if

$$|m| \leq l \tag{31.43}$$

which gives a relation between the magnetic and orbital quantum numbers.

We may further combine the solutions (31.29) and (31.42) to express the *spherical harmonics* $Y_{lm}(\theta,\varphi)$ as

$$Y_{lm}(\theta,\varphi) = N_{lm} P_l^{|m|}(\cos\theta) e^{\pm im\varphi} \tag{31.44}$$

where N_{lm} are normalization constants, $l = 0,1,2,\ldots$ and $m = 0,\pm 1,\pm 2,\ldots,\pm l$. The spherical harmonics are eigenfunctions of $\hat{L}^2$ and $\hat{L}_z$ and represent the angular part of the state function. The atomic states corresponding to the allowed values of *l* are successively labelled as *s* (for which $l = 0$), $p(l=1)$, $d(l=2)$, $f(l=3)$,... *states* in view of the *s*harp, *p*rincipal, *d*iffuse or *f*undamental aspect of their spectral lines.

EXAMPLE 31.1. The spatial quantization

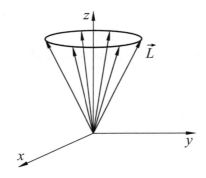

Figure 31.1. Spatial orientation of the orbital angular momentum vector

The probability density of the spherical harmonic eigenstates (31.44) is a function of θ only, so that it is determined by the inclination of the orbital angular momentum $\vec{L}$ to the *z*-axis, without any orientational restriction to the *x*- and *y*-axes. We illustrate this in Figure 31.1, where the vector observable $\vec{L}$ may lie anywhere on a cone about the *z*-axis. In an *s-state* the angular momentum is zero, because the quantum numbers are $l = 0$ and $m = 0$. The eigenfunction is a constant

$$Y_{00} = \frac{1}{\sqrt{4\pi}}$$

where the value of N_{00} was derived from a normalization condition analogous to Eq.(29.5).

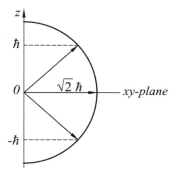

Figure 31.2. Quantization of the *p*-state orbital angular momentum

In a *p-state* we have $l=1$ and $m=0,\pm 1$ so that

$$L = \sqrt{l(l+1)}\hbar = \sqrt{2}\hbar \quad , \quad L_z = 0, \pm\hbar$$

and the spherical harmonics are

$$Y_{10} = \sqrt{\frac{3}{4\pi}}\cos\theta \quad , \quad Y_{1,\pm 1} = \mp\sqrt{\frac{3}{8\pi}}\sin\theta\, e^{\pm i\varphi}$$

The vector representation is given in Figure 31.2 on a polar diagram.

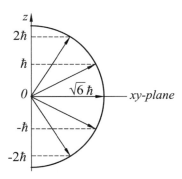

Figure 31.3. Quantization of the *d*-state orbital angular momentum

In a *d-state*, where $l=2$ and $m=0,\pm 1,\pm 2$, we obtain

$$L = \sqrt{6}\hbar \quad , \quad L_z = 0, \pm\hbar, \pm 2\hbar$$

and the corresponding eigenfunctions are

$$Y_{20} = \sqrt{\frac{5}{16\pi}}(3\cos^2\theta - 1) \quad , \quad Y_{2,\pm 1} = \mp\sqrt{\frac{15}{8\pi}}\cos\theta\sin\theta\, e^{\pm i\varphi} \quad , \quad Y_{2,\pm 2} = \sqrt{\frac{15}{32\pi}}\sin^2\theta\, e^{\pm 2i\varphi}$$

The allowed orientations of $\vec{L}$ about the z-axis are given on the polar diagram plotted in Figure 31.3. The vector representation of the quantum condition (31.28) is known as the *spatial quantization* of orbital angular momentum.

FURTHER READING

1. H. F. Hameka - QUANTUM MECHANICS: A CONCEPTUAL APPROACH, Wiley, New York, 2004.
2. A. C. Phillips - INTRODUCTION TO QUANTUM MECHANICS, Wiley, New York, 2004.
3. W. Greiner - QUANTUM MECHANICS: AN INTRODUCTION, Springer-Verlag, New York, 1994.

PROBLEMS

31.1. Find expressions for the operator $\hat{L}_x$, $\hat{L}_y$ and $\hat{L}_\pm = \hat{L}_x + i\hat{L}_y$ in spherical polar coordinates.

Solution

For dynamical systems with spherical symmetry, it is convenient to introduce the unit vectors $\vec{e}_r, \vec{e}_\theta$ and $\vec{e}_\varphi$ in the directions of increasing r, θ and φ, the spherical polar coordinates, given by $x = r\sin\theta\cos\varphi$, $y = r\sin\theta\sin\varphi$, $z = r\cos\theta$. This is the same as

$$r^2 = x^2 + y^2 + z^2, \quad \theta = \arctan\frac{\sqrt{x^2+y^2}}{z}, \quad \varphi = \arctan\frac{y}{x}$$

It follows that

$$\frac{\partial r}{\partial x} = \sin\theta\cos\varphi \qquad \frac{\partial \theta}{\partial x} = \frac{1}{r}\cos\theta\cos\varphi \qquad \frac{\partial \varphi}{\partial x} = -\frac{1}{r}\frac{\sin\varphi}{\sin\theta}$$

$$\frac{\partial r}{\partial y} = \sin\theta\sin\varphi \qquad \frac{\partial \theta}{\partial y} = \frac{1}{r}\cos\theta\sin\varphi \qquad \frac{\partial \varphi}{\partial y} = \frac{1}{r}\frac{\cos\varphi}{\sin\theta}$$

$$\frac{\partial r}{\partial z} = \cos\theta \qquad \frac{\partial \theta}{\partial z} = -\frac{1}{r}\sin\theta \qquad \frac{\partial \varphi}{\partial z} = 0$$

This yields

$$\frac{\partial}{\partial x} = \frac{x}{r}\frac{\partial}{\partial r} + \frac{1}{\sin\theta}\frac{zx}{r^3}\frac{\partial}{\partial \theta} - \frac{y/x^2}{(1+y^2/x^2)}\frac{\partial}{\partial \varphi}$$

and hence, in view of the previous expressions for x, y and z, one obtains

$$\frac{\partial}{\partial x} = \sin\theta\cos\varphi\frac{\partial}{\partial r} + \frac{\cos\theta\cos\varphi}{r}\frac{\partial}{\partial \theta} - \frac{\sin\varphi}{r\sin\theta}\frac{\partial}{\partial \varphi}$$

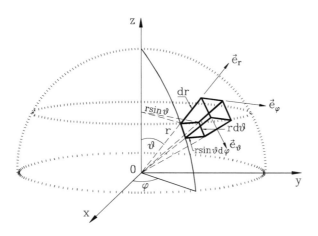

In a similar manner we have

$$\frac{\partial}{\partial y} = \sin\theta\sin\varphi\frac{\partial}{\partial r} + \frac{\cos\theta\sin\varphi}{r}\frac{\partial}{\partial \theta} + \frac{\cos\varphi}{r\sin\theta}\frac{\partial}{\partial \varphi} \quad , \quad \frac{\partial}{\partial z} = \cos\theta\frac{\partial}{\partial r} - \frac{\sin\theta}{r}\frac{\partial}{\partial \theta}$$

Substituting the expressions for the partial derivatives into Eqs.(29.16) for L_x, L_y yields

$$\hat{L}_x = i\hbar\left[\sin\varphi\frac{\partial}{\partial \theta} + \frac{\cos\varphi}{\tan\theta}\frac{\partial}{\partial \varphi}\right] \quad , \quad \hat{L}_y = i\hbar\left[-\cos\varphi\frac{\partial}{\partial \theta} + \frac{\sin\varphi}{\tan\theta}\frac{\partial}{\partial \varphi}\right]$$

and then, the operators $\hat{L}_+, \hat{L}_-$ assume the simple form

$$\hat{L}_+ = \hat{L}_x + i\hat{L}_y = i\hbar e^{i\varphi}\left(-i\frac{\partial}{\partial \theta} + \frac{\cos\theta}{\sin\theta}\frac{\partial}{\partial \varphi}\right) \quad , \quad \hat{L}_- = \hat{L}_x - i\hat{L}_y = i\hbar e^{-i\varphi}\left(i\frac{\partial}{\partial \theta} + \frac{\cos\theta}{\sin\theta}\frac{\partial}{\partial \varphi}\right)$$

31.2. If $\Psi(r) = Ne^{-\alpha r}$ is a state function of a system in a spherically symmetric potential energy $U(r)$, find an expression for $U(r)$ and calculate the corresponding magnitude of the angular momentum.

Solution

The Hamiltonian (31.14) should be independent of both θ and φ, and hence, the Schrödinger equation reduces to

$$-\frac{\hbar^2}{2m}\frac{1}{r^2}\frac{d}{dr}\left(r^2\frac{d\Psi}{dr}\right) + U(r)\Psi = \varepsilon\Psi(r)$$

Substituting $\Psi(r)$ yields

$$\left[\frac{\alpha\hbar^2}{mr} + U(r)\right] - \left(\varepsilon + \frac{\alpha^2\hbar^2}{2m}\right) = 0$$

and it follows that

$$U(r) = -\frac{\alpha \hbar^2}{mr}, \quad \varepsilon = -\frac{\alpha^2 \hbar^2}{2m}$$

The square of the magnitude of the angular momentum L^2 can be derived from Eq.(13.26), where

$$\xi(r) = Nre^{-\alpha r}, \quad \frac{d\xi(r)}{dr} = N(1-\alpha r)e^{-\alpha r}, \quad \frac{d^2\xi(r)}{dr^2} = N\alpha(\alpha r - 2)e^{-\alpha r}$$

Substituting into Eq.(31.26) yields

$$-\frac{\hbar^2}{2m}\alpha(\alpha r - 2) + \left(-\frac{\alpha \hbar^2}{mr} + \frac{L^2}{2mr^2}\right)r = \varepsilon$$

and this gives $L^2 = 0$.

31.3. Find the eigenstates and eigenvalues of a particle of mass m that rotates on a circle of radius r_0 in a spherically symmetric potential energy $U(r)$.

Solution

The problem is one dimensional, because the equation of the planar motion can be expressed in terms of polar coordinates in the form

$$x = r\cos\varphi, \quad y = r\sin\varphi$$

where $r = r_0$. We have

$$\nabla^2 = \frac{\partial^2}{\partial x^2} + \frac{\partial^2}{\partial y^2} = \frac{1}{r}\frac{\partial}{\partial r}\left(r\frac{\partial}{\partial r}\right) + \frac{1}{r^2}\frac{\partial^2}{\partial \varphi^2}$$

and the Schrödinger wave equation (6.4) reads

$$\frac{1}{r}\frac{\partial}{\partial r}\left(r\frac{\partial \psi}{\partial r}\right) + \frac{1}{r^2}\frac{\partial^2 \psi}{\partial \varphi^2} + \frac{2m}{\hbar^2}[E - U(r)]\psi = 0$$

Since $r = r_0$, it is apparent that $\partial \psi / \partial r = 0$ and $U(r_0) = U_0$ is a constant. Hence, we are left with an ordinary differential equation for the motion in a constant potential energy, which reads

$$\frac{d^2 \psi}{d\varphi^2} + k^2 \psi = 0, \quad k^2 = \frac{2m}{\hbar^2}r_0^2(E - U_0)$$

where $\psi = \psi(r_0, \varphi)$, so that we look for a solution of the form (6.11), that is,

$$\psi(r_0, \varphi) = A\sin(k\varphi + \delta).$$

The appropriate admissibility condition for wave functions, given in terms of angular coordinate variables, consists of the single-valueness requirement $\sin(k\varphi + \delta) = \sin[k(\varphi + 2\pi) + \delta]$, and this implies that $k = n$, where n is a positive integer $n = 1,2,3,\ldots$. The normalization condition

$$A^2 \int_0^{2\pi} \sin^2(n\varphi + \delta) d\varphi = 1$$

gives $A = 1/\sqrt{\pi}$, and hence the eigenstates are described by

$$\psi_n(r_0, \varphi) = \frac{1}{\sqrt{\pi}} \sin(n\varphi + \delta)$$

The corresponding eigenvalues form a discrete spectrum, given by $E_n = U_0 + \frac{n^2 \hbar^2}{2mr_0^2}$.

31.4. Find the expectation values $\langle \hat{L}_z \rangle$ and $\langle \hat{L}_z^2 \rangle$ in a state of a plane rotator described by the state function $\Psi(\varphi) = N \cos^2 \varphi$, where N is a real constant.

Solution

The normalizing condition yields

$$N^2 \int_0^{2\pi} \cos^4 \varphi \, d\varphi = N^2 \int_0^{2\pi} \left(\frac{1 + \cos 2\varphi}{2} \right)^2 d\varphi = \frac{N^2}{4} \left(2\pi + 2\int_0^{2\pi} \cos 2\varphi \, d\varphi + \int_0^{2\pi} \cos^2 2\varphi \, d\varphi \right) = 1$$

and hence, we have

$$\Psi(\varphi) = \frac{2}{\sqrt{3\pi}} \sin^2 \varphi = \frac{1}{\sqrt{3\pi}} \left(1 + \frac{1}{2} e^{2i\varphi} + \frac{1}{2} e^{-2i\varphi} \right) = \sqrt{\frac{2}{3}} \Phi_0 + \frac{1}{\sqrt{6}} \Phi_2 + \frac{1}{\sqrt{6}} \Phi_{-2}$$

where Φ_m stands for the eigenfunctions of $\hat{L}_z$, Eq.(31.29). The probability of finding the system in each of these eigenstates, Eq.(29.29), is given by

$$P_0 = \frac{2}{3}, \quad P_2 = P_{-2} = \frac{1}{6}$$

As $\hat{L}_z$ has eigenvalues $m\hbar$, Eq.(31.28), it follows that

$$\langle \hat{L}_z \rangle = \frac{2}{3} \times 0 + \frac{1}{6} \times 2\hbar + \frac{1}{6}(-2\hbar) = 0$$

and similarly, using the eigenvalues $l(l+1)\hbar^2$ of $\hat{L}^2$, Eq.(31.35), one obtains

$$\langle \hat{L}_z^2 \rangle = \frac{2}{3} \times 0 + \frac{1}{6} \times 4\hbar^2 + \frac{1}{6} 4\hbar^2 = \frac{4}{3} \hbar^2$$

31.5. Find the coordinate representation of the operator $\hat{p}_r$, associated with the radial component p_r of linear momentum.

Solution

The radial component of linear momentum is expressed in classical mechanics by the dot product of $\vec{e}_r$ and $\vec{p}$ as

$$p_r = \vec{e}_r \cdot \vec{p} = \frac{1}{r}\vec{r}\cdot\vec{p} = \frac{1}{r}(xp_x + yp_y + zp_z)$$

The classical Hamiltonian may be written in terms of the variables p_r and L, by using the identity

$$p^2 = p_x^2 + p_y^2 + p_z^2 = \frac{1}{r^2}\left[(xp_x + yp_y + zp_z)^2 + (yp_z - zp_y)^2 + (zp_x - xp_z)^2 + (xp_y - yp_x)^2\right]$$

$$= p_r^2 + \frac{1}{r^2}(L_x^2 + L_y^2 + L_z^2) = p_r^2 + \frac{1}{r^2}L^2$$

and hence, substituting into Eq.(29.15) gives

$$H = \frac{1}{2m}p_r^2 + \frac{1}{2mr^2}L^2 + U(r)$$

Comparison with Eq.(31.12) indicates the representation

$$\hat{p}_r^2 = -\hbar^2 \frac{1}{r^2}\frac{\partial}{\partial r}\left(r^2 \frac{\partial}{\partial r}\right)$$

Since, for any radial function $\Psi(r)$, we may write

$$\hat{p}_r^2 \Psi(r) = -\hbar^2 \left[\frac{\partial^2 \Psi(r)}{\partial r^2} + \frac{2}{r}\frac{\partial \Psi(r)}{\partial r}\right] = -\hbar^2 \left(\frac{\partial}{\partial r} + \frac{1}{r}\right)\left[\frac{\partial \Psi(r)}{\partial r} + \frac{\Psi(r)}{r}\right]$$

$$= -\hbar^2 \left(\frac{\partial}{\partial r} + \frac{1}{r}\right)\left(\frac{\partial}{\partial r} + \frac{1}{r}\right)\Psi(r)$$

it follows that we have

$$\hat{p}_r = -i\hbar\left(\frac{\partial}{\partial r} + \frac{1}{r}\right)$$

This shows that the rule of associating an operator of the form $-i\hbar\partial/\partial x_i$ to the linear momentum component conjugated to x_i is not applicable for the pair of conjugated variables r, p_r.

31.6. Find an expression for the potential energy $U(r)$, and calculate the energy eigenvalue ε and the magnitude of the angular momentum in a spherically symmetric state $\Psi(r) = Ne^{-\alpha r^2}$.

31.7. Prove the identities $\hat{L}^2 - \hat{L}_z^2 = \hat{L}_+\hat{L}_- - \hbar\hat{L}_z = \hat{L}_-\hat{L}_+ + \hbar\hat{L}_z$.

31.8. Calculate the expectation values $\langle \hat{L}_x \rangle$ and $\langle \hat{L}_x^2 \rangle$ in an eigenstate $Y_{lm}(\theta,\varphi)$ of $\hat{L}^2$ and $\hat{L}_z$.

31.9. Find the probability of obtaining the results $\hbar$, 0 and $-\hbar$, respectively, for a measurement of L_z, in a state described by the wave function

$$\Psi(\theta,\varphi) = \sqrt{\frac{3}{8\pi}}(\sin\theta\sin\varphi + i\cos\theta)$$

What is the probability of obtaining the result $2\hbar^2$ for a measurement of L^2 in the same state?

31.10. Find an expression for the transition energy $\Delta\varepsilon$ between two adjacent levels of the plane rotator described by the Hamiltonian $\hat{H} = \hat{L}^2/2I$, where I is the moment of inertia.

32. ONE-ELECTRON ATOMS

32.1. THE RADIAL EQUATION FOR ONE-ELECTRON ATOMS
32.2. THE HYDROGEN ATOM
32.3. THE ONE-ELECTRON ATOM IN AN EXTERNAL MAGNETIC FIELD

32.1. THE RADIAL EQUATION FOR ONE-ELECTRON ATOMS

Let us find the solution of the radial equation (31.26), where the eigenvalues of $\hat{L}^2$ expressed by Eq.(31.35) must be first inserted and then the form of the potential energy $U(r)$, must be given. We will take as a particular example the case of the one-electron atom which is a reference system for atomic physics. We will assume an atom with atomic number Z and all but one of its electrons removed, such as the helium ion He^+, lithium ion Li^{2+} and so on. We recall that their potential energy depends only upon the separation r of the particles $+Ze$ and $-e$ in the form

$$U(r) = -\frac{Ze_0^2}{r} \qquad (32.1)$$

so that the radial equation (31.26) becomes

$$\frac{d^2 \xi(r)}{dr^2} + \frac{2m_e}{\hbar^2}\left[\varepsilon + \frac{Ze_0^2}{r} - \frac{l(l+1)\hbar^2}{2m_e r^2}\right]\xi(r) = 0 \qquad (32.2)$$

where m_e stands for the electron mass. We restrict the present discussion to the bound states, which are the most significant for the atomic electron, where the discrete energy eigenvalues are negatives, that is $\varepsilon = -|\varepsilon|$. For large r, the equation reduces to

$$\frac{d^2 \xi(r)}{dr^2} + \frac{2m_e |\varepsilon|}{\hbar^2}\xi(r) = 0$$

32. ONE-ELECTRON ATOMS

The asymptotic condition $R(r) \to 0$ as $r \to \infty$ eliminates the divergent exponential solutions and indicates the acceptable asymptotic form as $\xi(r) \sim e^{-\lambda r}$, provided we set

$$\lambda = \left(\frac{2m_e|\varepsilon|}{\hbar^2}\right)^{1/2} \tag{32.3}$$

Therefore a trial solution to Eq.(32.2) is

$$\xi(r) = e^{-\lambda r} f(r) \tag{32.4}$$

and this leads to

$$\frac{d^2 f(r)}{dr^2} - 2\lambda \frac{df(r)}{dr} + \left[\frac{2m_e Z e_0^2}{\hbar^2 r} - \frac{l(l+1)}{r^2}\right] f(r) = 0 \tag{32.5}$$

Assuming a power series solution

$$f(r) = \sum_{k=0}^{\infty} a_k r^k \tag{32.6}$$

a procedure already used for the Hermite (30.20) and Legendre (31.32) equations, yields

$$\sum_k \left\{[k(k+1) - l(l+1)] a_{k+1} - 2\left(\lambda k - \frac{m_e Z e_0^2}{\hbar^2}\right) a_k\right\} r^{k-1} = 0$$

and hence, we obtain

$$a_{k+1} = \frac{2(\lambda k - m_e Z e_0^2 / \hbar^2)}{k(k+1) - l(l+1)} a_k \tag{32.7}$$

For large k, Eq.(32.7) becomes identical to the recurrence relation for the terms in the series expansion

$$e^{2\lambda r} = \sum_{k=0}^{\infty} \frac{(2\lambda)^k}{k!} r^k \quad \text{or} \quad \frac{a_{k+1}}{a_k} \to \frac{2\lambda}{k}$$

However, this form is not acceptable for $f(r)$ because the solution (32.4) will diverge like $e^{\lambda r}$ as r tends to infinity. Thus, the series must terminate after a finite number of terms, for instance $k = n$, so that from Eq.(32.7), where $a_n \neq 0$ and $a_{n+2} = 0$, we obtain

$$\lambda n = \frac{m_e Z e_0^2}{\hbar^2} \quad \text{or} \quad \frac{2m_e|\varepsilon|}{\hbar^2} n^2 = \frac{m_e^2 Z^2 e_0^4}{\hbar^4} \tag{32.8}$$

This is a quantum condition for the energy levels of one-electron atoms, of the form

$$\varepsilon_n = -\frac{Z^2}{n^2}\frac{m_e e_0^4}{2\hbar^2} = -\frac{Z^2}{n^2}Rh = -\frac{Z^2}{n^2}\frac{e_0^2}{2a_0} \tag{32.9}$$

where R is the Rydberg constant (28.21), and a_0 is the Bohr radius (28.17). This result was first derived in Eq.(28.18) from Bohr's model for one-electron atoms. The acceptable values of the principal quantum number n must obey

$$n \geq l+1 \tag{32.10}$$

as required by the recurrence relation (32.7). Otherwise, for $n=l$, a_{n+1} is infinite and so are the succeeding coefficients. From Eqs.(31.43) and (32.10) we may conclude that, for a one-electron atom, the three quantum numbers, required for the physically acceptable solutions of the time-independent Schrödinger equation, have the following allowed integer values

$$n = 1, 2, 3, \ldots \tag{32.11}$$

while

$$l = 0, 1, 2, \ldots, n-1 \tag{32.12}$$

and

$$-l \leq m \leq l$$

The radial eigenfunctions are obtained from Eqs.(31.25), (32.4), (32.6) and (32.8) as

$$R_{nl}(r) = \frac{1}{r}e^{-m_e Z e_0^2 r/n\hbar^2}\sum_{k=0}^{n}a_k r^k = \frac{1}{r}e^{-Zr/na_0}\sum_{k=0}^{n}a_k r^k \tag{32.13}$$

where the polynomials are solutions to Eq.(32.5), which is called the *associated Laguerre equation* and a_0 is the Bohr radius, as defined by Eq.(28.17). The expression (32.13) can be then combined with the appropriate spherical harmonics (31.44) to obtain the functions (31.17) in the form

$$\Psi_{nlm}(r,\theta,\varphi) = N_{lm}\left(\frac{1}{r}e^{-Zr/na_0}\sum_{k=0}^{n}a_k r^k\right)P_l^{|m|}(\cos\theta)e^{\pm im\varphi} \tag{32.14}$$

where N_{lm} is a normalization constant. These are the complete time-independent state functions of the bound states of one-electron atoms. They will be used to describe the case of the hydrogen atom, which corresponds to $Z=1$.

32.2. THE HYDROGEN ATOM

We first consider the *ground state* of the hydrogen atom, given by the lowest values of the quantum numbers: $n=1$, $l=0$, $m=0$. The eigenfunction (32.14) reads

$$\Psi_{100}(r) = N_{00} e^{-r/a_0} \tag{32.15}$$

and shows a complete spherical symmetry, as it is not θ- nor φ- dependent. According to the Born interpretation, formulated in Eq.(29.1) for one dimension, the probability of finding the electron in a given volume element of space $dV = r^2 dr \sin\theta\, d\theta\, d\varphi$ about a point (r,θ,φ) may be written as

$$P(r)dV = |\Psi_{100}|^2 dV = |N_{00}|^2 e^{-2r/a_0} r^2 \sin\theta\, d\theta\, d\varphi\, dr \tag{32.16}$$

The constant N_{00} is given by the normalization condition (29.2), which takes the form

$$1 = |N_{00}|^2 \int_0^\infty e^{-2r/a_0} r^2 dr \int_0^\pi \sin\theta\, d\theta \int_0^{2\pi} d\varphi = 4\pi |N_{00}|^2 \int_0^\infty r^2 e^{-2r/a_0} dr$$

The integral has the form (IV.15), Appendix IV, so that

$$1 = |N_{00}|^2 \pi a_0^3 \quad \text{or} \quad N_{00} = (1/\pi a_0^3)^{1/2} \tag{32.17}$$

and the ground state function of the hydrogen atom becomes

$$\Psi_{100}(r) = (1/\pi a_0^3)^{1/2} e^{-r/a_0} \tag{32.18}$$

In view of the spherical symmetry of the ground state, we may average the probability (32.16) over all the θ, φ values to obtain

$$P(r)dr = |N_{00}|^2 4\pi r^2 e^{-2r/a_0} dr = \frac{4}{a_0^3} r^2 e^{-2r/a_0} dr \tag{32.19}$$

which gives the radial probability of finding the hydrogen electron in a spherical shell of thickness dr at a distance r from the nucleus, which is at the origin. The probability density $P(r)$, plotted in Figure 32.1 as a function of r, is zero at the nucleus and in the asymptotic region, where $r \to \infty$, so that a radial position of maximum probability must exist, as given by

$$\frac{dP(r)}{dr} = 0 \quad \text{or} \quad e^{-2r/a_0}\left(2r - 2r^2 \frac{1}{a_0}\right) = 0 \quad \text{i.e.,} \quad r = a_0$$

Therefore, the maximum occurs at a distance from the origin equal to the Bohr radius a_0. However, we must emphasize that although the energy and the most probable radial position of the electron correspond to Bohr's results, a spherical distribution of the ground state probability is obtained, as illustrated by the angular probability density, given on a polar diagram in Figure 32.2, instead of a planar orbit predicted by Bohr's picture. The expectation radius of this spherical distribution is given by Eq.(29.8) as

$$\langle r \rangle = \int \Psi_{100}^* r \Psi_{100} dV = \frac{4}{a_0^3} \int_0^\infty r^3 e^{-2r/a_0} dr \qquad (32.20)$$

where we may use Eq.(IV.15), Appendix IV, to obtain

$$\langle r \rangle = \frac{3}{2} a_0 \qquad (32.21)$$

We may further obtain from Eqs.(29.8) and (IV.15) that

$$\langle r^2 \rangle = \frac{4}{a_0^3} \int_0^\infty r^4 e^{-2r/a_0} dr = 3a_0^2 \qquad (32.22)$$

which allows us to estimate of the standard deviation for the expectation electron radial position as

$$\Delta r = \sigma_r = \sqrt{<r^2> - <r>^2} = \frac{\sqrt{3}}{2} a_0 \qquad (32.23)$$

The expectation potential energy of the ground state can be calculated, starting from

$$\left\langle \frac{1}{r} \right\rangle = \frac{4}{a_0^3} \int_0^\infty \frac{1}{r} r^2 e^{-2r/a_0} dr = \frac{1}{a_0}$$

which, if substituted into Eq.(32.1), gives

$$\langle U(r) \rangle = -- = -\frac{Zm_e e_0^4}{\hbar^2} = 2\varepsilon_1 \qquad (32.24)$$

where the energy ε_1 was inserted according to Eq.(32.9). Since we have

$$\varepsilon_1 = \langle \hat{H} \rangle = \langle \hat{T} + U(r) \rangle$$

the relation between the expectation values of the kinetic and potential energy operators, in the ground state, given by

$$\langle \hat{T} \rangle = \varepsilon_1 - \langle U(r) \rangle = -\varepsilon_1 = -\frac{1}{2} \langle U(r) \rangle$$

obeys the classical virial theorem, Eq.(26.13), provided that the expectation values are interpreted as the average values of observables in the ground state.

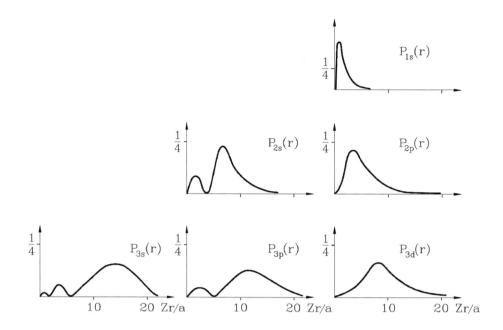

Figure 32.1. Radial probability distribution of the electron in the lowest energy states

Any state of a given $n > 1$, called an *excited state* of the hydrogen atom, is *degenerate*, which means that there are several electron eigenstates with different l and m which have the same energy. Since the energy (32.9) is not l-dependent, as would be expected because the orbital quantum number is a coefficient in the radial equation (32.2), to a given value of the principal quantum number there corresponds a set of n states with $l \leq n-1$. This degeneracy is called *accidental*. Furthermore, for a fixed l there are $2l+1$ values of the magnetic quantum number m, giving rise to a *spatial* degeneracy. As a result of both effects we have

$$\sum_{l=0}^{n-1}(2l+1) = n^2 \tag{32.25}$$

which means that the nth energy eigenstate is n^2-fold degenerate. In order to illustrate the concept of degeneracy it is common practice to use the correspondence between the electron states and the planar orbits of Sommerfeld's model. The accidental degeneracy is comparable with the case where different elliptical orbits have the same energy, provided they have the same major axis. The spatial degeneracy corresponds to the property of an orbit to preserve its energy during a rotation of its plane which is restricted by the spatial quantization of the orbital angular momentum.

For an arbitrary hydrogen state Ψ_{nlm}, as given by Eq.(32.14), the probability of finding the electron in a given volume element of space, written as

$$P(r)dV = R_{nl}^2 |Y_{lm}|^2 r^2 dr d\Omega = r^2 R_{nl}^2 dr |Y_{lm}|^2 d\Omega \tag{32.26}$$

contains a product of a radial probability density $r^2 R_{nl}^2(r)$ and an angular probability density $|Y_{lm}(\theta,\varphi)|^2$. Plots of the radial probability density for the s-states $\Psi_{100}, \Psi_{200}, \Psi_{300}$ of the hydrogen atom are represented in Figure 32.1.

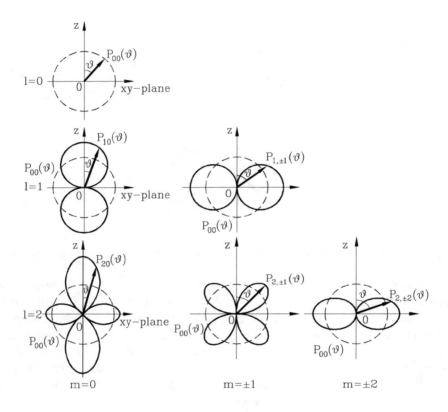

Figure 32.2. Polar representation of the angular probability density for electron states of lowest l

The probability of finding the electron within a unit solid angle element about the origin, $|Y_{lm}(\theta,\varphi)|^2 = |P_l^{|m|}(\cos\theta)|^2$, is plotted on polar diagrams in Figure 32.2 for s-, p-, and d-states. It is apparent that, as m increases, the angular probability density shifts from the z-axis towards the equatorial plane xy. When $|m|=l$ and l takes large values we approach the classical limit of Sommerfeld's planar orbits.

The degeneracy of the excited states can be removed by small interactions of the electron with external fields. Since some eigenfunctions of a degenerate eigenvalue are more affected than others, a splitting of the corresponding energy levels is expected to occur.

32.3. THE ONE-ELECTRON ATOM IN AN EXTERNAL MAGNETIC FIELD

The Hamiltonian of the one-electron atom must be modified in the presence of electromagnetic forces, as discussed in Chapter 14. We have shown that in this case the momentum $m\vec{v}$ of an electron of charge $-e$ and mass m_e must be replaced by a generalized momentum $\vec{p}$, defined by Eq.(14.18) which reads

$$\vec{v} = \frac{1}{m_e}\left(\vec{p} + e\vec{A}\right) \qquad (32.27)$$

and the potential energy U of the electron takes the form (14.19) which becomes

$$U = -eV + \frac{e}{m_e}\left(\vec{p} + e\vec{A}\right)\cdot\vec{A} \qquad (32.28)$$

Substituting Eqs.(32.27) and (32.28) into Eq.(24.17) we obtain the Lagrangian L as

$$L = \frac{1}{2m_e}\left(\vec{p} + e\vec{A}\right)^2 + eV - \frac{e}{m_e}\vec{A}\cdot\left(\vec{p} + e\vec{A}\right) \qquad (32.29)$$

In view of Eq.(24.29), the Hamiltonian must be written as

$$H = \sum_k p_k v_k - L = \frac{1}{m_e}\vec{p}\cdot\left(\vec{p} + e\vec{A}\right) - \frac{1}{2m_e}\left(\vec{p} + e\vec{A}\right)^2 + \frac{e}{m_e}\vec{A}\cdot\left(\vec{p} + e\vec{A}\right) - eV$$

It is convenient to denote the electrostatic part of the energy (32.28), having a central character, by $U(r) = -eV(r)$, so that we obtain

$$H = \frac{1}{2m_e}\left(\vec{p} + e\vec{A}\right)^2 + U(r) \qquad (32.30)$$

which is the desired form of the classical Hamiltonian. The appropriate Hamilton operator for an electron under a stationary electromagnetic field corresponds to Eq.(32.30), as required by Postulate 2, and allows us to put in evidence the removal of the spatial degeneracy produced by an external magnetic field. Because the Hamiltonian (32.30) is written in terms of the vector potential $\vec{A}$, it is convenient to use, instead of the external field $\vec{H}$, the external induction $\vec{B}_0 = \mu_0 \vec{H}$, where μ_0 stands for the permeability of free space. Since all the spatial derivatives of a *uniform field* vanish, it is common practice to take

$$\vec{A} = \frac{1}{2}\vec{B}_0 \times \vec{r} \qquad (32.31)$$

which is in a form suitable to obey the constraint (12.29) for the vector potential, that is,

$$\nabla \times \vec{A} = \nabla \times \left(\frac{1}{2}\vec{B}_0 \times \vec{r}\right) = \frac{1}{2}[(\nabla \cdot \vec{r})\vec{B}_0 - (\vec{B}_0 \cdot \nabla)\vec{r}] = \frac{1}{2}(3\vec{B}_0 - \vec{B}_0) = \vec{B}_0$$

where a standard vector identity, (I.15), Appendix I, was used. Substituting Eq.(32.31) into Eq.(32.30) the classical Hamiltonian takes the form

$$H = \frac{p^2}{2m_e} + U(r) + \frac{e}{2m_e}(\vec{p} \cdot \vec{A} + \vec{A} \cdot \vec{p}) + \frac{e^2}{2m_e}\vec{A}^2$$

$$= \frac{p^2}{2m_e} + U(r) + \frac{e}{2m_e}(\vec{r} \times \vec{p}) \cdot \vec{B}_0 + \frac{e^2}{2m_e}\left(\frac{1}{2}\vec{B}_0 \times \vec{r}\right)^2 \quad (32.32)$$

If we now make use of the orbital angular momentum definition, Eq.(3.1), the classical Hamiltonian can be rewritten as

$$H = \frac{p^2}{2m_e} + U(r) + \frac{e}{2m_e}\vec{B}_0 \cdot \vec{L} + \frac{e^2}{2m_e}\left(\frac{1}{2}\vec{B}_0 \times \vec{r}\right)^2 \quad (32.33)$$

The quadratic term in $\vec{B}_0$ may usually be dropped, since the magnetic fields experimentally available are weak, whereas the linear term will perturb slightly only the electron energy levels, as given by

$$H = \frac{p^2}{2m_e} + U(r) + \frac{g_l e}{2m_e}\vec{B}_0 \cdot \vec{L} = \frac{p^2}{2m_e} + U(r) - \vec{\mu}_l \cdot \vec{B}_0 = \frac{p^2}{2m_e} + U(r) + U_m \quad (32.34)$$

The physical observable $\vec{\mu}_l$, called the *orbital magnetic moment*, which has been defined by Eq.(28.27) for a circular electron path, is usually written in terms of the orbital angular momentum as

$$\vec{\mu}_l = -g_l \frac{e}{2m_e}\vec{L} \quad (32.35)$$

where, for convenience, a quantity $g_l = 1$, known as the *orbital giromagnetic factor*, can be formally introduced. Equation (32.34) allows us to obtain the orbital angular momentum, provided we know the energy U_m of the magnetic interaction, as

$$\vec{\mu}_l = -\frac{\partial U_m}{\partial \vec{B}_0} \quad (32.36)$$

If we take the z-direction to coincide with that of $\vec{B}_0$, $\vec{L}$ will be replaced by the operator $\hat{L}_z$ and the *Hamiltonian operator* $\hat{H}$, corresponding to Eq.(32.34), will be expressed as

$$\hat{H} = \frac{\hat{p}^2}{2m_e} + U(r) + \frac{g_l e}{2m_e}\vec{B}_0 \hat{L}_z = \hat{H}_0 + \frac{g_l e}{2m_e}\vec{B}_0 \hat{L}_z \quad (32.37)$$

where $\hat{H}_0$ stands for the central field Hamiltonian (31.1). Since $\hat{H}$ commutes with $\hat{H}_0$, they have a common set of eigenfunctions, so that we may find the eigenvalues of $\hat{H}$ in a uniform magnetic feld using the eigenfunctions (32.14) derived for $\hat{H}_0$, that is,

$$\hat{H}\Psi_{nlm} = \hat{H}_0 \Psi_{nlm} + \frac{g_l e B_0}{2m_e} \hat{L}_z \Psi_{nlm} = \left(\varepsilon_n + \frac{g_l e B_0}{2m_e} m\hbar\right) \Psi_{nlm} \qquad (32.38)$$

It follows that the energy levels ε_n, given by Eq.(32.9), are shifted according to

$$\varepsilon_{nm} = \varepsilon_n + m\mu_B B_0 g_l \qquad (32.39)$$

where the parameter μ_B is called the *Bohr magneton* and is related to the characteristic electron constants by

$$\mu_B = \frac{e\hbar}{2m_e} \qquad (32.40)$$

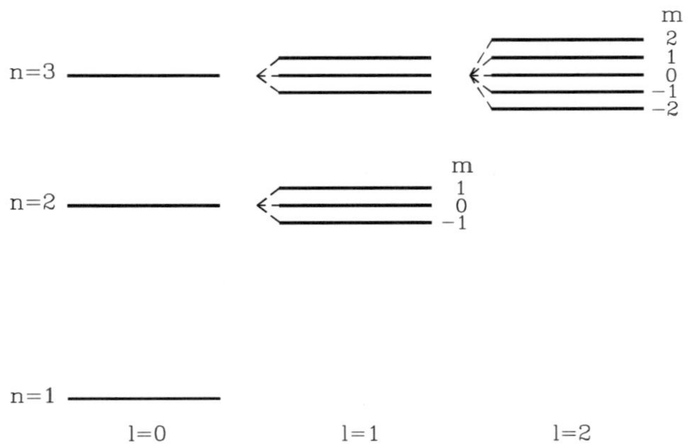

Figure 32.3. Normal Zeeman splitting of the energy levels for the first two excited states in hydrogen

Note that the coupling energy between the orbital magnetic moment and the external magnetic field, written as

$$\varepsilon_{nm} - \varepsilon_n = m\mu_B B_0 g_l = m\hbar\omega_L \qquad (32.41)$$

can be formally related to the classical picture of an angular momentum vector $\vec{L}$ precessing about the z-axis with Larmor frequency ω_L, provided that $\hbar\omega_L$ is interpreted to be the quantum of energy for the precession motion.

The interaction of the electron with a weak uniform external magnetic field, known as the *normal Zeeman effect* and illustrated in Figure 32.3, removes the spatial degeneracy according to Eq.(32.39). However, the accidental degeneracy, with respect to l, is not lifted.

Further Reading

1. B. H. Bransden, C. J. Joachain - PHYSICS OF ATOMS AND MOLECULES, Prentice Hall, Upper Saddle River, 2004.
2. G. C. Schatz, M. A. Ratner - QUANTUM MECHANICS IN CHEMISTRY, Dover Publications, New York, 2002.
3. J. Simons, J. Nichols - QUANTUM MECHANICS IN CHEMISTRY, Oxford University Press, Oxford, 1996.

Problems

32.1. Assuming that the bound states of one-electron atoms result from radial motion with negative energies corresponding to the classically allowed region, where $\varepsilon - U_{eff}(r) = p^2/2m_e > 0$, find the region where the electron will be found with highest probability in a stationary state defined by the quantum numbers n and l.

Solution

The electron is moving in the effective potential energy

$$U_{eff}(r) = -\frac{Ze_0^2}{r} + \frac{l(l+1)\hbar^2}{2\mu r^2}$$

and hence, the radial motion with negative energies corresponding to the classically allowed region can be defined by

$$\varepsilon - U_{eff}(r) = \varepsilon + \frac{Ze_0^2}{r} - \frac{l(l+1)\hbar^2}{2m_e r^2} \geq 0$$

The exact solution to the radial equation yields the energy eigenvalues ε_n given by Eq.(32.9), which upon substitution lead to the inequality

$$-\frac{Z^2}{n^2}\frac{e_0^2}{2a_0} + \frac{Z e_0^2}{r} - \frac{l(l+1)\hbar^2}{2m_e r^2} \geq 0$$

This is satisfied by the r values situated between the two roots of the equation

$$r^2 - 2\frac{a_0 n^2}{Z}r + \frac{a_0^2 n^2}{Z^2}l(l+1) = 0$$

which are

$$r_{1,2} = a_0 \frac{n}{Z}\left[n \pm \sqrt{n^2 - l(l+1)}\right]$$

It follows that, for a stationary state defined by the quantum numbers n and l, the electron will be found with highest probability in a region of width $a_0 n\sqrt{n^2 - l(l+1)}/Z$ about the value $r_n = a_0 n^2/Z$. For $n = 1$, $Z = 1$ we obtain $r_n = a_0$, and the electron moves in a spherical region defined by the Bohr radius, as expected. The width of the probability distribution decreases as l increases up to the maximum value of $n-1$, in a stationary state with a defined n value. When $r \to 0$, the repulsive potential energy dominates the Coulomb attraction, such that for small values of l, there will be a significant probability for the electron to get near to the nucleus, but this probability will decrease quickly as l increases up to its highest value $l = n-1$, and hence, the electron will be kept away from the nucleus.

32.2. Find an expression for the spherically symmetric potential of the hydrogen atom, assuming that the electron is in its ground state.

Solution

The electrostatic potential due to the electron charge distribution is given by Poisson's equation (11.14), which reads

$$\frac{1}{r^2}\frac{d}{dr}\left(r^2 \frac{dV_e}{dr}\right) = -\frac{\rho}{\varepsilon_0}$$

where the electron charge density in the ground state is given by

$$\rho = -e|\Psi_{100}|^2 = -\frac{e}{\pi a_0^3}e^{-2r/a_0} \quad \text{i.e.,} \quad \frac{d^2(rV_e)}{dr^2} = \frac{e}{\pi \varepsilon_0 a_0^3} r e^{-2r/a_0}$$

On integration with boundary conditions $V_e(0) = 0$ and $V_e(\infty) = 0$ one obtains the electron contribution as

$$rV_e = \frac{e}{4\pi\varepsilon_0}e^{-2r/a_0}\left(1+\frac{r}{a}\right) - \frac{e}{4\pi\varepsilon_0}$$

The potential of the one-electron atom includes the contribution from the point charge of the proton, so that one obtains

$$V = V_e + \frac{e}{4\pi\varepsilon_0} = \frac{e}{4\pi\varepsilon_0}\left(\frac{1}{r}+\frac{1}{a_0}\right)e^{-2r/a_0}$$

This reduces to the contribution of the point nuclear charge when $r \ll a_0$ and decays exponentially for $r \gg a_0$, due to the shielding effect of the electronic charge distribution.

32.3. Use the radial equation for one-electron atoms to calculate the expectation value $\langle 1/r^2 \rangle_{nl}$ in a stationary state of the hydrogen atom.

Solution

Because the spherical harmonic eigenfunctions of the rotational motion are normalized on the unit sphere, the expectation value of any power of r is obtained from

$$\langle r^k \rangle_{nl} = \langle \Psi_{nlm} | r^k | \Psi_{nlm} \rangle = \int_0^\infty r^k R_{nl}^2(r) r^2 dr = \int_0^\infty r^k \xi_{nl}^2(r) dr$$

where the functions $\xi_{nl}(r)$ are solutions to the radial equation (32.2). Substituting the energy eigenvalues ε_n from Eq.(32.9) gives

$$\frac{d^2 \xi_{nl}(r)}{dr^2} + \frac{2m_e}{\hbar^2}\left[-\frac{e_0^2}{2a_0 n^2} + \frac{e_0^2}{r} - \frac{l(l+1)\hbar^2}{2m_e r^2}\right]\xi_{nl}(r) = 0$$

or

$$\frac{d^2 \xi_{nl}(r)}{dr^2} + \left[-\frac{1}{a_0^2 n^2} + \frac{2}{a_0 r} - \frac{l(l+1)}{r^2}\right]\xi_{nl}(r) = 0$$

It is convenient to eliminate the Bohr radius a_0 by setting $r = a_0 \rho$, which gives

$$\frac{d^2 \xi_{nl}(\rho)}{d\rho^2} = \left[\frac{1}{n^2} - \frac{2}{\rho} + \frac{l(l+1)}{\rho^2}\right]\xi_{nl}(\rho)$$

The property of the radial functions givem by

$$\int_0^\infty \xi_{nl}^2 \frac{\partial}{\partial l}\left(\frac{\partial^2 \xi_{nl}/\partial \rho^2}{\xi_{nl}}\right) d\rho = \int_0^\infty \xi_{nl} \frac{\partial}{\partial l}\left(\frac{\partial^2 \xi_{nl}}{\partial \rho^2}\right) d\rho - \int_0^\infty \frac{\partial^2 \xi_{nl}}{\partial \rho^2} \frac{\partial \xi_{nl}}{\partial l} d\rho = 0$$

implies that

$$\int_0^\infty \xi_{nl}^2 \frac{\partial}{\partial l}\left[\frac{1}{n^2} - \frac{2}{\rho} + \frac{l(l+1)}{\rho^2}\right] d\rho = -\frac{2}{n^3}\int_0^\infty \xi_{nl}^2 d\rho + (2l+1)\int_0^\infty \frac{1}{\rho^2}\xi_{nl}^2 d\rho$$

$$= -\frac{2}{a_0 n^3}\int_0^\infty \xi_{nl}^2 dr + (2l+1) a_0 \int_0^\infty \frac{1}{r^2}\xi_{nl}^2 dr = 0$$

and this finally yields

$$\left\langle \frac{1}{r^2}\right\rangle_{nl} = \frac{2}{(2l+1)n^3}\frac{1}{a_0^2}$$

32.4. Use the radial equation for the hydrogen atom to derive the Kramers recurrence relation for the expectation values of the powers of r that, for $s > -2l-3$, is given by

$$\frac{s+1}{n^2}\langle r^s \rangle_{nl} - (2s+1)a_0 \langle r^{s-1}\rangle_{nl} + \frac{s a_0^2}{4}\left[(2l+1)^2 - s^2\right]\langle r^{s-2}\rangle_{nl} = 0$$

Solution

As shown in Problem 32.3, the radial equation can be put in the form

$$\frac{d^2\xi_{nl}(\rho)}{d\rho^2} = \left[\frac{1}{n^2} - \frac{2}{\rho} + \frac{l(l+1)}{\rho^2}\right]\xi_{nl}(\rho)$$

where $\rho = r/a_0$. It is convenient to multiply both sides by

$$\rho^s\left(\rho\frac{d\xi_{nl}}{d\rho} - \frac{s+1}{2}\xi_{nl}\right)$$

and then to integrate over ρ, which gives

$$\int_0^\infty \left\{\rho^s\left(\rho\frac{d\xi_{nl}}{d\rho}\frac{d^2\xi_{nl}}{d\rho^2} - \frac{s+1}{2}\xi_{nl}\frac{d^2\xi_{nl}}{d\rho^2}\right) + \left[-\frac{1}{n^2}\rho^{s+1} + 2\rho^s + l(l+1)\rho^{s-1}\right]\xi_{nl}\frac{d\xi_{nl}}{d\rho}\right\}d\rho$$

$$+ \frac{s+1}{2a_0}\left[\frac{1}{n^2}\langle\rho^s\rangle - 2\langle\rho^{s-1}\rangle + l(l+1)\langle\rho^{s-2}\rangle\right] = 0$$

In view of Eqs.(31.25), (32.10) and (32.13), it is apparent that $\xi_{nl} \sim \rho^{l+1}$, when $\rho \to 0$, and $\xi_{nl} \to 0$, if $\rho \to \infty$. Thus we have

$$\int_0^\infty \rho^{s+1}\frac{d\xi_{nl}}{d\rho}\frac{d^2\xi_{nl}}{d\rho^2}d\rho = \frac{1}{2}\int_0^\infty \rho^{s+1}\frac{d\xi_{nl}^2}{d\rho}d\rho = -\frac{s+1}{2}\int_0^\infty \rho^s\xi_{nl}^2 d\rho$$

and also

$$-\frac{s+1}{2}\int_0^\infty \rho^s\xi_{nl}\frac{d^2\xi_{nl}}{d\rho^2}d\rho = \frac{s+1}{2}\int_0^\infty \frac{d\xi_{nl}}{d\rho}\left(\rho^s\frac{d\xi_{nl}}{d\rho} + s\rho^{s-1}\xi_{nl}\right)d\rho$$

$$= \frac{s+1}{2}\int_0^\infty \rho^s\left(\frac{d\xi_{nl}}{d\rho}\right)^2 d\rho + \frac{s+1}{2}\int_0^\infty \rho^{s-1}\xi_{nl}\frac{d\xi_{nl}}{d\rho}d\rho$$

Substituting these results leads to

$$\frac{1}{2}\int_0^\infty\left\{-\frac{1}{n^2}\rho^{s+1} + 2\rho^s + \left[\frac{s(s+1)}{2} - l(l+1)\right]\rho^{s-1}\right\}\frac{d\xi_{nl}^2}{d\rho}d\rho$$

$$+ \frac{s+1}{2a_0}\left[\frac{1}{n^2}\langle\rho^s\rangle - 2\langle\rho^{s-1}\rangle + l(l+1)\langle\rho^{s-2}\rangle\right] = 0$$

Integrating by parts yields

$$\frac{s+1}{2a_0 n^2}\langle\rho^s\rangle - \frac{s}{a_0}\langle\rho^{s-1}\rangle - \frac{s-1}{2a_0}\left[\frac{s(s+1)}{2} - l(l+1)\right]\langle\rho^{s-2}\rangle$$

$$+ \frac{s+1}{2a_0 n^2}\langle\rho^s\rangle - \frac{s+1}{a_0}\langle\rho^{s-1}\rangle + \frac{(s+1)l(l+1)}{2a_0}\langle\rho^{s-2}\rangle = 0$$

Rearranging gives the Kramers recurrence relation, as required.

32.5. Solve the eigenvalue problem of an electron moving in a uniform magnetic field of induction $\vec{B}$.

Solution

The eigenvalue problem reads

$$\frac{1}{2m_e}(\hat{\vec{p}}+e\vec{A})^2 \Psi = \varepsilon \Psi$$

and we may take the direction of $\vec{B}$ to be the z-axis. If we choose the vector potential of the uniform field of the form given in Eq.(32.31), one obtains

$$\frac{1}{2}\vec{B}\times\vec{r} = \frac{1}{2}B(x\vec{e}_y - y\vec{e}_x) \quad \text{or} \quad A_x = -\frac{1}{2}By, \quad A_y = \frac{1}{2}Bx, \quad A_z = 0$$

It is, however, more convenient to take

$$A_x = -By, \quad A_y = 0, \quad A_z = 0$$

which also satisfies the restriction $\vec{B} = \nabla \times \vec{A}$ if $\vec{B}$ is along the z-axis. With this choice, we have $\nabla \cdot \vec{A} = 0$ and the eigenvalue problem simplifies to

$$-\frac{\hbar^2}{2m_e}\nabla^2\Psi + \frac{i\hbar}{m_e}eBy\frac{\partial\Psi}{\partial x} + \frac{e^2}{2m_e}B^2y^2\Psi = \varepsilon\Psi$$

The Hamiltonian operator commutes with $\hat{p}_x$ and $\hat{p}_z$, so that a trial solution has the form

$$\Psi(x,y,z) = e^{i(p_x x + p_z z)/\hbar}u(y)$$

Substituting into the eigenvalue equation and dropping the exponential factors gives

$$-\frac{\hbar^2}{2m_e}\frac{d^2u(y)}{dy^2} + \frac{e^2B^2}{2m_e}\left(y^2 - \frac{2yp_x}{eB} + \frac{p_x^2}{e^2B^2}\right)u(y) = \left(\varepsilon - \frac{p_z^2}{2m_e}\right)u(y)$$

We introduce the new variable $\rho = y - p_x/eB$ such that

$$-\frac{\hbar^2}{2m_e}\frac{d^2u(\rho)}{d\rho^2} + \frac{e^2B^2}{2m_e}\rho^2 u(\rho) = \left(\varepsilon - \frac{p_z^2}{2m_e}\right)u(\rho)$$

This equation can be reduced to the eigenvalue equation (6.62) of the linear harmonic motion, if we set

$$E = \varepsilon - \frac{p_z^2}{2m_e}, \quad \omega = \frac{eB}{m_e}$$

which gives

$$\frac{d^2 u(\rho)}{d\rho^2} + \frac{2m_e}{\hbar^2}\left(E - \frac{m_e \omega^2}{2}\rho^2\right) u(\rho) = 0$$

It follows that the energy eigenvalue of an electron in a uniform magnetic field can be written as

$$\varepsilon_{n,p_z} = \frac{p_z^2}{2m_e} + E_n = \frac{p_z^2}{2m_e} + \left(n + \frac{1}{2}\right)\hbar\omega = \frac{p_z^2}{2m_e} + \left(n + \frac{1}{2}\right)\hbar\frac{eB}{m_e}$$

where the first term has a continuous set of values, corresponding to free motion along the z-axis, and the second term has a discrete set of values associated with electron oscillation in a plane normal to $\vec{B}$. The corresponding eigenfunctions have the form (30.34), which reads

$$u_n(\xi) = N_n e^{-\xi^2/2} H_n(\xi)$$

where $H_n(\xi)$ are the Hermite polynomials and

$$\xi = \sqrt{\frac{m_e \omega}{\hbar}}\rho = \sqrt{\frac{eB}{\hbar}}\left(y - \frac{p_x}{eB}\right)$$

Finally we obtain

$$\Psi_{n p_x p_z}(x, y, z) = N_n\, e^{i(p_x x + p_z z)/\hbar} e^{-\xi^2/2} H_n(\xi)$$

and there is infinite degeneracy, because for values of n and p_z fixed by $\varepsilon_{n p_z}$, the function p_x can take a continuous set of values.

32.6. A hydrogen atom is in a state described by

$$\Psi(\vec{r}) = \frac{1}{\sqrt{14}}\left[2\Psi_{100}(\vec{r}) - 3\Psi_{200}(\vec{r}) + \Psi_{332}(\vec{r})\right]$$

Find the expectation values for energy, for L^2 and for L_z.

32.7. A one-electron atom is in a state given by

$$\Psi = \frac{1}{81}\sqrt{\frac{2}{\pi}}\left(\frac{Z}{a_0}\right)^{5/2} r\left(6 - \frac{Z}{a_0}r\right) e^{-Zr/3a_0} \cos\theta$$

Find the quantum numbers n, l and m_l of the state. If $Z = 1$, what is the most probable r value?

32.8. Use the Kramers recurrence relation to find expressions for $\langle 1/r \rangle_{nl}$, $\langle r \rangle_{nl}$ and $\langle r^2 \rangle_{nl}$ in a hydrogen atom.

32.9. Use the radial equation of one-electron atoms to find the radial functions and the energy eigenvalues of an electron trapped in a spherical box of radius a, defined by the potential energy

$$U(r) = 0 \quad r < a$$
$$= \infty \quad r \geq a$$

32.10. Find an expression for the radial function of a free particle, of energy $\varepsilon = (\hbar k)^2 / 2m$, assuming that $l \neq 0$.

33. MATRIX MECHANICS OF ANGULAR MOMENTUM

33.1. MATRIX REPRESENTATIONS
33.2. ANGULAR MOMENTUM MATRICES
33.3. SPIN ANGULAR MOMENTUM

33.1. MATRIX REPRESENTATIONS

The formulation of quantum mechanics in terms of differential operators and differential equations solved for state functions, known as *wave mechanics*, becomes more suitable for some applications if both the state functions and the operators are represented in an appropriate function space defined by a complete set of wave functions. For state functions of the form (29.4) the recipe is similar to that of representing a vector by a column matrix which gives its projections in a particular coordinate system

$$\Psi = \sum_{i=1}^{n} a_i \Psi_i \quad \text{or} \quad \Psi = \begin{pmatrix} a_1 \\ a_2 \\ \dots \\ a_n \end{pmatrix}, \quad \Phi = \sum_{i=1}^{n} b_i \Psi_i \quad \text{or} \quad \Phi = \begin{pmatrix} b_1 \\ b_2 \\ \dots \\ b_n \end{pmatrix}$$

The general form of a linear relation between the components a_i, b_i with respect to a given set of functions Ψ_i, written as

$$b_i = \sum_{k=1}^{n} f_{ik} a_k \qquad (33.1)$$

contains n^2 coefficients f_{ik} which form the elements of a matrix, so that Eq.(33.1) also reads

$$\begin{pmatrix} b_1 \\ b_2 \\ \cdots \\ b_n \end{pmatrix} = \begin{pmatrix} f_{11} & f_{12} & \cdots & f_{1n} \\ f_{21} & f_{22} & \cdots & f_{2n} \\ \cdots & \cdots & & \cdots \\ f_{n1} & f_{n2} & \cdots & f_{nn} \end{pmatrix} \begin{pmatrix} a_1 \\ a_2 \\ \cdots \\ a_n \end{pmatrix} \qquad (33.2)$$

We can represent the square matrix as a linear operator $\hat{f}$ which transforms one function into another, by writing Eq.(33.2) in the form

$$\Phi = \hat{f}\Psi \qquad (33.3)$$

The *unit* or identity matrix (see Eq.(II.9), Appendix II) will be represented as a linear operator $\hat{I}$ describing the multiplication of a function by unity

$$\Psi = \hat{I}\Psi$$

In view of the expansion (29.28), which holds for an arbitrary state function in the form

$$\Psi = \sum_{i=1}^{n} \langle \Psi_i | \Psi \rangle \Psi_i \quad \text{or} \quad |\Psi\rangle = \sum_{i=1}^{n} |\Psi_i\rangle\langle \Psi_i | \Psi \rangle$$

we may define the unit or *identity* operator as

$$\hat{I} = \sum_{i=1}^{n} |\Psi_i\rangle\langle \Psi_i | \qquad (33.4)$$

Conversely, for every linear operator $\hat{f}$ we can form a square matrix, in terms of a complete set of functions Ψ_i, by multiplying the left-hand side of Eq.(33.3) by Ψ_i^* and integrating, which gives

$$\int \Psi_i^* \Phi dV = \langle \Psi_i | \Phi \rangle = \langle \Psi_i | \hat{f} | \Psi \rangle = \langle \Psi_i | \hat{f}\hat{I} | \Psi \rangle$$

$$= \sum_{k=1}^{n} \langle \Psi_i | \hat{f} | \Psi_k \rangle \langle \Psi_k | \Psi \rangle = \sum_{k=1}^{n} \langle \Psi_i | \hat{f} | \Psi_k \rangle a_k$$

Since $\langle \Psi_i | \Phi \rangle = b_i$ this result is identical with that given by Eq.(33.1), provided we set

$$f_{ik} = \langle \Psi_i | \hat{f} | \Psi_k \rangle = \int \Psi_i^* \hat{f} \Psi_k dV \qquad (33.5)$$

which are called the *matrix elements* of the linear operator $\hat{f}$ in terms of the particular complete set of functions Ψ_i. It is a straightforward matter to show that the rules of matrix algebra for addition, multiplication, associativity and distributivity (see Appendix II) apply to the matrix elements (33.5), following from their definition. Two properties of matrix elements that are needed in quantum mechanics can be readily derived from Eq.(33.5). Using Eq.(29.19) we obtain

$$f_{ik} = \langle \Psi_i | \hat{f} | \Psi_k \rangle = \langle \Psi_k | \hat{f} | \Psi_i \rangle^* = f_{ki}^* \tag{33.6}$$

which means that *if $\hat{f}$ is a Hermitian operator, then its matrix representation in a complete set of functions is Hermitian*. Hence, the elements in the principal diagonal of a Hermitian matrix must be real. When all the elements are real, the matrix must be symmetric. From Eq.(29.20) we have

$$\langle \Psi_k | \hat{f} | \Psi_i \rangle = \lambda_i \langle \Psi_k | \Psi_i \rangle = \lambda_i \delta_{ik} \tag{33.7}$$

that is: *if the functions Ψ_i are the eigenfunctions of the Hermitian operator $\hat{f}$, the matrix representation is diagonal, and the matrix elements are the eigenvalues of the operator*. A quantum mechanical problem can be stated in a differential equation form or in a matrix form. In both cases the solution consists of finding the eigenvalues and the eigenfunctions of an operator associated with a physical observable. In view of a matrix representation in any complete set of functions, the eigenvalue equation (29.20) must be formulated as

$$\hat{f}\Psi = \lambda \Psi \tag{33.8}$$

where the arbitrary eigenfunction Ψ can be expanded, as given by Eq.(29.4), in terms of a complete set of functions Ψ_i. If we multiply Eq.(33.8) by Ψ_i^* and integrate, we obtain

$$\int \Psi_i^* \hat{f} \Psi dV = \lambda \int \Psi_i^* \Psi dV \quad \text{or} \quad \langle \Psi_i | \hat{f} \hat{I} | \Psi \rangle = \lambda \langle \Psi_i | \Psi \rangle$$

or also, in view of Eq.(33.4)

$$\sum_{k=1}^{n} \langle \Psi_i | \hat{f} | \Psi_k \rangle \langle \Psi_k | \Psi \rangle = \lambda \langle \Psi_i | \Psi \rangle$$

This can be rewritten as

$$\sum_{k=1}^{n} f_{ik} a_k = \lambda a_i$$

which reads

$$\begin{pmatrix} f_{11} & f_{12} & \cdots & f_{1n} \\ f_{21} & f_{22} & \cdots & f_{2n} \\ \cdots & \cdots & & \cdots \\ f_{n1} & f_{n2} & \cdots & f_{nn} \end{pmatrix} \begin{pmatrix} a_1 \\ a_2 \\ \cdots \\ a_n \end{pmatrix} = \lambda \begin{pmatrix} a_1 \\ a_2 \\ \cdots \\ a_n \end{pmatrix} \tag{33.9}$$

Equation (33.6) is called the *matrix eigenvalue equation* and consists of n linear homogenous equations for the a_k, given by

$$(f_{11} - \lambda)a_1 + f_{12}a_2 + ... + f_{1n}a_n = 0$$
$$f_{21}a_1 + (f_{22} - \lambda)a_2 + ... + f_{2n}a_n = 0 \quad (33.10)$$
$$...\quad ...\quad ...$$
$$f_{n1}a_1 + f_{n2}a_2 + ... + (f_{nn} - \lambda)a_n = 0$$

This system of simultaneous equations has a nontrivial solution for the a_k if and only if the characteristic determinant vanishes

$$\det(\hat{f} - \lambda\hat{1}) = \begin{vmatrix} (f_{11} - \lambda) & f_{12} & ... & f_{1n} \\ f_{21} & (f_{22} - \lambda) & ... & f_{2n} \\ ... & ... & & ... \\ f_{n1} & f_{n2} & ... & (f_{nn} - \lambda) \end{vmatrix} = 0 \quad (33.11)$$

If expanded, the determinant gives a polynominal of *n*th degree called the *characteristic polynominal* of $\hat{f}$ with *n* roots which are the desired eigenvalues. For each eigenvalue Eqs.(33.10) can be solved for the a_k giving the corresponding eigenfunction that is a column matrix.

33.2. ANGULAR MOMENTUM MATRICES

The orbital angular momentum operator was defined in quantum mechanics by Eqs.(29.16) which correspond, according to Postulate 2, to the classical definition (3.1). However, it is more convenient to consider the commutation relations (31.2), (31.3) and (31.7) as defining properties to be used for the matrix representation of angular momentum of any system, whether or not it possesses a classical analogue. Therefore, the components $\hat{J}_x, \hat{J}_y$ and $\hat{J}_z$ of any angular momentum are assumed to be Hermitian operators which satisfy the same commutation rules, namely,

$$\left[\hat{J}_x, \hat{J}_y\right] = i\hbar\hat{J}_z \quad , \quad \left[\hat{J}_y, \hat{J}_z\right] = i\hbar\hat{J}_x \quad , \quad \left[\hat{J}_z, \hat{J}_x\right] = i\hbar\hat{J}_y$$
$$\left[\hat{J}_x, \hat{J}^2\right] = \left[\hat{J}_y, \hat{J}^2\right] = \left[\hat{J}_z, \hat{J}^2\right] = 0 \quad (33.12)$$

The elements of the angular momentum matrices of $\hat{J}_x, \hat{J}_y, \hat{J}_z$ and $\hat{J}^2$ will be derived in the representation where $\hat{J}^2$ and $\hat{J}_z$, which correspond to compatible observables, are diagonal. In other words the matrix elements will be expressed in terms of the common set of eigenfunctions $\Psi_{\lambda m}$, with the eigenvalues $\lambda \hbar^2$ and $m\hbar$ introduced by

$$\hat{J}^2 \Psi_{\lambda m} = \lambda \hbar^2 \Psi_{\lambda m}$$
$$\hat{J}_z \Psi_{\lambda m} = m\hbar \Psi_{\lambda m} \qquad (33.13)$$

The obvious inequality satisfied by the expectation values

$$\langle \hat{J}^2 \rangle = \langle \hat{J}_x^2 \rangle + \langle \hat{J}_y^2 \rangle + \langle \hat{J}_z^2 \rangle \geq \langle \hat{J}_z^2 \rangle$$

which can be rewritten, using Eq.(29.8), as

$$\langle \Psi_{\lambda m} | \hat{J}^2 | \Psi_{\lambda m} \rangle \geq \langle \Psi_{\lambda m} | \hat{J}_z^2 | \Psi_{\lambda m} \rangle$$

yields the following restriction

$$\lambda \hbar^2 \geq m^2 \hbar^2 \qquad (33.14)$$

and this means that eigenfunctions $\Psi_{\lambda m}$ with $m \geq \lambda + 1$ should not exist.

The solution is usually obtained by a procedure similar to that used for the harmonic oscillator (see Problem 30.5). The difference of two squares $\hat{J}^2 - \hat{J}_z^2$ may be factorized into

$$\hat{J}^2 - \hat{J}_z^2 = (\hat{J}_x + i\hat{J}_y)(\hat{J}_x - i\hat{J}_y) - \hbar \hat{J}_z = (\hat{J}_x - i\hat{J}_y)(\hat{J}_x + i\hat{J}_y) + \hbar \hat{J}_z \qquad (33.15)$$

where use has been made of Eqs.(33.12). It is convenient to introduce the operators

$$\hat{J}_+ = \hat{J}_x + i\hat{J}_y \quad , \quad \hat{J}_- = \hat{J}_x - i\hat{J}_y \qquad (33.16)$$

so that Eq.(33.15) reads

$$\hat{J}^2 - \hat{J}_z^2 = \hat{J}_+ \hat{J}_- - \hbar \hat{J}_z = \hat{J}_- \hat{J}_+ + \hbar \hat{J}_z \qquad (33.17)$$

The physical significance of $\hat{J}_+$ can be demonstrated by applying it to one of the functions $\Psi_{\lambda m}$, which gives

$$\hat{J}_z (\hat{J}_+ \Psi_{\lambda m}) = \hat{J}_z (\hat{J}_x + i\hat{J}_y) \Psi_{\lambda m} = (\hat{J}_x + i\hat{J}_y)(\hat{J}_z + \hbar) \Psi_{\lambda m} = (m+1)\hbar (\hat{J}_+ \Psi_{\lambda m}) \qquad (33.18)$$

This means that if $\Psi_{\lambda m}$ is an eigenfunction of $\hat{J}_z$ with the eigenvalue $m\hbar$, $\hat{J}_+ \Psi_{\lambda m}$ is an eigenfunction of $\hat{J}_z$ with the eigenvalue $(m+1)\hbar$. In other words, $\hat{J}_+$ raises the eigenvalue $m\hbar$ of an eigenstate of angular momentum by $\hbar$. It represents a *raising operator*. Similarly, $\hat{J}_-$ lowers the eigenvalue of $\hat{J}_z$ by $\hbar$ and it represents a *lowering operator*. Because of the commutation relations (33.12), we have

$$\hat{J}^2(\hat{J}_+\Psi_{\lambda m}) = \hat{J}_+(\hat{J}^2\Psi_{\lambda m}) = \lambda\hbar^2(\hat{J}_+\Psi_{\lambda m}) \qquad (33.19)$$

and this shows that $\hat{J}_+$ and $\hat{J}_-$ leave the quantum number λ of $\Psi_{\lambda m}$ unchanged. There must be a limit to the number of times $\hat{J}_+$ can be applied in succession to $\Psi_{\lambda m}$ and this will be given by the highest value m, which can be set equal to an arbitrary integer $m = j$, such that

$$\hat{J}_+\Psi_{\lambda j} = 0 \qquad (33.20)$$

From Eqs.(33.13), (33.17) and (33.20) it follows that

$$(\hat{J}^2 - \hat{J}_z^2)\Psi_{\lambda j} = (\hat{J}_-\hat{J}_+ + \hbar\hat{J}_z)\Psi_{\lambda j} = j\hbar^2\Psi_{\lambda j} \quad \text{or} \quad \hat{J}^2\Psi_{\lambda j} = j(j+1)\hbar^2\Psi_{\lambda j} \quad (33.21)$$

In a similar manner, if $\hat{J}_-$ is successively applied to $\Psi_{\lambda m}$, there will be a lowest integer value of $m = k$, such that

$$\hat{J}_-\Psi_{\lambda j} = 0 \qquad (33.22)$$

so that, substituting Eq.(33.22), Eq.(33.17) gives

$$(\hat{J}^2 - \hat{J}_z^2)\Psi_{\lambda k} = (\hat{J}_+\hat{J}_- - \hbar\hat{J}_z)\Psi_{\lambda k} = -k\hbar^2\Psi_{\lambda k} \quad \text{or} \quad \hat{J}^2\Psi_{\lambda k} = k(k-1)\hbar^2\Psi_{\lambda k} \quad (33.23)$$

If compared to Eq.(33.21), Eq.(33.23) shows that the same eigenvalue $\lambda\hbar^2$ can be expressed in terms of either j or k, which yields

$$j(j+1) = k(k-1) \qquad (33.24)$$

Since, for a given eigenvalue $\lambda\hbar^2$, the eigenvalues of $\hat{J}_z$ are separated by integral multiples of $\hbar$, it follows that the difference between the highest and the lowest values of m must be some *positive* integer p, where $p = j - k$. Equation (33.24) gives

$$\begin{aligned} k &= -j \quad &\text{or} \quad p &= 2j \\ k &= j+1 \quad &\text{or} \quad p &= -1 \end{aligned} \qquad (33.25)$$

The latter solution is unacceptable, so that we must take $k = -j$, and this gives the eigenvalues of $\hat{J}^2$ as

$$\langle \Psi_{\lambda m} | \hat{J}^2 | \Psi_{\lambda m} \rangle = j(j+1)\hbar^2 \tag{33.26}$$

and the eigenvalues of $\hat{J}_z$ in the form

$$\langle \Psi_{\lambda m} | \hat{J}_z | \Psi_{\lambda m} \rangle = m\hbar \tag{33.27}$$

where $2j+1$ values for m are allowed, according to

$$-j \leq m \leq j \tag{33.28}$$

The quantum conditions (33.26) and (33.28) are analogous to those found for the angular momentum operator in Eqs.(31.35) and (31.43). However, Eq.(33.25) shows that, since p is an integer, *j can be either an integer or a half-integer*

$$j = 0, \tfrac{1}{2}, 1, \tfrac{3}{2}, \ldots$$

It is obvious that, if j assumes integer or half-integer values, so does m. The eigenstates in the common set for $\hat{J}^2$ and $\hat{J}_z$ are usually labelled by the quantum numbers j and m_j, rather than by λ and m, so that we write

$$\Psi_{jm_j} \equiv \Psi_{\lambda m}$$

The matrix elements of $\hat{J}^2$ and $\hat{J}_z$ follow from Eqs.(33.26) and (33.27) as

$$\left\langle \Psi_{j'm'_j} | \hat{J}^2 | \Psi_{jm_j} \right\rangle = j(j+1)\hbar^2 \delta_{jj'} \delta_{m_j m'_j}$$
$$\left\langle \Psi_{j'm'_j} | \hat{J}_z | \Psi_{jm_j} \right\rangle = m_j \hbar \delta_{jj'} \delta_{m_j m'_j} \tag{33.29}$$

since both matrices are diagonal. The complete set of functions Ψ_{jm_j} used for this representation consists of $2j+1$ eigenfunctions corresponding to different eigenvalues $m_j \hbar$, so that any two eigenfunctions corresponding to the same m_j must obey

$$\Psi'_{jm_j} = C_{m_j} \Psi_{jm_j}$$

where C_{m_j} is *real* and *positive*. Thus the only nonvanishing matrix elements in the $\hat{J}^2, \hat{J}_z$ representation follow from

$$\hat{J}_+ | \Psi_{jm_j} \rangle = | \Psi'_{j,m_j+1} \rangle = C_{m_j} | \Psi_{j,m_j+1} \rangle \tag{33.30}$$

and from the conjugate equation

$$\langle \Psi_{jm_j} | \hat{J}_- = C_{m_j} \langle \Psi_{j,m_j+1} | \qquad (33.31)$$

In view of Eq.(33.17), this leads to

$$\langle \Psi_{jm_j} | \hat{J}_- \hat{J}_+ | \Psi_{jm_j} \rangle = \langle \Psi_{jm_j} | \hat{J}^2 - \hat{J}_z^2 - \hbar \hat{J}_z | \Psi_{jm_j} \rangle = \left[j(j+1) - m_j(m_j+1) \right] \hbar^2 = C_{m_j}^2 \qquad (33.32)$$

Substituting Eq.(33.32), Eqs.(33.30) and (33.31) give the only nonvanishing matrix elements as

$$\langle \Psi_{jm_j+1} | \hat{J}_+ | \Psi_{jm_j} \rangle = C_{m_j} = \sqrt{j(j+1) - m_j(m_j+1)}\, \hbar$$

$$\langle \Psi_{jm_j-1} | \hat{J}_- | \Psi_{jm_j} \rangle = C_{m_j-1} = \sqrt{j(j+1) - m_j(m_j-1)}\, \hbar \qquad (33.33)$$

It follows in a straightforward manner, using Eq.(33.16), that the nonvanishing matrix elements of $\hat{J}_x$ and $\hat{J}_y$ have the form

$$\langle \Psi_{jm_j+1} | \hat{J}_x | \Psi_{jm_j} \rangle = i \langle \Psi_{jm_j+1} | \hat{J}_y | \Psi_{jm_j} \rangle = \frac{1}{2} C_{m_j} = \frac{1}{2}\sqrt{j(j+1) - m_j(m_j+1)}\, \hbar \qquad (33.34)$$

$$\langle \Psi_{jm_j-1} | \hat{J}_x | \Psi_{jm_j} \rangle = -i \langle \Psi_{jm_j-1} | \hat{J}_y | \Psi_{jm_j} \rangle = \frac{1}{2} C_{m_j-1} = \frac{1}{2}\sqrt{j(j+1) - m_j(m_j-1)}\, \hbar \qquad (33.35)$$

EXAMPLE 33.1. Orbital angular momentum matrices

Consider the particular case of an *orbital* angular momentum operator in a *p*-state, where $j = l = 1$ and $m_j = m = -1, 0, 1$. The matrix elements of $\hat{L}_x, \hat{L}_y, \hat{L}_z$ and $\hat{L}^2$ in the $\hat{L}^2, \hat{L}_z$ representation are given by Eqs.(33.29), (33.34) and (33.35), on substitution of the appropriate quantum numbers, as follows

$$
\begin{array}{c}
\quad m \quad\quad 1 \;\; 0 \;\; -1 \\
\begin{array}{c} 1 \\ 0 \\ -1 \end{array}
\hat{L}_z = \begin{pmatrix} 1 & 0 & 0 \\ 0 & 0 & 0 \\ 0 & 0 & -1 \end{pmatrix}\hbar, \;\;
\hat{L}^2 = \begin{pmatrix} 2 & 0 & 0 \\ 0 & 2 & 0 \\ 0 & 0 & 2 \end{pmatrix}\hbar^2, \;\;
\hat{L}_x = \begin{pmatrix} 0 & 1 & 0 \\ 1 & 0 & 1 \\ 0 & 1 & 0 \end{pmatrix}\frac{\hbar}{\sqrt{2}}, \;\;
\hat{L}_y = \begin{pmatrix} 0 & -i & 0 \\ i & 0 & -i \\ 0 & i & 0 \end{pmatrix}\frac{\hbar}{\sqrt{2}}
\end{array}
$$

$$(33.36)$$

It is easily confirmed that these matrices satisfy the commutation rules (31.2), (31.3) and (31.7). As the eigenfunctions of $\hat{L}^2$ and $\hat{L}_z$ are the spherical harmonics given by Eq.(31.44) as

$$\Psi_{1m} = Y_{1m}(\theta, \varphi)$$

the eigenfunctions of $\hat{L}_x$ and $\hat{L}_y$ can be derived in terms of the spherical harmonics

Y_{11}, Y_{10} and $Y_{1,-1}$ by means of their matrix representation. The matrix eigenvalue equation for $\hat{L}_x$ in a p-state reads

$$\frac{\hbar}{\sqrt{2}} \begin{pmatrix} 0 & 1 & 0 \\ 1 & 0 & 1 \\ 0 & 1 & 0 \end{pmatrix} \begin{pmatrix} a_1 \\ a_0 \\ a_{-1} \end{pmatrix} = m_x \hbar \begin{pmatrix} a_1 \\ a_0 \\ a_{-1} \end{pmatrix}$$

where

$$\Psi_{1m_x} = a_1 Y_{11} + a_0 Y_{10} + a_{-1} Y_{1,-1}$$

The eigenvalues are determined from the characteristic equation (33.11) which reduces to

$$\begin{vmatrix} -m_x & \frac{1}{\sqrt{2}} & 0 \\ \frac{1}{\sqrt{2}} & -m_x & \frac{1}{\sqrt{2}} \\ 0 & \frac{1}{\sqrt{2}} & -m_x \end{vmatrix} = -m_x^3 + m_x = 0$$

and this leads to the eigenvalues $0, \pm \hbar$ and then to the corresponding eigenfunctions

$$\Psi_{11} = \frac{1}{2}\left(Y_{11} + \sqrt{2} Y_{10} + Y_{1,-1}\right) \quad , \quad \Psi_{10} = \frac{1}{\sqrt{2}}\left(Y_{11} - Y_{1,-1}\right) \quad , \quad \Psi_{1,-1} = \frac{1}{2}\left(Y_{11} - \sqrt{2} Y_{10} + Y_{1,-1}\right)$$

If this set of functions is chosen to be a basis for the matrix representation, the matrix $\hat{L}_x$ becomes diagonal. Similar considerations apply to the case of $\hat{L}_y$.

33.3. SPIN ANGULAR MOMENTUM

The quantum operators have been introduced so far to be functions of coordinates and momenta derived from the classical expressions for the dynamical quantities according to Postulate 2. However, experiments have shown that there are physical observables that have no analogy in classical physics. Such quantities can be defined in terms of the characteristic properties of their associated operators.

Spin is an observable first detected for the electron which could not be explained in using classical concepts. If a beam of silver atoms is directed through an inhomogenous magnetic field of induction $\vec{B}$, whose direction provides the z-axis for spatial quantization, the coupling energy U_m between the magnetic moment $\vec{\mu}$ of the single outer electron of a silver atom and the magnetic field, as given by Eq.(32.34), allows us to express the net force (2.27) on the magnetic moment in the form

$$F_z = -\frac{\partial U_m}{\partial z} = -\frac{\partial}{\partial z}\left(-\mu_{lz}B_0\right) = \mu_{lz}\frac{\partial B_0}{\partial z} \qquad (33.37)$$

This force is the origin of the observed splitting of the beam into two components, as observed in the Stern-Gerlach experiment, illustrated in Figure 33.1.

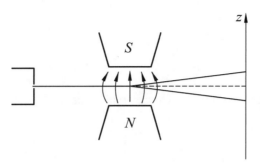

Figure 33.1. The Stern-Gerlach experiment

Since the atoms are in the ground state, with the outer electrons in an *s*-state, they have no orbital angular momentum, and hence, are not expected to exhibit any magnetic moment. The beam splitting into two components of equal intensity suggests, however, that all electrons have an intrinsic magnetic moment with the same magnitude and two possible orientations: parallel or antiparallel to the magnetic field. The origin of this observable can be understood if it is assumed that an electron has an intrinsic *spin angular momentum*, in addition to its orbital angular momentum. If the electron spin is characterized by two quantum numbers s and m_s, which are the analogues of l and m associated with the orbital angular momentum, the result of the experiment implies that the z-component of spin can take only two values $(2s+1=2)$, so that the quantum numbers are restricted to $s = \frac{1}{2}$ and $m_s = \pm\frac{1}{2}$. Therefore, the electron spin can be represented by an angular momentum operator $\hat{J} = \hat{S}$, defined in the special case $j = s = \frac{1}{2}$. Since there is no classical analogue of spin, its description may be introduced only as a matrix representation that gives the transformation properties of the eigenfunctions, in complete agreement with the experiment and with no concern about the analytical form of these functions.

As a manifestation of the spin as an angular momentum, the spin component operators $\hat{S}_x, \hat{S}_y$ and $\hat{S}_z$ and the total spin momentum operator $\hat{S}^2$ must satisfy the commutation relations (33.12), which read

$$\left[\hat{S}_i, \hat{S}_j\right] = i\hbar \hat{S}_k \quad , \quad \left[\hat{S}^2, \hat{S}_z\right] = 0 \qquad (33.38)$$

where i, j and k stand for x, y and z. In the $\hat{S}^2, \hat{S}_z$ representation, the eigenvalue equations to be solved read

$$\hat{S}^2 \chi = s(s+1)\hbar^2 \chi \quad , \quad \hat{S}_z \chi = m_s \hbar \chi \qquad (33.39)$$

where $s = \frac{1}{2}$ and $m_s = \pm\frac{1}{2}$. The spin matrices are obtained from Eqs.(33.29), (33.34) and (33.35), for $j = s = \frac{1}{2}$ and $m_j = m_s = \pm\frac{1}{2}$, as

$$\begin{array}{c} m_s \quad \tfrac{1}{2} \ -\tfrac{1}{2} \end{array}$$

$$\begin{array}{c} \tfrac{1}{2} \\ -\tfrac{1}{2} \end{array} \hat{S}_z = \begin{pmatrix} \tfrac{1}{2} & 0 \\ 0 & -\tfrac{1}{2} \end{pmatrix} \hbar = \tfrac{1}{2}\hbar \begin{pmatrix} 1 & 0 \\ 0 & -1 \end{pmatrix} = \tfrac{1}{2}\hbar\hat{\sigma}_z$$

$$\hat{S}_x = \begin{pmatrix} 0 & \tfrac{1}{2} \\ \tfrac{1}{2} & 0 \end{pmatrix} \hbar = \tfrac{1}{2}\hbar \begin{pmatrix} 0 & 1 \\ 1 & 0 \end{pmatrix} = \tfrac{1}{2}\hbar\hat{\sigma}_x \qquad (33.40)$$

$$\hat{S}_y = \begin{pmatrix} 0 & -\tfrac{1}{2}i \\ \tfrac{1}{2}i & 0 \end{pmatrix} \hbar = \tfrac{1}{2}\hbar \begin{pmatrix} 0 & -i \\ i & 0 \end{pmatrix} = \tfrac{1}{2}\hbar\hat{\sigma}_y$$

where $\hat{\sigma}_x, \hat{\sigma}_y$ and $\hat{\sigma}_z$ are known as the *Pauli matrices*. They obviously have the property (33.38), which reads

$$\left[\hat{\sigma}_i, \hat{\sigma}_j\right] = 2i\hat{\sigma}_k \qquad (33.41)$$

and also are *anticommuting*, that is

$$\hat{\sigma}_i\hat{\sigma}_j + \hat{\sigma}_j\hat{\sigma}_i = 0 \qquad (33.42)$$

Since

$$\hat{\sigma}_x^2 = \hat{\sigma}_y^2 = \hat{\sigma}_z^2 = 1 \qquad (33.43)$$

by combining Eqs.(33.42) and (33.43) we obtain the more compact form

$$\hat{\sigma}_i\hat{\sigma}_j + \hat{\sigma}_j\hat{\sigma}_i = 2\delta_{ij} \qquad (33.44)$$

Note that although Pauli matrices have been introduced here for the matrix representation of the spin operators, we have seen in Chapter 20 that they are also suitable for the description of other physical quantities that appear in two states only, such as the basic polarization states. The total spin is given by

$$\hat{S}^2 = \hat{S}_x^2 + \hat{S}_y^2 + \hat{S}_z^2 = \tfrac{1}{4}\hbar^2\left(\hat{\sigma}_x^2 + \hat{\sigma}_y^2 + \hat{\sigma}_z^2\right) = \tfrac{3}{4}\hbar^2 \begin{pmatrix} 1 & 0 \\ 0 & 1 \end{pmatrix} = \tfrac{3}{4}\hbar^2\hat{I} \qquad (33.45)$$

where $\hat{I}$ is the 2×2 unit matrix. Equations (33.40) and (33.45) show that the matrix equations for $\hat{S}^2$ and $\hat{S}_z$ have eigenvalues as postulated by Eqs.(33.39), that is,

$$s(s+1)\hbar^2 = \tfrac{1}{2}\left(\tfrac{1}{2}+1\right)\hbar^2 = \tfrac{3}{4}\hbar^2 \quad , \quad m_s\hbar = \pm\tfrac{1}{2}\hbar \qquad (33.46)$$

The common set of eigenfunctions consists of two component column vectors χ called *spinors*, which follow from Eq.(33.39) as

$$\hat{S}_z \chi = \tfrac{1}{2}\hbar \begin{pmatrix} 1 & 0 \\ 0 & -1 \end{pmatrix} \begin{pmatrix} a_+ \\ a_- \end{pmatrix} = \pm \tfrac{1}{2}\hbar \begin{pmatrix} a_+ \\ a_- \end{pmatrix} \qquad \text{i.e.} \qquad \begin{pmatrix} a_+ \\ -a_- \end{pmatrix} = \pm \begin{pmatrix} a_+ \\ a_- \end{pmatrix}$$

Hence, in the $\hat{S}^2, \hat{S}_z$ representation, the basis consists of the orthonormal spinors

$$\chi_+ = \begin{pmatrix} 1 \\ 0 \end{pmatrix} \quad , \quad \chi_- = \begin{pmatrix} 0 \\ 1 \end{pmatrix} \qquad (33.47)$$

and this defines the spatial quantization for electron spin illustrated in Figure 33.2, although the analytical form of the eigenfunctions is not specified.

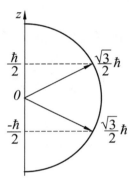

Figure 33.2. Quantization of the intrinsic spin angular momentum

An arbitrary spinor can be expanded using the complete set (33.47) in the form

$$\chi = \begin{pmatrix} a_+ \\ a_- \end{pmatrix} = a_+ \begin{pmatrix} 1 \\ 0 \end{pmatrix} + a_- \begin{pmatrix} 0 \\ 1 \end{pmatrix} = a_+ \chi_+ + a_- \chi_- = \sum_{m_s=-1/2}^{1/2} a_{m_s} \chi_{m_s} \qquad (33.48)$$

where, as a consequence of Postulate 3, $|a_{m_s}|^2$ can be interpreted to be the probability of obtaining the eigenvalue $m_s \hbar$ for $\hat{S}_z$ as a result of the measurement. The normalization condition reads

$$|a_+|^2 + |a_-|^2 = 1$$

EXAMPLE 33.2. Expectation value of spin components

The eigenfunctions of $\hat{S}_x$ with eigenvalues $\pm \tfrac{1}{2}\hbar$ are obtained from the matrix eigenvalue equation as

$$\hat{S}_x\chi = \tfrac{1}{2}\hbar \begin{pmatrix} 0 & 1 \\ 1 & 0 \end{pmatrix}\begin{pmatrix} a_+ \\ a_- \end{pmatrix} = \pm\hbar\begin{pmatrix} a_+ \\ a_- \end{pmatrix} \qquad \text{i.e.} \qquad \begin{pmatrix} a_- \\ a_+ \end{pmatrix} = \pm\begin{pmatrix} a_+ \\ a_- \end{pmatrix}$$

that is $a_+ = \pm a_-$. The normalization condition then gives $a_+ = a_- = 1/\sqrt{2}$ or $a_+ = -a_- = 1/\sqrt{2}$. The eigenfunctions follow from Eq.(33.48) as

$$\chi_+^x = \frac{1}{\sqrt{2}}(\chi_+ + \chi_-) = \frac{1}{\sqrt{2}}\begin{pmatrix} 1 \\ 1 \end{pmatrix} \quad , \quad \chi_-^x = \frac{1}{\sqrt{2}}(\chi_+ - \chi_-) = \frac{1}{\sqrt{2}}\begin{pmatrix} 1 \\ -1 \end{pmatrix}$$

Similarly, the eigenfunctions of $\hat{S}_y$ with eigenvalues $\pm\tfrac{1}{2}\hbar$ are

$$\chi_+^y = \frac{1}{\sqrt{2}}(\chi_+ + i\chi_-) = \frac{1}{\sqrt{2}}\begin{pmatrix} 1 \\ i \end{pmatrix} \quad , \quad \chi_-^y = \frac{1}{\sqrt{2}}(\chi_+ - i\chi_-) = \frac{1}{\sqrt{2}}\begin{pmatrix} 1 \\ -i \end{pmatrix}$$

The fact that $\chi_\pm^x, \chi_\pm^y$ and $\chi_\pm$ are all different means that no pair of $\hat{S}_x, \hat{S}_y$ and $\hat{S}_z$ observables are compatible, in accordance with the commutation rules (33.38) of the spin component operators. The expectation value of $\hat{S}_z$ is clearly equal to

$$\langle \hat{S}_z \rangle = |a_+|^2(\tfrac{1}{2}\hbar) + |a_-|^2(-\tfrac{1}{2}\hbar) = \tfrac{1}{2}\hbar(2|a_+|^2 - 1) = 0$$

in each of the eigenstates of $\hat{S}_x$ and $\hat{S}_y$ given above, which all have $a_+ = 1/\sqrt{2}$. This is consistent with the orientation of spin components along mutually perpendicular axes.

A rotation of spin coordinates through angle φ_0 about the z-axis that is in the xy-plane (see Figure 4.3) produces a new spin state in which the spin components $\hat{S}_x, \hat{S}_y, \hat{S}_z$ are replaced by $\hat{S}'_x, \hat{S}'_y, \hat{S}'_z$ such that, as in Eqs.(4.21), we obtain

$$\hat{S}_x = \hat{S}'_x \cos\varphi_0 - \hat{S}'_y \sin\varphi_0 \quad , \quad \hat{S}_y = \hat{S}'_x \sin\varphi_0 + \hat{S}'_y \cos\varphi_0 \quad , \quad \hat{S}_z = \hat{S}'_z \tag{33.49}$$

The eigenfunctions of $\hat{S}_x$ in terms of the new spin coordinates are obtained from the eigenvalue equation

$$\hat{S}_x \chi = (\hat{S}'_x \cos\varphi_0 - \hat{S}'_y \sin\varphi_0)\chi = \tfrac{1}{2}\hbar\begin{pmatrix} 0 & e^{i\varphi_0} \\ e^{-i\varphi_0} & 0 \end{pmatrix}\begin{pmatrix} a_+ \\ a_- \end{pmatrix} = \pm\tfrac{1}{2}\hbar\begin{pmatrix} a_+ \\ a_- \end{pmatrix}$$

which gives

$$\begin{pmatrix} a_- e^{i\varphi_0} \\ a_+ e^{-i\varphi_0} \end{pmatrix} = \pm\begin{pmatrix} a_+ \\ a_- \end{pmatrix}$$

so that the normalized eigenfunctions are given by

$$\chi_+^{\varphi_0} = \frac{1}{\sqrt{2}}\left(e^{i\varphi_0/2}\chi_+ + e^{-i\varphi_0/2}\chi_-\right) = \sum_{m_s=-1/2}^{1/2} e^{im_s\varphi_0} a_{m_s} \chi_{m_s}$$

$$\chi_-^{\varphi_0} = \frac{1}{\sqrt{2}}\left(e^{i\varphi_0/2}\chi_+ - e^{-i\varphi_0/2}\chi_-\right) = \sum_{m_s=-1/2}^{1/2} e^{im_s\varphi_0} a'_{m_s} \chi_{m_s}$$

(33.50)

It follows that the change of a spinor by rotation of spin coordinates through φ_0 about the z-axis can be expressed in terms of $\hat{S}_z$ as

$$\chi^{\varphi_0} = \sum_{m_s=-1/2}^{1/2} e^{im_s\varphi_0} a_{m_s} \chi_{m_s} = e^{i\hat{S}_z\varphi_0/\hbar} \chi \qquad (33.51)$$

In other words, the operator $e^{i\hat{S}_z\varphi_0/\hbar}$ is interpreted to be a spin rotation generator about the z-axis. This property is similar to that of the operator $e^{i\hat{L}_z\varphi_0/\hbar}$ which is used to express the change of spatial state functions $\Psi(x,y,z) = \Psi(r,\theta,\varphi)$ upon rotation of space coordinates through φ_0 about the z-axis. In this case, in view of Eqs.(4.21), we may use Taylor's expansion of the state function in the new coordinate system, that is

$$\Psi'(x,y,z) = \Psi(x'\cos\varphi_0 - y'\sin\varphi_0, x'\sin\varphi_0 + y'\cos\varphi_0, z)$$

$$= \Psi(x,y,z) + \sum_{n=1}^{\infty} \frac{\varphi_0^n}{n!}\left(\frac{\partial^n \Psi}{\partial \varphi_0^n}\right)_{\varphi_0=0}$$

where

$$\left(\frac{\partial \Psi}{\partial \varphi_0}\right)_{\varphi_0=0} = \left(\frac{\partial \Psi}{\partial x}\frac{\partial x}{\partial \varphi_0} + \frac{\partial \Psi}{\partial y}\frac{\partial y}{\partial \varphi_0}\right)_{\varphi_0=0} = \left(-y\frac{\partial \Psi}{\partial x} + x\frac{\partial \Psi}{\partial y}\right)_{\varphi_0=0} = \frac{i}{\hbar}\hat{L}_z\Psi$$

and use has been made of Eq.(29.16). It follows that

$$\Psi' = \left[1 + \sum_{n=1}^{\infty} \frac{1}{n!}(i\hat{L}_z\varphi_0/\hbar)^n\right]\Psi = e^{i\hat{L}_z\varphi_0/\hbar}\Psi \qquad (33.52)$$

which is analogous to Eq.(33.51). Note that a rotation through 2π about the z-axis, which is physically equivalent to the identity transformation, leaves any single-valued spatial eigenfunction of the orbital angular momentum $\Psi(r,\theta,\varphi)$ unchanged, whereas the rotation changes the sign of any spinor, as $e^{im_s 2\pi} = e^{\pm i\pi} = -1$. Although the spinors are double valued, no spin observable $S_i(i=x,y,z)$ is changed by a rotation through 2π, since the spinors that can be used to define the matrix elements of $\hat{S}_i$ either do not change sign or both change sign.

FURTHER READING

1. A. R. Edmonds - ANGULAR MOMENTUM IN QUANTUM MECHANICS, Princeton University Press, Princeton, 1996.
2. M. E. Rose - ELEMENTARY THEORY OF ANGULAR MOMENTUM, Dover Publications, New York, 1995.
3. M. Danos, V. Gillet - ANGULAR MOMENTUM CALCULUS IN QUANTUM PHYSICS, World Scientific, Singapore, 1990.

PROBLEMS

33.1. Solve the eigenvalue problem for the operator representing the component of the orbital angular momentum in a direction defined by its spherical polar coordinates θ and φ, given by

$$\hat{L}_r = \sin\theta\cos\varphi\,\hat{L}_x + \sin\theta\sin\varphi\,\hat{L}_y + \cos\theta\,\hat{L}_z$$

Solution

The eigenvalue problem reads

$$\hat{L}_r \psi_{l\lambda}(\theta,\varphi) = \lambda\hbar\,\psi_{l\lambda}(\theta,\varphi)$$

where $\psi_{l\lambda}$ can be expanded in terms of the spherical harmonics as

$$\psi_{l\lambda} = \sum_{m=-l}^{l} a_{lm} Y_{lm}(\theta,\varphi)$$

It is convenient to write the operator $\hat{L}_r$ in the form

$$\hat{L}_r = \frac{1}{2}\sin\theta\,e^{-i\varphi}\hat{L}_+ + \frac{1}{2}\sin\theta\,e^{i\varphi}\hat{L}_- + \cos\theta\,\hat{L}_z$$

and to use Eq.(33.33) which, for a given $j = l$ value, yields

$$\sum_{m=-l}^{l} a_m \left[\frac{1}{2}\sin\theta\,e^{-i\varphi}C_{lm}Y_{l,m+1} + \frac{1}{2}\sin\theta\,e^{i\varphi}C_{l,m-1}Y_{l,m-1} + (m\cos\theta - \lambda)Y_{lm}\right] = 0$$

where C_{lm} are given by Eq.(33.32). This can be rewritten as

$$\sum_{m=-l}^{l} \left[\frac{a_{m+1}}{2}\sin\theta\,e^{i\varphi}C_{lm} + a_m(m\cos\theta - \lambda) + \frac{a_{m-1}}{2}\sin\theta\,e^{-i\varphi}C_{l,m-1}\right] Y_{lm} = 0$$

where all the coefficients of the spherical harmonics must vanish, and this provides $2l + 1$ equations to be solved for the a_m. In the special case $l = 1$, for example, by taking $m = 1, 0, -1$ one obtains

$$a_1(\cos\theta - \lambda) + a_0 \frac{1}{\sqrt{2}} \sin\theta e^{-i\varphi} = 0$$

$$a_1 \frac{1}{\sqrt{2}} \sin\theta e^{i\varphi} - a_0 \lambda + a_{-1} \frac{1}{\sqrt{2}} \sin\theta e^{-i\varphi} = 0$$

$$a_0 \frac{1}{\sqrt{2}} \sin\theta e^{i\varphi} - a_{-1}(\cos\theta + \lambda) = 0$$

with the determinantal equation $-\lambda^3 + \lambda = 0$ which gives $\lambda = 0, \pm 1$. Hence the eigenvalues of $\hat{L}_r$ are $0, \pm\hbar$ and the corresponding normalized eigenfunctions are obtained by calculating a_1, a_0 and a_{-1} for each λ and then substituting into the expansion of $\psi_{1\lambda}$ which gives

$$\psi_{11} = \cos^2\frac{\theta}{2} e^{-i\varphi} Y_{11} + \frac{1}{\sqrt{2}} \sin\theta Y_{10} + \sin^2\frac{\theta}{2} e^{i\varphi} Y_{1,-1}$$

$$\psi_{10} = -\frac{1}{\sqrt{2}} \sin\theta e^{-i\varphi} Y_{11} + \cos\theta Y_{10} + \frac{1}{\sqrt{2}} \sin\theta e^{i\varphi} Y_{1,-1}$$

$$\psi_{1,-1} = -\sin^2\frac{\theta}{2} e^{-i\varphi} Y_{11} + \frac{1}{\sqrt{2}} \sin\theta Y_{10} - \cos^2\frac{\theta}{2} e^{i\varphi} Y_{1,-1}$$

33.2. Solve the eigenvalue problem for the operator

$$\hat{S}_r = \sin\theta\cos\varphi\,\hat{S}_x + \sin\theta\sin\varphi\,\hat{S}_y + \cos\theta\,\hat{S}_z$$

representing the spin component in a direction θ, φ.

Solution

It is convenient to use the representation (33.40), such that the matrix eigenvalue equation reads

$$\frac{\hbar}{2}\begin{pmatrix} \cos\theta & \sin\theta e^{-i\varphi} \\ \sin\theta e^{i\varphi} & -\cos\theta \end{pmatrix}\begin{pmatrix} a_+ \\ a_- \end{pmatrix} = \lambda\hbar\begin{pmatrix} a_+ \\ a_- \end{pmatrix}$$

and this is equivalent to the simultaneous equations

$$\left(\frac{\cos\theta}{2} - \lambda\right)a_+ + \frac{\sin\theta}{2}e^{-i\varphi}a_- = 0 \quad, \quad \frac{\sin\theta}{2}e^{i\varphi}a_+ - \left(\frac{\cos\theta}{2} + \lambda\right)a_- = 0$$

The characteristic equation immediately gives $\lambda = \pm 1/2$, as expected, such that the eigenvalues are $\pm\frac{1}{2}\hbar$, with corresponding spinors given by

$$\chi_+^r = \begin{pmatrix} \cos\frac{\theta}{2} e^{-i\varphi/2} \\ \sin\frac{\theta}{2} e^{i\varphi/2} \end{pmatrix}, \quad \chi_-^r = \begin{pmatrix} -\sin\frac{\theta}{2} e^{-i\varphi/2} \\ \cos\frac{\theta}{2} e^{i\varphi/2} \end{pmatrix}$$

33.3. A spin rotation generator through an angle α about the y-axis is represented by the operator $\hat{R}(\alpha) = e^{-i\alpha \hat{S}_y/\hbar}$. Find the matrix elements of this operator in the representation where $\hat{S}^2$ and $\hat{S}_z$ are diagonal.

Solution

Substituting the matrix form (33.40) one obtains

$$-\frac{i}{\hbar}\alpha \hat{S}_y = -\frac{i\alpha}{2}\begin{pmatrix} 0 & -i \\ i & 0 \end{pmatrix} = \frac{\alpha}{2}\begin{pmatrix} 0 & -1 \\ 1 & 0 \end{pmatrix} = \frac{\alpha}{2}\hat{A}$$

where $\hat{A}$ stands for a real matrix. Expanding the exponential gives

$$\hat{R}(\alpha) = e^{\alpha \hat{A}/2} = \hat{I} + \frac{\alpha}{2}\hat{A} + \frac{1}{2!}\left(\frac{\alpha}{2}\right)^2 \hat{A}^2 + \cdots + \frac{1}{n!}\left(\frac{\alpha}{2}\right)^n \hat{A}^n + \cdots$$

where $\hat{I}$ is the 2×2 unit matrix, and it is apparent that

$$\hat{A}^2 = \begin{pmatrix} 0 & -1 \\ 1 & 0 \end{pmatrix}^2 = -\begin{pmatrix} 1 & 0 \\ 0 & 1 \end{pmatrix} = -\hat{I}$$

This implies that

$$\hat{A}^{2n} = (-1)^n \hat{I} = (-1)^n \begin{pmatrix} 1 & 0 \\ 0 & 1 \end{pmatrix}, \quad \hat{A}^{2n+1} = (-1)^n \hat{A} = (-1)^n \begin{pmatrix} 0 & -1 \\ 1 & 0 \end{pmatrix}$$

It follows that

$$\hat{R}(\alpha) = \hat{I} + \sum_{n=0}^{\infty} \frac{1}{(2n)!}\left(\frac{\alpha}{2}\right)^{2n} \hat{A}^{2n} + \sum_{n=0}^{\infty} \frac{1}{(2n+1)!}\left(\frac{\alpha}{2}\right)^{2n+1} \hat{A}^{2n+1}$$

$$= \begin{pmatrix} 1 & 0 \\ 0 & 1 \end{pmatrix} + \begin{pmatrix} 1 & 0 \\ 0 & 1 \end{pmatrix}\sum_{n=0}^{\infty} \frac{(-1)^n}{(2n)!}\left(\frac{\alpha}{2}\right)^{2n} + \begin{pmatrix} 0 & -1 \\ 1 & 0 \end{pmatrix}\sum_{n=0}^{\infty} \frac{(-1)^n}{(2n+1)!}\left(\frac{\alpha}{2}\right)^{2n+1}$$

Thus the matrix elements have the form

$$R_{ii} = R_{jj} = \sum_{n=0}^{\infty} \frac{(-1)^n}{(2n)!}\left(\frac{\alpha}{2}\right)^{2n} = \cos\frac{\alpha}{2}, \quad R_{ij} = -R_{ji} = \sum_{n=0}^{\infty} \frac{(-1)^n}{(2n+1)!}\left(\frac{\alpha}{2}\right)^{2n+1} = \sin\frac{\alpha}{2}$$

and this provides the required matrix representation as

$$\hat{R}(\alpha) = \begin{pmatrix} \cos\dfrac{\alpha}{2} & -\sin\dfrac{\alpha}{2} \\ \sin\dfrac{\alpha}{2} & \cos\dfrac{\alpha}{2} \end{pmatrix}$$

33.4. For the operators $\hat{\sigma}_+$ and $\hat{\sigma}_-$, given by $\hat{\sigma}_\pm = (\hat{\sigma}_x \pm i\hat{\sigma}_y)/2$, prove that $(\hat{\sigma}_+\hat{\sigma}_-)^n = \hat{\sigma}_+\hat{\sigma}_-$ provided n is an integer.

Solution

It is straightforward to show that

$$\hat{\sigma}_+\hat{\sigma}_- = \frac{1}{4}(\hat{\sigma}_x + i\hat{\sigma}_y)(\hat{\sigma}_x - i\hat{\sigma}_y) = \frac{1}{4}\left[\hat{\sigma}_x^2 + \hat{\sigma}_y^2 + i(\hat{\sigma}_y\hat{\sigma}_x - \hat{\sigma}_x\hat{\sigma}_y)\right]$$

and hence, in view of Eqs.(33.41) and (33.43), one obtains

$$\hat{\sigma}_x\hat{\sigma}_y = \frac{1}{2}(1+\hat{\sigma}_z)$$

For $n = 2$ we have

$$(\hat{\sigma}_x\hat{\sigma}_y)^2 = \frac{1}{4}(1+\hat{\sigma}_z)^2 = \frac{1}{4}(1+2\hat{\sigma}_z+\hat{\sigma}_z^2) = \frac{1}{2}(1+\hat{\sigma}_z) = \hat{\sigma}_+\hat{\sigma}_-$$

and the required relation follows for an arbitrary n by mathematical induction.

33.5. Write down the matrix elements of the angular momentum operators $\hat{J}_x, \hat{J}_y$ and $\hat{J}_z$ in the $\hat{J}^2, \hat{J}_z$ representation where $j = \frac{3}{2}$. If the interaction on this angular momentum is described by the Hamiltonian

$$\hat{H} = A\hat{J}_x^2 + B\hat{J}_y^2 - (A+B)\hat{J}_z^2$$

find the energy eigenvalues.

Solution

There is a complete set consisting of $2j+1 = 4$ eigenfunctions

$$\Psi_{\frac{3}{2},\frac{3}{2}}, \Psi_{\frac{3}{2},\frac{1}{2}}, \Psi_{\frac{3}{2},-\frac{1}{2}}, \Psi_{\frac{3}{2},-\frac{3}{2}}$$

that correspond to the eigenvalues $m_j\hbar$ and should be used for this representation. In view of Eqs.(33.34) and (33.35), the matrices can be written as

$$\hat{J}_x = \begin{pmatrix} 0 & \sqrt{3}/2 & 0 & 0 \\ \sqrt{3}/2 & 0 & 1 & 0 \\ 0 & 1 & 0 & \sqrt{3}/2 \\ 0 & 0 & \sqrt{3}/2 & 0 \end{pmatrix}\hbar \quad , \quad \hat{J}_y = \begin{pmatrix} 0 & -i\sqrt{3}/2 & 0 & 0 \\ i\sqrt{3}/2 & 0 & -i & 0 \\ 0 & i & 0 & -i\sqrt{3}/2 \\ 0 & 0 & i\sqrt{3}/2 & 0 \end{pmatrix}\hbar$$

$$\hat{J}_z = \begin{pmatrix} 3/2 & 0 & 0 & 0 \\ 0 & 1/2 & 0 & 0 \\ 0 & 0 & -1/2 & 0 \\ 0 & 0 & 0 & -3/2 \end{pmatrix}\hbar$$

Squaring these matrices is straightforward and leads to the following Hamiltonian

$$\hat{H} = \begin{pmatrix} -\frac{3}{2}(A+B) & 0 & \frac{\sqrt{3}}{2}(A-B) & 0 \\ 0 & \frac{3}{2}(A+B) & 0 & \frac{\sqrt{3}}{2}(A-B) \\ \frac{\sqrt{3}}{2}(A-B) & 0 & \frac{3}{2}(A+B) & 0 \\ 0 & \frac{\sqrt{3}}{2}(A-B) & 0 & -\frac{3}{2}(A+B) \end{pmatrix}\hbar^2$$

In it now apparent that a more convenient form is obtained by rearranging the basis functions as

$$\Psi_{\frac{3}{2}\frac{3}{2}}, \Psi_{\frac{3}{2},-\frac{1}{2}}, \Psi_{\frac{3}{2},-\frac{3}{2}} \Psi_{\frac{3}{2}\frac{1}{2}}$$

so that the Hamiltonian can be written as

$$\hat{H} = \begin{pmatrix} -\frac{3}{2}(A+B) & \frac{\sqrt{3}}{2}(A-B) & 0 & 0 \\ \frac{\sqrt{3}}{2}(A-B) & \frac{3}{2}(A+B) & 0 & 0 \\ 0 & 0 & -\frac{3}{2}(A+B) & \frac{\sqrt{3}}{2}(A-B) \\ 0 & 0 & \frac{\sqrt{3}}{2}(A-B) & \frac{3}{2}(A+B) \end{pmatrix}\hbar^2$$

The characteristic equation immediately gives the energy levels

$$\varepsilon = \pm\hbar^2\sqrt{3(A^2 + B^2 + AB)}$$

which are two-fold degenerate.

33.6. Find the expectation value of the operator

$$\hat{J}_r = \sin\theta\cos\varphi\hat{J}_x + \sin\theta\sin\varphi\hat{J}_y + \cos\theta\hat{J}_z$$

in a common eigenstate Ψ_{jm_j} of $\hat{J}^2$ and $\hat{J}_z$.

33.7. If Ψ_{jm_j} is an eigenstate of both $\hat{J}^2$ and $\hat{J}_z$, find the m_j value that ensures the minimum uncertainty for the simultaneous determination of the J_x and J_y components of angular momentum.

33.8. Prove that if an operator commutes with two of the angular momentum component operators it will also commute with the third component operator.

33.9. Use the Pauli matrix representation of the spin operators to derive the magnitude of the square of the spin component in an arbitrary direction $\vec{e}_r$.

33.10. Prove the identity

$$\left(\hat{\vec{\sigma}}\cdot\hat{\vec{A}}\right)\left(\hat{\vec{\sigma}}\cdot\hat{\vec{B}}\right)=\left(\hat{\vec{A}}\cdot\hat{\vec{B}}\right)\hat{I}+i\hat{\vec{\sigma}}\cdot\left(\hat{\vec{A}}\times\hat{\vec{B}}\right)$$

where $\hat{\vec{A}}(\hat{A}_x,\hat{A}_y,\hat{A}_z)$ and $\hat{\vec{B}}(\hat{B}_x,\hat{B}_y,\hat{B}_z)$ are vectorial operators, $\hat{\vec{\sigma}}(\hat{\sigma}_x,\hat{\sigma}_y,\hat{\sigma}_z)$ denotes the Pauli matrices, and $\hat{I}$ is the 2×2 unit matrix.

34. THE SPINNING ELECTRON

34.1. ADDITION OF ANGULAR MOMENTA
34.2. SPIN-ORBIT INTERACTION
34.3. THE SPINNING ELECTRON IN EXTERNAL MAGNETIC FIELD

34.1. ADDITION OF ANGULAR MOMENTA

A total angular momentum can be associated with operators that represent a simultaneous rotation of both space and spin coordinates, in other words with the change of state functions that reflect both spatial and spin properties. The angular momentum states can be written in the form

$$\Phi_{nlmm_s} = \Psi_{nlm}(r,\theta,\varphi)\chi = \Psi_{nlm}a_+\chi_+ + \Psi_{nlm}a_-\chi_- = a_+\begin{pmatrix}\Psi_{nlm}\\0\end{pmatrix} + a_-\begin{pmatrix}0\\\Psi_{nlm}\end{pmatrix} \quad (34.1)$$

Since the spatial and spin parts of Eq.(34.1) obey Eqs.(33.51) and (33.52), respectively, the change of the total state function under a rotation through an angle φ_0 about the z-axis can be represented as

$$\Phi'_{nlmm_s} = \Psi_{nlm}(r,\theta,\varphi+\varphi_0)\chi^{\varphi_0} = e^{i(\hat{L}_z+\hat{S}_z)\varphi_0/\hbar}\Psi_{nlm}(r,\theta,\varphi)\chi = e^{i\hat{J}_z\varphi_0/\hbar}\Phi_{nlmm_s} \quad (34.2)$$

Let us introduce the *total angular momentum operator*, defined as

$$\hat{\vec{J}} = \hat{\vec{L}} + \hat{\vec{S}} \quad (34.3)$$

with components that satisfy the commutation relations (33.12), since $\hat{\vec{L}}$ depends on spatial coordinates of which $\hat{\vec{S}}$ is independent, so that they commute. Equation (34.3) shows that a rotation of both space and spin coordinates about the z-axis can be represented in terms of the $\hat{J}_z$ component of the total angular momentum. The angular momentum states of a system can be defined in terms of a set of common eigenfunctions

$$\Phi_{nlmm_s} \equiv \Phi_{mm_s}$$

corresponding to the eigenvalues $l(l+1)\hbar^2$, $m\hbar$, $s(s+1)\hbar^2$, $m_s\hbar$ of the four operators $\hat{L}^2, \hat{L}_z, \hat{S}^2, \hat{S}_z$ which represent compatible observables. There are $2l+1$ allowed values of m and two values of m_s, for given l and s, and hence, $2(2l+1)$ possible eigenfunctions Φ_{mm_s}. Squaring Eq.(34.3) gives

$$\hat{J}^2 = \hat{L}^2 + \hat{S}^2 + 2\hat{\vec{L}} \cdot \hat{\vec{S}} \tag{34.4}$$

which shows that the observables L^2, S^2, J^2 and J_z are also compatible. As a result, the states of the total angular momentum can also be represented in terms of the simultaneous eigenfunctions

$$\Psi_{nljm_j} \equiv \Psi_{jm_j}$$

where j and m_j are quantum numbers defined by the eigenvalue equations

$$\hat{J}^2 \Psi_{jm_j} = j(j+1)\hbar^2 \Psi_{jm_j}$$
$$\hat{J}_z \Psi_{jm_j} = m_j \hbar \Psi_{jm_j} \tag{34.5}$$

As there are $2j+1$ possible values of m_j for each value of j compatible with given l and s, the total number of eigenfunctions is

$$\sum_j (2j+1) = 2(2l+1) \tag{34.6}$$

since the number of angular momentum states must be the same in both representations. When the z-components of $\hat{\vec{L}}$ and $\hat{\vec{S}}$ are diagonal, so is that of $\hat{\vec{J}}$, and we may write

$$\hat{J}_z \Phi_{mm_s} = (m+m_s)\hbar \Phi_{mm_s}$$

and this, by consideration of Eq.(34.5), gives

$$m_j = m + m_s \tag{34.7}$$

Since the maximum values of m and m_s are l and s, the maximum m_j will be $l+s$, so that the condition $|m_j| \leq j$ yields

$$j_{max} = (m+m_s)_{max} = l+s \tag{34.8}$$

and the corresponding eigenfunction is

$$\Psi_{l+s,l+s} = \Phi_{ls} = R_n(r)Y_{ll}(\theta,\varphi)\chi$$

For $m_j = l+s-1$, there are two mutually orthogonal Φ_{mm_s} states written as

$$\Phi_{l-1,s} \quad , \quad \Phi_{l,s-1}$$

and these can be combined to give two linearly independent Ψ_{jm_j} states corresponding to the possible choices

$$j = l+s \quad , \quad m_j = l+s-1 = j-1 \quad : \quad \Psi_{l+s,l+s-1}$$

$$j = l+s-1 \quad , \quad m_j = l+s-1 = j \quad : \quad \Psi_{l+s-1,l+s-1}$$

For $m_j = l+s-2$ there will be three mutually orthogonal Φ_{mm_s} states denoted by

$$\Phi_{l,s-2} \quad , \quad \Phi_{l-1,s-1} \quad , \quad \Phi_{l-2,s}$$

from which three linear combinations can be built, each with a given m_j, of the form

$$j = l+s \quad , \quad m_j = l+s-2 = j-2 \quad : \quad \Psi_{l+s,l+s-2}$$

$$j = l+s-1 \quad , \quad m_j = l+s-2 = j-1 \quad : \quad \Psi_{l+s-1,l+s-2}$$

$$j = l+s-2 \quad , \quad m_j = l+s-2 = j \quad : \quad \Psi_{l+s-2,l+s-2}$$

This process may be continued until the minimum value of j, which also is the minimum value of $|m_j|$, is reached as given by

$$j_{\min} = l - s \tag{34.9}$$

so that the possible values of j are

$$l - s \leq j \leq l + s \tag{34.10}$$

The relation between the two representations of the total angular momentum eigenstates is usually given by linking the two sets of eigenfunctions Ψ_{jm_j} and Φ_{mm_s} in the form

$$\Psi_{jm_j} = \sum_m \sum_{m_s} C_{lsmm_s}^{jm_j} \Phi_{mm_s} \tag{34.11}$$

where the coefficients $C_{lsmm_s}^{jm_j}$ are called the *Clebsch-Gordan coefficients*.

EXAMPLE 34.1. Representation of angular momenta for $l=1$ and $s=\frac{1}{2}$

Consider a spinning electron with $s=\frac{1}{2}$ in a *p*-state of the one-electron atom where $l=1$. In a classical vector representation we may think of the electron as having both the orbital and spin angular momenta, with each producing a magnetic field due to the electron charge. We recall that the torque exerted by a magnetic field on a magnetic dipole moment will cause the axis of the angular momentum to precess around the field direction with a constant polar orientation. Thus the quantum condition (34.3) can be represented as in Figure 34.1, which shows the angular momenta $\vec{L}$ and $\vec{S}$ precessing about the common axis of their mechanical resultant $\vec{J}$. The opening angles of the precession cones are so adjusted that the resultant has the length corresponding to the possible values $j=l+\frac{1}{2}=\frac{3}{2}$ and $j=l-\frac{1}{2}=\frac{1}{2}$ given by Eq.(34.10).

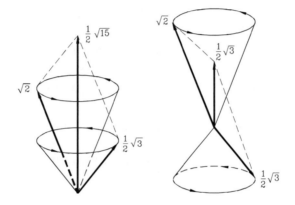

Figure 34.1. Precession of $\vec{L}$ and $\vec{S}$ and $\vec{J}$ for a spinning electron

There are six possible states of the angular momentum, as given by Eq.(34.6) which reads

$$\left(2\cdot\frac{3}{2}+1\right)+\left(2\cdot\frac{1}{2}+1\right)=2(2\cdot 1+1)=6$$

The six possible Φ_{mm_s} states of angular momentum

$$\Phi_{1,\frac{1}{2}},\Phi_{1,-\frac{1}{2}},\Phi_{0,\frac{1}{2}},\Phi_{0,-\frac{1}{2}},\Phi_{-1,\frac{1}{2}},\Phi_{-1,-\frac{1}{2}}$$

are related to the spatial quantization given in Figure 34.2, where the common axis of precession is taken to be the z-axis for both m and m_s. The quantum condition (34.7) allows us to give an alternative representation of the same states, in term of the six functions

$$\Psi_{\frac{3}{2}\frac{3}{2}},\Psi_{\frac{3}{2},-\frac{3}{2}},\Psi_{\frac{3}{2}\frac{1}{2}},\Psi_{\frac{3}{2},-\frac{1}{2}},\Psi_{\frac{1}{2}\frac{1}{2}},\Psi_{\frac{1}{2},-\frac{1}{2}}$$

which correspond to the spatial quantization of $\vec{J}$ in Figure 34.2. Equation (34.7) shows that

$$\Psi_{\frac{3}{2}\frac{3}{2}}=\Phi_{1,\frac{1}{2}}=R_n(r)Y_{11}(\theta,\varphi)\chi_+ \quad , \quad \Psi_{\frac{3}{2},-\frac{3}{2}}=\Phi_{-1,-\frac{1}{2}}=R_n(r)Y_{1,-1}(\theta,\varphi)\chi_-$$

whereas each of the remaining four states is given by a linear combination (34.11) of Φ_{mm_s}.

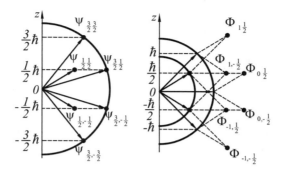

Figure 34.2. Spatial quantization of the total angular momentum

Using Eq.(33.33) we have

$$\hat{J}_-\Psi_{\frac{3}{2},\frac{3}{2}} = \sqrt{\tfrac{3}{2}(\tfrac{3}{2}+1) - \tfrac{3}{2}(\tfrac{3}{2}-1)}\,\hbar\Psi_{\frac{3}{2},\frac{1}{2}} = \sqrt{3}\hbar\Psi_{\frac{3}{2},\frac{1}{2}} \quad \text{or} \quad \Psi_{\frac{3}{2},\frac{1}{2}} = \frac{1}{\sqrt{3}\hbar}\hat{J}_-\Psi_{\frac{3}{2},\frac{3}{2}} = \frac{1}{\sqrt{3}\hbar}(\hat{L}_- + \hat{S}_-)\Phi_{1,\frac{1}{2}}$$

Now from Eq.(33.33) we also obtain

$$\hat{L}_-\Phi_{1,\frac{1}{2}} = \sqrt{1(1+1) - 1(1-1)}\,\hbar\Phi_{0,\frac{1}{2}} = \sqrt{2}\hbar\Phi_{0,\frac{1}{2}} \quad , \quad \hat{S}_-\Phi_{1,\frac{1}{2}} = \sqrt{\tfrac{1}{2}(\tfrac{1}{2}+1) - \tfrac{1}{2}(\tfrac{1}{2}-1)}\,\hbar\Phi_{1,-\frac{1}{2}} = \hbar\Phi_{1,-\frac{1}{2}}$$

and hence

$$\Psi_{\frac{3}{2},\frac{1}{2}} = \frac{1}{\sqrt{3}}\left(\Phi_{1,-\frac{1}{2}} + \sqrt{2}\Phi_{0,\frac{1}{2}}\right) = \frac{R_n(r)}{\sqrt{3}}\left(Y_{11}\chi_- + \sqrt{2}Y_{10}\chi_+\right)$$

Similarly, since

$$\hat{J}_-\Psi_{\frac{3}{2},\frac{1}{2}} = \sqrt{\tfrac{3}{2}(\tfrac{3}{2}+1) - \tfrac{1}{2}(\tfrac{1}{2}-1)}\,\hbar\Psi_{\frac{3}{2},-\frac{1}{2}} = 2\hbar\Psi_{\frac{3}{2},-\frac{1}{2}}$$

it follows that

$$\Psi_{\frac{3}{2},-\frac{1}{2}} = \frac{1}{2\sqrt{3}\hbar}\hat{J}_-\left(\Phi_{1,-\frac{1}{2}} + \sqrt{2}\Phi_{0,\frac{1}{2}}\right) = \frac{1}{\sqrt{3}}\left(\sqrt{2}\Phi_{0,-\frac{1}{2}} + \Phi_{-1,\frac{1}{2}}\right) = \frac{R_n(r)}{\sqrt{3}}\left(\sqrt{2}Y_{10}\chi_- + Y_{1,-1}\chi_+\right)$$

Under the restriction of $m + m_s = m_j$, Eq.(34.11) gives

$$\Psi_{\frac{1}{2},\frac{1}{2}} = C_1\Phi_{1,-\frac{1}{2}} + C_2\Phi_{0,\frac{1}{2}} \quad , \quad \Psi_{\frac{1}{2},-\frac{1}{2}} = C_1'\Phi_{0,-\frac{1}{2}} + C_2'\Phi_{-1,\frac{1}{2}}$$

Since $\Psi_{\frac{1}{2},\frac{1}{2}}$ is expressed in terms of the same Φ_{mm_s} states as $\Psi_{\frac{3}{2},\frac{1}{2}}$, their orthogonality requires that $C_1 + \sqrt{2}C_2 = 0$, and the normalization condition of $\Psi_{\frac{1}{2},\frac{1}{2}}$ gives $C_1^2 + C_2^2 = 1$. It follows that

$$\Psi_{\frac{1}{2},\frac{1}{2}} = \frac{1}{\sqrt{3}}\left(\sqrt{2}\Phi_{1,-\frac{1}{2}} - \Phi_{0,\frac{1}{2}}\right) = \frac{R_n(r)}{\sqrt{3}}\left(\sqrt{2}Y_{11}\chi_- - Y_{10}\chi_+\right)$$

Similarly $\Psi_{\frac{1}{2},-\frac{1}{2}}$, which is orthogonal to $\Psi_{\frac{3}{2},-\frac{1}{2}}$, is obtained as

$$\Psi_{\frac{1}{2},-\frac{1}{2}} = \frac{1}{\sqrt{3}}\left(\Phi_{0,-\frac{1}{2}} - \sqrt{2}\Phi_{-1,\frac{1}{2}}\right) = \frac{R_n(r)}{\sqrt{3}}\left(Y_{10}\chi_- - \sqrt{2}Y_{1,-1}\chi_+\right)$$

The representation of the total angular momentum eigenstates by two different sets of good quantum numbers can be understood in terms of two different schemes of coupling the orbital and spin angular momenta in a *vector model* of atoms. If we place the one electron atom into a magnetic field which is parallel to the z-direction, the precession of $\vec{L}$ and $\vec{S}$ or $\vec{J}$ around the z-axis results in a constant value of their z component, which must be one of the possible eigenvalues $(m+m_s)\hbar$ or $m_j\hbar$.

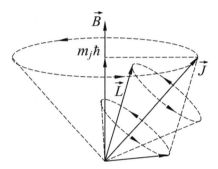

Figure 34.3. Precession of $\vec{J}$ around the field direction with $m_j\hbar$ as a constant of motion

The Ψ_{jm_j} representation of the total angular momentum implies that the operators $\hat{H}, \hat{L}^2, \hat{S}^2, \hat{J}^2$ and $\hat{J}_z$ commute, thus there is a set of common eigenfunctions with the following properties

$$\hat{H}\Psi_{jm_j} = \varepsilon_n \Psi_{jm_j} \quad , \quad \hat{L}^2\Psi_{jm_j} = l(l+1)\hbar^2\Psi_{jm_j} \quad , \quad \hat{S}^2\Psi_{jm_j} = s(s+1)\hbar^2\Psi_{jm_j}$$
(34.12)
$$\hat{J}^2\Psi_{jm_j} = j(j+1)\hbar^2\Psi_{jm_j} \quad , \quad \hat{J}_z\Psi_{jm_j} = m_j\hbar\Psi_{jm_j}$$

This description corresponds to an applied magnetic field that is weak compared to the extremely strong internal fields in the atom, so that the total angular momentum will precess as a whole and will adjust its polar orientation in such a way that its z component has one of the allowed values $m_j\hbar$. The good quantum numbers are l, s, j and m_j, since $m\hbar$ and $m_s\hbar$ vary in time as a result of precession, as illustrated in Figure 34.3, so that Ψ_{jm_j} are the suitable eigenfunctions.

For an applied field stronger than the internal fields inside the atom, the coupling interaction between $\vec{L}$ and $\vec{S}$ is negligible and each of them adjusts so that their z component becomes a constant of motion. This is shown in Figure 34.4 where $\vec{L}$ and $\vec{S}$

precess independently about the z-axis. This implies that only $\hat{L}^2, \hat{S}^2, \hat{L}_z$ and $\hat{S}_z$ are quantized for strong fields, and therefore, m and m_s are good quantum numbers.

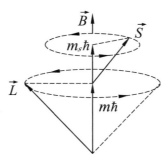

Figure 34.4. Independent precession of $\vec{L}$ and $\vec{S}$ around the field direction, with $m\hbar$ and $m_s\hbar$ as constants of motion

The eigenstates of the angular momentum can be described by the set of common eigenfunctions Φ_{mm_s} of the operators $\hat{H}, \hat{L}^2, \hat{S}^2, \hat{L}_z$ and $\hat{S}_z$ that obey

$$\hat{H}\Phi_{mm_s} = \varepsilon_n \Phi_{mm_s} \quad , \quad \hat{L}^2\Phi_{mm_s} = l(l+1)\hbar^2\Phi_{mm_s} \quad , \quad \hat{S}^2\Phi_{mm_s} = s(s+1)\hbar^2\Phi_{mm_s}$$
(34.13)
$$\hat{L}_z\Phi_{mm_s} = m\hbar\Phi_{mm_s} \quad , \quad \hat{S}_z\Phi_{mm_s} = m_s\hbar\Phi_{mm_s}$$

Note that in both cases described by Eqs.(34.12) or (34.13) $\hat{H}$ denotes the central force Hamiltonian, defined by Eq.(31.1).

34.2. Spin-Orbit Interaction

Because the electron is spinning, it will possess an intrinsic *spin magnetic moment*. Since, at present, we do not have a complete understanding of the structure of the electron, we must assume that, as in the case of orbital motion, the spin magnetic moment $\vec{\mu}_s$ is proportional to the spin angular momentum $\vec{S}$. Hence, we write

$$\vec{\mu}_s = -g_s \frac{e}{2m_e} \vec{S}$$
(34.14)

where the factor g_s is to be determined. If the spin were a classical angular momentum, g_s should have been unity, like the orbital giromagnetic factor introduced in Eq.(32.35).

However, the value of g_s which fits the experimental observation is $g_s = 2$, provided we neglect some very small corrections. In a magnetic field, the dipole moment (34.14) acquires an energy $U_m = -\vec{\mu}_s \cdot \vec{B}$ that has the same form as that found for the orbital magnetic moment in Eq.(32.34). Since the electron is moving through the electric field of the nucleus, as illustrated in Figure 34.5, it always experiences the effect of an internally generated magnetic field $\vec{B}_i$ of the atom.

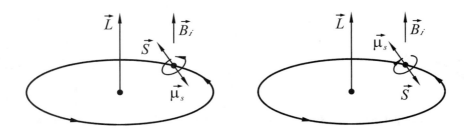

Figure 34.5. The spin and orbital motions of the electron

If $\vec{r}$ and $\vec{v}$ stand for the position and velocity of the electron with respect to the nucleus $+Ze$, the nucleus appears to the electron to be moving around with a velocity $-\vec{v}$, so that the magnetic field experienced by the electron is given by Eq.(12.17) as

$$\vec{B}_i = \frac{\mu_0}{4\pi} \frac{Ze(\vec{r} \times \vec{v})}{r^3} = \frac{Ze}{4\pi\varepsilon_0 m_e c^2} \frac{\vec{r} \times m_e \vec{v}}{r^3} = \frac{\vec{L}}{m_e ec^2} \frac{1}{r} \frac{d}{dr}\left(\frac{-Ze_0^2}{r}\right) = \frac{1}{m_e ec^2} \frac{1}{r} \frac{dU(r)}{dr} \vec{L}$$

(34.15)

where the potential energy $U(r)$ given by Eq.(32.1) has been inserted. The scalar product $-\vec{\mu}_s \cdot \vec{B}_i$ gives the increment in the potential energy of the spinning electron in the frame of reference where the electron is at rest. However, the interaction energy must be multiplied by a factor of one half, called the Thomas factor, if measured in the frame of reference where the nucleus is at rest, as a result of relativistic kinematical considerations. It follows that the interaction energy becomes

$$-\frac{1}{2}\vec{\mu}_s \cdot \vec{B}_i = -\frac{1}{2}\left(-\frac{e}{m_e}\vec{S}\right) \cdot \left(\frac{1}{m_e ec^2} \frac{1}{r} \frac{dU(r)}{dr}\vec{L}\right) = \frac{1}{2m_e^2 c^2} \frac{1}{r} \frac{dU(r)}{dr} \vec{S} \cdot \vec{L} \quad (34.16)$$

This result gives the *spin-orbit coupling* term of the spinning electron Hamiltonian, which becomes

$$\hat{H} = \hat{H}_0 + \frac{1}{2m_e^2 c^2} \frac{1}{r} \frac{dU(r)}{dr} \hat{\vec{S}} \cdot \hat{\vec{L}}$$

(34.17)

where $\hat{H}_0$ denotes the central force Hamiltonian (31.1) in the absence of spin. For the eigenstates of the Hamiltonian (34.17), m and m_s are not good quantum numbers, as the components of $\hat{\vec{L}}$ or $\hat{\vec{S}}$ are not compatible with the spin-orbit term. Instead, from Eq.(34.4), we have

$$\hat{\vec{S}} \cdot \hat{\vec{L}} = \frac{1}{2}\left(\hat{J}^2 - \hat{L}^2 - \hat{\vec{S}}^2\right)$$

and this gives

$$\left[\hat{\vec{S}} \cdot \hat{\vec{L}}, \hat{J}^2\right] = \frac{1}{2}\left(\left[\hat{J}^2, \hat{J}^2\right] - \left[\hat{L}^2, \hat{J}^2\right] - \left[\hat{\vec{S}}^2, \hat{J}^2\right]\right) = 0 \quad \text{and} \quad \left[\hat{\vec{S}} \cdot \hat{\vec{L}}, \hat{J}_z\right] = 0$$

This implies that the appropriate eigenfunctions for the Hamiltonian (34.17) are the Ψ_{jm_j}, Eqs.(34.12). Substituting into Eq.(34.17), the energy eigenvalues result from

$$\hat{H}\Psi_{jm_j} = \left[\hat{H}_0 + \frac{1}{2m_e^2 c^2}\frac{1}{r}\frac{dU(r)}{dr}\left(\hat{J}^2 - \hat{L}^2 - \hat{\vec{S}}^2\right)\right]\Psi_{jm_j} \tag{34.18}$$

which gives

$$\varepsilon_{nlj}\Psi_{jm_j} = \varepsilon_n \Psi_{jm_j} + \frac{1}{4m_e^2 c^2}\left[j(j+1) - l(l+1) - s(s+1)\right]\hbar^2 \frac{1}{r}\frac{dU(r)}{dr}\Psi_{jm_j}$$

Using Ψ_{jm}^* as a multiplying factor in front of each term and integrating gives

$$\varepsilon_{nlj} = \varepsilon_n + \frac{1}{2}\left[j(j+1) - l(l+1) - s(s+1)\right]\zeta_{nl} \tag{34.19}$$

where ζ_{nl} stands for the expectation value

$$\zeta_{nl} = \left\langle \Psi_{jm_j} \left| \frac{\hbar^2}{2m_e^2 c^2}\frac{1}{r}\frac{dU(r)}{dr} \right| \Psi_{jm_j} \right\rangle \tag{34.20}$$

Equation (34.20) reduces to an average over the radial eigenfunctions $R_{nl}(r)$, given by Eq.(32.13), so that ζ_{nl} depends on n and l only. If $l \neq 0$ there are two possible values of j for the spinning electron, namely, $j = l \pm \frac{1}{2}$. It follows that the spin-orbit interaction splits a state with given quantum numbers n and l and energy E_{nl} into two states with energies

$$\varepsilon_{nlj} = \varepsilon_n + \frac{1}{2}l\zeta_{nl} \quad , \quad j = l + \frac{1}{2}$$

$$= \varepsilon_n - \frac{1}{2}(l+1)\zeta_{nl} \quad , \quad j = l - \frac{1}{2} \tag{34.21}$$

States with $l = 0$ are not affected, but the splitting of all the other states is proportional to $2l + 1$. Thus the spin-orbit coupling completely removes the accidental degeneracy.

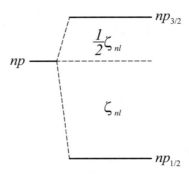

Figure 34.6. Fine structure of a np-state due to spin-orbit coupling

Eigenstates Ψ_{jm_j} of the spinning electron are labelled in the spectroscopic notation by n, l and j in the form

$$n(l\ \text{symbol})_j$$

with the usual symbols for the allowed values of l. An example is given in Figure 34.6, where the fine structure of a np-state (with $l=1$ for a spinning electron) is a result of splitting into a $np_{3/2}$ – state for which $l=1$ and $j=\frac{3}{2}$ and an $np_{1/2}$ – state for which $l=1$, $j=\frac{1}{2}$. When more electrons are present, the resultant orbital, spin and total angular momenta are labelled by capitals, and the notation of the atomic state is

$$^{2S+1}(L\ \text{symbol})_J$$

The superscript $2S+1$ denotes the multiplicity of the level of given L.

34.3. THE SPINNING ELECTRON IN EXTERNAL MAGNETIC FIELD

In the presence of an applied magnetic field $\vec{B}_0$, the energy of the spinning electron in central field is obtained by combining Eqs.(32.34) and (34.16), which give

$$H = H_0 - \frac{1}{2m_e ec^2}\frac{1}{r}\frac{dU(r)}{dr}\vec{\mu}_s \cdot \vec{L} - \vec{\mu}_l \cdot \vec{B}_0 - \vec{\mu}_s \cdot \vec{B}_0 \tag{34.22}$$

and yields the Hamiltonian operator

$$\hat{H} = \hat{H}_0 + \frac{1}{2m_e c^2}\frac{1}{r}\frac{dU(r)}{dr}\hat{\vec{S}} \cdot \hat{\vec{L}} + \frac{e}{2m_e}\left(\hat{\vec{L}} + 2\hat{\vec{S}}\right) \cdot \vec{B}_0 \tag{34.23}$$

By choosing $\vec{B}_0$ to be in the z-direction, Eq.(34.23) reduces to

$$\hat{H} = \hat{H}_0 + \frac{1}{2m_e^2 c^2} \frac{1}{r} \frac{dU(r)}{dr} \hat{\vec{S}} \cdot \hat{\vec{L}} + \frac{eB_0}{2m_e}\left(\hat{L}_z + 2\hat{S}_z\right) \quad (34.24)$$

There are two cases to consider: the *weak-field case*, where the energy due to the external magnetic field is small compared to the spin-orbit coupling so that the eigenfunctions Ψ_{jm_j} are required to describe the eigenstates of the spinning electron and the *strong-field case*, where the spin-orbit interaction becomes negligible and we can speak of a decoupling of $\vec{L}$ and $\vec{S}$ and of their separate coupling to the external field, such that the electron eigenstates will be described in terms of the Φ_{mm_s} set.

In the *weak-field case*, from Eq.(34.24) we obtain

$$\hat{H}\Psi_{jm_j} = \hat{H}_0 \Psi_{jm_j} + \frac{1}{2m_e^2 c^2} \frac{1}{r} \frac{dU(r)}{dr} \left(\hat{\vec{S}} \cdot \hat{\vec{L}}\right)\Psi_{jm_j} + \frac{eB_0}{2m_e}\left(\hat{J}_z + \hat{S}_z\right)\Psi_{jm_j}$$

and this gives

$$\varepsilon_{nlj}\Psi_{jm_j} = \varepsilon_n \Psi_{jm_j} + \varepsilon_{SL}\Psi_{jm_j} + \frac{eB_0}{2m_e} m_j \hbar \Psi_{jm_j} + \frac{eB_0}{2m_e}\hat{S}_z \Psi_{jm_j}$$

where ε_{SL} stands for the spin-orbit coupling energy given in Eq.(34.19). Using $\Psi_{jm_j}^*$ as a multiplying factor in front of each term and integrating gives

$$\varepsilon_{nlj} = \varepsilon_n + \varepsilon_{SL} + \mu_B B_0 m_j + \frac{eB_0}{2m_e}\left\langle \Psi_{jm_j} \middle| \hat{S}_z \middle| \Psi_{jm_j} \right\rangle \quad (34.25)$$

where μ_B is the Bohr magneton (32.40).

The calculation of the last matrix element requires explicit expressions for the Ψ_{jm_j} eigenfunctions. However, it reduces to the previously solved eigenvalue problem for $\hat{J}_z$, if the average of $\hat{S}_z$ is evaluated in terms of the vector model. Since the non-constant vector $\vec{S}$ precesses about the constant $\vec{J}$, as illustrated in Figure 34.3, only the component of $\vec{S}$ parallel to $\vec{J}$ is relevant, the other components averaging out. It follows that the average value of $\vec{S}$ can be represented in the vector model by

$$\langle \vec{S} \rangle = \left(\vec{S} \cdot \frac{\vec{J}}{J}\right)\frac{\vec{J}}{J} = \frac{\vec{S} \cdot \vec{J}}{J^2}\vec{J}$$

and this leads to

$$\langle S_z \rangle = \frac{\vec{S} \cdot \vec{J}}{J^2}J_z = (g_L - 1)J_z \quad \text{or} \quad \langle \hat{S}_z \rangle = \frac{\hat{\vec{S}} \cdot \hat{\vec{J}}}{\hat{J}^2}\hat{J}_z = (g_L - 1)\hat{J}_z \quad (34.26)$$

where g_L is called the *Landé factor*. It is obtained by squaring Eq.(34.3), which gives

$$\hat{\vec{S}}\cdot\hat{\vec{J}} = \frac{1}{2}\left(\hat{J}^2 + \hat{S}^2 - \hat{L}^2\right)$$

In terms of the quantum numbers j, s and l we then have

$$g_L = 1 + \frac{\hat{\vec{S}}\cdot\hat{\vec{J}}}{J^2} = 1 + \frac{j(j+1) + s(s+1) - l(l+1)}{2j(j+1)} \quad (34.27)$$

Using Eq.(34.26) it follows that

$$\left\langle \Psi_{jm_j} \left| \hat{S}_z \right| \Psi_{jm_j} \right\rangle = (g_L - 1)\left\langle \Psi_{jm_j} \left| \hat{J}_z \right| \Psi_{jm_j} \right\rangle = (g_L - 1)m_j \hbar$$

so that Eq.(34.25) gives

$$\varepsilon_{nlj} = \varepsilon_n + \varepsilon_{SL} + \mu_B B_0 g_L m_j \quad (34.28)$$

Therefore, the effect of a weak external magnetic field is to split each energy level of the fine structure, of energy $\varepsilon_{nl} + \varepsilon_{SL}$, into $2j+1$ levels labelled by the m_j values. This is known to be the *anomalous Zeeman effect*. The Landé factor (34.27) provides the constant relative separation $\mu_B B_0 g_L$ of the Zeeman levels.

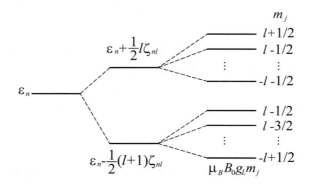

Figure 34.7. The anomalous Zeeman effect for the one electron atom

For one-electron atoms, Eq.(34.27) reads

$$g_L = 1 \pm \frac{1}{2l+1} \quad , \quad j = l \pm \tfrac{1}{2}$$

so that the complete expression for the Zeeman levels is obtained by combining Eqs.(34.21) and (34.28) as

$$\varepsilon_{nl, j=l+\frac{1}{2}} = \varepsilon_n + \frac{1}{2}l\zeta_{nl} + \mu_B B_0 \frac{2(l+1)}{2l+1} m_j$$

$$\varepsilon_{nl, j=l-\frac{1}{2}} = \varepsilon_n - \frac{1}{2}(l+1)\zeta_{nl} + \mu_B B_0 \frac{2l}{2l+1} m_j$$

(34.29)

The energy level splitting is represented in Figure 34.7.

In the *strong-field case*, where the spin splitting is small compared to the magnetic splitting of the energy levels, and hence, it can be neglected to a first approximation, the Hamiltonian (34.24) reduces to

$$\hat{H} = \hat{H}_0 + \frac{eB_0}{2m_e}\left(\hat{L}_z + 2\hat{S}_z\right) \quad (34.30)$$

It is convenient to write the energy eigenvalue equation in terms of the eigenfunctions Φ_{mm_s} as follows

$$\hat{H}\Phi_{mm_s} = \left[\hat{H}_0 + \frac{eB_0}{2m_e}\left(\hat{L}_z + 2\hat{S}_z\right)\right]\Phi_{mm_s} = \left[\varepsilon_n + \frac{eB_0}{2m_e}(m + 2m_s)\hbar\right]\Phi_{mm_s} \quad (34.31)$$

so that

$$\varepsilon_{nlmm_s} = \varepsilon_n + \mu_B B_0 (m + 2m_s) \quad (34.32)$$

It is seen that the effect of a strong external magnetic field consists of splitting a state with given quantum numbers n and l, and energy ε_n, into $2l+3$ states $(l \neq 0)$ with equally spaced energy levels. The separation $\mu_B B_0$ of the energy levels is the same as that obtained in Eq.(32.39) in case of the normal Zeeman effect.

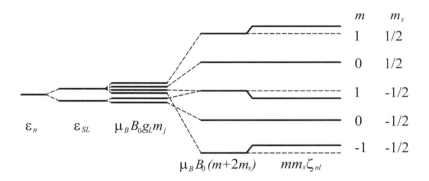

Figure 34.8. Transition from Zeeman to Paschen-Back levels of a *np*-state in the one electron atom

The small correction term to the Zeeman levels (34.32), due to the spin-orbit interaction, must be specified in terms of the n, l, m and m_s quantum numbers, instead of n, l, j and m_j that are used in Eq.(34.19). This can be readily achieved for those states with $j = l + \frac{1}{2}$ of the one-electron atoms corresponding to the highest $(m = 1, m_s = \frac{1}{2})$ and the lowest $(m = -1, m_s = -\frac{1}{2})$ Zeeman level. The spin-orbit coupling (34.21) in these states can be rewritten as

$$\varepsilon_{SL} = \frac{1}{2}l\zeta_{nl} = m_s m \zeta_{nl} \quad (34.33)$$

and it can be shown that Eq.(34.33) is valid for each Φ_{mm_s} state. From Eqs.(34.33) and (34.32), the strong-field energy levels can be written as

$$\varepsilon_{nlmm_s} = \varepsilon_n + \mu_B B_0 (m + 2m_s) + mm_s \zeta_{nl} \qquad (34.34)$$

This result is called the *Paschen-Back effect*. The transition from the weak-field Zeeman levels (34.29) to the strong-field Paschen-Back levels (34.34) is illustrated in Figure 34.8 for an np-state of the spinning electron.

FURTHER READING

1. R. Shankar - PRINCIPLES OF QUANTUM MECHANICS, Kluwer Academic, New York, 1997.
2. H. A. Bethe, R. Jackiw - INTERMEDIATE QUANTUM MECHANICS, Perseus Publishing, Philadelphia, 1997.
3. J. J. Sakurai - MODERN QUANTUM MECHANICS, University of Bangalore Press, Bangalore, 1997.

PROBLEMS

34.1. Solve the eigenvalue problem of the total angular momentum for the spinning electron if $l \neq 0$, and hence, $j = l \pm \frac{1}{2}$.

Solution

When $j = l \pm \frac{1}{2}$ it is convenient to solve the eigenvalue problem (34.5) in the 2×2 matrix representation

$$\hat{J}_z = \hat{L}_z + \hat{S}_z = \begin{pmatrix} \hat{L}_z + \hat{S}_z & 0 \\ 0 & \hat{L}_z - \hat{S}_z \end{pmatrix} = \begin{pmatrix} -i\hbar \frac{\partial}{\partial \varphi} + \frac{1}{2}\hbar & 0 \\ 0 & -i\hbar \frac{\partial}{\partial \varphi} - \frac{1}{2}\hbar \end{pmatrix}$$

$$\hat{J}^2 = \hat{L}^2 + \hat{S}^2 + 2(\hat{L}_x \hat{S}_x + \hat{L}_y \hat{S}_y + \hat{L}_z \hat{S}_z) = \begin{pmatrix} \hat{L}^2 + \frac{3}{4}\hbar^2 + \hbar \hat{L}_z & \hbar \hat{L}_- \\ \hbar \hat{L}_+ & \hat{L}^2 + \frac{3}{4}\hbar^2 + \hbar \hat{L}_z \end{pmatrix}$$

In view of Eq.(34.1), the eigenstates of $\hat{J}^2$ and $\hat{J}_z$ will be written in the general form

$$\Psi_{jm_j} = \Psi_+ \chi_+ + \Psi_- \chi_- = \begin{pmatrix} \Psi_+ \\ \Psi_- \end{pmatrix}$$

so that the eigenvalue equation for $\hat{J}_z$ reads

$$\begin{pmatrix} -i\hbar \dfrac{\partial}{\partial \varphi} + \tfrac{1}{2}\hbar & 0 \\ 0 & -i\hbar \dfrac{\partial}{\partial \varphi} - \tfrac{1}{2}\hbar \end{pmatrix} \begin{pmatrix} \Psi_+ \\ \Psi_- \end{pmatrix} = m_j \hbar \begin{pmatrix} \Psi_+ \\ \Psi_- \end{pmatrix}$$

and this is equivalent to

$$-i\hbar \frac{\partial \Psi_+}{\partial \varphi} = \left(m_j - \tfrac{1}{2}\right)\hbar \Psi_+, \qquad -i\hbar \frac{\partial \Psi_-}{\partial \varphi} = \left(m_j + \tfrac{1}{2}\right)\hbar \Psi_-$$

Because m_j is half-integer, like j, for the spinning electron, and Ψ_+ and Ψ_- should also be eigenfunctions of $\hat{L}^2$, their angular dependence will be given by the spherical harmonics $Y_{l,m_j \pm \frac{1}{2}}(\theta,\varphi)$, such that the eigenstates Ψ_{jm_j} become

$$\Psi_{jm_j} = C_+ R(r) Y_{l,m_j - \frac{1}{2}}(\theta,\varphi)\chi_+ + C_- R(r) Y_{l,m_j + \frac{1}{2}}(\theta,\varphi)\chi_-$$

The radial function $R(r)$ is a solution to Eq.(31.24) and C_+, C_- stand for the Clebsch-Gordan coefficients to be determined by the eigenvalue equation for $\hat{J}^2$ which reads

$$\begin{pmatrix} \hat{L}^2 + \tfrac{3}{4}\hbar^2 + \hbar \hat{L}_z & \hbar \hat{L}_- \\ \hbar \hat{L}_+ & \hat{L}^2 - \tfrac{3}{4}\hbar^2 - \hbar \hat{L}_z \end{pmatrix} \begin{pmatrix} C_+ RY_{l,m_j - \frac{1}{2}} \\ C_- RY_{l,m_j + \frac{1}{2}} \end{pmatrix} = j(j+1)\hbar^2 \begin{pmatrix} C_+ RY_{l,m_j - \frac{1}{2}} \\ C_- RY_{l,m_j + \frac{1}{2}} \end{pmatrix}$$

Since $\hat{L}_+$ and $\hat{L}_-$ must satisfy Eqs.(33.33) when j and m_j assume the integer values of the orbital and magnetic quantum numbers l and m, we obtain the simultaneous equations

$$C_+\left[l(l+1) - j(j+1) + m_j + \tfrac{1}{4}\right] + C_- \sqrt{l(l+1) - (m_j^2 - \tfrac{1}{4})} = 0$$

$$C_- \sqrt{l(l+1) - (m_j^2 - \tfrac{1}{4})} + C_-\left[l(l+1) - j(j+1) - m_j + \tfrac{1}{4}\right] = 0$$

The determinant provides the quantum numbers $j = l \pm \tfrac{1}{2}$. For $j = l + \tfrac{1}{2}$, we have

$$C_+ = \sqrt{l + m_j + \tfrac{1}{2}}, \qquad C_- = \sqrt{l - m_j + \tfrac{1}{2}}$$

Substituting into the expression for Ψ_{jm_j} yields the normalization condition $|C_+|^2 + |C_-|^2 = 2l + 1$, and therefore

$$\Psi_{l+\frac{1}{2},m_j} = \sqrt{\frac{l+m_j+\frac{1}{2}}{2l+1}}\, R(r)\mathrm{Y}_{l,m_j-\frac{1}{2}}\,\chi_+ + \sqrt{\frac{l-m_j+\frac{1}{2}}{2l+1}}\, R(r)\mathrm{Y}_{l,m_j+\frac{1}{2}}\,\chi_-$$

In the same way, for $j = l - \frac{1}{2}$, we find that

$$\Psi_{l-\frac{1}{2},m_j} = -\sqrt{\frac{l-m_j+\frac{1}{2}}{2l+1}}\, R(r)\mathrm{Y}_{l,m_j-\frac{1}{2}}\,\chi_+ + \sqrt{\frac{l+m_j+\frac{1}{2}}{2l+1}}\, R(r)\mathrm{Y}_{l,m_j+\frac{1}{2}}\,\chi_-$$

It follows that the Clebsch-Gordan coefficients are given by

$$C^{l+\frac{1}{2},m_j}_{l,m_j-\frac{1}{2},\frac{1}{2},\frac{1}{2}} = C^{l-\frac{1}{2},m_j}_{l,m_j+\frac{1}{2},\frac{1}{2},-\frac{1}{2}} = \sqrt{\frac{l+m_j+\frac{1}{2}}{2l+1}}\, , \quad C^{l+\frac{1}{2},m_j}_{l,m_j+\frac{1}{2},\frac{1}{2},-\frac{1}{2}} = -C^{l-\frac{1}{2},m_j}_{l,m_j-\frac{1}{2},\frac{1}{2},\frac{1}{2}} = \sqrt{\frac{l-m_j+\frac{1}{2}}{2l+1}}$$

34.2. Find the expectation values of the operators $\hat{S}_i$, $\hat{L}_i$ and $\hat{J}_i$ in the eigenstate of the total angular momentum in one-electron atoms.

Solution

Consider first the case $j = l + \frac{1}{2}$, where the eigenstates derived in Problem 34.1 are

$$\Psi_{l+\frac{1}{2},m_j} = R_{nl}(r)\begin{pmatrix}\sqrt{\dfrac{l+m_j+\frac{1}{2}}{2l+1}}\,\mathrm{Y}_{l,m_j-\frac{1}{2}}\\[6pt] \sqrt{\dfrac{l-m_j+\frac{1}{2}}{2l+1}}\,\mathrm{Y}_{l,m_j+\frac{1}{2}}\end{pmatrix} = R_{nl}\begin{pmatrix}\psi_+\\\psi_-\end{pmatrix}$$

where $R_{nl}(r)$ are the radial functions with no influence on the expectation values. We have

$$\langle\hat{S}_z\rangle = \int (\psi_+^*,\psi_-^*)\frac{\hbar}{2}\begin{pmatrix}1 & 0\\0 & -1\end{pmatrix}\begin{pmatrix}\psi_+\\\psi_-\end{pmatrix}d\Omega = \frac{\hbar}{2}\int\left(|\psi_+|^2 - |\psi_-|^2\right)d\Omega$$

$$= \frac{\hbar}{2}\frac{l+m_j+\frac{1}{2}}{2l+1}\int|\mathrm{Y}_{l,m_j-\frac{1}{2}}|^2\,d\Omega - \frac{\hbar}{2}\frac{l-m_j+\frac{1}{2}}{2l+1}\int|\mathrm{Y}_{l,m_j+\frac{1}{2}}|^2\,d\Omega = \frac{m_j\hbar}{2l+1} = \frac{m_j\hbar}{2j}$$

In the same way, using Eq.(33.40), one obtains

$$\langle\hat{S}_x\rangle = \frac{\hbar}{2}\int\psi_+^*\psi_-\,d\Omega - \frac{\hbar}{2}\int\psi_-^*\psi_+\,d\Omega = 0$$

because the spherical harmonics corresponding to different values of the magnetic quantum number are orthogonal. Similarly we have $\langle\hat{S}_y\rangle = 0$. For the operator $\hat{L}_z$ we can write

$$\langle\hat{L}_z\rangle = \int(\psi_+^*,\psi_-^*)\begin{pmatrix}\hat{L}_z & 0\\0 & \hat{L}_z\end{pmatrix}\begin{pmatrix}\psi_+\\\psi_-\end{pmatrix}d\Omega = \int(\psi_+^*,\psi_-^*)\begin{pmatrix}(m_j-\frac{1}{2})\hbar\psi_+\\(m_j+\frac{1}{2})\hbar\psi_-\end{pmatrix}d\Omega$$

or

$$\langle \hat{L}_z \rangle = \hbar(m_j - \tfrac{1}{2})\frac{l + m_j + \tfrac{1}{2}}{2l+1}\int |Y_{l,m_j-\tfrac{1}{2}}|^2 d\Omega + \hbar(m_j + \tfrac{1}{2})\frac{l - m_j + \tfrac{1}{2}}{2l+1}\int |Y_{l,m_j+\tfrac{1}{2}}|^2 d\Omega$$

Thus, we have

$$\langle \hat{L}_z \rangle = m_j\hbar - \frac{2m_j}{2l+1}\frac{\hbar}{2} = m_j\hbar\left(1 - \frac{1}{2j}\right)$$

and, in view of Eqs.(33.34) and (33.35), one obtains

$$\langle \hat{L}_x \rangle = \langle \psi_{jm_j} | \hat{L}_x | \psi_{jm_j} \rangle = 0, \quad \langle \hat{L}_y \rangle = 0$$

We observe that

$$\langle \hat{L}_z \rangle = m_j\hbar - \langle \hat{S}_z \rangle \quad \text{or} \quad \langle \hat{J}_z \rangle = m_j\hbar$$

as expected and, from Eqs.(33.34) and (33.35), we also have $\langle \hat{J}_x \rangle = \langle \hat{J}_y \rangle = 0$. For $j = l - \tfrac{1}{2}$ we similarly obtain that

$$\langle \hat{S}_z \rangle = -\frac{m_j\hbar}{2(j+1)}, \quad \langle \hat{L}_z \rangle = m_j\hbar\left[1 + \frac{1}{2(j+1)}\right], \quad \langle \hat{J}_z \rangle = m_j\hbar$$

34.3. Find the average magnetic moment for the spinning electron in one-electron atoms.

Solution

The total magnetic moment is given by

$$\hat{\vec{\mu}} = -\frac{\partial \hat{H}}{\partial \vec{B}} = -\frac{e}{2m_e}(\hat{\vec{L}} + 2\hat{\vec{S}})$$

where use has been made of Eq.(34.23) for the Hamiltonian in external magnetic field. The expectation values for the x and y components are equal to zero, since $\langle \hat{L}_x \rangle, \langle \hat{S}_x \rangle, \langle \hat{L}_y \rangle$ and $\langle \hat{S}_y \rangle$ are all zero, as shown in Problem 34.2. For the component along the z-axis we have

$$\langle \hat{\mu}_z \rangle = -\frac{e}{2m_e}\left[\langle \hat{L}_z \rangle + 2\langle \hat{S}_z \rangle\right] = -\frac{e}{2m_e}\left[\langle \hat{J}_z \rangle + \langle \hat{S}_z \rangle\right]$$

Using the results derived in Problem 34.2 we then obtain, for $j = l + \tfrac{1}{2}$ that

$$\langle \hat{\mu}_z \rangle = -\frac{e\hbar}{2m_e}\left(m_j + \frac{m_j}{2j}\right) = -\mu_B m_j \frac{2j+1}{2j} = -\mu_B m_j g_L(j = l + \tfrac{1}{2})$$

where g_L is the Landé factor for that particular j value. For $j = l - \tfrac{1}{2}$ we similarly obtain

$$\langle \hat{\mu}_z \rangle = -\frac{e\hbar}{2m_e}\left(m_j - \frac{m_j}{2(j+1)}\right) = -\mu_B m_j \frac{2j+1}{2(j+1)} = -\mu_B m_j g_L(j = l - \tfrac{1}{2})$$

Hence, in general, we can write $\langle \hat{\mu}_z \rangle = -\mu_B m_j g_L(j)$.

34.4. Evaluate the expectation value ζ_{nl} of the spin-orbit interaction energy in the hydrogen atom.

Solution

In view of Eq.(34.20), we have

$$\zeta_{nl} = \frac{\hbar^2}{2m_e^2 c^2}\left\langle \Psi_{jm_j}\left|\frac{1}{r}\frac{d}{dr}\left(-\frac{e_0^2}{r}\right)\right|\Psi_{jm_j}\right\rangle = \frac{e_0^2 \hbar^2}{2m_e^2 c^2}\left\langle \frac{1}{r^3}\right\rangle$$

where $\langle r^{-3}\rangle$ is given by the Kramers recurrence relation derived in Problem 32.4. Putting $s=-1$, substituting $\langle r^{-2}\rangle$ from Problem 32.3 and $\langle r^{-1}\rangle$ from Problem 32.8 yields

$$\langle r^{-3}\rangle = \frac{\langle r^{-2}\rangle}{a_0 l(l+1)} = \frac{2}{l(l+1)(2l+1)n^3}\frac{1}{a_0^3}$$

Thus, one obtains

$$\zeta_{nl} = \frac{e_0^2 \hbar^2}{m_e^2 c^2 a_0^3}\frac{1}{l(l+1)(2l+1)n^3}$$

34.5. Find the induction B_0 of the applied magnetic field in the configuration illustrated in Figure 34.7, if the energy of the lowest $np_{3/2}$ Zeeman level is the same as that of the highest $np_{1/2}$ Zeeman level.

Solution

The Zeeman levels are given by Eqs.(34.29), which can be written for $l=1$ as

$$\varepsilon_{nl,j+\frac{1}{2}} = \varepsilon_{ns} + \frac{1}{2}\zeta_{nl} + \frac{4}{3}\mu_B B_0 m'_j \quad,\quad \varepsilon_{nl,j-\frac{1}{2}} = \varepsilon_{ns} - \zeta_{nl} + \frac{2}{3}\mu_B B_0 m'_j$$

Thus, it is required that

$$\frac{1}{2}\zeta_{nl} + \frac{4}{3}\mu_B B_0 m'_{j\min} = -\zeta_{nl} + \frac{2}{3}\mu_B B_0 m'_{j\max} \quad \text{or} \quad \frac{3}{2}\zeta_{nl} = \frac{2}{3}\mu_B B_0\left(m'_{j\max} - 2m'_{j\min}\right)$$

where $m'_{j\max} = \frac{1}{2}$ and $m'_{j\min} = -\frac{3}{2}$. It follows that

$$\frac{3}{2}\zeta_{nl} = \frac{7}{3}\mu_B B_0 \quad \text{or} \quad B_0 = \frac{9}{14}\frac{\zeta_{nl}}{\mu_B}$$

34.6. If the angular momentum operator $\hat{\vec{J}} = \hat{\vec{J}}_1 + \hat{\vec{J}}_2$ is a resultant of two angular momentum vectors, it is required that $m = m_1 + m_2$, and hence, it follows that $j_{max} = (m_1 + m_2)_{max} = j_1 + j_2$. Find the minimum value of j.

34.7. Use the vector model of the angular momentum to find an expression for $\cos(\vec{S} \cdot \vec{L})$, where $\vec{S}$ and $\vec{L}$ are the observables of the electron in an one-electron atom.

34.8. Find an expression for the internally generated magnetic field B_i, corresponding to the fine structure of an np-state illustrated in Figure 34.6, in terms of the expectation value ζ_{nl} of the spin-orbit coupling.

34.9. Find the Landé factors for the ns_j, np_j and nd_j states of the one-electron atom.

34.10. A hydrogen atom in the ground state with $j = \frac{1}{2}, m_j = \frac{1}{2}$ is in a uniform magnetic field. If the field is suddenly rotated by $\theta = 60°$, what are the probabilities $P_{\pm\frac{1}{2}}$ associated with the electron being in states with $m'_j = \pm\frac{1}{2}$ relative to the new field immediately after the change in field.

35. SYSTEMS OF IDENTICAL PARTICLES

35.1. MANY-PARTICLE SYSTEMS
35.2. THE PAULI EXCLUSION PRINCIPLE
35.3. DISTRIBUTION LAWS FOR IDENTICAL PARTICLES

35.1. MANY-PARTICLE SYSTEMS

The postulates of quantum mechanics and their consequences, which have been developed to describe a single particle, are still valid when applied to a system of N particles. The particles can be described by a wave function Ψ which depends on the $3N$ coordinates of all the particles and on the time

$$\Psi = \Psi(\vec{r}_1,\ldots,\vec{r}_k,\ldots,\vec{r}_N,t) = \Psi(x_1,y_1,z_1,\ldots,x_k,y_k,z_k,\ldots,x_N,y_N,z_N,t) \quad (35.1)$$

If we consider $|\Psi(\vec{r}_k,t)|^2$ to be the probability density for finding, at time t, particle 1 at $\vec{r}_1,\ldots,$ particle k at $\vec{r}_k,\ldots$ and particle N at $\vec{r}_N$, a normalization condition analogous to Eq.(29.2) holds as

$$\int \Psi^* \Psi d\mathrm{V} = 1 \quad (35.2)$$

The integral is defined in a space with $3N$ dimensions called the *configuration space* of the system, where a configuration point is given by $3N$ position coordinates $x_1,y_1,z_1,\ldots,x_k,y_k,z_k,\ldots,x_N,y_N,z_N$, and the volume element is

$$d\mathrm{V} = d\mathrm{V}_1 \ldots d\mathrm{V}_k \ldots d\mathrm{V}_N = dx_1 dy_1 dz_1 \ldots dx_k dy_k dz_k \ldots dx_N dy_N dz_N$$

Physical observables of the system will be represented by Hermitian operators, obtained from the classical representation in terms of the position and momentum variables as indicated by Postulate 2. Thus the many-particle Hamiltonian results from Eq.(2.32) as

$$\hat{H} = \sum_{k=1}^{N} \left[-\frac{\hbar^2}{2m_k} \nabla_k^2 + U_k(\vec{r}_k, t) + \sum_{j \neq k} U_{kj}(\vec{r}_k, \vec{r}_j) \right] \quad (35.3)$$

where ∇_k only acts on the coordinates of the *k*th particle, of mass m_k. The one-particle potential energy $U_k(\vec{r}_k,t)$ is due to external fields where this particle moves and $U_{kj}(\vec{r}_k,\vec{r}_j)$ gives the mutual interaction between the particles *k* and *j*. It is assumed that the position and momentum operators of different particles commute, and so do all the operators describing single-particle observables if they refer to different particles.

The evolution of a many-particle system will be then formulated as a generalization of Eq.(30.6) in terms of Ψ and $\hat{H}$ given by Eqs.(35.1) and (35.3)

$$i\hbar \frac{\partial \Psi}{\partial t} = \hat{H}\Psi \quad (35.4)$$

If the Hamiltonian (35.3) does not depend explicitly on the time, it is usual to express the time dependence of the many-particle state function by means of the evolution operator $\hat{U}(t)$ defined as

$$\Psi(\vec{r}_1,\ldots,\vec{r}_k,\ldots,\vec{r}_N,t) = \hat{U}(t)\Psi_\alpha(\vec{r}_1,\ldots,\vec{r}_k,\ldots,\vec{r}_N) \quad (35.5)$$

which is formally derived from Eq.(35.4) as

$$i\hbar \frac{\partial}{\partial t}\left[\hat{U}(t)\Psi_\alpha\right] = i\hbar \frac{\partial \hat{U}(t)}{\partial t}\Psi_\alpha = \hat{H}\hat{U}(t)\Psi_\alpha$$

It follows that

$$\hat{U}(t) = e^{-i\hat{H}t/\hbar} \quad (35.6)$$

which is the analogue of Eq.(30.12). The evolution operator is *unitary*, Eq.(II.22), since

$$\hat{U}(t)\hat{U}^\dagger(t) = \hat{I} \quad (35.7)$$

Substituting Eq.(35.5), one obtains from Eq.(35.4) that the stationary states of the *N*-particle systems are given by the energy eigenvalue equation

$$\hat{H}\Psi_\alpha(\vec{r}_1,\ldots,\vec{r}_k,\ldots,\vec{r}_N) = E_\alpha \Psi_\alpha(\vec{r}_1,\ldots,\vec{r}_k,\ldots,\vec{r}_N) \quad (35.8)$$

The solution to the *N*-particle problem is facilitated if we take into account the symmetries of many systems of physical interest, which allow us to define constants of motion and criteria for simplifying the form (35.3) of the Hamiltonian operator. An alternative to complex explicit calculations is to associate with specific symmetries *time independent* unitary operators $\hat{U}$, which transform the state functions Ψ into $\hat{U}\Psi$, so that Eq.(35.4) gives

$$i\hbar\frac{\partial}{\partial t}(\hat{U}\Psi) = i\hbar\hat{U}\frac{\partial\Psi}{\partial t} = \hat{U}\hat{H}\Psi = (\hat{U}\hat{H}\hat{U}^\dagger)\hat{U}\Psi$$

This is again a Schrödinger equation for the state $\hat{U}\Psi$ with the new Hamiltonian

$$\hat{H}' = \hat{U}\hat{H}\hat{U}^\dagger$$

If the Hamiltonian remains unchanged by the transformation, we have

$$\hat{H} = \hat{U}\hat{H}\hat{U}^\dagger \quad\text{or}\quad [\hat{U},\hat{H}] = 0 \tag{35.9}$$

According to the evolution equation (29.43), this implies that $d\hat{U}/dt = 0$, so that U is a constant of motion. If the unitary transformation $\hat{U}$, which leaves the Hamiltonian invariant, is a function of an observable, its constancy implies that the given observable is a constant in any state of the system.

The *translation* of the system with a constant vector $\vec{r}_0$ is described by the associated operator

$$\hat{U} = e^{i\hat{\vec{P}}\cdot\vec{r}_0/\hbar} = \prod_{k=1}^{N} e^{i\hat{\vec{p}}_k\cdot\vec{r}_0/\hbar} = \prod_{k=1}^{N} e^{i(\hat{p}_{xk}x_0 + \hat{p}_{yk}y_0 + \hat{p}_{zk}z_0)/\hbar} \tag{35.10}$$

where $\vec{P}$ is the total momentum (2.19) of the system. In view of Eqs.(35.3) and (35.9), we must evaluate

$$\hat{U}x_k\hat{U}^\dagger = (x_k\hat{U} + [\hat{U},x_k])\hat{U}^\dagger = x_k - i\hbar\frac{\partial\hat{U}}{\partial p_{xk}}\hat{U}^\dagger = x_k + x_0$$

where the commutator corresponds to the Poisson bracket (24.37) where $U = f(p)$. It follows that

$$\hat{U}\vec{r}_k\hat{U}^\dagger = \vec{r}_k + \vec{r}_0$$

and this is a translation of the system through $\vec{r}_0$. If the potential energy is a function of the relative coordinates only, that is,

$$U(\vec{r}_1,\ldots,\vec{r}_k,\ldots,\vec{r}_N) = \sum_{k=1}^{N}\sum_{j\neq k} U_{kj}(\vec{r}_k,\vec{r}_j) = \sum_{k=1}^{N}\sum_{j\neq k} U_{kj}(\vec{r}_k - \vec{r}_j) \tag{35.11}$$

it is invariant under translation, and so is the Hamiltonian, since we can write

$$\hat{U}\hat{\vec{p}}_k\hat{U}^\dagger = \hat{\vec{p}}_k \quad\text{i.e.,}\quad \hat{U}\hat{H}\hat{U}^\dagger = \hat{H}$$

It follows that the transformation operator (35.10), and hence, the total momentum $\vec{P}$, are constants of motion, provided the condition (35.11) is satisfied. This is not the case in the presence of external fields, where the one-particle potential energies do not depend on relative coordinates only.

A *rotation* through a constant angle φ_0 about the z-axis is described by the associated operator

$$\hat{U} = e^{i\hat{L}_z \varphi_0 / \hbar} = \prod_{k=1}^{N} e^{i\hat{L}_{zk} \varphi_0 / \hbar} \qquad (35.12)$$

as it has been showed earlier for a single particle, Eq.(33.52). Since a rotation through φ_0 linearly transforms the coordinates and momentum components of all particles, Eq.(4.21), the operator (35.12), thus L_z, are constants of motion, if the Hamiltonian (35.3) can be written in terms of scalar products of the coordinate and momentum vectors only. This is valid for any component of $\vec{L}$, as the direction of the z-axis is arbitrary. In the presence of external fields having an axis of symmetry, only the component of $\vec{L}$ along this axis is a constant of motion.

Including the intrinsic spin of a particle, the rotation is described by the operator

$$\hat{U} = e^{i(\hat{L}_z + \hat{S}_z)\varphi_0 / \hbar} = \prod_{k=1}^{N} e^{i(\hat{L}_{zk} + \hat{S}_{zk})\varphi_0 / \hbar} \qquad (35.13)$$

where $\hat{S}_z$ denotes the total spin component along the z-axis. The wave function of a system of N particles, each with spin, must be modified in a similar manner to that of the spinning electron, Eq.(34.2), and will be a function of $3N$ position coordinates and N spin coordinates given by

$$\Phi = \Phi(\vec{r}_1 m_{s1}, \ldots, \vec{r}_k m_{sk}, \ldots, \vec{r}_N m_{sN}, t) = \Psi(\vec{r}_1, \ldots, \vec{r}_k, \ldots, \vec{r}_N, t) \chi(m_{s1}, \ldots, m_{sk}, \ldots, m_{sN}) \qquad (35.14)$$

where the interactions between the spin and translational motion of the particles have been neglected. The spin wave function χ has the usual meaning, that is $|\chi|^2$ gives the probability density for particle 1 to have spin orientation $m_{s1}, \ldots$, and for particle N to have spin orientation m_{sN}. This probability is independent of the state of translational motion described by the wave function (35.1).

35.2. THE PAULI EXCLUSION PRINCIPLE

N particles are called *identical* if any observable associated with them depends in the same way on the variables describing the particles. In other words, any observable is *symmetric* in the variables of N identical particles. Since a classical particle has a definite trajectory, it is possible to distinguish identical particles by following their paths. In quantum mechanics the uncertainty principle makes it impossible to define precisely the particle trajectory, so that a given particle, in a system of N identical particles, cannot be labelled. We will be able to determine the state of a system, but we will not be able to relate a state function Ψ_k with a particular particle k. This means that no change of the physical state of a system can be observed if two particles k and j are exchanged. The operation of exchanging the particle indices k and j is represented by a *unitary* operator $\hat{P}_{kj}$ defined as

$$\hat{P}_{kj}\Phi(\vec{r}_1 m_{s1},\ldots,\vec{r}_k m_{sk},\ldots,\vec{r}_j m_{sj},\ldots,\vec{r}_N m_{sN}) = \Phi(\vec{r}_1 m_{s1},\ldots,\vec{r}_j m_{sj},\ldots,\vec{r}_k m_{sk},\ldots,\vec{r}_N m_{sN})$$
(35.15)
$$= e^{i\alpha}\Phi(\vec{r}_1 m_{s1},\ldots,\vec{r}_k m_{sk},\ldots,\vec{r}_j m_{sj},\ldots,\vec{r}_N m_{sN})$$

where the normalized function Φ should change by a phase factor only, if particles are relabelled. A second exchange of the same two particles results in obtaining the original wave function, that is

$$\hat{P}_{kj}^2 \Phi = \hat{P}_{kj}(\hat{P}_{kj}\Phi) = \hat{P}_{kj}(e^{i\alpha}\Phi) = e^{2i\alpha}\Phi = \Phi$$

which gives $\alpha = n\pi$, that is $e^{i\alpha} = e^{in\pi} = \pm 1$. This means that the wave functions must have a *definite symmetry*, since they either remain unchanged or change sign upon the exchange of any two particles, according to

$$\hat{P}_{kj}\Phi_s(\vec{r}_1 m_{s1},\ldots,\vec{r}_k m_{sk},\ldots,\vec{r}_j m_{sj},\ldots,\vec{r}_N m_{sN}) = \Phi_s(\vec{r}_1 m_{s1},\ldots,\vec{r}_j m_{sj},\ldots,\vec{r}_k m_{sk},\ldots,\vec{r}_N m_{sN})$$
(35.16)
$$\hat{P}_{kj}\Phi_a(\vec{r}_1 m_{s1},\ldots,\vec{r}_k m_{sk},\ldots,\vec{r}_j m_{sj},\ldots,\vec{r}_N m_{sN}) = -\Phi_a(\vec{r}_1 m_{s1},\ldots,\vec{r}_j m_{sj},\ldots,\vec{r}_k m_{sk},\ldots,\vec{r}_N m_{sN})$$

where Φ_s is called *symmetric*, and Φ_a is called *antisymmetric* with respect to the exchange of any two particles.

Since the interaction between identical particles is always symmetric under their exchange, the Hamiltonian is left unchanged by the unitary operator $\hat{P}_{kj}$, that is

$$\hat{P}_{kj}\hat{H}\hat{P}_{kj}^\dagger = \hat{H} \quad \text{or} \quad [\hat{H},\hat{P}_{kj}] = 0 \tag{35.17}$$

Hence, the exchange operator is a constant of motion. In other words, *the symmetry or antisymmetry of a state function is permanent*. Experiment has shown that the particular state of symmetry of a system is directly related to the intrinsic spin of the identical particles. Empirically it has been demonstrated that
  ♦systems of identical particles with integer spins, called *bosons,* are described by symmetric wave functions.
  ♦systems of identical particles with half-integer spins, called *fermions,* are described by antisymmetric wave functions.
The symmetry requirements (35.16) imply that the antisymmetric wave function vanishes when two particles have the same set of coordinates $\vec{r}_k m_{sk} = \vec{r}_j m_{sj} = \vec{r} m_s$, since

$$\Phi_a(\vec{r}_1 m_{s1},\ldots,\vec{r} m_s,\ldots,\vec{r} m_s,\ldots,\vec{r}_N m_{sN}) = -\Phi_a(\vec{r}_1 m_{s1},\ldots,\vec{r} m_s,\ldots,\vec{r} m_s,\ldots,\vec{r}_N m_{sN})$$

or

$$\Phi_a(\vec{r}_1 m_{s1},\ldots,\vec{r} m_s,\ldots,\vec{r} m_s,\ldots,\vec{r}_N m_{sN}) = 0 \tag{35.18}$$

In other words, a state where two fermions having the same spin are occupying the same position cannot exist. This is one form of the **Pauli exclusion principle**: *two identical fermions can never simultaneously occupy the same state $\vec{r} m_s$*. This statement has a direct effect on the statistical distribution of fermions which, unlike that of bosons, is restricted by the condition of statistical repulsion: if one fermion is situated at a given position $\vec{r}$, it excludes all the others having the same spin state m_s from that position.

A special formulation of the Pauli exclusion principle, appropriate to the structure of multielectron atoms, can be derived for systems of N identical noninteracting particles. The particles are assumed to move independently in the field, which gives rise to the one-particle potential energies $U(\vec{r}_k)$, so that Eq.(35.3) reduces to

$$\hat{H} = \sum_{k=1}^{N} \hat{H}_k = \sum_{k=1}^{N}\left[-\frac{\hbar^2}{2m}\nabla_k^2 + U(\vec{r}_k)\right] \tag{35.19}$$

The solution to the Schrödinger equation $i\hbar\,\partial\Phi/\partial t = \hat{H}\Phi$ can be written in terms of stationary states Φ_α in a form analogous to Eq.(35.5), that is

$$\Phi = e^{-i\hat{H}t/\hbar}\Phi_a = e^{-iE_a t/\hbar}\Phi_a \tag{35.20}$$

where

$$(\hat{H}_1 + \ldots + \hat{H}_N)\Phi_\alpha(\vec{r}_1 m_{s1},\ldots,\vec{r}_N m_{sN}) = E_\alpha \Phi_\alpha(\vec{r}_1 m_{s1},\ldots,\vec{r}_N m_{sN}) \tag{35.21}$$

We assume that each single-particle eigenvalue problem has known solutions which can be considered to form an orthonormal set of stationary states

$$\hat{H}_k \varphi_{\varepsilon_k}(\vec{r}_k, m_{sk}) = \varepsilon_k \varphi_{\varepsilon_k}(\vec{r}_k, m_{sk}) \tag{35.22}$$

The Schrödinger equation (35.21) is then solved by the product of single-particle wave functions given by

$$\Phi_\alpha(\vec{r}_1 m_{s1},\ldots,\vec{r}_N m_{sN}) = \varphi_{\varepsilon_1}(\vec{r}_1 m_{s1})\ldots\varphi_{\varepsilon_N}(\vec{r}_N m_{sN}) \tag{35.23}$$

If, in the general case, we consider n_k particles to be in the state φ_{ε_k}, the total energy is

$$E_\alpha = \sum_k n_k \varepsilon_k \tag{35.24}$$

where

$$\sum_k n_k = N \tag{35.25}$$

The same energy eigenvalue E_α will, however, be obtained for $N!$ eigenfunctions (35.23), which can be generated by permutations of the N particles. None of these eigenfunctions is, in general, symmetric or antisymmetric with respect to particle interchange, but their sum is clearly symmetric, so that the symmetric wave function, for identical bosons, results as

$$\Phi_\alpha^s = \frac{1}{\sqrt{N!}} \sum_{n=1}^{N!} \hat{P}_{kj}^{n-1} \varphi_{\varepsilon_1}(\vec{r}_1 m_{s1}) \ldots \varphi_{\varepsilon_N}(\vec{r}_N m_{sN}) \qquad (35.26)$$

The antisymmetric state is formed by inserting a $\pm$ sign in Eq.(35.26), namely + for every permutation involving the exchange of an even number of pairs of identical particles and − for every permutation involving an odd number of exchanges, that is

$$\Phi_\alpha^a = \frac{1}{\sqrt{N!}} \sum_{n=1}^{N!} \left(-\hat{P}_{kj}\right)^{n-1} \varphi_{\varepsilon_1}(\vec{r}_1 m_{s1}) \ldots \varphi_{\varepsilon_N}(\vec{r}_N m_{sN}) \qquad (35.27)$$

An equivalent way of representing the state (35.27) of N identical fermions is the *Slater determinant* given by

$$\Phi_\alpha^a = \frac{1}{\sqrt{N!}} \begin{vmatrix} \varphi_{\varepsilon_1}(\vec{r}_1 m_{s1}) & \varphi_{\varepsilon_1}(\vec{r}_2 m_{s2}) & \ldots & \varphi_{\varepsilon_1}(\vec{r}_N m_{sN}) \\ \varphi_{\varepsilon_2}(\vec{r}_1 m_{s1}) & \varphi_{\varepsilon_2}(\vec{r}_2 m_{s2}) & \ldots & \varphi_{\varepsilon_2}(\vec{r}_N m_{sN}) \\ \ldots & \ldots & & \ldots \\ \varphi_{\varepsilon_N}(\vec{r}_1 m_{s1}) & \varphi_{\varepsilon_N}(\vec{r}_2 m_{s2}) & \ldots & \varphi_{\varepsilon_N}(\vec{r}_N m_{sN}) \end{vmatrix} \qquad (35.28)$$

as it can be checked by expanding the determinant. If two fermions of the same spin m_s are at the same position $\vec{r}$, two columns in Eq.(35.28) will be identical and the eigenfunction Φ_α^a will vanish, in agreement with the previous formulation of the Pauli exclusion principle. Furthermore, if two of the occupied one-particle states φ_{ε_j} and φ_{ε_k} are the same, the determinant (35.28) has two equal rows and again vanishes. This leads to the special formulation of the **Pauli exclusion principle:** *in any configuration, the completely defined one-electron states cannot be occupied by more than one fermion.* In other words, no two fermions can be assigned the same set of quantum numbers.

35.3. Distribution Laws for Identical Particles

A suitable representation of the state function for a system of N noninteracting identical particles is obtained by specifying the number of particles n_k, called the *occupation number*, in every one-particle state φ_{ε_k}. Equations (35.24) and (35.25) give the energy E_α and the number of particles N corresponding to a configuration of the system, described by the state function (35.23). We have seen that there is a so-called *exchange degeneracy*, which means that it is possible to rearrange the occupation numbers keeping the energy and the total number of particles fixed. In other words to a macroscopic state Φ_α, defined by the parameters E_α and N, there are many compatible microstates, defined by ε_k and n_k. If the expectation value of a many-particle operator $\hat{F}$ in an energy eigenstate Φ_α is written in a form similar to that given for a single particle, that is

$$\langle \hat{F} \rangle_\alpha = \langle \Phi_\alpha | \hat{F} | \Phi_\alpha \rangle \qquad (35.29)$$

the expectation value in an arbitrary state of the system, which can be expanded as

$$\Phi = \sum_\alpha C_\alpha \Phi_\alpha \qquad (35.30)$$

takes the form

$$\langle \hat{F} \rangle = \langle \Phi | \hat{F} | \Phi \rangle = \sum_\alpha |C_\alpha|^2 \langle \Phi_\alpha | \hat{F} | \Phi_\alpha \rangle \qquad (35.31)$$

We may introduce the *distribution function* f_α, associated with finding the system in the state Φ_α, as

$$f_\alpha = |C_\alpha|^2 \quad \text{or} \quad \sum_\alpha f_\alpha = 1 \qquad (35.32)$$

so that Eq.(35.31) reads

$$\langle \hat{F} \rangle = \sum_\alpha f_\alpha \langle \Phi_\alpha | \hat{F} | \Phi_\alpha \rangle \qquad (35.33)$$

This equation can be interpreted to be an ensemble average, Eq.(25.15), over the representative points in configuration space. Since the energy of the system and the number of particles are never precisely known, we must regard in general both E_α and N as variables of the system, with average values determined by external conditions. As a result, the system of N identical particles becomes an element of a grand canonical ensemble, with a grand distribution function (25.81), which reads

$$f_\alpha = \frac{1}{Z_G} e^{-E_\alpha/k_B T + \mu_0 N/k_B T} = \frac{1}{Z_G} e^{-\left(\sum_k n_k \varepsilon_k\right)/k_B T + \mu_0 \left(\sum_k n_k\right)/k_B T}$$

$$= \frac{1}{Z_G} \prod_k \left(e^{-\varepsilon_k/k_B T + \mu_0/k_B T} \right)^{n_k} = \prod_k f_k(n_k) \qquad (35.34)$$

and gives the probability density of finding n_1 particles in the one-particle state ε_1, n_2 particles in the state ε_2 and so on. The one-particle distribution function $f_k(n_k)$ stands for the probability density that n_k particles are in the state ε_k, irrespective of any other occupation number, and is defined as

$$f_k(n_k) = \text{const} \cdot \left(e^{-\varepsilon_k/k_BT + \mu_0/k_BT}\right)^{n_k} \tag{35.35}$$

where the constant results from the condition (35.32), which can then be rewritten in terms of $f_k(n_k)$ as

$$\sum_{n_k} f_k(n_k) = 1 \tag{35.36}$$

In view of Eq.(35.24), which reads

$$\langle \Phi_\alpha | \hat{H} | \Phi_\alpha \rangle = \sum_{\varepsilon_k} n_k \langle \varphi_{\varepsilon_k} | \hat{H} | \varphi_{\varepsilon_k} \rangle \tag{35.37}$$

the energy ensemble average (35.33) over many-particle states reduces to an average over one-particle states, as we have

$$\overline{E} = \langle H \rangle = \sum_\alpha f_\alpha \langle \Phi_\alpha | \hat{H} | \Phi_\alpha \rangle = \sum_{n_k} \sum_{\varepsilon_k} n_k f_k(n_k) \langle \varphi_{\varepsilon_k} | \hat{H} | \varphi_{\varepsilon_k} \rangle = \sum_{\varepsilon_k} f(\varepsilon_k) \langle \varphi_{\varepsilon_k} | \hat{H} | \varphi_{\varepsilon_k} \rangle \tag{35.38}$$

where $f(\varepsilon_k)$ is the distribution function for the average ocupation of the state φ_{ε_k}, given by

$$f(\varepsilon_k) = \sum_{n_k} n_k f_k(n_k) \tag{35.39}$$

Similarly the average particle number of the system is obtained as

$$\overline{N} = \langle N \rangle = \sum_{\varepsilon_k} f(\varepsilon_k) \tag{35.40}$$

For a system of N fermions, the occupation numbers n_k can only assume the values 0 or 1, according to the Pauli exclusion principle. Thus Eq.(35.36) reduces to

$$\text{const} \cdot \left[1 + e^{-(\varepsilon_k - \mu_0)/k_BT}\right] = 1 \tag{35.41}$$

so that

$$f_k(0) = \frac{1}{1 + e^{-(\varepsilon_k - \mu_0)/k_BT}} \quad , \quad f_k(1) = \frac{e^{-(\varepsilon_k - \mu_0)/k_BT}}{1 + e^{-(\varepsilon_k - \mu_0)/k_BT}} \tag{35.42}$$

From Eq.(35.39) we finally obtain

$$f(\varepsilon_k) = 0 \cdot f_k(0) + 1 \cdot f_k(1) = \frac{1}{e^{(\varepsilon_k - \mu_0)/k_BT} + 1} \tag{35.43}$$

This result is called the *Fermi-Dirac distribution* law for identical fermions, and $f(\varepsilon_k)$ is represented in Figure 35.1 for various temperatures.

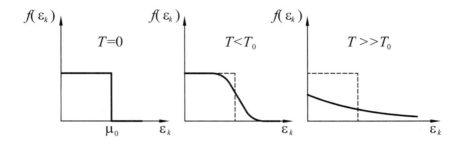

Figure 35.1. The Fermi-Dirac distribution

For a system consisting of N identical bosons, the occupation numbers can assume any value, so that Eq.(35.36) becomes

$$\sum_{n_k=0}^{\infty} f_k(n_k) = \text{const} \cdot \sum_{n_k=0}^{\infty} \left[e^{-(\varepsilon_k - \mu_0)/k_B T} \right]^{n_k} = \frac{\text{const}}{1 - e^{-(\varepsilon_k - \mu_0)/k_B T}} = 1$$

Substituting the value of this constant into Eq.(35.39) gives

$$f(\varepsilon_k) = \text{const} \cdot \sum_{n_k=0}^{\infty} n_k \left[e^{-(\varepsilon_k - \mu_0)/k_B T} \right]^{n_k} = \text{const} \cdot e^{-(\varepsilon_k - \mu_0)/k_B T} \sum_{n_k=0}^{\infty} n_k \left[e^{-(\varepsilon_k - \mu_0)/k_B T} \right]^{n_k - 1}$$

$$= \frac{\text{const} \cdot e^{-(\varepsilon_k - \mu_0)/k_B T}}{\left[1 - e^{-(\varepsilon_k - \mu_0)/k_B T} \right]^2} = \frac{e^{-(\varepsilon_k - \mu_0)/k_B T}}{1 - e^{-(\varepsilon_k - \mu_0)/k_B T}} = \frac{1}{e^{(\varepsilon_k - \mu_0)/k_B T} - 1} \tag{35.44}$$

where use has been made of the identity

$$\sum_{n_k=0}^{\infty} n_k x^{n_k - 1} = \frac{d}{dx} \sum_{n_k=0}^{\infty} x^{n_k} = \frac{d}{dx} \left(\frac{1}{1-x} \right) = \frac{1}{(1-x)^2} \tag{35.45}$$

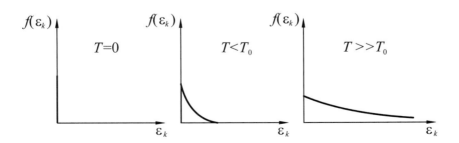

Figure 35.2. The Bose-Einstein distribution

Equation (35.44) is called the *Bose-Einstein distribution law* for identical bosons and is plotted in Figure 35.2. By combining Eqs.(35.43) and (35.44) we obtain the distribution function for the quantum average occupation as

$$f(\varepsilon) = \frac{1}{e^{(\varepsilon-\mu_0)/k_BT} \pm 1} \qquad (35.46)$$

Near absolute zero, the Fermi-Dirac distribution has one fermion in each state below a given energy μ_0 and none in the states situated above μ_0, while the Bose-Einstein distribution has most of the bosons in the one-particle state with $\varepsilon = 0$, a behaviour known as the Einstein condensation. As the average spacing between particles increases with temperature, we may expect that there exists a critical temperature T_0 where the exchange degeneracy disappears. For temperatures higher than T_0, Eq.(35.46) becomes

$$f(\varepsilon) \to e^{-(\varepsilon-\mu_0)/k_BT} = \text{const} \cdot e^{-\varepsilon/k_BT}$$

and this implies that both quantum distributions are reduced to the Boltzmann distribution law (26.25) for distinguishable identical particles.

FURTHER READING

1. J. M. Cassels - BASIC QUANTUM MECHANICS, Krieger, Melbourne, 1995.
2. D. S. Betts, R. E. Turner - INTRODUCTORY STATISTICAL MECHANICS, Addison-Wesley, Reading, 1992.
3. W. Pauli - GENERAL PRINCIPLES OF QUANTUM MECHANICS, Springer-Verlag, New York, 1990.

PROBLEMS

35.1. Given that, according to the principle of indistinguishability, the wave function of a system of two identical particles should predict no observable change of properties if the particles are interchanged, show that such a wave function should bear a definite symmetry under the exchange of the two particles.

Solution

A two-particle system is described by a wave function (35.14) which reads

$$\Phi = \Psi(\vec{r}_1, \vec{r}_2, t)\chi(m_{s1}, m_{s2})$$

If we exclude spin, the Hamiltonian of the system has the form (35.3) which reduces to

$$\hat{H}(\vec{r}_1, \vec{r}_2) = \left[-\frac{\hbar^2}{2m_e}\nabla_1^2 + U(\vec{r}_1, t)\right] + \left[-\frac{\hbar^2}{2m_e}\nabla_2^2 + U(\vec{r}_2, t)\right] + U_{12}(\vec{r}_{12})$$

Assuming that the interaction between particles gives a potential energy depending on their separation only, the symmetric Hamiltonian with respect to the exchange of the two particles is

$$\hat{H}(\vec{r}_1, \vec{r}_2) = \hat{H}(\vec{r}_2, \vec{r}_1)$$

The time-dependent Schrödinger equation (35.4) can be written either in terms of $\Psi(\vec{r}_1, \vec{r}_2, t)$ as

$$i\hbar \frac{\partial \Psi(\vec{r}_1, \vec{r}_2, t)}{\partial t} = \hat{H}(\vec{r}_1, \vec{r}_2)\Psi(\vec{r}_1, \vec{r}_2, t)$$

or in terms of $\Psi(\vec{r}_2, \vec{r}_1, t)$, in the form

$$i\hbar \frac{\partial \Psi(\vec{r}_2, \vec{r}_1, t)}{\partial t} = \hat{H}(\vec{r}_2, \vec{r}_1)\Psi(\vec{r}_2, \vec{r}_1, t) = \hat{H}(\vec{r}_1, \vec{r}_2)\Psi(\vec{r}_2, \vec{r}_1, t)$$

It follows that both wave functions satisfy the same evolution equation, and hence, we must take their linear combination to be a general solution given by

$$\Psi_{12}(\vec{r}_1, \vec{r}_2, t) = A\Psi(\vec{r}_1, \vec{r}_2, t) + B\Psi(\vec{r}_2, \vec{r}_1, t)$$

so that we have

$$\Psi_{12}(\vec{r}_2, \vec{r}_1, t) = B\Psi(\vec{r}_1, \vec{r}_2, t) + A\Psi(\vec{r}_2, \vec{r}_1, t)$$

Since no change of probability density is expected through interchange of the particles, that is,

$$|\Psi_{12}(\vec{r}_1, \vec{r}_2, t)|^2 = |\Psi_{12}(\vec{r}_2, \vec{r}_1, t)|^2$$

one obtains two conditions for the complex coefficients, written as

$$A^*A = B^*B \quad , \quad A^*B = B^*A$$

The first condition indicates that the only difference between A and B must be a phase factor, $A = Ce^{i\alpha}$, $B = Ce^{i\beta}$ say, and hence, the second condition becomes

$$e^{2i(\alpha-\beta)} = 1 \quad \text{or} \quad \alpha - \beta = n\pi$$

where n is an arbitrary integer, so that

$$B = Ce^{i(\alpha-n\pi)} = Ae^{-in\pi} = \pm A$$

Thus, there exist only one symmetric $(B = A)$ and one antisymmetric $(B = -A)$ linear combination, given by

$$\Psi_s(\vec{r}_1,\vec{r}_2,t) = A[\Psi(\vec{r}_1,\vec{r}_2,t) + \Psi(\vec{r}_2,\vec{r}_1,t)] = \Psi_s(\vec{r}_2,\vec{r}_1,t)$$

$$\Psi_a(\vec{r}_1,\vec{r}_2,t) = A[\Psi(\vec{r}_1,\vec{r}_2,t) - \Psi(\vec{r}_2,\vec{r}_1,t)] = -\Psi_a(\vec{r}_2,\vec{r}_1,t)$$

as required.

35.2. Find expressions for the symmetric and antisymmetric spatial wave functions of a system of two identical particles in terms of single-particle stationary states.

Solution

For the two-particle system considered in Problem 35.1, the wave function takes the form

$$\Psi(\vec{r}_1,\vec{r}_2,t) = e^{-iEt/\hbar}\Psi(\vec{r}_1,\vec{r}_2) = e^{-i(\varepsilon_1+\varepsilon_2)t/\hbar}\psi_{\varepsilon_1}(\vec{r}_1)\psi_{\varepsilon_2}(\vec{r}_2)$$

Exchange of the two particles gives

$$\Psi(\vec{r}_2,\vec{r}_1,t) = e^{-i(\varepsilon_1+\varepsilon_2)t/\hbar}\psi_{\varepsilon_1}(\vec{r}_2)\psi_{\varepsilon_2}(\vec{r}_1)$$

so that the symmetric and antisymmetric linear combinations become

$$\Psi^s(\vec{r}_1,\vec{r}_2) = A\left[\psi_{\varepsilon_1}(\vec{r}_1)\psi_{\varepsilon_2}(\vec{r}_2) + \psi_{\varepsilon_1}(\vec{r}_2)\psi_{\varepsilon_2}(\vec{r}_1)\right]$$

$$\Psi^a(\vec{r}_1,\vec{r}_2) = A\left[\psi_{\varepsilon_1}(\vec{r}_1)\psi_{\varepsilon_2}(\vec{r}_2) - \psi_{\varepsilon_1}(\vec{r}_2)\psi_{\varepsilon_2}(\vec{r}_1)\right]$$

We recall that spin is excluded, and we further assume that the one-particle stationary states form an orthonormal set, that is

$$\int \psi^*_{\varepsilon_1}(\vec{r}_1)\psi_{\varepsilon_2}(\vec{r}_1)dV_1 = \int \psi^*_{\varepsilon_1}(\vec{r}_2)\psi_{\varepsilon_2}(\vec{r}_2)dV_2 = \delta_{\varepsilon_1\varepsilon_2}$$

For $\varepsilon_1 \neq \varepsilon_2$, the normalization condition (35.2) for either Ψ^s or Ψ^a gives

$$\int (\Psi^s)^*\Psi^s dV_1 dV_2 = \int (\Psi^a)^*\Psi^a dV_1 dV_2 = 2A^2 = 1$$

so that the stationary wave functions become

$$\Psi^s = \frac{1}{\sqrt{2}}\left[\psi_{\varepsilon_1}(\vec{r}_1)\psi_{\varepsilon_2}(\vec{r}_2) + \psi_{\varepsilon_1}(\vec{r}_2)\psi_{\varepsilon_2}(\vec{r}_1)\right]$$

and

$$\Psi^a = \frac{1}{\sqrt{2}}\left[\psi_{\varepsilon_1}(\vec{r}_1)\psi_{\varepsilon_2}(\vec{r}_2) - \psi_{\varepsilon_1}(\vec{r}_2)\psi_{\varepsilon_2}(\vec{r}_1)\right] = \frac{1}{\sqrt{2}}\begin{vmatrix}\psi_{\varepsilon_1}(\vec{r}_1) & \psi_{\varepsilon_1}(\vec{r}_2)\\ \psi_{\varepsilon_2}(\vec{r}_1) & \psi_{\varepsilon_2}(\vec{r}_2)\end{vmatrix}$$

respectively. For $\varepsilon_1 = \varepsilon_2 = \varepsilon$, the normalization condition reads

$$\int (\Psi^s)^* \Psi^s dN_1 dN_2 = 4A^2 = 1$$

so that the symmetric wave function is

$$\Psi^s = \frac{1}{2}[\psi_\varepsilon(\vec{r}_1)\psi_\varepsilon(\vec{r}_2) + \psi_\varepsilon(\vec{r}_2)\psi_\varepsilon(\vec{r}_1)] = \psi_\varepsilon(\vec{r}_1)\psi_\varepsilon(\vec{r}_2)$$

and the antisymmetric wave function vanishes.

35.3. Find an expression for the antisymmetric wave function of a system of two noninteracting fermions with spin $s_1 = s_2 = \frac{1}{2}$, including both the spatial and the spin properties.

Solution

Using single-particle wave functions, which reflect both spatial and spin properties in the form

$$\Phi_\varepsilon(\vec{r}, m_s) = \psi_\varepsilon(\vec{r})\chi_{m_s}$$

where $m_s = \pm\frac{1}{2}$ and the corresponding χ_{m_s} have the form (33.47), the antisymmetric state of the two-electron system is given by Eq.(35.28) as

$$\Phi^a = \frac{1}{\sqrt{2}} \begin{vmatrix} \psi_{\varepsilon_1}(\vec{r}_1)\chi^{(1)}_{m_{s1}} & \psi_{\varepsilon_1}(\vec{r}_2)\chi^{(2)}_{m_{s1}} \\ \psi_{\varepsilon_2}(\vec{r}_1)\chi^{(1)}_{m_{s2}} & \psi_{\varepsilon_2}(\vec{r}_2)\chi^{(2)}_{m_{s2}} \end{vmatrix}$$

It follows immediately, in terms of the symmetric and antisymmetric spatial wave functions, as derived in Problem 35.2, Φ^a takes the form

$$\Phi^a = \frac{1}{\sqrt{2}}\left[\psi_{\varepsilon_1}(\vec{r}_1)\psi_{\varepsilon_2}(\vec{r}_2)\chi^{(1)}_{m_{s1}}\chi^{(2)}_{m_{s2}} - \psi_{\varepsilon_1}(\vec{r}_2)\psi_{\varepsilon_2}(\vec{r}_1)\chi^{(2)}_{m_{s1}}\chi^{(1)}_{m_{s2}}\right]$$

$$= \frac{1}{2\sqrt{2}}\left[\psi_{\varepsilon_1}(\vec{r}_1)\psi_{\varepsilon_2}(\vec{r}_2) + \psi_{\varepsilon_1}(\vec{r}_2)\psi_{\varepsilon_2}(\vec{r}_1)\right]\left[\chi^{(1)}_{m_{s1}}\chi^{(2)}_{m_{s2}} - \chi^{(2)}_{m_{s1}}\chi^{(1)}_{m_{s2}}\right]$$

$$+ \frac{1}{2\sqrt{2}}\left[\psi_{\varepsilon_1}(\vec{r}_1)\psi_{\varepsilon_2}(\vec{r}_2) - \psi_{\varepsilon_1}(\vec{r}_2)\psi_{\varepsilon_2}(\vec{r}_1)\right]\left[\chi^{(1)}_{m_{s1}}\chi^{(2)}_{m_{s2}} + \chi^{(2)}_{m_{s1}}\chi^{(1)}_{m_{s2}}\right] = \frac{1}{\sqrt{2}}(\Psi^s\chi^a + \Psi^a\chi^s)$$

The antisymmetric spin function is nonvanishing for $m_{s1} \neq m_{s2}$ only, so that we can take $m_{s1} = +\frac{1}{2}$, $m_{s2} = -\frac{1}{2}$, which gives

$$\chi^a = \frac{1}{\sqrt{2}}(\chi^{(1)}_+ \chi^{(2)}_- - \chi^{(2)}_+ \chi^{(1)}_-)$$

The symmetric spin function

$$\chi^s = \frac{1}{\sqrt{2}}(\chi^{(1)}_{m_{s1}}\chi^{(2)}_{m_{s2}} + \chi^{(2)}_{m_{s1}}\chi^{(1)}_{m_{s2}})$$

describes three possible spin states of the two electron system, corresponding to the two spins either parallel ($m_{s1}=m_{s2}=\frac{1}{2}$ and $m_{s1}=m_{s2}=-\frac{1}{2}$) or antiparallel ($m_{s1}=\frac{1}{2}, m_{s2}=-\frac{1}{2}$) to one another. The normalized symmetric spin functions read

$$\chi_1^s = \chi_+^{(1)}\chi_+^{(2)}, \qquad \chi_0^s = \frac{1}{\sqrt{2}}(\chi_+^{(1)}\chi_-^{(2)} + \chi_+^{(2)}\chi_-^{(1)}), \qquad \chi_{-1}^s = \chi_-^{(1)}\chi_-^{(2)}$$

35.4. Prove that the spin functions χ^a and $\chi_{m_s}^s$ for a two-fermion system with spins $s_1 = s_2 = \frac{1}{2}$ are eigenfunctions of the operator $\left(\hat{\vec{S}}_1 \cdot \hat{\vec{S}}_2\right)^n$, where n is an integer, and show that the eigenvalue problem can be reduced to that of the operator $\hat{\vec{S}}_1 \cdot \hat{\vec{S}}_2$.

Solution

It is straightforward to show that

$$\hat{S}^2 = \hat{S}_1^2 + \hat{S}_1^2 + 2\hat{\vec{S}}_1 \cdot \hat{\vec{S}}_2 \qquad \text{or} \qquad \hat{\vec{S}}_1 \cdot \hat{\vec{S}}_2 = \frac{1}{2}\left(\hat{S}^2 - \frac{3}{2}\hbar^2\right)$$

where the total spin is $S = 0$ in the antisymmetric state χ^a and $S = 1$ in a symmetric state $\chi_{m_s}^s$. Using the spin functions defined in Problem 35.3, it follows that

$$\left(\hat{\vec{S}}_1 \cdot \hat{\vec{S}}_2\right)\chi^a = -\tfrac{3}{4}\hbar^2 \chi^a, \qquad \left(\hat{\vec{S}}_1 \cdot \hat{\vec{S}}_2\right)\chi_{m_s}^s = \left(\hbar^2 - \tfrac{3}{4}\hbar^2\right)\chi_{m_s}^s = \tfrac{1}{4}\hbar^2 \chi_{m_s}^s$$

and hence, the required eigenvalue corresponding to χ^a and $\chi_{m_s}^s$ are

$$\left(\hat{\vec{S}}_1 \cdot \hat{\vec{S}}_2\right)^n \chi^a = (-3)^n \left(\tfrac{1}{2}\hbar\right)^{2n} \chi^a, \qquad \left(\hat{\vec{S}}_1 \cdot \hat{\vec{S}}_2\right)^n \chi_{m_s}^s = \left(\tfrac{1}{2}\hbar\right)^{2n} \chi_{m_s}^s$$

It is convenient to assume a linear relationship of the form

$$\left(\hat{\vec{S}}_1 \cdot \hat{\vec{S}}_2\right)^n = A + B\left(\hat{\vec{S}}_1 \cdot \hat{\vec{S}}_2\right)$$

and to choose A and B such that it is valid when acting on both χ^a and $\chi_{m_s}^s$. This gives

$$(-3)^n \left(\tfrac{1}{2}\hbar\right)^{2n} = A - 3\left(\tfrac{1}{2}\hbar\right)^2 B, \qquad \left(\tfrac{1}{2}\hbar\right)^{2n} = A + \left(\tfrac{1}{2}\hbar\right)^2 B$$

Thus one obtains

$$A = \frac{3+(-3)^n}{4}\left(\tfrac{1}{2}\hbar\right)^{2n}, \qquad B = \frac{1-(-3)^n}{4}\left(\tfrac{1}{2}\hbar\right)^{2(n-1)}$$

and the relationship can be written in the form

$$\left(\hat{\vec{S}}_1 \cdot \hat{\vec{S}}_2\right)^n = \frac{3+(-3)^n}{4}\left(\tfrac{1}{2}\hbar\right)^{2n} + \frac{1-(-3)^n}{4}\left(\tfrac{1}{2}\hbar\right)^{2(n-1)}\left(\hat{\vec{S}}_1 \cdot \hat{\vec{S}}_2\right)$$

which holds true for any spin function of the two-fermion system. In particular, for $n=2$, the eigenvalue problem reads

$$\left(\hat{\vec{S}}_1 \cdot \hat{\vec{S}}_2\right)^2 \chi = \tfrac{1}{4}\hbar^2\left[\tfrac{3}{4}\hbar^2 - 2\left(\hat{\vec{S}}_1 \cdot \hat{\vec{S}}_2\right)\right]\chi = \lambda_2 \chi$$

If we put $\left(\hat{\vec{S}}_1 \cdot \hat{\vec{S}}_2\right)\chi = \lambda_1 \chi$, this implies that

$$\lambda_2 = \tfrac{1}{4}\hbar^2\left(\tfrac{3}{4}\hbar^2 - 2\lambda_1\right)$$

Similarly, for $n=3$, one obtains

$$\lambda_3 = \left(\tfrac{1}{2}\hbar\right)^4\left(-\tfrac{3}{4}\hbar^2 + 7\lambda_1\right)$$

and hence, the eigenvalue problem can be reduced to that of the $\hat{\vec{S}}_1 \cdot \hat{\vec{S}}_2$ operator, as required.

35.5. Find an expression for the symmetric wave function of a system of two noninteracting bosons with spin $s_1 = s_2 = 1$, in terms of both the spatial and spin variables.

Solution

The symmetric state of the two-boson system, including both spatial and spin properties, can be written as

$$\Phi^s = \frac{1}{\sqrt{2}}\left[\psi_{\varepsilon_1}(\vec{r}_1)\psi_{\varepsilon_2}(\vec{r}_2)\chi^{(1)}_{m_{s1}}\chi^{(2)}_{m_{s2}} + \psi_{\varepsilon_1}(\vec{r}_2)\psi_{\varepsilon_2}(\vec{r}_1)\chi^{(2)}_{m_{s1}}\chi^{(1)}_{m_{s2}}\right]$$

$$= \frac{1}{2\sqrt{2}}\left[\psi_{\varepsilon_1}(\vec{r}_1)\psi_{\varepsilon_2}(\vec{r}_2) + \psi_{\varepsilon_1}(\vec{r}_2)\psi_{\varepsilon_2}(\vec{r}_1)\right]\left[\chi^{(1)}_{m_{s1}}\chi^{(2)}_{m_{s2}} + \chi^{(2)}_{m_{s1}}\chi^{(1)}_{m_{s2}}\right]$$

$$+ \frac{1}{2\sqrt{2}}\left[\psi_{\varepsilon_1}(\vec{r}_1)\psi_{\varepsilon_2}(\vec{r}_2) - \psi_{\varepsilon_1}(\vec{r}_2)\psi_{\varepsilon_2}(\vec{r}_1)\right]\left[\chi^{(1)}_{m_{s1}}\chi^{(2)}_{m_{s2}} - \chi^{(2)}_{m_{s1}}\chi^{(1)}_{m_{s2}}\right] = \frac{1}{\sqrt{2}}(\Psi^s \chi^s + \Psi^a \chi^a)$$

The normalized symmetric spin functions have the form

$$\chi^s_{m_{s1}m_{s2}} = \frac{1}{\sqrt{2}}(\chi^{(1)}_{m_{s1}}\chi^{(2)}_{m_{s2}} + \chi^{(2)}_{m_{s1}}\chi^{(1)}_{m_{s2}}), \quad m_{s1} \neq m_{s2} \quad \text{and} \quad \chi^s_{m_{si}m_{si}} = \chi^{(1)}_{m_{si}}\chi^{(2)}_{m_{si}}$$

where $m_{si} = -1, 0, 1$ and

$$\chi^{(i)}_1 = \begin{pmatrix}1\\0\\0\end{pmatrix}, \quad \chi^{(i)}_0 = \begin{pmatrix}0\\1\\0\end{pmatrix}, \quad \chi^{(i)}_{-1} = \begin{pmatrix}0\\0\\1\end{pmatrix}$$

35.6. Show that antisymmetric spin function χ^a of a two-electron system of spin $\hat{\vec{S}} = \hat{\vec{S}}_1 + \hat{\vec{S}}_2$ is an eigenfucntion of the operators $\hat{S}^2$ and $\hat{S}_z$ and find the corresponding eigenvalues.

35.7. Show that the symmetric spin functions χ^s of a two-electron system where $\hat{\vec{S}} = \hat{\vec{S}}_1 + \hat{\vec{S}}_2$ describe three possible spin states, and solve the eigenvalue problem $\hat{S}^2$ and $\hat{S}_z$.

35.8. Find the spin states described by the symmetric eigenfunctions of the total spin operators $\hat{S}^2$ and $\hat{S}_z$ for a two-boson system where $s_1 = s_2 = 1$.

35.9. Solve the eigenvalue problem for the total spin operators $\hat{S}^2$ and $\hat{S}_z$ using the antisymmetric eigenfunctions of a two-boson system where $s_1 = s_2 = 1$.

35.10. Find the number of symmetric one-particle states n_s and that of antisymmetric states n_a for a system of two identical particles of spin s.

36. MULTIELECTRON ATOMS

36.1. STATIONARY STATE PERTURBATION THEORY
36.2. THE HELIUM ATOM
36.3. THE CENTRAL-FIELD APPROXIMATION

36.1. STATIONARY STATE PERTURBATION THEORY

The determination of the stationary states of *N*-particle systems involves the solution of the energy eigenvalue equation (35.8). This can only be achieved for idealized systems of noninteracting particles, using known functions in closed form. In many physical situations where the interaction energy represents a small disturbance or a *perturbation* of an idealized system, for which the eigenvalue problem is soluble, perturbation theory provides a systematic method of obtaining approximate solutions.

The Hamiltonian $\hat{H}$ of a given system is compared with the Hamiltonian $\hat{H}_0$ of an *unperturbed* system, of which the eigenvalues ε_k^0 and the complete set of eigenfunctions Ψ_k^0 of the stationary states are already known

$$\hat{H}_0 \Psi_k^0 = \varepsilon_k^0 \Psi_k^0 \qquad (36.1)$$

It is assumed that $\hat{H}$ includes small effects neglected in $\hat{H}_0$ and represented by a perturbation operator $\hat{V}$, that is

$$\hat{H} = \hat{H}_0 + \hat{V} \qquad (36.2)$$

Approximate eigenvalues and eigenstates are derived for the Hamiltonian

$$\hat{H}(\lambda) = \hat{H}_0 + \lambda \hat{V} \qquad (36.3)$$

where λ, called the *smallness parameter*, satisfies $0 \leq \lambda \leq 1$, such that $\lambda = 0$ corresponds to the unperturbed system $\hat{H}_0$, and $\lambda = 1$ corresponds to the given system (36.2). The eigenvalues and eigenfunctions of $\hat{H}(\lambda)$, given by

$$\left(\hat{H}_0 + \lambda \hat{V}\right)\Psi_k = \varepsilon_k \Psi_k \tag{36.4}$$

are assumed to be analytic functions of λ, expressible as convergent power series

$$\Psi_k = \Psi_k^0 + \lambda \Psi_k^{(1)} + \lambda^2 \Psi_k^{(2)} + \ldots$$

$$\varepsilon_k = \varepsilon_k^0 + \lambda \varepsilon_k^{(1)} + \lambda^2 \varepsilon_k^{(2)} + \ldots \tag{36.5}$$

where $\Psi_k^{(1)}, \Psi_k^{(2)}$ and $\varepsilon_k^{(1)}, \varepsilon_k^{(2)}$ are the first- and second-order corrections to be determined. On substituting Eq.(36.5) into Eq.(36.4) and collecting the terms of like powers of λ one obtains

$$\left(\hat{H}_0 - \varepsilon_k^0\right)\Psi_k^0 + \left[\left(\hat{H}_0 - \varepsilon_k^0\right)\Psi_k^{(1)} + \left(\hat{V} - \varepsilon_k^{(1)}\right)\Psi_k^0\right]\lambda$$

$$+ \left[\left(\hat{H}_0 - \varepsilon_k^0\right)\Psi_k^{(2)} + \left(\hat{V} - \varepsilon_k^{(1)}\right)\Psi_k^{(1)} - \varepsilon_k^{(2)}\Psi_k^0\right]\lambda^2 + \ldots = 0 \tag{36.6}$$

Since λ can take any value between $\lambda = 0$ and $\lambda = 1$, the coefficients of the various powers of λ must be equal to zero. The eigenvalues and eigenfunctions of the zeroth order equation are assumed to be known. The higher order equations must be solved in succession for $\varepsilon_k^{(1)}, \Psi_k^{(1)}, \varepsilon_k^{(2)}, \Psi_k^{(2)}, \ldots$ since the nth order corrections are obtained providing all the lower order terms have been previously calculated.

The *first-order corrections* result from the term in Eq.(36.6) containing λ to first-order, which reads

$$\left(\hat{H}_0 - \varepsilon_k^0\right)\Psi_k^{(1)} + \left(\hat{V} - \varepsilon_k^{(1)}\right)\Psi_k^0 = 0 \tag{36.7}$$

If the eigenstates Ψ_k^0 are *nondegenerate*, that is, if the eigenvalues ε_k^0 are all distinct, they form a complete set in terms of which we can make the expansion

$$\Psi_k^{(1)} = \sum_{m=1}^{\infty} a_{km} \Psi_m^0 \tag{36.8}$$

On substitution into Eq.(36.7), one obtains

$$\left(\hat{H}_0 - \varepsilon_k^0\right)\sum_{m=1}^{\infty} a_{km}\Psi_m^0 + \left(\hat{V} - \varepsilon_k^{(1)}\right)\Psi_k^0 = 0 \quad \text{or} \quad \varepsilon_k^{(1)}\Psi_k^0 = \hat{V}\Psi_k^0 + \sum_{m \neq k}\left(\varepsilon_m^0 - \varepsilon_k^0\right)a_{km}\Psi_m^0 \tag{36.9}$$

Multiplying through $\left(\Psi_i^0\right)^*$ and integrating gives

$$\varepsilon_k^{(1)} \delta_{ik} = \left\langle \Psi_i^0 \left| \hat{V} \right| \Psi_k^0 \right\rangle + \left(\varepsilon_i^0 - \varepsilon_k^0 \right) a_{ki} \tag{36.10}$$

If we set $i = k$, the first-order eigenvalue correction to the unperturbed state Ψ_k^0 is obtained as the expectation value of the perturbation operator $\hat{V}$ in this state

$$\varepsilon_k^{(1)} = \left\langle \Psi_k^0 \left| \hat{V} \right| \Psi_k^0 \right\rangle \tag{36.11}$$

whereas for $i \neq k$, the coefficients a_{ki} can be obtained from Eq.(36.10) as

$$a_{ki} = \frac{\left\langle \Psi_i^0 \left| \hat{V} \right| \Psi_k^0 \right\rangle}{\varepsilon_k^0 - \varepsilon_i^0} \tag{36.12}$$

and the coefficients a_{kk} must be obtained in a different manner. Since the eigenfunction (36.5) is determined up to a phase factor, we may choose the coefficients a_{kk} to be real. The normalization of the eigenfunctions (36.5) to first order in λ gives

$$\left\langle \Psi_k | \Psi_k \right\rangle = \left\langle \Psi_k^0 + \lambda \Psi_k^{(1)} \middle| \Psi_k^0 + \lambda \Psi_k^{(1)} \right\rangle = 1 + \lambda \left\langle \Psi_k^0 \middle| a_{kk} \Psi_k^0 \right\rangle + \lambda \left\langle a_{kk} \Psi_k^0 \middle| \Psi_k^0 \right\rangle$$

$$= 1 + 2\lambda \operatorname{Re} a_{kk} = 1$$

so that the result is $a_{kk} = 0$. Substituting Eqs.(36.8), (36.11) and (36.12), Eq.(36.5) gives the first-order results for the system (36.2) under consideration, that is, for $\lambda = 1$, as

$$\Psi_k = \Psi_k^0 + \sum_{i \neq k} \frac{\left\langle \Psi_i^0 \left| \hat{V} \right| \Psi_k^0 \right\rangle}{\varepsilon_k^0 - \varepsilon_i^0} \Psi_i^0 \quad , \quad \varepsilon_k = \varepsilon_k^0 + \left\langle \Psi_k^0 \left| \hat{V} \right| \Psi_k^0 \right\rangle \tag{36.13}$$

If certain eigenstates Ψ_k^0 are *degenerate*, that is, if there are n independent eigenfunctions $\Psi_{k\alpha}^0$, $\alpha = 1, 2, \ldots, n$, which all correspond to the same eigenvalue ε_k^0, the formula (36.12) does not hold and one should proceed as follows. Assuming that the degenerate states $\Psi_{k\alpha}^0$ are orthonormal and allowing for degeneracy in our expansion (36.5) we obtain

$$\Psi_k = \sum_{\alpha=1}^n A_{k\alpha} \Psi_{k\alpha}^0 + \lambda \Psi_k^{(1)} + \ldots \tag{36.14}$$

If we substitute this expression into the eigenvalue equation (36.4), instead of the first-order equation (36.9), we obtain a set of n relations

$$\varepsilon_k^{(1)} \sum_{\alpha=1}^n A_{k\alpha} \Psi_{k\alpha}^0 = \hat{V} \sum_{\alpha=1}^n A_{k\alpha} \Psi_{k\alpha}^0 + \sum_{m \neq k} \left(\varepsilon_m^0 - \varepsilon_k^0 \right) a_{km} \Psi_m^0$$

where the expansion (36.8) must be written in terms of states Ψ_m^0 outside $\Psi_{k\alpha}^0$. In other words, Ψ_m^0 are states orthogonal to $\Psi_{k\alpha}^0$ for all α and $\varepsilon_m^0 - \varepsilon_k^0 \neq 0$. On multiplication by $\left(\Psi_{k\gamma}^0\right)^*$ and integration one obtains, using orthonormality, that

$$\varepsilon_k^{(1)} A_{k\gamma} = \sum_{\alpha=1}^n A_{k\alpha} \left\langle \Psi_{k\gamma}^0 \left| \hat{V} \right| \Psi_{k\alpha}^0 \right\rangle \tag{36.15}$$

which is a set of homogenous equations for the coefficients $A_{k\alpha}$. If we set

$$V_{\gamma\alpha} = \left\langle \Psi_{k\gamma}^0 \left| \hat{V} \right| \Psi_{k\alpha}^0 \right\rangle \tag{36.16}$$

we may rewrite Eq.(36.15) as

$$\sum_{\alpha=1}^n V_{\gamma\alpha} A_{k\alpha} = \varepsilon_k^{(1)} A_{k\gamma} \tag{36.17}$$

which is a matrix eigenvalue equation, analogous to Eq.(33.9), of the form

$$\begin{pmatrix} V_{11} & V_{12} & \cdots & V_{1n} \\ V_{21} & V_{22} & \cdots & V_{2n} \\ \cdots & \cdots & \cdots & \cdots \\ V_{n1} & V_{n2} & \cdots & V_{nn} \end{pmatrix} \begin{pmatrix} A_{k1} \\ A_{k2} \\ \cdots \\ A_{kn} \end{pmatrix} = \varepsilon_k^{(1)} \begin{pmatrix} A_{k1} \\ A_{k2} \\ \cdots \\ A_{kn} \end{pmatrix} \tag{36.18}$$

The condition for nonidentically vanishing $A_{k\alpha}$ is given by the characteristic equation

$$\det\left(\hat{V} - \varepsilon_k^{(1)}\hat{I}\right) = \begin{vmatrix} V_{11} - \varepsilon_k^{(1)} & V_{12} & \cdots & V_{1n} \\ V_{21} & V_{22} - \varepsilon_k^{(1)} & \cdots & V_{2n} \\ \cdots & \cdots & \cdots & \cdots \\ V_{n1} & V_{n2} & \cdots & V_{nn} - \varepsilon_k^{(1)} \end{vmatrix} = 0 \tag{36.19}$$

with n roots for $\varepsilon_k^{(1)}$, which are denoted by $\varepsilon_{k\alpha}^{(1)}$, $\alpha = 1, 2, \ldots, n$. If the roots are all distinct, it is said that the degeneracy has been completely removed due to the perturbation $\hat{V}$, otherwise it is said to be only partially removed. As the eigenvalues $\varepsilon_{k\alpha}^{(1)}$ give the first-order corrections to the energy, the formerly n-fold degenerate state splits, under the influence of perturbation, into n close-lying states with energies (36.5) which can be written as

$$\varepsilon_{k\alpha} = \varepsilon_k^0 + \varepsilon_{k\alpha}^{(1)} \tag{36.20}$$

Inserting each eigenvalue $\varepsilon_{k\alpha}^{(1)}$ into the matrix equation (36.18) yields a set of n coefficients $A_{k\alpha}$ that is restricted by the normalization condition

$$\sum_{\alpha=1}^{n} |A_{k\alpha}|^2 = 1 \qquad (36.21)$$

If *all* the eigenstates are degenerate, we no longer have states Ψ_m^0 outside $\Psi_{k\alpha}^0$ to make the expansion possible (36.8), so that Eq.(36.14) reduces to the first term and therefore the eigenfunction Ψ_k can be written as a linear combination of the degenerate states for each energy (36.20). Thus, the first-order results (36.13) of the perturbation theory for nondegenerate states are replaced by Eqs.(36.14) and (36.20) in the case of degenerate states, which are of special interest in experimental physics.

The second-order corrections can be obtained using a similar approach to the first-order case. We start by inserting the first-order corrections (36.13) into the second-order equation (36.6), and this gives

$$\varepsilon_k^{(2)} \Psi_k^0 = \left(\hat{H}_0 - \varepsilon_k^0\right)\Psi_k^{(2)} + \left(\hat{V} - \varepsilon_k^{(1)}\right)\Psi_k^{(1)} \qquad (36.22)$$

If we multiply Eq.(36.22) by $\left(\Psi_i^0\right)^*$ and integrate, one obtains for $i = k$ the second-order eigenvalue corrections as

$$\varepsilon_k^{(2)} = \left\langle \Psi_k^0 \left| \hat{V} \right| \Psi_k^{(1)} \right\rangle = \sum_{m \neq k} a_{km} \left\langle \Psi_k^0 \left| \hat{V} \right| \Psi_m^0 \right\rangle = \sum_{m \neq k} \frac{\left|\left\langle \Psi_k^0 \left| \hat{V} \right| \Psi_m^0 \right\rangle\right|^2}{\varepsilon_k^0 - \varepsilon_m^0} \qquad (36.23)$$

where use has been made of Eqs.(36.8) and (36.12). It follows that the second-order correction to ground state energy ε_k^0 is always negative. Inserting Eq.(36.23) into Eq.(36.5) provides, for $\lambda = 1$, the energy

$$\varepsilon_k = \varepsilon_k^0 + \left\langle \Psi_k^0 \left| \hat{V} \right| \Psi_k^0 \right\rangle + \sum_{m \neq k} \frac{\left|\left\langle \Psi_k^0 \left| \hat{V} \right| \Psi_m^0 \right\rangle\right|^2}{\varepsilon_k^0 - \varepsilon_m^0} \qquad (36.24)$$

The corresponding approximation to Ψ_k may be obtained making use of Eq.(36.22) for $i \neq k$. However, it is common practice to avoid terms beyond the first-order correction to the wave function and beyond the second-order correction in energy

36.2. THE HELIUM ATOM

The helium atom, consisting of a nucleus of charge $Ze = 2e$ and two orbiting electrons, is the simplest multielectron atom. If the nucleus is placed at the origin and the spin-interaction terms are ignored, the Hamiltonian operator may be written as

$$\hat{H} = -\frac{\hbar^2}{2m_e}\left(\nabla_1^2 + \nabla_2^2\right) - 2e_0^2\left(\frac{1}{r_1} + \frac{1}{r_2}\right) + \frac{e_0^2}{|\vec{r}_1 - \vec{r}_2|} = \hat{H}_1 + \hat{H}_2 + \hat{V} \quad (36.25)$$

where $\hat{V}$ denotes the mutual electrostatic interaction of the two electrons, which will be treated as a perturbation. The Hamiltonians $\hat{H}_1$ and $\hat{H}_2$ are single-particle operators for two individual electrons that satisfy the eigenvalue equations

$$\hat{H}_i \varphi_{n_i l_i m_i m_{si}} = \left(-\frac{\hbar^2}{2m_e}\nabla_i^2 - \frac{2e_0^2}{r_i}\right)\varphi_{n_i l_i m_i m_{si}} = \varepsilon_{n_i}\varphi_{n_i l_i m_i m_{si}} \quad (36.26)$$

where $i = 1, 2$. The eigenstates $\varphi_{n_i l_i m_i m_{si}}$ have the form (34.1) previously derived for the spinning electron of one-electron atoms, which is

$$\varphi_{n_i l_i m_i m_{si}}(\vec{r}_i) = R_{n_i l_i}(r) Y_{l_i m_i}(\theta, \varphi) \chi \quad (36.27)$$

and correspond to degenerate energy levels (32.9) or (28.18) given by

$$\varepsilon_{n_i} = -13.6\frac{Z^2}{n_i^2} = -\frac{54.4}{n_i^2} \text{ (eV)} \quad (36.28)$$

Thus, if the mutual electrostatic interaction $\hat{V}$ is ignored, the idealized helium atom can be regarded as a system of two identical fermions, of energy

$$E_\alpha = \varepsilon_{n_1} + \varepsilon_{n_2} = -54.4\left(\frac{1}{n_1^2} + \frac{1}{n_2^2}\right) \text{ (eV)} \quad (36.29)$$

with the corresponding eigenfunction represented by the Slater determinant (35.28), which reads

$$\Phi_\alpha = \frac{1}{\sqrt{2}}\begin{vmatrix} \varphi_{n_1 l_1 m_1 m_{s1}}(\vec{r}_1) & \varphi_{n_1 l_1 m_1 m_{s1}}(\vec{r}_2) \\ \varphi_{n_2 l_2 m_2 m_{s2}}(\vec{r}_2) & \varphi_{n_2 l_2 m_2 m_{s2}}(\vec{r}_2) \end{vmatrix} \quad (36.30)$$

The ground state configuration of this system contains both electrons in the lowest energy level, that is $n_1 = n_2 = 1$, $l_1 = l_2 = 0$, $m_1 = m_2 = 0$, $m_{s1} = \frac{1}{2}$, $m_{s2} = -\frac{1}{2}$. The ground state energy (36.29) is $E_g = -108.8$ eV and

$$\Phi_g = \frac{1}{\sqrt{2}}\begin{vmatrix} \psi_{100}(\vec{r}_1)\chi_+^{(1)} & \psi_{100}(\vec{r}_2)\chi_+^{(2)} \\ \psi_{100}(\vec{r}_1)\chi_-^{(1)} & \psi_{100}(\vec{r}_2)\chi_-^{(2)} \end{vmatrix} = \psi_{100}(\vec{r}_1)\psi_{100}(\vec{r}_2)\chi^S \quad (36.31)$$

The spatial part is symmetrical whereas the spin function χ^S, which is antisymmetrical under spin label interchange, as given by

$$\chi^S = \begin{vmatrix} \chi_+^{(1)} & \chi_+^{(2)} \\ \chi_-^{(1)} & \chi_-^{(2)} \end{vmatrix} = \frac{1}{\sqrt{2}} \left(\chi_+^{(1)} \chi_-^{(2)} - \chi_-^{(1)} \chi_+^{(2)} \right) \tag{36.32}$$

defines a *singlet* spin state, of total spin $S = 0$, such that $2S + 1 = 1$. As both the spatial and spin functions are unique, the ground state is nondegenerate.

The first excited state contains one electron in the ground-state level $n = 1$, $l = 0$, $m = 0$ having the spatial wave function ψ_{100} and another in the energy level defined by $n = 2$, with $l = 0$, $m = 0$ or $l = 1$, $m = 0$, ± 1 and corresponding spatial wave functions ψ_{2lm}. The configuration is fourfold degenerate because of both exchange and spin degeneracy, with total energy $E_e = -68 \text{ eV}$, as provided by Eq.(36.29). Two of the corresponding Slater determinants (36.30) are products of a spatial function and a spin function, both normalized and of definite symmetry, written as

$$\Phi_{lm} = \frac{1}{\sqrt{2}} \begin{vmatrix} \psi_{100}(\vec{r}_1)\chi_+^{(1)} & \psi_{100}(\vec{r}_2)\chi_+^{(2)} \\ \psi_{2lm}(\vec{r}_1)\chi_+^{(1)} & \psi_{2lm}(\vec{r}_2)\chi_+^{(2)} \end{vmatrix} = \psi_a \chi_+^{(1)} \chi_+^{(2)} \tag{36.33}$$

and

$$\Phi_{lm}^I = \frac{1}{\sqrt{2}} \begin{vmatrix} \psi_{100}(\vec{r}_1)\chi_-^{(1)} & \psi_{100}(\vec{r}_2)\chi_-^{(2)} \\ \psi_{2lm}(\vec{r}_1)\chi_-^{(1)} & \psi_{2lm}(\vec{r}_2)\chi_-^{(2)} \end{vmatrix} = \psi_a \chi_-^{(1)} \chi_-^{(2)} \tag{36.34}$$

where $\psi_a = [\psi_{100}(\vec{r}_1)\psi_{2lm}(\vec{r}_2) - \psi_{2lm}(\vec{r}_1)\psi_{100}(\vec{r}_2)]/\sqrt{2}$. For the other two determinants

$$\Phi_{lm}^{II} = \frac{1}{\sqrt{2}} \begin{vmatrix} \psi_{100}(\vec{r}_1)\chi_+^{(1)} & \psi_{100}(\vec{r}_2)\chi_-^{(2)} \\ \psi_{2lm}(\vec{r}_1)\chi_+^{(1)} & \psi_{2lm}(\vec{r}_2)\chi_-^{(2)} \end{vmatrix}, \quad \Phi_{lm}^{III} = \frac{1}{\sqrt{2}} \begin{vmatrix} \psi_{100}(\vec{r}_1)\chi_-^{(1)} & \psi_{100}(\vec{r}_2)\chi_+^{(2)} \\ \psi_{2lm}(\vec{r}_1)\chi_-^{(1)} & \psi_{2lm}(\vec{r}_2)\chi_+^{(2)} \end{vmatrix} \tag{36.35}$$

a linear combination will be needed, as shown below, in order to obtain eigenfunctions of definite symmetry.

The influence of the mutual electrostatic interaction $\hat{V}$ is taken into account by using the perturbation theory. For the nondegenerate ground state, due to the spin-independence of $\hat{V}$, the first-order energy correction (see Problem 36.2) reads

$$K_{10} = \langle \Phi_g | \hat{V} | \Phi_g \rangle = \int \psi_{100}^*(\vec{r}_1) \psi_{100}^*(\vec{r}_2) \frac{e_0^2}{|\vec{r}_1 - \vec{r}_2|} \psi_{100}(\vec{r}_1) \psi_{100}(\vec{r}_2) dV_1 dV_2 = \frac{5}{4} \frac{e_0^2}{a_0} \tag{36.36}$$

The positive energy shift K_{10} represents the repulsive electrostatic interaction between the two charge densities $\rho_1 = e|\psi_{100}(\vec{r}_1)|^2$ and $\rho_2 = e|\psi_{100}(\vec{r}_2)|^2$. The magnitude of K_{10} is 34 eV, and this yields $E_{10} = E_g + K_{10} = -108.8 + 34 = -74.8$ eV, a value which is closer to the experimental ground-state energy of $E_{10} = -79$ eV.

For the degenerate excited state of helium, the first-order corrections to the energy level E_e are solutions to the characteristic equation (36.19) where, in view of Eq.(36.33), one obtains

$$V_{11} = \langle \Phi_{lm} | \hat{V} | \Phi_{lm} \rangle = \int \psi_a^* \chi_+^{(1)} \chi_+^{(2)} \hat{V} \psi_a \chi_+^{(1)} \chi_+^{(2)} dV_1 dV_2$$

$$= e_0^2 \int \frac{|\psi_{100}(\vec{r}_1)|^2 |\psi_{2lm}(\vec{r}_2)|^2}{|\vec{r}_1 - \vec{r}_2|} dV_1 dV_2 - e_0^2 \int \frac{\psi_{100}^*(\vec{r}_1) \psi_{2lm}^*(\vec{r}_2) \psi_{2lm}(\vec{r}_1) \psi_{100}(\vec{r}_2)}{|\vec{r}_1 - \vec{r}_2|} dV_1 dV_2$$

$$= K_{2l} - J_{2l} \tag{36.37}$$

and similarly, from Eq.(36.34), we have

$$V_{22} = \langle \Phi_{lm}^I | \hat{V} | \Phi_{lm}^I \rangle = \int \psi_a^* \chi_-^{(1)} \chi_-^{(2)} \hat{V} \psi_a \chi_-^{(1)} \chi_-^{(2)} dV_1 dV_2 = K_{2l} - J_{2l} \tag{36.38}$$

since the total spin functions $\chi_+^{(1)} \chi_+^{(2)}$ $(S=1, M_S = 1)$ and $\chi_-^{(1)} \chi_-^{(2)}$ $(S=1, M_S = -1)$ are assumed to be orthogonal and normalized. The term K_{2l} corresponds to a repulsive electrostatic interaction between the charge distributions of the two electrons, analogous to that denoted by K_{10} in Eq.(36.36), while the second term J_{2l} is called the *exchange integral* and has no classical counterpart. The magnitude of the exchange integral depends on the product $\psi_{100} \psi_{2lm}$, that is, on the overlapping of the two wave functions. It can be shown, using Eqs.(36.35), that

$$\langle \Phi_{lm}^{II} | \hat{V} | \Phi_{lm}^{II} \rangle = \langle \Phi_{lm}^{III} | \hat{V} | \Phi_{lm}^{III} \rangle = K_{2l} \quad , \quad \langle \Phi_{lm}^{II} | \hat{V} | \Phi_{lm}^{III} \rangle = \langle \Phi_{lm}^{III} | \hat{V} | \Phi_{lm}^{II} \rangle = -J_{2l} \tag{36.39}$$

whereas all the other elements of the characteristic determinant vanish, because of the orthogonality of the spin functions. Thus, the characteristic equation (36.19) reduces to

$$\begin{vmatrix} K_{2l} - J_{2l} - E^{(1)} & 0 & 0 & 0 \\ 0 & K_{2l} - J_{2l} - E^{(1)} & 0 & 0 \\ 0 & 0 & K_{2l} - E^{(1)} & -J_{2l} \\ 0 & 0 & -J_{2l} & K_{2l} - E^{(1)} \end{vmatrix}$$

$$\tag{36.40}$$

$$= \left(K_{2l} - J_{2l} - E^{(1)} \right)^2 \begin{vmatrix} K_{2l} - E^{(1)} & -J_{2l} \\ -J_{2l} & K_{2l} - E^{(1)} \end{vmatrix} = \left(K_{2l} - J_{2l} - E^{(1)} \right)^3 \left(K_{2l} + J_{2l} - E^{(1)} \right) = 0$$

and has a triple root $E_T^{(1)}$ and a single root $E_S^{(1)}$ given by

$$E_T^{(1)} = K_{2l} - J_{2l} \quad , \quad E_S^{(1)} = K_{2l} + J_{2l} \tag{36.41}$$

Substituting $E_T^{(1)}$ into the matrix eigenvalue equation yields

$$\begin{pmatrix} K_{2l} & -J_{2l} \\ -J_{2l} & K_{2l} \end{pmatrix} \begin{pmatrix} A_1 \\ A_2 \end{pmatrix} = E^{(1)} \begin{pmatrix} A_1 \\ A_2 \end{pmatrix} \tag{36.42}$$

that is, $A_1 = A_2 = 1/\sqrt{2}$. Thus, the corresponding eigenfunction, which defines a state with $S=1$, $M_S = 0$, can be written as

$$\Phi_{lm}^{IV} = \frac{1}{\sqrt{2}}\left(\Phi_{lm}^{II} + \Phi_{lm}^{III}\right) = \psi_a \frac{1}{\sqrt{2}}\left(\chi_+^{(1)}\chi_-^{(2)} + \chi_-^{(1)}\chi_+^{(2)}\right) \qquad (36.43)$$

where ψ_a is the antisymmetric spatial function. By combining the space-antisymmetric, spin-symmetric wave functions (36.33), (36.34) and (36.43) we can represent the so-called *triplet* state $S=1$, $M_S = 0, \pm 1$ of helium as

$$\Phi_{lm}^T = \frac{1}{\sqrt{2}}\left[\psi_{100}(\vec{r}_1)\psi_{2lm}(\vec{r}_2) - \psi_{2lm}(\vec{r}_1)\psi_{100}(\vec{r}_2)\right]\chi^T = \psi_a\chi^T \qquad (36.44)$$

where χ^T is the triplet spin function, given by

$$\chi^T = \chi_+^{(1)}\chi_+^{(2)} \qquad (M_S = 1)$$

$$= \frac{1}{\sqrt{2}}\left(\chi_+^{(1)}\chi_-^{(2)} + \chi_-^{(1)}\chi_+^{(2)}\right) \qquad (M_S = 0)$$

$$= \chi_-^{(1)}\chi_-^{(2)} \qquad (M_S = -1) \qquad (36.45)$$

Similarly, by substituting $E_S^{(1)}$, Eq.(36.42) gives $A_{S1} = -A_{S2} = 1/\sqrt{2}$, so that there is a corresponding *singlet* state given by

$$\Phi_{lm}^S = \frac{1}{\sqrt{2}}\left(\Phi_{lm}^{II} - \Phi_{lm}^{III}\right) = \psi_s \frac{1}{\sqrt{2}}\left(\chi_+^{(1)}\chi_-^{(2)} - \chi_-^{(1)}\chi_+^{(2)}\right) = \psi_s\chi^S \qquad (36.46)$$

This is a space-symmetric, $\psi_s = \left[\psi_{100}(\vec{r}_1)\psi_{2lm}(\vec{r}_2) + \psi_{2lm}(\vec{r}_1)\psi_{100}(\vec{r}_2)\right]$, spin-antisymmetric wave function, analogous to the ground-state eigenfunction (36.31).

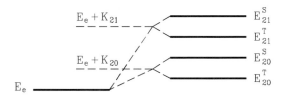

Figure 36.1. Ground level shift and first excited level splitting in helium

Therefore, if the interaction between electrons is taken into account, there is a shift K_{10} of the ground-state level, and the first excited states split into a triplet E_{2l}^T and a singlet E_{2l}^S as

$$E_{20}^T = E_e + K_{20} - J_{20} \quad , \quad E_{20}^S = E_e + K_{20} + J_{20}$$

$$E_{21}^T = E_e + K_{21} - J_{21} \quad , \quad E_{21}^S = E_e + K_{21} + J_{21}$$

where $E_e = -68$ eV. Since both K_{2l} and J_{2l} are positive, the energy shift of the triplet state level E_{2l}^T is lower than that of the singlet state level E_{2l}^S by an amount of $2J_{2l}$ (see Figure 36.1). In other words, for a given excited configuration of helium, the states of highest spin have the lowest energy. It also follows that, as a result of the interaction between the two electrons, the accidental degeneracy of E_e is removed.

36.3. THE CENTRAL-FIELD APPROXIMATION

The Hamiltonian for the motion of Z electrons about a nucleus of mass M and charge Ze has the form

$$\hat{H} = \sum_{i=1}^{Z}\left[-\frac{\hbar^2}{2m_e}\nabla_i^2 - \frac{Ze_0^2}{r_i}\right] + \frac{1}{2}\sum_{i=1}^{Z}\sum_{k\neq i}\frac{e_0^2}{|\vec{r}_i - \vec{r}_k|}$$

which includes the kinetic energy of the electrons, the electrostatic interaction between the nucleus and each of the Z electrons and the mutual interaction between each pair of electrons. The energy eigenvalue problem for this system of identical fermions can only be solved approximately, under the assumption that each electron i moves in an averaged central force field, which is due partly to the nucleus and partly to all the other electrons $k \neq i$, represented by *uniform* charge distributions $-e|\psi_k(\vec{r}_k)|^2$, as given by

$$\hat{H}_0 = \sum_{i=1}^{Z}\left[-\frac{\hbar^2}{2m_e}\nabla_i^2 - \frac{Ze_0^2}{r_i} + V_{eff}(r_i)\right] \tag{36.47}$$

where the part of the potential due to all the other electrons is assumed to be a function of the radial coordinate r_i of one electron only, that is

$$V_{eff}(r_i) = e_0^2 \sum_{k\neq i}\int\frac{|\psi_k(\vec{r}_k)|^2}{|\vec{r}_i - \vec{r}_k|}dV_k \tag{36.48}$$

The form (36.47) of the Hamiltonian is called the *central-field approximation*. In this approximation the energy eigenvalue problem, which reads

$$\hat{H}_0 \Phi_0 = E_0 \Phi_0 \tag{36.49}$$

can be separated into Z eigenvalue equations, assuming that

$$\Phi_0 = \prod_{i=1}^{Z} \varphi_i(\vec{r}_i, m_{si}) = \prod_{i=1}^{Z} \begin{pmatrix} \psi_{i+}(\vec{r}_i) \\ \psi_{i-}(\vec{r}_i) \end{pmatrix} \tag{36.50}$$

and

$$E_0 = \sum_{i=1}^{Z} \varepsilon_i^0 \tag{36.51}$$

Substituting Eqs.(36.50) and (36.51) into Eq.(36.49), the spatial wave functions ψ_{i+} and ψ_{i-} are found to satisfy the so-called *Hartree equations*, of the form

$$\left[-\frac{\hbar^2}{2m_e} \nabla_i^2 - \frac{Ze_0^2}{r_i} + V_{\mathit{eff}}(r_i) \right] \psi_{i\pm} = \varepsilon_i^0 \psi_{i\pm} \tag{36.52}$$

which are similar to the energy eigenvalue equation for one-electron atoms. However, these equations cannot be solved exactly because of the form of $V_{\mathit{eff}}(r_i)$ given by Eq.(36.48). As $V_{\mathit{eff}}(r_i)$ is spherically symmetric, the spatial wave functions can be characterized by l_i and m_i quantum numbers, and hence, the Hamiltonian commutes with both the orbital and spin angular momentum operators. A *self consistent* method of solving the set of Z simultaneous equations (36.52) consists of assuming an initial set of functions $\psi_{n_i l_i m_i}$ to obtain the potential energies $V_{\mathit{eff}}(r_i)$ and then solving numerically the Hartree equations for a new set of one-electron functions. This process is repeated until wave functions $\psi_{n_i l_i m_i}$ are found which give the same potential energy $V_{\mathit{eff}}(r_i)$ as that used in the equations solved for them.

The combination of the Pauli principle with the Hartree method gives an insight into the building up of atoms by the addition of more and more electrons to the appropriate nucleus, which provides the charge Ze. A single-electron state in central field is described by a wave function of the form (36.27), called an *orbital*, which is given by

$$\varphi_{n_i l_i m_i m_{si}}(\vec{r}_i) = R_{n_i l_i}(r) Y_{l_i m_i}(\theta, \varphi) \chi_{m_{si}} \tag{36.53}$$

and is completely defined by a set of four quantum numbers $nlmm_s$. Since the potential energy (36.52) no longer has the $1/r_i$ form, the accidental degeneracy of all the states with a given n is removed, so that the energy eigenvalues are functions of both n and l. The energy eigenvalue ε_{nl} increases as n increases but there will be a splitting for different l values for a given n, with the lowest level corresponding to $l = 0$ and an increasing energy as l increases. This is expected in view of the results obtained for radial motion in one-electron atoms, where a significant probability for the electron to approach the nucleus and to feel the full nuclear attraction was found for small l values only, while the electron is kept away from the nucleus by the centrifugal barrier as l increases. Although the l-dependent splitting is smaller than the splitting between different n values for low Z, it increases with increasing n, such that for larger Z values we see from Figure 36.2 that there are ns energy levels lower than $(n-1)d$ or $(n-2)f$ levels.

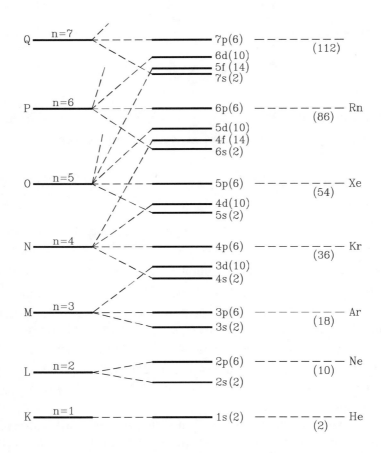

Figure 36.2. Energy levels for electrons in multielectron atoms

By listing the n and l values, in the order of increasing electron energy, we obtain the *electron configuration* of the multielectron atoms. The orbitals with the same value n are said to form a *shell*, and this is denoted by K, L, M, N, O, P, Q as n = 1, 2, 3, 4, 5, 6, 7 respectively. The orbitals with the same value for both n and l form a *subshell* and are called *equivalent*, as they all correspond to the same energy eigenvalue ε_{nl}. We have shown before, Eq.(34.6), that the maximum number of equivalent electrons, obtained by summing over either m and m_s or j and m_j angular momentum quantum numbers, is $2(2l+1)$. A subshell which contains the maximum number of electrons, namely two in a given s-subshell, six in a p-subshell, ten in a d-subshell, fourteen in an f-subshell, etc., is called a *closed* subshell. The maximum number of electrons in a shell, which is then also said to be closed, is $\sum_{l=0}^{n-1} 2(2l+1) = 2n^2$. Consequently, regularities in the shell model of multielectron atoms might be expected for $Z = 2, 10, 28, 60$ and 110. However, the experimentally observed regularities, illustrated in Figure 36.2, which are expressed by the so-called *atomic magic numbers*, which are $Z = 2, 10, 18, 36, 54$ and 86, are determined by the maximum number of equivalent electrons $2(2l+1)$ rather than by the closed shells.

However, the information provided by the electron configuration is insufficient for a complete description of the state of the atom, because it does not specify how the orbital and spin angular momenta of the individual electrons are combined to give the total angular momentum of the atom. Experiment shows that different orbitals corresponding to the same electron configuration are split, such that we have to consider a classification of the energy eigenstates according to the eigenvalues of angular momenta which are constants of motion. Neglecting the weak interaction, it can be assumed that the total orbital angular momentum $\vec{L}$ and the total spin angular momentum $\vec{S}$, defined by

$$\vec{L} = \sum_{i=1}^{Z} \vec{L}_i, \qquad \vec{S} = \sum_{i=1}^{Z} \vec{S}_i \qquad (36.54)$$

are separately conserved, which means that the operators $\hat{L}^2, \hat{L}_z, \hat{S}^2$ and $\hat{S}_z$ commute with the Hamiltonian. In this approximation, $\vec{L}$ and $\vec{S}$ are precessing independently about the z-axis, such that $M = \sum_{i=1}^{Z} m_i$ and $M_S = \sum_{i=1}^{Z} m_{si}$ are conserved, as shown before for one-electron atoms. Since the magnitudes $L(L+1)\hbar^2$ and $S(S+1)\hbar^2$ are well defined for a given energy level, this will be completely specified by both the electron configuration and the *LS* values. The spectroscopic description of an energy level of definite *L* and *S* is $^{2S+1}(L$ symbol), corresponding to a multiplet of $(2L+1)(2S+1)$ orbitals which are degenerate. The *L* symbols are $S(L = 0)$, $P(L = 1)$, $D(L = 2)$, $F(L = 3)$, and the superscript indicates the spin multiplicity which defines *singlet* ($S = 0$), *doublet* ($S = \frac{1}{2}$), *triplet* ($S = 1$) or *quadruplet* ($S = \frac{3}{2}$) spin states. Experiment shows that orbitals with different *L* and *S*, which belong to the same configuration, are split by fractions of an electron volt, and this is one order of magnitude less than the separation of a few electron volts between different electron configurations. The order of terms with different multiplicity was established earlier by the first two empirical *Hund rules* which state that

(i) *The lowest energy in a given configuration corresponds to the LS term with the largest S value.*

(ii) *For a given S, the LS term with the largest L value has the lowest energy.*

Both rules account for the increasing of the average electronic distance, so reducing the average repulsive interaction of the electrons, and hence, the energy eigenvalue, either by increasing *S* or by increasing *L* for a given *S*. If the spin-orbit coupling can be treated to be a small perturbation to the Hamiltonian (36.47), which is the situation found for most atoms, the interaction is called *normal* or *LS coupling*.

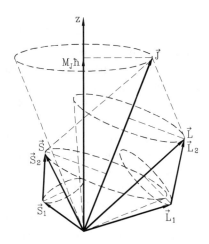

Figure 36.3. Vector representation of *LS* coupling

In a vector model representation, given in Figure 36.3 for the two-electron system, although the $\vec{L}_i \cdot \vec{S}_i$ interaction results in $\vec{L}_i$ and $\vec{S}_i$ precessing about their resultant, this precession is much slower than that of $\vec{L}_i$ about $\vec{L}$ and that of $\vec{S}_i$ about $\vec{S}$, as determined by stronger electrostatic interactions. Consequently, only the components of the $\vec{L}_i$ parallel to $\vec{L}$ and those of the $\vec{S}_i$ parallel to $\vec{S}$ are relevant, the other components averaging out. It follows that we may replace the independent $\vec{L}_i$ and $\vec{S}_i$ angular momenta by their average values along the $\vec{L}$ and $\vec{S}$ directions, and it is clear from Figure 36.3 that $\hat{\vec{L}} \cdot \hat{\vec{S}}$ commutes with $\hat{L}^2$ and $\hat{S}^2$, such that L and S still characterize the energy levels, but M and M_S vary in time as a result of precession. The constant vector is now the total angular momentum $\hat{\vec{J}} = \hat{\vec{L}} + \hat{\vec{S}}$, of magnitude $J(J+1)\hbar^2$, where the possible values of J are as given by Eq.(34.10), that is,

$$J = L+S,\ L+S-1, \ldots, |L-S| \tag{36.55}$$

Hence, in the presence of the spin-orbit interaction, which can be put in a form similar to that found for one-electron atoms, Eq.(34.19), the energy levels are completely specified by the *LSJ* values, in the form of a $^{2S+1}(L\,\text{symbol})_J$. The $(2L+1)(2S+1)$-fold degenerate *LS* multiplets are split into levels with defined J, given by Eq.(36.55), which are known as the *fine structure* of the energy spectrum. The separation between the fine structure levels is of the order of hundredths of an electron volt. The order of levels having a given J in a *LS* multiplet is established by a third empirical rule, derived from spin-orbit calculations, which states that

> (iii) *If the configuration is no more than half filled, then the lowest energy corresponds to the smallest J value, $J = |L - S|$. If the configuration is more*

than half filled then the lowest energy corresponds to the maximum value $J = L + S$.

In a further approximation we may consider the atomic magnetic moment of the atom, associated to the total angular momentum as discussed for one-electron atoms. The Landé factor for LS coupling is defined in the form

$$g_L = 1 + \left\langle \frac{\hat{\vec{S}} \cdot \hat{\vec{J}}}{\hat{J}^2} \right\rangle = 1 + \frac{J(J+1) + S(S+1) - L(L+1)}{2J(J+1)} \qquad (36.56)$$

and this implies, as discussed in the case of a single spinning electron, Eq.(34.28), that in the presence of a weak magnetic field B_0 each LSJ multiplet of the fine structure splits into $2J+1$ magnetic sublevels, with a constant relative separation $\mu_B B_0 g_L$. Thus, the Zeeman magnetic splitting completely removes any degeneracy.

FURTHER READING

1. A. F. Levi - APPLIED QUANTUM MECHANICS, Cambridge University Press, Cambridge, 2003.
2. W. A. Harrison - APPLIED QUANTUM MECHANICS, World Scientific, Singapore, 2000.
3. H. Kroemer - QUANTUM MECHANICS, Prentice Hall, Englewood Cliffs, 1994.

PROBLEMS

36.1. Derive the first-order effect of a perturbation operator $\hat{V}$ on a twofold degenerate stationary state, assuming that $\hat{V}$ is symmetric with respect to the unperturbed eigenfunctions.

Solution

Consider a twofold degenerate state of energy ε_k^0 for which the eigenvalue equation (36.17) reads

$$\begin{pmatrix} V_{11} & V_{12} \\ V_{21} & V_{22} \end{pmatrix} \begin{pmatrix} A_1 \\ A_2 \end{pmatrix} = \varepsilon^{(1)} \begin{pmatrix} A_1 \\ A_2 \end{pmatrix}$$

where we have deleted the index k because it always appears in the same way and $\alpha = 1, 2$. The characteristic equation (36.19) becomes

$$\begin{pmatrix} V_{11} - \varepsilon^{(1)} & V_{12} \\ V_{21} & V_{22} - \varepsilon^{(1)} \end{pmatrix} = 0$$

As a further simplification, we assume that the perturbation operator is symmetric with respect to the degenerate eigenfunctions Ψ_{k1}^0 and Ψ_{k2}^0, such that

$$K = V_{22} = V_{11} = \langle \Psi_{k1}^0 | \hat{V} | \Psi_{k1}^0 \rangle = \int (\Psi_{k1}^0)^* \hat{V} \Psi_{k1}^0 dV$$

$$J = V_{21} = V_{12} = \langle \Psi_{k1}^0 | \hat{V} | \Psi_{k2}^0 \rangle = \int (\Psi_{k1}^0)^* \hat{V} \Psi_{k2}^0 dV$$

It follows that $\varepsilon_1^{(1)} = K + J$, $\varepsilon_2^{(1)} = K - J$ and the degeneracy of the level ε_k^0 is lifted according to Eq.(36.20), which yields $\varepsilon_{k1} = \varepsilon_k^0 + K + J$ and $\varepsilon_{k2} = \varepsilon_k^0 + K - J$. If $\varepsilon_{k1}^{(1)} = K + J$ is inserted into Eq.(36.20), the normalized coefficients become $A_1 = A_2 = 1/\sqrt{2}$ so that the corresponding eigenfunction (36.14) can be written as

$$\Psi_{k1} = \frac{1}{\sqrt{2}} \left(\Psi_{k1}^0 + \Psi_{k2}^0 \right)$$

Similarly $A_1' = -A_2' = 1/\sqrt{2}$ corresponds to the $\varepsilon_{k2}^{(1)} = K - J$, and it follows from Eq.(36.14) that

$$\Psi_{k2} = \frac{1}{\sqrt{2}} \left(\Psi_{k1}^0 - \Psi_{k2}^0 \right)$$

36.2. Find the first-order energy correction K_{10} to the ground-state energy of the helium atom due to the electrostatic interaction between the two electrons.

Solution

Using $\Psi_{100}(\vec{r})$ as given by Eqs.(32.14), one obtains

$$K_{10} = \frac{e_0^2}{\pi^2} \left(\frac{2}{a_0} \right)^6 \int \frac{e^{-4(r_1+r_2)/a_0}}{|\vec{r}_1 - \vec{r}_2|} r_1^2 dr_1 d\Omega_1 r_2^2 dr_2 d\Omega_2$$

$$= \frac{1}{16\pi^2} \frac{e_0^2}{a_0} \int \frac{e^{-(\rho_1+\rho_2)}}{|\vec{\rho}_1 - \vec{\rho}_2|} \rho_1^2 d\rho_1 d\Omega_1 \rho_2^2 d\rho_2 d\Omega_2$$

where $\vec{\rho}_{1,2} = 4\vec{r}_{1,2}/a_0$ are dimensionless variables. If we choose the z-axis along the direction of $\vec{\rho}_1$, such that θ denotes the angle between $\vec{\rho}_1$ and $\vec{\rho}_2$, we can write

$$K_{10} = \frac{1}{16\pi^2} \frac{e_0^2}{a_0} \int e^{-\rho_1} I(\rho_1) \rho_1^2 d\rho_1 d\Omega_1 = \frac{1}{4\pi} \frac{e_0^2}{a_0} \int_0^\infty e^{-\rho_1} I(\rho_1) \rho_1^2 d\rho_1$$

where

$$I(\rho_1) = 2\pi \int_0^\infty \int_0^\pi \frac{e^{-\rho_2} \rho_2^2 d\rho_2 \sin\theta\, d\theta}{(\rho_1^2 - 2\rho_1\rho_2 \cos\theta + \rho_2^2)^{1/2}}$$

The integration over θ is straightforward and gives

$$\int_0^\pi \frac{\sin\theta\, d\theta}{(\rho_1^2 - 2\rho_1\rho_2 \cos\theta + \rho_2^2)^{1/2}} = \frac{1}{\rho_1\rho_2} \int_0^\pi \frac{d}{d\theta}(\rho_1^2 + \rho_2^2 - 2\rho_1\rho_2 \cos\theta)^{1/2} d\theta$$

$$= \frac{1}{\rho_1\rho_2}(\rho_1 + \rho_2 - |\rho_1 - \rho_2|) = \frac{2}{\rho_1} \quad \text{(if } \rho_2 < \rho_1\text{)} \quad \text{or} \quad = \frac{2}{\rho_2} \quad \text{(if } \rho_2 > \rho_1\text{)}$$

It follows that $I(\rho_1)$ can be written in the form

$$I(\rho_1) = 4\pi \left[\frac{1}{\rho_1} \int_0^{\rho_1} e^{-\rho_2} \rho_2^2\, d\rho_2 + \int_{\rho_1}^\infty e^{-\rho_2} \rho_2\, d\rho_2 \right] = \frac{4\pi}{\rho_1}\left[2 - e^{-\rho_1}(2 + \rho_1)\right]$$

and hence, we have

$$K_{10} = \frac{e_0^2}{a_0}\left[2\int_0^\infty e^{-\rho_1}\rho_1\, d\rho_1 - 2\int_0^\infty e^{-2\rho_1}\rho_1\, d\rho_1 - \int_0^\infty e^{-2\rho_1}\rho_1^2\, d\rho_1 \right] = \frac{5}{4}\frac{e_0^2}{a_0}$$

36.3. Find the energy level splitting of a p^2 configuration under applied magnetic field assuming *LS* coupling of angular momenta. What are the energy levels for this configuration if the electrons obey *jj* coupling, where the spin-orbit interaction of each electron dominates over its electrostatic interaction with the other electron.

Solution

For a p^2 configuration, we have $m_1, m_2 = 1, 0, -1$ and $m_{s1}, m_{s2} = \pm\frac{1}{2}$. Since the equivalent orbitals have to be specified by different m or m_s, the 36 possible combinations of the quantum numbers are restricted to 15 different states of the configuration. We find three cases: $M = \pm 2$, $M_S = 0$ which indicate a fivefold degenerate singlet term with $L = 2$, $S = 0$ denoted by ^1D; then $M = \pm 1$, $M_S = \pm 1$ which lead to $L = 1$ and $S = 1$ corresponding to the triplet state ^3P (ninefold degenerate); and finally $M = 0$, $M_S = 0$ which defines the singlet state ^1S.

The separation of the *LS* terms is given in Figure (a), where the order of the terms in increasing energy is ^3P, ^1D and ^1S, according to Hund's rules. Figure (b) illustrates the fine structure of the configuration, where the spin-orbit interaction breaks the ninefold degenerate multiplet ^3P with $L = 1$, $S = 1$ and produces the $^3P_0, ^3P_1$ and $^3P_2$ levels corresponding to the possible values $J = 0, 1, 2$. The order of these levels is established by the third empirical rule. In the presence of a weak magnetic field, each *LSJ* multiplet of the fine structure splits into $2J + 1$ magnetic sublevels, and hence the degeneracy is completely removed, as illustrated in Figure (c).

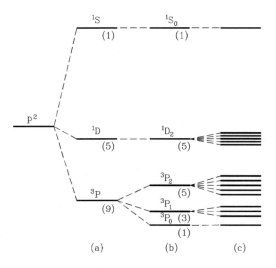

The vector model for *jj coupling* is represented below, in Figure (a), where the good quantum numbers which describe the electron orbitals are n_i, l_i, j_i and m_{ji} and the constants of motion for the atoms are given by

$$\hat{\vec{J}} = \sum \hat{\vec{J}}_i, \qquad \hat{J}_z = \sum \hat{J}_{iz}$$

For the two-electron configuration p^2 we have $j_1, j_2 = \frac{3}{2}, \frac{1}{2}$, and this leads to the energy levels $(j_1 j_2)$ illustrated in Figure (b). According to the exclusion principle, the configuration $(n1\frac{3}{2}m_{j1})(n1\frac{3}{2}m_{j2})$ is only allowed for different m_j values, such that $M_j = m_{j1} + m_{j2}$ can only assume the values 2, 1, 0, 0, −1, −2, and this means that $J = 0, 2$. Similarly we obtain the values $J = 1, 2$ for the configuration $(n1\frac{3}{2}m_{j1})(n1\frac{1}{2}m_{j2})$ and $J = 0$ for $(n1\frac{1}{2}m_{j1})(n1\frac{1}{2}m_{j2})$. As a result, the electrostatic perturbation breaks the spin-orbit multiplets into the fine structure levels, with defined J values, which are plotted in Figure (c). Each level of the fine structure is again split by the Zeeman interaction into $2J+1$ magnetic levels, as shown in Figure (d). It is clear that the same number of states is obtained assuming either *LS* or *jj* coupling.

36.4. Find the fine structure levels for an atom with p^3 electron configuration. If the atom is placed in a weak magnetic field B, how many Zeeman levels are obtained, and what is their relative spacing?

Solution

The LS terms in a p^3 configuration might have $L = 3, 2, 1, 0$ and $S = \frac{3}{2}, \frac{1}{2}$. The exclusion principle eliminates the $L = 3$ state, and this implies a symmetric spatial part of the wave function, and hence, a completely antisymmetric spin function, which is impossible to form with more than two spins $\frac{1}{2}$. The allowed combinations are $L = 2, S = \frac{1}{2}$ or ^2D, which involve $(2L+1)(2S+1) = 10$ states; $L = 1, S = \frac{1}{2}$ or ^2P (6 states); and $L = 0, S = \frac{3}{2}$ or ^4S (4 states). This gives a total of 20 orbitals. The spin-orbit interaction splits these terms into fine structure multiplets, according to the possible J values (36.55), each having $2J+1$ components, namely: $^2D_{\frac{5}{2}}$ (6 components), $^2D_{\frac{3}{2}}$ (4), $^2P_{\frac{3}{2}}$ (4), $^2P_{\frac{1}{2}}$ (2) and $^4S_{\frac{3}{2}}$ (4). The Zeeman effect completely removes the degeneracy of the fine structure terms. The magnetic separation $\mu_B B g$ is a function of the Landé g-factor, and hence, depends on L, S and J, as given by Eq.(36.56). We thus obtain

Fine structure levels:	$^2D_{\frac{5}{2}}$	$^2D_{\frac{3}{2}}$	$^2P_{\frac{3}{2}}$	$^2P_{\frac{1}{2}}$	$^4S_{\frac{3}{2}}$
Number of Zeeman levels:	6	4	4	2	4
Energy spacing:	$\frac{6}{5}\mu_B B$	$\frac{4}{5}\mu_B B$	$\frac{4}{3}\mu_B B$	$\frac{2}{3}\mu_B B$	$2\mu_B B$

36.5. Since the energy eigenvalues ε_{nl} of the electron in a multielectron atom are lower than the Bohr values ε_n, the energy levels ε_{nl} can be written for low Z values in the form

$$\varepsilon_{nl} = -\frac{Rh}{n^2} Z_{eff}^2$$

where the difference $Z - Z_{eff}$ is called the *shielding constant* and gives the effect of the other $Z - 1$ electrons in reducing the electrostatic attraction between the nucleus and the electron in the state given by nl. Calculate Z_{eff} for the electron configurations of the five lightest atoms.

Solution

Hydrogen $(Z = 1)$ has a single electron in the ground-state configuration $1s^1 (n = 1, l = 0)$. The energy required to remove the electron, known as the binding or ionization energy of the one-electron atom, is defined, in view of Eq.(32.9), by

$$\Delta \varepsilon_n = \varepsilon_\infty - \varepsilon_n = \frac{Rh}{n^2} Z^2$$

For hydrogen we obtain $\Delta \varepsilon_1 = Rh = 13.6$ eV in the ground-state configuration.

Helium $(Z=2)$ has the ground state configuration $1s^2$. The ionization energy in a multielectron atom, defined as the energy required to remove the most weakly bound electron, is

$$\Delta\varepsilon_{nl} = \varepsilon_\infty - \varepsilon_{nl} = \frac{Rh}{n^2} Z_{eff}^2$$

Since the total binding energy of the ground state of helium is $\Delta E_{10} = -E_{10} = 79$ eV and the binding energy of a single electron in a $1s$ orbit about a $Z=2$ nucleus is given by Eq.(36.54) as $\Delta\varepsilon_1 = Rh(2)^2 = 54.4$ eV, the ionization energy must be the difference between the two, namely $\Delta\varepsilon_{10} = 24.6$ eV. This gives $Z_{eff} = \sqrt{24.6/13.6} = 1.35$. In other words, each electron moves in the field of a central charge $Z_{eff} e = 1.35 e$. The shielding constant is $Z - Z_{eff} = 0.65$. The configuration of helium $1s^2$ represents the closed K shell.

Lithium $(Z=3)$ is obtained by adding one electron to the $1s^2$ configuration. The experimental ionization energy is $\Delta\varepsilon = 5.4$ eV so that $Z_{eff} = \sqrt{5.4/13.6} = 0.63$ for a third electron in the K shell and $Z_{eff} = 2\sqrt{5.4/13.6} = 1.26$ if it belongs to the L shell, where $n=2$. The first value implies a shielding factor $Z - Z_{eff} = 2.37$ which exceeds the charge of the two inner electrons in the configuration $1s^2$. Thus, we must assume that the electron configuration of the helium atom is $1s^2 2s^1$. The outer electron moves in the field of a central charge $Z_{eff} e = 1.26 e$. The shielding constant of the inner electrons is $Z - Z_{eff} = 1.74$.

Beryllium $(Z=4)$ has an additional electron which leads to the configuration $1s^2 2s^2$ and to an experimental value of 9.3 eV for the ionization energy. From Eq.(36.55) it follows that $Z_{eff} = 2\sqrt{9.3/13.6} = 1.65$, which gives the shielding constant as $Z - Z_{eff} = 2.35$. As 1.74 is the contribution of the inner configuration $1s^2$ to the shielding, it follows that the contribution of the second $2s$ electron is 0.61. In other words, each of the two equivalent electrons in the subshell $2s^2$ screens 0.61 of a proton charge, a value which is close to the shielding constant 0.65 of an electron in the subshell $1s^2$ of the helium atom.

Boron $(Z=5)$ has ionization energy of 8.3 eV. From Eq.(36.55) it follows that $Z_{eff} = 2\sqrt{8.3/13.6} = 1.56$, if the fifth electron belongs to the L shell and, if it belongs to the M shell, with $n=3$, then $Z_{eff} = 3\sqrt{8.3/13.6} = 2.34$. The latter value implies that an electron in the M shell, which is expected to be further away from the nucleus than an electron in the L shell, moves in the field of a larger central charge $Z_{eff} e = 2.34 e$ than that determined for electrons in the $2s$ subshell. It follows that the electron must have $n=2$. Since the exclusion principle forbids a $2s^3$ configuration, the suggestion is that the fifth electron makes the boron configuration to be $1s^2 2s^2 2p^1$.

36.6. Find the energy level splitting of the first excited state of hydrogen, due to an applied electric field E.

36.7. Find the degeneracy of the electronic configuration $2p\,3d$, and list the possible $^{2S+1}L_J$ values.

36.8. Find the possible $^{2S+1}L_J$ values for the electronic configuration $3d\,4d$, and determine its degeneracy.

36.9. Find the $^{2S+1}L_J$ terms of the fine structure for a d^2 electronic configuration.

36.10. Show that $L = S = J = 0$ for a full shell, and use this result to find the degeneracy and the $^{2S+1}L_J$ terms of the $3d^9$ electronic configuration.

37. ATOMIC RADIATION

37.1. TIME-DEPENDENT PERTURBATION THEORY
37.2. FERMI'S GOLDEN RULE
37.3. EMISSION AND ABSORPTION OF RADIATION
37.4. SPONTANEOUS EMISSION

37.1. TIME-DEPENDENT PERTURBATION THEORY

The interaction of either the electric or magnetic fields of electromagnetic radiation with the electric or magnetic moment of an atomic system of charged particles has a time-dependent perturbing effect on the atomic stationary states. The perturbation may result in the system undergoing transitions between stationary states, which are specified by the eigenvalues ε_k^0 and the complete set of eigenfunctions $\Psi_k^0(t)$ given by

$$i\hbar \frac{\partial \Psi_k^0(t)}{\partial t} = \hat{H}_0 \Psi_k^0(t) = \varepsilon_k^0 \Psi_k^0(t) \tag{37.1}$$

A general solution to Eq.(37.1) has the form (30.20) which reads

$$\Psi^0(t) = \sum_k a_k^0 \Psi_k^0(t) = a_k^0 e^{-i\varepsilon_k^0 t/\hbar} \Psi_k^0 \tag{37.2}$$

where the a_k^0 are independent of time and restricted by

$$\sum_k |a_k^0|^2 = 1 \tag{37.3}$$

During the presence of the radiation perturbation $\hat{V}(t)$, the Hamiltonian of the system becomes

$$\hat{H} = \hat{H}_0 + \hat{V}(t) \tag{37.4}$$

and, since the state of the system varies with time, we must adopt the time-dependent Schrödinger equation to be a basis for the perturbation theory, that is

$$i\hbar \frac{\partial \Psi(t)}{\partial t} = \left[\hat{H}_0 + \hat{V}(t)\right]\Psi(t) \tag{37.5}$$

We assume that Eq.(37.5) has normalized solutions which can be expanded in terms of the Ψ_k^0 as

$$\Psi(t) = \sum_k a_k(t)\Psi_k^0(t) = \sum_k a_k(t)e^{-i\varepsilon_k^0 t/\hbar}\Psi_k^0 \tag{37.6}$$

where the normalization condition on $\Psi(t)$ gives

$$\sum_k |a_k(t)|^2 = 1 \tag{37.7}$$

that is, $|a_k(t)|^2$ gives the probability that the system is in the state Ψ_k^0 at time t. Substituting Eq.(37.6) into Eq.(37.5) yields

$$i\hbar \sum_k \left[\frac{da_k(t)}{dt} - \frac{i\varepsilon_k^0}{\hbar}a_k(t)\right]e^{-i\varepsilon_k^0 t/\hbar}\Psi_k^0 = \sum_k \left[\varepsilon_k^0 + \hat{V}(t)\right]a_k(t)e^{-i\varepsilon_k^0 t/\hbar}\Psi_k^0$$

which reduces to

$$i\hbar \sum_k \frac{da_k(t)}{dt}e^{-i\varepsilon_k^0 t/\hbar}\Psi_k^0 = \sum_k \hat{V}(t)a_k(t)e^{-i\varepsilon_k^0 t/\hbar}\Psi_k^0 \tag{37.8}$$

In view of orthogonality of the stationary state eigenfunctions, multiplication by $(\Psi_n^0)^*$ followed by integration over the configuration space gives

$$i\hbar \sum_k \frac{da_k(t)}{dt}e^{-i\varepsilon_k^0 t/\hbar}\delta_{nk} = \sum_k a_k(t)e^{-i\varepsilon_k^0 t/\hbar}\langle\Psi_n^0|\hat{V}(t)|\Psi_k^0\rangle \tag{37.9}$$

Denoting the matrix elements of the perturbation operator between the stationary states n and k by

$$V_{nk}(t) = \langle\Psi_n^0|\hat{V}(t)|\Psi_k^0\rangle \tag{37.10}$$

we can rewrite Eq.(37.9) as

$$i\hbar \frac{da_n(t)}{dt} = \sum_k a_k(t)V_{nk}(t)e^{i(\varepsilon_n^0 - \varepsilon_k^0)t/\hbar} \tag{37.11}$$

which is a set of simultaneous differential equations of the first order for the coefficients $a_n(t)$. If we assume that the system is in a definite stationary state Ψ_i^0 at some particular time $t = 0$, and then it is allowed to interact with the radiation field, we obtain from Eq.(37.6) that

$$\Psi_i^0 = \Psi(0) = \sum_k a_k(0) \Psi_k^0$$

Multiplying by $\left(\Psi_k^0\right)^*$ and integrating gives

$$a_k(0) = \delta_{ki} \tag{37.12}$$

that is, $a_i(0) = 1$ and $a_k(0) = 0$ for $k \neq i$. If we consider a time interval t sufficiently short for the coefficients $a_k(t)$ in Eqs.(37.11) to be approximated by the form (37.12), which means that the term $k = i$ gives the greatest contribution to the sum and all other terms can be neglected, we have

$$i\hbar \frac{da_n(t)}{dt} = V_{ni}(t) e^{i(\varepsilon_n^0 - \varepsilon_i^0)t/\hbar} \quad , \quad i\hbar \frac{da_i(t)}{dt} = a_i(t) V_{ii}(t) \tag{37.13}$$

and on integration we obtain

$$a_n(t) = -\frac{i}{\hbar} \int_0^t V_{ni}(t) e^{i(\varepsilon_n^0 - \varepsilon_i^0)t'/\hbar} dt' \quad (n \neq i) \quad , \quad a_i(t) = 1 - \frac{i}{\hbar} \int_0^t V_{ii}(t') dt' \tag{37.14}$$

The probability (29.29) that the system is in the state Ψ_n^0 at time t, in other words, that $\Psi(t)$ coincides with Ψ_n^0, is given by

$$P_{in}(t) = \left|\langle \Psi_n^0 | \Psi(t) \rangle\right|^2 = \left|a_n(t)\right|^2 \tag{37.15}$$

and this is taken to be the *probability of direct transition* from state Ψ_i^0 to state Ψ_n^0, in a time t, induced by the presence of the radiation field. Assuming a transition where a photon is emitted, that is, $\varepsilon_i^0 > \varepsilon_n^0$, we recall the Bohr condition (28.19), which reads

$$\varepsilon_i^0 - \varepsilon_n^0 = \hbar \omega_{in} \tag{37.16}$$

so that Eqs.(37.14) can be rewritten as

$$a_n(t) = -\frac{i}{\hbar} \int_0^t V_{ni}(t') e^{-i\omega_{in} t'} dt' \quad (n \neq i)$$

$$\tag{37.17}$$

$$a_i(t) = 1 - \frac{i}{\hbar} \int_0^t V_{ii}(t') dt'$$

As we can see from the above, the time-dependent perturbation theory is concerned with transitions that result from perturbation, between the unperturbed levels of the system, while time-independent perturbation theory calculates the stationary shifts in the energy levels.

37.2. Fermi's Golden Rule

Consider the case of a *constant perturbation*, switched on at time $t=0$. The integrals (37.17) become straightforward, and we obtain

$$a_n(t) = \frac{V_{ni}}{\hbar \omega_{in}}\left(e^{-i\omega_{in}t} - 1\right) \quad (n \neq i)$$

$$a_i(t) = 1 - \frac{i}{\hbar} V_{ii} t$$

(37.18)

Substituting Eqs.(37.18) into Eq.(37.6) gives, to a first approximation, the wave function of the atomic system as

$$\Psi(t) = e^{-iV_{ii}t/\hbar}\Psi_i^0(t) + \sum_{n \neq i} \frac{V_{ni}}{\hbar \omega_{in}}\left(e^{-i\omega_{in}t} - 1\right)\Psi_n^0(t) \qquad (37.19)$$

The probability (37.15) of observing the system in the state Ψ_n^0 at time t, for the coefficients $a_n(t)$ given by Eq.(37.18), reduces to

$$P_{in}(t) = \frac{|V_{ni}|^2}{\hbar^2 \omega_{in}^2} 4\sin^2(\omega_{in} t / 2) \qquad (37.20)$$

which means that transitions from state Ψ_i^0 to states Ψ_n^0 only take place if the matrix elements V_{ni} do not vanish. The plot of $P_{in}(t)$ as a function of ω_{in}, given in Figure 37.1 for a perturbation which is on for a time t, suggests that the probability behaves like a Dirac δ-function as t approaches infinity. We may represent the δ-function as

$$\lim_{t \to \infty} \frac{\sin(\omega_{in} t/2)}{\pi \omega_{in}/2} = \lim_{t \to \infty} \frac{1}{2\pi} \int_{-t}^{t} e^{i\omega_{in} x/2} dx = \frac{1}{2\pi} \int_{-\infty}^{\infty} e^{i\omega_{in} x/2} dx = \delta(\omega_{in}/2) = 2\delta(\omega_{in}) \quad (37.21)$$

where use was made of a property derived from the normalization condition, written as

$$\int_{-\infty}^{\infty} \delta(u) du = \int_{-\infty}^{\infty} \delta(ku) d(ku) = 1 \quad \text{or} \quad \delta(ku) = \frac{1}{k}\delta(u) \qquad (37.22)$$

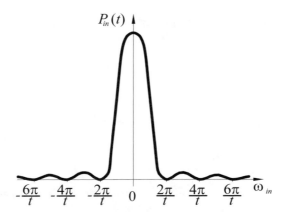

Figure 37.1. Dependence of the transition probability on frequency

It follows from Figure 37.1 that, in the large-t limit, we have $\omega_{in} \cong 0$, and hence $\left[2\sin(\omega_{in}t/2)/\omega_{in}t \right] \to 1$. Thus, one obtains

$$\lim_{t \to \infty} \frac{4\sin^2(\omega_{in}t/2)}{\omega_{in}^2 t} = \lim_{t \to \infty} \left[\frac{\sin(\omega_{in}t/2)}{\omega_{in}t/2} \pi \frac{\sin(\omega_{in}t/2)}{\pi \omega_{in}/2} \right] = \pi\delta(\omega_{in}/2) = 2\pi\delta(\omega_{in})$$

so that the transition probability becomes

$$\lim_{t \to \infty} P_{in}(t) = \frac{2\pi t}{\hbar^2}|V_{ni}|^2 \delta(\omega_{in}) = \frac{2\pi t}{\hbar}|V_{ni}|^2 \delta(\varepsilon_i^0 - \varepsilon_n^0) \qquad (37.23)$$

It follows that the transitions take place between states of equal unperturbed energies $\varepsilon_n^0 \to \varepsilon_i^0$. The transition probability is proportional to the time interval t during which the perturbation is on, that is $P_{in}(t)$ approaches infinity as $t \to \infty$. Since no experiment lasts for an infinite time, it is the transition probability per unit time, called the *transition rate* $P_{i \to n}$, which is of particular interest. Normally, ε_n^0 is part of a continuum, and hence, the transition might end in any state which belongs to a finite range of N final states, rather than in a particular state. We may then define the transition rate as

$$P_{i \to n} = \lim_{t \to \infty} \frac{1}{t} \int P_{in}(t)dN \qquad (37.24)$$

The integral over the final states is usually transformed into one over energy, if we define a *density of states* $\rho(\varepsilon)$, representing the number of continuum states per energy interval, by taking

$$dN = \rho(\varepsilon_n^0)d\varepsilon_n^0 \qquad (37.25)$$

so that, substituting Eqs.(37.23) and (37.25), Eq.(37.24) gives

$$P_{i\to n} = \lim_{t\to\infty} \frac{1}{t} \int P_{in}(t)\rho(\varepsilon_n^0)d\varepsilon_n^0 = \frac{2\pi}{\hbar}|V_{ni}|^2 \rho(\varepsilon_n^0)\delta(\varepsilon_i^0 - \varepsilon_n^0) \qquad (37.26)$$

This result is known as *Fermi's golden rule*, which states that *transitions may occur only between states of equal energies for which the matrix elements of the perturbation operator do not vanish*, and that *the transition rate is proportional to both the square modulus of these matrix elements and the density of final states*.

37.3. EMISSION AND ABSORPTION OF RADIATION

Consider the transitions between stationary states which are induced by a *harmonic perturbation* of the form

$$\hat{V}(t) = 2\hat{V}(\vec{r})A(\omega)\cos\omega t = \hat{V}(\vec{r})A(\omega)(e^{i\omega t} + e^{-i\omega t}) \qquad (37.27)$$

It is assumed that the harmonic perturbation corresponds to the interaction of either the electric or magnetic field components of the incoming radiation with the electric or magnetic moment of the atom. Thus, the coupling between the system and the perturbing field can be taken in the form of a field amplitude $A(\omega)$ times a displacement of the system $\hat{V}(\vec{r})$. On substitution of Eq.(37.27) into Eq.(37.10) one obtains

$$V_{ni}(t) = \langle \Psi_n^0 | \hat{V}(\vec{r}) | \Psi_i^0 \rangle A(\omega)(e^{i\omega t} + e^{-i\omega t}) = V_{ni}A(\omega)(e^{i\omega t} + e^{-i\omega t}) \qquad (37.28)$$

It follows from Eq.(37.17) that

$$a_n(t) = -\frac{i}{\hbar}V_{ni}A(\omega)\int_0^t \left[e^{-i(\omega_{in}-\omega)t'} + e^{-i(\omega_{in}+\omega)t'}\right]dt'$$

$$= \frac{V_{ni}A(\omega)}{\hbar}\left[\frac{e^{-i(\omega_{in}-\omega)t'}}{\omega_{in}-\omega} + \frac{e^{-i(\omega_{in}+\omega)t'}}{\omega_{in}+\omega}\right]_0^t = \frac{V_{ni}A(\omega)}{\hbar}\left[\frac{e^{-i(\omega_{in}-\omega)t}-1}{\omega_{in}-\omega} + \frac{e^{-i(\omega_{in}+\omega)t}-1}{\omega_{in}+\omega}\right] \qquad (37.29)$$

If the atomic system is bathed into a radiation field of frequency $\omega = \omega_{ni} = -\omega_{in}$, which implies that $\varepsilon_n^0 > \varepsilon_i^0$ according to the Bohr condition (37.16), the transition from state Ψ_i^0 to state Ψ_n^0 describes a process of *induced absorption*, where the energy of the system is increased by energy transfer from the radiation field to the atom. If $\omega = \omega_{ni} = -\omega_{in}$, the second term on the right-hand side of Eq.(37.29) grows without limit, so that the first term can be neglected.

If, however, $\omega_{in} > 0$, so that $\varepsilon_i^0 > \varepsilon_n^0$, the transition from state Ψ_i^0 to state Ψ_n^0 is accompanied by a decrease in the energy of the system. Such a process is termed *induced emission*, where energy is transferred to a radiation field of frequency $\omega = \omega_{in}$. Induced emission is described by the first term in Eq.(37.29), which has the dominant contribution to $a_n(t)$ for this value of ω.

If we consider the effect of each term in Eq.(37.29) on the transition probability between states, the only change which occurs in Eq.(37.18) is a shift of $\pm \omega$ in the resonance frequency, which reads

$$a_n^{\pm}(t) = \frac{V_{ni} A(\omega)}{\hbar(\omega_{in} \pm \omega)} \left[e^{-i(\omega_{in} \pm \omega)t} - 1 \right] \tag{37.30}$$

Proceeding as before, one obtains a transition probability similar to that given by Eq.(37.23), namely,

$$\lim_{t \to \infty} P_{in}^{\pm}(t) = \frac{2\pi t}{\hbar} |V_{ni}|^2 |A(\omega)|^2 \delta\left(\varepsilon_i^0 - \varepsilon_n^0 \pm \hbar\omega\right) \tag{37.31}$$

where $P_{in}^+(t)$, which vanishes unless the final energy ε_n^0 is greater than the initial energy ε_i^0 (in other words $\varepsilon_n^0 = \varepsilon_i^0 + \hbar\omega$) describes the *absorption* of energy by the system, whereas $P_{in}^-(t)$ represents the probability of energy *emission*, restricted by the condition $\varepsilon_i^0 = \varepsilon_n^0 + \hbar\omega$. The transition rate is then derived from Eq.(37.31) if we divide by t and integrate over the initial states, the final states, or the radiation frequencies, which may be elements of a continuum. The usual situation now is that of a transition between two discrete bound states induced by a perturbation continuous in ω, that is, by a radiation field represented by a wave packet in terms of conjugate parameters ω and t. Hence, we consider the absorption and emission processes to be induced by a perturbation

$$\hat{V}(t) = \hat{V}(\vec{r}) \int_{-\infty}^{\infty} A(\omega) e^{\pm i\omega t} d\omega \tag{37.32}$$

so that the transition rate is defined as

$$P_{i \to n} = \lim_{t \to \infty} \frac{1}{t} \int_{-\infty}^{\infty} P_{in}^{\pm}(t) d\omega \tag{37.33}$$

Substituting Eq.(37.31) into Eq.(37.33) one obtains

$$P_{i \to n} = \frac{2\pi}{\hbar^2} |V_{ni}|^2 \int_{-\infty}^{\infty} |A(\omega)|^2 \hbar \delta\left(\varepsilon_i^0 - \varepsilon_n^0 \pm \hbar\omega\right) d\omega = \frac{2\pi}{\hbar} |V_{ni}|^2 |A(\omega_{in})|^2 \tag{37.34}$$

In other words, the transition rate is the same for both the induced absorption and induced emission. The radiative transition can only be induced between stationary states for which the matrix elements V_{ni} do not vanish by radiation fields which contain the corresponding resonant frequency component $\omega = \omega_{in}$.

EXAMPLE 37.1. Selection rules for electric dipole transitions

Consider a one-electron atom exposed to electromagnetic radiation that is described by the electric vector (20.8), which reads

$$\vec{E} = \vec{E}_0(\omega)e^{-i\omega t}$$

We will evaluate the electric field at the centre of the atom, which we choose to be at the origin of x, y and z coordinates. For optical transitions we may neglect the spatial variation of the electric field, because the wavelengths are large as compared to the size of atoms. The electron acquires in the electric field a potential energy, which following Eq.(11.8), we can write as

$$U = e\int \vec{E} \cdot d\vec{r} = e\vec{E} \cdot \vec{r} = -\vec{p} \cdot \vec{E} \tag{37.35}$$

where $\vec{p} = -e\vec{r}$ is the atomic dipole moment. Equation (37.35) can be expressed either in Cartesian or spherical coordinates in the form

$$e\vec{r} \cdot \vec{E} = xE_x + yE_y + zE_z = r\sin\theta\left(E_x \cos\varphi + E_y \sin\varphi\right) + r\cos\theta E_z \tag{37.36}$$

$$= \frac{1}{2}r\sin\theta e^{i\varphi}\left(E_x - iE_y\right) + \frac{1}{2}r\sin\theta e^{-i\varphi}\left(E_x + iE_y\right) + r\cos\theta E_z = \sum_k r_k E_k = \sum_k r_k E_{0k}(\omega)e^{-i\omega t}$$

where k stands for the three components of the dipole interaction. For each component of the electric field the perturbation operator assumes the form $\hat{V}_k(t) = er_k E_{0k}(\omega)e^{-i\omega t}$, so that Eq.(37.28) reads

$$\left[\hat{V}_k(t)\right]_{ni} = \left\langle \Psi_n^0 \left| er_k \right| \Psi_i^0 \right\rangle E_{0k}(\omega)e^{-i\omega t}$$

Substituting into Eq.(37.34), where $A(\omega)$ is identified with $E_{0k}(\omega)$ and the displacement matrix elements are identified with those of the components of the atomic dipole moment, one obtains

$$P_{i \to n} = \frac{2\pi}{\hbar^2} \left|\left\langle \Psi_n^0 \left| er_k \right| \Psi_i^0 \right\rangle\right|^2 \left|E_{0k}(\omega_{in})\right|^2 \tag{37.37}$$

Transitions between states Ψ_i^0 and Ψ_n^0, for which the matrix elements $e\left\langle \Psi_n^0 \left| r_k \right| \Psi_i^0 \right\rangle$ vanish, are said to be *forbidden* for electric dipole radiation. We may derive some of the selection rules for electric dipole transitions between states described by the spatial wave functions (32.14), which read

$$\Psi_{nlm} = N_{lm} R_{nl}(r) P_e^m(\cos\theta) e^{im\varphi}$$

The electron spin is neglected for simplicity. In spherical coordinates, the matrix elements (37.37) take the form

$$\left\langle \Psi_{n'l'm'} \left| \frac{1}{2} r \sin\theta e^{i\varphi} \right| \Psi_{nlm} \right\rangle = \frac{1}{2} N_{l'm'} N_{lm} \int_0^\infty R_{n'l'}^* R_{nl} r^3 dr$$

$$\times \int_0^\pi P_{l'}^{m'}(\cos\theta) P_l^m(\cos\theta) \sin\theta d(\cos\theta) \int_0^{2\pi} e^{-i(m'-m-1)\varphi} d\varphi$$

$$\left\langle \Psi_{n'l'm'} \left| \frac{1}{2} r \cos\theta e^{-i\varphi} \right| \Psi_{nlm} \right\rangle = \frac{1}{2} N_{l'm'} N_{lm} \int_0^\infty R_{n'l'}^* R_{nl} r^3 dr$$

$$\times \int_0^\pi P_{l'}^{m'}(\cos\theta) P_l^m(\cos\theta) \cos\theta d(\cos\theta) \int_0^{2\pi} e^{-i(m'-m+1)\varphi} d\varphi$$

$$\left\langle \Psi_{n'l'm'} \left| r \cos\theta \right| \Psi_{nlm} \right\rangle = \frac{1}{2} N_{l'm'} N_{lm} \int_0^\infty R_{n'l'}^* R_{nl} r^3 dr$$

$$\times \int_0^\pi P_{l'}^{m'}(\cos\theta) P_l^m(\cos\theta) \cos\theta d(\cos\theta) \int_0^{2\pi} e^{-i(m'-m)\varphi} d\varphi$$

The selection rule on the magnetic quantum number results from the integrals over φ which vanish unless $m' - m = -1$, $m' - m = 1$ or $m' = m$, respectively. In other words, the allowed transitions for electric dipole radiation are restricted to a change in magnetic quantum number of

$$\Delta m = 0, \pm 1 \tag{37.38}$$

The selection rule on l can be derived from the integrals over θ, substituting the recurrence formulae involving the associated Legendre functions given by

$$(2l+1)\sin\theta P_l^m(\cos\theta) = P_{l+1}^{m+1}(\cos\theta) - P_{l-1}^{m+1}(\cos\theta)$$

$$(2l+1)\cos\theta P_l^m(\cos\theta) = (l-m+1)P_{l+1}^m(\cos\theta) + (l+m)P_{l-1}^m(\cos\theta)$$

The first and second integrals over θ vanish unless $l' = l \pm 1$, and this implies that the allowed transitions for electric dipole radiation are also restricted by

$$\Delta l = \pm 1 \tag{37.39}$$

As the angular momentum states, Eq.(34.1), can be written as a product of electronic wavefunctions for space and spin orbitals, the transition probability (37.37) can be put in the form

$$P_{i \to n} = \frac{2\pi}{\hbar^2} \left| \left\langle \Phi_n^0 \left| er_k \right| \Phi_i^0 \right\rangle \right|^2 \left| E_{0k}(\omega_{in}) \right|^2 = \frac{2\pi}{\hbar^2} \left| \left\langle \Psi_n^0 \left| er_k \right| \Psi_i^0 \right\rangle \right|^2 \left| \left\langle \chi_n^0 \left| \chi_i^0 \right\rangle \right|^2 \left| E_{0k}(\omega_{in}) \right|^2$$

and it follows that $P_{i \to n} = 0$, so that the transition is forbidden, unless

$$\Delta s = 0 \tag{37.40}$$

which is a spin selection rule for electric dipole transitions.

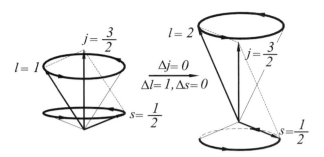

Figure 37.2. Angular momenta representation in an electric dipole transition where $\Delta j = 0$

If j and m_j are good quantum numbers, and hence, $\vec{L}$ and $\vec{S}$ are precessing around $\vec{J}$, as described by the vector model of one-electron atoms, Figure 34.3, from $j = l \pm s$ it follows that $\Delta j = \Delta l \pm \Delta s$ and, in view of Eqs.(37.39) and (37.40), one obtains $\Delta j = \pm 1$. However, it is also possible to preserve the value of j, that is to observe electric dipole radiation for $\Delta j = 0$, if the relative position of $\vec{L}$ and $\vec{S}$ changes in a transition where $\Delta l = \pm 1$ and $\Delta s = 0$, as shown in Figure 37.2. Therefore, the selection rules on j for the allowed transitions have the form

$$\Delta j = 0, \pm 1 \qquad (37.41)$$

The corresponding restrictions on m_j are similar to those found for the orbital magnetic quantum number, Eq.(37.38), that is

$$\Delta m_j = 0, \pm 1 \qquad (37.42)$$

There is no selection rule on n, and hence, any change Δn is allowed for a radiative transition.

37.4. Spontaneous Emission

Consider a system consisting of a large number of atoms in a radiation field of spectral energy density $u(\omega, T)$ in the frequency range ω to $\omega + d\omega$ given by

$$u_\omega = u(\omega, T) = \frac{1}{2}\left[\varepsilon_0 \langle E^2 \rangle + \frac{1}{\mu_0}\langle B^2 \rangle\right] = \varepsilon_0 \langle E^2 \rangle = \frac{1}{2}\varepsilon_0 |E_0(\omega)|^2 \qquad (37.43)$$

where the thermodynamic equilibrium at a temperature T is obtained by absorption and emission of radiation. In the electric dipole approximation (37.37), if the radiation is incident upon the atom from all directions and with random polarization, the transition rate must include independent contributions from the matrix element of each component

r_k, that is,

$$P_{i\to n} = \frac{2\pi e^2}{\hbar^2}|E_{0k}(\omega_{in})|^2 \sum_k |\langle\Psi_n^0|r_k|\Psi_i^0\rangle|^2$$

If r_k are the Cartesian components of $\vec{r}$, we also have

$$|E_{0x}(\omega_{in})|^2 = |E_{0y}(\omega_{in})|^2 = |E_{0z}(\omega_{in})|^2 = \frac{1}{3}|E_0(\omega_{in})|^2$$

This gives

$$P_{i\to n} = \frac{2\pi e^2}{3\hbar^2}|E_0(\omega_{in})|^2 \sum_k |\langle\Psi_n^0|r_k|\Psi_i^0\rangle|^2 = B_{i\to n}u(\omega_{in},T)$$

where $B_{i\to n}$ is defined as

$$B_{i\to n} = \frac{4\pi e^2}{3\varepsilon_0\hbar^2}\sum_k |\langle\Psi_n^0|r_k|\Psi_i^0\rangle|^2 \tag{37.44}$$

for the electric dipole transitions. If $\varepsilon_n^0 > \varepsilon_i^0$, $B_{i\to n}$ is associated with an absorption process and is called *Einstein's coefficient of absorption*.

If the system in state Ψ_n^0 makes a transition to a lower state Ψ_i^0, the transition rate may similarly be written as

$$P_{n\to i} = B_{n\to i}u(\omega_{in},T)$$

where $B_{n\to i}$ is called *Einstein's coefficient of induced emission*. We have showed earlier that there is the same transition rate for induced absorption and induced emission, and hence, we can write $B_{i\to n} = B_{n\to i}$. Let there be N_i atoms in the energy level ε_i^0 and N_n atoms in the energy level ε_n^0. Since no net energy is absorbed or emitted by the atoms, the spectral energy density $u(\omega_{in},T)$ remains constant, and this implies that

$$N_i B_{i\to n}u(\omega_{in},T) = N_n B_{n\to i}u(\omega_{in},T) = N_n B_{i\to n}u(\omega_{in},T)$$

However, this result is inconsistent with the occupation numbers of the energy levels given by the Boltzman distribution law (26.25) at thermal equilibrium between atoms and the radiation field, which should be related as

$$\frac{N_n}{N_i} = \frac{e^{-\varepsilon_n^0/k_B T}}{e^{-\varepsilon_i^0/k_B T}} = e^{-\hbar\omega_{in}/k_B T} \tag{37.45}$$

Thus, we must assume that there exists a process, independent of the presence of the radiation field $u(\omega_{in},T)$, which makes the lower energy levels more densely populated. The concept of *spontaneous emission* was formally introduced by Einstein in terms of an intrinsic transition rate defined as

$$P_{n \to i} = A_{n \to i} \tag{37.46}$$

and known as *Einstein's coefficient of spontaneous emission*. The equation which describes the balance between absorption and emission at thermal equilibrium becomes

$$N_i B_{i \to n} u(\omega_{in}, T) = N_i B_{n \to i} u(\omega_{in}, T) = N_n [B_{n \to i} u(\omega_{in}, T) + A_{n \to i}] \tag{37.47}$$

and may be solved for $u(\omega_{in}, T)$ which gives

$$u(\omega_{in}, T) = \frac{A_{n \to i}}{B_{n \to i}} \frac{1}{N_i / N_n - 1} = \frac{A_{n \to i}}{B_{n \to i}} \frac{1}{e^{\hbar \omega_{in} / k_B T} - 1} \tag{37.48}$$

As we expect this relation to obey the Planck radiation formula (27.36), one obtains a relationship between the spontaneous and induced emission coefficients in the form

$$A_{n \to i} = \frac{\hbar \omega_{in}^3}{\pi^2 c^3} B_{n \to i} \tag{37.49}$$

Therefore, the rate of spontaneous emission for electric dipole radiation results from Eqs.(37.42) and (37.49) as

$$A_{n \to i} = \frac{8e^2 \omega_{in}^3}{3\varepsilon_0 hc^3} \sum_k \left| \langle \Psi_n^0 | r_k | \Psi_i^0 \rangle \right|^2 \tag{37.50}$$

FURTHER READING

1. B. H. Bransden, C. J. Joachain - QUANTUM MECHANICS, Pearson Education, Boston, 2000.
2. J. C. Townsend - MODERN APPROACH TO QUANTUM MECHANICS, University Science Books, Sausalito, 2000.
3. F. Battaglia - NOTES IN CLSSSICAL AND QUANTUM MECHANICS, Blackwell Scientific, Oxford, 1990.

PROBLEMS

37.1. Use the time-dependent perturbation theory to find the probability of a $1s \to 2s$ transition in a one electron atom, due to a sudden change of its nuclear charge $Z \to Z \pm 1$ following a $\beta^\pm$ decay.

Solution

Because of the sudden change at time $t = 0$, when a constant perturbation $\hat{V} = \pm e_0^2/r$ is switched on, we can integrate by parts Eq.(37.13) and, for $n \neq i$, this gives

$$a_n(t) = -V_{ni} \left[\frac{e^{i(\varepsilon_n^0 - \varepsilon_i^0)t'/\hbar}}{\varepsilon_n^0 - \varepsilon_i^0} \right]_0^t + \int_0^t \frac{\partial V_{ni}(t')}{\partial t'} \frac{e^{i(\varepsilon_n^0 - \varepsilon_i^0)t'/\hbar}}{\varepsilon_n^0 - \varepsilon_i^0} dt'$$

The time dependence of the matrix elements $V_{ni}(t')$ accounts for the effect of suddenly applying the constant perturbation at $t = 0$ and can be described by

$$\frac{\partial V_{ni}(t')}{\partial t'} = V_{ni}\delta(t') \quad \text{i.e.} \quad V_{ni}(t) = V_{ni}\int_{-\infty}^t \delta(t')dt' = V_{ni} \text{ (if } t \geq 0\text{)} \quad \text{or} \quad V_{ni}(t) = 0, \text{ (if } t < 0\text{)}$$

This leads to

$$a_n(t) = \frac{V_{ni}}{\varepsilon_i^0 - \varepsilon_n^0}\left[e^{i(\varepsilon_n^0 - \varepsilon_i^0)t/\hbar} - 1\right] - \frac{V_{ni}}{\varepsilon_n^0 - \varepsilon_i^0}$$

The first term corresponds to a transition probability of the form

$$\frac{|V_{ni}|^2}{(\varepsilon_i^0 - \varepsilon_n^0)^2} 2\left(1 - \cos\frac{\varepsilon_n^0 - \varepsilon_i^0}{\hbar}t\right) = \frac{|V_{ni}|^2}{(\varepsilon_i^0 - \varepsilon_n^0)^2} 4\sin^2\left(\frac{\varepsilon_n^0 - \varepsilon_i^0}{2\hbar}t\right)$$

which, for large t values, behaves like the Dirac δ-function $\delta(\varepsilon_n^0 - \varepsilon_i^0)$. In other words, we will obtain in this limit a transition probability different from zero only for energies $\varepsilon_n^0 \to \varepsilon_i^0$. Neglecting this term, as we are interested in a first order transition to an excited state, and substituting $a_n(t)$ into Eq.(9.68), gives

$$P_{in}(t) = \frac{|V_{ni}|^2}{(\varepsilon_n^0 - \varepsilon_i^0)^2}$$

This result is valid when the perturbation energy $\hat{V} = \pm e_0^2/r$ is small, namely, for large Z values only and, in this case, we have $E_{2s} - E_{1s} = 3Z^2 e_0^2/8a_0$, which yields

$$P_{1s \to 2s} \approx \frac{16 a_0^2}{9Z^4 e_0^4} \left|\langle \Psi_{2s}| \pm \frac{e_0^2}{r}|\Psi_{1s}\rangle\right|^2 = \frac{2}{3}\frac{2^{11}}{9^4 Z^2}$$

37.2. An electron resonance transition between the Zeeman levels, occurring in the presence of a static field $\vec{B}_0$, is induced by a time-dependent perturbation of the form

$$\hat{V}(t) = 2\mu_B g B_1 \hat{J}_x \cos \omega t$$

which is produced by a magnetic field $B_x = B_1 \cos \omega t$ oscillating along the x direction. Use Fermi's golden rule to find an expression for the transition rate.

Solution

The time-dependent perturbation can be written as

$$\hat{V}(t) = 2\mu_B g B_1 \hat{J}_x \cos \omega t = \mu_B g B_1 \hat{J}_x (e^{i\omega t} + e^{-i\omega t})$$

and, by taking the matrix elements between two Zeeman angular momentum eigenstates ψ_{jm_j} and $\psi_{jm'_j}$, one obtains

$$\hat{V}_{m_j m'_j}(t) = \mu_B g B_1 \langle \psi_{jm'_j} | \hat{J}_x | \psi_{jm_j} \rangle (e^{i\omega t} + e^{-i\omega t})$$

which has a form similar to Eq.(37.28). It follows that the transition rate from state ψ_{jm_j} to state $\psi_{jm'_j}$ can be written in the form given by Eq.(37.31), which reads

$$P_{m_j \to m'_j} = \frac{2\pi}{\hbar^2} \mu_B^2 g^2 B_1^2 |\langle \psi_{jm'_j} | \hat{J}_x | \psi_{jm_j} \rangle|^2 \delta(\omega_{m_j m'_j} - \omega)$$

If we further use Eqs.(33.34) and (33.35) for the nonvanishing matrix elements of $\hat{J}_x$, and let $\gamma = \mu_B g/\hbar$, we obtain

$$P_{m_j \to m_j \pm 1} = \frac{\pi}{2} \gamma^2 B_1^2 [j(j+1) - m_j(m_j \pm 1)] \delta(\omega_{m_j, m_j \pm 1} - \omega)$$

These radiative transitions involve the perturbation produced by the magnetic dipole moment, and hence, are called *magnetic dipole transitions*. The selection rule on m_j, implied by the expression derived for the transition probability, is $\Delta m_j = \pm 1$.

37.3. Find the energy levels of the ^3F and ^3D multiplets of the fine structure, and use the selection rules to determine the allowed transitions $^3F \to {}^3D$.

Solution

The fine structure energy levels are given by Eq.(34.19) which, for multielectron atoms, reads

$$E_{SL} = \tfrac{1}{2}[J(J+1) - L(L+1) - S(S+1)]\zeta_{LS}$$

As J is given by Eq.(36.55), one obtains for the two LS multiplets that

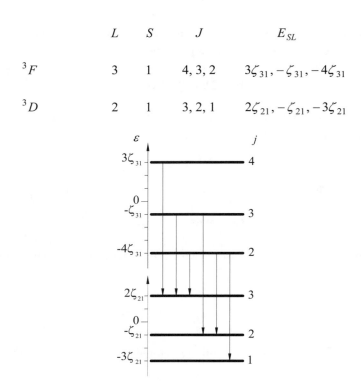

	L	S	J	E_{SL}
^3F	3	1	4, 3, 2	$3\zeta_{31}, -\zeta_{31}, -4\zeta_{31}$
^3D	2	1	3, 2, 1	$2\zeta_{21}, -\zeta_{21}, -3\zeta_{21}$

Using the selection rules on L and J, which have the form (37.39) and (37.41), respectively, leads to the six allowed radiative transitions illustrated in the figure.

37.4. If a system of atoms can make radiative transitions to the ground state, with a decay constant of the excited stated given by Einstein's coefficient of spontaneous emission $A_{n \to i}$, find an expression for the spectral intensity of the radiation.

Solution

The spontaneous emission of photons with a constant rate $A_{n \to i}$ results in a finite lifetime of the excited state ψ_n for the ensemble of atoms. If we assume that N_n atoms were certainly in the excited state at a particular time t, the probability per unit time that decay will occur after a short time is proportional to N_n and the rate of spontaneous emission is given by

$$\frac{dN_n}{dt} = -A_{n \to i} N_n \quad \text{or} \quad N_n(t) = N_n(0) e^{-A_{n \to i} t} = N_n(0) e^{-t/\langle t \rangle}$$

This shows that Einstein's coefficient $A_{n \to i}$ stands for the *decay constant* of the excited state, which is a measure of the *mean lifetime* $\langle t \rangle = 1/A_{n \to i}$. It follows that there is a time dependence of the radiation intensity of the form

$$|E|^2 = |E_0(\omega_{ni})|^2 e^{-A_{n \to i} t}$$

which makes varying any field component about the central frequency ω_{ni} of the transition as

$$E = E_0(\omega_{ni})e^{-i\omega_{ni}t - A_{n\to i}t/2} = \int_{-\infty}^{\infty} E_0(\omega)e^{-i\omega t}d\omega$$

The Fourier transform of this equation is

$$E_0(\omega) = \frac{E_0(\omega_{ni})}{2\pi}\int_0^\infty e^{i(\omega-\omega_{ni}) - A_{n\to i}t/2}d\omega = \frac{E_0(\omega_{ni})}{2\pi}\frac{1}{-i(\omega-\omega_{ni}) + \frac{1}{2}A_{n\to i}}$$

and this gives the spectral intensity in the so-called Lorentzian form

$$|E_0(\omega)|^2 = \frac{|E_0(\omega_{ni})|^2}{4\pi^2}\frac{1}{(\omega-\omega_{ni})^2 + \frac{1}{4}A_{n\to i}^2}$$

Thus, the natural linewidth in spontaneous emission is determined by Einstein's coefficient $A_{n\to i}$.

37.5. Assuming that a population inversion $N_n > N_i$, where $\varepsilon_n^0 > \varepsilon_i^0$, can be produced in an optical medium, show that the radiation of wavelength λ will be amplified as it passes through such a medium, and find an expression for the gain constant.

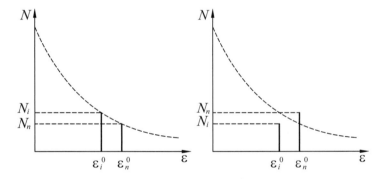

Solution

When a radiation beam propagates through an optical medium, each process of absorption by a single atom substracts a photon of energy $\hbar\omega_{in}$ from the beam, while each process of induced emission results in addition of the same energy, so that the net time rate of change of the spectral energy density is

$$\frac{du_\omega}{dt} = (N_n - N_i)P_{i\to n}\hbar\omega_{in} = (N_n - N_i)B_{i\to n}u_\omega\hbar\omega_{in}$$

If we make use of Eqs.(37.43), (17.30) and (17.23), this equation can be rewritten in terms of the average energy flow, given by the spectral intensity $I(\omega)$ as

$$\frac{dI(\omega)}{dx} = (N_n - N_i)B_{i\to n}\frac{\hbar\omega_{in}}{c}I(\omega) = \alpha I(\omega)$$

Integrating over the distance x travelled by the beam gives $I(\omega) = I_0(\omega)e^{\alpha x}$. The Boltzman distribution law (37.45) shows that $N_n = N_i e^{-\hbar\omega_{in}/k_B T} \ll N_i$, that is, $\alpha < 0$, and the radiation will be attenuated as it passes through the medium, because the loss due to absorption exceeds the gain due to the induced emission. A *population inversion* can be produced, however, if $N_n > N_i$, where $\varepsilon_n^0 > \varepsilon_i^0$, as illustrated in the figure. Substituting Eq.(37.49), Eq.(37.51) gives α, called the *gain constant,* as

$$\alpha = (N_n - N_i)\frac{\lambda^2}{4} A_{n \to i}$$

If the population inversion exists, the radiation will be amplified as it passes through the medium, as in laser systems.

37.6. If a perturbation $V(t) = -\alpha x e^{-\gamma^2 t^2}$ is applied to the ground state ψ_0 of energy $\varepsilon_0 = \hbar\omega/2$ of a simple harmonic oscillator of mass m and frequency ω, find the transition probability to the first excited state ψ_1 of energy $\varepsilon_1 = 3\hbar\omega/2$.

37.7. What is the number of allowed radiative transitions between the multiplets of the fine structure $^4D \to {}^4P$ and $^4P \to {}^4S$ in a multielectron atom?

37.8. Find the $^{2S+1}L_J$ terms for the fine structure of hydrogen corresponding to $n = 3$ and $n = 4$. What is the number of allowed radiative transitions between these fine structure multiplets?

37.9. Find the number of allowed radiative transitions $3d \to 2p$ between the Zeeman multiplets of the fine structure of hydrogen.

37.10. Find an expression for Einstein's coefficient of spontaneous decay of the hydrogen atom from the $2p$ excited state to the ground state.

38. Systems of Atoms

38.1. THE ADIABATIC APPROXIMATION
38.2. LINEAR LATTICE VIBRATIONS
38.3. PHONONS
38.4. LATTICE HEAT CAPACITY

38.1. The Adiabatic Approximation

We will consider a sufficiently large collection of atoms so that they exhibit the characteristic behaviour of the bulk material. In such a material the nucleus may only retain its more tightly bound electrons, whereupon the entity is known to be an *ion*. Hence, we will consider a system of ions of position vectors $\vec{R}_p$, masses M_p and charges $Z_p e$ and electrons $\vec{r}_j, m_e, -e$, described by the Hamiltonian operator

$$\hat{H} = -\frac{\hbar^2}{2m_e}\sum_j \nabla_j^2 - \sum_p \frac{\hbar^2}{2M_p}\nabla_p^2 + \frac{1}{2}\sum_j\sum_{k\neq j}\frac{e_0^2}{r_{jk}} - \sum_j\sum_p \frac{Z_p e_0^2}{\left|\vec{r}_j - \vec{R}_p\right|} + \frac{1}{2}\sum_p\sum_{q\neq p}\frac{Z_p Z_q e_0^2}{R_{pq}}$$

(38.1)

$$= -\frac{\hbar^2}{2m_e}\sum_j \nabla_j^2 - \sum_p \frac{\hbar^2}{2M_p}\nabla_p^2 + U_{ee}(r) + U_{ei}(r,R) + U_{ii}(R)$$

where p is a running index over all the ions, the index j extends over all the electrons, $U_{ee}(r)$, $U_{ei}(r,R)$ and $U_{ii}(R)$ are the potential energies of the electron-electron, electron-ion and ion-ion interaction, respectively. The electron motion may be separated from that of the ions, in the Schrödinger equation of the system, by considering first the Hamiltonian $\hat{H}_0$ describing the electron motion for a set of fixed ionic positions, denoted collectively by R, as

$$\hat{H}\Phi(r,R) = \left[\hat{H}_0 - \sum_p \frac{\hbar^2}{2M_p}\nabla_p^2\right]\Phi(r,R) = E\Phi(r,R) \qquad (38.2)$$

The eigenvalue problem for the electronic motion reads

$$\hat{H}_0 \Psi(r,R) = \left[-\frac{\hbar^2}{2m_e} \sum_j \nabla_j^2 + U_{ee}(r) + U_{ei}(r,R) + U_{ii}(R) \right] \Psi(r,R) = E(R)\Psi(r,R) \quad (38.3)$$

where $\Psi(r,R)$ and $E(R)$ are the eigenfunctions and eigenvalues of $\hat{H}_0$ which depend on the fixed parameter R only. The electron coordinates, denoted collectively by r, are the dynamical variables in Eq.(38.3). Assuming that the eigenfunctions $\Psi(r,R)$ are known, the wave functions $\Phi(r,R)$ which solve the Schrödinger equation (38.2) are usually approximated by the following expression

$$\Phi(r,R) = \Psi(r,R)\varphi(R) \quad (38.4)$$

and this is known to be the *adiabatic approximation*. Substituting Eq.(38.4), Eq.(38.2) becomes

$$-\sum_p \frac{\hbar^2}{2M_p}\left[\varphi(R)\nabla_p^2 \Psi(r,R) + 2\nabla_p \Psi(r,R) \cdot \nabla_p \varphi(R) + \Psi(r,R)\nabla_p^2 \varphi(R) \right]$$
$$+ E(R)\Psi(r,R)\varphi(R) = E\Psi(r,R)\varphi(R) \quad (38.5)$$

Multiplying to the left of each term by $\Psi^*(r,R)$ and integrating over all the electron positions, we obtain the equation for $\varphi(R)$ as

$$-\sum_p \frac{\hbar^2}{2M_p}\nabla_p^2 \varphi(R) + E(R)\varphi(R)$$
$$= E\varphi(R) + \sum_p \frac{\hbar^2 \varphi(R)}{2M_p} \int \Psi^*(r,R)\nabla_p^2 \Psi(r,R)dr + \sum_p \frac{\hbar^2}{M_p}\nabla_p \varphi(R) \cdot \int \Psi^*(r,R)\nabla_p \Psi(r,R)dr \quad (38.6)$$

The last two terms on the right-hand side vanish if the electrons can be considered to be perfectly free, because $\Psi(r,R)$ becomes independent of the ion coordinates R, so that $\nabla_p \Psi(r,R)$ is zero. Such an approximation can be used for metallic systems. Otherwise, the eigenfunctions $\Psi(r,R)$ are assumed to be slowly varying functions of the ionic coordinates R and the second and third terms on the right-hand side of Eq.(38.6) can be neglected to a first approximation. Under this assumption Eq.(38.6) reduces to the equation for ionic motion

$$-\sum_p \frac{\hbar^2}{2M_p}\nabla_p^2 \varphi(R) + E(R)\varphi(R) = E\varphi(R) \quad (38.7)$$

As a result of the adiabatic approximation to the many-body eigenvalue problem of a system of atoms, it is possible either to suppress the ionic interactions while considering the motion of the electrons, described by Eq.(38.3), or to neglect the electronic interactions when solving the eigenvalue problem (38.7) for the motion of the ions. In other words, ionic motion can be ignored when describing the binding of the electrons to the nucleus, whereas the chemical bond may be regarded as a parameter when considering the ionic motion. The state functions (38.4) are obtained by first solving the electronic problem for fixed ionic positions and then using the total electronic energy $E(R)$ as a potential function for the ionic motion. The potential energy $E(R)$ exhibits a set of minima, corresponding to the equilibrium positions R_0 of the ions. For R close to R_0 we may then expand as

$$E(R) = E(R_0) + \frac{1}{2}(R-R_0)^2 \left(\frac{\partial^2 E}{\partial R^2}\right)_{R=R_0} + \ldots \cong E(R_0) + \frac{1}{2}\kappa_0 (R-R_0)^2$$

where the first order derivatives are zero at equilibrium and κ_0 stands for the *coupling constant*, having the dimensions of an elastic constant. Therefore the effect of the electrons appears, to a first approximation, to be an elastic coupling of the ions, with force constants which are dependent on the electronic states.

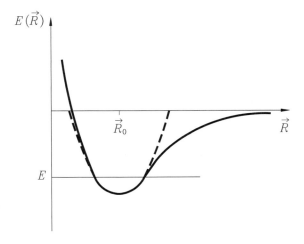

Figure 38.1. Electronic potential energy for the ionic motion and its harmonic approximation

Any change in the electronic states $\Psi(r, R)$ modifies the potential energy $E(R)$ felt by the ions, and so modifies their coupling. For many purposes it is sufficient to describe ion dynamics using the *harmonic approximation*, illustrated in Figure 38.1 by a broken line, which limits the treatment to harmonic motion of the ions.

38.1. LINEAR LATTICE VIBRATIONS

Since many of the interesting features of lattice harmonic motion do not depend on either the three-dimensional or quantum mechanical approach, we first consider a vibrating lattice in one dimension, described by the classical counterpart of the Hamiltonian (38.7) which reads

$$H = \frac{1}{2M}\sum_{p=1}^{N} P_p^2 + E(R_1, R_2, \ldots, R_N) \qquad (38.8)$$

where N identical ions of mass M and position coordinates R_p have been assumed. The ionic motion is described by the canonical equations (24.27), which take the form

$$\dot{R}_p = \frac{\partial H}{\partial P_p} \quad , \quad \dot{P}_p = -\frac{\partial H}{\partial R_p} \qquad (p = 1, 2, \ldots, N) \qquad (38.9)$$

The equilibrium position R_{p0} of the pth ion may be expressed in terms of the equilibrium spacing a of the ions as $R_{p0} = pa$, and hence, the actual position R_p can be written in terms of the displacement s_p from equilibrium as

$$R_p = R_{p0} + s_p \qquad (38.10)$$

If the interaction between ions is assumed to be conservative, the potential energy for the ionic motion takes the form (2.33), so that the Hamiltonian (38.8) becomes

$$H = \frac{1}{2M}\sum_p P_p^2 + \frac{1}{2}\sum_p \sum_{q \neq p} E_{pq}(R_p - R_q) \qquad (38.11)$$

It is useful to expand the potential energy in terms of s_p, s_q about the equilibrium separation $R_{p0} - R_{q0}$, up to the second order, and this gives

$$E_{pq}(R_p - R_q) = E_{pq}(R_{p0} - R_{q0}) + s_p \left(\frac{\partial E_{pq}}{\partial R_p}\right)_{R_{p0}, R_{q0}} + s_q \left(\frac{\partial E_{pq}}{\partial R_q}\right)_{R_{p0}, R_{q0}}$$

$$+ \frac{1}{2} s_p^2 \left(\frac{\partial^2 E_{pq}}{\partial R_p^2}\right)_{R_{p0}, R_{q0}} + \frac{1}{2} s_q^2 \left(\frac{\partial^2 E_{pq}}{\partial R_q^2}\right)_{R_{p0}, R_{q0}} + s_p s_q \left(\frac{\partial^2 E_{pq}}{\partial R_p \partial R_q}\right)_{R_{p0}, R_{q0}}$$

$$= E_{pq}(R_{p0} - R_{q0}) + \frac{1}{2}\left[\left(s_p \frac{\partial}{\partial R_p} + s_q \frac{\partial}{\partial R_q}\right)^2 E_{pq}\right]_{R_{p0}, R_{q0}} \qquad (38.12)$$

where the first-order derivatives vanish, because of the equilibrium conditions

$(\partial E_{pq}/\partial R_p)_{R_{p0},R_{q0}} = (\partial E_{pq}/\partial R_q)_{R_{p0},R_{q0}} = 0$ and the second-order derivatives, denoted by

$$\kappa_{pq} = \kappa_{qp} = \left[\frac{\partial^2 E_{pq}(R_p - R_q)}{\partial R_p \partial R_q}\right]_{R_{p0},R_{q0}} \tag{38.13}$$

are known as *coupling constants*. They have the dimensions of elastic constants, so that the force exerted on the pth ion, when the qth ion is displaced by a distance s_q, is $-\kappa_{pq} s_q$. If all the ions are displaced by a constant amount, the resulting force on each ion corresponding to that displacement, in essence a translation, must be zero, that is

$$\sum_{q \neq p} \kappa_{pq} = 0 \tag{38.14}$$

The total potential energy can now be written as

$$\frac{1}{2}\sum_p \sum_{q \neq p} E_{pq}(R_p - R_q) = \frac{1}{2}\sum_p \sum_{q \neq p} E_{pq}(R_{p0} - R_{q0}) + \frac{1}{2}\sum_p \sum_{q \neq p} \kappa_{pq} s_p s_q$$

The first term on the right-hand side is just a constant, and the zero of the potential energy can be chosen such that this constant is zero. Hence, the potential energy reduces, in the harmonic approximation, to a quadratic function of the displacements and the Hamiltonian (38.11) becomes

$$H = \frac{1}{2M}\sum_p P_p^2 + \frac{1}{2}\sum_p \sum_{q \neq p} \kappa_{pq} s_p s_q \tag{38.15}$$

Substituting Eq.(38.15) into Eqs.(38.9), and setting $\dot{R}_p = \dot{s}_p$ and $\partial/\partial R_p = \partial/\partial s_p$, according to Eq.(38.10), we obtain the equation of motion for the pth ion as

$$M\ddot{s}_p = -\sum_{q \neq p} \kappa_{pq} s_q \tag{38.16}$$

It is reasonable to assume *normal mode* trial solutions of the form

$$s_p = A_p e^{-i\omega t} \tag{38.17}$$

where A_p are considered to be time-independent. Substituting Eq.(38.17) into Eq.(38.16) gives

$$M\omega^2 A_p = \sum_{q \neq p} \kappa_{pq} A_q \tag{38.18}$$

A similar equation can be written for the $(p+1)$th ion as

$$M\omega^2 A_{p+1} = \sum_{q \neq p+1} \kappa_{p+1,q} A_q = \sum_{q+1 \neq p+1} \kappa_{p+1,q+1} A_{q+1} = \sum_{q \neq p} \kappa_{pq} A_{q+1} \qquad (38.19)$$

where we made the change $q \to q+1$ in the dummy variable of summation, and it is obvious, from the definition (38.13), that $\kappa_{p+1,q+1} = \kappa_{pq}$. Comparing Eqs.(38.18) and (38.19) we see that A_p and A_{p+1} can only differ by a phase factor, that is

$$A_{p+1} = e^{ika} A_p \qquad (38.20)$$

where a is the lattice spacing, and k is a constant to be determined from the boundary conditions. By applying p times Eq.(38.20), starting with $p=0$, we obtain

$$A_p = e^{ikpa} A_0 \qquad (38.21)$$

Note that the allowed k values must be real, in order to keep A_p finite when $p \to \infty$. The significance of k may be derived by using the periodic boundary conditions, which require that ions p and $p+N$ are one and the same

$$A_{p+N} = A_p \qquad (38.22)$$

so that from Eq.(38.21) we obtain

$$Nka = 2\pi l \qquad (38.23)$$

where l is an arbitrary integer, which labels the normal mode solutions. We may infer that k should be a wave number of the form $k = 2\pi l/Na$ which belongs to the one-dimensional k-space, where the separation of two adjacent modes is given by $\Delta k = 2\pi \Delta l/Na = 2\pi/Na$, since the smallest value of Δl is 1, as l is an integer. Hence, the density of modes $1/\Delta k$ in the one-dimensional k-space, for a linear chain of length L, is

$$\frac{Na}{2\pi} = \frac{L}{2\pi} \qquad (38.24)$$

It is obvious that A_p, given by Eq.(38.21), remains unchanged if we replace k by $k + 2\pi/a$, so that all the solutions of physical significance can be obtained by restricting k to lie within a range of wave numbers equal to $2\pi/a$, defined as

$$-\frac{\pi}{a} \leq k \leq \frac{\pi}{a} \qquad (38.25)$$

which is referred to as the *first Brillouin zone* in one dimension. For this range of k, the normal modes will be labelled by N consecutive integers, namely, $-N/2 \leq l \leq N/2$, except $l=0$ which corresponds to a translation of the crystal with zero frequency and without change in potential energy. The normal frequencies ω_k are obtained by inserting

Eq.(38.21) into Eq.(38.17), which becomes

$$s_p = A_0 e^{i(kpa-\omega_k t)} \tag{38.26}$$

and then substituting the displacements (38.26) into the equation of motion (38.16), which gives

$$M\omega_k^2 = \sum_{q \neq p} \kappa_{pq} e^{ika(q-p)} = \sum_h \kappa(h) e^{ikah} = \sum_{h>0} \kappa(h) \cos(kha) \tag{38.27}$$

where $h = q - p$ is an integer. From Eq.(38.13) we obtain $\kappa(h) = \kappa_{qp} = \kappa_{pq} = \kappa(-h)$. Equation (38.27) represents a *dispersion relation*, as it gives the dependence $\omega_k(h)$, which is clearly determined by the coupling constants $\kappa(h)$ contributing the potential energy. If we assume that only nearest neighbour forces are involved, $\kappa(h)$ are all zero, except $\kappa_0 = \kappa(0) = \kappa_{pp}$ and $\kappa(\pm 1) = \kappa_{p,p\pm 1}$. Substituting into Eq.(38.21), gives

$$\kappa_0 + \kappa(1) + \kappa(-1) = 0 \quad \text{or} \quad \kappa(1) = \kappa(-1) = -\frac{\kappa_0}{2} \tag{38.28}$$

In this case, the equation of motion (38.16) for the *p*th ion reduces to

$$M\ddot{s}_p = -\kappa_0 s_p + \frac{\kappa_0}{2} s_{p+1} + \frac{\kappa_0}{2} s_{p-1} = -\frac{\kappa_0}{2}(s_p - s_{p-1}) - \frac{\kappa_0}{2}(s_p - s_{p+1}) \tag{38.29}$$

which has the form of the elastic interaction between the nearest neighbours only, as in Figure 38.2, where $\kappa_0 / 2$ stands for the restoring force constant.

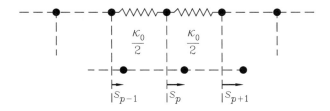

Figure 38.2. The linear monatomic chain model

Hence, the dispersion relation (38.27) reduces to

$$\omega_k^2 = \frac{1}{M}\left[\kappa_0 + \kappa(1)\cos ka + \kappa(-1)\cos(-ka)\right] = \frac{\kappa_0}{M}(1 - \cos ka)$$

and this reads

$$\omega_k = \sqrt{\frac{2\kappa_0}{M}} \left|\sin\frac{ka}{2}\right| = \omega_{max} \left|\sin\frac{ka}{2}\right| \tag{38.30}$$

Substituting this result into Eq.(38.26) shows that each ion may oscillate in N different normal modes, with normal frequencies ω_k corresponding to the allowed k values, given by the boundary condition (38.23). It follows that the ion displacement can be described as a superposition of the harmonic solutions (38.26) of the form

$$s_p = \sum_k A_k e^{i(kpa-\omega_k t)} \tag{38.31}$$

For a large one-dimensional lattice, we may assume an almost continuous distribution of normal frequencies, and hence, Eq.(38.30) can be represented by an almost continuous set of points that is by a dispersion curve (Figure 38.3).

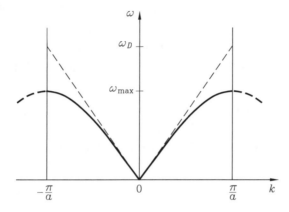

Figure 38.3. Dispersion curve for a linear monatomic lattice

It is convenient, in this case, to define a frequency distribution function $N(\omega)$ by the number of allowed frequencies between ω and $\omega+d\omega$, such that

$$\int N(\omega)d\omega = N = \int 2\left(\frac{Na}{2\pi}\right)dk \tag{38.32}$$

where the number of modes in the element dk of k-space is given by Eq.(38.23) to be $Na\,dk/2\pi$, and a factor of 2 comes from the fact that $\omega(k)=\omega(-k)$. In view of Eq.(38.30), it follows that

$$N(\omega)d\omega = \frac{Na}{\pi}\frac{dk}{d\omega}d\omega \quad \text{or} \quad \frac{dk}{d\omega} = \frac{d}{d\omega}\left[\frac{2}{a}\arcsin\left(\frac{\omega}{\omega_{max}}\right)\right] = \frac{2}{a\left(\omega_{max}^2 - \omega^2\right)^{1/2}}$$

We thus obtain the frequency distribution function as

$$N(\omega) = \frac{2N/\pi}{\left(\omega_{max}^2 - \omega^2\right)^{1/2}} \tag{38.33}$$

with a cut-off frequency ω_{max}, which is illustrated in Figure 38.4.

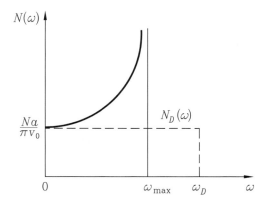

Figure 38.4. Frequency distribution for a linear monatomic lattice

---

EXAMPLE 38.1. The Debye approximation

Lattice vibrations can be described in terms of waves propagating along a linear chain, such that the amplitude of vibration A_p of the pth ion should be equal to the amplitude of vibration of the zeroth ion A_0, provided $pa = \lambda$. From Eq.(38.21) it follows that $e^{ikpa} A_0 = e^{ik\lambda} A_0 = A_0$ or $\lambda = 2\pi / k$. There are two values of k equal in magnitude and opposite in sign, corresponding to each ω, which describe the waves moving in opposite directions along the chain. The phase velocity of the waves can be derived from Eq.(38.30) as

$$v = \frac{\omega}{k} = a\sqrt{\frac{\kappa_0}{2M}} \frac{\sin(ka/2)}{ka/2} = v_0 \frac{\sin(ka/2)}{ka/2} \qquad (38.34)$$

In the limit where the wavelength is long compared to the lattice spacing a, the lattice may be regarded to be a continuous distribution of matter. As $k \to 0$, the limiting value of velocity is

$$v_0 = a\sqrt{\frac{\kappa_0}{2M}} \qquad (38.35)$$

because $2\sin(ka/2)/ka \to 1$. This result can be interpreted in terms of the elastic constants for longitudinal vibrations in a continuous line, if a finite section of the lattice of length a is regarded to be a uniform elastic string of cross sectional area S and density $\rho = M / Sa$. We can take the x-axis to lie along the string. The elastic force between nearest neighbours at the ends of such a finite section acts as a tensile force $\kappa_0 (s_p - s_{p-1}) = 2S\sigma_x$ which determines the dilatational strain $\varepsilon_x = (s_p - s_{p-1})/a$ (see Figure 38.2), so that from Eq.(15.18) we can define Young's modulus to be $E = \sigma_x / \varepsilon_x = a\kappa_0 / 2S$. Substituting into Eq.(38.35) gives

$$v_0 = \sqrt{\frac{E}{\rho}}$$

and this coincides with the velocity (15.23) of longitudinal waves propagating in a stretched string. Thus, in the *continuum* or *Debye approximation*, a linear lattice can sustain long wavelength elastic (acoustic) waves, which propagate with no dispersion, according to Eq.(38.35), which reads

$$\omega = v_0 k \tag{38.36}$$

The frequency distribution function (38.33) for a linear lattice, when $k \to 0$ and therefore $\omega/\omega_{max} \ll 1$, can be approximated by

$$N(\omega) = \frac{2N}{\pi \omega_{max}} \left(1 + \frac{1}{2} \frac{\omega^2}{\omega_{max}^2}\right) = \frac{Na}{\pi v_0} + \frac{Na^3}{8\pi v_0^3}\omega^2 \tag{38.37}$$

The relationships (38.36) and (38.37) are represented for comparison by dashed lines in Figures 38.3. and 38.4

38.2. Phonons

The N-body problem of lattice vibrations can be solved exactly if it is reduced to N one-body problems. This can be achieved by reformulating the Hamiltonian (38.15) as a function of *generalized coordinates* ξ_k and *generalized momenta* η_k, which are defined in terms of displacements s_p and momenta P_p by

$$\xi_k = \frac{1}{\sqrt{N}} \sum_p s_p e^{-ikpa} \quad , \quad \eta_k = \frac{1}{\sqrt{N}} \sum_p P_p e^{ikpa} \tag{38.38}$$

where $1/\sqrt{N}$ is a normalizing factor. Our linear lattice, with periodic boundary conditions (38.23), has the useful property

$$\frac{1}{N}\sum_k e^{ik(p-p')a} = \frac{1}{N}\sum_{l=0}^{N-1} e^{i2\pi(p-p')l/N} = \frac{1}{N}\frac{1-e^{i2\pi(p-p')}}{1-e^{i2\pi(p-p')/N}} = \frac{1}{N}\frac{1-1}{1-e^{i2\pi(p-p')/N}} = 0 \tag{38.39}$$

since $p - p'$ is an integer for $p \neq p'$. For $p = p'$ the sum gives N, so that we have

$$\frac{1}{N}\sum_k e^{ik(p-p')a} = \delta_{pp'} \quad , \quad \frac{1}{N}\sum_k e^{i(k-k')pa} = \delta_{kk'} \tag{38.40}$$

From Eq.(38.38) we then obtain

$$\frac{1}{\sqrt{N}}\sum_k \xi_k e^{ikpa} = \frac{1}{N}\sum_{p'} s_{p'} \sum_k e^{ik(p-p')a} = s_p, \quad \frac{1}{\sqrt{N}}\sum_k \eta_k e^{-ikpa} = \frac{1}{N}\sum_{p'} P_{p'} \sum_k e^{ik(p'-p)a} = P_p$$

(38.41)

Since both the displacements s_p and momenta P_p are real, we have

$$\xi_{-k} = \xi_k^* \quad \text{and} \quad \eta_{-k} = \eta_k^* \quad (38.42)$$

and substituting Eqs.(38.41) into Eq.(38.15) gives

$$H = \frac{1}{2M}\sum_k \sum_{k'} \eta_k \eta_{k'} \frac{1}{N}\sum_p e^{-i(k+k')pa} + \frac{1}{2}\sum_k \sum_{k'} \xi_k \xi_{k'} \sum_p \sum_{q \neq p} \kappa_{pq} e^{i(kp+k'q)a}$$

As we have

$$e^{i(kp+k'q)a} = e^{i(k+k')pa} e^{ik'(q-p)a} = e^{i(k+k')pa}(A_q / A_p)$$

where use has been made of Eq.(38.21), it follows, in view of Eq.(38.18), that

$$H = \frac{1}{2M}\sum_k \sum_{k'} \eta_k \eta_{k'} \delta_{k',-k} + \frac{1}{2}\sum_k \sum_{k'} \xi_k \xi_{k'} \delta_{k',-k} \left(\frac{1}{A_p}\sum_{q \neq p} \kappa_{pq} A_q\right)$$

$$= \sum_k \left[\frac{1}{2M}\eta_k \eta_k^* + \frac{1}{2}M\omega_k^2 \xi_k \xi_k^*\right] = \sum_k H_k \quad (38.43)$$

which is a Hamiltonian for N independent harmonic oscillators. Their generalized coordinates, obtained by substituting the ion displacement (38.31) into Eq.(38.41), as $\xi_k = \sqrt{N} A_k e^{-i\omega_k t}$ are solutions to N equations of the form $\ddot{\xi} + \omega_k^2 \xi_k = 0$, which describe simple harmonic motions with normal frequencies ω_k. The Hamiltonian H_k for each single mode may be factorized in such a way that the lattice vibrations can be described in terms of raising and lowering operators $\hat{a}_k^\dagger$ and $\hat{a}_k$ defined as

$$\hat{a}_k = \sqrt{\frac{M\omega_k}{2}}\hat{\xi}_k + i\frac{\hat{\eta}_{-k}}{\sqrt{2M\omega_k}}, \quad \hat{a}_k^\dagger = \sqrt{\frac{M\omega_k}{2}}\hat{\xi}_{-k} - i\frac{\hat{\eta}_k}{\sqrt{2M\omega_k}} \quad (38.44)$$

where $\hat{\xi}_k, \hat{\xi}_{-k}$ and $\hat{\eta}_k, \hat{\eta}_{-k}$ are considered to be operators which obey the commutation relations (29.32), postulated for conjugate variables. It follows that the commutation relations (30.39) hold for $\hat{a}_k^\dagger$ and $\hat{a}_k$. Equations (38.44) lead to a quantum mechanical Hamiltonian $\hat{H}_k$ for a single mode of the form

$$\hat{H}_k = \omega_k \hat{a}_k^\dagger \hat{a}_k + \frac{1}{2}\hbar\omega_k \quad (38.45)$$

similar to that discussed in Problem 30.5. In view of Eq.(38.43), Eq.(38.45) allows us to introduce the total Hamiltonian operator $\hat{H}$ as

$$\hat{H} = \sum_k \left(\omega_k \hat{a}_k^\dagger \hat{a}_k + \frac{1}{2} \hbar \omega_k \right) \tag{38.46}$$

with the energy eigenvalue

$$E = \sum_k \hbar \omega_k \left(n_k + \frac{1}{2} \right) \tag{38.47}$$

which is the sum of the separate harmonic oscillator energies, each having the form (30.33). The vibrational eigenstate of $\hat{H}$ is then defined by the number n_k of quanta of energy $\hbar \omega_k$ in each normal mode of the linear lattice. These quanta are called *phonons*. There can be any number of quanta in each mode, so that the average energy per normal mode can be defined from Eq.(38.47) by

$$\langle \varepsilon_k \rangle = \left(\langle n_k \rangle + \frac{1}{2} \right) \hbar \omega_k = \left[f(\omega_k) + \frac{1}{2} \right] \hbar \omega_k \tag{38.48}$$

where $f(\omega_k) = \langle n_k \rangle$ is the average value of n_k at thermal equilibrium at a temperature T. The mean energy can be expressed by an ensemble average over a canonical ensemble, using Eq.(26.65), which reads

$$\langle \varepsilon_k \rangle = \frac{\sum_{n_k=0}^{\infty} \varepsilon_k e^{-\varepsilon_k/k_B T}}{\sum_{n_k=0}^{\infty} e^{-\varepsilon_k/k_B T}} = \frac{\sum_{n_k=0}^{\infty} n_k \hbar \omega_k e^{-n_k \hbar \omega_k/k_B T}}{\sum_{n_k=0}^{\infty} e^{-n_k \hbar \omega_k/k_B T}} + \frac{1}{2} \hbar \omega_k$$

If we set $x = e^{-\hbar \omega_k/k_B T}$, this equation can be rewritten as

$$\langle \varepsilon_k \rangle = \left(\frac{\sum_{n_k=0}^{\infty} n_k x^{n_k}}{\sum_{n_k=0}^{\infty} x^{n_k}} + \frac{1}{2} \right) \hbar \omega_k = \left(\frac{x \sum_{n_k=0}^{\infty} n_k x^{n_k-1}}{\sum_{n_k=0}^{\infty} x^{n_k}} + \frac{1}{2} \right) \hbar \omega_k$$

$$= \left(\frac{x}{1-x} + \frac{1}{2} \right) \hbar \omega_k = \left(\frac{1}{e^{\hbar \omega_k/k_B T} - 1} + \frac{1}{2} \right) \hbar \omega_k = \varepsilon(\omega_k) \tag{38.49}$$

where use was made of the identity (35.45). Comparing Eqs.(38.49) and (38.48) gives

$$f(\omega_k) = \frac{1}{e^{\hbar \omega_k/k_B T} - 1} \tag{38.50}$$

which is essentially the Bose-Einstein distribution law (35.44) for identical bosons. As a result of quantization of the vibrational modes, the ionic lattice may be regarded to be a system of independent quantum oscillators. In other words, the ionic lattice may be replaced by a gas of noninteracting phonons of energies $\hbar\omega_k$.

38.3. Lattice Heat Capacity

Since the vibrational properties of the ionic lattice are equivalent to those of a set of independent quantum oscillators, the internal energy U, which is the total vibrational energy, can be obtained by summing over all the normal modes. As the normal frequencies lie close together, it is convenient to assume a continuous distribution of frequencies $N(\omega)$ and to replace the sum by the integral

$$U = \int_0^{\omega_{max}} \varepsilon(\omega) N(\omega) d\omega \tag{38.51}$$

where $\varepsilon(\omega)$ is given by Eq.(38.49), $N(\omega)d\omega$ is the number of oscillators with frequencies in the range between ω and $\omega + d\omega$, and ω_{max} is the highest frequency of any normal mode. Hence, using Eq.(7.19), we can express the lattice heat capacity as

$$C_V = \left(\frac{\partial U}{\partial T}\right)_V = \int_0^{\omega_{max}} k_B \left(\frac{\hbar\omega}{k_B T}\right)^2 \frac{e^{\hbar\omega/k_B T}}{\left(e^{\hbar\omega/k_B T} - 1\right)^2} N(\omega) d\omega \tag{38.52}$$

However, Eq.(38.52) can only be compared with experimental data on heat capacity if we consider the three-dimensional case of a system of N ions having $3N$ degrees of freedom. A lattice which consists of N ions is therefore equivalent to $3N$ independent quantum oscillators, and we should assume that

$$\int_0^{\omega_{max}} N(\omega) d\omega = 3N \tag{38.53}$$

At high temperatures, where $\hbar\omega \ll k_B T$, one obtains

$$C_V \to k_B \int_0^{\omega_{max}} N(\omega) d\omega = 3N k_B \tag{38.54}$$

This is the Dulong-Petit law (26.8), which can be understood in classical terms and appears to be justified in the high-temperature range only.

The temperature dependence of the lattice heat capacity, in agreement with the predictions (10.29) of the third law of thermodynamics, can be derived by making assumptions on the frequency distribution function $N(\omega)$.

The simplest model assumes that all the modes have the same frequency ω_E, as illustrated in Figure 38.5, so that Eq.(38.52) reduces to

$$C_V = 3Nk_B \left(\frac{\hbar\omega_E}{k_B T}\right)^2 \frac{e^{\hbar\omega_E/k_B T}}{\left(e^{\hbar\omega_E/k_B T}-1\right)^2} = 3Nk_B \left(\frac{\theta_E}{T}\right)^2 \frac{e^{\theta_E/T}}{\left(e^{\theta_E/T}-1\right)^2} \qquad (38.55)$$

where $\theta_E = \hbar\omega_E / k_B$ defines the so-called *Einstein temperature*. At high temperatures, where $T \gg \theta_E$, Eq.(38.55) approaches the Dulong-Petit value (38.54). Furthermore, the Einstein temperature, which is usually determined by the best fit of experimental data to be of the order of a few hundred degrees absolute, can be regarded to be the temperature below which one obtains substantial departure from the classical model. However, at low temperatures, where $T \ll \theta_E$, Eq.(38.55) becomes

$$C_V = 3Nk_B \left(\frac{\theta_E}{T}\right)^2 e^{-\theta_E/T} \qquad (38.56)$$

and this is in poor agreement with the experimental T^3 dependence. This is illustrated by a dashed line in Figure 38.6.

Since at low temperatures the average energy per mode (38.49) becomes small when ω_k is large, the significant contribution in this temperature range to the internal energy comes from the long wavelength modes only. It is reasonable to assume that vibrations of adjacent ions are no longer independent, in the long wavelength limit, so that the lattice can be treated as a homogenous solid where all the lattice waves travel with the same phase velocity. This is the continuum or *Debye approximation*, discussed before for the linear lattice, which assumes that ω is a linear function of k, as given by the dispersion relation (38.36). A cut-off frequency $\omega_{max} = \omega_D$ is defined by the condition (38.53).

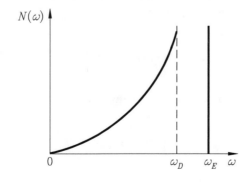

Figure 38.5. Frequency distribution for the Debye and Einstein models

For a three-dimensional lattice, the wave number k defined for a linear chain should be replaced by a wave vector $\vec{k}$ associated with each normal mode, such that the components of $\vec{k}$ should satisfy the boundary conditions derived for one dimension. Thus, the number of states per unit volume in $\vec{k}$-space will be given by

$$\frac{Pa}{2\pi}\frac{Qb}{2\pi}\frac{Sc}{2\pi} = \frac{V}{(2\pi)^3}$$

which is a straightforward generalization of Eq.(38.24) for the case of a three-dimensional lattice of volume V. Because these states are uniformly distributed, the number of modes with wave vector up to a given magnitude $|\vec{k}| = k$, which are contained in a spherical volume of radius k, is given by

$$N(k) = \left(\frac{4\pi}{3}k^3\right)\frac{V}{(2\pi)^3} = \frac{Vk^3}{6\pi^2} \qquad (38.57)$$

where V is the lattice volume. Thus, the number of modes with wave vector magnitude between k and $k + dk$ can be written as

$$N(k)dk = \frac{V}{2\pi^2}k^2 dk$$

and the frequency distribution function $N(\omega)$ follows as $N(\omega)d\omega = N(k)(dk/d\omega)d\omega$. Using the Debye approximation, Eq.(38.36), we obtain

$$N(\omega) = \frac{V\omega^2}{2\pi^2 v^3} \qquad (38.58)$$

It is assumed that to every lattice point we may assign one longitudinal and two transverse modes, with phase velocities v_l and v_t, respectively, so that for a given frequency ω, we must take

$$N(\omega) = \frac{V\omega^2}{2\pi^2}\left(\frac{1}{v_l^3} + \frac{2}{v_t^3}\right) = \frac{3V}{2\pi^2 v^3}\omega^2 \qquad (38.59)$$

where v stands for a mean phase velocity. The cut-off frequency is obtained by substituting Eq.(38.59) into Eq.(38.53), which gives

$$\omega_D = v\left(6\pi^2 \frac{N}{V}\right)^{1/3} \qquad (38.60)$$

so that Eq.(38.59) becomes

$$N(\omega) = \frac{9N}{\omega_D^3} \omega^2 \qquad (38.61)$$

Equation (38.54) is represented in Figure 38.5. Using the frequency distribution function (38.61), the lattice heat capacity (38.52) becomes

$$C_V = \frac{9Nk_B}{\omega_D^3} \int_0^{\omega_D} \left(\frac{\hbar\omega}{k_BT}\right)^2 \frac{\omega^2 e^{\hbar\omega/k_BT}}{\left(e^{\hbar\omega/k_BT}-1\right)^2} d\omega = 9Nk_B \left(\frac{T}{\theta_D}\right)^3 \int_0^{\theta_D/T} \frac{x^4 e^x}{\left(e^x-1\right)^2} dx \qquad (38.62)$$

where $x = \hbar\omega/k_BT$. The *Debye temperature* θ_D was defined as

$$\theta_D = \frac{\hbar\omega_D}{k_B} = \frac{\hbar}{k_B} v \left(6\pi^2 \frac{N}{V}\right)^{1/3} \qquad (38.63)$$

and can be estimated by an appropriate averaging of the sound velocities in the solid. It is seen from Eq.(38.62) that C_V increases with temperature, as shown in Figure 38.6.

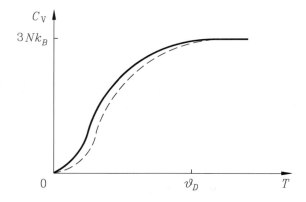

Figure 38.6. Debye heat capacity (solid line) and Einstein's description (dashed line)

For $\theta_D/T \ll 1$, the integral in Eq.(38.62) can be approximated by $(\theta_D/T)^3/3$ so that C_V approaches the Dulong-Petit value $3Nk_B$ expected at high temperatures. For $\theta_D/T \gg 1$, if the upper limit on the integral is made infinite, the integral is approximately $4\pi^4/15$, so that the low temperature limit for the heat capacity becomes

$$C_V = \frac{12\pi^4}{5} Nk_B \left(\frac{T}{\theta_D}\right)^3 \qquad (38.64)$$

and this is known as the *Debye T^3 law*. One would expect that θ_D for a given solid should be a constant, and this can be tested by considering it as an adjustable parameter which can be evaluated at different temperatures from the observed heat capacity.

FURTHER READING
1. H. Ibach, H. Lüth - SOLID STATE PHYSICS, Springer-Verlag, New York, 2003.
2. L. D. Landau, E. M. Lifschitz - QUANTUM MECHANICS, NON-RELATIVISTIC THEORY, Elsevier, Amsterdam, 1997.
3. S. M. McMurry - QUANTUM MECHANICS, Addison-Wesley, Wokingham, 1993.

PROBLEMS

38.1. Find the energy levels of a diatomic molecule of reduced mass μ, assuming that the electronic energy is described by the Morse potential $E(R) = E(R_0)\left[1 - e^{-\alpha(R-R_0)}\right]^2$, which reduces to a harmonic potential for small oscillations.

Solution

The eigenvalue equation of the two-atom system can be obtained from Eq.(31.26), by putting $L = 0$ and using the potential energy as represented by the Morse potential, that is

$$-\frac{\hbar^2}{2\mu}\frac{d^2\xi(R)}{dR^2} + E(R_0)\left[1 - e^{-\alpha(R-R_0)}\right]^2 \xi(R) = E\xi(R)$$

If we let $x = Ae^{-\alpha(R-R_0)}$, where $A^2 = 2\mu E(R_0)/\alpha^2\hbar^2$, the eigenvalue equation reduces to

$$\frac{d^2\xi}{dx^2} + \frac{1}{x}\frac{d\xi}{dx} + \left[\frac{E - E(R_0)}{E(R_0)}\frac{A^2}{x^2} + 2\frac{A}{x} - 1\right]\xi = 0$$

where the trial solution $\xi(x) = e^{-x}u(x)$ leads to

$$\frac{d^2u}{dx^2} + \left(\frac{1}{x} - 2\right)\frac{du}{dx} + \left[\frac{E - E(R_0)}{E(R_0)}\frac{A^2}{x^2} + \frac{2A-1}{x}\right]u = 0$$

This equation is solved by the power series $u(x) = \sum_{k=0}^{\infty} a_k x^{k+p}$, where $p \neq 0$ and $a_0 \neq 0$, otherwise $x = 0$ is a singularity. Substituting, one obtains

$$\sum_{k=0}^{\infty}\left\{\left[p^2 + \frac{E - E(R_0)}{E(R_0)}A^2\right]a_k x^{k+p-2} + \left[(k+1)(2p+1)a_{k+1} + (2A - 2p - 1)a_k\right]x^{k+p-1}\right.$$

$$\left. + (k+1)\left[(k+2)a_{k+2} - 2a_{k+1}\right]x^{k+p}\right\} = 0$$

and, since $a_0 \neq 0$, this implies that

$$p^2 + \frac{E-E(R_0)}{E(R_0)} A^2 = 0 \quad \text{or} \quad E = E(R_0) - \frac{E(R_0)}{A^2} p^2$$

It is convenient to redefine the index k, which gives

$$\sum_{k=0}^{\infty} \left\{ \left[p^2 + \frac{E-E(R_0)}{E(R_0)} A^2 + (k+1)(2p+1) + k(k+1) \right] a_{k+1} - (1-2A+2p+2k)a_k \right\} x^{k+p-1} = 0$$

and provides the recurrence relation

$$a_{k+1} = \frac{2(k+p-A)+1}{(k+p+1)^2 + [E-E(R_0)]A^2/E(R_0)} a_k \to \frac{2}{k} a_k$$

For large k values, this becomes identical to the recurrence relation of the expansion $e^{2x} = \sum_{k=0}^{\infty} (2x)^k/k!$ which is expected to be divergent as x tends to infinity. It follows that the series must terminate after a finite number of terms, for instance $k = n$, so that, by taking $a_n \neq 0$ and $a_{n+1} = 0$, we obtain the quantum condition

$$1 + 2(n+p-A) = 0 \quad \text{or} \quad p = A - \left(n + \frac{1}{2}\right)$$

and this gives

$$E_n = E(R_0) - \frac{E(R_0)}{A^2} \left[A^2 - 2A\left(n + \frac{1}{2}\right) + \left(n + \frac{1}{2}\right)^2 \right] = \left(n + \frac{1}{2}\right)\hbar \sqrt{\frac{2\alpha^2 E(R_0)}{\mu}} - \left(n + \frac{1}{2}\right)^2 \frac{\alpha^2 \hbar^2}{2\mu}$$

For harmonic molecular vibrations of frequency ω we can write

$$E(R) = E(R_0)\left[1 - e^{\alpha(R-R_0)}\right]^2 \approx E(R_0)\alpha^2 (R-R_0)^2 = \frac{\mu\omega^2}{2}(R-R_0)^2$$

and the required energy levels result as

$$E_n = \left(n + \frac{1}{2}\right)\hbar\omega - \left(n + \frac{1}{2}\right)^2 \frac{\alpha^2 \hbar^2}{2\mu}$$

38.2. If the electronic energy of a single ion pair is represented by $E(R) = \lambda e^{-R/\rho} - e_0^2/R$, where R_0 is the equilibrium separation, find an expression for the binding energy of a linear chain of N ion pairs with alternating charges.

Solution

For a system of N ion pairs the potential energy takes the form

$$U(R) = N\left(\sum_{j \neq i} \lambda_{ij} e^{-p_{ij}R/\rho} - \frac{e_0^2}{R} \sum_{j \neq i} \frac{\pm 1}{p_{ij}} \right)$$

where the distance $R_{ij} = p_{ij} R$ between the ions i and j is given in terms of the nearest neighbour separation R.

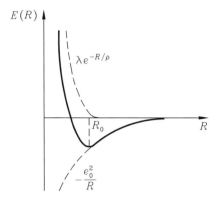

Because of the exponential dependence of the repulsive energy, we may consider, without loss of accuracy, only the nearest-neighbour separation term in the first series on the right-hand side. It is straightforward to evaluate the second series for an infinite linear chain of ions at separation R, and the result is known to be the Madelung constant, given by

$$A = \sum_{j \neq i} \frac{\pm 1}{p_{ij}} = 2\left(\frac{1}{1} - \frac{1}{2} + \frac{1}{3} - \frac{1}{4} + \ldots\right) = 2 \ln 2$$

where use has been made of the series expansion $\ln(1+x) = x - x^2/2 + x^3/3 - x^4/4 + \ldots$. The potential energy of the system of ions then reduces to

$$U(R) = N\left(\lambda e^{-R/\rho} - \frac{A e_0^2}{R}\right)$$

where ρ can be derived in terms of the equilibrium separation R_0 and isothermal compressibility k_T, Eq.(8.42). The static equilibrium condition ($p=0$) on the ionic lattice reads

$$\left(\frac{\partial U}{\partial R}\right)_{R=R_0} = 0 = N\left(\frac{A e_0^2}{R_0^2} - \frac{\lambda}{\rho} e^{-R_0/\rho}\right) \quad \text{or} \quad \lambda = \frac{\rho A e_0^2}{R_0^2} e^{R_0/\rho}$$

It follows that

$$\frac{1}{k_T} = \frac{1}{9N\alpha R_0}\left(\frac{\partial^2 U}{\partial R^2}\right)_{R=R_0} = \frac{1}{9\alpha R_0}\left(\frac{\lambda}{\rho^2} e^{-R_0/\rho} - \frac{2 A e_0^2}{R_0^3}\right) \quad \text{or} \quad \frac{R_0}{\rho} = 2 + \frac{9\alpha R_0^4}{A e_0^2 k_T}$$

Finally, the binding energy is obtained in the form

$$U(R_0) = -\frac{N A e_0^2}{R_0}\left(1 - \frac{\rho}{R_0}\right)$$

which is close to the total electrostatic energy at equilibrium separation, provided $\rho \ll R_0$.

38.3. Find the dispersion relation for a monatomic square array of lattice constant a, assuming atoms of mass M, restricted to move in the plane of the lattice, and considering nearest-neighbours interactions only.

Solution

In view of Eq.(38.29), the equation of motion for the atom p, q (see the figure) is

$$M\ddot{s}_{p,q} = \frac{\kappa_0}{2}(s_{p+1,q} + s_{p-1,q} - 2s_{p,q}) + \frac{\kappa_0}{2}(s_{p,q+1} + s_{p,q-1} - 2s_{p,q})$$

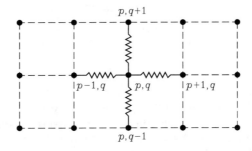

If all the atoms are considered to obey exactly the same equation of motion, the solutions have the harmonic form $s_{p,q} = A_0 e^{i(k_x pa + k_y qa - \omega t)}$ which reduces to Eq.(38.26) in the case of a linear lattice. This implies that a longitudinal mode is propagating with polarization vector parallel to the wave vector $\vec{k}$. Substituting into the equation of motion one obtains

$$M\omega^2 s_{p,q} = \frac{\kappa_0}{2}(2 - e^{ik_x a} - e^{-ik_x a})s_{p,q} + \frac{\kappa_0}{2}(2 - e^{ik_y a} - e^{-ik_y a})s_{p,q}$$

which yields the dispersion relation

$$\omega^2 = \frac{\kappa_0}{M}(2 - \cos k_x a - \cos k_y a) = \frac{2\kappa_0}{M}\left(\sin^2 \frac{k_x a}{2} + \sin^2 \frac{k_y a}{2}\right)$$

This reduces to Eq.(38.30) for $\vec{k}$ along the [10] or [01] directions, where $k_y = 0$ or $k_x = 0$, respectively, such that $k_{max} = \pi/a$, $\omega_{max} = \sqrt{2\kappa_0/M}$. The quantization of $\vec{k}$ is obtained by imposing periodic boundary conditions, as in the one-dimensional case

$$s_{p+N,q} = e^{ik_x Na} s_{p,q} = s_{p,q} \quad \text{or} \quad e^{ik_x Na} = 1$$

and this yields $k_x Na = 2\pi l$, which is equivalent to Eq.(38.23). Similarly, one obtains

$$s_{p,q+N} = e^{ik_y Na} s_{p,q} = s_{p,q} \quad \text{or} \quad e^{ik_y Na} = 1$$

which means that $k_y Na = 2\pi m$. As a result, the first two-dimensional Brillouin zone is defined by

$$-\frac{\pi}{a} \le k_x \le \frac{\pi}{a}, \quad -\frac{\pi}{a} \le k_y \le \frac{\pi}{a}$$

For $\vec{k}$ along the [11] direction, we have $k_x = k_y = k/\sqrt{2}$, and this gives a different dispersion curve described by

$$\omega = 2\sqrt{\frac{\kappa_0}{M}} \left| \sin \frac{ka}{2\sqrt{2}} \right|$$

where $k_{max} = \pi\sqrt{2}/a$ and $\omega_{max} = 2\sqrt{\kappa_0/M}$.

38.4. Use the Bose-Einstein distribution law for phonons to estimate the temperature dependence of the vibrational energy in a three-dimensional lattice.

Solution

From the graphical representation (see Figure 35.2) of $f(\omega_k)$ given by Eq.(38.50), it is apparent that only modes of frequency ω_k less than a critical value, given by $\hbar\omega_k = k_BT$, have an appreciable number of phonons, at a particular temperature T. In the continuum approximation, Eq.(38.36), this implies that the critical k value is proportional to T.

The high-frequency modes, $\hbar\omega_k \gg k_BT$, have occupation and average energy given by

$$f(\omega_k) = \frac{1}{e^{\hbar\omega_k/k_BT}-1} \to e^{-\hbar\omega_k/k_BT} \quad \text{i.e.,} \quad \langle \varepsilon_k \rangle \to \left(e^{-\hbar\omega_k/k_BT} + \frac{1}{2} \right)\hbar\omega_k$$

It is apparent that the average energy in each mode, Eq.(38.49), becomes small and tends toward the zero-point value $\hbar\omega_k/2$, as $f(\omega_k)$ tends to zero. Thus, the significant contribution to the vibrational energy comes from the low-frequency modes only, for which $\hbar\omega_k \ll k_BT$, such that $f(\omega_k) \to k_BT/\hbar\omega_k$, $\langle E_k \rangle \to k_BT$. This is the classical result, according to which a mode contributes k_BT to the vibrational energy. Assuming a uniform distribution of modes in $\vec{k}$-space, Eq.(38.57), the number of modes with wave vector lower than the critical value $|\vec{k}| = k$ will be proportional to the spherical volume $V = 4\pi k^3/3$, and thus to T^3. It follows that the total vibrational energy will be proportional to T^4, and this is found for many solids at low temperatures.

At sufficiently high temperatures, where $k_BT \gg \hbar\omega_k$ for all modes, irrespective of their frequency, the energy of every mode can be taken to be k_BT. In this case, the total number of modes is given by the number of degrees of freedom of the lattice, which is $3N$ for a set of N atoms, and therefore the vibrational energy is $3Nk_BT$, a result consistent with the Dulong-Petit law, Eq.(26.8).

38.5. Find expressions for the heat capacity of a linear monatomic lattice of Debye temperature θ_D in the low- and high-temperature limits.

Solution

The total vibrational energy is given by Eq.(38.51), which reads

$$U = \int_0^{\omega_{max}} \frac{\hbar\omega}{e^{\hbar\omega/k_B T}-1} N(\omega)d\omega$$

where the zero-point energy has been neglected, as it has no contribution to the heat capacity. Substituting the frequency distribution function for the linear monatomic lattice, Eq.(38.33), one obtains

$$U = \frac{2N\hbar}{\pi}\int_0^{\omega_{max}} \frac{\omega d\omega}{(e^{\hbar\omega/k_B T}-1)(\omega_{max}^2-\omega^2)^{1/2}}$$

At low temperatures ($\hbar\omega \ll k_B T$), we may assume that the highest frequency modes are effectively frozen such that $\omega_{max}/\omega \gg 1$, and hence

$$U = \frac{2N\hbar}{\pi}\int_0^{\omega_{max}} \frac{d\omega}{(e^{\hbar\omega/k_B T}-1)\left[(\omega_{max}/\omega)^2-1\right]^{1/2}} = \frac{2N\hbar}{\pi\omega_{max}}\int_0^{\omega_{max}} \frac{\omega d\omega}{e^{\hbar\omega/k_B T}-1}$$

Setting $x = \hbar\omega/k_B T$ and $\theta_D = \hbar\omega_{max}/k_B$, yields

$$U = \frac{2Nk_B}{\pi}\frac{T^2}{\theta_D}\int_0^{\theta_D/T} \frac{xdx}{e^x-1} = \frac{\pi Nk_B}{3\theta_D}T^2$$

where, as $\theta_D/T \gg 1$, the upper limit of the integral was made infinite, such that the integral is approximately $\pi^2/6$. Hence, the low-temperature heat capacity reads

$$C_V = \left(\frac{\partial U}{\partial T}\right)_V = \frac{2\pi}{3}Nk_B\left(\frac{T}{\theta_D}\right)$$

The high-temperature limit for the heat capacity directly follows from

$$U = \int_0^{\omega_{max}} \frac{\hbar\omega}{1+\frac{\hbar\omega}{k_B T}-1} N(\omega)d\omega = k_B T\int_0^{\omega_{max}} N(\omega)d\omega = Nk_B T$$

where the last integral represents the normalization relation for the frequency distribution function in a linear chain of N atoms. It follows that $C_V = Nk_B$, and this is the Dulong-Petit value, Eq.(26.8), in one dimension.

38.6. If the electronic interaction in a molecular solid is given by

$$E(R) = \frac{A}{R^n} - \frac{B}{R^m}$$

where R is the nearest-neighbour separation, show that the system is stable for $n>m$, and find the ratio of the attractive to repulsive energy terms at equilibrium.

38.7. Find an expression for the compression work W which is required to reduce the equilibrium separation in a linear ionic chain from R_0 to $R_0(1-\varepsilon)$.

38.8. Find the normal frequency ω_k for a linear chain of five atoms of mass M where the opposite end atoms are held fixed, assuming that only nearest-neighbours interact and the coupling constant is κ_0.

38.9. Find an expression for the heat capacity of a system of N atoms, each having two energy levels 0 and ε.

38.10. Find the heat capacity of a two-dimensional square lattice of N atoms, if $T \gg \theta_D$ and $T \ll \theta_D$, given the frequency distribution function

$$N_D(\omega) = \frac{4N}{\omega_D^2} \omega$$

in the Debye approximation.

39. STRUCTURE OF SOLIDS

39.1. THE CRYSTAL LATTICE
39.2. THE RECIPROCAL LATTICE
39.3. STRUCTURE DETERMINATION

39.1. THE CRYSTAL LATTICE

We now know that the properties of atoms are determined by their electronic configuration. We should therefore expect that the properties of solids, which consist of spatial arrangements of atoms, reflect both the electronic structure of the separate atoms and the existing order in the atomic arrangements. An understanding of the solid is simplified if it possesses a *crystalline structure* characterized by a regular arrangement of identical structural units that each contains one or more atoms. Each of these units is called a *basis* and if the basis position is represented by a point in space, we obtain a periodic array of points called a *lattice*. The concept of a lattice only provides information about the geometry of the structure and has no physical reality. The crystalline structure is obtained by first defining the atomic environment, the basis, of each lattice point, from which the structure can be built up by translation.

The position vector of a lattice point is defined in terms of one, two or three primitive translation vectors, for one-, two- or three-dimensional lattices respectively, as

$$\vec{R}_{pqs} = p\vec{a} + q\vec{b} + s\vec{c} \tag{39.1}$$

where the numbers p, q and s are integers, so that the lattice is invariant under any translation that consists of multiples of the individual primitive vectors $\vec{a}, \vec{b}, \vec{c}$. The primitive vectors may be omitted, and hence, the vectors (39.1) can be written by convention as $[pqs]$. It follows that the primitive vectors themselves may be denoted by $\vec{a} = [100]$, $\vec{b} = [010]$, $\vec{c} = [001]$. The lattice is generated by repeated arbitrary translation (39.1) of either a single lattice point or a *primitive cell* associated with it, as illustrated in Figure 39.1 for a two-dimensional lattice. A primitive cell always contains one lattice point only. Either there are points at the corners of the cell only (a crystallographic cell) or one lattice point is in the centre of a cell which is defined by the planes bisecting the lines joining the central point with the adjacent lattice points (a Wigner-Seitz cell). Figure 39.1 shows that the sides of the Wigner-Seitz cell are determined by the equation

$$-\vec{r} \cdot \vec{R} = \frac{1}{2}\vec{R}^2 \quad \text{or} \quad 2\vec{r} \cdot \vec{R} + \vec{R}^2 = 0 \quad \text{i.e.} \quad (\vec{r} + \vec{R})^2 = \vec{r}^2 \quad (39.2)$$

where $\vec{r}$ is the position vector with respect to the central lattice point.

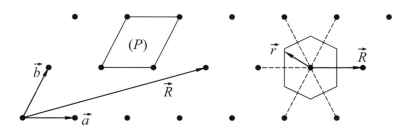

Figure 39.1. The generation of a two-dimensional lattice

Lattices are classified into groups or *systems* according to the shape of the *unit cell*, and the unit cell generates the entire lattice by translation. The shape of the unit cell is chosen for geometrical convenience and so may contain more than one lattice point. The unit cell is also a primitive (P) cell, if lattice points are located at the corners only. Otherwise it may be body-centred (I), with an extra point at the centre, face-centred (F), with an extra point at the centre of each face or base-centred (C), having a point at the centre of the base. The sides of the unit cell are called axes and are designated a, b and c. The angles between axes are α (between b and c), β (between a and c) and γ (between a and b). The faces of the unit cell which intersect the a-, b- and c-axes, respectively are called A, B and C. Symmetry considerations restrict the number of different possible lattices, known as *Bravais lattices*.

Consider two points P and P' separated by a primitive lattice translation, R say, in a two-dimensional lattice, as in Figure 39.2. Assuming that a rotation through an angle $\pm \varphi$ is allowed, we can generate the lattice points Q and Q' a distance R' apart and each at a distance R from P and P', respectively. The distance R' between the lattice points Q and Q' must be an integral multiple of R

$$R' = R - 2R\cos\varphi = nR$$

which yields

$$\cos\varphi = \frac{1-n}{2} \quad \text{or} \quad -1 \leq \frac{1-n}{2} \leq 1 \quad \text{i.e.,} \quad -1 \leq n \leq 3$$

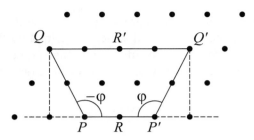

Figure 39.2. Arrangement of lattice points obtained by rotation

It follows that the only allowed rotations that are consistent with the translational invariance are

$$\varphi = \arccos\frac{1-n}{2} = 0, \frac{\pi}{3}, \frac{\pi}{2}, \frac{2\pi}{3}, \pi$$

This leads to four systems and five possible two-dimensional lattices, known as oblique (P), rectangular (P) and (C), square (P) and hexagonal (P), illustrated in Figure 39.3.

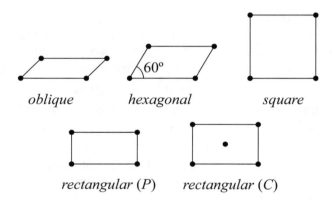

Figure 39.3. Two-dimensional Bravais lattices

The 14 possible three-dimensional Bravais lattices belong to one of the seven different systems, illustrated in Figure 39.4, each of which has its own basic symmetry given by

- ◆ cubic $\quad a = b = c, \quad \alpha = \beta = \gamma = 90° \quad (P), (I) \text{ and } (F)$
- ◆ tetragonal $\quad a = b \neq c, \quad \alpha = \beta = \gamma = 90° \quad (P) \text{ and } (I)$
- ◆ orthorhombic $\quad a \neq b \neq c, \quad \alpha = \beta = \gamma = 90° \quad (P), (I), (F) \text{ and } (C)$
- ◆ hexagonal $\quad a = b \neq c, \quad \alpha = \beta = 90°, \gamma = 120° \quad (P)$
- ◆ monoclinic $\quad a \neq b \neq c, \quad \alpha = \beta = 90° \neq \gamma \quad (P) \text{ and } (C)$
- ◆ triclinic $\quad a \neq b \neq c, \quad \alpha \neq \beta \neq \gamma \quad (P)$
- ◆ trigonal $\quad a = b = c, \quad \alpha = \beta = \gamma < 120° (\neq 90°) \quad (P)$

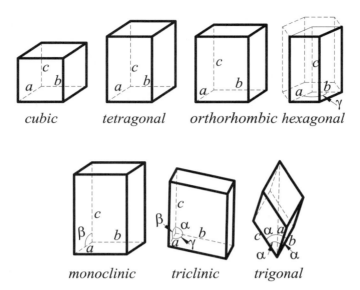

Figure 39.4. Unit cell of the seven crystal systems

A characteristic of each Bravais lattice is the volume per primitive unit cell, and therefore, the volume per lattice site. For any of the unit cells illustrated in Figure 39.4, it is apparent that the unit cell volume is given by

$$V_0 = \vec{c} \cdot (\vec{a} \times \vec{b}) = \vec{a} \cdot (\vec{b} \times \vec{c}) = \vec{b} \cdot (\vec{c} \times \vec{a}) \tag{39.3}$$

and it is invariant, whichever face is chosen as the base.

EXAMPLE 39.1. Simple crystalline structures

A crystalline structure is defined by a particular Bravais lattice and the basis. It is found that many crystalline structures, for example that of typical metals which is determined by the nondirectional character of the metallic bond, may be considered in terms of the packing together of atoms or ions in the form of hard spheres.

A two-dimensional layer of spheres of radius r may be most efficiently packed in a hexagonal arrangement, as shown in Figure 39.5 (a), where the intrinsic symmetry elements are specified. Any sphere has six nearest neighbours, and it is said that its *coordination number* is 6. The open spaces are called *interstitial* holes. A planar hole, represented in Figure 39.5 (b), has the coordination number 3, and the largest radius r_i of an interstitial atom which will fit into it is given by

$$r + r_i = \frac{\sqrt{3}}{3}(2r) \quad \text{or} \quad r_i = \left(\frac{2\sqrt{3}}{3} - 1\right)r = 0.155\,r$$

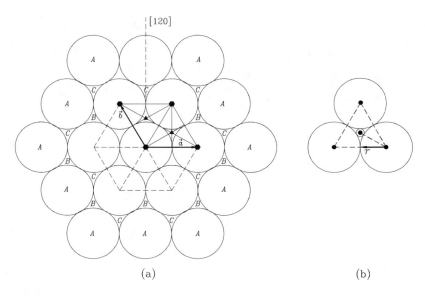

(a)   (b)

Figure 39.5. Hexagonal close-packing of spheres in two dimensions

In a three-dimensional structure we must consider additional close packed layers above and below, so that any central sphere will be surrounded by 12 nearest neighbours, three from the plane above and three from the plane below, in addition to the six already mentioned. The symmetry of the three-dimensional arrays of closed-packed spheres, illustrated by the unit cell in Figure 39.6, can be either hexagonal (hcp) or face-centred cubic (fcc), as a result of two non-equivalent stacking sequences of layers A and B with a hexagonal arrangement, represented as $ABAB$ and ABC, respectively.

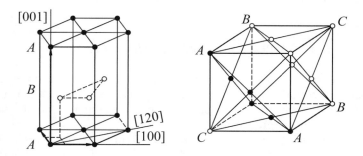

Figure 39.6. Hexagonal and cubic close-packing of spheres in three dimensions

If we assume that the basis of a Bravais lattice consists of an atom or ion represented by a hard sphere of radius r, we may evaluate the *packing factor f*, defined as the ratio of the volume of the atoms in the unit cell to the volume of the unit cell. For close-packed *fcc* structures there are four atoms of radius r per unit cell of volume a^3. Since $(4r)^2 = 2a^2$, as seen in Figure 39.7 (a), that is, $r = a\sqrt{2}/4$, the packing factor is

$$f_{fcc} = \frac{4(4\pi/3)(\sqrt{2}a/4)^3}{a^3} = \frac{\sqrt{2}\pi}{6} = 0.74$$

and this means that only 74% of the space is occupied by atoms in a close-packed cubic structure (and also in a closed-packed hexagonal structure).

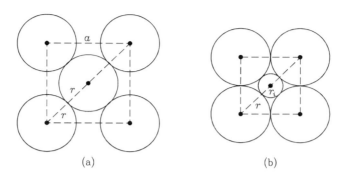

Figure 39.7. Magnified views for determination of the packing factor (a) and the radius of an octahedral interstitial atom (b) in an *fcc* structure

The interstitial spaces in a close-packed structure may have either an octahedral or a tetrahedral coordination. An octahedral hole is an open space surrounded by six lattice atoms, as in Figure 39.8 (a). If the atoms at the centres of the six faces of the cubic (F) cell are connected, they produce an octahedron with an octahedral hole at the centre of both the cube and the octahedron. Similarly, if atoms at four of the cube corners are connected, a tetrahedral hole with coordination number 4 appears at the centre of the tetrahedron and the cube. There are one octahedral and two tetrahedral holes for each primitive cell in both the cubic or hexagonal close-packed structures.

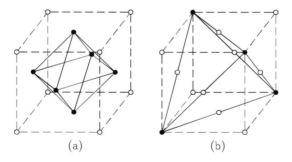

Figure 39.8. Octahedral (a) and tetrahedral (b) coordination of interstitial holes

The radius r_i of the interstitial atom which fits into an octahedral hole without forcing the atoms apart can be obtained from the magnified view in Figure 39.7 (b), which gives

$$(2r + 2r_i)^2 = (2r)^2 + (2r)^2$$

that is, $r_i = (\sqrt{2}-1)r = 0.414r$. A similar calculation for a tetrahedral hole gives $r_i = 0.225r$.

There are two other cubic structures where no close-packed layers are involved. The body-centred cubic (or *bcc*) structure, where each atom has only eight nearest neighbours (instead of 12 nearest neighbours involved in the *fcc* structure) is illustrated in Figure 39.9.

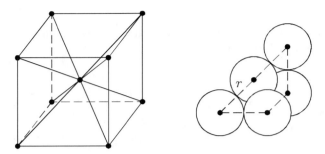

Figure 39.9. Body-centred cubic unit cell and a magnified view for determining the packing factor

The unit cell contains two atoms. The allowed atomic radius, r, is given by the restriction that the diagonal of the unit cell is $4r$, which gives $(4r)^2 = a^2 + (a\sqrt{2})^2 = 3a^2$, that is, $r = \sqrt{3}a/4$. The packing factor of the bcc structure can then be written as

$$f_{bcc} = \frac{2(4\pi/3)(\sqrt{3}a/4)^3}{a^3} = \frac{\sqrt{3}\pi}{8} = 0.68$$

The simple-cubic (or sc) structure is not often found in nature. Each atom in this structure has six nearest neighbours, and there is a single atom per unit cell (Figure 39.10). The radius of an atom must be $r = a/2$, so that the packing factor is

$$f_{sc} = \frac{(4\pi/3)(a/2)^3}{a^3} = \frac{\pi}{6} = 0.52$$

It is clear that as the number of the nearest neighbours, in a cubic structure, decreases the packing factor also decreases.

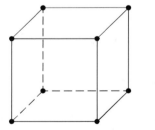
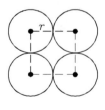

Figure 39.10. Simple cubic unit cell and a magnified view for determination of the packing factor

A cubic interstitial hole in a simple cubic structure has a coordination number of 8, and the radius r_i of the interstitial atom results from $(2r + 2r_i)^2 = a^2 + (a\sqrt{2})^2 = 3a^2 = 3(2r)^2$, so that

$$r_i = (\sqrt{3} - 1)r = 0.732r$$

The highest possible coordination of an interstitial hole is 12, as in the case of a vacancy in a close-packed structure, where $r_i = r$. It is apparent that there is a monotonic increase in the allowed radius of an interstitial atom from $0.155r$ to r, as its coordination increases from 3 to 12.

38.2. THE RECIPROCAL LATTICE

The experimental determination of the crystal structure of materials is usually performed in terms of the orientation of the sets of parallel planes that characterize the structure. We must therefore extend the vector representation of directions (39.1), associated with lattice points, to describe the lattice planes. The two-dimensional array of points in a lattice plane can be represented by the vector product of any two vectors in the plane. The magnitude of the vector product is equal to the area of the parallelogram defined by the two vectors. Consider an arbitrary plane which intersects the axes of the crystal lattice at the points $p\vec{a}$, $q\vec{b}$ and $s\vec{c}$ where p, q and s are integers (Figure 39.11). The triangular portion of the plane has sides given by

$$p\vec{a} - q\vec{b}, \quad q\vec{b} - s\vec{c}, \quad s\vec{c} - p\vec{a} \tag{39.4}$$

such that the vector product of any two of these yields

$$(p\vec{a} - q\vec{b}) \times (q\vec{b} - s\vec{c}) = qs(\vec{b} \times \vec{c}) + ps(\vec{c} \times \vec{a}) + pq(\vec{a} \times \vec{b}) \tag{39.5}$$

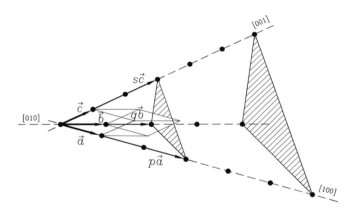

Figure 39.11. A set of parallel planes with integral intercepts on axes $\vec{a}, \vec{b}$ and $\vec{c}$ of the direct lattice

It is usual to set

$$h = qs, \quad k = ps, \quad l = pq$$

which leads to the ratios

$$h:k:l = qs:ps:pq = \frac{1}{p}:\frac{1}{q}:\frac{1}{s} \tag{39.6}$$

The vector representation (39.5) of an arbitrary plane is defined for convenience as

$$\vec{K}_{hkl} = \frac{2\pi}{V_0}\left[h\left(\vec{b}\times\vec{c}\right)+k\left(\vec{c}\times\vec{a}\right)+l\left(\vec{a}\times\vec{b}\right)\right] \tag{39.7}$$

where V_0 is the primitive unit cell volume (39.3). Equation (39.7) reduces to

$$\vec{K}_{hkl} = h\vec{a}^* + k\vec{b}^* + l\vec{c}^* \tag{39.8}$$

which gives the desired vector representation of the lattice planes in terms of the so-called *reciprocal* basis set defined as

$$\vec{a}^* = 2\pi\frac{\vec{b}\times\vec{c}}{\vec{a}\cdot\left(\vec{b}\times\vec{c}\right)}, \quad \vec{b}^* = 2\pi\frac{\vec{c}\times\vec{a}}{\vec{b}\cdot\left(\vec{c}\times\vec{a}\right)}, \quad \vec{c}^* = 2\pi\frac{\vec{a}\times\vec{b}}{\vec{c}\cdot\left(\vec{a}\times\vec{b}\right)} \tag{39.9}$$

It is a straightforward matter to show, from Eq.(39.9), that

$$\vec{a}^*\cdot\vec{a} = 2\pi, \quad \vec{a}^*\cdot\vec{b} = \vec{a}^*\cdot\vec{c} = 0$$

$$\vec{b}^*\cdot\vec{b} = 2\pi, \quad \vec{b}^*\cdot\vec{a} = \vec{b}^*\cdot\vec{c} = 0 \tag{39.10}$$

$$\vec{c}^*\cdot\vec{c} = 2\pi, \quad \vec{c}^*\cdot\vec{a} = \vec{c}^*\cdot\vec{b} = 0$$

which is a set of equations often treated as formal definitions of the reciprocal basis set. It provides both the magnitude of the components of the reciprocal basis set as

$$\left|\vec{a}^*\right| = \frac{2\pi}{|\vec{a}|}, \quad \left|\vec{b}^*\right| = \frac{2\pi}{|\vec{b}|}, \quad \left|\vec{c}^*\right| = \frac{2\pi}{|\vec{c}|} \tag{39.11}$$

and their orientation. The three components are normal to the faces *A*, *B* and *C* of the primitive unit cell, respectively.

It is apparent that Eq.(39.8) has the familiar form (39.1) of a lattice generating equation, with a different basis set $\vec{a}^*, \vec{b}^*$ and $\vec{c}^*$. Hence, the vectors (39.8) associated with arbitrary sets of parallel planes in the real lattice, define a lattice with the reciprocal basis set (39.9), called the *reciprocal lattice*. A geometrical procedure for constructing the two-dimensional reciprocal oblique lattice, using the normals to the unit cell sides, is illustrated in Figure 39.12.

The Wigner-Seitz cell in the reciprocal lattice is known as the first *Brillouin zone*. Its sizes are determined by a relation similar to Eq.(39.2), namely,

$$2\vec{k}\cdot\vec{K} + \vec{K}^2 = 0 \tag{39.12}$$

where $\vec{k}$ denotes some general vector in reciprocal or $\vec{k}$-space. In the reciprocal lattice representation, each direction $\vec{K}_{hkl}$ corresponds to a possible orientation of planes in the crystal lattice, and therefore to a set of parallel planes.

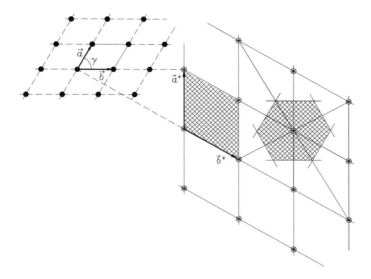

Figure 39.12. Geometric relationship between the direct and reciprocal oblique lattice

The components h, k, l of the reciprocal lattice vector $\vec{K}_{hkl}$ are called the *Miller indices* of a given orientation. It is seen from Eq.(39.6) and Figure 39.11 that they are the inverse ratio of the intercepts of a plane in the crystal lattice. It is common practice to rewrite the vectors (39.4), lying in an arbitrary plane of the real lattice, in terms of the Miller indices as

$$\frac{\vec{a}}{h} - \frac{\vec{b}}{k}, \quad \frac{\vec{b}}{k} - \frac{\vec{c}}{l}, \quad \frac{\vec{c}}{l} - \frac{\vec{a}}{h} \qquad (39.13)$$

A particular set of parallel planes in the real lattice is written by convention with round parentheses (hkl). If one indice is negative, the minus sign is written as a bar. Some planes (hkl) and directions $[pqr]$ in a simple-cubic lattice are shown in Figure 39.13.

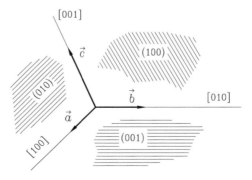

Figure 39.13. Lattice direction $[pqs]$ and plane orientations (hkl)

The orientation of the reciprocal lattice vector follows from Eqs.(39.8), (39.9) and (39.13), if we consider the identities

$$\vec{K}_{hkl} \cdot \left(\frac{\vec{a}}{h} - \frac{\vec{b}}{k}\right) = \left(h\vec{a}^* + k\vec{b}^* + l\vec{c}^*\right) \cdot \left(\frac{\vec{a}}{h} - \frac{\vec{b}}{k}\right) = 2\pi\left(\frac{h}{h} - \frac{k}{k}\right) = 0$$

and similarly

$$\vec{K}_{hkl} \cdot \left(\frac{\vec{b}}{k} - \frac{\vec{c}}{l}\right) = \vec{K}_{hkl} \cdot \left(\frac{\vec{c}}{l} - \frac{\vec{a}}{h}\right) = 0$$

In other words $\vec{K}_{hkl}$ is perpendicular to two vectors in the plane (hkl), and hence, it is perpendicular to the set of parallel planes (hkl) of the crystal lattice. If the spacing of these planes is d_{hkl}, which can be defined by

$$d_{hkl} = \frac{\vec{a}}{h} \cdot \vec{e}_{hkl} = \frac{\vec{b}}{k} \cdot \vec{e}_{hkl} = \frac{\vec{c}}{l} \cdot \vec{e}_{hkl} \qquad (39.14)$$

where $\vec{e}_{hkl}$ is the unit vector in the direction $\vec{K}_{hkl}$ of the normal to the planes, we have

$$d_{hkl} = \frac{\vec{a}}{h} \cdot \frac{\vec{K}_{hkl}}{|\vec{K}_{hkl}|} = \frac{\vec{a}}{h} \frac{h\vec{a}^* + k\vec{b}^* + l\vec{c}^*}{|\vec{K}_{hkl}|} = \frac{2\pi}{|\vec{K}_{hkl}|}$$

and this gives the magnitude of the reciprocal lattice vector in terms of the interplanar spacing as

$$\vec{K}_{hkl} = \frac{2\pi}{d_{hkl}} \vec{e}_{hkl} \qquad (39.15)$$

Thus, to every reciprocal lattice vector $\vec{K}_{hkl}$ there corresponds a set of direct lattice planes which are normal to it, whose spacing is inversely proportional to the shortest reciprocal lattice vector in the direction of $\vec{K}_{hkl}$. This defines a one-to-one correspondence between the possible orientations of planes in a real lattice and the possible directions of lines in its reciprocal lattice, such that the density of planes $1/d_{hkl}$ in one lattice is proportional to the density of lines in the other. By considering the scalar product given by

$$\vec{K} \cdot \vec{R} = \left(h\vec{a}^* + k\vec{b}^* + l\vec{c}^*\right) \cdot \left(p\vec{a} + q\vec{b} + s\vec{c}\right) = 2\pi\left(hp + kq + ls\right) \qquad (39.16)$$

which is an integral multiple of 2π for all $\vec{K}$ and $\vec{R}$, we obtain

$$e^{i\vec{K} \cdot \vec{R}} = 1 \qquad (39.17)$$

and this is a formal relationship, important for later use, between the real and reciprocal lattices.

39.3. STRUCTURE DETERMINATION

The X-ray region of the electromagnetic spectrum has wavelengths in a range which is suitable for the investigation of the structure of solids. Crystal structure is usually determined by diffraction of X-rays with wavelengths of about 10^{-10} m, and this is comparable to the distances between atoms in a crystal. The amplitude of scattered radiation from a crystal structure is given by the diffraction formula (19.17) which reads

$$\Psi = \Psi_0 \sum_{j=1}^{N} e^{ik\Delta(\vec{r}_j)} \qquad (39.18)$$

where the sum runs over all the atoms in the crystal.

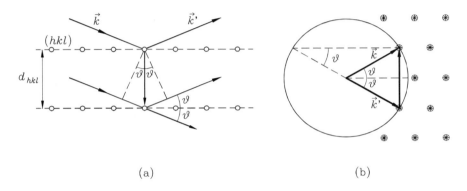

Figure 39.14. Diffraction from the crystal lattice (a) and the Ewald diagram in the reciprocal lattice (b)

For a point scatterer at a given position $\vec{r}_j$, the path difference $\Delta(\vec{r}_j)$ results from the simple geometric construction shown in Figure 39.14 (a) as

$$\Delta(\vec{r}_j) = \vec{r}_j \cdot (-\vec{e}_k) + \vec{r}_j \cdot \vec{e}_{k'} = \vec{r}_j (\vec{e}_{k'} - \vec{e}_k)$$

so that the phase shift can be written in the form

$$k\Delta(\vec{r}_j) = \vec{r}_j \cdot (\vec{k}' - \vec{k}) = \vec{r}_j \cdot \Delta\vec{k} \qquad (39.19)$$

where $\vec{e}_{k'}, \vec{e}_k$ are the unit vectors of the incident and diffracted beams, and $\Delta\vec{k}$ is called the *scattering vector*. Substituting Eq.(39.19), Eq.(39.18) becomes

$$\Psi = \Psi_0 \sum_{j=1}^{N} e^{i\vec{r}_j \cdot \Delta\vec{k}} \qquad (39.20)$$

If we assume the crystal to be a finite parallelepiped of dimensions $P\vec{a}$, $Q\vec{b}$ and $S\vec{c}$, and hence, containing $N = PQS$ identical atoms, one at every lattice point, the position vector $\vec{r}_j$ reads

$$\vec{r}_j = \vec{R}_j = p_j \vec{a} + q_j \vec{b} + s_j \vec{c} \tag{39.21}$$

and Eq.(39.20) takes the form

$$\Psi = \Psi_0 \sum_{p_j,q_j,s_j} e^{i(p_j \vec{a} + q_j \vec{b} + s_j \vec{c}) \cdot \Delta \vec{k}} = \Psi_0 \sum_{p_j=0}^{P-1} e^{i p_j \vec{a} \cdot \Delta \vec{k}} \sum_{q_j=0}^{Q-1} e^{i q_j \vec{b} \cdot \Delta \vec{k}} \sum_{s_j=0}^{S-1} e^{i s_j \vec{c} \cdot \Delta \vec{k}} \tag{39.22}$$

Each factor on the right-hand side can similarly be transformed as follows

$$\sum_{p_j=0}^{P-1} \left(e^{i\vec{a} \cdot \Delta \vec{k}}\right)^{p_j} = \frac{1 - e^{iP\vec{a} \cdot \Delta \vec{k}}}{1 - e^{i\vec{a} \cdot \Delta \vec{k}}} = \frac{e^{iP\vec{a} \cdot \Delta \vec{k}/2}}{e^{i\vec{a} \cdot \Delta \vec{k}/2}} \frac{\sin\left(P\vec{a} \cdot \Delta \vec{k}/2\right)}{\sin\left(\vec{a} \cdot \Delta \vec{k}/2\right)}$$

so that the intensity distribution of the diffraction pattern is

$$I = |\Psi|^2 = I_0 \frac{\sin^2\left(P\vec{a} \cdot \Delta \vec{k}/2\right)}{\sin^2\left(\vec{a} \cdot \Delta \vec{k}/2\right)} \frac{\sin^2\left(Q\vec{b} \cdot \Delta \vec{k}/2\right)}{\sin^2\left(\vec{b} \cdot \Delta \vec{k}/2\right)} \frac{\sin^2\left(S\vec{c} \cdot \Delta \vec{k}/2\right)}{\sin^2\left(\vec{c} \cdot \Delta \vec{k}/2\right)} \tag{39.23}$$

a form which is analogous to that obtained in Eq.(18.11) as a result of multiple-beam interference. Maxima of intensity

$$I_{max} = P^2 Q^2 S^2 I_0 = N^2 I_0 \tag{39.24}$$

occur whenever both the numerator and denominator tend to zero, that is

$$\vec{a} \cdot \Delta \vec{k} = 2\pi h, \quad \vec{b} \cdot \Delta \vec{k} = 2\pi k, \quad \vec{c} \cdot \Delta \vec{k} = 2\pi l \tag{39.25}$$

where h, k and l are arbitrary integers. Multiplying each of these conditions by p, q and s, respectively, and then taking the sum, we obtain

$$\left(p\vec{a} + q\vec{b} + s\vec{c}\right) \cdot \Delta \vec{k} = \vec{R} \cdot \Delta \vec{k} = 2\pi(ph + qk + sl) \tag{39.26}$$

By comparison with Eq.(39.16) it is seen that diffraction maxima occur if, and only if, the scattering vector $\Delta \vec{k}$ is one of the reciprocal lattice vectors (39.8), that is

$$\Delta \vec{k} = \vec{K}_{hkl} = h\vec{a}^* + k\vec{b}^* + l\vec{c}^* \tag{39.27}$$

The condition (39.27) on the scattering vector, which is equivalent to Eq.(39.26) or to the previous Eq.(39.17), is called the *Laue condition* for constructive interference in crystal diffraction. A simple geometrical interpretation of Eq.(39.27), known as the *Ewald diagram* is illustrated in Figure 39.14 (b). If we draw a sphere of radius $k = |\vec{k}|$ about the origin of the wave vector $\vec{k}$ of the incident radiation, where $\vec{k}$ is taken to pass through a point of the reciprocal lattice, it follows that wherever the sphere intersects any other point of the reciprocal lattice constructive interference is observed. It is convenient to square both sides of Eq.(39.27), written as

$$\vec{k}' = \vec{k} + \vec{K} \quad \text{or} \quad k^2 = k^2 + 2\vec{k} \cdot \vec{K} + \vec{K}^2$$

to obtain

$$2\vec{k} \cdot \vec{K} + \vec{K}^2 = 0 \tag{39.28}$$

By comparison with Eq.(39.12) it is seen that the Laue condition is satisfied on the faces of the first Brillouin zone.

The Laue condition shows that constructive interference is associated with particular vectors $\vec{K}_{hkl}$, that is with particular sets of parallel planes of the real lattice. Assuming that 2θ is the angle between the incident and scattered beams, Eq.(39.27) can be rewritten as

$$\left|\Delta \vec{k}\right| = 2k \sin\theta = \left|\vec{K}_{hkl}\right| = \frac{2\pi}{d_{hkl}}$$

and reduces to the familiar *Bragg's law* for constructive interference by reflection from planar arrays (hkl) in the real lattice

$$2 d_{hkl} \sin\theta = \lambda \tag{39.29}$$

where d_{hkl} is the interplanar spacing.

EXAMPLE 39.2. X-ray diffraction patterns

The *positions* of X-ray reflections from a given structure are functions of the size, shape and symmetry of the unit cell. For a simple cubic lattice, with primitive unit cell of size a, Eq.(39.9) shows that the reciprocal lattice is also cubic, with a unit cell of size $a^* = 2\pi/a$. From Eq.(39.8) it follows that the reciprocal lattice vector has the magnitude

$$K = \frac{2\pi}{a} \sqrt{h^2 + k^2 + l^2}$$

By comparison with Eq.(39.15) we obtain the interplanar spacing as

$$d_{hkl} = \frac{a}{\sqrt{h^2 + k^2 + l^2}} \tag{39.30}$$

so that the angular positions θ of X-ray reflections follow in the form

$$\sin\theta = \frac{\lambda}{a} \frac{\sqrt{h^2 + k^2 + l^2}}{2}$$

The diffraction pattern (39.23) from a *simple cubic crystal* is then expected to be similar to that plotted in Figure 18.6 for a grating with narrow slits, with modified reflection positions given by

$$\sin\theta = \frac{1}{2}\frac{\lambda}{a}, \quad \frac{\sqrt{2}}{2}\frac{\lambda}{a}, \quad \frac{\sqrt{3}}{2}\frac{\lambda}{a}, \quad \frac{\lambda}{a}, \quad \ldots$$

$$(hkl) = (100), \quad (110), \quad (111), \quad (200), \quad \ldots$$

(39.31)

The *intensities* of X-ray reflections from a given structure are dependent on the nature and position of atoms in the unit cell, where each atom can be specified by its coordinates x_i, y_i, z_i, that is by a position vector $\vec{\rho}_i = x_i\vec{a} + y_i\vec{b} + z_i\vec{c}$, and also by an *atomic scattering factor* f_i which is dependent on the number of electrons the atom contains. The diffraction formula (39.20) then becomes

$$\Psi = \Psi_0 \sum_{j=1}^{n}\sum_{i=1}^{m} f_i e^{i(\vec{R}_j+\vec{\rho}_i)\cdot\Delta\vec{k}} = \Psi_0 \sum_{j=1}^{n} e^{i\vec{K}\cdot\vec{R}_j} \sum_{i=1}^{m} f_i e^{i\vec{K}\cdot\vec{\rho}_i}$$

where it was assumed that there are m atoms in each of the n unit cells, so that $N = nm$, and use has been made of the Laue condition (39.27). A *structure factor* can be defined as

$$F(hkl) = \sum_{i=1}^{m} f_i e^{i\vec{K}\cdot\vec{\rho}_i} = \sum_{i=1}^{m} f_i e^{i2\pi(hx_i + ky_i + lz_i)}$$

(39.32)

so that the intensities (39.24) become

$$I_{\max} = n^2 |F(hkl)|^2 I_0$$

(39.33)

The structure factor is always nonvanishing for primitive Bravais lattices, where x_i, y_i and z_i are integers. However, $F(hkl)$ may vanish for certain fractional coordinates of the atomic positions in the body-centred (I), face-centred (F) and base-centred (A) or (B) or (C) unit cells, leading to the so-called *extinction rules* which state that certain X-ray reflections are forbidden.

A body-centred (I) lattice contains two identical atoms per unit cell, of coordinates (x_i, y_i, z_i) and $(x_i + \frac{1}{2}, y_i + \frac{1}{2}, z_i + \frac{1}{2})$, so that

$$F_{(I)}(hkl) = \left[1 + e^{i\pi(h+k+l)}\right] \sum_{i=1}^{m} f_i e^{i2\pi(hx_i+ky_i+lz_i)} = \left[1 + e^{i\pi(h+k+l)}\right] F_{(P)}(hkl)$$

Hence, a body-centred (I) lattice can only have reflections where the sum of h, k and l are even. For a *body-centred cubic lattice* this extinction rule reduces the reflections (39.31), to

$$\sin\theta = \qquad , \quad \frac{\sqrt{2}}{2}\frac{\lambda}{a}, \quad , \quad \frac{\lambda}{a}, \quad \ldots$$

$$(hkl) = \cancel{(100)}, \quad (110), \quad \cancel{(111)}, \quad (200), \quad \ldots$$

(39.34)

We similarly obtain for the base-centred unit cells that

$$F_{(A)}(hkl) = \left[1 + e^{i\pi(k+l)}\right] F_{(P)}(hkl)$$

$$F_{(B)}(hkl) = \left[1 + e^{i\pi(h+l)}\right] F_{(P)}(hkl)$$

$$F_{(C)}(hkl) = \left[1 + e^{i\pi(h+k)}\right] F_{(P)}(hkl)$$

which means that A-base-centred (A) lattices give only reflections with $k+l$ even, B-base-centred (B) with $h+l$ even and C-base-centred (C) with $h+k$ even. It follows for a face-centred (F) lattice that all these rules must be simultaneously satisfied, so that only reflections where h, k and l are all even or all odd will be allowed. Hence, in case of a *face-centred cubic lattice*, the allowed reflections (39.31) from a simple cubic lattice will be reduced to

$$\sin\theta = \qquad , \qquad , \quad \frac{\sqrt{3}}{2}\frac{\lambda}{a}, \quad \frac{\lambda}{a}, \quad \ldots$$

(39.35)

$$(hkj) = \cancel{(100)}, \quad \cancel{(110)}, \quad (111), \quad (200), \quad \ldots$$

FURTHER READING
1. W. Massa – CRYSTAL STRUCTURE DETERMINATION, Springer-Verlag, New York, 2004.
2. M. F. C. Ladd, R. A. Palmer – STRUCTURE DETERMINATION BY X-RAY CRYSTALLOGRAPHY, Kluwer Academic, New York, 2003.
3. J. J. Rousseau – BASIC CRYSTALLOGRAPHY, Wiley, New York, 1998.

PROBLEMS

39.1. Find a relation between the Miller indices (hkl) corresponding to the orientation of a macroscopic crystal face and the lattice directions $[pqs]$ of its edges.

Solution

If an edge $\vec{R}_{pqs}$, Eq.(39.1), belongs to face (hkl), it must be perpendicular to $\vec{K}_{hkl}$, Eq.(39.8), and hence Eq.(39.16) reduces to

$$hp + kq + ls = 0$$

This relation defines the condition for the face (hkl) to intersect some other face along edge $[pqs]$. As a set of faces parallel to a certain edge is called a zone, having the edge direction as zone axis, the relation is known as the Weiss zone law. It provides the condition satisfied by any lattice plane (hkl) which belongs to a zone $[pqs]$, and it is usually applied to find new faces and edges consistent with the observed ones. If the crystal faces are polygons, having at least two nonparallel edges $[pqs]$ and $[p'q's']$, we may combine the equations

$$hp + kq + ls = 0 \quad \text{and} \quad hp' + kq' + ls' = 0$$

to find the Miller indices of the face, in the form

$$h = qs' - sq', \quad l = sp' - ps', \quad k = qp' - pq'$$

In other words, two possible edges of a crystal determine a possible crystal face. In the same way, knowing the Miller indices of two crystal faces, the zone law provides the coordinates of a possible crystal edge.

39.2. For a hexagonal close-packed (hcp) structure, having a simple hexagonal Bravais lattice with a basis consisting of two atoms, one at 000 and the other at $\tfrac{1}{3}\tfrac{2}{3}\tfrac{1}{2}$, find the ratio of the height of the cell to the nearest-neighbour spacing c/a and the packing factor f_{hcp}.

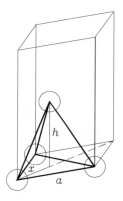

Solution

If the atoms are approximated by hard spheres of radius r, the conventional hexagonal unit cell shown in Figure 39.6 contains a number of $\tfrac{1}{6}12 + \tfrac{1}{2}2 + 3 = 6$ atoms, of total atomic volume $24\pi r^3/3$. The side of the hexagonal base is $a = 2r$, which yields a base area of $6r^2\sqrt{3}$. The height of the ideal cell can be derived using the tetrahedron of side a formed by three A atoms and a B atom, as shown in the figure, where $x = a/\sqrt{3}$ and $h = a\sqrt{2/3}$. Because $h = c/2$, this yields a value of $2\sqrt{2/3}$ which is called the ideal c/a ratio. Therefore, we have $c = 4r\sqrt{2/3}$, and the packing factor is

$$f_{hcp} = \frac{24\pi r^3/3}{24r^3\sqrt{2}} = \frac{\pi}{3\sqrt{2}} = 0.74$$

39.3. Find the reciprocal lattice of a face-centred cubic (*fcc*) lattice. What are the metric and angular relations of the reciprocal primitive cell?

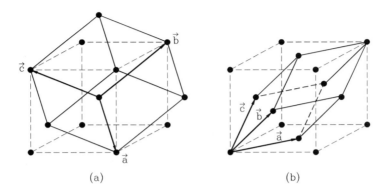

(a) (b)

Solution

If a is the period of the primitive cubic cell, defined by the primitive translations $\vec{a} = a\vec{e}_x$, $\vec{b} = a\vec{e}_y$ and $\vec{c} = a\vec{e}_z$, such that $V_{0sc} = a^3$, for a face-centred cubic lattice with the same period, the primitive cell is defined as in Figure (b) by

$$\vec{a}' = \frac{1}{2}(\vec{b} + \vec{c}) = \frac{a}{2}(\vec{e}_y + \vec{e}_z), \quad \vec{b}' = \frac{a}{2}(\vec{e}_z + \vec{e}_x), \quad \vec{c}' = \frac{a}{2}(\vec{e}_x + \vec{e}_y)$$

such that $V_{0fcc} = \vec{a}' \cdot (\vec{b}' \times \vec{c}') = a^3/4$. Thus the reciprocal basis set consists of

$$\vec{a}^* = \frac{2\pi}{a}(-\vec{e}_x + \vec{e}_y + \vec{e}_z), \quad \vec{b}^* = \frac{2\pi}{a}(\vec{e}_x - \vec{e}_y + \vec{e}_z), \quad \vec{c}^* = \frac{2\pi}{a}(\vec{e}_x + \vec{e}_y - \vec{e}_z)$$

and this yields the metric and angular relations of the reciprocal primitive cell

$$|\vec{a}^*| = |\vec{b}^*| = |\vec{c}^*| = \frac{2\pi}{a}\sqrt{3} \quad , \quad \cos\alpha^* = \cos\beta^* = \cos\gamma^* = -\frac{1}{3}$$

Thus, $V^*_{0bcc} = \vec{a}^* \cdot (\vec{b}^* \times \vec{c}^*) = (2\pi)^3(4/a^3) = (2\pi)^3/V_{0fcc}$ is the primitive volume in reciprocal space. This is a rhombohedral primitive cell which corresponds to a *body-centred cubic* lattice, as shown in Figure (a). It is convenient to write $|\vec{a}^*|$ in terms of the cube edge, l say, in the reciprocal lattice, as $|\vec{a}^*| = \sqrt{3}l/2$, and hence, the volume of the body-centred cube is

$$V^*_{bcc} = l^3 = (2\pi)^3\left(\frac{2}{a}\right)^3 = 8V^*_{0sc}$$

Hence, the reciprocal lattice of a face-centred cubic direct lattice has a period which is twice that constructed from a simple cubic direct lattice of equal size a. In the same way, the reciprocal lattice of a body-centred cubic lattice of period a is found to be face-centred cubic, having the primitive cell volume $V^*_{0fcc} = (2\pi)^3(2/a^3)$.

39.4. The diamond structure can be described as a cubic lattice with a base consisting of eight carbon atoms, two at 000 and $\frac{1}{4}\frac{1}{4}\frac{1}{4}$, and other six at the equivalent positions due to the (F) lattice. Find the extinction rules for the X-ray diffraction spectra from such a structure.

Solution

The base of the diamond structure consists of eight C atoms at sites

$$000; \ \tfrac{1}{2}\tfrac{1}{2}0; \ \tfrac{1}{2}0\tfrac{1}{2}; \ 0\tfrac{1}{2}\tfrac{1}{2}; \ \tfrac{1}{4}\tfrac{1}{4}\tfrac{1}{4}; \ \tfrac{3}{4}\tfrac{3}{4}\tfrac{1}{4}; \ \tfrac{3}{4}\tfrac{1}{4}\tfrac{3}{4}; \ \tfrac{1}{4}\tfrac{3}{4}\tfrac{3}{4}$$

If f_C is the atomic scattering factor, the structure factor, Eq.(39.32), reads

$$F_C(hkl) = \left[f_C + f_C e^{i\pi(h+k+l)/2} \right]\left[1 + e^{i\pi(h+k)} + e^{i\pi(h+l)} + e^{i\pi(k+l)} \right]$$

$$= f_C \left[1 + e^{i\pi(h+k+l)/2} \right]\left[1 + (-1)^{h+k} + (-1)^{h+l} + (-1)^{k+l} \right]$$

where the structure factor of a diatomic base at sites 000 and $\frac{1}{4}\frac{1}{4}\frac{1}{4}$ is multiplied by the geometric factor of the *fcc* lattice. Apart the systematic extinction introduced by the *fcc* lattice symmetry, Eq.(39.35), we obtain the following regularities:

- if $h+k+l$ is twice an even number: $F_C(hkl) = 8 f_C$
- when $h+k+l$ is twice an odd number: $F_C(hkl) = 0$
- for odd $h+k+l$: $F_C(hkl) = 4 f_C (1 \pm i)$

39.5. Use the intensity distribution of the diffraction peak in one dimension to derive the Scherrer formula for X-ray diffraction line broadening Γ in terms of wavelength λ and the average size b of the crystal grains.

Solution

The X-ray amplitude diffracted in a given direction θ produces a diffraction peak having the intensity distribution similar to that obtained from a single slit of width b, Eq.(19.28), that is

$$I = |\Psi|^2 = I_0 \left(\frac{\sin \gamma}{\gamma} \right)^2$$

where $\gamma = (\pi b / \lambda) \sin \theta$ is a function of the angular position of the diffraction peak. Consider, for clarity, that b is the size of a small crystal, in the direction normal to the (hkl) planes that produce a diffraction peak in the direction θ. The intensity distribution vanishes at $\theta \pm \delta\theta$, where $\Gamma = \delta\theta$ defines the half width of the broadened diffraction line. The first zero occurs for

$$\sin\left(\frac{\pi b \sin \theta}{\lambda} \right) = 0 \quad \text{or} \quad \frac{\pi b \sin(\theta \pm \delta\theta)}{\lambda} = \pm \pi$$

and this yields

$$b[\sin(\theta + \delta\theta) - \sin(\theta - \delta\theta)] = 2\lambda \quad \text{or} \quad 2b \cos\theta \, \delta\theta = 2\lambda$$

This result can be written as the Scherrer formula for diffraction line broadening in the form

$$\Gamma = \frac{\lambda}{b \cos \theta}$$

which provides the average size b of the crystal grains. The line broadening is observed for b ranging from 1 to 100 nm, whereas the effect is negligible if the crystal grains exceed 10^3 nm in size, where they can be treated as infinitely large.

39.6. Find an expression for the unit cell volume of a Bravais lattice in terms of its sides a, b and c and their angular relations specified by α, β and γ.

39.7. Find an expression for the angle between two translation vectors $[pqs]$ and $[p'q's']$ in a rectangular lattice where a, b and c are the sides of the unit cell.

39.8. Find the metric and angular relations in the primitive cell of a face-centred cubic lattice.

39.9. Calculate the packing factor for the diamond structure.

39.10. Prove that the direct and reciprocal Bravais lattices, defined by the primitive translations $\vec{a}, \vec{b}$ and $\vec{c}$ and $\vec{a}*, \vec{b}*$ and $\vec{c}*$, respectively, are reciprocal to each other.

39.11. Find the structure factor of a disordered A_3B metallic alloy of cubic structure.

40. FREE ELECTRONS IN SOLIDS

40.1. FREE-ELECTRON APPROXIMATION
40.2. FREE-ELECTRON THERMODYNAMIC FUNCTIONS
40.3. ELECTRON SPIN PARAMAGNETISM
40.4. ELECTRICAL CONDUCTION

40.1. FREE-ELECTRON APPROXIMATION

The energy eigenvalue problem (38.3) for a system of electrons is usually formulated in the *one-electron approximation*, where each electron is assumed to move in an average field due partly to the net positive charge of the ion cores and partly to all the other electrons. The ions can be treated as a uniform positive background charge and the electrons, for which $k \neq j$, can be represented by the uniform charge distribution $-e|\psi_k(\vec{r}_k)|^2$. Equation (38.3) reduces to

$$\hat{H}_0 \Psi(r,R) = \sum_j \left[-\frac{\hbar^2}{2m_e} \nabla_j^2 + U_{ee}(r_j) + U_{ei}(r_j, R) \right] \Psi(r,R) = E(R)\Psi(r,R) \quad (40.1)$$

where the Coulomb potential energy of interaction $U_{ii}(R)$ between the ions has been omitted since we will allow the electrons to move through an array of stationary ions, so that the interionic distance R can be treated as a constant parameter. If $\Psi(r,R)$ is assumed to be a product of one-electron functions $\psi_j(\vec{r}_j)$, each corresponding to one or other of two possible eigenvalues $m_{si}\hbar = \pm\frac{1}{2}\hbar$ of the spin component, Eq.(40.1) leads to the one-electron equations

$$\left[-\frac{\hbar^2}{2m_e} \nabla_j^2 + e_0^2 \sum_{k \neq j} \frac{|\psi_k(\vec{r}_k)|^2}{\vec{r}_j - \vec{r}_k} dV_k + U_{ei}(r_j, R) \right] \psi_j(r_j) = \varepsilon_j \psi_j(r_j) \quad (40.2)$$

Since the second term on the left-hand side of Eq.(40.2), $U_{ee}(r_j)$, depends on the functions of all the other electrons, a self-consistent procedure, similar to that outlined before for the Hartree equations (36.52) should, in general, be used. It is usual to rewrite Eq.(40.2) in the compact form

$$\left[-\frac{\hbar^2}{2m_e}\nabla^2 + U(\vec{r})\right]\psi(\vec{r}) = \varepsilon\psi(\vec{r}) \qquad (40.3)$$

where $U(\vec{r})$ represents the last two terms on the left-hand side of Eq.(40.2), that are assumed to be dependent on the position $\vec{r}$ of a given electron only. In other words, the N-electron system represented by Eq.(40.1) can be treated as N one-electron systems described by Eq.(40.3).

We will consider a metal to consist of positive ions located on lattice sites, bathed in a sea of conduction electrons, and take a particular example where N valence electrons move in a structure containing N monovalent ion cores. Then the second term in Eqs.(40.2), which describes the positive interaction between equal amounts of uniform electronic charge, and the third term, which represents the negative interaction between the uniform ionic charge and an equal amount of electronic charge, will cancel. We can therefore drop $U(\vec{r})$ in Eq.(40.3), in the zeroth order of approximation, to obtain the one-electron, free-electron Schrödinger equation as

$$-\frac{\hbar^2}{2m_e}\nabla^2\psi(\vec{r}) = \varepsilon\psi(\vec{r}) \qquad (40.4)$$

Although this *free-electron approximation* is a drastic assumption, Eq.(40.4) provides an excellent model for the physical properties of metals. The plane wave solutions of Eq.(40.4), which read

$$\psi(\vec{r}) = \frac{1}{\sqrt{V}}e^{i\vec{k}\cdot\vec{r}} \qquad (40.5)$$

are normalized with respect to the volume V of the crystal and correspond to energy eigenvalues given by

$$\varepsilon = \frac{\hbar^2 k^2}{2m_e} \qquad (40.6)$$

The momentum eigenvalues of the free electrons result from

$$\hat{p}\psi(\vec{r}) = \frac{\hbar}{i}\nabla\left[\frac{1}{\sqrt{V}}e^{i\vec{k}\cdot\vec{r}}\right] = \hbar\vec{k}\psi(\vec{r})$$

in the form

$$\vec{p} = \hbar\vec{k} \qquad (40.7)$$

If we assume the volume V as a finite parallelepiped of sides $P\vec{a}, Q\vec{b}$ and $S\vec{c}$, we obtain the *free-electron gas* model where, since the interaction between electrons is neglected, we only need to describe the motion of one electron at a time inside an impenetrable rectangular box. Although the physical situation requires that the wave function $\psi(\vec{r})$ vanishes at the walls of the box, we shall use instead the *periodic boundary condition* which requires that the solutions should be periodic over the distances Pa, Qb and Sc, respectively, and hence

$$\psi(\vec{r} + P\vec{a}) = \psi(\vec{r} + Q\vec{b}) = \psi(\vec{r} + S\vec{c}) = \psi(\vec{r})$$

Substituting Eq.(40.5), it follows that

$$P\vec{k} \cdot \vec{a} = 2\pi p \quad , \quad Q\vec{k} \cdot \vec{b} = 2\pi q \quad , \quad S\vec{k} \cdot \vec{c} = 2\pi s$$

where p, q and s are arbitrary integers. Comparing these equations with the definitions of the reciprocal base set (39.10), it is clear that $\vec{k}$ may be considered to be a vector in $\vec{k}$-space of the form

$$\vec{k} = \frac{p}{P}\vec{a}^* + \frac{q}{Q}\vec{b}^* + \frac{s}{S}\vec{c}^*$$

Note that, for a vector in $\vec{k}$-space, the components p/P, q/Q and s/S can be fractions of an integer. If they are integers, say $h = p/P$, $k = q/Q$ and $l = s/S$, $\vec{k}$ becomes a reciprocal lattice vector, as defined by Eq.(39.8) and denoted by $\vec{K}_{hkl}$. In the unit cell volume $V_0^* = \vec{a}^* \cdot (\vec{b}^* \times \vec{c}^*)$ of $\vec{k}$-space, defined as in Eq.(39.3) which gives V_0 for the direct lattice, Eq.(39.3), there are PQS $\vec{k}$-states of the free electron, so that the number of states per unit volume of $\vec{k}$-space is

$$\frac{PQS}{V_0^*} = \frac{PQS}{\vec{a}^* \cdot (\vec{b}^* \times \vec{c}^*)} = \frac{(PQS)V_0}{(2\pi)^3} = \frac{V}{(2\pi)^3} \tag{40.8}$$

where V is the volume of the crystal. Since the states in $\vec{k}$-space are uniformly distributed, the number of states with wave vector up to an arbitrary value of $|\vec{k}| = k$, that is, with energy less than $\varepsilon = \hbar^2 k^2 / 2m_e$, is

$$\left(\frac{4\pi}{3}k^3\right)\frac{V}{(2\pi)^3} = \frac{Vk^3}{6\pi^2} = \frac{V}{6\pi^2}\left(\frac{2m_e\varepsilon}{\hbar^2}\right)^{3/2} \quad \text{i.e.} \quad \frac{V}{3\pi^2}\left(\frac{2m_e\varepsilon}{\hbar^2}\right)^{3/2} \tag{40.9}$$

provided we include spin, because each $\vec{k}$-state corresponds to one of the two possible eigenvalues $\pm\frac{1}{2}\hbar$ of the spin component. Assuming a continuous distribution of free-electron states, it is convenient to define the *density of states* $N(\varepsilon)$ to be the number of states with energy between ε and $\varepsilon + d\varepsilon$, such that

$$\frac{V}{3\pi^2}\left(\frac{2m_e\varepsilon}{\hbar^2}\right)^{3/2} = \int_0^\varepsilon N(\varepsilon)d\varepsilon \qquad (40.10)$$

and it follows that

$$N(\varepsilon) = \frac{d}{d\varepsilon}\left[\frac{V}{3\pi^2}\left(\frac{2m_e\varepsilon}{\hbar^2}\right)^{3/2}\right] = \frac{V}{2\pi^2}\left(\frac{2m_e}{\hbar^2}\right)^{3/2}\varepsilon^{1/2} = C\varepsilon^{1/2} \qquad (40.11)$$

If we wish to accomodate a total of N free electrons, one by one, into the volume V of the real space, *at absolute zero*, only one electron will be permitted by the Pauli exclusion principle to be in a given state of the system, at a given time. This behaviour, where the occupation numbers n_k of the electron states can assume only the values 0 or 1, is described in terms of the Fermi-Dirac distribution.

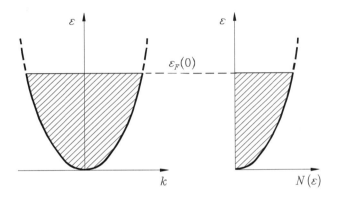

Figure 40.1. Energy band for free electrons at absolute zero

The lowest energy states will be progressively filled up to a maximum value $\varepsilon_F(0)$, called the *Fermi energy*, which is obtained by taking the number of states (40.9) to be N, and this gives

$$\varepsilon_F(0) = \frac{\hbar^2}{2m_e}\left(3\pi^2\frac{N}{V}\right)^{2/3} = \frac{\hbar^2 k_F^2}{2m_e} \qquad (40.12)$$

where the Fermi wave vector was introduced as

$$k_F = \left(3\pi^2\frac{N}{V}\right)^{1/3} \qquad (40.13)$$

There is one electron in each state below $\varepsilon_F(0)$ and none in the states situated above $\varepsilon_F(0)$, as illustrated in Figure 40.1 for $T = 0$. The result is a parabolic energy band for free electrons at absolute zero, plotted in Figure 40.1 as a function of both k, Eq.(40.6), and $N(\varepsilon)$, Eq.(40.11).

At temperatures *above absolute zero*, the average number of electrons per state is given by the Fermi-Dirac distribution function (35.43) which reads

$$f(\varepsilon) = \frac{1}{e^{(\varepsilon-\mu_0)/k_BT}+1} = \frac{1}{e^{(\varepsilon-\varepsilon_F)/k_BT}+1} \quad (40.14)$$

where we have identified the threshold energy μ_0 with ε_F. Equation (40.14) provides us with an alternative definition of the Fermi energy as the energy for which the average occupation of electron states is one half $f(\varepsilon_F) = 1/2$. It is convenient to introduce a temperature T_F that is equivalent to the Fermi energy at absolute zero, as given by $k_B T_F = \varepsilon_F(0)$, which is called the *Fermi temperature*. For most metals we have $T_F > 10^4 K$ so that $T \ll T_F$ and the departure of $f(\varepsilon)$ from its value at absolute zero is small for the usual range of temperatures, in the way plotted in Figure 40.2. There are also illustrated the fraction $\delta f(\varepsilon) = f(\varepsilon) - f_0(\varepsilon)$ of electrons thermally excited into unoccupied states, which may contribute to the measured properties of the electron gas, and the derivative of the distribution function

$$\frac{\partial f(\varepsilon)}{\partial \varepsilon} = -\frac{1}{k_BT}\frac{e^x}{(e^x+1)^2} \quad (40.15)$$

which has an appreciable value only where $x = (\varepsilon - \varepsilon_F)/k_BT$ is near zero, that is, where ε is near the Fermi energy.

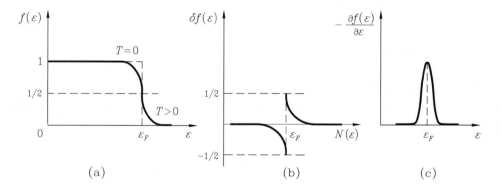

Figure 40.2. Free electron distribution for normal temperatures $T \ll T_F$: (a) $f(\varepsilon)$, (b) $\delta f(\varepsilon)$ and (c) $-\partial f(\varepsilon)/\partial \varepsilon$

Since the behaviour of any function of energy $F(\varepsilon)$ referring to a property of the electron gas is determined by a small region of the free electron distribution close to ε_F, we may expand $F(\varepsilon)$ in a Taylor series about the Fermi energy as

$$F(\varepsilon) = F(\varepsilon_F) + (\varepsilon - \varepsilon_F)\left[\frac{\partial F(\varepsilon)}{\partial \varepsilon}\right]_{\varepsilon=\varepsilon_F} + \frac{1}{2}(\varepsilon - \varepsilon_F)^2 \left[\frac{\partial^2 F(\varepsilon)}{\partial \varepsilon^2}\right]_{\varepsilon=\varepsilon_F} + \ldots \quad (40.16)$$

If $F(\varepsilon)$ is a function which vanishes when ε vanishes, it is common practice to introduce the integral

$$I = \int_0^\infty f(\varepsilon)\frac{\partial F(\varepsilon)}{\partial \varepsilon}d\varepsilon = \left[f(\varepsilon)F(\varepsilon)\right]_0^\infty - \int_0^\infty F(\varepsilon)\frac{\partial f(\varepsilon)}{\partial \varepsilon} = -\int_0^\infty F(\varepsilon)\frac{\partial f(\varepsilon)}{\partial \varepsilon}d\varepsilon \quad (40.17)$$

where $F(0)$ and $f(\infty)$ are zero. Substituting $F(\varepsilon)$ from Eq.(40.16), Eq.(40.17) becomes a series given by

$$I = I_0 F(\varepsilon_F) + I_1 \left[\frac{\partial F(\varepsilon)}{\partial \varepsilon}\right]_{\varepsilon=\varepsilon_F} + I_2 \left[\frac{\partial^2 F(\varepsilon)}{\partial \varepsilon^2}\right]_{\varepsilon=\varepsilon_F} + \ldots$$

where the coefficients are obtained, by using term-by-term integration, as

$$I_0 = -\int_0^\infty \frac{\partial f(\varepsilon)}{\partial \varepsilon}d\varepsilon = f(0) - f(\infty) = 1$$

$$I_1 = -\int_0^\infty (\varepsilon - \varepsilon_F)\frac{\partial f(\varepsilon)}{\partial \varepsilon}d\varepsilon = -k_B T \int_{-\infty}^\infty \frac{xe^x}{(e^x+1)^2}dx = 0$$

$$I_2 = -\frac{1}{2}\int_0^\infty (\varepsilon - \varepsilon_F)^2 \frac{\partial f(\varepsilon)}{\partial \varepsilon}d\varepsilon = \frac{(k_B T)^2}{2}\int_{-\infty}^\infty \frac{x^2 e^x dx}{(e^x+1)^2} = \frac{\pi^2}{6}(k_B T)^2$$

Substituting I from Eq.(40.17), one obtains the useful approximation

$$\int_0^\infty f(\varepsilon)\frac{\partial F(\varepsilon)}{\partial \varepsilon}d\varepsilon \cong F(\varepsilon) + \frac{\pi^2}{6}(k_B T)^2 \left[\frac{\partial^2 F(\varepsilon)}{\partial \varepsilon^2}\right]_{\varepsilon=\varepsilon_F} \quad (40.18)$$

We may now develop an expression for the temperature dependence of the Fermi energy $\varepsilon_F(T)$ of the free electron gas, from the condition that the integral over all the occupied states is equal to N, which can be written as

$$N = \int_0^\infty f(\varepsilon)N(\varepsilon)d\varepsilon \quad (40.19)$$

at temperatures $T > 0$, where the density of occupied states is $f(\varepsilon)N(\varepsilon)$. At absolute zero, it reduces to

$$N = \int_0^{\varepsilon_F(0)} N(\varepsilon)d\varepsilon = \frac{2}{3}C\varepsilon_F^{3/2}(0) \quad (40.20)$$

where use has been made of Eq.(40.11). The integral (40.19) has the form of the left-hand side of Eq.(40.18), provided we take

$$F(\varepsilon) = \int_0^\varepsilon N(\varepsilon)d\varepsilon = \frac{2}{3}C\varepsilon^{3/2} \qquad (40.21)$$

so that Eq.(40.19) may be rewritten, using Eq.(40.11), as

$$N \cong \int_0^\varepsilon N(\varepsilon)d\varepsilon + \frac{\pi^2}{6}(k_B T)^2 \left[\frac{\partial N(\varepsilon)}{\partial \varepsilon}\right]_{\varepsilon=\varepsilon_F} = \frac{2}{3}C\varepsilon_F^{3/2}(T) + \frac{\pi^2}{6}(k_B T)^2 \frac{C}{2\varepsilon_F^{1/2}(T)} \qquad (40.22)$$

Equating Eqs.(40.22) and (40.20) we have

$$\varepsilon_F^{3/2}(0) \cong \varepsilon_F^{3/2}(T) + \frac{\pi^2}{8}\frac{(k_B T)^2}{\varepsilon_F^{1/2}(T)} \quad \text{or} \quad \varepsilon_F^{3/2}(T) \cong \varepsilon_F^{3/2}(0)\left\{1 - \frac{\pi^2}{8}\left[\frac{k_B T}{\varepsilon_F(0)}\right]^2\right\}$$

if we let $\varepsilon_F(T) \cong \varepsilon_F(0)$. The last term is a small correction of order $(T/T_F)^2$, so that we may approximate $(1-\alpha)^{2/3} \cong 1 - 2\alpha/3$, and hence, we finally obtain

$$\varepsilon_F(T) \cong \varepsilon_F(0)\left\{1 - \frac{\pi^2}{12}\left[\frac{k_B T}{\varepsilon_F(0)}\right]^2\right\} \qquad (40.23)$$

which is the usual approximation for the temperature dependence of the Fermi energy at normal temperatures.

40.2. Free-Electron Thermodynamic Functions

The *internal energy* U of the free-electron gas is the total kinetic energy determined from

$$U = \int_0^\infty \varepsilon f(\varepsilon)N(\varepsilon)d\varepsilon \qquad (40.24)$$

Comparison with Eq.(40.18), where it is now convenient to take

$$F(\varepsilon) = \int_0^\varepsilon \varepsilon N(\varepsilon)d\varepsilon = \frac{2}{5}C\varepsilon^{5/2} \qquad (40.25)$$

allows us to write Eq.(40.24) as

$$U \cong \int_0^{\varepsilon_F} \varepsilon N(\varepsilon) d\varepsilon + \frac{\pi^2}{6}(k_B T)^2 \left[\frac{\partial [\varepsilon N(\varepsilon)]}{\partial \varepsilon}\right]_{\varepsilon=\varepsilon_F} = \frac{2}{5} C \varepsilon_F^{5/2}(T) + \frac{\pi^2}{6}(k_B T)^2 \frac{3}{2} C \varepsilon_F^{1/2} \quad (40.26)$$

where $F(\varepsilon_F)$ has been expressed from Eq.(40.25) and $N(\varepsilon)$ from Eq.(40.11). Substituting Eq.(40.23), the internal energy may be obtained, to a first approximation, in terms of $\varepsilon_F(0)$ in the form

$$U \cong \frac{2}{5} C \varepsilon_F^{5/2}(0) \left\{1 - \frac{\pi^2}{12}\left[\frac{(k_B T)}{\varepsilon_F(0)}\right]^2\right\}^{5/2} + \frac{\pi^2}{4} C (k_B T)^2 \varepsilon_F^{1/2}(0)$$

$$\cong \frac{2}{5} C \varepsilon_F^{5/2}(0) - \frac{\pi^2}{12} C (k_B T)^2 \varepsilon_F^{1/2}(0) + \frac{\pi^2}{4} C (k_B T)^2 \varepsilon_F^{1/2}(0)$$

$$= \frac{2}{5} C \varepsilon_F^{5/2}(0) + \frac{\pi^2}{6} C \varepsilon_F^{1/2}(0) (k_B T)^2 = C \varepsilon_F^{5/2}(0) \left[\frac{2}{5} + \frac{\pi^2}{6}\left(\frac{T}{T_F}\right)^2\right]$$

$$= N \varepsilon_F(0) \left[\frac{3}{5} + \frac{\pi^2}{4}\left(\frac{T}{T_F}\right)^2\right] \quad (40.27)$$

where T_F is the Fermi temperature, and C has been inserted as given by Eq.(40.20). The free-electron *heat capacity* is obtained by applying Eq.(7.19), which gives

$$C_V = \left(\frac{\partial U}{\partial T}\right)_V = \frac{\pi^2}{2} \frac{N \varepsilon_F(0)}{T_F^2} T = \gamma T \quad (40.28)$$

and it is interpreted as the electronic contribution to the heat capacity of metals. A more convenient form for γ is obtained by substituting $\varepsilon_F(0)$ in terms of T_F, and this yields

$$\gamma = \frac{\pi^2}{2} \frac{N \varepsilon_F(0)}{T_F^2} = \frac{\pi^2}{2} \frac{N k_B}{T_F} \quad (40.29)$$

Thus the heat capacity (40.28) can be written as

$$C_V = \frac{\pi^2}{2} N k_B \left(\frac{T}{T_F}\right) \quad (40.30)$$

which is a product of a term close to the value $3 N k_B / 2$, as derived in Eq.(26.7) for a perfect gas from the theorem of equipartition of energy, and the ratio T/T_F. This ratio is proportional to the small fraction $\delta f(\varepsilon)$ of electrons thermally excited into empty states, which thereby contribute to the heat capacity of the metal (see Figure 40.2). It is a common experimental procedure to separate the electron contribution to the heat capacity of metals at low temperatures, by plotting C_V/T versus T^2, as shown in Figure 40.3. For most metals the plot follows the straight line given by

$$\frac{C_V}{T} = \gamma + \delta T^2 \qquad (40.31)$$

where the second term on the right-hand side of Eq.(40.31) represents the contribution of lattice vibrations or phonons to the heat capacity, as derived in Chapter 38.

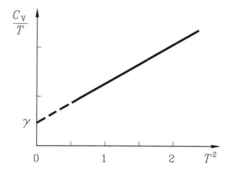

Figure 40.3. Plot of C_V/T versus T^2 for the determination of γ

For simple metals, such as alkali metals, there is a broad agreement between experimental values for γ and those calculated from Eq.(40.29). Substituting $\varepsilon_F(0)$ given by Eq.(40.12) into Eq.(40.29), where $T_F = \varepsilon_F(0)/k_B$, we have

$$\gamma = \frac{\pi^2}{2} \frac{Nk_B^2}{\varepsilon_F(0)} = \frac{\pi^2 Nk_B^2}{(\hbar k_F)^2} m_e \qquad (40.32)$$

from which we can see that the agreement between the measured and calculated values of γ may always be improved by making the electron mass an adjustable parameter. Exact agreement is obtained by assigning the electron an *effective mass* m_e^* defined as

$$m_e^* = m_e \frac{\gamma_{\exp}}{\gamma} \qquad (40.33)$$

The effective mass will be discussed in Chapter 41 and has its origin in the influence of the lattice of positive ions which prevents the electrons from being totally free.

Substituting the heat capacity (40.30) into Eq.(8.22), which reads

$$S = \int_0^T \frac{C_V}{T} dT$$

we obtain the entropy of the free-electron gas as

$$S = \frac{\pi^2}{2} Nk_B \left(\frac{T}{T_F}\right) \quad \text{or} \quad TS = \frac{\pi^2}{2} Nk_B T_F \left(\frac{T}{T_F}\right)^2 \quad (40.34)$$

Hence, the temperature dependence of both the *TS* product and the internal energy (40.27) involves $(T/T_F)^2$, which is of the order of $10^{-4} - 10^{-6}$ for most metals at normal temperatures. Since the thermodynamic potentials *H*, *F* and *G* are defined in terms of *U* and *TS*, it follows that all the thermodynamic properties of the free-electron gas can be regarded to be temperature-independent at normal temperatures.

40.3. ELECTRON SPIN PARAMAGNETISM

Consider the effect of an applied magnetic field of induction $\vec{B}_0 = \mu_0 \vec{H}$ on the spin magnetic moments $\vec{\mu}_s$ of free electrons. At absolute zero, in the absence of the field, the energy band for free electrons, represented by a plot of energy ε versus density of states $N(\varepsilon)$, exhibit two identical halves for electrons with spin parallel $(N_p(\varepsilon))$ and antiparallel $(N_a(\varepsilon))$ to the z-axis, see Figure 40.4 (a). If the direction of the field is taken to be the z-axis, the energy ε of the spinning electron is raised or lowered, dependending on the eigenvalue of its spin component, by

$$-\vec{\mu}_s \cdot \vec{B}_0 = g_s \frac{e}{2m_e} \vec{S} \cdot \vec{B}_0 = 2 \frac{e}{2m_e} m_s \hbar \vec{B}_0 = \pm \mu_B B_0$$

as obtained by combining Eqs.(34.22) and (34.14).

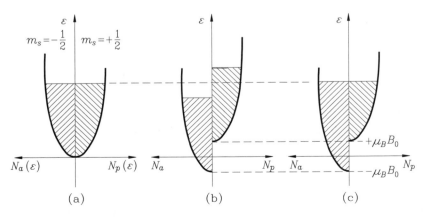

Figure 40.4. Energy bands for free electrons in applied magnetic field

The result is an instantaneous displacement of the two halves of the energy band, as in Figure 40.4 (b), resulting in a nonequilibrium configuration, so that the electrons with $m_s = +\frac{1}{2}$, which have higher energy than those with $m_s = -\frac{1}{2}$, fall into the empty states which they then occupy by reversal of their spins. Consequently in the equilibrium configuration in applied magnetic field, shown in Figure 40.4 (c), there will be more electrons with spin antiparallel (and magnetic moment parallel) to the magnetic field than electrons with parallel spin, as given by

$$N_p = \frac{1}{2} \int_{\mu_B B_0}^{\infty} f(\varepsilon) N_p(\varepsilon) d\varepsilon = \frac{1}{2} \int_0^{\infty} f(\varepsilon) N(\varepsilon - \mu_B B_0) d\varepsilon$$

(40.35)

$$N_a = \frac{1}{2} \int_{-\mu_B B_0}^{\infty} f(\varepsilon) N_a(\varepsilon) d\varepsilon = \frac{1}{2} \int_0^{\infty} f(\varepsilon) N(\varepsilon + \mu_B B_0) d\varepsilon$$

The net magnetic moment *parallel* to the magnetic field is then given by

$$M = \mu_B (N_a - N_p) = \frac{\mu_B}{2} \int_0^{\infty} f(\varepsilon) \left[N(\varepsilon + \mu_B B_0) - N(\varepsilon - \mu_B B_0) \right] d\varepsilon \quad (40.36)$$

where we may expand the density of states functions, by use of a Taylor series, as

$$\frac{1}{2} \left[N(\varepsilon + \mu_B B_0) - N(\varepsilon - \mu_B B_0) \right] \cong \mu_B B_0 \frac{\partial N(\varepsilon)}{\partial \varepsilon} \quad (40.37)$$

Substituting Eq.(40.37), Eq.(40.36) becomes

$$M \cong \mu_B^2 B_0 \int_0^{\infty} f(\varepsilon) \frac{\partial N(\varepsilon)}{\partial \varepsilon} d\varepsilon \quad (40.38)$$

On comparison with Eq.(40.18), where we put $F(\varepsilon) = N(\varepsilon) = C\varepsilon^{1/2}$, the total magnetic moment (40.38) results as

$$M \cong \mu_B^2 B_0 N(\varepsilon_F) = \mu_B^2 B_0 C \varepsilon_F^{1/2}(T) = \frac{3}{2} \frac{N \mu_B^2 B_0}{\varepsilon_F^{3/2}(0)} \varepsilon_F^{1/2}(T) \cong \frac{3N\mu_B^2}{2k_B T_F} B_0 \left[1 - \frac{\pi^2}{24} \left(\frac{T}{T_F} \right)^2 \right]$$

(40.39)

where use has been made of Eq.(40.23). The total magnetic moment (40.39) represents the *magnetization* of a free-electron gas, as defined by Eq.(13.7), if N is taken to be the number of electrons per unit volume. An induced magnetization in the direction of the applied field defines the phenomenon known as *paramagnetism*, which will be discussed in Chapter 44. In contrast to the common effect, where an applied field induces a strong temperature dependent magnetization, the observed paramagnetic behaviour of alkali metals, often called *Pauli spin paramagnetism* is a weak and almost temperature-independent effect. Equation (40.39) provides a reasonable approximation for the usual range of temperatures, where $T \ll T_F$. This result shows that the free-electron approximation has some validity in describing certain elementary aspects of metallic behaviour.

40.4. ELECTRICAL CONDUCTION

The electron flow in a conductor, under the influence of an external electric field, can be associated with a change of the free-electron distribution $f(\vec{k})$ in $\vec{k}$-space. We will consider the simplest possible situation where a small electric field $\vec{E}$ is applied in the x-direction and the force can be expressed to be the rate of change of the momentum, using Eq.(40.7), in the form

$$-e\vec{E} = \hbar \frac{d\vec{k}}{dt} \quad \text{or} \quad -eE = \hbar \frac{dk_x}{dt}$$

The rate of change of the distribution function $f(\vec{k})$, caused by the field, and therefore associated with electron *drift*, may be expressed as

$$\frac{df}{dt} = \frac{\partial f}{\partial t} + \left(\frac{\partial f}{\partial k_x}\right)\left(\frac{dk_x}{dt}\right) = \frac{\partial f}{\partial t} - \frac{eE}{\hbar}\left(\frac{\partial f}{\partial k_x}\right) \tag{40.40}$$

For small deviations of the distribution function from equilibrium, we may assume, in a real crystal, that $f(\vec{k})$ returns to its thermal equilibrium form $f_0(\vec{k})$ if the external field is removed, due to electron-lattice collisions, according to

$$(f - f_0)_t = (f - f_0)_{t=0} e^{-t/\tau}$$

where τ stands for the relaxation time of the system due to electron scattering by phonons. Alternatively, we can say that the rate of change of $f(\vec{k})$ caused by scattering is proportional to the deviation of the function from equilibrium given by

$$\left(\frac{\partial f}{\partial t}\right)_{scatt} = -\frac{f - f_0}{\tau} \tag{40.41}$$

A steady-state electron flow, where $\partial f / \partial t = 0$, is established as a result of the dynamic equilibrium between drift and scattering effects, which can be written, using Eqs.(40.40) and (40.41), as

$$\frac{df}{dt} = \left(\frac{\partial f}{\partial t}\right)_{scatt} \quad \text{or} \quad -\frac{eE}{\hbar}\left(\frac{\partial f}{\partial k_x}\right) = -\frac{f-f_0}{\tau} \qquad (40.42)$$

The distribution function then follows as

$$f = f_0 + \frac{e\tau E}{\hbar}\frac{\partial f}{\partial k_x} = f_0 + e\tau E \frac{\hbar k_x}{m_e}\frac{\partial f}{\partial \varepsilon} = f_0 + e\tau E v_x \frac{\partial f}{\partial \varepsilon} \qquad (40.43)$$

where use has been made of Eqs.(40.6) and (40.7), and v_x is the electron velocity in the x-direction. The second term on the right-hand side of Eq.(40.43) represents the perturbation of the equilibrium distribution f_0 and, provided the electric field E is weak, we may assume $\partial f / \partial \varepsilon \cong \partial f_0 / \partial \varepsilon$, which yields

$$f = f_0 + e\tau E v_x \frac{\partial f_0}{\partial \varepsilon} \qquad (40.44)$$

Since $f_0(\vec{k})$ is the Fermi distribution function and the states in $\vec{k}$-space are uniformly distributed, in view of Eq.(40.13) we can write

$$\int_0^{k_F} f_0(\vec{k}) d\vec{k} \cong \frac{4\pi}{3}k_F^3 = \frac{4\pi}{3}\left(3\pi^2 \frac{N}{V}\right) = 4\pi^3 \frac{N}{V} \quad \text{or} \quad n = \frac{N}{V} \cong \frac{1}{4\pi^3}\int_0^{k_F} f_0(\vec{k}) d\vec{k}$$

which gives the number of electrons per unit volume, in the absence of applied fields. In the presence of a weak external field we may replace $f_0(\vec{k})$ by $f(\vec{k})$, so that the electron concentration in the volume element $d\vec{k}$ of reciprocal space will be taken as

$$dn = \frac{dN}{V} = \frac{1}{4\pi^3} f(\vec{k}) d\vec{k} \qquad (40.45)$$

where $d\vec{k} = dk_x dk_y dk_z = k^2 \sin\theta \, d\theta \, d\varphi \, dk$. By definition (12.7) we know that for N electrons in a volume V, all moving with the same velocity $\vec{v}$, the current density is $\vec{j} = -(N/V)e\vec{v} = -ne\vec{v}$. If we assume a distribution of drift velocities, the current density j_x in the x-direction can be calculated by summing the electron contribution over all the carriers in $\vec{k}$-space, and this can be written as

$$j_x = -\int e v_x dn = -\frac{e}{4\pi^3}\int v_x f(\vec{k}) d\vec{k} \qquad (40.46)$$

The first term in Eq.(40.44) represents equilibrium and does not contribute the current density (40.46). Setting $v_x = v\cos\theta$ and integrating over θ and φ, Eq.(40.46) becomes

$$j_x = -\frac{e^2 E}{4\pi^3}\int v_x^2 \tau\left(\frac{\partial f_0}{\partial \varepsilon}\right)d\vec{k} = -\frac{e^2 E}{4\pi^3}\int_0^\infty \int_0^\pi \int_0^{2\pi}(v\cos\theta)^2 \tau\left(\frac{\partial f_0}{\partial \varepsilon}\right)k^2 \sin\theta\, d\theta\, d\varphi\, dk$$

$$= -\frac{e^2 E}{3\pi^2}\int_0^\infty v^2 \tau\left(\frac{\partial f_0}{\partial \varepsilon}\right)k^2 dk = -\frac{2e^2 E}{3\pi^2 m_e}\int_0^\infty \varepsilon\tau\left(\frac{\partial f_0}{\partial \varepsilon}\right)k^2 dk \tag{40.47}$$

where $v^2 = 2\varepsilon/m_e$. Using the parabolic energy dependence (40.6), Eq.(40.47) can be rewritten as

$$j_x = -\frac{2(2m_e)^{1/2} e^2 \tau_F E}{3\pi^2 \hbar^3}\int_0^\infty \varepsilon^{3/2}\frac{df_0(\varepsilon)}{d\varepsilon}d\varepsilon = \frac{(2m_e)^{1/2} e^2 \tau_F E}{\pi^2 \hbar^3}\int_0^\infty \varepsilon^{1/2} f_0(\varepsilon)d\varepsilon \tag{40.48}$$

where use has been made of Eq.(40.17), by taking $F(\varepsilon) = \varepsilon^{3/2}$. Note that, because $df_0(\varepsilon)/d\varepsilon$ has an appreciable value only where ε is close to the Fermi energy, as shown in Figure 40.2, τ_F can be interpreted to be the effective value of the relaxation time at the Fermi level. Substituting $N(\varepsilon)$ from Eq.(40.11) and then N given by Eq.(40.19), Eq.(40.48) reduces to

$$j_x = \frac{(2m_e)^{1/2} e^2 \tau_F E}{\pi^2 \hbar^3}\frac{2\pi^2}{V}\left(\frac{\hbar^2}{2m_e}\right)^{3/2}\int_0^\infty N(\varepsilon)f_0(\varepsilon)d\varepsilon = \frac{N}{V}\frac{e^2 \tau_F}{m_e}E \tag{40.49}$$

Comparing Eq.(40.49) with Eq.(12.10), the *electrical conductivity* follows as

$$\sigma = \frac{N}{V}\frac{e^2 \tau_F}{m_e} = \frac{ne^2 \tau_F}{m_e} \tag{40.50}$$

which is the same as the static conductivity (23.50), derived from the classical free-electron model. Since electron-lattice collisions occur near the Fermi energy only, the relaxation time $\tau = \tau_F$ is independent of energy, and therefore the classical treatment yields the same expression for σ as that obtained by the quantum approach.

FURTHER READING
1. W. A. Harrison - ELEMENTARY ELECTRONIC STRUCTURE, World Scientific, Singapore, 2004.
2. R. E. Hummel - ELECTRONIC PROPERTIES OF MATERIALS, Springer-Verlag, New York, 2000.
3. R. H. Bube - ELECTRONS IN SOLIDS: AN INTRODUCTORY SURVEY, Elsevier, Amsterdam, 1992.

PROBLEMS

40.1. Show that there is a change, induced by dimensionality, in the density of states of a free-electron gas with N-electrons, of Fermi energy $\varepsilon_F(0)$, and find expressions for $N_{3-D}(\varepsilon)$, $N_{2-D}(\varepsilon)$ and $N_{1-D}(\varepsilon)$.

Solution

For the free-electron states in 3-D we can combine Eqs.(40.10) and (40.9) in the form

$$N_{3-D}(\varepsilon)d\varepsilon = 2\frac{V}{(2\pi)^3}\int_{S_\varepsilon}d^3k = \frac{2V}{(2\pi)^3}\int_{S_\varepsilon}dS_\varepsilon dk = \frac{2V}{(2\pi)^3}\int_{S_\varepsilon}dS_\varepsilon \frac{dk}{d\varepsilon}d\varepsilon$$

where $dS_\varepsilon = 4\pi k^2$ and $dk/d\varepsilon = m_e/\hbar^2 k$, which leads to

$$N_{3-D}(\varepsilon) = \frac{2V}{(2\pi)^3}\int_{S_\varepsilon}dS_\varepsilon \frac{dk}{d\varepsilon} = \frac{2V}{(2\pi)^3}\frac{4\pi k^2 m_e}{\hbar^2 k} = \frac{V}{2\pi^2}\left(\frac{2m_e}{\hbar^2}\right)^{1/2}\varepsilon^{1/2}$$

and this is the same as given by Eq.(40.11). In view of Eq.(40.12), this can be written as

$$N_{3-D}(\varepsilon) = \frac{3N}{2\varepsilon_F^{3/2}(0)}\varepsilon^{1/2}$$

In a similar way, for 2-D systems, the number of electron states inside the surface element of two-dimensional reciprocal space, bounded by the contours of constant energy ε and $\varepsilon + d\varepsilon$, reads

$$N(\varepsilon)d\varepsilon = 2\frac{A}{(2\pi)^2}\int_{l_\varepsilon}d^2k = 2\frac{A}{(2\pi)^2}\int_{l_\varepsilon}dl_\varepsilon \frac{dk}{d\varepsilon}d\varepsilon$$

For the free-electron gas in two dimensions, $d\varepsilon/dk = \hbar^2 k/m_e$ is a constant along a circular path of radius k, and this gives a constant density of states of the form

$$N_{2-D}(\varepsilon) = 2\frac{A}{(2\pi)^2}\frac{2\pi k}{\hbar^2 k/m_e} = \frac{m_e A}{\pi\hbar^2}$$

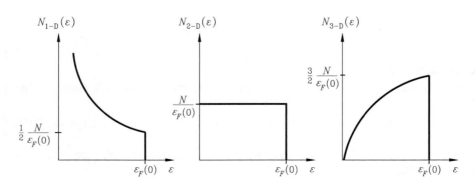

At absolute zero, the normalization condition yields

$$\int_0^{\varepsilon_F(0)} N_{2-D}(\varepsilon)\,d\varepsilon = N \quad \text{or} \quad N_{2-D}(\varepsilon) = \frac{N}{\varepsilon_F(0)}$$

In one dimension, the integration contour reduces to the points $\pm k$, and the free-electron density of states becomes

$$N_{1-D}(\varepsilon) = 2\frac{L}{2\pi}\frac{2}{\hbar^2 k/m_e} = \frac{2m_e L}{\pi\hbar^2}\frac{1}{k} = \frac{L\sqrt{2m_e}}{\pi\hbar}\varepsilon^{-1/2} = C_1\varepsilon^{-1/2}$$

The normalization condition at absolute zero yields

$$C_1 \int_0^{\varepsilon_F(0)} \frac{d\varepsilon}{\varepsilon^{1/2}} = N \quad \text{or} \quad C_1 = \frac{N}{2\varepsilon_F^{1/2}(0)} \quad \text{i.e.,} \quad N_{1-D}(\varepsilon) = \frac{N}{2\varepsilon_F^{1/2}(0)}\varepsilon^{-1/2}$$

The change in $N(\varepsilon)$ induced by dimensionality is apparent from the graphical representation.

40.2. Find the expressions for the Fermi energy and the heat capacity of a free-electron gas in one dimension at temperatures $T > 0$.

Solution

At temperatures $T > 0$, the normalization condition in one dimension reads

$$N = \int_0^\infty f(\varepsilon) N_{1-D}(\varepsilon)\,d\varepsilon = \frac{N}{2\varepsilon_F^{1/2}(0)}\int_0^\infty f(\varepsilon)\varepsilon^{-1/2}\,d\varepsilon$$

where use has been made of $N_{1-D}(\varepsilon)$, as derived in Problem 40.1. In view of the identity (40.18), it is convenient to choose $F(\varepsilon) = 2\varepsilon^{1/2}$, such that

$$2\varepsilon_F^{1/2}(0) = \int_0^\infty f(\varepsilon)\frac{\partial(2\varepsilon^{1/2})}{\partial\varepsilon}\,d\varepsilon = 2\varepsilon_F^{1/2}(T) - \frac{\pi^2}{6}(k_B T)^2\frac{1}{2\varepsilon_F^{3/2}(T)}$$

which approximately yields

$$\varepsilon_F(T) = \varepsilon_F(0)\left[1 - \frac{\pi^2}{24}\frac{(k_B T)^2}{\varepsilon_F^2(T)}\right]^{-2} = \varepsilon_F(0)\left[1 + \frac{\pi^2}{12}\left(\frac{T}{T_F}\right)^2\right]$$

if we let $\varepsilon_F(T) \approx \varepsilon_F(0) = k_B T_F$. The internal energy, Eq.(40.24), reduces in one dimension to

$$U = \int_0^\infty \varepsilon f(\varepsilon) N_{1-D}(\varepsilon)\,d\varepsilon = \frac{N}{2\varepsilon_F^{1/2}(0)}\int_0^\infty f(\varepsilon)\varepsilon^{1/2}\,d\varepsilon$$

By taking $F(\varepsilon) = (2/3)\varepsilon^{3/2}$, one obtains from Eq.(40.18) that

$$U = \frac{N}{2\varepsilon_F^{1/2}(0)} \left[\frac{2}{3} \varepsilon_F^{3/2}(T) + \frac{\pi^2}{6} (k_B T)^2 \frac{1}{2\varepsilon_F^{1/2}(T)} \right]$$

$$= \frac{N}{2\varepsilon_F^{1/2}(0)} \left[\frac{2}{3} \varepsilon_F^{3/2}(0) + \frac{\pi^2}{12} \frac{(k_B T)^2}{\varepsilon_F^{1/2}(0)} + \frac{\pi^2}{12} \frac{(k_B T)^2}{\varepsilon_F^{1/2}(T)} \right] = N \varepsilon_F(0) \left[\frac{1}{3} + \frac{\pi^2}{12} \left(\frac{T}{T_F} \right)^2 \right]$$

which gives the electronic contribution to the heat capacity as

$$C_V = \left(\frac{\partial U}{\partial T} \right)_V = \frac{\pi^2}{6} \frac{N \varepsilon_F(0)}{T_F^2} T = \frac{\pi^2}{6} N k_B \left(\frac{T}{T_F} \right)$$

40.3. Find expressions for the temperature dependence of the paramagnetic susceptibility for a free-electron gas in one, two and three dimensions.

Solution

The magnetization of the free-electron gas is given by Eq.(40.39) as

$$M = \frac{(N_a - N_p) \mu_B}{V} = \frac{\mu_B^2 B_0 N(\varepsilon_F)}{V}$$

where the density of states must be replaced as derived in Problem 40.1. It follows that the susceptibility $\chi = \mu_0 M / B_0$ takes, in one dimension, the form

$$\chi_{1-D} = \frac{\mu_0 \mu_B^2}{V} N_{1-D}(\varepsilon_F) = \frac{n}{2} \frac{\mu_0 \mu_B^2}{\varepsilon_F^{1/2}(0)} \varepsilon_F^{-1/2}$$

where $n = N/V$ is the electron concentration. By virtue of Eq.(40.23), one obtains

$$\chi_{1-D} = \frac{n}{2} \frac{\mu_0 \mu_B^2}{k_B T_F} \left[1 + \frac{\pi^2}{24} \left(\frac{T}{T_F} \right)^2 \right]$$

Similarly, in the two-dimensional case, the result is a constant susceptibility of the form

$$\chi_{2-D} = \frac{\mu_0 \mu_B^2}{V} N_{2-D}(\varepsilon_F) = n \frac{\mu_0 \mu_B^2}{\varepsilon_F(0)} = n \frac{\mu_0 \mu_B^2}{k_B T_F}$$

In view of Eq.(40.39), we may write

$$\chi_{3-D} = \frac{\mu_0 M}{B_0} = \frac{3}{2} \frac{n \mu_0 \mu_B^2}{k_B T_F} \left[1 - \frac{\pi^2}{24} \left(\frac{T}{T_F} \right)^2 \right]$$

and the effect of dimensionality becomes apparent at absolute zero where, assuming the same n and T_F, one has $\chi_{3-D} : \chi_{2-D} : \chi_{1-D} = 3 : 2 : 1$.

40.4. If a current is flowing along a metal strip in the presence of a normal magnetic field $\vec{B}_0$, show that a lateral drift of electrons, called the Hall effect, can be observed, and find an expression for the transverse voltage in terms of the electron concentration n.

Solution

The electron equation of motion (23.32) can be written as

$$m_e \frac{d\vec{v}}{dt} + \frac{m_e \vec{v}}{\tau} = -e(\vec{E} + \vec{v} \times \vec{B}_0)$$

where τ stands for the relaxation time due to electron scattering. Thus, the electron drift results as

$$\vec{v} = -\frac{e\tau}{m_e}(\vec{E} + \vec{v} \times \vec{B}_0)$$

Substituting the velocity components in the presence of a magnetic field of induction $\vec{B}_0 = (0, 0, B_0)$, one obtains

$$j_x = -nev_x = \frac{\sigma_0}{1 + \omega_c^2 \tau^2}(E_x - \omega_c \tau E_y) \quad, \quad j_y = \frac{\sigma_0}{1 + \omega_c^2 \tau^2}(\omega_c \tau E_x + E_y) \quad, \quad j_z = \sigma_0 E_z$$

where $\omega_c = eB_0/m_e$ is the cyclotron frequency. The electrical conductivity is given by a tensor

$$\begin{pmatrix} j_x \\ j_y \\ j_z \end{pmatrix} = \frac{\sigma_0}{1 + \omega_c^2 \tau^2} \begin{pmatrix} 1 & -\omega_c \tau & 0 \\ \omega_c \tau & 1 & 0 \\ 0 & 0 & 1 + \omega_c^2 \tau^2 \end{pmatrix} \begin{pmatrix} E_x \\ E_y \\ E_z \end{pmatrix}$$

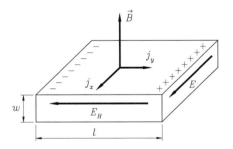

If the electric field is taken to be $\vec{E} = (E_0, 0, 0)$ along the metal strip, as shown in the figure, the magnetic field $\vec{B}_0$ causes a drift of electrons in the y-direction, called the Hall effect. Since the current cannot flow out of the strip, a transverse field E_H should occur. In the steady state, we have $j_y = 0$ and this implies that

$$\omega_c \tau E_0 + E_H = 0 \quad \text{or} \quad E_H = -\omega_c \tau E_0 = -\frac{eB_0 \tau}{m_e} E_0$$

whereas the current density along the applied electric field is just as if the magnetic field were absent, that is,

$$j_x = \frac{\sigma_0}{1+\omega_c^2 \tau^2}(E_0 + \omega_c^2 \tau^2 E_0) = \sigma_0 E_0$$

The Hall coefficient R_H is defined by the ratio

$$R_H = \frac{E_H}{j_x B_0} = -\frac{e\tau}{\sigma_0 m_e} = -\frac{1}{ne}$$

where n is the electron concentration. In a Hall effect experiment, the current $I = j_x A = j_x lw$ and Hall voltage $V_H = E_H l$ are measured rather than the current density and the field E_H. The observed voltage is given by $V_H = (B_0 I / w) R_H$, and this allows us to compute R_H, and hence, the electron concentration.

40.5. Use the Debye model for the lattice vibrations to find the temperature dependence of the electrical resistivity in a simple metal of Debye temperature θ_D.

Solution

The temperature dependence of resistivity can be described in terms of the scattering interaction between free electrons and phonons. The probability of scattering, $1/\tau$, along the path of an electron, depends only on the geometrical cross-section associated with the phonon, and this is measured by the mean-square displacement $\langle s^2 \rangle$ in a given direction. In the Debye model the total energy is just the sum of the mean energies over all frequencies, given by

$$U = \sum_k M\omega_k^2 \langle s^2(\omega_k) \rangle.$$

and, for a continuous distribution of frequencies, this becomes

$$\int_0^{\omega_D} E(\omega) N(\omega) d\omega = \int_0^{\omega_D} M\omega^2 \langle s^2(\omega) \rangle d\omega$$

Assuming, for simplicity, that $\langle s^2(\omega) \rangle$ is approximately a constant, yields

$$\langle s^2 \rangle = \frac{3}{M\omega_D^3} \int_0^{\omega_D} \frac{\hbar\omega}{e^{\hbar\omega/k_B T} - 1} N(\omega) d\omega$$

Because only a small fraction $\delta f(\varepsilon)$ of electrons with energies close to the Fermi surface are involved in the electron-phonon interactions, exchanging energy $\hbar\omega = k_B(T_F \pm T)$, the integral must fall off rapidly for interactions involving phonons with frequencies outside this range. This implies that the Debye distribution function, Eq.(39.61), must be multiplied by a factor $T/T_F = \hbar\omega/\varepsilon_F = \hbar\omega/(m_e v_F^2/2) = 2\hbar\omega/m_e v_F^2$, which is proportional to $\delta f(\varepsilon)$. This leads to

$$N(\omega) d\omega = \frac{18\hbar}{m_e v_F^2 \omega_D^3} \omega^3 d\omega$$

Thus, one obtains

$$\langle s^2 \rangle = \frac{54\hbar k_B T}{M m_e v_F^2 \omega_D^6} \int_0^{\omega_D} \frac{\hbar\omega/k_B T}{e^{\hbar\omega/k_B T}-1} \omega^3 d\omega = \frac{54\hbar k_B T}{M m_e v_F^2 \omega_D^2} \left(\frac{T}{\theta_D}\right)^4 \int_0^{\theta_D/T} \frac{x^4 dx}{e^x-1}$$

where $x = \hbar\omega/k_B T$ and $\theta_D = \hbar\omega_D/k_B$. It follows that

$$\rho(T) \sim \frac{T}{M\theta_D^2} \left(\frac{T}{\theta_D}\right)^4 \int_0^{\theta_D/T} \frac{x^4 dx}{e^x-1}$$

At high temperatures, where θ_D/T is small, the integral tends to $(\theta_D/T)^4$, and the result is a linear dependence of the resistivity on temperature, given by

$$\rho(T) \sim \frac{T}{M\theta_D^2}$$

40.6. Find the average energy at absolute zero for a free electron in one-, two- and three-dimensional metals of Fermi energy $\varepsilon_F(0)$.

40.7. Find the temperature dependence of the Fermi energy and the electron heat capacity of a two-dimensional metal.

40.8. Find expressions for the total heat capacity, including electron and phonon contributions, of one- and two-dimensional metals.

40.9. Find an expression for the magnetic flux $\Phi_n = BS_{xy}(n)$ through the allowed electron orbits, of energy ε_n, in a plane perpendicular to the uniform magnetic field B.

40.10. Find an expression for the temperature dependence of electrical resistivity assuming that the lattice vibrations are described by the Einstein model.

40.11. Find the lattice vibration contribution to the electrical resistivity at low temperatures in the Debye approximation.

41. Electronic Energy Bands

41.1. BLOCH WAVES
41.2. THE WEAK-BINDING APPROXIMATION
41.3. THE TIGHT-BINDING APPROXIMATION

41.1. Bloch Waves

Crystal lattices possess translational symmetry, and this requires that the potential energy $U(\vec{r})$ of the one-electron problem (40.3) has to be a periodic function, which can be written as

$$U(\vec{r}+\vec{R}) = U(\vec{r}) \tag{41.1}$$

where $\vec{R}$ is any lattice vector of the form (39.1) that connects equivalent points in two unit cells. We will show in this chapter how the periodic potential splits the allowed electron energies into bands separated by forbidden regions, called energy gaps. We start by substituting $\vec{r}+\vec{R}$ for $\vec{r}$ in Eq.(40.3), and this gives

$$\left[-\frac{\hbar^2}{2m_e}\nabla^2 + U(\vec{r}+\vec{R})\right]\psi(\vec{r}+\vec{R}) = \varepsilon\psi(\vec{r}+\vec{R})$$

which can be written, in view of Eq.(41.1), as

$$\left[-\frac{\hbar^2}{2m_e}\nabla^2 + U(\vec{r})\right]\psi(\vec{r}+\vec{R}) = \varepsilon\psi(\vec{r}+\vec{R}) \tag{41.2}$$

On comparison with Eq.(40.3), we may infer that $\psi(\vec{r}+\vec{R})$ and $\psi(\vec{r})$ correspond to the same eigenvalue ε, that is,

$$\psi(\vec{r}+\vec{R}) = C(\vec{R})\psi(\vec{r}), \quad \text{where} \quad |C(\vec{R})|^2 = 1 \tag{41.3}$$

In other words, under any translation $\vec{R}$ that leaves the lattice invariant the wave function is multiplied by a phase factor $C(\vec{R})$. Substituting $\vec{R}+\vec{R}'$ for $\vec{R}$ yields

$$\psi(\vec{r}+\vec{R}+\vec{R}') = C(\vec{R}+\vec{R}')\psi(\vec{r}) = C(\vec{R})\psi(\vec{r}+\vec{R}') = C(\vec{R})C(\vec{R}')\psi(\vec{r})$$

that is,

$$C(\vec{R}+\vec{R}') = C(\vec{R})C(\vec{R}') \quad \text{or} \quad C(\vec{R}) = e^{i\vec{k}\cdot\vec{R}} \qquad (41.4)$$

and hence, Eq.(41.3) reduces to the *Bloch condition* given by

$$\psi(\vec{r}+\vec{R}) = e^{i\vec{k}\cdot\vec{R}}\psi(\vec{r}) \qquad (41.5)$$

The significance of $\vec{k}$ may be derived by using periodic boundary conditions on the wave functions (41.5), which read

$$\psi(\vec{r}+P\vec{a}) = \psi(\vec{r}+Q\vec{b}) = \psi(\vec{r}+S\vec{c}) = \psi(\vec{r})$$

and this requires that

$$P\vec{k}\cdot\vec{a} = 2\pi p, \quad Q\vec{k}\cdot\vec{b} = 2\pi q, \quad S\vec{k}\cdot\vec{c} = 2\pi s$$

where p, q and s are arbitrary integers. Thus, $\vec{k}$ is a vector of $\vec{k}$-space of the form

$$\vec{k} = \frac{p}{P}\vec{a}^* + \frac{q}{Q}\vec{b}^* + \frac{s}{S}\vec{c}^* \qquad (41.6)$$

where the components p/P, q/Q and s/S are, in general, fractions of an integer. With each $\vec{k}$-value we may associate a one-electron wave function of the form

$$\psi_{\vec{k}}(\vec{r}) = e^{i\vec{k}\cdot\vec{r}} u_{\vec{k}}(\vec{r}) \qquad (41.7)$$

called a *Bloch wave* which, upon substitution into the Bloch condition (41.5), leads to

$$e^{i\vec{k}\cdot(\vec{r}+\vec{R})} u_{\vec{k}}(\vec{r}+\vec{R}) = e^{i\vec{k}\cdot\vec{R}} e^{i\vec{k}\cdot\vec{r}} u_{\vec{k}}(\vec{r}) \quad \text{or} \quad u_{\vec{k}}(\vec{r}+\vec{R}) = u_{\vec{k}}(\vec{r}) \qquad (41.8)$$

Equations (41.7) and (41.8) express the *Bloch theorem* which states that in a periodic potential (41.1) one can always choose the one-electron wave functions to be Bloch waves of the form (41.7), where $u_{\vec{k}}(\vec{r})$ has the periodicity (41.8) of the lattice.

Substituting Eq.(41.7) into the one-electron Schrödinger equation (40.3) we find an equation for $u_{\vec{k}}(\vec{r})$ in the form

$$-\frac{\hbar^2}{2m_e}\left[\nabla^2 + 2i\vec{k}\cdot\nabla - \vec{k}^2\right]u_{\vec{k}}(\vec{r}) + U(\vec{r})u_{\vec{k}}(\vec{r}) = \varepsilon u_{\vec{k}}(\vec{r}) \tag{41.9}$$

which can be rewritten as

$$\left[-\frac{\hbar^2}{2m_e}\nabla^2 + \frac{\hbar}{m_e}\vec{k}\cdot\vec{p} + U(\vec{r})\right]u_{\vec{k}}(\vec{r}) = \left(\varepsilon - \frac{\hbar^2 k^2}{2m_e}\right)u_{\vec{k}}(\vec{r}) \tag{41.10}$$

Considering the complex conjugate equation, yields $u_{\vec{k}}^*(\vec{r}) = u_{-\vec{k}}(\vec{r})$ and this leads to

$$\psi_{\vec{k}}^*(\vec{r}) = \psi_{-\vec{k}}(\vec{r}) \tag{41.11}$$

and also

$$\varepsilon(-\vec{k}) = \varepsilon(\vec{k}) \tag{41.12}$$

which means that any energy band is symmetric about the origin of $\vec{k}$-space.

EXAMPLE 41.1. Bloch waves in a one-dimensional periodic potential

We have seen so far that the influence of the lattice structure on the electron states consists, according to the Bloch theorem, of modulating the plane-wave functions of free-electron states by a function having the periodicity of the lattice. As an example, consider the one-dimensional periodic potential energy $U(x)$ plotted in Figure 41.1 where, to a first approximation, $U(x)$ may be taken to be a periodic array of delta functions

$$U(x) = -\sum_j V_0 \delta(x - ja)$$

where a is the lattice constant, $V_0 > 0$, and j is an integer. Considering the lowest energy state $(k = 0)$ for the electron motion, the Schrödinger equation (41.10) may be written in terms of $\psi_0(\vec{r}) = u_0(\vec{r})$ as

$$\left[-\frac{\hbar^2}{2m_e}\frac{d^2}{dx^2} + U(x)\right]u_0(x) = \varepsilon_0 u_0(x)$$

In the interval $0 < x < a$, where the potential vanishes, to a first approximation, a trial solution is

$$u_0(x) = Ae^{\alpha x} + Be^{-\alpha x}$$

The form of this function is determined by two conditions. The first condition is that this function should have the same periodicity as the lattice, and this, from Eq.(41.8), can be written as

$$u_0(x + a) = u_0(x)$$

For $x = 0$ we obtain

$$Ae^{\alpha a} + Be^{-\alpha a} = A + B \quad \text{or} \quad \left(Ae^{\alpha a/2} - Be^{-\alpha a/2}\right)\left(e^{\alpha a/2} - e^{-\alpha a/2}\right) = 0$$

which can be rewritten as

$$\alpha A e^{\alpha a/2} - \alpha B e^{-\alpha a/2} = \left[\frac{du_0(x)}{dx}\right]_{x=a/2} = 0$$

This means that the slope of the function $u_0(x)$ vanishes at $x = ja/2$, where j is an integer.

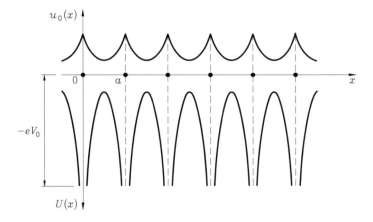

Figure 41.1. Bloch wave of lowest energy in a one-dimensional periodic potential

The second condition is that $u_0(x)$ should satisfy the Schrödinger equation at the lattice sites, and this gives

$$\varepsilon_0 u_0(x) + \frac{\hbar^2}{2m_e}\frac{d^2 u_0(x)}{dx^2} = -\sum_j V_0 \delta(x - ja) u_0(x)$$

Integrating both sides over a small interval about $x = 0$, we obtain

$$\varepsilon_0 \int_{-\rho}^{\rho} u_0(x)\,dx + \frac{\hbar^2}{2m_e}\left[\left(\frac{du_0(x)}{dx}\right)_{x=\rho} - \left(\frac{du_0(x)}{dx}\right)_{x=-\rho}\right] = -V_0 u_0(0)$$

The first term on the left-hand side vanishes for $\rho \to 0$, and the remaining terms show that $u_0(x)$ must be an even function, such that

$$\left[\frac{du_0(x)}{dx}\right]_{x=0} = -\frac{m_e V_0}{\hbar^2} u_0(0) \quad \text{or} \quad \alpha(A - B) = -\frac{m_e V_0}{\hbar^2}(A + B)$$

As a result of these two conditions on $u_0(x)$ and its derivative, the lowest energy Bloch wave may be represented as in Figure 41.1. The two homogenous equations for A and B have non-identically vanishing solutions provided that

$$e^{\alpha a} = \frac{\alpha + m_e V_0/\hbar^2}{\alpha - m_e V_0/\hbar^2} \quad \text{or} \quad \frac{m_e V_0}{\hbar^2} = \alpha \tanh\frac{\alpha a}{2}$$

This equation can be graphically solved and provides α in terms of V_0 and a that define the periodic potential energy $U(x)$.

41.2. THE WEAK-BINDING APPROXIMATION

If we think of the potential of the ionic cores as being rather small and acting as a slight perturbation on the free electrons in the solid, we obtain the *weak-binding approximation*, also called the *nearly-free electron model*. Such an approximation is valid for electrons in conductors and semiconductors. It is convenient to use the Fourier expansion in terms of reciprocal lattice vectors $\vec{K}$ for the periodic potential energy $U(\vec{r})$, which again has the translational symmetry of the lattice as

$$U(\vec{r}) = \sum_{\vec{K}} U(\vec{K}) e^{i\vec{K}\cdot\vec{r}} \qquad (41.13)$$

and also for the periodic part $u_{\vec{k}}(\vec{r})$ of the Bloch waves, which reads

$$u_{\vec{k}}(\vec{r}) = \sum_{\vec{K}} a(\vec{K}) e^{i\vec{K}\cdot\vec{r}} \qquad (41.14)$$

From Eq.(41.7) it follows that

$$\psi_{\vec{k}}(\vec{r}) = \sum_{\vec{K}} a(\vec{K}) e^{i(\vec{k}+\vec{K})\cdot\vec{r}} \qquad (41.15)$$

Substituting Eqs.(41.13) and (41.15) into the one-electron Schrödinger equation (40.3) yields

$$\sum_{\vec{K}} \left[\varepsilon - \frac{\hbar^2(\vec{k}+\vec{K})^2}{2m_e} \right] a(\vec{K}) e^{i(\vec{k}+\vec{K})\cdot\vec{r}} = \sum_{\vec{K}',\vec{K}''} U(\vec{K}') a(\vec{K}'') e^{i(\vec{k}+\vec{K}'+\vec{K}'')\cdot\vec{r}} \qquad (41.16)$$

If, in the zeroth order approximation, we consider a constant potential (41.13), where $U(0) \neq 0$ and $U(\vec{K}) = 0$ for all $\vec{K} \neq 0$, Eq.(41.16) reduces to

$$\left[\varepsilon - \frac{\hbar^2(\vec{k}+\vec{K})^2}{2m_e} \right] a(\vec{K}) = U(0) a(\vec{K})$$

and this gives the energy shift of the free electron of wave vector $\vec{k}+\vec{K}$ as

$$\varepsilon = U(0) + \frac{\hbar^2(\vec{k}+\vec{K})^2}{2m_e} = U(0) + \varepsilon_{\vec{k}+\vec{K}} \qquad (41.17)$$

Thus, the free electron eigenvalues are shifted, for $\vec{K} = 0$, by

$$\varepsilon = U(0) + \frac{\hbar^2 \vec{k}^2}{2m_e} = U(0) + \varepsilon_{\vec{k}} \qquad (41.18)$$

In the next approximation, we consider an *almost* constant potential (41.13), where $U(\vec{K}) \ll U(0)$ for all $\vec{K} \neq 0$, so that Eq.(41.16) becomes

$$\left[\varepsilon - U(0) - \varepsilon_{\vec{k}+\vec{K}}\right] a(\vec{K}) - U(\vec{K}) a(0) = \sum_{\vec{K}' \neq 0, \vec{K}} U(\vec{K}') a(\vec{K} - \vec{K}') \qquad (41.19)$$

We may drop the sum on the right-hand side of Eq.(41.19), provided that the $a(\vec{K}-\vec{K}')$ coefficients are negligible for $\vec{K}' \neq \vec{K}$, that is $a(\vec{K}-\vec{K}') \ll a(0)$. Since $\vec{K}$ is an arbitrary vector of the reciprocal lattice, the validity of this assumption can be understood from the reduced equation

$$\left[\varepsilon - U(0) - \varepsilon_{\vec{k}+\vec{K}}\right] a(\vec{K}) - U(\vec{K}) a(0) = 0 \qquad (41.20)$$

if ε is approximated by the free electron shifted eigenvalues (41.18), which gives

$$\left(\varepsilon - \varepsilon_{\vec{k}+\vec{K}}\right) a(\vec{K}) - U(\vec{K}) a(0) = 0$$

As, in general, we have $\varepsilon_{\vec{k}+\vec{K}} - \varepsilon_{\vec{k}} = \hbar^2 \left(2\vec{k}\cdot\vec{K} + \vec{K}^2\right)/2m_e \neq 0$, one obtains

$$a(\vec{K}) = \frac{U(\vec{K})}{\varepsilon_{\vec{k}} - \varepsilon_{\vec{k}+\vec{K}}} a(0) = -\frac{2m_e U(\vec{K})}{\hbar^2 \left(2\vec{k}\cdot\vec{K}+\vec{K}^2\right)} a(0) \qquad (41.21)$$

but we have assumed $U(\vec{K})$ to be negligible, and hence, $a(\vec{K}) \ll a(0)$ for all $\vec{K} \neq 0$.

Since $U(\vec{K})$ is a Fourier coefficient of the potential energy expansion (41.13), it can be written as the Fourier transform of $U(\vec{r})$, Eqs.(16.61), and this can be regarded as a matrix element of $U(\vec{r})$ which joins two free-electron states $\psi_{\vec{k}}^0$ and $\psi_{\vec{k}+\vec{K}}^0$ differing in their wave number by a reciprocal lattice vector $\vec{K}$, that is,

$$U(\vec{K}) = \int e^{-i\vec{K}\cdot\vec{r}} U(\vec{r}) dV = \int e^{-i(\vec{k}+\vec{K})\cdot\vec{r}} U(\vec{r}) e^{i\vec{k}\cdot\vec{r}} dV = \langle \psi_{\vec{k}+\vec{K}}^0 | U(\vec{r}) | \psi_{\vec{k}}^0 \rangle \quad (41.22)$$

Substituting Eqs.(41.22) and (41.21), the Bloch wave (41.15) takes a form similar to that derived from the first-order perturbation theory in Eq.(36.13), of the form

$$\psi_{\vec{k}}(\vec{r}) = \psi_{\vec{k}}^0 + \sum_{\vec{K}} \frac{\langle \psi_{\vec{k}+\vec{K}}^0 | U(\vec{r}) | \psi_{\vec{k}}^0 \rangle}{\varepsilon_{\vec{k}} - \varepsilon_{\vec{k}+\vec{K}}} \psi_{\vec{k}+\vec{K}}^0 \quad (41.23)$$

Standard perturbation treatment, developed in Chapter 36, leads to the second-order expression (36.24) for energy, which reads

$$\varepsilon = \frac{\hbar^2 k^2}{2m_e} + U(0) + \sum_{\vec{K}} \frac{|\langle \psi_{\vec{k}+\vec{K}}^0 | U(\vec{r}) | \psi_{\vec{k}}^0 \rangle|^2}{\varepsilon_{\vec{k}} - \varepsilon_{\vec{k}+\vec{K}}} = \frac{\hbar^2 k^2}{2m_e} + U(0) - \frac{2m_e}{\hbar^2} \sum_{\vec{K}} \frac{|U(\vec{K})|^2}{2\vec{k}\cdot\vec{K} + \vec{K}^2} \quad (41.24)$$

For each given value of $\vec{K}$, the wave vectors $\vec{k}$ which span the boundary of the corresponding *Brillouin zone* in $\vec{k}$-space satisfy Eq.(39.12), that is

$$2\vec{k}\cdot\vec{K} + \vec{K}^2 = 0 \quad \text{or} \quad \varepsilon_{\vec{k}+\vec{K}} = \varepsilon_{\vec{k}} \quad (41.25)$$

Thus, if $\vec{k}$ is near a Brillouin zone boundary, defined by a particular value of $\vec{K}$, one of the $a(\vec{K})$ coefficients becomes large, according to Eq.(41.21), and therefore, all the $a(\vec{K} - \vec{K}')$ coefficients in Eq.(41.19) become negligible either for $\vec{K}' \neq \vec{K}$, in which case $a(\vec{K} - \vec{K}') \ll a(0)$, or for $\vec{K}' \neq 0$, which implies that $a(\vec{K} - \vec{K}') \ll a(\vec{K})$. In these conditions, there are two valid approximations only for Eq.(41.19), which form a set of simultaneous homogenous equations for $a(\vec{K})$ and $a(0)$ given by

$$\left[\varepsilon - U(0) - \varepsilon_{\vec{k}+\vec{K}}\right] a(\vec{K}) - U(\vec{K}) a(0) = 0$$
$$\left[\varepsilon - U(0) - \varepsilon_{\vec{k}}\right] a(0) - U(-\vec{K}) a(\vec{K}) = 0 \quad (41.26)$$

where $U(-\vec{K}) = U^*(\vec{K})$ for any Fourier coefficient of the *real* potential function $U(\vec{r})$.

The free-electron states $\psi_{\vec{k}}$ and $\psi_{\vec{k}+\vec{K}}$ appear in this case, from Eq.(41.25), to be degenerate, so that the Bloch wave (41.15) should be written in terms of the two coefficients $a(0)$ and $a(\vec{K})$ as

$$\psi_{\vec{k}}(\vec{r}) = a(0) e^{i\vec{k}\cdot\vec{r}} + a(\vec{K}) e^{i(\vec{k}+\vec{K})\cdot\vec{r}}$$

which may be interpreted to be a linear combination of degenerate states. The matrix eigenvalue equation for the degenerate case is obtained from Eq.(41.26) as

$$\begin{pmatrix} U(0)+\varepsilon_{\vec{k}} & U(\vec{K}) \\ U^*(\vec{K}) & U(0)+\varepsilon_{\vec{k}+\vec{K}} \end{pmatrix} \begin{pmatrix} a(0) \\ a(\vec{K}) \end{pmatrix} = \varepsilon \begin{pmatrix} a(0) \\ a(\vec{K}) \end{pmatrix} \qquad (41.27)$$

and the characteristic equation is

$$\left[\varepsilon - U(0) - \varepsilon_{\vec{k}}\right]\left[\varepsilon - U(0) - \varepsilon_{\vec{k}+\vec{K}}\right] - \left|U(\vec{K})\right|^2 = 0 \qquad (41.28)$$

This provides two energy eigenvalues of the form

$$\varepsilon_\pm = U(0) + \frac{\varepsilon_{\vec{k}} + \varepsilon_{\vec{k}+\vec{K}}}{2} \pm \sqrt{\left(\frac{\varepsilon_{\vec{k}} - \varepsilon_{\vec{k}+\vec{K}}}{2}\right)^2 + \left|U(\vec{K})\right|^2} \qquad (41.29)$$

For $\left|U(\vec{K})\right| \ll \left|\varepsilon_{\vec{k}} - \varepsilon_{\vec{k}+\vec{K}}\right|$, Eq.(41.29) reduces to the solutions (41.18) and (41.17) which give the first order energy shifts for the free-electron states $\vec{k}$ and $\vec{k}+\vec{K}$. At each zone boundary in $\vec{k}$-space, where $\varepsilon_{\vec{k}} = \varepsilon_{\vec{k}+\vec{K}}$, there is an energy gap of magnitude

$$\Delta\varepsilon = \varepsilon_+ - \varepsilon_- = 2\left|U(\vec{K})\right| \qquad (41.30)$$

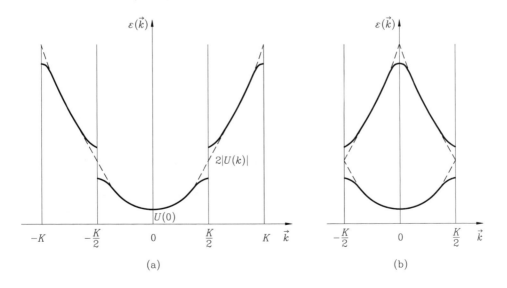

Figure 41.2. Energy-band structure in the weak binding approximation (a) and its representation reduced to the first Brillouin zone (b)

The energy spectrum of nearly-free electrons exhibits allowed energy bands $\varepsilon(\vec{k})$ at successively higher energies, with forbidden energy gaps between adjacent energy bands, as illustrated in Figure 41.2 (a) for the one-dimensional case. The significance of the energy gaps in the ε versus $\vec{k}$ spectrum, at $\vec{k}$-values which are said to define zone boundaries in $\vec{k}$-space, can be given a simple interpretation in terms of Bragg reflection. Bloch waves with wave numbers which satisfy Eq.(41.25), and hence, the Laue condition (39.28), are Bragg reflected and do not propagate, giving rise to a gap in energy. In this respect, the effect on the Bloch waves of the potential energy $U(\vec{r})$, having the periodicity of the lattice, is similar to that of a diffraction grating on the propagation of plane waves. It follows that we must expect that a "bands and gaps" structure of the energy spectrum will occur for any periodic potential. Since the Bloch wave functions have the periodicity of the reciprocal lattice given by

$$\psi_{\vec{k}+\vec{K}}(\vec{r}) = e^{i(\vec{k}+\vec{K})\cdot\vec{r}} u_{\vec{k}+\vec{K}}(\vec{r}) = e^{i\vec{k}\cdot\vec{r}} u_{\vec{k}}(\vec{r}) = \psi_{\vec{k}}(\vec{r})$$

their substitution into the Schrödinger wave equation shows that the energy bands for values of $\vec{k}$ differing by a reciprocal lattice vector must be exactly the same

$$\varepsilon(\vec{k} + \vec{K}) = \varepsilon(\vec{k})$$

and this means that $\varepsilon(\vec{k})$ is also a periodic function of $\vec{k}$ having the translational periodicity of the reciprocal lattice. Hence, the energy levels of the electron in a periodic potential can be described by a set of functions $\varepsilon_n(\vec{k})$, where $\vec{k}$ is restricted to the first Brillouin zone, and this is referred to as the *energy-band structure* of the crystal. Each $\varepsilon(\vec{k})$, specified by the band index n, can be represented as a multiple-valued function of $\vec{k}$, see Figure 41.2 (b), where the lowest branch of $\varepsilon(\vec{k})$ is called the *first band*, the next branch above it is called the *second band*, and so on.

EXAMPLE 41.2. The effective mass

We now consider the dynamics of an electron in an allowed energy band under the influence of both an external force field $\vec{F}$ and the periodic field of the lattice. Provided the influence of the periodic lattice field, which is not treated explicitly, is represented by an effective mass m_e^* instead of the usual inertial mass m_e, we may use Eq.(2.21) for a free particle, that is

$$\frac{d\varepsilon}{dt} = \vec{F} \cdot \vec{v} = \sum_i F_i v_i \qquad (41.31)$$

where $i = x, y, z$. As a relationship between ε and $\vec{k}$ can be interpreted as a dispersion relation for Bloch waves, it is convenient to define the electron velocity components in terms of energy by

$$v_i = \frac{1}{\hbar}\frac{d\varepsilon}{dk_i} \qquad \text{or} \qquad \vec{v} = \frac{1}{\hbar}\nabla_{\vec{k}}\varepsilon \qquad (41.32)$$

and this corresponds to the concept of group velocity, given by Eq.(16.66). Equation (41.32) may also be derived from the energy-velocity relation (5.3), by substituting the momentum (40.7) of a free electron. An equation completely analogous to the classical equation of motion (2.10), which reads

$$\frac{d\vec{v}}{dt} = \frac{1}{m_e}\vec{F} \tag{41.33}$$

can be formulated in terms of the velocity components as

$$\frac{dv_i}{dt} = \frac{1}{\hbar}\frac{\partial}{\partial k_i}\left(\frac{d\varepsilon}{dt}\right) = \frac{1}{\hbar}\frac{\partial}{\partial k_i}\left(\frac{1}{\hbar}\sum_j F_j v_j\right) = \frac{1}{\hbar^2}\left(\sum_j \frac{\partial^2 \varepsilon}{\partial k_i \partial k_j}\right)F_j \tag{41.34}$$

Comparing Eqs.(41.34) and (41.33), we may infer that the effective mass may be taken to be a tensor of components

$$\frac{1}{m^*_{ij}} = \frac{1}{\hbar^2}\frac{\partial^2 \varepsilon}{\partial k_i \partial k_j} \tag{41.35}$$

where $i, j = x, y, z$. This tensor of second-rank corresponds to three-dimensional anisotropic energy surfaces, and it becomes symmetric if ε is a quadratic function of $\vec{k}$. In this case, three mutually orthogonal principal directions always exist, and the effective mass tensor reduces to three principal components of the form

$$\frac{1}{m^*_i} = \frac{1}{\hbar^2}\frac{\partial^2 \varepsilon}{\partial k_i^2} \quad \text{or} \quad \varepsilon = \frac{\hbar^2}{2}\sum_i \frac{k_i^2}{m^*_i} \tag{41.36}$$

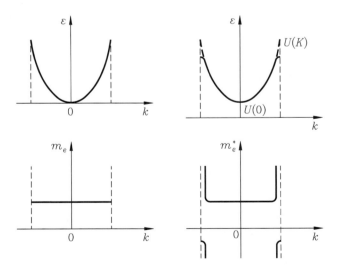

Figure 41.3. Dispersive relation and effective mass as a function of k in an energy band

An isotropic effective mass of the electron will be represented by a scalar quantity which is related to the curvature of the energy band as

$$\frac{1}{m_e^*} = \frac{1}{\hbar^2} \frac{\partial^2 \varepsilon}{\partial k^2} \qquad (41.37)$$

so that narrow energy bands lead to large effective masses. The variation of m_e^* throughout an energy band is represented in qualitative form in Figure 41.3 for a free electron, where $m_e^* = m_e$, and for a nearly-free electron, where $m_e^* \neq m_e$. It is seen that effective mass, which is positive in the lower part of the energy band, becomes negative close to the zone boundary, as the band is concave down. A negative effective mass is a result of the Bragg reflection of Bloch waves at the zone boundary, arising from the strong electron-lattice interaction during reflection, which results in a negative change in the electron momentum. The electron behaves as if it had a negative mass.

41.3. THE TIGHT-BINDING APPROXIMATION

We also can demonstrate the existence of electronic energy bands, with gaps between neighbouring bands, by solving the one-electron problem (40.3) in the *tight-binding approximation*, where we think of electrons as bound to separate atoms in the solid. Such an approximation is valid for electrons in insulators. In the configuration illustrated in Figure 41.4 it is convenient to separate the contribution to the potential energy of an electron in the atom, given by the ionic field $U_0(\vec{r} - \vec{R}_j)$, from that given by the average field of all the other electrons $U(\vec{r})$. The one electron problem (40.3) can then be formulated as

$$\left[-\frac{\hbar^2}{2m_e} \nabla^2 + U_0(\vec{r} - \vec{R}_j) + U(\vec{r}) - U_0(\vec{r} - \vec{R}_j) \right] \psi(\vec{r})$$

$$= \left[\hat{H}_0 + U(\vec{r}) - U_0(\vec{r} - \vec{R}_j) \right] \psi(\vec{r}) = \varepsilon \psi(\vec{r}) \qquad (41.38)$$

where $\hat{H}_0$ is the free-atom Hamiltonian, with eigenfunctions $\psi_0(\vec{r} - \vec{R}_j)$ obtained from

$$\hat{H}_0 \psi_0(\vec{r} - \vec{R}_j) = \left[-\frac{\hbar^2}{2m_e} \nabla^2 + U_0(\vec{r} - \vec{R}_j) \right] \psi_0(\vec{r} - \vec{R}_j) = \varepsilon_0 \psi_0(\vec{r} - \vec{R}_j) \qquad (41.39)$$

The tight-binding approach to the one-electron problem (41.38) consists of writing the crystal wave function to be a superposition of atomic wave functions $\psi_0(\vec{r} - \vec{R}_j)$ as

$$\psi(\vec{r}) = \sum_{j=1}^{N} a_j \psi_0(\vec{r} - \vec{R}_j) = \sum_{j=1}^{N} e^{i\vec{k} \cdot \vec{R}_j} \psi_0(\vec{r} - \vec{R}_j) \qquad (41.40)$$

where N is the number of lattice ions, located at $\vec{R}_j$, and the form of the coefficients a_j was chosen such that $\psi(\vec{r})$ satisfies the Bloch condition (41.5).

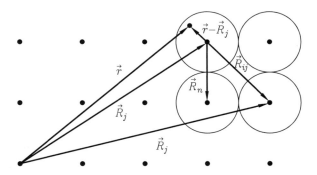

Figure 41.4. Coordinates of a tightly bound electron

The energy eigenvalues ε of the electron in the crystal can be evaluated from Eq.(41.38), if we multiply by $\psi^*(\vec{r})$ and integrate, which gives

$$\sum_{j=1}^{N}\sum_{i=1}^{N}e^{i\vec{k}\cdot(\vec{R}_j-\vec{R}_i)}\int\psi_0^*(\vec{r}-\vec{R}_i)\left[U(\vec{r})-U_0(\vec{r}-\vec{R}_j)\right]\psi_0(\vec{r}-\vec{R}_j)dV$$

$$=(\varepsilon-\varepsilon_0)\sum_{j=1}^{N}\sum_{i=1}^{N}e^{i\vec{k}\cdot(\vec{R}_j-\vec{R}_i)}\int\psi_0^*(\vec{r}-\vec{R}_i)\psi_0(\vec{r}-\vec{R}_j)dV \quad (41.41)$$

Neglecting any overlap of the atomic wave functions at neighbouring lattice sites, we may approximate the integral on the right-hand side by δ_{ij}, so that

$$\sum_{j=1}^{N}\sum_{i=1}^{N}e^{i\vec{k}\cdot(\vec{R}_j-\vec{R}_i)}\int\psi_0^*(\vec{r}-\vec{R}_i)\psi_0(\vec{r}-\vec{R}_j)dV \cong \sum_{j=1}^{N}\sum_{i=1}^{N}e^{i\vec{k}\cdot(\vec{R}_j-\vec{R}_i)}\delta_{ij} = N \quad (41.42)$$

It follows from Eq.(41.41) that

$$\varepsilon = \varepsilon_0 + \frac{1}{N}\sum_{j=1}^{N}\sum_{i=1}^{N}e^{i\vec{k}\cdot(\vec{R}_j-\vec{R}_i)}\int\psi_0^*(\vec{r}-\vec{R}_i)\left[U(\vec{r})-U_0(\vec{r}-\vec{R}_j)\right]\psi_0(\vec{r}-\vec{R}_j)dV$$

$$\cong \varepsilon_0 + \sum_{i}e^{-i\vec{k}\cdot\vec{R}_{ij}}\int\psi_0^*(\vec{r}-\vec{R}_{ij})\left[U(\vec{r})-U_0(\vec{r})\right]\psi_0(\vec{r})dV \quad (41.43)$$

where we have assumed that the terms of summation over the lattice positions $\vec{R}_j$ are identical. Therefore, we can express the energy as a function of the separation $\vec{R}_{ij}=\vec{R}_i-\vec{R}_j$ of neighbouring atoms from a given ion core $\vec{R}_j$. If we restrict ourselves to the nearest-neighbour contributions to energy and assume that $\psi_0(\vec{r})$ are isotropic s-electron states, such that all these contributions are approximately identical, the sum on the right-hand side of Eq.(41.43) can be written in terms of two parameters α and γ only. The parameter α defines the atomic interaction as

$$\alpha = -\int \psi_0^*(\vec{r})\left[U(\vec{r})-U_0(\vec{r})\right]\psi_0(\vec{r})dV \qquad (41.44)$$

while γ describes the nearest-neighbour interaction in the form

$$\gamma = -\int \psi_0^*(\vec{r}-\vec{R}_n)\left[U(\vec{r})-U_0(\vec{r})\right]\psi_0(\vec{r})dV \qquad (41.45)$$

where $\vec{R}_n$ denotes the nearest-neighbour separation from a given lattice site. Both parameters α and γ are positive, since we have assumed that the average field of all the electrons is weaker than the atomic field, so that $U(\vec{r})-U_0(\vec{r})$ is always negative. Substituting Eqs.(41.44) and (41.45), the energy (41.43) becomes

$$\varepsilon = \varepsilon_0 - \alpha - \gamma \sum_n e^{-i\vec{k}\cdot\vec{R}_n} \qquad (41.46)$$

The second term on the right-hand side shifts each atomic level by a constant amount, while the last term broadens each discrete energy level $\varepsilon_0 - \alpha$ into an energy band. In other words, when the atoms are brought close together to form the solid, their mutual interaction gives rise to the energy bands. If bands originate from discrete levels ε_0, where the energy difference between neighbouring levels is larger than the broadening of levels, we are left with an energy gap between adjacent bands. The width of a particular band depends on the electron wave vector $\vec{k}$ and on the number and the separation of nearest neighbours, corresponding to the type of Bravais lattice.

EXAMPLE 41.3. Energy bands in a simple cubic lattice

We have seen that each atom in a simple cubic structure, illustrated in Figure 38.10, has six nearest neighbours, at positions $(\pm a,0,0)$, $(0,\pm a,0)$ and $(0,0,\pm a)$ say, so that, for this particular case, Eq.(41.46) becomes

$$\varepsilon = \varepsilon_0 - \alpha - 2\gamma(\cos k_x a + \cos k_y a + \cos k_z a) \qquad (41.47)$$

Each electron state of energy ε_0 in the free atom is now confined to an energy band with limits $\pm 6\gamma$ which depend directly upon the overlap γ between wave functions of adjacent atoms in the solid. If there are N atoms in the solid, each band which corresponds to a nondegenerate state $\psi_0(\vec{r})$ contains $2N$ states (including spin). For small values of k we may expand the cosines up to the second order, and this gives

$$\varepsilon \cong \varepsilon_0 - \alpha - 2\gamma\left[\left(1-\frac{k_x^2 a^2}{2}\right)+\left(1-\frac{k_y^2 a^2}{2}\right)+\left(1-\frac{k_z^2 a^2}{2}\right)\right] = \varepsilon_0 - \alpha - 6\gamma + \gamma k^2 a^2 \quad (41.48)$$

Thus, near the bottom of the band, the electron energy depends upon the square of k in agreement with the results for both free and nearly free electrons. Since the effective mass (41.37) depends on the overlap γ of adjacent orbitals as

$$m_e^* = \frac{\hbar^2}{2\gamma a^2} \quad (41.49)$$

it follows that narrow bands are characterized by large effective masses, as expected.

FURTHER READING

1. J. Singleton - BAND THEORY AND ELECTRONIC PROPERTIES OF SOLIDS, Oxford University Press, Oxford, 2001.
2. M. N. Rudden, J. Wilson - ELEMENTS OF SOLID STATE PHYSICS, Wiley, New York, 1993.
3. J. R. Hook, H. E. Hall - SOLID STATE PHYSICS, Wiley, New York, 1993.

PROBLEMS

41.1. Find the first four Brillouin zones for a two-dimensional square lattice with a primitive cell of edge a.

Solution

The reciprocal lattice vector, Eq.(39.7), can be written in two dimensions as

$$\vec{K}_{hl} = h\vec{a}^* + l\vec{b}^* = \frac{2\pi}{a}(h\vec{e}_x + l\vec{e}_y)$$

and hence, the Brillouin zone boundaries will be defined by Eq.(39.12), where $\vec{k} = k_x \vec{e}_x + k_y \vec{e}_y$, such that

$$hk_x + lk_y = \frac{\pi}{a}(h^2 + l^2)$$

The first Brillouin zone is defined by the lines

$$k_x = \pm \frac{\pi}{a} \quad (h = \pm 1, l = 0), \quad k_y = \pm \frac{\pi}{a} \quad (h = 0, l = \pm 1)$$

The region enclosed between the first zone and the four intersecting lines

$$k_x \pm k_y = \frac{2\pi}{a} \quad (h = 1, l = \pm 1), \quad k_x \pm k_y = -\frac{2\pi}{a} \quad (h = -1, l = \pm 1)$$

is referred to as the second Brillouin zone. It contains all the electron states in the second energy band. It has the same area of $(2\pi/a)^2$ as the first zone. The region outside the second zone and within the area enclosed by the four lines

$$k_x = \pm \frac{2\pi}{a} \quad (h = \pm 2, l = 0), \quad k_y = \pm \frac{2\pi}{a} \quad (h = 0, l = \pm 2)$$

is much larger than that of the first and second zones.

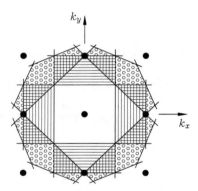

The third Brillouin zone is defined by additional boundaries $k_x = \pm \pi/a$ and $k_y = \pm \pi/a$ which run through the region outside the first two zones and divide the space such as to encompass an area of $(2\pi/a)^2$. The fourth zone is similarly obtained by using the next boundary lines given by

$$\pm k_x \pm 2k_y = \frac{5\pi}{a} \quad (h = \pm 1, l = \pm 2), \quad \pm 2k_x \pm k_y = \frac{5\pi}{a} \quad (h = \pm 2, l = \pm 1)$$

A graphical representation of the results is given in the figure.

41.2. Use the nearly-free electron model to find the energy band structure for a linear monatomic lattice of parameter a.

Solution

In a linear lattice, the Bragg condition, Eq.(41.25), reduces to

$$k = \frac{1}{2}K = n\frac{\pi}{a} \quad (n = \pm 1, \pm 2, \cdots)$$

and hence, given the periodicity $U(x) = U(x+a)$, the only nonvanishing matrix elements associated with the electron scattering at the Brillouin zone boundary are

$$U(K) = \langle \psi_{-k} | U(x) | \psi_k \rangle = \frac{1}{a} \int_{-a/2}^{a/2} U(x) e^{2i(n\pi/a)x} dx$$

and its complex conjugate. In other words, there is two-fold degeneracy of the electron states in one dimension. The energy eigenvalue problem, which reads

$$-\frac{\hbar^2}{2m_e} \frac{d^2\psi}{dx^2} + U(x)\psi(x) = \varepsilon\psi(x)$$

should be solved by a linear combination of two independent free-electron Bloch waves corresponding to $\pm k$, that is,

$$\psi(x) = \frac{1}{\sqrt{a}} (A_1 e^{ikx} + A_2 e^{-ikx})$$

This leads, by substitution, to the matrix eigenvalue equation

$$\begin{pmatrix} 0 & U(K) \\ U^*(K) & 0 \end{pmatrix} \begin{pmatrix} A_1 \\ A_2 \end{pmatrix} = \left(\varepsilon - \frac{\hbar^2 k^2}{2m_e}\right) \begin{pmatrix} A_1 \\ A_2 \end{pmatrix}$$

with the characteristic equation given by

$$\begin{vmatrix} \varepsilon - \frac{\hbar^2 k^2}{2m_e} & U(K) \\ U^*(K) & \varepsilon - \frac{\hbar^2 k^2}{2m_e} \end{vmatrix} = 0$$

so that degeneracy is lifted at the zone boundary according to

$$\varepsilon_\pm = \frac{\hbar^2 k^2}{2m_e} \pm |U(K)|$$

This shows that electron scattering is accompanied by an energy gap $2|U(K)|$. By insertion of the energy eigenvalue into the matrix equation, one obtains the normalized state functions as

$$\psi_+(x) = \sqrt{\frac{2}{a}} \cos kx, \qquad \psi_-(x) = \sqrt{\frac{2}{a}} \sin kx$$

41.3. Find an expression for the change in the free-electron mass m_e, induced by the translational periodicity of the lattice, using the weak-binding approximation.

Solution

We may assume that the influence of the periodic lattice on the Bloch electron velocity can be represented in terms of the effective mass in the form $\vec{v} = \hbar\vec{k}/m_e^*$, and hence, an infinitesimal variation of the crystal momentum from $\hbar\vec{k}$ to $\hbar(\vec{k}+\delta\vec{k})$ implies a change in velocity

$$\delta\vec{v} = \frac{\hbar}{m_e^*}\delta\vec{k} \quad \text{or} \quad \delta v_i = \sum_j \frac{\hbar}{m_{ij}^*}\delta k_j$$

where $i, j = x, y, z$. The components of the effective mass tensor must be introduced whenever $\delta\vec{v}$ is not parallel to $\delta\vec{k}$. The shape of the energy bands will affect the dynamics of Bloch electrons through their mean velocity in a given energy band

$$\vec{v} = \frac{\langle \hat{\vec{p}} \rangle}{m_e} = -\frac{i\hbar}{m_e}\langle \psi_{\vec{k}} | \nabla \psi_{\vec{k}} \rangle = -\frac{i\hbar}{m_e}\int \psi_{\vec{k}}^* \nabla \psi_{\vec{k}}\, dV$$

Substituting the Bloch electron state, Eq.(41.15), one obtains

$$\vec{v} = \frac{1}{V}\sum_{\vec{K}}\sum_{\vec{K}'}\int a_{\vec{k}}^*(\vec{K})e^{-i(\vec{k}+\vec{K})\cdot\vec{r}}\frac{\hbar}{m_e}(\vec{k}+\vec{K}')a_{\vec{k}}(\vec{K}')e^{i(\vec{k}+\vec{K}')\cdot\vec{r}}\,dV$$

$$= \frac{1}{V}\sum_{\vec{K}}\sum_{\vec{K}'}a_{\vec{k}}^*(\vec{K})a_{\vec{k}}(\vec{K}')\frac{\hbar}{m_e}(\vec{k}+\vec{K}')\int e^{i(\vec{K}'-\vec{K})\cdot\vec{r}}\,dV = \sum_{\vec{K}}\sum_{\vec{K}'}a_{\vec{k}}^*(\vec{K})a_{\vec{k}}(\vec{K}')\frac{\hbar}{m_e}(\vec{k}+\vec{K}')\delta_{\vec{K}\vec{K}'}$$

$$= \sum_{\vec{K}}|a_{\vec{k}}(\vec{K})|^2\frac{\hbar}{m_e}(\vec{k}+\vec{K}) = \frac{\hbar}{m_e}\vec{k} + \sum_{\vec{K}}|a_{\vec{k}}(\vec{K})|^2\frac{\hbar}{m_e}\vec{K}$$

Hence, δv_i can be expressed in the form

$$\delta v_i = \frac{\hbar}{m_e}\delta k_i + \frac{\hbar}{m_e}\sum_{\vec{K}}\sum_j K_j \frac{\partial}{\partial k_j}|a_{\vec{k}}(\vec{K})|^2 \delta k_j = \sum_j \frac{\hbar}{m_{ij}^*}\delta k_j$$

and this means that the components of the effective mass tensor are related to m_e by

$$\frac{m_e}{m_{ij}^*} = \delta_{ij} + \sum_{\vec{K}} K_i \frac{\partial}{\partial k_i}|a_{\vec{k}}(\vec{K})|^2$$

Since $|a_{\vec{k}}(\vec{K})|^2$ is the probability to find the Bloch electron in the nearly-free state with crystal momentum $\hbar(\vec{k}+\vec{K})$, it is apparent that the second term represents the required change in the free-electron mass m_e, as induced by the translational symmetry of the lattice.

41.4. Find the dispersion relation for tightly-bound electrons in a body-centred cubic lattice with the primitive cell of sides a.

Solution

The dispersion relation has the form (41.46), where the sum runs over the lattice vectors $\vec{R}_n$ that connect the origin with its nearest neighbours. For a body-centred cubic crystal there are eight nearest neighbours at $\vec{R}_n = \frac{1}{2}a(\pm 1, \pm 1, 1)$, $\frac{1}{2}a(\pm 1, \pm 1, -1)$ and we may choose $\vec{k} = (k_x, k_y, k_z)$ such that

$$\sum_n e^{-i\vec{k}\cdot\vec{R}_n} = e^{-ia(k_x+k_y+k_z)/2} + e^{ia(k_x+k_y+k_z)/2} + e^{-ia(k_x-k_y+k_z)/2} + e^{ia(k_x-k_y+k_z)/2}$$

$$+ e^{-ia(k_x+k_y-k_z)/2} + e^{ia(k_x+k_y-k_z)/2} + e^{-ia(-k_x+k_y+k_z)/2} + e^{ia(-k_x+k_y+k_z)/2}$$

$$= 2\left(\cos\frac{k_x+k_y+k_z}{2}a + \cos\frac{k_x-k_y+k_z}{2}a + \cos\frac{k_x+k_y-k_z}{2}a + \cos\frac{-k_x+k_y+k_z}{2}a\right)$$

This yields

$$\sum_n e^{-i\vec{k}\cdot\vec{R}_n} = 4\cos\frac{k_y a}{2}\left(\cos\frac{k_x+k_y}{2}a + \cos\frac{k_x-k_z}{2}a\right) = 8\cos\frac{k_x a}{2}\cos\frac{k_y a}{2}\cos\frac{k_z a}{2}$$

which means that the dispersion relation (41.46) reads

$$\varepsilon = \varepsilon_0 - \alpha - 8\gamma\cos\frac{k_x a}{2}\cos\frac{k_y a}{2}\cos\frac{k_z a}{2}$$

41.5. Find expressions for the density of states $N(\varepsilon)$ of nearly-free and tightly-bound electrons in a simple cubic lattice.

Solution

The density of states for nearly-free electrons is given by Eq.(40.11), provided m_e is replaced by the effective mass

$$N(\varepsilon) = \frac{V}{2\pi^2}\left(\frac{2m_e^*}{\hbar^2}\right)^{3/2}(\varepsilon - \varepsilon_\Gamma)^{1/2}$$

where ε_Γ stands for the bottom of the energy band. For tightly-bound electrons, near the bottom of the energy band, where ε has the form (41.48), which reads

$$\varepsilon = \varepsilon_\Gamma + \gamma k^2 a^2$$

one has $\partial\varepsilon/\partial k = 2\gamma a^2 k$, and hence, on substitution into Eq.(40.11), yields

$$N(\varepsilon) = \frac{2V}{(2\pi)^3}\frac{4\pi k^2}{2\gamma a^2 k} = \frac{V}{2\pi^2}\frac{(\varepsilon-\varepsilon_\Gamma)^{1/2}}{(\gamma a^2)^{3/2}} = \frac{V}{2\pi^2}\left(\frac{2m_e^*}{\hbar^2}\right)^{3/2}(\varepsilon-\varepsilon_\Gamma)^{1/2}$$

This is the same as the parabolic density of states found for nearly-free electrons. In a similar manner, close to the corners of the Brillouin zone we may expand

$$\cos\left(\frac{\pi}{a} - \delta k_i\right)a = \cos(\pi - \delta k_i a) = -\cos(\delta k_i a) = -1 + \frac{(\delta k_i)^2 a^2}{2} \quad (i = x, y, z)$$

such that the energies near the top of the band, in view of Eq.(41.47), are given by

$$\varepsilon = \varepsilon_0 - \alpha - 2\gamma\left(\cos k_x a + \cos k_y a + \cos k_z a\right) = \varepsilon_0 - \alpha + 6\gamma - \gamma(\delta k)^2 a^2 = \varepsilon_\Gamma + 12\gamma - \gamma(\delta k)^2 a^2$$

This immediately leads to a density of state function which is vanishing at the top of the band, where the energy is $\varepsilon_\Gamma + 12\gamma$, and has a parabolic increase for decreasing energy, given by

$$N(\varepsilon) = \frac{V}{2\pi^2}\left(\frac{2m_e^*}{\hbar^2}\right)(\varepsilon_\Gamma + 12\gamma - \varepsilon)^{1/2}$$

41.6. For a two-dimensional square lattice, find the ratio of the free-electron energy at the corners to that at the mid boundaries of the first Brillouin zone. What is the corresponding result in the case of a simple cubic lattice?

41.7. Find the Bragg condition for the zone boundaries of a two-dimensional rectangular lattice in which the side ratio in the primitive cell is $n:1$, where n is an integer, and construct the first four Brillouin zones.

41.8. Find the energy gap at the zone boundary for nearly-free electrons in a linear lattice with potential energy $U(x) = U_0 \cos(2\pi x/a)$. .

41.9. Find an expression for the effective mass of tightly-bound electrons in a linear lattice with atomic separation a.

41.10. Find the dispersion relation for tightly-bound electrons in a face-centred cubic lattice with conventional cell of side a.

42. SEMICONDUCTOR PHYSICS

42.1. FREE CHARGE CARRIERS
42.2. INTRINSIC SEMICONDUCTORS
38.3. IMPURITY SEMICONDUCTORS

42.1. FREE CHARGE CARRIERS

The existence of nonmetallic solids is a direct consequence of the energy gaps that separate bands of quasicontinuous energy eigenvalues, for both nearly-free and tightly bound electrons. We have seen that the location of energy gaps in $\vec{k}$-space is determined by the translational symmetry of the lattice through the Laue condition (39.28). The magnitude of the gaps depends on the crystal potential strength, Eq.(41.30).

If N is the number of primitive cells in a solid of volume V, which is assumed to be the same as the total number of atoms, the number of $\vec{k}$-states contained in a unit cell of reciprocal space of volume V_0^*, is given by

$$N(k)V_0^* = \frac{NV_0}{(2\pi)^3} V_0^* = N$$

where the density of states $N(k)$ was obtained from Eq.(40.8). In other words, since each allowed band is associated with a Brillouin zone, the number of electron states in each energy band, including spin, is $2N$. If there is one valence electron per atom, the N electrons will fill only half of the states available in the band, as in Figure 42.1. Since nearly-free electrons can be accelerated by an applied electric field, conduction will take place, in this case, as if the electrons were free. The material will exhibit metallic behaviour with an electrical conductivity defined by Eq.(40.50). Note that, if there are two valence electrons per atom, overlap between adjacent bands is predicted by the weak-binding approximation. The Fermi energy ε_F is such that the first band is partially empty and the second band is partially filled, and hence, the nearly-free electrons behave much as if they were free electrons, and the solid is also a metal.

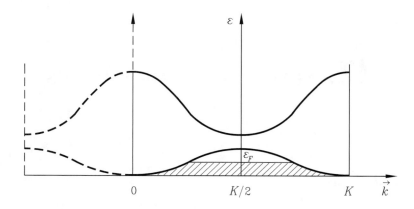

Figure 42.1. Energy bands for nearly-free electrons in metals with one valence electron per atom

Overlap between adjacent bands can also be described in terms of the tight-binding approximation where the Fermi energy lies within the energy gap designated by $\varepsilon_g = \varepsilon_C - \varepsilon_V$ as in Figure 42.2. The first band, completely filled at $T = 0$ because each atom has two valence electrons, is termed the *valence* band. The second band, completely empty at $T = 0$, is called the *conduction* band. A solid with such a band structure is an *insulator* if $\varepsilon_g \gg k_B T$ or an *intrinsic semiconductor* if $\varepsilon_g \geq k_B T$. For a *semimetal*, defined by $\varepsilon_g < k_B T$, there is overlap at $T = 0$, where $\varepsilon_g < 0$.

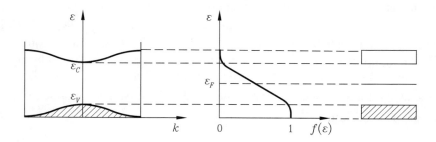

Figure 42.2. Energy bands for tightly bound electrons

An insulator is characterized by narrow energy bands separated by large energy gaps. We may assign therefore, to a first approximation, the same energy ε_C to all the allowed electron states in the conduction band, and the same energy ε_V to all the states in the valence band. As a result, the average number of electrons per state in the two bands will be given by the Fermi-Dirac distribution function (40.14) as

$$f(\varepsilon_C) = \frac{1}{e^{(\varepsilon_C - \varepsilon_F)/k_B T} + 1} \tag{42.1}$$

and

$$f(\varepsilon_V) = \frac{1}{e^{(\varepsilon_V - \varepsilon_F)/k_B T} + 1} \tag{42.2}$$

As the energy gap ε_g is large in insulators, it is reasonable to restrict the possible electron states to those contained in the first two energy bands. In other words, we will assume that $f(\varepsilon_C) + f(\varepsilon_V) = 1$, and this yields

$$f(\varepsilon_V) = 1 - \frac{1}{e^{(\varepsilon_C - \varepsilon_F)/k_B T} + 1} = \frac{1}{e^{(\varepsilon_F - \varepsilon_C)/k_B T} + 1}$$

Substituting this result into Eq.(42.2) gives

$$\varepsilon_C + \varepsilon_V - 2\varepsilon_F = 0 \quad \text{or} \quad \varepsilon_F = \frac{1}{2}(\varepsilon_C + \varepsilon_V) \tag{42.3}$$

It is seen in Figure 42.2 that the Fermi energy for insulators is in the middle of the band gap at any temperature.

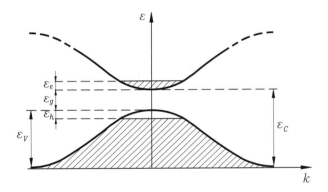

Figure 42.3. Energy bands in semiconductors and semimetals

In a pure or intrinsic semiconductor, because $\varepsilon_g \sim k_B T$, thermal energy is sufficient to excite electrons from the valence band into the conduction band, producing empty states in the former, as shown in Figure 42.3. A state in the valence band that is not occupied by an electron is considered to be occupied by a *hole*. The motion of a hole being that of a missing electron under the influence of an electric or magnetic force $\vec{F}$, the acceleration components defined before in Eq.(41.34) remains unchanged, and may formally be expressed as

$$\frac{dv_i}{dt} = \frac{1}{\hbar^2}\left(\sum_j \frac{\partial^2 \varepsilon}{\partial k_i \partial k_j}\right) F_j = \frac{1}{\hbar^2}\left[\sum_j \frac{\partial^2 (-\varepsilon)}{\partial k_i \partial k_j}\right](-F_j) = \sum_j \left[\frac{1}{\hbar^2}\frac{\partial^2 (-\varepsilon)}{\partial k_i \partial k_j}\right](-F_j) \tag{42.4}$$

The minus sign of the applied force means that a charge opposite in sign to that of the electron must be assigned to a hole. The effective mass tensor of a hole, of components

$$\left(\frac{1}{m_h^*}\right)_{ij} = \frac{1}{\hbar^2}\frac{\partial^2 (-\varepsilon)}{\partial k_i \partial k_j} = -\left(\frac{1}{m_e^*}\right)_{ij} \tag{42.5}$$

can be rewritten in terms of the energy $\varepsilon_h = \varepsilon_V - \varepsilon$ measured from the top of the valence band as

$$\left(\frac{1}{m_h^*}\right)_{ij} = \frac{1}{\hbar^2}\frac{\partial^2(-\varepsilon)}{\partial k_i \partial k_j} = \frac{1}{\hbar^2}\frac{\partial^2(\varepsilon_V - \varepsilon)}{\partial k_i \partial k_j} = \frac{1}{\hbar^2}\frac{\partial^2 \varepsilon_h}{\partial k_i \partial k_j}$$

Since a hole and a missing electron move with the same velocity under the influence of a given applied force $\vec{F}$, Eq.(41.32) may be written as

$$\vec{v} = \frac{1}{\hbar}\nabla_{\vec{k}}\varepsilon = \frac{1}{\hbar}\nabla_{-\vec{k}}(-\varepsilon) = \frac{1}{\hbar}\nabla_{-\vec{k}}(\varepsilon_h) \qquad (42.6)$$

It follows that the motion of a hole must be described by a wave vector

$$\vec{k}_h = -\vec{k} \qquad (42.7)$$

which corresponds to the total momentum of the valence band with one missing electron

$$\hbar \vec{k}_h = \sum_i \hbar \vec{k}_i - \hbar \vec{k} = -\hbar \vec{k}$$

because the total momentum of a completely filled band is zero. Hence, the overall behaviour of electrons in a nearly-filled valence band can be described in terms of holes. Since the missing electrons are likely to be from states close to the top of the valence band, it is expected that holes behave under the force $\vec{F}$ as if they had a positive effective mass m_h^* and a positive charge. Note that not all the empty electron states act as holes, but only those which would normally be occupied by electrons of negative effective mass. In the case of intrinsic semiconductors, electronic conduction must be considered in terms of free charge carriers in both bands: electrons in the conduction band and holes in the valence band. Conventionally, the concentration of electrons in the conduction band and that of holes in the valence band are denoted by n and p, respectively.

EXAMPLE 42.1. Carrier effective mass

The effective mass may be derived for electrons in the conduction band and for holes in the valence band using the electron energy eigenvalues (41.29), in the configuration represented in Figure 42.3. In the conduction band, just above the first energy gap, we may choose $\vec{k} = -\vec{K}/2 - \Delta\vec{k}$ so that $\vec{k} + \vec{K} = \vec{K}/2 - \Delta\vec{k}$, and we have

$$\frac{1}{2}\left(\varepsilon_{\vec{k}} + \varepsilon_{\vec{k}+\vec{K}}\right) = \frac{1}{2}\frac{\hbar^2}{2m_e}\left[2\left(\frac{K}{2}\right)^2 + 2(\Delta\vec{k})^2\right] = \varepsilon_{\vec{K}/2} + \frac{\hbar^2}{2m_e}(\Delta\vec{k})^2$$

Expanding $\left(\varepsilon_{\vec{k}} - \varepsilon_{\vec{k}+\vec{K}}\right)^2$ to first order gives

$$\left(\frac{\varepsilon_{\vec{k}}-\varepsilon_{\vec{k}+\vec{K}}}{2}\right)^2 = \left\{\frac{1}{2}\left[\frac{\hbar^2}{2m_e}\left(4\frac{\vec{K}}{2}\cdot\Delta\vec{k}\right)\right]\right\}^2 = 4\frac{\hbar^2}{2m_e}\left(\frac{\vec{K}}{2}\right)^2\frac{\hbar^2}{2m_e}(\Delta\vec{k})^2 = 4\varepsilon_{\vec{K}/2}\frac{\hbar^2}{2m_e}(\Delta\vec{k})^2$$

On substitution into Eq.(41.29), it follows that

$$\varepsilon_+ = U(0) + \varepsilon_{\vec{K}/2} + \frac{\hbar^2}{2m_e}(\Delta\vec{k})^2 + |U(\vec{K})|\left[1 + \frac{4\varepsilon_{\vec{K}/2}}{|U(\vec{K})|}\frac{\hbar^2}{2m_e}(\Delta\vec{k})^2\right]^{1/2}$$

$$\cong U(0) + \varepsilon_{\vec{K}/2} + |U(\vec{K})| + \frac{\hbar^2}{2m_e}(\Delta\vec{k})^2\left(1 + \frac{2\varepsilon_{\vec{K}/2}}{|U(\vec{K})|}\right)$$

Hence, the kinetic energy of electrons in the conduction band is

$$\Delta\varepsilon = \varepsilon_+ - \left[U(0) + \varepsilon_{\vec{K}/2} + |U(\vec{K})|\right] \cong \frac{\hbar^2}{2m_e}\left(1 + \frac{2\varepsilon_{\vec{K}/2}}{|U(\vec{K})|}\right)(\Delta\vec{k})^2$$

where both $\Delta\varepsilon$ and $\Delta\vec{k}$ are defined with respect to the bottom of the conduction band. A comparison with Eq.(41.37) yields

$$\frac{1}{m_e^*} = \frac{1}{m_e}\left(1 + \frac{2\varepsilon_{\vec{K}/2}}{|U(\vec{K})|}\right) \tag{42.8}$$

indicating a positive effective mass smaller than m_e.

For an electron in the valence band, just below the first energy gap, we must take $\vec{k} = -\vec{K}/2 + \Delta\vec{k}$, so $\vec{k} + \vec{K} = \vec{K}/2 + \Delta\vec{k}$ and, from Eq.(41.29), we similarly obtain

$$\varepsilon_- \cong U(0) + \varepsilon_{\vec{K}/2} - |U(\vec{K})| - \frac{\hbar^2}{2m_e}(\Delta\vec{k})^2\left(1 + \frac{2\varepsilon_{\vec{K}/2}}{|U(\vec{K})|}\right)$$

The kinetic energy is

$$\Delta\varepsilon = \varepsilon_- - \left[U(0) + \varepsilon_{\vec{K}/2} - |U(\vec{K})|\right] = -\frac{\hbar^2}{2m_e}\left(1 + \frac{2\varepsilon_{\vec{K}/2}}{|U(\vec{K})|}\right)(\Delta\vec{k})^2$$

so that the electron effective mass is a negative quantity given by

$$\frac{1}{m_e^*} = -\frac{1}{m_e}\left(1 + \frac{2\varepsilon_{\vec{K}/2}}{|U(\vec{K})|}\right)$$

Hence, it is convenient to consider the overall motion of electrons near the top of the valence band in terms of holes, of effective mass

$$\frac{1}{m_h^*} = -\frac{1}{m_e^*} = \frac{1}{m_e}\left(1 + \frac{2\varepsilon_{\vec{K}/2}}{|U(\vec{K})|}\right) \qquad (42.9)$$

which has the advantage of being positive.

42.2. INTRINSIC SEMICONDUCTORS

Under the influence of an external electric field $\vec{E}$, the current density in a pure semiconductor is obtained as a sum of contributions from both the electrons and holes, each written according to Eq.(12.10), which gives

$$\vec{j} = (\sigma_e + \sigma_h)\vec{E} = (ne\mu_e + pe\mu_h)\vec{E} \qquad (42.10)$$

where the electron and hole conductivities are given by Eq.(40.50), in terms of a relaxation time τ, as

$$\sigma_e = \frac{ne^2\tau_e}{m_e^*} \quad \text{and} \quad \sigma_h = \frac{pe^2\tau_h}{m_h^*} \qquad (42.11)$$

Therefore the *mobility* of electrons and holes, respectively, can be defined as

$$\mu_e = \frac{\sigma_e}{ne} = \frac{e\tau_e}{m_e^*} \quad \text{and} \quad \mu_h = \frac{\sigma_p}{pe} = \frac{e\tau_p}{m_h^*} \qquad (42.12)$$

The carrier concentrations n and p at a given temperature may be derived by treating both the electrons and holes as free particles with effective masses m_e^* and m_h^*. Under this assumption, the densities of states $N_C(\varepsilon_e)$ near the bottom of the conduction band and $N_V(\varepsilon_h)$ near the top of the valence band take the form (40.11) derived in the free-electron approximation. In the configuration represented in Figure 42.3, we have $\varepsilon_e = \varepsilon - \varepsilon_C$ and $\varepsilon_h = \varepsilon_V - \varepsilon$, and this, substituted into Eq.(40.11), yields

$$N_C(\varepsilon) = \frac{V}{2\pi^2}\left(\frac{2m_e^*}{\hbar^2}\right)^{3/2}(\varepsilon - \varepsilon_C)^{1/2} \quad , \quad N_V(\varepsilon) = \frac{V}{2\pi^2}\left(\frac{2m_h^*}{\hbar^2}\right)^{3/2}(\varepsilon_V - \varepsilon)^{1/2} \quad (42.13)$$

The total number of carriers in the two bands is expressed by Eq.(40.19), using the average number of electrons per state given at thermal equilibrium by the Fermi-Dirac distribution function (40.14). In the conduction band, where $\varepsilon - \varepsilon_F > k_BT$, the exponential becomes very large compared to unity, and thus

$$f_e(\varepsilon) = e^{-(\varepsilon-\varepsilon_F)/k_BT} \tag{42.14}$$

which has the form of the Boltzmann distribution law (26.25). In the nearly-filled valence band we may approximate

$$f_e(\varepsilon) = \frac{1}{e^{(\varepsilon-\varepsilon_F)/k_BT}+1} \cong 1 - e^{(\varepsilon-\varepsilon_F)/k_BT}$$

so that the probability of finding a missing electron, that is, a hole, at an energy ε is

$$f_h(\varepsilon) = 1 - f_e(\varepsilon) = e^{(\varepsilon-\varepsilon_F)/k_BT} \tag{42.15}$$

Thus the hole distribution follows the same classical statistical law. Substituting Eqs.(42.13), (42.14) and (42.15) into Eq.(40.19) gives the carrier concentrations as

$$n = \int_{\varepsilon_C}^{\infty} f_e(\varepsilon) N_C(\varepsilon) d\varepsilon = \frac{1}{2\pi^2} \left(\frac{2m_e^*}{\hbar^2}\right)^{3/2} e^{\varepsilon_F/k_BT} \int_{\varepsilon_C}^{\infty} (\varepsilon-\varepsilon_C)^{1/2} e^{-\varepsilon/k_BT} d\varepsilon$$

$$= \frac{1}{2\pi^2} \left(\frac{2m_e^*}{\hbar^2}\right)^{3/2} (k_BT)^{3/2} e^{(\varepsilon_F-\varepsilon_C)/k_BT} \int_{\varepsilon_C}^{\infty} \left(\frac{\varepsilon-\varepsilon_C}{k_BT}\right)^{1/2} e^{-(\varepsilon-\varepsilon_C)/k_BT} d\left(\frac{\varepsilon-\varepsilon_C}{k_BT}\right) \tag{42.16}$$

and

$$p = \int_{-\infty}^{\varepsilon_V} f_h(\varepsilon) N_V(\varepsilon) d\varepsilon = \frac{1}{2\pi^2} \left(\frac{2m_h^*}{\hbar^2}\right)^{3/2} e^{(\varepsilon_V-\varepsilon_F)/k_BT} \int_{\varepsilon_C}^{\infty} (\varepsilon_V-\varepsilon)^{1/2} e^{-(\varepsilon_V-\varepsilon)/k_BT} d(\varepsilon_V-\varepsilon)$$

$$= \frac{1}{2\pi^2} \left(\frac{2m_h^*}{\hbar^2}\right)^{3/2} (k_BT)^{3/2} e^{(\varepsilon_V-\varepsilon_F)/k_BT} \int_0^{\infty} \left(\frac{\varepsilon_V-\varepsilon}{k_BT}\right)^{1/2} e^{-(\varepsilon_V-\varepsilon)/k_BT} d\left(\frac{\varepsilon_V-\varepsilon}{k_BT}\right) \tag{42.17}$$

Making obvious substitutions, both integrals on the right-hand side of Eqs.(42.16) and (42.17) take the standard form evaluated in Appendix IV, Eq.(IV.4), as

$$\int_0^{\infty} u^{1/2} e^{-u} du = 2\int_0^{\infty} x^2 e^{-x^2} dx = \sqrt{\pi}/2$$

Upon substitution into Eqs.(42.16) and (42.17), we obtain the concentration of electrons in the conduction band in the form

$$n = 2\left(\frac{m_e^* k_BT}{2\pi\hbar^2}\right)^{3/2} e^{(\varepsilon_F-\varepsilon_C)/k_BT} = N_C e^{(\varepsilon_F-\varepsilon_C)/k_BT} \tag{42.18}$$

and the concentration of holes in the valence band as

$$p = 2\left(\frac{m_h^* k_B T}{2\pi \hbar^2}\right)^{3/2} e^{(\varepsilon_V - \varepsilon_F)/k_B T} = N_V e^{(\varepsilon_V - \varepsilon_F)/k_B T} \qquad (42.19)$$

where N_C and N_V are known to be the *effective densities of states* of the two bands. If we take the product of Eqs.(42.18) and (42.19), we find

$$np = 4\left(\frac{k_B T}{2\pi \hbar^2}\right)^3 \left(m_e^* m_h^*\right)^{3/2} e^{-\varepsilon_g / k_B T} \qquad (42.20)$$

This result is known as the *law of mass action* which states that the product np is a constant for a given semiconductor, with a specified value of ε_g, at a given temperature T. Being independent of the Fermi energy, Eq.(42.20) may be used in discussing both the intrinsic and extrinsic, or impurity, semiconductors.

For an intrinsic semiconductor, in which electrons are excited directly from the valence band to the conduction band, we have the condition $n_i = p_i$, where the subscript i indicates intrinsic concentrations. It follows that

$$np = n_i^2 \qquad (42.21)$$

and hence, using Eq.(42.20), one obtains

$$n_i = 2\left(\frac{k_B T}{2\pi \hbar^2}\right)^{3/2} \left(m_e^* m_h^*\right)^{3/4} e^{-\varepsilon_g / 2k_B T} = \left(N_C N_V\right)^{1/2} e^{-\varepsilon_g / 2k_B T} \qquad (42.22)$$

The Fermi energy is obtained by equating n and p given by Eqs.(42.18) and (42.19), respectively, and then solving for ε_F, which gives

$$\varepsilon_F = \varepsilon_{Fi} = \frac{\varepsilon_C + \varepsilon_V}{2} + \frac{3}{4} k_B T \ln\left(\frac{m_h^*}{m_e^*}\right) = \frac{\varepsilon_C + \varepsilon_V}{2} + \frac{1}{2} k_B T \ln\left(\frac{N_V}{N_C}\right) \qquad (42.23)$$

At $T = 0$, the Fermi energy of an intrinsic semiconductor ε_{Fi} is in the middle of the energy gap. This will also be true for finite temperatures, provided $m_e^* \approx m_h^*$, which is usually assumed for many semiconductors.

The intrinsic conductivity $\sigma = \sigma_e + \sigma_h$, Eq.(42.10), may now be written, using Eq.(42.22), as

$$\sigma_i = n_i e(\mu_e + \mu_h) = 2e\left(\frac{k_B T}{2\pi \hbar^2}\right)^{3/2} \left(m_e^* m_h^*\right)^{3/4} (\mu_e + \mu_h) e^{-\varepsilon_g / 2k_B T} \qquad (42.24)$$

or

$$\ln \sigma_i = -\frac{\varepsilon_g}{2k_B T} + \ln\left[2e\left(\frac{k_B T}{2\pi\hbar^2}\right)^{3/2} \left(m_e^* m_h^*\right)^{3/4} \left(\mu_e + \mu_h\right)\right] \qquad (42.25)$$

Thus, the magnitude of the band gap ε_g can be obtained from the linear decrease of $\ln \sigma_i$ with reciprocal temperature.

42.3. IMPURITY SEMICONDUCTORS

If impurity atoms are added to an intrinsic semiconductor, they may either donate electrons to the conduction band (donor impurities) or bind additional electrons, and hence, causing holes in the valence band (acceptor impurities). The change in the number of carriers affects both the Fermi energy, which must adjust such that overall charge neutrality is maintained and the conduction properties, and this will define the so-called *extrinsic* or *impurity semiconductivity*.

EXAMPLE 42.2. Doped semiconductors

A semiconductor whose conductivity depends on donor impurities is called an *n*-type semiconductor. For simplicity consider a pentavalent impurity atom such as As substituted for a host atom of a pure semiconductor, such as Si. Four of the As valence electrons will match the covalent bonds of neighbouring atoms, leaving one valence electron to move in the electrostatic potential of the impurity ion As^+. The configuration illustrated in Figure 42.4 (a) is similar to that of a one-electron atom, so that we expect the electron energy to be given by Eq.(32.9), provided we substitute the effective mass of the electron and the dielectric constant $\varepsilon_r = 16$, which accounts for the dielectric shielding of the host crystal and is not to be confused with an energy level. Taking $Z = 1$ for As^+ and $n = 1$ for the ground state, Eq.(32.9) becomes

$$\varepsilon_e = -\frac{m_e^* e^4}{32\pi^2 \varepsilon^2 \hbar^2} = \frac{m_e^*/m_e}{\varepsilon_r^2} \varepsilon_1 = -0.053 \frac{m_e^*}{m_e} \quad (\text{eV})$$

where use has been made of Eq.(28.18) which gives $\varepsilon_1 = -13.6$ eV. The experimental value $\varepsilon_e = -0.013$ eV indicates that the effective mass of the electron is less than that of a free electron, as indicated by Eq.(42.8). The discrete one-electron states associated with pentavalent *donor atoms* are situated at energies $\varepsilon_d = |\varepsilon_e|$ below the bottom of the conduction band, where the electron can be promoted by thermal excitation, to become a free carrier. The donor levels $\varepsilon_D = \varepsilon_C - \varepsilon_d$ are indicated in Figure 42.4 (b).

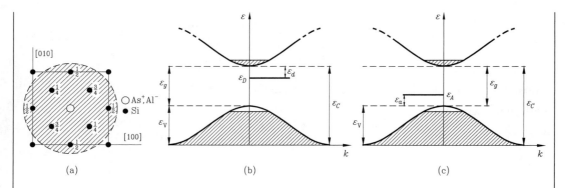

Figure 42.4. Donor and acceptor levels

Let N_d be the number of donor atoms per unit volume, which is usually smaller than the effective density of states of the conduction band, that is, $N_d < N_C$. If we assume complete ionization of the donors $N_d = N_d^+$, the carrier concentrations, always related by Eq.(42.21), will obey the electric neutrality condition

$$n = p + N_d^+ \tag{42.26}$$

Solving for n the simultaneous Eqs.(42.21) and (42.26) yields

$$n = \frac{N_d^+}{2}\left\{1 + \left[1 + \left(2n_i / N_d^+\right)^2\right]^{1/2}\right\} \tag{42.27}$$

and hence, $n > n_i$, as expected for a donor-doped material. If the number of donors greatly exceeds the number of intrinsic carriers $N_d^+ \gg n_i$, the root expansion gives

$$n = N_d^+ + n_i\left(\frac{n_i}{N_d^+}\right) \tag{42.28}$$

so that from Eq.(42.26) we obtain

$$p = p_i\left(\frac{n_i}{N_d^+}\right) \ll p_i \tag{42.29}$$

In other words, if we heavily dope with donors, the electrons become *majority carriers*, as their number is increased, while the number of holes, referred to as *minority carriers*, is decreased.

A semiconductor whose conductivity depends on acceptor impurities, such as an Si host containing trivalent impurities like Al, is known as a *p*-type semiconductor. An aluminium impurity can satisfy the bonding requirement of the lattice by accepting an electron from the host crystal and, hence, producing a positive hole in the crystal valence band. The calculation of the energy of the hole states may be carried out in a similar manner to that of the electron states. The discrete hole states associated with trivalent *acceptor atoms* will be situated at energies $\varepsilon_A = \varepsilon_a + \varepsilon_V$ above the top of the valence band, as indicated in Figure 42.4 (c). The valence band electrons can be promoted to these *acceptor levels* by thermal excitation, producing free positive holes in the valence band. Assuming that there are N_a acceptor levels per unit volume, completely occupied by electrons from the valence band $\left(N_a = N_a^-\right)$, the electric neutrality equation reads

$$p = n + N_a^- \tag{42.30}$$

Substituting n from the law of mass action (42.21) yields

$$p = \frac{N_a^-}{2}\left\{1 + \left[1 + \left(2n_i/N_a^-\right)^2\right]^{1/2}\right\} \tag{42.31}$$

For $N_a^- \gg n_i$ it follows that the holes are the majority carriers, since

$$p = N_a^- + n_i\left(\frac{n_i}{N_a^-}\right) \tag{42.32}$$

and the concentration of electrons, given by Eq.(42.30) as

$$n = n_i\left(\frac{n_i}{N_a^-}\right) \ll n_i \tag{42.33}$$

becomes negligible.

A more general problem is to consider the possibility of a semiconductor containing both donor and acceptor impurities. Let N_d and N_a be the concentration of donor and acceptor atoms. If N_d^0 and N_a^0 are the concentrations of neutral donors and acceptors, whereas N_d^+ and N_a^- are the ionized concentrations, we can write

$$N_d = N_d^0 + N_d^+ \quad \text{and} \quad N_a = N_a^0 + N_a^- \tag{42.34}$$

The concentration of neutral donors must equal the concentration of electrons in the donor states with energy $\varepsilon_D = \varepsilon_C - \varepsilon_d$, which is

$$N_d^0 = \frac{N_d}{\frac{1}{2}e^{(\varepsilon_D - \varepsilon_F)/k_BT} + 1} = \frac{N_d}{\frac{1}{2}e^{(\varepsilon_C - \varepsilon_d - \varepsilon_F)/k_BT} + 1} \tag{42.35}$$

The Fermi-Dirac distribution function is modified by the presence of the factor $\frac{1}{2}$ before the exponential term in the denominator due to the spin degeneracy of the donor states that can be occupied by either a spin-up or a spin-down electron (see Problem 42.5). From Eq.(42.34), we then obtain

$$N_d^+ = \frac{N_d}{2e^{(\varepsilon_F - \varepsilon_C + \varepsilon_d)/k_BT} + 1} \tag{42.36}$$

The concentration of electrons in the acceptor states with energy $\varepsilon_A = \varepsilon_a + \varepsilon_V$ equals the concentration of ionised acceptors, which is similarly written as

$$N_a^- = \frac{N_a}{2e^{(\varepsilon_A - \varepsilon_F)/k_B T} + 1} = \frac{N_a}{2e^{(\varepsilon_a + \varepsilon_V - \varepsilon_F)/k_B T} + 1} \qquad (42.37)$$

Substituting Eqs.(42.18), (42.19), (42.36) and (42.37) for n, p, N_d^+ and N_a^-, respectively, into the condition for electrical neutrality, which reads

$$n + N_a^- = p + N_d^+ \qquad (42.38)$$

one obtains

$$N_C e^{(\varepsilon_F - \varepsilon_C)/k_B T} + \frac{N_a}{2e^{(\varepsilon_a + \varepsilon_V - \varepsilon_F)/k_B T} + 1} = N_V e^{(\varepsilon_V - \varepsilon_F)/k_B T} + \frac{N_d}{2e^{(\varepsilon_F - \varepsilon_C + \varepsilon_d)/k_B T} + 1} \qquad (42.39)$$

The set of simultaneous equations (42.39), (42.18) and (42.19) can be used to determine the Fermi energy and the electron and hole concentrations. It is convenient to discuss the extrinsic behaviour by making certain assumptions on the concentration of carriers.

Consider first the model with no acceptors present, where $N_a = 0$, and suppose that we have doped heavily enough with donors so that p has become very small, according to Eq.(42.21), and hence it can be neglected. In this limit only electrons are important, and also $n \gg n_i$. Equation (42.39) reduces to

$$N_C e^{(\varepsilon_F - \varepsilon_C)/k_B T} = \frac{N_d}{2e^{(\varepsilon_F - \varepsilon_C + \varepsilon_d)/k_B T} + 1}$$

which can be solved for ε_F in the form

$$2e^{2\varepsilon_F/k_B T}\left[e^{(-2\varepsilon_C + \varepsilon_d)/k_B T}\right] + e^{\varepsilon_F/k_B T}\left[e^{-\varepsilon_C/k_B T}\right] - \frac{N_d}{N_C} = 0$$

and this gives

$$\varepsilon_F = \varepsilon_C - \varepsilon_d + k_B T \ln \frac{1}{4}\left[-1 + \left(1 + 8\frac{N_d}{N_C}e^{\varepsilon_d/k_B T}\right)^{1/2}\right] \qquad (42.40)$$

At low temperatures, where $8(N_d/N_C)e^{\varepsilon_d/k_B T} \gg 1$, we obtain

$$\varepsilon_F = \varepsilon_C - \frac{\varepsilon_d}{2} + \frac{k_B T}{2} \ln \frac{N_d}{2N_C} = \varepsilon_C - \frac{\varepsilon_d}{2} - \frac{k_B T}{2} \ln \frac{2N_C}{N_d} \qquad (42.41)$$

so that the Fermi level lies between the bottom ε_C of the conduction band and the donor levels of energy $\varepsilon_C - \varepsilon_d$. The Fermi energy tends to $\varepsilon_C - \varepsilon_d/2$ as the temperature

$T \to 0$ and begins to fall as T increases, as shown in Figure 42.5, because the donor levels become ionized ($N_d \ll 2N_C$ and the logarithm on the right-hand side of Eq.(7.32) is positive). Substitution of Eq.(42.41) into Eq.(42.18) gives the number of conduction electrons at low temperatures as

$$n = N_C e^{-\varepsilon_d/2k_BT - (1/2)\ln(2N_C/N_d)} = \left(\frac{N_C N_d}{2}\right)^{1/2} e^{-\varepsilon_d/2k_BT} \quad (42.42)$$

The form of Eq.(42.42) becomes identical to that for the intrinsic situation, Eq.(42.22), if $N_d/2$ is replaced by N_V and the band gap ε_d is replaced by ε_g. The linear decrease in $\ln n$ with reciprocal temperature, given by

$$\ln n = -\frac{\varepsilon_d}{2k_BT} + \frac{1}{2}\ln\left(\frac{N_C N_d}{2}\right) \quad (42.43)$$

is determined by the magnitude of the donor band gap ε_d below the bottom ε_C of the conduction band (see Figure 42.5).

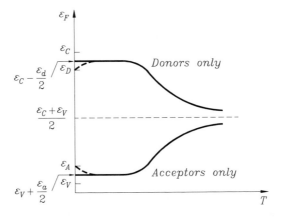

Figure 42.5. Temperature dependence of the Fermi level in an extrinsic semiconductor

At higher temperatures, where $8(N_d/N_C)e^{\varepsilon_d/k_BT} \ll 1$, we may use the binomial expansion of the root in Eq.(42.40), which gives

$$\varepsilon_F = \varepsilon_C + k_BT \ln\frac{N_d}{N_C} = \varepsilon_C - k_BT \ln\frac{N_C}{N_d} \quad (42.44)$$

The Fermi energy moves down as in Figure 42.5, with a steeper negative slope than that indicated for low temperatures by Eq.(42.41). Substitution of Eq.(42.44) into Eq.(42.18) yields a constant number of carriers equal to the number of donors, according to

$$n = N_C e^{\ln(N_d/N_C)} = N_d \quad (42.45)$$

This region of temperature is designated to be the *saturation* or *exhaustion region* in which all the donor levels have been emptied. The carrier concentration is a constant, as in Figure 42.6, and can be controlled by varying N_d in practical devices. By combining Eqs.(42.44) and (42.23) we obtain for the saturation region

$$\varepsilon_F - \varepsilon_{Fi} = \frac{\varepsilon_g}{2} - k_B T \ln\left[\left(N_C N_V\right)^{1/2} / N_d\right]$$

which can be rewritten in terms of n given by Eq.(42.45) and n_i given by Eq.(42.22) as

$$\varepsilon_F - \varepsilon_{Fi} = k_B T \ln\left(\frac{n}{n_i}\right)$$

It follows that the carrier concentration can be written in the form

$$n = n_i e^{(\varepsilon_F - \varepsilon_{Fi})/k_B T} \qquad (42.46)$$

Note that, for a *p*-type semiconductor, from Eq.(42.46) and the law of mass action (42.21), we obtain

$$p = n_i e^{(\varepsilon_{Fi} - \varepsilon_F)/k_B T} \qquad (42.47)$$

At still higher temperatures the electrons are excited from the valence band into the conduction band, so that the N_d term becomes small compared to the hole term in Eq.(42.39). In this limit, under the assumption of no acceptors $(N_a = 0)$, Eq.(42.39) reduces to

$$N_C e^{(\varepsilon_F - \varepsilon_C)/k_B T} = N_V e^{(\varepsilon_V - \varepsilon_F)/k_B T}$$

which leads to values for ε_F expressed by Eq.(42.23), that is the Fermi level drops to the centre of the intrinsic energy gap, as plotted in Figure 42.5, where a similar behaviour is indicated for *p*-type semiconductors, where $N_d = 0$.

In other words, intrinsic behaviour predominates at high temperatures, where the conductivity is due to electrons and holes and will change with temperature as given by Eq.(42.25). This so-called *intrinsic region* is seen in Figure 42.6 above the saturation region. The temperature at which intrinsic behaviour takes over increases with ε_g and usually represents the maximum working temperature of a semiconductor. It follows that, since ε_g is higher in *Si* than in *Ge*, it is possible to operate silicon semiconductors at higher temperatures than germanium semiconductors.

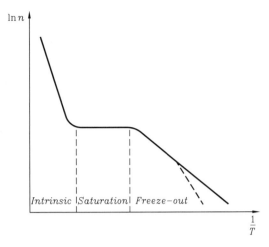

Figure 42.6. Temperature dependence of the carrier concentration for an extrinsic semiconductor

We now consider the model of a real semiconductor, where there are always a small number of acceptor states $N_a \ll N_d$ present, so that the electrons are the majority carriers. We may assume that all the acceptors are ionized since some of the electrons in the donor levels will lose energy and fill the acceptor sites, such that $N_a^- = N_a$. In the limit $n_i \ll n$, Eq.(42.39) can be written as

$$N_C e^{(\varepsilon_F - \varepsilon_C)/k_B T} + N_a = \frac{N_d}{2e^{(\varepsilon_F - \varepsilon_C + \varepsilon_d)/k_B T} + 1} \tag{42.48}$$

It is convenient to write the Fermi level in terms of n, by means of Eq.(42.18), as

$$\varepsilon_F = \varepsilon_C + k_B T \ln \frac{n}{N_C}$$

Substituting this expression for ε_F into Eq.(42.48), we readily obtain

$$\frac{n(n + N_a)}{N_d - (n + N_a)} = \frac{N_C}{2} e^{-\varepsilon_d / k_B T} \tag{42.49}$$

At very low temperatures it is reasonable to assume that there are only a few electrons in the conduction band, so that $n \ll N_a$, such that Eq.(42.49) becomes

$$n = \frac{N_C}{2} \left(\frac{N_d}{N_a} - 1 \right) e^{-\varepsilon_d / k_B T} \tag{42.50}$$

Substituting Eq.(42.50) into Eq.(42.49) gives

$$\varepsilon_F = \varepsilon_C - \varepsilon_d + k_B T \ln\left[\frac{1}{2}\left(\frac{N_d}{N_a} - 1\right)\right] \qquad (42.51)$$

which means that, at absolute zero, the Fermi level lies exactly at the donor level $\varepsilon_D = \varepsilon_C - \varepsilon_d$. This change in the position of the Fermi level in the presence of residual acceptor states, which occurs because some of the donor levels are empty due to compensation of the acceptor levels, is represented by a dashed line in Figure 42.5. Equation (42.50) can be rewritten as

$$\ln n = -\frac{\varepsilon_d}{k_B T} + \ln\left[\frac{N_C}{2}\left(\frac{N_d}{N_a} - 1\right)\right] \qquad (42.52)$$

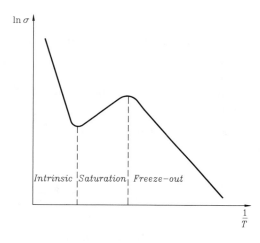

Figure 42.7. Temperature dependence of conductivity for an extrinsic semiconductor

Comparing Eq.(42.52) to Eq.(42.43) we can see that the presence of acceptors also changes the temperature dependence of the majority carrier concentration at very low temperatures, as plotted by a dashed line in Figure 42.6. The slope is proportional to ε_d or $\varepsilon_d/2$ depending on the presence of acceptors and their relative concentration. This range of temperature is known to be the *freeze-out region*.

At a higher temperature, we may assume that the number of conduction electrons increases, such that $N_a \ll n \ll N_d$ and Eq.(42.49) reduces to the form (42.42), that is, to the model with no acceptors. Hence, in the saturation region the majority carrier concentration will be expressed by Eqs.(42.46) and (42.47).

The temperature dependence of electrical conductivity for an extrinsic semiconductor, plotted in Figure 42.7, indicates that carrier mobility is also affected by the doping concentration. Since n is constant in the saturation region, the conductivity $\sigma = ne\mu$ is dominated by the mobility, which decreases with increasing temperature. Experiment shows that conductivity can be controlled by the addition of impurities to cover the whole range from insulating to metallic behaviour.

FURTHER READING

1. L. Solymar, D. Walsh - ELECTRICAL PROPERTIES OF MATERIALS, Oxford University Press, Oxford, 2004.
2. K. Seeger - SEMICONDUCTOR PHYSICS: AN INTRODUCTION, Springer-Verlag, New York, 2004.
3. R. P. Pierret - ADVANCED SEMICONDUCTOR FUNDAMENTALS, Prentice Hall, Upper Saddle River, 2002.

PROBLEMS

42.1. Find an expression for the change $\Delta\varepsilon_g$ in the energy gap of an intrinsic semiconductor induced by the presence of a magnetic field of induction B_0.

Solution

Assuming, for simplicity, that the effective mass is isotropic for both electrons and holes, the dispersion relation for electrons and holes reads

$$\varepsilon_e = \varepsilon_C + \frac{\hbar^2 k^2}{2m_e^*}, \quad \varepsilon_h = \varepsilon_V - \frac{\hbar^2 k^2}{2m_h^*}$$

as shown in Figure 41.3. In the presence of a magnetic induction, B_0, applied along the z-axis, the carrier energy in each band can be written in the form derived in Problem 32.5, where $p_z = \hbar k_z$. The carrier spin should also be included through the energy change due to spin alignment with the magnetic field $\hat{\vec{\mu}}_s \cdot \vec{B}_0 = \hat{\mu}_{sz} B_0 = \pm \mu_B B_0$, and hence, we have

$$\varepsilon_e(k_z, n) = \varepsilon_C + \frac{\hbar^2 k_z^2}{2m_e^*} + \left(n + \tfrac{1}{2}\right)\hbar\omega_C \pm \mu_B B_0$$

$$\varepsilon_h(k_z, n) = \varepsilon_V - \frac{\hbar^2 k_z^2}{2m_h^*} - \left(n + \tfrac{1}{2}\right)\hbar\omega_C' \mp \mu_B B_0$$

Substituting the cyclotron frequencies eB_0/m_e^* and $-eB_0/m_h^*$ for electrons and holes, respectively, and assuming that $k_z = 0$ and $n = 0$, the new energy gap, $\varepsilon_g(B_0)$, is given by

$$\varepsilon_g(B_0) = \varepsilon_e(0,0) - \varepsilon_h(0,0) = \varepsilon_g + \frac{e\hbar B_0}{2}\left(\frac{1}{m_e^*} + \frac{1}{m_h^*}\right) - 2\mu_B B_0 = \varepsilon_g + \frac{e\hbar B_0}{2}\left(\frac{1}{m_e^*} + \frac{1}{m_h^*} - \frac{2}{m_e}\right)$$

where $\varepsilon_g = \varepsilon_C - \varepsilon_V$ stands for the energy gap in the absence of the applied field. It follows that

$$\Delta\varepsilon_g = \frac{e\hbar B_0}{2}\left(\frac{1}{m_e^*}+\frac{1}{m_h^*}-\frac{2}{m_e}\right)$$

depends on the carrier effective mass values. Semiconductor-semimetal transitions can be induced if the sign of the energy gap is changed from $\varepsilon_g > 0$ to $\varepsilon_g < 0$, as found in semimetals at $T = 0$.

42.2. Find the carrier concentration in an intrinsic semiconductor where the constant energy surfaces about the conduction and valence band edges are given by

$$\varepsilon = \varepsilon_C + \frac{\hbar^2}{2}\left(\frac{k_x^2}{m_x}+\frac{k_y^2}{m_y}+\frac{k_z^2}{m_z}\right), \qquad \varepsilon = \varepsilon_V - \frac{\hbar^2}{2}\left(\frac{k_x^2}{m_x}+\frac{k_y^2}{m_y}+\frac{k_z^2}{m_z}\right)$$

for electrons and holes, respectively.

Solution

It is convenient to introduce new variables

$$\kappa_x = \sqrt{\frac{m}{m_x}}\,k_x, \quad \kappa_y = \sqrt{\frac{m}{m_y}}\,k_y, \quad \kappa_z = \sqrt{\frac{m}{m_z}}\,k_z$$

such that the constant energy surface becomes spherical in $\vec{\kappa}$-space, and hence

$$\varepsilon - \varepsilon_C = \frac{\hbar^2\kappa^2}{2m} \quad \text{and} \quad \varepsilon_V - \varepsilon = \frac{\hbar^2\kappa^2}{2m}$$

The density of states reads

$$N(\varepsilon)d\varepsilon = \frac{V}{(2\pi)^3}\int d^3k = \left(\frac{m_x m_y m_z}{m^3}\right)^{1/2}\frac{V}{(2\pi)^3}\int d^3\kappa$$

Using Eq.(40.11) for the density of states associated with spherical surfaces of constant energy, in $\vec{\kappa}$-space, one obtains, respectively, that

$$N(\varepsilon) = \frac{V}{2\pi^2}\left(\frac{2}{\hbar^2}\right)^{3/2}(m_x m_y m_z)^{1/2}(\varepsilon - \varepsilon_C)^{1/2}, \quad N(\varepsilon) = \frac{V}{2\pi^2}\left(\frac{2}{\hbar^2}\right)^{3/2}(m_x m_y m_z)^{1/2}(\varepsilon_V - \varepsilon)^{1/2}$$

In view of Eq.(47.13), the effective mass m^* is, in general, defined as $m^* = (m_x m_y m_z)^{1/3}$, and it is called the density-of-states effective mass. In most materials, two of the effective mass parameters are equal, and hence, m^* is written, in terms of a longitudinal (m_l) and a transverse (m_t) effective mass with respect to the symmetry axis, as $m^* = (m_l m_t^2)^{1/3}$.

42.3. Find an expression for the Fermi level in an impurity semiconductor assuming the presence of completely ionized donors, $N_d^+ = N_d$, and acceptors, $N_a^- = N_a$.

Solution

The condition for electrical neutrality, Eq.(47.39), reduces to

$$N_V e^{(\varepsilon_V - \varepsilon_F)/k_B T} - N_C e^{(\varepsilon_V - \varepsilon_F)/k_B T} + N_d - N_a = 0$$

and it is convenient to set $x = e^{\varepsilon_F/k_B T}$, $c = e^{-\varepsilon_C/k_B T}$, $v = e^{\varepsilon_V/k_B T}$, which leads to the equation

$$x^2 - \frac{N_d - N_a}{N_C c} x - \frac{N_V v}{N_C c} = 0$$

Thus, the Fermi level is

$$\varepsilon_F = k_B T \ln x = k_B T \ln \left\{ \frac{N_d - N_a}{2 N_C c} + \left[\left(\frac{N_d - N_a}{2 N_C c} \right)^2 + \frac{N_V v}{N_C c} \right]^{1/2} \right\}$$

Using the inverse hyperbolic sine function given by $\sinh^{-1}(\alpha/u) = \ln\left(\alpha + \sqrt{\alpha^2 + u^2}\right) - \ln u$, and setting $u = (N_V v / N_C c)^{1/2} = (m_h^*/m_e^*)^{3/4} e^{(\varepsilon_C + \varepsilon_V)/2k_B T}$, such that

$$\frac{\alpha}{u} = \frac{N_d - N_a}{2(N_C N_V c v)^{1/2}} = \frac{N_d - N_a}{2(N_C N_V)^{1/2} e^{-(\varepsilon_C - \varepsilon_V)/2 k_B T}} = \frac{N_d - N_a}{2 n_i}$$

where n_i is the intrinsic concentration, Eq.(47.20), one obtains

$$\varepsilon_F = \frac{1}{2}(\varepsilon_C + \varepsilon_V) + \frac{3}{4} k_B T \ln \frac{m_h^*}{m_e^*} + k_B T \sinh^{-1}\left(\frac{N_d - N_a}{2 n_i}\right) = \varepsilon_{Fi} + k_B T \sinh^{-1}\left(\frac{N_d - N_a}{2 n_i}\right)$$

It follows that the Fermi level is adjusted above ε_{Fi}, Eq.(47.23), in n-type semiconductors, where $N_d > N_a$, and below ε_{Fi} in p-type semiconductors.

42.4. If an intrinsic semiconductor having $N_C(T_0)$ at room temperature is doped with $N_d = 10^{-3} \times N_C(T_0)$ donors, find the temperature T at which the highest value of the Fermi level is reached.

Solution

As indicated in Figure 42.5, the Fermi level is expected to reach its highest value, situated between the donor energy and the bottom of the conduction band, in the low temperature range, and hence, the suitable approximation for ε_F is that given by Eq.(42.41). Since from Eq.(42.18) it follows that $N_C = (T/T_0)^{3/2} N_C(T_0)$, one obtains

$$\varepsilon_F(T) = \varepsilon_C - \frac{\varepsilon_d}{2} - \frac{k_B T}{2}\left[\ln \frac{2 N_C(T_0)}{N_d} + \frac{3}{2} \ln \frac{T}{T_0}\right]$$

and the condition of maximum on $\varepsilon_F(T)$ gives the required temperature as

$$\ln\frac{2N_C(T_0)}{N_d}+\frac{3}{2}\ln\frac{T}{T_0}+\frac{3}{2}=0 \quad \text{or} \quad T=\left[\frac{N_d}{2N_C(T_0)}\right]^{2/3}\frac{T_0}{e}$$

The numerical result is $T = 3.16$ K.

42.5. Find an expression for the Fermi-Dirac distribution over the impurity levels that accounts for the spin degeneracy of the impurity states.

Solution

There are two quantum states associated with each donor level, corresponding to the spin-up and spin-down orientations of the electron. However, double occupation is precluded, since the valency requirements of the donor ion are satisfied by one electron of either spin only. In other words, the average occupation is determined by considering the empty state, with no contribution to energy, and two single electron states, with probabilities given by Eqs.(35.37) and (35.38) as

$$f_n(0)=\frac{1}{1+2e^{-(\varepsilon_D-\varepsilon_F)/k_BT}}, \quad f_n(1)=\frac{e^{-(\varepsilon_D-\varepsilon_F)/k_BT}}{1+2e^{-(\varepsilon_D-\varepsilon_F)/k_BT}}$$

which yields the occupation probability

$$f_n(\varepsilon_D)=0\cdot f_n(0)+2\cdot f_n(1)=\frac{2e^{-(\varepsilon_D-\varepsilon_F)/k_BT}}{1+2e^{-(\varepsilon_D-\varepsilon_F)/k_BT}}=\frac{1}{\frac{1}{2}e^{(\varepsilon_D-\varepsilon_F)/k_BT}+1}$$

The concentration $N_d^0 = N_d f_n(\varepsilon_D)$ of unionized donors follows as given by Eq.(42.35). We may also derive the average occupation with electrons of the acceptor level ε_A, which can either be singly occupied by a spin-up or a spin-down electron, or doubly occupied. The empty state would correspond to two holes localized at the acceptor impurity, and hence it is prohibited. According to Eqs.(35.37) and (35.38), one obtains

$$f_n(1)=\frac{e^{-(\varepsilon_A-\varepsilon_F)/k_BT}}{2e^{-(\varepsilon_A-\varepsilon_F)/k_BT}+e^{-2(\varepsilon_A-\varepsilon_F)/k_BT}}, \quad f_n(2)=\frac{e^{-2(\varepsilon_A-\varepsilon_F)/k_BT}}{2e^{-(\varepsilon_A-\varepsilon_F)/k_BT}+e^{-2(\varepsilon_A-\varepsilon_F)/k_BT}}$$

which yields the mean number of electrons at the acceptor level as

$$f_n(\varepsilon_A)=2\cdot f_n(1)+2f_n(2)=\frac{e^{(\varepsilon_F-\varepsilon_A)/k_BT}+1}{\frac{1}{2}e^{(\varepsilon_F-\varepsilon_A)/k_BT}+1}=\langle n_a\rangle$$

The mean number of holes is $\langle p_a\rangle = 2-\langle n_a\rangle$, where 2 is the maximum number of electrons on one level ε_A, and hence

$$\langle p_a\rangle=\frac{1}{\frac{1}{2}e^{(\varepsilon_F-\varepsilon_A)/k_BT}+1} \quad \text{and} \quad N_a^0=N_a\langle p_a\rangle=\frac{N_a}{\frac{1}{2}e^{(\varepsilon_F-\varepsilon_A)/k_BT}+1}$$

The concentration of ionized acceptors, Eq.(42.37), follows using Eq.(42.34).

42.6. Given that the carrier concentration n_i is found to be independent of the variation with temperature of the gap width ε_g in an intrinsic semiconductor, find an expression for ε_g as a function of temperature.

42.7. Calculate the energy gap ε_p in an intrinsic semiconductor in terms of the carrier concentrations n_0 and $n = \alpha n_0$ measured at room temperature T_0 and at $T = \beta T_0$, if $\alpha = 30$, $\beta = 7/6$.

42.8. A Ge semiconductor of electron concentration $N_C(T_0)$ at room temperature T_0 is doped with $N_d = 3 \times 10^{-3} N_C(T_0)$ donors. If the Fermi level reaches the donor level energy ε_D at a temperature $T = \beta T_0$, find the magnitude of the donor band gap ε_d.

42.9. Show that the carrier concentration in the conduction and valence bands, in the presence of both the donor and acceptor impurities, depends on the excess of ionizable impurity flaws $\Delta N = N_d - N_a$ only, and find expressions for n and p in terms of n_i and ΔN.

43. Solid State Electronics

43.1. CARRIER TRANSPORT PHENOMENA
43.2. THE *pn* JUNCTION
43.3. THE JUNCTION TRANSISTOR

43.1. Carrier Transport Phenomena

Semiconductors are technologically important because the mechanism for electrical conductivity is different from that found in a perfect conductor. When an electric field is applied to a semiconductor, charge transport can be described by current density equations and by the requirement of continuity for electrons and holes. The current density (42.10) for a pure semiconductor, which represents the drift of free electrons and holes in the external field $\vec{E}$, is modified by the presence of a concentration gradient, which can be established by the chemical addition of impurities. We must in addition take into account a new mechanism of charge transport, which is not significant in metals, consisting of the *diffusion* of carriers by thermal motion and occuring whenever a concentration gradient exists. The current density caused by diffusion can be written for electrons in the form

$$j_e(x) = eD_e \frac{dn}{dx} \tag{43.1}$$

and for holes as

$$j_h(x) = -eD_h \frac{dp}{dx} \tag{43.2}$$

where D_e and D_h are the *diffusion constants* for electrons and holes, respectively. We have limited ourselves to the one-dimensional model with a simple distance parameter x, which is found to be sufficient for technologically important semiconductors with cubic structure and isotropic conduction properties.

The difference in sign between the two diffusion currents arises because, although diffusion takes place down the concentration gradient regardless of charge, the resulting current is in opposite directions for electrons and holes. Hence, the electron and hole current densities in a given direction consist of a drift component caused by an applied field and a diffusion component due to the carrier concentration gradient, that is

$$j_e = e\left(n\mu_e E + D_e \frac{dn}{dx}\right) \quad (43.3)$$

$$j_h = e\left(p\mu_h E - D_h \frac{dp}{dx}\right) \quad (43.4)$$

where the electric field is applied parallel to the *x*-axis. A relationship between mobility and the diffusion constant for both electrons and holes may be derived from the thermal equilibrium condition, assuming no applied electric field and no current flow. Since j_e and j_h must separately be equal to zero, Eq.(43.3) reduces to

$$n\mu_e E = -D_e \frac{dn}{dx} \quad (43.5)$$

This implies that diffusion results in an internal electric field which produces an electrostatic potential $V(x)$, given by Eq.(11.9) as $E = -dV(x)/dx$. We have seen in Chapter 42 that the energy distribution of carriers follows the Boltzmann distribution law (26.25), so that the variation of *n* with *x*, as a result of the motion of electrons in the electrostatic potential, will be given by

$$n = N_C e^{eV(x)/k_B T} \quad (43.6)$$

or

$$\frac{dn}{dx} = N_C \frac{e}{k_B T} \frac{dV(x)}{dx} e^{eV(x)/k_B T} = n \frac{e}{k_B T} \frac{dV(x)}{dx} \quad (43.7)$$

Hence, Eq.(43.5) yields

$$-n\mu_e \frac{dV(x)}{dx} = -D_e n \frac{e}{k_B T} \frac{dV(x)}{dx} \quad \text{or} \quad D_e = \frac{k_B T}{e}\mu_e \quad (43.8)$$

Similarly, from Eq.(43.4) we may obtain

$$D_h = \frac{k_B T}{e}\mu_h \quad (43.9)$$

Equations (43.8) and (43.9) are known to be the *Einstein relations* which, on substitution into Eqs.(43.3) and (43.4), lead to

$$j_e = e\mu_e \left(nE + \frac{k_B T}{e} \frac{dn}{dx} \right) \tag{43.10}$$

$$j_h = e\mu_h \left(pE - \frac{k_B T}{e} \frac{dp}{dx} \right) \tag{43.11}$$

In the absence of an external field E there will be no net electron or hole current flow, so that, substituting for n from Eq.(42.46) in Eq.(43.10), we obtain

$$j_e = 0 = \mu_e k_B T \frac{d}{dx}\left[n_i e^{(\varepsilon_F - \varepsilon_{Fi})/k_B T} \right] = \mu_e n \frac{d\varepsilon_F}{dx} \tag{43.12}$$

and similarly, from Eqs.(42.47) and (43.11), we have

$$j_h = 0 = \mu_h p \frac{d\varepsilon_F}{dx} \tag{43.13}$$

In other words, the Fermi energy is constant throughout an inhomogenous semiconductor in thermal equilibrium as

$$\frac{d\varepsilon_F}{dx} = 0 \tag{43.14}$$

EXAMPLE 43.1. Energy-band diagrams

It is often convenient to represent inhomogeneous semiconductors by energy-band diagrams which visualize the electrical behaviour described by the carrier transport equations. Most phenomena can be understood if we use a one-dimensional effective energy-band model (Figure 43.1), where ε_C is the lowest energy in the conduction band, and ε_V is the highest energy in the valence band for any possible crystallographic direction. In other words $\varepsilon_g = \varepsilon_C - \varepsilon_V$ is the minimum thermal energy required to excite electrons from the valence band into the conduction band and the x-axis represents an averaged direction through the crystal. This convention is illustrated in Figure 43.1 for homogenous semiconductors.

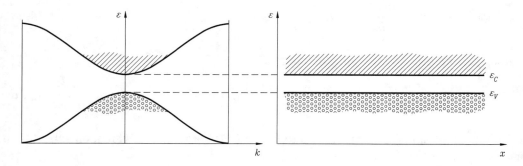

Figure 43.1. Conventional representation of the band structure in homogenous semiconductors

For inhomogeneous semiconductors at thermal equilibrium, according to Eq.(43.14), the Fermi energy is always represented in such diagrams by a horizontal line (Figure 43.2). Any local increase or decrease in the carrier potential energy with respect to the average effective value will be represented by an upward or downward bending of the bands. From Eqs.(43.6) and (42.18) it follows that

$$\varepsilon_C = \varepsilon_F - eV(x) \quad \text{or} \quad \frac{d\varepsilon_C}{dx} = -e\frac{dV(x)}{dx} = eE \quad (43.15)$$

A similar relation holds for ε_V, so that we can see that the slope of the band edges gives the magnitude of the electric fields, whatever their source and allows us to determine the forces which cause the carrier motion. It is apparent that, in an energy-band representation, the electrons move towards the bottom of the conduction band, while the holes move towards the top of the valence band.

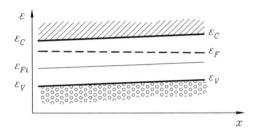

Figure 43.2. The energy-band diagram for inhomogeneous semiconductors at thermal equilibrium

The equation of continuity (12.15) should be modified, for carrier flow, due to generation and recombination of electrons and holes, which are continuous processes at thermal equilibrium. The excess concentration of carriers above their equilibrium value, generated by promotion of electrons from the valence band to the conduction band, is denoted by Δn and Δp, respectively, so that

$$n = n_0 + \Delta n \quad \text{and} \quad p = p_0 + \Delta p \quad (43.16)$$

where we will assume that $\Delta n = \Delta p$ to maintain neutrality. It is obvious that we must be concerned with the influence of generation and recombination on the *minority carriers* in extrinsic semiconductors, because the concentration of majority carriers is sufficiently large as to be unaffected. If the probability that an excess pair recombines in a time interval dt is taken to be dt/τ, then the rate at which excess carriers disappear will be given by

$$d(\Delta n) = -\Delta n \frac{dt}{\tau} \quad \text{or} \quad \frac{d(\Delta n)}{dt} = -\frac{\Delta n}{\tau} \quad (43.17)$$

from which it follows that

$$\Delta n(t) = \Delta n(0) e^{-t/\tau} \quad (43.18)$$

It is seen that the decay is exponential, at a rate governed by τ, which is the average *lifetime* of the minority carriers, in this case, electrons. Similar equations hold for holes. If the pair generation rate is denoted by g, the rate of change of the excess minority carrier concentration in a volume element will be determined by generation and recombination as given by Eq.(43.17), and the divergence of the current density will be expressed by Eq.(12.15). For the one-dimensional flow of a small pulse consisting of Δn excess minority electrons we obtain

$$\frac{d(\Delta n)}{dt} = g_e - \frac{\Delta n}{\tau_e} + \frac{1}{e}\frac{dj_e}{dx}$$

and, substituting j_e from Eq.(43.3), we obtain the equation of motion

$$\frac{d(\Delta n)}{dt} = g_e - \frac{\Delta n}{\tau_e} + (\Delta n)\mu_e \frac{dE}{dx} + \mu_e E \frac{d(\Delta n)}{dx} + D_e \frac{d^2(\Delta n)}{dx^2} \qquad (43.19)$$

The continuity requirement for the excess hole motion, given by

$$\frac{d(\Delta p)}{dt} = g_h - \frac{\Delta p}{\tau_h} - \frac{1}{e}\frac{dj_h}{dx}$$

provides a similar equation for the motion of the excess hole concentration Δp as

$$\frac{d(\Delta p)}{dt} = g_h - \frac{\Delta p}{\tau_h} - (\Delta p)\mu_h \frac{dE}{dx} - \mu_h E \frac{d(\Delta p)}{dx} + D_h \frac{d^2(\Delta p)}{dx^2} \qquad (43.20)$$

EXAMPLE 43.2. Steady-state diffusion of excess minority carriers

For the steady-state motion of excess holes in a *n*-type semiconductor we must take $d(\Delta p)/dt = 0$ in Eq.(43.20). If we assume $g_h = 0$ and the applied field is $E = 0$, Eq.(43.20) reduces to the diffusion equation

$$\frac{d^2(\Delta p)}{dx^2} - \frac{\Delta p}{D_h \tau_h} = 0 \qquad (43.21)$$

where it is usual to define a *diffusion length* for holes as

$$L_h = \sqrt{D_h \tau_h} \qquad (43.22)$$

Equation (43.21) has the general solution

$$\Delta p = a e^{x/L_h} + b e^{-x/L_h} \qquad (43.23)$$

where *a* and *b* are arbitrary constants. In a similar manner the steady-state diffusion equation for excess electrons in a *p*-type semiconductor may be derived from Eq.(43.19) as

$$\frac{d^2(\Delta n)}{dx^2} - \frac{\Delta n}{D_e \tau_e} = 0 \tag{43.24}$$

The solution is

$$\Delta n = c e^{x/L_e} + d e^{-x/L_e} \tag{43.25}$$

where $L_e = \sqrt{D_e \tau_e}$ is the diffusion length for electrons, and c and d are arbitrary constants. The constants a, b, c, d in Eqs.(43.23) and (43.25) are determined, as we shall show later on, by the boundary conditions required for semiconductor-device operation.

43.2. THE *pn* JUNCTION

The *pn* junction, which is essential to the operation of most semiconductor devices and integrated circuits, consists of an abrupt discontinuity or a graded distribution of the two types of doping impurities across a region of a particular semiconductor sample. Although no surface states are involved in the boundary between the *p*-type region and the *n*-type region, the electrostatic conditions at the junction can be obtained if we consider the special case where a piece of *p*-type material is brought into intimate contact with a piece of *n*-type material. The energy-band diagram at the moment of contact is given in Figure 43.3 (a). Owing to the concentration gradient, electrons diffuse from the conduction band of the *n*-type region into that of the *p*-type region and recombine with the free holes by dropping into the valence band. The space charge, produced by the negatively charged acceptors in the *p*-type side and the positively charged donors left behind in the *n*-type side, exerts a repulsive force on further charges crossing the junction and inhibits further flow. In terms of the energy-band diagram the Fermi energy of the *p*-type region is raised with respect to that of the *n*-type region until the Fermi levels become equal, as in Figure 43.3 (b), according to Eq.(43.14). The transferred charge which terminates on the immobile impurity atoms in the transition region produces an electric field which is directed from the *n*-type side to the *p*-type side. This *built-in potential barrier* V_{int} corresponds to the difference between the Fermi levels on the *n* and *p* sides before contact was made. We will assume that the electric field associated with V_{int} is so high that, to a first approximation, it excludes any thermally generated mobile carriers from the transition region, which is then referred to as the *depletion* region.

Let N_d and N_a be the concentrations of charged donor and acceptor impurities and x_d, x_a the thicknesses of the space charge regions on each side of the junction, respectively. For an *abrupt junction*, N_a and N_d are constants, independent of x, that is,

$$N_d(x) = 0, \quad N_a(x) = N_a \quad \text{for} \quad -x_a < x < 0$$

$$N_d(x) = N_d, \quad N_a(x) = 0 \quad \text{for} \quad 0 < x < x_d$$

and we may assume that all the impurities are ionized, so that the thermal equilibrium carrier densities at the edges of the space charge regions will be taken as $n_0(x_d) = N_d$ and $p_0(-x_a) = N_a$.

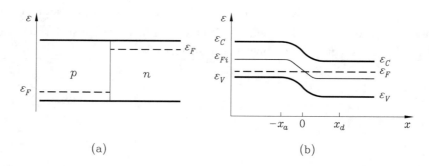

(a)　　　　　　　　　　　　(b)

Figure 43.3. Energy-band diagram for an abrupt *pn* junction (a) before contact (b) in thermal equilibrium

The built-in potential barrier results from the thermal equilibrium condition (43.5), if the internal field is expressed in terms of the electrostatic potential by Eq.(11.9) as

$$\frac{dV}{dx} = \frac{k_B T}{e} \frac{1}{n} \frac{dn}{dx}$$

where use has been made of Eq.(43.8) for D_e. On integration over the transition region, from $x = -x_a$ to $x = x_d$, one obtains

$$V_{int} = V(x_d) - V(-x_a) = \frac{k_B T}{e} \ln \frac{n_0(x_d)}{n_0(-x_a)} \tag{43.26}$$

We can take $n_0(x_d) = N_d$ and $n_0(-x_a) = n_i^2 / p_0(-x_a) = n_i^2 / N_a$. Hence, the thermal equilibrium potential barrier, whose *n*-type side is positive with respect to the *p*-type side, can be written as

$$V_{int} = \frac{k_B T}{e} \ln \frac{N_d N_a}{n_i^2} \tag{43.27}$$

The variation of the potential in the depletion region is obtained by solving Poisson's equation (11.14), written in terms of the dielectric constant ε_s as

$$\frac{d^2V}{dx^2} = \frac{eN_a}{\varepsilon_s} \quad \text{for} \quad -x_a < x < 0 \quad , \quad \frac{d^2V}{dx^2} = -\frac{eN_d}{\varepsilon_s} \quad \text{for} \quad 0 < x < x_d \qquad (43.28)$$

subject to the boundary conditions of zero field at both extremities, which read

$$\left(\frac{dV}{dx}\right)_{x=-x_a} = \left(\frac{dV}{dx}\right)_{x=x_d} = 0 \qquad (43.29)$$

Since the *pn* junction is made of separate pieces of *p*-type and *n*-type material, we must have the same charge per unit area on each side of the transition region, due to the charge separation, that is

$$N_d x_d = N_a x_a \qquad (43.30)$$

On the *p*-type side of the junction $(-x_a < x < 0)$, integrating Eq.(43.28) and bearing in mind the boundary condition (43.29), we obtain the field distribution

$$\frac{dV}{dx} = \frac{eN_a}{\varepsilon_s}(x + x_a) \qquad (43.31)$$

Integrating Eq.(43.31) gives

$$V(x) = \frac{eN_a}{2\varepsilon_s}(x + x_a)^2 + \text{const}$$

where the arbitrary constant is eliminated by defining the zero of potential at $-x_a$ as

$$V(x) = \frac{eN_a}{2\varepsilon_s}(x + x_a)^2 \quad \text{for} \quad -x_a < x < 0 \qquad (43.32)$$

that is,

$$V(-x_a) = 0 \quad \text{and} \quad V(0) = \frac{eN_a x_a^2}{2\varepsilon_s} \qquad (43.33)$$

Similarly, on the *n*-side $(0 < x < x_d)$ we have

$$V(x) = -\frac{eN_d}{2\varepsilon_s}(x_d - x)^2 + \text{const}$$

and the continuity at $x = 0$ allows us to evaluate the constant, giving

$$V(x) = \frac{eN_d x_d^2}{2\varepsilon_s} + \frac{eN_a x_a^2}{2\varepsilon_s} - \frac{eN_d}{2\varepsilon_s}(x_d - x)^2 \quad \text{for} \quad 0 < x < x_d \qquad (43.34)$$

From Eq.(43.34) it follows that

$$V(x_d) = \frac{eN_d x_d^2}{2\varepsilon_s} + \frac{eN_a x_a^2}{2\varepsilon_s}$$

The built-in potential barier $V_{int} = V(x_d) - V(-x_a) = V(x_d)$ can be written, using the condition (43.30), in the form

$$V_{int} = \frac{eN_d x_d}{2\varepsilon_s}(x_d + x_a) \qquad (43.35)$$

From Eq.(43.30), taking $w = x_d + x_a$ for the total width of the depletion region, we have

$$\frac{x_d}{w} = \frac{x_d}{x_d + x_a} = \frac{N_a}{N_d + N_a} \qquad (43.36)$$

On substitution into Eq.(43.35) we obtain

$$V_{int} = \frac{eN_d N_a}{2\varepsilon_s (N_d + N_a)} w^2 \qquad (43.37)$$

so that the width of the depletion region at thermal equilibrium can be written as

$$w = \left(\frac{2\varepsilon_s}{e} \frac{N_d + N_a}{N_d N_a} V_{int} \right)^{1/2} \qquad (43.38)$$

A *pn* junction is in dynamic equilibrium which can be described, in terms of the energy-band diagram of Figure 43.3, as a balance between majority carriers, moving from the *n*-type to the *p*-type side and climbing the potential barrier, and an equal number of minority carriers, moving down the barrier from the *p*-type to the *n*-type side.

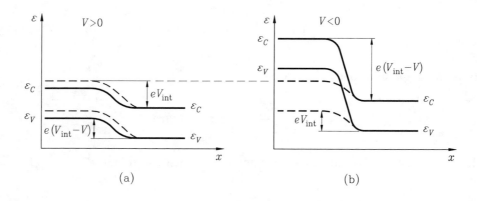

Figure 43.4. Energy-band diagrams for a *pn* junction (a) forward bias (b) reverse bias

When a potential difference V is applied across the junction, so that the p-type side is biased positively with respect to the n-type side ($V > 0$ or *forward bias*), the barrier to diffusion is lowered, as in Figure 43.4 (a), resulting in large forward currents. Reversing the polarity of the applied voltage ($V < 0$ or *reverse bias*), the barrier to diffusion is raised, as in Figure 43.4 (b), so that the minority carriers near the transition region only that produce a small reverse current.

In the presence of an applied voltage V across the junction, the potential barrier (see Figure 43.4) becomes $V_{int} - V$, if we adopt the convention $V > 0$ for forward bias and $V < 0$ for reverse bias, and Eq.(43.26) can be written as

$$V_{int} - V = \frac{k_B T}{e} \ln \frac{n(x_d)}{n(-x_a)} \qquad (43.39)$$

Substituting V_{int} from Eq.(43.26) we have

$$V = \frac{k_B T}{e} \ln \frac{n(-x_a)}{n(x_d)} + \frac{k_B T}{e} \ln \frac{n_0(x_d)}{n_0(-x_a)} \qquad (43.40)$$

It is assumed that the majority carrier concentration will only slightly be disturbed by the bias voltage, so that we can take $n(x_d) = n_0(x_d) = N_d$ and $p(-x_a) = p_0(-x_a) = N_a$, so that Eq.(43.40) reduces to

$$V = \frac{k_B T}{e} \ln \frac{n(-x_a)}{n_0(-x_a)} \qquad (43.41)$$

or

$$n(-x_a) = n_0(-x_a) e^{eV/k_B T} \qquad (43.42)$$

In a similar manner we can express

$$p(x_d) = p_0(x_d) e^{eV/k_B T} \qquad (43.43)$$

Equations (43.42) and (43.43) show that *minority* carrier concentrations are modified by a factor of $e^{eV/k_B T}$ at the edges of the transition region when applying a bias potential V. The excess minority carrier concentration Δn at the p-edge can be obtained from Eq.(43.42) as

$$\Delta n = n(-x_a) - n_0(-x_a) = n_0(-x_a)\left(e^{eV/k_B T} - 1\right) = n_a \left(e^{eV/k_B T} - 1\right) \qquad (43.44)$$

and Δp at the n-edge is given by Eq.(43.43) as

$$\Delta p = p(x_d) - p_0(x_d) = p_0(x_d)\left(e^{eV/k_B T} - 1\right) = p_d \left(e^{eV/k_B T} - 1\right) \qquad (43.45)$$

The steady-state diffusion of excess electrons through the transition region is described by the solution (43.25) to the diffusion equation (43.24), where the coefficients c and d are determined by the boundary conditions

$$\Delta n = 0 \quad \text{for} \quad x = -\infty \quad (43.46)$$

$$\Delta n = n_a \left(e^{eV/k_B T} - 1\right) \quad \text{for} \quad x = -x_a$$

and this gives

$$\Delta n = n_a \left(e^{eV/k_B T} - 1\right) e^{(x+x_a)/L_e} \quad (43.47)$$

If we assume that the excess minority electron current at $x = -x_a$ is entirely due to diffusion, we obtain the current density from Eq.(43.1) as

$$j_e = eD_e \frac{d(\Delta n)}{dx}\bigg|_{x=-x_a} = \frac{eD_e n_a}{L_e}\left(e^{eV/k_B T} - 1\right) \quad (43.48)$$

On the n-type side of the junction, at $x = x_d$, we similarly obtain from Eqs.(43.2) that

$$j_h = -eD_h \frac{d(\Delta p)}{dx}\bigg|_{x=x_d} = \frac{eD_h p_d}{L_h}\left(e^{eV/k_B T} - 1\right) \quad (43.49)$$

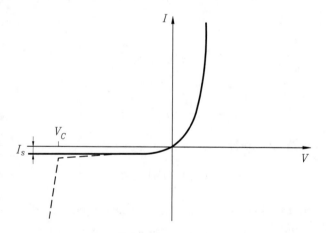

Figure 43.5. The static characteristic of the *pn* junction

The total current density can then be written as

$$j = j_e + j_h = \left(\frac{eD_e n_a}{L_e} + \frac{eD_h p_d}{L_h}\right)\left(e^{eV/k_B T} - 1\right) = j_s \left(e^{eV/k_B T} - 1\right) \quad (43.50)$$

This result is called the *Shockley equation* for the ideal *pn* junction. Hence, the net current flow from the *p*-region to the *n*-region is

$$I = Aj_s = I_s \left(e^{eV/k_B T} - 1\right) \quad (43.51)$$

where A stands for the area of the junction. Equation (43.51) is usually represented in a current against voltage plot as in Figure 43.5, called a *static characteristic* and shows the nonlinear behaviour of the *pn* junction. Under forward bias the large currents are due to the increased probability of *majority* carriers climbing the potential barrier $V_{int} - V$. The small currents obtained under reverse bias arise from *minority* carriers sliding down the barrier. This explains the use of junctions as rectifiers, in which they pass current when biased in the forward direction and block current when biased in the reverse direction. In real devices there is a gradual increase in reverse current with increasing reverse voltage, until a critical voltage V_C is reached, as indicated by the dashed line in Figure 43.5.

43.3. THE JUNCTION TRANSISTOR

The junction transistor consists of two *pn* junctions which have been formed in the same crystal in one of two possible configurations *npn* or *pnp*. Since their operation is similar, provided the roles of electrons and holes are interchanged and the polarities of the bias potentials reversed, we will consider the *npn* transistor. The *npn* structure is conventionally represented as a sequence of three sections called *emitter*, *base* and *collector*. The arrangement shown in Figure 43.6 (a) is known to be the *common-base* configuration with the base connection common to both the input circuit, which biases the *emitter* junction in the forward direction and the output circuit, which biases the *collector* junction in the reverse direction. The symbol of the *npn* transistor is shown in Figure 43.6 (b) where the arrow always points in the direction of positive current, in this case away from the base, since electrons are the majority carriers in the emitter.

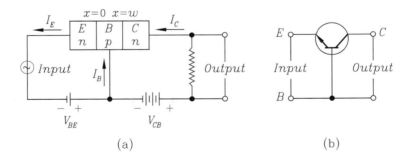

Figure 43.6. Structure and symbol for the *npn* junction transistor in the common base configuration

The operation of the device can be understood in terms of the energy band diagrams drawn in Figure 43.7 with electron energy increasing upwards and hole energy increasing downwards. If the input circuit is not connected, there is only a flow of minority carriers across the reverse-biased collector junction, electrons from the base and holes from the collector. The result is a small current I_{C0} flowing from collector to base. When the emitter junction is biased in the forward direction, the current I_E across it from base to emitter is due to electron flow from the emitter and hole flow from the base. Electrons that were majority carriers in the emitter become *minority* carriers into the base, where the hole concentration is only slightly disturbed by the bias voltage V_{BE} but the electron concentration is increased by a factor of e^{eV_{BE}/k_BT}, as discussed before for the *pn* junction. Excess electrons are then to be *injected* by the emitter into the base.

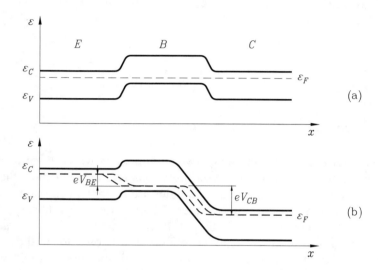

Figure 43.7. Energy-band diagram for a *npn* transistor (a) in equilibrium (b) in the common-base configuration

The fraction of the current I_E that is due to excess electron injection is called the *emitter efficiency* γ and is defined as

$$\gamma = \frac{I_{eE}}{I_E} = \frac{I_{eE}}{I_{eE} + I_{hE}} = \frac{1}{1 + I_{hE}/I_{eE}} \qquad (43.52)$$

where I_{eE} and I_{hE} are the electron and hole components of the emitter current. The emitter efficiency can be estimated using the electron and hole current densities, defined by Eq.(43.48) and (43.49), which on substitution into Eq.(43.52) yield

$$\gamma = \left(1 + \frac{D_h L_e p_d}{D_e L_h n_a}\right)^{-1} = \left(1 + \frac{D_{hE} L_{eB} N_{aB}}{D_{eB} L_{hE} N_{dE}}\right)^{-1} \qquad (43.53)$$

where $p_d = p_0(x_d) = n_i^2/N_d$ and $n_a = n_0(-x_a) = n_i^2/N_a$ as discussed before and the subscripts E and B indicate the emitter and the base, respectively. Thus, the emitter efficiency can be controlled by the ratio of the doping concentrations in the emitter and base and approaches unity if the emitter is heavily doped as compared to the base.

Excess minority electrons injected by the emitter diffuse across the base under the influence of their own concentration gradient. They reach the collector junction if the base region is sufficiently narrowed compared to the diffusion length L_{eB}, to prevent recombination with majority holes. The *base transport efficiency* is defined by

$$\delta = \frac{I_{eC}}{I_{eE}} \tag{43.54}$$

where I_{eC} is the electron component of the collector current. The steady-state diffusion of excess electrons through the base is described by the current density (43.1) as

$$j_e(x) = eD_{eB} \frac{d(\Delta n)}{dx}$$

where Δn is the solution of the form (43.25) to the diffusion equation (43.24). The coefficients c and d are determined by two boundary conditions. The first condition is written at the emitter junction, where $x = 0$ (Figure 43.6) and Δn assumes a value Δn_0 as a result of electron injection into the base, and hence

$$\Delta n(0) = c + d = \Delta n_0$$

The second boundary condition is taken at the collector junction $(x = w)$, where the recombination rate is high, such that

$$j_e(w) = eD_{eB} \frac{d(\Delta n)}{dx}\bigg|_{x=w} = \frac{e\Delta n(w)}{\tau} \quad \text{or} \quad \frac{D_{eB}}{L_{eB}}\left(ce^{w/L_{eB}} - de^{-w/L_{eB}}\right) = \frac{1}{\tau}\left(ce^{w/L_{eB}} + de^{-w/L_{eB}}\right)$$

Assuming a short average lifetime τ before recombination, $1/\tau \gg D_{eB}/L_{eB}$, we obtain

$$\Delta n = (\Delta n_0)\left[e^{x/L_{eB}} - e^{(2w-x)/L_{eB}}\right]/\left(1 - e^{2w/L_{eB}}\right)$$

that is,

$$j_e(x) = \frac{e(\Delta n_0)D_{eB}}{L_{eB}}\left[e^{x/L_{eB}} + e^{(2w-x)/L_{eB}}\right]/\left(1 - e^{2w/L_{eB}}\right)$$

The base transport efficiency δ can be expressed to be the ratio of the electron currents at $x = w$ and $x = 0$ in the form

$$\delta = \frac{j_e(w)}{j_e(0)} = \text{sech}\frac{w}{L_{eB}} \cong 1 - \frac{1}{2}\frac{w^2}{L_{eB}^2} \tag{43.55}$$

where use has been made of the assumption $w \ll L_{eB}$. Equation (43.55) shows that δ approaches unity by decreasing the base width w. From Eqs.(43.52) and (43.54) we obtain the total collector current as

$$I_C = \gamma\delta I_E + I_{C0} = \alpha I_E + I_{C0} \tag{43.56}$$

where α is called *current gain* and is almost unity in the conditions specified by Eqs.(43.53) and (43.55). The current directions selected in Figure 43.6 imply that

$$I_E = I_C + I_B \tag{43.57}$$

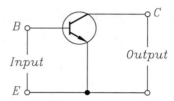

Figure 43.8. Common-emitter configuration for a *npn* transistor

The *common-emitter* configuration, represented in Figure 43.8, is useful for achieving current gain. It has a static characteristic similar to that of the common-base configuration, where the input current is I_B rather than I_E. From the basic transistor equations (43.56) and (43.57) we then obtain

$$I_C = \frac{\alpha}{1-\alpha} I_B + \frac{1}{1-\alpha} I_{C0} = \beta I_B + I'_{C0} \tag{43.58}$$

It is seen that the current gain

$$\beta = \frac{\alpha}{1-\alpha} = \frac{1}{1/\alpha - 1} \tag{43.59}$$

is much greater than in the common-base configuration, and it is clearly important that $\alpha = \gamma\delta$ should be as near unity as possible. Substituting Eqs.(43.53) and (43.55) into Eq.(43.59) and neglecting products of the small quantitites, we obtain

$$\beta \cong \left(\frac{D_{hE} L_{eB} N_{aB}}{D_{eB} L_{hE} N_{dE}} + \frac{w^2}{2L_{eB}^2} \right)^{-1} \tag{43.60}$$

In most transistors, the dominant term is the first term on the right-hand side, which is related to the emitter efficiency. The second term shows the need for a long diffusion length L_{eB}, which restricts the range of semiconductor materials that may be used for junction devices.

FURTHER READING

1. J. P. Colinge, C. A. Colinge - PHYSICS OF SEMICONDUCTOR DEVICES, Kluwer Academic, New York, 2002.
2. D. A. Neaman - SEMICONDUCTOR PHYSICS AND DEVICES, McGraw-Hill, New York, 2002.
3. S. M. Sze - SEMICONDUCTOR DEVICES: PHYSICS AND TECHNOLOGY, Wiley, New York, 2001.

PROBLEMS

43.1. Assuming that direct electron-hole recombination may occur through vertical transitions in which the electron drops across the gap between the conduction and the valence band, find the average lifetime of the minority carriers in such a semiconductor.

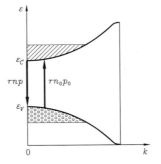

Solution

Suppose, for convenience, that every electron near the conduction band edge has the same probability of vertical transition, and every hole near the valence band edge has the same probability of being filled, the rate at which excess carriers disappear can be taken to be $-rnp$, where r is a constant called the recombination coefficient, and n, p are given by Eq.(43.16). Thermal excitation of electrons from the valence to the conduction band is always present, at a rate R_0 which obviously is the same as that of recombination at thermal equilibrium, where

$$\frac{dn}{dt} = -rn_0 p_0 + R_0 = 0 \quad \text{or} \quad R_0 = rn_0 p_0$$

It follows that the net rate of recombination away from equilibrium is given by

$$\frac{dn}{dt} = -rnp + rn_0 p_0$$

Substituting n and p from Eq.(43.16) and making the approximation that $\Delta n \ll n_0 + p_0$, which is justified when excess carrier concentrations are small, yields

$$\frac{d(\Delta n)}{dt} = -r(n_0 + p_0)\Delta n - r(\Delta n)^2 \approx -r(n_0 + p_0)\Delta n$$

This equation has the same form as Eq.(43.17), and thus gives the reciprocal average time spent by an electron in the conduction band, before recombining, to be $\tau^{-1} = r(n_0 + p_0)$.

43.2. Assuming an increase $\Delta\sigma$ in the electrical conductivity of an intrinsic semiconductor, if it is illuminated for a long period of time, find the pair generation rate g due to light absorption.

Solution

The electrical conductivity, Eq.(42.10), can be written as

$$\sigma = ne\mu_e + pe\mu_h = (n_0 + \Delta n)e\mu_e + (p_0 + \Delta p)e\mu_h$$

$$= e(n_0\mu_e + p_0\mu_h) + e(\mu_e + \mu_h)\Delta n = \sigma_0 + e(\mu_e + \mu_h)\Delta n$$

where use has been made of Eqs.(43.16). It is assumed that $\Delta n = \Delta p$ at any time, since a hole is created for every electron excited across the gap, and that every electron that drops across the gap recombines with a hole. If g is the optical generation rate, the time dependence of the excess carrier concentration is given by

$$\frac{d(\Delta n)}{dt} = g - \frac{\Delta n}{\tau} \qquad \text{or} \qquad \Delta n(t) = \tau g(1 - e^{-t/\tau})$$

In the long-time limit, the steady value τg is reached, and this gives

$$\sigma - \sigma_0 = e(\mu_e + \mu_h)\tau g = \frac{e(\mu_e + \mu_h)}{r(n_0 + p_0)} g$$

such that the optical generation rate results as $g = \dfrac{r(n_0 + p_0)}{e(\mu_e + \mu_h)} \Delta\sigma$.

43.3. Show that a *pn* junction behaves like a capacitor under reverse-bias voltage, and find an expression for its depletion capacitance.

Solution

Since the transition region remains depleted of free carriers if a reverse-bias voltage is applied across the junction, the built-in potential barrier V_{int} can be replaced by $V_{int} - V$ in Eq.(43.38), which becomes

$$w = \left(\frac{2\varepsilon_s}{e} \frac{N_d + N_a}{N_d N_a}\right)^{1/2} (V_{int} - V)^{1/2}$$

Substituting this result into Eq.(43.36), the charge-per-unit area in the transition region is obtained from Eq.(43.30) as

$$Q = eN_d x_d = \frac{eN_d N_a}{N_d + N_a} w = \left(\frac{2e\varepsilon_s N_d N_a}{N_d + N_a}\right)^{1/2} (V_{int} - V)^{1/2}$$

so that the capacitance of the *pn* junction can be written as

$$C = \left|\frac{dQ}{dV}\right| = \left[\frac{e\varepsilon_s N_d N_a}{2(V_{int} - V)(N_d + N_a)}\right]^{1/2} = \frac{C_0}{(1 - V/V_{int})^{1/2}}$$

Thus the *pn* junction is equivalent to a capacitor with spacing between plates equal to the transition region width and dielectric constant ε_s.

43.4. Find the light-generated current in a *pn* junction where the electron-hole pairs are generated at a constant rate g.

Solution

The diffusion equation (43.24) for excess electrons in a *p*-type semiconductor must be written as

$$D_e \frac{d^2(\Delta n)}{dx^2} - \frac{\Delta n}{\tau_e} + g = 0$$

and we may assume steady-state injection at $x = -x_a$, such that the solution is

$$\Delta n = \left[\Delta n(-x_a) - \tau_e g\right] e^{(x+x_a)/L_e} + \tau_e g$$

where $L_e = \sqrt{D_e \tau_e}$ is the diffusion length for electrons. The electron current density, Eq.(43.1), becomes

$$j_e(x) = eD_e \frac{d(\Delta n)}{dx} = \frac{eD_e}{L_e}\left[\Delta n(-x_a) - \tau_e g\right] e^{(x+x_a)/L_e}$$

Substituting $\Delta n(-x_a)$ as given by Eq.(43.44) at the left side of the transition region $(x = -x_a)$, one obtains

$$j_e = j_e(-x_a) = \frac{eD_e n_a}{L_e}(e^{eV/k_B T} - 1) - eL_e g$$

In a similar manner, the hole current density at the right side of the transition region $(x = x_d)$ is

$$j_h = \frac{eD_h p_d}{L_h}(e^{eV/k_B T} - 1) - eL_h g$$

and hence, the total current density in the junction becomes

$$j = \left(\frac{eD_e n_a}{L_e} + \frac{eD_h p_d}{L_h}\right)(e^{eV/k_B T} - 1) - e(L_e + L_h)g$$

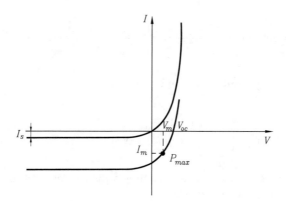

If A denotes the area of the junction, the current is $I = I_s(e^{eV/k_BT} - 1) - eA(L_e + L_h)g$, and it is apparent that, when the *pn* junction is illuminated, its static characteristic given by Eq.(43.51) is shifted downward.

43.5. If a sinusoidal signal is applied to a *npn* junction transistor, find an expression for the common-base current gain α as a function of the *ac* frequency ω.

Solution

In the common-base configuration shown in Figure 43.6 (b), the concentration of excess electrons injected into the base, which is given by Eq.(43.19) as

$$\frac{d(\Delta n)}{dt} = -\frac{\Delta n}{\tau_e} + D_e \frac{d^2(\Delta n)}{dx^2}$$

will vary sinusoidally in the form $\Delta n = f(x)e^{i\omega t}$, and hence, we have

$$\left(\frac{L_{eB}^2}{1+i\omega\tau_e}\right)\frac{d^2 f(x)}{dx^2} - f(x) = 0$$

This is the same as Eq.(43.24) for Δn, provided we introduce a complex diffusion length $L_{eB}^* = L_{eB}/(1+i\omega\tau_e)^{1/2}$. It is convenient to set $(1+i\omega\tau_e)^{1/2} = \xi + i\eta$, where

$$\xi, \eta = \frac{1}{\sqrt{2}}\left(\sqrt{1+\omega^2\tau_e^2} \pm 1\right)^{1/2}$$

such that Δn takes the form $\Delta n = e^{i\omega t}\left[c e^{\xi x/L_{eB}} e^{i\eta x/L_{eB}} + d e^{-\xi x/L_{eB}} e^{-i\eta x/L_{eB}}\right]$, similar to that given by Eq.(43.25). It follows that, under *ac* conditions, the diffusion length L_{eB} should be replaced by

$$\frac{L_{eB}}{\xi} = \frac{L_{eB}\sqrt{2}}{\left(\sqrt{1+\omega^2\tau_e^2}+1\right)^{1/2}}$$

which is a function of ω that reduces to the steady-state value L_{eB} in the low frequency limit and decreases with increasing ω. Hence, L_{eB}/ξ should be substituted for L_{eB} in the base transport efficiency, Eq.(43.55), such that one obtains the *ac* current gain as

$$\alpha(\omega) = \gamma\left(1 - \frac{w^2}{2L_{eB}^2}\xi^2\right) = \gamma\left[1 - \frac{w^2}{4L_{eB}^2}\left(\sqrt{1+\omega^2\tau_e^2}+1\right)\right] = \alpha + \frac{\gamma w^2}{4L_{eB}^2}\left(1 - \sqrt{1+\omega^2\tau_e^2}\right)$$

where α stands for the common-base current gain at zero frequency.

43.6. Calculate the light-generated current I_g and the open circuit voltage V_{oc} that can be obtained across a uniformly illuminated *pn* junction.

43.7. Find a relation between the potential V_m across a *pn* junction operating as a solar cell that corresponds to its maximum power output, and the maximum voltage V_{oc} at zero output current.

43.8. If avalanche breakdown occurs in a *pn* junction, where a carrier acquires energy ε_g in traversing a distance of the order of the mean free path λ and an electron-hole pair is created by impact ionization of covalent bonds, find the critical voltage required for the onset of breakdown.

43.9. Find the emitter doping N_{dE} that yields a maximum current gain β for an *npn* junction transistor operating in the common-emitter configuration, given that a rise in N_{dE} results in narrowing the energy gap by $\Delta\varepsilon_g = \alpha\sqrt{N_{dE}}$.

43.10. Find an expression for the collector current I_C of a *pnpn* device operating in the open-base configuration, assuming that α_p and α_n are the current gains for the *pnp* and *npn* sections, respectively.

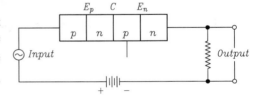

44. SOLID STATE MAGNETISM

44.1. DIAMAGNETISM
44.2. PARAMAGNETISM
44.3. FERROMAGNETISM
44.4. ANTIFERROMAGNETISM
44.5. FERRIMAGNETISM

44.1. DIAMAGNETISM

The magnetic state of a solid is characterized by the magnetic moment per unit volume or magnetization $\vec{M}$ given by Eq.(13.7), and its response to an applied magnetic field $\vec{H}$ is represented by the magnetic susceptibility χ_m defined by Eq.(13.14). It is common practice to introduce, instead of the external field $\vec{H}$, an external magnetic induction $\vec{B}_0 = \mu_0 \vec{H}$, where μ_0 is the permeability of free space, so that Eq.(13.14) reads

$$\vec{M} = \chi_m \frac{\vec{B}_0}{\mu_0} \qquad (44.1)$$

The influence of an applied field on a solid has been discussed in Chapter 40, where the solid was treated as an assembly of free electrons. Experiments have shown, however, that solids where the electrons are localized on isolated atoms and described by the tight-binding approximation display larger magnetic effects than those with nearly-free electrons, described by the weak-binding model. This suggests that magnetism is of atomic origin and can be understood by treating the solid as a system of isolated atoms. If any interaction between atoms is negligible, the solids exhibit just diamagnetic or paramagnetic behaviour. If there is a cooperative interaction between the atoms, the solids may be in a state that shows spontaneous magnetization, corresponding to ferromagnetism, antiferromagnetism or ferrimagnetism.

Consider a system of identical non-interacting atoms, arranged in a lattice such that $\vec{B}_0$ and χ_m are the same for all the atoms, and the magnetization is given by $\vec{M} = N\vec{\mu}$, where $\vec{\mu}$ is the atomic magnetic moment, and N stands for the number of atoms per unit volume. From Eq.(44.1) it follows that

$$\chi_m \vec{B}_0 = N\mu_0 \vec{\mu} \tag{44.2}$$

Using the central-field approximation, discussed in Chapter 36, the Z-electron problem for an atom can be reduced to Z one-electron problems, described by the Hamiltonian (36.47). The effect of an external magnetic field is to quantize the electron energy with respect to its motion normal to $\vec{B}_0$, and this is described as in Eq.(32.33). Hence, in the presence of the field, if we consider the electron spin, the Hamiltonian is given by

$$\hat{H} = \sum_{i=1}^{Z} \left[-\frac{\hbar^2}{2m_e}\nabla_i^2 - \frac{Ze_0^2}{r_i} + V_{\mathit{eff}}(r_i) + \frac{e}{2m_e}\vec{B}_0 \cdot \left(\hat{\vec{L}}_i + g_s\vec{S}_i\right) + \frac{e^2}{2m_e}\left(\frac{1}{2}\vec{B}_0 \times \vec{r}_i\right)^2 \right]$$

$$= \hat{H}_0 + \sum_{i=1}^{Z} \left[\frac{e}{2m_e}\vec{B}_0 \cdot \left(\hat{\vec{L}}_i + g_s\vec{S}_i\right) + \frac{e^2}{2m_e}\left(\frac{1}{2}\vec{B}_0 \times \vec{r}_i\right)^2 \right] \tag{44.3}$$

where $g_s = 2$. If the direction of $\vec{B}_0$ is taken to be the z-axis, we have

$$\frac{1}{2}\vec{B}_0 \times \vec{r}_i = \frac{1}{2}B_0\left(x_i\vec{e}_y - y_i\vec{e}_x\right) \quad \text{i.e.,} \quad \left(\frac{1}{2}\vec{B}_0 \times \vec{r}_i\right)^2 = \frac{B_0^2}{4}\left(x_i^2 + y_i^2\right)$$

and Eq.(44.3) reduces to

$$\hat{H} = \hat{H}_0 + \frac{eB_0}{2m_e}\sum_{i=1}^{Z}\left(\hat{L}_{zi} + 2\hat{S}_{zi}\right) + \frac{e^2 B_0^2}{8m_e}\sum_{i=1}^{Z}\left(x_i^2 + y_i^2\right) \tag{44.4}$$

where we assume that the energy eigenvalues (36.51) of $\hat{H}_0$, given by the Hartree equations (36.52), are known. The last two terms in Eq.(44.4) can be treated as perturbations. The first term is a paramagnetic component, which originates in the angular momentum of the electron, and the second one is a diamagnetic component. Using the first-order perturbation theory, Eq.(36.11), the energy in the presence of the applied magnetic field is

$$E = E_0 + \frac{eB_0}{2m_e}\sum_{i=1}^{Z}\langle\psi_i|\hat{L}_{zi} + \hat{S}_{zi}|\psi_i\rangle + \frac{e^2 B_0^2}{8m_e}\sum_{i=1}^{Z}\langle\psi_i|x_i^2 + y_i^2|\psi_i\rangle \tag{44.5}$$

It can be shown that a nonvanishing expectation value of the angular momentum is obtained for unfilled shells only. Transition metals, rare earth metals and actinides have unfilled 3d, 4f and 5f shells, respectively, and are all therefore expected to exhibit paramagnetic behaviour in the solid state. For closed shells the sum of the z-components of the angular momentum is zero, so that the first term in Eq.(44.5) vanishes, and the second term is responsible for the diamagnetic behaviour. Assuming for simplicity that ψ_i are spherically symmetric, we can write

$$\langle \psi_i | x_i^2 + y_i^2 | \psi_i \rangle = \frac{2}{3} \langle \psi_i | r_i^2 | \psi_i \rangle$$

so that the diamagnetic interaction energy per atom is obtained from Eq.(44.5) as

$$E_d = \frac{e^2 B_0^2}{12 m_e} \sum_{i=1}^{Z} \langle \psi_i | r_i^2 | \psi_i \rangle = \frac{Z e^2 B_0^2}{12 m_e} \langle r^2 \rangle$$

where $\langle r^2 \rangle$ is the mean square distance from the nucleus. By comparing this result with Eq.(32.36), it is seen that the expectation value of the magnetic moment is

$$\mu_d = -\frac{\partial E_d}{\partial B_0} = -\frac{Z e^2 \langle r^2 \rangle}{6 m_e} B_0 \quad \text{or} \quad \vec{\mu}_d = -\frac{Z e^2 \langle r^2 \rangle}{6 m_e} \vec{B}_0$$

From Eq.(44.2), the diamagnetic susceptibility χ_d for a solid which consists of atoms or ions with closed shells can be written as

$$\chi_d = -\frac{N Z e^2 \mu_0}{6 m_e} \langle r^2 \rangle \tag{44.6}$$

It is temperature independent, and the negative sign follows from the fact that the induced magnetic moments are in opposite direction to the applied field. In other words, the induced magnetization of a diamagnetic solid always results in a decrease in the external magnetic induction, which becomes $\vec{B} = \vec{B}_0 + \mu_0 \vec{M}$. A perfect diamagnetic solid is characterized by

$$\vec{B} = \vec{B}_0 + \mu_0 \vec{M} = 0 \tag{44.7}$$

which means that $\chi_d = -1$ or the relative permeability $\mu_r = 0$. A contribution to susceptibility from atoms or ions with closed shells can be calculated for every solid, and hence, the susceptibility always has a diamagnetic component (44.6), although it may be negligible with respect to the contributions from other magnetic effects.

44.2. PARAMAGNETISM

In a solid described as a system of isolated atoms, at the origin of paramagnetism are the unpaired electrons in unfilled shells. An external magnetic field interacts with both the orbital magnetic moment (32.35) and the spin magnetic moment (34.14) of these atomic electrons. If we consider, for simplicity, a system of N noninteracting one-electron atoms per unit volume, and neglect the small diamagnetic effect, each single atom is subjected, in the presence of a field $\vec{B}_0$, to the interaction (44.4) which reads

$$\hat{H} = \hat{H}_0 + \frac{eB_0}{2m_e}\sum_{i=1}^{Z}(\hat{L}_{zi} + 2\hat{S}_{zi}) = \hat{H}_0 + \frac{eB_0}{2m_e}(\hat{L}_z + 2\hat{S}_z) = \hat{H}_0 + \frac{eB_0}{2m_e}(\hat{J}_z + \hat{S}_z) \quad (44.8)$$

Following the same steps used to derive the energy levels (34.28) of the Hamiltonian (34.24) for one-electron atoms, it is straightforward to show that the effect of weak external fields is to split each atomic energy level of the fine structure into $2J+1$ magnetic sublevels, with a constant separation $\mu_B B_0 g_L$, that is

$$\varepsilon = \varepsilon_0 + \mu_B B_0 g_L m_J = \varepsilon_0 - \mu_{m_J} B_0 = \varepsilon_0 - \varepsilon_{m_J} \quad (44.9)$$

where J is given by Eq.(36.55) and g_L by Eq.(36.56). The levels labelled by their m_J values are assigned, using Eq.(32.36), to effective magnetic moments given by

$$\mu_{m_J} = -\frac{\partial \varepsilon_{m_J}}{\partial B_0} = -\mu_B g_L m_J = -g_L \frac{e}{2m_e}\langle \psi_{Jm_J}|\hat{J}_z|\psi_{Jm_J}\rangle \quad (44.10)$$

The μ_{m_J} can therefore be regarded as the components, along the external field direction, of the magnetic moment

$$\vec{\mu}_J = -g_L \frac{e}{2m_e}\vec{J} \quad (44.11)$$

Equation (44.11) has the same form as those obtained for the orbital and spin magnetic moments. It is common practice to introduce an *effective magnetic moment* of magnitude

$$\mu_{eff}^2 = g_L^2 \left(\frac{e}{2m_e}\right)^2 \langle \psi_{Jm_J}|\hat{J}^2|\psi_{Jm_J}\rangle = g_L^2 \mu_B^2 J(J+1) \quad (44.12)$$

If each atom is assumed to be an element of a canonical ensemble that is in equilibrium at a temperature T under the applied magnetic field, the average magnetic moment per atom can be written, using Eq.(22.65) for the ensemble average, as

$$\langle \mu \rangle = \frac{\sum_{m_J=-J}^{J}\mu_{m_J}e^{\mu_{m_J}B_0/k_BT}}{\sum_{m_J=-J}^{J}e^{\mu_{m_J}B_0/k_BT}} = \frac{\sum_{m_J=-J}^{J}(-\mu_B g_L m_J)e^{-\mu_B B_0 g_L m_J/k_BT}}{\sum_{m_J=-J}^{J}e^{-\mu_B B_0 g_L m_J/k_BT}}$$

$$= \mu_B g_L \frac{\sum_{m_J=-J}^{J}(-m_J)e^{(-m_J)x}}{\sum_{m_J=-J}^{J}e^{(-m_J)x}} = \mu_B g_L \frac{\sum_{m_J=-J}^{J}m_J e^{m_J x}}{\sum_{m_J=-J}^{J}e^{m_J x}} \quad (44.13)$$

where a dimensionless parameter $x = \mu_B B_0 g_L / k_B T$ has been defined to be the ratio of the magnetic interaction to the thermal energy. It follows that

$$\langle \mu \rangle = \mu_B g_L \frac{d}{dx}\left[\ln\left(\sum_{m_J=-J}^{J} e^{m_J x}\right)\right] = \mu_B g_L \frac{d}{dx}\left\{\ln\left[e^{Jx}\frac{1-e^{-(2J+1)x}}{1-e^{-x}}\right]\right\}$$

$$= \mu_B g_L \frac{d}{dx}\left[\ln\left(\frac{\sinh[(2J+1)x/2]}{\sinh(x/2)}\right)\right]$$

which can be written as

$$\langle \mu \rangle = \mu_B g_L J\left[\frac{2J+1}{2J}\coth\left(\frac{2J+1}{2J}Jx\right) - \frac{1}{2J}\coth\left(\frac{Jx}{2J}\right)\right] = \mu_B g_L J B_J(Jx) \quad (44.14)$$

where $B_J(Jx)$ stands for the *Brillouin function*. Substituting Eq.(44.14) into Eq.(44.2), the paramagnetic susceptibility of the solid can formally be written as

$$\chi_p = \frac{N\mu_0 \mu_B g_L J}{B_0} B_J(Jx) \quad (44.15)$$

Note that, in deriving Eq.(44.15), we have assumed that all the atoms are in the same state of quantum number J. We recall that this assumption is valid in the weak field approximation only. For $x \ll 1$, that is, for high temperatures, we have

$$\coth x \cong \frac{3+x^2}{3x} \quad \text{or} \quad B_J(Jx) \cong \left(\frac{J+1}{3J}\right)Jx \quad (44.16)$$

and Eq.(44.15) takes the form

$$\chi_p = \frac{N\mu_0 \mu_B g_L J}{B_0}\frac{J+1}{3}\frac{\mu_B B_0 g_L}{k_B T} = \frac{N\mu_0}{3k_B T}\left[g_L^2 \mu_B^2 J(J+1)\right] = \frac{N\mu_0 \mu_{\text{eff}}^2}{3k_B T} = \frac{C}{T} \quad (44.17)$$

which is known as the *Curie law*, where the *Curie constant* C is given by

$$C = \frac{N\mu_0 \mu_{\text{eff}}^2}{3k_B} \quad (44.18)$$

For $x \gg 1$, that is, for low temperatures, where $\coth x \cong 1$ or $B_J(Jx) \cong 1$, Eq.(44.15) reduces to

$$\chi_p = \frac{N\mu_0 \mu_B g_L J}{B_0} \quad \text{or} \quad M_0 = N\mu_B g_L J \cong N\mu_{\text{eff}} \quad (44.19)$$

where use has been made of Eqs.(44.1) and (44.12). Thus, the magnetization M_0 is a constant for given N and J and represents a saturation value where every atom is contributing its maximum possible magnetic moment. Saturation is shown in Figure 44.1 by a plot of the Brillouin function as a function of Jx, where, for any J state, $B_J(Jx)$ asymptotically approaches a constant value of unity, resulting in magnetic saturation.

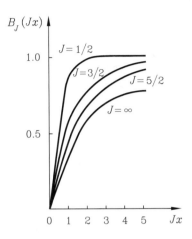

Figure 44.1. Brillouin function for one-electron states with various J values

44.3. FERROMAGNETISM

At the origin of the spontaneous magnetization, which is exhibited by ferromagnetic solids in the absence of an applied field, is the *exchange interaction* between nearly-localized electrons of isolated atoms. We first consider, for simplicity, the effect of exchange in bringing about the alignment of the electron spins in a two-atom system, which we can regard as a rudimentary magnet.

EXAMPLE 44.1. Exchange interaction in a two-electron system

For a hydrogen molecule, the eigenvalue problem for the electronic motion, Eq.(33.3), reads

$$\hat{H}_0 \Psi(r,R) = -\frac{\hbar^2}{2m_e}\nabla_1^2\Psi - \frac{\hbar^2}{2m_e}\nabla_2^2\Psi + \left(\frac{e_0^2}{r_{12}} - \frac{e_0^2}{r_{p1}} - \frac{e_0^2}{r_{p2}} - \frac{e_0^2}{r_{q1}} - \frac{e_0^2}{r_{q2}} + \frac{e_0^2}{R}\right)\Psi = E(R)\Psi(r,R)$$

where we have used 1 and 2 to designate the electrons, whereas p and q designate the protons, as illustrated in Figure 44.2. If the Schrödinger equations describing noninteracting hydrogen atoms where the electrons 1 or 2 are associated to the protons p or q, respectively, have the form

$$\hat{H}_{p1}\Psi_p(1)=E_p\Psi_p(1)\ ,\ \hat{H}_{q1}\Psi_q(1)=E_q\Psi_q(1)\ ,\ \hat{H}_{p2}\Psi_p(2)=E_p\Psi_p(2)\ ,\ \hat{H}_{q2}\Psi_q(2)=E_q\Psi_q(2)$$

the two-electron wave function can be described as a linear combination of single-electron states by $\Psi(r,R)=A\Psi_p(1)\Psi_q(2)+B\Psi_q(1)\Psi_p(2)$, and this gives

$$A\left[\frac{e_0^2}{r_{12}}-\frac{e_0^2}{r_{p2}}-\frac{e_0^2}{r_{q1}}-E(R)+E_p+E_q+\frac{e_0^2}{R}\right]\Psi_p(1)\Psi_q(2)$$

$$+B\left[\frac{e_0^2}{r_{12}}-\frac{e_0^2}{r_{p1}}-\frac{e_0^2}{r_{q2}}-E(R)+E_p+E_q+\frac{e_0^2}{R}\right]\Psi_q(1)\Psi_p(2)=0$$

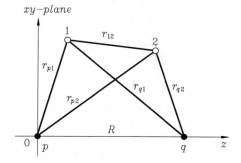

Figure 44.2. Relative positions of electrons and protons in the hydrogen molecule

If we set for convenience $\varepsilon=E(R)-E_p-E_q-e_0^2/R$, multiplying to the left by $\Psi_p^*(1)\Psi_q^*(2)$ and integrating over the two-electron positions yields

$$A\left[K(R)-\varepsilon\right]+B\left[J(R)-\varepsilon S^2(R)\right]=0$$

where $K(R)$ represents the mutual electrostatic interaction of the electron charge densities $e|\Psi_p(1)|^2$ and $e|\Psi_q(2)|^2$ and their interaction with the other proton, that is,

$$K(R)=e_0^2\left[\int\frac{|\Psi_p(1)|^2|\Psi_q(2)|^2}{r_{12}}dN_1dN_2-\int\frac{|\Psi_p(1)|^2}{r_{q1}}dN_1-\int\frac{|\Psi_q(2)|^2}{r_{p2}}dN_2\right]$$

and $J(R)$ defines the *exchange interaction*, which depends on the overlapping of both electron wave functions in the form

$$J(R)=e_0^2\int\frac{\Psi_p^*(1)\Psi_q^*(2)\Psi_q(1)\Psi_p(2)}{r_{12}}dN_1dN_2-e_0^2\int\frac{\Psi_p^*(1)\Psi_q(1)}{r_{p1}}dN_1\int\Psi_q^*(2)\Psi_p(2)dN_2$$

$$-e_0^2\int\frac{\Psi_q^*(2)\Psi_p(2)}{r_{q2}}dN_2\int\Psi_p^*(1)\Psi_q(1)dN_1$$

The *overlap integral* $S(R)$ for each electron wave function, defined as

$$S(R) = \int \Psi_p^*(1) \Psi_q(1) dV_1 = \int \Psi_q^*(2) \Psi_p(2) dV_2$$

is a function of the proton separation. It tends to unity as $R \to 0$, since p and q become indistinguishable, and vanishes as $R \to \infty$. Similarly, by multiplying to the left of each term by $\Psi_q^*(1)\Psi_p^*(2)$ and integrating gives

$$A\left[J(R) - \varepsilon S^2(R)\right] + B\left[K(R) - \varepsilon\right] = 0$$

where the same values are obtained for $K(R)$ and $J(R)$ if the indices p, q and 1, 2 of their arguments are simultaneously interchanged. Thus, the matrix eigenvalue equation is

$$\begin{pmatrix} K(R) & J(R) - \varepsilon S^2(R) \\ J(R) - \varepsilon S^2(R) & K(R) \end{pmatrix} \begin{pmatrix} A \\ B \end{pmatrix} = \varepsilon \begin{pmatrix} A \\ B \end{pmatrix}$$

and has the characteristic equation

$$\begin{vmatrix} K(R) - \varepsilon & J(R) - \varepsilon S^2(R) \\ J(R) - \varepsilon S^2(R) & K(R) - \varepsilon \end{vmatrix} = \left[K(R) - \varepsilon\right]^2 - \left[J(R) - \varepsilon S^2(R)\right]^2 = 0$$

This gives the energy corrections to $E(R)$ as $\varepsilon_\pm = \dfrac{K(R) \pm J(R)}{1 \pm S^2(R)}$ and the energy levels $E_\pm(R)$ are

$$E_\pm(R) = E_p + E_q + \frac{K(R) \pm J(R)}{1 \pm S^2(R)} + \frac{e_0^2}{R}$$

where $E_+(R)$ represents a singlet state and corresponds to antiparallel spins, while $E_-(R)$ represents a triplet state and corresponds to parallel spins.

The energy levels of a two-electron system can be put in the convenient form

$$E_\pm(R) = E_p + E_q + \frac{e_0^2}{R} + \frac{K(R) - J(R)S^2(R)}{1 - S^4(R)} \pm \frac{J(R) - K(R)S^2(R)}{1 - S^4(R)} = E_0(R) \pm J_{12}(R) \tag{44.20}$$

where $J_{12}(R)$ is called the *exchange constant* and is related to the energy separation of the energy levels for parallel (or *ferromagnetic*) and antiparallel (or *antiferromagnetic*) spin configurations in the form

$$J_{12}(R) = \frac{1}{2}\left[E_+(R) - E_-(R)\right] = \frac{J(R) - K(R)S^2(R)}{1 - S^4(R)} \tag{44.21}$$

It is apparent that $J_{12}(R)$ depends, in general, on the atomic separation $R = |\vec{R}_p - \vec{R}_q|$ and reduces to the exchange integral $J(R)$ for $R \to \infty$, because in this case the overlap integral $S(R)$ vanishes. The energy $E_-(R)$ is always lower than $E_+(R)$, and hence, the ferromagnetic state of a system is stable if $J_{12}(R) > 0$. In other words, electrons with a positive $J_{12}(R)$ favour parallel alignment of their spins (or magnetic moments), while electrons with a negative $J_{12}(R)$, like those in the hydrogen molecule, favour antiparallel alignment. Since the orientation of the electron spins relative to one another is closely related to the sign of the exchange integral, it is convenient to show this formally in Eq.(44.20), by introducing a spin coupling energy, which is expected to be proportional to their scalar product $\hat{\vec{S}}_1 \cdot \hat{\vec{S}}_2$, as in the case of the spin-orbit coupling energy (34.16). The eigenvalues of $\hat{\vec{S}}_1 \cdot \hat{\vec{S}}_2$ can be formed by using

$$\hat{\vec{S}}_1 \cdot \hat{\vec{S}}_2 = \frac{1}{2}\left[\left(\hat{\vec{S}}_1 + \hat{\vec{S}}_2\right)^2 - \hat{\vec{S}}_1^2 - \hat{\vec{S}}_2^2\right]$$

where, in view of Eq.(33.39), the eigenvalue of $\left(\hat{\vec{S}}_1 + \hat{\vec{S}}_2\right)^2$ is $S(S+1)\hbar^2$. This reduces to $2\hbar^2$ in the triplet state where $S = 1$ and to zero in the singlet state. Similarly $s_1(s_1+1)\hbar^2 = s_2(s_2+1)\hbar^2 = \frac{3}{4}\hbar^2$, since $s_1 = s_2 = \frac{1}{2}$, so that $\hat{\vec{S}}_1 \cdot \hat{\vec{S}}_2$ has eigenvalues $\frac{1}{4}\hbar^2$ in the ferromagnetic state and $-\frac{3}{4}\hbar^2$ in the antiferromagnetic state. If we now set $J_{12}(R) = J_{12}\hbar^2$, from Eq.(44.20) we obtain

$$E_+(R) = E_0(R) + J_{12}\hbar^2 - \left(3J_{12}\hbar^2/2 + 2J_{12}\hat{\vec{S}}_1 \cdot \hat{\vec{S}}_2\right) = E_0(R) + J_{12}(R)/2 - 2J_{12}\hat{\vec{S}}_1 \cdot \hat{\vec{S}}_2$$

(44.22)

where it is obvious that the sum between the brackets is zero in the singlet state and can be subtracted without altering the result. In a similar manner, for the triplet state we can derive a formally identical relation that reads

$$E_-(R) = E_0(R) - J_{12}\hbar^2 + \left(J_{12}\hbar^2/2 - 2J_{12}\hat{\vec{S}}_1 \cdot \hat{\vec{S}}_2\right) = E_0(R) - J_{12}(R)/2 - 2J_{12}\hat{\vec{S}}_1 \cdot \hat{\vec{S}}_2$$

The first two terms do not depend on the relative spin orientation and can be dropped if we consider the relative and not the absolute energy; whereas the last term represents the desired form of the spin coupling energy, appropriate for both the parallel and antiparallel alignment. The effect of the exchange interaction in a solid can be evaluated using a system of N atoms per unit volume, each having one unpaired electron with zero orbital angular momentum, $\hat{\vec{J}}_p = \hat{\vec{S}}_p$, so that

$$\hat{H}_{ex} = -2\sum_{p}\sum_{q} J_{pq}\hat{\vec{S}}_p \cdot \hat{\vec{S}}_q = -2J_{ex}\sum_{p}\sum_{r}\hat{\vec{S}}_p \cdot \hat{\vec{S}}_{p+r} \qquad (44.23)$$

where the index p runs over all the atoms and $q = p + r$. Because the interaction between nearest neighbours only has a significant contribution to the exchange energy, the index r runs over all the nearest neighbours of a given atom. We also assume that the exchange constant is the same for each nearest neighbour. Equation (44.23) can be further simplified by using the mean-field approximation, where the spin operators $\hat{\vec{S}}_{p+r}$ are replaced by their expectation value, which can be taken to be the same for all the atoms, so that Eq.(44.23) becomes

$$\hat{H}_{ex} = -\sum_{p}\hat{\vec{S}}_p \cdot \left(2J_{ex}\sum_{r=1}^{n}\left\langle\hat{\vec{S}}_{p+r}\right\rangle\right) = -\sum_{p}\hat{\vec{S}}_p \cdot \left(2nJ_{ex}\left\langle\hat{\vec{S}}\right\rangle\right) = \left(\frac{2nJ_{ex}}{g_s e/2m_e}\left\langle\hat{\vec{\mu}}_s\right\rangle\right)\cdot\sum_{p}\hat{\vec{S}}_p \qquad (44.24)$$

where n is the number of nearest neighbours, and $\langle\hat{\vec{\mu}}_s\rangle$ is the average atomic magnetic moment given by Eq.(34.14).

In the presence of an external magnetic field which interacts with the spin magnetic moment of each atom, it follows from Eq.(44.8), by taking $\vec{J} \equiv \vec{S}$, and hence, $\hat{J}_z \equiv \hat{S}_z$, in view of Eq.(44.24), that the magnetic interaction can be written as

$$\hat{H} = \frac{e}{2m_e}\vec{B}_0 \cdot \sum_{p}\left(\hat{\vec{J}}_p + \hat{\vec{S}}_p\right) + \hat{H}_{ex} = \frac{2e}{2m_e}\vec{B}_0 \cdot \sum_{p}\hat{\vec{S}}_p + \hat{H}_{ex}$$

$$= \frac{g_s e}{2m_e}\vec{B}_0 \cdot \sum_{p}\hat{\vec{S}}_p + \frac{2nJ_{ex}}{g_s e/2m_e}\left\langle\hat{\vec{\mu}}_s\right\rangle \cdot \sum_{p}\hat{\vec{S}}_p = \frac{g_s e}{2m_e}\left(\vec{B}_0 + \vec{B}_m\right)\cdot\sum_{p}\hat{\vec{S}}_p \qquad (44.25)$$

where $\vec{B}_m$ represents the mean field, often called *molecular field*, that is produced at each atomic site by the exchange interaction and has the form

$$\vec{B}_m = \frac{2nJ_{ex}}{(g_s e/2m_e)^2}\left\langle\hat{\vec{\mu}}_s\right\rangle = \frac{2nJ_{ex}}{N(g_s e/2m_e)^2}\vec{M} = \gamma\vec{M} \qquad (44.26)$$

In the mean (or molecular) field approximation, the exchange interaction gives rise to a local field that is proportional to the magnetization with a *molecular field constant* γ as the constant of proportionality, first introduced by Weiss as a phenomenological parameter. In the *molecular field approximation* (44.26) the magnetic interaction (44.25) is formally identical to that of system of N isolated atoms which experience an effective magnetic induction $\vec{B}_{eff} = \vec{B}_0 + \gamma\vec{M}$, parallel to $\vec{B}_0$. Thus, the expression for the average magnetic moment given by Eq.(44.14) can be applied to ferromagnetic materials by incorporating $\vec{B}_{eff}$, instead of $\vec{B}_0$, and hence, along the external field direction we have

$$M = N\langle\mu\rangle = N\mu_B g_s SB_S\left[\frac{S\mu_B g_s}{k_B T}(B_0 + \gamma M)\right] = N\mu_B g_s SB_S(Sy) \quad (44.27)$$

where $B_S(Sy)$ is the Brillouin function, and S is the atomic spin quantum number. Since the Brillouin function increases with increasing y from zero at $y = 0$ up to a saturation value of unity (see Figure 44.1), it is convenient to define the saturation magnetization

$$M_0 = N\mu_B g_s S \quad (44.28)$$

and this is analogous to that defined by Eq.(44.19), so that y has a linear dependence on M/M_0, which is called the *reduced magnetization*

$$y = \frac{\mu_B g_s}{k_B T}B_0 + \left(\frac{\gamma\mu_B g_s M_0}{k_B T}\right)\frac{M}{M_0} \quad \text{or} \quad \frac{M}{M_0} = \left(\frac{k_B T}{\gamma S\mu_B g_s M_0}\right)Sy - \frac{B_0}{\gamma M_0} \quad (44.29)$$

Thus, Eq.(44.27) represents an implicit equation for M/M_0 of the form

$$\frac{M}{M_0} = B_S(Sy) \quad (44.30)$$

A graphical solution can be found by a simultaneous plot of Eqs.(44.30) and (44.29) in a rectangular coordinate system with axes Sy and M/M_0, as in Figure 44.3.

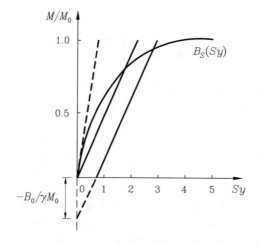

Figure 44.3. Graphical solution of the implicit equation for the reduced magnetization

The ordinate of intersection of the straight line (44.29) with the $B_S(Sy)$ curve gives M/M_0. The term $-B_0/\gamma M_0$ corresponds to an intercept on the vertical axis, from which an estimate of the molecular field constant γ can be made. At constant temperature (constant slope $k_B T/\gamma S\mu_B g_s M_0$) a decrease in the external field B_0 shifts the straight line, producing a decrease in the reduced magnetization, since the field experienced by the magnetic dipoles is now lower. With $B_0 = 0$, the straight line passes through the origin, and the resulting reduced magnetization M/M_0 obtained from the intercept with $B_S(Sy)$, known as the reduced *spontaneous magnetization*, defines the existence of the *ferromagnetic state* of the solid. The temperature dependence of the reduced spontaneous magnetization is obtained by recording the changing intercept as the slope of the straight lines increases with decreasing temperature. The resulting curve is plotted in Figure 44.4 for a ferromagnetic solid with $S = \frac{1}{2}$.

For the straight line to intersect the curve $B_S(Sy)$, in Figure 44.3, in order to obtain a solution for M/M_0 when $B_0 = 0$, its slope must be smaller than that of the Brillouin function (which is given by Eq.(44.16) near $Sy = 0$), that is

$$\frac{k_B T}{\gamma S \mu_B g_s M_0} \leq \frac{dB_S(Sy)}{d(Sy)} = \frac{S+1}{3S} \qquad (44.31)$$

The temperature T_C at which the spontaneous magnetization M vanishes is called the *Curie temperature* and corresponds to the condition of equal slopes for the straight line and the Brillouin function, which is illustrated by a dashed straight line in Figure 44.3. The Curie temperature can be obtained from Eq.(44.31) which, for equal slopes, gives

$$T_C = \frac{\gamma \mu_B g_s M_0}{k_B} \frac{S+1}{3} = \frac{N \mu_B^2 g_s^2 S(S+1)}{3 k_B} \gamma \qquad (44.32)$$

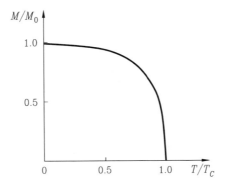

Figure 44.4. Reduced spontaneous magnetization as a function of the reduced temperature for a ferromagnetic solid

The Curie temperature defines a *phase transition*, such that spontaneous magnetization occurs for all temperatures below T_C, in the ferromagnetic state, and is zero above T_C, in the paramagnetic state. In the absence of the exchange interaction, when $\gamma = 0$, we have $T_C = 0$, and there is no phase transition. The magnetic susceptibility can be derived above T_C using Eqs.(44.1) and (44.27), which give

$$\chi_f = \mu_0 \frac{dM}{dB_0} = N\mu_0\mu_B g_s S \frac{dB_S(Sy)}{d(Sy)} \frac{S\mu_B g_s}{k_B T}\left(1+\gamma\frac{dM}{dB_0}\right)$$

$$= N\mu_0\mu_B g_s S\left(\frac{S+1}{3S}\right)\frac{S\mu_B g_s}{k_B T}\left(1+\frac{\gamma}{\mu_0}\chi_f\right) = \frac{\mu_0 T_C}{\gamma T}\left(1+\frac{\gamma}{\mu_0}\chi_f\right) \quad (44.33)$$

where we have substituted the slope of the Brillouin function near $Sy=0$, Eq.(44.31), and T_C as defined by Eq.(44.32). Solving Eq.(44.33) for χ_f we obtain

$$\chi_f = \frac{\mu_0 T_C/\gamma}{(T-T_C)} = \frac{C}{T-T_C} \quad (44.34)$$

which is known as the *Curie-Weiss law*. In view of Eq.(44.32), the Curie-Weiss constant

$$C = \mu_0 \frac{T_C}{\gamma} = \frac{N\mu_0\mu_B^2 g_s^2 S(S+1)}{3k_B} = \frac{N\mu_0\mu_{effB}^2}{3k_B} \quad (44.35)$$

is identical to the Curie constant (44.18), since it was assumed that only the spin term in Eq.(44.8) contributed to the total angular momentum $(J \equiv S)$. Since T_C is proportional to the molecular field constant γ, we can see that the difference in the expression for susceptibility from the Curie law (44.17) to the Curie-Weiss law, Eq.(44.34), illustrated in Figure 44.5, is determined by the exchange interactions that are still present in the paramagnetic state of ferromagnetic solids.

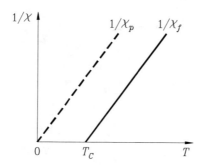

Figure 44.5. The Curie-Weiss behaviour of a ferromagnet in the paramagnetic state. (The Curie law is plotted by a dashed line.)

44.4. ANTIFERROMAGNETISM

If the exchange constant J_{ex} is negative, there will be a tendency to produce cooperative order where the spin pairs line up antiparallel. Consider the simple case of a solid whose magnetic structure can be divided into two sublattices 1 and 2, such that the nearest neighbours of an ion in one sublattice are ions in the other sublattice. *Identical interpenetrating magnetic sublattices*, where the negative J_{ex} is only appreciable between nearest neighbours can readily lead to *antiferromagnetic* order where the spin magnetic moments of all the nearest neighbours have antiparallel alignment, and therefore yield no net spontaneous magnetization.

The behaviour of antiferromagnetic solids can be treated in terms of the molecular field approximation (44.26), assuming a negative molecular field constant $\gamma = -|\gamma|$. From considerations similar to those leading to Eq.(44.30), there are two simultaneous equations for the reduced magnetization M_1 and M_2 for each sublattice

$$\frac{M_1}{M_0} = B_S\left(\frac{S\mu_B g_s}{k_B T}B_0 + \frac{\gamma S\mu_B g_s M_0}{k_B T}\frac{M_2}{M_0}\right) = B_S(Sy_1)$$

(44.36)

$$\frac{M_2}{M_0} = B_S\left(\frac{S\mu_B g_s}{k_B T}B_0 + \frac{\gamma S\mu_B g_s M_0}{k_B T}\frac{M_1}{M_0}\right) = B_S(Sy_2)$$

where M_0 has the form (44.28) with N taken to be the ion concentration in each sublattice. In the absence of an applied field $(B_0 = 0)$ the spontaneous magnetizations are opposite in sign, as required by the symmetry of the two equations with a negative γ. This results in no net spontaneous magnetization, as

$$M_1 = -M_2 \quad \text{or} \quad M = M_1 + M_2 = 0 \quad (44.37)$$

Substituting Eq.(44.37), Eqs.(44.36) reduce to

$$\frac{M_1}{M_0} = B_S\left(-\frac{\gamma S\mu_B g_s M_0}{k_B T}\frac{M_1}{M_0}\right) = B_S\left(\frac{|\gamma| S\mu_B g_s M_0}{k_B T}\frac{M_1}{M_0}\right)$$

(44.38)

$$\frac{M_2}{M_0} = B_S\left(\frac{|\gamma| S\mu_B g_s M_0}{k_B T}\frac{M_2}{M_0}\right)$$

which are both applicable to a ferromagnet for $B_0 = 0$. Following the development leading to Eq.(44.32), we obtain that both M_1 and M_2 vanish at the critical temperature

$$T_N = \frac{|\gamma|\mu_B g_s M_0}{k_B}\frac{S+1}{3} = \frac{N\mu_B^2 g_s^2 S(S+1)}{3k_B}|\gamma| \quad (44.39)$$

where T_N is known as the *Néel temperature* and defines a phase transition from an antiferromagnetic state below T_N to a paramagnetic state above T_N. The magnetic susceptibility in the paramagnetic state, near T_N, is obtained from Eq.(44.36) as

$$\chi_a = \mu_0 \frac{d(M_1 + M_2)}{dB_0} = \mu_0 M_0 \frac{dB_S(Sy_1)}{d(Sy_1)} \frac{S\mu_B g_s}{k_B T}\left(1 + \gamma \frac{dM_2}{dB_0}\right)$$

$$+ \mu_0 M_0 \frac{dB_s(Sy_2)}{d(Sy_2)} \frac{S\mu_B g_s}{k_B T}\left(1 + \gamma \frac{dM_1}{dB_0}\right) \quad (44.40)$$

where the slope of both Brillouin functions can be approximated by $(S+1)/3S$, not far from T_N, so that we have

$$\chi_a = \frac{\mu_0 T_N}{|\gamma| T}\left(2 + \frac{\gamma}{\mu_0}\chi_a\right) \quad (44.41)$$

Solving for χ_a gives

$$\chi_a = \frac{2\mu_0 T_N}{|\gamma|}\frac{1}{(T+T_N)} = \frac{2C}{T+T_N} \quad (44.42)$$

where C is the Curie-Weiss constant, defined by Eq.(44.35). The susceptibility χ_a has a temperature dependence similar to the Curie-Weiss law (44.34), but with the opposite sign for T_N inferring that at the Néel temperature the susceptibility remains finite, $\chi_a = \mu_0/|\gamma|$. Below T_N there is no net spontaneous magnetization and, if an external field $\vec{B}_0$ is applied, we must differentiate between the cases of parallel and perpendicular orientation of $\vec{B}_0$ relative to the direction of spins. If the external field is applied parallel to the spins of sublattice 1, it gives rise to an increase in magnetization from M_1 to $M_1 + dM_1$, which will be nearly equal to the decrease in magnetization in sublattice 2, from $M_2 = -M_1$ to $-(M_1 - dM_1) = -M_1 + dM_1$. Thus the external field applied parallel to the direction of spins results in a net magnetization

$$M_{||} = (M_1 + dM_1) + (-M_1 + dM_1) = 2dM_1 \quad (44.43)$$

The parallel susceptibility for the antiferromagnetic state is obtained by combining Eqs.(44.43) and (44.36) which yields

$$\chi_{||} = \mu_0 \frac{dM_{||}}{dB_0} = 2\mu_0 \frac{dM_1}{dB_0} = 2\mu_0 M_0 \frac{dB_S(Sy_1)}{d(Sy_1)} \frac{S\mu_B g_s}{k_B T}\left(1 + \gamma \frac{dM_1}{dB_0}\right) \quad (44.44)$$

because $dM_2/dB_0 = dM_1/dB_0$. Equation (44.44) is similar to Eq.(44.40) and leads to

$\chi_\parallel = \chi_a = \mu_0/|\gamma|$ at $T = T_N$ and shows that $\chi_\parallel$ decreases with temperature below T_N, as given by the slope of the Brillouin function. It follows that at very low temperatures, where $Sy_1 \gg 1$ and $B_S(Sy_1) \cong 1$, $\chi_\parallel$ is zero (see Figure 44.6).

If the field is applied perpendicular to the direction of spins, M_1 and M_2 will rotate through some small angle φ, resulting in a component $M_\perp$ of magnetization parallel to the field $M_\perp = M_1 \sin\varphi + M_2 \sin\varphi$. The molecular fields that are proportional to these components must oppose and balance the external field. Thus

$$B_0 + \gamma M_1 \sin\varphi + \gamma M_2 \sin\varphi = 0 \quad \text{or} \quad \gamma M_\perp = -B_0$$

and the perpendicular susceptibility is given by

$$\chi_\perp = \mu_0 \frac{dM_\perp}{dB_0} = \frac{\mu_0}{|\gamma|} \tag{44.45}$$

which is temperature independent and is equal to the value of χ_a at the Néel temperature. The temperature dependence of the magnetic susceptibility for an antiferromagnetic solid is represented in Figure 44.6.

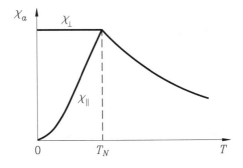

Figure 44.6. Variation with temperature of the antiferromagnetic susceptibility

44.5. FERRIMAGNETISM

Ferrimagnetism can be described in terms of two interpenetrating magnetic sublattices with unequal magnetic moments, interacting through a negative J_{ex}, which is assumed appreciable between nearest neighbours only. Thus antiparallel alignment of spins will be compatible with a net spontaneous magnetization, which can be written as the resultant magnetization of the two sublattices in the form

$$M = M_1 - M_2 \tag{44.46}$$

For ferrimagnetic solids, Eqs.(44.36) read

$$\frac{M_1}{M_{01}} = B_{S_1}\left(\frac{S_1\mu_B g_s}{k_B T}B_0 + \frac{\gamma_1 S_1 \mu_B g_s M_{02}}{k_B T}\frac{M_2}{M_{02}}\right)$$

$$\frac{M_2}{M_{02}} = B_{S_2}\left(\frac{S_2\mu_B g_s}{k_B T}B_0 + \frac{\gamma_2 S_2 \mu_B g_s M_{01}}{k_B T}\frac{M_1}{M_{01}}\right) \tag{44.47}$$

where the added subscripts 1 and 2 indicate the possibility of having different spin quantum numbers S_1, S_2, molecular field constants $\gamma_1 = -|\gamma_1|, \gamma_2 = -|\gamma_2|$ and saturation magnetizations given by

$$M_{01} = N_1\mu_B g_s S_1 \quad , \quad M_{02} = N_2\mu_B g_s S_2 \tag{44.48}$$

for the two sublattices. In each case the difference may result in ferrimagnetic behaviour, which is illustrated below a transition temperature T_C. Figure 44.7 shows a plot of the temperature dependence of the saturation magnetization M, obtained from Eqs.(44.47) and (44.46) for $B_0 = 0$, using the graphical method discussed earlier. The separate sublattice magnetizations are represented by dashed lines.

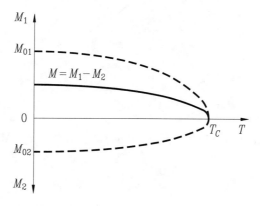

Figure 44.7. Temperature dependence of the ferrimagnetic spontaneous magnetization

For temperatures $T > T_C$ the solid is in a paramagnetic state, where only the linear portion of the Brillouin functions B_{S_1} and B_{S_2} will be considered, as in the cases of ferromagnetic and antiferromagnetic solids. In view of Eq.(44.16), Eqs.(44.47) can be written as

$$M_1 = N_1\mu_B g_s S_1 \left(\frac{S_1+1}{3S_1}\right) \frac{S_1\mu_B g_s}{k_B T}(B_0 + \gamma_1 M_2) = \frac{C_1}{\mu_0 T}(B_0 + \gamma_1 M_2)$$

$$M_2 = N_2\mu_B g_s S_2 \left(\frac{S_2+1}{3S_2}\right) \frac{S_2\mu_B g_s}{k_B T}(B_0 + \gamma_2 M_1) = \frac{C_2}{\mu_0 T}(B_0 + \gamma_2 M_1)$$

(44.49)

where

$$C_1 = \frac{N_1\mu_0\mu_B^2 g_s^2 S_1(S_1+1)}{3k_B} \quad , \quad C_2 = \frac{N_2\mu_0\mu_B^2 g_s^2 S_2(S_2+1)}{3k_B} \quad (44.50)$$

define a Curie-Weiss constant for each sublattice, similar to that introduced by Eq.(44.35). The two simultaneous equations (44.49) can be rewritten as

$$M_1\mu_0 T + |\gamma_1|C_1 M_2 = C_1 B_0$$

$$|\gamma_2|C_2 M_1 + M_2\mu_0 T = C_2 B_0$$

(44.51)

The condition for nonidentical vanishing spontaneous sublattice magnetizations M_1 and M_2, when $B_0 = 0$, is given by the characteristic equation

$$\begin{vmatrix} \mu_0 T & |\gamma_1|C_1 \\ |\gamma_2|C_2 & \mu_0 T \end{vmatrix} = 0 \quad (44.52)$$

which defines the Curie temperature for a ferrimagnetic solid as

$$T_C = \frac{1}{\mu_0}\left(|\gamma_1||\gamma_2|C_1 C_2\right)^{1/2} \quad (44.53)$$

For $T > T_C$, the simultaneous solution to Eqs.(44.51) gives

$$M_1 = \frac{C_1\mu_0 T - |\gamma_1|C_1 C_2}{\mu_0^2(T^2 - T_C^2)} B_0 \quad , \quad M_2 = \frac{C_2\mu_0 T - |\gamma_2|C_1 C_2}{\mu_0^2(T^2 - T_C^2)} B_0 \quad (44.54)$$

so that, for a ferrimagnetic solid, Eq.(44.1) becomes

$$\chi = \mu_0 \frac{d(M_1 + M_2)}{dB_0} = \frac{(C_1 + C_2)T - C_1 C_2(|\gamma_1| + |\gamma_2|)/\mu_0}{T^2 - T_C^2} \quad (44.55)$$

Thus the magnetic susceptibility is not a linear function of temperature in the paramagnetic state of a ferrimagnetic solid. This is illustrated in Figure 44.8 where the inverse susceptibility of an antiferromagnet which follows a linear law in the paramagnetic state is plotted for comparison.

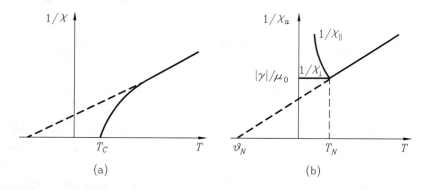

Figure 44.8. Inverse susceptibility of a ferrimagnet (a) and an antiferromagnet (b)

FURTHER READING

1. E. T. de Lacheisser, D. Gignoux, M. Schlenker - MAGNETISM, Springer-Verlag, New York, 2004.
2. N. A. Spaldin - MAGNETIC MATERIALS: FUNDAMENTALS AND DEVICE APPLICATIONS, Cambridge University Press, Cambridge, 2003.
3. R. C. O'Handley - MODERN MAGNETIC MATERIALS: PRINCIPLES AND APPLICATIONS, Wiley, New York, 1999.

PROBLEMS

44.1. Derive the Langevin equation for the average magnetic moment in a paramagnet, under a magnetic field $\vec{B}_0$ at a temperature T, assuming a net permanent magnetic moment $\vec{\mu}$ per atom.

Solution

The potential energy of the moment $\vec{\mu}$ in a magnetic field $\vec{B}_0$ is given by $\varepsilon = -\vec{\mu} \cdot \vec{B}_0 = -\mu B_0 \cos\theta$, where θ is the angle between $\vec{\mu}$ and $\vec{B}_0$. Thermal energy tends to randomize the alignment of the moments, thus the expectation value is $\langle \mu \rangle = \mu \langle \cos\theta \rangle$, where $\langle \cos\theta \rangle$ is the average orientation. This can be calculated as the Boltzmann thermal average, if the moments are assumed to be noninteracting, which yields

$$\langle \cos\theta \rangle = \frac{\int_0^\pi \int_0^{2\pi} \cos\theta\, e^{-\varepsilon/k_B T} \sin\theta\, d\theta\, d\varphi}{\int_0^\pi \int_0^{2\pi} e^{-\varepsilon/k_B T} \sin\theta\, d\theta\, d\varphi} = \frac{\int_0^\pi e^{-\mu B_0 \cos\theta/k_B T} \cos\theta \sin\theta\, d\theta}{\int_0^\pi e^{-\mu B_0 \cos\theta/k_B T} \sin\theta\, d\theta}$$

If we put $x = \cos\theta$, $dx = -\sin\theta\, d\theta$, we have

$$\langle \cos\theta \rangle = \frac{\int_{-1}^1 e^{\mu B_0 x/k_B T} x\, dx}{\int_{-1}^1 e^{\mu B_0 x/k_B T}\, dx} = \coth\frac{\mu B_0}{k_B T} - \frac{1}{(\mu B_0/k_B T)} = L(\mu B_0/k_B T)$$

where $L(\mu B_0/k_B T)$ is known as the Langevin function. It follows that $\langle\mu\rangle = \mu L(\mu B_0/k_B T)$. In normal conditions we have $\mu B_0 \ll k_B T$, and hence, we may approximate $L(\mu_0 B/k_B T) \approx \mu B_0/3k_B T$ so that the Langevin equation reduces, in view of Eq.(44.2), to the Curie law

$$\langle\mu\rangle = \frac{\mu^2 B_0}{3k_B T} \quad \text{or} \quad \chi_p = \frac{N\mu_0}{B_0}\langle\mu\rangle = \frac{N\mu_0 \mu^2}{3k_B T}$$

At low temperatures and high magnetic fields, where $\mu B_0 \gg k_B T$, the Langevin function tends to unity, and this corresponds to all the dipoles aligned with the field, such that $\langle\mu\rangle = \mu$.

44.2. For a paramagnetic salt where the magnetic moment per ion is $\mu_{m_J} = \pm\mu_B$, find an expression for the magnetic contribution to the heat capacity C_V. .

Solution

According to Eq.(44.10), each ion has one unpaired electron in an s-state ($L = 0$), such that we can take $J = S = \frac{1}{2}$, and hence, $g_L = 2$. There are two allowed states, one parallel and one antiparallel to the applied field, with energies $\pm\mu_B B_0$. The average magnetic moment is given by Eq.(44.13) as

$$\langle\mu\rangle = \frac{\mu_B e^{\mu_B B_0/k_B T} - \mu_B e^{-\mu_B B_0/k_B T}}{e^{\mu_B B_0/k_B T} + e^{-\mu_B B_0/k_B T}} = \mu_B \tanh\frac{\mu_B B_0}{k_B T}$$

and thus the magnetic energy per unit volume reads

$$\varepsilon = -N\vec{\mu}\cdot\vec{B}_0 = -N\mu_B B_0 \tanh\frac{\mu_B B_0}{k_B T}$$

It follows that the magnetic contribution to the heat capacity has the form

$$C_V = V\left(\frac{\partial\varepsilon}{\partial T}\right)_V = VNk_B\left(\frac{\mu_B B_0}{k_B T}\right)^2 \operatorname{sech}^2\frac{\mu_B B_0}{k_B T}$$

44.3. If the low-field permeability can be represented by $\mu(H) = \mu(0) + \nu H$, where ν is a constant, find an analytical representation for the corresponding hysteresis loop, and calculate the hysteresis loss in alternating fields.

Solution

The hysteresis loop has a parabolic dependence of B on H, given by

$$B = \mu(H)H = \mu(0)H + \nu H^2$$

where the two terms can be assigned to reversible and irreversible magnetization, respectively.

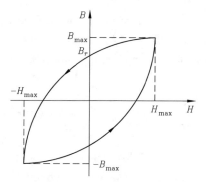

This loop, referred to as the Rayleigh hysteresis loop, has a lower branch described by

$$B + B_{max} = \mu(0)(H + H_{max}) + \nu(H + H_{max})^2$$

such that, at (H_{max}, B_{max}), it reduces to

$$B_{max} = \mu(0)H_{max} + 2\nu H_{max}^2$$

The upper branch can be represented, if (H_{max}, B_{max}) is at the origin, by

$$B - B_{max} = \mu(0)(H - H_{max}) - \nu(H - H_{max})^2$$

Substituting $B_{max}(H_{max})$ into the two equations for $B(H)$ yields

$$B = \left[\mu(0) + 2H_{max}\right]H \pm \nu(H^2 - H_{max}^2)$$

When an alternating field is applied, the area of the hysteresis loop defines the loss for each complete cycle of the field (see Problem 13.5), and hence, the hysteresis loss in the low-field region results as

$$U = V \int_\Gamma H dB = V \int_{-B_{max}}^{B_{max}} H dB + V \int_{B_{max}}^{-B_{max}} H dB = \frac{8}{3}\nu V H_{max}^3$$

Note that the parabolic approximation of the hysteresis curve breaks down if the amplitude H_{max} of the applied magnetic field is increased.

44.4. Find an expression for the magnetization predicted by the Curie-Weiss theory for a ferromagnetic solid with $J = S = \frac{1}{2}$, at low temperatures.

Solution

For $J = S = \frac{1}{2}$, Eq.(44.30) reduces to

$$M = M_0 B_{\frac{1}{2}}\left(\frac{y}{2}\right) = N\mu_B \tanh\left(\frac{y}{2}\right)$$

where use has been made of Eq.(44.28) for M_0. We assume that $B_0 = 0$ in the ferromagnetic state, such that, according to Eq.(44.29), one obtains $y = \gamma\mu_B M/k_B T$. At low temperatures, where $M/M_0 \to 1$ as shown in Figure 44.4, y can be approximated by

$$y \approx \frac{\gamma\mu_B M_0}{k_B T} = \frac{\gamma N\mu_B^2}{k_B T} = \frac{2T_C}{T}$$

where T_C is given by Eq.(44.32), and it is large, so that one finally obtains

$$M = N\mu_B\left(1 - 2e^{-y}\right) = N\mu_B\left(1 - 2e^{-2T_C/T}\right)$$

44.5. Find an expression for the parallel susceptibility $\chi_\parallel$ for an antiferromagnetic solid where $S = \frac{1}{2}$.

Solution

The parallel susceptibility is given by Eq.(44.44), which reads

$$\chi_\parallel = 2\mu_0 \frac{dM_1}{dB_0} = 2\mu_0 N\mu_B g_s SB'_S(Sy_1)\frac{S\mu_B g_s}{k_B T}\left(1 - \frac{|\gamma|}{2\mu_0}\chi_\parallel\right)$$

where M_0 was substituted by Eq.(44.28), and Sy_1 has been defined as in Eq.(44.36). This yields

$$\frac{1}{\chi_\parallel} = \frac{|\gamma|}{2\mu_0} + \frac{k_B T}{2\mu_0 N\mu_B^2 g_s^2 S^2 B'_S(Sy_1)}$$

Close to $T = T_N$, the slope of the Brillouin function is $B'_S(Sy_1) = (S+1)/3S$, and, according to Eqs.(44.39) and (44.45), one obtains $1/\chi_\parallel \to |\gamma|/\mu_0 = 1/\chi_\perp$. At very low temperatures, where $B'_S(Sy_1) \to 0$ ($y_1 \gg 1$), $1/\chi_\parallel \to \infty$, and hence, $\chi_\parallel \to 0$. For $S = \frac{1}{2}$, we have $g_s S = 1$ and also

$$B_{\frac{1}{2}}\left(\frac{y_1}{2}\right) = \tanh\left(\frac{y_1}{2}\right) \quad \text{and} \quad B'_{\frac{1}{2}}\left(\frac{y_1}{2}\right) = \text{sech}^2\left(\frac{y_1}{2}\right)$$

such that the required expression is

$$\frac{1}{\chi_\parallel} = \frac{|\gamma|}{2\mu_0} + \frac{k_B T}{2\mu_0 N \mu_B^2} \cosh^2\left(\frac{y_1}{2}\right)$$

44.6. Calculate the diamagnetic susceptibility for a mole of He atoms in the ground state which is described by the wave-function

$$\psi(r_1, r_2) = \frac{Z^3}{\pi a_0^3} e^{-Z(r_1 + r_2)/a_0}$$

44.7. Show that the Brillouin function $B_J(Jx)$ reduces to the Langevin function $L(Jx)$ in the classical limit of $J \to \infty$, whereas, for $J = \frac{1}{2}$, it reduces to $\tanh(Jx)$.

44.8. Show that the susceptibility of a paramagnetic solid can be determined by measuring the force exerted on a specimen of volume V by a field gradient applied along a given direction.

44.9. Use the low-temperature magnetization derived in Problem 44.4 to find an expression for the heat capacity C_V of a ferromagnetic solid where $J = S = \frac{1}{2}$.

44.10. Find an expression for the maximum energy product $(BH)_{max}$ of an ideal ferromagnetic solid that is a measure of its capacity to retain the magnetization M.

45. SUPERCONDUCTIVITY

45.1. THE SUPERCONDUCTING STATE
45.2. COOPER PAIRS

45.1. THE SUPERCONDUCTING STATE

The basic property of superconductivity is the existence of a persistent current or a zero electrical resistance on cooling the sample below a critical temperature T_c. Superconductivity is exhibited by elements, alloys and oxides with a range of compositions and structures. The sharp transition between the normal and superconducting state is illustrated in Figure 45.1 for a metal, where the expected normal behaviour below T_c is represented by the dashed line.

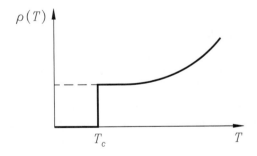

Figure 45.1. Temperature dependence of resistivity $\rho = 1/\sigma$ for a superconducting metal

The condition of perfect conductivity can be described in terms of a finite current density $\vec{j}$ which flows indefinitely in the superconducting state, such that Ohm's law (12.10) gives

$$\vec{E} = (1/\sigma)\vec{j} = \rho\vec{j} = 0 \qquad (45.1)$$

Substituting this result, Faraday's law (13.23) shows that, inside a perfect conductor, $\vec{B}$

cannot change with time

$$\frac{\partial \vec{B}}{\partial t} = -\nabla \times \vec{E} = 0 \quad \text{or} \quad \vec{B} = \text{const} \qquad (45.2)$$

Inside an electrical conductor which is cooled below T_c in an external magnetic field $\vec{B}_0$ and then becomes a perfect conductor, one would expect the magnetic flux through the sample to be maintained even after switching off the external field, due to induced persistent currents. However, it was found experimentally that, if a superconductor is cooled below T_c in an applied magnetic field, the magnetic flux is expelled rather than frozen, as in Figure 45.2. In other words, in the superconducting state it is always required that

$$\vec{B} = 0 \qquad (45.3)$$

This result, which is not predicted by Eq.(45.2), is called the *Meissner effect*, and can be described in terms of the *perfect diamagnetism* defined by Eq.(44.7).

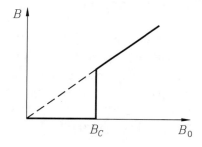

Figure 45.2. Behaviour of a superconductor and of a perfect conductor (dashed line) which are cooled below T_c in an external magnetic field B_0

The two properties of perfect conductivity and perfect diamagnetism for the superconducting state represent the phenomenological basis of the *London theory* or the *two-fluid model* of superconductivity. It assumes that a superconductor contains two types of electrons, superconducting and normal, with densities n_s and n_n, and velocities v_s and v_n, respectively. The normal electrons obey the familiar equations (12.7) and (12.10) which give the current density as

$$\vec{j}_n = -n_n e \vec{v}_n = \sigma_n \vec{E} \qquad (45.4)$$

while the superconducting electrons are considered subjected to no scattering by the lattice, so that the damping force can be dropped from their equation of motion (12.1), and this reduces to

$$m_e \frac{d\vec{v}_s}{dt} = -e\vec{E} \qquad (45.5)$$

If the current density of superconducting electrons takes a form similar to that given by Eq.(45.4), that is,

$$\vec{j}_s = -n_s e \vec{v}_s \qquad (45.6)$$

by combining Eqs.(45.5) and (45.6) we obtain the *first London equation* for a perfect conductor as

$$\frac{d\vec{j}_s}{dt} = \frac{n_s e^2}{m_e} \vec{E} \qquad (45.7)$$

If we take the *curl* of this equation, we have

$$\nabla \times \frac{d\vec{j}_s}{dt} = \frac{n_s e^2}{m_e} \nabla \times \vec{E} = -\frac{n_s e^2}{m_e} \frac{\partial \vec{B}}{\partial t} \quad \text{or} \quad \frac{\partial}{\partial t}\left[\nabla \times \vec{j}_s + \frac{n_s e^2}{m_e} \vec{B}\right] = 0 \qquad (45.8)$$

where use has been made of Eq.(13.23). Equation (45.8) is consistent with a perfect conductor, where the magnetic flux through the sample remains constant, as derived earlier from Eq.(45.2). Integrating this equation, the result can be made consistent with the Meissner effect if the constant of integration is set to zero, which yields

$$\nabla \times \vec{j}_s = -\frac{n_s e^2}{m_e} \vec{B} \qquad (45.9)$$

This is known as the *second London equation* for a perfect diamagnet. Under the assumption that the contribution of normal electrons can be neglected at low temperatures, where $\vec{j} = \vec{j}_s$, from Eq.(45.9) and Ampère's circuital law (12.27), we have

$$\nabla \times (\nabla \times \vec{B}) = \mu_0 (\nabla \times \vec{j}_s) = -\frac{\mu_0 n_s e^2}{m_e} \vec{B} = -\frac{1}{\lambda_L^2} \vec{B}$$

$$\nabla \times (\nabla \times \vec{j}_s) = -\frac{n_s e^2}{m_e}(\nabla \times \vec{B}) = -\frac{\mu_0 n_s e^2}{m_e} \vec{j}_s = -\frac{1}{\lambda_L^2} \vec{j}_s \qquad (45.10)$$

where λ_L is called the *London penetration depth*, defined as

$$\lambda_L = \left(\frac{m_e}{\mu_0 n_s e^2}\right)^{1/2} \qquad (45.11)$$

The two equations (45.10) can be rewritten as

$$\nabla^2 \vec{B} = \frac{1}{\lambda_L^2} \vec{B} \quad , \quad \nabla^2 \vec{j}_s = \frac{1}{\lambda_L^2} \vec{j}_s \qquad (45.12)$$

and this gives a correct description of the Meissner effect. Figure 45.3 represents the penetration of an external magnetic field $\vec{B}_0$ into a superconductor. To the right of the plane $x = 0$, in an external magnetic field $\vec{B}_0 = B_0 \vec{e}_z$, Eqs.(45.12) reduce to

$$\frac{\partial^2 B(x)}{\partial x^2} = \frac{1}{\lambda_L^2} B(x) \quad , \quad \frac{\partial^2 j_s(x)}{\partial x^2} = \frac{1}{\lambda_L^2} j_s(x) \tag{45.13}$$

where $\vec{j}_s = -(\partial B / \partial x) \vec{e}_y$, as required by Ampère's circuital law (12.27). Equations (45.13) have the solutions

$$B(x) = B_0 e^{-x/\lambda_L} \quad , \quad j_s(x) = \left(\frac{\partial B}{\partial x}\right)_{x=0} e^{-x/\lambda_L} \tag{45.14}$$

where $B(0) = B_0$ and $j_s(0) = (\partial B / \partial x)_{x=0}$ is the surface superconducting current density which screens the sample against external fields. The exponential decay $B(x)$ with distance into the sample is the Meissner effect again.

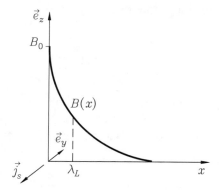

Figure 45.3. Penetration of an external magnetic field into a superconductor

It is found that superconductivity is destroyed by a sufficiently high external magnetic field or a current sufficient to generate a critical field B_c at the surface. The temperature dependence of B_c follows a parabolic relation

$$B_c(T) = B_c(0)\left[1 - \left(\frac{T}{T_c}\right)^2\right] \tag{45.15}$$

which leads to the phase diagram represented in Figure 45.4.

Since the transition from superconducting to normal state is found to be reversible, it can be discussed in terms of the equilibrium condition (10.16) for a one component system. As a result of the Meissner effect, the external field energy density

$B_0^2/2\mu_0$, Eq.(13.34), is absent over the volume of the sample since $B=0$ in the superconducting state. Hence, the expulsion of the magnetic flux by diamagnetic magnetization requires an energy of $B_0^2/2\mu_0$, which can be covered by the difference in the Gibbs function between the normal and superconducting states.

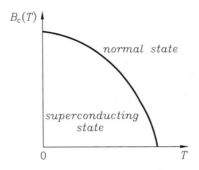

Figure 45.4. Phase diagram of a superconductor in applied magnetic field

The situation is illustrated in Figure 45.5 (a), where the plot of $G_s(B_0,T)$ for the superconducting state, at a given temperature $T<T_c$, follows the equation

$$G_s(B_0,T)=G_s(T)+\frac{V}{2\mu_0}B_0^2 \qquad (45.16)$$

while $G_n(B_0,T)=G_n(T)$ for the normal state is independent of B_0. For a critical value $B_0=B_c(T)$, the transition occurs if the equilibrium condition (10.16) is satisfied. Hence

$$G_s(B_c,T)=G_s(T)+\frac{V}{2\mu_0}B_c^2(T)=G_n(T) \qquad (45.17)$$

or

$$G_s(T)=G_n(T)-\frac{V}{2\mu_0}B_c^2(T) \qquad (45.18)$$

which is valid in the absence of an external magnetic field and implies that $B_c(T)$ is a measure of the stability of the superconducting state. Using Eq.(8.27) for the entropy, we obtain from the Clapeyron equation (10.18) that there is no latent heat λ in the transition, in zero external magnetic field, that is,

$$\lambda=T(S_s-S_n)_{T=T_c}=-T\frac{d}{dT}(G_s-G_n)_{T=T_c}=\frac{VT}{\mu_0}\left(B_c\frac{dB_c}{dT}\right)_{T=T_c}=0 \qquad (45.19)$$

since $B_c=0$ at $T=T_c$. However, there is a discontinuity in the heat capacity C_V that can be calculated by combining Eqs.(45.19) and (8.22) to obtain

$$C_{Vs} - C_{Vn} = T\frac{d}{dT}(S_s - S_n) = T\frac{d}{dT}\left(\frac{VB_c}{\mu_0}\frac{dB_c}{dT}\right)_{T=T_c}$$

$$= \frac{VT}{\mu_0}\left[\left(\frac{dB_c}{dT}\right)^2 + B_c\frac{d^2B_c}{dT^2}\right]_{T=T_c} = \frac{VT_c}{\mu_0}\left(\frac{dB_c}{dT}\right)^2 \qquad (45.20)$$

The change in the heat capacity at the superconducting transition, illustrated in Figure 45.5 (b), is characteristic for a second-order phase transition, which is associated with a discontinuity in the internal order but does not affect the lattice structure.

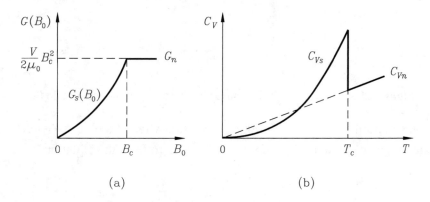

Figure 45.5. Continuity of the Gibbs function (a) and discontinuity of the heat capacity (b) at the superconducting transition

45.2. Cooper Pairs

The origin of superconductivity has been assigned to a mechanism involving interaction among electrons which leads to a new phase of the electron gas, the phase of paired electrons. It is postulated in the existence of *Cooper pairs*, which are sets of two electrons formed by an effective attractive interaction due to the presence of the positive ion lattice. In a rather qualitative picture, the lattice deformation produced by the motion of an electron is associated with an increased density of positive charge due to the ion cores, which can attract a second electron. In other words, the electron-lattice deformation coupling produces a weak attraction between pairs of electrons. Since the positive ions are more massive than the conduction electrons and bound in the lattice structure, we can think of an ion as remaining displaced for a time, while the electron which has attracted it moves out of the area. As a result, the two electrons related by the lattice deformation have a separation of the order of many lattice spacings. The large size of the Cooper pairs is consistent with the existence of an effective attraction between two electrons in spite of the fact that their Coulomb repulsion is completely screened out over such distances.

We have seen that a lattice ion displacement can be represented by a superposition of normal modes of vibration, Eq.(38.31). In quantum mechanical terms, a lattice vibrational state was defined by the number n_k of phonons of energy $\hbar\omega_k$ in each normal mode. We can think of phonons to be emitted by the coupling between the electron and the lattice deformation and then absorbed, to ensure energy conservation, after a time interval $\tau_k = 1/2\omega_k$, which is their lifetime estimated from the uncertainty relation (28.48). In other words, Cooper pairs are formed because of the exchange of phonons between electrons.

Consider two electrons in a free-electron system which are in states $\vec{k}_1$ and $\vec{k}_2$, with energies $\hbar^2 \vec{k}_1^2 / 2m_e$, $\hbar^2 \vec{k}_2^2 / 2m_e$ and total momentum $\vec{K} = \vec{k}_1 + \vec{k}_2$. Since electron pairs of opposite spin have a lower energy, Cooper pairs will be formed by electrons of opposite spin, and hence, their spatial wave function must be symmetric under the permutation of electrons and can be written as

$$\Psi_0(\vec{r}_1,\vec{r}_2) = A e^{i(\vec{k}_1 \cdot \vec{r}_1 + \vec{k}_2 \cdot \vec{r}_2)} = A e^{i(\vec{K} \cdot \vec{R} + \vec{k} \cdot \vec{r})} \tag{45.21}$$

where A is a normalization constant, $\vec{R} = (\vec{r}_1 + \vec{r}_2)/2$ denotes the position of the centre of mass of the pair, $\vec{r} = \vec{r}_1 - \vec{r}_2$ denotes the relative position of the two electrons and $\vec{k} = (\vec{k}_1 - \vec{k}_2)/2$. The energy of the pair corresponding to the state (45.21) results as

$$E_0 = \frac{\hbar^2}{2m_e}\left(\vec{k}_1^2 + \vec{k}_2^2\right) = \frac{\hbar^2}{m_e}\left(\frac{\vec{K}^2}{4} + \vec{k}^2\right) \tag{45.22}$$

An electron pair *at rest* is described by taking $\vec{K} = 0$, and this implies electron pairs with equal and opposite wave vectors $\vec{k} = \vec{k}_1 = -\vec{k}_2$ and energy $E_0 = \hbar^2 \vec{k}^2 / m_e$, which have the state functions

$$\Psi(\vec{k},-\vec{k}) = \Psi_{\vec{k}}(\vec{r}_1 - \vec{r}_2) = \Psi_{\vec{k}}(\vec{r}) = A e^{i\vec{k}\cdot\vec{r}} \tag{45.23}$$

In the presence of a weak attractive interaction $V(\vec{r}_1 - \vec{r}_2) < 0$ between the two electrons, the wave function of a pair must satisfy the energy eigenvalue equation

$$\left[-\frac{\hbar^2}{2m_e}\nabla_1^2 - \frac{\hbar^2}{2m_e}\nabla_2^2 + V(\vec{r}_1 - \vec{r}_2)\right]\Psi(\vec{r}_1 - \vec{r}_2) = E\,\Psi(\vec{r}_1 - \vec{r}_2) \tag{45.24}$$

which reduces to

$$\left[(E_0 - E) + V(\vec{r})\right]\Psi(\vec{r}) = 0 \tag{45.25}$$

where $\vec{r} = \vec{r}_1 - \vec{r}_2$, and E_0 is the energy of the pair in the absence of interaction. A trial solution to Eq.(45.25) is a linear combination of unperturbed pair states (45.23) given by

$$\Psi(\vec{r}_1 - \vec{r}_2) = \sum_{\vec{k}} a_{\vec{k}} A e^{i\vec{k}\cdot\vec{r}} \qquad (45.26)$$

Since the lectron-electron interaction is postulated to be an exchange of phonons, where a phonon can assume a maximum energy $\hbar\omega_D$ limited by the Debye frequency ω_D, Eq.(38.60), the summation must be restricted in Eq.(45.26) to $\vec{k}$-values in the range

$$\varepsilon_F < \frac{\hbar^2 \vec{k}^2}{2m} < \varepsilon_F + \hbar\omega_D \quad \text{or} \quad 2\varepsilon_F < \frac{\hbar^2 \vec{k}^2}{m_e} < 2(\varepsilon_F + \hbar\omega_D) \qquad (45.27)$$

This restriction is consistent with the situation where all the one-electron states with wave vector $\vec{k}$ within the Fermi sphere are filled at $T=0$, and all the states of energy higher than $\varepsilon_F(0)$ are empty. The states which do not comply with the restriction (45.27), where $\varepsilon_F \cong \varepsilon_F(0)$ for low temperatures, are considered not to be directly involved in the superconducting transition. Note that usually $\hbar\omega_D \ll \varepsilon_F$. Substituting Eq.(45.26) into Eq.(45.25), multiplying by $e^{-i\vec{k}'\cdot\vec{r}}$ and integrating over the normalization volume V yields

$$(E_0 - E)a_{\vec{k}} + \sum_{\vec{k}'} a_{\vec{k}'} \int V(\vec{r}) e^{i(\vec{k}-\vec{k}')\cdot\vec{r}} dV = 0 \qquad (45.28)$$

The electron-electron interaction $V(\vec{r})$ scatters the electron pair from a state $\Psi(\vec{k},-\vec{k})$ to a state $\Psi(\vec{k}',-\vec{k}')$. The scattering matrix element can be replaced by an average value $-V_0$ which is independent of $\vec{k}$, for $\vec{k}$-states within the range (45.27). This is the so-called *weak-interaction approximation*, which allows us to rewrite Eq.(45.28) as

$$(E_0 - E)a_{\vec{k}} = V_0 \sum_{\vec{k}'} a_{\vec{k}'} = C \qquad (45.29)$$

where C is independent of $\vec{k}$. On summing Eq.(45.29) over $\vec{k}$ we obtain

$$\sum_{\vec{k}} a_{\vec{k}} = C \sum_{\vec{k}} \frac{1}{E_0 - E} \quad \text{or} \quad \frac{1}{V_0} = \sum_{\vec{k}} \frac{1}{E_0 - E} \qquad (45.30)$$

The sum over $\vec{k}$, which is extended over all the electron pair states compatible with the restriction (45.27), can be replaced in Eq.(45.30) by an integral

$$\frac{1}{V_0} = \int_{2\varepsilon_F}^{2(\varepsilon_F + \hbar\omega_D)} \frac{f(E_0) dE_0}{E_0 - E} \qquad (45.31)$$

where $f(E_0)$ is the density of pair states. Since the range of interaction is small as compared to ε_F, one may replace $f(E_0)$ by $f(2\varepsilon_F)$, and this can be taken out of the integral, so that Eq.(45.31) becomes

$$\frac{1}{V_0 f(2\varepsilon_F)} = \int_{2\varepsilon_F}^{2(\varepsilon_F + \hbar\omega_D)} \frac{dE_0}{E_0 - E} = \ln\frac{2\varepsilon_F + 2\hbar\omega_D - E}{2\varepsilon_F - E} = \ln\left(1 + \frac{2\hbar\omega_D}{2\varepsilon_F - E}\right) \quad (45.32)$$

It is common practice to introduce the binding energy of the pair, at $T = 0$, defined by

$$2\Delta = 2\varepsilon_F - E \quad (45.33)$$

in terms of the density of one-electron states at the Fermi level, $N(\varepsilon_F)$, which is the same as the density of pair states at $2\varepsilon_F$, $f(2\varepsilon_F)$. It follows from Eq.(45.32) that

$$2\Delta = \frac{2\hbar\omega_D}{e^{1/V_0 N(\varepsilon_F)} - 1} \quad (45.34)$$

and this shows that the pair state $\Psi(\vec{r}_1 - \vec{r}_2)$, Eq.(45.26), corresponds to a pair of electrons with an energy E which is lower than $2\varepsilon_F$ by 2Δ. In other words, pairs of electrons with antiparallel spins form bound states near the Fermi energy. In the weak-interaction approximation we have $V_0 N(\varepsilon_F) \ll 1$ so that

$$\Delta \cong \hbar\omega_D e^{-1/V_0 N(\varepsilon_F)} \quad (45.35)$$

By taking $\Delta \sim k_B T_c$ and $\hbar\omega_D = k_B \theta_D$, as given by Eq.(38.63), we obtain an estimate of the critical temperature in terms of the Debye temperature of a superconducting solid as

$$T_c \cong \theta_D e^{-1/V_0 N(\varepsilon_F)} \quad (45.36)$$

By inserting the limiting value of the phase velocity v_0 of lattice waves for a linear lattice, Eq.(38.35), into Eq.(38.63) it can be shown that the Debye temperature is inversely proportional to the square root of the atomic mass M. Hence, Eq.(45.36) yields

$$T_c \sim M^{-1/2} \quad (45.37)$$

and this is called the *isotope effect* indicating at macroscopic scale that the lattice vibrational motion plays an essential role in the interaction mechanism of superconductivity. The pairing interaction produces a gap in the density of one-electron states $N(\varepsilon)$, as illustrated in Figure 45.6 (a), which should be compared with the similar plot given for free electrons at absolute zero in Figure 40.1. Since the energy gap 2Δ corresponds to the binding energy of a Cooper pair, an excitation across the gap that breaks up a pair to produce two particles is often interpreted to be similar to the production of an electron and a hole in an intrinsic semiconductor. However, the energy gap 2Δ is a function of temperature, as shown in Figure 45.6 (b), with $2\Delta(T)=0$ at $T=T_c$, and hence, it is different from the almost temperature-independent band gap ε_g of an intrinsic semiconductor.

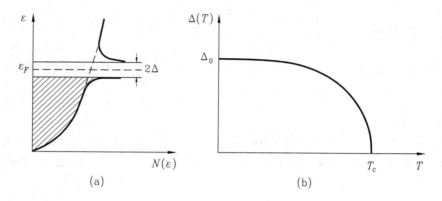

Figure 45.6. Density of one-electron states in a superconductor at absolute zero (a) and the temperature dependence of the superconducting gap parameter Δ (b)

The concept of a gap 2Δ in the energy spectrum allows us to understand how a superconductor can sustain a persistent current. When a current flows, the Cooper pair as a whole has a total momentum $\vec{K} \neq 0$, so that it is in a state $\Psi(\vec{k}+\vec{K}/2,-\vec{k}+\vec{K}/2)$, given by Eq.(45.21), and this is entirely equivalent to the superconducting rest state $\Psi(\vec{k},-\vec{k})$, Eq.(45.23), as the additional phase factor $e^{i\vec{K}\cdot\vec{R}}$ does not affect the measurable probability density. In other words, $\vec{K} \neq 0$ corresponds to an equivalent superconducting state for electrons drifting with a velocity $\vec{v}=\hbar\vec{K}/2m_e$. If n_s is the concentration of superconducting electrons, the superconducting current density is given by Eq.(45.6) as

$$\vec{j}_s = -\frac{n_s e\hbar}{2m_e}\vec{K} \qquad (45.38)$$

The scattering of individual electrons, which gives rise to the electrical resistance in a normal conductor, can only occur if the increase in the wave vector $\vec{k}$ by $\vec{K}/2$, for each electron of a Cooper pair, is higher than the energy 2Δ required for excitation across the gap, so that, for bound pairs, we have

$$\frac{\hbar^2(\vec{k}+\vec{K}/2)^2}{m_e} - \frac{\hbar^2\vec{k}^2}{m_e} = \frac{\hbar^2}{m_e}\vec{k}\cdot\vec{K} + \frac{\hbar^2\vec{K}^2}{4m_e} \leq 2\Delta$$

Since the one-electron states involved in superconductivity are situated in the vicinity of the Fermi energy, we can take $k \cong k_F$, and hence we can neglect the second-order term in $\vec{K}$, because $|\vec{K}| \ll k_F$, to obtain an upper value for K, which is compatible with the existence of the bound pair states, in the form

$$K \leq \frac{2m_e \Delta}{\hbar^2 k_F}$$

Substituting it into Eq.(45.38), gives an estimate for the critical current density as

$$j_c = -\frac{n_s e \Delta}{\hbar k_F} \qquad (45.39)$$

A persistent superconducting current will be decreased by scattering only if it exceeds the critical value given by Eq.(45.39).

EXAMPLE 45.1. High-temperature superconductivity

Conventional superconductors with T_c below 24 K, ranging from ordinary metals and alloys such as Nb ($T_c = 9.25$ K) and Nb$_3$Ge ($T_c = 23.2$ K) to oxides like BaPb$_{1-x}$Bi$_x$O$_3$ ($T_c = 13$ K) have been known for many years. To take advantage of their superconducting behaviour these conventional materials need to be cooled to the temperature of liquid helium, 4.2 K. A T_c of about 24 K appears to be an upper limit as far as metals and alloys are concerned. However, it was discovered that oxide superconductors of the Ba-La-Cu-O system show a transition temperature of about 80 K. Superconducting oxides with chemical formula MBa$_2$Cu$_3$O$_{7-x}$ where the metal M is Y or most of the rare-earth elements, have been reported to exhibit superconductivity above liquid nitrogen temperatures, with a T_c above 90 K, allowing liquid helium to be replaced by liquid nitrogen, which is a cheaper and more effective coolant. Almost all the macroscopic superconducting properties of the new copper-oxide-based materials are similar to those of conventional superconductors. They exhibit the classical characteristics of superconductivity, namely, the zero electrical resistance and the Meissner effect. However, the mechanism of superconductivity in these materials still needs to be explained.

The predicted *critical temperature* (45.36) can be influenced by varying three microscopic parameters V_0, $N(\varepsilon_F)$ and θ_D. An increase in either the electron-phonon coupling strength V_0 or the density of electronic states $N(\varepsilon_F)$ at the Fermi level is expected to produce a higher T_c, but not in excess of 40 K. It has been suggested that the electron-phonon interaction must be augmented or replaced by an interaction of particles with lower mass than the lattice ions, which could explain the observed high transition temperatures in view of a much higher ω_D, and hence, θ_D.

There is *spatial anisotropy* in high$-T_c$ compounds, as illustrated in Figure 45.7 for the temperature dependence of their electrical resistivity $\rho = 1/\sigma$ in the normal phase. The anisotropy is produced by the layered crystal structure sketched in Figure 45.7 (b), where the *ab*-planes of Cu - O, to which the superconductivity and charge transport are confined, are separated from each other by planes of other oxides and metals. It is seen that in the *ab*-plane the resistivity ρ_{ab} varies linearly with temperature, while along the *c*-axis ρ_c varies as $1/T$.

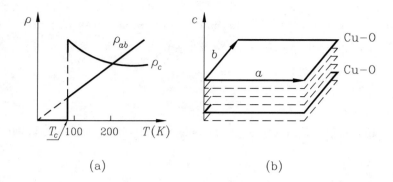

Figure 45.7. Temperature dependence of the resistivity ρ_{ab}, parallel to the Cu - O planes, and ρ_c along the c-axis (a) and the corresponding layered crystal structure (b)

As a result of anisotropy effects, the highest critical current densities at 77 K, $j_c \sim 10^6$ A/cm^2, required for practical devices, have only been reported for single-crystal thin films. This places a significant restriction on the possible range of applications for oxide superconductors.

FURTHER READING
1. M. Tinkham - INTRODUCTION TO SUPERCONDUCTIVITY, Dover Publications, New York, 2004.
2. V. L. Ginzburg, E. A. Andryushin - SUPERCONDUCTIVITY, World Scientific, Singapore, 2004.
3. J. B. Ketterson, S. Song - SUPERCONDUCTIVITY, Cambridge University Press, Cambridge, 1999.

PROBLEMS

45.1. Find an expression for the magnetization $M(x)$ under external magnetic field $\vec{B}_0 = B_0 \vec{e}_z$ for a superconducting slab of thickness $a \ll \lambda_L$, where λ_L is the London penetration depth.

Solution

A trial solution to Eq.(45.13), which describes the penetration of the external field into the superconducting slab is

$$B(x) = \alpha e^{-x/\lambda_L} + \beta e^{x/\lambda_L}$$

with the boundary conditions given by $B(-a/2) = B(a/2) \approx B_0$, since $a \ll \lambda_L$. This yields

$$\alpha = \beta = \frac{B_0}{2\cosh(a/2\lambda_L)} \quad \text{or} \quad B(x) = B_0 \frac{\cosh(x/\lambda_L)}{\cosh(a/2\lambda_L)}$$

The magnetization of the superconducting plate results in a decrease of the applied magnetic induction, according to $B(x) = B_0 + \mu_0 M(x)$, such that

$$\mu_0 M(x) = B(x) - B_0 = B_0 \left[\frac{\cosh(x/\lambda_L)}{\cosh(a/2\lambda_L)} - 1 \right]$$

For $x/\lambda_L \ll 1$, we may approximate $\cosh(x/\lambda_L) \approx 1 + x^2/2\lambda_L^2$, and this yields the required magnetization as

$$M(x) \approx \frac{B_0}{\mu_0} \frac{4x^2 - a^2}{8\lambda_L^2 + a^2} \approx \frac{B_0}{8\mu_0 \lambda_L^2}(4x^2 - a^2)$$

45.2. Show that a thermally isolated superconducting specimen at a temperature $T = \alpha T_c \, (\alpha < 1)$ will cool if an external field $B > B_c(T)$ is applied, and find the corresponding drop in temperature.

Solution

The entropy of the thermally isolated specimen at $T = \alpha T_c$ should remain constant when superconductivity is suppressed. According to Eq.(45.19), the transition implies an increase in entropy from the superconducting to the normal state, given by

$$(S_n - S_s)_{T=\alpha T_c} = -\frac{V}{\mu_0}\left(B_c \frac{dB_c}{dT}\right)_{T=\alpha T_c} = 2\alpha(1-\alpha^2)\frac{VB_c^2(0)}{\mu_0 T_c}$$

where use has been made of Eq.(45.15) for $B_c(T)$. As a result, the temperature must drop in the normal state from αT_c to a final temperature T_f, such as to yield a change in entropy of the form

$$\Delta S_n = \int_{\alpha T_c}^{T_f} \frac{C_V}{T} dT = \int_{\alpha T_c}^{T_f} \frac{\gamma T}{T} dT = \gamma(T_f - \alpha T_c)$$

where the heat capacity C_V of the normal metal, in the low temperature range, was approximated by the electronic contribution given by Eq.(40.28). Since $S_n(\alpha T_c) - S_s(\alpha T_c) + \Delta S_n = 0$, it follows that

$$\Delta T = T_f - \alpha T_c = -2\alpha(1-\alpha^2)\frac{VB_c^2(0)}{\mu_0 \gamma T_c}$$

and the specimen will cool, as expected.

45.3. Find the maximum value of the magnetic induction on the surface of a spherical superconducting specimen of radius R under external magnetic field $\vec{B}_0 = B_0 \vec{e}_z$.

Solution

If the magnetic induction is derived from a scalar magnetostatic potential ϕ, as given by Eq.(13.3), then ϕ satisfies the Laplace equation

$$\nabla \cdot \vec{B} = -\nabla^2 \phi = 0$$

which has been solved, in spherical coordinates, for the electrostatic potential $V(r)$, Eq.(11.19). Hence, we have

$$\phi(r) = \alpha r \cos\theta + \beta \frac{\cos\theta}{r^2}$$

where α and β are determined by suitable boundary conditions. If the superconducting sphere is placed at the origin of a Cartesian coordinate system, with negligible effect on the uniform field at large distances, the field components are obtained using the asymptotic form $\phi(\infty) = -B_0 r \cos\theta$ in the form

$$B_r(\infty) = -\frac{\partial \phi}{\partial r} = B_0 \cos\theta, \quad B_\theta(\infty) = -\frac{1}{r}\frac{\partial \phi}{\partial \theta} = -B_0 \sin\theta$$

and this means that $\phi(\infty) = \alpha r \cos\theta = -B_0 r \cos\theta$, or $\alpha = -B_0$. Thus, one obtains

$$B_r(r) = -\frac{\partial \phi}{\partial r} = B_0 \cos\theta + \frac{2\beta}{r^3}\cos\theta \quad , \quad B_\theta(r) = -\frac{1}{r}\frac{\partial \phi}{\partial \theta} = -B_0 \sin\theta + \frac{\beta}{r^3}\sin\theta$$

Due to flux exclusion from the interior of the superconducting sphere, for $r = R$, along the z-axis ($\theta = 0$) we have

$$B(R) = B_r(R) = B_0 + \frac{2\beta}{R^3} = 0 \quad \text{or} \quad \beta = -\frac{B_0 R^3}{2}$$

and it follows that

$$B_r(r) = B_0 \cos\theta \left(1 - \frac{R^3}{r^3}\right), \quad B_\theta(r) = -B_0 \sin\theta \left(1 + \frac{R^3}{2r^3}\right)$$

The magnetic induction on the surface of the sphere has the components

$$B_r(R) = 0, \quad B_\theta(R) = -\frac{3}{2} B_0 \sin\theta$$

and hence, a maximum value of $3B_0/2$ is found in the equatorial plane.

45.4. Find an expression for the current density produced by a Cooper pair, in the presence of an applied magnetic field $\vec{B} = \nabla \times \vec{A}$ in a homogeneous superconductor, and show that the spatial dependence of the wave function can be expressed through its phase φ only and not through its module.

Solution

If a Cooper pair, which is a carrier of charge $-2e$ and mass $m = 2m_e$, is described in terms of a wave function $\Psi(\vec{r})$, one obtains a statistical distribution of its position with a probability density $P_s = |\Psi(\vec{r})|^2$, and hence, the superconducting charge density ρ_s is $\rho_s = -2e|\Psi(\vec{r})|^2$. From the continuity equation for probability, Eq.(30.23), which reads

$$\frac{\partial P_s}{\partial t} + \nabla \cdot \vec{j} = 0$$

where $\vec{j}$ is the probability current density, it follows that the Cooper pair current density $\vec{j}_s$ should be taken as

$$\vec{j}_s = -2e\vec{j} = \frac{ie\hbar}{2m_e}(\Psi^*\nabla\Psi - \Psi\nabla\Psi^*)$$

In the presence of an applied magnetic field $\vec{B} = \nabla \times \vec{A}$, we make the usual substitution

$$\vec{p} \to \vec{p} + 2e\vec{A} \quad \text{or} \quad \nabla \to \nabla + \frac{2ie}{\hbar}\vec{A}$$

which yields the current density in the form

$$\vec{j}_s = \frac{ie\hbar}{2m_e}(\Psi^*\nabla\Psi - \Psi\nabla\Psi^*) - \frac{2e^2}{m_e}\vec{A}|\Psi|^2$$

If we write $\Psi(\vec{r}) = |\Psi|e^{i\varphi}$, because $|\Psi|^2$ is proportional to the density of superconducting electrons which is a constant at any given temperature below the critical temperature in a homogeneous superconductor, it is reasonable to assume that the spatial dependence of the wave function is expressed through its phase φ only, and not through its module $|\Psi|$. Thus, the current density follows as

$$\vec{j}_s = -\frac{e\hbar}{m_e}\nabla\varphi|\Psi|^2 - \frac{2e^2}{m_e}\vec{A}|\Psi|^2$$

Since $\nabla \times \nabla\varphi \equiv 0$, taking the *curl* of this expression leads to the second London equation, Eq.(45.9), provided we let

$$n_s = 2|\Psi|^2 \quad \text{or} \quad \Psi(\vec{r}) = \sqrt{\tfrac{1}{2}n_s}\, e^{i\varphi}$$

and hence, $|\Psi|$ should be a constant, as required. Note that this choice is the same as that made for ρ_s.

45.5. Find the current density j_s produced by a Cooper pair tunneling through a Josephson junction, which consists of two homogeneous superconductors with Cooper pair functions $|\Psi_1|e^{i\varphi_1}$ and $|\Psi_2|e^{i\varphi_2}$ separated by an oxide layer of thickness a.

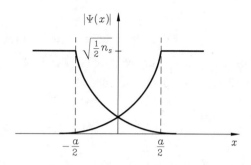

Solution

It is apparent that the wave functions overlap each other, and hence, in the tight-binding approximation, the total wave function through the oxide layer is the sum of those for the separated superconductors, with amplitudes corresponding to the exponential decay of the order parameter

$$\Psi(x) = C\left[e^{-(x+a/2)/\xi}e^{i\varphi_1} + e^{(x-a/2)/\xi}e^{i\varphi_2}\right]$$

where ξ stands for the so-called coherence length, which is a measure of the spread of the wave function associated with a Cooper pair, whose localization is limited to distances of the order of ξ. At the boundaries, this function reduces to

$$\Psi(-a/2) = C\left(e^{i\varphi_1} + e^{-a/\xi}e^{i\varphi_2}\right) \approx |\Psi_1|e^{i\varphi_1}, \quad \Psi(a/2) = C\left(e^{-a/\xi}e^{i\varphi_1} + e^{i\varphi_2}\right) \approx |\Psi_2|e^{i\varphi_2}$$

as expected. Substituting $\Psi(x)$ into the current density j_s (see Problem 45.4) gives

$$j_s = -\frac{\hbar k e n_s}{m_e \xi}e^{-a/\xi}\sin\Delta\varphi = j_0 \sin\Delta\varphi$$

where $\Delta\varphi = \varphi_2 - \varphi_1$. It follows that there will be an electron flow into the second superconductor, and hence, current will flow in the opposite direction. This is called the Josephson current and corresponds to a Cooper pair tunnelling through the oxide layer. The current density j_0 is a characteristic parameter for a given junction, which acts as a *dc* current generator in what is known as the *dc* Josephson effect.

If we further assume that a constant voltage difference $V = V_1 - V_2$ is applied to the junction, each electron potential energy shift is associated with a phase factor $e^{-i\varepsilon_j t/\hbar} = e^{2ieV_j t/\hbar}$ ($j = 1, 2$), such that the phase difference becomes

$$\varphi_2 - \varphi_1 = \Delta\varphi + \frac{2e(V_2 - V_1)t}{\hbar} = \Delta\varphi - \frac{2eVt}{\hbar}$$

Substituting into j_s one obtains

$$j_s = j_0 \sin\left(\Delta\varphi - \frac{2eVt}{\hbar}\right)$$

If V is a constant, the phase advances through the junction at a constant rate, and we observe an alternating current with a frequency $2eV/\hbar$ that corresponds to 483.6 MHz for an applied potential of 1 μV. This is known as the *ac* Josephson effect.

45.6. Show that the critical field B_f of a superconducting plate of thickness $a \gg \lambda_L$ is always greater than the critical field B_c of a bulk sample, due to field penetration, and find the B_f / B_c ratio.

45.7. Find an expression for the B_f / B_c ratio for a superconducting slab of thickness $a \ll \lambda_L$.

45.8. Find the magnetic flux contained in a superconducting ring under applied magnetic field $\vec{B}$ normal to its plane.

45.9. Show that a persistent superconducting current can be described by taking the Cooper pair wave function to be a plane wave $\Psi(\vec{r}) = e^{i\vec{K}\cdot\vec{r}}$.

45.10. If a magnetic field is applied normal to the plane of a symmetrical circuit containing two identical Josephson junctions assembled in parallel, show that the observed current density j_s depends periodically on the magnetic flux Φ enclosed by the loop.

46. Nuclear Structure

46.1. SEMICLASSICAL MODELS OF THE NUCLEUS
46.2. THE SHELL MODEL OF THE NUCLEUS

46.1. Semiclassical Models of the Nucleus

Using a logical extension of our picture of the atom, the atomic nucleus can be treated as a many-particle system which contains almost all the mass of the atom and a charge of $+Ze$. Nuclear masses are measured in terms of the *atomic mass unit, u*, defined in Chapter 1, which is close to the rest mass m_H of a hydrogen atom in its ground state. The *mass number A* of a nucleus is the integer nearest to the ratio between the nuclear mass and the atomic mass unit. Except for hydrogen, A is always greater than the *atomic number Z*. It was suggested by Heisenberg that the nucleus is a system of Z *protons* of mass m_p and charge $+e$, and $N = A - Z$ *neutrons* of mass m_n almost equal to that of a proton (see Chapter 1) and zero charge. Hence, to a first approximation, the mass of the nucleus is given by the sum of masses of the *nucleons* (protons and neutrons), and its charge is equal to the total charge of the protons. A nuclear species, or *nuclide*, is therefore specified by the symbol $^A_Z X$, where X is the chemical symbol, or simply by $^A X$ since, given X, the atomic number Z is known, and we can always find N as $A - Z$. Nuclides of the same Z but different N are called *isotopes*, those of the same N and different Z are known as *isotones* and those of the same mass number A, but different Z and N, are said to be *isobars*.

Specific approximate descriptions, called *models*, have been developed in the case of nuclei, each of them only being appropriate for a limited range of nuclear properties. A general understanding of systematic trends found for the time-independent properties, such as mass, size, charge and energy, can be obtained from the *semiclassical* or *particle models*, which give a description in terms of energy, with no insight about details inside the nucleus.

Since the nucleons are held together by strong attractive forces, it is convenient to introduce the binding energy $B(A, Z)$ of a nucleus that represents the energy released in bringing the nucleons together or, conversely, the work necessary to dissociate the nucleus into separate nucleons. It is common practice to define the nuclear binding energy according to the energy-inertia relation (5.9) in terms of atomic rather than nuclear masses, as

$$B(A,Z) = \left[Zm_H + (A-Z)m_n - M(A,Z) \right] c^2 = \Delta M c^2 \quad (46.1)$$

where $M(A,Z)$ is the atomic mass of the isotope, and ΔM is the so-called mass defect. Since the state of infinite separation of the nucleons is taken to be the zero level of energy, the total energy of the ground state of a nucleus is $-B(A,Z)$. Nuclear energies are conveniently measured in $\text{MeV}\,(1\,\text{MeV} = 10^6\,\text{eV})$, where the electron-volt is the common atomic unit of energy. Equation (5.9) gives the energy equivalent of an atomic mass unit as $1u = 931.44\,\text{MeV}$, so that one can work either with masses or energies as convenient. The binding energy per nucleon

$$f = \frac{B(A,Z)}{A} \quad (46.2)$$

is often called the *binding fraction*, and its plot as a function of the mass number, given in Figure 46.1, can be used for a systematic study of nuclear binding energy.

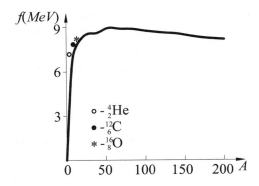

Figure 46.1. Binding fraction as a function of mass number

It is apparent that the binding fraction of most nuclei is about 8 MeV, and it is approximately independent of A, except for the lightest nuclei. This implies that a nucleon in a large nucleus is not bound to more nucleons than in a small one; in other words, there is a saturation of the interaction between nucleons. Hence, the nuclear forces have a range which is of the order of the diameter of one nucleon. Saturation indicates effects which keep nucleons apart from each other, which can be understood if, in addition to mass and charge, we consider the *spin angular momentum* of the nucleons. It is found experimentally that electrons, protons and neutrons have an intrinsic angular momentum $\tfrac{1}{2}\hbar$, so that the Pauli exclusion principle forbids two nucleons of the same kind to occupy states with identical quantum numbers, and this leads to the saturation of nuclear forces. The nuclei are most tightly bound near $A \approx 60 (\text{Fe}, \text{Ni})$, where f is a maximum. In light nuclei, a single nucleon is attracted by a few other nucleons, thus the nucleon separation is large and the stability is reduced. In heavy nuclei the decrease of stability is due to the Coulomb repulsion between protons, which becomes important for large Z and hence large A. It is therefore apparent that energy can be released using nuclei situated at both ends of the curve, either by combining light nuclei into heavier nuclei (nuclear *fusion*) or by breaking heavy nuclei into lighter nuclei (nuclear *fission*).

Liquid Drop Model for the Nucleus

Saturation properties of the nuclear forces are very similar to the properties of intermolecular forces in a liquid. A formula for the binding energy of the nucleus can be derived on the basis of the liquid drop analogy for nuclear matter. The parallel with the liquid drop implies the assumptions that the nuclear force is identical for every nucleon, proton or neutron, and that the nucleus consists of incompressible matter. If the nucleus is regarded as a spherical assembly of A nucleons, its volume must be proportional to A, that is

$$\frac{4\pi}{3} R^3 \sim A \quad \text{or} \quad R = R_0 A^{1/3} \qquad (46.3)$$

where the experimental value of the nuclear unit radius is $R_0 \cong 1.2 \times 10^{-15}$ m. This *liquid drop model* of the nucleus, first proposed by Bohr, allows us to predict the form of the most significant components of the binding energy.

A *volume energy* proportional to the number of nucleons A is expected to be released when nucleons combine together into a nucleus, under the influence of nuclear forces. This is the most obvious contribution to include in estimating $B(A, Z)$, because it corresponds to a constant binding fraction, which is found for most nuclei to be

$$B_V = a_1 A \qquad (46.4)$$

Since the number of pairs of nucleons is $A(A-1)/2$, Eq.(46.4) accounts for saturation of nuclear forces. It is, however, an overestimate of the binding energy because the nucleons near the surface, surrounded by fewer neighbours, are less tightly bound than those inside the volume. A *surface energy* term, proportional to the nuclear drop surface area, must be then subtracted from $B(A, Z)$, that is,

$$B_S = -a_2 A^{2/3} \qquad (46.5)$$

where a_2 can be estimated in terms of the volume interaction constant a_1. Since the average cubic volume per nucleon results from Eq.(46.3) as $4\pi R^3 / 3A = 4\pi R_0^3 / 3 = l^3$, it is convenient to separate a surface layer of thickness $l = R_0 \sqrt[3]{4\pi/3}$ from the inner region of the nuclear liquid drop, as in Figure 46.2 (a). The surface layer contains a number of nucleons given by

$$\frac{4\pi R^2 l}{4\pi R_0^3 / 3} = 3\sqrt[3]{4\pi/3} A^{2/3} \cong 4.8 A^{2/3}$$

for which the nuclear forces are only effective on their inner surface.

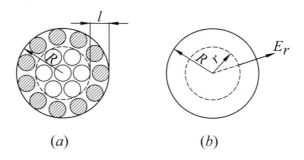

Figure 46.2. The nuclear liquid drop

As a result, the binding energy per nucleon must be reduced by a fraction ranging from 1/6 for a cubic volume to 1/4 for a spherical volume assigned to a nucleon. By taking the average fraction, which is 5/24, we obtain the surface energy correction as

$$B_S \approx \left(-\frac{5}{24}\right) a_1 4.8 A^{2/3} = -a_1 A^{2/3}$$

and this, compared to Eq.(46.5), shows that $a_2 \approx a_1$. Since the binding fraction of medium and heavy nuclei is about 12 MeV if the Coulomb repulsion energy is allowed for, it can be estimated that $a_1 \approx a_2 \approx 16$ MeV.

The continuous charge distribution of the nuclear liquid drop has a repulsion potential energy that diminishes the binding energy. This can be obtained from Eq.(11.30), where we must specify $\rho(\vec{r})$ and $V(\vec{r})$. The charge density is assumed to be uniform over the nuclear volume, as given by

$$\rho = \frac{Ze}{4\pi R^3 / 3} = \frac{3Ze}{4\pi R^3} \tag{46.6}$$

Instead of solving Poisson's equation (11.14) for $V(\vec{r})$ inside the nucleus (i.e., for $r < R$), it is more convenient to use Eq.(11.8) which derives the electrostatic potential from the radial component of the electric field, given by Gauss's law (11.4), in the configuration of Figure 46.2 (b), as

$$4\pi r^2 E_r = \frac{1}{\varepsilon_0} \rho \frac{4\pi}{3} r^3 \quad \text{or} \quad E_r = \frac{Ze}{4\pi \varepsilon_0 R^3} r \tag{46.7}$$

Substituting Eq.(46.7), Eq.(11.8) yields

$$V(r)-V(R) = -\int_R^r \vec{E}\cdot d\vec{r} = -\frac{Ze}{4\pi\varepsilon_0 R^3}\int_R^r r\,dr = \frac{Ze}{4\pi\varepsilon_0 R^3}\frac{R^2-r^2}{2} \tag{46.8}$$

Since outside the nuclear boundary ($r > R$) the electrostatic potential of the nuclear charge is that of a point charge at the origin, the boundary condition on Eq.(46.8) is

$$V(R) = \frac{Ze}{4\pi\varepsilon_0 R}$$

which, on substitution into Eq.(46.8), gives

$$V(r) = \frac{Ze}{4\pi\varepsilon_0 R}\left(\frac{3}{2} - \frac{r^2}{2R^2}\right) \tag{46.9}$$

It is straightforward to show that, by inserting Eqs.(46.6) and (46.9), Eq.(11.30) gives the Coulomb potential energy as

$$U = \frac{3}{5}\frac{(Ze)^2}{4\pi\varepsilon_0 R} = \frac{3}{5}\frac{(Ze_0)^2}{R} \tag{46.10}$$

Therefore, in order to take into account the Coulomb repulsion, the binding energy $B(A,Z)$ must be reduced by

$$B_C = -a_3 \frac{Z^2}{A^{1/3}} \tag{46.11}$$

where the theoretical estimation of the constant a_3 is

$$a_3 = \frac{3}{5}\frac{e_0^2}{R_0} \approx 0.72 \text{ MeV} \tag{46.12}$$

Gas Model of the Nucleus

For a realistic description of stable nuclei, the expression for the binding energy must also include correction terms which involve the quantum nature of the nucleons. A simple approach is to neglect the forces between pairs of nucleons and to simulate the overall force on each nucleon by requiring that all nucleons should be contained in a sphere of radius $R = R_0 A^{1/3}$. Since nucleons move freely inside the sphere, the nucleus can be treated as a gas, obeying Fermi-Dirac distribution law, because the particles have spin $\frac{1}{2}\hbar$. The Pauli exclusion principle, which is valid for noninteracting particles, requires that no two nucleons of one kind (protons or neutrons) have the same quantum numbers. In other words there will be at most two protons and two neutrons in each $\vec{k}$-state. The assumptions of the *gas model* for the nucleus are similar to those of the free-electron model discussed in Chapter 40. Hence, the number of states available to particles of one kind in the gas, with wave vector up to an arbitrary value of k, is $Vk^3/3\pi^2$, as given by Eq.(40.9) for the free-electron states, where the volume V of the crystal must be replaced by $V = 4\pi R_0^3 A/3$. The maximum k-value for protons and neutrons is obtained by taking the number of states equal to Z or N, which gives

$$k_{p\,max} = \frac{1}{R_0}\left(\frac{9\pi}{4}\frac{Z}{A}\right)^{1/3} \quad , \quad k_{n\,max} = \frac{1}{R_0}\left(\frac{9\pi}{4}\frac{N}{A}\right)^{1/3} \quad (46.13)$$

Hence, the maximum energies for protons and neutrons, which play the same role as the Fermi energy for the free-electron gas, are

$$\varepsilon_{p\,max} = \frac{\hbar^2 k_{p\,max}^2}{2m_p} = C\left(\frac{Z}{A}\right)^{2/3} \quad , \quad \varepsilon_{n\,max} = C\left(\frac{N}{A}\right)^{2/3} \quad (46.14)$$

where $C = \hbar^2(9\pi/4)^{2/3}/(2m_p R_0^2) \cong \hbar^2(9\pi/4)^{2/3}/(2m_n R_0^2) \cong 95$ MeV. The total energy for protons and neutrons can be expressed in terms of $\varepsilon_{p\,max}$ and $\varepsilon_{n\,max}$, respectively, in a form analogous to that derived in Eq.(40.27) for the internal energy of the free-electron gas in terms of ε_F, that is

$$E_p = \frac{3}{5}Z\varepsilon_{p\,max} = \frac{3}{5}CZ\left(\frac{Z}{A}\right)^{2/3} \quad , \quad E_n = \frac{3}{5}CN\left(\frac{N}{A}\right)^{2/3} \quad (46.15)$$

It follows that the kinetic energy of the nucleus, which produces a decrease in the binding energy, can be written as

$$E = \frac{3}{5}C\left[Z\left(\frac{Z}{A}\right)^{2/3} + N\left(\frac{N}{A}\right)^{2/3}\right] \quad (46.16)$$

It is convenient to introduce the *neutron excess* $\Delta = N - Z = A - 2Z$ so that substituting $Z = (A-\Delta)/2$ and $N = (A+\Delta)/2$ into Eq.(46.16) gives

$$E = \frac{3}{5}C\frac{A}{2^{5/3}}\left[\left(1-\frac{\Delta}{A}\right)^{5/3} + \left(1+\frac{\Delta}{A}\right)^{5/3}\right]$$

where, because Δ is small for any real nucleus, it is convenient to use the expansion $(1+x)^{5/3} = 1 + (5/3)x + (5/9)x^2 + \ldots$, which gives

$$E \cong \frac{3\sqrt[3]{2}}{20}C\left(2A + \frac{10}{9}\frac{\Delta^2}{A}\right) = \frac{3\sqrt[3]{2}}{10}CA + \frac{\sqrt[3]{2}}{6}C\frac{(A-2Z)^2}{A} \quad (46.17)$$

The last term is called the *asymmetry energy*, and it is seen that it vanishes for nuclei where $N = Z = A/2$, which are found to be more stable than their isobars with $N \ne Z$. As a result, we must consider a correction to the binding energy that is mostly important for light nuclei, and is written as

$$B_a = -a_4 \frac{(A-2Z)^2}{A} \qquad (46.18)$$

where $a_4 = C\sqrt[3]{2}/6 \approx 19$ MeV.

Under the assumption that nuclear forces between neutrons are identical to those between protons, the energy states of neutrons and protons predicted by the gas model must be identical. However, the Coulomb interaction between the protons results in rising the bottom of the proton well by $eV(r)$, where $V(r)$ is the electrostatic potential inside the nucleus, given by Eq.(46.9), which reads

$$eV(r) = \frac{Ze_0^2}{R_0 A^{1/3}} \left(\frac{3}{2} - \frac{r^2}{2R_0^2 A^{2/3}} \right) \qquad (46.19)$$

This represents a potential energy barrier for protons at the nuclear boundary given by

$$eV(R) = \frac{Ze_0^2}{R_0 A^{1/3}} \qquad (46.20)$$

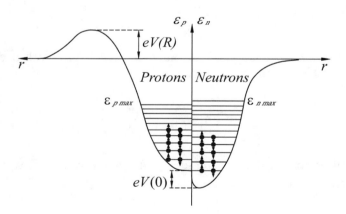

Figure 46.3. The gas model of a nucleus

There is no Coulomb barrier for neutrons, as illustrated in Figure 46.3, which shows a simplified energy level diagram for protons and neutrons, which is consistent with the gas model of the nucleus but is modified to account for the electrostatic interaction. It is seen that the spins of pairs of similar nucleons are antiparallel, as required by the Pauli exclusion principle. The gas model predicts that the kinetic energy of the nucleus does not increase smoothly with Z or N, as assumed by the liquid drop model, but rather in steps. If there is an odd number of either protons or neutrons in the nucleus, the highest energy state is only half-filled (see Figure 46.3) and so the next nucleon of the same kind will go into it. There are four possible types of nuclei: even-

even, odd-odd, even-odd and odd-even, where the first word designates the number of protons and the second the number of neutrons. Compared to the smooth increase in the kinetic energy, an even-even nucleus will have less and an odd-odd nucleus more energy, so that the most stable nuclei are of the even-even type and the least stable are of the odd-odd type. This step-like increase of the kinetic energy is represented in the binding energy formula by a *pairing energy* term of the form

$$B_p = \delta(A,Z) = \lambda \varphi(A) \qquad (46.21)$$

where $\lambda = 1$ for even-even nuclei, $\lambda = -1$ for odd-odd nuclei and $\lambda = 0$ otherwise. The parameter $\varphi(A)$ covers the overestimation of the maximum kinetic energy (46.14) for nucleons in an even-even nucleus (N,Z) as compared to the nuclei $(N-1,Z)$ or $(N,Z-1)$. Hence, $\varphi(A)$ can be estimated from

$$\varphi(A) = \varepsilon_{n\max}(N) - \varepsilon_{n\max}(N-1) = C\left(\frac{N}{A}\right)^{2/3}\left[1 - \left(1 - \frac{1}{N}\right)^{2/3}\right] \cong \frac{2}{3}C\left(\frac{N}{A}\right)^{2/3}\frac{1}{N} \approx \frac{2\sqrt[3]{2}}{3}\frac{C}{A} \qquad (46.22)$$

The best experimental fit indicates, however, a slightly different form for $\varphi(A)$, which is currently considered to be given by

$$\varphi(A) = \frac{a_5}{A^{3/4}} \qquad (46.23)$$

where $a_5 \cong 34$ MeV.

On combining Eqs.(46.4), (46.5), (46.11), (46.18), (46.21) and (46.23) we obtain an expression for the nuclear binding energy

$$B(A,Z) = a_1 A - a_2 A^{2/3} - a_3 \frac{Z^2}{A^{1/3}} - a_4 \frac{(A-2Z)^2}{A} + \lambda \frac{a_5}{A^{3/4}} \qquad (46.24)$$

called *von Weizsäcker's formula*. The contribution of the volume, surface and Coulomb energies, which originate from the liquid drop model and of the asymmetry energy, as predicted by the gas model, are represented as a function of the nucleon number A in Figure 46.4, which shows that the addition of successive terms improves the agreement with the experimental values, plotted in Figure 46.1. It is found that Eq.(46.24) agrees with experimental data to better than 1 percent, for $A > 15$.

Substituting Eq.(46.24) into Eq.(46.1) we obtain the *semiempirical mass formula* which, for constant A, has a parabolic dependence on Z

$$M(A,Z) = Zm_H + (A-Z)m_n - \frac{a_1}{c^2}A + \frac{a_2}{c^2}A^{2/3} + \frac{a_3}{c^2}\frac{Z^2}{A^{1/3}} + \frac{a_4}{c^2}\frac{(A-2Z)^2}{A} - \frac{\lambda}{c^2}\frac{a_5}{A^{3/4}}$$

$$= uA + vZ + wZ^2 \qquad (46.25)$$

where

$$u = m_n + (a_4 - a_1)/c^2 + a_2 A^{-1/3}/c^2 - \lambda a_5 A^{-7/4}/c^2$$

$$v = m_H - m_n - 4a_4/c^2 \cong -4a_4/c^2$$

$$w = a_3 A^{-1/3}/c^2 + 4a_4 A^{-1}/c^2$$

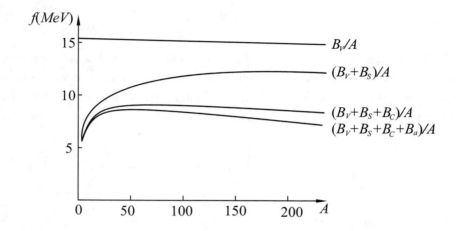

Figure 46.4. The influence of the various terms in von Weizsäcker's formula B_V = volume energy, B_S = surface energy, B_C = Coulomb energy, B_a = asymmetry energy

For odd A, we have $\lambda = 0$ and the plot of $M(A,Z)$ versus Z is *one* parabola, illustrated in Figure 46.5. The smallest mass, which corresponds to the most stable nucleus among isobars of given A, is obtained by taking $\partial M(A,Z)/\partial Z = 0$ in Eq.(46.25). The result is

$$Z = Z_A = -\frac{v}{2w} \cong \frac{A/2}{1 + (a_3/4a_4)A^{2/3}} \tag{46.26}$$

which is not usually an integer. There is *one* stable nucleus lying close to the minimum of the parabola, while the other nuclei on either side are radioactive, and they can decay into the stable nucleus, as in Figure 46.5, by beta decay (β^-) or electron capture (β^+) as we shall discuss in Chapter 47.

For even A the presence of $\lambda = \pm 1$ gives two parabolas, one which is lower in energy for even-even nuclei and another, higher in energy, for odd-odd nuclei, both plotted in Figure 46.5 by dashed lines. There may be up to three stable even-even nuclei, while no odd-odd stable nuclei are known for $A > 15$ (the few stable ones are ^2_1H, ^6_3Li, $^{10}_5B$, $^{14}_7N$).

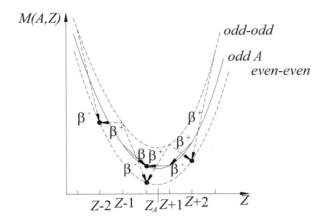

Figure 46.5. Plot of $M(A,Z)$ versus Z for isobars

EXAMPLE 46.1. Nuclear fission

If the binding energy of a nucleus is smaller than the total binding energy of any fragment into which it can disintegrate, the nucleus is unstable and fission becomes possible, and is accompanied by the release of energy which appears as kinetic energy of the fragments. Consider the fission of a nucleus of mass number A into two fragments with mass numbers xA and $(1-x)A$ and charge numbers yZ and $(1-y)Z$, respectively, according to

$$^A_Z X \to ^{xA}_{yZ} X' + ^{(1-x)A}_{(1-y)Z} X''$$

The difference in the binding energies released in this process, given by

$$\Delta B = B(xA, yZ) + B[(1-x)A, (1-y)Z] - B(A,Z)$$

can be calculated by applying Eq.(46.24) which yields

$$\Delta B = a_2 A^{2/3}\left[1 - x^{2/3} - (1-x)^{2/3}\right] + a_3 \frac{Z^2}{A^{1/3}}\left[1 - y^2 x^{-1/3} - (1-y)^2 (1-x)^{-1/3}\right]$$

$$+ a_4 \frac{4Z^2}{A}\left[1 - y^2 x^{-1} - (1-y)^2 (1-x)^{-1}\right]$$

where the difference between the pairing energies is small and has been neglected. It is a straightforward matter to show that ΔB is a maximum when $x = y = 1/2$, so that

$$\Delta B_{max} = a_2 A^{2/3}\left(1 - \sqrt[3]{2}\right) + a_3 \frac{Z^2}{A^{1/3}}\left(1 - \frac{1}{\sqrt[3]{4}}\right)$$

$$\cong A^{2/3}\left(-0.26 a_2 + 0.37 a_3 \frac{Z^2}{A}\right) \cong A^{2/3}\left(-4.3 + 0.26 \frac{Z^2}{A}\right)$$

where the numerical values of a_2 and a_3 estimated from the liquid drop model, have been inserted. Fission is possible if $\Delta B_{max} \geq 0$, or

$$Z^2 \geq \frac{4.3}{0.26} A \cong 16A$$

In other words, a symmetric fission with release of energy is predicted for all the nuclei having $Z > 35$ and $A > 80$. Since fission consists of formation of two nuclear liquid drops from one larger drop, it implies an increase in the surface area accompanied by a decrease in Coulomb repulsion. The overall effect is equivalent to a potential barrier through which the fission products must pass. It can be shown that fission will be spontaneous for nuclei for which the surface energy effect is equal to or less than the Coulomb repulsion. This happens for

$$Z^2 > \frac{2a_2}{a_3} A \cong 50A$$

that is for nuclei having $Z > 140$ and $A > 380$. In the well known case of ^{238}U we have $Z^2 = 36A$ and the fission is usually accelerated by means of slow neutron capture, which releases the amount of energy $(\sim 7 \text{ MeV})$ required to overcome the fission potential barrier.

46.2. THE SHELL MODEL OF THE NUCLEUS

All the evidence indicates that there are periodicities in the nuclear binding energies which might be due to a shell structure similar to the atomic shell structure. For neutrons, closed shells have been associated with the *magic numbers* $N = 2, 8, 20, 28, 50, 82$ and 126 and, for protons, with $Z = 2, 8, 20, 50$ and 82, because the nuclei with one or both of the magic numbers show a particular stability when compared with other nuclei. The magic numbers can be derived using the *single-particle shell model* where the many-nucleon problem is reduced to a single-particle problem, under the assumption that, despite the strong overall attraction between nucleons, which provides the nuclear binding energy, each nucleon moves in a nuclear potential due to all the other nucleons. In other words, the short-range nuclear forces should average out to produce the nuclear potential, so that any coupling between nucleons can be neglected. If the nuclear potential energy is represented, in the simplest case, by a spherically symmetric function $U(r)$, which is dependent on r only, the situation becomes similar to the one-electron atom problem, where the electron motion under the electrostatic interaction (32.1) gives rise to a series of electron shells with stable closed configurations.

The nuclear states for single nucleons are thus obtained by solving the Schrödinger equation (31.13) which can be separated, for any shape of the spherical function $U(r)$, into an angular equation (31.20), having the spherical harmonics as solutions, and a radial equation (31.26). If an orbital angular momentum $\hat{\vec{L}}$, specified by a quantum number l according to the quantum condition (31.35), is assigned to the nucleon, the radial equation (31.26) takes the form

$$\frac{d^2\xi(r)}{dr^2} + \frac{2m}{\hbar^2}\left[\varepsilon - U(r) - \frac{l(l+1)\hbar^2}{2mr^2}\right]\xi(r) = 0 \tag{46.27}$$

where $\xi(r) = rR(r)$ and m stands for either m_p or m_n. Since the nuclear forces are of short range, the nucleons experience an average potential which is nearly constant inside the nucleus and falls sharply to zero near the nuclear surface. The simplest forms of $U(r)$ simulating such behaviour (see Figure 46.6) are the harmonic oscillator potential

$$U(r) = -V_0 + \frac{1}{2}m\omega^2 r^2 \tag{46.28}$$

and the square well potential, given by

$$U(r) = -V_0 \quad , \quad r \leq R$$
$$= 0 \quad , \quad r > R \tag{46.29}$$

Substituting Eq.(46.28), the radial equation takes the form

$$\frac{d^2\xi(r)}{dr^2} + \left[\frac{2m}{\hbar^2}\left(\varepsilon + V_0 - \frac{m\omega^2}{2}r^2\right) - \frac{l(l+1)}{r^2}\right]\xi(r) = 0 \tag{46.30}$$

For $r \to 0$, Eq.(46.30) can be approximated by

$$\frac{d^2\xi(r)}{dr^2} - \frac{l(l+1)}{r^2}\xi(r) = 0 \quad \text{i.e.,} \quad \xi(r) \sim r^{l+1} \tag{46.31}$$

For $r \to \infty$, Eq.(46.30) reduces to

$$\frac{d^2\xi(r)}{dr^2} + \frac{2m}{\hbar^2}\left(\varepsilon + V_0 - \frac{m\omega^2}{2}r^2\right)\xi(r) = 0$$

which is similar to the equation of a quantum oscillator (30.25). Thus the asymptotically valid solutions must take the form (30.28) which, if we let $s = m\omega/\hbar$, reads

$$\xi(r) \sim e^{-sr^2/2} \tag{46.32}$$

A trial solution, valid for all r, must include the asymptotic solutions (46.31) and (46.32) and can be written as

$$\xi(r) = r^{l+1} e^{-sr^2/2} \eta(r) \tag{46.33}$$

On substitution into Eq.(46.30) and multiplication by $\eta(r)/\xi(r)$ one obtains

$$\frac{d^2\eta(r)}{dr^2}+2\frac{d\eta(r)}{dr}\left(\frac{l+1}{r}-sr\right)-s\eta(r)\left[2l+3-\frac{2}{\hbar\omega}(\varepsilon+V_0)\right]=0$$

or

$$\frac{d^2\eta(r)}{dr^2}+\frac{d\eta(r)}{dr}\left(\frac{2q-1}{r}-2sr\right)-4ps\eta(r)=0 \qquad (46.34)$$

where $2l+3=2q$ and $2l+3-2(\varepsilon+V_0)/\hbar\omega=4p$. We look for $\eta(r)$ in the form of a power series $\eta(r)=\sum_k a_k r^k$ which, on insertion into Eq.(46.34), yields the recurrence relation

$$(k+2)(k+1)a_{k+2}+(2q-1)(k+2)a_{k+2}-2ska_k-4psa_k=0 \text{ or } a_{k+2}=\frac{2s(k+2p)}{(k+2q)(k+2)}a_k$$

$$(46.35)$$

It follows that the power series assumes the form of the confluent hypergeometric function, defined as

$$\eta(r)=F(p,q,sr^2)=1+\frac{p}{q}(sr^2)+\frac{p(p+1)}{q(q+1)}\frac{(sr^2)^2}{2!}+\frac{p(p+1)(p+2)}{q(q+1)(q+2)}\frac{(sr^2)^3}{3!}+\cdots$$

which must terminate after a finite number of terms, so that the asymptotic behaviour can be satisfied. This means that p must be equal to a negative integer, say $-i$, that is,

$$2l+3-\frac{2}{\hbar\omega}(\varepsilon+V_0)=-4i$$

This gives the energy eigenvalues of the form

$$\varepsilon_n=-V_0+\hbar\omega\left(2i+l+\frac{3}{2}\right)=-V_0+\hbar\omega\left(n+\frac{3}{2}\right) \qquad (46.36)$$

where $n=2i+l$ is a *quantum number* such that, if n is even, $l=n, n-2,\ldots, 0$ and, if n is odd, $l=n, n-2,\ldots, 1$. The energy levels (46.36) are equally spaced. The ground state $(n=0)$, of energy

$$\varepsilon_0=-V_0+\frac{3}{2}\hbar\omega \qquad (46.37)$$

is not degenerate, since it can only be formed with $i=0, l=0$.

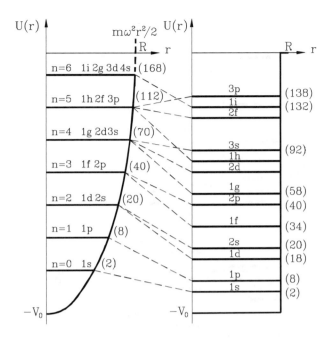

Figure 46.6. Energy levels for protons or neutrons in an oscillator potential and in an infinite square well potential

As there are $2l+1$ values of the magnetic quantum number in a given l state, the higher energy levels are degenerate. If we include spin, the total number of protons or neutrons in a given l state is $2(2l+1)$ so that the degeneracy of the level n is $(n+1)(n+2)$. A nucleon state is normally specified using the appropriate spectroscopic symbol $s, p, d, f, g, \ldots$ for its l value and a number in front of the letter, which is 1 if the symbol following appears for the first time, 2 if it appears for the second time, and so on. Hence, the sequence of the energy levels from the bottom of the parabolic well is $1s\ (n=0, l=0)$, $1p\ (n=1, l=1)$, $1d, 2s\ (n=2, l=2, 0)$, $1f, 2p\ (n=3, l=3, 1)$, $1g, 2d, 3s$ $(n=4, l=4, 2, 0)$, $1h, 2f, 3p\ (n=5,\ l=5, 3, 1)$, $1i, 2g, 3d, 4s\ (n=6, l=6, 4, 2, 0)$, as shown in Figure 46.6. Note that the number in front of the spectroscopic symbol is not the quantum number n, as in the designation of atomic spectral lines. It is seen from Figure 46.6 that the oscillator potential function (46.28) leads to the magic numbers 2, 8, 20, 40, 70, 112, 168 which are not identical with the nuclear magic numbers mentioned earlier. Use of the square well potential (46.29) changes the level spacing and slightly changes the order of levels, as indicated in Figure 46.6, predicting the shell closure at 2, 8, 18, 20, 34, 40, 58, 92, 132, 138 which still does not correspond to the experimental evidence.

A more realistic scheme takes into account the possible splitting of each energy level into two, according to whether the orbital and spin angular momenta $\hat{\vec{L}}$ and $\hat{\vec{S}}$ of the nucleon in the level are in the same or opposite directions.

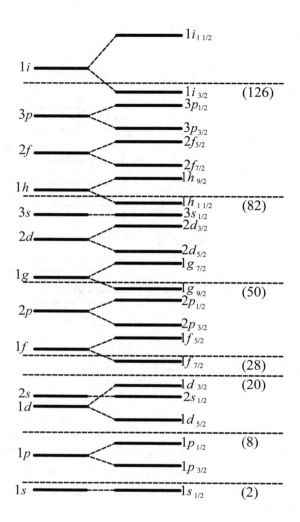

Figure 46.7. Nucleon levels in the single particle shell model

Such additional energy could arise from a spin-orbit coupling involving the unpaired nucleon, which is postulated to be analogous to that introduced for electrons in an atom, Eq.(34.17), as

$$-\frac{\lambda}{2m^2c^2}\frac{1}{r}\frac{dU(r)}{dr}\hat{\vec{S}}\cdot\hat{\vec{L}} \qquad (46.38)$$

where m stands for either m_p or m_n, and $U(r)$ is the nuclear potential energy. Note that, since an interaction energy of the form used in Eq.(34.17) is too small and also has the opposite sign with respect to the expected splitting, the sign was changed and a dimensionless constant λ was introduced in Eq.(46.38), to fit the data. This means that the origin of the spin-orbit coupling of nucleons might be different from the electromagnetic interaction discussed in Chapter 34, and hence, it is left unspecified.

Taking into account the spin-orbit coupling, we obtain the splitting of the energy levels ε_n, as derived in Eq.(34.21), given by

$$\varepsilon_{nl} = \varepsilon_n - \frac{1}{2}l\zeta \quad , \quad j = l + \tfrac{1}{2}$$

$$\varepsilon_{nl} = \varepsilon_n + \frac{1}{2}(l+1)\zeta \quad , \quad j = l - \tfrac{1}{2}$$

(46.39)

where ε_n are the energy eigenvalues in the absence of spin-orbit interaction, and ζ is a parameter having the dimensions of energy

$$\zeta = \frac{\lambda \hbar^2}{2m^2 c^2} \frac{1}{r} \frac{dU(r)}{dr} \tag{46.40}$$

The level splitting (46.39) is such that the level with total angular momentum $j = l + \tfrac{1}{2}$ lies below the $j = l - \tfrac{1}{2}$ level, and this is the opposite to what happens for electrons but agrees with the experimental evidence. The resulting states are specified by adding the half-integral values j as subscripts in the spectroscopic symbols.

The correct magic numbers for large l can be obtained if we count the levels with largest l, and $j = l + \tfrac{1}{2}$, together with the group of lower energy. The procedure is illustrated in Figure 46.7, using the energy levels of the square well potential shown in Figure 46.6.

FURTHER READING

1. S. S. Wong - INTRODUCTORY NUCLEAR PHYSICS, Wiley, New York, 2004.
2. H. A. Enge, R. P. Redwine - INTRODUCTION TO NUCLEAR PHYSICS, Addison-Wesley, San Francisco, 1999.
3. K. Heyde – FROM NUCLEONS TO THE ATOMIC NUCLEUS, Springer-Verlag, New York, 1997.

PROBLEMS

46.1. Show that the nuclear unit radius R_0 can be estimated in terms of the difference ΔB in binding energy between two mirror nuclei that are a pair of nuclei where the number of protons and neutrons are interchanged.

Solution

Since $N = A - Z$, the difference in binding energy is obtained from Eq.(46.24) as

$$\Delta B = B(A,Z) - B(A, A-Z) = \frac{a_3}{A^{1/3}}\left[(A-Z)^2 - Z^2\right] = a_3 A^{2/3}(A - 2Z)$$

where a_3 is given by Eq.(46.12). It follows that

$$\Delta B = \frac{3}{5}\frac{e_0^2}{R_0} A^{2/3}(A - 2Z) \quad \text{or} \quad R_0 = \frac{3}{5}\frac{e_0^2}{\Delta B} A^{2/3}(A - 2Z)$$

46.2. Assuming that a neutron star can be described in terms of a spherical nucleus $^A_0 X$, use von Weizsäcker's formula to estimate the size of such a star.

Solution

The Coulomb interaction should be replaced in Eq.(46.24) by a gravitational energy term of the form

$$U = \frac{3}{5} G \frac{(Am_n)^2}{R}$$

which is similar to that derived in Eq.(46.10). It follows that

$$B(A,0) = (a_1 - a_4)A - a_2 A^{2/3} + \lambda a_5 A^{-3/4} + \frac{3}{5}\frac{G}{R_0 A^{1/3}}(Am_n)^2$$

For large A values we may neglect both the surface and the pairing energy terms, and hence, stability is obtained if

$$B(A,0) = (a_1 - a_4)A + \frac{3}{5}\frac{G}{R_0 A^{1/3}}(Am_n)^2 \geq 0$$

and this happens for

$$A^{2/3} \geq \frac{5}{3}\frac{(a_4 - a_1)R_0}{Gm_n^2}$$

Substituting m_n and the gravitational constant G as given in Chapter 1, and using the empirical values for R_0, a_1 and a_4, one obtains $A \approx 5 \times 10^{55}$, and this implies a size R of 4.3 km for a stable neutron star.

46.3. Solve the eigenvalue problem for the $^3S_1$ ground state of the deuteron, which consists of a proton and a neutron with parallel spins, assuming that their interaction is described by the square well potential.

46. NUCLEAR STRUCTURE

Solution

As the ground state of deuteron is $^3S_1$, we have $l_1 = l_2 = 0$, $s_1 = s_2 = \frac{1}{2}$, and hence, $S = 1$ for the two parallel spins. The radial equations (31.26) for the two particles reduce to

$$\frac{d^2\xi(r)}{dr^2} + \frac{2\mu}{\hbar^2}\left[\varepsilon - U(r)\right]\xi(r) = 0$$

where $\varepsilon = -|\varepsilon|$ is of about 2.2 MeV in the bound state, $\xi(r) = rR(r)$ and $\mu = m_p m_n/(m_p + m_n) \approx m_p/2$ stands for the reduced mass of the two nucleons. It follows that

$$\frac{d^2\xi(r)}{dr^2} + \frac{2\mu}{\hbar^2}(V_0 - |\varepsilon|)\xi(r) = 0 \quad \text{or} \quad \frac{d^2\xi(r)}{dr^2} + k^2\xi(r) = 0 \quad, \quad r \leq R$$

$$\frac{d^2\xi(r)}{dr^2} - \frac{2\mu}{\hbar^2}|\varepsilon|\xi(r) = 0 \quad \text{or} \quad \frac{d^2\xi(r)}{dr^2} - \alpha^2\xi(r) = 0 \quad, \quad r > R$$

where

$$k^2 = \frac{2\mu}{\hbar^2}(V_0 - |\varepsilon|) \quad, \quad \alpha^2 = \frac{2\mu}{\hbar^2}|\varepsilon|$$

A convenient trial solution is

$$\xi(r) = A\sin kr \quad (r \leq R) \quad, \quad \xi(r) = Be^{-\alpha(r-R)} \quad (r > R)$$

so that the boundary condition for $\xi(r)$ and its first derivative at $r = R$, given by

$$\frac{1}{\xi(r)}\frac{d\xi(r)}{dr}\bigg|_{r\leq R} = \frac{1}{\xi(r)}\frac{d\xi(r)}{dr}\bigg|_{r>R}$$

takes the form

$$k\cot(kR) = -\alpha \quad \text{or} \quad \frac{1 - \sin^2(kR)}{\sin^2(kR)} = \frac{\alpha^2}{k^2}$$

Since $\xi(r)$ is continuous at $r = R$, that is, $A\sin(kR) = B$, one also obtains

$$\frac{1 - (B/A)^2}{(B/A)^2} = \frac{\alpha^2}{k^2} \quad \text{or} \quad k^2 A^2 = (k^2 + \alpha^2)B^2$$

The normalization condition on $\xi(r)$ reads

$$4\pi \int_0^\infty \xi^2(r)dr = 1 \quad \text{or} \quad A^2\int_0^R \sin^2(kr)dr + B^2\int_R^\infty e^{-2\alpha(r-R)}dr = \frac{1}{4\pi}$$

and provides a second relation between A and B in the form

$$\frac{A^2}{2k}\left[2kR - \sin(2kR)\right] + \frac{B^2}{\alpha^2} = \frac{1}{2\pi}$$

Solving for B the two equations, yields

$$\frac{B^2}{2k}\left(1 + \frac{\alpha^2}{k^2}\right)\left[2kR - \sin(2kR)\right] + \frac{B^2}{\alpha^2} = \frac{1}{2\pi}$$

and this can be approximated as

$$\frac{\alpha}{2\pi} = B^2\left\{1 + \frac{\alpha}{2k}\left(1 + \frac{\alpha^2}{k^2}\right)\left[2kR - \sin(2kR)\right]\right\}$$

$$= B^2\left\{1 + \alpha R\left(1 + \frac{\alpha^2}{k^2}\right)\left[1 - \frac{\sin(2kR)}{2kR}\right]\right\} = B^2(1 + \alpha R + \cdots)$$

It follows that

$$B = \sqrt{\frac{\alpha}{2\pi}}(1 + \alpha R + \cdots)^{-1/2} \approx \sqrt{\frac{\alpha}{2\pi}}\left(1 - \frac{\alpha R}{2}\right) \approx \sqrt{\frac{\alpha}{2\pi}}e^{-\alpha R/2}$$

and hence $A = B(1 + \alpha^2/k^2)^{1/2} \approx B(1 + \alpha^2/2k^2) \approx B$. This finally gives

$$\xi(r) = \sqrt{\frac{\alpha}{2\pi}}e^{-\alpha R/2}\sin kr \quad \text{i.e.,} \quad R(r) = \sqrt{\frac{\alpha}{2\pi}}e^{-\alpha R/2}\frac{\sin kr}{r} \quad (r \le R)$$

$$\xi(r) = \sqrt{\frac{\alpha}{2\pi}}e^{\alpha R/2}e^{-\alpha r} \quad \text{i.e.,} \quad R(r) = \sqrt{\frac{\alpha}{2\pi}}e^{\alpha R/2}\frac{e^{-\alpha r}}{r} \quad (r > R)$$

The energy eigenvalue is obtained in terms of the interaction potential V_0 from the boundary condition, which reads

$$\cot(kR) = -\frac{\alpha}{k} = -\left(\frac{|\varepsilon|}{V_0 - |\varepsilon|}\right)^{1/2}$$

As it is found that V_0 is of the order of 30-60 MeV, and hence, $V_0 \gg |\varepsilon|$ to a first approximation, we can put $\cot(kR) \approx 0$ or $kR \approx \pi/2$. This implies that

$$k^2 \approx \frac{\pi^2}{4R^2} \quad \text{or} \quad (V_0 - \varepsilon)R^2 \approx \frac{\pi\hbar^2}{8\mu}$$

Thus, a bound state of the deuteron is obtained if $V_0 R^2 \ge \frac{\pi^2 \hbar^2}{8\mu}$.

46.4. Find the energy split between the $^1S_0$ and $^3S_1$ states of the deuteron, assuming that the interaction of the two nucleon spins is given by $U = -V_0(a + b\vec{\sigma}_1 \cdot \vec{\sigma}_2)$, where $\vec{\sigma}_1$ and $\vec{\sigma}_2$ are the Pauli spin matrices.

Solution

In view of Eqs. (33.40), we have $\vec{S}_1 = \frac{1}{2}\hbar\vec{\sigma}_1$, $\vec{S}_2 = \frac{1}{2}\hbar\vec{\sigma}_2$ for the two fermions, and hence, using the results derived in Problem 35.4, one obtains for the two states

$$(\hat{\vec{\sigma}}_1 \cdot \hat{\vec{\sigma}}_2)\chi^a = -3\chi^a \quad , \quad (\vec{\sigma}_1 \cdot \vec{\sigma}_2)\chi^s = \chi^s$$

where χ^a and χ^s are the antisymmetric ($S = 0$) and symmetric ($S = 1$) states of the deuteron, respectively. It follows that

$$\varepsilon_{^1S_0} = -V_0(a - 3b) \quad , \quad \varepsilon_{^3S_1} = -V_0(a + b)$$

and this gives the required energy split as $\Delta\varepsilon = 4bV_0$.

46.5. Calculate the magnetic moment for odd A nuclides assuming that it is due entirely to the last unpaired nucleon. The spin g-factors were found experimentally to be $g_s = 2.79$ for protons and $g_s = -1.91$ for neutrons.

Solution

By analogy with the magnetic moment for one-electron atoms we can define

$$\hat{\vec{\mu}} = \hat{\vec{\mu}}_l + \hat{\vec{\mu}}_s = \frac{\mu_N}{\hbar}(g_l \hat{\vec{l}} + 2g_s \hat{\vec{s}})$$

where $g_l = 1$ for a proton and 0 for a neutron (which makes no contribution to the magnetic moment by virtue of its orbital motion, as it has zero electric charge). Since only the components of $\vec{l}$ and $\vec{s}$ which are parallel to $\vec{j}$ are relevant, all the others averaging out, we have

$$\hat{\vec{\mu}} = \frac{\mu_N}{\hbar}\left(g_l \left\langle \frac{\hat{\vec{l}} \cdot \hat{\vec{j}}}{\hat{j}^2} \right\rangle + 2g_s \left\langle \frac{\hat{\vec{s}} \cdot \hat{\vec{j}}}{\hat{j}^2} \right\rangle\right)\hat{\vec{j}}$$

where

$$\left\langle \frac{\hat{\vec{l}} \cdot \hat{\vec{j}}}{\hat{j}^2} \right\rangle = \frac{j(j+1) + l(l+1) - s(s+1)}{2j(j+1)} \quad , \quad \left\langle \frac{\hat{\vec{s}} \cdot \hat{\vec{j}}}{\hat{j}^2} \right\rangle = \frac{j(j+1) + s(s+1) - l(l+1)}{2j(j+1)}$$

Because $j = l \pm \frac{1}{2}$, we immediately obtain

$$\hat{\vec{\mu}} = \frac{\mu_N}{\hbar}\left[g_l + \frac{2g_s - g_l}{2j}\right]\hat{\vec{j}}, \quad j = l + \tfrac{1}{2} \quad \text{and} \quad \hat{\vec{\mu}} = \frac{\mu_N}{\hbar}\left[g_l - \frac{2g_s - g_l}{2(j+1)}\right]\hat{\vec{j}}, \quad j = l - \tfrac{1}{2}$$

If the magnetic moment is determined by the last unpaired proton (odd A and odd Z), this gives

$$\hat{\vec{\mu}} = \frac{\mu_N}{\hbar}\left(1 + \frac{2.29}{j}\right)\hat{\vec{j}} \quad (j = l + \tfrac{1}{2}), \qquad \hat{\vec{\mu}} = \frac{\mu_N}{\hbar}\left(1 - \frac{2.29}{j+1}\right)\hat{\vec{j}} \quad (j = l - \tfrac{1}{2})$$

whereas, for an odd number of neutrons (odd A and even Z) we obtain

$$\hat{\vec{\mu}} = \frac{\mu_N}{\hbar}\left(-\frac{1.91}{j}\right)\hat{\vec{j}} \quad (j = l + \tfrac{1}{2}), \qquad \hat{\vec{\mu}} = \frac{\mu_N}{\hbar}\left(\frac{1.91}{j+1}\right)\hat{\vec{j}} \quad (j = l - \tfrac{1}{2})$$

46.6. Use von Weizsäcker's formula to find the element for which the binding fraction f is a maximum, assuming a stable nucleus with $Z = N = A/2$ and neglecting the pairing energy.

46.7. Use the definition of the binding energy $B(A, Z)$ to find expressions for the separation energies $S_n(A, Z)$ needed to remove a neutron and a proton, respectively, from the nucleus ${}_Z^A X$.

46.8. Assuming for simplicity that the ground state of the deuteron can be described by the wave function $\xi(r) = rR(r) = (\alpha/2\pi)^{1/2} e^{-\alpha r}$ from $r = 0$ to $r \to \infty$, calculate the average range $\langle r \rangle$ of the nuclear interaction.

46.9. Use the Heisenberg uncertainty relations to estimate the range Δr of the nuclear interaction in the bound state of the deuteron.

46.10. If the magnetic moments of two nucleons are defined in terms of the Pauli spin matrices as $\mu\vec{\sigma}_1$ and $\mu\vec{\sigma}_2$, show that their magnetic interaction can be written in the form

$$U = V(r)\left[3\frac{(\hat{\vec{\sigma}}_1 \cdot r)(\hat{\vec{\sigma}}_2 \cdot r)}{r^2} - \hat{\vec{\sigma}}_1 \cdot \hat{\vec{\sigma}}_2\right]$$

and find an expression for $V(r)$.

47. NUCLEAR DYNAMICS

47.1. RADIOACTIVE DECAY LAW
47.2. ALPHA DECAY
47.3. BETA DECAY

47.1. RADIOACTIVE DECAY LAW

Dynamic nuclear properties are associated with transitions from some initial to some final system of nucleons. The transitions will occur spontaneously through *radioactive decay*, if the total energy of the final system is less than that of the initial system. They can also be produced through *nuclear reactions*, if energy must be furnished to the initial system. In the approximation that the interaction energy involved in the transition is small and can be treated as constant perturbation, which is on for a time t, the transition probability from an initial state Ψ_i into a final state Ψ_f can be expressed by Fermi's Golden Rule (37.26), which reads

$$P_{i \to f} = \frac{2\pi}{\hbar} \left| \left\langle \Psi_f \left| \hat{V} \right| \Psi_i \right\rangle \right|^2 \frac{dn}{d\varepsilon} \tag{47.1}$$

where $\hat{V}$ is the operator associated with the interaction, and $dn/d\varepsilon$ is the number of final states in the interval ε to $\varepsilon + d\varepsilon$ of the total energy (the density of final states). Equation (47.1) holds for both nuclear decays and nuclear reactions, which therefore bear a strong theoretical similarity.

The three radioactive decay modes, known as α-rays (or α-particles, the nuclei of $^{4}_{2}\text{He}$, consisting of two protons and two neutrons), β-rays (or β-particles that are electrons) and γ-rays (electromagnetic radiation), have a common time dependence, which can be derived if it is assumed that the decay is of statistical nature. In other words the probability that a *parent* (initial) nuclear state decays is of the same magnitude at any time for any nucleus. If N nuclei are in a given nuclear state at time t and λdt is the probability of decay in the time interval dt, the decrease in the number N in the short time dt is given by

$$-dN = N\lambda dt \qquad (47.2)$$

which, on integration, yields the radioactive decay law as

$$N(t) = N_0 e^{-\lambda t} = N_0 e^{-t/\tau} \qquad (47.3)$$

where $N(t)$ is the number of nuclei present at time t in the given state, $N_0 = N(0)$, λ is the *decay constant*, and τ is the *mean lifetime* of the state, given by

$$\tau = \frac{1}{N_0} \int_0^{N_0} t \, dN = \frac{1}{N_0} \int_\infty^0 t\left(-N_0 e^{-\lambda t} \lambda\right) dt = \int_0^\infty \lambda t e^{-\lambda t} dt = \frac{1}{\lambda} \qquad (47.4)$$

Instead of counting the number of nuclei $N(t)$, it is easier to count the rate at which decays occur in the radioactive sample, and this is called the *activity*, defined as

$$\left|\frac{dN}{dt}\right| = \lambda N_0 e^{-\lambda t} = \lambda N(t) \qquad (47.5)$$

and follows the same exponential decay law (47.3). The SI unit for activity is the *Becquerel* (Bq), equal to one decay per second, and the common unit is the *Curie* (Ci), defined as $1\,\text{Ci} = 3.7 \times 10^{10}\,\text{Bq}$. It is also convenient to introduce the *half-life* $t_{1/2}$ as the time interval in which the activity is reduced to one-half

$$t_{1/2} = \frac{\ln 2}{\lambda} = \tau \ln 2 \qquad (47.6)$$

The probability of radioactive decay through the direct transition of the nucleus from an initial state Ψ_i to a final state Ψ_f, in time t, is given by

$$P_{if}(t) = |a_f(t)|^2 \qquad (47.7)$$

where the wave function amplitudes $a_f(t)$ are obtained from Eq.(37.11), which can be approximated by

$$i\hbar \frac{da_f(t)}{dt} = a_i(t) V_{fi}(t) e^{i(\varepsilon_f - \varepsilon_i)t/\hbar} \qquad (47.8)$$

Since the probability of finding our decaying system in the initial state Ψ_i must decrease with time according to the radioactive decay law (47.3), which gives

$$|\Psi_i(t)|^2 = |\Psi_i|^2 e^{-\lambda t} \qquad (47.9)$$

we have to choose

$$a_i(t) = e^{-\lambda t/2} \tag{47.10}$$

rather than $a_i = 1$, as considered in Eq.(37.13). Substituting this result into Eq.(47.8), and assuming a constant perturbation which is switched on at time $t = 0$, yields

$$i\hbar \frac{da_f(t)}{dt} = V_{fi} e^{-\lambda t/2 + i(\varepsilon_f - \varepsilon_i)t/\hbar} \tag{47.11}$$

Integrating, we get an oscillatory behaviour of the wave function amplitude, given by

$$a_f(t) = \frac{\hbar V_{fi}\left[e^{i(\varepsilon_f - \varepsilon_i + i\hbar\lambda/2)t/\hbar} - 1\right]}{(\varepsilon_f - \varepsilon_i) + i\hbar\lambda/2} \tag{47.12}$$

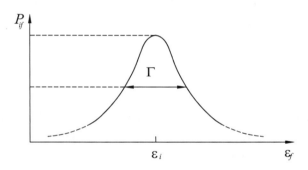

Figure 47.1. Energy spectrum of a decaying state of width Γ

At the end of the nuclear transition we should consider the amplitude corresponding to $t \to \infty$, which reads

$$a_f(\infty) = \frac{\hbar V_{fi}}{(\varepsilon_f - \varepsilon_i) + i\hbar\lambda/2} \tag{47.13}$$

and this yields the transition probability (47.7) in the form

$$P_{if}(\infty) = \frac{\hbar^2 |V_{fi}|^2}{(\varepsilon_f - \varepsilon_i)^2 + \frac{1}{4}\Gamma^2} \tag{47.14}$$

Hence, the probability to observe the nucleus with energy ε_f in the vicinity of ε_i follows a Lorentzian distribution, illustrated in Figure 47.1. By analogy with the atomic transitions, $\Gamma = \hbar\lambda = \hbar/\langle t\rangle$ gives the natural linewidth of the emitted radiation, which is the same as the *width* of the decaying nuclear state. Nuclear states which are populated in ordinary decays typically have lifetimes greater than 10^{-12} sec, corresponding to

$$\Gamma(\text{eV}) = 0.66 \times 10^{-15}/\tau \tag{47.15}$$

which is of the order of 10^{-3} eV. Since this width is small compared with the energy spacing of nuclear levels, which is of the order of 10^{-3} MeV, the nuclear decaying states may be regarded to be discrete quasi-stationary states.

EXAMPLE 47.1. Nuclear resonance

Each nuclear state can be assigned a nuclear angular momentum $\vec{I}$, called *nuclear spin*, representing the sum of the orbital and intrinsic angular momenta of all the nucleons in the nucleus. The nuclear spin obeys the same rules of quantization as the total angular momentum $\vec{J}$ of the electron states in the outer atom, Eq.(34.5), and this can be written as

$$\left|\vec{I}\right| = \sqrt{I(I+1)}\hbar \quad , \quad I_z = m_I \hbar \quad (m_I = -I, -I+1, \ldots, I-1, I)$$

The *magnetic dipole moment* is introduced in formal analogy with Eq.(32.35) as

$$\vec{\mu}_I = g_I \frac{e}{2m_p} \vec{I}$$

where the giromagnetic factor g_I varies from nucleus to nucleus and is unpredictable, unlike the atomic Landé factor, since the particle coupling inside the nucleus is not known. Under the influence of an external magnetic field $\vec{B}_0$, there is a change in energy

$$U_m = -\vec{\mu}_I \cdot \vec{B}_0 = -\mu_I B_0 \cos\theta$$

where θ is the angle between $\vec{\mu}_I$ and the z-axis, which in this case is taken to coincide with the field direction, that is

$$\cos\theta = \frac{I_z}{\left|\vec{I}\right|} = \frac{m_I}{\sqrt{I(I+1)}}$$

It follows that

$$U_m = -g_I \frac{e\hbar}{2m_p} B_0 m_I = -g_I \mu_N B_0 m_I$$

and this is an expression for the *nuclear Zeeman effect*, where the *nuclear magneton* μ_N is introduced as

$$\mu_N = \frac{e\hbar}{2m_p}$$

The Zeeman splitting of the lowest nuclear levels in $^{57}_{26}\text{Fe}$ is illustrated in Figure 47.2.

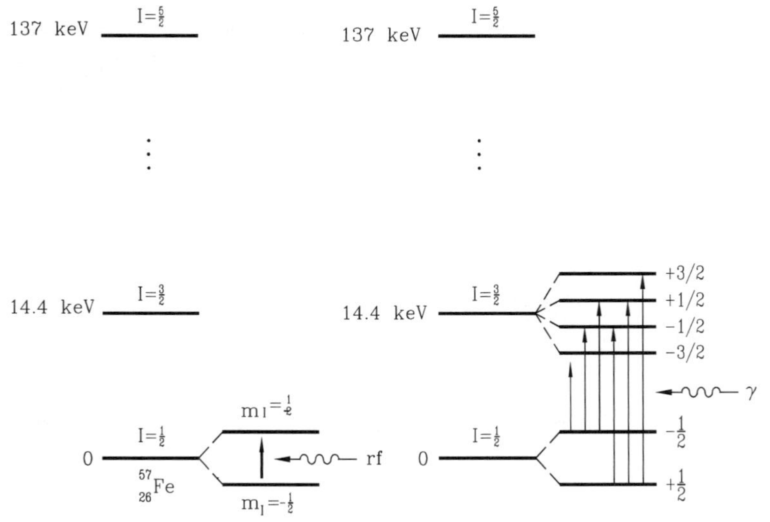

Figure 47.2. Resonance transitions between nuclear Zeeman levels in $^{57}_{26}$Fe

It can be assumed, as in the similar atomic problem, Eq.(28.19), that the energy of electromagnetic radiation emitted and absorbed in nuclear transitions, is given by the energy difference between nuclear levels as

$$\hbar \omega_{if} = \varepsilon_i - \varepsilon_f$$

Transitions between the ground Zeeman levels of $I = \frac{1}{2}$ may be produced by absorbing radiofrequency power, if the *nuclear magnetic resonance* condition is satisfied, that is

$$\hbar \omega_{if} = E(m_I = -\tfrac{1}{2}) - E(m_I = \tfrac{1}{2}) = g_I \mu_N B_z$$

A sharp resonance line is observed either by sweeping the magnetic field through the resonance value at a fixed frequency ω_{if}, or by sweeping the frequency at a given static field B_z.

It might be expected that nuclear resonant absorption should also occur for *gamma radiation*, emitted when nuclei in excited states lose their energy by radiative transitions, provided the wavelength is large compared to the size of the emitting system

$$kR = \frac{2\pi}{\lambda} R = \frac{\omega R}{c} = \hbar \omega \frac{R}{\hbar c} = (\varepsilon_i - \varepsilon_f) \frac{R}{\hbar c} \ll 1$$

which means that $\varepsilon_i - \varepsilon_f \ll \hbar c / R \cong 20$ MeV. However, the effect of the recoil momentum, which can be neglected for atomic radiation, becomes dominant for γ- radiation, because of its much higher energy. For emission or absorption of radiation from free atoms, we have a recoil energy $R = p_R^2 / 2M = p_\gamma^2 / 2M = E_\gamma^2 / 2Mc^2$, where M stands for the nuclear mass, and the recoil momentum is equal to that of the emitted photon $p_\gamma = E_\gamma / c$. This recoil energy prevents the observation of nuclear resonance absorption in free atoms. However, it has been shown that nuclear gamma resonance may occur in some nuclides if the emitting and absorbing atoms are bound in a solid lattice, such that they are no longer able to recoil individually. The effect observed in $^{57}_{26}$Fe is illustrated in Figure 47.2. The energy spectrum consists of six sharp lines, corresponding to the allowed transitions, according to the selection rule $\Delta m_I = 0, \pm 1$ between the nuclear Zeeman levels.

47.2. ALPHA DECAY

Alpha decay is the process of spontaneous emission of an α-particle ($^4_2$He), represented by the equation

$$^A_Z X \to ^{A-4}_{Z-2} X' + \alpha \tag{47.16}$$

where the atomic number, and hence, the chemical nature of the residual nucleus $^{A-4}_{Z-2}X'$ is different from that of its parent $^A_Z X$. The *separation energy* S_α needed to remove an α-particle from the parent nucleus is

$$S_\alpha = B(A,Z) - B(A-4, Z-2) - B(\alpha) = Af(A) - (A-4)f(A-4) - B(\alpha)$$

$$= (A-4)\left[f(A) - f(A-4)\right] + 4f(A) - B(\alpha)$$

where $f(A)$ is the binding fraction which, for $A > 60$, varies smoothly enough with A for its derivative with respect to A to exist, so that

$$S_\alpha = 4(A-4)\frac{df}{dA} + 4f(A) - B(\alpha) \tag{47.17}$$

The binding energy of the α-particle is $B(\alpha) = 28$ MeV and, for heavy nuclei, $f(A) \cong 7.5$ MeV, so that $4f(A) - B(\alpha) \cong 2$ MeV. Since, for large A, we have $df/dA < 0$ (see Figure 46.1) we find that S_α becomes negative for $A > 150$ and, consequently, many heavy nuclei are alpha radioactive. Experiment shows that alpha emitters with large decay energies $Q_\alpha = -S_\alpha$ have short half-lives $t_{1/2}$. The converse is also true, and this result is known as the *Geiger-Nutall law* which reads

$$\ln \lambda = a - bQ_\alpha^{-1/2} = a - \frac{b'}{v_\alpha} \tag{47.18}$$

where v_α is the velocity with which the α-particle leaves the nucleus, related to the decay energy, in the nonrelativistic case, by $Q_\alpha = m_\alpha v_\alpha^2/2$.

The mechanism of alpha decay has been explained using a single-particle model which assumes that the α-particles exist inside the nucleus and move in the field of the residual nucleus. The corresponding potential energy is given by the superposition of the nuclear and electrostatic potentials, resulting in the Coulomb barrier shown in Figure 47.3, which can be represented, to a good approximation, by

$$U(r) = -V_0 \quad , \quad r < R \tag{47.19}$$

$$= \frac{zZ'e_0^2}{r} \quad , \quad r \geq R$$

where $z = 2$ for an α-particle, $Z' = Z - 2$ is the atomic number of the residual nucleus, and $r = R$ defines the boundary of the parent nucleus.

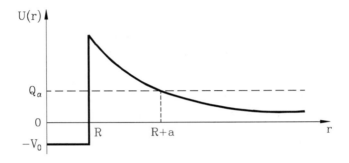

Figure 47.3. Coulomb barrier for alpha decay

We have seen that a particle can penetrate a square potential barrier of finite width, a say, and height V_0 with a probability given by the transmission coefficient $T \sim e^{-2\gamma a}$, where γ has the form (30.24). A good approximation for alpha particle tunnelling through the barrier, shown in Figure 47.3, is to replace the true potential barrier by a sequence of infinitesimal square barriers of height $U(R)$, as given by Eq.(47.19) for $r > R$, so that the probability to penetrate the complete barrier can be defined to be

$$P = e^{-2G} \tag{47.20}$$

where, in view of Eq.(30.24), the so-called *Gamov factor G* is given by

$$G = \int_R^{R+a} \gamma(r)\,dr = \int_R^{R+a} \sqrt{\frac{2m_\alpha}{\hbar^2}\left[U(r) - Q_\alpha\right]}\,dr = \int_R^{R+a} \sqrt{\frac{2m_\alpha}{\hbar^2}\left(\frac{2Z'e_0^2}{r} - Q_\alpha\right)}\,dr \tag{47.21}$$

It is clear from Figure 47.3 that the width a is specified by the condition

$$\frac{2Z'e_0^2}{R+a} = Q_\alpha \tag{47.22}$$

so that it is convenient to introduce the energy ratio

$$\frac{1}{x} = \frac{2Z'e_0^2}{rQ_\alpha} = \frac{R+a}{r} \tag{47.23}$$

Equation (47.21) may then be rewritten as

$$G = \sqrt{\frac{2m_\alpha Q_\alpha}{\hbar^2}} \frac{2Z'e_0^2}{Q_\alpha} \int_{R/(R+a)}^{1} \sqrt{\frac{1}{x}-1}\, dx = \sqrt{\frac{2m_\alpha}{\hbar^2 Q_\alpha}} 2Z'e_0^2 \left[\sqrt{x(1-x)} - \arccos\sqrt{x}\right]_{R/(R+a)}^{1}$$

$$= \sqrt{\frac{2m_\alpha}{\hbar^2 Q_\alpha}} 2Z'e_0^2 \left[\arccos\sqrt{\frac{R}{R+a}} - \sqrt{\frac{R}{R+a}\left(1-\frac{R}{R+a}\right)}\right] \tag{47.24}$$

For small energies Q_α we have $R \ll a$ (see Figure 47.3), and we can approximate the last term in Eq.(47.24) by a series expansion which gives

$$G = \sqrt{\frac{2m_\alpha}{\hbar^2 Q_\alpha}} RU(R)\left(\frac{\pi}{2} - 2\sqrt{\frac{R}{R+a}}\right) = \sqrt{\frac{2m_\alpha}{\hbar^2 Q_\alpha}} RU(R)\left[\frac{\pi}{2} - 2\sqrt{\frac{Q_\alpha}{U(R)}}\right]$$

$$= \frac{\pi RU(R)}{\hbar v_\alpha} - \frac{2R}{\hbar}\sqrt{2m_\alpha U(R)} \tag{47.25}$$

Substituting Eq.(47.25), the penetration probability (47.20) is thus obtained in terms of v_α, R and Z'. The decay constant λ of an α-emitter is given by the product of the penetration probability P and the frequency v_i/R with which the α-particle hits the barrier, that is

$$\lambda \cong \frac{v_i}{R} P \tag{47.26}$$

where v_i denotes the velocity of the α-particle inside the nucleus. By combining Eqs.(47.26), (47.25) and (47.20) we then obtain

$$\ln \lambda \cong \ln \frac{v_i}{R} - \frac{2\pi RU(R)}{\hbar v_\alpha} + \frac{4R}{\hbar}\sqrt{2m_\alpha U(R)} \tag{47.27}$$

and this correctly reproduces the Geiger-Nuttal law (47.18) to within a factor of 44. Due to the strong dependence of λ on R, Eq.(47.27) can be used to evaluate R. The available data are consistent with the relation (46.3) and yield R_0 in the range of 1.4 to 1.5×10^{-15} m, which represents an overestimate, due to the fact that the size of the alpha particles has not been considered.

These results hold for heavy nuclei in their ground state $l = 0$. For $l > 0$ a centrifugal potential should be added to the potential energy $U(r)$, as shown before in Eq.(46.27). As a result, the height of the potential barrier (47.19) and its effective width a are both increased. Instead of Eq.(47.21), the Gamov factor must be expressed as

$$G = \int_R^{R+a} \sqrt{\frac{2m_\alpha}{\hbar^2}\left[\frac{2Z'e_0^2}{r} + \frac{l(l+1)\hbar^2}{2m_\alpha r^2} - Q_\alpha\right]}\, dr$$

However, it is clear that, for large Z, the influence of the centrifugal term is small, because the ratio of the centrifugal to the Coulomb barrier height is

$$\frac{l(l+1)\hbar^2}{4m_\alpha Z'e_0^2 R} \cong 0.002\, l(l+1)$$

and hence, λ does not depend strongly on the quantum number l. The simplifying assumptions of the one-particle model seem therefore to be justified by the small influence of the neglected factors on the decay constant λ.

47.3. Beta Decay

The term β-decay describes processes in which a nucleus makes an isobaric transition where the mass number A remains constant and the atomic number Z increases or decreases by one. These processes are the emission of electrons e^- (negative beta decay, β^-) or positrons e^+ (positive beta decay, β^+). The second of these two processes is always simultaneous with the capture by the nucleus of an orbital atomic electron (electron capture). Although positive or negative electrons are emitted in β-decay, there is strong evidence that they cannot be a constituent part of nuclei. We are then led to the assumption that electrons are created at the moment the nucleus decays. Since the only change between the initial and final nucleus is that a neutron has been changed into a proton, in a β^--decay, or conversely, in a β^+-decay, the basic decay processes can be assigned to nucleons as

$$_Z^A X \rightarrow {_{Z+1}^A} X' + e^- \quad \text{or} \quad {_0^1 n} \rightarrow {_1^1 p} + e^-$$

$$_Z^A X \rightarrow {_{Z-1}^A} X' + e^+ \quad \text{or} \quad {_1^1 p} \rightarrow {_0^1 n} + e^+$$

$$_Z^A X + e^- \rightarrow {_{Z-1}^A} X' \quad \text{or} \quad {_1^1 p} + e^- \rightarrow {_0^1 n}$$

Such two-body processes would predict that each β-particle should be emitted with a well-defined energy. However, it is observed experimentally that electrons in β-decay have a continuous energy distribution. To account for the variable energy of β-particles, it must be postulated that another particle is emitted simultaneously, taking up the remaining energy and momentum, and this should be written as

$$_0^1n \to {}_1^1p + e^- + \bar{\nu}$$

$$_1^1p \to {}_0^1n + e^+ + \nu \qquad (47.28)$$

$$_1^1p + e^- \to {}_0^1n + \nu$$

The additional particle, called a *neutrino* ν or an *antineutrino* $\bar{\nu}$, must be neutral, because the charge is already conserved without taking the neutrino into account, and must have zero or almost zero rest mass, because in all β-decays the maximum electron energy observed is practically equal to the total energy available. Since protons, neutrons and electrons each have spin $\frac{1}{2}\hbar$, the conservation of the angular momentum requires the neutrino to have spin $\frac{1}{2}\hbar$. The theory of β-decay is based on the analogy of the decay processes (47.28) with photon emission, where a photon is created during a transition caused by a perturbation (weak interaction) of the stationary states. Hence, Fermi's golden rule (47.1) can be used to obtain the probability per unit time $N(p)dp$ for the emission of a β-particle within the momentum range p to $p+dp$, which reads

$$N(p)dp = \frac{2\pi}{\hbar} \left| \langle \psi_e \psi_\nu \Psi_f | \hat{V} | \Psi_i \rangle \right|^2 \frac{dn}{dQ} \qquad (47.29)$$

where Q is the decay energy, and Ψ_i, Ψ_f are the time-independent wave functions of the nuclear system in the initial and final states. The wave functions ψ_e, ψ_ν of the escaping electron and neutrino can be taken in the free-particle form, normalized within a volume V as

$$\psi_e(\vec{r}) = \frac{1}{\sqrt{V}} e^{i\vec{p}\cdot\vec{r}/\hbar} \quad , \quad \psi_\nu(\vec{r}) = \frac{1}{\sqrt{V}} e^{i\vec{q}\cdot\vec{r}/\hbar} \qquad (47.30)$$

Since the interaction occurs at the position of nucleus only, and $\vec{p}\cdot\vec{r} \ll \hbar$, $\vec{q}\cdot\vec{r} \ll \hbar$ over the nuclear volume, we may expand the exponentials in power series and keep only the first term as

$$e^{i\vec{p}\cdot\vec{r}/\hbar} = 1 + i\vec{p}\cdot\vec{r}/\hbar + \ldots \cong 1$$
$$\qquad (47.31)$$
$$e^{i\vec{q}\cdot\vec{r}/\hbar} = 1 + i\vec{q}\cdot\vec{r}/\hbar + \ldots \cong 1$$

To this approximation, the interaction matrix element becomes independent of the energies of both the electron and neutrino, and it is responsible for the so-called *allowed transitions*. It is usually expressed in terms of a constant g of the weak interaction as

$$\langle \Psi_f | \hat{V} | \Psi_i \rangle = \frac{g}{V} \langle \Psi_f | \hat{M} | \Psi_i \rangle = \frac{g}{V} M_{if} \tag{47.32}$$

where M_{if} is called the nuclear matrix element. The form of the $\hat{M}$ operator is left unspecified, because the weak interaction is not known. Hence, the probability (47.29) of an allowed transition depends on the electron and neutrino energies through the density of the final states only. The number of states available for a free particle in the wave vector interval dk can be derived from Eq.(40.9), which gives

$$dn = \frac{d}{dk}\left(\frac{Vk^3}{6\pi^2}\right)dk = \frac{Vk^2}{2\pi^2}dk$$

if it is assumed that the spin direction remains fixed. Thus, an electron of momentum $\vec{p} = \hbar \vec{k}$ has a number of final states in the momentum interval dp given by

$$dn_e = V \frac{4\pi p^2}{(2\pi \hbar)^3} dp \tag{47.33}$$

and similarly, for the neutrino of momentum $\vec{q} = \hbar \vec{k}$, the number of final states in the momentum interval dq is

$$dn_\nu = V \frac{4\pi q^2}{(2\pi \hbar)^3} dq \tag{47.34}$$

It follows that the number of final states which have simultaneously an electron and a neutrino with appropriate momenta can be written as

$$dn = \frac{16\pi^2 V^2}{(2\pi \hbar)^6} p^2 q^2 dp dq \tag{47.35}$$

The kinetic energy of the neutrino can be derived from the decay energy Q, as given by

$$Q = T_e + T_\nu \quad \text{or} \quad T_\nu = Q - T_e$$

so that its relativistic momentum (5.12) reads

$$q = \frac{T_\nu}{c} = \frac{Q - T_e}{c} \tag{47.36}$$

where it was assumed that the neutrino has zero rest mass. For a fixed p, that is, at constant T_e, we have $dq = dQ/c$ and Eq.(47.35) becomes

$$\frac{dn}{dQ} = \frac{16\pi^2 V^2}{(2\pi\hbar)^6 c^3}(Q-T_e)^2 p^2 dp \qquad (47.37)$$

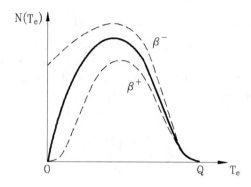

Figure 47.4. Electron energy distribution in β-decay. The dashed lines show the effect of the Coulomb interaction on the spectrum shape for negative and positive β-decays

Substituting Eqs.(47.37) and (47.32) into Eq.(47.29) gives

$$N(p)dp = \frac{g^2}{2\pi^3 \hbar^7 c^3}|M_{if}|^2(Q-T_e)^2 p^2 dp \qquad (47.38)$$

This momentum distribution can be transformed into an energy distribution, using Eq.(5.11), as

$$c^2 p\, dp = E_e dE_e = (T_e + m_e c^2)dT_e$$

so that Eq.(47.38) may assume the form

$$N(T_e)dT_e = \frac{g^2}{2\pi^3 \hbar^7 c^3}|M_{if}|^2(Q-T_e)^2(T_e^2 + 2T_e m_e c^2)^{1/2}(T_e + m_e c^2)dT_e \qquad (47.39)$$

A plot of Eq.(47.39) is shown in Figure 47.4. The distribution vanishes at the minimum energy $T_e = 0$ and also at the endpoint $T_e = Q$, which is the maximum energy of the electron. At low kinetic energies the Coulomb effect of the nucleus on the emitted β-particles, which has been neglected so far, becomes significant. The electrons are slowed down by the attractive Coulomb potential, as they leave the nucleus and the β^--spectrum is shifted toward lower energies. The positrons are accelerated by electrostatic repulsion, and there will be a deficiency of low energy positrons in the β^+-spectrum (see Figure 47.4). The Coulomb correction is applied by multiplying Eqs.(47.38) and (47.39) by the Fermi function $F(Z',p)$ and $F(Z',T_e)$, respectively, where Z' is the atomic number of the residual nucleus.

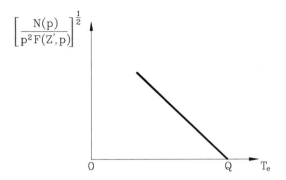

Figure 47.5. Kurie plot for allowed β transitions

The momentum distribution (47.38) may then be rewritten as

$$N(p)dp = \text{const} \cdot F(Z',p)(Q-T_e)^2 p^2 dp \qquad (47.40)$$

or

$$\left[\frac{N(p)}{p^2 F(Z',p)}\right]^{1/2} = \text{const} \cdot (Q-T_e) \qquad (47.41)$$

If the function on the left-hand side of Eq.(47.41) is plotted versus T_e, a straight line should be obtained for the allowed transitions, which makes an intercept Q on the energy axis, as in Figure 47.5. Such a representation, called a *Kurie plot*, gives an accurate way of measuring Q and provides a simple experimental test of the theory.

The total decay probability λ is obtained by integrating Eq.(47.40) over the momentum range, which yields

$$\lambda = \int_0^{p_{max}} N(p)dp = \text{const} \cdot \int_0^{p_{max}} F(Z',p)(Q-T_e)^2 p^2 dp$$

$$= \text{const} \cdot m_e^5 c^7 f\left(Z', \frac{Q}{m_e c^2}\right) = Cf\left(Z', \frac{Q}{m_e c^2}\right) \qquad (47.42)$$

where

$$C = \frac{m_e^5 c^4}{2\pi^3 \hbar^7} g^2 |M_{if}|^2 \qquad (47.43)$$

Since values for $F(Z',p)$ and $f(Z',Q/m_e c^2)$ have been calculated using experimental nuclear data, Eq.(47.42) can be written in terms of the half-life $t_{1/2}$ of the decay as

$$t_{1/2} = \frac{\ln 2}{\lambda} = \frac{\ln 2}{Cf} \quad \text{or} \quad ft_{1/2} = \frac{\ln 2}{C} \qquad (47.44)$$

The quantity $ft_{1/2}$ is called the *ft-value* for the transition, and this can be calculated knowing $t_{1/2}$ and Q (thus f) given by experiment. By combining Eqs.(47.44) and (47.43) we can estimate the magnitude of the β-decay interaction, which is of the order of 10^{-13} weaker than the so-called strong interaction between nucleons.

FURTHER READING
1. W. N. Cottingham, D. A. Greenwood - AN INTRODUCTION TO NUCLEAR PHYSICS, Cambridge University Press, Cambridge, 2001.
2. J. S. Lilley - NUCLEAR PHYSICS, Wiley, New York, 2001.
3. R. J. Blin-Stoyle - NUCLEAR AND PARTICLE PHYSICS, Chapman and Hall, London, 1991.

PROBLEMS

47.1. Find an expression for the transition probability of a nuclear transition observed when the detection time t_0 is short compared to the mean lifetime $\langle t \rangle$ of the decaying state.

Solution

If observations are made up to a given time t_0, Eq.(47.12) becomes

$$a_f(t_0) = \frac{\hbar V_{fi}}{(\varepsilon_f - \varepsilon_i) + i\Gamma/2}\left[e^{-\Gamma t_0/2\hbar} e^{i(\varepsilon_f - \varepsilon_i)t_0/\hbar} - 1 \right]$$

and therefore, we obtain the transition probability as

$$P(\varepsilon) = \frac{\hbar^2 |V_{fi}|^2}{(\varepsilon - \varepsilon_i)^2 + \frac{1}{4}\Gamma^2}\left[1 + e^{-\Gamma t_0/\hbar} - 2e^{-\Gamma t_0/2\hbar} \cos\left(\frac{\varepsilon - \varepsilon_i}{\hbar} t_0\right) \right]$$

For $t_0 \ll \langle t \rangle = \hbar/\Gamma$ or $\Gamma t_0/\hbar \ll 1$, it follows that the observed line profile can be written in the form

$$P(\varepsilon) \approx 2P_0 \frac{1 - \cos\left(\frac{\varepsilon - \varepsilon_i}{\hbar} t_0\right)}{(\varepsilon - \varepsilon_i)^2} = P_0 \frac{\sin^2\left(\frac{\varepsilon - \varepsilon_i}{2\hbar} t_0\right)}{(\varepsilon - \varepsilon_i)^2}$$

The linewidth is approximately given by $\hbar/t_0$, which is much larger than Γ, and hence, the spread of the observed energy distribution is much increased.

47.2. Find the nuclear spin I predicted by the nuclear shell model for the following odd A nuclides: $^{7}_{3}\text{Li}$, $^{15}_{7}\text{N}$, $^{39}_{19}\text{K}$, $^{59}_{27}\text{Co}$ and $^{23}_{11}\text{Na}$.

Solution

The diagram given in Figure 46.7, where each nuclear level $n(l\text{ symbol})_j$ may contain $2j+1$ nucleons, is valid for nucleons of one kind, and hence it is convenient to separate the total number of neutrons and protons accommodated in the scheme. All the involved nuclides have an even number of neutrons that combine in pairs with antiparallel spins, and this gives a resultant zero angular momentum. In the case of $^{7}_{3}\text{Li}$, for example, the closed shell $1s_{\frac{1}{2}}$ contains two neutrons and the other two are accommodated in the $1p_{\frac{3}{2}}$ state. Hence we are concerned with the occupation numbers for protons only. The proton configuration is obtained from Figure 46.7 as

$$^{7}_{3}\text{Li}: \quad (1s_{\frac{1}{2}})^2 (1p_{\frac{3}{2}})^1$$

$$^{15}_{7}\text{N}: \quad (1s_{\frac{1}{2}})^2 (1p_{\frac{3}{2}})^4 (1p_{\frac{1}{2}})^1$$

$$^{39}_{19}\text{K}: \quad (1s_{\frac{1}{2}})^2 (1p_{\frac{3}{2}})^4 (1p_{\frac{1}{2}})^2 (1d_{\frac{5}{2}})^6 (2s_{\frac{1}{2}})^2 (1d_{\frac{3}{2}})^3$$

$$^{59}_{27}\text{Co}: \quad (1s_{\frac{1}{2}})^2 (1p_{\frac{3}{2}})^4 (1p_{\frac{1}{2}})^2 (1d_{\frac{5}{2}})^6 (2s_{\frac{1}{2}})^2 (1d_{\frac{3}{2}})^4 (1f_{\frac{7}{2}})^7$$

The nuclear spin is determined by the angular momentum of the last unpaired proton, such that $I = \frac{1}{2}$ for $^{15}_{7}\text{N}$, $I = \frac{3}{2}$ for $^{7}_{3}\text{Li}$ and $^{39}_{19}\text{K}$, and $I = \frac{7}{2}$ for $^{59}_{27}\text{Co}$, as experimentally measured.

Note that the prediction based on the nuclear shell model breaks down in a few cases, for example for $^{23}_{11}\text{Na}$, which has the proton configuration given by

$$^{23}_{11}\text{Na}: \quad (1s_{\frac{1}{2}})^2 (1p_{\frac{3}{2}})^4 (1p_{\frac{1}{2}})^2 (1d_{\frac{5}{2}})^3$$

It would be expected to have a spin $I = \frac{5}{2}$, but the measured value is $I = \frac{3}{2}$. In this case the pairing is broken, and the three protons in the $1d_{\frac{5}{2}}$ level combine their angular momenta to produce a nuclear spin of $\frac{3}{2}$.

47.3. Use the expression of the Gamov factor G to calculate the nuclear unit radius R_0 in terms of Z, A, Q_α and $t_{1/2}$ values for a given α-emitting nuclide.

Solution

It is convenient to use the Gamov factor, Eq.(47.25), written as

$$G = \sqrt{\frac{2m_\alpha}{\hbar^2 Q_\alpha}} \, 2Z'e_0^2 \left(\frac{\pi}{2} - 2\sqrt{\frac{R}{R+a}}\right) = \frac{e^2\sqrt{2m_\alpha}}{4\varepsilon_0 \hbar} \frac{Z'}{\sqrt{Q_\alpha}} - \sqrt{\frac{4m_\alpha e^2}{\pi\varepsilon_0 \hbar^2}} \sqrt{Z'R}$$

Given that the energies of the α-particles range from 4 to 9 MeV, and the nuclear radius is of the order of femtometers (fm) or 10^{-15} m, this formula simplifies to

$$G = 1.97 \frac{Z'}{\sqrt{Q_\alpha}} - 1.49 \sqrt{Z'R}$$

if Q_α is measured in MeV and R in fm. The corresponding velocities of the emitted α-particles are of about 10^7 m/s, and hence, assuming velocities v_i of the same order inside the nucleus, we obtain

$$\log t_{1/2} = \log \frac{\ln 2}{\lambda} \approx \log \frac{v_i}{R} - \frac{1}{2.3}\left(3.94 \frac{Z'}{\sqrt{Q_\alpha}} - 2.98\sqrt{Z'R}\right)$$

or

$$R_0 = \left(\frac{17 + 0.77 \log t_{1/2}}{\sqrt{Z-2}} + 1.32\sqrt{\frac{Z-2}{Q_\alpha}}\right)^2 A^{-1/3}$$

if $t_{1/2}$ is measured in seconds. The available data yield R_0 values in the range of 1.4 to 1.5 fm, which represents, however, an overestimate, due to the fact that the size of the α-particles has not been considered. However, it is obvious that variations of several orders of magnitude in the half-life of α-emitting isotopes, which is in the range of 10^{-7} s to 10^9 years, result in a small change in the R_0 value.

47.4. Find the kinetic energy T_α of an α-particle emitted by a nucleus of mass number A, given the decay energy Q_α.

Solution

The conservation of energy for the alpha decay process, Eq.(47.16), can be written as

$$M_X c^2 = M_{X'} c^2 + M_\alpha c^2 + T_\alpha + T_{X'}$$

where M stands for the nuclear mass. It follows that

$$Q_\alpha = M_X c^2 - (M_{X'} + M_\alpha) c^2 = T_\alpha + T_{X'}$$

where, since $|\vec{p}_{X'}| = |\vec{p}_\alpha|$, the recoil energy $T_{X'}$ is given by

$$T_{X'} = \frac{p_{X'}^2}{2M_{X'}} = \frac{M_\alpha}{M_{X'}}\left(\frac{p_\alpha^2}{2M_\alpha}\right) = \frac{M_\alpha}{M_{X'}}T_\alpha$$

It follows that

$$Q_\alpha = T_\alpha \frac{M_{X'}+M_\alpha}{M_{X'}} \approx T_\alpha \frac{(A-4)+4}{A-4} = T_\alpha \frac{A}{A-4}$$

and hence

$$T_\alpha = Q_\alpha\left(1-\frac{4}{A}\right)$$

a value which is close to Q_α for heavy nuclei.

47.5. Calculate the average energy $\langle T_e \rangle$ of β-decay, assuming that $Q < m_e c^2 = 0.51$ MeV.

Solution

If $Q < m_e c^2$ (for instance in the β-decay of ^{14}C, where $Q = 0.156$ (MeV)), it follows that $T_e \ll m_e c^2$, and hence, the energy distribution can be approximated as $N(T_e)dT_e = T_e^{1/2}(Q-T_e)^2 dT_e$. The average energy is given by

$$\langle T_e \rangle = \frac{\int_0^Q T_e N(T_e)dT_e}{\int_0^Q N(T_e)dT_e} = \frac{\int_0^Q (Q-T_e)^2 T_e^{3/2} dT_e}{\int_0^Q (Q-T_e)^2 T_e^{1/2} dT_e}$$

$$= \frac{Q^2 \int_0^Q T_e^{3/2} dT_e - 2Q \int_0^Q T_e^{5/2} dT_e + \int_0^Q T_e^{7/2} dT_e}{Q^2 \int_0^Q T_e^{1/2} dT_e - 2Q \int_0^Q T_e^{3/2} dT_e + \int_0^Q T_e^{5/2} dT_e} = \frac{Q^{9/2}(2/5-4/7+2/9)}{Q^{7/2}(2/3-4/5+2/7)} = \frac{Q}{3}$$

47.6. Assuming that, at time $t = 0$, we have N_0 atoms of a radioactive nuclide X_1 which decays through the series

$$X_1 \xrightarrow{\lambda_1} X_2 \xrightarrow{\lambda_2} X_3 \xrightarrow{\lambda_3} \cdots$$

show that the number of atoms $N_i(t)$ present at time $t > 0$, where $i = 1, 2, 3, \ldots$, is given by $N_i = \sum_{j=1}^{i} A_{ij} e^{-\lambda_j t}$, and find a recurrence relation for the A_{ij}.

47.7. Show that the transition probability for the gamma radiation emitted by an ensemble of radioactive nuclides of mass M at temperature T follows a Gaussian distribution, and find the linewidth Δ of the emitted radiation.

47.8. Find the scattering angle θ from a free ^{12}C nucleus at rest for an alpha particle, assuming an elastic collision in which the particle loses $f = 20\%$ of its initial speed.

47.9. Show that the β-decay energy can be written as

$$Q_{\beta^\pm} = 2wc^2\left[\pm(Z-Z_A)-\frac{1}{2}\right]-2m_ec^2\delta$$

where Z_A corresponds to the most stable nucleus among the isobars of given A, and w is the coefficient of the parabolic representation for the semiempirical mass formula. What is the value of the parameter δ for the $\beta^\pm$-decay?

47.10. Show that the ratio of the nuclear recoil energy T_R to the maximum energy T_e of the electron in β^--decay is given by

$$\frac{T_R}{T_e} = \frac{Q+2m_ec^2}{Q+2m_Rc^2}$$

Appendix I. Vector Calculus

❶ A function $\Psi(x, y, z)$ that describes a scalar physical property at all points in space, by a single value for each point, is called a *scalar field*. Vector physical quantities are described by *vector fields* $\vec{A}(x, y, z)$ with three components for each point. The derivative of a vector field is defined by analogy with ordinary calculus as the vector sum of the derivatives of its components, that is,

$$\frac{d\vec{A}}{d\xi} = \vec{e}_x \frac{dA_x}{d\xi} + \vec{e}_y \frac{dA_y}{d\xi} + \vec{e}_z \frac{dA_z}{d\xi}$$

so that the usual rules for differentiation are followed.

The spatial change of a scalar field associated with coordinate changes at some point (x, y, z) is given by the elementary equation

$$d\Psi = \frac{\partial \Psi}{\partial x} dx + \frac{\partial \Psi}{\partial y} dy + \frac{\partial \Psi}{\partial z} dz$$

that can be written as the scalar product of a certain vector $\vec{A}$ with $d\vec{r}$ in the form

$$d\Psi = \left(\vec{e}_x \frac{\partial \Psi}{\partial x} + \vec{e}_y \frac{\partial \Psi}{\partial y} + \vec{e}_z \frac{\partial \Psi}{\partial z} \right) \cdot \left(\vec{e}_x dx + \vec{e}_y dy + \vec{e}_z dz \right) = \vec{A} \cdot d\vec{r}$$

The vector $\vec{A}$ is called the *gradient* of Ψ, denoted by $\nabla \Psi$, and gives the spatial rate of change of Ψ as

$$\nabla \Psi = \vec{e}_x \frac{\partial \Psi}{\partial x} + \vec{e}_y \frac{\partial \Psi}{\partial y} + \vec{e}_z \frac{\partial \Psi}{\partial z}$$

where the *gradient* or *del* operator ∇ is defined as

$$\nabla = \vec{e}_x \frac{\partial}{\partial x} + \vec{e}_y \frac{\partial}{\partial y} + \vec{e}_z \frac{\partial}{\partial z} \tag{I.1}$$

Hence, in terms of the gradient operator, $d\Psi$ is equally well expressed by

$$d\Psi = \nabla \Psi \cdot d\vec{r}$$

The various vector operations of ∇ follow directly from its definition. The *divergence* of a vector field $\vec{A}$ at some point, denoted by $\nabla \cdot \vec{A}$, is a scalar given by

$$\nabla \cdot \vec{A} = \left(\vec{e}_x \frac{\partial}{\partial x} + \vec{e}_y \frac{\partial}{\partial y} + \vec{e}_z \frac{\partial}{\partial z} \right) \cdot \left(\vec{e}_x A_x + \vec{e}_y A_y + \vec{e}_z A_z \right) = \frac{\partial A_x}{\partial x} + \frac{\partial A_y}{\partial y} + \frac{\partial A_z}{\partial z} \quad (I.2)$$

The value of the divergence of $\vec{A}$ at any given point represents the total outgoing flux of the vector field $\vec{A}$ per unit volume, that is,

$$\nabla \cdot \vec{A} = \lim_{V \to 0} \frac{\Phi}{V}$$

where the flux of $\vec{A}$ through the surface element $d\vec{S}$ is

$$d\Phi = \vec{A} \cdot d\vec{S}$$

To extend this result to a finite volume it is necessary to sum the individual fluxes of a large number of infinitesimal volume elements that yield the *divergence* or *Gauss theorem* in the form

$$\int_V (\nabla \cdot \vec{A}) dV = \int_S \vec{A} \cdot d\vec{S} \quad (I.3)$$

where the integral on the left-hand side involves values of $\nabla \cdot \vec{A}$ throughout the volume, V, whereas the right-hand integral involves values of $\vec{A}$ on the closed surface S bounding the volume only.

A vector field $\vec{A}$ has a *curl* denoted by $\nabla \times \vec{A}$ which is a vector of the form

$$\nabla \times \vec{A} = \begin{vmatrix} \vec{e}_x & \vec{e}_y & \vec{e}_z \\ \partial/\partial x & \partial/\partial y & \partial/\partial z \\ A_x & A_y & A_z \end{vmatrix} = \vec{e}_x \left(\frac{\partial A_z}{\partial y} - \frac{\partial A_y}{\partial z} \right) + \vec{e}_y \left(\frac{\partial A_x}{\partial z} - \frac{\partial A_z}{\partial x} \right) + \vec{e}_z \left(\frac{\partial A_y}{\partial x} - \frac{\partial A_x}{\partial y} \right)$$

(I.4)

The value of $\nabla \times \vec{A}$ at any given point represents the line integral around a closed path of the vector field $\vec{A}$ per unit surface

$$\nabla \times \vec{A} = \lim_{S \to 0} \frac{C}{S}$$

where the line integral of $\vec{A}$ along an elementary path $d\vec{l}$ is

$$dC = \vec{A} \cdot d\vec{l}$$

If this result is added up for all the elements $d\vec{S}$ making up a finite surface, we obtain a relationship between the flux of $\nabla \times \vec{A}$ through this surface and the line integral around the perimeter Γ of the surface of the form

$$\int_S (\nabla \times \vec{A}) \cdot d\vec{S} = \int_\Gamma \vec{A} \cdot d\vec{l} \tag{I.5}$$

which is known as *Stokes's theorem*. If the line integral of a certain vector field $\vec{A}$ is zero around any closed path, the vector field is called a *conservative* field. From Stokes's theorem (I.5) it follows that for a conservative field we have $\nabla \times \vec{A} = 0$.

The divergence of the gradient of a scalar field, denoted by $\nabla^2 \Psi$, is obtained as

$$\nabla^2 \Psi = \nabla \cdot (\nabla \Psi) = \left(\vec{e}_x \frac{\partial}{\partial x} + \vec{e}_y \frac{\partial}{\partial y} + \vec{e}_z \frac{\partial}{\partial z} \right) \cdot \left(\vec{e}_x \frac{\partial \Psi}{\partial x} + \vec{e}_y \frac{\partial \Psi}{\partial y} + \vec{e}_z \frac{\partial \Psi}{\partial z} \right)$$

$$= \vec{e}_x \frac{\partial \Psi}{\partial x} + \vec{e}_y \frac{\partial \Psi}{\partial y} + \vec{e}_z \frac{\partial \Psi}{\partial z}$$

where ∇^2 is called the *Laplacian* operator, which is defined by

$$\nabla^2 = \frac{\partial^2}{\partial x^2} + \frac{\partial^2}{\partial y^2} + \frac{\partial^2}{\partial z^2} \tag{I.6}$$

❷ The following vector identities, where Ψ and Φ are scalar fields and $\vec{A}$ and $\vec{B}$ are vector fields, may be proved in a straightforward manner by expanding the left- and right-hand sides and comparing the components:

$$\nabla(\Psi + \Phi) = \nabla\Psi + \nabla\Phi \tag{I.7}$$

$$\nabla(\Psi\Phi) = \Psi\nabla\Phi + \Phi\nabla\Psi \tag{I.8}$$

$$\nabla \times (\nabla\Psi) = 0 \tag{I.9}$$

$$\nabla \cdot (\Psi\vec{A}) = \vec{A} \cdot \nabla\Psi + \Psi\nabla \cdot \vec{A} \tag{I.10}$$

$$\nabla \times (\Psi\vec{A}) = (\nabla\Psi) \times \vec{A} + (\nabla \times \vec{A}) \tag{I.11}$$

$$\nabla(\vec{A} + \vec{B}) = \nabla\vec{A} + \nabla\vec{B} \tag{I.12}$$

$$\nabla \times (\vec{A} + \vec{B}) = \nabla \times \vec{A} + \nabla \times \vec{B} \tag{I.13}$$

$$\nabla \cdot (\vec{A} \times \vec{B}) = \vec{B} \cdot (\nabla \times \vec{A}) - \vec{A} \cdot (\nabla \times \vec{B}) \tag{I.14}$$

$$\nabla \times (\vec{A} \times \vec{B}) = (\vec{B} \cdot \nabla)\vec{A} - \vec{B}(\nabla \cdot \vec{A}) - (\vec{A} \cdot \nabla)\vec{B} + \vec{A}(\nabla \cdot \vec{B}) \tag{I.15}$$

$$\nabla(\vec{A}\cdot\vec{B}) = (\vec{B}\cdot\nabla)\vec{A} + (\vec{A}\cdot\nabla)\vec{B} + \vec{B}\times(\nabla\times\vec{A}) + \vec{A}\times(\nabla\times\vec{B}) \tag{I.16}$$

$$\nabla\cdot(\nabla\times\vec{A}) = 0 \tag{I.17}$$

$$\nabla\times(\nabla\times\vec{A}) = \nabla(\nabla\cdot\vec{A}) - \nabla^2\vec{A} \tag{I.18}$$

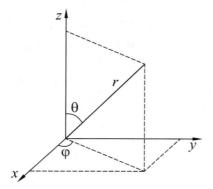

Figure I. Connection between Cartesian and spherical polar coordinates

❸ The identities (I.7) to (I.18) are independent of the type of coordinate system. However, the forms of the gradient, divergence, *curl* and Laplacian depend on the kind of coordinate system used. In spherical polar coordinates, where $\vec{e}_r$, $\vec{e}_\theta$ and $\vec{e}_\varphi$ are unit vectors in the directions of increasing r, θ and φ, as illustrated in Figure I, we have

$$x = r\sin\theta\cos\varphi \quad , \quad y = r\sin\theta\sin\varphi \quad , \quad z = r\cos\theta \tag{I.19}$$

Transformation of Eqs.(I.1), (I.2), (I.4) and (I.6) to the spherical polar coordinate system yields

$$\nabla\Psi = \vec{e}_r\frac{\partial\Psi}{\partial r} + \vec{e}_\theta\frac{1}{r}\frac{\partial\Psi}{\partial\theta} + \vec{e}_\varphi\frac{1}{r\sin\theta}\frac{\partial\Psi}{\partial\varphi} \tag{I.20}$$

$$\nabla\cdot\vec{A} = \frac{1}{r^2}\frac{\partial}{\partial r}(r^2 A_r) + \frac{1}{r\sin\theta}\frac{\partial}{\partial\theta}(\sin\theta A_\theta) + \frac{1}{r\sin\theta}\frac{\partial A_\varphi}{\partial\varphi} \tag{I.21}$$

$$\nabla\times\vec{A} = \frac{\vec{e}_r}{r\sin\theta}\left(\frac{\partial}{\partial\theta}(\sin\theta A_\varphi) - \frac{\partial A_\theta}{\partial\varphi}\right) + \frac{\vec{e}_\theta}{r}\left[\frac{1}{\sin\theta}\frac{\partial A_r}{\partial\varphi} - \frac{\partial}{\partial r}(rA_\varphi)\right]$$

$$+ \frac{\vec{e}_\varphi}{r}\left[\frac{\partial}{\partial r}(rA_\theta) - \frac{\partial A_r}{\partial\theta}\right] \tag{I.22}$$

$$\nabla^2\Psi = \frac{1}{r^2}\frac{\partial}{\partial r}\left(r^2\frac{\partial\Psi}{\partial r}\right) + \frac{1}{r^2\sin\theta}\frac{\partial}{\partial\theta}\left(\sin\theta\frac{\partial\Psi}{\partial\theta}\right) + \frac{1}{r^2\sin^2\theta}\frac{\partial^2\Psi}{\partial\varphi^2} \tag{I.23}$$

In cylindrical coordinates, where $\vec{e}_r$, $\vec{e}_\theta$ and $\vec{e}_z$ are unit vectors in the directions of increasing r, θ and z, we have

$$x = r\cos\theta \quad , \quad y = r\sin\theta \quad , \quad z \qquad (I.24)$$

and therefore

$$\nabla\Psi = \vec{e}_r \frac{\partial \Psi}{\partial r} + \vec{e}_\theta \frac{1}{r}\frac{\partial \Psi}{\partial \theta} + \vec{e}_z \frac{\partial \Psi}{\partial z} \qquad (I.25)$$

$$\nabla \cdot \vec{A} = \frac{1}{r}\frac{\partial}{\partial r}(rA_r) + \frac{1}{r}\frac{\partial A_\theta}{\partial \theta} + \vec{e}_z \frac{\partial A_z}{\partial z} \qquad (I.26)$$

$$\nabla \times \vec{A} = \vec{e}_r \left(\frac{1}{r}\frac{\partial A_z}{\partial \theta} - \frac{\partial A_\theta}{\partial z} \right) + \vec{e}_\theta \left(\frac{\partial A_r}{\partial z} - \frac{\partial A_z}{\partial \theta} \right) + \vec{e}_z \left[\frac{1}{r}\frac{\partial}{\partial r}(rA_\theta) - \frac{1}{r}\frac{\partial A_r}{\partial \theta} \right] \qquad (I.27)$$

$$\nabla^2 \Psi = \frac{1}{r}\frac{\partial}{\partial r}\left(r\frac{\partial \Psi}{\partial r} \right) + \frac{1}{r^2}\frac{\partial^2 \Psi}{\partial \theta^2} + \frac{\partial^2 \Psi}{\partial z^2} \qquad (I.28)$$

FURTHER READING
1. P. V. O'Neil - ADVANCED ENGINEERING MATHEMATICS, Thomson Learning, Stamford, 2002.
2. D. G. Zill, M. R. Cullen - ADVANCED ENGINEERING MATHEMATICS, Johnes & Bartlett, Sudbury, 1999.
3. M. Greenberg - ADVANCED ENGINEERING MATHEMATICS, Pearson Education, Boston, 1998.

APPENDIX II. MATRICES

❶ The components A_i of a vector $\vec{A}$ with respect to a Cartesian coordinate system, where i stands for x, y or z, can be written in the form of a column

$$\vec{A} = \begin{pmatrix} A_x \\ A_y \\ A_z \end{pmatrix} \tag{II.1}$$

which is the representative of the vector in our particular coordinate system. A *transformation* of the vector $\vec{A}$ into another vector $\vec{B}$, which in general changes both the direction and the magnitude of $\vec{A}$, is said to be *linear* if any component of the representative of $\vec{B}$ is a linear combination of the components of the representative of $\vec{A}$ of the form

$$\sum_j T_{ij} A_j = B_i \quad \text{or} \quad \hat{T}\vec{A} = \vec{B} \tag{II.2}$$

The constant coefficients T_{ij}, having two indices, form a set of nine components $(i, j = x, y, z)$ of a so-called second-rank Cartesian *tensor*, which can be arranged in a square array, called a *matrix*

$$\hat{T} = \begin{pmatrix} T_{xx} & T_{xy} & T_{xz} \\ T_{yx} & T_{yy} & T_{yz} \\ T_{zx} & T_{zy} & T_{zz} \end{pmatrix} \tag{II.3}$$

which is the representative of the linear transformation in the given coordinate system. *Tensor* is a general name given to quantities that transform in prescribed ways when the coordinate system is rotated. A scalar Ψ which is independent of the coordinate system is a zeroth-rank tensor. A vector, whose components A_i transform as do the coordinates of a point, is a first-rank tensor. A second-rank tensor has components T_{ij} which transform as do products of a pair of vector components $A_i B_j$. From Eq.(II.2) it follows that the operation of $\hat{T}$ on the sum of two vectors $\vec{A}$ and $\vec{C}$ can be written as

$$\sum_j T_{ij}(A_j + C_j) = \sum_j T_{ij} A_j + \sum_j T_{ij} C_j \quad \text{or} \quad \hat{T}(\vec{A} + \vec{C}) = \hat{T}\vec{A} + \hat{T}\vec{C}$$

and that, for any scalar n, we have

$$\sum_j T_{ij}(nA_j) = n\sum_j T_{ij}A_j \quad \text{or} \quad \hat{T}(n\vec{A}) = n\hat{T}\vec{A}$$

Because of these two properties, a second-rank tensor is known to be a *linear operator*.

Substituting Eqs.(II.1) and (II.2), Eq.(II.2) shows that a square matrix may be multiplied by a column matrix to yield another column matrix, using the rule

$$\begin{pmatrix} T_{xx} & T_{xy} & T_{xz} \\ T_{yx} & T_{yy} & T_{yz} \\ T_{zx} & T_{zy} & T_{zz} \end{pmatrix} \begin{pmatrix} A_x \\ A_y \\ A_z \end{pmatrix} = \begin{pmatrix} T_{xx}A_x & T_{xy}A_y & T_{xz}A_z \\ T_{yx}A_x & T_{yy}A_y & T_{yz}A_z \\ T_{zx}A_x & T_{zy}A_y & T_{zz}A_z \end{pmatrix} = \begin{pmatrix} B_x \\ B_y \\ B_z \end{pmatrix}$$

If the vector $\vec{A}$ itself has been created from a vector $\vec{C}$ by another linear transformation, that is

$$A_j = \sum_k T'_{jk} C_k$$

the connection between B_i and C_k is obtained from Eq.(II.2) as

$$B_i = \sum_j T_{ij} \left(\sum_k T'_{jk} \right) C_k = \sum_k \left(\sum_j T_{ij} T'_{jk} \right) C_k = \sum_k P_{ik} C_k$$

The P_{ik} coefficients are given by

$$\sum_j T_{ij} T'_{jk} = P_{ik} \tag{II.4}$$

and this shows that a square matrix can be multiplied by a square matrix to produce another square matrix in the form

$$\begin{pmatrix} T_{xx} & T_{xy} & T_{xz} \\ T_{yx} & T_{yy} & T_{yz} \\ T_{zx} & T_{zy} & T_{zz} \end{pmatrix} \begin{pmatrix} T'_{xx} & T'_{xy} & T'_{xz} \\ T'_{yx} & T'_{yy} & T'_{yz} \\ T'_{zx} & T'_{zy} & T'_{zz} \end{pmatrix} = \begin{pmatrix} P_{xx} & P_{xy} & P_{xz} \\ P_{yx} & P_{yy} & P_{yz} \\ P_{zx} & P_{zy} & P_{zz} \end{pmatrix}$$

The element P_{ik} of the product square is formed by multiplying the elements of the ith row of the first square onto those of the kth column of the second square and summing up. It is easily verified from Eq.(II.4) that the multiplication of square matrices is in general noncommutative, that is

$$\sum_j T_{ij} T'_{jk} \neq \sum_j T'_{ij} T_{jk}$$

The sum of two matrices is defined by

$$T_{ij} + T'_{ij} = S_{ij} \tag{II.5}$$

If the matrix (II.3) is modified by interchanging the rows and the columns, and hence, by symmetrically exchanging the positions of all the elements with respect to the main diagonal, it is said to be transposed. The new matrix $\hat{\tilde{T}}$ is known to be the *transpose* of the original matrix, written in terms of its elements as

$$\tilde{T}_{ij} = T_{ji} \tag{II.6}$$

If $\hat{T}$ is its own transpose, and therefore $T_{ij} = T_{ji}$, then $\hat{T}$ is called a *symmetrical* matrix. If $T_{ij} = -T_{ji}$, the matrix $\hat{T}$ is called *antisymmetrical* and all its diagonal terms must be zero. Any matrix can always be expressed as the sum of a symmetrical and an antisymmetrical part, using the identity

$$T_{ij} \equiv \frac{1}{2}(T_{ij} + T_{ji}) + \frac{1}{2}(T_{ij} - T_{ji}) \tag{II.7}$$

where it is clear that the first term is symmetric, the second antisymmetric. The sum of the elements on the main diagonal of any matrix is called the *trace*

$$Tr(\hat{T}) = \sum_i T_{ii} \tag{II.8}$$

The unit matrix, symbolized by $\hat{I}$, is a matrix for which all the elements on the main diagonal are unity and all others are zero

$$\hat{I} = \begin{pmatrix} 1 & 0 & 0 \\ 0 & 1 & 0 \\ 0 & 0 & 1 \end{pmatrix} \quad \text{or} \quad T_{ij} = \delta_{ij} \tag{II.9}$$

where δ_{ij} is the *Kronecker delta*, which is defined as being 1 for $i = j$ and 0 for $i \neq j$. If $\det(\hat{T}) \neq 0$, where $\det(\hat{T})$ is the determinant of the array of elements T_{ij}, the square matrix $\hat{T}$ has an inverse matrix, which is also square and is denoted by $\hat{T}^{-1}$. The relationship between the two is

$$\hat{T}^{-1}\hat{T} = \hat{T}\hat{T}^{-1} = \hat{I} \tag{II.10}$$

If the direction of a vector $\vec{A}$ is not affected by the operation of a square matrix on it, and hence, if we have

$$\hat{T}\vec{A} = \lambda\vec{A} \tag{II.11}$$

where λ is a real or complex number, $\vec{A}$ is called an *eigenvector* of the square matrix, and λ is the corresponding *eigenvalue*. Equation (II.11) can be rewritten as

$$\sum_j T_{ij} A_j = \lambda A_i = \lambda \sum_j \delta_{ij} A_j \quad \text{or} \quad \sum_j (T_{ij} - \lambda \delta_{ij}) A_j = 0 \quad \text{(II.12)}$$

which is a set of three homogenous equations

$$(T_{xx} - \lambda) A_x + T_{xy} A_y + T_{xz} A_z = 0$$

$$T_{yx} A_x + (T_{yy} - \lambda) A_y + T_{yz} A_z = 0 \quad \text{(II.13)}$$

$$T_{zx} A_x + T_{zy} A_y + (T_{zz} - \lambda) A_z = 0$$

The condition for solutions other than the trivial one, $\vec{A} = 0$, to exist is known to be the secular equation, written as

$$\det(\hat{T}) = \begin{vmatrix} T_{xx} - \lambda & T_{xy} & T_{xz} \\ T_{yx} & T_{yy} - \lambda & T_{yz} \\ T_{zx} & T_{zy} & T_{zz} - \lambda \end{vmatrix} = 0$$

It can be shown that such an equation has three *real* roots λ_n, where $n = 1$, 2 and 3, if T_{ij} is symmetric. If we assume that λ_n are all distinct, there is an independent eigenvector $\vec{A}_n$ for each root, with components A_{jn} obtained by solving a set of three equations (II.13) for each n. It is convenient to form a square matrix $\hat{A}$ with columns A_{jn} and a diagonal matrix $\hat{\Lambda}$ with elements λ_k, which read

$$\hat{A} = \begin{pmatrix} A_{x1} & A_{x2} & A_{x3} \\ A_{y1} & A_{y2} & A_{y3} \\ A_{z1} & A_{z2} & A_{z3} \end{pmatrix}, \quad \hat{\Lambda} = \begin{pmatrix} \lambda_1 & 0 & 0 \\ 0 & \lambda_2 & 0 \\ 0 & 0 & \lambda_3 \end{pmatrix} \quad \text{(II.14)}$$

so that the set of satisfied eigenvalue equations can be written as

$$\hat{T}\hat{A} = \hat{A}\hat{\Lambda}$$

For independent eigenvectors $\vec{A}_n$, the inverse of $\hat{A}$ always exists, so that we have

$$\hat{A}^{-1}\hat{T}\hat{A} = \hat{A}^{-1}\hat{A}\hat{\Lambda} = \hat{\Lambda} \quad \text{(II.15)}$$

In other words, a symmetric matrix $\hat{T}$ can be transformed, using the eigenvector matrix $\hat{A}$, into a matrix $\hat{\Lambda}$ which has diagonal elements only. It can be shown that the unit vectors $\vec{e}_n$ along the directions of the independent eigenvectors $\vec{A}_n$ are mutually perpendicular, if the λ_n are all distinct. They define the so-called *principal axes* of the symmetric tensor $\hat{T}$. The diagonal form (II.14) of a symmetric tensor can always be written as

$$\hat{T} = \begin{pmatrix} T_x & 0 & 0 \\ 0 & T_y & 0 \\ 0 & 0 & T_z \end{pmatrix} \tag{II.16}$$

where T_x, T_y and T_z are the components of the tensor along its principal axes.

❷ All the previous results are still valid if we consider an *n*-dimensional space, where vectors are expressed in terms of their components $A_1, A_2, \ldots, A_n$ with respect to n different coordinate axes, such that the representative of the vector is a column matrix

$$\begin{pmatrix} A_1 \\ A_2 \\ \vdots \\ A_n \end{pmatrix} \tag{II.17}$$

Hence, the representative of a linear operator in this coordinate system will be an $n \times n$ square matrix of the form

$$\hat{T} = \begin{pmatrix} T_{11} & T_{12} & \cdots & T_{1n} \\ T_{21} & T_{22} & \cdots & T_{2n} \\ \cdots & \cdots & \cdots & \cdots \\ T_{n1} & T_{n2} & \cdots & T_{nn} \end{pmatrix} \tag{II.18}$$

If the T_{ij} elements of such a matrix are complex, as we frequently encounter, the complex conjugate $\hat{T}^*$ of the matrix $\hat{T}$ is obtained by taking the complex conjugate of each element. The Hermitian adjoint $\hat{T}^\dagger$ of the matrix (II.18) is defined to be the complex conjugate of the transpose of $\hat{T}$, that is

$$\hat{T}^\dagger = \tilde{\hat{T}}^* \tag{II.19}$$

or

$$\begin{pmatrix} T_{11} & T_{12} & \cdots & T_{1n} \\ T_{21} & T_{22} & \cdots & T_{2n} \\ \cdots & \cdots & \cdots & \cdots \\ T_{n1} & T_{n2} & \cdots & T_{nn} \end{pmatrix}^\dagger = \begin{pmatrix} T_{11}^* & T_{21}^* & \cdots & T_{n1}^* \\ T_{12}^* & T_{22}^* & \cdots & T_{n2}^* \\ \cdots & \cdots & \cdots & \cdots \\ T_{1n}^* & T_{2n}^* & \cdots & T_{nn}^* \end{pmatrix} \tag{II.20}$$

If a matrix is equal to its Hermitian adjoint, it is said to be *Hermitian*, according to

$$\hat{T}^\dagger = \hat{T} \quad \text{or} \quad T^*_{ji} = T_{ij} \qquad (\text{II.21})$$

In other words, the diagonal elements of a Hermitian matrix are real and the off-diagonal elements which form complex conjugate pairs are symmetrically disposed with respect to the main diagonal. If a matrix $\hat{T}$ satisfies

$$\hat{T}^\dagger \hat{T} = \hat{T}\hat{T}^\dagger = \hat{I} \quad \text{or} \quad \hat{T}^\dagger = \hat{T}^{-1} \qquad (\text{II.22})$$

it is said to be *unitary*.

FURTHER READING
1. D. A. Danielson - VECTORS AND TENSORS IN ENGINEERING AND PHYSICS, Perseus Publishing, Philadelphia, 2003.
2. B. Spain - TENSOR CALCULUS: A CONCISE COURSE, Dover Publications, New York, 2003.
3. E. C. Young - VECTOR AND TENSOR ANALYSIS, Marcel Dekker, New York, 1992.

APPENDIX III. SOME PROPERTIES OF PARTIAL DERIVATIVES

❶ A function $F(x, y, z)$ that, for any positive λ, satisfies the equation

$$F(\lambda x, \lambda y, z) = \lambda^n F(x, y, z)$$

is said to be a *homogeneous function* of nth order with respect to the variables x and y. The differential of $F(\lambda x, \lambda y, z)$ with respect to λ can be written as

$$\frac{\partial F(\lambda x, \lambda y, z)}{\partial \lambda} = \frac{\partial F(\lambda x, \lambda y, z)}{\partial(\lambda x)}\frac{\partial(\lambda x)}{\partial \lambda} + \frac{\partial F(\lambda x, \lambda y, z)}{\partial(\lambda y)}\frac{\partial(\lambda y)}{\partial \lambda} = n\lambda^{n-1} F(x, y, z)$$

or

$$x\frac{\partial F(\lambda x, \lambda y, z)}{\partial(\lambda x)} + y\frac{\partial F(\lambda x, \lambda y, z)}{\partial y} = n\lambda^{n-1} F(x, y, z)$$

For $\lambda = 1$, this result is known as *Euler's theorem* for homogeneous functions of nth order, written as

$$x\frac{\partial F}{\partial x} + y\frac{\partial F}{\partial y} = nF \tag{III.1}$$

❷ If the three variables x, y and z are related by

$$F(x, y, z) = 0$$

we can rearrange the equation to express z as a function of x and y as $z = z(x, y)$. The total differential of z is defined to be

$$dz = \left(\frac{\partial z}{\partial x}\right)_y dx + \left(\frac{\partial z}{\partial y}\right)_x dy \tag{III.2}$$

We can equally express x in terms of the other two variables as

$$x = x(y, z) \quad \text{or} \quad dx = \left(\frac{\partial x}{\partial y}\right)_z dy + \left(\frac{\partial x}{\partial z}\right)_y dz$$

APPENDIX III. SOME PROPERTIES OF PARTIAL DERIVATIVES

and, if we substitute this expression for dx in Eq.(III.2), one obtains

$$\left[\left(\frac{\partial z}{\partial x}\right)_y \left(\frac{\partial x}{\partial y}\right)_z + \left(\frac{\partial z}{\partial y}\right)_x\right] dy + \left[\left(\frac{\partial z}{\partial x}\right)_y \left(\frac{\partial x}{\partial z}\right)_y - 1\right] dz = 0 \qquad \text{(III.3)}$$

If we choose $dy = 0$, which implies that $dz \neq 0$, because y and z are independent variables, Eq.(III.3) reduces to

$$\left(\frac{\partial z}{\partial x}\right)_y \left(\frac{\partial x}{\partial z}\right)_y = 1 \quad \text{or} \quad \left(\frac{\partial z}{\partial x}\right)_y = \left(\frac{\partial x}{\partial z}\right)_y^{-1} \qquad \text{(III.4)}$$

If alternatively we choose $dz = 0$ and $dy \neq 0$, Eq.(III.3) yields

$$\left(\frac{\partial z}{\partial x}\right)_y \left(\frac{\partial x}{\partial y}\right)_z + \left(\frac{\partial z}{\partial y}\right)_x = 0 \quad \text{or} \quad \left(\frac{\partial z}{\partial x}\right)_y \left(\frac{\partial x}{\partial y}\right)_z \left(\frac{\partial y}{\partial z}\right)_x = -1 \qquad \text{(III.5)}$$

where use has been made of Eq.(III.4). Equation (III.5) is known as the *cyclical rule*.

❸ An expression of the form

$$A(x, y)dx + B(x, y)dy \qquad \text{(III.6)}$$

is called an *exact differential* if some function z exists such that

$$dz = A(x, y)dx + B(x, y)dy \qquad \text{(III.7)}$$

A finite change in z, when x changes from x_i to x_f and y from y_i to y_f, is given by

$$\Delta z = \int_{x_i}^{x_f}\int_{y_i}^{y_f} dz = z(x_f, y_f) - z(x_i, y_i)$$

Since the result depends on the values of z at the limit points (x_i, y_i) and (x_f, y_f) only, it is said that the integral is *path independent*, and therefore $z(x, y)$ is called a *function of state*. Comparing Eqs.(III.7) and (III.2) we obtain $A(x, y) = (\partial z / \partial x)_y$ and $B(x, y) = (\partial z / \partial y)_x$ so that the condition for an exact differential is

$$\left(\frac{\partial A}{\partial y}\right)_x = \left(\frac{\partial B}{\partial x}\right)_y \qquad \text{(III.8)}$$

It can be shown that Eq.(III.8) is not only a necessary condition for the differential expression (III.6) to be exact, but it is also a sufficient condition.

FURTHER READING
1. G. B. Folland - ADVANCED CALCULUS, Prentice Hall, Upper Saddle River, 2001.
2. A. E. Taylor, W. R. Mann - ADVANCED CALCULUS, Wiley, New York, 1999.
3. W. A. Kosmala - ADVANCED CALCULUS: A FRIENDLY APPROACH, Pearson Education, Boston, 1998.

APPENDIX IV. EVALUATION OF SOME INTEGRALS

❶ Integrals containing Gaussian functions, such as

$$I = \int_{-\infty}^{\infty} e^{-\alpha x^2} dx \qquad (IV.1)$$

and others of a similar form are called *Gaussian integrals*. To evaluate the integral (IV.1) we can multiply by I to obtain

$$I^2 = \int_{-\infty}^{\infty} e^{-\alpha x^2} dx \int_{-\infty}^{\infty} e^{-\alpha y^2} dy = \int_{-\infty}^{\infty}\int_{-\infty}^{\infty} e^{-\alpha(x^2+y^2)} dxdy$$

Transforming to polar coordinates r and θ, where $r^2 = x^2 + y^2$, the element of area is $rdrd\theta$, and I^2 can be rewritten as

$$I^2 = \int_0^\infty\int_0^{2\pi} e^{-\alpha r^2} rdrd\theta = 2\pi\int_0^\infty e^{-\alpha r^2} rdr = \frac{\pi}{\alpha}\int_0^\infty e^{-\alpha r^2} d(\alpha r^2) = \frac{\pi}{\alpha}\int_0^\infty e^{-u} du = \frac{\pi}{\alpha}\left(-e^{-u}\right)_0^\infty = \frac{\pi}{\alpha}$$

It follows that

$$I = \int_{-\infty}^{\infty} e^{-\alpha x^2} dx = \sqrt{\frac{\pi}{\alpha}} \qquad (IV.2)$$

and, because the Gaussian function $e^{-\alpha x^2}$ is symmetrical about $x = 0$, one obtains

$$I_0 = \int_0^\infty e^{-\alpha x^2} dx = \frac{1}{2}\sqrt{\frac{\pi}{\alpha}} \qquad (IV.3)$$

Successive differentiation of Eq.(IV.3) with respect to α yields

$$I_2 = \int_0^\infty e^{-\alpha x^2} x^2 dx = -\frac{dI_0}{d\alpha} = \frac{\sqrt{\pi}}{4}\alpha^{-3/2} \qquad (IV.4)$$

$$I_4 = \int_0^\infty e^{-\alpha x^2} x^4 dx = -\frac{dI_2}{d\alpha} = \frac{3\sqrt{\pi}}{8}\alpha^{-5/2} \qquad (IV.5)$$

$$I_6 = \int_0^\infty e^{-\alpha x^2} x^6 dx = -\frac{dI_4}{d\alpha} = \frac{15\sqrt{\pi}}{16}\alpha^{-7/2} \qquad (IV.6)$$

It is a straightforward matter to show that

$$I_1 = \int_0^\infty e^{-\alpha x^2} x\,dx = \frac{1}{2\alpha}\int_0^\infty e^{-\alpha x^2} d(\alpha x^2) = \frac{1}{2\alpha} \qquad (IV.7)$$

By differentiation of I_1 with respect to α we then find

$$I_3 = \int_0^\infty e^{-\alpha x^2} x^3 dx = -\frac{dI_1}{d\alpha} = \frac{1}{2\alpha^2} \qquad (IV.8)$$

$$I_5 = \int_0^\infty e^{-\alpha x^2} x^5 dx = -\frac{dI_3}{d\alpha} = \frac{1}{2\alpha^3} \qquad (IV.9)$$

It is obvious that in general we will obtain

$$\int_{-\infty}^\infty e^{-\alpha x^2} x^n dx = 2I_n \qquad (n \text{ even})$$

$$= 0 \qquad (n \text{ odd}) \qquad (IV.10)$$

❷ Some integrals can be evaluated using the Riemann zeta function

$$\zeta(z) = \sum_{r=1}^\infty \frac{1}{r^z} \qquad (z>1) \qquad (IV.11)$$

which it is shown to have the values

$$\zeta(2) = \frac{\pi^2}{6}, \qquad \zeta(4) = \frac{\pi^4}{90} \qquad (IV.12)$$

For instance we have

$$\frac{1}{e^x - 1} = e^{-x}\frac{1}{1-e^{-x}} = e^{-x}\sum_{n=0}^\infty e^{-nx} = \sum_{n=0}^\infty e^{-(n+1)x}$$

and also, integrating by parts

$$\int_0^\infty e^{-(n+1)x} x\,dx = -\frac{1}{n+1}\left[xe^{-(n+1)x}\right]_0^\infty + \frac{1}{n+1}\int_0^\infty e^{-(n+1)x} dx = \frac{1}{(n+1)^2}$$

Hence, we can write

$$\int_0^\infty \frac{x\,dx}{e^x - 1} = \sum_{n=0}^\infty \int_0^\infty e^{-(n+1)x} x\,dx = \sum_{n=0}^\infty \frac{1}{(n+1)^2} = \frac{\pi^2}{6} \qquad \text{(IV.13)}$$

In a similar manner, one obtains

$$\int_0^\infty \frac{x^3\,dx}{e^x - 1} = \sum_{n=0}^\infty \int_0^\infty e^{-(n+1)x} x^3\,dx = \sum_{n=0}^\infty \frac{6}{(n+1)^4} = \frac{\pi^4}{15} \qquad \text{(IV.14)}$$

❸ Starting from the simple integral

$$\int_0^\infty e^{-\alpha x}\,dx = \frac{1}{\alpha}$$

we obtain by differentiation with respect to α that

$$\int_0^\infty e^{-\alpha x} x^n\,dx = (-1)^n \frac{d^n}{d\alpha^n} \int_0^\infty e^{-\alpha x}\,dx = (-1)^n \frac{d^n}{d\alpha^n}\left(\frac{1}{\alpha}\right) = \frac{n!}{\alpha^{n+1}} \qquad \text{(IV.15)}$$

If we now set $\alpha = 1$, we find

$$n! = \int_0^\infty e^{-x} x^n\,dx = \int_0^\infty e^{n \ln x - x}\,dx$$

The exponent is a function

$$f(x) = n \ln x - x$$

which has a maximum for $x = n$, where $(df/dx)_{x=n} = 0$. Hence, if we expand $f(x)$ in a Taylor series about this point, we obtain

$$f(x) = f(n) + (x-n)\left(\frac{df}{dx}\right)_{x=n} + \frac{1}{2}(x-n)^2\left(\frac{d^2 f}{dx^2}\right)_{x=n} + \ldots$$

$$= f(n) + \frac{1}{2}(x-n)^2\left(\frac{d^2 f}{dx^2}\right)_{x=n} + \ldots = n \ln n - n - \frac{1}{2n}(x-n)^2 + \ldots$$

It follows that $n!$ can be approximated by

$$n! \cong e^{n \ln n - n} \int_0^\infty e^{-(1/2n)(x-n)^2}\,dx = e^{n \ln n - n} \int_{-n}^\infty e^{-(1/2n)y^2}\,dy$$

where $y = x - n$. For large n the lower limit of the integral can be replaced by $-\infty$, and therefore, in view of Eq.(IV.2), we obtain

$$n! \cong e^{n\ln n - n} \int_{-n}^{\infty} e^{-(1/2n)y^2} dy = e^{n\ln n - n} (2\pi n)^{1/2}$$

or

$$\ln n! \cong n\ln n - n + \frac{1}{2}\ln(2\pi n) \cong n\ln n - n \tag{IV.16}$$

This result is known as *Stirling's approximation* for the asymptotic form of $\ln n!$, which is valid for large n.

FURTHER READING
1. B. Das - MATHEMATICS FOR PHYSICS WITH CALCULUS, Prentice Hall, Upper Saddle River, 2004.
2. R. Snieder - MATHEMATICAL METHODS FOR THE PHYSICAL SCIENCES, Cambridge University Press, Cambridge, 2001.
3. G. B. Arfken, H. J. Weber - MATHEMATICAL METHODS FOR PHYSICISTS, Elsevier, Amsterdam, 2000.

Appendix V. Glossary of Symbols

Because there are not enough ordinary symbols for all the physical quantities that have been discussed, and because several accepted conventions use the same symbol for quite different things, an attempt has been made to coordinate the choice of symbols used in order to preserve most of the conventions, while reducing the duplication of symbols as much as possible. The list of symbols given here includes those used in more than a single chapter together with their usual meaning or meanings. A number of other symbols, which are either related to one particular subject only or appear briefly in the text to simplify the discussion, are defined in each case.

a	linear lattice spacing
a, b	semimajor and semiminor axes of the ellipse
a	width parameter
a_0	Bohr radius
$\vec{a}$	acceleration
$\vec{a}, \vec{b}, \vec{c}$	base lattice vectors
$\vec{a}^*, \vec{b}^*, \vec{c}^*$	base reciprocal lattice vectors
$\hat{a}, \hat{a}^+$	lowering and raising operators
A	mass number of the nucleus
A	oscillation and wave amplitude
A	complex amplitude
$A_{n \to i}, B_{n \to i}, B_{i \to n}$	Einstein's coefficients
A_μ	four-vector
$\vec{A}$	vector potential
B	binding energy of the nucleus
B_j	Brillouin function
$\vec{B}$	magnetic induction
$\vec{B}_0$	applied magnetic field
$\vec{B}_c$	critical field
c	velocity of light
c_V, c_p	specific heat, at constant volume and pressure
C	Curie constant
C_V, C_p	heat capacity, at constant volume and pressure
D	phase density
$\vec{D}$	electric displacement

D_e, D_h	diffusion constant, electrons and holes
e	electron charge
$\vec{e}_x, \vec{e}_y, \vec{e}_z, \vec{e}_r, \vec{e}_\theta, \vec{e}_\varphi, \vec{e}_n, \vec{e}_s$	unit vectors
$\hat{e}$	rate of strain tensor
E	total energy
E	Young's modulus
$\vec{E}$	electric field
f	focal length
f	distribution function
f	binding fraction
F	Helmholtz function
$\vec{F}$	force
$\vec{F}_e$	external force
$\vec{F}_L$	Lorentz force
g	specific Gibbs function
g_l, g_s, g_L, g_I	gyromagnetic factors
G	Gibbs function
G	gravitation constant
$\vec{G}$	torque of force
h	Planck's constant, $\hbar = h/2\pi$
h	specific enthalpy
H	enthalpy
$H, \hat{H}$	Hamiltonian
$\vec{H}$	magnetic field
i	square root of minus one
i	angle of incidence
I	electric current
I	wave intensity
I	quantum number of the nuclear angular momentum
$I_\omega, I(\omega)$	spectral intensity
$\vec{I}$	nuclear angular momentum
j	quantum number of the total angular momentum
$\vec{j}, \vec{j}_e, \vec{j}_s, \vec{j}_n, \vec{j}_h$	current density, electrons and holes
J	Jones vector
J	action integral
J	exchange integral
$\vec{J}, \hat{J}^2$	total angular momentum
k	wave number
k_B	Boltzmann constant
k_T, k_S	compressibility, isothermal and adiabatic
$\vec{k}$	wave vector

$\vec{k}$	vector in $\vec{k}$-space
K	bulk modulus
$\vec{K}$	reciprocal lattice vector
l	length
l	orbital quantum number
L	Lagrangian
L_e, L_n	diffusion length, electrons and holes
$\vec{L}, \hat{L}^2$	orbital angular momentum
m	mass
m	order of interference and diffraction
m, m_s, m_j, m_I	magnetic quantum numbers
m^*	effective mass
m_e, m_h	carrier mass, electrons and holes
m_e^*, m_h^*	effective mass, electrons and holes
M	atomic and ionic mass
$\hat{M}$	Jones matrices
$\vec{M}$	magnetization
n	refractive index
n	principal quantum number
n, N	number of entities
n, n_s, n_n, n_i	electron concentration
N	density of particles
$N(\varepsilon)$	density of states
N_A	Avogadro number
p	pressure
p	concentration of holes
p_k	generalized momentum
p_C, V_C, T_C	critical constants
p_r, V_r, T_r	reduced parameters
p_μ	four-momentum
$\vec{p}$	linear momentum
$\vec{p}$	electric dipole moment
P	power
$P_{i \to n}$	transition rate
P_l	Legendre polynomials
$P_l^{|m|}$	associated Legendre function
$\vec{P}$	total linear momentum
$\vec{P}$	polarization
q	electric charge

$q_k, \dot{q}_k$	generalized coordinate and velocity
Q	quantity of heat
Q	decay energy
r	angle of reflection
$\vec{r}, \vec{R}$	position vectors
R	radius, sphere and cylinder
R	Rydberg's constant
R	gas constant
R	reflectivity
R	resistance
RP	resolving power
$R(r)$	spatial function
s	spin quantum number
s	specific entropy, entropy density
S	entropy
S	principal function of Hamilton
S, dS	surface, surface element
$S(R)$	overlap integral
S, S'	reference frames
SWR	standing wave ratio
$\vec{S}$	Poynting vector
$\vec{S}, \hat{S}^2$	spin angular momentum
t	time variable
t	temperature on the Celsius scale
$t_{1/2}$	half-life
T	period of oscillation
T	transmissivity
T	kinetic energy
T	absolute temperature
T_c	critical temperature
T_C	Curie temperature
T_F	Fermi temperature
T_{tr}	triple point of water
u, u_{mec}, u_E, u_H	energy density, mechanic, electrostatic and magnetic
$u_\omega, u(\omega)$	spectral energy density
U	internal energy
U	potential energy
U_ω	spectral energy
$\hat{U}$	evolution operator
v	specific volume
v_μ	four-velocity
$\vec{v}$	velocity

$\vec{v}_g$	group velocity of waves or particles
V	electrostatic potential
V, dV	volume, volume element
V_0	unit cell volume
V_0^*	unit cell volume in $\vec{k}$-space
$\hat{V}$	perturbation operator
w	width of the depletion region
$\vec{w}$	ray velocity
W	work
W	thermodynamic probability
x	extensive variable
x, x', y, y', z, z'	displacement Cartesian coordinates
X	intensive variable
X, X', X''	nuclear species
Y	Airy function
Z	atomic number
Z	characteristic impedance
Z, Z_G	partition function, canonical and grand canonical
Z_0	characteristic impedance per unit area
$\vec{Z}$	Hertz vector
α	atomic or molecular polarizability
α, β, γ	direction cosines
$\beta = v/c$	relativistic constant, $\gamma = 1/\sqrt{1-\beta^2}$
γ	adiabatic exponent
γ	molecular field constant
γ	degree of coherence
Γ	closed path
Γ	linewidth
$\delta(x)$	Dirac delta function
δ_{ij}	Kronecker delta
δ	variation symbol
Δ	volume expansion
2Δ	superconductor energy gap
ε	dielectric constant
ε	energy
ε_0	permittivity of free space
ε_r	relative permittivity
ε_F	Fermi energy
$\hat{\varepsilon}$	strain tensor
$\hat{\varepsilon}_r$	dielectric tensor
ζ	spin-orbit coupling constant

η	coefficient of viscosity
θ	spherical angle
θ	empirical temperature
θ_D, θ_E	Debye and Einstein temperatures
λ	wavelength
λ	latent heat
λ	decay constant
λ, μ	Lamé constants
λ_L	London penetration depth
Λ_0	Compton wavelength
μ	permeability
μ	shear modulus
μ	linear density
μ, μ_0	molar chemical potential
μ_0	permeability of free space
μ_B, μ_N	magneton, Bohr and nuclear
μ_e, μ_n	mobility, electrons and holes
μ_r	relative permeability
$\vec{\mu}, \vec{\mu}_l, \vec{\mu}_s, \vec{\mu}_j, \vec{\mu}_I$	magnetic moment, orbital, spin, total and nuclear
ν	frequency
$\nu, \overline{\nu}$	neutrino, antineutrino
υ	Poisson's ratio
$\vec{\xi}$	displacement vector
$\vec{\pi}$	wave momentum density
ρ	density
ρ	charge density
ρ	amplitude coefficient of reflection
$\rho = 1/\sigma$	resistivity
$\rho(\varepsilon)$	density of states
σ	surface charge density
σ	standard deviation
$\sigma, \sigma_e, \sigma_h$	electrical conductivity, electrons and holes
$\hat{\sigma}_x, \hat{\sigma}_y, \hat{\sigma}_z$	Pauli matrices
Σ, Σ'	frames of reference
τ	decay time
τ	proper time
τ, τ_l, τ_h	relaxation time between collisions
φ	gravitational potential
φ	spherical angle
ϕ	magnetostatic scalar potential

Φ	magnetic flux
Φ	wave function
Ψ, ψ	wave functions
Ψ, Φ	Fourier transforms
χ	spinor
χ, χ_e, χ_m	susceptibility, electric and magnetic
$\omega = 2\pi\nu$	angular frequency, oscillations and waves
ω_p	plasma frequency
ω_D	Debye frequency
Ω	solid angle
Ω	volume in phase space
$\vec{A}$	vector quantity symbol
$\times$	vector multiplication of two vectors
$\cdot$	scalar multiplication of two vectors
$\hat{A}$	linear operator, tensor and matrix symbol
A^*	complex conjugate of A
$\det \hat{A}$	determinant
$\mathrm{Im} A$	imaginary part of A
$\mathrm{Re} A$	real part of A
$\prod$	product symbol
$\sum$	summation symbol
∂	partial derivative symbol
$\oint_\Gamma$	integral over a closed path
∇	del operator
∇^2	Laplacian

Problem Hints

H 2.8. A constant α implies that the same acceleration exists for the horizontal translation motion of both M and m.

H 2.9. The ball should strike normally the face of the incline.

H 2.10. The string remains stretched as long as there is a spring deformation $x \geq 0$.

H 2.11. The kinetic energy is conserved and, since no external forces act on the system, the momentum is also conserved.

H 2.12. Both the momentum and the kinetic energy are conserved when the ball m is pulled upward.

H 2.13. Put $p_2^2 = p_{2x}^2 + p_{2y}^2$ in both the energy and momentum conservation laws and then eliminate p_1^2.

H 2.14. The tension T can be evaluated using the virtual work done by the system to rise his centre of mass.

H 3.8. The linear velocities $\vec{v}_i$ of the particles the body is formed of, can be written as $\vec{v}_i = \vec{v}_0 + \vec{v}_i'$, where $\vec{v}_0$ is the velocity of the centre of mass, and $\vec{v}_i'$ are the velocities relative to a reference system that is connected with the centre of mass.

H 3.9. The spool rolls without sliding, and hence, $a = R d\omega / dt$.

H 3.10. The acceleration is the same for the translation motion down the plane of the two bodies.

H 3.11. For a parabolic orbit, the velocity at the point at infinity is equal to zero, and this corresponds to the zero value of the total energy, Eq.(3.25).

H 3.12. Use a reference system moving in an orbit round the Sun, like the Earth.

H 4.5. Use the transformation of velocity components, Eqs.(4.19), and then $\tan(\alpha/2) = v_y / v_x$.

H 4.6. Consider $\vec{v}'(v'_x, v'_y, v'_z) = \vec{v}_1(v_{1x}, v_{2x}, v_{3x})$ and $\vec{v}(v,0,0) = -\vec{v}_2(-v_2,0,0)$ in Eqs.(4.19), and then $v^2 = (\vec{v}_1 - \vec{v}_2)^2$.

H 4.7. Write down the Lorenz contraction, Eq.(4.13), and the time dilatation, Eq.(4.15), in differential form.

H 5.6.	Write down $\rho = dm/dV$ and consider the transformation laws for inertial mass and volume.
H 5.7.	See Problem 4.1.
H 5.8.	Use the vector form of the equation of motion given in Problem 5.1.
H 5.9.	The directions of the force and acceleration coincide if we put either $\vec{F}\cdot\vec{v}=0$ or $\vec{F}\cdot\vec{v}=Fv$ into the equation of motion of the particle.
H 5.10.	Use the conservation laws for momentum and energy.
H 6.6.	Represent both the stress and strain tensors as a sum of two tensors, one involving only the normal stresses (strains) and the other including only the shear stresses (strains).
H 6.7.	Write down Hooke's law for a section dl long, of elastic constant $k(l/dl)$, from which hangs a mass dependent on its position. On integration one obtains the new length.
H 6.8.	Represent a given displacement of the load in terms of the resulting changes in length of the two springs and relate the corresponding linear forces on springs with the restoring force acting on the load.
H 6.9.	Use the conservation of energy, including the work of the sliding friction.
H 6.10.	Look at the assembly when the spring length is a minimum. The velocity of its centre of mass is given by the conservation of momentum, and its energy is equal to that of the ball.
H 6.11.	The relationship can be obtained by differentiating the condition of the constancy of the mass flux along the stream tube, using $dp/d\rho = v_S^2$, see Eq.(15.32) in Chapter 15.
H 6.12.	The reaction force is given by $\rho Q v$, where the volume outflow rate Q is obtained by integrating Poiseuille's law over the cross section of the pipe.
H 6.13.	A free surface of water of inclination α above the horizontal indicates a gradient of pressure due to the accelerated motion of the vessel, and this gives $a_0 = g\sin\alpha$.
H 7.6.	Use the definition $C = \delta Q/dT$, Eq.(7.15), where δQ is given by the first law, Eq.(7.18).
H 7.7.	Write down the first law, Eq.(7.18), in terms of C and C_V and integrate the equation.
H 7.8.	Compare the amount of work done by the system in each process.
H 7.9.	Use $x = (|Q_1|+|Q_2|)/Q_m$ and the definitions for the efficiency η and the coefficient of performance ε_p.
H 7.10.	The amount of heat absorbed in process $ikli$ is the same as that delivered to the surroundings in the other process.

H 8.6. Write down the equation of state and the equation of the reversible process in differential form and put $dS = 0$ in the first TdS equation.

H 8.7. Use $V = V(T, p)$ in differential form and the definitions of β_p, Eq.(8.41), and k_T, Eq.(8.42).

H 8.8. First derive an expression for U from Eq.(7.19) and then for S, using Eq.(8.13). The thermodynamic potentials F, G and H follow from their definitions.

H 8.9. Write down dU using p and V as independent variables, and then substitute $(\partial U / \partial p)_V$ and $(\partial U / \partial V)_p$ as given by the first law.

H 8.10. Derive $(\partial S / \partial V)_p$ in terms of C_p and β_p.

H 9.6. Use Eq.(7.9) for the efficiency and $Q = T\Delta S$, see Problem 9.1.

H 9.7. Use the condition $dS = 0$ in Eq.(8.13), where dV is given by the internal energy equation (8.40) and note that $(\partial p / \partial T)_V = \beta_p / k_T$.

H 9.8. See Problem 9.4.

H 9.9. Use the general relationship (8.43) and the equation of state (9.29).

H 9.10. Use the critical conditions, see Problem 9.5.

H 10.6. Write down the constant of integration of the Clapeyron equation, derived in Problem 10.1, in terms of p_0 and T_0.

H 10.7. The latent heat needed to produce a mass $-dm$ of vapour is obtained by cooling down the two liquids by dT.

H 10.8. Put $\theta = T - 273$ and substitute p into the Clapeyron equation at low pressure, obtained in Problem 10.1.

H 10.9. Since the water temperature changes from T to T_0, consider that the Carnot refrigerator operates in infinitesimal cycles for the heat transfer to be reversible.

H 10.10. Use the Maxwell relation (8.36) and the third law, Eq.(10.27).

H 11.7. Calculate the electric field due to the charges $\pm q_1$ of one capacitor at a large distance from it, and then the contribution of $\pm q_2$ toward the electrostatic force on the other capacitor.

H 11.8. Write down the electric field due to the line of charge, see Problem 11.2, and then find the force by integrating over the entire dielectric wire.

H 11.9. The force is equal and opposite to that felt by the hemisphere, see Problem 11.3.

H 11.10. Write down Eq.(11.31), where the electric field is given by Gauss' law, Eq.(11.4).

H 12.6. For the equivalent RC circuit, the capacitor is discharging with a time constant $\tau = RC$, where the resistance of the dielectric material can be written in terms of its conductivity σ.

H 12.7. Calculate the current density at a radius $r < R$ at time $t = 0$, and then use the continuity equation (12.15) to find its time dependence.

H 12.8. Add the contributions from all the conducting sides of finite length.

H 12.9. See Problem 12.4.

H 12.10. Integrate a series of contiguous circular current loops.

H 13.6. Use Eq.(13.28) to find the current due to the change of flux linked by the loop, and then the equation of vertical motion.

H 13.7. Use both Faraday's law and the energy conservation law.

H 13.8. Determine the circular arch of the electron path inside the solenoid, in view of the Lorentz force law, Eq.(13.30).

H 13.9. Write down the energy conservation law in the LC circuit at the instant τ.

H 13.10. Show first that $L_1 I_1 = L_2 I_2$, and then use energy conservation.

H 14.6. Calculate the time-varying fields $\vec{E}$, Eq.(13.25) and $\vec{B}$, Eq.(12.29), and then identify their sources as given by Eqs.(14.2).

H 14.7. Integrate the equation of continuity over a spherical surface surrounding the radioactive solid to obtain the rate of the charge decrease, and then write down the displacement current density.

H 14.8. Substitute the constitutive relations (14.4) into Maxwell's field equations (14.2).

H 14.9. Substituting V and the components A_r and A_θ into Eqs.(13.25) and (12.29), written in spherical polar coordinates, one obtains $E_\varphi = 0$, $E_r \cong 0$ (neglecting terms of order $1/r^2$ and higher) and $B_r = B_\theta = 0$. The Poynting vector is given by $E_\theta B_\varphi / \mu_0$.

H 14.10. Write down the components of Faraday's law and Ampère-Maxwell's law in S and S'.

H 15.6. Use Eq.(15.26) where the tension F in the ring is supplying the necessary centripetal acceleration.

H 15.7. Use the result of Problem 15.3 and the continuum approximation where $a \ll \omega/v$.

H 15.8. Integrate the equation of motion for a volume element of the fluid.

H 15.9. Express dp/p from the adiabatic equation (9.10).

H 15.10. Calculate the energy stored per unit volume of dielectric due to the axial current, and use Eq.(13.32) to find L_0.

H 16.7. Consider that the phase difference at time t is zero at some position x and it is 2π at the next position x', and put $x'-x = a$.

H 16.8. By Fermat's principle all of the ray trajectories from a point O of the optical axis, which intercept the lens ending up to a point I of that axis, are equally favored and take the same time.

H 16.9. Follow the deviation by refraction of a light ray approaching the glass sphere parallel to its optical axis.

H 16.10. Use $n = c/v(y)$ to show that the ray path is the same as that of a particle of velocity $v(y)$ moving in a harmonic potential.

H 16.11. Write down the Fourier transform (17.13) of the wave function $A(\xi)e^{ik_0\xi}$.

H 16.12. Substitute E_z into the electromagnetic wave equation (15.52) to obtain $v = \omega/k_x$ and then $v_g = \partial\omega/\partial k_x$.

H 17.6. See Problem 15.10.

H 17.7. Use Ampère's law to determine the magnetic flux density, and then write down the flux per unit length linked by the cross section area, in order to find L_0.

H 17.8. There is an almost total reflection $(R = 1)$ of the sound wave energy at an air-water interface.

H 17.9. An isotropic sound wave propagates outward from its source as given by Eq.(16.30).

H 17.10. Stationary waves are obtained by superposition of an incident wave $A\cos(kx - \omega t)$ and a reflected wave $\rho A\cos(kx + \omega t)$.

H 18.6. Determine the path difference at the central fringe due to some displacement of the point source with respect to the axis of the instrument.

H 18.7. Write down the radii of curvature R_1 and R_2, respectively, of the two spherical surfaces of the lens in terms of r_m and r_m'.

H 18.8. Find the wavelength λ matching the condition for destructive interference at the position of the spectrometer and put $\lambda_m < \lambda < \lambda_M$.

H 18.9. The intensities I_{max} and I_{min} at the maxima and minima of the fringe pattern should be expressed by Eq.(19.5), where I_1 and I_2 are related Eq.(18.52).

H 18.10. Consider the superposition of three waves $A\cos\omega t$, $A\cos(\omega t + \delta)$ and $A\cos(\omega t - \delta)$, where $\delta = kay/x$.

H 19.6. The intensity given by Eq.(19.28) is a maximum wherever $dI/d\gamma = 0$.

H 19.7. Put $N = 2$ in Eq.(19.24), where the condition for an interference maximum $\delta = n\pi$ should be satisfied at the same position θ as the condition for a diffraction minimum $\gamma = m\pi$.

H 19.8. Write down Eq.(19.35) for both the first and the highest order of diffraction.

H 19.9. The critical angular separation is given by the angular size of the Airy disk derived in Problem 19.3.

H 19.10. Write down the equation obeyed by the zone radius ρ_1 in order to achieve constructive interference at a distance $R_0 + r_0$.

H 20.6. Integrate $dt = -dW / \langle P \rangle$, where dW is the energy loss and $\langle P \rangle$ is the time averaged radiated power, over the decay time τ, and put $\delta = 2\pi / \omega\tau$.

H 20.7. See Problem 20.2, as the electric vector of the wave can always be resolved into two orthogonal harmonic components E_x and E_y with a phase difference of $\pi/2$.

H 20.8. Solve the eigenvalue equation $\hat{M}J = \lambda J$ as shown in Problem 20.3.

H 20.9. Use Eq.(23.63) to describe the effect of the polarizer, and put $I = A_x^2 + A_y^2 = J^+ J$ for both the incident and transmitted beams.

H 20.10. See Problem 20.4.

H 20.11. The coherency matrix can always be reduced to the diagonal form $\hat{C}_{diag} = \begin{pmatrix} \lambda_1 & 0 \\ 0 & \lambda_2 \end{pmatrix}$.

H 21.7. For a parallel-sided plate, the incident and emergent rays are parallel.

H 21.8. The reflected beam is circularly polarized when $\Delta\varphi = \varphi_{TM} - \varphi_{TE} = \pi/2$.

H 21.9. Follow the steps in the derivation of the transfer matrix at normal incidence, using Eqs.(21.23) for *TE* polarization.

H 21.10. Calculate the transfer matrix for a bilayer consisting of two quarter-wave films of different refractive indices n and n'.

H 21.11. Calculate the transfer matrix of the trilayer, solve the matrix equation for $\rho = E_\rho / E_i$ and put $\rho = 0$.

H 21.12. The transfer matrix of a half-wave layer is $M = \begin{pmatrix} -1 & 0 \\ 0 & -1 \end{pmatrix}$.

H 22.6. Write down the phase matching conditions derived in Problem 22.1 for $k = k_o$ and $k = k_e$, and consider $k_\tau = \omega / c$ for air.

H 22.7. If I_n is the intensity of natural light in the incident beam, the intensity of the emergent beam has a maximum of $I_e + I_n / 2$ and a minimum of $I_n / 2$.

H 22.8. Find a relation between the orientations of $\vec{D}$ and $\vec{E}$ with respect to the optic axis.

H 22.9. The phase difference between the ordinary and extraordinary waves upon traveling a distance d through the medium is $\Delta\varphi = \dfrac{2\pi}{\lambda_0}(n_e - n_o)d$.

H 22.10. Use $\delta(\lambda_0)$ given in Problem 22.3, where ε_i has the form derived in Problem 22.4.

H 23.6. Use Eq.(23.22) and the magnitude of the impedance found in Problem 23.1.

H 23.7. Write down boundary conditions (14.3) assuming finite σ_f and j_f in the conductor.

H 23.8. The ionic contribution to the dispersive relation (23.45) should be written in terms of a free-ion plasma frequency $\Omega_p^2 = \omega_p^2 \mu / m_e$.

H 23.9. The relation is obtained by differentiating the given condition for minimum deviation.

H 23.10. Rearrange the differential equation separating the variables ω and k and then integrate it.

H 24.6. Use Eq.(24.14).

H 24.7. See Problem 24.1.

H 24.8. Set $q = \theta$, which is the angle between the pendulum bob and the vertical, and use Cartesian coordinates with respect to the centre of the loop.

H 24.9. The constraint condition is $r\theta - x = 0$.

H 24.10. Use the Lagrangian derived in Problem 24.4 to find p_i and $\dot{q}_i = v_i$, and then the definition (24.22).

H 24.11. Write down the Hamiltonian given in Problem 24.10 in terms of $\partial S / \partial x_i = p_i$ and take the square of Eq.(24.46).

H 25.7. Show that the partition function can be factored as $Z = z^3$.

H 25.8. Use Eq.(25.72), where $Z = z^N$.

H 25.9. Use Eq.(12.6) for the electrical conductivity, where n is the number of point defects at a given temperature as derived in Problem 25.1.

H 25.10. Use Eq.(25.70).

H 25.11. Follow the steps given in Problems 25.3 and 25.4, using Eq.(8.42) for the adiabatic compressibility k_S.

H 26.6. Use Eq.(26.6) and the result of Problem 25.6.

H 26.7. Set $dn/n = f(v)dv = f(\varepsilon)d\varepsilon$ in Eq.(26.30).

H 26.8. Use the Gaussian probability distribution given by Eq.(26.28) to calculate $\sigma^2(\vec{v}) = 3\sigma^2(v_k) = 3(\langle v_k^2 \rangle - \langle v_k \rangle^2)$.

H 26.9. The relative motion of two molecules can be described to be the motion of a single particle of reduced mass μ, as given by Eq.(3.5).

H 26.10. Solve for p the equation (9.29) and then expand in powers of b/V, which is less than one, using the series expansion $\dfrac{1}{1-x} = 1 + x + x^2 + x^3 + \cdots$.

H 27.6. Derive $pV = f(U)$ for thermal radiation from Eqs.(27.4) and (27.5), and for a perfect gas from Eqs.(26.19).

H 27.7. Use Eq.(27.36) where the energy density within a range of frequencies should be the same as the energy density within the corresponding range of wavelengths, that is $u_\omega |d\omega| = u_\lambda |d\lambda|$.

H 27.8. The partition function of the N-oscillator system is $Z = z^N$, where z has been derived in Problem 27.2.

H 27.9. Use the single-oscillator partition function z derived in Problem 27.3 to obtain the N-oscillator partition function given by $Z = z^N$.

H 27.10. Calculate dU_ω / dT using Eq.(27.36), and substitute it into the expression for the variance of energy, derived in Problem 25.6.

H 28.7. The photoelectron should be emitted with a kinetic energy higher than the electrostatic attraction Ne_0^2 / R.

H 28.8. Write down the components of the law of conservation of momentum given in Problem 28.2.

H 28.9. Set $p_\varphi = \partial E_n / \partial \omega = I\omega$ in Eq.(28.22).

H 28.10. The limiting value v_0 is that corresponding to the inelastic collision.

H 28.11. Solve Eq.(5.6) for $\beta = v/c$, and then substitute E from Eq.(5.10) and p from Eq.(28.36).

H 28.12. Use the uncertainty relation $\langle p \rangle \langle r \rangle \sim \hbar$ to express the total energy E_g of the two-electron system in terms of position only, and then minimize E.

H 29.6. Solve the eigenvalue equation by introducing a new function $\Phi = x\Psi$.

H 29.7. Use the eigenvalue equation $\hat{\Pi}\Psi(x) = \pi \Psi(x)$ and the property $\hat{\Pi}^2 \Psi(x) = \Psi(x)$ to find the eigenvalues. The eigenfunctions should have a definite parity.

H 29.8. Use the commutation relations to show that $\hat{S}(a)$ always commutes with the kinetic energy $\hat{p}_x^2/2m$ and that $[U(x), \hat{S}(a)] = 0$ provided $U(x)$ is a periodic function with period a.

H 29.9. Write down the function $f(\hat{A})$ and its derivative as power series expansions of the form

$$f(\hat{A}) = \sum_{n=0}^{\infty} c_n \hat{A}^n \quad , \quad \frac{df(\hat{A})}{d\hat{A}} = \sum_{n=0}^{\infty} n c_n \hat{A}^n$$

H 29.10. Substitute the Hamiltonian of the simple harmonic oscillator into the evolution equations for $\hat{x}$ and $\hat{p}_x$, which can be reduced to the classical equation of motion.

H 30.6. It is straightforward to separate a factor proportional to the probability density $P = \Psi^*\Psi$ into Eq.(30.22).

H 30.7. Substitute $\psi(x)$ into the time-independent Schrödinger equation in one dimension, Eq.(30.17).

H 30.8. Follow the same steps as in Problem 30.1.

H 30.9. Use Eq.(29.8) to derive the average values.

H 30.10. Since $U(\vec{r}) = U(x) + U(y) + U(z)$, the Schrödinger equation can be separated into three eigenvalue equations in one dimension.

H 31.6. Follow the steps used to solve the Problem 31.2.

H 31.7. Use the definition of the operators $\hat{L}_\pm$ and the commutation relation (31.2).

H 31.8. Since L^2 and L_z are known, it follows that $\langle \hat{L}_x \rangle = \langle \hat{L}_y \rangle$ and also $\langle \hat{L}_x^2 \rangle = \langle \hat{L}_y^2 \rangle$.

H 31.9. Expand $\Psi(\theta, \varphi)$ in terms of $Y_{lm}(\theta, \varphi)$.

H 31.10. The eigenvalue problem reduces to that of $\hat{L}^2$.

H 32.6. Derive the probabilities of finding the atom in each eigenstate, and use Eq.(29.29) to calculate the expectation values.

H 32.7. Use Eq.(32.13) to find n and l by direct inspection and Eq.(31.27) to determine the value of m. The most probable value of r occurs when the radial probability density $r^2 R_{nl}^2$ is a maximum.

H 32.8. Put $s = 0$ in the Kramers recurrence relation and use $\langle r^{-2} \rangle_{nl}$ as derived in Problem 32.3, to find $\langle r^{-1} \rangle_{nl}$. Then put $s = 1$ to derive $\langle r \rangle$, and $s = 2$ to obtain $\langle r^2 \rangle$.

H 32.9. The radial equation reduces to that of a harmonic oscillation inside the box. The energy eigenvalues result from the boundary condition $R(a) = 0$.

H 32.10. For $l = 0$, the solution is similar to that given in Problem 32.9, $R_{k0}(r) = \sqrt{\dfrac{2}{\pi}} \dfrac{\sin kr}{r}$.

For $l \neq 0$, assume that $R_{kl}(r) = r^l u_{kl}(r)$.

H 33.6. Use Eqs.(33.29), (33.34) and (33.35).

H 33.7. Show that $(\Delta J_x)^2 = \langle \hat{J}_x^2 \rangle$ and $(\Delta J_y)^2 = \langle \hat{J}_y^2 \rangle$, and then use Eqs.(33.29).

H 33.8. Combine the two commutation relations with that satisfied by the corresponding angular momentum component operators, Eqs.(33.12).

H 33.9. Write down $\left(\vec{e}_r \cdot \hat{\vec{S}} \right)^2$, where $\vec{e}_r = a_x \vec{e}_x + a_y \vec{e}_y + a_z \vec{e}_z$ $\left(a_x^2 + a_y^2 + a_z^2 = 1 \right)$ and the components of $\hat{\vec{S}}$ are given by Eqs.(33.40).

H 33.10. Use the property $\hat{\sigma}_i \hat{\sigma}_j = i \hat{\sigma}_k$ of the Pauli matrices, which can be derived in a straightforward way from Eqs.(33.41), (33.42) and (33.43).

H 34.6. Use the fact that the total number $(2j_1 + 1)(2j_2 + 1)$ of angular momentum states must be the same in both representations.

H 34.7. Use Eq.(34.4) to find the allowed values of $\hat{\vec{S}} \cdot \hat{\vec{L}}$ in the eigenstates Ψ_{jm_j}.

H 34.8. Use Eq.(34.16) and the energy levels (34.21).

H 34.9. Use Eq.(34.27) where $j = l \pm \tfrac{1}{2}$.

H 34.10. Immediately after the field is rotated, the expectation value of $\hat{\vec{J}} \cdot \vec{e}_z' = \hat{J}_z \cos\theta$ should remain unchanged.

H 35.6. Use χ^a as defined in Problem 35.3, and put the operators in the form

$$\hat{S}^2 = \tfrac{3}{2} \hbar^2 + \hat{S}_{1+} \hat{S}_{2-} + \hat{S}_{1-} \hat{S}_{2+} + 2\hat{S}_{1z} \hat{S}_{2z} \quad , \quad \hat{S}_z = \hat{S}_{1z} + \hat{S}_{2z}$$

H 35.7. Use χ^s as introduced in Problem 35.3.

H 35.8. Write down the spin operators as $\hat{S}^2 = 4\hbar^2 + \hat{S}_{1+} \hat{S}_{2-} + \hat{S}_{1-} \hat{S}_{2+} + 2\hat{S}_{1z} \hat{S}_{2z}$, $\hat{S}_z = \hat{S}_{1z} + \hat{S}_{2z}$, and the symmetric spin function as derived in Problem 35.5.

H 35.9. Use the antisymmetric spin function introduced in Problem 35.5.

H 35.10. The states in which both particles are in the same state are always symmetric.

H 36.6. Write down the characteristic equation (36.19) using the matrix elements $\langle \Psi_{2lm} | -eEr\cos\theta | \Psi_{2lm'} \rangle$.

H 36.7. Each 2p state may be combined with a 3d state and an antisymmetric function is formed from the combination.

H 36.8. The two electrons are in different shells so that any of their combinations is possible and hence the degeneracy is $10 \times 10 = 100$.

H 36.9. See Problem 36.3.

H 36.10. For a set of electrons one short of a full shell, the L, S, J values are the same as if there is only one electron in shell.

H 37.6. Use Eqs.(9.66) and (9.68), where $\langle \psi_0 | x | \psi_1 \rangle = \sqrt{\hbar/2m\omega}$.

H 37.7. See Problem 37.3.

H 37.8. Write down the energy levels (34.21) for $l = 0, 1, \ldots, n-1$ and then use the selection rules (37.39) and (37.41).

H 37.9. Write down the energy levels (34.29) for the $\varepsilon_{3d_{5/2}}$, $\varepsilon_{3d_{3/2}}$, $\varepsilon_{2p_{3/2}}$ and $\varepsilon_{2p_{1/2}}$ Zeeman multiplets, and then use the selection rules on j, Eq.(37.41), and on m_j, Eq.(37.42).

H 37.10. Use Eq.(37.50) and calculate the matrix elements between the Ψ_{21m} and Ψ_{100} states.

H 38.6. The separation R_0 at equilibrium is obtained from $dE(R)/dR = 0$, and the system is stable if $\left[d^2E(R)/dR^2 \right]_{R=R_0} > 0$.

H 38.7. The work is given by the increase in the potential energy $U(R)$ described in Problem 38.2.

H 38.8. There are only three vibrating atoms, each motion being described by Eq.(38.27).

H 38.9. Find the internal energy U of the system, and then use Eq.(7.19).

H 38.10. See Problem 38.5, using the numerical approximation $\int_0^\infty \frac{x^2 dx}{e^x - 1} \cong 2.4$.

H 39.6. Write down Eq.(39.3) as a determinant.

H 39.7. The angle θ is defined by $\cos\theta = \dfrac{\vec{R}_{pqs} \cdot \vec{R}_{p'q's'}}{|\vec{R}_{pqs}||\vec{R}_{p'q's'}|}$.

H 39.8. Use the primitive translations $\vec{a}'$, $\vec{b}'$ and $\vec{c}'$ defined in Problem 39.3.

H 39.9. Use the atom distribution in the unit cell described in Problem 39.4.

HINTS

H 39.10. Use Eqs.(39.9) to show that the reciprocal primitive vectors of $\vec{a}^*, \vec{b}^*$ and $\vec{c}^*$ coincide with $\vec{a}, \vec{b}$ and $\vec{c}$, respectively.

H 39.11. The A_3B cubic alloy has six A atoms on the faces of the unit cell and eight B atoms at the corners, and hence, there is a face-centred cubic structure.

H 40.6. Use Eqs.(40.24) and (40.20) for U and N to calculate $\langle \varepsilon \rangle = U/N$ by substituting the densities of states $N_{1-D}(\varepsilon)$, $N_{2-D}(\varepsilon)$ and $N_{3-D}(\varepsilon)$.

H 40.7. See Problem 40.2 using $N_{2-D}(\varepsilon)$ as derived in Problem 40.1.

H 40.8. The contribution of lattice vibrations has been derived in Problems 38.5 and 38.10.

H 40.9. The energy eigenvalues of the planar motion of an electron in a uniform field B_0, derived in Problem 32.5, can be interpreted as $\dfrac{p_x^2}{2m_e} + \dfrac{p_y^2}{2m_e} = \left(n+\dfrac{1}{2}\right)\hbar\dfrac{eB}{m_e}$, where $S_{p_x p_y}(n) = \pi(p_x^2 + p_y^2)$. The Lorentz force law implies that $S_{p_x p_y}(n) = (eB_0)^2 S_{xy}(n)$.

H 40.10. See Problem 40.5, taking into account that all the vibration modes have the same frequency given by $\hbar\omega_E = k_B \theta_E$.

H 40.11. See Problem 40.5, where the integral tends to a constant if the upper limit is made infinite at low temperatures, where $T \ll \theta_D$.

H 41.6. Use Eq.(41.6), where $\vec{k}$ should be expressed in terms of its components given in Problem 41.1.

H 41.7. Use Eq.(39.12), where $\vec{a}^* = (2\pi/a)\vec{e}_x$ and $\vec{b}^* = (2\pi/na)\vec{e}_y$.

H 41.8. Use Eq.(41.30), where $U(K)$ for a linear lattice has the form given in Problem 41.2.

H 41.9. For a linear lattice, the dispersion relation (41.47) reduces to $\varepsilon = \varepsilon_0 - \alpha - 2\gamma\cos\alpha$.

H 41.10. See Problem 41.4.

H 42.6. Use Eq, (42.22) for n_i, assuming that ε_g has a polynomial dependence on temperature.

H 42.7. Write down Eq.(42.22) at T_0 and βT_0.

H 42.8. Let $\varepsilon_F = \varepsilon_D$ in Eq.(42.40), and solve it for ε_d.

H 42.9. Use the condition for electrical neutrality (42.38), assuming completely ionized donors and acceptors.

H 43.6. Use the statistic characteristic given in Problem 43.4 to find the short-circuit current at zero output voltage and the maximum voltage at zero output current.

H 43.7. V_m corresponds to the point of maximum power P_m (shown on the static characteristic given in Problem 43.4) where $dP/dV = 0$.

H 43.8. Use Eqs.(43.28) and (43.26) to calculate the field, and then relate the width w of the depletion region with V_C as in Problem 43.3.

H 43.9. Write down Eqs.(43.59) and (43.60) for β, using $n_E = n_i e^{\alpha \sqrt{N_{dE}/k_B T}}$.

H 43.10. Use Eq.(43.56) to express the total collector current.

H 44.6. Use Eq.(44.6) for χ_d, where $\langle r^2 \rangle = \langle r_1^2 \rangle + \langle r_2^2 \rangle$ is calculated from Eq.(29.8).

H 44.7. Use Eq.(44.14) for the Brillouin function. The Langevin function is defined in Problem 44.1.

H 44.8. Write down the force F_x on a specimen, Eq.(2.27), where $U = -V\vec{M} \cdot \vec{B}$.

H 44.9. Use Eq.(7.19), where $U = V\left(-\vec{B}_{ef} \cdot \vec{M}\right)_{B_0=0}$.

H 44.10. Use H given by Eq.(13.12), and write down the maximum condition on the energy product.

H 45.6. Use Eq. 45.17, where B_f should be associated with a volume which extends over the paramagnetic region.

H 45.7. The energy per unit volume associated with flux exclusion becomes a function of position given by $B_f [B_f - B(x)]/2\mu_0 = -B_f M(x)/2$, where $M(x)$ is as derived in Problem 45.1.

H 45.8. Use Eq.(13.18) for the magnetic flux, where $\vec{A}$ is obtained by integrating $\vec{j}_s$ (derived in Problem 45.4) over closed path inside the ring along which the current density is zero everywhere.

H 45.9. Use the expression obtained in Problem 45.4 for the current density $\vec{j}_s$ of a Cooper pair.

H 45.10. The total current density splits into two parts, so that $j_s = 2j_0 \sin \Delta\varphi$.

H 46.6. Write down the maximum condition on $f = B(A,Z)/A$, where $B(A,Z)$ is given by Eq.(56.24).

H 46.7. Use the energy-inertia relation (59) which, if a neutron is removed, reads

$$M(A,Z)c^2 + S_n(A,Z) = M(A-1,Z)c^2 + m_n c^2$$

H 46.8. The expectation radius is given by Eq.(29.8).

H 46.9. Use the condition for a stable bound state of the deuteron derived in Problem 46.3, where $p^2/2\mu \leq V_0$.

H 46.10. Put $U = -\mu\vec{\sigma}_2 \cdot \vec{B}$, where the magnetic induction due to the other nucleon has the form

$$\vec{B} = -\nabla\left(\frac{\mu_0}{4\pi} \frac{\mu\vec{\sigma}_1 \cdot \vec{r}}{r^3}\right)$$

given by Eqs.(13.3) and (13.6).

H 47.6. Solve the equations $dN_k = \lambda_{k-1} N_{k-1} dt - \lambda_k N_k dt$ which describe the decay processes.

H 47.7. Write down the Boltzmann distribution law for the translational motion of the nucleus in one dimension, along the direction in which the γ-radiation is emitted. The observed frequencies ω are different from ω_{if} due to the Doppler effect.

H 47.8. Use Eq.(47.25) for $M(A,Z)$.

H 47.9. Use the conservation laws for energy and momentum.

H 47.10. Calculate the energy levels $U_Q(m_I)$ for $m_I = \pm\frac{1}{2}, \pm\frac{3}{2}$ $\left(I = \frac{3}{2}\right)$ given by Eq.(47.64).

Answers

A 2.8. $m = M \sin\alpha / (1 - \sin\alpha)$

A 2.9. $\tan\alpha = 2\tan\varphi + \cot\varphi$

A 2.10. $D = mg/k$

A 2.11. $u \leq (M-m)\sqrt{\dfrac{2gH}{M(m+M)}}$

$v \leq 2m\sqrt{\dfrac{2gH}{M(m+M)}}$

A 2.12. $h = l\left[\dfrac{4}{3} - \dfrac{1}{4(v_0^2/2gl - 1)}\right]$

$t = \dfrac{1}{\sqrt{v_0^2 - 2gl}}$

A 2.13. $\left(p_{2x} - \dfrac{p_0}{1+A}\right)^2 + p_{2y}^2 = \left(\dfrac{p_0}{1+A}\right)^2$

A 2.14. $T = \dfrac{3}{2}mg$

A 3.9. $a = \dfrac{g}{1 + 2I/mR^2}$, $a = 0$

A 3.10. $\tan\beta = \dfrac{2}{5}\tan\alpha$

A 3.11. $v_{cr} = \sqrt{Rg} \approx 7.39$ km/s,

$v_{par} = \sqrt{2Rg} \approx 11.2$ km/s

A 3.12. $v_{sev}^2 = v_0^2\left(\sqrt{2}-1\right)^2 + v_{par}^2$

A 4.5. $\tan\left(\dfrac{\alpha}{2}\right) = \sqrt{\dfrac{1}{\beta^2} - 1}$

A 4.6. $v^2 = \dfrac{(\vec{v}_1 - \vec{v}_2)^2 - (\vec{v}_1 \times \vec{v}_2)^2/c^2}{(1 - \vec{v}_1\cdot\vec{v}_2/c^2)^2}$

A 5.6. $\rho' = \dfrac{\rho}{1 - \beta^2}$

A 5.7. $\vec{p}' = \vec{p} + (\gamma - 1)\dfrac{\vec{p}\cdot\vec{v}}{v^2}\vec{v} - \gamma\dfrac{E}{c^2}\vec{v}$

$E' = \gamma(E - \vec{v}\cdot\vec{p})$

A 5.8. $\vec{a} = \dfrac{1}{m_0}\sqrt{1 - \dfrac{v^2}{c^2}}\left[\vec{F} - \left(\dfrac{\vec{v}}{c}\cdot\vec{F}\right)\dfrac{\vec{v}}{c}\right]$

$\vec{a} = \dfrac{q}{m_0}\sqrt{1 - \dfrac{v^2}{c^2}}\left[\vec{E} + \vec{v}\times\vec{B} - \left(\dfrac{\vec{v}}{c}\cdot\vec{E}\right)\dfrac{\vec{v}}{c}\right]$

A 5.9. $\begin{pmatrix}F_\perp \\ F_\parallel\end{pmatrix} = m_0\gamma\begin{pmatrix}1 & 0 \\ 0 & \gamma^2\end{pmatrix}\begin{pmatrix}a_\perp \\ a_\parallel\end{pmatrix}$

A 5.10. $M = \left(m_1^2 + m_2^2 + \dfrac{2m_1 m_2}{\sqrt{1 - v_0^2/c^2}}\right)^{1/2}$

$\vec{v} = \vec{v}_0 \dfrac{m_1}{m_1 + m_2\sqrt{1 - v_0^2/c^2}}$

A 6.7. $x = \dfrac{mg}{2k}$

A 6.8. $T = 2\pi\sqrt{5m/k}$

A 6.9. $v_0 = \mu g\sqrt{15m/k}$

A 6.10. $l_{\min} = l - v_0\sqrt{m/2k}$,

$l_{\max} = l + v_0\sqrt{m/2k}$

A 6.12. $F = \pi r^6 \rho\left(\dfrac{\rho g h}{8\eta\ell}\right)^2$

A 6.13. $\mu = \dfrac{\rho S v^2}{mg\cos\alpha}$

A 7.6. $C = C_V + \dfrac{R}{2}$

A 7.7. $VT^{(C_V-C_0)/R} e^{(C_0-C)2R} = \text{const}$

A 7.8. $Q_{ij} < Q_{ik}$

A 7.9. $T_0 = \dfrac{x T_M T_m}{T_M + (x-1) T_m}$

A 7.10. $\dfrac{\eta'}{\eta} = \dfrac{4\gamma + 1}{3\gamma + 2}$

A 8.6. $T = \dfrac{\gamma a^2}{Rb(\gamma+1)^2}$

A 8.7. $p^{3/4}(V-b) = \text{const} \cdot T$

A 8.8. $F = F_0 + \nu a T + \dfrac{\nu b}{2} T^2$
$\qquad -\nu a T \ln T - \nu R T \ln V$

$G = F_0 + \nu a T + \dfrac{\nu b}{2} T^2$
$\qquad -\nu a T \ln T - \nu R T (\ln V - 1)$

$H = U_0 + \nu(a+R)T - \dfrac{\nu b}{2} T^2$

A 8.10. $\left(\dfrac{\partial S}{\partial V}\right)_p = \dfrac{C_p}{VT} \dfrac{1}{\beta_p}$

A 9.6. $\dfrac{\eta'}{\eta} = \dfrac{2x}{x+1}$

A 9.7. $\left[\dfrac{1}{a}\left(\dfrac{p}{p_0} + bV - 1\right)\right]^{C_V} e^{\beta_p V/k_T} = \text{const}$

A 9.8. $\mu = -b/C_p$

A 9.9. $C_p - C_V =$
$\left(p + \dfrac{a}{V^2}\right) \dfrac{R}{p - a(V-2b)/V^3}$

A 9.10. $p_c = \dfrac{a}{16b(4b)^{2/3}}$

$V_c = 4b$

$T_c = \dfrac{15a}{16R(4b)^{2/3}}$

$p_c V_c = \dfrac{4}{15} R T_c$

A 10.6. $T \cong T_0 - \left(1 - \dfrac{p}{p_0}\right) \dfrac{RT_0^2}{\lambda}$

A 10.7. $f = \left(1 + \dfrac{m_0 c_0}{mc}\right) e^{-c\theta/\lambda} - \dfrac{m_0 c_0}{mc}$
$\cong 1 - \left(1 + \dfrac{m_0 c_0}{mc}\right) \dfrac{c\theta}{\lambda}$

A 10.8. $\lambda = 4 \dfrac{RT^2}{T - 273}$

A 10.9. $W = C_p T \ln \dfrac{T}{T_0} - C_p(T - T_0)$
$\qquad + \lambda\left(\dfrac{T}{T_0} - 1\right)$

A 11.7. $F = \dfrac{3 q_1 q_2 d^2}{2\pi \varepsilon_0 D^4}$

A 11.8. $F = \dfrac{q\lambda}{2\pi \varepsilon_0 l} \ln\left(1 + \dfrac{l}{R}\right)$

A 11.9. $F = \dfrac{qQ}{2\pi^2 \varepsilon_0 R^2}$

A 11.10. $U = \dfrac{q^2}{8\pi\varepsilon}\left(\dfrac{1}{R_1} - \dfrac{1}{R_2}\right)$

A 12.6. $\sigma = \dfrac{\varepsilon}{\tau} \ln n$

A 12.7. $j_r = \dfrac{\sigma}{3\varepsilon} \rho_0 r e^{-\sigma t/\varepsilon}$

A 12.8. $B = n\dfrac{\mu_0 I}{2R}\tan\left(\dfrac{\pi}{n}\right)$

A 12.9. $B_r = \dfrac{\mu_0 I}{2\pi r}\dfrac{R_2^2 - r^2}{R_2^2 - R_1^2}$

A 12.10. $B_x = \dfrac{\mu_0 q\omega}{2\pi R^2}\left(\dfrac{R^2 + 2x^2}{\sqrt{R^2 + x^2}} - \dfrac{2x^2}{\sqrt{x^2}}\right)$

A 13.6. $\omega = \dfrac{\alpha a^2}{\sqrt{mL}}$

A 13.7. $F = \dfrac{B^2 l^3}{4R}\omega$

A 13.8. $\tau = \dfrac{2m_e}{\mu_0 e N_0 I}\tan^{-1}\left(\dfrac{\mu_0 e N_0 IR}{m_e v}\right)$

A 13.9. $Q_0 = \dfrac{CE}{2}\sqrt{2 - \cos^2\omega\tau}$, $\omega = \sqrt{\dfrac{2}{LC}}$

A 13.10. $I_1 = V_0\sqrt{\dfrac{L_2 C}{L_1(L_1 + L_2)}}$,

$I_2 = V_0\sqrt{\dfrac{L_1 C}{L_2(L_1 + L_2)}}$

A 14.6. Point charge $q = -\alpha$, $\vec{j} = 0$

A 14.8. $\rho_P = -\nabla\cdot\vec{P}$, $\vec{j}_P = \dfrac{\partial\vec{P}}{\partial t}$,

$\vec{j}_M = \nabla\times\vec{M}$

A 14.9. $E_\theta = \dfrac{k^2 p_0}{4\pi\varepsilon_0}\dfrac{e^{i(kr-\omega t)}}{r}\sin\theta$,

$B_\varphi = -\dfrac{\mu_0 c k^2 p_0}{4\pi}\dfrac{e^{i(kr-\omega t)}}{r}\sin\theta$

$S = \dfrac{ck^4 p_0^2}{16\pi^2\varepsilon_0}\dfrac{\sin^2\theta}{r^2}\cos^2(kr - \omega t)$

A 14.10. $E'_x = E_x$

$E'_y = \gamma(E_y - vB_z)$

$E'_z = \gamma(E_z + vB_y)$

$B'_x = B_x$

$B'_y = \gamma\left(B_y + \dfrac{v}{c^2}E_z\right)$

$B'_z = \gamma\left(B_z - \dfrac{v}{c^2}E_y\right)$

A 15.6. $v = R\omega$

A 15.7. $\omega_{\lim} = 2\omega_0$, $v_{\lim} = a\omega_0$

A 15.8. $\Delta p = \rho v^2 s$

A 15.10. $L_0 = \dfrac{\mu}{2\pi}\ln\left(\dfrac{R_2}{R_1}\right)$

$C_0 = \dfrac{2\pi\varepsilon}{\ln(R_2/R_1)}$, $v = \dfrac{1}{\sqrt{\varepsilon\mu}}$

A 16.7. $\lambda_1 = a\dfrac{v_1 - v_2}{v_0 + v_2}$, $\lambda_2 = a\dfrac{v_1 - v_2}{v_0 + v_1}$

A 16.8. $\dfrac{1}{f} = (n-1)\left(\dfrac{1}{R_1} + \dfrac{1}{R_2}\right)$

A 16.9. $f = R\dfrac{n}{2(n-1)}$

A 16.10. $y = L\cos i_0 \sin\left(\dfrac{x}{L\sin i_0}\right)$

A 16.11. $\Phi(k) = \dfrac{AL}{2\pi}\dfrac{\sin\alpha}{\alpha}$,

$\alpha = \dfrac{(k_0 - k)L}{2}$, $\Delta\xi\Delta k = 4\pi$

A 16.12. $v = \dfrac{c}{\sqrt{1 - \pi^2 c^2/a^2\omega^2}} = c\dfrac{k}{k_x}$

$v_g = c\dfrac{k_x}{k}$

A 17.6. $Z_0 = \dfrac{1}{2\pi}\sqrt{\dfrac{\mu}{\varepsilon}}\ln\dfrac{R_2}{R_1}$

A 17.7. $L_0 = \mu\dfrac{a}{b}$, $C_0 = \varepsilon\dfrac{b}{a}$,

$Z_0 = \dfrac{a}{b}\sqrt{\dfrac{\mu}{\varepsilon}}$

A 17.8. $p_r = 2\dfrac{I}{v_S}$

A 17.9. $\gamma = \dfrac{\ln(I_1 r_1^2 / I_2 r_2^2)}{2(r_2 - r_1)}$

A 17.10. $\rho = (SWR-1)/(SWR+1)$

A 18.6. $v = \dfrac{\lambda h}{x}\nu$

A 18.7. $f = \dfrac{4r_m^2}{5m\lambda(n-1)}$

A 18.8. $\lambda = 20\lambda_M/(2m+1)$

$m = 10, 11, 12, 13, 14, 15, 16, 17$

A 18.9. $V = \dfrac{2(1-R)}{2(1-R)+R^2}$

A 18.10. $I = I_0 \dfrac{\sin^2(3\delta/2)}{\sin^2(\delta/2)}$

A 19.6. $N = 2m+1 = 2\left[\dfrac{b}{\lambda}\right]$

A 19.7. $n = m\dfrac{a}{b}$

A 19.8. $m = [1/\sin\theta_1] = 6$

A 19.9. $RP = 0.82\dfrac{D}{\lambda}$

A 19.10. $f = \rho_1^2/\lambda$

A 20.6. $\delta = \dfrac{\mu_0 e^2 \omega}{6m_e c}$

A 20.7. $\langle P \rangle = \dfrac{\mu}{6\pi c}\dfrac{e^4 E_0^2 \omega^4}{m_e^2 (\omega_0^2 - \omega^2)^2}$

A 20.8. $\lambda_1 = 1$, $J_x = \begin{pmatrix}1\\0\end{pmatrix}$;

$\lambda_2 = i$, $J_y = \begin{pmatrix}0\\1\end{pmatrix}$

A 20.9. $\theta = \pi/4$, $I_f/I_i = 9/10$

A 20.10. $I = 36$ V²/m², $Q = 4$ V²/m²,

$U = 8\sqrt{15}$ V²/m², $V = 8\sqrt{5}$ V²/m²;

$a = 5.8$, $b = 1.54$, $\tan 2\Psi = 2\sqrt{15}$

A 21.7. $\tau_1 \tau_2 + \rho_2^2 = 1$

A 21.8. $n \approx 1 + \dfrac{3}{2}\tan\left(\dfrac{\pi}{8}\right) \approx 1.47$

A 21.9. $\beta = kd\cos\theta$, $\gamma = n\cos\theta$,

$\gamma_{i,\tau} = n_{i,\tau}\cos\theta_{i,\tau}$

A 21.10. $R = \left[\dfrac{(n'/n)^N - 1}{(n'/n)^N + 1}\right]^2$

A 21.11. $\dfrac{n_1 n_3}{n_2} = \sqrt{n_\tau}$

A 21.12. $R = \left(\dfrac{n_\tau n_1^2 - n_3^2}{n_3^2 + n_\tau n_1^2}\right)$

A 22.6. $\dfrac{1}{n_o} < \sin\theta < \dfrac{1}{n_e}$

A 22.7. $P = \dfrac{1}{2}$

A 22.8. $\tan\alpha = \dfrac{n_o^2 - n_e^2}{2n_o n_e}$

A 22.9. $\alpha = \pi/4$, $d = \lambda_0 / 4(n_e - n_o)$

A 22.10. $\delta(\lambda_0) = \dfrac{\pi e \omega_p^2 \omega}{\lambda_0 n_o m_e (\omega_0^2 - \omega^2)^2} B$

A 23.6. $\dfrac{\langle u_E \rangle}{\langle u_H \rangle} = \dfrac{\sigma}{\varepsilon\omega}$

A 23.7. $\sigma_f = 2\varepsilon_0 E_0 \sin\theta$,

$j_f = 2\sqrt{\dfrac{\varepsilon_0}{\mu_0}} E_0$

A 23.8. $\dfrac{\mu}{m_H} = 1840 \dfrac{B\lambda_2^4}{C\lambda_1^4} = 19.5$

A 23.9. $\dfrac{d\delta(\lambda)}{d\lambda} = \dfrac{2\sin(A/2)}{\sqrt{1-n^2\sin^2(A/2)}} \dfrac{dn}{d\lambda}$

A 23.10. $n^2 = 1 + \dfrac{\text{const}}{\omega^2}$

A 24.7.

$\theta = C\cos\left(\sqrt{\dfrac{g}{l}}t + \varphi\right) + \dfrac{A\omega^2}{l\omega^2 - g}\cos\omega t$

A 24.8.

$H = \dfrac{ml^2}{2}\left[\dfrac{p^2}{ml^2} - \dfrac{R\omega}{l}\cos(\theta - \omega t)\right]^2$

$- \dfrac{mR^2\omega^2}{2} - mg(l\cos\theta + R\cos\omega t)$

A 24.9. $\ddot{x} = \dfrac{g\sin\alpha}{2}$, $\ddot{\theta} = \dfrac{g\sin\alpha}{2r}$

A 24.10. $H = \sqrt{c^2(\vec{p}-q\vec{A})^2 + m_0^2 c^4} + qV$

A 24.11.

$\left(\dfrac{\partial S}{\partial t} - qV\right)^2 = c^2(\nabla S - q\vec{A})^2 + m_0^2 c^4$

A 25.7.

$F = -3k_B T \ln\left(e^{-\varepsilon_1/k_B T} + e^{-\varepsilon_2/k_B T} + e^{-\varepsilon_3/k_B T}\right)$

A 25.8. $U = N\varepsilon \dfrac{3e^{-\varepsilon/k_B T}}{2 + 3e^{-\varepsilon/k_B T}}$

A 25.9. $\dfrac{R(T)}{R(T_0)} = e^{\Delta(T_0 - T)/k_B T_0 T}$

A 25.11. $\sigma^2 = \dfrac{k_B T}{k_s V}$

$P(p) = \sqrt{\dfrac{k_s V}{2\pi k_B T}} e^{-k_s V(p-p_0)^2 / 2k_B T}$

A 26.6. $\dfrac{\langle (\Delta E)^2 \rangle}{\langle E \rangle^2} = \dfrac{2}{f}$

A 26.7. $f(\varepsilon) = \dfrac{2\pi}{(\pi k_B T)^{3/2}} \varepsilon^{1/2} e^{-\varepsilon/k_B T}$

$\varepsilon_p = \dfrac{1}{2} k_B T \neq \dfrac{1}{2} m v_p^2$

A 26.8. $\dfrac{\sigma^2(v)}{\sigma^2(\vec{v})} = 1 - \dfrac{8}{3\pi}$

A 26.9. $\bar{v}_R = 4\left(\dfrac{k_B T}{\pi m}\right) = \sqrt{2}\bar{v}$

A 26.10. $B = b - \dfrac{a}{RT}$, $C = b^2$

A 27.6. $pV = \dfrac{1}{3}U$, $\gamma = 4/3$ (radiation)

$pV = \dfrac{2}{3}U$, $\gamma = 5/3$ (gas)

A 27.7. $u_\lambda(x) = \dfrac{8\pi k_B^5 T^5}{h^4 c^4} \dfrac{x^5}{e^x - 1}$

$x = hc/\lambda k_B T$

$\lambda_0 T = \dfrac{hc}{4.97 k_B}$

A 27.8. $F = G = Nk_B T \ln\left(\dfrac{\hbar\omega}{k_B T}\right)$

$S = Nk_B \left[1 - \ln\left(\dfrac{\hbar\omega}{k_B T}\right)\right]$

$U = Nk_B T$

A 27.9. $F = G = Nk_B T \ln\left(1 - e^{-\hbar\omega/k_B T}\right)$

$S = Nk_B \left[\dfrac{\hbar\omega/k_B T}{e^{\hbar\omega/k_B T} - 1} - \ln\left(1 - e^{-\hbar\omega/k_B T}\right)\right]$

$U = \dfrac{N\hbar\omega}{e^{\hbar\omega/k_B T} - 1}$

A 27.10. $\sigma^2 = \hbar\omega U_\omega + \dfrac{c^3 \pi^2}{V\omega^2 d\omega} U_\omega^2$

A 28.7. $N = \dfrac{Rch}{e_0^2}\left(\dfrac{1}{\lambda} - \dfrac{1}{\lambda_0}\right)$

A 28.8. $\tan\varphi = \left(\cot\dfrac{\theta}{2}\right) / \left(1 + \dfrac{h}{m_e c\lambda}\right)$

A 28.9. $E_n = n^2 \dfrac{\hbar^2}{2I}$

A 28.10. $v_0 = \sqrt{\dfrac{3\varepsilon_1}{M}}$

A 28.11. $v = c / \sqrt{1 + \left(\dfrac{m_e c a \sin\theta}{h}\right)^2}$

A 28.12. $E_g = -\dfrac{49}{8}\dfrac{m_e e_0^2}{2\hbar^2} \approx -83\text{ eV}$

A 29.6. $\lambda = -\alpha^2$, $\Psi(x,\alpha) = N\dfrac{\sin\alpha r}{r}$

for any real α

A 29.7. $\pi = \pm 1$

$\psi(x,1) = \dfrac{1}{\sqrt{2}}[\psi(x) + \psi(-x)]$

$\psi(x,-1) = \dfrac{1}{\sqrt{2}}[\psi(x) - \psi(-x)]$

A 29.10. $\hat{x}(t) = x\cos\omega t + \dfrac{p_x}{m\omega}\sin\omega t$

$\hat{p}_x(t) = p_x \cos\omega t - m\omega x \sin\omega t$

A 30.6. $\vec{v} = \dfrac{i\hbar}{2m} \sum_{i=x,y,z} \vec{e}_i \dfrac{\partial}{\partial x_i}\left[\ln\left(\dfrac{\Psi^*}{\Psi}\right)\right]$

A 30.7. $U(x) = \dfrac{\hbar^2}{2m}\dfrac{n}{x}\left(\dfrac{n-1}{x} - 2\gamma\right)$,

$\varepsilon = -\dfrac{\hbar^2}{2m}\gamma^2$

A 30.8.

$\dfrac{1}{2m}\sum_i \left(\dfrac{\partial S_0}{\partial x_i}\right)^2 + U = H\left(x_i, \dfrac{\partial S_0}{\partial x_i}\right) = \varepsilon$

$\sum_i \dfrac{\partial}{\partial x_i}\left(A_0^2 \dfrac{1}{m}\dfrac{\partial S_0}{\partial x_i}\right) = \sum_i \dfrac{\partial}{\partial x_i}(A_0^2 v_i)$

$= \nabla \cdot (P\vec{v}) = 0$

A 30.9. $\langle(\Delta x)^2\rangle = \dfrac{a^2}{2}$, $\langle(\Delta p)^2\rangle = \dfrac{\hbar^2}{2a^2}$

A 30.10. $\varepsilon_n = \hbar\omega\left(n + \dfrac{3}{2}\right)$, $n = n_1 + n_2 + n_3$

Degeneracy of ε_n is $\dfrac{1}{2}(n+1)(n+2)$

A 31.6. $U(r) = \dfrac{2\alpha^2 \hbar^2}{m}r^2$, $\varepsilon = \dfrac{\alpha\hbar^2}{m}$,

$L^2 = 0$

A 31.8. $\langle \hat{L}_x \rangle = 0$,

$\langle \hat{L}_x^2 \rangle = [l(l+1) - m^2] \dfrac{\hbar^2}{2}$

A 31.9. $P(0) = \dfrac{1}{2}$, $P(\pm\hbar) = \dfrac{1}{4}$, $P(2\hbar^2) = 1$

A 31.10. $\Delta\varepsilon = \varepsilon_{l+1} - \varepsilon_l = \dfrac{l(l+1)\hbar^2}{I}$

A 32.6. $\langle \hat{H} \rangle = -0.45 \dfrac{e_0^2}{2a_0}$, $\langle \hat{L}^2 \rangle = \dfrac{3}{7}\hbar^2$,

$\langle \hat{L}_z \rangle = \dfrac{1}{7}\hbar$

A 32.7. $n = 3$, $l = 1$, $m = 0$, $r = 12a_0$

A 32.8. $\left\langle \dfrac{1}{r} \right\rangle_{nl} = \dfrac{1}{n^2 a_0}$,

$\langle r \rangle_{nl} = \dfrac{1}{2}[3n^2 - l(l+1)]a_0$,

$\langle r^2 \rangle_{nl} = \dfrac{n^2}{2}[5n^2 + 1 - 3l(l+1)]a_0^2$

A 32.9. $R_n(r) = \sqrt{\dfrac{2}{a}} \dfrac{\sin(n\pi r/2a)}{r}$,

$E_n = n^2 \dfrac{h^2}{8m_e a^2}$

A 32.10.

$R_{kl}(r) = \sqrt{\dfrac{2}{\pi}} (-1)^l r^l \left(\dfrac{1}{kr} \dfrac{d}{dr} \right)^l \dfrac{\sin kr}{r}$

A 33.6. $\langle \hat{J}_r \rangle = m\hbar \cos\theta$

A 33.7. $|m_j| = j$

A 33.9. $(\vec{e}_r \cdot \hat{\vec{S}})^2 = \dfrac{\hbar^2}{4}$

A 34.6. $j_{\min} = |j_1 - j_2|$

A 34.7.

$\cos(\vec{S}\cdot\vec{L}) = \dfrac{j(j+1) - l(l+1) - s(s+1)}{2\sqrt{l(l+1)s(s+1)}}$

A 34.8. $B_i = \sqrt{l(l+1)} \dfrac{\zeta_{nl}}{\mu_B} = \sqrt{2} \dfrac{\zeta_{nl}}{\mu_B}$

A 34.9. $g_L = 2$; $g_L = \dfrac{2}{3}, \dfrac{4}{3}$; $g_L = \dfrac{4}{5}, \dfrac{6}{5}$

A 34.10. $P_{+\frac{1}{2}} = \cos^2\left(\dfrac{\theta}{2}\right) = \dfrac{3}{4}$,

$P_{-\frac{1}{2}} = \sin^2\left(\dfrac{\theta}{2}\right) = \dfrac{1}{4}$

A 35.6. $S^2 = 0$, $S_z = 0$

A 35.7. $S^2 = 2\hbar^2$, $S_z = 0, \pm\hbar$ $(S = 1)$

A 35.8. $S^2 = 0, 6\hbar^2$,

$S_z = 0, \pm\hbar, \pm 2\hbar$ $(S = 0, 2)$

A 35.9. $S^2 = 2\hbar^2$, $S_z = 0, \pm\hbar$ $(S = 1)$

A 35.10. $n_s = (s+1)(2s+1)$, $n_a = s(2s+1)$

A 36.6. $\varepsilon_2 = -\dfrac{e_0^2}{8a_0}, -\dfrac{e_0^2}{8a_0} \pm 3eEa_0$

A 36.7.

$60: {}^3F_4, {}^3F_3, {}^3F_2, {}^3D_3, {}^3D_2, {}^3D_1, {}^3P_2, {}^3P_1, {}^3P_0,$
${}^1F_3, {}^1D_2, {}^1P_1$

A 36.8.

$100: {}^3G_5, {}^3G_4, {}^3G_3, {}^3F_4, {}^3F_3, {}^3F_2, {}^3D_3, {}^3D_2, {}^3D_1,$
${}^3P_2, {}^3P_1, {}^3P_0, {}^3S_1, {}^1G_4, {}^1F_3, {}^1D_2, {}^1P_1, {}^1S_0$

A 36.9.

${}^3F_4, {}^3F_3, {}^3F_2, {}^3P_2, {}^3P_1, {}^3P_0, {}^1G_4, {}^1D_2, {}^1S_0$

A 36.10. $10: {}^2D_{5/2}, {}^2D_{3/2}$

A 37.6. $P_{01}(\infty) = \dfrac{\pi \alpha^2}{2\hbar m \omega \gamma^2} e^{-\omega^2/2\gamma^2}$

A 37.7. $^4D \to {}^4P$: 8 lines
$^4P \to {}^4S$: 3 lines

A 37.8.
$4f_{7/2}, 4f_{5/2}, 4d_{5/2}, 4d_{3/2}, 4p_{3/2}, 4p_{1/2}, 4s_{1/2}$
$3d_{5/2}, 3d_{3/2}, 3p_{3/2}, 3p_{1/2}, 3s_{1/2},$ 13 lines

A 37.9. $^2d_{5/2} \to {}^2p_{3/2}$: 12 lines,
$^2d_{5/2} \to {}^2p_{3/2}$: 10 lines, $^2d_{3/2} \to {}^2p_{1/2}$: 6 lines

A 37.10. $A_{2p \to 1s} = \dfrac{2^{10}}{3^8} \dfrac{e_0^8}{\hbar^4 c^3 a_0}$

A 38.6. $\left[\dfrac{E_{attractive}}{E_{repulsive}}\right]_{R=R_0} = -\dfrac{n}{m}$

A 38.7. $W_{R_0 \to R_0(1-\varepsilon)} = \dfrac{NAe^2}{2\rho} \varepsilon^2$

A 38.8. $\omega_k^2 = \dfrac{\kappa_0}{M}, \dfrac{\kappa_0}{M}\left(1 \pm \dfrac{\sqrt{2}}{2}\right)$

A 38.9. $C_V = Nk_B \left(\dfrac{\varepsilon}{k_BT}\right)^2 \dfrac{e^{\varepsilon/k_BT}}{\left(1+e^{\varepsilon/k_BT}\right)^2}$

A 38.10. $C_V = 28.8\, Nk_B \left(\dfrac{T}{\theta_D}\right)^2, T \ll \theta_D$
$C_V = 2Nk_B, T \gg \theta_D$

A 39.6.
$V_0 = abc\sqrt{1 - \cos^2 \alpha - \cos^2 \beta - \cos^2 \gamma + 2\cos \alpha \cos \beta \cos \gamma}$

A 39.7.
$\cos \theta = \dfrac{pp'a^2 + qq'b^2 + ss'c^2}{\sqrt{\left(p^2 a^2 + q^2 b^2 + s^2 c^2\right)\left(p'^2 a^2 + q'^2 b^2 + s'^2 c^2\right)}}$

A 39.8. $a' = b' = c' = \dfrac{\sqrt{2}}{2} a,$
$\alpha' = \beta' = \gamma' = 60°$

A 39.9. $f = \dfrac{\sqrt{3}\pi}{16} = 0.34$

A 39.11.
$F_{A_3B} = f_{A_3B}\left[1 + e^{i\pi(k+l)} + e^{i\pi(l+h)} + e^{i\pi(k+h)}\right]$

A 40.6. $\langle \varepsilon \rangle_{1-D} = \varepsilon_F(0)/3,$
$\langle \varepsilon \rangle_{2-D} = \varepsilon_F(0)/2,$
$\langle \varepsilon \rangle_{3-D} = 3\varepsilon_F(0)/5$

A 40.7. $C_V = \dfrac{\pi^2 Nk_B}{3} \dfrac{T}{T_F}$

A 40.8. $1-D : \dfrac{C_V}{T} = \dfrac{2\pi Nk_B}{3\theta_D}\left(1 + \dfrac{\pi}{4}\dfrac{\theta_D}{T_F}\right)$
$2-D : \dfrac{C_V}{T} = Nk_B\left(\dfrac{\pi^2}{3T_F} + \dfrac{28.8}{\theta_D^2} T\right)$

A 40.9. $\Phi_n = \left(n + \tfrac{1}{2}\right)\dfrac{h}{e}$

A 40.10. $\rho(T) \sim \dfrac{T}{M\theta_E^2}$

A 40.11. $\rho_{phonon}(T) \sim T^5$

A 41.6. $\dfrac{\varepsilon_c}{\varepsilon_m} = 2$ (square),
$\dfrac{\varepsilon_c}{\varepsilon_m} = 3$ (simple cubic)

A 41.7. $hk_x + \dfrac{l}{n} k_y = \dfrac{\pi}{a}\left(h^2 + \dfrac{l^2}{n^2}\right)$

A 41.8. $\Delta \varepsilon = U_0$

A 41.9. $\dfrac{1}{m_e^*} = \dfrac{a^2}{\hbar^2}(\varepsilon_0 - \alpha - \varepsilon)$

A 41.10. $\varepsilon = \varepsilon_0 - \alpha - 4\gamma \cos\dfrac{k_x a}{2}\cos\dfrac{k_y a}{2}$
$-4\gamma \cos\dfrac{k_y a}{2}\cos\dfrac{k_z a}{2} - 4\gamma\cos\dfrac{k_z a}{2}\cos\dfrac{k_x a}{2}$

A 42.6. $\varepsilon_g = \varepsilon_{g_0} - \alpha T$

A 42.7.
$\varepsilon_g = \dfrac{2k_B T_0 \beta}{\beta - 1}\left(\ln\alpha + \dfrac{3}{2}\ln\beta\right) = 1.16\text{ eV}$

A 42.8. $\varepsilon_d = \dfrac{3}{2}k_B T_0 \beta \ln(100\beta)$

A 42.9. $n, p = \left[n_i^2 + \left(\dfrac{\Delta N}{2}\right)^2\right]^{1/2} \pm \dfrac{\Delta N}{2}$

A 43.6. $I_g = -eA(L_e + L_h)g$,
$V_{oc} = \dfrac{k_B T}{c}\ln\left(\dfrac{I_g}{I_s} + 1\right)$

A 43.7. $V_{oc} = V_m + \dfrac{k_B T}{e}\ln\left(1 + \dfrac{eV_m}{k_B T}\right)$

A 43.8. $V_C = \dfrac{\varepsilon_s \varepsilon_g^2}{2e^3 \lambda}\left(\dfrac{1}{N_a} + \dfrac{1}{N_d}\right)$

A 43.9. $N_{dE} = \left(\dfrac{2k_B T}{\alpha}\right)^2$

A 43.10. $I_C = \dfrac{I_{C0}}{1 - (\alpha_n + \alpha_p)}$

A 44.6. $\chi_d = -\dfrac{N_A e^2 a_0^2}{m_e Z^2}$

A 44.8. $\chi = \dfrac{2\mu_0 F_x}{V(\partial B^2/\partial x)}$

A 44.9. $C_V = Nk_B V\left(\dfrac{2T_c}{T}\right)^2 e^{-2T_c/T}$

A 44.10. $(BH)_{\max} = \dfrac{\mu_0}{4}M^2$

A 45.6. $\dfrac{B_f}{B_c} = 1 + \dfrac{\lambda_L}{a}$

A 45.7. $\dfrac{B_f}{B_c} = 2\sqrt{3}\,\dfrac{\lambda_L}{a}$

A 45.8. $\Phi = m\dfrac{h}{2e}$ (m integer)

A 45.9. $\vec{j}_s = -\dfrac{n_s e\hbar \vec{K}}{2m_e}$

A 45.10. $j_s = 2j_0 \cos^2\dfrac{e\Phi}{\hbar}$

A 46.6. $Z = \dfrac{a_2}{a_3} \approx 26$ (Fe)

A 46.7. $S_n(A,Z) = B(A,Z) - B(A-1,Z)$
$S_p(A,Z) = B(A,Z) - B(A-1,Z-1)$

A 46.8. $\langle r \rangle = \dfrac{1}{2\alpha}$

A 46.9. $\Delta r \sim \dfrac{2}{\hbar}R$

A 46.10. $V(r) = -\dfrac{\mu_0}{4\pi}\dfrac{\mu^2}{r^3}$

A 47.6. $A_{ij} = \dfrac{\lambda_{i-1}}{\lambda_i - \lambda_j}A_{i-1,j}$

A 47.7. $\Delta = \dfrac{\omega_{if}}{c}\sqrt{(2k_B \ln 2)\dfrac{T}{M}}$

A 47.8. $\delta = 1(\beta^+),\ 0(\beta^-)$

A 47.10. $\Delta U_Q = \tfrac{1}{2}eQV_{zz}$

INDEX

Absorption, induced, 735
Acceleration, 12
  centripetal, 13
  Coriolis, 13
  due to gravity, 10
Action integral, 477, 566
Activity, 932
Adiabatic exponent, 161
Air breakdown, 222
Air friction, 29
Airy disk, 382
d'Alembertian, 80
d'Alembert's formula, 301
d'Alembert's principle, 471
Ampère, 2
Ampère's circuital law, 237
  modified by Maxwell, 266
Amplitude, 18, 301
  complex, 303
  division of, 351, 356
Amplitude-matching
  condition, 413
Anemometer, 360
Angle,
  Brewster, 416, 423
  critical, 419
  ellipticity, 398
  of emergence, 353
  of incidence, 352
  polarization, 416, 422
  of shear, 106
  of twist, 125
Angular momentum, 40
  nuclear, 934
  orbital, 619
  spin, 662
  total, 673
Anisotropy, spatial, 903
Antiferromagnetism, 883
Antineutrino, 940
Antinode, 311
Antisymmetry, 696

Aperture, 369
  circular, 381
  rectangular, 380
  uniform, 370
Apodisation, 376
Approximation,
  adiabatic, 748
  central-field, 718
  continuum, 756
  Debye, 755, 760
  free-electron, 791
  harmonic, 749
  molecular field, 879
  one-electron, 790
  quasi-classical, 613
  scalar, 348
  Stirling's, 536, 966
  tight binding, 820
  weak binding, 814
  weak-interaction, 900
Atom,
  acceptor, 838
  donor, 837
  helium, 713
  hydrogen, 639
  multielectron, 709
  one-electron, 636
Atomic mass unit, 10, 910
Attenuation, 336
Average,
  ensemble, 490, 506, 522
  time, 490
Axis,
  Cartesian, 68
  optic, 435, 440
  principal, 108, 285, 430, 958
  ray, 440

Band,
  conduction, 830
  energy, 810, 822
  valence, 830

Bandwidth theorem, 324
Barometric equation, 530
Base, 771
Basis, 770
Becquerel, 932
Bernoulli equation, 121
Bias,
  forward, 859
  reverse, 859
Binding fraction, 911
Biot-Savart law, 233
Birefringence, 442
Black body, 544
Bloch condition, 811
Bloch theorem, 811
Bohr radius, 565
Boltzmann's principle, 500
Boltzmann relation, 501
Boson, 696
Boundary condition, 311, 338
  no-slip, 119
  periodic, 792
Boyle's law, 168
Bragg's law, 783
Brewster window, 417
Brillouin zone, 816
  first, 752, 778

Candela, 2
Capacitance,
  electric, 3
  depletion, 866
Capacitor, 222, 267
  parallel-plate, 223
Capture,
  electron, 939
  neutron, 920
Carnot cycle, 138
Carnot engine, 143
Carnot theorem, 137
Carrier, 829
  majority, 838

minority, 838
Carrier concentration, 842
Cavity, 545
Cell,
   base-centred, 771
   body-centred, 771
   face-centred, 771
   primitive, 770
   unit, 771
   Wigner-Seitz, 771
Centre of mass, 23
Characteristic, static, 861
Charge,
   electric, 3, 204
   electron, 9
   point, 204
   polarization, 459
Circulation law, 207
Clapeyron equation, 191
Clausius-Mosotti equation, 226
Clausius's theorem, 150
Coefficient,
   absorption, 452
   air friction, 32
   Clebsch-Gordan, 675
   Einstein's, 740, 741
   Fourier, 815
   of friction, 32
   Joule, 182
   Joul-Kelvin, 183
   of performance, 140
   Sellmeyer, 464
   transmission, 608, 615
   virial, 168, 542
   of viscosity, 116
Coherent, 349
Collector, 861
Collision, 35, 457
   elastic, 86
   electron-lattice, 801
Commutation relations, 591
Commutator, 591
Compatibility, 592
Compressibility,
   adiabatic, 161
   isothermal, 161, 202, 765
Condensation, 289
Conductance, electric, 3
Conductivity,
   complex, 450
   electrical, 228, 803
   frequency-dependent, 460

   intrinsic, 836
   perfect, 893
   static, 460
   thermal, 7, 344
Conductor, 213
   good, 453, 461
   poor, 453
Configuration,
   common-base, 861
   common-emitter, 864
   electron, 720
   spin, 878
Conic section, 49
Conservation,
   of angular momentum, 41
   of energy, 27
   of momentum, 24
Constant, physical, 10
   Avogadro, 9
   Boltzmann, 9, 501, 527
   coupling, 749, 751
   critical, 177
   Curie, 874, 932
   damping, 457
   decay, 744, 932
   diffusion, 850
   elastic, 285, 749
   exchange, 878
   gain, 746
   gas, 170
   gravitational, 9
   Lamé, 114
   Madelung, 765
   molecular field, 879
   of motion, 478, 679
   phase, 18, 301, 303
   Planck, 9, 554
   Rydberg, 563, 638
   shielding, 728
   spring, 32, 124
Constraint, 468
   forces of, 469
   rheonomous, 469
   scleronomous, 469
Continuum, 105
Contrast, 357
Cooling energy ratio, 139
Cooper pair, 898
Coordinates,
   Cartesian, 11
   cylindrical, 953
   generalized, 473, 756
   polar, 13

   polar spherical, 630
   spin, 665
Correct Boltzmann counting, 536
Correction,
   first-order, 710
   second-order, 713
Cotton-Mouton effect, 448
Coulomb, 3
Coulomb's inverse-square law, 205
Coupling,
   LS, 721
   jj, 726
   spin-orbit, 664, 680, 924
Crystal,
   biaxial, 435, 448
   nonactive, 443
   simple cubic, 784
   uniaxial, 436, 441
Curie law, 874
Curie-Weiss law, 882
Curl, 950
Curve,
   dispersion, 754
   inversion, 184
   isenthalpic, 184
Cyclic process, 135
   bithermal, 135
   monothermal, 136
   reversible, 136, 138

Damping, 456
Debye T^3 law, 762
Deformation, 105
Degeneracy,
   accidental, 641
   exchange, 699
   spatial, 641
Degree of freedom, 469
Density, 109, 129
   charge, 205
   current, 228, 802
   displacement current, 266
   electron number, 466
   free-charge, 216
   linear, 287
   magnetic flux, 236
   magnetization current, 251
   momentum, 335, 336
   phase, 491
   polarized charge, 213
   probability current, 606

spectral, 325
superconducting current, 902
surface current, 251
Density of states, 506, 734
   effective, 836
Depth,
   London penetration, 895
   skin, 454
Deuteron, 926
Diagram,
   BM, 139
   energy-band, 852
   Ewald, 781
   free-body, 28
   phase, 897
   pV, 138, 153
   pVT, 176
   spacetime, 75
   TS, 153, 172
Diamagnetism, 870
   perfect, 894
Dielectric, 213
   anisotropic, 445
   isotropic, 215, 447
   lossy, 242, 451
Dielectric constant, 225
   principal, 431
Differential, exact, 961
Diffraction, 366
   Fraunhofer, 372
   Fresnel, 372
   Fresnel theory of, 367
   Kirchhoff formulation of, 366
   linear approximation of, 370
   order of, 377
   X-ray, 783
Diffraction pattern, 372
   Fraunhofer, 375
   Fresnel, 382
   grating, 377
   X-ray, 783
Diffusion, 850
   steady-state, 854
Dilatation, 108
Dimension, 4
Dimensional analysis, 5
Dimensional formula, 3
Dipole, electric, 208
Dirac δ-function, 344, 590
Dirac notation, 583
Dispersion, 317

angular, 378
anomalous, 459
normal, 458
wave, 315
Dispersion relation, 316, 458, 753
Displacement, 129, 288
   electric, 216
   vector, 292
   virtual, 469
Displacement current, 266
Distribution function, 495, 699
   Bose-Einstein, 701
   canonical, 509
   Fermi-Dirac, 701
   frequency, 760
   grand, 513
   microcanonical, 505
Distribution law,
   Boltzmann, 530, 702
   Bose-Einstein, 702, 767
   Fermi-Dirac, 701
   Maxwell, 532
   Maxwell-Boltzmann, 529
Disturbance, 284
Divergence, 950
   four-, 278
Dulong-Petit law, 524, 759

Efficiency,
   base transport, 863
   of a Carnot cycle, 138
   emitter, 862
Ehrenfest equations, 193
Ehrenfest theorem, 600, 611
Eigenfrequency, 312
Eigenfunction, 312, 588
   degenerate, 322
   radial, 638
Eigenstate,
   degenerate, 711
   nondegenerate, 710
Eigenvalue, 312, 588, 957
   nondegenerate, 592
Eigenvalue equation, 622
   matrix, 655
Eigenvector, 957
   Jones, 406
Eikonal, 306
Eikonal equation, 307
Einstein relations, 851
Electric current, 2, 227

Electron, 227
   bound, 457
   free, 790
   nearly-free, 814
   spinning, 673
Electronics, solid state, 850
Elliptic pendulum, 483
Emission,
   induced, 736
   spontaneous, 739
Emitter, 861
Energy, 2
   asymmetry, 915
   binding, 728, 911
   centrifugal potential, 45
   Coulomb repulsion, 913
   electromagnetic, 270
   electrostatic, 216
   effective potential, 45, 625
   equipartition of, 523
   Fermi, 793
   internal, 132, 154
   ionization, 728
   kinetic, 24
   pairing, 917
   potential, 24, 327
   relativistic, 85
   rest, 88
   separation, 936
   surface, 912
   total, 25, 88
   transition, 635
   volume, 912
   zero point, 611
Energy density,
   electromagnetic, 271
   electrostatic, 217
   magnetostatic, 258
   mechanical, 271
   spectral, 546, 561
   total, 327
   wave, 338
Energy-inertia relation, 88
Ensemble, 490
   canonical, 509, 527
   grand canonical, 514, 528
   microcanonical, 504, 525
   stationary, 494
Enthalpy, 153, 155, 175
Entropy, 2, 151, 501
   at absolute zero, 197
   microcanonical, 506
   of mixing, 535

radiation, 546
spectral, 561
statistical, 516
of the van der Waals gas, 179
Equal *a priori* probability, 498
Equation,
angular, 623
canonical, 475
characteristic, 17, 712
constitutive, 252
continuity, 109, 230, 276
of corresponding states, 178
determinantal, 668
diffusion, 343
elastic wave, 115
electromagnetic wave, 269
evolution, 478, 594, 598
Hertz vector, 275
isobaric, 173
isochoric, 173
of motion, 89, 111
plane wave, 429, 431, 452
polytropic, 170
potential, 273
radial, 623, 636
of a ray, 308
of state, 131, 170
of telegraphy, 337
of wave motion, 281, 283
Equation of state, 131, 170
Dieterici, 176
van der Waals, 176
Equilibrium,
conditions for, 25
dynamic, 801
macroscopic, 499
stability of, 26
thermal, 129
thermodynamic, 129, 498
Equipartition theorem, 523
Equivalence principle, 93
Ergodic hypothesis, 497
Ether, 66
Euler's equations, 120
Euler's theorem, 960
Event, 75
Exchange integral, 716
Expansion,
adiabatic, 547
coefficient of, 2
free, 169
isobaric, 161, 163

Joule, 182
Joule-Kelvin, 184
Taylor, 570
virial, 168, 541
Expansion theorem, 583
Expectation value, 589
Experiment,
Cockroft-Walton, 93
Davisson and Germer, 571
Franck-Hertz, 564
Michelson-Morley, 67
Pound and Rebka, 95
Stern-Gerlach, 662
Young's, 350, 379
Extinction index, 451, 461

Factor,
atomic scattering, 784
Boltzmann, 533
Gamov, 937
integrating, 151
Landé, 683
obliquity, 370
packing, 774
structure, 784
weighting, 590
Farad, 3
Faraday's law, 253
Fermi's golden rule, 735
Fermion, 696
Ferrimagnetism, 885
Ferromagnetism, 875
Field,
current density, 227
effective, 224
electric, 204
electromagnetic, 272
magnetic, 237, 251, 643
molecular, 879
scalar, 949
solenoidal, 236
strong, 685
uniform, 643
vector, 949
weak, 683
Fission, nuclear, 912, 919
Flow,
current, 227
energy, 329
laminar, 119
steady-state, 120
tube of, 120
viscous, 116

Fluctuation, 518, 541
Fluid, 116
incompressible, 118, 492
perfect, 119
viscous, 118
Flux,
electric, 206
energy, 330
luminous, 3
magnetic, 3, 236
Flux line, 205
Force, 2, 14, 129
body, 109
central, 40
coercivity, 263
conservative, 25
contact, 109
damping, 18, 227
driving, 330
electromotive, 253
external, 23
four-, 91
generalized, 474
harmonic, 17
internal, 23
inverse-square, 46, 60
Lorentz, 104, 257
nuclear, 916
random, 540
Fourier coefficient, 314
Fourier integral formula, 314
Fourier's law, 343
Fourier series, 313
Fourier spectrum, 315, 344
Fourier transform, 315
Four-scalar, 79
Four-vector, 89
current density, 279
momentum-energy, 90
position, 101
potential, 279
wave, 101
Frame, reference, 11
centre of mass, 92
inertial, 15, 68
laboratory, 92
noninertial, 28
rest, 86
Fraunhofer diffraction
formula, 373
Frequency, 2
angular, 301
cut-off, 761

INDEX

cyclotron, 446
forcing, 457
fundamental, 313
Larmor, 569
natural, 19, 446
normal, 549
plasma, 447, 457, 461
resonance, 22, 459
spatial, 373
Fresnel equations, 416, 418
Fresnel integrals, 383
Fresnel-Kirchhoff diffraction formula, 370
Fresnel zones, 385
Fringe, 350
Fizeau, 353
Haidinger, 351
localized, 353
nonlocalized, 351
ft value, 944
Function,
Airy, 357
aperture, 375
associated Legendre, 628
Bessel, 381
Brillouin, 874
confluent hypergeometric, 922
correlation, 361
Fermi, 942
Gibbs, 154, 156, 175
Hamilton principal, 480
Helmholz, 153, 155, 175
homogenous, 960
Langevin, 889
Planck, 157
Riemann zeta, 964
of state, 129, 961
thermodynamic, 172, 796
thermodynamic potential, 174
Fusion, nuclear, 912

Galileo's principle of relativity, 18
Gap, energy, 817, 901
Gas,
Dieterici, 184
free-electron, 460, 791
ideal, 168
perfect, 168, 525
real, 168
van der Waals, 175

Gauge, 240
Coulomb, 241, 273
Lorentz, 274
Gauss's law, 206
Gauss's theorem, 950
Gaussian curve, 345
Gaussian integral, 963
Gaussian probability distribution, 531
Geiger-Nuttal law, 936
Gibbs-Helmholtz equations, 157
Gibbs phase rule, 195
Giromagnetic factor,
nuclear, 934
orbital, 644
spin, 679
Glan prism, 448
Gradient, 949
four-, 80
pressure, 119
Grating, 354
diffraction, 376
Grating formula, 377
Gravitational red shift, 95
Graviton, 89
Green's theorem, 367

Half-life, 932
Hall effect, 807
Hamiltonian, 475
Hamilton-Jacobi equation, 481, 488, 614
Hamilton's principle, 472
Harmonic oscillator, 477, 609
Planck's, 558
quantum, 609, 616, 759
Harmonics, 313
Hartree equations, 719
Heat, 2
Joule, 455
latent, 191, 200
specific, 133
Heat capacity, 2, 133, 160, 523
free electron, 797
lattice, 759
Heat conduction, 343
Heat engine, 135, 145
Heat pump, 140
Heat reservoir, 507
Heisenberg inequality, 593
Heisenberg picture, 595, 614

Heisenberg uncertainty principle, 574
Henry, 3
Hertz, 2
Hole, 831
interstitial, 773
Hooke's law, 113
Huygens's principle, 370
Hydrogen molecule, 876
Hypothesis,
de Broglie, 571
Planck's, 554
quasi-ergodic, 497
statistical, 495, 496, 498
Hysteresis loop, 262
Rayleigh, 890

Illumination, 3
Impedance,
characteristic, 330, 331
complex, 463
intrinsic, 390
Impurity, 837
Inductance, 3, 255
mutual, 255
self-, 256
Insulator, 830
Intensity,
electric field, 204
electromagnetic wave, 390
luminous, 2
wave, 333
Interaction,
exchange, 876
repulsive, 715
spin-orbit, 679
strong, 944
weak, 900
Interface, 338, 412
Interference, 348
multiple beam, 355
order of, 351
Interference pattern, 352, 358
Interferometer,
amplitude-splitting, 359
two-slit, 361
Internal energy equation, 159
Interval,
four-, 79
proper time, 72
spacelike, 77
timelike, 77
Interval squared, 76

Isobar, 910
Isotone, 910
Isotope, 910
Isotope effect, 901

Jacobi's identity, 479
Josephson effect,
  ac, 909
  dc, 908
Joule, 2
Joule's law,
  for an electric current, 229
  for a perfect gas, 168
Junction, pn, 855
  abrupt, 855
  collector, 861
  emitter, 861
  Josephson, 908

Kelvin, 2
Kelvin statement, 162
Kepler's law,
  first, 50
  second, 52
  third, 52
Kepler's problem, 47
Kilogram, 2
Kirchhoff's integral theorem, 369
Kirchhoff's rules, 231
Kramers recurrence relation, 648
Kronecker delta, 956
Kurie plot, 943

Lagrange's equations, 474
Lagrangian, 472
Laguerre equation, associated, 638
Langevin equation, 888
Laplace's equation, 211
Larmor's formula, 405
Laser, 746
Lattice, 750, 770
  body-centred cubic, 784
  Bravais, 771, 772
  cubic, 782
  face-centred cubic, 785
  hexagonal, 772
  ionic, 765
  linear, 750
  monatomic, 755
  monoclinic, 772

  orthorhombic, 772
  reciprocal, 777
  tetragonal, 772
  triclinic, 772
  trigonal, 772
Laue condition, 782
Law of increase of entropy, 152
Law of mass action, 836
Law of thermodynamics,
  first, 132
  second, 135
  third, 195
  zeroth, 130
Lecher wires, 342
Legendre equation, 627
  associated, 627
Length, 2
  diffusion, 855
  focal, 352
  proper, 71
Lenz's law, 253
Level, energy, 611, 923
  acceptor, 838
  donor, 837
  excited, 718
  Fermi, 841
  ground, 718
  nucleon, 924
  Zeeman, 684, 743
Lifetime, 854
  mean, 744, 932
Light cone, 77
Light, natural, 408
Linewidth, 745, 933
Liouville's theorem, 492
Logarithmic decrement, 19
London equations, 895
Lorentz contraction, 71
Lorentz force law, 257, 335
Lumen, 3
Lux, 3

Magnetic flux law, 265
Magnetic induction, 3, 139, 232, 244
Magnetization, 139, 250
  reduced, 880
  saturation, 875
  spontaneous, 881
Magneton,
  Bohr, 645
  nuclear, 934

Mass, 2
  effective, 798, 818
  inertial, 13, 93
  gravitational, 93
  reduced, 43
  rest, 9, 86
Matrix, 954
  angular momentum, 656
  coherency, 407
  correlation, 395
  Hermitian, 959
  inverse, 438
  Jones, 400
  orbital angular momentum, 660
  spin, 663
  strain, 107
  stress, 110
  transfer, 426
  transpose, 956
  unitary, 959
Matrix element, 654
Maxwell's equations, 267
Maxwell relations, 158
Mayer relation, 170
Medium,
  anisotropic, 430, 433
  biaxial, 435
  conducting, 450
  dispersive, 315, 453
  double refracting, 447
  homogenous, 268
  isotropic, 268
  linear, 268
  neutral, 450
  nonactive, 431
  nonconducting, 389
  optically active, 431
  uniaxial, 435
Meissner effect, 894
Metre, 2
Miller indices, 779
Mobility, 834
  electron, 851
  hole, 851
Model,
  Bohr, 579
  Debye, 760
  Einstein, 760
  free-electron, 791
  gas, 914
  liquid drop, 912
  Lorentz, 457

nearly-free electron, 460, 814
particle, 910
shell, 920
Sommerfeld's, 568, 641
two-fluid, 894
vector, 678
Modulus,
 bulk, 113
 shear, 113
 Young's, 285, 755
Mole, 2, 3
Molecular speed,
 average, 533
 most probable, 533
 root-mean-square, 534
Molecule,
 diatomic, 763
 hydrogen, 876
Moment,
 effective magnetic, 873
 electric dipole, 208, 213
 of inertia, 53
 magnetic dipole, 249
 orbital magnetic, 644
 spin magnetic, 679
Momentum, 13, 87
 four-, 90, 91
 generalized, 272, 474, 756
 relativistic, 85
 total, 23
 wave, 334
Momentum-energy relation, 89
Motion,
 bound, 47
 Brownian, 540
 damped harmonic, 18
 electron, 567
 forced damped harmonic, 21
 orbital, 680
 oscillatory, 17
 in phase space, 493
 planar, 44
 planetary, 50
 rolling, 56
 simple harmonic, 17
 steady-state, 459
 unbound, 46, 47
 under a central force, 42
 under a constant force, 15

Navier-Stokes equations, 118

Nernst-Simon statement, 195
Neutrino, 940
Neutron, 910
Neutron excess, 915
Newmann's formula, 256
Newton, 2
Newton's law,
 first, 13
 second, 14, 227
 third, 14
Newton's rings, 353
Nodal line, 322
Node, 310
Normal mode, 329, 549, 751
 degenerate, 550
 electromagnetic, 550
Nuclear unit radius, 925
Nucleon, 910
Nucleus, 910
 even-even, 917
 odd-odd, 917
Nuclide, 910
Number,
 atomic, 910
 coordination, 773
 magic, 721, 920
 mass, 910

Observable, physical, 496
 compatible, 620
Occupation number, 698
Ohm, 3
Ohm's law, 228
Operator, 584
 adjoint, 587
 angular momentum, 586
 del, 949
 deviation, 589
 evolution, 603
 gradient, 949
 Hamiltonian, 586
 Hermitian, 587
 identity, 600, 654
 kinetic energy, 586
 Laplacian, 951
 linear, 654, 955
 lowering, 617, 658
 momentum, 586
 perturbation, 709
 position, 586
 quad, 278
 quantum, 594
 raising, 617, 658

spin angular momentum, 662
total angular momentum, 673
unitary, 693, 696
Optical activity, 442
Orbital, 719
Oscillation, steady-state, 457
Overlap integral, 877

Paradox,
 Gibbs, 534
 twin, 103
Paramagnetism, 872
 Pauli spin, 800
Parameter,
 conjugate, 324
 constitutive, 450, 457
 critical, 178, 184
 reduced, 178
 smallness, 709
 statistical, 503
 Stokes, 401
Particles, identical, 536, 695, 698
Partition function,
 canonical, 509
 grand, 512
 microcanonical, 505
Pascal, 2
Paschen-Back effect, 686
Pauli exclusion principle, 697, 698
Pauli spin matrices, 403, 663
Penetration, 454, 896
Permeability,
 differential, 263
 of free space, 10, 232
 relative, 252
Permittivity,
 complex, 458
 of free space, 10
 relative, 216
Perturbation, 709
 constant, 733
 harmonic, 735
Perturbation theory,
 stationary state, 709
 time-dependent, 730
Phase, 476
Phase difference, 349, 361
Phase matching condition, 413

Phase point, 490
Phase shift, 351, 421
Phase space, 476, 490
Phase transition, 187, 881
  first order, 190
  second order, 192
Phonon, 758
Photoelectron, 575
Photon, 89, 563
Planck's radiation formula, 555
Plasma, 461
Poincaré sphere, 402
Point,
  critical, 176
  lattice, 772
  triple, 177, 191
Poiseuille's law, 119
Poisson's brackets, 478
  fundamental, 479
Poisson bright spot, 385
Poisson's equation, 210
Poisson's ratio, 121, 285
Poisson's theorem, 480
Polarizability, 225
Polarization of a dielectric, 213, 446, 457
Polarization of waves, 387
  circular, 399
  complete, 394
  degree of, 409
  elliptical, 396
  linear, 395
  partial, 408
  by reflexion, 416
  state of, 393
  transverse electric, 413
  transverse magnetic, 413
Polarization ellipse, 397
Polynomial,
  characteristic, 656
  Hermite, 612
  Legendre, 627
Polytropic exponent, 171
Population inversion, 746
Postulate,
  Bohr's, 563
  of discrete transitions, 565
  of quantum mechanics, 583, 585, 589, 592, 602
  of special relativity, 67
  of stationary states, 563
Potential,

  chemical, 194, 503
  electric, 2
  electrostatic, 208, 210
  magnetic vector, 239, 245
  magnetostatic scalar, 239
  Morse, 763
  periodic, 812
  q-, 514
  thermodynamic, 153
Potential barrier, 607
  built-in, 855
  Coulomb, 937
Power, 2, 330
  average, 332
  radiated, 404
  specific rotatory, 445, 465
Poynting's theorem, 271
Precession, 676
  Larmor, 569
Pressure, 2, 117
  radiation, 336, 545
  standard, 10
Principle,
  of correspondence, 566
  of indistinguishability, 702
  of Newtonian mechanics, 11
  of special relativity, 65
  of statistical mechanics, 490
  of superposition, 17, 204, 282, 582
Probability,
  thermodynamic, 498
  transition, 734, 736
Probability current density, 606
Probability density, 581, 615
Process,
  adiabatic, 174
  bithermal, 137
  cyclic, 135
  irreversible, 152
  isentropic, 151
  isobaric, 156, 174
  isochoric, 156, 174
  isothermal, 156, 173
  monothermal, 136, 162
  polythermal, 149
  polytropic, 170, 180
  reversible, 152
  throttling, 183
Proton, 910
Pulse,
  Gaussian, 324, 375

  harmonic, 323
  rectangular, 325

Quantization, 625
  spatial, 568, 628
Quantum number, 922
  good, 679
  magnetic, 568, 625
  orbital, 627
  principal, 567, 638

Radiation,
  atomic, 730
  black body, 544
  electromagnetic, 544
  gamma, 935
  thermal, 544
Radioactive decay, 931
  alpha, 936
  beta, 939
Radioactive decay law, 932
Rate,
  optical generation, 866
  pair generation, 854
  transition, 735
Rayleigh-Jeans law, 551
Reciprocity theorem, 158
Recombination, electron-hole, 865
Reflection, 339, 414
  amplitude coefficient of, 340
  Bragg, 818
  external, 416
  internal, 416
  total internal, 420
Reflectivity, 340, 418
Refraction, 414
  double, 442
Refractive index, 295, 451
  complex, 451
  generalized, 306
  Maxwell's relation for, 460
  principal, 431
Refrigerator, 139
Region,
  classically allowed, 646
  depletion, 855
  extinction, 784
  freeze-out, 844
  intrinsic, 842
  phase, 493
  saturation, 842

transparent, 464
ultraviolet, 456
visible, 456
X-ray-, 781
Relaxation time, 388
Remanence, 263
Representation,
  coordinate, 572
  Eulerian, 105
  matrix, 653, 655
  momentum, 573
  polar, 642
  principal-axis, 431
Reservoir, heat, 507
Resistance, electric, 3
Resistivity, 808, 904
Resolving power, 364, 378
Resonance, 22
  nuclear, 934
  nuclear gamma, 935
  nuclear magnetic, 935
Ritz combination principle,
  563
Rotation, 695
  allowed, 772
  Faraday, 449
Rule, 583, 604
  cyclical, 961
  extinction, 784
  Hund, 721, 722
  quantization, 566, 568
  selection, 737, 738, 739
Sackur-Tetrode equation, 537
Scattering, electron, 801
Scherrer formula, 788
Schrödinger equation,
  time-dependent, 602
  time-independent, 605
Schrödinger picture, 601, 614
Second, 2
Sellmeyer's equation, 460
Semiconductor, 829
  extrinsic, 837
  impurity, 837
  inhomogenous, 852
  intrinsic, 830, 834
  n-type, 837
  p-type, 838
Semiempirical mass formula,
  917
Semimetal, 830
Shell,
  closed, 721

energy, 497
Shockley equation, 860
SI, 2
Siemens, 3
Simultaneity, 70
Single-valuedness condition,
  623
Skin effect, 454
Slater determinant, 698
Snell's law, 308, 414
Solenoid, 237
Space, 11
  configuration, 692
  Euclidean, 96
  function, 582
  $\vec{k}$-, 432, 761
  n-dimensional, 958
  phase, 476
  Riemann, 96
  $\vec{w}$-, 439
  world, 78
Spacetime, 78
Spacing,
  equilibrium, 750
  interplanar, 780
Speed,
  of light, 65
  of sound, 128
Sphere, 773
Spherical harmonics, 623
Spin,
  electron, 661
  nuclear, 934
Spinor, 934
Standard deviation, 573
Standing wave ratio, 347
State,
  bound, 605, 609
  d-, 629
  doublet, 721
  excited, 565
  ferromagnetic, 881
  ground, 565
  normal, 897
  p-, 629
  paramagnetic, 882
  quadruplet, 721
  singlet, 715
  superconducting, 893
  triplet, 717
  unbound, 605
Stefan-Boltzmann law, 546

Steradian, 3
Stokes's theorem, 207, 951
Strain, 106
  dilatational, 106, 755
  principal, 107
  rate of, 108
  shear, 106
Streamline, 120
Stress, 109
  compressive, 110
  dilatational, 110
  principal, 114
  shear, 110
  tensile, 110
Structure,
  close-packed, 774
  crystalline, 770
  energy-band, 818
  fine, 682
  nuclear, 910
Subshell, closed, 720
Superconductivity, 893
  high-temperature, 903
Surface,
  antinodal, 350
  nodal, 350
  pVT, 176
  ray-velocity, 439
  wave-vector, 432
Susceptibility,
  antiferromagnetic, 884
  diamagnetic, 872
  electric, 215, 458
  ferrimagnetic, 887
  ferromagnetic, 882
  magnetic, 252, 850
  parallel, 884
  paramagnetic, 806
  perpendicular, 885
Symmetry, 696
System,
  crystal, 773
  many-particle, 692
  two-electron, 876

Taylor's criterion, 364
Taylor series, 965
TdS equation,
  first, 159
  second, 159
Temperature, 2, 130
  absolute, 132, 138
  Boyle, 168

critical, 903
Curie, 881
Debye, 762
Einstein, 760
empirical, 131
Fermi, 794
Néel, 884
standard, 10
Temperature scale, 131
  Celsius, 131
  Kelvin, 132, 138
Tensor, 954
  dielectric, 431, 442
  strain, 107
  stress, 111
Thermal contact, 129, 499
Thermal isolation, 154
Thermometer, 540
Time, 2
  dilation, 72
Toroid, 238
Torque, 40
Trace, 956
Trajectory, 476
  virtual, 471
Transformation,
  Galilean, 16, 66
  linear, 68, 954
  Lorentz, 67, 70
  of velocity, 72
Transistor, junction, 861
Transition, 730
  allowed, 739, 941
  electric dipole, 739
  forbidden, 737
  magnetic dipole, 743
  radiative, 737
  superconducting, 898
Translation, 694, 770
Transmission, 339
  amplitude coefficient of, 340
Transmission line, 297, 299
Transmissivity, 340, 418
Tunnel effect, 607
Tunnelling, 609

Uncoherent, 350
Unit, 1
  base, 1
  derived, 1

Vacuo, 455

Vaporisation, 198
Variable,
  extensive, 129
  intensive, 129
  specific, 129
Variance, 519
Variation, 472
Vector, 949
  angular momentum, 628
  bra, 583
  Hertz, 274
  Jones, 399
  ket, 583
  position, 11
  Poynting, 9, 271, 332
  primitive translation, 770
  reciprocal lattice, 780
  scattering, 781
  state, 582
  stress, 109
Velocity, 12
  areal, 52
  drift, 802
  four-, 79
  generalized, 473
  group, 316, 462, 570
  of light, 9, 294
  parabolic, 64
  phase, 316, 437, 462
  radial, 13
  ray, 437
  solar escape, 64
  of sound, 290
  tangential, 13
Vibration, lattice, 750
Virial, 524
Virial theorem, 524
Viscosity, 540
  coefficient of, 116
Visibility, 362
Volt, 3

Watt, 2
Wave, 280
  Bloch, 810, 812
  compressional, 288
  continuous, 313
  dilatational, 286
  elastic, 284
  electromagnetic, 294, 331
  extraordinary, 436
  harmonic, 297, 301, 332
  line, 282

  longitudinal, 115, 292, 299, 331
  matter, 571
  monochromatic, 314, 349
  ordinary, 436
  plane, 283
  quasi-monochromatic, 392
  reflected, 421
  scalar, 366
  shear, 287
  sound, 288, 299, 331
  spherical, 305, 366
  stationary, 309, 334
  tidal, 296
  transmitted, 420
  transverse, 115, 288, 389
  travelling, 281, 331
Waveform, 281
Wavefront, 282, 306, 394
  division of, 354
Wave function, 281
  antisymmetric, 696
  scalar, 298
  symmetric, 696
Wavelength, 302
  de Broglie, 577
  Compton, 577
Wave number, 301
  complex, 452
Wave packet, 315, 569
  Gaussian, 570
Wave vector, 304
  complex, 451
  Fermi, 793
Weber, 3
Weiss zone law, 786
von Weizsäcker's formula, 917
Wien formula, 553
Wien law, 548
Wien's displacement law, 548
Work, 24, 134
  virtual, 470
Worldline, 75

Young's ratio, 121

Zeeman effect,
  anomalous, 684
  normal, 645
  nuclear, 934